DISEASES IN ANTIQUITY

DISEASES IN ANTIQUITY

A Survey of the Diseases, Injuries and Surgery
of Early Populations

Compiled and Edited by

DON BROTHWELL
British Museum (Natural History)
London, England

and

A. T. SANDISON
Pathology Department
University of Glasgow
Western Infirmary
Glasgow, Scotland

With a Foreword by

WARREN R. DAWSON

CHARLES C THOMAS • PUBLISHER
Springfield • Illinois • U.S.A.

Published and Distributed Throughout the World by
CHARLES C THOMAS • PUBLISHER
BANNERSTONE HOUSE
301-327 East Lawrence Avenue, Springfield, Illinois, U.S.A.
NATCHEZ PLANTATION HOUSE
735 North Atlantic Boulevard, Fort Lauderdale, Florida, U.S.A.

This book is protected by copyright. No part of it may be reproduced in any manner without written permission from the publisher.

© 1967, *by* CHARLES C THOMAS • PUBLISHER
Library of Congress Catalog Card Number: 66-16786

With THOMAS BOOKS *careful attention is given to all details of manufacturing and design. It is the Publisher's desire to present books that are satisfactory as to their physical qualities and artistic possibilities and appropriate for their particular use.* THOMAS BOOKS *will be true to those laws of quality that assure a good name and good will.*

Printed in the United States of America

CONTRIBUTORS OF ORIGINAL PAPERS

V. ALEXANDERSEN, *Universitets Anthropologiske Laboratorium, Institute of Anatomy, København, Denmark.*

J. LAWRENCE ANGEL, *Division of Physical Anthropology, Smithsonian Institution, U.S. National Museum, Washington, D.C., U.S.A.*

J. B. BOURKE, *The London Hospital Medical College, London, England.*

H. BRABANT, *Professor, Clinique Stomatologique, Hospital Universitaire Saint-Pierre, Bruxelles, Belgium.*

DON BROTHWELL, *British Museum (Natural History), London, England.*

CYRIL B. COURVILLE, *Professor, Cajal Laboratory of Neuropathology, Los Angeles, California, U. S. A.*

WARREN R. DAWSON, *Simpson House, Simpson, Bletchley, Bucks, England.*

C. W. GOFF, *Hartford, Connecticut, U. S. A.*

P. H. K. GRAY, *British Museum (Natural History), London; and St. Thomas' Hospital, Surrey, England.*

C. J. HACKETT, *Institute of Orthopaedics, London W.1., England.*

H. HAMPERL, *Professor and Direktor des Pathologischen Instituts der Universität Bonn, Bonn, Germany.*

RONALD HARE, *Laboratory Animals Centre, Medical Research Council Laboratories, Carshalton, Surrey, England.*

FRANK L. LAMBRECHT, *Maun, Republic of Botswana, Africa.*

F. FILCE LEEK, *Hemel Hempstead, Herts, England.*

F. P. LISOWSKI, *Professor, Institute of Medical Sciences, Haile Selassie I University, Addis Ababa, Ethiopia.*

THOMAS LODGE, *Department of Radiology, The Royal Hospital, Sheffield, England.*

LU GWEI-DJEN, *Lucy Cavendish College, Cambridge, England.*

EDWARD L. MARGETTS, *Professor, Department of Psychiatry, Faculty of Medicine, University of British Columbia, Vancouver, B. C., Canada.*

WILLIAM McKENZIE, *Harley St., London, England.*

VILHELM MØLLER-CHRISTENSEN, *Professor, Københavns Universitets Medicinsk Historiske Museum, København, Denmark.*

DAN MORSE, *Director, Peoria Municipal Tuberculosis Sanitarium, Peoria, Illinois, U. S. A.*

GERALD C. MOSS, *Royal Perth Hospital, Perth, Western Australia.*

JOSEPH NEEDHAM, F. R. S., *Gonville & Caius College, Cambridge, England.*

ADAM PATRICK, *Emeritus Professor, Department of Medicine, University of St. Andrews, St. Andrews, Scotland.*

A. W. PIKE, *Commonwealth Bureau of Helminthology, St. Albans, Herts, England.*

IVAN V. POLUNIN, *Department of Social Medicine and Public Health, University of Singapore, Singapore.*

J. THOMPSON ROWLING, *The Royal Hospital, Sheffield, England.*

PHILIP SALIB, *Department of Orthopaedic Surgery, Massachusetts General Hospital, Boston, Massachusetts, U. S. A.*

A. T. SANDISON, *Pathology Department, University of Glasgow, Western Infirmary, Glasgow, Scotland.*

MAX SUSSMAN, *Department of Bacteriology, The Royal Infirmary, Cardiff, Wales.*

CALVIN WELLS, *Mulbarton, Norwich, Norfolk, England.*

J. V. KINNIER WILSON, *Institute of Oriental Studies, Cambridge, England.*

Foreword

THE TERM Palaeopathology was first suggested by the late Sir Marc Armand Ruffer (1859-1917) when he was professor of bacteriology at the Cairo Medical School, as a convenient label for the study of disease in ancient human remains.[1] He himself made many important contributions to the subject, and others before and after him have described many particular cases that came under their notice. Clear evidences of disease, dating from very remote periods, have been found in the skulls and other bones of some of the fossil hominids that flourished long before the emergence of *Homo sapiens,* or were contemporary with him. The Lower Pleistocene australopithecine group of southern Africa displays evidence of caries and enamel hypoplasia, and the *Homo erectus* specimens from China and Java showed clear evidence of trauma (perhaps as a result of group combat). However, perhaps the most famous diseased fossil specimen is the cranium of the Rhodesian Man, discovered in 1921, which was found to have been affected for a long period with chronic sepsis. The teeth displayed dental caries and alveolar abscesses leading to general oral sepsis. This Pleistocene African also had suppurative middle ear disease involving the mastoid process with an abscess which broke through the cortex at its base. In addition to this, the tibia showed signs of arthritic changes.[2]

Skulls and skeletons provide evidence only of diseases that leave their mark upon the bones: many affections that involve only the tissues are thus excluded from investigation. Most of the peoples of antiquity inhumed their dead; a smaller number used cremation (a method that robs the pathologist of most of his data); a still smaller number practised embalming. The resulting mummies, preserving more or less completely the soft tissues, have provided a vast field for research into the traces of disease. The dried and shrunken tissues of mummies can, by appropriate laboratory treatment, be rendered in a state that admits of histological investigation. Mummification was first practised in Egypt, but was adopted by other peoples and has remained in vogue until comparatively recent times over a wide geographical range.[3] Egypt and Nubia have provided by far the greatest contingent of material for the study of palaeopathology. This is

---

1. *J. Path. Bact., 18:* (1913) 149.

2. Macleod Yearsley in *Rhodesian Man and Associated Remains,* (British Museum, Natural History, 1928), pp. 59-63.

3. I have dealt with mummification in countries other than Egypt in a series of memoirs. Australia and America: *J. Roy Anthrop. Inst., 58:* (1928), 115-138, pls. 8-13; *Man,* 1928, 73-74, pl. E.; Canary Islands: *Proc. roy Soc. Med., 20:* (1927), 832-842; Torres Straits: *ibid.* 850-851; *Ann. Arch. Anthrop.* 11, (1924), 87-94, pls. 10-12.

mainly due to the fact that the late Sir Grafton Elliot Smith, F.R.S., while he was professor of anatomy in Cairo, was called upon to examine the remains in various degrees of completeness of some 30,000 skeletons and mummies brought to light by the excavations of archaeologists in Egypt, and during the Archaeological Survey of Nubia. The cemeteries and tombs from which these remains emanated, dating from Predynastic times, through the long period of the thirty historic dynasties, then through the Ptolemaic and Roman periods, down to early Christian times, cover a total span of over 4,000 years. Not only did he place on a scientific basis the definition of the technique of embalming and its variations of method from time to time, but he encountered in the course of his investigations numerous cases of pathological interest which he usually referred for investigation to his colleagues, for he was, as he himself often emphasized, an anatomist and not a pathologist.[4] There is no need here to summarise the pathological results obtained, as they have already been placed on record.[5]

The Archaeological Survey of Nubia has produced the largest mass of material from a single area hitherto brought together. The number of specimens of each period was large enough to provide data for statistical and general conclusions. Elliot Smith and such colleagues as Frederic Wood Jones and Douglas Derry, who assisted him, have ably recorded not only the data for the physical anthropologist, but also all the anatomical anomalies and pathological changes revealed by the thousands of skeletons and mummies brought to light by the excavations of the archaeologists. All the traces of disease and of wounds and fractures have been fully and interestingly recorded in their report, a large folio volume accompanied by forty-nine plates and numerous text-figures.[6]

Many Egyptian mummies were brought to England in the first few decades of the nineteenth century. Some of these were unwrapped and examined by the surgeons of the day—Sir Everard Home, Sir Benjamin Brodie, and others. As most of these mummies were destined to be museum specimens, the examinations of them had perforce to be limited to external observations so as to preserve the integrity of the specimens and consequently no dissection was permissible. A notable exception is provided by a mummy of the Persian period of Egypt (XXVII Dynasty, B.C. 525-402) which came into the hands of Augustus Bozzi Granville, F.R.S., in 1820. He made a thorough examination of the body, and not only completely dissected it, but also made numerous preparations which were exhibited at the Royal Society. This mummy had not been eviscerated and had not the customary incision in the left flank. Most of the viscera were *in situ;* all were present

---

4. References to such cases abound in his published works, notably in *The Royal Mummies,* Cairo, 1912; *Contributions to the study of Mummification* in Mem. Inst. Eg. Cairo, 1906, 5, 1-53, pls. 1-19.

5. Elliot Smith, G. and Dawson, W. R.: *Egyptian Mummies,* London, 1924, 154-162; Rowling, J. T.: *Proc. roy. Soc. Med.:* 54 (1961), 409-414.

6. *The Archaeological Survey of Nubia: Report on the Human Remains,* Cairo, 1910.

except the lower part of the intestines, the right kidney and the liver. These three organs had been removed by the embalmer *per anum* by incising and enlarging the orifice. The left kidney, with its capsule and ureter attached, as well as the uterus and its appendages, showed evidence of having been in a diseased state for some time previous to death. Granville's conclusions were that the woman died at the age of fifty to fifty-five years, and that the cause of her death was ovarian dropsy attended by structural derangement of the uterine system generally. I have mentioned this case because it is the first instance known to me of a detailed scientific and anatomical examination of a mummy. Granville's paper[7] deserves to be read in its entirety, and it compares honourably with many later and more superficial publications. His preparations from the viscera were examined by several well-known contemporary anatomists—Matthew Baillie, Wilson, Carpue and Joshua Brookes—who confirmed his conclusions. Dr. Granville may be truly considered a worthy pioneer of palaeopathology.

The improvements in scientific technique since Granville's time, particularly in histology, have provided later investigators with improved and more accurate facilities for research. Thus Ruffer was enabled to make investigations that considerably advanced the science of palaeopathology: he investigated cases of calculi, arterial disease, lesions of the lungs and kidneys, and bilharzia infection, of which he found ova in the tissues of mummies. His contributions, published in various scientific journals between 1910 and 1914, have been conveniently collected into a volume, edited posthumously by Roy L. Moodie.[8]

Radiology in more recent years has been a valuable aid to research in palaeopathology. The first mummy to be X-rayed was that of the Pharaoh Tuthmosis IV (XVIII Dynasty, *c.* 1413-1405, B.C.) in 1904. At that time the only apparatus in Cairo was that in a private nursing-home (a primitive one by modern standards), and Elliot Smith described to me very amusingly how he, assisted by Howard Carter, conveyed the rigid Pharaoh in a cab for the purpose. The result was interesting, because the condition of the epiphyses enabled Elliot Smith to estimate with some precision the age of the king at the time of his death. During 1931, a large series of Egyptian and Peruvian mummies in the Field Museum of Chicago have been radiographed, and the results published.[9] During 1963 and 1964, over sixty mummies in the British Museum, as well as a number of others in various museums, have been radiographed (as part of a new survey of material in Europe.) Among these, cases of pathological interest have been found.

I have confined this sketch mainly to Egypt, because it is the material

---

7. *Phil. Trans.*: *115*, (1825), 269-319, pls. 18-23.

8. *Studies in the Palaeopathology of Egypt,* Chicago, 1921.

9. Roy L. Moodie: *Roentgenologic Studies of Egyptian and Peruvian Mummies.* Chicago, Field Museum of Natural History. Anthropological Series, Vol. 3, 1931.

from that country that I have principally studied, and because it has been my good fortune to be closely associated with Elliot Smith for many years until the time of his death in 1937, and with the recent radiological investigations. The bibliography of palaeopathology is already extensive and it is rapidly growing. Throughout the world contributions have been made by a wide variety of specialists, and one of the important features of this book is the demonstration of the interest of anatomists, pathologists, surgeons, radiologists, biologists and medical historians in this fascinating study of disease through space and time. There is clearly much which can be contributed by these various disciplines, although it is to be hoped that the continuing efforts by all such workers will eventually establish palaeopathology as a subject *per se,* a subject worthy of academic standing and just as acceptable from the point of view of specialist employment and research grants as serology and cytogenetics are today. Although mummified remains are relatively uncommon, human skeletal material continues to appear at a steady rate from the numerous excavations undertaken every year in different parts of the world. This ensures an ever broadening knowledge of ancient disease, and with continually growing numbers of diseased specimens, we can look forward to more exact statistical comparisons of pathological data in the future. With further discoveries, present conclusions and theories on ancient disease will, of course, have to be modified, just as recent work reported in this book has to some extent modified the findings of those dealing with human remains during the first three decades of this century. This is both inevitable and desirable.

Although no work of this size can hope to present a fully comprehensive survey of this broad subject, this book nevertheless goes a considerable way to satisfying this need. It is to be welcomed as the first reference work for some time which has considered such a wide variety of topics in such detail, and is certainly the first symposium to benefit from such a large number of specialist contributors.

WARREN R. DAWSON

Editorial Prolegomenon: The Present and Future

THIS BOOK originates from a feeling among students of early disease, that the time has come for some form of palaeopathological stock-taking and pooling of recently collected data. Although during the early part of this century and continuing into the 1930's there was much interest shown in the medical biology of ancient populations, the past three decades have seen but small advances. Admittedly there has been a number of introductory works considering ancient disease, and it is customary for general works on medical history to include a section on disease in antiquity. But there was clearly a need for something beyond this elementary level—and by far the best way of achieving this end and at the same time guaranteeing a comprehensive and mature coverage of the subject, was to gather a series of specialist contributions. This is what we have attempted, and although we do not pretend that it has been possible to get complete coverage of this broad and complex field of study, most of the major topics have received attention. Some of these subjects are controversial, and in some cases diagnosis has of necessity been highly speculative, but this is to be expected and although we hope the subject has been generally approached with a critical outlook, it would have been wrong at times not to enter into hypothesis and speculation.

Although the majority of contributions to this volume have been written specifically for this work, there seemed every advantage in including a few "classic" papers. This has double value in not only adding to the content of this work and perhaps helping to bridge some gaps in the coverage, but also it is a pleasure to include contributions by earlier authors who would not otherwise be represented in this volume.

Some studies on the various undeveloped aboriginal populations of today have a bearing on the study of the health of earlier peoples, and because of this they have not been neglected here. Such studies have had, of course, to be limited in a book of this size, and some aspects of modern primitive groups—for example their mental health—have had to be left out entirely.

A prime reason for this symposium is to stimulate yet further interest in the fascinating problem of human disease viewed through time. There is certainly evidence that there is a healthier outlook to such studies on the part of museums and other institutions in which human material is preserved. Not only is access to such material becoming easier, but curators are more prepared than ever before to permit specimens to be submitted to new techniques, and to be moved and manipulated for X-raying. There is also some evidence that specialists in rather esoteric lines of medicine and biology are becoming increasingly interested in the temporal dimensions of disease. Improved techniques in radiology are being brought to bear on exploring the internal contents of mummy packs and the pathological changes taking place within bones. It is possible that slab radiographs of bones might also be informative. Improved histological techniques and staining methods including histochemistry are permitting a much more detailed study of preserved ancient soft tissues. Palaeobiochemistry, initiated before World War II by serological investigations on ancient mummy and skeletal tissue, has extended in the last decade to amino-acid determinations on the nitrogenous residues to be found in some prehistoric bones. One wonders if tissue antigens, other than those of blood groups, might eventually be identifiable and, if so, whether anything could ever be found out about immune diseases? Although excavated human remains will probably never be ideal for electron microscopy, which demands rapid and excellent fixation, some results can be hoped for, and indeed the study of early hair—to give an example—has revealed information on the melanin granules by this means.

There is great need at the present time for a reappraisal of the ancient classical medical writers. Far too much literature on early writers is itself becoming rather dated, and we have need of more studies along the lines of the recent collaboration between Chadwick and Mann (1950) on the medical works of Hippocrates. One would, for instance, like to see re-examined in this way the remainder of *Corpus Hippocraticorum* and the works of such writers as Galen, Celsus, Paulus, Aretaeus and Soranus of Ephesus.

Two serious deterrents to the progress of palaeopathology seem worthy of special mention here, and one can only feel a little puzzled that no attempt has been made in the past to rectify these matters. First and foremost is the generally miserable nature of comparative collections of recent bone pathology. Admittedly, radiographic data is more plentiful, but this does not exclude the urgent need for microscopic and macroscopic data on bone pathology, which is also well catalogued with accurate case histories. Already the conquest of antibiotics has rendered some forms of bone reaction very uncommon, at least in fairly advanced societies, and clearly it is now or never even in the case of undeveloped societies (for they will not remain thus for much longer). In some instances, as for example in yaws and leprosy, there is still every hope of obtaining worthwhile comparative specimens, and thus the needs of the palaeopathologist here need emphasizing. Specialists working on such diseases in the living cannot be expected to know the value of amputation material and autopsy specimens to the study of ancient disease.

The second deterrent we refer to is that of the distribution of specimens showing evidence of disease, and of skeletal collections which it would be worth while examining (or re-examining) for evidence of pathological change. Taking the British Isles as an example, there are dozens of museums which contain skeletal material excavated from local sites, dated to between Neolithic and Medieval times; to visit all these museums is both costly and time consuming. In the same way, visiting collections in other parts of Europe presents the same difficulty, but on a more expensive scale. Clearly the problem will be no different in other parts of the world. The only workable answer we can see to this is to have two or three major study centres where, not only will original palaeopathological material be easily available for research, but where casts of specimens in less accessible museums may be referred to. Also, in association with these original and cast specimens will be X-rays of the bone disease, and wherever possible sections of the involved region. This may seem rather ambitious, but is the only reasonable answer to the wide scattering of important specimens. The same type of centralization is of course needed for data on mummies and naturally preserved bodies, where at least duplicate X-rays and perhaps tissue samples could be made available. At the British Museum (Natural History) a move has already been made to collect together for research purposes not only original specimens (including some of the famous diseased Nubian specimens, until recently at the Royal College of Surgeons), but also X-rays and casts of specimens in other museums. The results so far are very promising, and there has been good cooperation with other museums so far approached.

Finally, mention might be made of some of the more archaeological aspects of this discipline. Directors of excavations are becoming increasingly aware that the study of early human disease is not only of value to the human biologist and medical historian, but that some diseases could affect and mould the activities and development of early societies. Moreover, the early town or city consisted not only of streets, mansions, cottages and temples, but also of drains, cesspools, wells and other sources of drinking water (not alway clean), and perhaps even carelessly discarded food waste. These latter details are not only useful supporting evidence, but could at times even directly contribute to the study of disease. Thus, for example, during recent excavations at Roman Fishbourne and Medieval Winchester, structures were revealed which may have been cesspools or latrine pits. Careful analyses of the soils from such structures may in some instances reveal evidence of intestinal parasites (particularly in water-logged or highly desiccated conditions).

There is also a need for further systematic excavations specifically with a view to finding pathological changes in skeletons. In Europe, the question of the antiquity, frequency and distribution of leprosy and syphilis can only be answered by purposeful searching for these diseases. Professor Møller-Christensen has set a high standard in this respect, not only in his detailed studies of leper skeletons, but also in his careful research into the location and distribution of Danish leprosaria, and in directing so ably the excavation of some of these sites. Similar records and sites are available in other parts of Europe, and clearly this opportunity should not be missed (and indeed, especially at this time when so much town and city redevelopment is covering, perhaps for all time, Medieval foundations and original street layout). In some regions it is possible that the history of syphilis might be investigated along similar lines. To give but one instance, it is recorded in *The History and Antiquities of the County of Surrey* by Manning and Bray (1814) that "stews" in a certain part of Southwark were in evidence as early as 1162, being suppressed by Henry VIII in 1546. Also, they state that prostitutes living in such stews were not permitted Christian burial, but were interred at a special site called The Single Woman's Burial Ground. Clearly historical evidence of this nature demands careful consideration, followed by excavation whenever a suitable site can be fixed in relation to recent plans.

We have perhaps digressed somewhat in parts of this editorial preface. It is to be hoped, however, that in so doing we have helped to underline not only present developments and problems, but also future possibilities in the challenging multi-disciplined field of palaeopathology.*

DON BROTHWELL
A. T. SANDISON

---

*Addendum: Since going to press, D. A. Rokhlin has published *Diseases of Ancient Man* (1965, In Russian, Publishing House 'Nauka', Moscow) and an American symposium *Human Palaeopathology* (Saul Jarcho, Editor. Yale University Press, 1966) have become available for reference. Although too late to bring their findings into contributions here, it would nevertheless seem worth mentioning them at this point.

EDITORIAL ACKNOWLEDGMENTS

Although the acknowledgments of individual contributions are given separately at the end of the relevant papers, we should like to take this opportunity of thanking various individuals and institutions who have helped us during the compilation and preparation of this book.

Miss Mary Scott Ross and Mrs. Irene Copeland have given patient secretarial assistance. Mr. George Kerr has similarly helped with photographic needs; and Miss Theya Molleson and Miss Rosemary Powers with certain illustrations and other work.

Various colleagues, and especially Drs. J. S. Beck, Mary Catto, David Charles and S. E. P. Miller, have kindly given help.

The Wellcome Historical Medical Museum and Library, London, kindly and promptly supplied a number Xerox copies of articles we needed.

For advice an Egyptological aspects we must thank Cyril Aldred of the Royal Scottish Museum. Similarly, for advice regarding copyright we wish to acknowledge the help of the Copyright Office of the Library of Congress, Washington.

In the painful matter of proof checking, and in the preparation of an index, we have been helped by Patricia Brothwell.

In the case of contributions to this book which have already been published elsewhere, full details of the original place of publication are given in the list of contents to the volume. We would, however, like to take this opportunity to thank the editors of the *American Anthropologist, Journal of Pathology and Bacteriology, Bulletin of the History of Medicine, Archives of Pathology, Chinese Medical Journal,* and *British Medical Journal.* Also to the Executive Managing Editor of the American Medical Association, and the Periodicals Manager of The Johns Hopkins Press. In the case of the now extinct *Annals of Medical History,* we should like at least to acknowledge the important part this journal played in the furtherance of palaeopathology during the first three decades of this century.

Finally, we wish to thank Macmillan of London for permitting us to use parts of the Jones book on malaria in early Greece and Rome: and to Dr. E. A. Underwood and Oxford University Press for permitting republication of the Dawson paper.

D.B.
A.T.S.

Contents

	Page
Foreword—*Warren R. Dawson*	vii
Editorial Prolegomenon: The Present and Future	xi
Editorial Acknowledgments	xv

SECTION I
INTRODUCTORY STUDIES

Chapter

1. Pseudopathology—*Calvin Wells* 5
2. Calcinosis Intervertebralis, with Special Reference to Similar Changes Found in Mummies of Ancient Egyptians—*P. H. K. Gray* 20
3. General Considerations of the Evidences of Pathological Conditions Found among Fossil Animals—*Roy L. Moodie* 31
4. Notes on Diseases and Healed Fractures of Wild Apes—*Adolph H. Schultz** 47
5. The Bio-cultural Background to Disease—*Don Brothwell* 56
6. Health and Disease in Contemporary Primitive Societies—*Ivan V. Polunin* 69
7. The Egyptian Medical Papyri—*Warren R. Dawson* 98

SECTION II
CONTRIBUTIONS TO PARASITOLOGY

8. The Antiquity of Diseases Caused by Bacteria and Viruses, A Review of the Problem from a Bacteriologist's Point of View—*Ronald Hare* 115
9. Trypanosomiasis in Prehistoric and Later Human Populations, a Tentative Reconstruction—*Frank L. Lambrecht* 132
10. The Human Treponematoses—*C. J. Hackett* 152
11. The Prevalence of Malaria in Ancient Greece—*W. H. S. Jones** 170
12. Note on the Presence of "Bilharzia Haematobia" in Egyptian Mummies of the Twentieth Dynasty (1250-1000 b.c.)—*Marc Armand Ruffer** 177
13. Parasitic Diseases—*A. T. Sandison* 178
14. The Recovery of Parasite Eggs from Ancient Cesspit and Latrine Deposits: an Approach to the Study of Early Parasite Infections—*A. W. Pike* 184

xvii

SECTION III
GEOGRAPHIC STUDIES

Chapter		Page
15.	Organic Diseases of Ancient Mesopotamia—*J. V. Kinnier Wilson*	191
16.	Diseases in the Bible and the Talmud—*Max Sussman*	209
17.	Records of Diseases in Ancient China—*Lu Gwei-Djen* and *Joseph Needham*	222
18.	Disease in Antiquity: Ancient Greece and Rome—*Adam Patrick*	238

SECTION IV
SOMATIC DISEASES (INDIVIDUAL)

19.	Tuberculosis—*Dan Morse*	249
20.	Paraplegia—*J. Thompson Rowling*	272
21.	Syphilis—*C. W. Goff*	279
22.	Evidence of Leprosy in Earlier Peoples—*Vilhelm Møller-Christensen*	295
23.	Evidence on the Palaeopathology of Yaws—*T. D. Stewart* and *Alexander Spoehr**	307
24.	The Evidence for Neoplasms—*Don Brothwell*	320
25.	An Eruption Resembling that of Variola in the Skin of a Mummy of the Twentieth Dynasty (1200-1100 b.c.)—*Marc Armand Ruffer* and *A. R. Ferguson**	346
26.	Evidence of Endemic Calculi in an Early Community—*Don Brothwell*	349
27.	A Review of the Palaeopathology of the Arthritic Diseases—*J. B. Bourke*	352
28.	Osteitis Fibrosa in a Skeleton of a Prehistoric American Indian—*Henri Stearns Denninger**	371
29.	Porotic Hyperostosis or Osteoporosis Symmetrica—*J. Lawrence Angel*	378
30.	A New Approach to Palaeopathology: Harris's Lines—*Calvin Wells*	390
31.	Thinning of the Parietal Bones in Early Egyptian Populations and Its Aetiology in the Light of Modern Observations—*Thomas Lodge*	405
32.	Biparietal Thinning in Early Britain—*Don Brothwell*	413
33.	Historical Notes on Some Vitamin Deficiency Diseases in China—*T'ao Lee**	417
34.	Major Congenital Anomalies of the Skeleton: Evidence from Earlier Populations—*Don Brothwell*	423
35.	Hernia in Egypt—*J. Thompson Rowling*	444

SECTION V
SOMATIC DISEASE (REGIONAL AND SYSTEMIC)

Chapter		Page
36.	Diseases of the Skin—*A. T. Sandison*	449
37.	Diseases of the Eyes—*A. T. Sandison*	457
38.	Diseases in the Ear Region—*William McKenzie* and *Don Brothwell*	464
39.	Degenerative Vascular Disease—*A. T. Sandison*	474
40.	Respiratory Disease in Egypt—*J. Thompson Rowling*	489
41.	Disease of the Alimentary System in Egypt—*J. Thompson Rowling*	494
42.	Diseases of the Reproductive System—*A. T. Sandison* and *Calvin Wells*	498
43.	Endocrine Diseases—*A. T. Sandison* and *Calvin Wells*	521
44.	Urology in Egypt—*J. Thompson Rowling*	532
45.	Palaeostomatology—*H. Brabant*	538
46.	The Pathology of the Jaws and the Temporomandibular Joint—*V. Alexandersen*	551

SECTION VI
ACCIDENTAL TRAUMA AND SURGICAL INTERVENTION

47.	Trauma and Disease of the Post-cranial Skeleton in Ancient Egypt—*Philip Salib*	599
48.	Cranial Injuries In Prehistoric Man—*Cyril B. Courville*	606
49.	The Evidence for Injuries to the Jaws—*V. Alexandersen*	623
50.	The Osteological Consequences of Scalping—*H. Hamperl*	630
51.	Primitive Surgery—*Erwin H. Ackerknecht**	635
52.	Prehistoric and Early Historic Trepanation—*F. P. Lisowski*	651
53.	Trepanation of the Skull by the Medicine-men of Primitive Cultures, with Particular Reference to Present-day Native East African Practice—*Edward L. Margetts*	673
54.	Reputed Early Egyptian Dental Operation, An Appraisal—*F. Filce Leek*	702

SECTION VII
MENTAL ABNORMALITY

55.	Mental Disorder in Antiquity—*Gerald C. Moss*	709
56.	Mental Diseases of Ancient Mesopotamia—*J. V. Kinnier Wilson*	723
57.	Sexual Behaviour in Ancient Societies—*A. T. Sandison*	734
	Index	757

*Contributions marked with an asterisk are in our opinion some of the "classics" in the field of palaeopathology (Eds).

DISEASES IN ANTIQUITY

SECTION I
INTRODUCTORY STUDIES

Chapter 1

Pseudopathology

CALVIN WELLS

It is no trespass, I hope, on editorial privilege to point out that this volume is designed to be much more than a collection of essays in the form of a reference book. It is also calculated to attract new workers to the field of palaeopathology and to inspire a livelier awareness of the ancient evidences of disease and of what they reveal about early patterns of living. But the interpretation of archaic pathology is often a task of extreme complexity. Pitfalls abound for the unwary. Precipitate judgments and hasty conclusions are the surest road to errors of diagnosis and interpretation. As R. L. Moodie (1928) once wrote: "It sometimes takes years of painstaking comparison to be sure a diagnosis of an ancient lesion is correct." A corollary of this is, as I have said elsewhere (Wells, 1964) that "in palaeopathology the best opinions are usually tentative opinions." One of the commonest sources of error is to be misled into diagnosing abnormality where none exists. Experienced clinicians and archaeologists alike often fail to recognize the many conditions which can mimic disease and overlook the changes, artifacts and simulations which their material may present. Innumerable examples of this "pseudopathology" could be quoted. The purpose of this chapter is to alert workers in this field to the existence of these pitfalls and to describe a few of the types most commonly encountered.

Inorganic Agents: Mechanical

The vast majority of skeletal remains are inhumations recovered from a great diversity of burial places. It is important, therefore, to know how bones can be affected by the soil that covers them and to what extent pathological states can be imitated.

Mechanical forces are very common and usually exert themselves in one of two ways: (a) by simple pressure on the bone or (b) by erosion of its cortex. Pressure may result in various deformations. When the skull is affected it is easily mistaken for deliberate moulding (Fig. 1). As long ago as 1853 Akerman described a deformed skull with tilted sphenoid wings from an Anglo-Saxon cemetery at Harnham, near Salisbury. Davis (1862) strongly urged that this was artificially contrived but Dingwell (1931) refuted the suggestion and regarded it as post-mortem warping. Lagotala (1922) noted a similar appearance of deliberate artificial moulding in a group of eleven crania from a dolmen at Guiry. These, too, were almost certainly the result of soil pressure and few long series of skulls are found that do not have examples of this kind. If only a single specimen is found it may sometimes be difficult to decide whether it shows intentional deformation or merely post-mortem warping, but usually the diagnosis is helped by the fact that where ethnic head

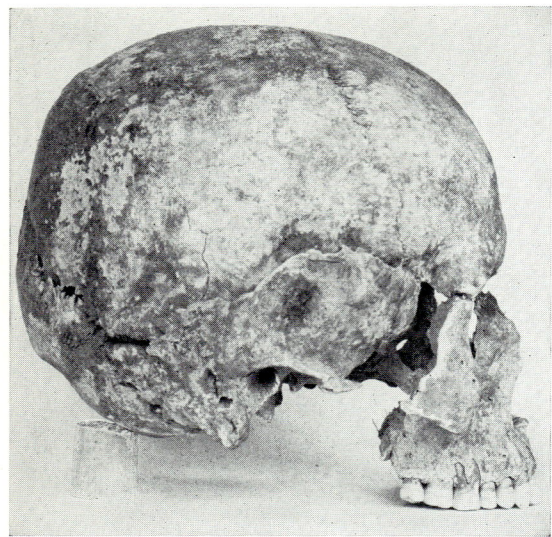

FIGURE 1. An Anglo-Saxon calvarium from Thorpe St. Catherine, Norfolk. It is a strongly planoccipital skull and highly brachy-cranial. (Cranial Index 95.2). This is entirely the result of soil pressure. In life it was probably dolichocephalic with a well-rounded occiput and tuber occipitale. The destruction of the alveolus around the first molar tooth is due to soil erosion, not disease. Author's collection.
Photograph Hallam Ashley.

moulding occurs it will be applied to many or all of the skulls in a cemetery, as at Pecos Pueblo (Hooton, 1930). Moreover, posthumous deformation is usually much more asymmetrical than that produced during life and tends not to fall into any of the recognized standard types: it is difficult to see, for instance, how it can produce conical forms. But unintentional obliquity from cradle decubitus is sometimes simulated by posthumous plagiocrany.

Skulls on which the weight of earth has led to much lateral compression may resemble a naturally occurring anomaly: scaphocephaly. This is easy to exclude if the sagittal suture is unfused but more care is needed when the parietals have united. In true scaphocephaly compensatory elongation is present together with other changes in cranial architecture. In young children the thin, ununited bones of the cranial vault often bulge outward as a result of pressure from soil which has seeped inside the skull. This may mimic hydrocephaly but careful inspection of the whole vault and base should suffice to show the real nature of the condition. In some cases of true hydrocephaly the sella turcica is enlarged but this does not occur under conditions of mechanical postmortem pressure.

In many soils, especially where drainage is poor and acid conditions lead to decalcification, the long bones may become warped by earth pressure. This has commonly led to a diagnosis of rickets. It can be excluded by the absence of such typical features as expanded metaphyses (including ribs), an S-shaped tibial curve, and cranial bossing. It should be remembered, too, that under appropiate conditions these simulations of rickets may be very common whereas the disease itself is rare before late Mediaeval times. Unequal subsidence from the infilling of the grave may fracture various bones. Ribs are especially vulnerable but mid-shaft breaks in the tibia and fibula are also common. If these accidents occur soon after burial it may be impossible to distinguish them from fractures inflicted shortly before death. Exceptionally, an ante-mortem blood stain may still be visible on bones long after burial. Wood Jones (1908) found many such cases in his Nubian material and one has been described in a Scottish stone cist (Waterston, 1927). When present this distinguishes ante-mortem from post-inhumation injury. Pressure may sometimes be minutely localized by the action of a small flint or similar rock. If the teeth are affected they may suffer a post-mortem chipping not unlike an ethnic mutilation. Jackson (1914) quotes a neolithic tooth

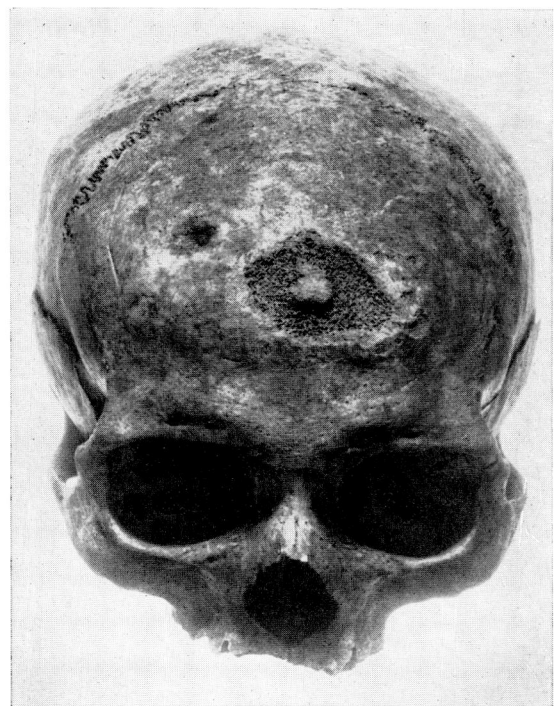

FIGURE 2. An Anglo-Saxon calvarium from Caister-on-Sea, Norfolk. The frontal bone shows two "lesions:" a small healed depressed fracture above the right orbit and a circinate erosion above the glabella. The latter is due to soil erosion but a thin deposit of mineral salts over exposed diploë and around the margin of the "lesion" above the left orbit gives an appearance of osteitic reaction and early proliferation of new bone. Castle Museum, Norwich. Acc. No. 343.957. Photograph Ministry of Public Building and Works.

"filing" from Belas Knapp long barrow that was possibly due to this kind of injury.

Inorganic Agents: Chemical

So far I have discussed the purely mechanical effects of soil pressure but other forms of pseudopathology result from chemical erosion, though the two agents are often combined. When the compact cortex of long bones is lightly affected by solution processes a "grained" appearance may be produced that closely resembles periostitis. On the cranial vault and in the orbit a more pitted effect should not be mistaken for cribra parietalis or orbitalia. Chemical osteolysis which bites more deeply into the bone can imitate osteitis and even osteomyelitis if it erodes through to the marrow cavity. Absence of osseous reaction, which is found in all but the most rapidly lethal osteomyelitis, will avert misdiagnosis. A more patchy and circinate type of erosion on skull or limb bones has not infrequently been assessed as syphilitic gummatous ulceration or yaws (Fig. 2).

Bodies buried in chalk, such as the many Bronze Age round barrow interments of Wiltshire and Dorset, typically have very light, friable bones. In the absence of enough comparative material to establish that this is a chemical change it is easy to suppose that the condition is due to an intra-vitam osteoporosis. Leprosy, frostbite, syringomyelia and other forms of peripheral gangrene are sometimes diagnosed as a result of finding eroded metacarpals, metatarsals and phalanges. All these diseases have precise pathological changes, a knowledge of which will prevent these mistakes being made.

Even tiny areas of cortical destruction may simulate specific lesions. Two common examples are the erosion of the postero-lateral surface of the mastoid process (Fig. 3) or a thin layer of the maxilla to expose the root of the canine tooth. The first may be mistaken for a discharging fistula of mastoiditis, the second for a paradontal abscess. When much of the jaw is involved in this erosive change it may simulate chronic alveolar infection or scurvy, especially if associated with loss of teeth. Occasionally an appearance of dental caries is given by a combination of postmortem chipping and chemical erosion but in general the teeth are by far the most stable elements in the dead skeleton. If a narrow zone of anterior alveolus is eroded

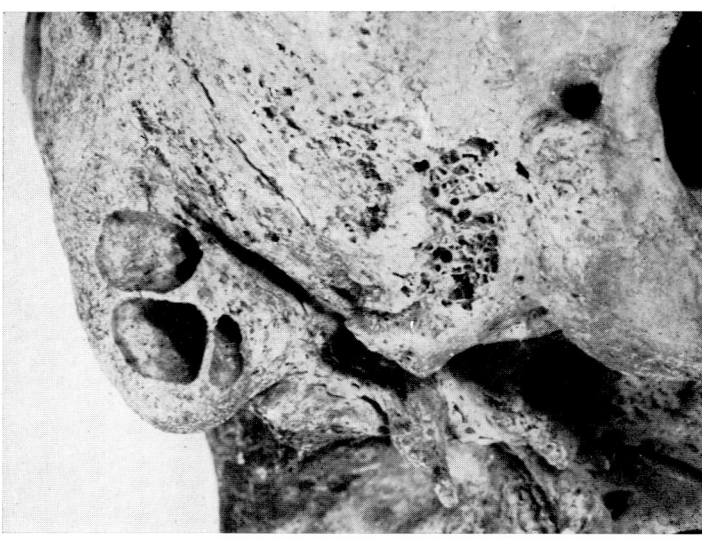

FIGURE 3. An Anglo-Saxon skull from Thorpe St. Catherine, Norfolk. The left mastoid process has three "fistulae" on the postero-inferior surface. Destruction of septa within the bone has taken place. These openings are strongly suggestive of discharging sinuses in chronic mastoiditis but there is no doubt that in fact they are due to physico-chemical erosion. This process has also attacked the adjacent part of the occipital bone. Castle Museum, Norwich. Acc. No. 15.953/41.

with posthumous loss of incisors, it may at times be necessary to consider the possibility of accidental loss or ritual ablation during life.

In many fully fossilized specimens a peculiar chemical change known as pyrites disintegration may be found. It is common in Jurassic remains such as ammonites and belemnites, also in ichthyosaurs from the Kimmeridge clay in which it may resemble the changes produced by chronic periostitis and other lesions. It may sometimes be detected by the strongly sulphurous smell which emanates from it.

In addition to pressure and chemical erosion on interred bones further pseudo-pathological changes may be produced by physico-chemical deposition. This takes many forms, but films of precipitated carbonates are common and when found on teeth may be mistaken for tartar (which, indeed, they chemically resemble) although their shape and distribution are quite different from the deposits formed in a living mouth. Sometimes these accretions may be coloured brown by ferruginous infiltration or greenish from copper salts and other matter. They can then be mistaken for the similarly coloured deposits which are a frequent result of coca or betel chewing (Leigh, 1937).

Fine alluvium may percolate into the maxillary antrum or other sinuses and, adhering to the floor, produce a granular appearance suggestive of chronic sinusitis. Post-mortem damage to these cavities often permits a view of their interior and workers should beware of diagnosing antral infection in these circumstances. In fossilized material, especially when it is coated with a smooth mineral deposit or embedded in a closely investing matrix, an illusion of congenital or pathological synostosis is sometimes given. This can be most deceiving in the small bones of hands and feet, and in the vertebrae. Careful laboratory preparation of the specimen usually reveals the true situation without difficulty. Sometimes radiography is the simplest way of recognizing the condition.

Miscellaneous Agents

Physical factors other than soil action can play a part in pseudopathological changes. Fluctuations of temperature, particularly alternate freezing and thawing, are important; wind and rain likewise. In Eskimo graves, many of which are simple stone cists, the body is exposed to the vagaries of a harsh climate. Under these conditions various factors combine to demineralize teeth as well as bones. Shrinkage, especially of the dentine, takes place, and when the tooth dries out the enamel may flake off to give the effect of functional attrition or ante-mortem dental fracture (Pedersen, 1949). A peculiar perforation of the skull is common in Eskimo skulls. It is usually bilateral in the occipital squama and might be mistaken for congenital lacunae, trephination, or other anomaly (Pales et al., 1952). It appears to be due to a combination of physicochemical causes including humidity, high winds and the weight of the skull itself.

Living Agents: Bacteria and Moulds

Bacteria are known to invade bone and to set up changes which may be mistaken for disease. Renault (1896) long ago thought that he could detect fossilized micro-organisms in his material and, although it was partly a post-mortem putrefaction process which he observed, later workers (Moodie, 1923) have left little doubt that bacteria can be recognized from exceedingly archaic times—some would say from pre-Cambrian deposits. Their effect on human skeletons is variable. Periostitis may be mimicked and there is some reason to believe that they play a part in producing an extensive destruction of the vertebral bodies which must not be confused with a tuberculous or pyogenic infection. When the sternum is attacked an appearance of congenital perforation may result.

Fungi, too, have been noted as ancient pathogens—at least as far back as the Eocene. But they are also important agents of pseudopathological change. Their mode of action seems to be that the fungal mycelia penetrate the bone and live on its organic matter, producing acids which dissolve the mineral content. Solution channels (*canaux de forage*) are produced (Morgenthaler *et al.*, 1957) and may appear in the guise of osteoporotic lesions. The essential nature of this process was recognized by Wedl (1864) and was further studied by Roux (1887) and by Schaffer, (1889) who named them "*Bohrkanäle*. Sognnaes (1950) has made a special study of this process as it affects teeth by invasion of the dentine and cementum. The large irregular channels which result should not be confused with intra-vitam decay.

Higher Plants

Plant roots often entwine themselves around bones and score the compact surface. These roots may be of any size from hair-line upwards. Typically they form a fine reticulate mesh on the cortex (Wells, 1963) which may be mistaken for minute vascular channels indicative of a hyperaemic condition such as periostitis (Fig. 4). Larger rootlets, up to 2 or 3 mm in diameter, have been mistaken for anomalous blood vessels. Either of these conditions, if present at all, tends to affect many bones of a skeleton or many bodies in a cemetery. It is common to find roots growing through cranial or other foramina. As the root enlarges it may split the bone and

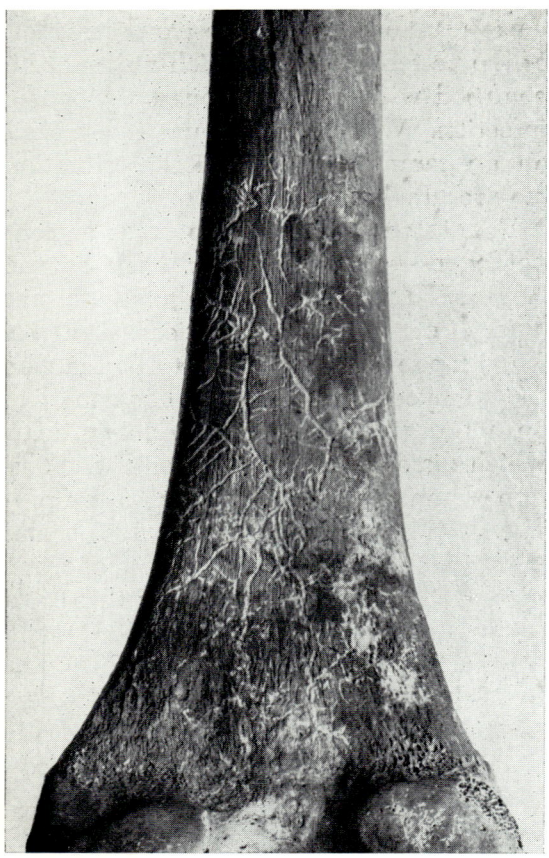

FIGURE 4. The distal end of an Anglo-Saxon femur from Burgh Castle, Suffolk. Two post-mortem appearances are seen: (a) thin, reticulate and branching lines which are the result of a web of roots eroding the cortex of the bone; (b) a fine longitudinal striation of parallel lines due to soil erosion which emphasizes the natural striated texture of the femur. Neither condition should be mistaken for the effects of periostitis or other hyperaemic conditions. Author's collection. Photograph Hallam Ashley.

make it look as though it had sustained an ante-mortem fracture. Misdiagnosis is likely here if the root has long since died and left no trace other than a soil stain from its carbonized remains. Fine roots may also encircle the necks of teeth and set up chemical solution of the dental tissue which has been interpreted as the effect of using tooth picks. (This is not to be confused with a somewhat similar chemical solution occurring in the living mouth.) (Hartweg, 1945; Brothwell, 1963) The pseudopathology noted above as affecting Eskimo teeth seems to be aggravated by contact with the decaying mosses which commonly invade the burial cairns.

In some soils, for example certain fine well drained gravels, matted collections of root fibrils can invade the medulla and decay into a deep black mass which stains most of the bone. This can readily be mistaken for an inefficient cremation or even for ante-mortem burning, the more so as it is often associated with an unusual brittleness of the cortex. A similar appearance can occur in peat burials. In these cases tannins seem to augment the effect.

Animal Agents

Animals of many different sizes attack exposed or interred bones and produce curious imitations of disease. Lortet (1907) noted an ancient skull from Roda in Upper Egypt with serpiginous ulceration of the left parietal and other areas. He claimed that it was syphilitic but it seems likely that the appearance was due to post-mortem gnawing by beetles (Gangolphe, 1913) since it is well recognized that various coleoptera attack recently buried remains. Fouquet (1897) spoke of these creatures as *les travailleurs de la mort*. Negro and other African crania are very often found with marks due to gnawing by rats and similar rodents. Porcupines are said to be vigorous offenders. The commonest area to be attacked is the upper margin of the eye sockets (Fig. 5). The result may be mistaken for ante-mortem injury or osteitis. Pales (1952) has reported a trephined Gallo-Roman skull in which the margin of the hole had been retouched by *un grignotage de souris*

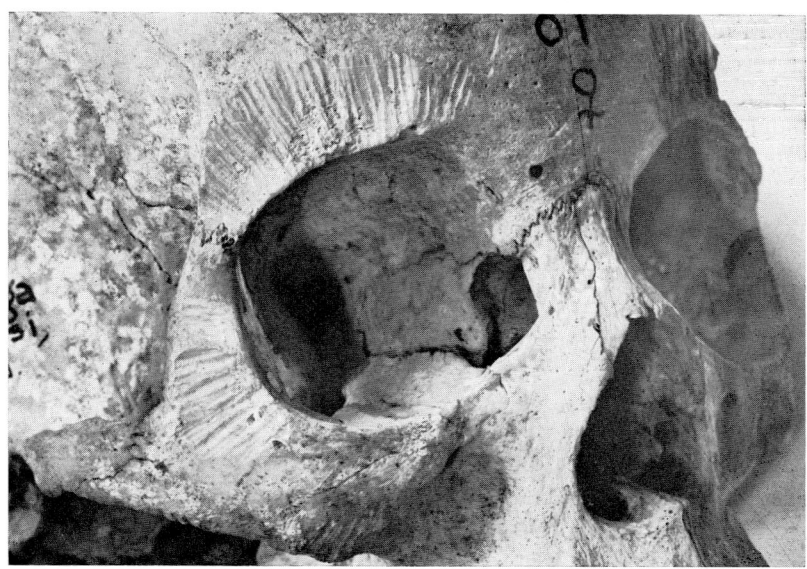

FIGURE 5. A Haya skull, Musira Island, Bukoba, Africa. The parallel grooves and sharpened edge around the superior and lateral margin of the right orbit are typical of gnawing by small rodents. They are also on the lower border of the zygomatic arch. Cambridge University, Duckworth Laboratory, Reg. No. Af. 23.0.199. Photograph Hallam Ashley.

—mouse nibbling. Larger carnivores leave larger traces. Dubois (1927) alleged that certain marks (not the exostosis) on the original Pithecanthropus femur were made by the teeth of a crocodile; grooves on a mammoth's acetabulum from Suffolk may have been caused by a wolf; and the hole in the left temporal squama of the Rhodesian skull, which is probably related to its gross mastoiditis, has been attributed to a leopard or other large carnivore. Hughes (1954) has discussed the relationship of hyaena feeding habits to what has been found at Makapansgat and other Australopithecine sites. Janssens (1963) illustrates possible fox bite on a bone from Furfooz.

To the direct attacks that these animals make on bones we can add the indirect effects of disturbance from burrowers. Rabbits, prairie dogs and similar creatures may greatly disturb inhumations, and at the excavation of these burials strange anomalies can be found. Phalanges and metatarsals may be removed, so that loss by gangrene or trauma has to be considered. I once found a well-articulated, apparently undisturbed body which seemed to have only six cervical vertebrae. The missing bone was found two metres away beside the neighbouring inhumation and had, presumably, been cleanly removed from the first by the misguided activity of a mole.

Excavation Hazards

This leads on to the consideration of hazards arising from incompetent and unvigilant excavation. All too often, even now, the first sight that the "archaeologist" has of a body is when his pick sinks into its skull. The freshness of the broken edges of bone should never deceive anyone into mistaking this for an ancient death wound, still less a trephination. Nor should the numerous excavation fractures which are produced under difficult field conditions in ribs, long bones, etc. present any ambiguity. But in crowded burial grounds careless excavation or inadequate recording may lead to confusion. All physical anthropologists are well used to opening a box of recently recovered bones only to find a two headed or three legged monster within—a situation which is seldom difficult to unravel. But if, especially in damaged incomplete bodies, a less dramatic substitution occurs the chimaera may not be easy to resolve. A small ulna and radius wrongly ascribed to a heavily built skeleton can be mistaken for the disuse atrophy of poliomyelitis; straying vertebrae may suggest supernumerary segments in the column (and this cannot always be excluded by articulation if the bones are much eroded by soil action); misplaced pelvic fragments can lead to errors of sexing or age estimation and thus vitiate studies in the vital statistics of the group; hasty excavation may confuse the relationship between a newborn child and its mother—if the child has been laid on the woman it implies a post-partum death from haemorrhage, puerperal fever, etc., if the foetus is clearly shown to be in the pelvis we may suspect an obstetric death from obstructed labour.

Laboratory Techniques

Mistakes due to muddle in the field are paralleled by errors perpetrated in the laboratory. Skeletons may be inextricably mixed and pose problems similar to those noted above. This is most likely to happen in teaching units or where several people handle the material. Reference to excavation photographs will only rarely enable the confusion to be disentangled. The moral is that every bone should be labelled *before* any work is done on it—and even this will be unrewarding unless the labelling is done accurately.

Various laboratory techniques also lead to pseudopathological appearances. Although mercury is hardly ever used now in the estimation of cranial volume many old skulls exist in which it has been used. Radiography of these calvaria may show minute speckles of the pocketed metal. If they are sufficiently small and the skull itself is more than usually radio-opaque an appearance of patchy osteopetrosis results which is not easy to interpret. It is probable that the future will see great progress in palaeoserology including the identification of skeletal blood groups and even perhaps the detection of abnormal haemoglobins or specific diagnostic reactions for treponeme and other infections. If these tests are to be valid, extreme care must be taken to record precisely what physicochemical treatment is given to the bones, such as the use of waxes, shellacs, Alvar 1570 and other preparations. Finally, the Piltdown episode serves as a reminder that even deliberate deceptions may, on rare occasions, be perpetrated.

Morphological Uncertainty

Another cause of pseudopathological diagnosis is our inadequate knowledge of normal morphology. This can even arise with human material. Is the Peruvian skull from Chilca, with a volume of only 490 cc, that of a normal midget as Hrdlička (1943) believed or does it transgress the bounds of normality and qualify as a pathological microcephalic? Is the Piltdown skull within the normal limits of thickness or should it be interpreted as an abnormal response to some blood dyscrasia? In palaeolithic and still more in archaic non-human forms these problems have commonly arisen.

More than a hundred years ago Robert Knox (1863) discussed whether or not the newly found Neandertal skull was artificially deformed. Hamy (1870) had already decided that the "sword blade" tibia of the Cro-Magnons was an ethnic distinction but many early workers believed it to be syphilitic or rachitic (Hanseman, 1904). The first Pithecanthropus erectus skull stirred similar controversy as to whether it was normal or diseased. Here, too, rickets was diagnosed. Sherwood Moore (1955) has recently re-examined an old problem and asks "Did the tall Cro-Magnon people finally succumb to gigantism?" Brown (1917) discusses ossified vertebral tendons in the dinosaur *Trachodon annectens*. So, too, does Moodie (1928) who also considers crumpling of the edge of the mantle in a Miocene clam (*Venus tridachnoides*) (1923) and rickets in a Pleistocene wolf and a small hawk from La Brea. Hatcher (1901) and Romer (1924) refer to an exostosis commonly occurring on the radius of an Oligocene cat (*Daphoenus*). Yet is is possible that each of these has been judged to be pathological merely because we do not know enough about the normal range of variation in these animals.

Mummies

Mummies are particularly fertile in pseudopathology owing to the extreme difficulty in interpreting their dried and distorted soft tissues. In addition, the ancient embalmers were often content to include parts of different bodies or even animal remains in the mummy pack. One of the most clearly recognized lesions from Egyptian material is atherosclerosis which often affected the aorta and other large vessels. Sandison (1962) has described an associated splitting of the wall in some atheromatous arteries and warns that such

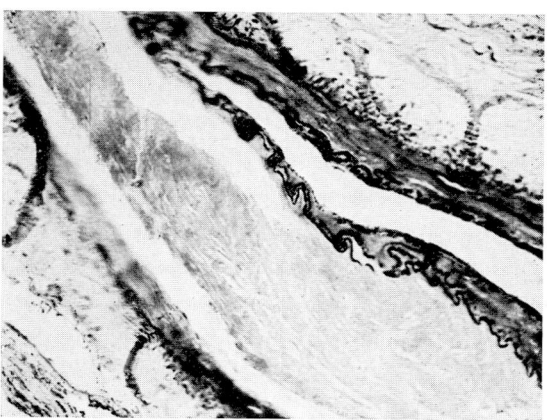

FIGURE 6. Tibial artery from an elderly female Egyptian mummy. The appearances suggest a dissecting aneurysm but are in fact the result of splitting of an atheromatous intimal plaque. The amorphous material is sudanophilic lipid. Heidenhain's iron-haematoxylin X 260. Preparation and photograph Dr. A. T. Sandison.

sectoral clefts are artifacts which must not be mistaken for dissecting aneurysms (Fig. 6). He also suspects (1963) that moulds may simulate leucocytes, giving rise to a false impression of an inflammatory lesion, and notes the danger of accepting a prolific post-mortem growth of bacteria as evidence of a heavy invasion during life. The whole problem of interpreting embalmed remains is one that we need to know much more about. Changes in the brain and viscera are peculiarly difficult to unravel.

Closely allied to naturally dessicated mummies are other instances of surviving soft tissue such as bog-burials and the frozen corpses of the Siberian Altai. Tissue autolysis in some bog-bodies can simulate degenerative conditions occurring during life and putrefactive changes in muscles may resemble ante-mortem infection by gas gangrene (Clostridia spp.).

Cremation

Cremation has been the chosen method of burial in many places and periods. Its fragmented remains are a fruitful source of pseudopathology. Two outstanding changes are found in cremated bones: severe warping and multiple fragmentation. The first is not to be mistaken for the various diseases, such as rickets or osteogenesis imperfecta, which cause distortion of living bones. The second must not be confused with fractures that have occurred before death. This can easily happen when confronted with a tooth from which the crown has been cracked off by burning and the exposed dentine stained by ferruginous gravel from the pyre and even glazed by fused silicaceous matter to give an appearance closely resembling functional attrition. A curious effect is sometimes seen on fragments of the cranial vault. It consists of small pits, a few millimetres across, eroded into the convex table of the skull and may even resemble the erosion of secondary malignant deposits. They are not pathological but are merely the depressions which contained the glandulae Pacchionii. They appear on the convex surface of the vault when an extreme degree of warping has turned a fragment of it "inside out"—a common event.

Some of these fragments of cremated parietal may be almost circular and are not easy to distinguish from rondelles excised by trephination (Messeri, 1959). Grave goods may complicate the interpretation of cremated remains. In an Anglo-Saxon I found what seemed to be a good example of arthritic synostosis of two cervical vertebrae: only the closest inspection revealed that the joint surfaces were welded together by a fused glass bead (the remains of a necklace) to which pulverized bone adhered, giving an effect remarkably like a small osteophyte. Also in Anglo-Saxons in England, but apparently not on the Continent, tiny fragments of ivory bag handles are easily mistaken for splintered and deformed tooth roots. Parts of domestic animals are often found mixed with human cremations and when they are small and warped the unwary examiner can readily mistake them for some kind of human pathology.

Miscellaneous

There is almost no limit to the pseudopathology that freakish circumstances can produce. A single example must suffice.

At Tell Duweir (Lachish) a female skull was discovered which gave evidence of having had a dental filling inserted into a cavity on the occlusal surface of the mandibular right second molar. Close inspection showed that this was illusory. The woman had bitten on a tiny metallic fragment of some kind which had become embedded in a normal pit on the crown of the tooth and remained in place after death. It measured 1 x 3 mm in area and was 1 mm deep—a very exact simulation of a small dental stopping (Risdon, 1939).

Cannibalism

Although this subject is not strictly within the compass of pseudopathology it is worth a brief mention owing to the frequency with which it is diagnosed on wholly inadequate evidence. It is often assumed that unnatural scattering of the skeleton, especially when combined with splintered long bones and a damaged cranial base, means that the body has been eaten—the bones cracked for marrow and the skull opened for its brains. This is reckless abuse of evidence. Bones may be scattered and broken from many causes and disintegration of the base of the skull is so common that few long series of inhumations fail to show many such specimens (Brothwell, 1961; Wells, 1964).

Trephination

This is another condition that is often diagnosed on inadequate evidence, whilst Moodie (1929) even went so far as to suggest successful bone transplantation in one pre-Columbian Peruvian: a fantastically improbable event. I have already referred to holes in Eskimo skulls which result from weathering and to "rondelles" as artifacts of cremation. These are imitations of lesions where none exist. There are, however, many genuinely pathological conditions which produce what we may call pseudo-trephination and they merit a brief mention here in view of the confusion they cause. Perforations of the cranial vault can be due to six types of lesion: congenital (Parry, 1928), injury (Cornwall, 1954), infection (Pales et al., 1937), tumours (Derry, 1914), haemopoietic diseases and osseous dysplasias (Wertheimer et al., 1956). Each of these groups offers a variety of lesions and trephination should not be accepted until all other possibilities have been excluded, especially in skulls from areas where it occurs but rarely.

Radiography

This is an indispensable part of the investigation of ancient bones, whether normal or pathological. It is important, therefore, to be alert to the many ways in which radiographs can simulate abnormal conditions. Broadly speaking they can be deceptive in two ways: (a) by giving a false appearance as to the morphology of a bone; (b) by misleading as to its texture. Examples of the first are minute grains of sand which may insinuate themselves between the roots of molar teeth close to the crown: these are easily mistaken for pulp stones. Chemical erosion may break down some of the septa inside an otherwise intact mastoid process: in a radiograph this must be distinguished from a somewhat similar appearance given by mastoiditis. Alluvial or crystalline deposits within the frontal sinus must be clearly recognized as such, otherwise the reduced size of the cavity can lead to a misdiagnosis of the sex of a skull or the deposits may even be mistaken for evidence of chronic sinusitis or an osteoma. Post-mortem fractures are not to be confused with those present during life: the same applies to processes of erosion (Fig. 7). Fine silt can percolate into long bones where it may occasionally lie in thin wisps of radio-opaque shadows across the shaft. This gives a false appearance of Harris's lines of arrested growth. Finally, when radiographing mummies countless intrusive objects may be found which demand the greatest circumspection on the part of the radiologist.

Changes of texture are even more perplexing. They take many forms and are ill understood. Irregular areas of increased density, imitating osteopetrosis, occur in intact long bones and have been attributed to the deposition of silicon compounds (Decker et al., 1939). I have elsewhere noted the frequent appearance of fluorosis or heavy metal poisoning given by many bones recovered from graves in chalk (Wells, 1963). Local areas of radiolucency in long bones may mimic a Brodie's abscess or similar destructive lesion whilst overall rarefaction may be mistaken for osteoporosis occurring during life. Ambiguous changes of texture such as these are very common: an excellent example is the radiographic simulation of Paget's disease by the Gamble's cave skull. It is a golden rule of palaeopathology that the specimen itself should always be inspected when its radiographs are being scanned.

Artistic Representations

Works of art of many different kinds provide an important source of evidence

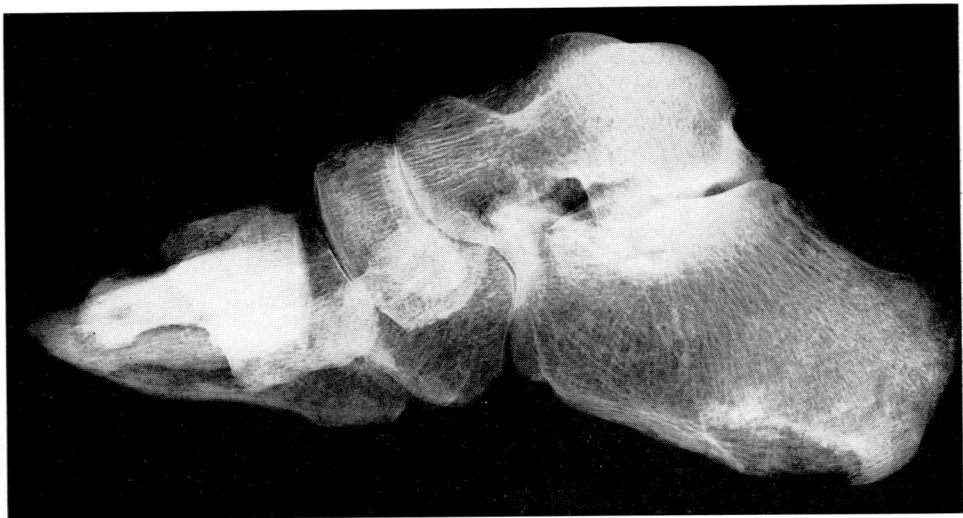

FIGURE 7. Lateral radiograph of left foot of an Early Saxon from Beckford, Gloucestershire. The missing phalanges and metatarsal heads are the result of destruction of the toes by leprosy. The inferior surface of the calcaneus appears to be eroded by a trophic ulcer which would be compatible with the disease. Only on inspection of the specimen itself can it be seen that this is a modern defect due to damage at the time of excavation or subsequently in the laboratory. Author's collection. Radiograph Dr. Brian Maxwell.

of ancient disease. Unfortunately their interpretation may be extremely difficult. Errors of diagnosis are commonly made here in two ways: (a) by seeing pathology where none exists; (b) by interpreting at its face value a pathological appearance which is only the expression of an artistic convention. To avoid the first it is necessary to have a clear view of the object (which may not be possible in the case, for example, of a disintegrating palaeolithic painting); to avoid the second a high level of artistic knowledge and sensitivity is required.

Many examples of diagnosing non-existent pathology could be quoted. In the Cesnola collection of Cypriot antiquities is an *ex-voto* in the form of a woman's torso. An irregularity below the right breast has been described as a cancer of that organ (Haddow, 1936), whereas it is merely the remains of a weathered bunch of grapes—a common fertility symbol. (Wells, 1964). Similar erosion of the stone appears to have led to another classical statue having been diagnosed as portraying the same disease (Meyer-Steineg *et al.*, 1921; Long, 1928). From seeing a photograph I recently thought that a Lenapé (Delaware) Indian mask was intended to show a pathological dilatation of one pupil (Fig. 8). Inspection of the "orbit" revealed that this appearance was due to loss of a silver disc which had originally been present in the eye. The much diagnosed trachoma in the left eye of the painted limestone statue of Queen Nephretiti (Berlin Museum) is equally due to unintentional damage to the statue.

It is more interesting and more difficult to assess the intention of the artist towards features which he completed in a frankly pathological form. Many Peruvian pots of the Chimú period seem to be excellent

FIGURE 8. A Lenapé (Delaware) mask of "Old Solid Face." The artistic style is highly conventionalized but the inequality in the size of the pupils, as shown in the photograph, might suggest that the artist was seeking to portray an organic paralysis. Inspection of the specimen shows that the right eye is filled with a silver disc drilled to represent a small pupil. This has been lost from the left eye. New York, Museum of the American Indian. No. 2/1067. Photograph Museum of the American Indian.

representations of double hare lip. Yet this is perhaps nothing more than an artistic convention derived from a feline motif, despite the fact that in general these *huacos* are an outstandingly naturalistic form of art. The appearance of artificial head moulding in Egyptian sculpture of the el Amarna period is certainly a convention divorced from reality, just as the curious representation of shoulders is (Madsen, 1905). So, too, are the many simulations of facial paralysis and asymmetry found in Iroquois masks of the "False Face" Society.

Indications of flat feet in anthropomorphic pots and statues from many areas are nothing but an artist's device to impart stability to his work. The puffy, oedematous eyelids of innumerable African masks and figurines are also a conventional way of carving these structures but the problem posed by enlarged genitalia and prominent navels is less easily solved. Some examples seem to be undoubted portrayals of elephantiasis and umbilical hernia.

Finally, aesthetic preferences may be responsible for deviations which verge on the pathological. Many early Renaissance painters depicted their madonnas with arachnodactyly (Bruno, 1959), whilst the bloated and obese infants and Venuses of the Reubens studio could well be interpreted as endocrine dyscrasias.

Textual Errors

Because the study of early disease in the historic period is often dependent on literary evidence, it should always be borne in mind that textual sources are a prolific source of error in many ways. The document may be indecipherable either because it is physically defective, as is common in many of the Akkadian cuneiform records; or because of our present inability to identify many of the linguistic usages, as in early Chinese manuscripts. It may be a corrupt text because it is a mixed recension of earlier documents, as is found in parts of the Ebers papyrus; or because of a scribe's error of transcription. Even when there is no doubt about the transliteration of a word its precise range of meaning may be problematical. No better example could be found than the Greek λεπρα, commonly translated leprosy." But in the Classical period it certainly embraced a wide range of skin diseases such as psoriasis, eczema, dermatoses, etc. and possibly never true leprosy. In mediaeval usage it undoubtedly extended its range to include leprosy but there are strong reasons for believing that it often referred to syphilis also (Holcomb, 1941). The common Egyptian *âaâ* is another term of great ambiguity which has been the subject of extensive discussion (Ghalioungui,

1963). Doubt about the precise date of a document may sometimes be crucial for its content. There is a 1412 manuscript in the Laurentian library, Florence. which suggests a pre-Columbian reference to syphilis. However, it is just possible that the relevant passage may have been added after March 1493 (Thorndike, 1942). Old documents are often factually wrong. Recent exhumation of Tamerlaine's body has disproved early records that he was an albino (Froggatt, 1962). Modern errors and misprints can initiate a very tenacious pseudopathology. A recent popular account of ancient medicine illustrates the Egyptian use of splints on a fractured forearm but the note to the plate describes it as "both shinbones" (Thorwald, 1963). The same book prints the classic Peruvian trephined skull found by Squier (Broca, 1867; Squier, 1877) from a reversed photo: this makes it appear that the lesion was on the right of the frontal bone not the left. Gejvall (1960) discusses and illustrates part of a femur previously described by Sjovall, apparently believing it to be a tibia. His reproduction of a very small length of the shaft is not sufficient to let the reader see that a mistake has been made. Other inaccuracies such as the use of a wrong scale or faulty numbering of plates and diagrams (Thorwald, 1963) may also be the progenitors of mistaken pathology.

When we remember the many ways in which a pseudopathological appearance can be produced—or a genuine lesion obscured—it no longer seems extraordinary that palaeopathologists occasionally make a wrong diagnosis. The wonder is that we ever make a right one.

REFERENCES

AKERMAN, J. W., 1853: An account of excavations in an Anglo-Saxon burial ground at Harnham Hill, near Salisbury. *Archaeologia*, 25:259.

BROCA, P. 1867: Cas singulier de trépanation chez les Incas. *Bull. Soc. Anthrop.*, 2ᵉ sér., 2:403.

BROTHWELL, D. R., 1961: Cannibalism in Early Britain. *Antiquity*, 35:304.

BROTHWELL, D. R., 1963: The macroscopic dental pathology of some earlier human populations. *Dental Anthropology*: Pergamon Press, 271.

BROWN, B., 1917: A complete skeleton of the Horned Dinosaur *Monoclonius*, and description of a second skeleton showing skin impressions. *Bull. Amer. Mus. Nat. Hist.*, 36:Art. 10, 281.

BRUNO, G., 1959: Malformazioni delle dita della mano nella patologia e nell'arte. *Minerva Med.*, 50:No. 92, 3685.

CORNWALL, I. W., 1954: The human remains from Sutton Walls. *Arch. J., London*, 110:66.

DAVIS, J. B., 1862: Notes on the distortions which present themselves in the crania of Ancient Britons. *Nat. Hist. Rev.*, n.s., 2:No. 7, 290.

DECKER, F. H. and BOHROD, M. G., 1939: Medullary artifacts in prehistoric bones. *Amer. J. Roentgen.*, 42:374.

DERRY, D. E., 1914: Parietal perforation accompanied with flattening of the skull in an ancient Egyptian. *J. Anat. Physiol.*, 48:417.

DINGWALL, E. J., 1931: *Artificial Cranial Deformation*. London, Bale and Danielson.

DUBOIS, E., 1927: Über die Hauptmerkmale des Femur von Pithecanthropus erectus. *Anthrop. Anz.*, 4:131.

FOUQUET, D., 1897: Recherches sur les crânes de l'époque de la pierre taillée en Égypte. Appendix to J. de Morgan, *Recherches sur les origines de l'Égypte*, 2:269.

FROGGATT, P., 1962: The albinism of Timur, Zāl and Edward the Confessor. *Med. Hist.*, 6:328.

GANGOLPHE, M., 1913: Syphilis osseuse préhistorique. *Mém. Acad. Sci Belles-Lettres Arts*, Lyon, 3ᵉ sér., 3:131.

GEJVALL, N-G., 1960: *Westerhus*. Lund, Ohlssons.

GHALIOUNGUI, P., 1963: *Magic and Medicine in Ancient Egypt*. London, Hodder and Stoughton, pp. 53-56.

HADDOW, A., 1936: Historical notes on cancer from the MSS. of Louis Westenra Sambon. *Proc. Roy. Soc. Med.*, 29:1015.

HAMY, E. T., 1870: *Précis de Paléontologie Humaine*. Paris, Baillière.

HANSEMAN, D. v., 1904: Über die rachitischen Veranderungen des Schadels. *Zeit. Ethnol.*, 36:373.

HARTWEG, R., 1945: Remarques sur la denture et statistiques sur la carie en France aux époques préhistorique et protohistorique. *Bull. Mém. Soc. Anthrop.*, 9ᵉ sér., 6:71.

HATCHER, J. B., 1901: Diplodocus Marsh, its osteology, taxonomy and probable habits, with a restoration of the skeleton. *Mem. Carnegie Mus., Pittsburg*, 1:85.

HOLCOMB, R. C., 1941: The antiquity of congenital syphilis. *Bull. Hist. Med.*, 10:148.

HRDLIČKA, A., 1943: Skull of a midget from Peru, Amer. J. Phys. Anthrop., n.s. 1:77.

HUGHES, A. R., 1954: Hyaenas versus Australopithecines as agents of bone accumulation. Amer. J. Phys. Anthrop., 12:467.

JACKSON, J. W., 1914: Dental mutilation in Neolithic human remains. J. Anat. Phys., 49:72.

JANSSENS, P., 1963: La race de Furfooz: son âge sa pathologie. Bull. Soc. Belg. Anthrop. Préhist., 73:45.

KNOX, R., 1863: On the deformation of the human cranium, supposed to be produced by mechanical means. Anthrop. Rev., 1:271.

LAGOTALA, H., 1922: Au sujet de quelques crânes déformés provenant du dolmen de Guiry. C-R. Ass. Fran. Av. Sci., 45e Sess., Rouen, 1921, 784-787.

LEIGH, R. W., 1937: Dental morphology and pathology of Pre-Spanish Peru. Amer. J. Phys. Anthrop., 22:267.

LONG, E. R., 1928: *A History of Pathology.* Baltimore, Williams and Wilkins.

LORTET, L. C., 1907: Crâne syphilitique et nécropoles préhistoriques de la Haute Égypte. Bull. Soc. Anthrop. Lyon, 26:211.

MADSEN, H., 1905: Ein künstlerisches Experiment im alten Reiche. Z. Ägyptische Sprache Altertumskunde, 42:65.

MESSERI, P., 1959: Rondelle ossee non intenzionali in reperti del Bronzo italiano. Arch. Antrop. Etnol. Firenze, 89:241.

MEYER-STEINEG, T. and SUDHOFF, K., 1921: *Geschichte der Medizin.* Jena.

MOODIE, R. L., 1923: *Paleopathology. An introduction to the study of ancient evidences of disease.* Urbana, Univ. Illinois Press.

MOODIE, R. L., 1928: The histological nature of ossified tendons found in dinosaurs. Amer. Mus. Novitates, No. 311, New York.

MOODIE, R. L., 1929: Surgery in precolumbian Peru. Ann. Med. Hist., n.s. 1:698.

MOORE, S., 1955: *Hyperostosis cranii.* Springfield, Thomas.

MORGENTHALER, P. W. and BAUD, A. C., 1957: Sur une cause d'altération des structures dans l'os humain fossile. Bull. Schweiz. Gesell. Anthrop. Ethnol., 33:9.

PALES, L., 1952: Le crâne perforé de la sépulture Gallo-Romaine de Camp-Grand (Lot). Bull. Mém. Soc. Anthrop. 10e sér. 3:110.

PALES, L. and BOTREAU-ROUSSEL, 1937; Faut-il réviser les trépanations préhistoriques? Rev. anthrop., 47:296.

PALES, L., FALCK, E. and LUTROT, J., 1952: Les perforations posthumes naturelles des crânes Eskimo du Groenland. Bull. Mém. Soc. Anthrop., 10e sér., 3:229.

PARRY, T. W., 1928: Holes in the skull of prehistoric Man and their significance. Arch. J., London, 85:91.

PEDERSEN, P. O., 1949: *The East Greenland Eskimo dentition.* Copenhagen, Reitzels.

RENAULT, B., 1896: Recherches sur les Bactériacées fossiles. Ann. Sci. Nat. Bot., (VIII), 2:275.

RISDON, D. L., 1939: A study of the cranial . . . remains from . . . Tell Duweir (Lachish). Biometrika, 31:99.

ROMER, A. S. A., 1924: A radial exostosis in the fossil Canid Daphoenus. Amer. J. Sci., 8:235.

ROUX, W., 1887: Über eine im Knochen lebende Gruppe von Fadenpilzen (Mycelites ossifragus). Z. Wiss. Zool., 45:227.

SANDISON, A. T., 1962: Degenerative vascular disease in the Egyptian mummy. Med. Hist., 6:77.

SANDISON, A. T., 1963: The study of mummified and dried human tissues. In Science in Archaeology. Ed Brothwell, D. R. and Higgs, E. Chapter 40:413-425.

SCHAFFER, J., 1889: Über den feineren Bau fossiler Knochen. Sitzgsber. Akad. Wiss. Wien, Math.-naturwiss. Kl, (III), 98:319.

SOGNNAES, R. F., 1950: Histological studies of ancient and recent teeth with special regard to differential diagnosis between intra-vitam and post-mortem characteristics. (Abst. in) Amer. J. Phys. Anthrop., 8:269.

SQUIER, E. G., 1877: *Incidents of travel and exploration in the land of the Incas.* New York.

THORNDIKE, L., 1942: A possible reference to syphilis before the discovery of America. Bull. Hist. Med., 11:474.

THORWALD, J., 1963: *Science and Secrets of Early Medicine.* London, Thames and Hudson.

WATERSTON, D., 1927: A stone cist and its contents found at Piekie Farm, near Boarhills, Fife. Proc. Soc. Antiq. Scot., 61:30.

WEDL, C., 1864: Über einen im Zahnbein und Knochen keimenden Pilz. Sitzgsber. Akad. Wiss. Wien, Math.-naturwiss. Kl (1), 50:171.

WELLS, CALVIN, 1963: The radiological examination of human remains. In Science in Archaeology, Brothwell, D. R. and Higgs, E. (Eds.) Chapter 39:401-412.

WELLS, CALVIN, 1963: Cortical grooves on the tibia. Man, 63:227.

WELLS, CALVIN, 1964: *Bones, Bodies and Disease.* London, Thames and Hudson.

WERTHEIMER, P., AVET, J., LEVY, A. and JENOT, J., 1956: Les lacunes osseuses de la voûte cranienne. Presse Méd. 68:1556.

WOOD JONES, F., 1908: The post-mortem staining of bone produced by the ante-mortem shedding of blood. Brit. Med. J., 1:734.

Chapter 2

Calcinosis Intervertebralis, with Special Reference to Similar Changes Found in Mummies of Ancient Egyptians

P. H. K. GRAY

The intervertebral discs are composed of three structures, the nucleus pulposus, the annulus fibrosus and the cartilaginous end plate of the adjacent vertebrae. Under normal conditions, these components are radiolucent. Calcification of the disc structures is not uncommon in adult life, but is relatively rare in childhood.

Calcification of the Disc Structures in Adults

This is often seen in the aged or middle-aged and is probably the result of normal degeneration. Trauma, no doubt, also plays its part. Calcification of the discs in the ochronotic stage of alkaptonuria is well known, and also occurs in the late stage of chondro-osteodystrophy and ankylosing spondylitis (Golding, 1959).

In alkaptonuria, the spinal changes are seldom found before the fourth decade (Golding, 1960). However, Umber and Buerger (1913) reported marked disc calcification in four out of eight alkaptonuric children whose father suffered from the disease.

Calcification of the nucleus alone has been frequently recorded, and is probably the result of trauma, but Kerley (1936) states that it may be associated with any disease which produces abnormality of calcium metabolism.

Calcification of the Disc Structures in Children

The incidence of disc calcification in children is low, but case reports are becoming more frequent. Köhler (1935) reports a case of a twelve-year-old child who fell ill with pyrexia and a partial fixation of the spine. A radiograph revealed some calcification of two of the intervertebral discs at the junction of the dorsal and lumbar vertebrae. This calcification disappeared within a year. Sandstroem (1951) reported transient disc calcification as the result of vitamin D poisoning. Further cases of transient disc calcification in children are recorded by Weens (1945), Walker (1954) and Newton (1958). In 1963, Melnick and Silverman produced a most comprehensive study of this problem. The youngest child in their series was seven days old and had the largest number of disc calcifications. In a summary of their article in the *Year Book of Radiology, 1936/64,* (Year Book Medical Publishers, Chicago) the Editors state:

"Intervertebal disk calification in children appears to be a self limited disease. It may be postulated that such calcification results from unknown causes in early childhood, disappears by maturity, is occasionally associated with local pain and differs from the degenerative calcification of senescence."

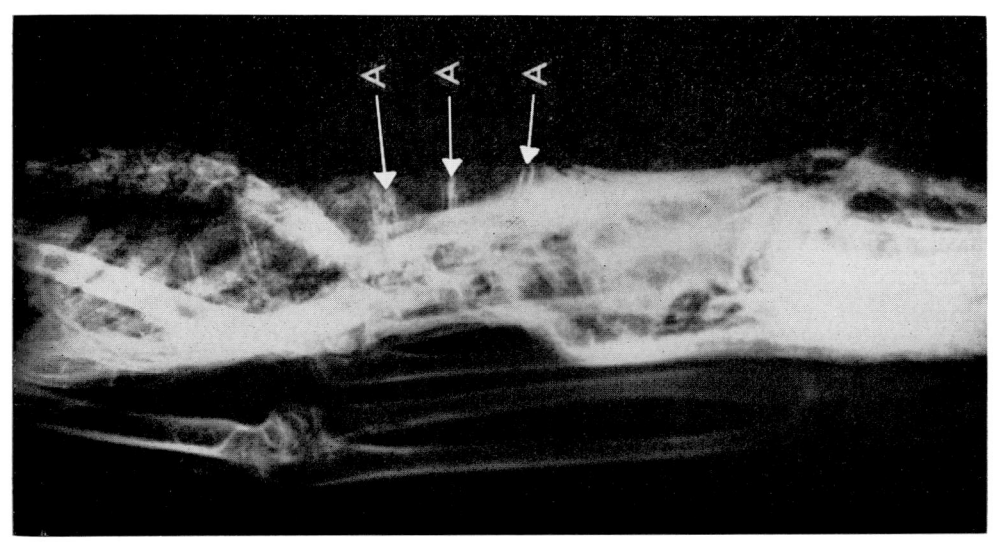

FIGURE 2. Lateral view of Figure 1 again showing the calcification (A).

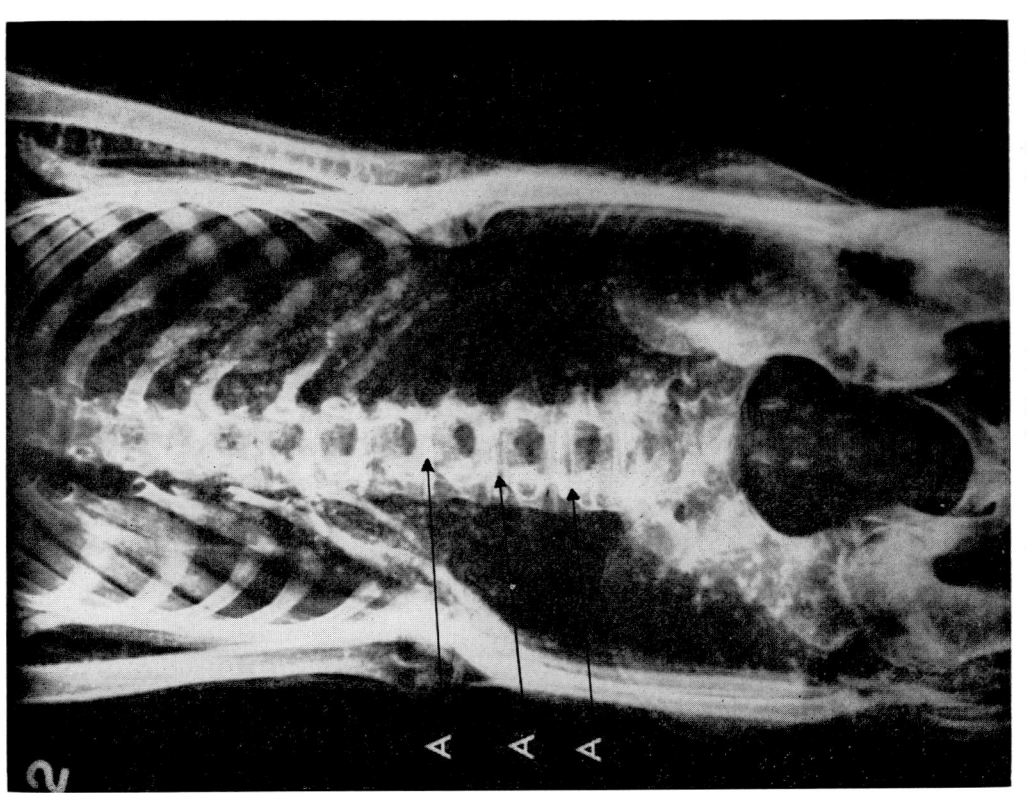

FIGURE 1. Mummy of a young child of the Roman period (? sex). Note the dense calcification of the intervertebral discs (A). British Museum Reg. No. 30362.

Calcification of the Disc Structures in Ancient Egyptian Mummies

It is apparent that a definite aetiology of calcification of disc structures, particularly in children, has not been established, and it is hoped that the fact that similar shadows can be demonstrated in the intervertebral discs of ancient Egyptian mummies, both adults and children, may stimulate further research into this matter.

Frequent mention of ankylosing spondylitis is made by the earlier writers on the spinal pathology of ancient Egyptians, so much so that Zorab (1961), who was studying the history of the disease, decided to investigate the matter further. He states:

"The more one reads about 'ankylosing spondylitis' in ancient Egyptians the more difficult it becomes to be sure what the various writers meant by the term and how frequently the disease did occur. I thought it worth while, therefore, to try to form a first hand opinion by examining the available remains in the Egyptology Department of the British Museum."

Radiographs of the spine and sacro-iliac joints of seven mummies were obtained by Zorab. No case of ankylosing spondylitis, using the term in its modern sense, was noted. However, the intervertebral discs of one of the mummies (a child of the Roman period) showed changes which were considered consistent with those of the ochronotic stage of alkaptonuria and the case was reported as such by Simon and Zorab (1961). This interesting article no doubt stimulated Wells and Maxwell (1962) to X-ray the two mummies in the Castle Museum, Norwich. The first, an adult female of the XXVI Dynasty, showed identical changes with those found in the spine of the child in the British Museum. The second mummy was still in its cartonnage case and the views of the spine were, to some extent, obscured by the packing material within the body cavity of the specimen. However, they state: "Parts of the spine were sufficiently clear to show that considerable calcification of the intervertebral discs is present." Early in 1964, a further case of presumptive alkaptonuria in an ancient Egyptian mummy was recorded by Adams who arranged to have the specimen in the Royal Albert Memorial Museum, Exeter, X-rayed.

It is not within the scope of a paper of this nature to go into the details of alkaptonuria. These can be found in any standard text book of medicine. Indeed, Golding (1960) goes as far as to state "it is better known than many conditions which are common." It will be sufficient to say that it is a very rare inborn error of metabolism, in which homogentisic acid is excreted in the urine, and the final diagnosis rests upon the identification of this substance. Dent (1964) estimates it to have frequency of one in a million. Clinically it is characterised by the passage of dark urine and a tendency to arthritic changes in later life. Radiologically the spinal changes are as follows:

a. A wide spread calcification within the discs.
b. A *marked* loss of the disc spaces associated with marginal "lipping."
c. Some rarefaction of the adjacent vertebrae.

This paper is based on the results of a radiographic skeletal survey of 102 ancient Egyptian mummies at present in museums in Britain. In this survey, apart from other observations of medical and archaeological interest, the incidence of intervertebral disc calcification both in adults and children, was found to be far greater than

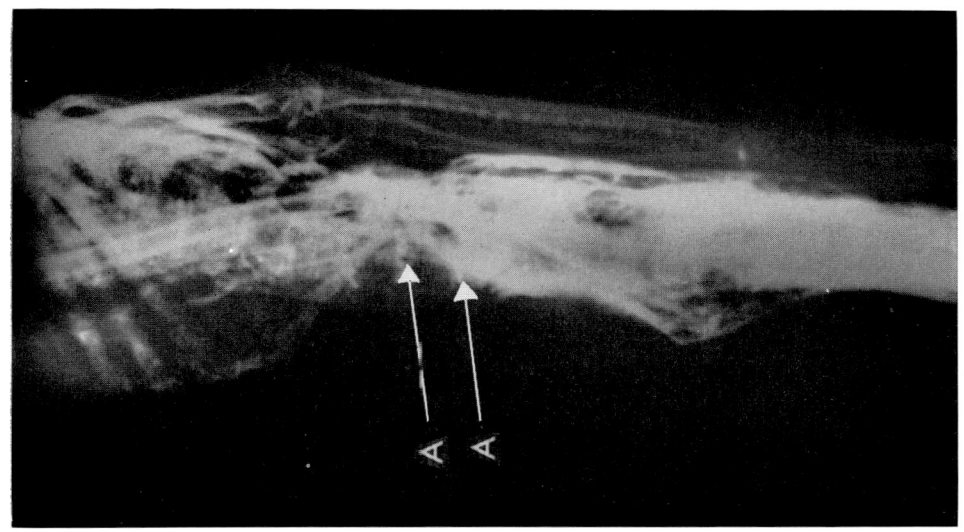

FIGURE 4. Lateral view of Figure 3 again showing the calcification (A).

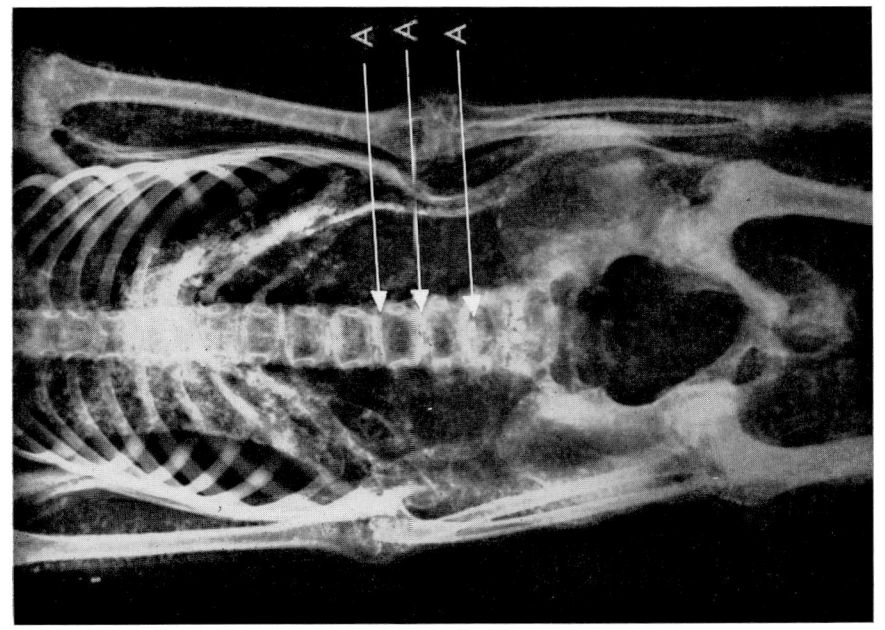

FIGURE 3. Mummy of a young male child of the Roman Period. Note dense calcification of the intervertebral discs (A). British Museum Reg. No. 30364.

would be expected, and was such as to cast considerable doubt on the previous suggestions incriminating alkaptonuria as the cause. Of the mummies radiographed, sixty-four satisfied the basic criteria of having a known date and provenance and intervertebral discs suitable to pass a radiological opinion.

TABLE SHOWING THE INCIDENCE OF DISC CALCIFICATION IN ANCIENT EGYPTIAN MUMMIES

Era	Positive	Possible	Negative	Total
a. Adults				
Predynastic and early dynastic naturally desiccated mummies.	Nil	Nil	7	7
Dyn. 11	Nil	Nil	1	1
Dyn. 17	Nil	Nil	1	1
Dyn. 21-25	7	1	Nil	8
Dyn. 25-26	2	2	5	9
Dyn. 27-30	Nil	Nil	5	5
Ptolemaic	1	1	7	9
Roman	Nil	1	7	8
				48
b. Children				
Predynastic and early dynastic naturally desiccated mummies	Nil	Nil	1	1
Dyn. 21-25	1	Nil	Nil	1
Dyn. 27-30	1	Nil	Nil	1
Roman	6	Nil	7	13
				16
Total No. of mummies				64

From the above table, it will be seen that intervertebral disc calcification in the mummies occurs in no less than eighteen out of sixty-four, an incidence of under one in four. This represents an enormous difference from the present day frequency of one in 1,000,000, and it is most unlikely that the incidence of alkaptonuria could change so drastically even over periods of eighteen to thirty centuries. Other evidence that the disc changes are not likely to be due to alkaptonuria is as follows:

1. Today, spinal changes in alkaptonuria are rare, in this already rare disease, before the fourth decade. It will be noted from the table that these changes were seen in no less than eight out of sixteen children, including the very young.

2. The radiographic findings themselves are not entirely typical.
   a. Allowing for post mortem changes, the disc spaces themselves are virtually within normal limits.
   b. There is no associated marginal "lipping."
   c. The end plates of the mummified discs show a marked increase in density. This feature is not so apparent in modern cases.

Allowing for the extremely faint possibility that these changes are, in fact, due to alkaptonuria, the absence of changes in the spines of the predynastic and early dynastic naturally desiccated mummies seem to imply that the disease became almost universal amongst the wealthy Egyptians of the XXI Dynasty. The disease was uncommon in the later dynasties, virtually absent during the Ptolemaic period but not uncommon during the Roman times, especially amongst the children.

Correlation of the Findings with the Mummification Process

The essential process of artificial mummification in ancient Egypt consisted of dehydration of the body and subsequent desiccation. The whole process, including bandaging and preparation, from time of death to interment is said to have lasted seventy days. It is generally agreed that the dehydrating agent was natron (*Pure natron being composed of sodium carbonate and sodium bicarbonate*), but there is a difference in opinion whether it was employed in a dry or liquid state. (For details concerning the use of natron, see Lucas, 1932, and Sandison, 1963.)

The fact that the tissues of many of the mummies are found to be infested with insects suggests that the body may have been laid out in the open to be dried by the heat of the sun.

It is also accepted that the art of embalming reached its peak during the XXI Dynasty, and the new and elaborate techniques as described by Smith (1906) tended to survive until the XXV Dynasty. But thereafter a decline set in and more reliance was placed upon the use of resin and less upon natron. Dawson (1964) states:

"In the Ptolemaic period well made mummies are occasionally found, and these had probably been subjected to the traditional natron process, but generally the treatment was very careless. Most of them are abundantly covered with resin applied hot in a molten state when it invades every crevice of the cavity and even the cancellous structure of the bone, . . . but in Roman times natron was often used again and the lavish use of resin discontinued."

It will be noticed that the radiographic changes appear to coincide closely with the use of natron, that is, nearly all the mummies of the XXI Dynasty show the disc changes; these are rarely found in mummies of the Ptolemaic period when resin was mainly used, but can be demonstrated in specimens of the Roman era when natron was used once more.

In their article concerning the Norwich mummy, Wells and Maxwell expressed their doubts that the calcification is due to the embalming process because in their specimen the cervical discs were normal and that the changes had not been demonstrated in every mummy. With regard to the former point, it is believed that the head and neck of the body received special attention and did not come into contact with the natron. Concerning the latter, there were several variations employed in the process of mummification in ancient Egypt.

1. Herodotus and Diodorus Siculus inform us that in their times three methods were used depending upon the wealth of the relatives.
2. In some periods natron was used and in others resin.
3. The time factor for the period of dehydration and desiccation also appears to have varied. Zaki and Iskander (1943), after having made analyses of material taken from various parts of a mummy, came to the following conclusion: "It seems for some political reason or other, the relatives of the dead were afraid of leaving the body of the General Imntfnht for the required seventy days in the mummifying laboratory. They therefore cut the proceedings very short indeed."

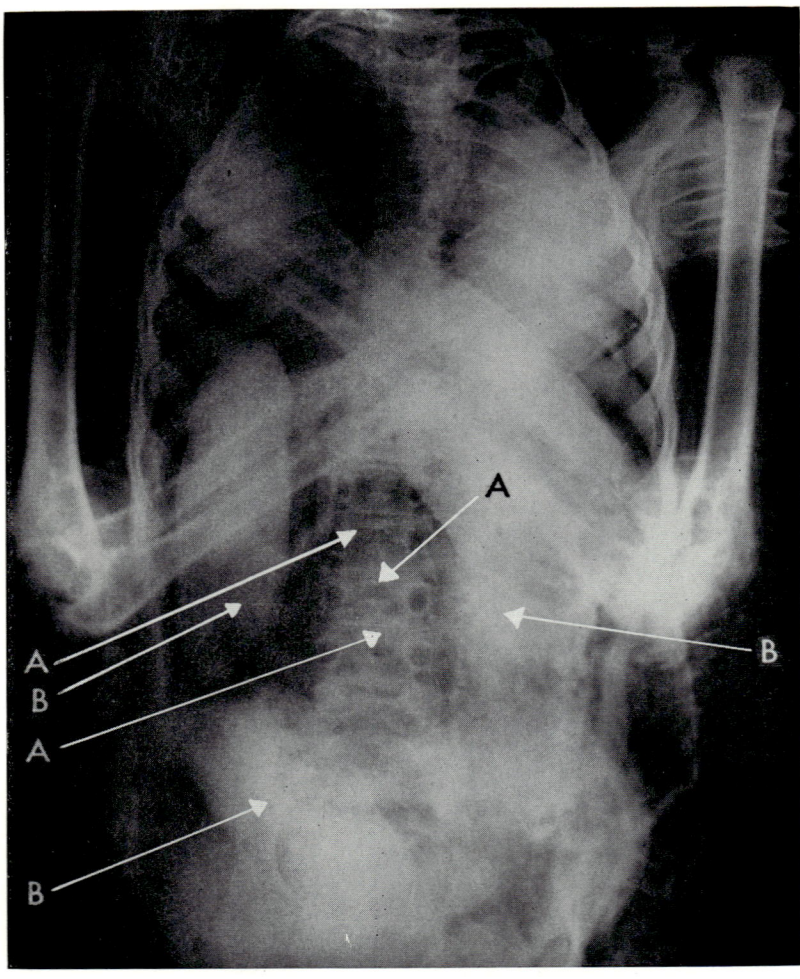

Figure 5. Mummy of a child (? sex) of the XXX Dynasty. Note changes in lumbar and dorsal disc spaces. The shadows are not so dense as Figures 1 and 3. The cylindrical objects (B) are bundles of linen soaked in resin. They probably represent a ritual recalling of the customs of the XXI Dynasty when the viscera were replaced into the body cavity in four separate linen wrapped bundles, each with its protective deity. British Museum Reg. No. 6699B.

In 1964, Darlow soaked the spine of a monkey in a saturated solution of sodium chloride, carbonate, bicarbonate and sulphate for seventy days, and then dried the specimen. No radiological changes were noted in the intervertebral discs before or after the experiment. Similar experiments are being performed elsewhere using dry natron, and it is hoped that when these results are available they will help solve this problem. The ancient Egyptians almost certainly used the encrusted crude natron from the shores of the Wadi Natrun and other sites, full of impurities and with varying proportions of the component salts. Many dumps of the ancient embalmers' refuse and natron have been found and analysed (Lucas, 1932 and 1960, and

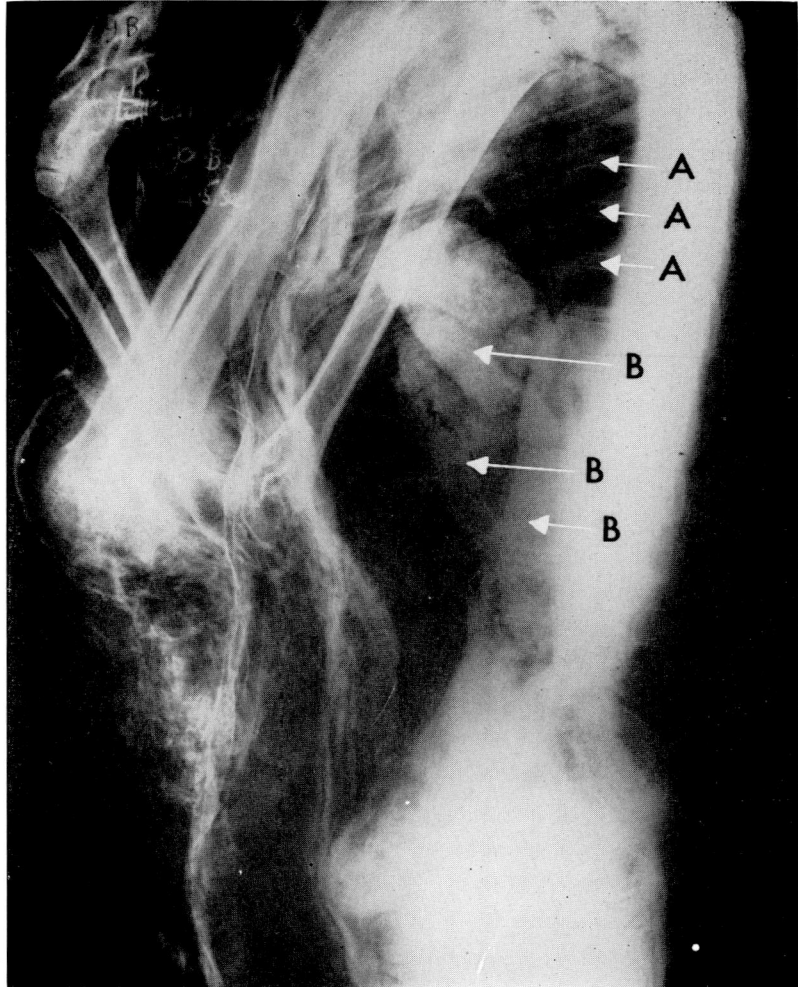

FIGURE 6. Lateral view of Figure 5 showing the changes in the spine (A) and the bundles (B).

Lauer and Iskander, 1955). These show an enormous variation in their constituents, and the presence of many impurities such as silicon and the salts of calcium and iron in varying amounts.

As disc calcification in the mummies of ancient Egyptians is most unlikely to be due to disease, it is suggested that the dense shadows within the intervertebral discs may represent deposits of impurities contained in the crude natron. If this be so, the question now arises, how did these impurities find their way into the cartilaginous structures? It would seem possible that they became permeated with these impurities whilst they were in contact with crude natron when in a fluid state. It is, however, quite obvious that the matter is still in a very speculative stage, and the nature of the opacities can only be determined by dissection and analysis. Most mummies have been considered as mere

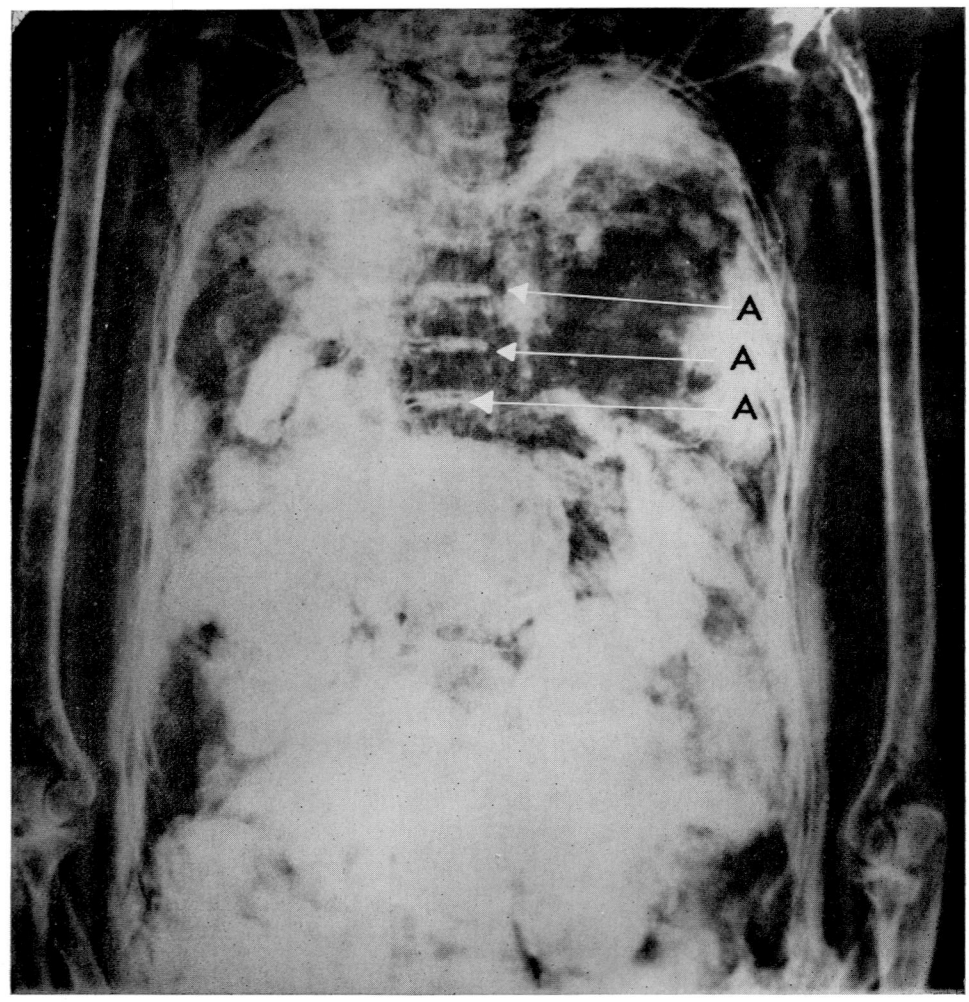

FIGURE 7. View of the thorax of a female mummy of Dyn. XXI to XXIV. Note the dense nature of the dorsal intervertebral discs (*A*). Note also the visceral packs, typical of this period. *N. B. This radiograph is reproduced by kind permission of C. V. A. Adams, Esq., of the Prince Albert Memorial Museum, Exeter.*

curios and far too few have been scientifically investigated. It is hoped that this paper will stimulate others to help achieve this end.

SUMMARY AND CONCLUSIONS

Over one hundred ancient Egyptian mummies have been subjected to a radiographic skeletal survey.

In sixty-four mummies of a known date and provenance, eighteen cases of disc calcification were observed. It is suggested that the dense shadows within the intervertebral discs of the mummies are not the result of alkaptonuria, but may represent a deposition of impurities derived from crude natron which was used in the embalming process.

Other items of medical and archaeological interest were observed in this survey,

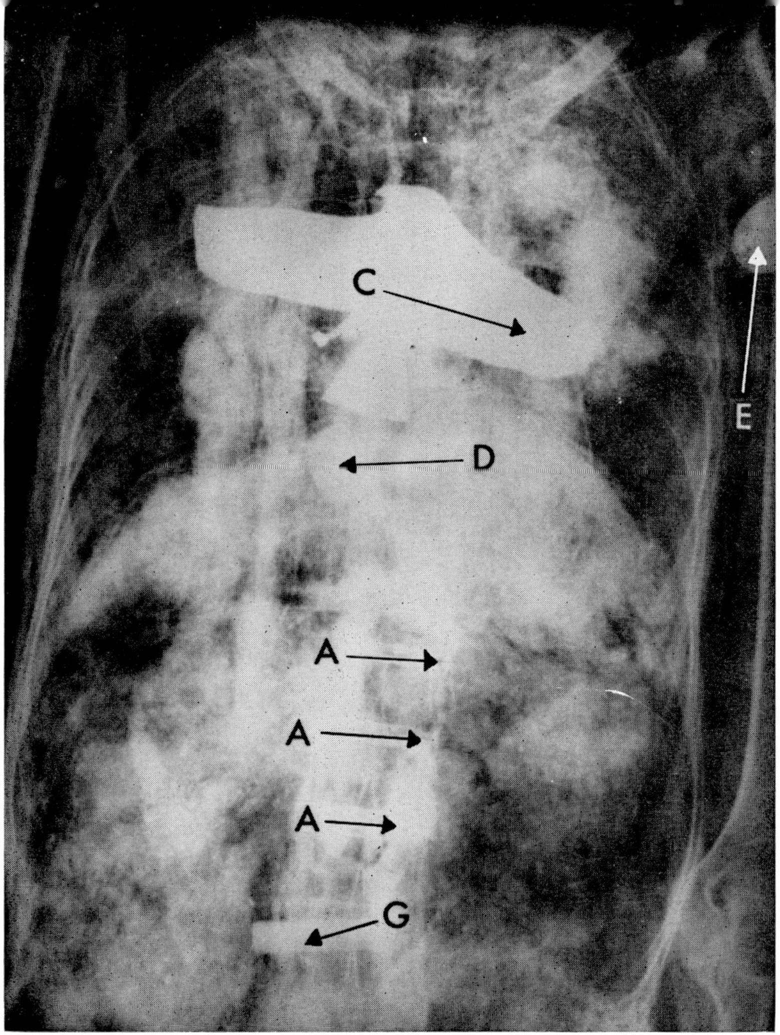

FIGURE 8. View of the thorax of a female mummy of the XXI Dynasty. Note the changes in the spine (A). The winged pectoral (C), heart scarab (D) and the linear amulet (G) are either on the body surface or within the innermost bandages. Note also the visceral packs, similar to Plate 7. The opacity (E) in the left axilla probably represents a *bennu* bird amulet. The mummy is British Museum Reg. No. 22939.

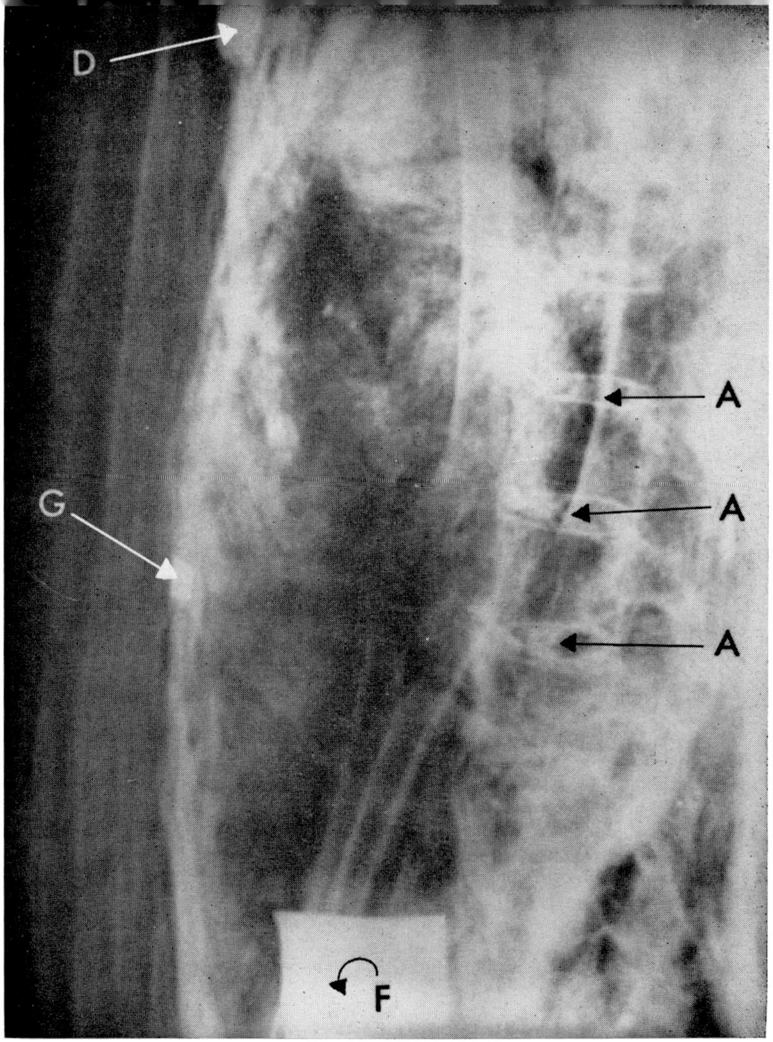

FIGURE 9. Lateral view of Figure 8. Note the changes (A) in the spine. The heart scarab (D) and the linear amulet (G) are again demonstrated. The rectangular opacity (F) is the flank plate. This covers the embalmer's incision in the left flank, and on it would be engraved the symbolic Eye.

and it is further suggested that if a radiographic study of more mummies of a known date is made, it is possible that the findings may lead to a harmless method of dating specimens, and furthermore, throw some light upon disease in antiquity.

ACKNOWLEDGMENTS

The writer would like to express his thanks to the following for allowing him to X-ray their collection of mummies:

Cyril Aldred, Esq., The Royal Scottish Museum, Edinburgh.
Dr. I. E. S. Edwards, Department of Egyptian Antiquities, British Museum, London.
A. Jewell, Esq., Haslemere Educational Museum, Haslemere.
Dr. O. W. Samson, The Horniman Museum, Forest Hill, London.
P. S. Rawson, Esq., Gulbenkian Museum of Oriental Art, Durham.
A. M. Tynan, Esq., The Hancock Museum, Newcastle-upon-tyne.

He would also like to express his thanks to the following for the use of their Dark Room:

Dr. J. Blewett, King's College Hospital, London.
Dr. G. T. Holroyd, Dryburn Hospital, Durham .
Dr. W. N. Thomson, City Hospital, Fairmilehead, Edinburgh.
Dr. C. K. Warrick, Royal Victoria Infirmary, Newcastle-upon-Tyne.

His thanks are also due to E. Elston Esq., of Messrs. Associated Electrical Industries for his invaluable help and loan of "a portable X-Ray unit" for part of the survey.

His special thanks must go to J. D. Dain Esq., and his Staff of Medical Sales Division, Kodak Limited, Kingsway, London, who most generously gave, and processed the majority of the radiographs.

REFERENCES

Adams, C. V. A., 1964: The songstress of Amen-re. *Medical News.* March 27.
Darlow, H. M., 1964: Surgeon Commander, Royal Navy. Microbiological Research Establishment, Porton, Wilts. *Personal Communication.*
Dawson, Warren R., 1964: *Personal Communication.*
Dent, C. E., 1964: *Personal Communication.*
Golding, F. C., 1959: Static and paralytic lesions: the intervertebral discs. *A Text Book of X-Ray Diagnosis,* Shanks, S. C. and Kerley P. (Eds.). London, Lewis.
Golding, F. C., 1960: Rare diseases of bone. *Modern Trends in Diagnostic Radiology.* 3rd Series, J. W. McLaren (Ed.) Butterworths, London.
Kerley, P., 1936: *Recent Advances in Radiology,* 2nd Edition. Churchill, London.
Köhler, A., 1935: *Röntgenology. The borderlands of the normal and early pathological in the skiagram,* 2nd Edition, Turnbull A. (Trans. and Ed). London, Baillière, Tindall and Cox.
Lauer, J. R. and Iskander, Z., 1955: Données nouvelles sur le momiefication. *Annales du Service,* 53:181.
Lucas, A., 1932: The use of natron in mummification. *J. Egypt. Archaeol.,* 18:125.
Lucas, A., 1960: *Ancient Egyptian Materials and Industries,* 4th Edition. London, Arnold.
Melnick, C., and Silverman, F. N., 1963: Intervertebral disc calcification in childhood. *Radiology,* 80:399.
Newton, T. H., 1958: Cervical intervertebral-disc calcification in children. *J. Bone Joint Surg.,* 40-A:107.
Sandison, A. T., 1963: The use of natron in mummification in ancient Egypt. *J. Near East. Studies,* 22:259.
Sandstroem, C., 1951: Calcifications of the intervertebral discs and the relationship between the various types of calcifications in the soft tissues of the body. *Acta Radiol. (Stockh.),* 36:217.
Simon, G. and Zorab, P. A., 1961: The radiographic changes in alkaptonuric arthritis. *Brit. J. Radiol.,* 34:384.
Smith, G. E., 1906: A contribution to the study of mummification in Egypt, with special reference to the methods adopted during the time of the XXIst Dyn. for moulding the form of the body. *Mem. Inst. Égypte.* T. V. Cairo.
Umber, F. and Buerger, M., 1913: Alkaptonurie mit Ochronose und Osteoarthritis deformans. *Deutsch. Med. Wschr.* 39:2337.
Walker, C. S., 1954: Calcification of the intervertebral discs in children. *J. Bone Joint Surg.,* 36-B:601.
Weens, H. S., 1945: Calcification of the intervertebral discs in childhood. *J. Pediat.,* 26:178.
Wells, C. and Maxwell, B. M., 1962: Alkaptonuria in an Egyptian mummy. *Brit. J. Radiol.,* 35:679.
Zorab, P. A., 1961: The Historical and prehistorical background of ankylosing spondylitis. *Proc. Roy. Soc. Med. (Hist. Sect.)* 54:415.
Zaki, A. and Iskander, Z., 1943: Materials and method used for mummifying the body of Amentefnekht, Saqqara, 1941. *Annales du Service,* 42:247.

Chapter 3

General Considerations of the Evidences of Pathological Conditions Found Among Fossil Animals

ROY L. MOODIE*

DEFINITION AND SCOPE OF PALAEOPATHOLOGY

The study of the evidences of disease among ancient man and fossil animals is known as *palaeopathology*, the term having first been applied by Sir Marc Armand Ruffer[1] in 1913[2] to methods he had developed in studying the pathological anatomy of the ancient Egyptian mummies. He defined it as follows:

Palaeopathology is the science of the diseases which can be demonstrated in human and animal remains of ancient times.

The significance of the term has been dwelt upon by Klebs,[3] and a further extension of its meaning to include, not only the diseases of the ancient Egyptians but those of prehistoric man and fossil animals as well, has been suggested by the writer.[4] The field thus involved includes the resources of anthropology and paleontology, as well as some details contained in archeological studies.

The present paper deals mainly with the so-called prehistoric,[5] and especially prehuman, evidences of disease (prior to 500,000 B.C.) of the extinct vertebrates. It is interesting to note that the history of disease, from the first geological evidences at present obtainable down to the historical data contained in August Hirsch's *Handbook of Geographical and Historical Pathology* (c. 600 B.C. to 1875 A.D.), will be seen as a series of consecutive events from the introduction of diseased conditions among animals and plants down to the present time. There can be no doubt that many of the diseases existing today are of very great antiquity, having a history extending back into geological time for many millions of years.

It is not necessary nor pertinent to review in this place the studies of Ruffer, Elliot Smith, Wood Jones, Rietti, Fouquet and other writers on the pathological anatomy of the ancient Egyptian mummies, since their results are so readily accessible. Their material might be regarded, from a certain standpoint, as *fossil*, meaning something 'dug up." The term *fossil*, however, as used in this paper refers to material which is thoroughly petrified, the age of which must be reckoned by geological standards. The studies of the above-mentioned writers have been briefly reviewed and summarized by Garrison,[6] Klebs,[7] and Sudhoff,[8] and will be extensively referred to elsewhere by the writer.

* It has proved difficult to select a publication to represent Moodie, because he produced more than one contribution which could be regarded as classic. In the end we have edited and reduced one of his more general articles as it fits particularly well in our Introductory section. The larger original study appeared in *Ann. Med. Hist.* 1:374, 1917 (Eds).

The studies of Ales Hrdlička and Langdon[9] on the pathological anatomy of the North American Indians, and of Hrdlička, Eaton[10] and other writers on the ancient Peruvians, must also be neglected, as well as the meager details of fossil man as they are recounted in the various works on anthropology. The subject of the diseases of ancient human races has never been systematically studied. The writer will present a consideration of this subject at some future time.

PALEONTOLOGICAL EVIDENCES OF DISEASE

The study of palaeopathology is still in its initial stages, and especially is the application of pathological methods to fossil lesions a new field. But the comparative scantiness of facts so far brought out and the difficulties of research should not hinder its successful prosecution. What the final results may be remains to be seen. The immediate results are certain to bring attention to the presence of characteristic lesions of disease far back in geological time, and it is very interesting, if not important, to find in past geological ages evidences of pathological processes which are so familiar to us today. If we can trace the known lesions to any definite cause among the extinct animals it will be a step toward the erection of the newest branch of pathology, dealing with the oldest aspects of that science.

In regard to the importance of this branch of study, Klebs[11] says:

We need only consider what definite influence diseases exert in our individual lives, what profound social upheavals were brought about through the incidence of epidemics, less perceptibly perhaps but none the less strongly, through wide-spread chronic ailments, through professional diseases, how whole districts and countries are forsaken because disease made them uninhabitable, how disease affecting early childhood and others producing sterility led to the gradual extinction of whole peoples. . . . For the grasp of such problems, the study of disease as it appears to us now does not suffice; the traces left during immense periods of time have to be taken into account and it is in just such questions, not approachable by other methods, that palaeopathology in time to come may furnish important solutions.

The attitude of students of paleontology toward this subject has been negative. Even men like Leidy, a trained anatomist and an eminent medical man, paid scant attention to the subject, although he did describe an example of caries in a mastodon tooth[12] from Florida. Cuvier too, eminent as he was in the field of comparative anatomy, failed to recognize the importance of this phase of paleontology. His discussions of the few lesions he recognized were meager and inadequate. He has described a fractured skull of a Pleistocene *Hyaena* and a fractured femur of *Anoplotherium*.

Paleontology lends considerable light to the study of the antiquity of disease. The study of the lesions so far known among fossil animals indicates nothing new in the nature of pathological processes but simply extends our knowledge of disease to a vastly earlier period than had previously been known. It seems quite probable that some of the diseases exhibited by the extinct vertebrates went out of existence with the race of animals which were afflicted. If this proves to be true it will be an interesting opportunity to study the details of lesions of extinct diseases. There seems to be little possibility of determining the fundamental cause of disease other than is already known; for disease is apparently one of the manifestations of life, and has followed the same lines of evolution as have plants and animals, and is possibly directed by the same factors. Such a study as the present may, however, throw light on the origin of many of the diseases to which the human race is a prey. A knowledge of the pathological processes which have taken place in animals of geological antiquity will aid in an understanding of the general nature of disease.

The literature of vertebrate paleontology contains a number of incidental references to the diseased nature of the fossilized bones of fishes, reptiles, birds, and mammals, the lesions described indicating a variety of diseases, some of which are not uncommon today. It is manifestly impossible to diagnose correctly, on the basis of our modern knowledge of recent diseases, all of the lesions which are preserved in a fossil condition. In the extinction of the ancient races of animals, certain diseases, without doubt, became extinct with them,[13] and it is partly the purpose of this paper to inaugurate an inquiry into the nature of the diseases of fossil vertebrates. No one has yet made a study of the evidences of disease among fossil animals, since these conditions, whenever noted, have been referred to only in an incidental way, by writers on paleontological subjects.

Geological evidences of the diseased state of animals are necessarily restricted to pathological lesions on the hard parts of fossil animal remains. Soft parts[14] are seldom fossilized, and the few specimens known have not been subjected to disease. Since the pathological changes which affect the hard parts of animals today are relatively few when compared to the diseases which afflict the body as a whole, it is to be supposed that the paleontological evidences of disease are but partial indications of the prevalence of pathological conditions in geological time. The following account, too, must be read in the light of the paucity of evidence available for discussion. The details are meager, but since they are all we have, they may be deemed worthy of consideration.

It will be clearly evident, after a consideration of geological matters, that all paleontological evidence is of relative value, since such small portions of the ancient faunas and flora are preserved in the rocks. However, we are safe in stating, from such evidence as we have, the probabilities of the occurrence of numerous diseases among extinct animals, just as it is safe for us to state, on the basis of a single tooth in a definite geological horizon, that such and such an animal existed at the time the formation was being deposited, provided, of course, the deposit is a primary one and the fossil was not moved by shifting in a secondary deposition.

All that we know of the earliest land vertebrates, prior to the Pennsylvanian, for instance, is a single footprint from the Devonian, and a few series of footprints from the Mississippian. On the basis of these footprints we are able to say definitely that there existed in North America a diversified fauna of vertebrates, probably amphibian, which preceded the well-known amphibian faunas of the great Coal Period.

DEFINITION OF DISEASE AS USED IN THIS STUDY

Disease, as the term is used in this study, may be defined as any deviation from the healthy or normal state of the body which has left a visible impress upon the fossilized skeleton. The evidence may take the form of broken bones, tumors, necroses, hyperplasias and arthritides of various kinds. Only the diseases of animals have been considered. This is done with a full realization of the enormous domain of phytopathology and is a confession of a limitation to a restricted field. Some of the paleobotanical literature has been read, but apparently no attempt has been made to trace the rise and progress of phytopathology from fossil material.

This is doubtless due to the unsatisfactory condition of fossil plant material which is usually quite fragmentary. Some idea of the nature of plant diseases of the past may be had from the following brief summary[15] for which I am indebted to Professor Edward W. Berry:

Bacterial and fungus activity are known in Carboniferous plants, and would probably be detectable at much earlier horizons if petrified material of greater age were available for study, since the bacteria appear to be the earliest forms of life. Material preserved as impressions at all horizons, more especially the post-Paleozoic ones, show abundant leaf-spot fungi, and such remains from the Cretaceous and Tertiary show abundant insect galls and leaf cutting by caterpillars or bees; but this class of material is usually more or less indefinite. Whenever one handles much petrified material, one is struck with traces of fungal ravages and bacterial action.

EVIDENCES OF DISEASE IN FOSSIL PLANTS

It is often difficult to decide whether the ravages of fungi and bacteria are pre-mortem or post-mortem. The agents of decay are well known to have existed early in geological time. During the Carboniferous there existed conditions which were especially favorable to the growth of a mycological flora, and much of it was probably on dead plant material.

Professor Berry writes further[16] concerning the primitive fungi:

Among the relics of former vegetation that carry the record back many millions of years the remains of fungi are so rarely found that their presence is always exceptional, although it is obvious that many times during the long history of the earth the environment has offered optimum conditions for their abundant development. To mention but one such occasion, that of the formation of the Coal Measures must have witnessed an exceedingly abundant mycological flora. That these plants were present thus early is indicated by the abundance of hyphae, and other traces of fungal activity such as butyric fermentation, in the tissues of Carboniferous vascular plants, and the scarcity of described forms must be attributed to the perishable nature of most fungal tissues and to the lack of systematic work by experienced mycologists on the more or less obscure material available. To be sure, quite a considerable number of fossil forms referred to Fungi have been recorded from various geologic horizons but the vast majority of these are leaf-spot types based upon real or fancied resemblances, and found on impressions of foliage and without definite botanical characters. Some doubtless represent fungal ravages, others are due to insects, some are glandular, and others are purely imaginary.

Professor Berry refers to A. Meschinelli's *Fungorum Fossilium Omnium Iconographia* (1902, 144 pp., 31 plates), for a rather complete illustrated list of all of the forms referred to down to the year 1900. Other and more complete studies on the bacteria and fungi of the Coal Measures of France particularly have been made by Van Tieghem and Renault. A fairly complete list of their numerous papers is to be found in Smith's bibliography.[17] Other information may be gleaned from the memoirs and textbooks dealing with Paleobotany.

METCHNIKOFF ON DISEASES IN REMOTE EPOCHS

The possible presence of disease among animals of remote epochs of the earth's history was suggested by Élie Metchnikoff in the following words:

Diseases in general and infective diseases in particular were developed on the earth at a very remote epoch. Far from being peculiar to man, animals and the higher plants, they attack inferior forms and are widely distributed among unicellular organisms, Infusoria and Algae. Diseases undoubtedly play an important *rôle* in the history of life on our planet, and it is very probable that they have contributed in a marked degree to the extinction of certain species.[18] When we observe the ravages produced by parasitic Fungi among the young fish which we are trying to rear, or the destruction of cray-fish in certain countries in consequence of the rapid increase of epizoötic germs, we are involuntarily led to the conclusion that pathogenic micro-organisms must have brought about the disappearance of certain animal and vegetable species.[19]

It would be interesting in this connection to know Metchnikoff's sources of information relative to the presence of diseases at remote epochs. Virchow's studies on the cave bears of Europe[20] were well known, and his characterization of the arthritic lesions of the fossorial animals as the "Höhlengicht," was certainly famous at the time Metchnikoff wrote. The studies of Mayer[21] on the lesions of the cave bears and cave lions of Europe as well as the writings of Cuvier (1820), Esper (1774), G o l d f u s s (1810), Walther (1825), Schmerling (1835), Owen (1842), Schaafhausen (1858), Newton and Parker (1870), Graff (1885) and Leidy (1886) may, any or all of them, have been known to Metchnikoff. They all suggest the pathology of ancient times and some deal entirely with the pathological anatomy of fossil remains. None, however, are studies which deal with remains older than the middle Tertiary, and to a paleontologist the term "remote epoch" hardly applies, when compared to lesions known from the Carboniferous, for example. I am, therefore, forced to conclude that Metchnikoff simply forecast what would be discovered, on the basis of what he knew in modern plants and animals. All of the literature in paleontology dealing with pathological evidences of any great antiquity, prior to the mid-Tertiary, has appeared since 1900. The literature is meager and unsatisfactory. Paleontological studies seldom deal specifically with diseased conditions, so that the field is still to be explored. The studies in paleontology dealing with pathological evidences among fossil vertebrates have been reviewed by Abel,[22] and a much fuller review is planned for a memoir on paleopathology, now in preparation.

APPARENT ABSENCE OF DISEASE AMONG EARLY PALEOZOIC ANIMALS

A careful study of the literature of paleontology shows that, so far as present observations are concerned, the animals of the earlier periods of the earth's history were free from disease. Although bacteria[23] are known to have occurred in the algal deposits of the Newland limestone, a formation of the Beltian series of Algonkian rocks in central Montana, they are not known to have been of a pathogenic nature, but rather are supposed to have been active in the deposition of limestones, together with the algae with which they were associated. No lesions due to accident or to infection have been described among either the vertebrates or invertebrates of the early geological periods, prior to the Carboniferous. This lack of knowledge concerning the evidence of disease may be due to ignorance on our part, for the lesions may have been seen but were not deemed worthy of description. Or, it may be due to the fact that the invertebrates of the Proterozoic and Paleozoic, which were the predominant types of animal life during these periods, were free from disease which affected the skeleton, as are, in general, the invertebrates of today, although many of the recent forms are highly parasitized and are occasionally subject to disease. It appears probable that vertebrates have been more liable to diseases which afflict the hard parts than have the invertebrates, and this liability to pathologic processes has been increased with the passage of geologic time.

IMMUNITY IN MODERN INVERTEBRATES

The greater immunity of early Paleozoic animals to disease, based on the evidences of paleontological material, is probably not a true index to actual conditions, though it may be so. It is probably not safe to conclude from present-day conditions what the state of Paleozoic animals may have been as regards disease. At any rate the paleontological evidences are not wholly substantiated by conditions found in modern forms. Metchnikoff[24] has called attention to the occurrence of epidemics of a severe nature among protozoa, such as diseases in *Amoebae* caused by the *Microsphaera* and the disease in *Actinophrys* attributed to Fungi allied to the genus *Pythium*. Pasteur's studies on the *pébrine* and *flâcherie* of the silkworms will be remembered as instances of severe epidemics in an invertebrate group. Molluscs, however, are apparently largely immune to infection, and since the molluscous animals formed such a large percentage of the preserved faunas of the early periods of the earth's history we may attribute our ignorance of the presence of disease to this factor, in part at least. The immunity of many intermediate hosts to infection[25] is well known, and the classical example of the mosquito-borne infections will suffice, although it is well known that insects of many kinds are subject to fatal diseases. Kowalevsky has discussed the anthrax of crickets and many other students have studied the problem. The entire question of immunity in its relation to all forms of extinct animals is of course a new and unsolved, probably an insolvable, problem. But it seems certain that if the early animals were diseased, the ensuing pathology was of such a nature as to leave no impress upon the fossilized part; or else we have not yet learned to recognize these lesions.

Phagocytosis[26] doubtless began very early in the history of animal life, and it is probable that the natural immunity of the early animals was sufficiently strong to resist the invasion by any pathogenic organisms in sufficient numbers to produce disease. The breaking down of this immunity may possibly be correlated with the development of senescence[27] among the early races of animals, which reached a climax in some forms—the trilobites, for instance,—at about the time when we find the first indications of disease among fossil animals. The breaking down of the immunity, due to the development of race senescence and the introduction of disease, doubtless was of very great importance in the extinction of the trilobites and other great groups of animals which have disappeared from the earth.[28]

I do not intend to assert that senility or senescence is a disease, but that age weakens the organism and the race and allows the ingress of disease. Minot has stated:

> Old age is not a disease and cannot be cured; it is an accumulation of changes which begin during earliest youth and continue throughout the entire life of the individual.

It may be said that disease in the past has often attacked the races of animals which showed senescence. Many of the virile races of animals in the past were also subject to disease. The paleontological indications of senescence are the reduction in size, the loss of vigor and the production of apparently useless spines as seen in the races of animals which have become reduced or extinct, such as the crinoids, trilobites, brachiopods, ammonites and the dinosaurs. Other examples of senescence may be seen among some of the Permian reptiles which assumed bizarre forms. The tendency of many races of animals to acquire spinous and other useless excrescences of the hard parts shortly before the extinction of the group is noteworthy, and this tendency has been regarded by paleontologists as an indication of senescence.

LESIONS OF PARASITISM IN CARBONIFEROUS CRINOIDS

Our knowledge of the history of disease, as it is based on paleontological evidence, begins with the Carboniferous, when certain crinoids were afflicted in their stems with tumor-like lesions, possibly due to the parasitic action of myzostomids such as commonly attack crinoid stems today. A careful description of the enlarged stems of recent crinoids and the parastic action of the myzostomids is to be found in the reports of the Challenger Exploring Expedition. A comparison of the ancient and recent lesions on the stems of crinoids leads one to accept the enlargements of fossil crinoid stems as due to the parasitic action of the myzostomids or some similar form.

The evidences for such a conclusion are, apparently, incontrovertible, and have been established by a number of writers on fossil crinoids. Parasitized crinoid stems are known from the Carboniferous of Scotland, Germany and the Keokuk beds of North America. Graff[29] found the carbonized remains of the parasite in one of the enlargements which he studied and which he referred to as the fossilized integument of the myzostomid. The presence of this soft-bodied animal so early in the geological history of the world is not surprising, since from the researches of Walcott[30] we know that jellyfishes, sea cucumbers, many types of annulates, and soft-bodied crustaceans lived during the Cambrian, many millions of years earlier. The parasitism of animals during the Carboniferous was preceded by partial parasitism or commensalism of the earlier periods, and is known to have occurred among fossil corals of the Devonian. The intimate association of animals and the origin of parasitism and commensalism during the early part of the Paleozoic has been studied by Clarke.[31] The reader is referred to his paper for further details.

The remains of the early vertebrates prior to the Permian have shown no noteworthy pathological lesions. There may have been disease among these early forms, but the lesions have not yet been discovered. We find, to be sure, certain laterally compressed fishes preserved in the attitude of the opisthotonos and pleurothotonos in horizons prior to the Permian. These attitudes may have been due to spastic distress induced by cerebrospinal infections or to some form of poisoning. Since this subject will be more fully treated elsewhere nothing more need be said than that these attitudes possibly represent diseased conditions of the central nervous system.

PATHOLOGY OF THE PERMIAN VERTEBRATES

Several pathological conditions are indicated among the vertebrates of the Permian. Renault[32] has described caries of certain fish bones preserved in coprolites from the Autun basin. He concludes that this type of caries is due to several types of bacteria which he has described and figured. A left radius of *Dimetrodon*, a primitive reptile, from the Permian of Texas shows an incompletely healed fracture with abundant osteosclerosis and some intermediary callus. *This is the oldest known case of fracture.* It was a simple fracture cutting the bone at right angles, and the healing process has taken place with very little shortening. The bone has no medullary cavity, so that attempts to study the nature of the fracture by means of the *x*-ray have been a failure. The Texas red beds, from which the bone comes, are impregnated with iron, and the radius reacts to the *x*-rays much as a bar of iron would. A fractured rib with an old callus is also known from the Permian of Texas. A description of this lesion with illustrations is to be found in *The Surgical Clinics of Chicago*, April, 1918. Von Huene has described the skull of a phytosaur from the Triassic of Germany, showing a fractured snout with many necrotic sinuses.

DISCUSSION OF GRAPH SHOWING INCREASE OF DISEASE IN GEOLOGICAL TIME

It is not necessary at this time to go into further details concerning the progress of disease, since the details are to be given in a later paper. The accompanying graph (Fig. 1) will show how, according to present evidences, disease has progressed during the geological history of the earth. The twenty-five divisions on the base line a-d (Fig. 1) represent as many periods of the earth's history. The divisions on the vertical line d-b represent the approximate number of diseases present in each period, as indicated by the known fossil lesions. The time intervals in the graph are shown as of equal value, but the geological periods are not at all of equal duration nor of equal character. This should be kept in mind in studying the graph.

At the point "a" we may say that organic life is first known. It will be seen that the line "a-b," representing the history of disease, follows a base level for the first twelve periods of the earth's history. Then the curve gradually rises until, during the Cretaceous, at "c," diseases and accidents —such as caries, osteoperiostitis, deforming arthritides, necroses, hyperostosis, osteophytes, osteomata, fractures—and many infective processes, reached a maximum of development among the dinosaurs, mosasaurs, crocodiles, plesiosaurs, and turtles.

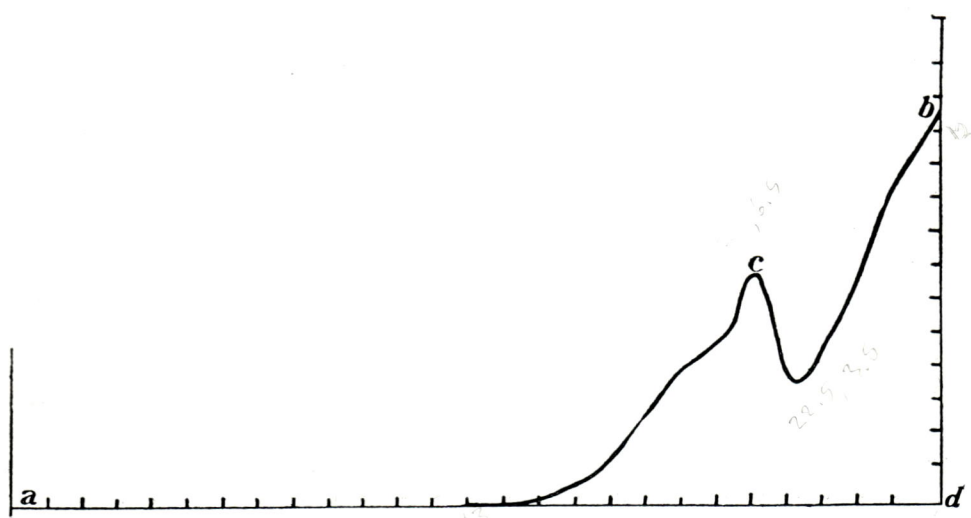

FIGURE 1. Graph showing increase of disease in geological time. The divisions on the base line, *a-d*, represent geological periods, and on the vertical line prevalence of pathological conditions.

The curve suddenly and sharply descends from "c." For with the close of the Cretaceous and the sudden extinction of large groups of the giant reptiles, the incidence of disease also decreased. It seems quite probable that many of the diseases which afflicted the dinosaurs and their associates became extinct with them.

The mammals of the Cretaceous and early Tertiary periods do not seem to have been so generally afflicted with disease as were the preceding groups of giant reptiles, nor as were the later mammals. The ascending curve therefore is not so abrupt as one might expect. Certain processes of disease seem to have been acquired by the mammals from preceding forms, for *caries* and other primitive diseases are evident among early Tertiary mammals. The curve rises rapidly, however, and reaches the highest point at "b," indicating that disease is much more prevalent at the present time than ever before in the history of the world.

The geological development of disease has certain curious characteristics which parallel facts in the evolution of animals and plants. Huxley many years ago called attention to certain persistent types of animals which had existed almost unchanged from early geological periods down to the present. Among the known diseases of geological antiquity a few can certainly be called persistent or primitive types which have remained the same since the close of the Paleozoic. Other diseases arose and became extinct, but some of them have retained the same characteristics, as seen in the resulting changes of structure.

According to present evidences, disease is, from the geological standpoint, of relatively recent origin and has afflicted the inhabitants of the earth for only the last one-quarter of the earth's history. Future discoveries will doubtless modify our present conceptions, but the above outline is a summary of our present knowledge of the rise and development of disease among animals.

TABULATION OF GEOLOGICAL EVIDENCES

The table given below will show at a glance the antiquity of pathological evidences in geological history. The estimates of time are based upon the relative thickness of the pre-Cambrian and post-Cambrian rocks, after Walcott and Schuchert, as given by Osborn in his *Origin and Evolution of Life*.[33] The estimates of the duration of the geological periods vary greatly. The duration of the Proterozoic was as great, probably, as all post-Cambrian time, which has been estimated as high as 500,000,000 years. A study of radioactive substances gives estimates as high as 3,000,000,000 years for the duration of the Archeozoic.

FOSSIL PATHOLOGICAL LESIONS

The following annotated list and illustrations of fossil lesions will indicate the extent of diseases among fossil vertebrates. The study of these lesions is by no means complete, and other pathological processes will doubtless be indicated as the study of them progresses.

CARIES is very common among fossil vertebrates and has been described by Renault as occurring among Permian fishes. A large marine reptile, from Belgium, one of the Cretaceous mosasaurs, according to Abel, shows in the left mandibular ramus extensive evidences, of the ravages of this disease. In an early Tertiary species of the three-toed horse, the mandible has been affected by caries and possibly also by actinomycosis, as well

GEOLOGICAL EVIDENCES OF PALAEOPATHOLOGY*

Age in Millions of Years	Eras	Geological Periods	Chief Animal Groups	Evidences of Pathology
	CENOZOIC	Quaternary	Age of Man	Abundant lesions on fossil and sub-fossil human remains
		Tertiary	Age of Mammals	Numerous diseases represented on animal remains from the deposits of the period
70				
	MESOZOIC	Cretaceous	Age of Reptiles	Lesions on the bones of mosasaurs, dinosaurs, plesiosaurs, turtles, crocodiles, phytosaurs and other reptiles representing diseases similar to the modern forms of periostitis, hemangioma, necrosis, caries, pyorrhea alveolaris, arthritides, fracture with callus, pachyostosis, osteoma, opisthotonos and other lesions which cannot be interpreted.
140		Comanchian		
		Jurassic		
		Triassic		
195				
220	PALEOZOIC	Permian	Age of Amphibians	The lesions known represent dental caries, pyorrhea alveolaris, fracture, callus and parasitism. These periods witnessed the beginnings of disease. Bacteria and fungi were abundant.
		Pennsylvanian		
275		Mississippian		
		Devonian	Age of Fishes	No evidences of disease are known from these periods. Beginning of dependent life.
		Silurian		
350		Ordovician	Age of Invertebrates	
520		Cambrian		
	PROTEROZOIC	Keweenawan	First known fossils	Bacteria (non-pathogenic)
		Animikian		
		Huronian		
		Algomian		
		Sudburian		
	ARCHEOZOIC	Laurentian	No life known	
		Paleolaurentian		
3000				

* Age estimates of strata are recent ones and not those originally given by Moodie. (Eds).

as some necrotic process which has resulted in the exposure of the roots of the teeth and the absorption of the alveolar margins, similar to the results of pyorrhea alveolaris. Caries has been noted also in the tooth of a mastodon, and in the early cave bears of Europe. The early races of men were singularly free from this disease as evidenced by the fossil remains.

PYORRHEA ALVEOLARIS, or some similar pathologic process, is especially evident in the absorbed alveolar margins and in the loosened teeth of a three-toed horse from the Miocene of North America. It is also extensively indicated in the mandibles of the European cave bears, and in a Cretaceous mosasaur from France.

DEFORMING ARTHRITIDES are fairly common among fossil vertebrates and indicate a variety of pathologic conditions. Arthritides are especially common in Pleistocene mammals. The most prominent case of a deformed joint is the case of two caudal vertebrae of a large dinosaur, the interarticular surfaces of which have been extensively deformed and surrounded by a huge exostosial growth.

one of the vertebrae measures 27 cm, and the same measurement around the middle of the tumor-like mass is 38.5 cm. The lesion has involved a length of 12 cm. Its surface generally is rather deeply pitted. There is an unusual ventral growth, which is shown in its normal condition in Figure 2 at "A." This bony process, "the chevron," which served to protect the caudal vein and artery, is commonly present in the tail of these reptiles. The growth of the diseased portion is unequal and has involved more of the vertebrae on one side than on the other; likewise, the

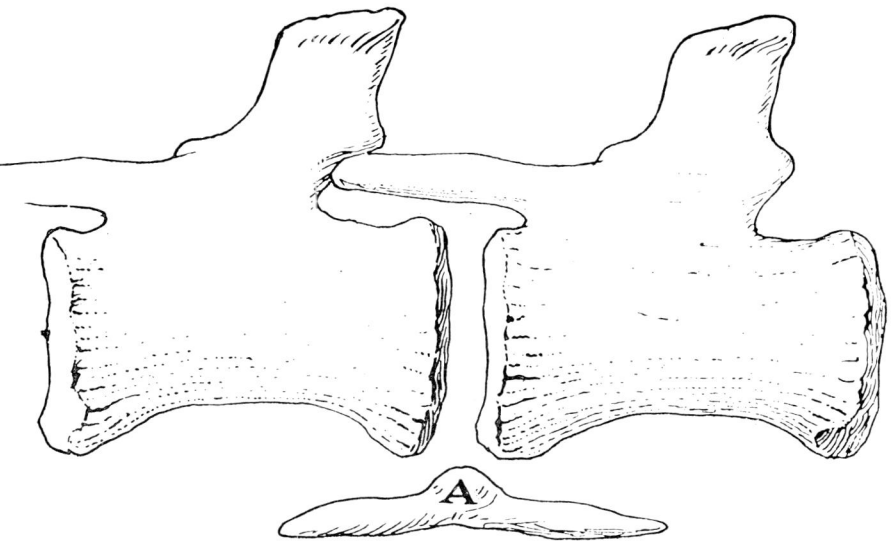

FIGURE 2. Outline sketch showing normal appearance of the two vertebrae, based on *Diplodocus* and *Apatosaurus*. A=chevron. This process in the pathological specimen has been shoved far ventralward and involved in the tumor-like mass.

The mass resembles closely the tumor-like masses seen on oak trees. It entirely encircles the vertebrae and has involved fully half of the two bones. All evidences of separate structures are obliterated, and the two vertebrae are fused into a single mass. The specimen has a length of 26.5 cm and a weight of 5.1 kg. The circumference of the normal articular surface of

growth has attained greater lateral dimensions on one side.

The enlargement is somewhat suggestive of the lesion of chronic osteomyelitis. It may be a callous growth, possibly due to a fracture of the caudal vertebrae; or it may be a bone tumor. The character of the lesion is naturally problematic, but it is interesting that pathological growths in the early geological periods so closely resemble the lesions of today. Section of the tumor mass shows the presence of numerous vascular spaces, so that in this respect it resembles a haemangioma. Microscopic study of the periphery shows the presence of well-developed Haversian systems of osseous lamellae.

The bones exhibiting these interesting indications of Mesozoic pathology are the caudal vertebrae of a huge land reptile, one of the sauropodous dinosaurs, possibly *Apatosaurus*. The sauropodous dinosaurs were the most gigantic of all land vertebrates, although not nearly so large as some of the modern whales. The largest of these reptiles attained a length of nearly seventy feet and an estimated weight of thirty-nine tons. The head was approximately the size of that of a modern draft horse and the contained brain was no larger than one's fist. The lumbar intumesence, however, was ten times the size of the cephalic portion of the nervous system, or at least the subdural space was. Whether the nervous material filled the entire cavity or not is unknown. The animals lived, possibly, in the swamps and low-lying rivers, feeding on the succulent vegetation, and are said to have been capable of attaining the ripe age of 1,000 years. Diseases are rarely seen on fossil dinosaur bones, in spite of the great abundance of their remains.

The tail in some of these large animals was very long and slender, and it may have been used in swimming, as a muskrat uses his today. The terminal caudals in some species were reduced to mere slender rods of bone, so that a fracture or injury of any kind in this region could easily occur. Aside from possible blows from the head, the dinosaur to which the above described vertebrae belonged was entirely defenseless. The tail, for example, might be seized by one of the carnivorous dinosaurs and vigorously chewed for some time before the owner of the tail was able to turn his huge body and knock the offender away.

Lesions of similar nature, but not so well developed, are known to occur in the tail of *Cetiosaurus leedsi*, an English dinosaur; and Hatcher has described the same lesions in the tail of *Diplodocus*. A fuller discussion of these lesions is reserved for another time.

The nature of the above-described lesion is such that it may have been due to bacterial activity, and suggests, at any rate, the presence of pathogenic bacteria in the early part of the Cretaceous period. Bacteria and primitive fungi have, indeed, been described from much older periods. The best account of their occurrence is contained in "Microorganismes des combustibles fossiles," by B. Renault.[34] Renault has described and figured many forms of bacteria and fungi in the fossilized feces (coprolites) of fishes, in fossil wood and in coal. He has also discovered in the teeth of some ancient fishes what he regards as indications of the activity of organisms which have produced results similar to *caries*. He shows in one of his plates photomicrographs of fossil bone from the petrified feces in which the ravages of the bacteria, *Micrococcus* are evident in the canaliculi and the bone corpuscles, which appear in various stages of destruction.

Other deforming arthritides are represented by the arthritic condition sometimes spoken of as rheumatoid arthritis which has been noted by Virchow in the cave bears, by other observers in certain fossil human skeletons, in the famous Lansing man of Kansas,[35] and it is probably indicated in the Cretaceous mosasaurs, where a well-developed osteoma accompanied the arthritic inflammation.

OSTEOMYELITIS is probably indicated in the dinosaurian caudals and in certain phalangeal elements of a giant wolf from the Pleistocene of California.

EXOSTOSES due to trauma, indicated as callous growths around fractures of ribs and limb bones, or as outgrowths due to chronic irritation or infection, are fairly common among fossil vertebrates. Healed fractures are very common among mammals and are occasionally seen among fossil reptiles. Dinosaurs exhibiting broken ribs, vertebrae, and horn cores attest the accidents or fights which caused these traumatic conditions, and have led Abel to infer that the males of these animals contested during the breeding season for the female. An exostosis which is especially clearly marked is evident on the inner or visceral surface of a dinosaur scapula, where it takes the form of a hook-like process, evidently due to chronic irritation. An exact duplicate of this lesion may be seen on a recent human femur. One of the most perfect exostoses is seen in a mosasaur from the Cretaceous of Kansas where there is a decided lump at the articular surface between the third and fourth dorsal vertebrae, resulting in what is probably the only known fossil osteoma. Curious exostoses which are bilaterally symmetrical occur on the radii of an Oligocene dog, the skeleton of which is in the Carnegie Museum of Pittsburgh.

OSTEOSARCOMATA have not been positively identified among extinct animals, but the condition is suggested in several instances. Esper, in 1774, described what he thought was an osteosarcoma in the femur of a cave bear, but Mayer, who studied the specimen later, suggested that it might have been a fracture with callus and necrosis.

FISTULAE are evident in the lower jaw of an ancient and primitive whale from the Eocene of Egypt, and an enlargement of the mandible of a three-toed horse from the Miocene of North America indicates the presence of a fistula, possibly due to actinomycosis, in its early stages. Dental fistulae are occasionally seen among the known remains of fossil man, often resulting in the loss of teeth.

RICKETS is indicated, according to Abel, among the apes which are found mummified in the old Egyptian graves.

NECROSES, due possibly to a variety of causes, and attributed by certain French writers to tuberculosis, are fairly common among fossil vertebrates. A marked necrosis of the ilium of a large dinosaur, accompanied by expansion and thickening of the bone, is evident in the mounted skeleton of *Camptosaurus* on exhibition at the National Museum in Washington. A mosasaur bone from the Cretaceous of Kansas and certain crocodile limb bones from the Jurassic of England show lesions of a necrotic nature. The assignment of any of the lesions to a definite cause is manifestly impossible, and while tuberculosis has been suggested as a possible cause, the diagnosis is so uncertain as to be nearly worthless. In the crocodile skeleton, above referred to, there is abundant evidence that the infection, the focus of which was in the pelvis, was carried by metastasis to the bones of the palate which were also involved, as well as other parts of the body.

HYPEROSTOSIS OR PACHYOSTOSIS, which is similar to the enlargement of the bones in gigantism, is indicated as thickened and enlarged portions of the skeleton. This condition has been detected in certain fossil Paleozoic fishes and Mesozoic reptiles, some of them of great geological antiquity. A genus of fossil whales, known as *Pachycanthus*, has the neural, vertebral spines very greatly enlarged and swollen.

A similar condition is seen in the skeleton of a Triassic nothosaur. We are not justified in stating on these evidences the presence of pituitary disturbances in ancient animals, but further studies in this line may add very interesting data.

OSTEOPERIOSTITIS or some similar disturbance is the result seen in the arm bones of a mosasaur from the Cretaceous of Kansas. The articular surfaces are very greatly roughened and the surfaces of the bones are covered with smooth, somewhat flattened excrescences, possibly due to a subperiosteal irritation. The lesions have been observed in no other instances, so that no comparative statements can be made. Microscopic study of the peripheral lesions reveals many interesting histological details. One area shows typical osteoid tissue, similar in all essential respects to osteoid tissue developed in a human humerus in a case of osteomyelitis. Other areas show perforating fibers of Sharpey, as seen in the dark bundles, and the nature of the osseous lacunae. The whole section is filled with vascular spaces. An especially large one, filled with calcite crystals, is seen in the upper portion of the picture. There are no apparent Haversian systems or canals. Whether this is due to the pathology of the bone or whether it is an occurrence in normal bone of the mosasaurs will be determined later by microscopic study of the normal tissues.

OPISTHOTONOS and the allied phenomena, pleurothotonos and emprosthotonos, are quite frequently seen among fossil vertebrates. It has been suggested elsewhere that these attitudes represent possible cerebrospinal infections or other neurotoxic conditions, and they must be considered in connection with the study of disease among fossil animals.[36] The skeleton of the small dinosaur, *Struthiomimus altus*, described by Osborn,[37] shows a very well-developed condition of opisthotonos, with the the head thrown sharply back, the tail strongly flexed, and the toes contracted and appressed. The whole attitude strongly suggests a spastic distress, possibly brought on by some form of poisoning of the central nervous system, from infection or the deglutition of some poisonous substance.

OSTEOMALACIA is evidently the cause of the hypertrophy of the bones of *Limnocyon potens,* an early carnivore from the Washakie Eocene of Wyoming, nearly 70,000,000 years old.

MATERIALS AND METHODS

The material described in the present paper has been loaned the writer for description by the Field Museum of Chicago, by the American Museum of Natural History of New York City, by Walker Museum of the University of Chicago, and by the University of Kansas Natural History Museum. A beautiful specimen of an osteoma, the only one known so far, on the vertebra of a Kansas Cretaceous mosasaur, was given the writer by Dr. J. M. Armstrong of St. Paul. The writer expresses his obligations to the gentlemen connected with the above-mentioned institutions and to Dr. Armstrong.

The methods used are a combination of procedures in the various lines involved. Microscopic sections, which can be made thin enough for immersion lens study, are made by the well-known petrographic methods so common in all geological laboratories. The diagnoses, where they are attempted, are made from comparisons of the material with similar lesions in recent human material; but strict diagnosis has not been attempted. We must have some name for the lesions, so the terms used must be regarded as suggestive rather than an accurate statement of conditions. The interpretation of the lesions in the fossil material is a matter of experience with fossil remains. The author feels that twelve years experience in the study of fossils should be sufficient to avoid most of the usual pitfalls.

NOTES

1. Sir Marc Armand Ruffer, distinguished for his work in preventive medicine and for his studies on the pathology of the Egyptian mummies, lost his life while engaged in Red Cross work. See: "Memorial Notice of Sir Marc Armand Ruffer," by F. H. GARRISON, *Ann. Med. Hist.* 1917, vol. i, No. 2, pp. 218-220, with portrait.

2. RUFFER: Studies in palæopathology in Egypt, *J. Path. Bact.* 1913, vol. 18, p. 149.

3. KLEBS: *Bull. Hopkins Hosp.* 1917, vol. 28, pp. 261-266.

4. MOODIE: *Amer. J. Sc.* 1916, vol. 41, pp. 530-531; *Science*, N. S. 1916, vol. xliii, p. 425

5. The term *prehistoric*, of course, usually refers to events prior to the details of recorded human history, and is variously designated according to the region under discussion. Thus in Egypt any grave earlier than the time of the first dynasty is often called prehistoric. This implies an age of 6,000 years or more. In France LeBaron defines the prehistoric period as closing at about 222 B.C., and several centuries later in Algeria. To the paleontologist the term is meaningless. Klebs has said: "The adjective 'prehistoric,' used so often, would seem a misnomer, because the distinction of a history read in written records from one seen and studied in equally characteristic objects, chronologically determinable, is purely arbitrary and artificial and it would do no harm to drop it altogether."

6. GARRISON: History of Medicine 1917, 2nd ed., p. 50; *Ann. Med. Hist.,* 1917, vol. I, No. 2, p. 219.

7. KLEBS: *Bull. Hopkins Hosp.* 1917, vol. 28, pp. 261-266.

8. SUDHOFF: J. L. Pagel's Einführung in die Geschichte der Medizin, p. 33.

9. LANGDON: The Madisonville prehistoric cemetery. Anthropological notes. *J. Cincin. Soc. Nat. Hist.* 1880, vol. iii, p. 40; 1881, vol. iv, p. 250, Figs. 1-22. Good discussion of pathology of early North American Indians.

10. EATON: The collection of osteological material from Machu Picchu" *Mem. Connect. Acad. Arts & Sc.* May, 1916, vol. v.

11. KLEBS: *Bull. Hopkins Hosp.,* 1917, vol. 28, pp. 261-266.

12. LEIDY: *Proc. Acad. Nat. Sc.,* Philadelphia, 1886, p. 38.

13. Among diseases which have become extinct within historical times may be mentioned the *sweating sickness* described by Hecker in "Epidemics of the Middle Ages," 1846, pp. 181-353.

14. The soft parts of fossil vertebrates have been discussed by a number of writers. Our knowledge of the entire subject is reviewed in the author's paper, A new fish brain from the Carboniferous of Kansas, with a review of other fossil brains, *J. Comp. Neurol.,* April, 1915, vol. 25, No. 2, where an annotated bibliography of fifty papers will enable the interested reader to see just how meager is our knowledge of the soft parts of extinct vertebrates. Many of the softer structures are represented by impressions on the stone.

15. BERRY: Letter to author, Jan. 4, 1918.

16. BERRY: Remarkable fossil fungi, *Mycologia,* 1916, vol. 8, No. 2, pp. 73-78, plates 180-182, containing sixteen figures.

17. ERWIN F. SMITH: Bacteria in Prehistoric Times. In *Bacteria in Relation to Plant Diseases,* 1905, vol. 1, p. 262.

18. The question of extinction is still one of the unsolved problems of paleontology. The importance of those diseases which leave an impress on the skeleton has been referred to by the author in the following words:

"It is not my intention to contend that disease has not been influential in the extinction of races (or species); it probably has been; but those diseases which have left an impress on the fossilized skeleton certainly cannot be regarded as among those diseases which would produce widespread extinction. Some other has been the dominant factor. The present results of the study of fossil pathology indicate the early appearance in geological time and widespread distribution of diseases of many kinds, but none of them, so far as the fossil lesions may be interpreted, were sufficiently severe to have played a part in the extinction of any of the known groups of fossil vertebrates. They are to be regarded rather as chronic infectious or constitutional diseases which may have played a part in extinction, but there must have been some other and more powerful ally which is at present unknown." (The Influence of disease in the extinction of races. *Science,* N. S., Jan. 19, 1917, vol. xlv., No. 1151, pp. 63-64.

19. ÉLIE METCHNIKOFF: *Immunity in Infective Diseases,* 1915. Translated from the French by Francis G. Binnie, p. 8.

20. RUDOLF VIRCHOW: Ueber einen Besuch der westfällischen Knochenhöhle. *Z. Ethnol.* 1870, Bd. 2, p. 365, footnote; "Knochen vom Höhlenbären mit krankhaften Veränderungen," *Ibid.,* 1895, Bd. 27, pp. 706-708, figs. 1-4; Beitrag zur Geschichte der Lues, *Dermat. Z.* 1896, Bd. 3, p. 4.

21. MAYER: Ueber krankhafte Knochen vorweltlicher Thiere. In Nova Acta Leopoldina (Novorum Actorum Academia Cæsareæ Leopoldino-Carolinæ Naturæ Curiosum), Bd. xxiv, pt. II, pp. 673-689, pl. 30.

22. O. ABEL: Grundzüge der Paleobiologie der Wirbelthiere, 1912; Spuren von Kämpfen, pp. 88-91; Knochenerkrankungen, pp. 91-95.

23. These bacteria are described and figured by C. D. WALCOTT and H. F. OSBORN.

C. D. WALCOTT: Discovery of Algonkian bacteria. *Proc. Nat. Acad. Sc.,* April 1915, p. 256; Evidences of primitive life. *Smithsonian Rep. for* 1915, pp. 235-255, illustrated.

H. F. OSBORN: Origin and evolution of life, 1917, p. 153; *Science,* N. S., 1917, vol. 46, No. 1192, pp. 432-434.

Their discovery was forecasted by WALCOTT in his Pre-Cambrian algal flora. *Smithsonian Misc. Collect,* 1914, vol. 64, No. 2, p. 95.

24. METCHNIKOFF: *Immunity in Infective Diseases,* translated from the French by Francis G. Binnie, 1905, p. 18; also Chap. iii.

25. EDWARD HINDLE: *Flies in Relation to Disease (Blood sucking Flies),* 1914.

G. S. GRAHAM-SMITH: *Flies in Relation to Disease (Non-Blood sucking Flies),* 1914.

26. ELIAS METCHNIKOFF: "Die Lehre von den Phygocyten und deren experimentelle Grundlagen." In *Kolle und Wassermann's Handbuch der pathogenen Mikroorganismen,* 1913, Bd. ii, erste Hälfte, pp. 655-731, with an excellent bibliography.

27. The studies of Charles Emerson Beecher (1856-1904), an American paleontologist, upon evolutionary phases of the early fossil brachiopods and trilobites are especially important to consider in connection with the question of race senescence and the extinction of animal groups. His papers have been collected into a volume: *Studies in Evolution,* New York, 1901.

The entire subject of senescence in the recent lower animals is discussed by Child in *Senescence and Rejuvenescences,* University of Chicago Press, 1915.

28. This suggestion has been discussed by RENÉ LARGER in his paper "La contre-évolution oú dégénérescence par l'hérédité pathologique cause naturelle de l'extinction des groupes animau. Essai de paleopathologique générale comparée," 1916, *Bull. Mèm. Soc. anthrop. Par.*

29. GRAFF: Paleontographica, 1885, Bd. 31, pp. 183-192, Taf. xvi.

30. C. D. WALCOTT: Evidences of Primitive Life. *Smithsonian Rep. for* 1915, pp. 235-255, with plates.

31. JOHN M. CLARKE: The beginnings of dependent life. *Fourth Ann. Rep., Director of Science Div., New York State Education Dept.,* 1908, pp. 1-28. Pl. 1-13.

32. B. RENAULT: "Microorganismes des combustibles fossiles," *Bull. Soc. Industrie minérale Saint-Etienne,* 1899-1900, Ser. III, with atlas of plates.

33. OSBORN: *Origin and Evolution of Life,* p. 153.

34. B. RENAULT: Microorganismes des combustibles fossiles. *Bull. Soc. Industrie minérale Saint-Etienne,* Paris, 1899-1900, Tomes 13-14, with folio atlas of 20 plates of photomicrographs.

35. CHARLES A. PARKER: Evidences of Rheumatoid Arthritis in the Lansing Man. *Amer. Geol.,* 1904, vol. xxxiii. pp. 39-42. fig. I.

36 This subject has been discussed at length by the writer, in "Opisthotonos and Allied Phenomena among Fossil Vertebrates." *American Naturalist,* 1918.

37. OSBORN: *Bull. Amer. Mus. Natl. Hist.,* 1917, vol. 35, p. 733, pl. 28.

In wishing to modify as little as possible of the paper, we have retained the original references, but have modified dates according to recent views. We have also left in certain minor statements with which we are not personally in agreement. Eds.

Chapter 4

Notes on Diseases and Healed Fractures of Wild Apes

ADOLPH H. SCHULTZ*

There exists little hope that we may ever discover much direct evidence of the diversity and prevalence of diseases and injuries among the various forms of early man. Even though remains of fossil man have come to light in recent years with unexpected and most encouraging frequency, the finds are naturally restricted to skeletal parts and teeth alone and represent at best only very few samples of a population including rarely old individuals who are the ones most likely to manifest pathological conditions.

In their attempts to advance our understanding of man, physical anthropology, anatomy, embryology, neurology, physiology and psychology have already greatly benefited by investigations on man's nearest animal relations, the apes and monkeys. In the study of the antiquity of human disease, likewise, we can expect some helpful information from data appertaining to pathological processes among primates in general and the man-like apes in particular. Most of the morphological and physiological conditions of the latter resemble strikingly those of man, as has been amply proved. In regard to the protozoan parasites of primates Hegner (1928) concluded that those of monkeys, apes and man "belong for the most part to the same species or are so similar in their structure, lifecycles, and host-parasite relations as to be practically indistinguishable." That the nutritional, social and most other environmental factors of the recent anthropoid apes cannot be very far removed from the conditions having prevailed in the ages during and after which man separated from the common ancestral stock of all higher primates appears to be a fully justifiable assumption. It follows that the conditions for the acquisition of and reaction to diseases and injuries in early man were in all probability also quite similar to those persisting among the apes. In speculating, therefore, on the prevalence and diversity of diseases and injuries in the earliest history of man we can and should consider the evidence of the corresponding conditions among primates besides man.

Quite commonly one encounters more or less vague statements to the effect that the diseases of man became much more frequent with advance in civilization and that the incidence of disease in modern man is equalled only by that in animals under domestication or, possibly, in cap-

---

* It may be noted that a more recent and detailed paper on diseases in the primates has been written by Professor Schultz (The occurrence and frequency of pathological and teratological conditions and of twinning among non-human primates, *Primatologia*, 1956 I: 965). However, the paper of his which we include represents the first of its kind, and deserves to be regarded as a classic. The complete original paper appeared in *Bull. Hist. Med.*, 7:571, 1939 (Eds.).

tivity. In a wild state, it is assumed, animals suffer from disease only rarely and then fall speedily a prey to their enemies or fail to survive in the competitive struggle for existence.[1] The following notes will show that these claims are far too generalized and often do not at all agree with the known facts at least in regard to the animals of greatest interest in this connection, the wild monkeys and apes.

This paper is intended chiefly as a tentative, brief and quite incomplete survey of the literature on the subject of some pathological conditions in wild primates and as a preliminary report on some of the diseases and injuries in a new series of anthropoid apes which the writer has at present under investigation mostly for other purposes. The majority of the latter material was obtained on the Asiatic Primate Expedition of 1937[2] of which the author was a member.

A tremendous number and variety of both harmless and pathogenic parasites occurring in primates have been catalogued by Stiles, Hassall and Nolan (1929) and it appears that parasitic infections have already been recorded for the great majority of the many different species of monkeys and apes. It must be noted, however, that this useful paper includes observations on wild as well as captive specimens and, though furnishing welcome information regarding the diversity of parasites, does not indicate the frequency with which animals of the same species have become infected. The latter problem is elucidated in a report by Clark (1930) who studied the wild monkeys of Panama. According to this publication Spider monkeys contained tertian-like malaria parasites, accompanied by marked splenic enlargement, in 29 per cent of the cases and Capuchin monkeys quartan-like malaria parasites in 10 per cent of the cases. Microfilaria was found in the former series in 92 per cent, in the latter in 87 per cent and among Titi monkeys in 91 per cent of the specimens. A large trypanosome occurred in 42 per cent of Capuchin monkeys and in 26 per cent of Titi monkeys. It may be specially mentioned here also that in chimpanzees and gorillas of West Africa malarial parasites have been discovered which correspond in their various stages exactly to all three of the species known in man (Reichenow, 1920 and Blacklock and Adler, 1922). The gibbons (*Hylobates lar*) examined by the writer in one locality of Northern Siam were heavily infected with filariid worms. These parasites (*Tetrapetalonema digitata*, determined by Sandground, 1938) were found in twelve of nineteen young apes and in seventy-eight of eighty adults. Among the latter the peritoneal cavity was usually crammed full with these parasites and in some cases they had also invaded the thoracic cavity. In two adults the ovaries were cystically enlarged with filariid worms penetrating into these organs. In the same series of one hundred Siamese gibbons intestinal macroscopic parasites of as yet undetermined species occurred in at least thirty-nine cases. In four of eight orang-utans in Borneo the writer found *Dirofilaria pongoi* (three times intermuscularly and once, apparently, intramuscularly). In view of the

---

1. For instance, Pearl (1928) states: "The wild animals that happen to be killed by hunters are the survivors of a steady process of natural selection, and are therefore apt to be in an excellent state of health. The sick, weak and defective have long since died off, under circumstances practically always making impossible post-mortem examination. Again wild animals probably only rarely die a natural death due to internal pathological changes. When an animal in a state of nature falls appreciably below par in health, sooner or later something kills it."

2. This expedition was organized by Mr. H. J. Coolidge, Jr. The Carnegie Institute of Washington generously contributed a grant towards the author's expenses on this expedition.

fact that parasitic infections are known from all recent, wild man-like apes it seems more justifiable to assume that early man, too, had suffered from parasitic infestations than to claim that he differed radically in this respect from all his nearest cousins.

Of other diseases in wild primates we possess as yet even less information. Fox (1939) has recently described many cases of chronic arthritis in mammals. Among wild primates he finds marked arthritic changes in a baboon,[3] in fifteen of eighty-nine gorillas and in two of forty-one orang-utans. Among one-hundred-eighteen wild adult gibbons of various species (mostly *Hylobates lar*), recently surveyed by the writer, there are six cases of unmistakable arthritis of the spinal column. One of these cases is shown as No. 7 in plate 2. In addition this series contained at least two skeletons with evidently arthritic changes of some of the phalanges and wrist bones. In this same material were found eight specimens with diseased sterno-clavicular joints (six times bilaterally) and three specimens with diseased mandibular joints (three times bilaterally). In all these cases the particular joint surfaces were enlarged by exostoses, quite irregular, rough, and partly porous. The author is not competent to decide whether these grossly abnormal conditions were caused by some form of arthritis or are due to some local infection.[4] It may be mentioned here also that among the wild, adult gibbons, studied by the writer, there are two in which one lower extremity is strikingly atrophied, the circumference of the affected limb being less than half that of the normal limb, chiefly on account of the enormous difference in the thickness of the leg muscles. In another wild-shot *Hylobates lar* the right temporal muscle was atrophied to one fourth the size of the left one. Still another adult specimen, finally, possessed an enormous umbilical hernia. The consequences of infections of the maxillary sinus were observed by the author (1939) in three of thirty-five adult chimpanzees. Large drainage holes developed in these cases: once through the outer wall of the maxillary bone, once into the nasal cavity and once through the floor of the orbit. The last condition is present also in an adult wild gibbon.

Diseases affecting the dentition are known to occur among all the major groups of primates as has been shown chiefly by Sir Frank Colyer (1936) and for some aspects by the author (1935). Alveolar abscesses are particularly common in old specimens with advanced attrition of the teeth. The writer found alveolar abscesses among the latter in 62 per cent of orang-utans, in 59 per cent of gorillas and in 58 per cent of chimpanzees. Among gibbons they occurred in only 20 per cent of the oldest specimens. Individually as many as 18 of the permanent teeth may show alveolar abscesses (in one wild, old, male gorilla) and 14 to 17 abscesses have been counted in a considerable number of wild, old apes. Carious cavities in permanent teeth become also much more frequent with advance in age. For instance, in the writer's (1935) series of 95 orang-utans the percentage of specimens with caries amounts to 0 in young animals, to 4.4 in adult ones and to 28.5 in really old ones. In another series of 194 adult wild orangutans Selenka (1898) recorded caries in

---

3. Moodie (1931, plate LXIV) has pictured extensive arthritic changes in the spinal column of a mummified baboon of ancient Egypt.

4. An exactly corresponding condition in the mandibular joint of an adult wild chimpanzee is illustrated in a recent paper by the author (1939) and has also been found in several wild orang-utans, as will be reported later. No pus could be noticed in opening the affected joints of any of the gibbons.

ten specimens (= 5.2 per cent), while Colver (1936) found it in only 2.4 per cent of orang-utans, but fails to give the ages in his extensive material. A typical case of carious cavities in the dentition of an adult, wild orang-utan is shown in plate 3. In the chimpanzee caries is even more common than in the orang-utan. According to the writer's observations caries occurred in 6.7 per cent of adult chimpanzees and in 29.2 per cent of old ones. Colyer gives a much lower value for his series of 465 chimpanzees, namely 4.5 per cent, but his material may have contained many young specimens. In the gorilla, finally, caries is quite rare, having been found in only 0.7 per cent of the cases according to the latter author and in only slightly over one per cent, even in old animals, according to the writer. That wild apes can outlive the usefulness of their dentition is shown, e.g., by the case of an old chimpanzee with only fourteen of the misnamed *permanent* teeth left, the other eighteen having become lost and their alveolar processes completely resorbed (Schultz, 1939).

We possess direct evidence in support of the conclusion that dental disease did exist in fossil man. Thus the early Rhodesian man suffered from extensive caries. According to Hrdlička (1930), who examined the original skull, "at least nine of the teeth had advanced decay, in half of the cases nothing remaining but a small shell of the tooth. The destruction is such that there is no other explanation" (than caries). "In addition there were some root abscesses and probably some pyorrhea." In the typical Neanderthal skull of La Chapelle-aux-Saints only two of the permanent teeth are left and one of these has an open pulp cavity. All the other teeth had become lost *intra vitam* most likely through infection, as there are traces of alveolar abscesses. That such abscesses as well as caries were quite common in some ancient human populations has been shown, e. g. by Sir Armand Ruffer (1920) for Egyptians (including predynastic periods) and by McCurdy (1923) for Peruvians.

It remains to consider bony fractures in wild apes, particularly their incidence and their repair without any medical assistance. Healed fractures in anthropoid apes have been described by several authors. Duckworth (1904) discussed such cases in orang-utans and Holland (1924) one case in an old gorilla. Korschelt and Stock (1928) and Korschelt (1930) reported a considerable number of at times very well healed fractures in all three great apes. The question of the frequency of old, healed fractures among adult, wild great apes has been considered by the writer (1937) in another paper. In the humerus, radius, femur, tibia and clavicle alone such fractures occurred among 262 specimens in nine per cent of the cases.

In the already mentioned series of 118 wild, adult gibbons there are 42, or 36 per cent, with healed fractures. Twenty-six of the latter have only one fracture each and in the remaining sixteen skeletons there are two to four fractures in each. These healed fractures are distributed as follows:

Humerus	= 11 times	Metatarsi	= 2 times
Radius	= 3 "	Phalanges	= 9 "
Ulna	= 5 "	Clavicle	= 4 "
Femur	= 12 "	Scapula	= 2 "
Tibia	= 5 "	Pelvis	= 1 "
Fibula	= 3 "	Sternum	= 1 "
Patella	= 1 "	Ribs	= 1 "
Wrist or ankle	= 3 "	Orbital rim	= 2 "
Total			= 65 fractures

This is an astonishingly high rate of fractures, all of which have become repaired

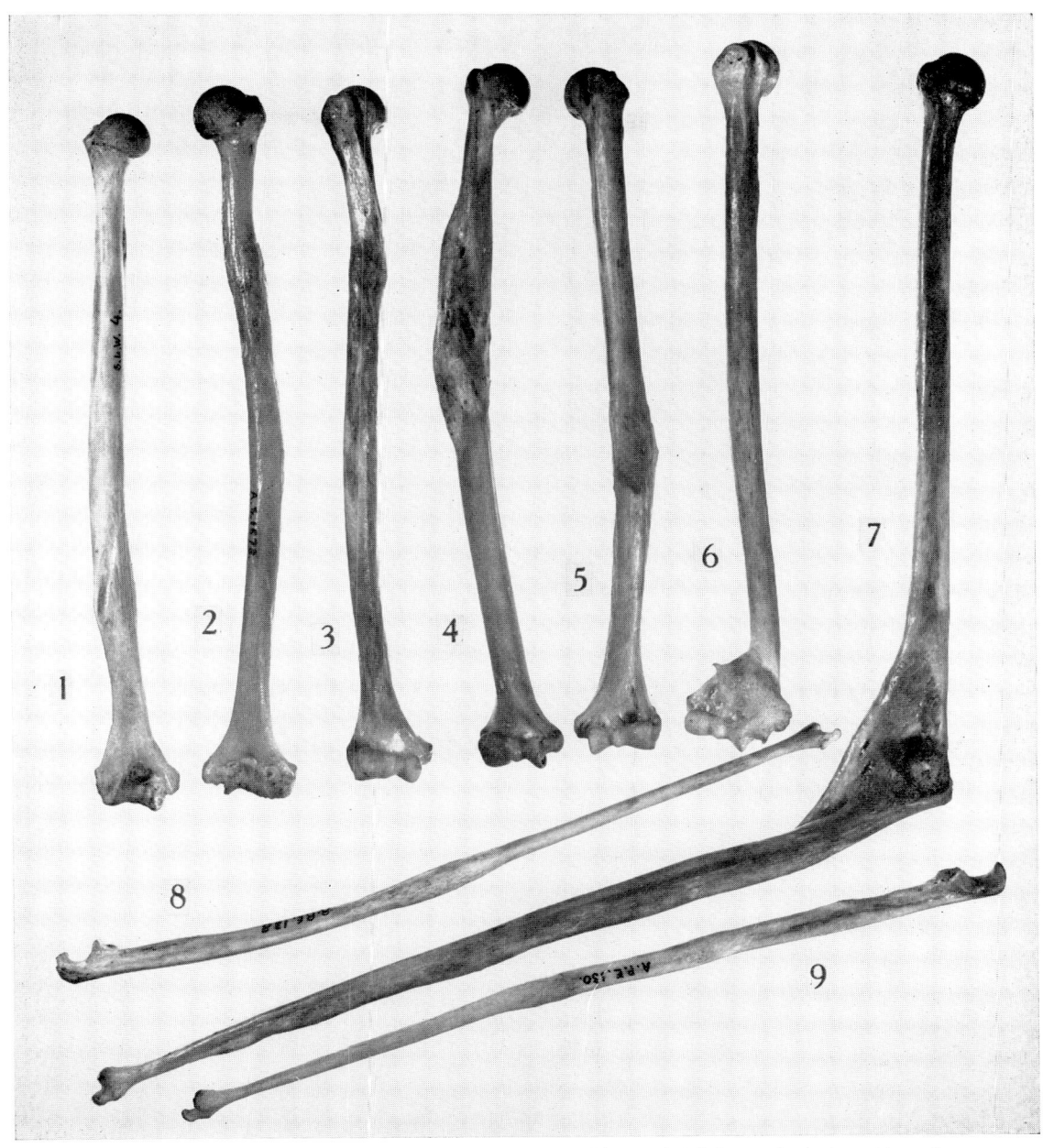

Figure 1. Healed fractures in adult wild gibbons (*Hylobates lar*), 1 to 7 = humerus, 8 and 9 = ulna, 7 = bony ankylosis of humerus and ulna.

in widely varying degrees of perfection. It seems that not many fractures can end fatally and that they do not, as a rule, incapacitate the apes sufficiently to lead to death by starvation or through capture by their many enemies, since gibbons were found to be very abundant in the districts visited by the writer, yet their fertility is rather low, as will be fully discussed in a later paper. A number of healed fractures

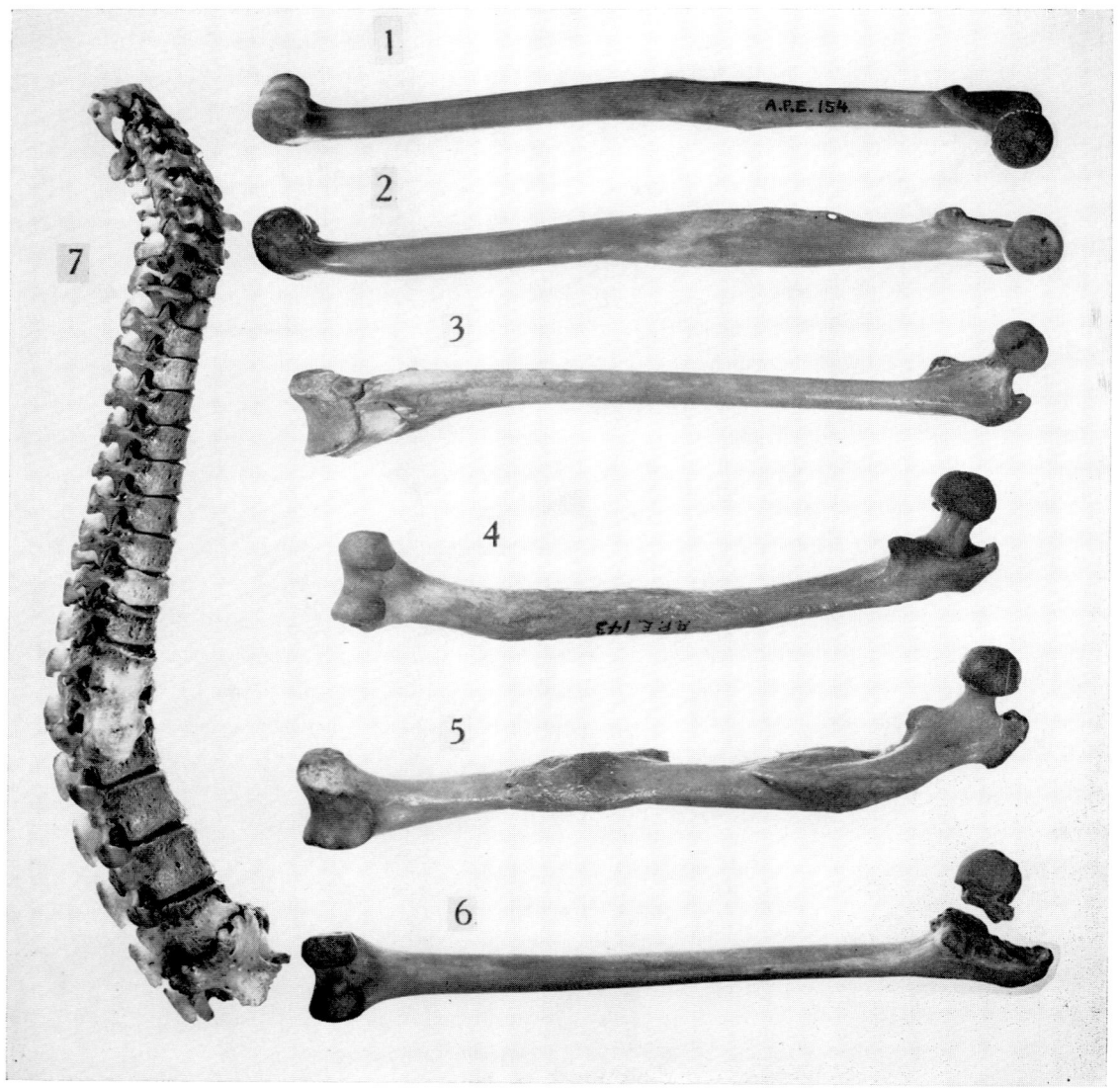

FIGURE 2. 1 to 5 = healed fractures of femora in adult wild gibbons (*Hylobates lar*), 6 = unrepaired fracture of neck of femur in an adult wild gibbon (*H. lar*), 7 = arthritis of spinal column affecting 10th, 12th and 13th thoracic and 1st, 4th and 5th lumbar verterbrae in an adult wild gibbon (*Hylobates moloch*).

in the limb bones of adult gibbons are shown in the Figures 1 and 2. Most of these have been selected to show successful repair of the injuries, accompanied by proportionately little shortening of the affected bone. The following tabulation compares the lengths of the fractured bones, shown in the Figures 1 and 2, with the lengths of the corresponding, normal bones:

No. in figure		Length (mm) of bone: fractured	normal	Difference in % of length of normal bone
1	Figure 1	221	224	− 1.3
2	" "	228	228	.0
3	" "	223	219	(both fractured)
4	" "	226	238	− 5.0
5	" "	220	228	− 3.5
6	" "	225	236	− 4.7
7	" "	236	235	+ 0.4
8	" "	249	250	− 0.4
9	" "	262	266	− 1.5
1	" 2	200	199	+ 0.5
2	" "	199	206	− 3.4
3	" "	197	205	− 3.9
4	" "	172	201	−14.4
5	" "	188	205	− 8.3

In several publications Jäger (1907 and 1909) has described many healed fractures in skeletal remains of prehistoric and early historic man. These he grouped in two categories, namely *well healed fractures,* having most likely received artificial help ("Gut geheilte—Wohl durch Kunsthilfe"), and *badly healed fractures* which supposedly had been left to themselves. In this way Jäger attempted to demonstrate that the art of setting and holding broken bones is very old. The latter statement may be true, but it is certainly not at all proved or even rendered probable by the mere finding of well-healed fractures.[5] Among the skeletons of wild gibbons, examined by the writer, there are many with fractures fully as well healed and this with at least as little shortening and axial distortion as in most the "well-healed" fractured bones of ancient man, described and illustrated by Jäger as having benefited by artificial aid.

In conclusion it can be stated that, judging by the example of the gibbons specially discussed here, anthropoid apes in a state of nature show pathological conditions in an unexpectedly high proportion of the cases. In addition these gibbons are very commonly subject to developmental disturbances such as polydactyly, brachydactyly, spina bifida occulta, incomplete formation of an entire limb or of ribs, impacted and congenitally lacking teeth, fusion of atlas with occiput, and cryptorchism,[6] yet in spite of these manifold afflictions they manage to survive. All this in populations of apes in their natural environment gives a picture which casts serious doubt on the often assumed prevalence of health and normality in nature and on the efficacy of natural selection.

It seems impossible to find any valid reason why early man, having developed from the same ancestral stock, should have been very different from man-like apes in these respects.

---

5. Jäger (1907) mentions, e.g., that the "well-healed" fracture of the humerus of a juvenile individual had shortened the bone by 3 cm. Since a juvenile human humerus is hardly over 30 cm. long this would mean a relative shortening of the bone by about 10 per cent or much more than in most of the fractured long bones of wild gibbons, as shown by the above tabulation.

6. Three cases of unilateral cryptorchism in wild gibbons have been described by the author (1938) in another paper. The other abnormalities mentioned here will be discussed in detail in a later publication.

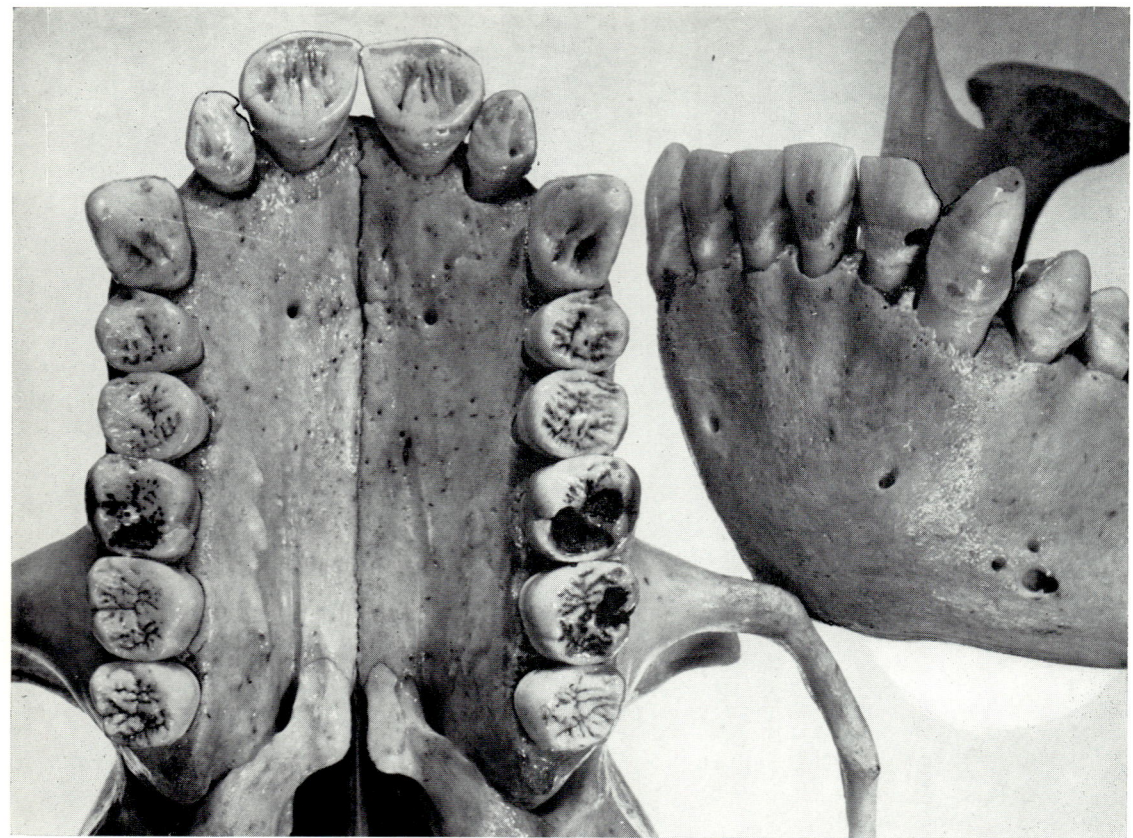

FIGURE 3. Dentition of an adult, female, wild orang-utan showing carious cavities in crowns of upper first molars and left second molar and in both lower lateral incisors.

REFERENCES

BLACKLOCK, B. and S. ADLER, 1922: A parasite resembling *Plasmodium falciparum* in a chimpanzee. *Ann. Trop. Med. & Parasit.*, 16:99.

CLARK, H., 1930: A preliminary report on some parasites in the blood of wild monkeys of Panama. *Amer. J. Trop. Med.*, 10:25.

COLYER, F., 1936: Variations and disease of the teeth of animals. London.

DUCKWORTH, W. L. H., 1904: Os fracturés des Orangoutans. *Studies from the Anthropological Laboratory, The Anatomical School Cambridge*, 51-53. Cambridge.

FOX, H., 1939; Chronic arthritis in wild mammals. *Transact. Amer. Philos. Soc., New Ser.*, 31:71.

HEGNER, R., 1928: The evolutionary significance of the protozoan parasites of monkeys and man. *Quart. Review of Biol.*, 3:225.

HOLLAND, W. J., 1924: Account of a skeleton of a gorilla remarkable because showing recovery from gunshot wounds. *Annals Carnegie Mus.*, 15:293.

HRDLIČKA, A., 1930: The skeletal remains of early man. *Smithsonian Misc. Coll.*, LXXXIII (entire volume).

JAGER, K., 1907: *Beiträge zur frühzeitlichen Chirurgie* (with Atlas). Wiesbaden.

———, 1909; Beiträge zur prähistorischen Chirurgie (Paläochirurgie). *Deutsch. Z. Chirurgie*, 102, 109.

KORSCHELT, E. and H. STOCK, 1928. Geheilte Knochenbrüche bei wildebenden und in Gefangenschaft gehaltenen Tieren. Berlin.

KORSCHELT, E., 1930: Weitere Beobachtungen an geheilten Knochembrüchen wildlebender Tiere. *Sitzungsber. Ges. Z. Beförder. Naturwiss.*, Marburg, 64:111.

MCCURDY, G. G., 1923: Human skeletal remains from the highlands of Peru. *Amer. J. Phys. Anthrop.* 6: 217.

MOODIE, R. L., 1931: Roentgenologic studies of Egyptian and Peruvian mummies. *Field Mus. Nat. Hist., Anthrop. Mem.*, III.

PEARL, R., 1928: Evolution and mortality. *Quart. Rev. Biol.*, 3:271.

REICHENOW, E., 1920: Ueber das Vorkommen der Malaria Parasiten des Menschen bei den afrikanischen Menschenaffen. *Centrabl. Bakter. Parasit., 85*:207.

RUFFER, A., 1920: Study of abnormalities and pathology of ancient Egyptian teeth. *Amer. J. Phys. Anthrop. 3*:335.

SANDGROUND, J. H., 1938: A redescription of Tetrapetalonema digitata (Chandler, 1929) comb. nov., a filariid parasite of Gibbon apes, with an enumeration of its congeners. *Bull. Mus. Comp. Zoology, 85*:49.

SCHULTZ, A. H., 1935: Eruption and decay of the permanent teeth in primates. *Amer. J. Phys. Anthropol., 19*:489.

———, 1937: Proportions, variability and asymmetries of the long bones of the limbs and the clavicles in man and apes. *Human Biol., 9*:281.

———, 1938: The relative weight of the testes in primates. *Anatom. Rec., 72*:387.

———, 1939: Growth and development of the chimpanzee. Contrib. to *Embryology*, Carnegie Inst. Wash. Publication (in press).

SELENKA, E., 1898: Rassen, Schädel und Bezahnung des Orang-utan. Studien über Entwicklungsgesch., 6. H., Menschenaffen, 1. Lief., 1-91.

STILES, C. W., A. HASSALL and M. O. NOLAN, 1929; Key-catalogue of parasites reported for primates (monkeys and lemurs) with their possible public health importance and key-catalogue of primates for which parasites are reported. U. S. Treasury Dept., Public Health Service, Hygienic Labor. Bull. No. 152.

Chapter 5

The Bio-cultural Background to Disease

DON BROTHWELL

ALTHOUGH A detailed survey of man's cultural development and ever changing environment would be out of place here, it is important not to exclude a review of these subjects in view of their undoubted repercussions on the patterns of human disease through time.

Four major stages can be distinguished in human evolution and cultural development (Fig. 1). The first is the differentiation of the early hominids, generally called australopithecines. Continuing discoveries show this group to be quite variable, and there is no doubt that at least three distinct varieties are represented. Their divergence from a "pre-hominid" stock may well have taken place over 1,500,000 years ago (on the evidence of potassium-argon results). Some australopithecines were responsible for the purposeful manufacture of the first recognizable—but still extremely crude—stone tools and although the worked edges were blunt, well aimed blows to the head could have produced considerable injury. Possibly certain bones were split into useful shapes and used in this way (although I doubt the complexities of early "osteodontokeratic cultures" elaborated by some workers). Also, of course, wood might have been utilized, and indeed, this possibility is one major objection to using the evidence for stone tools in taxonomic decisions, for it cannot be verified that perishable material was not used for defence and attack by these early groups.

It is likely that the diet of the australopithecines varied according to variety (?species) and ecological background, and in view of the dental caries and hypoplasia which have been noted, the food may at times have been unsatisfactory in quantity and quality. The majority could well have be omnivorous to some extent, and what may have begun as simple insect eating and carrion feeding seems likely to have developed into intentional hunting, singly or in bands (enough is now known of similar activities in baboons to make this very feasible). This period of "experimental" hunting, with or without weapons, may have been critical for the evolution of the erect posture. With the emergence of bipedalism, the hand was completely freed for new pursuits, and it was probably this new challenge which made increased brain size and efficiency so advantageous.

By 500,000 B.C., *Homo erectus* (*Pithecanthropus*) had become fully differentiated. In this group the brain capacity was well beyond the range found in the Great Apes, and overlapped with the range known in modern man (Fig. 2). The trend to facial recession continued, and from the fragmentary post-cranial remains, modern body proportions and stature were more or less attained. These pithecanthropines were widely distributed in the Old World

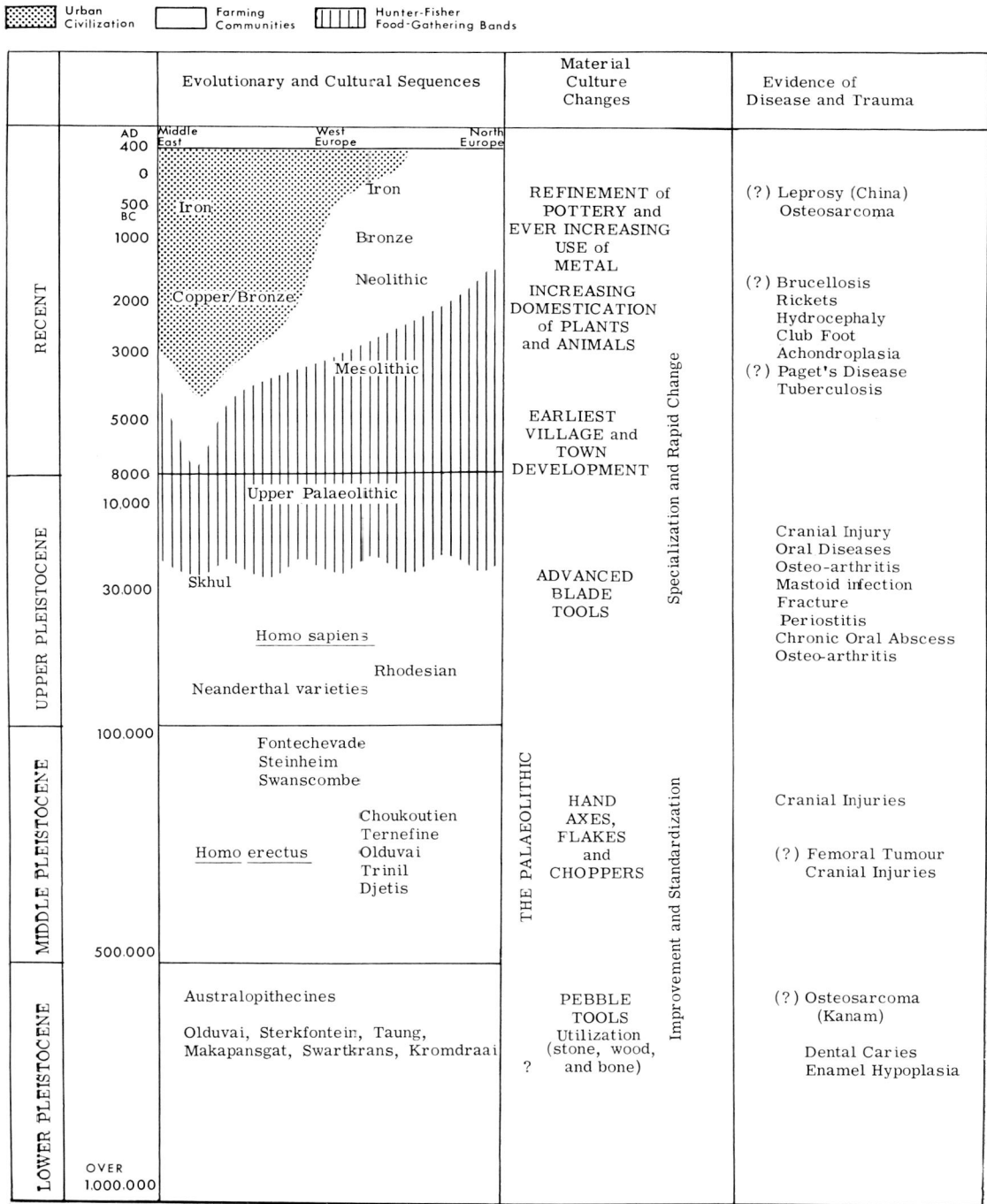

FIGURE 1. Some relationships between human evolution, cultural change, and diseases and trauma known to have occurred. (The last ten thousand years is with special reference to Europe and the Middle East).

from Java (Djetis, Trinil), China (Choukoutien), Africa (Olduvai, Ternefine) and through into Europe (Heidelberg) although of course the specimens so far discovered are by no means contemporary. (Coon, 1963, gives a useful survey of the extent of hominid fossils so far described.) As might be expected, t o o l m a k i n g achieved a higher standard than previously seen, although it was still very crude in comparison with later artifacts. These were probably multi-purpose tools—a fact not usually stressed or even suggested by prehistorians—and whenever hunting and collecting bands came into conflict, the artifacts would become useful weapons. Weidenreich (1939), in his descriptions of cranial injuries in this and other groups, shows clearly the type of injury which might have resulted from such weapons. (See also Courville, p. 606.)

More advanced hominids (which can be referred to by the group term hominines) had emerged by later Middle Pleistocene times (c. 200,000 years B.C.). The most well known specimens representing early hominines are the Steinheim and Swanscombe skull fragments. These show a morphological stage very near the Upper Pleistocene varieties of man. As seen in the handaxes from Furze Platt, near Maidenhead, (Fig. 5,a) artifacts could vary very considerably in size (the larger of the two being nearly 30cm in length). In times of group conflict, such tools—especially if methods were available for fixing them in wooden shafts—and the wooden spear (probably with fire hardened tip), would have been very effective weapons. It is interesting that on the external surface of the two Swanscombe parietals are three well defined but shallow depressions (Brothwell, 1964) w h i c h could be interpreted as healed minor injuries.

Regarding man's use of fire, there is evidence that *Homo erectus* did so, if only sporadically, and that Upper Pleistocene men not only commonly used fire, but were fire-*makers* (Oakley, 1955). From a dietary point of view, fire probably made possible a greater use of flesh as valuable nourishment for young infants and senile individuals.

With the development of advanced Palaeolithic techniques for tool making, followed by the emergence of Mesolithic industries, groups with such material culture were at a distinct advantage over less advanced culture—not only in terms of combative weapons but also from the point of view of competitive hunting and fishing. It was during these times also, that waves of early mongoloids slowly penetrated down into the New World, taking diseases with them which were then to evolve independently of Old World influence for thousands of years. (Dr. Hackett, elsewhere in this book considers treponemal infection from this point of view.)

There is a growing variety of dental evidence to show that, as a result of the coarseness of Palaeolithic and later stone age diets, very considerable demands were sometimes made on the teeth. In a number of fossil specimens there is extreme wear, and in some cases pulp exposure has resulted from the insufficient deposition of secondary dentine (Fig. 6a). This has caused pulp infection and apical abscess formation in one Shanidar Neandertaler (Stewart, 1962, and Brothwell, 1963 for X-ray), and probably the antemortem tooth loss evident in the Gibraltar I and La Chapelle Neandertalers is at least partly the result of this process. In the case of tooth loss, caries cannot of course be ruled out, and the Upper Pleistocene Rhodesian skull displays a number of abscesses result-

The Bio-cultural Background to Disease

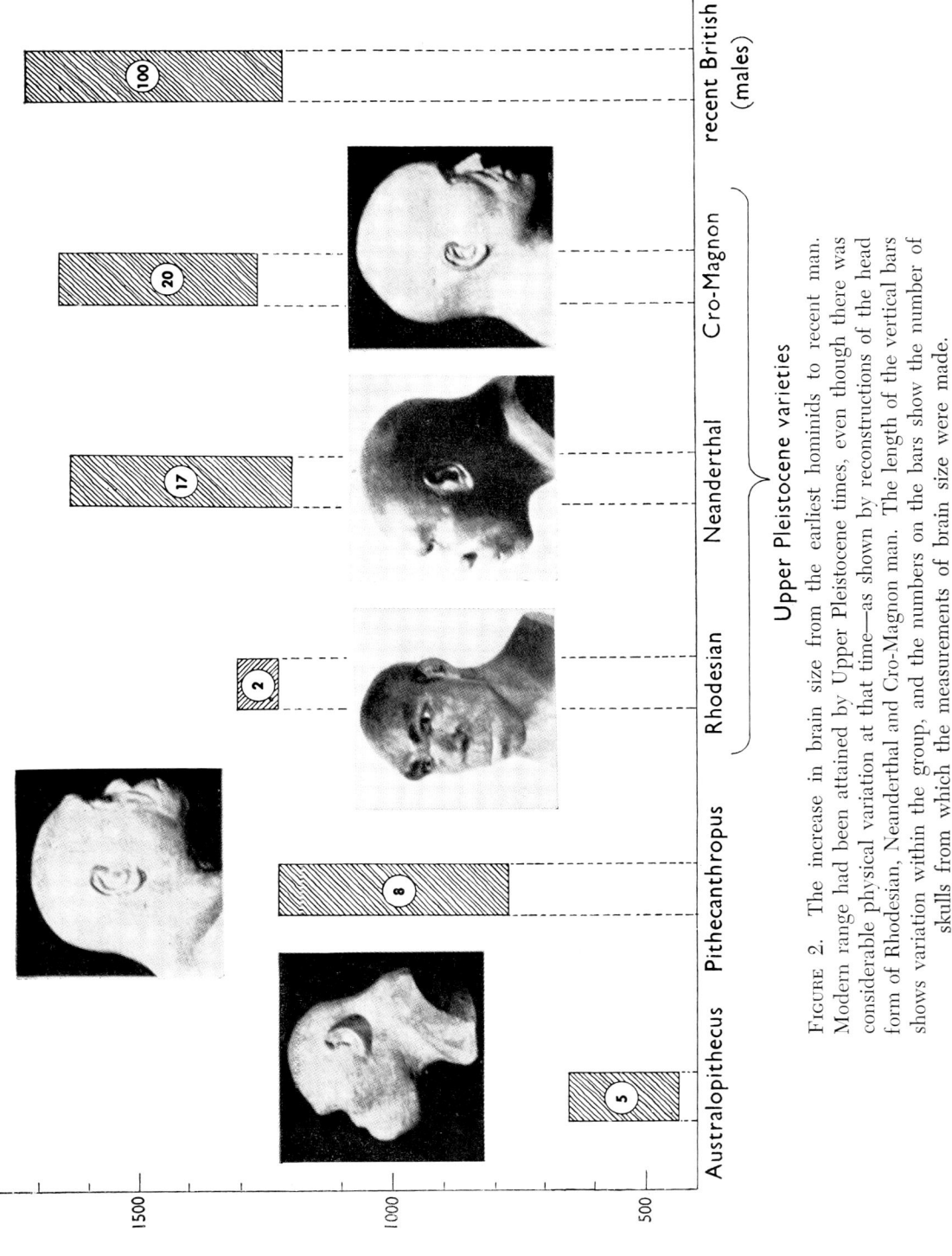

FIGURE 2. The increase in brain size from the earliest hominids to recent man. Modern range had been attained by Upper Pleistocene times, even though there was considerable physical variation at that time—as shown by reconstructions of the head form of Rhodesian, Neanderthal and Cro-Magnon man. The length of the vertical bars shows variation within the group, and the numbers on the bars show the number of skulls from which the measurements of brain size were made.

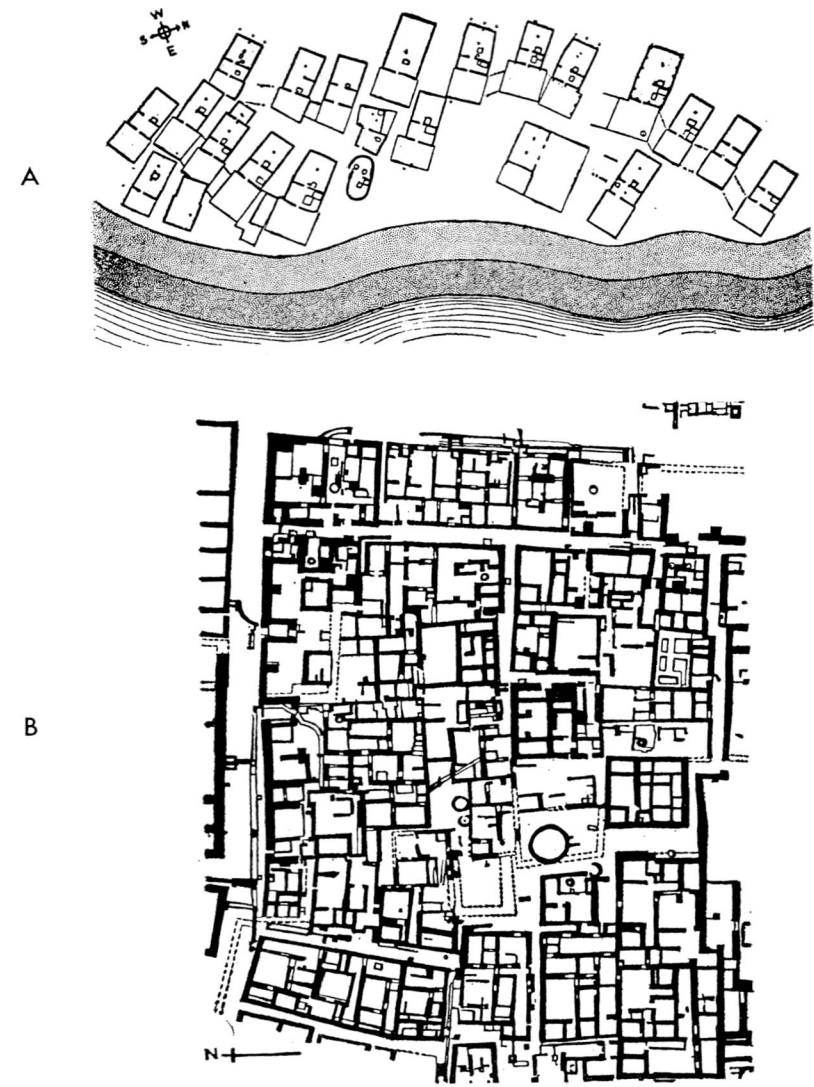

FIGURE 3. A. Plan of the Neolithic lake-side village of Aichbühl. After Schmidt. B. The ancient city of Mohenjo-daro. After Mackay.

ing from extensive cavities, proof that dental decay is by no means a "disease of civilization."

The slender harpoons manufactured in quantity by some Mesolithic communities raises the very interesting question of the versatility of such implements. Clearly they had considerable economic value, especially in fishing, but it seems impor- tant not to underestimate any such tool as a potentially efficient weapon in times of conflict. The harpoons from the Mesolithic Yorkshire site of Star Carr (Fig. 5b), for example, could have produced extensive wounds in combat (provided the hafting was securely fixed). This point seems worth stressing since comment has been made in the past of the apparent "paci-

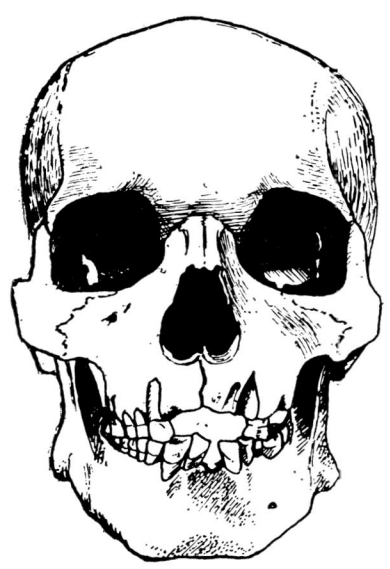

FIGURE 4. A typical case of tooth evulsion in a Mesolithic male skull from North Africa (Mechta 8), after Cabot Briggs. Drawing by Rosemary Powers.

fism" of some earlier communities considering the scarcity of special weapons for combat. Thus, for instance, Keith (1948) writes: "Herzfeld and other students of the village settlements of ancient Iran have been impressed by the absence from them of warlike equipment. The villages were open and unwalled; stone mace-heads and axes were found in them; there were sling stones, but no arrow-heads or spearheads. The villagers were pacific in nature; they were not big-boned, big-bodied, warlike folk" (p. 285).

Although the Neolithic Revolution, beginning some eight thousand years ago, was a major culture change which had considerable repercussions throughout the world, its beginnings were nevertheless modest and the spread of this new era took some millennia to achieve (Fig. 1). Neolithic and later developments depended entirely upon the new practice of cultivation and stock rearing, and hand in hand with this went domestication. Environment dictated to some extent what could be planted or reared, and by experimentation, man created further domestic varieties as the centuries went by (see Zeuner, 1963). The result of this profound change in economy was that families could band together into more extended and permanent communities (and indeed, it was beneficial for them to do so). Further concentration of people, perhaps aided by a population spurt as a result of the more guaranteed food supply throughout the year (and the consequent survival of more individuals into reproductive life), encouraged early village and town development. The concentration of dwellings at Tell es-Sultan, Jericho, was even larger than village size as early as 6,500 B.C. with well-defined town walls (Fig. 6b). The village of Qalat Jarmo began about 7,000 B.C., and Neolithic sites of a more recent date are also known in Europe; for example, the lake-side village of Aichbühl, Württemberg (Fig. 3a). By the third millennium B.C. cities had developed as far east as the Indus valley, and at Mohenjo-daro (Fig. 3b) there is clear evidence that the community was well aware of the public health aspects of city buildings (at least in the well-to-do quarters). Of the better houses, Piggott (1950) writes: "Rubbish-shoots occur, however, running out through the wall into a brick-built rectangular bin outside, presumably cleared by order of the municipal authorities, who must similarly have been responsible for the elaborate drainage system which ran under the streets and into which house drains communicated. The main drains could be cleared by lifting large, specially made brick "manhole-covers," and the whole conception shows a remarkable concern for sanitation and health without parallel in the Orient in

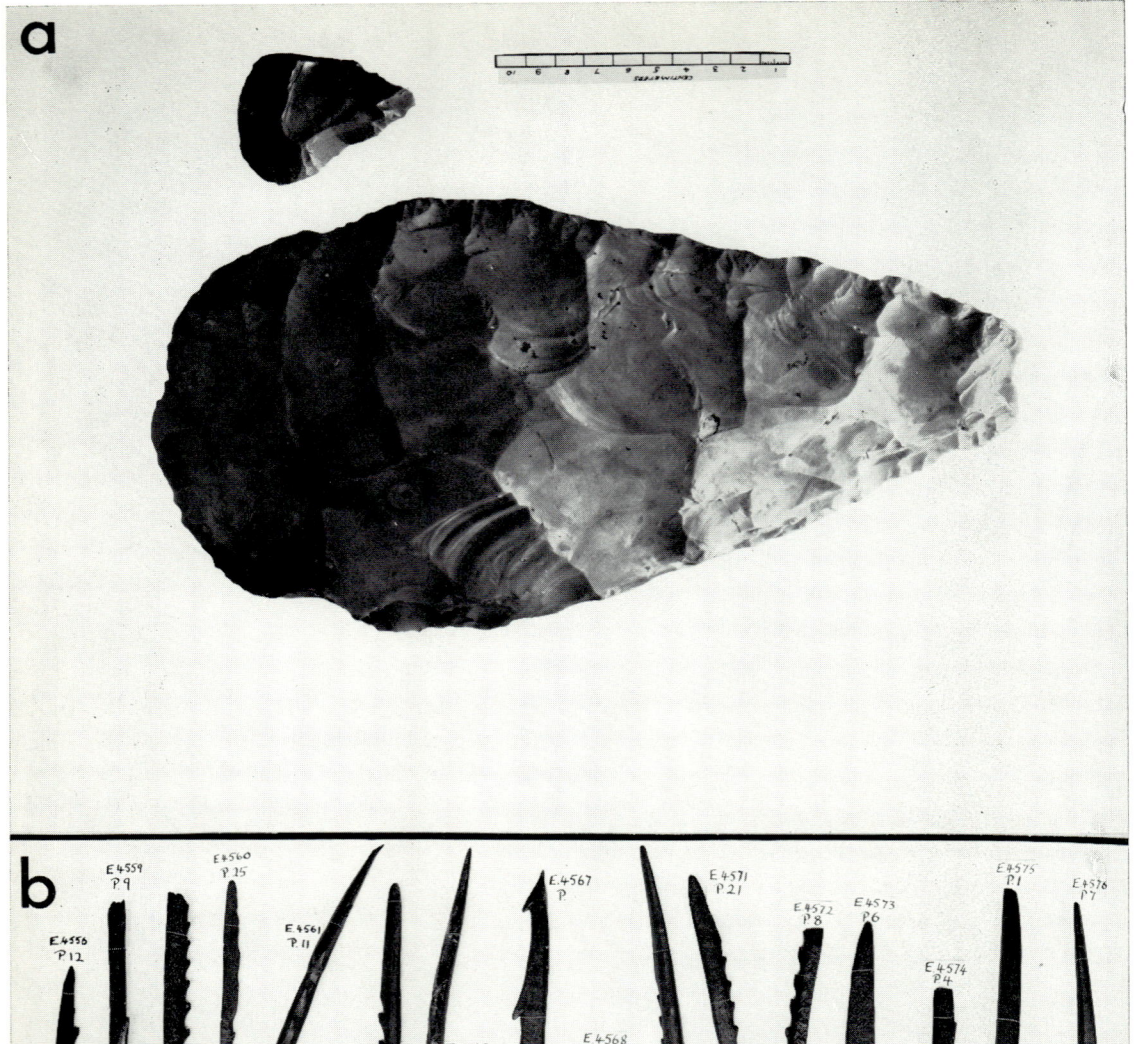

FIGURE 5. Palaeolithic and Mesolithic artifacts which could have been used as weapons, although primarily of economic value. (a) Acheulian hand-axes from Furze Platt, near Maidenhead. After K. P. Oakley.
(b) Harpoon points from the Mesolithic hunter-fisher camp of Star Carr in Yorkshire.

the prehistoric past or at the present day. Soak-pits took the eventual sewage" (p. 168).

The considerable affect which man had upon his surrounding environment during late prehistoric and early historic times could well have influenced the patterns of disease to a marked degree. Polgar (1964) summed up the situation nicely when he wrote:

Nomadic or semi-nomadic hunters and gatherers change the environment very little. When man settled down into permanent villages, he began to alter the landscape around him in many important ways. These alterations created opportunities for some animals and plants to establish new niches for themselves closely associated with human habitations. Certain species of insects and rodents domiciled this way have an important role in the transmission of many diseases. A type of mosquito, breeding in small bodies of water collected in man-made receptacles, is mainly responsible for the transmission of the viruses of yellow fever and dengue.

The clearings created in the West African tropical forest by the introduction of agriculture were probably a causal factor in making malaria an endemic disease there by allowing the proliferation of another kind of mosquito which does not breed in the dark shade of tall trees. Murine typhus is mostly a disease of wild rodents, but domestic rats can become infected and man can get the disease if bitten by the rat flea. Deforestation by man also leads to the establishment of new breeding places for field rodents and their mites, which are implicated in many parts of Asia in the transmission of scrub typhus. When farmers begin to cultivate land once allowed to run to waste, they will become exposed to the bites of these insects and thus catch the disease. The housefly is often involved in the spread of diseases associated with fecal matter. The milk of domestic animals as well as their skin, hair, tissues or even the dust in the places where they are kept are also vehicles for passing on to man a number of diseases like anthrax, Q. fever, Brucellosis and tuberculosis (pp. 4-5).

One might add that the construction of irrigation channels, artificial dams and lakes, no doubt similarly encouraged disease carrying vectors in some regions. As well as considering domestic animals and disease, one must not forget that plant domestication has also brought troubles in its wake. There are various comments on rust diseases dating back to over 700 B.C. (suggesting severely lowered crop yields and resulting famine) and there is possible recorded evidence of ergot contamination of crops as early as *ca.* 2,500 B.C.

The over consumption (relative to other foods) of cereals must have resulted at times in pellagra in the New World and kwashiorkor in the Old World. A final general point regarding the Neolithic Revolution, the common use of pottery—especially in unglazed forms—was a potentially serious new source of infection. In underdeveloped countries today, unclean food containers are probably responsible for a considerable amount of gastroenteritis particularly in young infants who depend on feeding bottles; and there is no reason to think that this hazard was unlikely in earlier cultures. Feeding bottles were certainly in use in some communities by as early as Neolithic times, and efficient cleaning would have been far more difficult than today.

The Neolithic gave way to the Metal Ages, and the growth of the Greek and Roman Empires, again showing temporal and spatial variations in their impact upon Old World communities. With the slowly growing world population and the increasing contact between communities, dispersion of infective diseases was ever more possible. It seems reasonable to assume that some intrusive diseases were previ-

FIGURE 6. (a) The anterior part of the palate of the Gibraltar I skull, showing considerable dental wear and some indication of pulp exposure.
(b) Neolithic town walls of Jericho. Courtesy of K. M. Kenyon.
(c) The forearm stump from Sedment, Egypt.

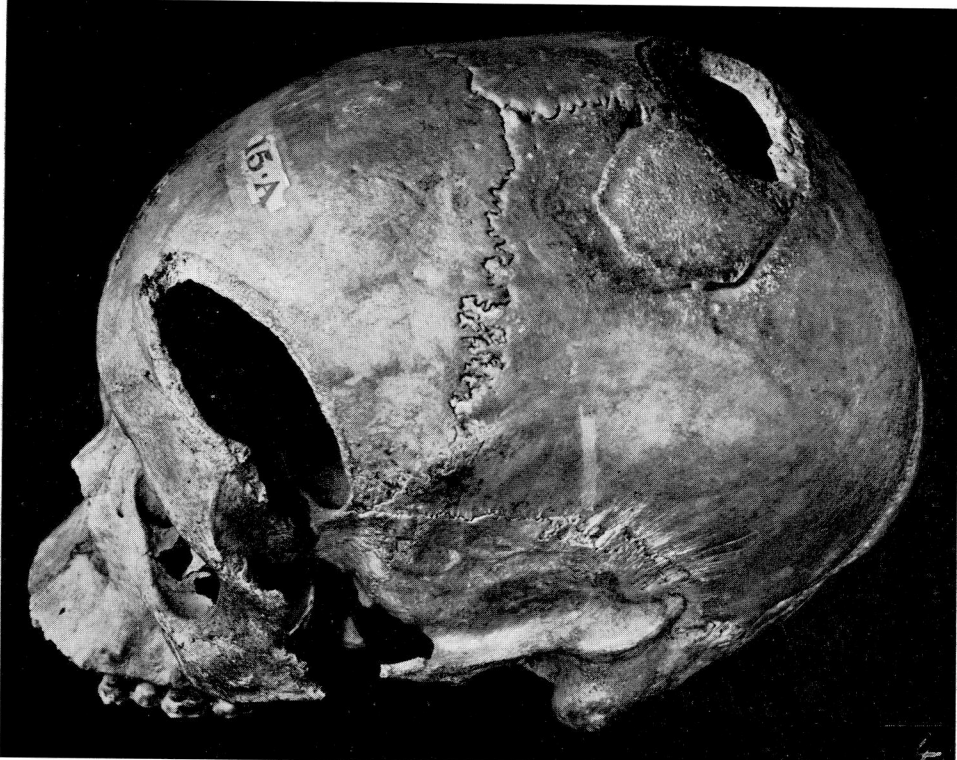

FIGURE 7. The male skull from grave 121 of the Nubian Cemetery 92.

ously uncommon or unknown in the communities which they now affected. Moreover, the impact of some newly intrusive diseases, may at times have been quite different centuries or millennia ago and the pathological changes following infection need not be known today. Furthermore, as in the case of leprosy in Mediaeval Europe, we can expect some earlier diseases to have reached serious epidemic proportions in populations which today show little evidence of them. Any appreciation of human population genetics and world patterns of disease, must not ignore this fact, for to consider possible relationships between the two without taking account of the dimension of time is a serious error.

One of the few recorded cases of ancient disease being transmitted for the first time into a community and reaching epidemic proportions, is described by *Pliny* in his *Natural History*. His description does not permit firm diagnosis, but the painless and disfiguring nature of the disease (of group of diseases?) is particularly suggestive of a mycotic or leprous condition. He writes:

The face of man has also been afflicted with new diseases, unknown in past years not only to Italy but also to almost the whole of

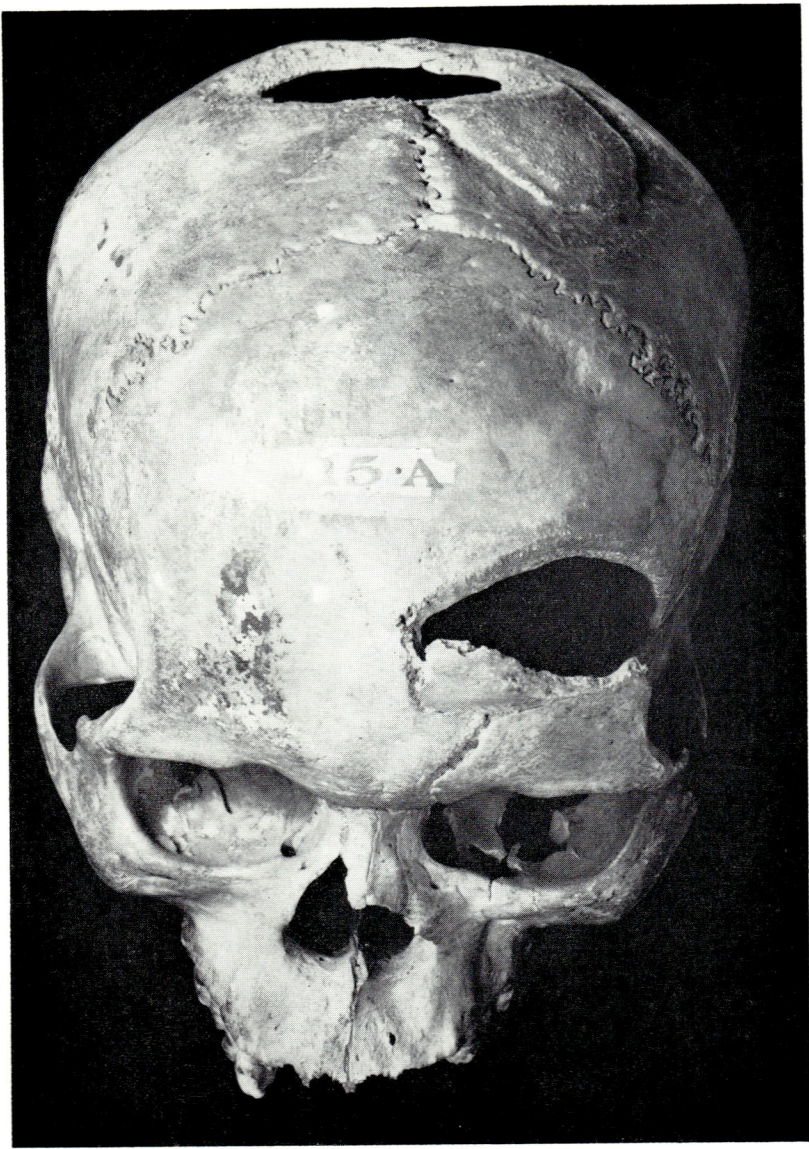

FIGURE 7 (cont'd.)

Europe, and even then they did not spread all over Italy, or through Illyricum, the Gauls, and the Spains to any great extent, or in fact anywhere except in and around Rome. Though they are painless and without danger to life, yet they are so disfiguring that any kind of death would be preferable. The most severe of these they called by a Greek name lichens. . . . . . . The disease seized in many cases at least the whole of the face, with the eyes only unaffected, but passed down however also to the neck, chest, and hands, covering the skin with a disfiguring, scaly eruption. This plague was unknown to our fathers and forefathers (Book XXVI).

The nature of war injuries clearly changed with the growing use of metal axes and swords after about 2,000 B.C.

TABLE I

DISTRIBUTION OF INJURIES TO THE POST-CRANIAL SKELETON IN BRITISH AND NUBIAN SERIES

	British cases, Neolithic-Anglo-Saxon	% of total	Nubian cases*	% of total	Recent London cases*	% of total
Ribs	1	2.6	9	6.9	121	8.3
Femur	2	5.1	20	15.3	93	6.4
Tibia and Fibula	6	15.4	16	12.2	272	18.7
Clavicle	7	17.9	22	16.8	133	9.2
Humerus	2	5.1	11	8.4	145	10.0
Radius and Ulna	19	48.7	50	38.2	266	18.3
Hand	2	5.1	3	2.3	421	29.0
Total	39**		131		1451	

* Data from Elliot Smith and Wood Jones (1910).

** Data from Brothwell (1961).

Such weapons were far more likely to produce well defined cuts than the crushing and splitting action of blunter stone weapons. Some of the injuries described in earlier specimens are remarkable for their severity in individuals who clearly survived the traumatic experience. One such case, a male skull from Grave 121 of the Nubian Cemetery 92 (X-group; Byzantine Pagan-Period) was described originally by Elliot Smith and Derry (1910). Of the three vault injuries (Fig. 7), two show much healing and of these the one in the left frontal area is a huge gaping opening, there being no doubt at all that in life the prefrontal region of the cerebral hemisphere must have been severely damaged.

With regard to frequency and distribution of fractures, there is clearly a need for more statistics on earlier populations, although the little data so far available suggests that noticeable differences can be expected, a fact which is not surprising in view of the varied, sometimes stormy histories of earlier peoples. In Table I, data is given on post-cranial injuries in two pre-industrial series, and a comparatively recent London sample. Without expanding further on these figures, it can nevertheless be seen, that differences do occur and at least some of them are likely to be significant, although two of the samples are small, reflecting changes for a cultural nature (including warlike propensities).

Although injury as a result of combat is by no means unique to man, he alone has invented methods of modifying physique by non-combative means. Much will be said in three later contributions to this book about primitive surgery. Although nearly all evidence of surgical intervention is of Neolithic or more recent date, it could well have a much more ancient history, and the missing right hand (certainly lost before death) in the Neandertal Shanidar I individual (Stewart, 1959) might conceivably be interpreted as removal following extensive injury or disease.

But intentional modification of body form has been undertaken for other reasons. A forearm stump (Fig. 6c) of IX Dynasty date from Sedment, Egypt, raises the interesting questions of to what extent hands were removed for the purpose of keeping score of prisoners as well as the dead (Brothwell and Møller-Christensen, 1963a), although this procedure may have been completely restricted to counting the dead (Aldred, 1964). As a punishment, the removal of hands and feet was certainly being practised by Saxon times and there is evidence of this in a seventh century British skeleton (Brothwell and Møller-Christensen, 1963b). As regards tooth evulsion, this was being practised in the Mediterranean area (especially North Africa) by Mesolithic and Neolithic times (Fig. 4).

One of the most controversial aspects of Upper Palaeolithic cave art is presented by the several outlines of mutilated hands. These have given rise to much speculation, and various diseases have been enumerated to account for the varying loss of phalanges. One possible explanation, to my mind far more likely than any other, is that we have in these deformities, evidence of cultural mutilation still practised today in parts of New Guinea and which might well have been far more widespread in earlier Stone Age times.

This is but a brief review of the complex background into which any consideration of the palaeopathology of ancient populations must be fitted. I hope, however, that it is in sufficient detail to suggest the many ways in which man's environment and cultural development have affected his health in the past.

REFERENCES

ALDRED, C., 1964: A possible case of amputation. *Man*, 64:56.

BROTHWELL, D. R., 1961: The palaeopathology of early British man: an essay on the problems of diagnosis and analysis, *J. Roy. Anthrop. Inst.*, 91:318.

BROTHWELL, D. R., 1963: The macroscopic dental pathology of some earlier human populations. In: *Dental Anthropology*, London, Pergamon Press, pp. 271-288.

BROTHWELL, D. R., 1964: Further comments on the right parietal from Swanscombe; anomalies and endocranial features. In: *The Swanscombe Skull. A Survey of Research on a Pleistocene Site.* (Ed. C. D. Ovey). Roy. Anthrop. Inst., London. pp. 173-4.

BROTHWELL, D. R., and MØLLER-CHRISTENSEN, V., 1963a: A possible case of amputation, dated to c. 2000 B.C. *Man*, 63: 192.

BROTHWELL, D. R. and MØLLER-CHRISTENSEN, V., 1936b: Medico-historical aspects of a very early case of mutilation. *Danish Med. Bull.*, 10:21.

COON, C. S., 1963: *The Origin of Races*. London, Cape.

ELLIOT SMITH, G. and DERRY, D. E., 1910: Anatomical report. *Archaeo. Survey of Nubia*, Bull. 5, 11. National Printing Department, Cairo.

ELLIOT SMITH, G. and WOOD JONES, F., 1910: Report on the human remains. *Archaeo. Survey of Nubia. Report of 1907-1908*, 11:375pp. Cairo.

KEITH, A., 1948: *A New Theory of Human Evolution*. London, Watts.

OAKLEY, K. P., 1955: Fire as Palaeolithic tool and weapon. *Proc. Prehist. Soc.*, 21:36.

POLGAR, S., 1964: Evolution and the ills of mankind. *Voice of America. Forum Anthropology Series*, 20:1.

STEWART, T. D., 1959: The restored Shanidar I skull. *Smithsonian Report for 1958*, Washington, pp. 473-480.

STEWART, T. D., 1962: Personal communication.

WEIDENREICH, F., 1939: The duration of life of fossil man in China and the Pathological lesions found in his skeleton. *Chin. Med. J.*, 55:34.

ZEUNER, F. E., 1963: *A History of Domesticated Animals*. London, Hutchinson.

Chapter 6

Health and Disease in Contemporary Primitive Societies

IVAN V. POLUNIN

INTRODUCTION

There exist in various parts of the world groups of people who are usually called primitive. This term is unsatisfactory as Mednick (1960) has pointed out, for it implies that the primitives were the first people to inhabit the area they occupy, and with the spread of civilization it is becoming more offensive and inappropriate. The word is used here merely to describe simple ways of living, free from implications of genetic or cultural evolution.

Primitive peoples tend to be far removed, both geographically and in their ways of life, from the more developed societies. Their societies appear to have changed much more slowly, until recently at least, and there has been little of the scientific and technological development which has so enormously changed the life of the people in advanced countries. Development has not proceeded to a similar degree in all directions, and it is not surprising that certain types of knowledge and ability are highly developed in primitives. Thus, knowledge of useful plants is much greater among Negritos in the Philippines than in typical members of advanced societies (Fox, 1952). Social attributes of less immediately obvious practical importance, such as language or religious ceremonies, may be extremely complex in otherwise primitive societies. Simplicity and slight degree of development is, however, true of the broad range of their culture.

Conditions of life have changed dramatically for some groups, because of contacts with other peoples. Some contacts have involved armed conflict, while others involved increasing dependency on relief aid, or a change to a cash economy. Contacts have usually involved the acquisition of cultural characteristics from the intruding groups, as well as pathogens and sometimes genes.

ANCIENT MEN AND CONTEMPORARY PRIMITIVE MEN

The justification for discussing disease in primitive societies in this book is that modern primitive men may be more similar to ancient peoples than are modern civilised men, from whom most of our knowledge of disease is obtained. If this is true, then a brief review of the disease pattern in present-day primitive peoples should give us some indication of situations which can be studied only by limited methods in ancient peoples.

All primitive men have in common the fact that they live very different lives from that of the sophisticated reader. Viewed a little closer, different groups show a remarkable variety in the type of environment they inhabit, in their means of livelihood and in their social organization. Furthermore, no s p e c i a l characteristic exists which divides primitive societies from the others, though possession of a written script perhaps comes nearest to this. It would seem inevitable that this variety of circumstances should be reflected in marked differences in disease pattern. A similar variation must have existed among different examples of "ancient men." These include groups who lived in areas and in a way not greatly different from present day primitive people. Others like the Ancient Greeks and Romans shared with modern men a tendency to urban living and the production of written records, while the Ancient Egyptians, with their sophisticated technology concerning the dead, have produced the most useful materials for study by palaeopathologists.

To sum up, it may be said that the ancient men whose remains survive should not be considered as similar to modern primitive men, nor should primitive men be regarded as surviving examples of ancient men. However, it can be stated with some confidence that present day primitive men live under conditions which are more like those which were widespread in ancient days than those in advanced communities, and their disease patterns are probably more like those of ancient men, than those of modern sophisticated men.

The dividing line between civilised and non-civilised disease pattern comes usually in a very different place to that between primitive and civilised peoples. It is probably true that most of the ancient civilisations were primitive as far as public health measures and disease pattern were concerned, and most of the so-called "diseases of civilisation" are only being reported to occur very commonly in the last generation or two in the most developed communities. "Diseases associated with modern life" might be a better term.

Some Characteristics of Primitive Populations

Settlement Size and Isolation

Primitive settlements tend to be small, for several reasons. Hunting, gathering and to a lesser extent shifting cultivation are methods which exploit the environment extensively, not intensively. A little use is made of a wide geographical area, and a large settlement uses so large an area that it is more convenient to split into smaller groups, particularly when the population is sparse, and where transport is by man-carrying, rather than by more efficient methods such as boats, pack animals and wheeled vehicles (Polunin, 1961). Where the land is unproductive as in the Australian and Kalahari deserts, small bands of hunter-gatherers may be the only groups that can survive and it takes many square miles to support a man. In small societies there tends to be a lack of specialisation between individuals and an impoverishment of the total range of ideas and skills available in the community. The adverse effects of living in small endogamous groups, which have been ascribed to the genetic effects of in-breeding, may be mainly due to what could be loosely called "cultural in-breeding." It is not surprising that many of the most distinctive primitive populations studied by anthropologists, such as the Negritos, the Bushmen and the Veddahs, are small.

Small populations are usually isolated, with few outside contacts allowing a flow of pathogenic propagules, ideas and genes.

(The word "propagule" is used to include any self-reproducing agent such as virus particle, bacterium, spore, egg or infective stage larva.) It is not surprising under such conditions that striking differences exist in the prevalence of infectious diseases, ideas and genes. In the case of pathogenic propagules which spread from man to man, and genes, their disappearance from the population will usually lead to a permanent loss, unless they are reintroduced from outside. Unusual gene frequencies, with complete absence of commonly occurring genes, are not uncommon in small genetically isolated populations.

Environmental Effects

It is characteristic of primitive groups, that their activities lead to only minor disturbance of the environment. This is least in hunting and gathering but even shifting cultivation usually interferes with the forest only temporarily, provided the intensity of exploitation is not too great. This is to be contrasted with the more massive changes associated with permanent crop husbandry, irrigation and urban development. These activities lead to a decrease in the number of species and an increase

FIGURES 1 & 2. Carvings made by members of the Jah Hut Tribe, Central Malaya.

FIGURE 1. *Bes Chemak N'tang* (The Sharp Eared Spirit). It lives in trees of the strangling fig, and its sharp ears pierce the victim's ear and may give rise to ear disease. If magical treatment is delayed, it may remove the soul through the external auditory meatus with fatal results.

FIGURE 2. *Bes Penajin*. A forest spirit which drinks the milk of lactating mothers causing loss of weight in the mother and stoppage of milk flow, which can lead to infant death. If no other milk is available, Bes Penajin will drink her own milk.

of population density of those species which are represented. These may be useful to men and are hence encouraged to reproduce, or are "weed" species of animal or plant which are able to exploit the new conditions in competition with men. In South-East Asian tropical rain forest, felling may lead to the replacement of a mosaic of several hundred species of trees with a regenerating thicket where one species is heavily dominant; Audy and Harrison (1951) have shown smaller but similar changes in rodent numbers.

This encouragement of certain species can be expected to lead to changes in human disease patterns, particularly when they concern animals of medical interest. These may be the reservoir hosts of pathogens, as rodents are for scrub typhus and plague bacilli, or domestic ungulates for Brucella. They may be vectors of disease, such as the snail vectors of schistosomiasis which has increased greatly in numbers due to modern irrigation practices in parts of Africa, and the insect vectors of such diseases as the African trypanosomiases (Langridge *et al.*, 1963) and malaria which have been greatly influenced by modifications of the vegetation as well as by the presence of Man himself. Dense human settlement eliminates the tsetse vectors of trypanosomiasis.

Nutritional Aspects

It is to be expected that the primitive hunter-gatherer would eat a wide range of animal and plant species, and that he would tend to eat the whole of the edible portion. Because of the large number of nutrients necessary for normal bodily function and their uneven distribution in foods, such a person would be expected to have a fairly well balanced diet. The introduction of subsistence agriculture may be expected to reduce the number of species eaten.

FIGURE 3. Starvation in a baby due to failure of lactation in the Mother (Semai-Senoi tribe, Central Malaya).

The situation can become dangerous where a population has moved over to cash-crop agriculture or wage-earning and where refined foodstuffs such as highly milled cereal products and sugar form a major part of the diet. It appears that beri-beri in the Far East, and kwashiorkor in Africa have increased in prevalence for reasons such as these. Trowell (1950) stated that kwashiorkor is uncommon in really primitive tribes normally practising mixed farming, but is common in more sophisticated areas where there is a cash-crop economy and population pressure, and Hirsch (1883) noted the recent appearance of pellagra. Collis, Dema and

Omololu (1962) described the bad effects which can follow the sudden shift to a cash-crop economy by Nigerian cocoa growers. Beri-beri has been most severe among wage earners in South-East Asia eating little else but highly milled rice and is rare in areas where home-pounded rice is eaten (*Rice and Rice Diets*, 1954). In advanced countries increases in the availability of different foodstuffs and the application of nutritional knowledge have led to great improvements in nutrition in spite of the increased use of purified foodstuffs.

All primitives do not have a well balanced diet, however. The Papuan highlanders are heavily dependent on the sweet potato (Oomen *et al.*, 1961), to the extent of 90 per cent of calorie and 50 per cent of protein needs. Little animal food is available even though the people eat a wide range of animals. Not surprisingly, kwashiorkor is endemic (Couvée, 1962a; Bailey and Whiteman, 1963). Scurvy epidemics have been reported in Australian Aborigines particularly during droughts (Basedow 1932). The quantity of food in the diet of primitive men is often inadequate. There may be hungry seasons of the year, or famine due to climate, epidemics or hostilities.

Specialisation in Primitive Groups

Primitive societies usually show a low degree of specialisation. Characteristic activities of their members are largely those conferred by membership of biological groups determined by sex and age, although the activities of such groups are largely prescribed by custom. It is com-

FIGURE 4. *Lanoh* Negrito lean-to shelter in North Malaya. There is close bodily contact in a well-ventilated dwelling which offers little protection from bad weather.

FIGURE 5. Part of a well-ventilated home. The raised floor of the house is easily cleaned as it is made of convex bamboo slats, while numerous slits in the flattened bamboo stem wall also provide ventilation. Murut Tribe, North Borneo.

mon to find only two types of specialists in a community, the leader and those persons concerned with magic, religion and health. We would expect from this that diseases would tend to be rather evenly distributed within such a biological group, in contrast to the situation in complex societies where disease patterns often differ markedly in different social strata and occupational groups.

Fixity and Mobility of Settlements

The settlements inhabited by primitives tend to be less restricted to one place than are more advanced communities. Shifting cultivation encourages shifting habitation. This can have several different types of epidemiological effect. Houses or shelters which have to be made at frequent intervals are often very cramped, and overcrowding leads to an increased spread of contact infections especially within the family, in spite of an extremely low density of population. On the other hand temporary dwellings, such as those shown in Figures 4 and 5, are usually very well ventilated, which lessens the risk of airborne infections. The igloo, a poorly ventilated temporary dwelling, is a notable exception.

There is also the possibility that nomadic groups may travel too rapidly to favour the transmission of certain infections. Thus de Zulueta (1956) has found that the nomadic Punan of Sarawak had less malaria than their neighbours who inhabited permanent longhouses. The Punan may not have stayed long enough in one place to render the local vector mosquitoes infective to man, which takes about ten days in Sarawak. Polunin (1961) thought that the low prevalence of *Ascaris lumbricoides* in several populations in South East Asia could be due to frequent shifts of house, which took place before heavy enough "seeding" of the soil could take place with *Ascaris* eggs to effect a high level of transmission. Another possible explanation is that *Ascaris* has not yet become established in certain remote populations.

The Demographic Pattern of Primitive Populations

Most primitive people marry early, and tend to have high fertility rates though these may not approach the maximum potential fertility rates for various reasons such as a taboo on sexual intercourse during a prolonged lactation which is almost universal in East and Central Africa (Foy et al., 1954), and abortion in Melanesia (Baker, 1928), and Yap (Schneider, 1955) or damage to the reproductive organs due to infections.

Death rates at all ages are usually very high in primitive populations, particularly at the youngest ages, giving a broad based population pyramid, with a high percentage of young people and a low percentage of old people. This influences the disease pattern in the population, and we should expect it to lead to a low prevalence of the malignant growths and degenerative diseases which are characteristic of old people. Macdonald (1951) considers that a population with high fertility can maintain itself in spite of the death of half the children before the age of fifteen years.

It is difficult to provide any precise information on demographic processes in the absence of censuses and birth and death registration. The most accurate data available concern the survival of children born to surviving mothers. Child wastage rates are usually high, as shown by the data in Table I. The typical primitive population shows a slow increase of population rate, with a high birth rate nearly balanced by a high death rate. The rate of increase from year to year would be higher but for disasters, such as epidemics, famines or wars which may cause periodic diminution in the population. The small size and wide separation of most existing primitive communities supports the conclusion that population growth is not rapid. It also fits in with all estimates made for the growth of the world's human population during the periods when the majority of people were uncivilised.

Primitive populations sometimes have an excess of elderly males over females (for examples see Scragg, 1957; and Schofield, 1962). This is the reverse of what is usually seen in advanced populations, and it may be due to poorer food, the hard and unprotected life of such women and to maternal mortality.

Depopulation

There are many instances of primitive populations undergoing a serious decline in numbers, or even becoming extinct like the Tasmanians and Fuegians, and wide areas have been depopulated.

Such changes have probably occurred at all stages of history, as is shown by the human fossil record, but they have been particularly important in recent times, when they have usually followed abrupt contact with outside peoples. They are

TABLE I
PERCENTAGE MORTALITY RATES FOR CHILDREN FROM BIRTH TO VARIOUS AGES

Ethnic Group	Country	Reference	Percentage
Moï	South Vietnam	Farinaud and Prost (1939)	30% Infant Mortality Rate (under age 1 year).
Various	N. Netherlands New Guinea	Van der Hoeven (1956)	10 to 35% Infant Mortality Rate.
Navaho	United States	Hadley (1956)	13.9% Infant Mortality Rate.
Various	Uganda	McFie (1959)	25% Infant Mortality Rate.
	Imesi W. Nigeria (1957	Morley (1963)	29.5% Infant Mortality Rate.
	Uganda	McFie (1959)	30% die before age 5 years.
	Imesi W. Nigera (1957)	Morley (1963)	69.7% die before age 5 years.
	N. Netherlands New Guinea	Van der Hoeven (1956)	41 to 58% die before age 15 years.
Kha and Xos	Central Laos	Chesneau (1930)	35.3% die before age 18 years.
Phou Thai	Central Laos	Chesneau (1930)	33.0% die before age 18 years.
Primitive Seks	Central Laos	Chesneau (1930)	31% die before age 18 years.
Primitive Nhos	Central Laos	Chesneau (1930)	26.1% die before age 18 years.
Gola	Liberia	Poindexter (1959)	80% Conceptuses die before age 18 years.
Various tribes	Malaya	Polunin (1953-4)	28.7 to 59.5% Predecease their surviving mothers.
Kapaukus	Netherlands New Guinea Highlands	Couvée (1962b)	38.7 to 47.2% Predecease their surviving mothers.
Muruts	North Borneo	Jones (1962)	33.8% Predecease their surviving mothers.

somewhat similar to the decline and extinction of certain animal species which have occurred in prehistoric times, and which are continuing at the present time because of human influences. In general it is the populations which do not adapt to the changing circumstances which suffer most. In man where rapid adaptability is possible by social means, the decline in numbers following outside contacts leads to extinction or it may stop and be followed by an increase in the population (Cook, 1945). Such a "contact cycle of depopulation" has occurred among North American Indians, the Maori (Price, 1950) and the Muruts of North Borneo (Jones, 1962).

If we ignore social factors such as migration or change in ethnic identity, a change in the size of a population depends entirely on the balance between births and deaths. Any factor influencing births or deaths will influence population size and is relevant in explaining depopulation. The causes of depopulation must therefore always be multiple. Nevertheless certain factors may have aroused notice because of their overwhelming importance, their recent appearance or their unusual nature.

Hostilities between natives and settlers have led to depopulation in several areas, though only a part of this depopulation is due to deaths by violence. Epidemics due to newly introduced pathogens have been a very important factor, and Buxton (1928) has pointed out that they can be introduced by local travellers, and can precede the entry of foreigners from outside.

Violence and killing infectious diseases are easily comprehensible direct causes of depopulation, but it is likely that peaceable incursions by immigrant populations have other effects such as interference with livelihood which may indirectly lead to depopulation. Rivers (1920, 1922), in essays which aroused wide interest, suggested that depopulation in Melanesia was due to the loss of interest in life and of the "will to live," which followed the destruction of traditional ways of life.

Many physicians and their patients believe that the outcome of an illness may depend considerably on the patient's mental attitude. Among Australian Aborigines death sometimes appears to occur following "bone-pointing" witchcraft. The victim is convinced that he is going to die and there is usually no obvious physical cause of the death (Cleland, 1928b). Davidson (1954) states that sick Africans, who believe they are victims of witchcraft, may lie down and die if they are not satisfied that the evil spirit has left them. Schofield and Parkinson (1963) described a New Guinea victim of sorcery who was in a hysterical rigid state and who did not eat or drink. But for their intervention he might have died of dehydration and of hypostatic pneumonia. In the gorilla, bereavement can lead to loss of appetite and death without obvious external cause (Annotation, *Lancet* 1939). However, the idea that depopulation occurs merely because the people do not want to live has not been generally acceptable, perhaps because physiological processes necessary to life are generally under automatic control. Baker (1928) in criticism of Rivers' ideas states that deaths among the Melanesians he observed were due to recognisable diseases, such as influenza and dysentery, and that demoralisation was the effect, rather than the cause of depopulation.

The indirect effect of psychological factors has been argued by Polunin (1959). Mental states and processes are determinants of behaviour, and man's environment depends greatly on his behaviour. Most killing disease among primitives is due to unfavourable environmental conditions, many of which could be improved even in the absence of special medical or hygienic measures, and it is reasonable to suppose that states of mind influence morbidity by such indirect means. Observers of many primitive groups have noted a falling off of standards of traditional arts and technology following outside contacts. It is likely that there is also a falling off of standards of less obvious activities which maintain the sanitary adequacy of the environment. Many authors (Rivers, 1920; Buxton, 1928; Hercus and Faine, 1951) have described a worsening of the domestic environment when traditional housing is superseded by modern housing, without an appropriate change in habits. In North Borneo the traditional slatted bamboo floor of the temporary house (Fig. 5) tends to be replaced by a rough sawn floor of planks placed close together. The smooth convex bamboo floor with adequate space between slats is easy to clean and is self draining of water and dust, while the wooden floor which is not washed or cleaned collects and retains most pathogenic propagules deposited in the room. This is particularly important for those which survive for long periods, such as roundworm eggs, tubercle bacilli, streptococci and tetanus spores. The wearing of copious European clothes which were allowed to stay wet and dirty on the body have often been mentioned as a cause of diseases. Baker (1928), Dubos (1959), Motulsky (1960) and others mention epidemics of communicable diseases which killed a majority of an affected primitive group. Waddy (1959) points out the importance of more insidious infections in leading to depopulation, such as trypanosomiasis, malaria, tuberculosis and venereal diseases, and onchocerciasis which leads to people abandoning villages for fear of blindness.

Failure to produce children has been shown to be an important factor in some depopulating groups. Most infertile women examined by Scragg (1957) in New Ireland had blocked Fallopian tubes, which he ascribed to gonococcal infection, and De Zulueta and Lachance (1956) ascribed depopulation to gonorrhoea in the Tinjar area of Sarawak. Polunin and Saunders (1958) in North Borneo found that infertile women had abnormalities of the uterus or Fallopian tubes incompatible with childbearing, which they considered were due to postabortal or puerperal nongonococcal infections.

Epidemiology of Communicable Diseases

Epidemics in Primitive Populations

Diseases transmitted from man to man can be divided into those conferring a high and prolonged degree of immunity and those conferring little or no immunity. In the former, the size of settlement and amount of contact between settlements is of great importance, as infected individuals are rendered insusceptible and are "used up" while reintroduction of the infection depends on contact with neighbouring populations. In primitive populations which tend to be small and isolated such epidemics tend to be infrequent and hence to occur in populations with a high proportion of susceptibles. This situation is seen in its most extreme form in small populations on islands, but small isolated inland populations can be looked on as islands of people in a sea of uninhabited country. Close contacts (within the group) and lack of measures to isolate infected persons during an epidemic lead to most of the susceptibles becoming infected. The devastating effect of what in sophisticated populations are usually mild childhood infections has been noticed, in primitive groups exposed to a disease apparently for the first time, by Hirsch (1883), Cleland (1928a), Perla and Marmorston (1941) and many others. Adels and Gajdusek (1963) reported a mortality of 10 to 20 per cent from a measles epidemic in New Guinea where they had previously demonstrated serologically that the under-twenty-four-year-olds had not previously suffered from measles. They describe measles in such areas as following a "snake like pattern of progression," sweeping through populations and affecting all ages in some villages and bypassing others. The great influenza pandemic of 1918 had features in common with these primitive epidemics, except that the susceptibility of the population was probably due to the emergence of a new type of virus.

It has been suggested that prolonged

exposure of a population to a pathogen may increase the resistance of the population by weeding out susceptibles (Cleland, 1928a). However, an increase in the resistance of the population as shown by falling mortality rates has in some instances been rapidly developed within one or two generations. For example, measles case fatality rates in Fiji fell from a figure of over 20 per cent in 1875 (Christensen et al., 1952) to 6 per cent in 1907 (Corney, 1913) and tuberculosis mortality rate in the Canadian Indians of Qu'Appelle Valley fell from almost 10 per cent per annum in the 1890's to 0.2 per cent forty years later (Ferguson 1955). Such a marked and rapid change if due to natural selection would involve a rapid change of gene frequency, and this would only occur if susceptibility depended on a single dominant or sex-linked recessive gene. We would then expect two clinical types of the disease with greatly different mortality rates, with characteristic distribution patterns in kin groups, and this has not been demonstrated. Genetic determination of susceptibility to most infections depends on several genes, in which case natural selection would only produce a slow effect on gene frequencies. High mortality rates may be partly due to low standards of care of the sick, which are made worse by the disorganization to which an epidemic may lead, and low resistance due to malnutrition. There are several instances quoted by Hirsch (1883), during severe killing epidemics of this type, of low and conventional death rates among those individuals who were adequately cared for, and the first epidemic of measles to effect Greenland had a death rate of 1.8 per cent probably because of the good care the patients received (Christensen et al., 1953). "Virgin soil" epidemics in natives of Ungava Bay and Baffin Land had case mortality rates of 7 per cent and 2 per cent, the difference being ascribed to differences in environmental conditions and care (Peart and Nagler, 1954).

Another possible explanation concerns the age at which persons are attacked. Many diseases show marked variations in severity at different ages. The commonest situation in most childhood infections is a death rate which is higher in babyhood and adulthood, and lower in childhood (Perla and Marmorston, 1941). A common endemic infection conferring prolonged immunity characteristically affects children. It will be uncommon during babyhood as babies usually have restricted opportunities to acquire such an infection; and most susceptibles become infected and hence immune before they reach adulthood. Such an explanation for the apparent resistance of Africans to yellow fever has been advanced by Boyce (1910, 1911).

A similar situation may occur where immunity is partial. A population whose members receive frequent malaria infections has a high degree of resistance. There may be no very obvious signs of malaria in the population except for large spleens and parasitaemia. In spite of this a mortality estimated to be usually about 1 to 2 per cent per year is thought to occur in babies before resistance is acquired at the age of about nineteen months (Wilson, et al., 1950). These children are likely to die at home without being seen by health workers. In addition, malaria is probably a contributing cause of death in a much greater proportion of small children (Bruce-Chwatt, 1952).

Genetic Factors and Infections

In spite of these complicating factors there is evidence that genetically determined variations in susceptibility to infections do exisit in man. Carefully con-

FIGURE 6. *Lanoh* Negrito with deep burn of thigh from a fall onto a fire during an epileptic convulsion.

trolled studies on domesticated mammals leave no doubt of the role of genetic factors in resistance to infectious disease (Motulsky, 1960). However, human populations are genetically more like "wild" populations in that artificial selection and intensive in-breeding are not practised and the evidence for genetically determined variations in susceptibility is not so striking in man (see review by Motulsky, 1960).

The clearest example of a gene influencing susceptibility to an infection is provided by the haemoglobin S. gene, which is an allele of the gene for normal haemoglobin. Allison (1954) has shown that children heterozygous for haemoglobin S have a better chance of surviving infection with malignant tertian malaria than have normal homozygotes.

The haemoglobin S. gene is prevented from reaching a frequency of 100 per cent in malarious populations where it occurs, because the homozygotes develop a serious hemolytic anaemia which is usually fatal. Red cell glucose 6-phosphate dehydrogenase deficiency, thalassaemia which interferes with normal haemoglobin synthesis, and the other haemoglobinopathies such as haemoglobin C trait and disease are single-gene characters which are common in some populations and which can give rise to serious disease, and evidence is accumulating that they also protect against malaria (Annotation, *Lancet*, 1962).

Where the advantage to haemoglobin S. heterozygotes has been abolished in populations not exposed to malaria, the haemoglobin S. frequency will be lowered by the elimination of heterozygotes, as has occurred in Surinam (Jonxis, 1959) and probably in United States Negroes (Allison, 1959). It is generally true that man has evolved for most of his history under "primitive" conditions, and we would expect him to be reasonably well adapted to them and less well adapted to life in societies of advanced technology. Genes influencing susceptibility to "diseases of civilisation" will in future be the subject of selection, providing they affect individuals before they have completed their families. The primitive environments are selectors of genes, some of which are no longer favourable, much in the same way that aggressive patterns of behaviour which developed during "primitive" conditions pose a danger to men living in the modern mechanised world.

Man and Other Mammals and Their Infections

There are many human infections which are only known to occur in man, or in which man is an obligatory host. These include viral, bacterial, fungal, protozoal, and helminth infections, as well as arthropod ectoparasites. Perla and Marmorston (1941) give an incomplete list of twenty-six such conditions. Garnham (1958) has pointed out that all human infections must have been zoonoses at the time of the evolutionary emergence of Man. It is reasonable to suppose that when human populations first evolved, man was by no means the dominant species he is now, and his effect on the environment was not greatly different from that of other large mammals. Early men, like the contemporary African villagers described by Carmichael (1952), lived in contact with a large number of wild and domestic animals.

In early times it was likely that the number of exclusively human infections was much smaller than it is now. With the closer contact of early man with other mammalian species, there may have been many "sylvatic" infections which affected man. Evolution of pathogens in man would tend to occur in a different direction than in the other hosts where the internal environment is different. In time, given conditions in which transmission took place without infecting mammals other than man, a pathogen might be evolved which was so highly specialised for life in man that it could no longer inhabit other mammals.

There are several diseases such as yellow fever, plague and relapsing fever (Garnham, 1958), where a single pathogen can cause epidemiologically or even clinically different patterns owing to differences in contacts between man and reservoir hosts and vectors. Thus yellow fever virus causes an infection in monkeys and *Aedes africanus* mosquitoes in the jungle canopy, which later may occasionally infect man by biting him in the forest at night. Monkeys raiding garden clearings in the forest may infect men via *Aedes simpsoni*, and an infected man may travel to town and initiate an epidemic of yellow fever transmitted by urban *Aedes* species (Garnham 1952). The situation is similar for South American trypanosomiasis. A disease of forest mammals transmitted by reduviid bugs, it may infect the peridomestic opossum and hence dogs, bugs and men in houses, where it may become a purely man-bug-man infection (Garnham 1958). Malayan filariasis (Wilson, 1961) found in man and transmitted by insect vectors is a zoonosis and loiasis (Gordon et al., 1950) and quartan malaria (Garnham, 1958) may be. In all of these diseases the primitive forest dweller is more likely to acquire the infection from a forest vector than is the case with more civilised people, where greater chances exist for acquisition from reservoir hosts such as man, domesticated animals or animal "weeds."

Some of the diseases just discussed appear to be on the way to becoming exclusively human infections, although it is unlikely that modern men will ever allow this to happen.

Communicable Diseases Which Do Not Confer Immunity

The epidemiological effects on settlement size and isolation are rather different for diseases which do not confer a significant degree of immunity. Random factors may lead to rapid changes in prevalence rates of diseases in small populations while they will tend to cancel each other out in larger populations. This is somewhat analagous to genetic drift, which is expected to be marked in small breeding populations.

The smaller a settlement the greater the chance that a purely human pathogen will be absent from it; and the greater the isolation from outside human contacts, the smaller the chance of it entering. The chance of a pathogen being absent is greater when its general prevalence is low. Thus it is probable that a high percentage of small, isolated "primitive" communities have no leprosy, and it is certain that many do not have the easily recognisable ringworm, Tinea imbricata, even though similar neighbouring communities have these diseases. It is also commonly found that where such an infection occurs, its prevalence is high. This is because of close contact between individuals, poor standards of hygienic practice and the absence of effective therapy.

Heinbecker and Irvine-Jones (1928) have described the situation among the Polar and Baffin Land Eskimos, who were exceptionally isolated. They found an absence of upper respiratory tract infections except after visits from outsiders which regularly led to epidemics of such diseases, and a remarkable tendency for wounds to heal without signs of infection. However, they found evidence of circulating diphtheria antitoxin and streptococcus antitoxin.

Mann (1961) reported an almost complete absence of *Staphylococcus aureus* from nasal and ocular cultures in Australian Aborigines from Groote Eyelandt, together with an apparent absence of styes and blepharitis.

Health Promoting Practices, Personal and Environmental Hygiene and Care of the Sick

All peoples and even higher animals undertake some practices which might be called hygienic in that they promote health, by design or accident. In primitives these measures are usually carried out with insufficient care, and they are often inappropriate in the light of what we know of the maintenance of health and the spread of propagules.

Personal cleanliness depends largely on the beliefs and practices of the people, the climate and the availability of water. Most peoples living at low altitudes in the tropics with abundant water bathe frequently and thoroughly. If the people live far from water or in colder climates, washing tends to be infrequent.

Practices which can all be regarded as improvements on the worst possible hygienic standards are spitting through gaps in the floor, throwing refuse out of the house, defaecating away from the house, the perianal toilet after defaecation by means of water, stick or leaves, the cleaning up of faeces deposited on the floor by small children, the washing of clothes, the placing of food in leaves, bamboo tubes, gourds, baskets or modern containers, particularly if lidded, the cooking of food, the separation of domestic animals from people in the house and excreta disposal by coprophagous pigs and dogs.

American Indians (Hirsch, 1883) abandoned and burned their homes during epidemics.

In general, primitive standards of hygiene are extremely low. Defaecation is commonly on the ground in shady places, where there is privacy and this helps in the preservation of propagules contained in the faeces. Defaecation into water is dangerous if the water is stagnant or if flow is small relative to the population density and it can lead to water born infections, and to the maintenance of helminth infections acquired while wading or by eating raw foods obtained from stagnant water.

Modern hygienic practices depend on detailed knowledge of the aetiology of certain diseases which have often only been obtained by highly technical means. Much of this knowledge runs counter to what might be called common sense. Thus it is not obvious that multiple vaginal examinations, practised by Muruts of North Borneo, (Polunin and Saunders, 1958) are dangerous to women at childbirth, but they may be a major cause of infertility, while the connection between infected faeces and some bowel diseases is often poorly understood. Primitive beliefs on the causation of disease have been discussed by Sigerist (1951). Contact with spirits, the offending of spirits or failing to appease them are considered to be important causes of disease among animistic peoples (Fig. 1, 2). Disease producing spirits are thought of as having attributes similar to the human being rather than to micro-organisms. Another type of belief prevalent in New Guinea (Schofield and Parkinson, 1963; Van Loghem, 1960) and in Africa is that disease is due to harmful practices carried out by people of ill will, by methods which we would call magical, and often involving manipulation of body products such as hair or faeces. In South and East Asia disease may be explained by the wandering of souls away from the body, and the breaking of taboos is widely believed to be a cause of disease. Animistic theories of disease lead to what we would call magical practices which do not usually lead to effective prevention or treatment of diseases, though they may be important in promoting mental health and curing psychogenic illness.

Aesthetic considerations also play a part in promoting hygienic standards. Thus there is a tendency to put faeces and refuse where it cannot be seen or smelt or where it may not cause a nuisance by the flies it attracts. Care of the skin and hair may be more for aesthetic than for hygienic reasons.

Little effort may be spent in maintaining hygienic standards. It is common to see food ready for eating left uncovered for long periods, and no attempt made to remove spoiled food or to clean utensils. This appears to be part of a general passivity when dealing with matters which are not considered to be of great interest, and is to be contrasted with the much more vigorous, time consuming and costly exertions in fields the people consider important, such as magical means of treatment.

Methods used for the physical care of the sick are usually little if at all more effective than the measures which are instinctively followed by sick persons or animals; such as rest of the whole body, or of those parts which are the site of pain, drinking to satisfy thirst, stopping eating when there is a loss of appetite, sitting up with orthopnoea and other such measures.

Actions taken to deal with sickness may themselves be harmful. Thus sparganosis among traditionalist Chinese is due to the application of raw frog poultices to inflamed eyes, and opium addiction among the Meo (Miaou) often follows opium smoking for the relief of symptoms of disease. The Muruts of North Borneo and the Fijians were said to treat fevers by plunging into cold water and the diets of many peoples are seriously restricted during the dangerous periods of pregnancy and the puerperium when needs are increased. The reasons for such practices may be highly logical but they are based on erroneous premises. Schofield and Parkinson (1963) point out that beliefs with consequences harmful to health may nevertheless play a valuable part in promoting social cohesion and should not be countered by direct opposition.

The Characteristics of Available Data on Primitive Populations

Data on hospital admissions are limited or non-existent for most primitive populations, as the people tend to be remote from hospitals and reluctant to enter them, and the findings are likely to be lost among the statistics for more advanced groups. Necropsy data are even more scanty. Information on hospital patients gives an inaccurate picture of the disease situation in a population, unless all the sick come to hospital.

Much of the information on primitives has been obtained as a result of surveys, as it is often possible to muster a large proportion of a population for a medical examination. Such a survey describes only the disease situation at the time of examination, though some information can be obtained by observing the residual signs of disease, such as scars and deformities, or the serological evidence for previous infections. Such surveys are of little use in studying diseases of short duration, which include many important causes of sickness and death.

Another limitation of such surveys is that unless enormous numbers are studied, it will not be possible to establish prevalence rates for rare conditions. The impression may be gained that uncommon diseases are less common than they really are. It is often said that the malignant neoplasms are rare in primitives but it is doubtful whether there are adequate grounds for making such a statement, when the age structure of their populations is considered, though there is evidence (Davies, et al, 1962) that cancer incidence rates do not show their usual increase with age in Ugandans.

These limitations are not as serious as might be expected, as a large proportion of the total morbidity is often due to a small number of diseases which are present in a large proportion of the population. This is one of the more striking characteristics of the disease pattern of primitives, and is to be contrasted with the situation in sophisticated populations where different diseases each contribute a small part to the total morbidity, as the formerly common diseases are largely under control.

As many persons will be harbouring several infections and perhaps dietary deficiencies, the clinical picture is often a complicated and confused one. Thus the relative importance of hookworm infections, malaria and dietary inadequacy is difficult to assess in a lactating woman with a hypochromic anaemia or protein-calorie malnutrition.

DISEASES CHARACTERISTIC OF PRIMITIVE WAYS OF LIFE

When we allow for the effects of the age distribution of the population, marked differences still occur between primitives and modern civilized peoples. Though there is great variability between groups the general "disease load" usually appears to be heavier in primitives than in more advanced peoples.

The question arises as to whether there are diseases which are characteristic of people living in a more primitive manner. It is exceptional for a disease to be confined to primitives, which is not surprising, as there is no special environmental feature which distinguishes primitive ways from other ways of living.

Kuru

Kuru is a progressive degenerative disease of the central nervous system with distinctive clinical, pathological and epidemiological features which only occurs among the Fore peoples of eastern New Guinea, where it has been intensively studied by Gajdusek (1963) and others. It is slightly commoner in girls than in boys, and occurs in women but only rarely in men. It occurs in Fore people living outside the tribal area under modern conditions, and has not appeared among outsiders who have moved into Fore territory. It has been suggested that kuru is due to a gene which produces the disease in homozygous children and heterozygous women. It causes high death rates (up to 2 per cent per annum) among people of reproductive potential and this would lead us to expect that kuru would be reduced to rarity by natural selection. It could even threaten the survival of the Fore population, though this was increasing in 1962. No biological advantage has been demonstrated for carriers of the supposed kuru gene which could explain its high frequency. This and the apparent recent origin of the disease has led to the suggestion that another factor besides a gene has recently become operative and is needed to produce kuru. Extensive searches for environmental factors which would play a part in the causation of kuru have so far been negative. The age and sex distribution of the disease, its apparent exacerbation by sex hormones and the beneficial effects of pregnancy point to a possible endocrine factor. On the other hand similarities between kuru and scrapie, an indolent virus infection of sheep occurring in genetically susceptible animals, has led to suggestions that a similar factor may operate in kuru.

Most diseases which are commoner in primitive groups are those which need environmental conditions of a type which are generally avoided or eradicated in more sophisticated groups and several of these will now be considered.

Malaria

Tropical primitive populations vary from those where malaria is holoendemic, all members of settlements receiving several reinfections every year, to populations which are free of malaria, such as the inhabitants of parts of the Pacific.

Generally speaking, malaria prevalence appears to be rather higher in primitive than in more sophisticated village groups. However, the factors responsible for the malariometric status of a settlement are so complex, and depend so much on highly specialised features of the environment determining the distribution of vector *Anopheles* that it is not surprising that no characteristic pattern can be found in the primitives.

The environmental characteristics associated with primitive ways of life may or may not promote breeding by malaria vectors. Where environmental conditions are not significantly disturbed by man, and where these conditions support a population of malaria vectors, a primitive way of life is associated with much malaria, because of the people's close association with an unchanged environment.

Human interference with the environment, which, generally speaking, increases with density of population and increasing civilisation may promote or eliminate vector Anophelines. In certain types of terrain in Malaya, generally speaking hilly land with granite rock, clearing the forest often encourages the breeding of the vector *A. maculatus*. Clearing of forest is a typical activity of shifting cultivators, and this may explain why spleen rates are often high in Malayan hill tribes (Polunin, 1953-4). Other effects of denser and more complex settlement, such as the pollution of water, will hinder the breeding of anophelines.

There are several factors which will often operate in primitive groups, which tend to increase malaria prevalence. Open type dwellings allow free access to mosquitoes, as does the generally scanty clothing, and the absence of mosquito nets, and repellents apart from smoke. Groups with few domestic animals may receive more infective bites because animals divert zoophilous mosquitoes from biting man, and where the human population is small each person receives more bites. De Zulueta (1956) considered that such factors operated in farm huts of the Sarawak Dayaks where Colbourne, *et al* (1959) demonstrated the anthropophilous feeding habits of a largely zoophilous malaria vector in farm huts, and an association between the use of farm huts and high malaria prevalence.

Yaws

Yaws is a disease which is mainly spread by contact between people which must be fairly direct as the causative organism cannot withstand drying, while treatment readily makes the sufferer nonineffective. Inaccessibility of treatment facilities, toleration of open lesions, scanty clothing and

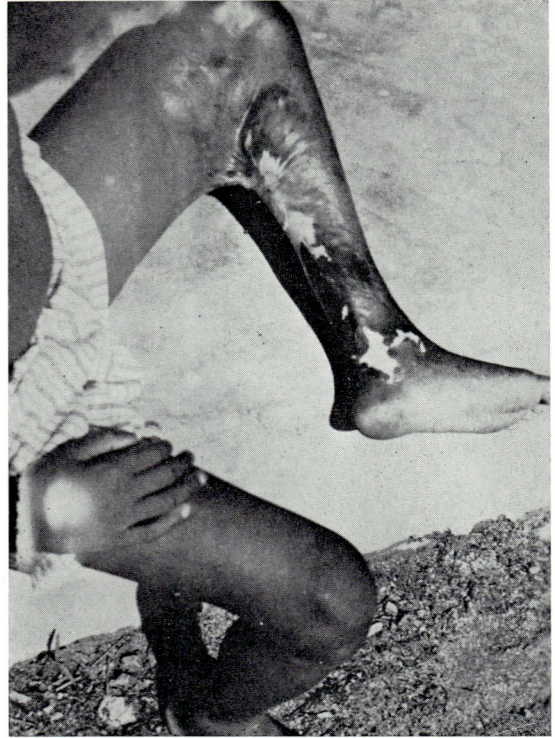

FIGURE 7. Legs of a native of the south coast of western New Guinea with gross limitation of movements of the left knee due to scar tissue. Both late yaws ulceration and burning from rolling onto the fire while sleeping apparently played a part in the development of the condition.
Photo: Dr. J. W. L. Kleevens.

close contacts between children are all causes of the high prevalence of yaws in many primitive groups in the humid tropics. The long-lasting partial cross-immunity which an attack of yaws confers against syphilis has probably succeeded in preventing the spread of syphilis to many areas such as Fiji, where gonorrhoea is said to be common.

Tinea Imbricata

Tinea imbricata, due to infection with the fungus *Trichophyton concentricum* is an infection which in South-East Asia is largely confined to the more primitive peoples, though people of any race can be infected. There is some evidence that it was formerly common in Malays, though this is no longer true (Polunin, 1952). It has not been found growing naturally except in human skin.

Prevalence is always far below 100 per cent, though all members of some communities must be exposed to infection. Infection usually develops during childhood and lasts a lifetime. The disease is difficult to cure, so it is unlikely that its disappearance is due to treatment. The conditions under which people can become infected are clearly of importance.

Schofield, et al (1963) have thrown much light on the epidemiology of the disease in Maprik, in eastern New Guinea. They found that the risk of infection depended on the closeness of contact with an infected person, though some children with both parents affected escaped the disease. It was nearly always contracted in the first two years of life and spontaneous remission was rare except among people who went to work on plantations where they were provided with an adequate diet and gained much weight. People who are cured, spontaneously or following treatment, usually develop the disease again when they return to their original way of life. Schofield, et al. (1963) found that Tinea imbricata often developed in children after a period of failure to gain weight. Of the few people who developed the disease during adulthood the onset in most of the men followed a weight-losing illness, while in women it was always associated with pregnancy or lactation which causes loss of weight in Maprik women.

Endemic Goitre

This disease due to iodine deficiency is common in many primitive groups, together with endemic cretinism and deaf-mutism. Gajdusek (1962) has described the recently contacted Neolithic people of the Mulia region of West New Guinea, many of whom had large goitres. Over 12 per cent were deaf mute or mentally defective, and all appeared to be at least slightly mentally subnormal.

Iodine concentration in each type of foodstuff usually shows extremely wide variations, depending on the availability of iodine in the area. All sea foods contain appreciable amounts of iodine, and endemic goitre is commonly found away from the sea, and where sea food is not an important item of diet. In those inland areas where the iodine content of soil water is low, as in many demineralised hilly areas in the tropics, iodine intake is inadequate for normal human requirements, and compensatory hypertrophy of the thyroid followed by colloid retention may be found in the majority of people. As many 'primitive' groups live in such areas, it is not surprising that endemic goitre is common among them. Iodization of salt entering the area is a simple and effective preventive measure which has largely prevented the condition in some areas, but it is less likely to be provided for primitive groups.

Hypochromic Anaemias

These are common and severe in many primitive groups. Thus Jelliffe et al. (1962) found mean haemoglobin values in small Lugbara children in Uganda of 6.3 grams/100 ml or well under half the normal value, and Polunin (1953-4) found two aboriginal groups in the Malay Peninsula with mean haemoglobin concentrations (all ages) of 10.6 grams/100 ml.

The low values of haemoglobin in primitives have complex causes. Inadequate iron absorption due to inadequate dietary iron intake, chronic haemorrhage due to hookworm infection, increased needs in women of reproductive age and infections such as malaria probably all play a part.

Other Nutritional Conditions

Unusual patterns of plasma proteins have often been reported. Total plasma proteins have usually been on the normal or high side, but albumin values may be low, suggesting inadequate protein intake or impaired liver function. The gamma globulin is often high (Couvée, 1962a; Mann, et al., 1962), which is of interest as antibodies are gamma globulins. This pattern is typical of African Negroes (Trowell, 1960).

Various conditions of possible nutritional significance have been noted to be common among many primitive groups, such as separation of the rectus abdominis muscles particularly in children, and enlargement of the parotid glands (Polunin, 1953-4, Couvée 1962a). Parotid gland enlargement may be a simple work hypertrophy due to eating a copious dry starchy diet, or due to protein deficiency (Raoult, et al., 1957), while separation of the rectus muscles appears to be due to poor muscle tone and is often associated with "pot belly."

DISEASES CHARACTERISTICALLY UNCOMMON AMONG PEOPLE LIVING PRIMITIVELY

Several diseases have been recognised with greatly increased frequency during the last few decades in technologically advanced countries and it is reasonably certain that these have not been entirely due to changes in age distribution of the population, improvements in medical coverage or improvement of diagnostic methods. As the living conditions of modern men are becoming less and less primitive, it is likely that these conditions are uncommon in primitives, though this may be difficult to demonstrate.

Carcinoma of the Lung

Carcinoma of the lung has recently shown a greatly increased incidence in many advanced countries. This is highest in areas of marked atmospheric pollution by smoke and fumes, and among cigarette smokers, and it is probably uncommon in primitives.

Myocardial Infarction

It is astonishing that this condition, which has only been recognised during the last few decades should now be a major cause of death and disability in many of the advanced countries. A vast literature has grown up around the subject. As the underlying condition is usually atheroma of the coronary arteries, a great deal of interest has centered around these subintimal deposits of cholesterol, although there is less evidence that atheroma is more widespread nowadays than formerly. Plasma cholesterol values in people who have had myocardial infarcts appear to be higher than those in other members of the population, and peoples with little myocardial infarction generally have low mean plasma cholesterol levels. These values appear to be influenced by a great variety of dietary and other factors and are notably increased by a diet rich in saturated (animal) fats which also promotes thrombosis and inhibits fibrinolysis and appears to be an important determinant of infarction.

Low plasma cholesterol values have been found in the East and West of New Guinea (Luyken and Jansen, 1960; Goldrick and Whyte, 1959) and exceptionally low values have been found by Mann et al. (1962) in Congo Pygmies.

Whyte (1958) quotes Price as finding death attributable to coronary artery diseases only once in 600 autopsies in natives of New Guinea, together with a comparative freedom from atheroma. Mann et al. (1962) failed to find any definite electrocardiographic evidence of myocardial infarction in 248 pygmies. There is much evidence, summarised by Trowell (1960) that atheromatous coronary disease is rare in African Negroes coming to autopsy, and serum cholesterol values tend to be lower and show a slower rise with age than is commonly found in Europeans. Coronary thrombosis appeared to be rarer in Negroes than in Whites in the U.S.A. (Lewis, 1942).

Shaper, et al (1961) found low plasma cholesterol and phospholipid values and little electrocardiographic evidence of ischaemic heart disease in Samburu men in Kenya who lived mainly on milk and meat but who walk great distances. Shaper (1962) thought fat intake would have to be continually high to produce a rise of lipid values.

Essential Hypertension

The condition of benign essential hypertension is characterised by a rise of arterial blood pressure above an arbitrarily defined level together with associated changes, for which no cause can be found. It occurs so commonly among the older members of civilised populations, that it is sometimes thought of as one of the normal concomitants of ageing. Mean values rise with age particularly after the age of about forty years. However, a small percentage of such people have low values in old age, and the situation is dramatically different in many of the primitive populations which have been studied, as can be seen from the reviews of Bays and Scrimshaw (1953), and Lowenstein (1961). In some of these groups hypertension appears to be practically absent, and there is little or no rise with age, or even a fall. However, some groups living under fairly primitive conditions have shown high arterial pressures, such as the semicivilised Australian Aborigines reported by Casley-Smith (1959), while the Congo Pygmies studied by Mann et al. (1962) had blood pressures which did not differ significantly from Americans.

The factors responsible for these differences are not clear, but appear to operate in populations differing widely in race and geographical location. That environmental factors are important is shown by the numerous reports that hypertension is commoner in American Negroes than in Whites (Kean, 1941), unlike the situation among Negroes living primitively (Donnison, 1929). The lack of obesity may play a part, as falls in blood pressure have been noted in advanced peoples during periods of famine (Lups and Francke, 1947).

Dental Caries

Most primitive peoples are remarkably free of dental caries. Thus Klatzky (1948) found a low caries rate in 4,000 skulls from forty-six different racial stocks in a museum collection, and a complete absence of caries in those of aborigines of British Columbia and Eskimos. He also found much attrition of occlusal surfaces together with lower jawbones which had well developed ridges where muscles were attached. In the living subject, East Greenland Eskimos (Pedersen, 1947), New Guinea Natives (Sinclair, et al, 1950), Australian Aborigines (Campbell, 1938) and many others have shown similar characteristics.

The effects of ways of life on oral diseases have been reviewed by Davies

(1963). No consistent association between dental caries and any known nutrient has been established, though numerous surveys have established a direct relationship between the prevalence of dental caries and the frequency with which fermentable carbohydrate in a sticky form is consumed. Many authors who have conducted surveys in underdeveloped countries claim that the physical nature of the food is the most significant factor in the initiation of dental caries. The diets of primitive people are often abrasive, giving rise to attrition of occlusal surfaces of the teeth which appears to protect them from caries. It has often been found that resistance to caries is lost as soon as the people change to a modern diet. Chewing of the betel-nut quid which is widespread among traditionalist populations from India to the Pacific, leads to considerable attrition of teeth (Fig. 8), and is associated with a lower caries rate in Malayan aborigines (Polunin, 1953-4).

Obesity

Obesity, though it occurs, is uncommon in primitive people, presumably because they do not suffer from the physical inactivity of mechanised men, nor do they have available excessive quantities of appetising foods. Brock (1963) has pointed out that most unsophisticated diets give a feeling of fullness because of their high cellulose content. It is probable that this lack of obesity is partly responsible for the low prevalence of hypertension and coronary heart disease.

The changes in weight with advancing age in adults show some differences. Mean body weights in civilized groups tend to rise until the fifth or sixth decade. In some primitive groups loss of weight begins at a much earlier age (Polunin, 1953-4).

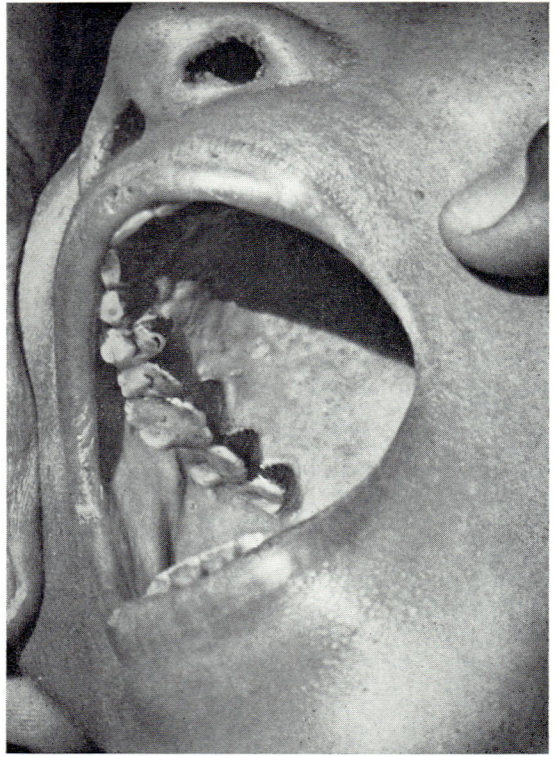

FIGURE 8. Marked attrition, absence of dental caries, regular dental arches and staining of teeth in a betel chewer. *Lanoh* Negrito estimated age forty years, North Malaya.

Schofield (1962) presents data from Sepik, West New Guinea which show that women progressively lose weight from the twenties onwards, and that one cause for this is prolonged lactation on a diet devoid of food of animal origin, and Mann *et al.* (1962) describe loss of subcutaneous tissues in Congo Pygmy women. Evidence from usually well nourished populations with restricted food intake (Neuprez, 1945; Smith and Woodruff, 1951) suggests that the loss of weight with age is due to undernutrition.

Growth and Development

Geber and Dean (1957) found that newborn Ugandan children had reached a significantly higher stage of behavioural development than newborn Europeans. Growth rates appear to be slow after the first few months, probably due to the lack of breast milk, though Jelliffe (1955) in his review of nutrition in the tropics and subtropics, found that breast feeding commonly continues for years.

Deficiency of muscle and fat is often seen in primitive children. The factors responsible are probably complex, and include insufficiency of total calories and essential amino acids; and infections by helminths, bacteria and protozoa. Maturation as measured by mean age at menarche, takes place at fourteen and one-half years in Plateau Pagans of North Nigeria, and aborigines of Central India, which is later than in advanced communities (Wilson and Sutherland, 1953). The growth curves of Congo Pygmies published by Mann et al. (1962) suggest that maturation is early as shown by the slowing of growth rates at an early age. It was unfortunate that ages had to be guessed. Mann et al. believed that small stature is largely due to heredity, and mentions the lack of evidence of unusual growth among Pygmy orphans in Mission Stations with apparently adequate diets.

It is characteristic of primitive populations that most of them are of lower stature than neighbouring groups of villagers. Genes may in part be responsible but it is difficult to escape the conclusion that hostile environmental factors have been important. Price et al. (1963) have found a higher prevalence of all intestinal parasites in the forest Pygmies than in their taller neighbours. However, it would seem that the primitive groups have not yet started to show that continuous increase of mean height which has been and continues to be characteristic of civilized peoples. Comparison of stature of various groups of Malayan aborigines at the beginning of the century and fifty years later offers little evidence that mean height is changing significantly (Polunin, 1953-4).

MENTAL DISEASES

No clear picture emerges of the prevalence of mental diseases in primitive peoples. Various opinions have been expressed on their mental health reflecting somewhat the writer's general view of the nature of man. On the one hand, there is the idea of the healthy mental state of the "noble savage" living in a "Garden of Eden." Thus Manning (quoted by Cleland, 1928c) thought that the Australian Aborigines in the wild state with a "simple and uneventful existence without wear or strain" were originally largely free of mental illnesses, which were increasing because of the evils of civilization. At the other extreme is the attitude of Christian missionaries who aimed to lift their charges out of the fear and darkness of savagery.

Cleland (1928c) has given case histories of Australian Aborigines admitted to mental hospitals. They showed a wide range of mental abnormalities. Much psychosis was ascribed to physical causes, such as senility, syphilis and alcohol, but depressive and delusional states occurred.

Carothers (1953) quotes observations which suggest that mental illness is rare in traditionalist African populations and commoner in groups undergoing change, but this may be an appearance due to the greater ability of traditional societies to cope with the mentally sick without recourse to hospitals. Schizophrenia was the commonest psychosis and affective psychoses were usually manic and only rarely depressive. Anxiety states centered on witchcraft and acute maniacal outbursts were also found. Carothers ascribed the rarity of depressive illness to cultural

TABLE II

DISEASES SHOWING DIFFERENCES IN CASE FREQUENCIES BETWEEN NEGROES AND WHITES IN MANY AREAS OF AFRICA SOUTH OF THE SAHARA

Data from Trowell (1960) by kind permission of Author and Publisher

Common in Many Negroes Uncommon in Whites	Uncommon in Many Negroes Common in Whites
Liver and gall-bladder	
Fatty changes of infantile malnutrition and kwashiorkor. Periportal reactions and fibrosis. Cirrhosis. Primary carcinoma of liver.	Gall-stones Cholecystitis Carcinoma of gall-bladder
Cardio-vascular	Severe degrees of atheroma of coronary artery. Coronary thrombosis Angina of effort Simple (postural) thrombophlebitis Pulmonary embolus
Alimentary	
Parotosis	Dental caries Carcinoma of stomach
Intussusception of large bowel	Diverticulosis Haemorrhoids
Volvulus of colon	Carcinoma of colon and rectum
Neurological	Cerebral vascular accidents Disseminated sclerosis Cerebral tumour, especially glioma Prolapsed lumbar disc Cervical spondylosis Simple tics and habit spasms
Metabolic and endocrine	
Adrenal cortical hypoplasia. Gynaecomastia in males	Thyrotoxicosis Diabetes mellitus
Blood	
Low Haemoglobin (outside area of iron-overload in S. Africa). Low neutrophil counts. Nutritional megaloblastic anaemia. High ESR.	Pernicious anaemia Leukaemia and certain reticuloses Polycythaemia vera
Other systems	
Keloid formation	Carcinoma of bronchus Eclampsia
Kaposi's sarcoma	Rheumatoid arthritis Osteitis deformans (Paget) Psoriasis
Nutrition	
Infantile malnutrition Kwashiorkor	Overnutrition and obesity
Cancer	
The incidence at many sites differs significantly from that of US Negroes and Europeans of comparable age and sex.	

TABLE III
DISEASES SHOWING DIFFERENCES IN CASE FREQUENCIES BETWEEN NEGROES AND WHITES IN CERTAIN COUNTRIES OF AFRICA SOUTH OF THE SAHARA
Date from Trowell (1960) by kind permission of Author and Publisher

Common in Negroes, Uncommon in Whites

Siderosis*	Multicentric sarcoma of jaw
Porphyria*	Endemic goitre
Idiopathic Cardiomyopathy*	Thrombocytopenic purpura (onyalai)
Abnormal Electrocardiographic findings*	Acquired haemolytic anaemia
Endomyocardial fibrosis	Splenomegaly of obscure cause
Multiple thrombophlebitis	Pyomyositis
Carcinoma of oesophagus	Pellagra and riboflavin deficiency
Retrobulbar neuropathy	Scurvy
Demyelinating neuropathy	Vitamin A deficiency
Encephalopathy states*	

Uncommon in Negroes, Common in Whites

Peptic ulcer	Appendicitis

*As in S. A. Bantu

causes, pointing out that self reproach was uncommon among Africans.

Brain (1963) mentions the rarity of suicide in West Africa. It is, however, common among the Muruts of North Borneo.

Benedict and Jacks (1954) state that mental disorders in pre-literate peoples are preponderantly hysterical-like excitements with a relative absence of severe depressions. Parker (1961) comparing Ojibwa Indians and Eskimos noted that depressive symptoms predominated in the Ojibwa who are brought up to be individualistic and suspicious, while the Eskimos who live in cooperative and supportive societies, are prone to temporary outbursts of hysterical excitement.

Some Findings on Negro People

Trowell (1960) has reviewed non-infective diseases of African Negroes. The case frequencies for many diseases show striking differences between Negroes and Whites in Africa and lists of these diseases are given in Tables II and III. The tables are based on hospital data and hence tend to exclude peoples living primitively. In spite of this, it is likely that this African disease pattern provides some valuable pointers to the situation in primitive groups, and several diseases not discussed in the present account have been reported in typical African frequency in primitive populations in various parts of the world. On the whole the disease pattern of Negroes in the United States is more like that of Europeans than Negro Africans, which suggests that it is mainly environmentally determined. Diseases associated with genetic polymorphisms are a notable exception to this tendency.

Trowell also points out that the concentrations of many important substances in the blood show characteristic differences between Negroes and Whites in Africa. Such changes are found in the mild syndrome of infantile malnutrition and still more in the similar but more serious condition, kwashiorkor, and Trowell believes that the period of malnutrition following weaning may if prolonged lead to permanent changes in the internal environment and homeostatic control. He suggests tentatively that this may have a significant effect in determining the pattern of diseases in African Negroes in later life. Brock (1955) has suggested that protein malnutrition may be a cause of liver cirrhosis, primary carcinoma of the liver and other conditions common in the tropics.

CHANGES IN THE GEOGRAPHICAL DISTRIBUTION OF HUMAN DISEASES

An interesting characteristic of many human infections has been the changes which have occurred in their geographical distribution. Two main trends can be mentioned. The first is a tendency for diseases of limited geographical distribution to be spread to new areas by the increases in communications and in population density which leads to a breakdown of isolation between peoples. This is a subject which is difficult to study because contacts sufficient to lead to the spread of diseases have usually taken place before epidemiological investigations have been possible. There is much evidence that many infections have been introduced from the Old World to the New World (Faust 1955). The opposite process, a shrinkage of endemic areas, has taken place for many diseases. Thus several diseases usually considered as tropical diseases were prevalent in temperate Europe centuries ago. The shrinkage in endemic areas has been due to a general improvement in living conditions, to specific control measures such as vaccination against smallpox and residual house spraying for the malarias, and to improved therapy as in the case of yaws. There is a tendency for such diseases to persist in populations where living conditions are still primitive.

REFERENCES

ADELS, B. R., and GAJDUSEK, D. C., 1963: Survey of Measles patterns in New Guinea, Micronesia and Australia. *Amer. J. Hyg.*, 77:317.

ALLISON, A. C., 1954: Protection afforded by the sickle cell trait against subtertian malarial infection. *Brit. Med. J.* 1:290.

ALLISON, A. C., 1959: In *Medical Surveys and Clinical Trials*, Ed. Witts, L.J. London, Oxford University Press.

ANNOTATION, 1962: Genes affecting susceptibility to disease. *Lancet*, 1:626.

ANNOTATION, 1939: Death of a gorilla. *Lancet*, 2:210.

AUDY, J. R. and HARRISON, J. L., 1951: A review of investigations on mite typus in Burma and Malaya, 1945-50. *Trans. Roy. Soc. Trop. Med. Hyg.*, 44:371.

BAILEY, K. V. and WHITEMAN, J., 1963: Dietary studies in the Chimbu (New Guinea Highlands). *Trop. Geogr. Med.*, 15:377.

BAKER, J. R., 1928: Depopulation in Espiritu Santo, New Hebrides. *J. Roy. Anthrop. Inst.*, 58:279.

BASEDOW, H., 1932: Diseases of the Australian Aborigines. *J. Trop. Med. Hyg.*, 35:247.

BAYS, R. P. and SCRIMSHAW, N. S., 1953: Facts and fallacies regarding the blood pressure of different regional and racial groups. *Circulation*, 8:655.

BENEDICT and JACKS, 1954: Cited by Parker (1961).

BOYCE, H. W., 1910-11: Cited by Perla and Marmorston, 1941.

BRAIN, W. R., 1963: Nigeria in the solar plexus. *Lancet*, 2:1325.

BROCK, J. F., 1955: Chronic protein malnutrition. *Nutr. Rev.*, 13:1.

BROCK, J. F., 1963: Sophisticated diets and man's health, pp. 35-56. In: *Man and his Future*, Wostenholme, G. (Ed.) London. Churchill.

BRUCE-CHWATT, L. J., 1952: Malaria in African infants and children in Southern Africa. *Ann. Trop. Med. Parasit.*, 46:173.

BUXTON, P. A., 1928: Researches in Polynesia and Melanesia. *Mem. Lond. Sch. Hyg. Trop. Med.*, No. 2.

CAMPBELL, T. D., 1938: Observations on the teeth of Australian Aborigines. *Aust. J. Dent.* 42:41.

CARMICHAEL, J., 1952: Animal-man relationship in tropical diseases in Africa. *Trans. Roy. Soc. Trop. Med. Hyg.*, 46:385.

CAROTHERS, J. C., 1953: *The African Mind in Health and Disease*. Geneva, W.H.O. Monograph Series.

CASLEY-SMITH, J. R., 1959: Blood pressures in Australian Aborigines. *Med. J. Aust.*, 1:627.

CHESNEAU, P., 1930: Natalité et mortalité infantile au Cammon, province du Moyen Laos. *Bull. Soc. Med Chir. de l'Indochine*, 16th year:2.

CHRISTENSEN, P. E., SCHMIDT, H., JENSEN, O., BANG, H. O., ANDERSEN, V. and JORDAL, B., 1952: An Epidemic of Measles in Southern Greenland, 1951. *Acta Med. Scand.*, 144:313.

CHRISTENSEN, P. E., SCHMIDT, H., BANG, H. O., ANDERSEN, V., JORDAL, B. and JENSEN, O., 1953: Epidemic of measles in Southern Greenland, 1951, Part 2. *Acta. Med. Scand.* 144:430.

CLELAND, J. B., 1928a: Disease amongst the Australian Aborigines, Part I. *J. Trop. Med. Hyg.*, 31:53.

CLELAND, J. B., 1928b: Disease amongst the Australian Aborigines, Part 4 (continued). *J. Trop. Med. Hyg.*, 31:232.

CLELAND, J. B., 1928c: Disease amongst the Australian Aborigines, Part 5. *J. Trop. Med. Hyg.* 31:262.

COLBOURNE, M. J., HUEHNE, W. H. and LACHANCE, F. de, 1959: The Sarawak anti-malaria project. *Sarawak Mus. J.,* 9:215.

COLLIS, W. R. F., DEMA, I. and OMOLOLU, A., 1962: On the Ecology of Child Nutrition and Health in Nigerian Villages. II. *Trop. Geogr. Med.,* 14:201.

COOK, S. F., 1945: Demographic consequences of European contact with primitive peoples. *Ann. Amer. Acad. Polit. Soc. Sci.,* 237:107.

CORNEY, B. G., 1913: Notes on an epidemic of measles in Rotuma, 1911. *Proc. roy. Soc. Med.,* 6:138.

COUVÉE, L. M. J., 1962a: The nutritional condition of the Kapauku in the Central Highlands of West New Guinea. II. *Trop. Geogr. Med.,* 14:314.

COUVÉE, L. M. J., 1926b: Marriage, obstetrics, and Infant mortality among the Kapauku in the Central Highlands, West New Guinea. *Trop. Geogr. Med., 14:325.*

DAVIDSON, L. S. P., 1954: African Journey. *Lancet,* 1:614.

DAVIES, G. N., 1963: Social customs and habits and their effect on oral disease. *J. Dent. Res.,* 42:209.

DAVIES, J. N. P., WILSON, B. A., and KNOWELDEN, J. 1962: Cancer incidence of the African Population of Kyadondo (Uganda). *Lancet,* 2:328.

DE ZULUETA, J., 1956: Malaria in Sarawak and Brunei. *Bull. W.H.O.* 15:651.

DE ZULUETA, J. and LACHANCE, F., 1956: A malaria-control experiment in the interior of Borneo. *Bull. W.H.O.* 15:673.

DONNISON, C. P., 1929: Blood pressure in the African native. *Lancet,* 1:6.

DUBOS, R., 1959: *Mirage of Health.* New York, Harper & Brothers. Cited by Motulsky: 1960.

F.A.O. Nutritional Studies No. 1, 1954 *Rice and Rice Diets.* Rome, F.A.O.

FARINAUD, E. and PROST, E., 1939: Recherches sur les modalités de l'impaludation en milieu Moï et en milieu Annamite. *Bull. Soc. Path. Exot.,* 32:762.

FAUST, E. C., 1955: History of human parasitic infections. *Public Health Rep.,* 70:958.

FERGUSON, R. G., 1955: *Studies in Tuberculosis.* Toronto, University of Toronto Press. Quoted by Motulsky: 1960.

FOX, R. B., 1952: The Pinatubo Negritos. Their useful plants and material culture. *Philipp. J. Sci., 81:* 173.

FOY, H., KONDI, A., TIMMS, G. L., BRASS, W., and BUSHRA, 1954: Variability of sickle-cell rates in the tribes of Kenya and the Southern Sudan. *Brit. Med., J.* 1:294.

GAJDUSEK, D. C., 1962: Congenital defects of the central nervous system associated with hyperendemic goiter in a neolithic highland society of Netherlands New Guinea. I. Epidemiology. *Paediatrics,* 29:345.

GAJDUSEK, D. C., 1963: Kuru. *Trans. Roy. Soc. Trop. Med. Hyg.,* 57:151.

GARNHAM, P. C. C. 1952: Discussion. *Trans. Roy. Soc. Trop. Med. Hyg.,* 46:397.

GARNHAM, P. C. C., 1958: Zoonoses of infections common to man and animals. *Trop. Med. Hyg.,* 61:92.

GEBER, M. and DEAN, R. F. A., 1957: The state of development of newborn African children. *Lancet, 1:*1216.

GOLDRICK, R. B. and WHYTE, H. M., 1959: A study of blood clotting and serum lipids in natives of New Guinea and Australians. *Aust. Ann. Med.,* 8:238.

GORDON, R. M., KERSHAW, W. E., CREWE, W., and OLDROYD, H., 1950: The problem of loiasis in West Africa with special reference to recent investigations at Kumba in the British Cameroons and at Sapele in Southern Nigeria. *Trans. Roy. Soc. Trop. Med. Hyg.,* 44:11.

HADLEY, J. N., 1956: Health Conditions among Navajo Indians. *Public Health Rep.* 70:831.

HEINBECKER, P., and IRVINE-JONES, E., 1928: Susceptibility of Eskimos to the common cold and a study of their natural immunity to diphtheria, scarlet fever, and bacterial filtrates. *J. Immun.,* 15:395.

HERCUS, C. E., and FAINE, S., 1951: The Rarotongan Villager's Environment. *Trans. Roy. Soc. Trop. Med. Hyg.,* 45:353.

HIRSCH, A., 1883: *Handbook of Geographical and Historical Pathology.* 3 vols. London, New Sydenham Society.

JELLIFFE, D. B., 1955: Infant nutrition in the tropics and subtropics. *W.H.O. Monograph Series No. 29.* Geneva, World Health Organization.

JELLIFFE, D. B., BENNETT, F. J., WHITE R. H. R., CULLINAN, T. R., and JELLIFFE, E. F. P., 1962: The children of the Lugbara. *Trop. Geogr. Med.* 14:33.

JONES, L. W., 1962: *North Borneo. Report on the Census of Population taken on 10th August, 1960.* Kuching, Government Printing Office.

JONXIS, J. H. P., 1959: The frequency of haemoglobin S and haemoglobin C carriers in Curacao and Surinam. In Jonxis, J. H. P. and Delafresnaye, J. F. (Eds), *Abnormal Haemoglobins.* Oxford, Blackwell Scientific Publications.

KEAN, B. H., 1941: Blood pressure studies on West Indians and Panamanians living on the Isthmus of Panama. *Arch. Intern. Med.,* 68:466.

KLATSKY, M., 1948: Studies in the dietaries of contemporary primitive peoples. *J. Amer. Dent. Ass.,* 36: 385.

LANGRIDGE, W. P., KERNAGHAN, R. J., and GLOVER, P. E., 1963: Review of recent knowledge of the ecology of the main vectors of trypanosomiasis. *Bull. W.H.O.* 28:671.

LEWIS J. H., 1942: *Biology of the Negro.* Chicago, The University of Chicago Press.

LOWENSTEIN, F. W., 1961: Blood pressure in relation to

age and sex in the tropics and subtropics. *Lancet, 1:* 389.

Lups, S. and Francke, C., 1947: On the changes in blood pressure during the period of starvation and after the liberation in Utrecht, Holland. *Acta Med. Scand., 126:*449.

Luyken, R. and Jansen, A. A. J., 1960: The cholesterol level in the blood serum of some population groups in New Guinea. *Trop. Geogr. Med., 12:*145.

Macdonald, G., 1951: Community aspects of immunity to malaria. *Brit. Med. Bull., 8:*33.

McFie, J., 1959: Malnutrition in Uganda. *Lancet, 1:*91.

Mann, I., 1961: Climate, culture and eye disease. *Trans. Ophthal. Soc., 81:*261.

Mann, G. V., Roels, O. A., Price, D. L. and Merrill, J. M., 1962: Cardiovascular disease in African Pygmies. *J. Chronic Dis., 15:*341.

Mednick, L., 1960: Memorandum on the word primitive. *Curr. Anthrop., 1:*441.

Morley, D., 1963: Medical service for children under five years of age in West Africa. *Trans. Roy. Soc. Trop. Med. Hyg., 57:*79. and correspondence *Ibid.* 392.

Motulsky, A. G., 1960: Metabolic polymorphisms and the role of infectious diseases in human evolution. *Hum. Biol., 32:*28.

Neuprez, R., 1945: Cited by Keys, A., *et al.*, 1950: in *The Biology of Human Starvation.* Minneapolis, the University of Minnesota Press.

Oomen, H. A. P. C., Spoon, W., Heesterman, J. E., Ruinard, J., Luyken, R. and Slump, P., 1961: The sweet potato as the staff of life of the highland Papuan. *Trop. Geogr. Med., 13:*55.

Parker, S., 1961: Depression and hysteria in two nonliterate societies. *Proc. 3rd Wld. Congr., Psychiatry, 2:*1298.

Peart, A. F. W., and Nagler, F. P., 1954: Measles in the Canadian Arctic, 1952. *Canad. J. Public Health, 45:*146.

Pedersen, P. O., 1947: Dental investigations of Greenland Eskimos. *Proc. Roy. Soc. Med. 40:*726.

Perla, D., and Marmorston, J., 1941: *Natural Resistance and Clinical Medicine.* Boston, Little, Brown.

Poindexter, H. A., 1953: Epidemiological survey among the Gola tribe in Liberia. *Amer. J. Trop. Med. Hyg., 2:*30.

Polunin, I., 1952: Tinea imbricata in Malaya. *Brit. J. Derm., 64:*378.

Polunin, I., 1953-4: The medical natural history of Malayan aborigines. *Med. J. Malaya, 8:*55.

Polunin, I., and Saunders, M. 1958: Infertility and depopulation, a study of the Murut tribes of North Borneo. *Lancet, 2:*1005.

Polunin, I., 1959: The Muruts of North Borneo and their declining population. *Trans. Roy. Soc. Trop. Med. Hyg., 53:*312.

Polunin, I., 1961: The effects of shifting agriculture on human health and disease, in *Proceedings of the Symposium on Humid Tropics Vegetation*, held in Goroka in 1961.

Price, A. G., 1950: *White Settlers and Native Peoples.* Cambridge, University Press.

Price, D. L., Mann, G. V., Roels, O. A., and Merrill, J. M., 1963: Parasitism in Congo Pygmies. *Amer. J. Trop. Med. Hyg., 12:*383.

Raoult, A., Thomas, J., Thiery, G., Perrin, G., and Perrellon., G., 1957: Les parotidoses de malnutrition en A.O.F. *Bull. Méd. de l'Afrique Occidentale Française* (ns. 2, 5-72) Abstract in *Trop. Dis. Bull., 54:*1454.

Rice and Rice Diets 1954: F.A.O. Nutritional Studies No. 1. Rome, Food and Agriculture Organization of the United Nations.

Rivers, W. H. R., 1920: The Dying out of Native Races. *Lancet 1:*42, 109.

Rivers, W. H. R., 1922: *Essays on the Depopulation of Melanesia.* Cambridge University Press.

Schneider, D. M., 1955: Abortion and depopulation on Yap Island, in *Health, Culture and Community.* Paul, B. D. (Ed.), New York, Russell Sage Foundation.

Schofield, F. D., 1962: Differences in palpable liver and spleen rates between men and women of the Sepik district, New Guinea. *Trans. Roy. Soc. Trop. Med. Hyg., 56:*60.

Schofield, F. D. and Parkinson, A. D., 1963: Social medicine in New Guinea: beliefs and practices affecting health among the Abelam and Wam Peoples of the Sepik district. *Med. J. Aust., I:* 1, 29.

Schofield, F. D., Parkinson, A. D., and Jeffrey, D., 1963: Observations on the epidemiology, effects and treatment of Tinea imbricata. *Trans. Roy. Soc. Trop. Med. Hyg., 57:*214.

Scragg, R. F. R., 1957: *Depopulation in New Ireland, a study of Demography and Fertility.* pp. 144. M.D. Thesis, University of Adelaide. Port Moresby, Administration of Papua and New Guinea.

Shaper, A. G., 1962: Cardiovascular studies in the Samburu tribe of northern Kenya. *Amer. Heart. J. 63:* 437.

Shaper, A. G., Jones, M., and Kyobe, J., 1961: Plasmalipids in an African tribe living on a diet of milk and meat. *Lancet, 2:*1324.

Sigerist, H. E., 1951: *A History of Medicine.* Vol. 1. *Primitive and Archaic Medicine.* New York, Oxford University Press.

Sinclair, B., Cameron, D. A., and Goldsworthy, N. E., 1950: Some observations on dental conditions in Papua-New Guinea (1947) with special reference to dental caries. *Dent. J. Aust., 22:*120.

Smith, D. A., and Woodruff, M. F. A., 1951: Deficiency diseases in Japanese prison camps. *Spec. Rep. Med. Res. Coun. Lond.* No. 274.

TROWELL, H. C., 1950: Problems raised by kwashiorkor. *Nutr. Rev.* 8:161.

TROWELL, H. C., 1960: *Non-Infective Disease in Africa.* London, Edward Arnold.

VAN DER HOEVEN, J. A., 1956: Some demographic data from Netherlands New Guinea. *Docum. Med. Geogr. Trop,* 8:303.

VAN LOGHEM, J. T., 1960: Sickness and death in the primitive world especially in New Guinea. *Trop. Geogr. Med.* 12:196.

WADDY, R. B., 1959: Discussion in *Trans. Roy. Soc. Trop. Med. Hyg.,* 53:322.

WHYTE, H. M., 1958: Body fat and blood pressure of natives in New Guinea: reflections on essential hypertension. *Aust. Ann. Med.,* 7:36.

also DE WOLFE, M. S. and WHYTE, H. M. *ibid.* 47. Cited by Anon. Middle age in New Guinea. *Lancet,* 2: 407.

WILSON, D. B., GARNHAM, P. C. C., and SWELLENGREBEL, N. H., 1950: A review of hyperendemic malaria. *Trop. Dis. Bull.* 47:677.

WILSON, D. C., and SUTHERLAND, I. 1953: The age of menarche in the tropics. *Brit. Med. J.,* 2:607.

WILSON, T., 1961: Filariasis in Malaya, a general review. *Trans. Roy. Soc. Trop. Med. Hyg.,* 55:107.

Chapter 7

The Egyptian Medical Papyri*

WARREN R. DAWSON

EDITORIAL COMMENT

We are privileged to reprint this important paper by Warren R. Dawson, the doyen of medical history and of the study of diseases in ancient societies. Elsewhere in this volume there are numerous references to Ancient Egypt and to its medical papyri; Dawson's paper brings together much valuable material on these subjects.

We also feel that this contribution helps to demonstrate the various problems encountered in the literary study and translation of early medical documents; such difficulties are often overlooked in superficial reviews and it is salutary that medical historians and palaeopathologists be reminded of them.

The medicine of ancient Egypt undoubtedly occupies an important place in the history of the science, but it has generally been unfortunate in the manner of its treatment by medical historians. In most of the standard medical histories, as well as in innumerable separate pamphlets, lectures and papers, general accounts of Egyptian medicine are to be found, but most of them give a very erroneous notion of the real nature of the medical knowledge and practice of the Egyptians. The reason for this state of affairs is not far to seek: medical historians, not being Egyptologists, have been unable to consult the original documents for themselves, and have been obliged to collect their data at second or third hand from various old and pre-existing published sources, most of them written many years ago when the knowledge of the ancient writing of the hieroglyphs (or in manuscripts, its derivative, hieratic) was in a far less advanced state than it is to-day. Moreover, the nature of Egyptian magic, which is the parent of medicine, was so completely misunderstood as to give a quite false picture of the real state of affairs. Most of the information imparted by medical historians, and the writers upon whom they drew, has been based upon a single document—the Ebers Papyrus—of which, unfortunately for science, a very incorrect and unsatisfactory German "translation" appeared as long ago as 1890.[1] It is to this well-intended but unfortunate work that most of the errors and misconceptions of

* Reprinted from *Science, Medicine, and History: Essays on the Evolution of Scientific Thought and Medical Practice*, written in honour of Charles Singer, Oxford U.P. 1953. Vol I, pp 47-60; with kind permission of Dr. E. A. Underwood (Editor) and the publishers.

1. *Papyros Ebers. Das älteste Buch über Heilkunde . . . übersetzt von Dr. H. Joachim*. Berlin, 1890.

later historians can be traced. Even so learned an Egyptologist as Professor Ebers, a former owner of the document (who named it after himself), so entirely mistook its nature and purport as to call it a "Hermetic Book."[2] An attempt will be made in the following pages to indicate the true nature of the Ebers and the other so-called Medical Papyri, but in the meantime sufficient has been said to show the unsatisfactory nature of the principal materials hitherto available to historians. For over forty years, I have made a special study of Egyptian medicine directly from the texts themselves, and have endeavoured to form conclusions uninfluenced by the statements of other writers. I am well aware that in order to handle the subject really adequately we must await the advent of a qualified medical man who is also a competent Egyptologist, and whilst this combination of elements has not so far appeared in one individual, medical historians have been obliged to borrow their Egyptological data more or less untested; and conversely, Egyptologists writing on so highly technical a subject must necessarily depend upon borrowed plumes on the purely medical side. If I may lay some claim to Egyptological knowledge, I can certainly lay no claim to medical qualifications; but I have been at pains to acquire some acquaintance with the elements of anatomy and physiology, and hope I may claim, without undue immodesty, to know enough of these subjects to enable me to appreciate the purport, if not the actual words, of the ancient medical writings. I mention these facts because it has frequently been assumed, from my previously-published writings on Egyptian medicine, that I am a medical man; but I wish to make it quite clear that I have no academic connection either with Egyptology or with medicine, and am most anxious that I should not be thought to sail under false colours.

The ancient Egyptians have always enjoyed a great reputation for their medical knowledge, especially in classical times. In the *Odyssey*, for instance, it is stated that the physicians of Egypt were skilled beyond those of all other lands, and Herodotus several times mentions the medical practitioners of Egypt, each of whom, he says, was a specialist, applying himself to the study of a particular branch. The same writer relates that Cyrus sent to Egypt for an oculist, and that Darius held that the Egyptians enjoyed the highest reputation for their medical skill: elsewhere similar references are to be found.[3] The "wisdom of the Egyptians" is indeed proverbial, and although they were incapable of that characteristic of the Greek mind—true philosophy and a b s t r a c t thought—there is no doubt that they were a very highly gifted people, with a great capacity for practical achievement. There can no longer be any reasonable doubt that the foundations of medical science were laid in Egypt more than fifty centuries ago, and although many modern writers have credited the Egyptians with medical and scientific knowledge of profound extent, others have denied this claim almost to the point of its non-existence,[4] though the writers on both sides of the question usually display a lack of true knowledge of the facts. The truth, of course, lies between these extremes; and indeed a nation which had acquired sufficient knowledge and skill to plan and carry

---

2. *Papyros Ebers. Das hermetische Buch uber die Ärzeneimittel der alten Ägypter in hieratischer Schrift.* 2 vols. Leipzig, 1875.

3. In 1924, I published an article on Herodotus as a Medical Writer in the *Ann. Med. Hist.*, vi, 357-60.

4. e.g., T. Wingate Todd, *Amer. Anthrop.*, 1921, xxiii, 460-70.

out feats of engineering and architecture, such as the Pyramids, as early as the fourth millenium before Christ, and whose mathematical knowledge, whilst wholly practical in aim, included highly complex calculations involving the principles of angles, cubic capacity, the square-root and fractional notation was obviously far ahead of its contemporaries in intellectual capacity. That this real knowledge evolved and developed amidst the meshes of a complex web of magic and superstition, does not in the least detract from its value, but rather enhances it. It may appear to be a sweeping generalization, but I believe it is nevertheless quite true, that magic was the basic motive that prompted almost all the practical achievements of the Egyptians. The real purpose which inspired the erection of the gigantic pyramids and temples, for the preservation of the dead and the vast and valuable equipment with which they were provided, was essentially *magic*, and the impressive and multifarious remains of the ancient civilization which extend from the Delta to the Cataracts and far into Nubia, are but the translation of the magical motive into practical fact. In medicine, as we shall see, magic played the leading part, and time has spared for us a considerable number of written documents—the so-called Medical Papyri—in which Egyptian ideas on magic and medicine are formulated, and which, as the oldest body of medical literature in the world, are the foundation upon which our knowledge of Egyptian medicine and its derivatives must necessarily be based. It will be convenient, before proceeding further, to enumerate and comment on these documents.

The contents of these papyri fall into two groups: those which may claim to be called medical books, and those which are mainly magical in purport or are collections of popular recipes. These two groups I will indicate respectively by the letters A and B. The contents of some of the papyri fall wholly in group A, a few wholly or mainly in group B, and others again contain elements of both groups, more or less indiscriminately combined.

1. *The Ebers Papyrus.* This document is placed first in the list (although in point of age it deserves no such priority), because it is the longest, most complete, and most famous of all these documents. It was found, together with the Edwin Smith Papyrus (below, No. 5), in 1862 by Edwin Smith (1822-1906), an American resident in Egypt, who traded in antiquities.[5] Smith sold it a few years later to Professor Georg Ebers of Leipzig, who named it after himself and published a sumptuous facsimile of it.[6] In this edition, Ebers did not attempt a translation, but he gave an introductory account of the manuscript, the nature of which, as already mentioned, he wholly misconceived. It also contains an Egyptian-Latin Glossary by Ludwig Stern, which is still useful as an index of words, but as a glossary is almost valueless, as it is full of mere guesswork. A transcript of the hieratic text into hieroglyphic characters was published without letterpress in 1913 by the late Dr. Walter Wreszinski.[7] This convenient edition of the text is that most generally used by Egyptologists, although frequent reference must still be made to Ebers's facsimile in order to control Wreszinski's readings. This editor has conveniently divided the 110 pages of the original manuscript into 877 numbered sections. A courageous attempt

---

5. Concerning Edwin Smith and his connection with Egyptology, see my *Life of Charles Wycliffe Goodwin.* Oxford, 1934, pp. 111 ff.

6. *op. cit.* (2).

7. *Der Papyrus Ebers. I Teil: Umschrift.* Leipzig, 1913. No further parts have appeared.

at a full translation into English was made by Dr. B. Ebbell a few years ago, but, whilst this is an immense improvement on Joachim's translation of 1890, it cannot be considered as altogether satisfactory.[8]

Physically, the Ebers Papyrus is in perfect condition and is easily legible from beginning to end. It contents, written on the recto, are medical and magical throughout, but on the verso, and quite unrelated to the recto, is written a short calendar which has been of great importance in the study of the difficult problems of Egyptian chronology. The rest of the verso is uninscribed. The document was written at the beginning of the XVIIIth Dynasty, but there is abundant evidence, based on philological and other grounds, that it was copied from a series of books many centuries older. It is not a book in the proper sense of the word, but a miscellany of extracts, recipes and jottings collected from at least forty different sources, and it is in this respect exactly analogous to the collections of household and medical recipes of Europe in later times. To speak of the Ebers Papyrus, as many writers have done, as an authoritative Egyptian medical treatise from a temple library, is manifestly absurd. The text consists mainly of a large collection of prescriptions for numerous ailments, most of which are named but not diagnosed, specifying the drugs to be used, the measures of each, and the method of preparing and administering them. A few of the sections are extracts from a more serious general medical treatise belonging to the group A, specified above, and of which treatise some portions have come down to us in the Edwin Smith and Kahûn Papyri. The extracts in the Ebers Papyrus just referred to relate to diseases of the stomach, to the action of the heart and its vessels, and to the surgical treatment of cysts, boils, carbuncles and similar conditions. All these excerpts can be readily identified by the fact that in them alone symptoms are described and diagnoses made, and that certain distinctive formulae occur in them which are absent from the rest of the text. Freely interspersed amongst these elements are magical spells and incantations.

2. *The Ramesseum Papyri.* These three papyri, and that to be next described, are the oldest of the known medical papyri. They belong to the Middle Kingdom, and were written in the XIIth or XIIIth Dynasties. The name that has attached itself to them is therefore misleading, for the Ramesseum (the great mortuary temple of the famous Pharaoh Ramesses II) dates of course from the XIXth Dynasty, some centuries later. Actually they were found, together with many other fragmentary papyri, in a Middle Kingdom tomb beneath the foundations of the Ramesseum, the existence of which was not known to the builders of the latter.. The discovery was made in 1896 by the late Mr. J. E. Quibell. Some of the literary papyri from this important find have long since been published, but many of the others, including the medical have only more recently been made known.[38] The three medical texts I transcribed in 1923 from photographs kindly lent to me by the late Sir Alan Gardiner. They are fragmentary and full of lacunae, but sufficient is preserved to establish their nature. Nos. 1 and 2 are written in vertical col-

---

8. *The Papyrus Ebers: the greatest Egyptian medical document.* Copenhagen and London, 1937. I have examined this publication in *J. Egyptian Archaeology*, xxiv, 250, and have there stated the reasons for my opinion.

38. The texts of these have lately been published, but without translation. J. W. B. Barns, *Five Ramesseum Papyri*, Oxford, 1956. Ram. Pap. No. 3, pp. 15-23; No. 4, pp. 24-29: No. 5, pp. 30-34; Pl. 11-23.

umns of hieratic writing, and from the surviving fragments I can distinguish parts of eighteen different recipes for the protection of mothers and their babies.[9] The remedies are sometimes incantations, sometimes prescriptions of drugs. Amongst the recipes may be mentioned one for ascertaining whether a new-born infant will live or die, and one for contraception. Some passages of this text occur also in the Kahûn Papyrus (below, No. 3). The third Ramesseum Papyrus contains a collection of prescriptions for rheumatoid or arthritic complaints, and many of them closely resemble the similar recipes in the Ebers and Hearst Papyri. This document is also written in vertical columns, but in a form of script known as linear hieroglyphic or semi-hieratic. The larger five fragments contain the remains of twenty-six prescriptions. Both papyri belong wholly to group B.

3. *The Kahûn Papyrus.* This was discovered at Lâhûn in the Fayuum district of Lower Egypt in 1889. Like those just described, it is of Middle Kingdom date. Although fragmentary, it contains the remains of thirty-four sections, all of which are gynaecological. Most of the sections are extracts from the general medical treatise of which the Edwin Smith and parts of the Ebers Papyri are excerpts (group A), but there are also some medico-magical recipes and incantations of group B, doubtless brought together because of their subject-matter. The prescriptions are for affections of the uterus and vagina, and there are also nostrums for ascertaining pregnancy and the sex of unborn children, these sections occurring in three other papyri (below, Nos. 7, 9, 12). The text was published many years ago with a translation and commentary.[10]

4. *The Hearst Papyrus* was found at Deir el Ballâs in Upper Egypt in 1899, and is now preserved in the University of California. A photographic facsimile, with an introduction and index of words, but without translation, was published by the late Dr. G. A. Reisner a few years later.[11] The outermost folds of the roll are fragmentary, but otherwise the document is in good condition and consists of fifteen almost undamaged columns, which Dr. Wreszinski, who has published a hieroglyphic transcript, has divided into 250 sections.[12] The papyrus is very similar to the Ebers, of which it contains a number of duplicate passages. It is somewhat later in date than the Ebers, and may be assigned to the time of Tuthmosis III (XVIIIth Dynasty).

5. *The Edwin Smith Papyrus.* This manuscript was found in Egypt, together with the Ebers Papyrus and other documents, in 1862. Edwin Smith did not sell it, and it remained in his possession until his death. His daughter presented it to the New York Historical Society in 1906. In 1930, the late Professor J. H. Breasted of Chicago, having devoted several years to a close study of it, published the papyrus in a sumptuous edition, consisting of a photographic facsimile and transcript in a folio volume, and a stout quarto containing a translation, commentary and glossary.[13] The greater part of this valuable

---

9. The text is similar to, but not a duplicate of, a later papyrus (Berlin 3027) published by Erman as *Zauberspruch für Mutter und Kind.* Berlin, 1901.

10. *Hieratic Papyri from Lahûn and Gurob,* by F. Ll. Griffith. London, 1898, 5-11, and Pls. V, VI.

11. *The Hearst Medical Papyrus.* Leipzig, 1905.

12. *Der Londoner Medizinische Papyrus und der Papyrus Hearst.* Leipzig, 1912, 1-333 (text, German translation, and commentary).

13. *The Edwin Smith Surgical Papyrus.* Chicago, Univ. Press, 1930. The late Sir D'Arcy Power published two interesting articles on this papyrus in *Brit. J. Surg.,* 1933, xxi, 1-4; 385-7.

document belongs to our group A, and is devoted to the treatment of wounds and fractures, extracted from the same general medical treatise as the before-mentioned sections of the Ebers Papyrus and the Kahûn Papyrus. It contains forty-eight long sections, each dealing with a particular case, i.e. the affection of a particular organ or region of the body. In addition to these rationally and almost scientifically described cases, the papyrus contains thirteen medico-magical incantations and prescriptions, as well as some explanatory glosses. The medico-magical elements fall into group B.

6. *The Chester-Beatty Papyri.* Eighteen of this valuable collection of nineteen hieratic papyri of Ramesside period were presented to the British Museum by Sir Chester and Lady Beatty and edited by Sir Alan Gardiner in 1935.[14] The series contains literary and magical documents, two of which are entirely on medical subjects. The recto of Papyrus No. VI (Brit. Mus. 10686) may be described as a fragment of a treatise on proctology, as it is concerned exclusively with the treatment of affections of the anus and rectum. Although the prescriptions of drugs resemble the familiar pattern of those in the Ebers, Hearst and other papyri, this document differs in an important respect from all the others, and this difference, which will be indicated later in this paper, entitles the papyrus (in my opinion) to be placed in our group A. The verso contains spells and incantations for epilepsy, and these are to be placed in group B. Papyrus No. X (Brit. Mus. 10690) is a book of aphrodisiacs consisting of spells and drugs against impotency, etc. Several of the longer papyri in the Chester-Beatty Collection are filled with magical incantations which are medical in so far as their purport is the cure and prevention of disease.

7. *The Berlin Medical Papyrus.* This document, of the XIXth Dynasty (Berlin Museum 3038), came to Europe early in the nineteenth century with the Passalacqua Collection, and is said to have been found at Sakkara. A lithographic facsimile of it, not very satisfactory, was published many years ago by Brugsch,[15] and more recently a photographic facsimile, with a hieroglyphic transcript, German translation and glossary, appeared under the editorship of Dr. Wreszinski, who divided the text into 204 sections.[16] The contents are very similar to those of the Ebers and Hearst Papyri, of which it contains several duplicate passages. The text is extremely faulty and corrupt, often to the pitch of unintelligibility. On the verso occur the nostrums for the ascertainment of pregnancy, etc., which occur in several other papyri. Nearly all the contents must be relegated to group B, but there is one extract of group A.

8. *The London Medical Papyrus.* A fragmentary papyrus of the latter part of the XVIIIth Dynastry: it is a badly written palimpsest, many traces of an earlier erased text being discernible. It consists entirely of medico-magical recipes and prescriptions, all of group B. A photographic facsimile with a hieroglyphic transcript, German translation and glossary has been published by Dr. Wreszinski.[17] This document was formerly in the library of the Royal Institution, but how it came there is not known. It was presented to

---

14. *Hieratic Papyri in the British Museum. 3rd series: Chester-Beatty Gift.* 2 vols. London, 1935.

15. *Recueil de monuments égyptiens.* Ed. by H. Brugsch and J. Dümichen. Leipzig, 1862-85, vol. ii, Pls. 85-107.

16. *Der grosse Medizinische Papyrus des Berliner Museums.* Leipzig, 1909.

17. *Der Londoner Medizinische Papyrus und der Papyrus Hearst.* Leipzig, 1912, 137-237.

the British Museum in 1860 (Brit. Mus. 10059).

9. *The Carlsberg Papyrus.* The papyrus known as Carlsberg No. VIII is a mere fragment, but it is of considerable interest. The recto contains the remains of prescriptions for the eyes, duplicating a passage of the Ebers Papyrus, and the verso is filled with the birth-prognoses known to us in three other Egyptian papyri (Nos. 3, 7, 12), and is the prototype of many popular birth-prognoses which appear in popular medical (or rather, pseudo-medical) literature over a period of many centuries.[18]

10. *Magical Papyri.* The museums of London, Paris, Leyden, Turin, Budapest, Rome (Vatican), Berlin, and other cities contain considerable numbers of magical papyri, which, although not generally therapeutic in character, are medical in so far as their purpose is the treatment and cure of disease, injury and the attacks of venomous animals. The Chester-Beatty Collection, as already mentioned, contains several documents of this class, but the richest stores are those of the museums of Leyden and Turin. All these, fantastic as they often are, have a place in the development of medicine, and they well exemplify the *modus operandi* of the magician.[19]

11. *Ostraca.* All the medical texts described in the foregoing paragraphs are collections of prescriptions, some of which are large and all of which, when complete, were probably extensive. Isolated prescriptions are rare, but a few have been found on ostraca, i.e. limestone flakes or potsherds. Thus an ostracon in the Louvre (3255) contains two prescriptions for the ear; Milne Ostracon C.I. is a spell for the protection of the limbs on the principle that "like influences like"; Ramesseum Ostracon 35 is a protection against snake-bite; Medineh Ostracon 1091 is a prescription for a "cure-all" in any part of the body.

12. *Later documents.* The texts enumerated above are all of the Pharaonic Period, that is to say of the Middle and New Kingdoms, but in addition to these we have some documents of Ptolemaic and later date. An important magical papyrus,[20] part of which is in the British Museum and part in Leyden, written in the demotic script and of the third century A.D., contains a good deal of medical matter, though all of it belongs to group B. Of Coptic material, the great medical papyrus of Mashâykh, discovered in 1892, is the most important item. This long document contains 237 prescriptions, and is, in the main, very similar to the medical papyri of Pharaonic times. Although it was written as late as the ninth or tenth century A.D. it closely adheres to the traditional forms, albeit some Greek and Arabic elements have obtruded themselves. It was published *in extenso* in a bulky volume in 1921.[21] Other Coptic medical texts, much smaller in extent, exist in the British Museum;[22] the John Rylands Library at

---

18. *Papyrus Carlsberg No. VIII, with some remarks on the Egyptian Origin of some popular Birth Prognoses,* by Erik Iversen. (Kgl. Danske Videnskobernes Selskab, Hist-Filologiske Meddeleser XXVI. 5.) Copenhagen, 1939.

19. See my Frazer Lecture, *Folk-Lore,* 1936, xlvii, 234-62 on the subject of the magical papyri generally.

20. *The Demotic Magical Papyrus of London and Leiden,* by F. Ll. Griffith and [Sir] Herbert Thompson. 3 vols. London, 1904-9.

21. *Un Papyrus Medical Copte,* by Emile Chassinat. Cairo, 1921. See also my paper in *Proc. roy. Soc. Med.,* 1924, xvii (Sect. Hist. Med.), 51-7.

22. MS. Or. 4920(3), W. E. Crum, *Cat. Coptic MSS. Brit. Mus.* No. 527, p. 255; an ointment for the eyes, Ostr. 27422, H. R. Hall, *Coptic and Greek Texts,* 1905, p. 64.

Manchester;[23] the Vatican;[24] the Berlin Museum[25] and the University of Michigan,[26] and Paris.[27]

From various sites in Egypt, chiefly in the Fayyum, has come to light a series of medical papyri wholly Egyptian in character, though written in Greek. These are now dispersed in various museums. Most of them follow the ancient pattern and are obviously of native origin. Thus we find in one fragment eleven prescriptions for the ear,[28] in a second, five prescriptions for the eyes,[29] and in a third, fourteen prescriptions for a variety of ailments.[30] Three fragments deal with the making of lozenges (τροχίσκοι) and tooth-powder (ὀδοντότριμμα).[31] But in addition to these collections of popular recipes, medical papyri not only written in Greek, but also of Greek origin, have been found in Egypt. Thus the Golenischef Papyrus, a Greek medical fragment of the third century A.D., although too fragmentary to be translated as it stands, contains a text closely resembling a passage from the *Gynaecology* of Soranus of Ephesus.[32] Another document, the Cattaui Papyrus, contains two columns of twenty-seven lines each, a fragment of a surgical treatise of the third century A.D.[33] A papyrus in the John Rylands Library describes the omens and significance of the quivering or twitching of every part of the body. It begins with the abdomen and enumerates forty-one organs or parts of the body from that point to the soles of the feet. Presumably, when complete, it began at the top of the head.[34] A Greek medical papyrus from Egypt of late first or early second century A.D. deals with luxation of the jaw.[35]

Such, in outline, is the documentary material available for the study of Egyptian medicine, and it is at the same time voluminous and f r a g m e n t a r y. Tradition ascribed to various gods, to certain early kings and to sages such as Imhotep (the Imouthes of the Greeks) the authorship of medical treatises. Whilst we have no indication of the authenticity of such attributions, we do know that one or two general treatises did exist, by whomsoever they were compiled, and that fragments of them have come down to us in several papyri.

---

23. W. E. Crum, *Cat. Coptic MSS. John Rylands Library*, Nos. 102, 104, 106-10.

24. G. Zoëga, *Catalogus Codicum Copticorum*. Rome, 1810, pp. 626-30. (Contains 45 prescriptions, written on leaves of a vellum book, numbered 241-244.) This book, when complete, must have contained nearly 2000 prescriptions up to p. 244 if the average number per page revealed by the surviving fragment held good throughout. This assumes, of course, that the whole of its contents consisted of prescriptions, but the earlier leaves may have contained other matter.

25. Ostracon 2173, *Z. äg. Sprache*, 1878, xvi, Taf. i (prescription for spitting blood); Ostr. P. 4984, *Koptische Urkunden*, No. 27 (preparation of a drug); Ostr. P. 8109, *op. cit.*, i. p. 24 (prescriptions for insomnia, palpitation, menstruation, etc.).

26. Coptic MSS. 593-603. W. H. Worrell, 'A Coptic wizard's hoard.' *Amer. J. Semitic Languages*, 1930, xlvi, 239-62.

27. Urbain Bouriant, 'Fragment d'un Livre de Médecine en Copte.' *Compt. rend. Acad. Inscr. Belles-Lettres*, 1887, xv, 374. A single leaf from a vellum book containing eleven prescriptions for the breasts. The pages are numbered 214, 215.

28. Grenfell and Hunt, *Oxyrhynchus Papyri*, 1899, ii, 134, No. 234 (2-3 century A.D.).

29. *Tebtunis Papyri*, 1907, ii, 22, No. 273 (late second or early third century A.D.).

30. *Oxyrhynchus Papyri*, 1911, viii, 110, No. 1088 (early first century A.D.).

31. A. S. Hunt, *Cat. Greek Papyri, John Rylands Library*, 65-9, Nos. 29, 29A, 29B.

32. A. Bäckström, *Archiv für Papyrusforschung*, 1906, iii, 157-62. The passage from Soranus is in ed. Huber, Munich, 1894, 148 ff.; ed. Rose, ii, 31, 85.

33. Jules Nicole, *Archiv für Papyrusforschung*, 1908, iv, 269-71, with commentary by Johannes Ilberg, *ibid.*, 271-83; E. Chassinat, *Bull. Inst. Franc. Arch. Orientale*, 1910, viii, 111-12 and 1 Pl.

34. Rylands Greek Papyrus, No. 28. A. S. Hunt, *Cat. Greek Papyri, John Rylands Library*, i, 56-65.

35. Brit. Mus. Pap. Gr. 155, purchased from the Rev. Greville Chester, who obtained it in Egypt. F. G. Kenyon, *Cat. Greek Papyri*, 1898, ii, xiv, 144.

As already mentioned, magic played a very prominent part in the social and religious life of the Egyptians: it affected not only the relations of men with their living fellows, but also with the dead and with the gods.[36] By the Egyptian, magic was believed to be a sure means of accomplishing all his necessities and desires, and of performing indeed everything which the common procedure of daily life was inadequate to bring about. It was, theoretically at least, the private faith of the magician in his own omnipotence, his *credo quia impossibile*. The Egyptians, naturally a gifted and practical people, were incapable of abstract thought, yet in very early times they so far recognized the existence and importance of the mystical power of magic upon which they placed so much reliance, as to conceive it as an entity and to name it. In very early texts we meet with the word *hīke*, 'magic', a mystic power that was soon personified as a god, and it was by virtue of this *hīke* that they carried out throughout their long history the complex series of rites, customs and beliefs which we to-day describe as magical. The power of magic was co-extensive with the whole range of human activity and desire, and it was employed explicitly or implicitly for almost every conceivable purpose.

The magic art was most often exercised for defensive, protective, preventive, productive and prognostic purposes. The services of the magician are most commonly met with in the prevention and cure of sickness, injury, the bites or stings of noxious animals and other similar calamities befalling the individual. These medical applications of the magic art, besides being the most numerous, well exemplify the method of procedure of the practitioner, and enable us to discern the steps by which the germs of rational scientific ideas grew out of mere magical jargon.

In the numerous medico-magical texts which have come down to us, the leading idea is possession, for diseases are treated as if personified and are harangued and addressed by the magician. It is generally stated or implied that disease or suffering is due to the actual presence in the patient's body of the demon itself, but frequently it is implied that the suffering is due to some poison or other evil emanation that the demon has projected into the body of his victim. The possessing spirit is usually conceived as a god or a goddess, a dead man or a dead woman, an enemy male or female, or a pain male or female. This collocation of words with endless variations is an oft-recurring formula in the magical spells.

Once installed, the demon made the patient ill and the business of the magician was primarily to eject the intruder. The simplest method of procedure was the recitation of a spell in which the demon was summarily ordered to quit, or the poison to flow forth from the infected body. These spells, of varying length and elaboration, are full of references to the gods and are often of the highest mythological interest. Some of the more elaborate spells embody threats and exorcisms of the most daring character. In these simplest cases, the magician operated by word of mouth only, but in most spells the spoken words are accompanied by a ritual—by gestures or by the use of amulets or other objects. These two essential parts of the magician's art have been very aptly termed by Sir Alan Gardiner the *oral rite* and the *manual rite* respectively. It is usual in the magico-medical texts to find a rubric at the end of the spoken formula (or oral rite) giving directions as to the performance of an ac-

---

36. On Egyptian Magic generally, see my paper in *Folk-Lore*, 1936, xlvii, 234-62.

companying manual rite. The manual rite often took the form of reciting the spell over an image, a string of beads, a knotted cord, a strip of inscribed linen, an amulet, a stone, or some other object. These articles, thus magically charged, the magician is directed to lay upon or attach to some part of the patient's body. In cases of illness or injury, the manual rite often takes the form of repeating the formula over a mixture of substances which were then given to the patient to swallow, or were applied to his body externally as ointments or poultices, the medicine so given being thus rendered efficacious by the recitation of the oral rite. The medical papyri, which consist for the most part of prescriptions of drugs, are interspersed with magical spells, the object of which was to impart efficacy to the prescriptions which follow them. Such spells are the oral rites belonging to each group of prescriptions, the preparation and administration of each of which constitute the corresponding manual rite. Many of the doses contain noxious or offensive ingredients (coprotherapy being especially frequent), and their object is manifestly to make them as unpalatable as possible to the possessing spirit so as to give it no inducement to linger in the patient's body.

It is characteristic of the magician at all times that he should have more than one string to his bow, for if one remedy fails, another may succeed, and at all costs his prestige must be maintained. Thus it is that we find in the medical papyri numerous alternative prescriptions for each complaint, and in the magical texts alternative spells to be recited with the same object. Some of the remedies contain drugs that are really appropriate and beneficial, and such prescriptions, actually accomplishing their purpose, would tend to survive their more fantastic fellows. By such means, more and more reliance would come to be placed on the drugs themselves (i.e. upon the manual rite) and less and less upon the recited spells (i.e. the oral rite). The magicians, therefore, who would be most in request in case of sickness would be those who were skilled in the knowledge of the preparation of drugs and manipulative treatment. Such men were no longer mere magicians, but were becoming physicians—and thus out of magic grew medicine. But the evolution of the physician did not extinguish magic. It is rare in human experience for any new order to supersede the old completely. The first physicians kept magic as a stand-by: magical methods continued to be employed side by side with the more rational procedure, as the medical papyri of Pharaonic times clearly show. Moreover, the existence of many magical papyri dating from Ptolemaic times and later, written in demotic Egyptian, Coptic and Greek, reveal unmistakably that magical practices for the cure of disease were in active operation long after the influence of scientific medicine, which was mainly due to the Greeks, had made itself felt. Magic maintained powerful sway throughout the early centuries of the Christian Era and throughout the Middle Ages: it persisted into the sixteenth, seventeenth and eighteen centuries, and is by no means extinct to-day, even among highly civilized nations. The magician has survived: he has merely changed his role from time to time, becoming successively the palmer, the quack and the advertiser of medicines. The ancient magician, when, *malgré lui*, he had become physician, refused to part with the mysticism of his craft, and he often disguised his more rational treatment under a veneer of mystery, a method which has been followed throughout the ages by his successors.

The very multiplicity of the prescriptions in the medical papyri is of itself a confession of their purely arbitrary and unscientific character: the very fact that numerous alternative remedies are provided for one and the same complaint implies that if one fails, another should be tried until, perchance, an effective one was found. The procedure according to most of the medical papyri really amounts to this—Try X, *or* Y, *or* Z, and so on. A definite advance on this haphazard method is manifested by the Chester-Beatty Papyrus VI, where the plan is: Do X, *then* Y, *then* Z, etc. That is to say the prescriptions for each case were to be *all* employed successively, and not merely selected at will from many alternatives. It is for this reason that I would place the papyrus in question in the group A, as defined above.

Had the Egyptians any knowledge of anatomy and physiology, or was their medical lore confined to therapeutic treatment? There cannot be any doubt that they did indeed possess such knowledge in a far greater measure that any of their contemporaries. They had opportunities for acquiring such knowledge which elsewhere were entirely lacking, for the custom of embalming the dead, involving as it did the removal and handling of the viscera, had a profound influence upon the growth of medicine, although it was not carried out by physicians but was a religious observance. Not only did the custom of mummification familiarize the Egyptians with the appearance, nature and mutual positions of the internal organs of the body—opportunities that were denied to all people who interred or cremated their dead—but it also made them acquainted with the preservative properties of the salts and resins they employed in the process. Mummification thus provided also for the first time in human history opportunities for observations in comparative anatomy, for it enabled its practitioners to perceive the analogies between the human viscera and those of animals, the latter long familiar from the time-honoured custom of cutting up beasts for food or sacrifice. It is worthy of note that the various hieroglyphic signs representing parts of the body, and especially the internal organs, are pictures of the organs of mammals and not of human beings. This proves that the Egyptians' knowledge of mammalian anatomy is older than their knowledge of that of man, and that they recognized the essential identity of the two by devising signs based on the organs of mammals, and using them unaltered when referring to the corresponding organs of the human body.

It is significant that the knowledge of a people in respect of any technical subject can be gauged by the richness or paucity of its terminology. In the ancient Egyptian language there are considerably over one hundred anatomical terms, and this fact of itself proves that the Egyptians were able to differentiate and name a great many organs and organic structures, that a less enlightened people would have failed to perceive. For instance, every part of the alimentary canal had its separate name. Whilst, however, their terminology for the gross anatomy of the body is accurate, they entirely failed to understand the nerves, muscles, arteries and veins. They had but one word to denote all these structures, and appear to have regarded them all as parts of a single system of branching and radiating cables forming a network over all parts of the body. The word used for the blood-vessels communicating with the heart is the same as that employed for the muscles and nerves in the prescriptions for stiff joints and rheumatoid complaints.

I need not recapitulate the well-known passage in the Ebers Papyrus which describes the heart and its 'vessels', a passage which is duplicated in the Berlin Medical Papyrus, and triplicated by a fragmentary copy in the Edwin Smith Papyrus. Nothing like a system of physiology can be reconstructed from this corrupt and garbled text, but one fact clearly emerges—the appreciation that the heart was the centre of the vascular system and that all the vessels were dependent upon it. The Egyptians certainly regarded the heart as the most important organ of the body, and they attached no importance at all to the brain. The heart was held to be the seat of intelligence and of all emotions, and its presence in the body was so important that it was not ablated during mummification, but was carefully left, together with its great vessels, in its place in the thorax, although all the other viscera were removed. The 'vessels' were believed to be the vehicles, not only of blood, but also of air, water, mucus, semen, and other secretions. This erroneous conclusion doubtless arose from the post-mortem manipulations of the embalmers, and could not have been derived from the functional vessels of a living body. The exposure of organs by wounds and fractures also afforded opportunities for observation. It was observed, for instance (as we learn from the Edwin Smith Papyrus), that the brain is enclosed in a membrane and that its hemispheres are patterned with convolutions; that injury to the spinal column may cause priapism, and that such an injury may also cause meteorism. A passage in the Ebers Papyrus dealing with affections of the stomach, and other in the Kahûn Papyrus dealing with uterine and other female disorders, introduce a novel feature in that they describe symptoms and give a diagnosis. In nearly all the other medical texts the diagnosis is assumed and only treatment is provided. These passages, together with the greater part of the Edwin Smith Papyrus, and the concluding part of the Ebers Papyrus, are evidently drawn from one source, quite different from that of the bulk of the medical papyri, and belong to the type which has been designated above as Group A.

Space compels me to pass over in silence the subjects of pathology, therapeutics and materia medica, and indeed all these have been fully dealt with elsewhere,[39] but it is necessary to refer again to the "scientific" texts of Group A. These are clearly extracts from one and the same book, and their form and arrangement is far in advance of those of the greater part of the medical papyri, which consist merely of prescriptions. These surgical and physiological texts are drawn up with definite formulae in a fivefold form: (i) title; (ii) examination (symptoms); (iii) diagnosis, (iv) opinion (i.e., whether curable or not), and (v) treatment. In many cases glosses are added which help us to understand the obscurities and idioms of the text. From the rational and almost methodical manner in which these texts are presented, the late Professor Breasted claimed that the former belief in the magical origin of medicine is no longer tenable, that there is now evidence that anatomy was studied for its own sake, and that the Edwin Smith Papyrus is in the true sense a scientific book. He failed to recognize, however, that this particular papyrus is only a part of a larger body of texts into which magic enters to no small extent, as it does, indeed, into that very

---

39. See my papers in Z. äg. Sprache, 62 (1926) 20-23; Ann. Med. Hist. 9, (1927) 315-326; J. Roy. Asiatic Soc. (1927) 497-503; J. Eg. Arch., 18, (1932) 150-154; 19, (1933), 133-137; 20, (1934) 41-46; 185-188; 21, (1935), 37-40; etc.

papyrus itself. It does not in the least detract from the interest and importance of this text to prefer the opinion that, whilst it undoubtedly affords evidence that an attempt was being made to understand the structure of the body, yet it must be remembered that it deals only with wounds and fractures—injuries of palpable and intelligible origin—and not with diseases, the causes of which, to the ancients, were impalpable, invisible and unknown. A wound or injury caused by a fall, or by a weapon or tool, was well understood and generally treated by rational means; but the causes of headache and fever, of skin eruptions and swellings, and of countless other maladies, were wholly mysterious and attributed to possession. In the collections of remedies, in the papyri of Group B, each prescription is headed by a title in which, instead of the simple phrase 'prescription for curing' such and such a disease, we have 'prescription for banishing,' 'driving-out,' 'terrifying' or 'killing' such disease. In such phraseology the idea of possession is manifest, and when in the Edwin Smith Papyrus, in cases where the physician's opinion is favourable as to the possibility of a cure, the formula used is 'it is a malady that I will contend with,' or 'wrestle with,' words which clearly imply the ancient belief in magic, it is evident that the physician was still a magician at heart. As has already been indicated, even after rationalism began to pervade Egyptian medicine, the magical and the medical elements marched side by side, and the same age produced both the Edwin Smith and Ebers Papyri with their widely differing contents. Indeed, in the Edwin Smith Papyrus there is a magical incantation in the body of the surgical text itself, and on the back of it there is written a collection of charms and prescriptions of the familiar type that fills page after page of the once so-called medical papyri. The ancient owner of the Edwin Smith Papyrus saw no incongruity in copying into the same note-book elements that appear to us to-day as absolutely antagonistic in nature and content.

Finally, having considered the foregoing statements, we may ask—what is the place in medical history that Egypt may justly claim? I think the answer must be: a place of high priority. From Egypt we have the earliest medical books, the first observations in human and comparative anatomy, the first experiments in surgery and pharmacy, the first use of splints, bandages, compresses and other appliances, and the first anatomical and medical vocabulary, and that an extensive one. In general terms it may be said that the popular medicine of almost every country of Europe and the Near East largely owes its origin to Egypt, and in its various migrations it has preserved its ancestral form and its very words and phrases almost intact throughout the ages. Not only were many well-known drugs of universal vogue first used by the Egyptians, but in addition to the more obvious examples, many of the drugs, as well as the properties and traditions ascribed to them by the Egyptians, that occur in the works of Pliny, Dioscorides, Galen, and even in the Hippocratic Collection itself, are clearly borrowed from Egypt. These later writers, and others who followed them, are the sources from which the compilers of herbals and books of popular medicine mainly drew for their materials, and the works of classical writers are often merely the stepping-stones by which much of the ancient medical lore of Egypt reached Europe, apart from direct borrowings. Early medical books in Arabic, Hebrew and other Semitic languages have also drawn largely from the same stock. When a drug really possesses the virtues attributed to it and is an effective remedy, its survival into modern times is natural enough, but the fact that many quite fantastic and arbitrary remedies have been carried on almost to our own times is definite proof of the slavish copying from the works of one writer to the works of another, in a continuous line that originated many centuries ago on the banks of the Nile.[37] The use of certain arbitrary preparations, the use of the same formulae, idioms and colophons in the popular medical literature of many countries through many centuries, are all indications with an unmistakable interpretation.

In two other ways, most important of all, Egypt has served the history of medical science. First, through the distinctive custom of mummification, aided by favourable climatic conditions, hundreds of actual bodies, many of them accurately dateable, have carried down to us the earliest actual cases of the effects of disease. The history of the incidence of many diseases and morbid conditions can be thrust farther back into antiquity from the evidence provided by Egyptian mummies and skeletons: of calculi, bilharzia, arterial diseases, tuberculosis, arthritis and other bone diseases, as well as many inflammatory and other conditions. Nor is the evidence confined to that afforded by the bones, for many affections of the soft parts and viscera are likewise recognizable, such as pleural and appendicular adhesions, tumours, cysts, eruptive conditions of the skin and other manifestations of disease. And secondly, and most important of all, the Egyptians, by that same custom of mummification, had the greatest of all influences on the history of medicine—for that practice had reconciled the popular mind for more than twenty centuries to the idea of cutting open the dead human body. It was in Egypt that it became possible for the Greek physicians and anatomists of the Ptolemaic age to practise for the first time openly and unhampered the systematic dissection of the human body, which religious and popular odium and prejudice forbade in their own country and in all other parts of the world. To this one fact above all others the true science of medicine ultimately owes its origin, and the possibility of its development through the sister-science of anatomy. Making every allowance for the magical and more fantastic elements in Egyptian medicine, which were the necessary forerunners of more rational methods, it can be justly claimed that the earliest dawn of medical science broke over the Valley of the Nile.

---

37. Many instances might be quoted, but one may here suffice. "The milk of a woman who has borne a male child" occurs as a drug twelve times in the Ebers Papyrus, three times in the London Medical Papyrus, twice in the Berlin Medical Papyrus, and once each in four other papyri. It occurs in Coptic, in the Hippocratic Collection, in Dioscorides, in Pliny and in other classical writers, as well as in countless European medical works, both MS. and printed, from the Middle Ages until the eighteenth century. See my *Leechbook* (1934), p. 14.

SECTION II
CONTRIBUTIONS TO PARASITOLOGY

Chapter 8

The Antiquity of Diseases Caused by Bacteria and Viruses, A Review of the Problem from a Bacteriologist's Point of View

RONALD HARE

A GREAT many species of bacteria and viruses can cause disease in human beings and there seems no doubt that some at least were doing so in remote antiquity. There remain however a great many more that almost certainly did not begin to infect human beings until comparatively recent times. It is, nevertheless, very difficult to follow most of these diseases very far back because few of them leave traces on early human skeletons, the only remains usually available for examination. The antiquity of some may be ascertained by examination of bodies preserved by mummification, desiccation or immersion in bogs or water of high salt content, but such material is scanty and comes from only a few localities. Written records, the only remaining source of information are even more recent and frequently very difficult to interpret.

Nevertheless, much may be deduced about the antiquity of many diseases due to micro-organisms by an indirect method which takes into account what is now known about the mechanisms by which pathogenic micro-organisms were probably evolved, the sources from which they generally come and the effects that their size, density and social behaviour may have on the ability of a human population to act as suitable hosts for the establishment and indefinite survival of the different species.

Information obtained by these two methods, taken in conjunction with what is known of the mode of life led by human beings at different stages of their evolution, can provide clues as to which organisms might and, perhaps of greater importance, those that could not have been causing infection in societies with Palaeolithic, Neolithic and more recent cultures.

In this, much depends on the reservoirs or natural habitats from which the organisms come—whether they are generally derived from an inanimate environment, from lower animals or from Man himself— and it is on this basis that the antiquity of these diseases can most profitably be discussed.

ORGANISMS WHOSE NORMAL HABITAT IS INANIMATE NATURE

Under normal conditions, these organisms are not parasites, being able to multiply and survive quite successfully in soil, water and rotting vegetable or animal matter. In so far as Man is concerned, the most important are the *Clostridia*. These organisms have been found in soil from many parts of the world and were therefore probably evolved long before human beings. Since two of these produce infection by entering open wounds, it is extremely probable that such infections have occurred ever since Man appeared.

Direct evidence of this is, however, comparatively recent. The earliest reference to tetanus, caused by *Cl. tetani* is in the Edwin Smith Papyrus (Ghalioungui, 1963) probably compiled in the second millennium B.C. Four cases were described in Book IV and three in Book VII of the *Epidemics* of Hippocrates (Adams, 1849) which probably dates from the sixth century B.C. Aretaeus (Adams, 1846) early in the Christian era was also well acquainted with it.

Gas gangrene, similarly caused by the entry of other species of *Clostridia* into wounds is probably of equal antiquity and, certainly, one of its chief characteristics, gas formation in the tissues following injury, is referred to in the Papyrus Ebers (Ebbell, 1937; Ghalioungui, 1963).

Botulism for which another species of *Clostridium (Cl. botulinum)* is responsible, is due to the absorption of toxins produced by the growth of the organisms in food before it is eaten. As a general rule, they cannot do so unless the food is canned or pickled, but suitable conditions may arise if masses of meat or fish undergo decomposition. Indians and Eskimos do not object to such food and suffer accordingly (Dolman and Iida, 1963). It is therefore reasonable to suppose that Palaeolithic communities whose dietary habits may have been equally uncritical, were also subject to it.

ORGANISMS WHOSE NORMAL HABITAT IS A HOST OTHER THAN MAN

Such organisms fall into three separate categories.

Organisms Derived From Wild Vertebrates

Very few of the bacteria and viruses whose normal hosts are wild animals can cause human disease. The virus of rabies and the so-called B virus of monkeys can do so but, on the whole, such infections are very unusual.

Much more important is *Pasteurella pestis,* responsible for plague. It parasitizes many species of rodents but most human infections are caught from the black rat, *Rattus rattus,* because it tends to infest human habitations and its fleas, which convey the disease, bite human beings. Such infections are frequently sporadic. Epidemics amongst human beings are unlikely to occur unless the disease first becomes epizootic amongst the rodents that constitute the reservoir. The factors controlling such occurrences are not yet clearly understood but it is probably because of this that the incidence of the disease in human beings has waxed and waned very considerably over the centuries.

It is quite possible that epidemics occurred amongst human beings during the Pleistocene but there is no evidence that the disease occurred at any time during the pre-Christian era. There was, however, an outbreak in Syria and North Africa during the first century A.D. which, from its description by Rufus of Ephesus may well have been plague (Hirsch, 1883). There is no record of human infections in the succeeding centuries. But there is no doubt that it was responsible for the Plague of Justinian in the sixth century A.D. Procopius (Dewing, 1919) gives an admirable description of the disease. Although it spread all over the Near East, northern Africa and southern Europe, it disappeared and no further outbreaks occurred until the Black Death in the fourteenth century following which there was a series of smaller outbreaks during the next 400 years. Then came another plague-free period until the disease re-appeared in Canton in 1894.

Spirochaetal jaundice is a second disease that may be, and frequently is, acquired from the rat. It is due to *Leptospira* excreted in its urine. Nothing whatever is known about the antiquity of the disease in animals or Man but in view of the fact that several different but closely related species of *Leptospira* can cause the disease (Alston and Broom, 1958) and that their distribution is world wide, it is probable that they were evolved a very long time ago. Man may therefore have been subject to infection ever since his evolution.

Organisms From Animals That Have Become Domesticated

Although susceptible to infection by them, animals living wild are not parasitized by certain species of organisms that can become established in an animal population when animal to animal transmission is facilitated by close contact in herds and even more so if they are stalled in sheds or byres. If then human beings are susceptible, they may acquire the organisms in their turn.

The most important organism in this category is the bovine variety of *Mycobacterium tuberculosis*. It has never been isolated from wild animals (Griffith, 1928) nor has it ever become established as a human parasite. But it has infected dairy herds since before the Christian era (Francis, 1947). It is usually transmitted in the milk to cause glandular, abdominal and skeletal tuberculosis.

Because Palaeolithic societies did not domesticate cattle, it is improbable that this organism caused infection at that time, but it may well have done so in Neolithic and more recent societies. Except for the possibility that it may have been responsible for some cases of skeletal tuberculosis in the Neolithic period and referred to in more detail later, very little is, in fact, known about the antiquity of human disease caused by this organism. Nevertheless since milk seems to have been drunk in India in the pre-Christian era (Wise, 1860) and also in Egypt (Ghalioungui [1963] gives an illustration of a child suckling direct from the udder) while the Papyrus Ebers (Ebbell, 1937) refers to two varieties of enlarged cervical glands, a conditon that is frequently tuberculous and generally caused by bovine strains, it is possible that this organism was causing human infection at that time.

Although the evidence is less conclusive, it seems probable that domestication also favoured the establishment of *Brucella abortus*, *Bacillus anthracis* and the *Salmonella* of food poisoning in cattle, pigs or poultry and with it, the risk of the corresponding diseases in the human population.

Organisms Derived From Invertebrates

One of the most important organisms in this category is the virus of yellow fever. Its principal hosts are mosquitoes (various species of *Anopheles* and *Haemagogus*) but monkeys and Man himself can act as secondary hosts. It was probably evolved during the Pleistocene somewhere in central Africa (Mattingly, 1960). Human cases of the disease may therefore have occurred there ever since but the first record of the disease (in Nigeria) is as recent as 1750 (Findlay, 1948).

It is probable that it was taken from Africa to America some time after Columbus (Mattingly, 1960) and, certainly, the first human cases in the New World were seen in Barbadoes in 1647. It soon became widespread in the southern United States, the Caribbean and central America (Carter, 1931.)

The virus of dengue is also a parasite of mosquitoes *(Anopheles)* and although monkeys may be involved, its most important secondary host is Man. It probably originated in South East Asia (Mattingly, 1960) but the first human infections of which there is any record were in Cairo in 1779 and a year later, in Philadelphia (Hirsch, 1883). It has since become widespread in tropical and subtropical climates.

Of the several species of *Rickettsiae* that cause typhus, that which is responsible for endemic typhus *(R. mooseri)* may be the oldest. Its principal host is the rat flea and Baker (1943) has suggested that those at the eastern end of the Mediterranean were the first to be parasitized. Although it is not known when this occurred it is of interest that Murchison (1862) considered that Case 15 of Book III of the *Epidemics* of Hippocrates may have been typhus. But the first recorded cases clearly recognizeable as typhus occurred amongst the troops besieging Granada in 1489. Baker is also of the opinion that the parasite was then carried to America after the Spanish conquest where it still persists, particularly in Mexico.

A second species, *R. prowazeki* responsible for epidemic typhus is probably a mutant of *R. mooseri* that ceased to infect the rat flea and became a parasite of the human body louse (Zinsser, 1935). This probably occurred in the sixteenth century. The first outbreaks were in Italy in 1505 and 1528 (Hecker, 1844) since when the disease has become endemic in eastern Europe. Epidemic typhus is therefore a comparatively modern disease.

The spirochaete responsible for relapsing fever may likewise have first become parasitic in ticks, probably in the eastern Mediterranean. Some were carried south into Africa to survive as *Borrelia duttoni*. Others went north into Europe (Baker, 1943) where they ultimately underwent mutation into *Borrelia recurrentis* and became parasitic in lice. When this change of host occurred is not known but cases closely resembling relapsing fever were described by Hippocrates in Book I of the *Epidemics* (Adams 1849).

ORGANISMS WHOSE NORMAL HABITAT IS MAN HIMSELF

Man is the only known host for most of his pathogenic organisms. Their ability to survive once they have left his body varies according to the species but, in general, few are able to do so for more than a day or two. Even fewer can actually multiply in the inanimate substances commonly present in the human environment. In consequence, the survival of these organisms depends to a very large extent on the facility with which they can be transported from person to person. The methods available for this depend on the species and vary greatly in efficiency. But in general, they necessitate fairly close contact between donor and recipient.

This utter dependence on human beings for their nurture limits the activities of these organisms very considerably. It is accordingly very doubtful whether any of the organisms responsible for the more severe infections can have come with sapient Man from his Pleistocene ancestors. Too few human beings would have been available for many millenia to have permitted this; and even if the attempt had been made, Man himself would not have survived.

Additional evidence in favour of this view is provided by the fact that although other members of the Order Primates are susceptible to many human pathogens, and when in contact with human beings can contract such diseases as pneumococcal pneumonia, bacillary dysentery, tuberculosis (Ruch, 1959) and even smallpox (Bleyer, 1922), there is no evidence that when living wild and out of contact with human beings, they are either infected by or act as carriers of these or, indeed, most human pathogens.

If, therefore, Man acquired his pathogenic organisms some time after his appearance, this would have entailed their evolution as a result of mutation or more drastic alterations in genetic constitution by transformation, transduction or conjugation from a precursor such as a commensal, a now extinct species or, what was probably more usual, a related animal pathogen.

There seems no reason to doubt that evolution in this way, of organisms potentially pathogenic for Man, is going on all the time. They die out because they fail to become established as permanent human parasites. This establishment is an essential pre-requisite for the production of disease. Yet it is, almost certainly, a distinct and entirely separate step whose success largely depends on the mode of attack of the parasite itself and the constitution of the human population at the time of its evolution.

For these reasons, an organism that remains alive in its host for only a short time, producing a disease with a high mortality and to which any survivors are immune for the rest of their lives, would require a large population with a high birth rate if it is to establish itself at all, whereas this would not be necessary for another species which can survive for longer because it causes infections that become chronic or persists on the mucous membranes of symptomless carriers. Still more successful would be a commensal which can continue to live in the tissues for a whole lifetime, without as a rule, causing clinical infection.

These variations are of considerable importance in any discussion on the antiquity of these diseases in view of the fact that for over 98 per cent of his time on earth, Man lived in Palaeolithic societies as a hunter and food gatherer, few in numbers, and widely dispersed in groups of families that were constantly on the move. (Childe, 1942). This wide dispersion would have rendered him almost useless as a host for some organisms but quite adequate for others (Carr Saunders, 1922). Thus some of these diseases may well be very much older than others (Hare, 1954). On this basis, it is possible to divide them into four categories.

Acute Infections in Which the Organisms Disappear When Recovery or Death Occurs; Measles, Smallpox, Chickenpox and Herpes Zoster, Influenza, Mumps, Cholera and Whooping Cough

Whether the patient dies or recovers from these diseases, the organisms are usually available for transfer to other persons for no longer than the acute phase of the disease, that is, about seven to fourteen days, and except for the virus of smallpox which can remain alive for some months, those that have reached the external environment usually die within a day or so.

A second disadvantage faced by these organisms is that they are so readily transmitted from person to person that if they reach a non-immune community virtually everyone goes down with the disease within a very short time. An epidemic of measles in Greenland in 1951 for example, lasted only thirty-six days in Julianhaab and a little over forty-three days at Narsask and Stetten (Christensen *et al.*, 1952-53).

A third impediment is that with the possible exception of influenza those who recover are not only immune for the rest of their lives but cannot even act as symptomless carriers of the organisms.

Thus, survival and permanent establishment of the organisms responsible for these diseases will only occur if the population is large enough to ensure a constant supply of susceptible children and is not so widely dispersed that person to person transmission is difficult.

It is therefore highly improbable that any of these organisms would have become established in a scattered community with a Palaeolithic culture. And, certainly, none of them were present in the Americas, Australia or New Zealand before the first contacts were made with Europeans. It is therefore a legitimate conclusion that Siberia and South-East Asia from which they were colonized, were similarly free of them. So too was the remainder of the Old World, so long as its culture was Palaeolithic. But when the urban revolution came and large, stable populations became available in the Nile valley, The Mesopotamian area and possibly in China, the necessary conditions were for the first time provided for the permanent establishment in human beings of the organisms under consideration.

Nevertheless, it is probable that there was a long delay before most of these diseases eventually appeared. This may have been the case with smallpox. Lesions in the skin of mummies dating from 1555-1096 B.C. (XVIIIth and XXth Dynasties) which may have been smallpox were, for example, reported by Ruffer and Ferguson (1911) and Ruffer (1914). But some authorities, notably Unna were extremely doubtful about the diagnosis; and it is of interest that there is nothing suggestive of the disease in the Egyptian papyri. Nor, as Adams pointed out, is it mentioned in the Hippocratic corpus (1849).

It is also extremely doubtful that the disease was prevalent in India as early as this. The *Sushruta Samhita* (probably compiled in the sixth century B.C.) refers to eruptions (Bhishagratna, 1907) that Wise (1860) considered may have been smallpox, but the accompanying symptoms bear little resemblance to those of smallpox.

Nor is it at all certain that is was present in China. Indeed, the only evidence is the statement attributed to the Jesuit missionary, Cibot, that he had seen a book called *Treatise from the Heart on Smallpox* in the Imperial palace in Peking in which references were made to the disease first appearing at the time of the Tsche-U Dynasty, 1122-249 B.C. (Moore, 1815). But Wong (1929) doubted whether the disease was smallpox.

Whatever its previous history, there seems no doubt that smallpox definitely became a human disease early in the Christian era. It seems to have been prevalent in China during the reign of Chien Wu, that is, in A.D. 49 (Smith, 1871; Wong and Wu, 1932.) And, a little later, the alchemist Ko Hung who probably lived during the Chin Dynasty (A.D. 265-313) gives an unmistakable description of the disease, which he said came from the Huns, in his *Handbook of Prescriptions for Emergencies* (Wong, 1929).

In the meanwhile, western Asia, the Mediterranean littoral and Europe seem to have been free of it. It is however possible that there may have been an incursion during the second century A.D. known as the Antonine pestilence. It seems to have first appeared at Seleucia on the Tigris and was eventually taken to Rome, Italy and even reached the German tribes north of the frontier. It is by no means certain that the disease was smallpox, but its behaviour, the description by Galen of the signs and symptoms (Haeser, 1882) and the fact that Rhazes in the tenth century (Greenhill, 1848) recognized the disease from his description, would suggest that it may have been smallpox.

Nevertheless, if it was smallpox it probably died out after infecting all the susceptibles because writers in the early years of the Christian era such as Celsus (Spencer, 1935-38), Aretaeus (Adams, 1846) or Paulus Aegineta (Adams, 1844) do not mention it, an omission that in view of its high infectivity and death rate, its severity and its sequelae is of some considerable importance.

There seems however no doubt that it appeared in western Asia half way through the sixth century A.D. First prevalent in Syria (Hirsch, 1883) it seems to have reached Mecca some time between 569 and 571 A.D. Here, it infected a raiding party of Abyssinians accompanied by one elephant who were besieging the city. They had to retreat "continually falling by the wayside, dying miserably by every waterhole" (Guillaume, 1955). Their leader, *Abraha*, himself died of it.

There is no doubt that from that time onward the disease became permanently established in Arabia and the neighbouring countries. The Caleph *Yezid* who died in 683 A.D. was said to be pitted because of it and the Caleph *Abul-abas Alaffah* died of it in A.D. 753 (Woodville, 1796). Rhazes (Greenhill, 1848) too, in the tenth century A.D. refers, in the *Liber Continens*, to predecessors such as Bachtishwa of Irak and Georgius of Persia who were well acquainted with the disease in the eighth century. He also refers to the Pandects of Abrun who seems to have recognized the disease in Alexandria in the seventh century.

Whether or not it invaded Europe as early as this is doubtful. Epidemics described by Gregory of Tours and Marius of Avenches in the sixth century A.D. and by Bede in Britain in the seventh (Creighton, 1891) may have been smallpox but it would seem more probable that it was as a result of the Arab invasion of north Africa and ultimately of Spain in the eighth century that the disease reached Europe.

Nothing whatever is known about the early history of chickenpox. But its virus is either identical with or very closely related to that which causes herpes zoster. And there is no doubt about the antiquity of this disease. Scribonius Largus for example (Pusey, 1933; Beswick, 1962) in the first century A.D., referred to the characteristic distribution of the eruption when on the trunk—in the form of a narrow band—which was so reminiscent of the belt or girdle worn by women that the same name, *zona* was given to the disease. The term *zoster* was also used by the elder Pliny about the same time, for what would appear to have been the same disease (Mettler, 1947).

Unless, therefore, the behaviour of the virus was different from that of the present day, it can be assumed that chickenpox occurred as well. That the disease is not mentioned by any of the Greek or Roman writers may have been due to its mildness. Indeed, it was not until the sixteenth century that the disease was described and differentiated from the milder forms of smallpox by Ingrassias (Rolleston, 1937).

The early history of measles is also very obscure. Although it is, nowadays, a very characteristic disease, in that it usually attacks children, producing upper respiratory symptoms and an easily identified rash there is in fact, no mention of such a syndrome by Hippocrates or any of the later Greek and Roman writers. This would suggest that, like smallpox, it had not been evolved.

Nevertheless it seems to have been prevalent in Arabia and the neighbouring countries from the sixth century A.D. onwards because the Arabian physicians seem to have had difficulty in distinguishing it from some forms of smallpox. Indeed, it was not until the tenth century that measles was recognized as a distinct disease entity by Rhazes (Greenhill, 1848). Moreover the fact that in the *Divisio Morborum,* he mentions that "measles are more to be dreaded than the smallpox, except in the eye" would suggest that it was a comparatively new disease for which Man had not had time enough to acquire much innate resistance.

There is, however, very little doubt about the antiquity of mumps. It was described and attention drawn to the swelling of the testes as a complication in Book I, Section I, of the *Epidemics* of Hippocrates (Adams, 1848).

Influenza can be mimicked so successfully by many other diseases that it is difficult to ascertain its antiquity from what records are available. Nevertheless, few of its imitators produce short lived epidemics with a high attack rate and a tendency to cause pneumonia. This renders it probable that the outbreak in Perinthos described by Hippocrates in Book IV of the *Epidemics* was in fact influenza and that it was this disease that occurred from time to time in Europe in the succeeding centuries. Nevertheless, one of the first outbreaks to be clearly identifiable as influenza was that of 1570 (Short, 1749).

Although cholera can (and did on many occasions in the nineteenth century) cause human infections when imported into countries with temperate climates, its organism, *Vibrio cholerae* cannot become established as a permanent human parasite unless the terrain is low lying, with a high water table, the temperature and humidity are constantly high and the human population has extremely low standards of hygiene. For these reasons, it is improbable that the disease has ever been endemic in places other than the Delta of the Ganges and the 'El Tor' district of Sinai the two foci now responsible for epidemics of the disease.

Little is however known about its antiquity. There is no record of the disease anywhere prior to 1517 when diarrhoeal diseases, some of which may have been cholera, began to occur in India, Ceylon or the East Indies. But the first outbreak of undoubted cholera did not occur until 1817. This was at Jessore in the delta of the Ganges. (Hirsch, 1883; Macpherson, 1886-87).

Little is also known about the antiquity of whooping cough prior to 1519 (Goodall, 1934). Sydenham in 1676 was well acquainted with the disease and was responsible for calling it pertussis.

Acute Infections in Which the Organisms Can Persist in Convalescent or Contact Carriers. Scarlet Fever, Tonsillitis, Diphtheria, Pneumonia, Cerebrospinal Meningitis, Typhoid, Paratyphoid, Dysentery, Poliomyelitis and Infective Hepatitis

The organisms under consideration in this section produce acute infections lasting from about a week to a month, but a proportion of those who recover, become convalescent carriers with the organisms persisting in the nose, throat or intestinal canal according to the disease. Normal persons, too, may act as carriers. The carrier state may last several months; indeed, permanent carriers of *Salmonella typhi* are known but are very unusual.

The transmission of these organisms from person to person is a very much less efficient process than is that of any of those discussed in the preceding section so that casual contacts of the day may not suffice and it is usually in the home, dormitory or class room, for example, that transmission of the respiratory organisms generally occurs (Glover and Griffith, 1931). Conveyance to others of those excreted from the intestinal canal in the faeces is also by no means an invariable occurrence.

Nevertheless, this inefficiency, combined with the possibility that the organisms may persist for long periods in carriers, renders them much more capable of surviving in a scattered population such as that of a Palaeolithic society than any of those discussed in the preceding section.

It is possible that *Streptococcus pyogenes* and *Diplococcus pneumoniae* succeeded in doing so because one or other of these organisms is generally responsible for acute infections of the mastoid following an extension of an infection in the throat to the middle ear. Since pre-Columbian skulls have been found in America showing evidence of this type of infection (Hooton, 1930), it would seem probable that one or both of the organisms had become established in the Old World and were carried to America before the land bridge over the Bering Strait was covered as a result of melting ice about 8000-6500 B.C.

Traces of such infections have also been detected in pre-dynastic skulls from Egypt and Nubia (Derry, 1909; Smith and Dawson, 1924) while the advice given by Galen to scrape away carious bone behind the ear undoubtedly refers to the same condition (Guthrie, 1940). A second complication generally due to these organisms is cerebral abscess and this too is mentioned by Hippocrates.

It is therefore probable that the other diseases caused by these organisms are of equal antiquity. *Strep. pyogenes*, for example, also causes erysipelas. This is referred to in the Papyrus Ebers (Ebbell, 1937) and its occurrence as a complication of infections of the throat and of wounds were both described by Hippocrates in Book III of the *Epidemics* (Section III, Constitution 2) and in Book IV. He also describes five cases of puerperal fever which were so severe as to suggest a streptococcal infection (*Epidemics*, Book I, Cases 4 and 11, Book III, Section 1, Cases 11 and 12, Section 2, Case 14). Whether infections of this type occurred during the Pleistocene and so reached the Americas, Australia and New Zealand, is quite unknown, but it is at least a possibility.

Much less is known about the antiquity of scarlet fever, another streptococcal infection but due to a variant that produces a specific toxin. It was certainly not endemic in America or Australia at the time of their discovery. Indeed, Ingrassias in the seventeenth century, seems to have been the first to describe it (Rolleston, 1937).

Dip. pneumoniae may, as already pointed out, have become established as a human pathogen during the Pleistocene. If this be so, its most typical pathogenic activity, the production of lobar pneumonia might well have occurred during that period in both the Old and the New Worlds. There is no direct evidence of this but lobar pneumonia may be of considerable antiquity. Ruffer (1910) noted possible hepatization of the lung in which there were gram-positive cocci in a mummy of the XXth Dynasty (1200-1090 B.C.). Aretaeus (Adams, 1846) too gives an admirable description of the disease.

Although the epidemiology of diphtheria resembles that of streptococcal tonsillitis so closely that if the organism had been evolved it might have survived in a Palaeolithic and even more so in a Neolithic society, there is no evidence of this. This disease was certainly not carried to the Americas, Australia or New Zealand by early man.

No bony lesions are caused by it nor is there evidence of its presence in mummies, while its identification from surviving medical records is bedeviled by its close resemblance to streptococcal tonsillitis. Nevertheless, two complications are typical. One is suffocation by the false membrane. This is referred to by Aretaeus (Adams, 1846) and in Talmudic medicine (A.D. 200-600) (Snowman, 1935). The second is paralysis of the soft palate which allows food or drink to get into the nose. This was observed by Hippocrates (Case 7 in Section I of Book III of the *Epidemics*), by Aretaeus (Adams, 1846) and by Aetius (Andrewes *et al.*, 1923). However, it was not until it caused an outbreak of "Garotillo" in Spain in the sixteenth century that its real history begins.

Cerebrospinal meningitis, despite its effect on the brain, is fundamentally another infection of the respiratory tract and the organism responsible for the disease is transmitted in much the same way as are other respiratory organisms. It is however most improbable that it caused epidemics during the Pleistocene resembling in any way those sometimes seen nowadays. These generally occur when large numbers of susceptible young men are congregated together in dormitories of some kind (Glover, 1918). It may therefore have plagued armies since the Bronze Age; but this is no more than conjecture because the disease cannot be recognized in records and it was not until the early years of the nineteenth century that it was differentiated.

A number of organisms infect the intestinal tract itself or, as in poliomyelitis and infective hepatitis, are present in it even though the main focus of the disease is in another organ. They are excreted in the faeces of both cases and carriers and their transmission to others depends on this. There seems, therefore, no reason to doubt that if the organisms had been evolved, they could have become established in Palaeolithic societies and might even have been taken to the Americas, Australia and New Zealand. None of this can be proved but all were causing disease in the Old World before the Christian era.

Dysentery is by far the most common disease in this group, and both Greek and Roman writers refer to it. Typhoid and paratyphoid on the other hand, are more difficult to identify in the literature but there seems little doubt that this was the diagnosis in at least six cases in the Hippocratic collection, that is, Cases 6, 7 and 10 in *Epidemics* Book I, Case 3 in Section I and Cases 6 and 12 in Section 3 of Book III.

The acute phase of poliomyelitis involving fever with muscular paralyses is not described in any of the old medical texts. But certain deformities are frequently a late result of the disease. One is shortening and atrophy of a limb. This is portrayed on a stele of a priest of Ruma who lived in Memphis during the XVIIIth Dynasty (Hamburger, 1911). Unilateral shortening of the femur was also found in a Pre-Dynastic skeleton (Mitchell, 1900). Contracture of unopposed muscles can similarly produce deformity such as that of the foot of the Pharaoh Siptah of the

XIXth Dynasty (Ruffer, 1914).

Infective hepatitis is much more difficult to identify because jaundice, its most typical manifestation, may occur in many other diseases. But it is probable that it was responsible for the outbreak described in Book IV of the *Epidemics* of Hippocrates in which many of the cases had ocular signs.

There seems little doubt, therefore, that all these intestinal infections are of very considerable antiquity.

Diseases in Which the Organisms Persist in the Lesions of Chronic Infections. Syphilis, Yaws, Pinta, Tuberculosis, Leprosy, Gonorrhoea and Trachoma

The organisms responsible for these diseases are present in sputum, pus or other discharges and may continue to reach the outside world in this way for years. Except for the fact that, in a food gathering or hunting community, the patient may be so ill that he may soon die of neglect or starvation (Carr Saunders, 1922; Cummins, 1939) and may be unable to survive long distance travel, the organisms responsible for such diseases are much more likely to become permanent inhabitants of a sparse, widely dispersed human community such as that of a Palaeolithic society than any of those discussed above.

There is reason to believe that the closely related spirochaetes causing the three diseases, syphilis, yaws and pinta were able to do so. According to Hackett (1963), *Treponema pinta* causing the non-venereal infection pinta was probably the first of the three spirochaetes to be evolved and became established in Palaeolithic tribes in Central Asia about 15,000 B.C. It was then carried to America before the land bridge over the Bering Strait became impassable.

Some time after this, a mutant, *Trep. pertenue* the cause of yaws, probably appeared somewhere in the Old World. It did not reach America but did get to Australia.

About 7,000 B.C., *Trep. pallidum* was evolved in the Old World to cause the so-called endemic or non-venereal syphilis. This was largely confined to the warm arid climates of North Africa, South West Africa, and ultimately, Australia. The same organism then began to cause venereal syphilis in a mild form about 3,000 B.C. facilitated in this by the change in mode of life resulting from the urban revolution. It continued to behave in this way but in the early years of the sixteenth century was supplanted by a more virulent mutant which produced the widespread epidemic of severe syphilis in Europe frequently considered to have been imported from America by Columbus. This mutant eventually reached every country in the world.

The *Mycobacteria* are similarly responsible for chronic infections such as tuberculosis and leprosy. *Myo. tuberculosis* occurs in two forms, the human and bovine. The former can infect any organ but the pulmonary form of the disease is by far the most common and the most important. The bovine variety on the other hand only seldom causes pulmonary tuberculosis; it is more likely to infect lymph nodes, the abdomen and the skeleton. The two varieties of the organism are so closely related that it is extremely probable that one is a mutant of the other.

The potentialities of the bovine variety have already been described in that it does not apparently infect wild animals—only those that have been domesticated—and, secondly, that it has never established itself as a human parasite.

The human variety on the other hand, is now permanently established as a human parasite with Man as its only host.

Although its chief pathogenic activity is the causation of pulmonary tuberculosis and this disease tends to be commoner amongst people living in confined spaces than in sparse widely dispersed populations, there is no reason to suppose that it could not have occurred amongst Palaeolithic communities.

It is however open to question whether it did so. Indeed, the earliest unequivocal evidence of the occurrence of pulmonary tuberculosis is late in the second millenium before Christ. In India for example, the *Ordinances of Manu* (Puccinotti, 1850; Flick, 1925) of about 1300 B.C. warned prospective bridegrooms to beware of marriage into a family prone to the disease. Some time later, probably in the sixth century B.C. the *Sushruta Samhita* (Bhishagratna, 1907) refers to what is almost certainly pulmonary tuberculosis and because it was incurable, a physician with any regard for his professional reputation was advised not to treat it. There are also records of the disease in China about the same time (Puccinotti, 1850). In Egypt too, spinal lesions suggestive of tuberculosis were found in the mummy of a young priest of Amen of the XXI Dynasty (1090-945 B.C.). They were complicated by a typical psoas abscess (Smith and Ruffer, 1910). Long (1931) in a very short paper without illustrations has also described areas of alleged well capsulated caseation of the lungs in a mummy of an old woman of the XXI Dynasty (1095-945 B.C.). Finally, the Greek and Roman medical writers from the time of Hippocrates were obviously well acquainted with the disease.

Its antiquity prior to 1,300 B.C. is, however, not at all certain. Bones with lesions suggestive of tuberculosis have been found in widely dispersed localities in the Old World. Some from the Nile Valley may date as far back as 3,700 B.C. (Morse *et al.*, 1964).

Since human strains of *Myco. tuberculosis* can cause bony lesions it has been assumed that pulmonary tuberculosis was also prevalent in these populations at that time. Nevertheless, it must be pointed out that a great many diseases can produce changes in the bones that resemble very closely those of tuberculosis. These include osteomyelitis, traumatic and other forms of arthritis and various fungal diseases (Morse, 1961). It must not therefore be too readily assumed that these bony changes were the result of tuberculosis.

Furthermore, even if they were tuberculous, it is possible that they may have been caused by the bovine variety of the organism which is a common cause of skeletal, but virtually never pulmonary, tuberculosis.

It must also be mentioned that although the viscera from Egyptian mummies may be either missing or valueless as pathological specimens, the fact remains that pulmonary tuberculosis was not found in any of the many thousand bodies from Egypt and Nubia examined by Elliot Smith, Wood Jones, Derry, Dawson, or Ruffer (Smith and Wood Jones, 1910; Derry, 1909; Smith and Dawson, 1924; Ruffer, 1921). Finally Burke (1938) has pointed out that there is nothing suggestive of the disease in the Papyrus Ebers.

It is therefore doubtful whether pulmonary tuberculosis had become established as a human disease in the Old World before the second millennium B.C. If this be so, it is even less probable that it had been prevalent in Palaeolithic and early Neolithic societies. Confirmation of this comes from the fact that the disease had evidently not been carried to Australia, New Zealand and the Pacific Islands since it was unknown before their discovery. Whether or not it was taken to America by settlers from Siberia before the Bering Strait land bridge was closed in about 8,000 B.C. is however more controversial. There are, for example, statements by Jesuit missionaries that pulmonary tuberculosis was prevalent amongst North American Indians in Canada before the arrival of the white man (Maher 1929) but clinical diagnosis by amateurs must always be viewed with suspicion and the disease was certainly not prevalent in most parts of North America before its discovery. It has also been suggested that the hunchbacks portrayed in Peruvian pottery may have had tuberculosis of the spine. But Morse (1961) has pointed out that there are other causes of kyphosis than tuberculosis.

On the whole therefore, it would seem probable that pulmonary tuberculosis caused by human strains of *Myco. tuberculosis* did not make its appearance anywhere until about the second millennium B.C. If this be so, it is possible that the human strain is a mutant of the bovine which had previously become established in the dairy cattle that were domesticated during the Neolithic period.

A second species of *Mycobacterium* (*Myco. leprae*) is responsible for leprosy. It is an unusual organism in that it has never been cultivated *in vitro*. Since this is also a peculiarity of another species of *Mycobacterium* that causes a leprosy-like disease in rats, it seems probable that *Myco. leprae* may be a mutant.

There is no evidence to suggest that this mutation occurred in the Pleistocene. The disease was certainly not carried to Australia, New Zealand or the Americas during the migrations. It may however, have developed in eastern Asia in the first millennium before Christ. A disease is, for example, described in the *Charaka Samhita,* an Indian compilation of 600-400 B.C. which seems to have involved disturbances in the sweat reflex, development of black spots in the skin and what is very characteristic of leprosy, anaesthetic areas on the skin of the limbs (Gupta, 1909). The use of chaulmoogra oil, the only successful treatment for the disease until recently, also dates from that time. The *Sushruta Samhita* of about the same date also has references to something resembling the disease (Bhishagratna, 1907).

In China the *Nei Ching* probably written about 250 B.C. (but may be very much earlier) describes a disease that bears close resemblance to leprosy, involving loss of hair, the formation of nodules, together with ulceration and, again anaesthetic areas on the skin (Wong and Wu, 1932).

During the period referred to above, it is extremely improbable that the disease was present in western Asia, or Europe. The many references to leprosy in the Old Testament are probably a consequence of mistranslation of the term $\lambda\varepsilon\pi\rho\alpha$ in the Greek texts (MacArthur, 1953), a series of conditions such as leucoderma, ringworm, impetigo and pyogenic infections being more probable. Nowhere are the typical lepromatous or tuberculoid forms of the disease described.

Nor is there evidence that the disease was present in early Egypt. No traces have been found in pre-Christian bones or mummies. The papyri do not mention it: an incurable swelling of the limbs referred to in the papyrus Ebers is certainly not diagnostic of the disease; and a jar of about 1211 B.C. with a face said to depict lepromatous leprosy (Yoeli 1955) could with equal probability be no more than a caricature. Lastly, there is nothing resembling leprosy in the Hippocratic collection.

If then it be assumed that leprosy first developed in the East there seems no doubt that it reached western Asia, Africa and Europe some time after Christ. Aretaeus (Adams, 1846) in the *Therapeutics of Common Affections*, describes a disease that very closely resembles it. Galen too seems to have been acquainted with it. A Coptic mummy of the sixth century from Nubia shows clear evidence of the disease in the hands and feet (Smith and Derry, 1910, Smith and Dawson, 1924) and skeletal leprosy has also been detected in remains of the first millennium A.D. from Europe (Møller-Christensen, 1961; Brothwell, 1961; also the contribution by Møller-Christensen in this book).

Trachoma is the result of infection of the eye by a virus. It can lead to total blindness and is still common in many tropical and sub-tropical countries. There is no evidence that it developed in the Palaeolithic societies of the Old World or that it was taken to the Americas, Australia or New Zealand. But it seems to have been present in Egypt at the time the Papyrus Ebers was compiled (Ebbell, 1937) and is referred to by Hippocrates in Constitution 2 of Book III of the *Epidemics*, by Celsus (Spencer, 1935-38) and Paulus Aegineta (Adams, 1844).

The antiquity of gonorrhoea is of much the same order. It evidently did not reach America, Australia or New Zealand but was known in Mesopotamia long before Christ (Sigerist, 1951). It is also probable that it was one of the diseases referred to in the Mosaic Law. It was also well known to the Romans, for example, Celsus (Spencer, 1935-38).

Diseases Caused by Commensal Organisms. Boils, Carbuncles, Pyorrhoea, Abscesses of the Teeth and Jaws, Pyelitis and Cystitis, Appendicitis and Peritonitis

A great many species of organisms are permanent inhabitants of the skin and mucous membranes. Although generally harmless they can however cause infection when, as sometimes happens, there is a breakdown in the local defense mechanisms. Acquired by the baby from his mother or members of the family, and remaining with him until he dies, they would have no difficulty in surviving in a Palaeolithic community.

They have almost certainly accompanied man from his pre-hominid ancestor; non-human primates, for example, have a very similar commensal flora today; and remote, isolated human tribes have the same commensals as those living in well populated areas. *Staphylococcus aureus* and *Staph. albus* were, for example, found in the nasal flora of a tribe living in the remote highlands of New Guinea by Rountree (1956). Moreover, the phage types of the former organism were very much the same as those of strains prevalent today in Europe and America; some were even resistant to penicillin. The vaginal flora too may be similar, Hare and Polunin (1961) having found that the women of an equally remote tribe in North Borneo were vaginal carriers of a particular variety of anaerobic streptococci which is also a commensal of women in Europe and America.

There seems no reason to doubt that the diseases caused by commensal organisms have plagued human beings ever since man evolved and spread throughout the world, and that they were as prevalent in the Americas, Australia and New Zealand as in Europe, Asia or Africa.

A great many infections come into this category. Some of the most important are caused by *Staph. aureus* whose normal habitat is the nose. But it can cause osteomyelitis and traces of this disease have been observed in skeletal remains of considerable antiquity in both the New and Old Worlds (Roney, 1959; Hooton, 1930; Ruffer, 1914). This organism also infects open wounds and such complications are referred to by both Greek and Roman writers: paronychiae and carbuncles of the neck, too, are referred to in the Ebers Papyrus (Ebbell, 1937). More unusual is abscess of the breast. Two cases are mentioned in the Edwin Smith Papyrus (Cases 39 and 46) (Dawson, 1932) while that of Queen Atossa of Persia in the fifth century B.C. (Sandison, 1959) may well have been due to the same organism.

Commensals normally present in the mouth or throat may similarly produce infections of the nasal sinuses that are sufficiently severe to cause changes in the bones. These have been detected in very old skulls from as far afield as Egypt, Britain, France, Peru and Queensland (Wells, 1964). But pyorrhoea and dental abscesses are even more common and equally widespread. They have certainly been found in wild Primates (Colyer, 1936; Schultz, 1939) so that it is not surprising that somewhat similar lesions have been detected in Rhodesian man (Hrdlička, 1930) in Neandertal man (Schultz, 1939) as well as in Egyptian mummies (Smith and Dawson, 1924; Ruffer, 1914) and in places as far apart in the New World as California (Hooton, 1930; Roney, 1959) Peru (Bandelier, 1904) and Hawaii (Snow, 1961).

Infections of the lungs, particularly bronchopneumonia may also be caused by commensal organisms from the throat. The earliest probable recorded instance of this disease was found in a mummy of the XVIII Dynasty (1555-1350 B.C.) (Shaw, 1938) but there seems no reason to doubt that it has been a cause of death ever since Man evolved.

The organisms normally present in the intestinal canal can also produce infection. Thus *Escherichia coli* which is also a common animal parasite, is the usual cause of cystitis and pyelitis, and it is of interest that apparent multiple abscesses in the kidney in which there were gram negative organisms were found in an Egyptian mummy by Ruffer (1910). The same organism would undoubtedly have caused peritonitis whenever the bowel was pierced but it could cause or play a part, in spontaneous infections as well. And certainly traces of old appendicitis have been detected in mummies (Smith and Wood Jones, 1910) while an old pelvic abscess found in that of an aged priestess of the XXI Dynasty in Egypt (Smith and Dawson, 1924), may have been caused by such organisms.

CONCLUSION

Although the evidence on which to base an appraisal of the antiquity of disease caused by bacteria and viruses is scanty and in many respects unsatisfactory, there seems no doubt that some infections are as old as Man himself, others only made their appearance when he had provided the necessary conditions by becoming numerous and adopting an urban way of life and still others only began to infect him when in the course of evolution, insect or animal hosts themselves became parasitized.

There is also evidence that although many of these organisms infected the inhabitants of Europe, North Africa and Asia long ago, those living in the Americas, Australia, New Zealand and the Pacific Islands remained free until contact was made with Europeans.

Thus, the antiquity of these diseases depends on place as much as on time.

REFERENCES

ADAMS, F., Trans. 1844: *Paulus Aegineta, The Seven Books.* London, The Sydenham Society.

ADAMS, F., Trans. 1846: *Aretaeus the Cappadocian.* London, The Sydenham Society.

ADAMS, F., Trans. 1849: *Hippocrates. The Genuine Works.* London, The Sydenham Society.

ALSTON, J. M. and BROOM, J. C., 1958: *Leptospirosis in Man and Animals.* Edinburgh, Livingstone.

ANDREWES, F. W., BULLOCH, W., DOUGLAS, S. R., DREYER, G., GARDNER, A. D., FILDES, P., LEDINGHAM, J. C. G. and WOLF, C. G. L., 1923: *Diphtheria.* London, H.M.S.O.

BAKER, A. C., 1943: The typical epidemic series. *Amer. J. Trop. Med.,* 23:559.

BANDELIER, A. F., 1904: Aboriginal trephining in Bolivia. *Amer. Anthrop.,* 6:440.

BESWICK, T. S. L., 1962: The origin and use of the word herpes. *Med. Hist.,* 6:214.

BHISHAGRATNA, K. K. L., 1907: *The Sushruta Samhita.* Calcutta, Pub. by the translator.

BLEYER, J. C., 1922: Ueber auftreten von Variola unter affen der genera *Mycetes* und *Cebus* bei vordringen einer Pockenepidemie in urwaldiebete on den Nebenflüssen des alto Uraguay in Südbrasilen, *Münch. Med. Wchsr.,* 69:1009.

BROTHWELL, D., 1961: The palaeopathology of early British man, *J. Roy. Anthrop. Inst.,* 91:318.

BURKE, R. M., 1938: *A Historical Chronology of Tuberculosis,* Springfield, Thomas.

CARR SAUNDERS, A., 1922: *The Population Problem.* Oxford, Clarendon Press.

CARTER, H. R., 1931: *The Early History of Yellow Fever.* Baltimore, Williams and Wilkins.

CHILDE, G., 1942: *What Happened in History.* Harmondsworth, Pelican Books.

CHRISTENSEN, P. E., SCHMIDT, H., JENSEN, O., BANG, H. O., ANDERSEN V. and JORDAL, B., 1952-53: An epidemic of measles in southern Greenland, *Acta. Med. Scand.,* 144:313, 430.

COLYER, F., 1936: *Variations and Diseases of the Teeth of Animals.* London, John Bale.

CREIGHTON, C., 1891: *A History of Epidemics in Great Britain.* London, Cambridge University Press.

CUMMINS, S. L., 1939: *Primitive Tuberculosis.* London, John Bale.

DAWSON, W. R., 1932: The earliest surgical treatise. *Brit. J. Surg.,* 20:34.

DERRY, D. E., 1909: *Archaeological survey of Nubia,* Bull. No. 3, pp. 32, 44, 51.

DEWING, H. B., Trans. 1919: *Procopius, History of the Wars and Anecdota.* Loeb Classical Library, London, Heinemann.

DOLMAN, C. E. and IIDA, H., 1963: Type E botulism; its epidemiology, prevention and specific treatment. *Canad. Pub. Health J.,* 54:293.

EBBELL, B. L., 1937: *The Papyrus Ebers.* London, Oxford University Press.

FINDLAY, G. M., 1948: John Williams and the early history of yellow fever. *Brit. Med. J.* 2:474.

FLICK, L. F., 1925: *Development of Our Knowledge of Tuberculosis,* Philadelphia.

FRANCIS, J., 1947: *Bovine Tuberculosis.* London, Staples Press.

GHALIOUNGUI, P., 1963: *Magic and Medical Science in Ancient Egypt.* London, Hodder and Stoughton.

GLOVER, J. A., 1918: The cerebro-spinal fever epidemic of 1917 at X depot. *J. Hyg.,* 17:350.

GLOVER, J. A. and GRIFFITH, F., 1931: Acute tonsillitis and some of its sequels. *Brit. Med. J.* 2:521.

GOODALL, E. W., 1934: *A Short History of the Epidemic Infectious Diseases.* London, John Bale.

GREENHILL, W. A., Trans. 1848: *A Treatise on the Smallpox and Measles* by Abu Beer Mohammed Ibn Zacariyá Ar-Rázi (commonly called Rhazes). London, the Sydenham Society.

GRIFFITH, S. A., 1928: Tuberculosis in captive wild animals. *J. Hyg.,* 28:198.

GUILLAUME, A., 1955: *The Life of Muhammed.* London, Oxford University Press.

GUPTA, K. N. N. S., 1909: *The Ayurvedic System of Medicine.* Calcutta, Chatterjee.

GUTHRIE, D., 1940: The history of otology, *J. Laryng. and Otol.,* 55:473.

HACKETT, C. J., 1963: On the origin of the human treponematoses. *Bull. W.H.O.,* 29:7.

HAESER, H., 1882: *Lehrbuch der Geschichte der Krankheiten.* Jena, Gustav Fischer.

HAMBURGER, O., 1911: Un cas de paralysie infantile dans l'antiquité *Bull. Franc. Hist. Med.,* 10:407.

HARE, R., 1954: *Pomp and Pestilence.* London, Gollancz.

HARE, R. and POLUNIN, I., 1961: Anaerobic cocci in the vagina of native women in British North Borneo. *J. Obstet. Gynaec. Brit. Emp.,* 67:985.

HECKER, H. F. C., 1844: *Epidemics of the Middle Ages.* Trans. B. G. Babington, London, The Sydenham Society.

HIRSCH, A., 1883: *Handbook of Geographical and Historical Pathology.* Trans. C. Creighton, London, The Sydenham Society.

HOOTON, E. A., 1930: *The Indians of Pecos Pueblo.* New Haven, Yale University Press.

HRDLIČKA, A., 1930: *The Skeletal Remains of Early Man,* Smithsonian Misc. Coll., No. 83.

Long, A. R., 1931: Cardiovascular renal disease. Report of a case of three thousand years ago. *Archiv. Path.*, 12, 92.

MacArthur, W., 1953: Mediaeval leprosy in the British Isles. *Leprosy Rev.*, 24:8.

Macpherson, J., 1886-87: The history of cholera in the east. *Trans. Epid. Soc.*, 6:46.

Maher, S. J., 1929: Tuberculosis among the American Indians, *Amer. Rev. Tuberc.*, 19:407.

Mattingly, P. F., 1960: Ecological aspects of the evolution of mosquito-borne virus diseases, *Trans. Roy. Soc. Trop. Med.*, 54:97.

Mettler, C. A., 1947: *History of Medicine*, Philadelphia, Blakiston.

Mitchell, J. K., 1900: Study of a mummy affected with poliomyelitis, *Trans. Assoc. Amer. Phys.*, 15:134.

Møller-Christensen, V., 1961: *Bone Changes in Leprosy.* Copenhagen, Munksgaard.

Moore, J., 1815: *The History of the Smallpox*, London, Longmans.

Morse, D., 1961: Prehistoric tuberculosis in America, *Amer. Rev. Resp. Dis.*, 83:489.

Morse, D., Brothwell, D. R. and Ucko, P. J., 1964: Tuberculosis in Ancient Egypt. *Amer. Rev. Resp. Dis.*, 90:524.

Murchison, C., 1862: *A Treatise on the Continued Fevers of Great Britain.* London, Parker, Son and Bourn.

Puccinotti, F., 1850: *Storia Della Medicina*, Livorno, Presso Massimiliano.

Pusey, W. A., 1933: *The History of Dermatology.* Springfield, Thomas.

Rolleston, J. D., 1937: *The History of the Acute Exanthemata.* London, Heinemann.

Roney, J. G., 1959: Palaeopathology of a Californian archaeological site. *Bull. Hist. Med.*, 33:97.

Rountree, P. M., 1956: Staphylococci harboured by people in the western highlands of New Guinea. *Lancet*, 1:719.

Ruch, T. C., 1959: *Diseases of Laboratory Primates.* Philadelphia, Saunders.

Ruffer, M. A., 1910: Remarks on the histology and pathological anatomy of Egyptian mummies. *Cairo Sci. J.* 4:1. (Reprinted in Ruffer, 1921)

Ruffer, M. A., 1914: Pathological notes on the royal mummies of the Cairo museum. *Mittheil. z. Gesch. Med. und der Naturwissensch.*, 13:239. (Reprinted in Ruffer, 1921)

Ruffer, M. A., 1921: *Studies in the Palaeopathology of Egypt.* Chicago, University of Chicago Press.

Ruffer, M. A. and Ferguson, A. R., 1911: An eruption resembling that of variola in the skin of a mummy of the twentieth dynasty (1200-1100 B.C.). *J. Path. Bact.*, 15:1. (Reprinted in Ruffer, 1921)

Sandison, A. T., 1959: The first recorded case of inflammatory mastitis—Queen Atossa of Persia and the Physician Democêdes. *Med. Hist.*, 3:317.

Schultz, A. D., 1939: Notes on diseases and healed fractures of wild apes. *Bull. Hist. Med.*, 7:571.

Shaw, A. F. B., 1938: A histological study of the mummy of Har-Mose, the singer of the eighteenth dynasty (Circa 1490 b.c.). *J. Path. Bact.*, 47:115.

Short, T., 1749: *A General Chronological History of the Air, Weather, Seasons, Meteors etc.*, Quoted from Annals of Influenza. T. Thompson (Ed.), 1832, London, The Sydenham Society.

Sigerist, H. E., 1951: *A History of Medicine*, Vol 1. New York, Oxford University Press.

Smith, F. P., 1871: Small-pox in China. *Med. Times Gaz.* ii:277.

Smith, G. Elliot and Dawson, W. R., 1924: *Egyptian Mummies.* London, Allen and Unwin.

Smith, G. Elliot and Derry, D. E., 1910: Arch. Survey of Nubia Bull. No. 6, p. 29.

Smith, G. Elliot and Jones, F. Wood, 1910: The Arch. Survey of Nubia. Report for 1907-08, Vol. II, Cairo, National Printing Dept.

Smith, G. Elliot and Ruffer, M.A., 1910: Pott'sche Krankheit an einer Äegypteschen Mumie aus der Zeit der 21 Dynastie (um 1000 V, Chr.). *Historischen Biologie Krankheitserreger*, Heft 3, Giessen. (Reprinted in Ruffer, 1921)

Snow, C. E., 1961: An old Hawaiian population on Oahu. *Amer. J. Phys. Anthrop.*, 20:69.

Snowman, J., 1935: *A Short History of Talmudic Medicine.* London, John Bale.

Spencer, W. G., Trans. 1935-38: *Celsus, De Medicina.* London, Heinemann.

Wells, C., 1964: *Bones, Bodies and Disease.* London, Thames and Hudson.

Wise, T. A., 1860: *Commentary on the Hindu System of Medicine.* Calcutta, Thacker.

Woodville, W., 1796: *The History of the Inoculation of the Smallpox.* London, J. Phillips.

Wong, K. C., 1929: China's contribution to the science of medicine. *Chin. Med. J.*, 43:1193.

Wong, K. C. and Wu, L. T., 1932: *History of Chinese Medicine.* Tientsin, The Tientsin Press.

Yoeli, M., 1955: A "facies leontina" of leprosy on an ancient Canaanite jar. *J. Hist. Med.*, 10:331.

Zinsser, H., 1935: *Rats, Lice and History.* London, Routledge.

Chapter 9

Trypanosomiasis in Prehistoric and Later Human Populations, a Tentative Reconstruction

FRANK L. LAMBRECHT

INTRODUCTION

In the "Historical Sketch" which forms the prologue to Darwin's *On the Origin of Species,* he cited the lecture in 1813 of W. C. Wells before the Royal Society. It contained this remarkable passage:

Of the accidental varieties of man, which would occur among the first few and scattered inhabitants of the middle regions of Africa, some one would be better fitted than the other to bear the diseases of the country. This race would consequently multiply, while the others would decrease, not only from their inability to sustain the attacks of the disease, but from their incapacity of contending with their more vigorous neighbours.

This is an extraordinary statement, considering that it was made at a time when ideas of evolution had just begun to emerge. First, it points to disease as an important selective factor; and second, the influence of this factor is projected on a human race of "middle Africa" well before fossil evidence had shown this to be the probable region of man's origin.

It is important to bear in mind not only that natural selection, which includes the important elements of parasitism and disease, operates on existing populations but also that they in turn are the results of previous selective factors. Africa seems to have been the main region of early hominid evolution. As such it is the environmental background against which man's origins and evolution must be reconstructed. Moreover, it is in the effects of parasitic diseases which may have existed during those early times that important selective factors may be sought. Infectious diseases transmitted by direct contact can be eliminated as possible selective agents because they could not have circulated easily among the sparsely distributed human populations of early times. We must look for insect-borne diseases, in which the invertebrate vector acts as a transportation agent, thus increasing the range of the disease and the number of possible vertebrate hosts. An even more effective agent would be a parasite capable of circulating in alternate animal hosts and thereby surviving for long periods during man's absence from the area. Such diseases, of which there are many examples, are known as "zoonoses."

Important insect-borne diseases which fit this description are malaria, yellow fever, and trypanosomiasis. Trypanosomiasis is the most important for this discussion. Malaria and yellow fever are both mosquito-borne. The vector moves freely from arboreal primate reservoirs to man, a situation which must have existed since early times. It tends to confer cross-immunity, and chances are that the parasite slowly becomes less pathogenic to populations in regular contact with it. The human malaria parasites, *Plasmodium vivax* and *P. malariae* can be said to induce rather mild diseases; *P. falciparum* is more virulent. Identical or very similar plasmodia are found in certain African subhuman primate species, e.g., gorilla and chimpanzee. The high sickle-cell trait in certain human populations is a classic example of genetic selection in response to the challenge of the rather virulent *falciparum* infections in man. Significantly, no indication of abnormal sickling has been found in red cells of African apes, in which *falciparum*-like plasmodia induces only mild infections.

Tsetse-borne trypanosomiasis is confined to the specific ground-level habitat of the flies belonging to the genus *Glossina*. As such there is no transmission of trypanosomes to, nor chance of cross-immunity in, arboreal primates from animals living at lower levels. African tree-dwelling monkeys are all very susceptible to tsetse-borne trypanosomes, except the baboon, a ground-dweller. The essential point is that at some time during the late Miocene or early Pliocene, certain primates began to spend more time in the expanding open wood-and-grass land; they exposed themselves to parasitic microorganisms, among them the tsetse-borne trypanosomes, against which they had no native resistance. The discovery of *Homo habilis* recently described (Napier, 1964) from Bed I, Olduvai Gorge, provides some evidence that tool-making hominid ancestors had already evolved at the close of the Pliocene.

Protozoan flagellates are known to be of extreme antiquity. That they are found today in almost every form of vertebrate animal may indicate that they began their parasitic mode of life as early as the Palaeozoic, while free-living forms may have evolved even during pre-Cambrian times.

Glossina flies are today confined to the African continent. However, fossil impressions of *Glossina* flies have been found in Miocene beds in Colorado, U.S.A. Since connections between the American continent and Africa by way of Europe must have been disrupted since the early Tertiary or before, it can therefore be assumed that *Glossina* were present in Africa at least since the Miocene and that, in all probability, trypanosomes carried by tsetse flies were circulating. This means that the species of ground-dwelling vertebrates living in the areas of fly distribution had to be resistant to the trypanosome strains. As a result the selection of land animals differed from that in other parts of the world where tsetse were absent, and the unoccupied ground habitats, resulting from elimination of certain groups of animals, became available to other animal groups. It may be postulated here that these opportunities were exploited by certain primates which left the canopy of the forests in response to evolutionary urges of adaptive radiation. Those that succeeded were the ones that possessed or acquired resistance or tolerance to the then-present trypanosome strains. At this point a precise selection was made, extremely important to later primate evolution and especially to subsequent hominid types from which *Homo sapiens* would eventually emerge. The same element (tsetse-borne trypanosomiasis) which provided the unoccupied ecological niches and triggered the venture for certain primates to become ground-dwellers was also responsible for the genotypes that were to succeed.

SLEEPING SICKNESS IN EARLY MAN

After certain primate groups had separated from their arboreal life to become true ground-dwellers, they probably became rapidly differentiated among themselves. The trypanosome-resistant animals are the ones most likely to have been most successful during the first stages of ground-niche occupation, and other types may have evolved later with variable results. The baboon, a true ground-dwelling primate, survives in tsetse-infested areas being resistant to trypanosome infections, whereas man, a non-resistant primate type, must have had other characteristics of survival such as, for instance, superior knowledge making him less dependent on restricted (and maybe tsetse-infested) territories, or perhaps was related to primate types that had migrated outside trypanosome contact or were geographically isolated in tsetse-free areas. In more advanced hominid types, the voluntary avoidance of annoying high-density fly-belts would have decreased contact with the parasite, and thus the chances for build-up of immunity, but may also have saved them from extinction. The distribution of the early hominids was perhaps much influenced by fly-belt shape and density. The scattered and uneven dispersal of early hominid populations resulted in the formation of a partially man-adapted trypanosome strain, derived from a wild animal strain, which later would give rise to a second human strain even better adapted to man. This evolution can be represented schematically in a simplified way as follows:

called the polymorphic trypanosomes of the *brucei*-subgroup (Hoare, 1948; 1955; 1957). The development of the two trypanosomes responsible for human sleeping sickness, *T. rhodesiense* and *T. gambiense*, was the outcome of their evolution in two distinct and ecologically separated biotopes, one in the forest and one in the savannah. The same factors are still responsible for keeping the two trypanosomes apart as biologically distinct species (van den Berghe and Lambrecht, 1963; Lambrecht, 1964).

T. rhodesiense is maintained in certain wild animals of the savannahs, transmitted by *Glossina morsitans*-group flies. This trypanosome is infective to man, causing a very acute type of sleeping sickness. If untreated, the disease may cause death a few months after the onset of infection. Under normal conditions man is infected only occasionally because flies belonging to the *morsitans*-group take their bloodmeals mainly from non-human hosts. These flies disperse, at the same time as the animals on which they feed, when man interferes with the natural biotopes through agriculture and other pursuits.

T. gambiense is transmitted by *G. palpalis* and allied flies of the forest areas and forest galleries along rivers and lakes. It is essentially a man-to-man transmitted parasite, as the result of largely overlapping habitats of the flies with human settlements or gathering places along river banks and lake edges, and the avidity of this group of flies for human blood. It causes a chronic disease which, if un-

Trypanosoma brucei	→	*Trypanosoma rhodesiense*	→	*Trypanosoma gambiense*
(animals)		(animals and man)		(man)

These three morphologically similar trypanosomes are certainly closely related phylogenetically. They belong to what is treated, ends in death after one or more years. There seems to be no important wild animal reservoir of *T. gambiense*.

However, this requires further investigation.

T. brucei is found in many species of wild animals, causing no apparent ill effect to them. In most domestic animals, however, this trypanosome is very pathogenic. It does not produce infections in man.

It is assumed that *T. rhodesiense* evolved from *T. brucei* as a result of selective mutation through circulation in certain early hominid species. The better man-adapted *T. gambiense* may have evolved much later, after closer fly-man-trypanosome contact was established as a consequence of the overlapping habitats of human hosts and man-feeding tsetses.

The poorly man-adapted *T. rhodesiense* is clearly the result of discontinuing contact which evolved under circumstances of scattered host populations, unspecific vectors and "foreign" trypanosomes, circumstances experienced today in savannah belts.

Though most of their daylight activities, principally hunting, took place in the savannah, the early hominids must have spent a certain amount of time along forest edges and especially along river banks and lake shores. Visits for drinking, resting and bathing may have been more frequent at certain particularly favoured spots. With increasing organisation and higher density, hominid groups must have formed semi-permanent gathering places along the water edges. This brought them into contact with flies of the *palpalis* group —*G. palpalis* and *G. tachinoides*—for which the riverine vegetation was the habitat. Contact with riverine tsetse became even more intimate when hominids began to settle permanently along rivers and lakes and began to spend time on the water for fishing and travelling. In the absence of man, *G. palpalis* flies feed mainly on crocodiles and it is to be assumed that reptiles were its main source of blood before man invaded its habitat and provided the fly with an alternative host. Under these circumstances, trypanosomes of the *rhodesiense* type were brought into *G. palpalis* habitats. Through the effect of almost continuous circulation of the parasite through the same *Glossina* and host species, a parallel adaptation occurred, in which genetic antibody resistance of the host resulted in the selection of less virulent trypanosomes, producing the relatively mild strain known today as *T. gambiense*. Because of possible previous geographical isolation, one would expect to find *T. gambiense* strains of differing virulence.

Another possibility would be that both *T. gambiense*- and *T. rhodesiense*-like trypanosomes arose independently from a stock *T. brucei*-like parasite. *T. gambiense* could have evolved directly from such a stock strain through the survival of less pathogenic trypanosomes in the human host. In the case of *T. rhodesiense* the selection for human adapted trypanosomes was less specific, the parasite circulating mostly in wild animals through the agency of *morsitans*-group flies.

It has often been shown that the ecology of the vector is of tremendous importance in the epidemiology of insect-transmitted parasites. It is at the base of their distribution, their host range, their virulence, and a factor in the natural selection of both the parasite and the host.

Fly Ecology in Prehistoric Africa

The ecology of tsetse flies is related to climate, patterns of vegetation, distribution and density of suitable hosts, and, more recently, to interference by man. The above subtitle therefore seems a bit presumptuous. The physiological and ecological requirements of the various *Glos-*

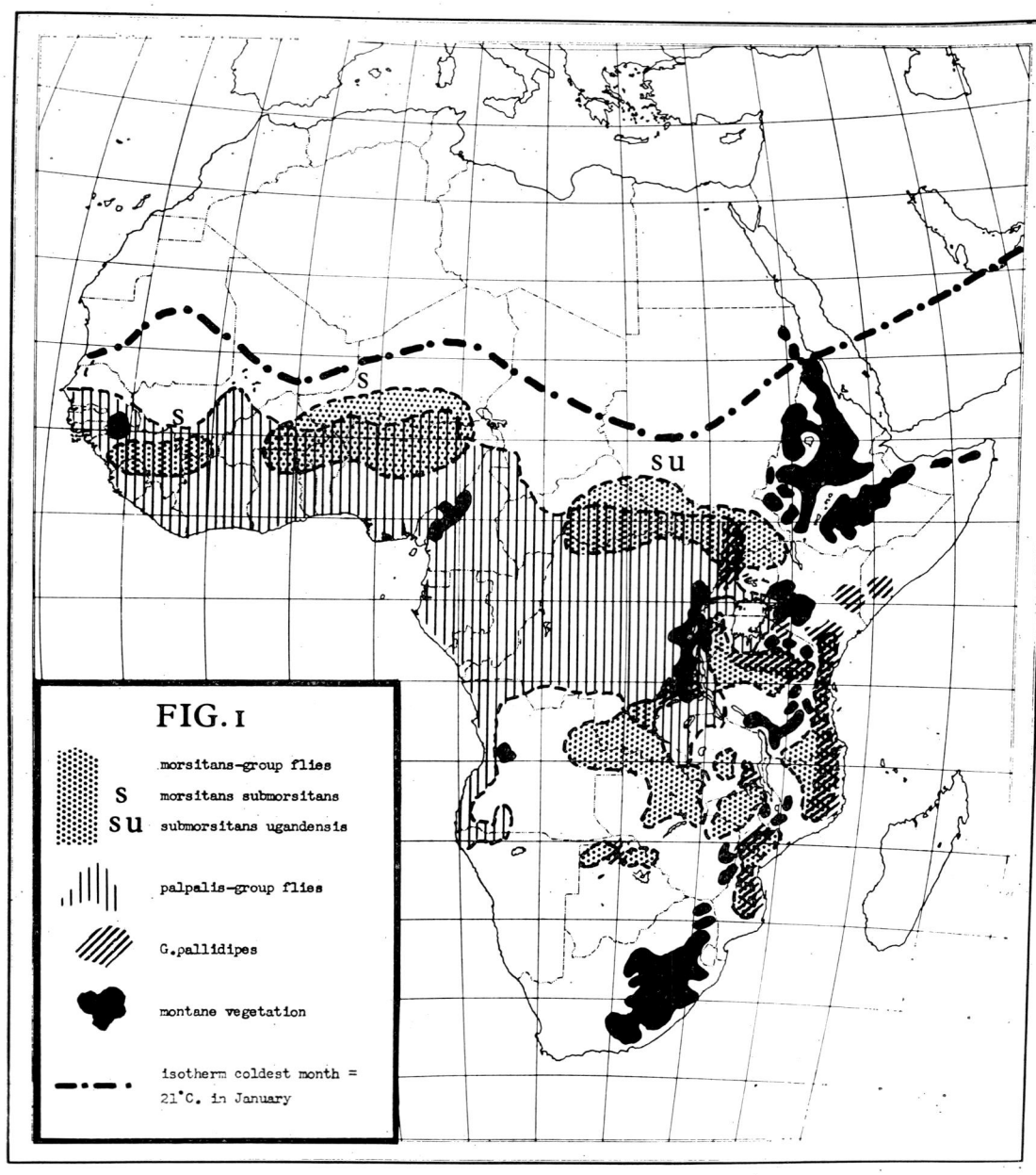

FIG. 1

- morsitans-group flies
- s morsitans submorsitans
- su submorsitans ugandensis
- palpalis-group flies
- G. pallidipes
- montane vegetation
- — · — isotherm coldest month = 21°C. in January

sina species, however, are sufficiently well known so that their survival potentialities under various climatic conditions can be predicted, at least in a general way, assuming that the fly's requirements have changed little within the last couple of million years.

Before discussing the ecology of tsetse flies under the varying conditions of the Pleistocene, let us examine some basic facts. Twenty-one species of *Glossina* are known and, in addition, ten subspecies have been described. They can be roughly classified as falling into two major eco-

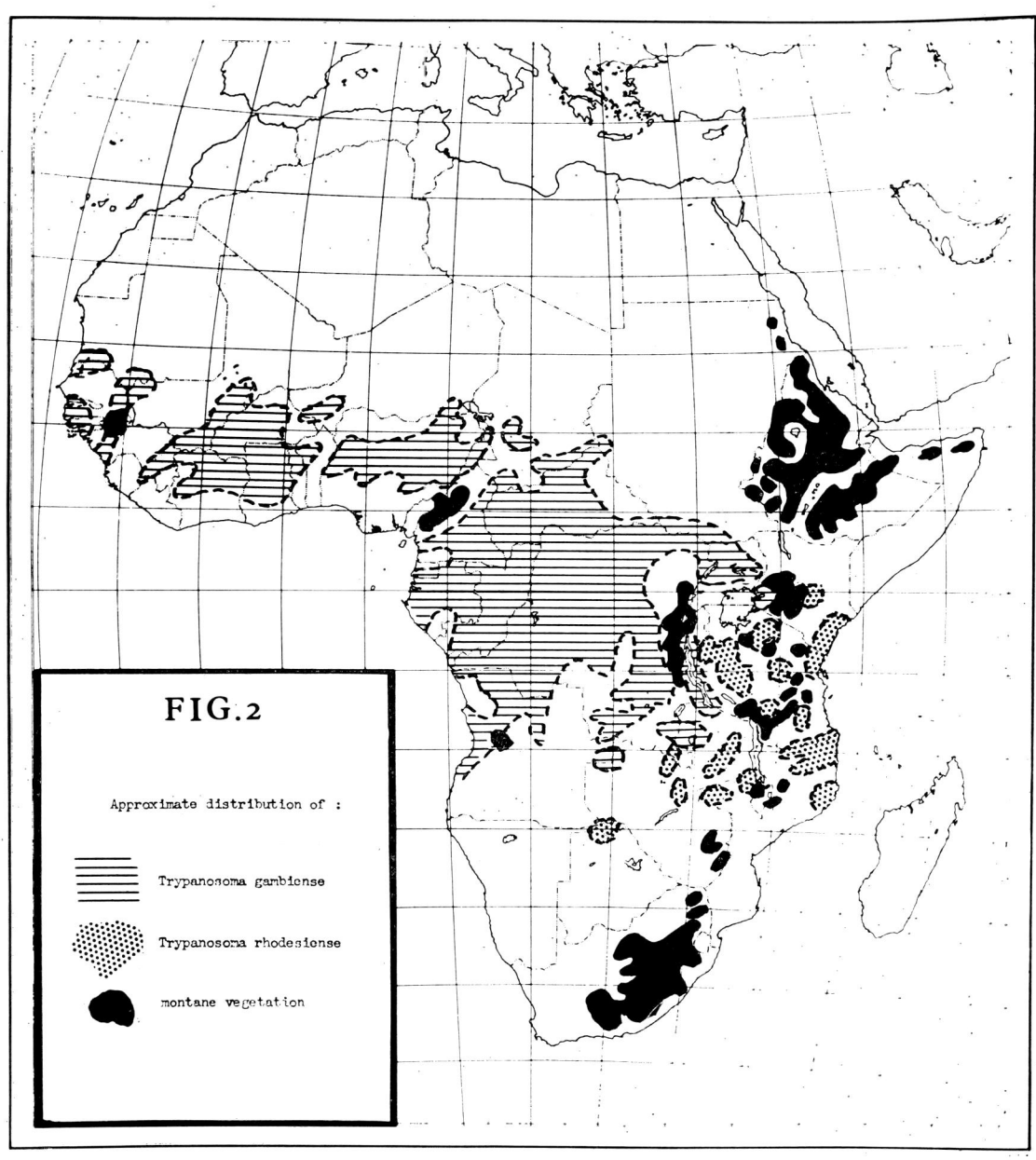

FIG. 2

Approximate distribution of:

≡ Trypanosoma gambiense

⋰ Trypanosoma rhodesiense

■ montane vegetation

logical groups: the *palpalis* and *fusca* group in the forest and dense vegetation of riverine galleries, and the *morsitans* group of the savannahs. All tsetse flies are viviparous, which is unusual but not unique. The egg hatches in the uterus; after about eleven days the larva is extruded; it burrows into the soil to a depth of about one inch where it hardens quickly into a black pupa. On the average it takes about thirty-five days for the adult fly to emerge. The length of time depends on soil temperatures and the fly species, varying from about three to thirteen weeks

(Glasgow, 1963). Adult flies live from one to four months, depending on the climate, especially on the length of the dry season, a long dry season shortening the life-span considerably. Each species shows a preference for certain vegetation types and this is especially marked in the savannah flies, but habitats may be largely overlapping in many instances. In woodland savannah, *G. morsitans* is found inside or at the edge of tree clusters; *G. pallidipes* is associated with dense thicket vegetation, while *G. swynnertoni* survives in rather open patches with short shrubs. *Glossina* species are found only in Africa, between latitudes 15° N and 29° S, in regions with temperatures of 20°-28° C, a relative humidity from 50 to 80 per cent, and rainfall between 25 and 60 inches. A low rainfall on heavy impermeable soil is more harmful, however, than a heavier rainfall on sandy soil.

It is clear that the climatic variations during the Pleistocene must have been of the greatest importance to fly distribution. Four major pluvials are generally recognized in Africa in various parts of the continent during that period, but further study is needed to establish the amplitude of the earlier three. Moreau (1963) has described the climatic variations during the late Pleistocene, and their influence on flora and fauna. It would seem that the four pluvials are the result of worldwide climatic disturbances coinciding with four glaciations during that period in the northern hemisphere. During maximum glaciation the temperature of the tropical areas was reduced by as much as 5° C. The lowering of the temperature reduced evaporation considerably and resulted in periods of heavier rainfall than today in most parts of Africa. The reduced temperature had a major effect on the vegetation. Moreau estimates that the reduction of 5° C would lower the mountain forest limit, at present at about 1,500 meters (4,900 feet) to 700 or even 500 meters (2,280 to 1,640 feet). When this information is plotted on a topographic map of Africa, the mountain biomes occupy, in one continuous block, the area from Abyssinia to the Cape, with a western extension to Mt. Cameroon. Lowland biomes are reduced to isolated areas of coastal strips, the Congo Basin, West Africa, North Africa, the Nile Valley, and the southeast coastal area along the Limpopo and Zambesi deltas (Fig. 3). Moreau found that the hypothetical montane biome idea was in accord with the present-day "island distribution" of certain faunas, inexplicable unless previous continuous distribution patterns are assumed. Geological evidence, such as glacial moraine levels, also indicates lower temperatures at certain periods of the Pleistocene.

Unsuitable conditions at higher levels limit the present-day altitude distribution of tsetse flies at about 1,500 meters. The lowering of this limit to 700 or 500 meters in response to a colder climate would inevitably restrict possible fly belts to coastal strip areas of the Congo Basin, Somaliland, West Africa, parts of North Africa, the Nile Valley, and the Limpopo and Zambesi deltas. It would perhaps be more correct to use the word "shift" instead of "restrict." Indeed, while the cold-wet climate would exclude tsetse from most parts of the African plateaus, it would at the same time provide suitable fly habitats by the development of savannah vegetation in areas at present covered by deserts: the Sahara, the Kalahari, the Nile Valley, Somaliland. However, suitable vegetation would not automatically provide suitable tsetse habitats. The map in Figure 1 shows the present-day isotherm of the average temperature during the coldest month of the year

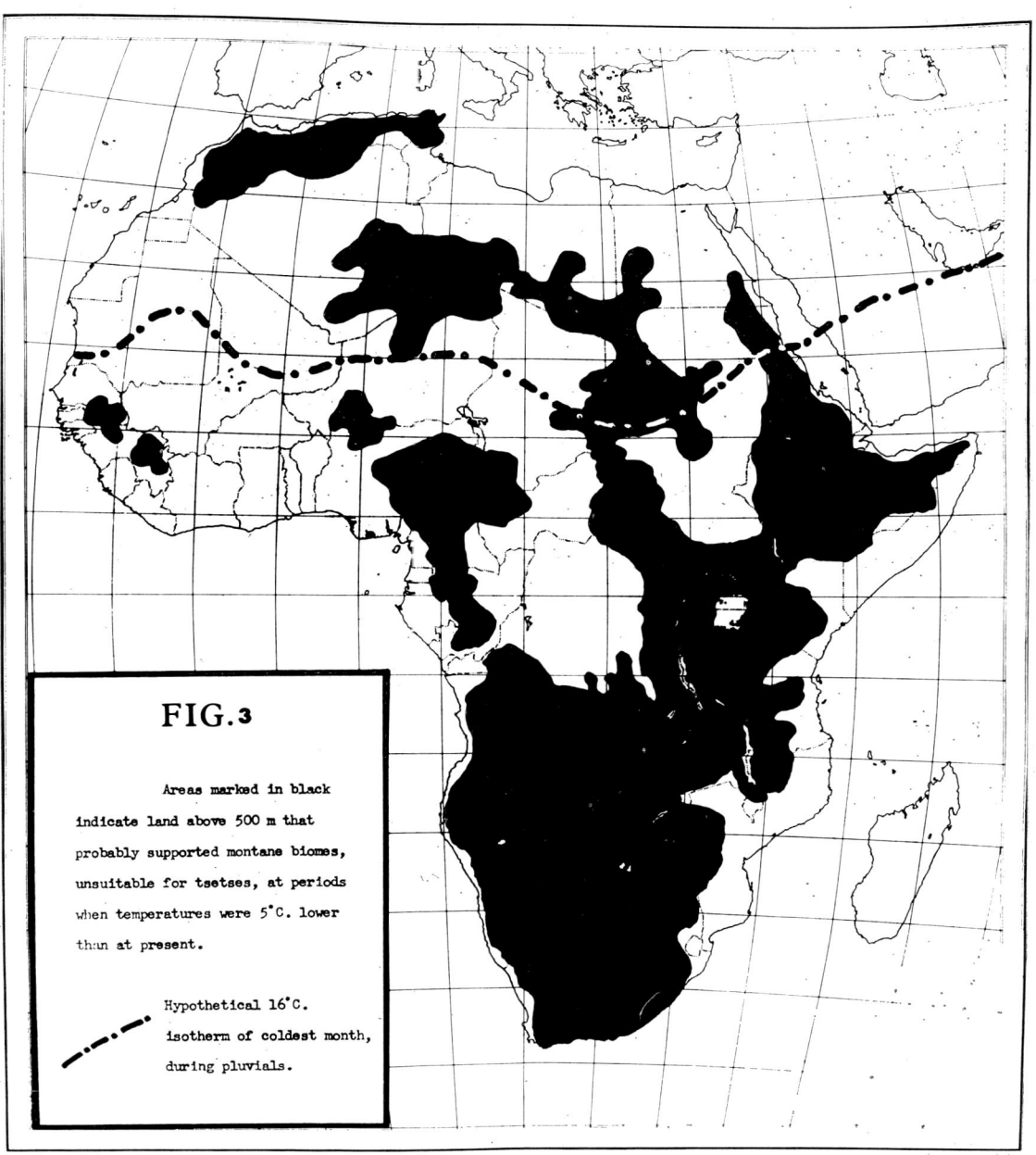

FIG. 3

Areas marked in black indicate land above 500 m that probably supported montane biomes, unsuitable for tsetses, at periods when temperatures were 5°C. lower than at present.

▬ ▪ ▬ ▪ Hypothetical 16°C. isotherm of coldest month, during pluvials.

(January), which is 70° F (21° C). The northern limit of today's fly belts stays just south of this line, except where the Ethiopian highlands force this line much further south. While this isotherm is not the reason *per se* for the tsetse boundary, it marks the northern limits of suitable climatic conditions and vegetation. During the pluvials, suitable vegetation would have extended much farther north over large parts of the Sahara. However, the lowering of the temperature by 5° C, which brought about these conditions, would also have restricted the fly's extension to a cer-

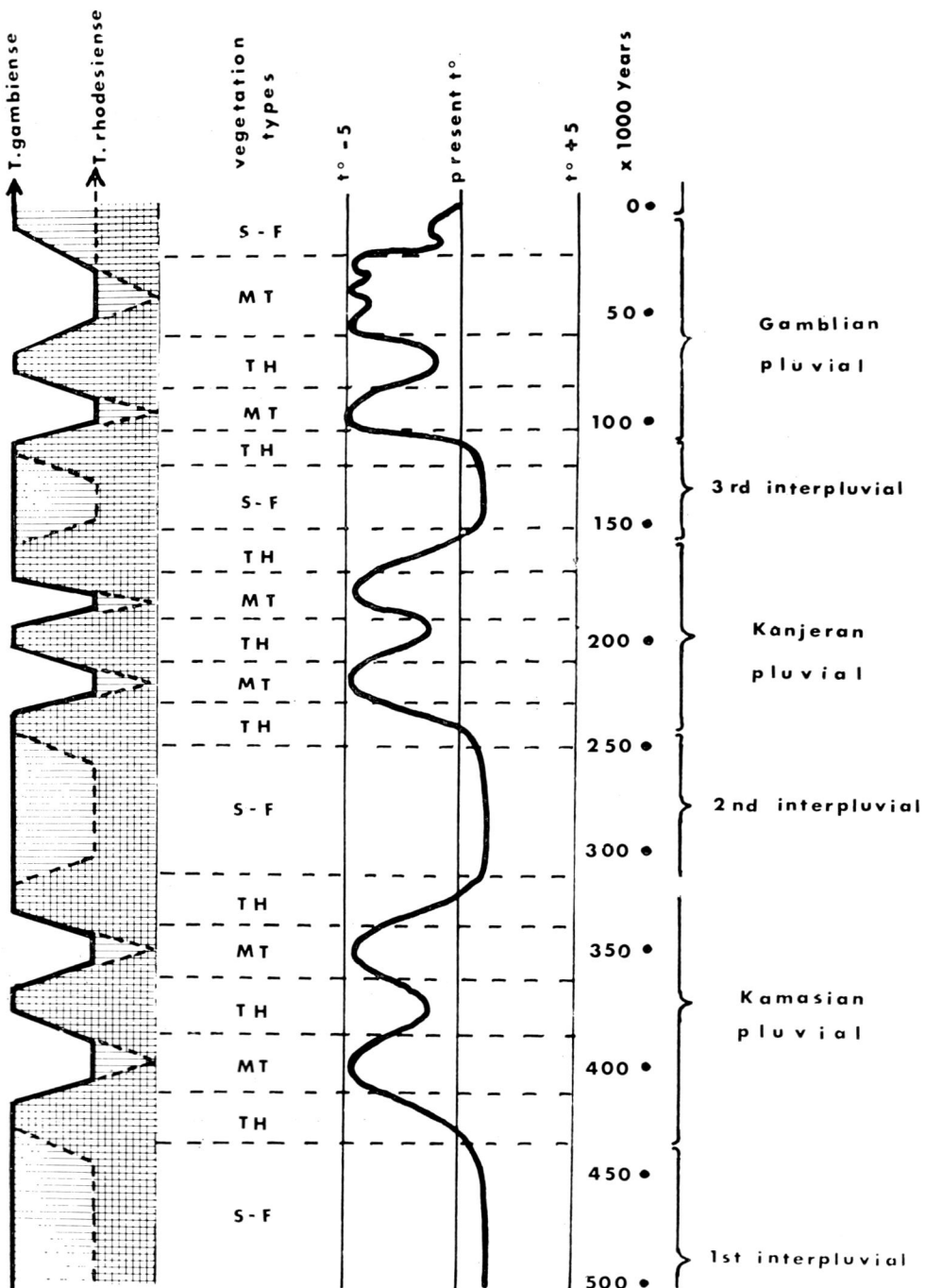

FIG. 4

tain latitude, where the long winter season would prevent normal breeding and the hatching of the pupae. It is doubtful that tsetse flies ever extended much farther north than their present limits. Indeed, during the pluvials, the present isotherm of 21° C was equivalent to that of about 16° C (Fig. 3) representing the limit of tsetse distribution, if only from the point of view of the fly's physiology. It is possible that, with a rise in temperature, and vegetation still being suitable, flies would have penetrated slightly more northward for certain periods of time.

The widely expansive Lake Chad, which at times reached the Tibesti Mts. and thus the mountain biomes in the north, may have sealed off western fly-belts from those in the east. Extension of lowland rain-forests well beyond their present limits may have further reduced savannah areas suitable for *morsitans* flies and *T. rhodesiense* transmission. Taking all these factors into account, Figure 4 has been prepared to illustrate possible climatic fluctuations, up to 500,000 years ago, and their possible influences upon fly distribution and circulation of Gambian and Rho-

←——— FIGURE 4. Relative areas of distribution of sleeping sickness during climatic fluctuations of the Upper and Middle Pleistocene.

This diagram shows the theoretical curve of temperature fluctuations during the last 500,000 years, and the generally accepted and recognized pluvial and interpluvial periods related to them. These factors are compared with the dominant vegetation type that presumably resulted and with the corresponding areas of suitable *T. gambiense* and *T. rhodesiense* transmissions.

These correlations are highly hypothetical but, in a broad sense, give some indication of how biotopes, and thus the ecology of glossinas and sleeping sickness, may have fluctuated along with climatic variations during these and earlier Pleistocene periods.

In each case the *T. gambiense* (=) curve overlaps that of *T. rhodesiense* (||). During periods of extensive montane biotope formation (MT) it is assumed that thickets, *G. pallidipes*, bushbuck and duiker would be at their greatest disadvantage, indicated by *T. rhodesiense* reaching the bottom of the curve during these periods. The transition periods between pluvials and interpluvials are assumed to be those of greatest *T. rhodesiense* transmission opportunities because they coincide with the greatest extension of thicket, *G. pallidipes*, bushbuck and duiker distribution, as shown by the peak of the curve at TH (thicket). A sort of intermediate situation is reached at the interpluvials, similar to present-day conditions (S-F : savannah + forests).

In the case of *T. gambiense* there could have been no period in which opportunities for transmission were lacking, providing man was present to serve as host. Its lowest point occurred at periods of extensive montane biome formation but at no time did it reach a point of no transmission. Even during periods of greatest reduction enough lowland forest would have been left, and especially extensive riverine vegetation along permanent rivers and lakes, to insure suitable *T. gambiense* transmission habitats. Periods of severe droughts may have resulted in a closer concentration of human hosts, *palpalis* flies and *T. gambiense* along riverine vegetation. Periods of transition and of savannah and forest formation provide conditions of maximum Gambian sleeping sickness. These conditions, we assume, are present today.

Comparison of the two curves lends additional support to the theory that *T. gambiense* had greater opportunity to become man-adapted than *T. rhodesiense* through longer periods of optimum transmission possibility.

desian sleeping sickness. Changes in fly and trypanosomiasis distribution corresponding to the changes above would be expected with each climatic fluctuation.

In addition to restricting favourable tsetse biotopes, the lower temperatures of the pluvials would also have influenced the transmission of trypanosomes. Lower temperatures would have retarded, if not stopped, the rate of development of the trypanosomes in the fly and resulted in much lower proportions of infected flies. This would not only have depended on the temperature to which the adult fly was submitted, but also on the temperature to which the pupae were exposed during their incubation time in the soil. Prolonged rains would have curtailed fly populations because water-logged soil causes high mortality in the pupae. At 15° C, the fly becomes sluggish and feeding and mating stops.

While the cold-wet climate was unfavourable to fly and trypanosome distribution because of their confinement to a number of restricted areas, this isolation may have been at the origin of speciation and strain differentiation. *T. rhodesiense* may have become "sealed off" in one of the isolated "pockets" at a period of extensive montane biomes, until "released" when climatic changes once more favoured the spread of savannah flies into other areas (Fig. 3).

Little information is available on environmental conditions during the interpluvial periods. According to Emiliani (1958), world temperatures may have been one degree, or perhaps two degrees, Centigrade, higher during those periods than at present. Summers (1960), who has studied climatic changes of the past in Southern Rhodesia, states that rainfall was reduced to half today's averages between the Middle Stone Age and the First Intermediate Stone Age. This would bring desert-like conditions to the southwestern part of that country, restricting savannah woodlands to the northeastern corner.

Between these two climatical extremes, an "interstage" is recognized during which the glacial era is interrupted. No information is available on what the environment in Africa might have been during those relatively short periods. However, it is logical to expect them to have been intermediate to the extremes discussed above. Moreover, it might be expected that during the interstages, and also during the periods leading to or receding from the pluvials, vegetation types must have been mixed, diversified, possibly with many large patches of dense thicket vegetation within more open woodland. If this were the case, then these periods would have been most favourable for tsetse fly dispersion and trypanosome transmission. This would apply more specifically to *T. rhodesiense* because the thicket vegetation is a favourable habitat for *G. pallidipes*, a fly of the *morsitans* group, proven to be a very effective transmitter of *T. rhodesiense*. Equally important, the thicket cover encourages the spread and numbers of bushbuck (*Tragelaphus scriptus*) and the common duiker (*Sylvicapra grimmia*), two antelopes known to be very effective *T. rhodesiense* reservoirs. By bringing together in a single habitat both a very suitable reservoir host and an equally suitable vector, thicket vegetation provides circumstances of maximum *T. rhodesiense* circulation. Fossil records indicate that forms of both the bushbuck and the duiker became widely distributed in South Africa during the Middle Pleistocene. Fossils from Sterkfontein (Early Pleistocene) reveal that the fauna differs only slightly in general aspects from that of the present (Wells, 1962). In East Africa, *T. scriptus* is recorded in Olduvai Bed IV, which would date it at Late Middle or Early Upper Pleistocene. From the above, one is tempted to speculate that periods of close contact between *Tragelaphus* sp., (and possibly *Sylvicapra* sp.), *G. pallidipes* and hominid hosts are at the origin of trypanosome infections in early man.

To summarize the discussion in this section:

1. *G. palpalis* and *T. gambiense* were largely confined to the same western regions throughout the climatic changes of the Pleistocene. This may have been a very important factor in the evolution of this parasite.

2. *G. morsitans* and *T. rhodesiense* belts were broken up and seriously reduced during the climatic fluctuations. This disruptive process must have seriously hampered the stability of the strain in regard to its adaptation to man. Isolated pockets may have been responsible for strain variations.

3. The partitioning into isolated ecological entities, over long periods of time must have had important consequences in the speciation of a number of organisms. It is possible that the separation of the *morsitans* fly belts during the pluvials, in a large western belt and an eastern branch, is responsible for the evolution of subspecies *submorsitans* in the western biomes.

4. The diagram in Figure 4 is a very hypothetical curve of "intensity of trypanosome circulation" plotted against vegetation type extensions during various climatic fluctuations. This suggests the possibility of better *T. gambiense* circulation for most of the periods and may have been yet another factor favourable for improved adaptation to man.

Trypanosomiasis and Early Food-gatherers

The reconstruction of the influence of trypanosomes on early human populations can only be hypothetical because information about the environment and composition of these populations is too scant. From other evidence, however, it can be surmised that tsetse-borne trypanosomes circulated in Africa at the time of the earliest evolution of ground-dwelling hominoids. Admittedly, it may have taken some time before a man-infecting trypanosome evolved, but this is probably negligible compared to the length of time hominoid, and later hominid, species were in contact with the vectors. Moreover, it could be that trypanosomes already having *T. rhodesiense*-like characteristics were present in wild animals long before certain primates became ground-dwellers and that these trypansomes were infective in later evolving hominid species, genetically susceptible to strains bearing those characteristics.

"It is possible that early savannah-dwelling hominids began to recognize that certain vegetations were irritating to man because of biting flies. Though they may not have connected this with the fatal disease that followed, they may nevertheless have avoided places where fly concentrations were particularly high. If they did, this may well have saved them. Camp sites may be expected to be found outside the main fly-belts for each of the periods to which they belong" (Lambrecht, 1964).

Other diseases, such as malaria, of course influenced human settlements and destiny. The very high incidence of the sickle-cell trait, reaching up to 40 per cent of the local population in certain parts of Africa, is the result of long-standing selection pressure by hyperendemic malaria infections in favour of a trait which is otherwise somewhat deleterious to the individual in heterozygous conditions, and mostly lethal when carried under homozygous conditions. The high proportion of such a selective trait in the presence of the disease is proof of how intimately it was tied with the evolution of these populations. It is also proof that the affected populations remained in contact with the *Plasmodium* parasite, causing the disease for very prolonged and continuous periods of time. Consequently one wonders whether human population movements, settlements and distribution were much influenced by malaria (Lambrecht, 1964).

Sleeping sickness, however, is a very acute disease at all ages and rapidly fatal in the case of *T. rhodesiense*. Places where the disease was prevalent were probably avoided as much as possible. Extensive fly belts with particular high fly densities may have been a serious handicap to human occupation and a barrier to the dispersal of human populations. As vegetation patterns and fly distribution changed during the slowly evolving climates of the Pleistocene, so may have the sites of hominid occupations (Lambrecht, 1964).

While in the relatively small populations of early Pleistocene hominids the environment may have been one of the major factors in natural selection, leading perhaps to species differentiation in the case of isolated groups, the later formation of agglomerates of larger groups resulted in a change of direction of natural selection. As aggregation intensified, the importance of infectious diseases increased and consolidated the relationship between parasite and man. Intestinal helminth parasitism must have become quite common. Increased malaria transmission occurred when sporozoite rates went up in *Anopheles* mosquitoes adapting to domestic breeding places created by man's local activities (Fig. 5).

The change of the human sedentary mode of life had a different effect on the trypanosomiasis of the forest from that of the savannah. Occupation of *G. palpalis* habitats had no adverse effect on *G. palpalis* but, on the contrary, increased contact between man and fly and promoted *T. gambiense* transmissions. In contrast, the agglomeration of large human groups in the savannah would have depleted both game and fly populations, thus decreasing the chances of *T. rhodesiense* transmission. The increase of human populations in the forest would result in *T. gambiense* becoming better man-adapted, but in the savannah would bring no change in the *T. rhodesiense*-man relationship. The difference in behaviour and ecology of their respective vectors maintained the two strains as separate biological races (Lambrecht, 1964).

During times of human aggregation into larger agglomerations, the total human populations must have been subject to two opposing forces: one an increase owing to improved hunting and tool-making techniques, the other a decrease due to higher mortality at all levels of the population, brought about by parasitic diseases. Whatever the immediate result may have been, it must have led to the selection of populations genetically, and perhaps phenotypically, different from previous populations.

Trypanosomiasis and the Cattle-owners

It is surprising to note that during the Mesolithic, after human groups in Africa had condensed into technically more advanced communities, they reached a level of relative non-development, losing initiative as leaders of human cultural evolution to those who in the meantime had popu-

SCHEMATIC REPRESENTATION OF THE EVOLUTION OF HUMAN TRYPANOSOMIASIS

FOREST BIOMES	SAVANNAH BIOMES
	PLIOCENE
Arboreal primates have no contact with Glossinas.	Circulation of tsetse-borne trypanosomes in animals other than primates.
	LOWER PLEISTOCENE 
Exploiting the opportunity of unoccupied ecological niches, certain primates become ground dwellers.	Contact between ground-dwelling protohominids and tsetse-borne trypanosomes.
	MIDDLE PLEISTOCENE 
Contact between forest tsetses and trypanosome-carrying hominids through overlapping areas of activities.	Trypanosomes of the *rhodesiense* type evolve in hominid species.
	UPPER PLEISTOCENE 
Establishment of man-adapted *T. gambiense* strains through continuous contact in common forest habitats of parasite, forest tsetses and man.	*T. rhodesiense* maintained in animals by game flies. Occasional return in man under conditions of optimum opportunities.

Fig. 5

lated regions of Eurasia. Clark (1960) notes that from Later Pleistocene times onwards, human culture south of the Sahara seems to have received more than it gave, possibly due to lack of those environmental stimuli present in more northerly latitudes.

Cattle were introduced and developed in northern Africa around 5,000 B.C. (Ford, 1960). Along with other domestic animals, they were, however, very susceptible to the trypanosomes commonly carried without ill effect in many species of African wild animals: *T. brucei, T. congolense* and *T. vivax*. Pastoralists migrating south must at first have sustained severe losses in stock, due to "nagana," the name by which animal trypanosomiasis is known. The main route for dispersion was down the high ridge country on either side of the central Rift Valley (Clark, 1962). The reason for this is likely to have been the stock-owners' need to find a tsetse-free route. The struggle of cattle-raising people in tsetse-fly country is well documented in later African history (Fuller, 1923; Dicke, 1932), as is that of early western explorers trying to penetrate the interior of the African continent with men and equipment of the expedition transported on pack-horses (Anderson, 1857).

Fly belts have more effectively channelled the movements of pastoralists than of other people. The southward migrating cattle owners, in their endeavor to avoid *nagana*, began to follow planned routes through fly-free corridors. These migration patterns became important in later economic development, establishing connecting links between trading-posts and other permanent settlements. In addition to natural fly-free corridors, other fly-free areas resulted from clearings made by agricultural tribes, established mainly along forest fringes because of fertile soil and adequate water supply. They provided a buffer strip between fly habitats of the forest and those of the savannah. Such areas were exploited by the pastoralists not only as migration routes, but also often as permanent pastures for their cattle. The occupation of the same area by tribes of very dissimilar cultural background led to a sort of feudal relationship in which the pastoralists imposed their social structure upon the agriculturalists. Such a system was in use for some 500 years in Rwanda-Burundi, until recent political events, following independence from the Belgian administration, ended the pastoral Batutsi overlordship in bloody uprisings of the Bahutu. In the wake of such events, previous pasture lands or agricultural fields were often abandoned and reverted to pyrophyllic woodland savannah and later to possible suitable habitats for tsetse flies (Ford and Hall, 1947; van den Berghe and Lambrecht, 1956; *Ibid*, 1962). Pastoral migration was responsible for certain fundamental changes in the economy of the African people. This population dispersion, moving along fly-free corridors, played an important part in the history of the development of Africa, especially during pre-colonial times.

TRYPANOSOMIASIS IN HISTORICAL TIMES

A. *Trypanosoma gambiense*

And it shall come to pass in that day, that the Lord shall hiss for the fly that is in the uttermost part of the rivers of Egypt. . . . And they shall come, and shall rest all of them in the desolate valleys, and in the holes of the rocks, and upon all thorns, and upon all bushes.

Isaiah vii, 18-19.

This quotation from the Bible may well be the earliest record of the implication of (tsetse) flies in unpleasant occurrences.

A tsetse-borne trypanosome was first described by Bruce (1895) in wild and domestic animals in Zululand. The strain was sent to England and described under the name of *Trypanosoma brucei* (Plimmer and Bradford, 1899).

The first known man-infecting trypanosome was found by Forde in the blood of the captain of a steamboat plying the Gambia River. It was described by Dutton (1902) under the name of *Trypanosoma gambiense*. In 1903, Castellani and Bruce and Nabarro were the first to describe the trypanosome and to identify it as the infective organism of sleeping sickness with cyclic transmission through the tsetse vector, *G. palpalis*.

The oldest written historical text which mentions sleeping sickness with certainty is by the great Arabian historian Ibn Khaldun of the XVI century (Meyerhof, 1941). He speaks of an informer, Judge Abu Abdallah ibn Wansul, who lived for a while in the Mali Kingdom, and relates that: "Sultan Jatah was overtaken by the sleeping sickness which is a disease that frequently befalls the inhabitants of those countries, and particularly their chieftains. The sufferer is attacked repeatedly and at any time by profound sleep, so that it is hardly possible to awake him, but for a short time only. It is harmful to the sufferer, and this disease continues until he dies. The king's disease remained in his constitution for two years until he died in the year 775 (1373-1374)."

This narrative discloses that sleeping sickness was a well-known and common disease in West Africa, and was present at least from the fourteenth century on. The two-year duration of the disease indicates that it was caused by *T. gambiense*. Early records of sleeping sickness on the west coast of Africa are listed in great detail by Duggan (1962). The first detailed clinical description is that in "The Navy Surgeon" by John Atkins in 1743 (Hoeppli and Lucasse, 1964). In 1592 there was recorded the story of a certain captain in the service of the Sultan of Morocco, who was forced to abandon his campaign along the Niger and return to Timbuctu after all his horses had been killed by tsetse flies. All earlier sleeping sickness records are from West Africa, but it should be remembered that this part of tropical Africa was bettter known than other regions because of the ivory, gold and slave trade. Furthermore, earlier events have been preserved by Arabic historians in literature. It should not be presumed from this that Gambian sleeping sickness originated in West Africa and then spread out when the continent was opened up by western explorers. Well-travelled trade routes criss-crossed to and from the interior to coastal areas many centuries before modern explorations (Vansina, 1962). It is probably true that dispersion of the disease increased during the intensification of caravan movements in the nineteenth and twentieth centuries. Also, the maps presented in Figures 1, 2 and 3 would seem to indicate that *T. gambiense* always had a western distribution, even at the height of climatic changes.

It was evident that Gambian sleeping sickness occurred elsewhere on the African continent when those regions became better known. Early reports from Guerin (1869), Mense (1909) and Rodhain et al. (1913) leave no doubt that the disease existed in the Congo Basin, causing terrifying epidemics. Rodhain found whole villages decimated by the disease (van den Berghe and Lambrecht, 1963). Bloss (1960) estimates at about half a million the number of people who died from sleeping sickness in that area between 1896 and 1906. The great epidemic of Gambian sleeping sickness north of Lake Victoria between the years 1902 and 1905 caused the death of 200,000 people, out of a total population of 300,000.

Sleeping sickness, among other causes of hardship and impaired health, must be considered one of the great factors which retarded penetration and development of Africa by western technology during the ages of exploration and mechanical revolution. The full impact of western civilisation on the interior of Africa started only after mechanized equipment like the railroad, automobile and aircraft freed transportation from its dependency on human

B. Trypanosoma rhodesiense

The story of sleeping sickness caused by *Trypanosoma rhodesiense* is different from that of *T. gambiense*. The strain was first isolated in 1908 from a white person who fell ill after having travelled through the Lwangwa Valley in southern Rhodesia. The strain was studied and described under the name of *T. rhodesiense* (Stephens and Fantham, 1910). It was later found to be morphologically identical with *T. gambiense* (and *T. brucei*), but because of its much more acute clinical manifestations and its behaviour in experimental animals, its specific name has properly been retained.

The behaviour of the disease, appearing in sudden unexplained epidemics in widely separated pockets, and occurring in savannah areas, pointed to the possibility of an animal reservoir. Positive proof of this was supplied by the studies of Heisch et al. (1958), when they isolated *T. rhodesiense* from wild bushbuck (*Tragelaphus scriptus*) in Kenya, using human volunteers. Moreover, *T. rhodesiense* strains were also isolated from wild *G. palpalis* from the same region. Van den Berghe and Lambrecht (1963) discussed reasons for the separation of *T. gambiense* and *T. rhodesiense* in nature and expressed their surprise at the fact that *T. rhodesiense* has not invaded the forest environment. An explanation has been proposed in what is called the "filtering-out mechanism" based on the difference in ecology and physiology between *palpalis* and *morsitans* flies, and the fact that *T. rhodesiense* is maintained successfully in animal reservoirs, while *T. gambiense* is not (Lambrecht, 1964).

However, this ecological barrier can be bypassed under certain circumstances, as predicted by van den Berghe and Lambrecht (1963), and as seen in certain areas east of Lake Victoria in Kenya, where *T. rhodesiense* is now being transmitted by *G. palpalis fuscipes* (= *G. fuscipes*) (Willett et al, 1964). The transfer of *T. rhodesiense* from one ecological entity into another can occur where *G. pallidipes* habitats overlap those of *G. palpalis*-group flies. Exchange of circulating trypanosomes is made possible because *G. pallidipes* will feed both on man and on wild animals, some good reservoirs of *T. rhodesiense*. Subsequent constant circulation in the local human population of *G. pallidipes*-introduced trypansome strains is then assured if *G. palpalis*-group flies are present in the same area, because of the latter close contact with man.

From the place of its first isolation in southern Rhodesia in 1908, *T. rhodesiense* seems to have spread northward very erratically. It appeared in Tanganyika Territory around 1924 and by 1930 had produced over 100,000 cases of sleeping sickness a year (Ormerod, 1961). Moving northwest around Lake Victoria, it made a sudden westward thrust into Rwanda-Burundi in 1954, causing a severe epidemic in these territories, parts of which had been invaded by *G. morsitans* possibly only ten years before. It was suggested by van den Berghe and Lambrecht (1963) that, by reducing its maintaining host through anti-warthog campaigns, *G. morsitans* had been forced to rely more on bushbuck for its food supply, thereby increasing contact with a good *T. rhodesiense* reservoir and becoming a more potential carrier of this trypanosome.

Rhodesian sleeping sickness seems to have conveniently started its migration from a single southern locus at a time when it could be witnessed by western observers. We should be suspicious of this coincidence. Isolations of trypanosome strains "with very pronounced *T. rhodesiense* characteristics" were recorded in parts of Africa well outside the alleged *T. rhodesiense* advance, or before this advance could have reached those parts. Lester (1933) described such *T. rhodesiense*-like strains from Nigeria. Cases of sleeping sickness of the Rhodesian type were mentioned by Archibald (1922) and Archibald and Riding (1926) from the Tembura district in the southern Sudan and by Duggan (1962) around Lake Chad. Sleeping sickness was reported as early as 1909 from Lake Ngami, about 800 miles southeast of the Lwangwa Valley. Though few details of these early reports are known, the isolation of trypanosomes of later cases from the same area confirmed the diagnosis of *T. rhodesiense* infections (Ormerod, 1961).

Probably *T. rhodesiense* strains were already present in areas of three widely separated regions at the beginning of the present century: In the north, (1) Lake Chad and Nigeria, and (2) the upper reaches of the White Nile; in the south, (3) the Zambesi River system: the Lwangwa Valley and Okovango River. The map in Figure 3 shows that during the pluvials, savannah flies and *T. rhodesiense* strains might indeed have become isolated in areas of the Zambesi River system. During those times another pocket could have been formed around the upper Nile Valley. The Nigerian and Chad strains may have been later spill-overs, or strains that simply became separated from the eastern main body. They may present slightly different characteristics from the typical *T. rhodesiense*.

During the pluvials, savannah flies must have been absent from the Congo basin and the coastal areas of West Africa where lowland rain-forest must have covered areas even larger than at present. As mentioned before, the northern limit of savannah flies during those periods may have been at about 20° North latitude. It is doubtful that even at the height of the pluvials Somaliland areas ever became covered with vegetation suitable for tsetses.

Confirming the habitat separation between western and eastern fly belts, which might have occurred at certain stages of the pluvials, is the fact that the *G. morsitans* of the western belt has been found morphologically distinct enough from the typical fly to have been described as a separate subspecies: *G. morsitans submorsitans* Newstead, as distinct from the classical *G. morsitans morsitans* Westwood.

The lowering of tsetse fly limits from 1,500 meters to about 600 meters during the cold pluvials of course also affected *palpalis*-group flies. They were confined to the lowland areas beneath 600 meters occupying riverine vegetation and rain-forests. They spread into the Congo Basin and other lowland rainforests in West Africa and wherever water-banks provided vegetation of suitable density. Present-day *T. gambiense* infections are found in these general areas, with extensions south and east, no doubt in the wake of the retreat of montane biomes which set in after the last pluvials.

The map in Figure 3 clearly indicates a solid ecological barrier of montane vegetation between eastern and western lowland areas during the pluvials. One wonders whether this was responsible for keeping *palpalis-gambiense* cycles in the western biomes, and the *morsitans-rhodesiense* cycles in the eastern biomes. If the lowland areas are expanded to present-day size, then the distribution of *T. gambiense* and *T. rhodesiense* become logical.

CONCLUSION

This essay began with the speculation that the exclusion of trypanosome-susceptible ground-dwelling fauna provided unoccupied ecological niches which were subsequently exploited by certain primate groups. In the new environment, the ground-dwelling primates rapidly evolved and eventually gave rise to the hominid branch. In the evolution of other animal forms similar mechanisms are not uncommon; on the contrary, they are indeed the regular patterns which organic evolution follows. Why ground-dwelling primates evolved on the African continent rather than in other tropical regions must be sought in opportunities offered by the unique composition of Africa's fauna. Faunal compositions are the result of migration, natural selection and geographic isolation. Tsetse-borne trypanosomiasis was one of the important factors that operated in Africa but was absent from other continents at the time of hominid evolution. Directly or indirectly trypansomiasis may have played an important role in the evolution of man. It could have been, however, only one of the many mechanisms by which natural selection is achieved.

The development of trypanosome-susceptible genotypes in the process of evolution of later hominid species may be due to certain other survival advantages such as higher birth rates, improved hunting ability, increased intelligence, and the like.

The baboon may be an archaic form of Primate belonging to one of the trypanosome-resistant ancestral groups. Though it has the advantage of being trypanosome-resistant, it clearly lacks some other advantages of the higher Primates.

Once a susceptible hominid species evolved, the struggle for survival forcibly included trypanosomiasis as a factor. Selection of genotypes may have been influenced by the disease. During the Pleistocene climatic fluctuations, the hominids, tsetse flies and trypanosomes, like all the other organisms, were subject to the vicissitudes of the changing environment. What exists today is what the extraordinarily complicated selection factors have sifted out during the untold centuries. To analyse in every detail the importance of each single factor is likewise extremely complicated, if not impossible. Within the content of this presentation, only one single aspect of human evolution has been discussed.

It is realized that infectious diseases induce very important changes in animal populations: in a direct way by causing high mortality and even extinction; indirectly by influencing genotype composition. Certain human genotypes, though deleterious in themselves, have been found to bestow resistance against certain infectious diseases. This is the case of the sickle-cell anemia trait found in inhabitants of highly malarious regions. Other high incidences of unusual, and sometimes deleterious, genotypes are thought to have evolved in populations possibly as a result of certain resistance to diseases in the past (Motulsky, 1960).

By eliminating the challenge of infectious diseases, a population showing genotypic resistance to them may become genotypically different from a previous (exposed) population. The control of malaria in high sickle-cell-trait populations will doubtless show this when later generations from these areas are examined.

REFERENCES

Anderson, A. J., 1857: *Lake Ngami: Four Years Wandering in the Wilds of South Western Africa (1850-1854)*. London, Hurst and Blackett.

Archibald, R. G., 1922: *Trypanosoma rhodesiense* in a case of sleeping sickness from the Sudan. Ann. Trop. Med. Parasit., 16:339.

ARCHIBALD, R. G. and RIDING, D., 1926: A second case of sleeping sickness caused by *Trypanosoma rhodesiense*. *Ann. Trop. Med. Parasit.*, 16:161.

VAN DEN BERGHE, L., LAMBRECHT, F. L. and CHRISTIAENSEN, A. R., 1956: Etude biologique et écologique des Glossines dans la région du Mutara, Ruanda. *Acad. Roy. Sci. Col.*, Memoir Tome IV, fasc. 2:1.

VAN DEN BERGHE, L., and LAMBRECHT, F. L., 1962: Étude biologique et écologique de *Glossina morsitans* dans le Bugesera (Ruanda-Urundi). *Acad. Roy. Sci. Outre-Mer*, Classe Sci. Nat. & Med., Memoir 13 (4):1.

VAN DEN BERGHE, L. and LAMBRECHT, F. L., 1963: The epidemiology and control of human trypanosomiasis in *Glossina morsitans* fly-belts. *Amer. J. Trop. Med. Hyg.*, 12:129.

BLOSS, J. F. E., 1960: The history of sleeping sickness in the Sudan. *Proc. Roy. Soc. Med.*, LIII:421.

BRUCE, D., 1895: *Preliminary Report on the Tsetse Fly Disease or Nagana, in Zululand.* Durban, Bennett and Davis.

BRUCE, D., and NABARRO, D., 1903: Progress reports on sleeping sickness in Uganda. *Refs. Sleep. Sickn. Comm. Roy. Soc.*:1-VII.

CASTELLANI, A., 1903: Presence of Trypanosoma in sleeping sickness. *Roy. Soc., Refs. Sleep. Sickn. Comm. 1:*1.

CLARK, J. D., 1960: Human ecology during Pleistocene and later times in Africa south of the Sahara. *Current Anthropology* 1:307.

CLARK, J. D., 1962: The spread of food production in in Sub-Saharan Africa. *J. Afr. History* 3:211.

DARWIN, C., 1872: *On the Origin of Species*, 6th ed. London, John Murray.

DICKE, B. H., 1932: The tsetse fly's influence on South African history. *S. Afr. J. Sci.* 29:792. Capetown, S.A.

DUGGAN, A. J., 1962: A survey of sleeping sickness in northern Nigeria from the earliest times to the present day. *Trans. Roy. Soc. Trop. Med. Hyg.* 56:439.

DUTTON, J. E., 1902: Note on a Trypanosoma occurring in the blood of man. *Brit. Med. J.*, 2:881.

EMILIANI, C., 1958: Ancient temperatures. *Sci. Amer.*, 198:54.

FORD, J., 1960: Distribution of African cattle. *Proc. 1st. Sci. Congress*, Salisbury, S. Rhodesia, pp. 357.

FORD, J. and HALL, R. DE Z. 1947: The history of the Karagwe, Bukoba District. *Tanganyika Dates and Records*, 24:3.

FULLER, C., 1923, 1924: Tsetse in the Transvaal and surrounding Territories: An historical review. *Union of S. Africa Dept. of Agriculture, Entomology Memoir No. 1*, 68 pp.; *Trop. Dis. Bull.* 21, 785.

GLASGOW, J. P., 1963: *The Distribution and Abundance of Tsetse.* New York, Macmillan.

GUERIN, P. M. A., 1869: *De la Maladie du Sommeil.* Thesis 201, Paris, Parent.

HEISCH, R. B., MCMAHON, J. P., and MANSON-BAHR, P. E. C., 1958: The isolation of *Trypanosoma rhodesiense* from a bushbuck. *Brit. Med. J.* (2):1203.

HOARE, C. A., 1948: The relationship of the haemoflagellates. *Proc. 4th. International Congr. Trop. Med. Mal., Washington*, pp. 1110.

HOARE, C. A., 1955: Intraspecific biological groups in pathogenic Protozoa. *Refuah Veterinarith*, 12:263.

HOARE, C. A., 1957: The classification of Trypanosomes of veterinary and medical importance. *Vet. Rev. Annot.*, 3:1-13.

HOEPPLI, R., and LUCASSE, C., 1964: Old ideas regarding cause and treatment of sleeping sickness in West Africa. *J. Trop. Med. Hyg.*, 67:60.

LAMBRECHT, F. L., 1964: Aspects of evolution and ecology of tsetse flies and trypanosomiasis in prehistoric African environment. *J. Afr. Hist.* 5:1.

LESTER, H. M. O., 1933: The characteristics of some Nigerian strains of the polymorphic trypanosomes. *Ann. Trop. Med. Parasit.*, 27:361.

MENSE, C., 1909: *Handb. Trop. Krankh.*, Leipzig, 3:628.

MEYERHOF, M., 1941: An early mention of sleeping sickness in Arabic chronicles. *J. Egypt. Med. Ass.*, 24:284.

MOREAU, R. E., 1963: Vicissitudes of the African biomes in the Late Pleistocene. *Proc. Med. Soc. London*, 141:395.

MOTULSKY, A. G., 1960: Metabolic polymorphism and the role of infectious diseases in human evolution. *Human Biol.* 32:28.

NAPIER, J. R., 1964: Profile of early man at Olduvai. *New Scientist*, 22 (386):86.

ORMEROD, W. E., 1961: The epidemic spread of Rhodesian sleeping sickness, 1908-1960. *Trans. Roy. Soc. Trop. Med. Hyg.*, 55:525.

PLIMMER, H. G., and BRADFORD, J. R., 1899: *Proc. Roy. Soc.* 65:274.

RODHAIN, J., PONS, C., VANDENBRANDEN, F., and BECQUART, J., 1913: *Rapport sur les Travaux de la Mission Scientifique du Katanga, 1910-1912.* Brussels, Hayez.

STEPHENS, J. W. W., and FANTHAM, H. B., 1910: On the peculiar morphology of a trypanosome from a case of sleeping sickness and the possibility of its being a new species (*T. rhodesiense*). *Ann. Trop. Med. Parasit.*, 4:343.

SUMMERS, R., 1960: Environment and culture in southern Rhodesia: A study in the 'personality' of a land-locked country. *Proc. Amer. Philos. Soc.* 104:266.

VANSINA, J., 1962: Long-distance trade-routes in Central Africa. *J. Afr. History* 3:375.

WELLS, L. H., 1962: Pleistocene faunas and the distribution of mammals in southern Africa. *Ann. Cape. Prov. Mus.*, 11:37.

WILLETT, K. C., MCMAHON, J. P., ASHCROFT, M. T., and BAKER, J. R., 1964: Trypanosomes isolated from *Glossina palpalis* and *G. pallidipes* in Sakwa, Kenya. *Trans. Roy. Soc. Trop. Med. Hyg.*, 58:391.

Chapter 10

The Human Treponematoses*

C. J. HACKETT

INTRODUCTION

THE interpretation of descriptions of disease in early writings is difficult and may be inaccurate; it must rely largely upon the descriptions of symptoms often with little relation to the course of the disease. However, reliable evidence of disease in the past may be sought in bony changes and in the soft tissues of mummies and dried bodies.

The geographical distribution of a disease or group of allied diseases at some not too remote and more or less stable period may assist in an understanding of them and perhaps other diseases in antiquity. The usefulness of such a distribution depends upon recognition of the diagnostic inaccuracies which often occurred when diagnoses were clinical and specific methods had not been discovered.

The human treponematoses offer good opportunities for such studies although each aspect needs more precision, but already scientific interest in many of these aspects has waned.

Four forms of human treponematoses are recognized:

1. pinta, a Central American disease with skin changes;

2. yaws, a disease of skin and bones occurring characteristically in rural populations of the humid tropics;

3. endemic syphilis, a similar disease found only in arid warm climates, and

4. venereal syhilis, which has no climatic restrictions and may affect any tissue of the body, including internal organs, particularly the brain and heart.

Each of the four treponemes and the four diseases related to them may for the present be regarded as different entities. Hudson (1946, 1958, 1963) has, however, strongly contended that all the treponemal infections are caused by one treponema, *T. pallidum,* and are one disease, treponematosis, "which presents different clinical patterns under different climatic and sociological conditions."

Certain aspects of these diseases were discussed in a paper on the origin of the treponematoses (Hackett, 1963) from which the maps below are taken and from which liberal quotation is made with permission of the World Health Organization. That paper arose from the consideration of the presence of endemic syphilis in Bechuanaland and possibly in Central Australia, despite its remoteness from similarly infected populations in comparable

---

* The chapter is based upon a paper published in 1963 in Bulletin WHO vol. 29, pp. 7-41, and was prepared while the author was a staff member of WHO, Geneva.

conditions in North Africa, Arabia through to central Asia. These northern and southern endemic syphilis areas are everywhere separated by large areas where the climate is humid and warm and the characteristic treponematosis is yaws (Maps 1 and 2, Figs. 2 and 3).*

The following summary of the paper referred to above is an outline of the suggested sequence, and the possible development of the treponemes and the infections they cause.

ORIGIN OF THE HUMAN TREPONEMATOSES

The probable distribution of the treponematoses at the beginning of the present century is given in Map 1 and, within broad limits, is probably moderately accurate. It is based upon reports from about 1900 with the omission of statements that can now be reasonably regarded as errors of diagnosis or interpretation. Until the present century "disgusting" or "revolting" disease manifestations were often reported as syphilis by explorers and even by doctors (Hackett, 1936b). The *Treponema pallidum* of venereal syphilis was then unknown, and diagnosis depended upon clinical examination. That yaws was not well known is clear from descriptions of the time (Charlouis, 1897; Manson, 1898) and also by the fact that the late destructive changes were not generally accepted until the 1920s (van Nitsen, 1944), despite the able descriptions by Rat (1891) and others of the "tertiary period" in yaws. Although Moseley (1800) said that yaws "ends in shocking nodes and destruction of the bones," Nicholls (1899) thought that "the sequels of yaws" and "the tertiary symptoms" resulted from treatment with mercury. There was much confusion between yaws and venereal syphilis in tropical countries. Endemic syphilis as seen in arid lands was not clearly recognized until the third decade of this century (Lacapère, 1923) and pinta was regarded as a mycosis until after 1930.

The geographical distribution of the treponematoses in Columbus's day would be of great assistance in considering their origin and Map 2 (Fig. 3) is an attempt to provide this.

The close relation of the causal organism of the treponemes of yaws, endemic syphilis and venereal syphilis and the intermediate position of endemic syphilis to the others is stressed by the laboratory studies of Turner and Hollander (1957). The treponeme of pinta, although microscopically identical with the other treponemes, has never been transmitted to animals, although such transmission is relatively easy with the other treponemes. Moreover, pinta remains infectious for many years, which does not occur in the other treponematoses.

The relationship of the treponeme of pinta to those of the other human treponematoses needs more precise definition. The closer this relationship, the more likely it is that the four human treponemes are a consecutive series starting with that of pinta (see hypothesis A in the accompanying diagram.) The less close this relationship, the more likely it is that the treponeme of pinta is an earlier and independent branch from the treponeme of an ancestral animal infection (see hypothesis B in the diagram). In this discussion the former has been proposed because of the occurrence of somewhat similar depigmen-

---

*The maps attempt to show the distribution of the four human treponematoses in the indigenous populations, except for yaws in the Americas. The distribution of venereal syphilis is that where other treponematoses are absent. No account is taken of syphilis in cities and towns and its extension therefrom into areas where another treponematosis might reasonably be expected to have been the earlier infection.

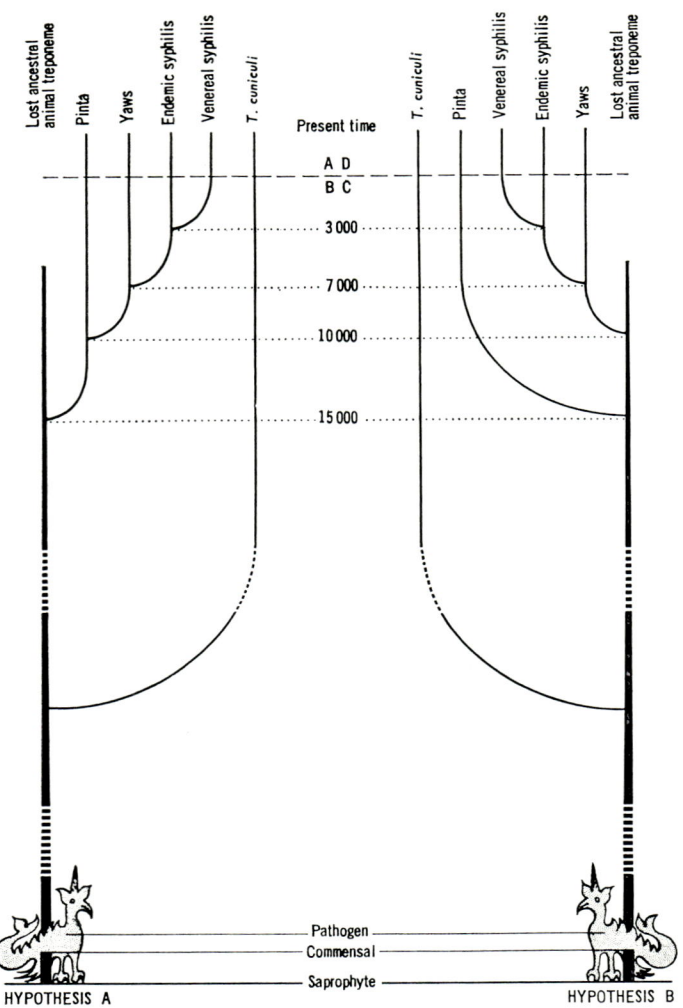

FIGURE 1. Two hypotheses as to the possible evolution of the syphilis-yaws treponemes

tation in pinta and in the other treponematoses. For this reason an independent origin for pinta in the Americas and another for the other human treponematoses in the Euro-Afro-Asian land mass has not been proposed.

A natural infection of rabbits with a treponeme, *T. cuniculi,* has been known for many years. Recently, however, Fribourg-Blanc *et al* (1963) have demonstrated specific serological reactions in baboons from a district in Guinea where yaws is prevalent in the human population. These workers have since found (personal communication Dr. Fribourg-Blanc, 1966) treponemes of the syphilis/yaws type in the popliteal lymph glands of apparently healthy baboons from Guinea (in press, *Bull. Soc. Path. exat.* 1966). The results of inoculation into hamsters are awaited. Reactive sera were not obtained from Kenyan or Cambodian monkeys. It is well known that nonpathogenic treponemes do not give rise in animals to reactions to the

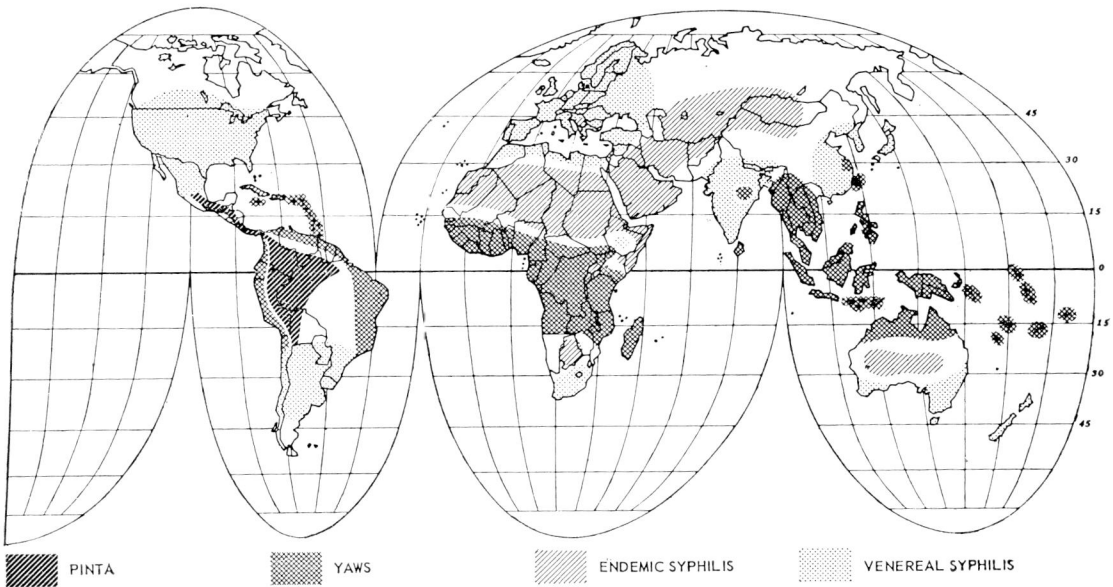

FIGURE 2. Map 1. Probable geographical distribution of the treponematoses about A.D. 1900. See footnote on page 153.

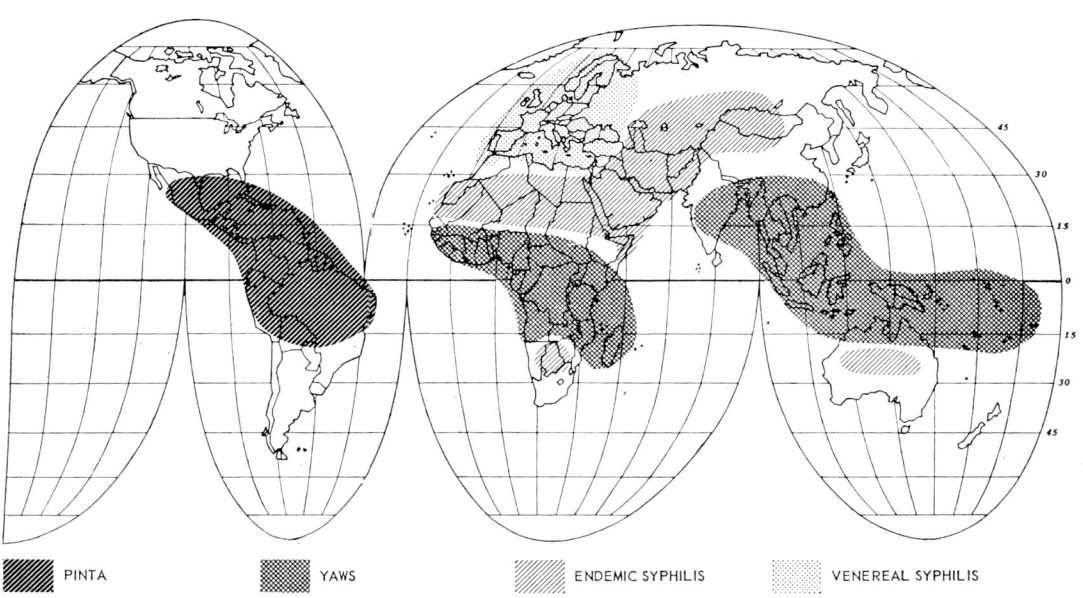

FIGURE 3. Map 2. Probable geographical pattern of the treponematoses at about the time of European expansion in the 16th century A.D.

specific antigens that these workers used. These important and interesting findings suggest that the human pathogenic treponemes arose more directly from an ancient animal infection than from a saprophyte or commensals.

Veldkamp (1960) and Bruijn (1961) have recently reported a saprophyte treponeme, *T. zuelzerae*, possibly related to the human treponemes. The treponeme was isolated by Veldkamp from mud in a small fresh water creek running through a small town, Pacific Grove, to Monterey Bay, California. Professor Veldkamp, however, in a personal communication (1965), says that treponemes of this type are very common in black muds in canals in Dutch towns and elsewhere. He has isolated similar treponemes from marine mud on a shallow coast near the Friesian Islands.

Step 1—Pinta (Map 3, Fig. 4)

Pinta has some characteristics of a primitive infection, such as a long duration of infectiousness (Burnet, 1962) which would maintain it in small population groups, even in families, which suggest that it is the early human treponematosis. Another characteristic that must be taken into account in considering its origin is its present isolation in the Amerindian populations in humid areas of the Americas (Map 1, Fig. 2).

The proposed original treponematosis, pinta, might have extended from Africa and Asia into the Americas with the migration of man and other animals over the ice-free Bering Strait land bridge which probably last existed about 15,000 to 10,000 B.C. (Zeuner, 1958). This would have been during the last part of the last (Wisconsin) glaciation before the ocean levels had risen from the melting of the polar ice-caps (Haag, 1962). The end of this glaciation is now generally regarded as being about 8,000 B.C. and probably little came out of the Americas thereafter until Columbus's discovery in the late fifteenth century.

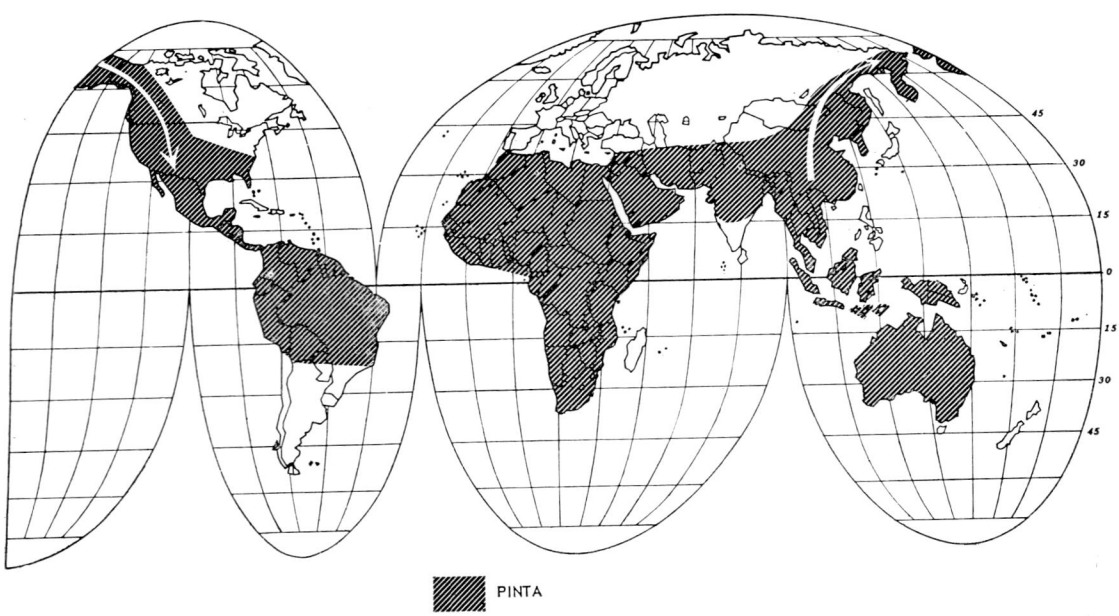

Figure 4. Map 3. Probable extent of pinta about 15,000 B.C.

Had pinta—like the primitive marsupials now isolated in the Americas and Australia (Mathew, 1939)—been worldwide, pockets of it would be expected in isolated and especially peripheral communities such as the Australian aborigines or the pygmies and negritos in remote mountain fastnesses, but no report of this is known. Pigmentary changes do occur in the other human treponematoses although they are not identical with those of pinta.

In attempting to date the spread of an infection from the Euro-Afro-Asian land mass to the peripheral Americas and Australia the absence of genes for the blood groups A and B from the Indians of South America and of that for B from the Australian Aborigines (Mourant et al, 1958) might be useful.

After taking the above points into consideration, it is proposed that pinta arose from an animal infection in the Euro-Afro-Asian land mass, perhaps before about 20,000 B.C. and had extended through the then accessible world by about 15,000 B.C. (Map 3, Fig. 4).

Step 2—Yaws (Map 4, Fig. 5)

Pinta is assumed to have extended throughout the world by about 15,000 B.C. and to have been cut off in the Americas from the main land masses after about this time. Maps 1 and 2 indicate the more recent patterns of yaws, endemic syphilis and venereal syphilis, at the beginning of the twentieth and sixteenth centuries.

With the exception of the small areas of endemic syphilis in southern Africa and possibly in Central Australia, the most peripheral infection of the three remaining infections is yaws. On Map 2 it will be seen that yaws occupies the tropical equatorial belt. This may mean that it was the first of the three to develop. It is therefore proposed that in Afro-Asia a humid, warm environment favoured mutants of the pinta treponemes which were more invasive and destructive of tissue and which caused yaws. This might have occurred about 10,000 B.C. Yaws then extended through the Afro-Asian area, but did not reach the Bering Strait before it was finally flooded. The last glaciation was extensive and many

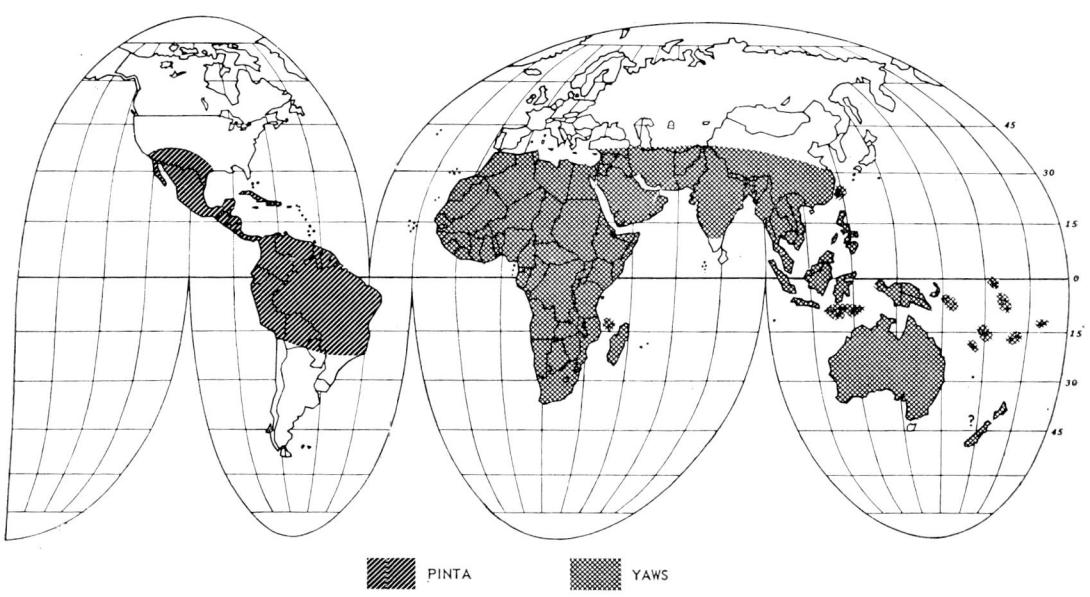

FIGURE 5. Map 4. Probable extent of pinta and yaws about 10,000 B.C.

of the now arid geographical areas probably enjoyed good rainfall and a more uniformly warm climate. Thus yaws might have extended throughout Africa and throughout Africa and southern Asia where the environment was favourable to reach Australia and the Pacific islands if or when they were inhabited, but not America. Mr. N. B. Tindale of the Department of Anthropology, South Australian Museum, Adelaide (personal communication, 1962) says that man was almost certainly in Australia for part of the last half of the last glaciation, perhaps before 20,000 B.C. and that three separate racial groups came into Australia from Asia. The earliest known were negritos, who were short in stature, dark-skinned with crisp, curly hair and of a different race from later arrivals who displaced them. Their most obvious survivors were the Tasmanians (Tindale, 1960). No pathological changes suggesting treponematoses have been described in the few Tasmanian bones that have survived.

Of mutation among the treponemes little is known and little may be learnt before they can be cultivated. Because drug resistance has not yet developed in treponemes, mutation affecting other characters is not excluded. The treponemes would be expected to undergo any particular mutation at a rate comparable to that found in other multiplications, that is about one in every million generations, but only those mutants would survive which changing climatic, social or other conditions favoured. Treponemes divide about once every thirty hours (Turner and Hollander, 1957).

Only *T. pallidum* of venereal syphilis flourishes in a wide range of climates; the others appear rather strictly limited to warm climates either humid (yaws and pinta) or arid (endemic syphilis).

In the Pacific islands yaws was apparently present when the first Europeans arrived. Stewart and Spoehr (1952), by radiocarbon dating, placed some skulls from the Mariana Islands with lesions probably due to treponematoses at the mid ninth century A.D. There is also considerable evidence of yaws in Hawaii (Samwell, 1786) although mistaken for "the venereal disease," and also in Australia (Hackett, 1936b) before the arrival of Europeans.

In India, yaws has survived in Orissa, Madhya Pradesh and Hyderabad in rather isolated aboriginal tribes in inaccessible localities. Saxena and Prasad (1963) believe that yaws was introduced from Ceylon in 1887 into Assam and thence into India within the past half century.

Kynsey (1881) and Spittel (1923) state that yaws was introduced into Ceylon, where it was known as *parangi*, meaning "the foreign disease," by slaves brought by the Portuguese from Africa. Such a means of introduction may also explain the focus of *S. haematobium* bilharziasis in Gimvi near Bombay.

In Africa, yaws has so flourished that some observers have thought that it was the cradle of the disease. The prevalence and severity of yaws in tropical Africa during the first thirty years of this century were probably much greater than anywhere else (Botreau-Roussel, 1962). More than one observer regarded yaws then as a bigger burden to some African populations than all other infections put together (van Nitsen, 1944). This high prevalence was almost certainly more directly related to the standards of living then than to other factors.

Thus yaws had the capacity in suitable conditions to become a widely spread, non-killing disease affecting most of the population by adult life.

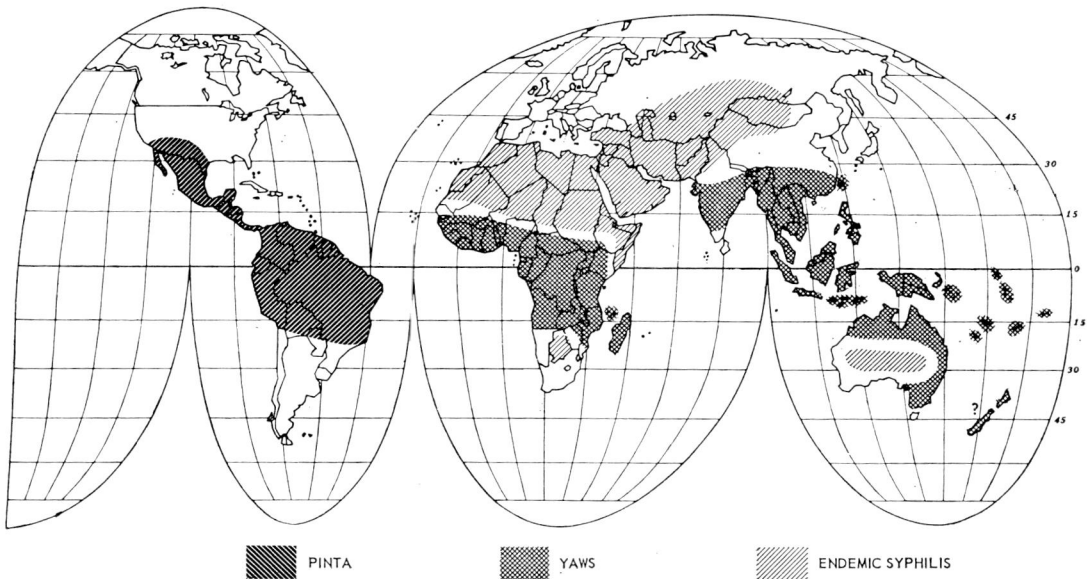

FIGURE 6. Map 5. Probable extent of pinta, yaws and endemic syphilis about 7,000 B.C.

To summarize, in Afro-Asia by about 10,000 B.C. conditions favoured certain mutants of the pinta treponeme that gave rise to yaws. This had then extended throughout the tropics except in the Americas, which were by then isolated by the reappearance of the Bering Strait.

This leaves unexplained the lack of evidence of survival of mutants of pinta treponemes in the Americas as it is proposed happened elsewhere. If the differences of colours of the pinta lesions reported from different South American countries are real, these may be due to mutants; perhaps the environment may have been too constant to favour mutants. The Americas differ from the other continents in that desert zones are much less extensive because both North and South America are so narrow in the latitudes concerned.

Step 3—Endemic Syphilis (Map 5, Fig. 6)

On Map 2 an uninterrupted block of endemic syphilis extends from Africa through western into central Asia, comprising deserts and semi-deserts, often with nomadic populations. There were two smaller but perhaps epidemiologically no less important areas of endemic syphilis in similar but southern latitudes, in Bechuanaland until recently (Murray et al., 1956) and perhaps in Central Australia in the past (Hackett, 1936a, 1936b). How endemic syphilis could have occurred in Bechuanaland is difficult to understand if it did not evolve from yaws already there.

In Australia the treponematosis pattern probably resembled that in Africa in the same southern latitudes. Bone lesions and serological reactions indicate that endemic treponematoses occurred among aborigines in Central Australia and further south (Hackett, 1936a). While yaws was the infection in the tropical northern Australian coastal area, the recently increased knowledge about endemic syphilis suggests that possibly the treponematosis, *irkintja* (Hackett, 1936a), in semi-arid central and southern Australia might have been endemic syphilis. Professor J. B. Cleland of Adelaide (personal communication, 1962; and Black and Cleland, 1938) has long proposed that *irkintja* was a non-venereal treponematosis other than yaws. The aboriginal population and its cultural introductions came into Australia in several waves from the north and north west. The nearest endemic syphilis area lies far to the north-west; thus yaws coming into northern Australia from Asia may have been the precursor of endemic syphilis in the semi-arid steppes of Central Australia.

Hudson (1963) has suggested that as the Australian aborigines moved south from the humid northern coasts "into the dry interior their yaws skin lesions dried up, and they had endemic syphilis." McKay (1938), on what would now be regarded as inadequate grounds, said that some of the lesions on aboriginal bones found in south eastern Australia, and probably pre-dating the arrival of Europeans, were most likely due to venereal syphilis; but they were probably due to endemic syphilis.

Since endemic syphilis is at present mostly restricted to arid, warm climates these may be assumed to favour the selection of mutants that gave rise to it. Such conditions would have developed in the present geographical distribution of endemic syphilis following the end of the last glaciation about 8,000 B.C. and have been maintained, with some variations, ever since.

The small area of endemic syphilis in East Africa (Map 1) in the Buganda Province of Uganda (Davies, 1956) may have been venereal syphilis with non-venereally acquired sporadic transmission, rather than endemic syphilis, and this may have arrived there much more recently. The endemic syphilis of Bosnia (Grin, 1953) was probably originally introduced from Anatolia with the Turkish occupation which began in the fifteenth century.

The effects of climate on the skin lesions of yaws have been reported by Lopez-Rizal and Sellards (1926), by Ramsay (1927) and also by Fitzgerald and Dey (1931). One of these was the frequency of buccal mucosal papules in cooler climates; such mucosal lesions are characteristic of endemic syphilis. In arid, intra-continental climates, the summers are hot and the winters cold but on the average the climate is warm. Turner and Hollander (1957) have shown that maintaining rabbits inoculated with treponemes at cool air temperatures resulted in higher infection rates, shorter incubation periods and more severe lesions than did maintaining them at higher air temperatures. The optimal temperature for treponemes is about $35°$-$37°C$ and the body temperature of the rabbit is about $39°C$; above $40°C$ treponemes are destroyed. Turner and Hollander say that one can only speculate about the effects of "environmental temperature on the treponematoses in human beings" or on the treponemes themselves.

Some observers have reported a higher prevalence of yaws in the cooler rainy season than in the hot dry season (Harding, 1949). However, the temperatures at the cool season in tropical climates would be much higher than those of the cool season in semi-arid, intra-continental ones, which at night may fall below $0°C$.

Turner and Hollander (1957) also report changes in behaviour in experimental animals of a few strains of the yaws treponeme towards that of venereal syphilis, especially when they were kept under favourable cool air conditions ($18°$-$21°C$).

It is thus proposed that mutants of the yaws treponeme were selected by arid, warm climates resulting from the retreat of the last glaciation and that by about

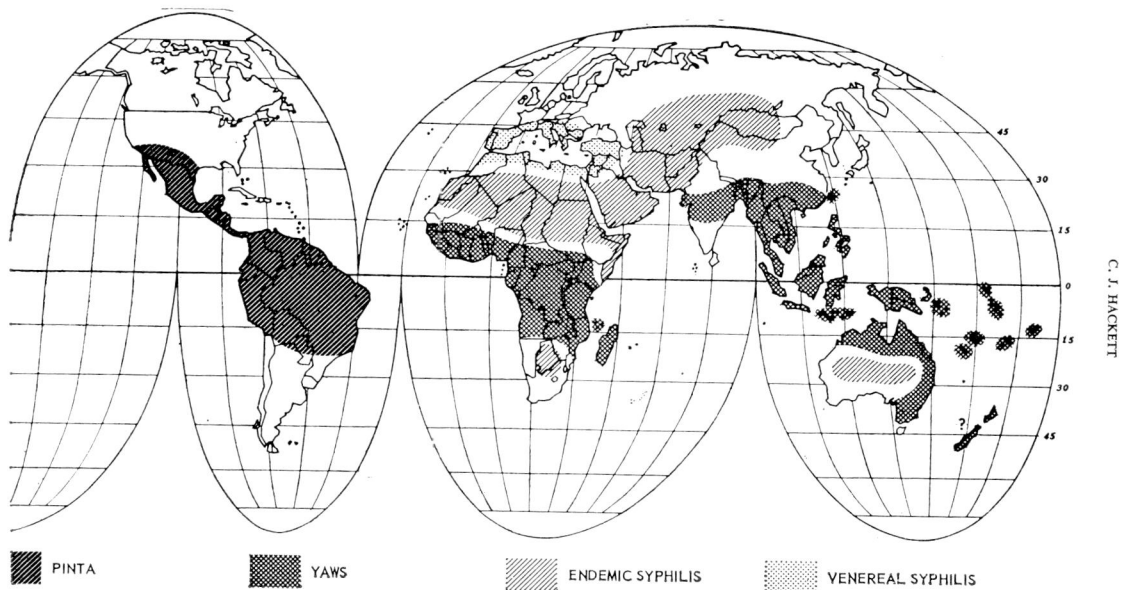

FIGURE 7. Map 6. Probable extent of pinta, yaws, endemic syphilis and venereal syphilis from about 3000 B.C. to the 1st century B.C.

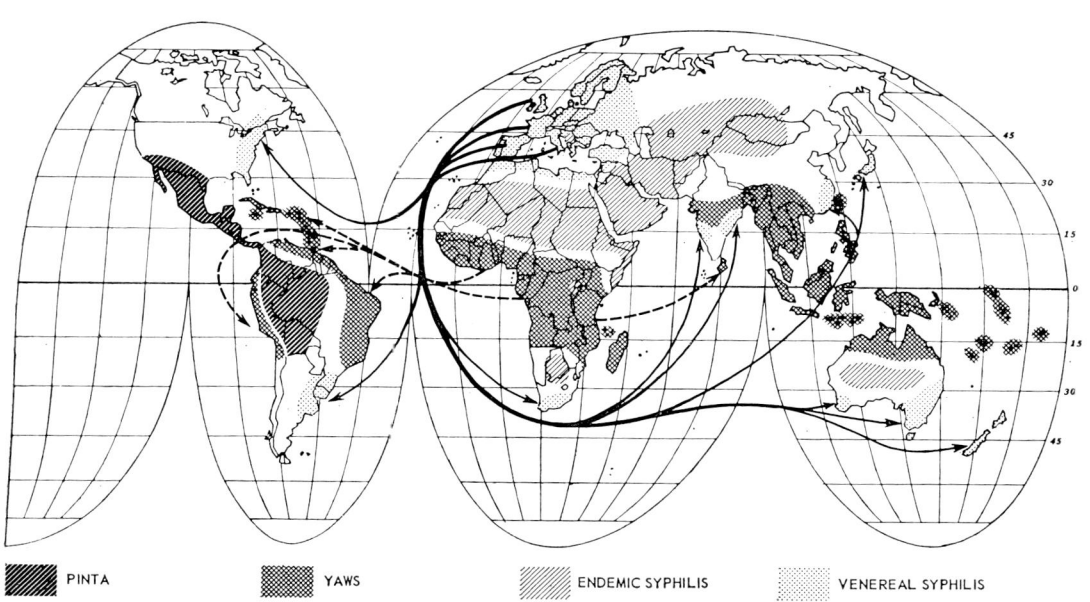

FIGURE 8. Map 7. Spread of venereal syphilis throughout Europe at about the 1st century A.D. and spread of venereal syphilis through the world and of yaws to the Americas and Ceylon from the 16th century A.D.

7,000 B.C. yaws in such areas had changed into endemic syphilis, while in humid warm climates yaws has continued unchanged.

Step 4—Venereal Syphilis (Maps 6 and 7, Figs. 7 and 8)

In Map 2, venereal syphilis is shown as occurring in Europe and the Mediterranean countries. Of the origin of venereal syphilis in Europe as in Persia (Elgood, 1951), India (Reddy, 1936), China (Wong, 1918; Eckstein, 1928) and Japan, there are two main explanations.

One is that venereal syphilis appeared within about ten years before or after the year A.D. 1500. Thus Bloch (1908), Hutchinson (1908), Capper (1925), Power (1934), Hamlyn (1939) and Harrison (1959) attribute venereal syphilis in Europe to its introduction by the crews of Columbus's ships returning in 1493 from the discovery of America; they cite the writings of the sixteenth century to support their contention. Their main points are the absence of references in writings before the end of the fifteenth century to a disease like venereal syphilis as known at present, statements that it was then a new disease, the absence from Europe of pre-Columbian bones with lesions resembling those of venereal syphilis and the alleged presence of such bones in America. However, diagnoses in some papers on "pre-Columbian syphilis in America" such as "most likely due to syphilis," "almost certain evidence of luetic infection," "changes regarded as pathognomonic of syphilis" and "changes consistent with, although not diagnostic of, syphilis" (Cole et al., 1955) leave much to be desired.

References by early writers, in the other countries mentioned above, to syphilis have been thought to have been mistaken; venereal syphilis in these countries was believed to have been recent and to have appeared about the time of its supposed introduction into Europe by Columbus's crews. Venereal syphilis was often known by such names as *phirangi roja* (India), indicating its foreign origin. The opinion of Elgood (1951; personal communication, 1962) that syphilis first appeared in Persia at the beginning of the sixteenth century is based upon wide personal examination of the original texts. However, as far as syphilis in Europe is concerned, the evidence for its origin in America is neither impressive nor precise. The history of venereal syphilis in Asia needs further study.

The other explanation (Buret, 1895; Sudhoff, 1925; Holcomb, 1940; Castiglione, 1947; Campbell, 1954; Singer and Underwood, 1962) is that venereal syphilis was present in Europe but was not identified from among a number of diseases, including some epidemic diseases, referred to *en masse* as "leprosy." They speak of the alleged venereal transmission of this "leprosy" which disappeared from the writings after the identification of syphilis in the sixteenth century and its naming by Fracastorius in 1530. They also refer to writings and edicts (Sudhoff, 1925) relating to the avoidance of venereal diseases of wider clinical range than gonorrhoea, either before the return of Columbus or within the following decades. Harrison (1959), after careful study, has called attention to the misdating of the earliest Paris edict relating to *gross vérole* (Holcomb, 1934) as 25 March, 1493, whereas the correct date is 6 March, 1497. Singer and Underwood (1962) however, suggest that perhaps a new strain of *T. pallidum* was brought to Europe from America, which changed the character of syphilis completely during the closing years of the fifteenth century.

Lancereaux (1868) believed that most of the manifestations now known as syphilis were known in Europe long before the Naples "epidemic." Only after that outbreak were many lesions of the various stages of the disease related to each other to give the picture of venereally transmitted syphilis. Previously some had been separately described and others were confused with other diseases.

Hirsch (1885) stated that there was little doubt that venereal syphilis occurred in antiquity in Greece and Rome and in the Middle Ages in Europe. He also stated that syphilis occurred before the end of the fifteenth century in other parts of the world such as India, China and Japan, but that it was taken to America and the Pacific islands by Europeans. European contact with Japan, however, was very restricted until about a century ago.

Dujardin (1949) suggests that the American origin of syphilis was introduced in the early sixteenth century to support the value of the American herb guaiac in the treatment of the disease, a value depending upon the belief that where a disease occurred there would be found its cure. This herb has no place in present day treatment. Dujardin alludes to the commercial implications of guaiac at the time and to the monoply of the influencial Fuggers for its importation into Europe. He is not satified that the lack of descriptions of a disease recognizable as syphilis in Greek and Roman writings is evidence of its absence because, he says, these writings do not contain a single description of scabies which he was convinced was present. The absence of recognizable descriptions might in part at least be accounted for by masking by the various theories of the causation of disease at different times, e.g., astrological, humoral, etc. and in part by its various manifestations being either regarded as separate diseases or confused by being grouped in artificial syndromes. It might, under these conditions, not have been until many patients were seen in a short time that the natural course of the disease would have been recognized and these errors corrected. He has little confidence in the American origin of syphilis but is more prepared to accept its presence in Europe, India and China at least, perhaps in a hidden form, long before the end of the fifteenth century, when it suddenly became a contagious venereal disease at Naples in 1495 and rapidly spread throughout Europe to reach the British Islands in 1497.

Cole *et al.* (1955) state: "Syphilis is a disease of antiquity. It was undoubtedly present from the earliest times among the Arabians, Greeks, and Romans in the Dark Ages, and it probably came from China, Japan and India."

A leading article in the British Medical Journal (Anon, 1962) typifies the confusion and lack of precision. It states that "syphilis continues to be the most controversial disease of all, there still being inadequate material to prove conclusively whether it originated in the New World or the Old." Later it is suggested that "there is, in fact, no good reason why both continents should not have been affected during pre-Columbian times,"* but no good reason is given why they should.

The question has also been raised why mercury, one of the first effective substances in the treatment of venereal syphilis, was of such high repute in Europe during the first fifteen centuries of the

---

* The possibility that venereal syphilis was brought to Europe by invaders from Asia in the thirteenth century needs investigation. Rat (1891) says that if syphilis did not originate in Europe, it might have been brought to Spain from Africa by the Moors. The last Moorish kingdom in Spain (Granada) did not surrender until 1492.

Christian era, if there were no venereal syphilis. This apparent absence of syphilitic bone lesions in Europe during this period still remains to be understood, but its absence is found to-day in Ethiopia.

Williams (1932, 1936) studied possible syphilitic lesions in pre-Columbian bones and stressed the difficulties of the diagnosis of venereal syphilis in ancient bones and also the difficulty of dating them. Williams concludes that there is no certain evidence of venereal syphilis in pre-Columbian bones in Egypt or Europe but he accepts some of the osteological evidence of venereal syphilis in American Indians before Columbus's arrival. Holcomb (1940) comments on William's (1932) conclusions. However, all this needs consideration, especially since Hrdlička (1932) (Lawrence, 1941) a leading anthropologist of the Amerindian, has stated that "from evidence of thousands of Indian skulls and skeletons predating the arrival of Columbus, there is, as yet, not a single instance of thoroughly authenticated pre-Columbian syphilis."

Some writers think that the epidemic of Naples in 1493, like others before it, was not due to venereal syphilis but to an outbreak of one of the epidemic communicable diseases that are known today. Others have pointed out that the American origin of the "Naples epidemic" was not raised until twenty-five years after Columbus's return.

In considering the absence of clear descriptions of syphilis before the early years of the sixteenth century one should remember that syphilis and gonorrhoea were not generally recognized as separate diseases until the second half of the eighteenth century (Bell in 1793). Even such an able investigator as John Hunter (1728-1793) did not accept this and carried out an experiment upon himself by which he erroneously satisfied himself that they were the same disease. Locomotor ataxia (tabes dorsalis) was not clearly described in the medical literature (Duchenne in 1858) nor its syphilitic origin recognized (Fournier in 1876) until the nineteenth century, although in its later stages it can be readily diagnosed at a distance by the gait of the sufferer. McGeoch (1960) reports that no reference to neurosyphilis has been recognized in the writings of Shakespeare (1564-1616). However, Scott (1962) in a study of *Shakespeare's melancholics* devotes a chapter to "Timon, the general paralytic" (*Timon of Athens*, 1604-1606).

Before and for some time after the fifteenth century the diagnosis of syphilis doubtless included many conditions, for example, some forms of leprosy and some skin diseases, now known not to be syphilitic. Something similar happened a generation ago with "influenza," which was a generic rather than a specific diagnosis before laboratory methods for diagnosis were developed.

This third and last suggested step in the evolution of the treponematoses would involve the change from a non-venereal disease (endemic syphilis) of rural children to a venereal disease (venereal syphilis) of urban adults. This might begin with the decrease of childhood infections because venereal transmission would not be expected in a population immunized by childhood with an endemic treponematosis. If childhood endemic syphilis infection decreased then more adults would be susceptible and a few "last cases" of delayed condylomatous relapses might provide the opportunity for venereal transmission to uninfected and susceptible partners. Guthe and Luger (1957) suggest this change in Syria. Denser settlement, more clothing and increasing sexual laxity as well as decreasing endemic syphilis might well have assisted climatic change in the selection of mutants towards venereal syphilis treponemes.

Davies (1956) has suggested that venereal syphilis was introduced into the lake areas of Uganda in the last half of the nineteenth century and since the turn of the century has replaced an already existing endemic syphilis. The increase in the amount of clothing worn in these districts during the twentieth century may have hastened this. However, the change, which would accompany the improvement of standards of living, might rather have been the reduction of non-venereally transmitted venereal syphilis. The affected population made up an organized kingdom in contrast to the organizational disorder of their relatively naked neighbours among whom yaws was highly endemic (Hackett 1947). The climate of the affected province, Buganda, is not arid nor semi-arid but humid.

If endemic syphilis was in Buganda, how it came is unknown. There is the same problem of venereal syphilis in Ethiopia, where it may have been introduced by visitors, slavers or other traders.

The change to venereal transmission might have occurred with the growth of big cities, together with increased clothing, in the now arid countries of the eastern Mediterranean and south-western Asia (Map 6). This might have been about 3000 B.C. during the post-glacial climatic optimun which culminated between 5000 B.C. and 2000 B.C. when the climatic zones were displaced towards the higher latitudes and temporarily the equatorial/monsoon belts were widened and the desert zones less severe. However, from about 3000 B.C. desiccation of the previously arid areas returned and has remained ever since (Brooks, 1949; Lamb, 1963) and was probably maximal during the twentieth century. Venereal syphilis was probably carried to the Mediterranean littoral by shipping and later, especially after the first century B.C. with the Roman conquests, throughout Europe, which was a treponemally uncommitted area. In rural populations in Syria, Saudi Arabia, Iran and Iraq the indigenous treponematosis is still endemic syphilis. In rural Syria (Hudson, 1946) venereal syphilis was known as *franji*, meaning "foreigner." This is also reported of parts of India (Reddy, 1936).

The absence of lesions resembling those of venereal syphilis in bones in Europe before the end of the fifteenth century and in early Egyptian burials (Rowling, 1961) needs careful consideration. However, the disease which, it is here suggested, evolved from endemic syphilis about 3000 B.C., may have still been a "mild" one, until a mutant appeared at the end of the fifteenth century in Europe which caused a more grave disease (Shrewsbury, quoted by Harrison, 1959; and *Med. J. Aust.*, 1961).

The social conditions, customs and habits of the congested urban populations of Europe in the fifteenth and sixteenth centuries (Taylor, 1953) would have assisted in spreading venereal syphilis both non-venereally and venereally. Venereal syphilis in Europe is generally supposed to have become less destructive since its first recognition at the end of the fifteenth century. At present (in America) venereal syphilis is regarded as being more gummatous in patients of African descent and more neuropathogenic in patients of European stock. However, any variations in its clinical manifestations might be due to other than treponemal factors.

One of Hudson's (1958) conclusions is that yaws arose in Africa from a saprophytic treponeme several hundred thousand years ago and spread eastwards to India and Oceania about 100,000 years ago. He also thinks yaws extended into America. As yaws left the tropics and entered cooler climates it became, he says, endemic syphilis and about 10,000 years ago in towns and cities on the Nile and Mesopotamia it became venereal syphilis.

Hudson (1962) at the end of an assessment of *las buvas* of Villalobos (1472-1556) states:

"It is more likely that in the latter decades of the fifteenth century the venereal form of syphilis was coming to be recognized as it emerged from a widely disseminated non-venereal endemic form of the disease in Europe. The process of transition in the epidemiology of syphilis from non-venereal to venereal, and endemic to sporadic, had been going on in that continent for many years before 1493. Columbus could have had nothing to do with it."

He has written again on this question (Hudson, 1963) starting with "the assumption of the African origin of treponematosis at some point in Paleolithic time, an assumption that few today would regard as fanciful." "There was first the transition from yaws to endemic syphilis, when man the hunter migrated from humid Central Africa to the drier [northern and southern] zones of savannah and desert. This process then extended to the rest of the restricted world of Palaeolithic man. Second, endemic syphilis was established in the early villages of the Neolithic period. This process then extended to all the world's villages; they constitute the reservoir of endemic treponematosis today. Third, there was a gradual transition from non-venereal to venereal syphilis when towns and cities came into being in the Neolithic period. The superior hygiene of the city gradually eliminated the childhood infection, and the increasing sophistication of the adult population provided opportunity for the spread of venereal infection."

He also states that "medical history provides many illustrations of the conversion of yaws into endemic syphilis and of endemic syphilis into venereal syphilis. Sometimes the process has been sudden, sometimes more deliberate . . ." in the former "in response to appropriate change in climatic conditions" and in the latter "when there is appropriate change in man's social organization and habits. Furthermore, medical history provides illustrations of the reverse processes." He states "each stage of human development would have the kind of treponemal infection appropriate to it."

"The differences between the treponemes are *strain* and not specific differences, since they still remain biologically susceptible and responsive to changes in human and physical environment. The treponemes may diverge, but they never lose their ability to regain their former state." "One must assume that endemic syphilis of the northern temperate zone of North America" (brought in by the hunters who "crossed the Bering bridge as the last glacial period was ending") "became yaws in the American tropics and that this, in turn, reverted to endemic syphilis in the southern latitudes." "It would be expected that venereal syphilis would develop in those advanced civilizations of Central and South America that lead to the growth of towns and cities. In Central America appeared a form of endemic treponematosis now called pinta. . . . This infection is obviously related closely to yaws and endemic syphilis." "There is nothing in the picture to support the belief, however, that it was either the original or the oldest form of treponemal infection in the New World."

Cockburn (1961) presents ideas of a "philosophical nature" about the characteristics and origin of the treponematosis. He considers (Cockburn, 1963) "the relationship of the discovery of America and the first appearance of syphilis would not be the cause and effect" but both were "a product of the Renaissance." He suggests that "pinta is the result of geographical isolation and of a unique "human host and his reaction to his parasite." Willcox (1960) stresses environmental factors in the "evolutionary cycle of the treponematosis."

Since venereal syphilis is geographically and pathologically more closely related to endemic syphilis than to yaws its origin from the former seems likely. Probably yaws was not present in the Americas when Columbus arrived; pinta is the ancient American endemic treponematosis. Venereal syphilis was said to have been unknown in North Canadian Indians before European contacts (Holcomb, 1940) and the limited studies of Medina (1954) of the resistance of pinta patients to the yaws treponeme would suggest possible resistance to syphilis. This in turn is further, though weak, support for the absence of syphilis in the Americas.

As stated above, it is suggested here in the cities, perhaps in the eastern Mediterranean or south-west Asia, conditions developed which favoured treponeme mutants tending towards venereal syphilis. This resulted in the replacement of endemic syphilis by venereal syphilis in urban populations, and later Europe became gradually but finally generally infected. Thence it or a further mutant causing a grave disease was probably carried during and after the sixteenth century to the Americas, southern Africa (Sax, 1952), India, south Asian countries, China and Australia (Map 7).

Dr. H. H. Lamb of the Meteorological Office, Bracknell, Berks., England (personal communication, 1962) has pointed out that in Europe during the fifteenth century the climate was deteriorating (cooling)—in some decades, particularly in the 1430s and 1490s, deteriorating sharply—though it temporarily recovered somewhat between 1500 and 1550. The fifteenth century was one of considerable difficulty for the rural people, particularly of northern Europe, because of the worsening of the climate and the grave results of the Black Death.

CONCLUSION

Among the many uncertainties in the above scheme is the presence or absence of syphilis in Europe and the Americas before the end of the fifteenth century. It is very doubtful if the problem can be answered by historical texts alone. One helpful source of evidence might come from the critical study of lesions in bones predating Columbus's discovery of America. Up to the present the diagnosis of such lesions depends largely on macroscopical inspection and radiographical examination. Newer and finer techniques are being used in the study of bones for medical purposes and these should be combined with macroscopical examinations for the establishment of widely acceptable criteria for the recognition of syphilitic and other lesions in ancient bones, especially from Europe and America.

There is a broader implication of the suggested evolution of the human treponematoses perhaps of relevance to other diseases in antiquity; this is the direct and indirect influence of climatic changes in the past on the environment and diet of man as well as on the selection of mutants of his pathogenic organisms. Thus, in any more precise study of the bone lesions of the treponematoses, other causes of contemporary bone lesions cannot be neglected both in the differentiation of treponemal from other lesions and also in the recognition of any evolution in them that might parallel that in the treponematoses.

REFERENCES

Anon., 1962: Diseases of ancient man. *Brit. Med. J.*, 1:852.
Black, E. C., and Cleland, J. B., 1938: Pathological lesions in Australian aborigines, Central Australia (Granites), and Flinders Range. *J. Trop. Med. Hyg.*, 41:69.
Bloch, I., 1908: History of syphilis. In: Power, D'A. and Murphy, J. K., (Eds.), *A System of Syphilis*. London, Hodder and Stoughton, p. 1.
Botreau-Roussel, J. M., 1962: Discussion en cours sur la lésion primaire du pian. *Bull. Soc. Path. Exot.*, 55:288.
Brooks, C. E. P., 1949: *Climate Through the Ages*. London, Ernest Benn.
Bruijn, J. H. de, 1961: Serological relationship between *Treponema zuelzerai* and the Reiter strain of *Treponema pallidum*. *Antonie v. Leeuwenhoek*, 27:98.
Buret, F., 1895: *Syphilis Today and Among the Ancients*. Philadelphia, vol. 2, p. 113.
Burnet, F. M., 1962: *Natural History of Infectious Diseases*, 3rd ed. Cambridge, University Press.
Campbell, D. J., 1954: The veneral diseases. In: Bett, W. R. (ed.), *The History and Conquest of Common Diseases*, Norman, University of Oklahoma Press, p. 178.
Capper, A., 1925: A epitome of the history of syphilis. *Arch. Derm. Syph. (Chicago)*, 12:509.
Castiglioni, A., 1947: *A History of Medicine*. New York, Knopf.
Charlouis, M., 1897: On polypapilloma tropicum (framboesia). In: *Selected Essays and Monographs*. London, New Sydenham Society, p. 285.
Cockburn, T. A., 1961: The origin of the treponematoses. *Bull. W. H. O.*, 24:221.
Cockburn, T. A., 1963: *The Evolution and Eradication of Infectious Diseases*. Baltimore, Johns Hopkins Press. p. 152.
Cole, N. H., Harkin, J. C., Kraus, B. S., and Moritz, A. R., 1955: Pre-Columbian osseous syphilis. Skeletal remains found at Kinishba and Vandal cave, Arizona, and some comments on pertinent literature. *Arch. Derm. (Chicago)*, 71:231.
Davies, J. N. P., 1956: The history of syphilis in Uganda. *Bull. W. H. O.*, 15:1041.
Dujardin, B., 1949: *Propos sur la Syphilis et son Histoire*. Union Chimique Belge S. A., Bruxelles.
Eckstein, O., 1928: Syphilis in ancient China. *Bull. Soc. Med. Hist. Chicago*, 4:71.

Elgood, C., 1951: *A Medical History of Persia*, Cambridge University Press.
Fitzgerald, G. H., and Dey, N. C., 1931: The manifestations of chronic yaws. *Indian med. Gaz.*, 66:425.
Fribourg-Blanc, A., Neil, G. and Mollaret, N. H. 1963: Note sur quelques aspects immunologiques du cynocephale africain. *Bull. Soc. Path. exot.* 56:474.
Grin, E. I., 1953: Epidemiology and control of endemic syphilis, Geneva (*World Health Organization*: Monograph Series, No. 11).
Guthe, T., and Luger, A., 1957: Epidemiological aspects of non-venereal "endemic" syphilis. *Dermatologica*, 115:248.
Haag, W. G., 1962: The Bering Strait land bridge. *Sci. Amer.*, 206:112.
Hackett, C. J., 1936a: Boomerang leg and yaws in Australian aborigines, London *Royal Society of Tropical Medicine and Hygiene, Monograph No. 1*.
Hackett, C. J., 1936b: Critical survey of some references to syphilis and yaws among Australian aborigines. *Med. J. Aust.*, 1:733.
Hackett, C. J., 1947: Incidence of yaws and of venereal diseases in Lango (Uganda). *Brit. Med. J.*, 1:88.
Hamlyn, H., 1939: The geography of treponematosis. *Yale J. Biol. Med.*, 12:29.
Harding, R. D., 1949: A yaws campaign in Sierra Leone. *Trans. Roy. Soc. Trop. Med. Hyg.*, 42:348.
Harrison, L. W., 1959: The origin of syphilis. *Brit. J. Vener. Dis.*, 35:1.
Hirsch, A., 1885: *Handbook of Geographical and Historical Pathology*. London, New Sydenham Society, vol. 2, p. 59.
Holcomb, R. C., 1934: Christopher Columbus and the American origin of Syphilis. *U.S. Nav. Med. Bull.*, 32:401.
Holcomb, R. C. 1940: Syphilis of the skull, among Aleuts, and the Asian and North American Eskimo about Bering and Arctic Seas. *U.S. Nav. Med. Bull.*, 38:177.
Hrdlička, A. 1932: Disease, medicine and surgery among the American aborigines. *JAMA*, 99:1661.
Hudson, E. H., 1946: *Treponematosis*. New York, Oxford Loose-leaf Medicine.
Hudson, E. H., 1958: *Non-venereal syphilis*. Edinburgh and London, Livingstone.
Hudson, E. H., 1962: Villalobos and Columbus. *Amer. J. Med.*, 32:578.
Hudson, E. H., 1963: Treponematosis and anthropology. *Amer. J. Med.*, 58:1037.
Hutchinson, J., 1908: Introduction. In: Power, D'A., and Murphy, J. K., (Eds.), *A System of Syphilis*. London, Hodder and Stoughton, p. xxiv.
Kynsey, W. R., 1881: *Report on the "parangi disease" of Ceylon*, Colombo.
Lacapère, G., 1923: *La Syphilis Arabe (Maroc, Algérie, Tunisie)*, Paris, Doin.
Lamb, H. H., 1963: *On the nature of certain climatic epochs which differed from the modern (1900-*

LANCEREAUX, E., 1868: *A treatise on syphilis.* London, New Sydenham Society, vol 1, p. 60.

LAWRENCE, E., 1941: Syphilis: as an anthropologist sees it. *Med. Press,* 205:300.

LOPEZ-RIZAL, L. and SELLARDS, A. W. 1926: A clinical manifestation of yaws observed in patients living in mountainous districts. *Philipp. J. Sci.,* 30:497.

MACKAY, C. V., 1938: Some pathological changes in Australian aboriginal bones. *Med. J. Aust.,* 2:537.

MANSON, P., 1898: *Tropical diseases; a manual of the diseases of warm countries.* London, Cassel, p. 423.

MATHEW, W. D., 1939: *Climate and evolution.* New York, New York Academy of Sciences, p. 93.

MCGEOCH, A., 1960: Shakespeare the syphilologist. *Med. J. Aust.,* 1:348.

Med. J. Aust., 1961: The origin of syphilis. 2:603.

MEDINA, R., 1954: Reacciones producidas en enfermos de pinta, buba o sifilis por inoculacion de *Treponema pertenue,* Castellani 1905.—Su posible aplicacion al diagnostico de curacion de esta treponematoses. *Arch. Venez. Pat. Trop.,* 2, No. 2, 51.

MOSELEY, B., 1800: On yaws. In: *Medical Tracts,* 2nd Ed. London, Cadell and Davies, p. 184.

MOURANT, A. E., KOPEC, A. C., and DOMANIEWSKA-SOBCZAK, K., 1958: *The ABO blood group; comprehensive tables and maps of world distribution,* Oxford, Blackwell.

MURRAY, J. F., MERRIWEATHER, A. M., FREEDMAN, M. J., and VILLIERS, D. J. DE, 1956: Endemic syphilis in the Bakwena Reserve of the Bechuanaland Protectorate. A report on mass examination and treatment. *Bull. W. H. O.,* 15:975.

NICHOLLS, H. A., 1899: In: *Twentieth Century Practice of Medicine,* Vol. 16. New York, Wood, p. 305.

NITSEN, R. VAN, 1944: Le pian. *Mém. Inst. colon. Belge Sci. nat.,* 13, No. 7.

POWER, D.'A., 1934: The venereal diseases. In: Bett, W. R. (Ed.), *A Short History of Some Common Diseases.* London, Oxford University Press, p. 34.

RAMSAY, C. G., 1927: The origin of yaws in Assam. *Trans. Roy. Soc. Trop. Med. Hyg.,* 20:506.

RAT, J. N., 1891: *Yaws: Its Nature and Treatment.* London, Waterlow, p. 41.

REDDY, D. V. S., 1936: Antiquity of syphilis (venereal diseases) in India. *Indian J. Vener. Dis.,* 2:103.

ROWLING, J. T., 1961: Pathological changes in mummies. *Proc. Roy. Soc. Med.,* 54:409.

SAMWELL, D., 1786: *A narrative of the death of Captain James Cook, to which are added some particulars concerning his life and character and observations respecting the introduction of the venereal disease into the Sandwich Islands,* London, Robinson.

SAX, S., 1952: The introduction of syphilis into the Bantu peoples of South Africa. *Afr. Med. J.,* 26:1037.

SAXENA, W. B., and PRASAD, B. G., 1963: An epidemiological study of yaws in Madhya Pradesh; i, historical and geographical. *Ind. J. Med. Res.,* 51:768.

SCOTT, W. I. D., 1962: *Shakespeare's Melancholics.* London, Mills and Boon, p. 108.

SINGER, C. J., and UNDERWOOD, E. A., 1962: *A Short History of Medicine,* 2nd ed. Oxford, Clarendon Press, pp. 106, 108.

SPITTEL, R. L., 1923: *Framboesia Tropica (Parangi of Ceylon).* London, Baillière, Tindall and Cox.

STEWART, T. D., and SPOEHR, A., 1952: Evidence on the paleopathology of yaws. *Bull. Hist. Med.,* 26:538.

SUDHOFF, K. F. J., 1925: *The Earliest Printed Literature on Syphilis. Being ten tractates from the years 1495-1498,* Florence, R. Lier and Co.

TAYLOR, G. R., 1953: *Sex in History.* London, Thames.

TINDALE, N. B., 1960: Man of the hunting age. *Colorado Quart.,* 8:229.

TURNER, T. B., and HOLLANDER, D. H., 1957: *Biology of the Treponematoses.* Geneva (World Health Organization: Monograph Series, No. 35)

VELDKAMP, H., 1960: Isolation and characteristics of Treponema zuelzerai, nov. spec., an anaerobic, free-living spirochaete. *Antonie v. Leeuwenhoek,* 26:103.

WILLCOX, R. R., 1960: Evolutionary cycle of the treponematoses. *Brit. J. Vener. Dis.,* 36:78.

WILLIAMS, H. U., 1932: The origin and antiquity of syphilis; the evidence from diseased bones. A review, with some new material from America. *Arch. Path. (Chicago),* 13:779, 931.

WILLIAMS, H. U., 1936: The origin of syphilis; evidence from diseased bones. Supplementary report. *Arch. Derm. Syph.,* 33:783.

WONG, K. C., 1918: *Chin. Med. J.,* 22:349.

ZEUNER, F. E., 1958: *Dating the Past,* 4th ed., London, Methuen.

Chapter 11

The Prevalence of Malaria in Ancient Greece*

W. H. S. JONES

Four points must be fully discussed in the present inquiry:

1. Did malaria exist in Greece?
2. If so, to what extent was it prevalent?
3. When was it introduced, or when did it become common?
4. Is there any ancient evidence of its effect upon character?

All these aspects of the question are important. Nevertheless it must be noticed that the precise date of the introduction of malaria is by no means so vital a point as to determine the period when it became widely extended. It may have lurked in corners without doing much harm; but its prevalence would necessarily bring about a decline in vigour and a change of character.

Greek Terms for "Fever"

Before going on to inquire into the prevalence of malaria in ancient Greece, it will be necessary to discuss the equivalence of the English word "fever."

The most common general term is πυρετός. Derived from πῦρ, fire, it is probably used in the sense of "heat" in *Iliad* xxii. 31 (of the dogstar):

Φέρει πολλὸν πυρετὸν δειλοῖσι βροτοῖσιν

The ancients seem to have taken the word to mean "heat," as is plain from the Latin poets Vergil, Lucan and Statius, who appear to have this line in mind when talking of the "burning dogstar." On the other hand a scholiast remarks that the word might mean "fever" as well as "heat."

It is a fact that the sense of "fever" does fit the passage. Summer and autumn are the seasons when fever was most prevalent in Greece. It seems likely, however, to judge from the usage of later non-medical writers, that either νόσος or the plural would have been used if a disease had been meant. In any case, even if "fever" be the meaning here, it is not necessarily malaria. It might very well be typhoid. In the present inquiry that meaning will be assumed to be true which tells most against the writer's own theory.

After this solitary instance in Homer there is a large gap. Hesiod does not appear to use πυρετός, although he might well have been expected to do so. The present writer cannot find that the word occurs again before Aristophanes. Herodotus does not use it, nor does Thucydides.

---

* This paper by W. H. S. Jones is an edited and reduced chapter from his book: *Malaria: A Neglected Factor in the History of Greece and Rome,* published in 1907. It is one of the earliest, and certainly one of the best, studies of the possible impact of disease on a protohistoric society, and it was therefore thought worthwhile including, even though space limitations have demanded a considerable reduction in the original text. We are most grateful to Macmillan of London for permission to reprint this extract from his work. (Eds)

It is remarkable that when the latter, in his description of the plague, wishes to express "feverishness," he seems to avoid the word πυρετός, and uses instead καῦμα or θέρμη.[1] On the other hand, when Galen is describing the same plague and is roughly quoting the words of Thucydides, he employs πυρετός twice within a few lines.

The places in Aristophanes, where πυρετός seems to occur for the first time after Homer, are interesting. In the *Wasps* (date 422 B.C.) occurs the following passage:

And he says that he attacked last year the shivers and fevers which by night strangled your fathers and throttled your grandsires, etc. 1037-1041.

It is sufficient to notice here that from this time onwards πυρετός is a fairly common word, while the verb πυρέσσω (first, apparently, in Eurip. *Cycl.* 228 and Pherecrates in Athen. III. 75), "I have an attack of fever," also frequently occurs.

In all the Greek medical writings, which date from 400 B.C. onwards, πυρετοί are divided into two classes, (a) continuous (συνεχεῖς) and (b) intermittent (διαλείποντες). The second class is again subdivided according to periodicity, the simpler forms of which give (1) quotidians (ἀμφημερινοί), (2) tertians (τριταῖοι) and (3) quartans (τεταρταῖοι). The first mention of this division which is to be found in non-professional writings occurs in the *Timaeus*[2] of Plato. This passage is to the effect that a body produces:

1. c o n t i n u o u s burnings (ξυνεχῆ καύματα) and πυρετοί, when suffering from excess of fire (πῦρ);
2. quotidian πυρετοί, when suffering from excess of air;
3. tertian πυρετοί, when suffering from excess of water;
4. quartan πυρετοί, when suffering from excess of earth.

In the popular speech, then, there is a tendency to limit πυρετοί to definite fevers, namely, to those exhibiting a certain periodicity. It cannot be said that this tendency is present in the professional writers, though even there πυρετοί usually means intermittents.

The present discussion will be occupied with the intermittents, and in the next section their symptoms, as given by the ancient medical writers, will be fully described. In the meantime two other names must receive attention.

1. καῦσος. This fever ("the burning disease") is very clearly described by Hippocrates. The chief symptoms are bodily ache and lassitude, intense thirst, sleeplessness and (sometimes) delirium. The tongue is rough, dry and very black. There gnawing pains about the bowels. The alvine discharges are watery and yellow. Major Ross says that this disease must be typhoid only, so that it will be neglected in the present inquiry.

2. ἠπίαλος. This curious word first occurs in Theognis. It is used twice by Aristophanes. Galen gives a brief account of the disease. It was a protracted quotidian, and of such a kind that the patient felt fever and shivering at one and the same time and in every part of the body. He adds that some Attic writers use the word to denote the shivers which precede an attack of fever. Major Ross, who has given his opinion after examining such evidence as the present writer could put before him, inclines to the belief that ἠπίαλος (the disease) was malaria or typhoid, though it might possibly be Malta fever. Of course, it is also possible that the disease was one which does not now exist.

Did Malaria Exist in Greece?

If tertian and quartan fevers existed among the Greeks, they certainly suffered from malaria. But it will be useful to apply the confirmatory test, splenomegaly, and to enter more fully into the symptoms and variations of the intermittent fevers which are described by the ancient medical writers.

The evidence that enlargement of the spleen was common is copious, but practically confined to professional works. This is only to be expected, especially as the Greeks had the excellent sense not to talk overmuch about their ailments. In a curious passage of the *Timaeus* Plato describes the spleen as a receptacle for the purgations of the liver, and accounts in this way for splenic enlargement. When it is remembered that the Greeks held that tertians and quartans were caused by bile, the words of Plato at once become full of meaning. Hippocrates says that men who drink marsh water get enlarged spleens. The phenomenon Hippocrates really observed was that dwellers by marshy places suffer from enlarged spleens. His interpretation of this phenomenon is incorrect. The enlargement was in all probability caused by malaria conveyed from one person to another by mosquitoes bred in the marshes.

Splenic enlargement is also caused by typhoid, and it may be remarked in passing that there is a tendency among modern physicians to diagnose most of the fever cases described in the Hippocratic writings as some form of this disease. This may be correct. Malaria and typhoid are sometimes extremely difficult to distinguish in ancient writings, especially when the former is of a complicated type.[3] But typhoid will not account for the many other fevers, mentioned in the same writings, which have a definite tertian or quartan periodicity. These must be malarial; and no one who reads the few passages about them which occur in the non-professional writings, or the accurate descriptions given by Hippocrates and Galen, will fail to come to the conclusion that they were among the commonest of the diseases with which the Greeks were afflicted. But the risk of confusing malaria and typhoid must make the historian cautious. Many fevers are described in the ancient writers which are, in all probability, though not certainly, some form of malaria. In the present inquiry no stress will be laid upon these, in order that there may be a firm foundation of fact upon which to build.

For these reasons it seems desirable not to discuss at length the vast number of cases in which splenomegaly is mentioned in the old medical writers. It will be sufficient to state that in a great number of instances it is allied with other symptoms which evidently show that it was not caused by any disease so serious as typhoid. Pyaemic fevers other than malaria may, of course, be meant, but the probability is that the latter disease is the one described. One or two examples must suffice. Hippocrates states that in autumn quartan fevers and splenic diseases are very common. The same writer says that bilious persons who have enlargement of the spleen are evil-complexioned, ulcerous and emaciated, and suffer from foul breath and constipation. These are most certainly the symptoms of malarial cachexia.

Malarial Fevers in Ancient Medical Writers

There are four kinds of malaria *parasites*, quartan, mild tertian, malignant tertian, and quotidian.

QUARTAN. Fever lasts on the average 9 hours and recurs every third day.

MILD TERTIAN. Fever last 11 hours and recurs every other day.

MALIGNANT TERTIAN. Fever lasts up to 40 hours—rises slowly, halts for hours, declines a little and then rises again to a greater height; and lastly falls. Recurs every other day.

QUOTIDIAN. Fever lasts 6 to 12 hours and comes on every day.

But a quotidian *fever* may be produced by (1) three parallel generations of quartan parasites; or (2) two generations of tertians, or (3) one generation of quotidian parasites.

There are also mixed infections, due to different parasites together, and double quartans.

To discuss fully all the different accounts of quartans, tertians and quotidians, as described in the Greek writers, would occupy a large treatise. It is certain, therefore, that they were constantly prevalent, and that they were more common diseases than other kinds of fevers. But it is not the object of the present inquiry to enter into details. The main point is to identify these fevers with malaria. The best description is to be found in Galen, who seems to have made a special study of periodicity. Although Galen is a late writer, who lived in Rome in the middle of the second century A.D., there is no risk in accepting his evidence. His account agrees in all essentials with the testimony of older writers, and there is no evidence that Galen describes diseases unfamiliar to his predecessors.

In the treatise περὶ τύπων Galen divides intermittents into (1) quotidians, with a daily access; (2) tertians, with an access every other day; (3) quartans, with an access every third day.

Quintans, and even less frequently recurring fevers, are also mentioned. There is also distinct recognition, both in this treatise and in others, of "mixed" and "double" infections. One instance only shall be quoted here. In the book περὶ περιόδων he remarks that a fever with attacks recurring every day is liable to be diagnosed by the uninitiated as a quotidian. But if a man take pains and have a genuine interest in medicine, he will not forget that the same effect can be produced by two tertians or three quartans.

There was also a fever which he calls "semi-tertian." It is regarded as a mixture of the tertian and a *continuous* quotidian. It was a dangerous disease,[7] attacked usually men in the prime of life, and especially in the autumn, and was marked by the length of the attack.[4] There seems to have been much irregularity in the length and severity of the paroxysms.[5] This fever was probably some variety of the tertian type, whether mild or malignant, produced by mixed or double infection.[6]

As an example of intermittent fevers, Galen gives a full account of the tertian. It begins with rigor and finishes with sweat and vomiting of bile. In some cases the intermission is short. Such fevers he calls "protracted" tertians (παρεκτείνοντες). Occasionally the fever lasts for forty hours, or even longer.

To descend to details, tertian fevers begin with shivering, and chill in the extremities. The pulse is hard and contracted. Gradually the chill is superseded by fever, and the pulse becomes quicker and larger. The patient often feels internal fever while the limbs are still chilled. The fever gradually increases until it has spread over the whole body. Then it subsides little by little, the d e c r e a s e being usually (τοῖς πλείστοις) accompanied by sweating.

The references in the Hippocratic writings to tertians are very numerous, and nothing would be gained by quoting them in full. The same applies to quartans.

Special mention, however, may be made of a passage in the first book of the *Epidemics,* where, among other interesting remarks, it is stated that the quartan is the longest but least severe type of intermittents, while the semi-tertian is the most deadly.[7] From other medical writers are quoted two passages of Oribasius, dealing with tertians and quartans respectively. For reasons given above, no further stress is laid upon quotidians.

Extent of the Prevalence of Malaria

The preceding section has not only shown that malaria existed in ancient Greece; it has also proved it to be widely prevalent. Even if all fevers except tertians and quartans be disregarded, these are mentioned so frequently, and by such a diversity of writers, as to leave no doubt whatsoever. From the year 400 B.C. onwards there is a vast quantity of evidence which points to the unmistakable conclusion that Greece was constantly in the clutches of an insidious and demoralising foe. Plato, Aristotle,[8] the Hippocratic writings, the long line of evidence represented by the works of Galen and Oribasius, all tell the same story. There is even a reference in an inscription.[9] That references to tertians and quartans do not occur more frequently in non-professional literature is not surprising. The Greeks were not in the habit of talking about their ailments. At any rate, when occasion arose to mention a fever, it would rarely be necessary to distinguish one kind from another, τριταῖος from ἀμφημερινός, and so forth. Especially would this be the case if the various forms of malaria were so common as to be designated in the popular speech by πυρετοί without further qualification. Though medical writers do not observe such a limitation, there is some evidence that the people did, as a general rule, limit πυρετοί to malaria. Thucydides seems particularly careful to avoid the word πυρετός in describing the feverish symptoms of the plague. The words he uses are καῦμα and θέρμη. Plato, also, speaks of ξυνεχῆ καύματα, and probably applies the term πυρετοί to continuous fevers because he conceived them to be due to excess of πῦρ. The use of καῦμα by Plato and Thucydides is a remarkable coincidence. Unfortunately there is not sufficient evidence to warrant a positive conclusion. But nevertheless, the frequent mention of malaria in the medical writers, combined with the remarkable passage from the *Timaeus,* makes it extremely likely that malaria was often called in the popular speech by the simple name of "fever." If this be so, whenever the words πυρετός, πυρέσσω, occur in non-medical writers, there is a strong presumption that malaria is meant. In any case, no doubt whatever can be entertained of its wide extension.

Owing to the incompleteness of the evidence, due entirely to the fact that few Greek states have left us any literature, it is impossible to say for certain how far malaria spread. Attica was certainly attacked, as it is attacked now, and of course it was prevalent in the districts which came under the observation of the medical writers. It is clear, from the Hippocratic treatise on *Airs, Waters and Places,* that:

1. the writer had been able to collect evidence about malaria from many districts;

2. the most he could say was that certain districts were less liable to malaria than others. Without going to the extreme of saying that no district was immune, there is every reason for supposing that malaria was widely spread.

When was Malaria Introduced?

Up to the present the inquiry has had a firm foundation of indisputable facts. It is easy to prove that malaria was present in Greece; it is difficult to find out when it first made its appearance, or when it became endemic. It is proverbially hard to prove a negative statement, and the present writer readily admits that it is impossible to show that there was no malaria in Greece before a fixed date. This does not mean that there is no evidence. On the contrary, the evidence, with respect to Attica at least, is very strong. But it is cumulative, and depends for its full force upon a due consideration of many lines of indirect testimony.

What of Its Effect Upon Character?

Many attacks of malaria are mild in character. The Greeks themselves observed that quartans were generally not severe. This fact explains why we have no definite mention in ancient writers of a time when it first came, or when it first assumed endemic form. Many a Greek must have been smitten with malaria without feeling any symptoms other than those he could express by the term πυρέσσω. But the permanent influence of malaria is not to be estimated by its mildness. A severe epidemic, such as one of small-pox, creates much stir at the time and causes many deaths. But it does not last long, and its victims are comparatively few. In all probability such epidemics do not lower the physical efficiency of a people. Even endemic diseases like measles, which cause such trouble to modern children, are transient, and in the great majority of cases do not permanently injure the health. But with malaria the case is different. Often not at all severe, it recurs again and again. Childhood may be one long sickness, the effects of which the adult carries to his grave. His faculties are dulled and he is less efficient generally. Experience proves that if malaria be endemic among a people, there must be a decline, physical, intellectual and moral.

CONCLUSION

Malaria was certainly prevalent in many parts of Greece, including Attica, during the fourth century B.C., though Greece was not "highly infected" in the technical sense of the words as used by Sir Patrick Manson. The evidence of language, and the fact that older people were frequently attacked, suggest that the disease had been but recently introduced. The use of the word μελαγχολία and its cognates shows that the Greeks themselves noticed the effect of malaria upon character. The change which gradually came over the Greek character from 400 B.C. onwards, was one which would certainly have been aided, and was in all probability at least partially caused, by the same disease.

The evidence given in the preceding pages is, from the nature of the case, chiefly cumulative. Many, but certainly not all, of the arguments brought forward might be attacked by a clever opponent. But taken together they are very strong. And it must not be forgotten that a vast amount of testimony, far exceeding that which has been offered, might have been cited if the writer had not wished to exclude as far as possible all cases and symptoms which might imply either malaria or diseases of the typhoid type. It is probable that many, it is certain that some of these were malaria. All this should be borne in mind in passing a judgment upon the question. If any one is still in doubt as to the devastating effects of malaria upon character, he should consult a specialist in tropical diseases, or have a few words with one who has himself suffered from the disease. His doubts will then vanish. Scepticism on the point is only possible in a land in which, happily, malaria is no longer prevalent.

---

1. THUCYD. II. 49.
2. Date uncertain; probably written between 380 and 360 B.C.
3. The difficulty may be best understood by considering a particular case. Suppose that outbreaks of malaria and typhoid occurred together, as they are wont to do in autumn. A Greek physician would almost certainly not distinguish between them, and his description of the epidemic would be a combination of the symptoms of both diseases. A modern physician, who usually has a predisposition to think typhoid the more probable disease, naturally hesitates to give a definite opinion.
4. *Ibid.* 467, 468.
5. KÜHN, VII. 435. Galen says here that this disease was very common in Rome.
6. Probably malignant tertian, which, when double, produces continuous fever with a tertian exacerbation.
7. KÜHN, III. 408, 409.
8. See *e.g. Problemata*, I. 57.
9. DITTENBERGER, *Sylloge*, 890. The inscription is Athenian, but late.

Chapter 12

Note on the Presence of "Bilharzia Haematobia" in Egyptian Mummies of the Twentieth Dynasty
[1250-1000 B.C.]

MARC ARMAND RUFFER*

In a previous note I described a process by which mummified tissues could be prepared for histological examination. I ventured to predict that it was highly probable that, by this method, one would be able to recognize pathological changes, such as cirrhosis, cancer, etc.

Thanks to the kindness of Professor Elliot Smith, Professor Flinders Petrie, and Professor Keatinge, I have obtained several organs from mummies of the XVIII to the XX Dynasty, and I may state at once that such diseases as atheroma, pneumonia, renal abscesses and cirrhosis of the liver are plainly recognizable. In the renal abscesses and in other lesions I have stained microorganisms with methylene blue, fuchsin, haematoxylin, and even by Gram's method.

At the present time there is perhaps no disease more important to Egypt than that caused by the *Bilharzia haematobia*. So far no evidence has been produced to show how long it has existed in this country, although medical papyri contain prescriptions against one of its most prominent symptoms—namely, haematuria. The lesions of this disease are best seen in the bladder and rectum, but unfortunately these are just the two mummified organs which I have not been able to obtain so far. Nevertheless, in the kidneys of two mummies of the twentieth dynasty I have demonstrated in microscopic sections a large number of calcified eggs of *Bilharzia haematobia*, situated, for the most part, among the straight tubules. Although calcified, these eggs are easily recognizable and cannot be mistaken for anything else. I may add that I showed some of my sections to Professors Looss and Ferguson, whose paramount authority on such a subject cannot be disputed, and both confirmed my diagnosis.

I have examined microscopically the kidneys of six mummies. The kidneys of two were apparently healthy; the left kidney of another was congenitally atrophied; those of the fourth contained multiple abscesses with well staining bacteria and other lesions, which so far I have not diagnosed; those of the fifth and sixth showed bilharzia eggs, and the latter had other lesions as well, which, owing to the shrunken state of the organ, I am unable to define accurately as yet.

Renal disease, therefore, was not infrequent among Egyptians living over three thousands years ago.

---

* Although a short note, it is nevertheless to be regarded as an important contribution to early palaeopathological studies. It was first published in the *Brit. Med. J.*, i:16, 1910. (Eds).

Chapter 13

Parasitic Diseases

A. T. SANDISON

In preparing this note I have drawn heavily on the publications of R. Hoeppli (1956, 1957, 1959) which constitute the most valuable single source for the student. Hoeppli redefines a parasite as "an organism which lives temporarily or permanently within or upon the body of another organism from which it takes its food, both parasite and host being either an animal or a plant."

We have no direct evidence that prehistoric man was aware of his infestation by parasites but Hoeppli points out that tribes in Sarawak and Borneo are aware of tapeworms, roundworms, threadworms, fleas, lice, bed bugs and maggots. It may, therefore, be reasonably deduced that prehistoric man would similarly be likely to recognize his own parasitic infestation.

Watson (1960) suggests that the change to an omnivorous habit and a terrestrial habitat exposed early man to helminthic infection to a much greater degree than his arboreal ancestors. His discovery of primitive weapons permitted the killing of animals and catching of fish with the consequent risk of meatborne and fishborne infestations which followed the introduction of flesh and fish into his diet. The domestication of animals also encouraged infestation by contact. As man settled down to an established agricultural existence intestinal round worm and hook worm infestation would constitute a further risk.

Intestinal worms are among the commonest endoparasites. Cuneiform texts from Mesopotamia indicate that intestinal worms occurred but were not clearly defined into types; Kinnier Wilson (page 194) indicates that Ascaris infestation was well known. Sung (1940) suggests that the worm mentioned in the Chinese *Book of Plain Questions* (2nd or 3rd cent. B.C.) was probably *Ascaris lumbricoides*. The Ancient Egyptians were certainly well aware of parasites and indeed exaggerated their importance in the production of disease. Papyrus Ebers (Ebbell, 1937) indicates that round worms and tapeworms were recognized. Much later Pliny mentioned the frequency of tapeworm infestation in the Egyptians. Since pork was not eaten *Taenia saginata* and not *T. solium* was implicated. It should be remembered, however, that the taxonomic distinction was not made until Linnaean times. It is not clear if threadworms were recognized; the worm *Herxetef* has been so interpreted. It has been suggested on very slender evidence that hookworm occurred and caused anaemia. Craig and Faust (1945) remind us that pomegranate infusion was used as a vermifuge. J. Thompson Rowling showed me sections he had prepared from the gut of a Ptolemaic mummy in which we saw unusual structures which were considered possibly to be helminth ova. However, these slides were sent to the Schools of Tropical Medicine in Liverpool and London and no positive identification was made (personal communications). These were, therefore, probably artefacts.

The Babylonian Talmud mentions the commoner human endoparasites and it has been suggested that Hebrew food taboos were perhaps intended to prevent parasitic infestations.

The *Corpus Hippocraticorum* contains frequent references to intestinal worms. Round worms are mentioned in *Aphorisms*, *Epidemics* and *Coan Prognosis*. The common infestation of children was known and in *Maladies des femmes* there is a reference to threadworm involvement of the vulva. Aristotle clearly distinguished between large and flat worms (*Taenia*), cylindrical worms (*Ascaris*) and thin worms (*Enterobius*). Fantham et al. (1916) discuss the nomenclature of intestinal worms during the Classical period. The Greek termed tapeworms ἕλμινθες πλατεῖαι more rarely χηρια. The Romans called them *taenia*, *tinea*, *taeniola* and later *lumbrici* usually with the addition *lati* to distinguish them from the *Lumbrici teretes* (*Ascaridae*). The proglottids were called *Vermes cucurbitini* and the cysticerci χάλαζαι ("hailstones") and later hydatids. The Greeks termed *Ascaris lumbricoides* ελμινς στρογγύλη; Pliny used the phrase *Tinea rotunda*. Later it was named *Lumbricus teres*. The ἄσκαρις of the Greeks and *ascaris* of the Romans is our *Enterobius vermicularis*.

Celsus and Pliny frequently spoke only of flat and round worms but Galen was well aware of the three main types and knew that round worms and tapeworms infested the small intestine while threadworms were present in the distal large bowel. Celsus described the occasional evacuation of round worms through the mouth or nose. Theophrastus was aware that certain populations harboured more tapeworms than others. Johannes Actuarius (14th Cent. A.D.), a Byzantine physician, refers to *Trichocephalus trichiuris* (*Trichuris trichiura*) as well as to the commoner intestinal worms. Lucretius (c. 50 B.C.) had earlier commented on the pallor of mine workers and it has been suggested that this pallor may have evidenced anaemia produced by hook worms.

The Ancient Hindu *Atharvaveda* contains spells against worms and the *Ayurveda* also mentions them. The Hindus at this time appear to have recognized round worms, threadworms and tapeworms. The Mediaeval Arabs also recognized round worms and threadworms but regarded segments of tapeworms (*cucurbitini*) as a separate species; true tapeworms were also recognized. In Ancient China clinical hookworm disease was described; round worm was known as early as 300 B.C. Threadworm and tapeworm infestation (probably both *T. saginata* and *T. solium*) also occurred.

From pre-Columbian Mexico and pre-conquest Peru come evidence that round worm was known. Some authorities believe that hook worm disease may have existed. Antihelminthics, including *chenopodium*, were certainly known (Hare, 1954).

Antihelminthics are indeed known to have been widely used in early societies. From China and India there are references to *areca*, *catechu*, *betel* and *ping-lang* and from Egypt to *pomegranate*. Pomegranate was also used by the Romans for tapeworms. *Filix mas* has long been known to be effective and was prescribed by Theophrastus as early as 300 B.C. (Culbertson, 1942). Pliny also mentions its use. Dioscorides used *santonin* for round worms (Hare, 1954) and Greek and Roman physicians also used derivatives of *Artemisia* species.

Hydatid disease was also known to the Ancient authors including the Ancient Mesopotamians; *Corpus Hippocraticorum* refers to hydatid cyst of man and animals. Operative removal was suggested. It is, however, clear that the relation of hydatids to *Taenia* was not known (Culbertson 1942). Galen was quite familiar with the appearance of cysticerci in the liver of slaughtered animals. Aristotle had earlier compared *Cysticercus cellulosae* with hailstones and Aristophanes in *Knights* refers to the examination of the tongues of pigs for cysticerci.

It should be mentioned here that parasites of animals were also well known in antiquity. Hippocrates and later Galen were aware of threadworm infestation of horses (*Oxyuris equi*) and Aristotle knew that both dogs and fish harboured worms. Columella (1st cent. A.D.) noted helminths in calves and horses. Galen knew that horses harboured round worms and tapeworms as well as threadworms and he also noted worms in the red mullet. Vegetius (4th cent. A.D.) also described round worms in the horse. Much later, in the thirteenth century, Demetrius Papagomenus described worms under the nictitating membrane of falcons; Myiasis of animals had, however, long been known. Aristotle described myiasis of the tongue of the deer and the Ancient Chinese knew of ocular myiasis of the horse and camel.

Infestation by *Dracunculus medinensis*, a macrofilarial type of organism, appears to have been recognized in early Mesopotamian societies and in Ancient Egypt. The *Fiery Serpents* of the Holy Bible which molested the Israelites by the Red Sea are also identified as *dracunculae*. Plutarch certainly believed *dracunculus* infestation to be common in Mesopotamia and Egypt. The disease was also mentioned by Agatharchides of Cnidus (2nd cent. B.C.) as common on the Red Sea littoral. The condition was known to Pliny, Galen, Soranus, Paulus Aegineta, Aetius of Amidea and later by the Mediaeval Arabs, although the latter interpreted the worm as a corrupted filament of the body. Agatharchides termed the worm Δρακόντιον and Galen called the disorder *dracontiasis*.

Elephantiasis arabum is said to have occurred in Old Kingdom Egypt; certainly Lucretius Carus (1st cent. B.C.) described elephantiasis as a common disease in Egypt. Elephantiasis was well known to the Ancient Hindus and later to Rhazes and Avicenna. It is not clear whether the *elephantiasis* of the Graeco-Roman writers did not contain other disorders, for example leprosy.

The important disease *Schistosomiasis* probably occurred in Ancient Mesopotamia since we know that a condition characterised by urethral bleeding was common. Lu and Needham (page 231) note that schistosomiasis and hepatic distomiasis also occurred in Ancient China. A somewhat similar condition disease occurred frequently in Ancient Egypt and the papyri mention haematuria. That schistosomiasis indeed occurred was decisively shown by Ruffer (1910) when he demonstrated a large number of calcified eggs of *Bilharzia haematobia* (*Schistosoma haematobium*) in the kidneys of two XXth Dynasty mummies. These eggs were situated among the straight tubules.

Leeches were widely used by physicians in antiquity. The *Corpus Hippocraticorum* has been interpreted as mentioning temporary endoparasitism in the human throat;

the translation, however, remains in doubt and some authors believe that varicose veins are described and not leeches (Hoeppli, 1959). The activities of the leech were familiar to the Hebrews (Proverbs 30: 15). Watson (1960) has suggested that the story of Gideon's army (Judges 7: 6) may be interpreted as preventing ingestion of leeches but this seems rather dubious. Mitchell (1951), however, in reporting three cases of endoparasitism by the leech in Arab troops in Palestine believes that "Gideon wisely chose an army of those who could take elementary precautions to avoid it." Pliny mentions leeches in the trachea and Solinus (3rd cent. A.D.) leeches in the eyes and ears of elephants. Herodotus described leeches in the mouth of the Nile crocodile. The Mediaeval Arabs were certainly aware of temporary endoparasitism by leeches in man.

Arthropod parasites and pests can also be traced in the ancient literature. The fly symbol of Nergal, god of Disease in Ancient Mesopotamia, suggests that the relation between flies and disease was already suspected. Assyro-Babylonian physicians treated pediculosis and scabies using sulphur for the latter. The Papyrus Ebers mentions flies, fleas and possibly lice and prescribes for mosquito bites. Ruffer (1914) noted nits on the scalp hairs of some Egyptian mummies. In Ancient Palestine flies and mosquitoes were prevalent: lice were suspected of transmitting "leprosy." Some authorities believe that Biblical leprosy may have included cases of scabies.

Aristotle described ticks and lice and recognized the existence of nits. The bedbug is mentioned by Aristotle, Aristophanes (in *Clouds*), Pliny and Petronius (in *Satyricon*). Shrewsbury (1964), however, throws some doubt on the assertions that Aristophanes and Aristotle were aware of bedbugs, although he agrees that they were certainly known to the Romans. Bedbugs were known to the Greeks as κάριο and to the Romans as *Cimex*. Fleas were also mentioned by Aristophanes. Mange of domesticated animals appears to have been known from the earliest times and was described by Marcus Porcius Cato (2nd cent. B.C.). According to Craig and Faust (1945) it was treated with sulphur. Scabies was known also to the Greeks and Romans although the mites were never visualized. Pliny and Columella discuss mosquitoes.

From Ancient India we have accounts of mosquitoes and of head and pubic lice: the Mediaeval Arabs also described fleas, lice, bedbugs and flies. Avenzoar described the *Sarcoptes scabei* and the Persian *Videvedad* mentions scabies. The Arab physicians used mercury in the treatment of scabies (Hare, 1954). The Ancient Chinese also recorded head and pubic lice, fleas, bedbugs and scabies; it is even possible that the actual scabies mite itself was recognized. The Mexicans of pre-Columbian times have left accounts of lice and also of infestation by *Tunga penetrans*: the latter also appears to have been noted in Peru. Dried bodies from Peru have shown nits and also lice; Brothwell and Spearman (1963) illustrate a specimen of *Pediculus humanus* found on a pre-Columbian dried body from Ancon, Peru.

If we turn now to myiasis we find that one of Job's afflictions (Job 7: 5) and one of the features of the death of Herod Agrippa (Acts 12: 23) have been interpreted as myiasis. Flavius Josephus in *The Jewish War* describes the death of Herod the Great: the latter appears to have progressed to cardio-renal failure and to have suffered terminal myiasis. Herodotus (Bk. 4:205) describes the death of Pheretime of Cyrene who was consumed by worms whilst still alive. Aristotle described myiasis of the diseased skin and of the gut. Plutarch gives an account of the extremely rare condition of urinary myiasis. The Ancient Hindus knew of infestation of diseased ears and noses and Papyrus Ebers contains a possible reference to myiasis of wounds of the extremities.

Anderson (1927) gives a useful account of malaria in antiquity. He believes that malaria may be readily recognized in ancient records and suggests that the Ancient Egyptians suffered an annual fever, probably malaria, as evidenced by an inscription at Denderah. Herodotus states that the Egyptian marsh dwellers were much troubled by gnats and protected themselves at night with nets. Ruffer (1913) described splenomegaly in Egyptian bodies of the Roman period from Fayoum and also of the Coptic period. Intermittent fevers also occurred in Ancient Mesopotamia.

Malaria is believed to have been rare in Greece during the Heroic period but became prevalent after the forests were felled and swamps began to form. Greek medical literature contains numerous references to quotidian, tertian and quartan fevers. Splenomegaly was noted, as was "phrenitis" which has been interpreted as cerebral malaria. It has been suggested that the decline in brilliance of the Ancient Greeks was due to increased prevalence of malaria. Craig and Faust (1945) attribute the decline in the Greek colonies of Southern Italy and Sicily to malaria.

Galen recognized multiple infections and Celsus also wrote at length on intermittent fevers. We know that Athens was plagued by mosquitoes and that mosquitoes were also common in Rome and its marshy environs. Dioscorides recommended the therapeutic administration of bedbugs in the treatment of malaria.

The Ancient Hindus described intermittent fevers in the *Ayurveda*: these were of tertian and quartan types. Susruta described the prominent symptoms of malaria. Some ancient cities in Ceylon are believed to have declined because of the prevalence of malaria. In Ancient China malaria also occurred and was associated with splenomegaly (Craig and Faust 1945). Hare (1954) states that antimalarial therapy was used. Probably malaria also occurred in pre-Columbian America; Hare indicates that the Incas used quinine.

Uta or American leishmaniasis appears to have been not uncommon in pre-Columbian America. Mochica period pots frequently show the destruction of the lips and tips of the nose characteristic of this malady. Wells (1964) points out, however, that some of these *huacos* may represent another disease, *South American blastomycosis*. Another pre-Columbian disease, occurring in Peru, of which we have some evidence is Oroya Fever, caused by an organism *Bartonella bacilliformis* (Wells, 1964). The taxonomic position of this organism remains uncertain.

Theodorides (1956) interprets certain conditions described by Celsus and Pliny as mycotic and due to a *Ctenomyces* organism. Theodorides also suggests that Alexander of Tralles (6th cent. A.D.) saw cases of amoebic dysentery. Chadwick and Mann (1950) interpret passages from *Epidemics, Prognosis* and *Coan Prognosis* of *Corpus Hippocraticorum* as possible evidence of hepatic and pulmonary amoebiasis. Some of these attributions are reasonable, for example that from *Coan Prognosis* xxii, 439. Elsewhere in this volume Patrick suggests that Celsus was also familiar with amoebic hepatitis. Hare points out that ipecacuanha was long used by South American Indians for chronic diarrhoea; ipecacuanha is now known to contain emetine which is effective in amoebiasis.

SUMMARY

It is clear that parasitic diseases may be recognized in the ancient medical records. There are also occasional instances of concrete evidence of such diseases, e.g., schistosomiasis and pediculosis capitis which have been demonstrated in mummies, and of intestinal parasites as reported in the following paper.

Indeed, it is not improbable that the relative incidence of parasitic diseases was greater in antiquity than in modern societies in which high standards of hygiene and sanitation obtain.

REFERENCES

ANDERSON, W. K., 1927: *Malarial Psychoses and Neuroses*. London, Oxford Medical Publications.

BROTHWELL, D., and SPEARMAN, R., 1962: The Hair of Earlier Peoples. In: *Science in Archaeology*. D. Brothwell and E. Higgs (Eds.), London, Thames and Hudson.

CHADWICK, J., and MANN W. N., 1950: *The Medical Works of Hippocrates*. Oxford, Blackwell.

CRAIG, C. F., and FAUST E. C., 1945: *Clinical Parasitology*. London, Kimpton.

CULBERTSON, J. T., 1942: *Medical Parasitology*. New York, Columbia University Press.

EBBELL, B., 1937: *The Papyrus Ebers: the Greatest Egyptian Medical Document*. Copenhagen, Levin and Munksgaard.

FANTHAM, H. B., STEPHENS, J. W. W., and THEOBALD, F. V., 1916: *The Animal Parasites of Man*. London, Bale, Sons and Danielsson.

HARE, R., 1954: *Pomp and Pestilence: Infectious Disease, its origins and Conquest*. London, Gollancz.

HOEPPLI, R., 1956: The knowledge of parasites and parasitic infections from ancient times to the 17th century. *Exp. Parasit.*, 5:398.

HOEPPLI, R., 1957: Knowledge of errors in ancient views on parasites and parasitic infections. *Elixir*. Winter, 21.

HOEPPLI, R., 1959: *Parasites and Parasitic Infections in Early Medicine and Science*. Singapore, University of Malaya Press.

MITCHELL, J. F. O., 1951: The leech as an endo-parasite. *J. Laryng.* 65:370.

RUFFER, M. A., 1910: Note on the presence of "Bilharzia haematobia" in Egyptian mummies of the Twentieth Dynasty (1250-1000 B.C.). *Brit. Med. J.*, 1:16.

RUFFER, M. A., 1913: On pathological lesions found in Coptic bodies (400-500 A.D.). *J. Path. Bact.*, 18:149.

RUFFER, M. A., 1914: Pathological notes on the royal mummies of the Cairo Museum. *Mitt. Gesch. Med. Naturw.*, 13:239.

SHREWSBURY, J. F. D., 1964: *The Plague of the Philistines and other Medical Historical Essays*. London, Gollancz.

SUNG, T. Y., 1940: History of gastro-intestinal diseases in China. *Chin. Med. J.*, 58:324.

THEODORIDES, J., 1956: La parasitologie chez les Byzantins. Essai de Comparaison avec les Arabes. *Arch. Ibamer. Hist. Med.*, 8:207.

WATSON, J. M., 1960: *Medical Helminthology*. London, Bailliere Tindal, Cox.

WELLS, C., 1964: *Bones, Bodies and Disease*. London, Thames and Hudson.

Chapter 14

The Recovery of Parasite Eggs from Ancient Cesspit and Latrine Deposits: an Approach to the Study of Early Parasite Infections

A. W. PIKE

THE literature on parasitic infections reviewed in the previous paper clearly shows that documentary evidence for the existence of parasites in early populations, both human and other animal, is extensive; but what evidence is available from preserved remains to support these accounts? And what new light can be thrown on the presence of parasites in early communities from this preserved evidence? This review is concerned mainly with the literature on this aspect of ancient parasitic diseases.

Death and subsequent decay of the host normally results in a similar decomposition of the parasites contained, except for arthropod parasites in which the exoskeleton is preserved, so it is unlikely that opportunities would arise for the study of intact parasites. Even the mummified or frozen bodies found on various occasions have not, apparently, yielded any recognizable worms, presumably because internal postmortem changes are not conducive to their preservation. Of course there is no reason to assume that hard parts of parasites, such as the hooks and spines, cannot survive in a recognizable form. Evidence of parasitic infections in the form of tissue damage should be available in mummified tissues, however, just as lesions caused by other diseases are known from them. A more profitable line of search for parasite remains exists in the examination for eggs or cysts of the parasites. The available literature entirely confirms that these can survive in preserved bodies, coprolites and cesspit and other deposits, often in exceptionally well-preserved conditions that have enabled some workers to suggest specific identities for the eggs and cysts recovered.

The first record of parasites from ancient remains is that of Ruffer (1910) who recognized eggs of *Schistosoma haematobium* in the kidney tubules of two mummified Egyptians dated between 1250 and 1000 B.C. More indirect evidence that S. *haematobium* existed in early populations comes from the work of Zakaria (1959). This author examined the ruins of a number of historic sites in Iraq and discovered shells of the fresh-water mollusc *Bulinus* in the mud bricks of the buildings, the oldest of which was four to six thousand years. The author states that "It appears from the findings of large numbers of subfossil shells of *Bulinus* in historical sites as described in the text that the snail must have been widely spread in irrigation canals in central Iraq in early times." Since it has been shown by other workers that *Bulinus* is today the intermediate host of S. *haema-*

tobium in Iraq, Zakaria suggests that the disease was probably common in historic if not pre-historic times. It would be interesting, as Zakaria comments, to confirm the existence of the disease in Iraq from an examination of mummified bodies, but so far none has been found.

Szidat (1944) reported finding helminth eggs in the gut contents of the preserved bodies of a girl aged twelve to fourteen years and a man, found in a bog in East Prussia. The body of the girl was dated at 600 B.C. and the gut contents included eggs of *Ascaris* and *Trichuris*. The man's body which was of more recent origin, 500 A.D., contained eggs of *Ascaris*, *Trichuris* and structures resembling *Diphyllobothrium latum* eggs, in the gut.

More recently there have been a number of reports from various parts of the world which indicate that a wide range of helminth eggs and protozoan cysts can be preserved. Pizzi and Schenone (1954) examined the body of an eight- or nine-year-old Inca child from a tomb in the Andes and found eggs of *Trichuris trichiura* and specifically unidentifiable cysts of the protozoan *Entamoeba* in the rectum; the approximate age of the body was given as four hundred and fifty years.

Taylor (1955) examined soil from a woodlined Medieval pit at Winchester in Britain, and found large numbers of eggs of *Ascaris lumbricoides*, *Trichuris trichiura* and of the trematode *Dicrocoelium dendriticum*. He discusses the possible origin of these eggs which may have been human or porcine and concludes that possibly the evidence is weighted in favour of the former.

The corpse of a man recovered from a peat bog at Grauballe, in Denmark, and dated as between the third and fifth centuries A.D. contained identifiable remains in the stomach which are described in detail by Helbaek (1958); they included eggs of *Trichuris trichiura*. Helbaek also mentions that a similar body recovered at Tollund, also in Denmark, contained eggs of the same nematode species.

Grzywiński (1962) examined one hundred and sixty-seven samples of faeces from cultural soil layers in a Slav settlement in the Opole District of Poland, dated from the tenth to thirteenth centuries A.D. and found that seven of them contained eggs believed to be those of the liver-fluke, *Fasciola hepatica*. The faecal samples containing the eggs were said to belong to large and small ruminants. This work was an extension of that reported earlier (Grzywiński, 1955 and 1961). Two coprolites of human origin found in a cave in the Nahal-Mishmar valley near the Dead Sea and corresponding to a layer eighteen hundred years old contained eggs of *Trichuris trichiura* and cysts of the protozoa, *Entamoeba histolytica*, *E. coli*, *Giardia lamblia* and *Chilomastix mesnili* (Witenberg, 1961). The first three species were said to be abundant whereas the last two species were fewer in number, but all were well preserved. Witenberg also mentions that all of these parasites exist in Israel at the present time.

Numerous species of helminth eggs were recovered by Jansen and Over (1962) from terp material in north-west Germany, the age of which was between 100 B.C. and 500 A.D. They recognized eggs of *Ascaris lumbricoides*, *Toxocara canis*, *Oxyuris equi*, *Trichuris trichiura*, *Trichuris ovis* or *Trichuris globulosa*, *Fasciola hepatica*, *Taenia saginata* or *Taenia solium* and *Diphyllobothrium latum*. In another report Over and Jansen (1962) describe the eggs of *Fasciola hepatica* and shells of the molluscan intermediate host of this species, *Lymnaea (Galba) truncatula* from

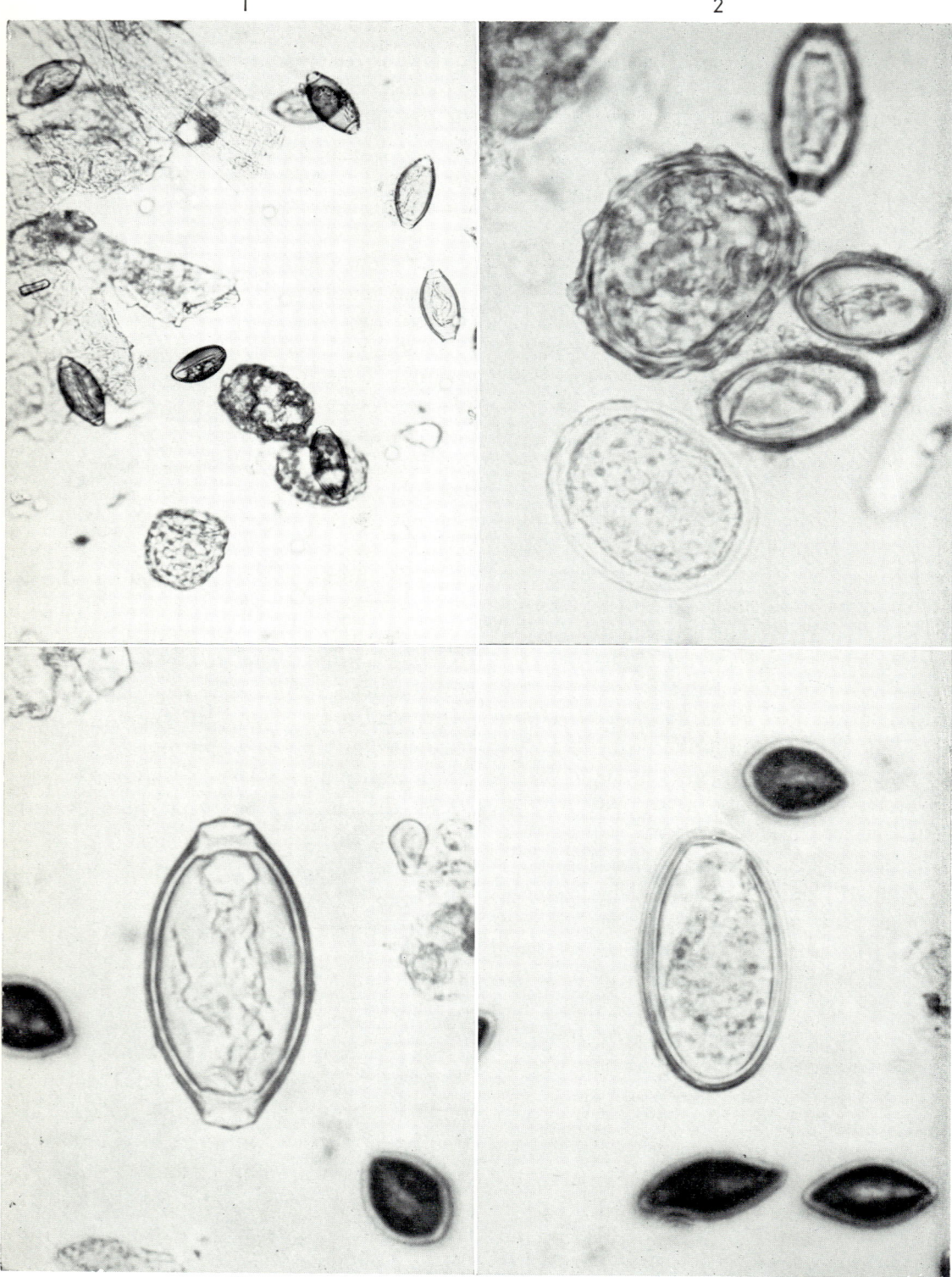

TABLE I
NUMBER OF EGGS PER GRAM OF DEPOSIT
(AVERAGES ARE OF TWELVE COUNTS)

	Range	Average
Ascaris sp.	200-1200	450
Trichuris sp.	1400-3200	2300
D. dendriticum	200-800	216

terp material eighteen hundred to two thousand years old.

Recently I had the opportunity of examining soil from a woodlined Medieval pit at Winchester, which was dated as approximately one thousand years old. The excavations from which the material was obtained were very near to those where Taylor (1955) collected his sample. The material, which was dark green-brown in colour, yielded eggs of *Ascaris* sp., *Trichuris* sp. and *Dicrocoelium dendriticum* thus confirming what Taylor had earlier found. All of the eggs were in a good state of preservation (Figs. 1-4) and occurred in very large numbers, Table I. The measurements of these eggs are given in Table II. A full account of this work is in preparation as a joint paper with Mr. M. Biddle, Director of Excavations at Winchester and will be published elsewhere.

Standard helminthological techniques were used for examination of the material from Winchester, the most convenient being the zinc sulphate flotation method for concentration of eggs. A sample of the material was washed several times with water and centrifuged. The sediment was then thoroughly mixed in a centrifuge tube with a solution of zinc sulphate (specific gravity 1.3) in which most parasite eggs will float and the latter removed from the surface layer with a glass coverslip. Semipermanent preparations can be made by ringing the coverslip on a glass microscope slide with a suitable slide mountant (e.g. Gurr's London, Glyceel or nail polish). Since some of the more delicate eggs may collapse in the zinc sulphate solution, samples of soil should also be treated by simple sedimentation techniques to recover these eggs.

The above methods were possible because the material was still moist and apparently had remained so almost since the

TABLE II
MEASUREMENTS OF HELMINTH EGGS IN MM. (AVERAGES ARE OF TEN SPECIMENS)

	Range	Average
Ascaris sp.	0.0546-0.0624 x 0.0702-0.0819	0.0593 x 0.0737
Trichuris sp.	0.0273-0.0312 x 0.0546-0.0858	0.0281 x 0.0608
D. dendriticum	0.0234-0.0273 x 0.0390-0.0468	0.0269 x 0.0425

FIGURE 1. Photomicrograph of a slide preparation showing the three types of eggs and plant tissue fragments. X200

FIGURE 2. Above, embryonated *Ascaris* egg; below, decorticated *Ascaris* egg. X610

FIGURE 3. *Trichuris* sp. egg, showing terminal plugs intact. X857

FIGURE 4. *Dicrocoelium dendriticum* egg, with several unidentified black, spindle-shaped structures. X857

time of deposition. Coprolites would, however, require some rehydration before eggs could be easily recognized and for this the method of Callen and Cameron, 1960 as given by Callen (1963) would probably be suitable. They suggest soaking coprolites in a 0.5% aqueous solution of trisodium phosphate for seventy-two hours or more. Witenberg (1961) used Triton solution (manufactured by Rohm and Haas, U.S.A.) to treat the coprolites he examined. However, he does not specify which one of the formulations of this detergent he used. Similar compounds may prove to be equally suitable for this purpose.

Besides the problems of recognizing eggs which have been preserved for long periods of time there is also the difficulty of determining the origin of the eggs in many instances. Unless the material is found in a recognizable coprolite the origin of which can be determined, it is very difficult to determine whether some eggs are of human origin or not. Such was the case with material from the Winchester excavations. If these problems can be overcome the possibility of learning more about the origin and distribution of parasitic infections could be realized.

REFERENCES

CALLEN, E. O., 1963: Diet as revealed by Coprolites. In: Science in Archaeology, Brothwell, D. R. and Higgs, E. (Eds.). London, Thames and Hudson.

GRZYWIŃSKI, L., 1955: [Parasite eggs in faeces originating from the XI, XII and XIII centuries.] Streszcz. ref. IV Zjazdu. [Summaries of papers of the IV meeting of the Polish Society of Parasitology, 97. (As quoted in Grzywiński, 1962)].

GRZYWIŃSKI, L., 1961: Analysis of feces from the Middle Age Period. Zoologica Pol., 10:195.

GRZYWIŃSKI, L., 1962: [Parasitologic analyses of excrements found in excavations.] Wiad. Parazyt., 8:543.

HELBAEK, H., 1958: [The last meal of Grauballe Man.] Kuml: 83.

JANSEN, JR., J. and OVER, H. J., 1962: Het voorkomen van parasieten in terpmateriaal uit Noordwest Duitsland. Tijdschr. Diergeneesk., 87:1377.

OVER, H. J. and JANSEN, JR., J. 1962: Het voorkomen van fascioliasis rond het begin van onze jaartelling in de omgeving van de 'Feddersen Wierde'. Tijdschr. Diergeneesk., 87:1440.

PIZZI, T. and SCHENONE, H., 1954: Hallazho de huevos de Trichuris trichiura en contenido intestinal de un cuerpo arqueológico incaico. Bol. Chile. Parasit., 9: 73.

RUFFER, M. A., 1910: Note on the presence of Bilharzia haematobia in Egyptian Mummies of the twentieth Dynasty [1250-1000 B.C.]. Brit. Med. J., 1:16.

SZIDAT, L., 1944: Uber die Erhaltungsfähigkeit von Helmintheneiern in Vor- und Frühgeschichtlichen Moorleichen. Z. Parasitenk., 13:265.

TAYLOR, E. L., 1955: Parasitic helminths in mediaeval remains. Vet. Rec., 67:216.

WITENBERG, G., 1961: [Human Parasites in Archeological Findings.] Bull. Israel Explor. Soc., 25:86.

ZAKARIA, H., 1959: Historical study of Schistosoma haematobium and its intermediate host, Bulinus truncatus, in central Iraq. J. Fac. Med., Baghdad, 1:2.

SECTION III
GEOGRAPHIC STUDIES

Chapter 15

Organic Diseases of Ancient Mesopotamia

J. V. KINNIER WILSON

THE source materials for conclusions reached in this and the following chapter are, for the most part, many hundreds of cuneiform tablets unearthed in the course of a century of excavation at many of the early cities of Babylonia and Assyria. This material, sometimes well preserved, sometimes much broken, is continually becoming better understood as more and more duplicates can be fitted into place and as the frontiers of Assyriology slowly advance in accurately determining the meanings of the technical terms used.

From the standpoint of disease these tablets fall into three groups. There are firstly those therapeutical "medical texts" —the ancient title is still unknown—which as Oppenheim, 1962, has convincingly shown were the traditional heritage in first millennium times of a vast corpus of knowledge on disease and its treatment which was first set down in the Old Babylonian period and never appreciably altered in later centuries. The second group consists of a single long work of forty tablets, anciently called *Sakīkū* or, approximately, "The Symptoms," which is concerned only with diagnosis and prognosis. Of central importance though this work is, many of the entries are no more than fragmented syndromes, or observations relating to individual parts of the body, and its information thus often disappoints the modern analyst. In the third group may be placed texts of a miscellaneous character which may be used to-day in furthering knowledge of disease in ancient Mesopotamia even if the original purpose of such documents was altogether different. In so far as the so-called physiognomical omens and birth omens may be included in this category it is from this section that one may draw for some insight into certain incurable and congenital disorders. To give detailed acknowledgement for every trespass upon this material would greatly overburden the present statement: where for purposes of verification and control references have been given, these are cited in accordance with standard abbreviations which are explained at the end of the chapter.

We need delay further only to emphasize two points. It is an obvious first duty of the chapter to synthesise the findings of older studies where these have attempted actual diagnosis. At the same time it is proper to stress that Babylonian medicine belongs, in the first instance, to the history of tropical medicine, and it is to this somewhat neglected field that particular attention is invited in the discussions of the following pages. In this connection the opportunity may be taken to express the extent of personal obligation owed to the modern Iraq medical journals. Their relevance will be at once clear from titles mentioned in the end references, and these may be thought the more welcome since the journals are not generally accessible outside of specialised libraries.

The second point must be to emphasise that interpretive difficulties are great and no purpose is served by underrating them. The whole of Babylonian medicine was written in days before pathology and one may easily suspect that the demarcation lines between disease and disease were often drawn in the wrong places; the differential diagnosis of the text-books is a constant reminder of the possibility of analytical errors; and the necessary linguistic standard is one in which the technical terms are understood with near precision. Thus what is in fact offered in the following account is a personally selected sample of diseases which are distinctive in the sense that they are not liable to be confused with anything else—although occasionally this rule will be broken when a group concept is itself important—and which at the same time are described in terms of a known vocabulary, exceptions under this heading being assigned a question mark. The selection attempts to be representative, but in no sense is it complete. No statement on the diseases of Babylon ever will be.

DEFICIENCY DISEASES

The fight for an adequate diet in a land of wars, famines and erratic rainfall, and which is not well supplied with citrus fruits, means that deficiency diseases must have occurred in ancient times almost within the natural order of things. "I removed the carcases," wrote Ashurbanipal in his Annals (iv 79), "of those whom the pestilence had struck down and whose bodies, devoured of dogs and (wild) pigs (sic), were obstructing the streets and open spaces of Babylon, even of those who had lost their lives in the terrible famine." No single sentence tells more about the privations of the times.

In the less dramatic atmosphere of clinical practice some observations concerning the eye may serve as a first pointer to deficiency disease. Thus by the definition of AMT 16, 3+ ii 8-9: "If a man can see nothing by day but everything by night, this is called *sillurmâ*: or if he can see everything by day but nothing by night, it is (again) called *sillurmâ*," it is evident that the one word served to describe both day- and night-blindness (*cf.* Thompson, 1926). A condition where the eyes "become dry" will doubtless be xerophthalmia or xerotic keratitis, found commonly throughout the East in times of famine and deprivation (Thompson, 1924; Krause, 1934); the use of an onion mentioned in one prescription for the condition—not that it would have been effective—underlines the point that, in such cases, there was diminished secretion of lacrymal fluid. For want of a better place we may also include here the condition *iṣṣanundā*, or "moving from side to side" of the eyes, which may confidently be taken to describe nystagmus, while verbal forms taken with Kraus, 1936, and against a recent expression of opinion to the contrary, to belong to a verb *uzuzzu*, "to stand or remain fixed," will, for eye associations, be descriptive of ophthalmoplegia. It should be explained that the justification for placing both of the latter terms under the present heading is the consideration that deficiency states must often have been a cause of the paralysis involved, not that the texts themselves directly express the connection. The identification of ophthalmoplegia finds its best support from a case considered below in the section on Poisons.

As to paralysis in the wider sense, it is now generally accepted that *šimmatu* was the normal word. The term denotes paralysis irrespective of aetiology, but particularly where it is given as a subject for treatment without further mention of other symptoms, it must often have been that a nutritional neuritis or polyneuritis lay at the root of the trouble. It may be noted that the word itself differs from the modern concept in that it actually denotes a substance, "paralysis-poison" or the like, and as I have elsewhere suggested (1956), instances where the body is said to "contain some *šimmatu*" probably express the condition of paresis. Closely related is a word *rimûtu* denoting a "looseness of the limbs" from a verb *ramû*, "to be loose." At least seven different translations have been offered for this word, but none is better than ataxia. This suggestion, although sensible in itself, is also somewhat controlled by references where *šimmatu* and *rimûtu* occur together, in which case we would see the condition as that of ataxic paraplegia.

The general picture becomes even more suggestive when the word *nuppuhu* is translated "oedema." This word comes from an onomatopoeic base *puh* meaning "puff," and as a verb will then basically mean "to be puffy." This is itself suggestive of oedema (and so, rather than "inflammation" which was Labat's translation in TDP) and it is of interest that Arabic physicians used the cognate verb in their own language when translating Galen's οἴδημα. All doubt is dispelled when one learns from TDP 178 19 that the *libbu* (or approximately "internal system"), the borders of the eyes and the ankles might become oedematous at the same time. Here, of course, it is not certain that we are still in the realm of deficiency disease for the observation could be equally characteristic of Bright's disease. The same difficulty in fact arises over the word *aganutillû*, known since the last century to mean "dropsy" since it was a condition in which the body filled with "water," for it must always remain uncertain whether cardio-renal disease, famine dropsy, wet beriberi(?), or the like was being described in a given case. However *nuppuhu* is of too common occurrence in the diagnostic texts for the oedema not to have been of malnutritional origin in many instances. The case of TDP 78 68, where face, trunk, arms and legs are all described as oedematous was obviously serious and is suggestive of general anasarca.

Against this background one disease stands out in strong relief, namely *bu'šānu* or "the stinking disease." Although not hitherto identified all the available clues point to *bu'šānu* as being the ancient word for scurvy. Here the first point that is important is one of placing, for, in the therapeutic texts, the prescriptions and incantations for *bu'šānu* are found in the second of the two Tablets which concern the teeth. This is altogether suitable since in scurvy the gums and teeth are both specifically involved, the gums becoming swollen, "spongy" and covered in fungating sores, and the teeth loosening and, in advanced cases, falling out. The seriousness of the condition may be gauged by the fact that three of the four columns of the Tablet are concerned with it, and one of the incantations, addressed to the personified disease, describes how he has attacked the glands of the mouth and the tongue, whereupon the demon finally centralises his attack "between the teeth." The name itself, from a verb *ba'āšu*, "to be evil-smelling," is an additional pointer to the diagnosis, and every support comes from the fact that a "medicine," also called *bu'šānu*, was prescribed as a cure for the disease of the same name. Since this second *bu'šānu* may be identified through the Hebrew *bĕʾušim* and the metathesised post-biblical *ʾŭbšin* both of which mean "wild grapes," and since this finding cross-checks with evidence of AMT 25,7 ii 9 and 13 that, in pharmaceutical use, the juice of *buʾšānu* was "pressed out" like wine, it will follow that we may declare the medicine of service for the disease. Abdulnabi *et al.*, 1961, in fact give the vitamin C content of black grapes in Iraq as 5.2 mg per 100 gm, and while this is not especially high the obvious merit of grape juice of any variety is that it will keep through the winter. It is also relevant to add that, in southern Iraq, many citrus fruits grow either badly or not at all, and the ubiquitous date-palm is of little help, containing vitamin C only in the proportion of 0.9 mg per 100 gm.

Scurvy, however, and for that matter anaemia also, will not have been the same disease in Assyria, for new texts coming from the Assyrian capital of Calah—the modern Nimrud—indicate that, at least from the times of Ashurnasirpal II (883 B.C.), all service personnel received a daily wine issue throughout the year to the order of three and a half pints for a unit of ten men. In Babylonia wine, being for the most part imported, was largely for the privileged class, and the national drink of beer can never have been the same thing from the medical point of view. But, to see scurvy at its worst one must look further afield. The scene in 597 B.C. when Nebuchadnezzar II laid siege to Jerusalem is graphically described in the Book of Lamentations (iv. 4ff):

Those who had fed on dainties went scavenging in the streets;
Those brought up in purple embraced the refuge-dumps...
They went unrecognised in the streets, their faces blacker than soot,
Their skin shrivelled upon their bones and dry as (tinder-)wood.
Happier were they slain of the sword than the victims of hunger
Who but slowly expired, stricken through for want of the fruits of the field.

In this passage the one diagnostic clue of the black skin may easily suggest the massive subcutaneous haemorrhage which characterizes scurvy in its terminal stages. It may thus be that, in the last line, the ancient poet wrote wiser than he knew.

BOWEL AND BLADDER

Disorders which fall under this headline occupy much of a physician's time in tropical countries, and it is not likely that ancient Mesopotamia was an exception.

Of helminthic disorders, which may be taken first, the one disease which stands out is ascariasis (for the modern picture in Iraq see, for instance, Bailey, 1955 and 1957; Baqir, 1960). It is even possible to see the responsible parasite, *Ascaris lumbricoides*, on an amulet from Nippur and probably datable to the first millennium B.C., bought locally from a dealer by Prof. Th. J. Meek of the University of Toronto by whose permission it is here published for the first time (Fig. 1). The amulet, numbered D. 1247, is now in the Royal Ontario Museum of Archaeology. Somewhat crudely engraved like many of its kind, it shows the demoness Lamastu surrounded by some animals of the lower orders, but the object of interest are the three crescent-shaped objects shown to the right of the left leg, and issuing from a point which may be thought to correspond to the anus. This placing, and the relative size of the worms, if such they be, could well suggest the ascaris which varies between 15 and 35 cm in length. This evidence apart, there would be a good linguistic claimant for the ascaris in the

FIGURE 1. The Ascaris (?) on an amulet from Nippur.

word *išqippu*, which is known both to mean an "earthworm" which ascaris is in fact said to resemble, and also an intestinal worm. The best evidence for the latter comes from a small incantation tablet published by Nougayrol, 1947, where the patient is stated to be "weary, sleepless and exhausted," and in whose body there is said to be both *ahhāzu*, "jaundice," and *išqippu*, "ascaris." The incantation is in fact to be recited over one of the worms. In that these may enter the pancreatic and bile ducts and give rise both to jaundice and abscess of the liver the diagnosis looks as if it should be hepatic ascariasis (cf. Muazzam et al., 1960).

There appears as yet to be no direct evidence in support of any suggestion that ancylostomiasis, or hookworm disease, existed in Babylonia, although notice of a severe pain in the epigastrium, with passing of blood and prognosis of death (TDP 114 42) at least simulates the symptoms of a bad and neglected case. In AMT 58,1 8 + 56,5 8, where the patient "has a sore anus and scratches it constantly," the symptoms are those of pruritus ani, and the additional indication of "wind" makes it the more likely that intestinal parasites were responsible.

The subject of dysentery is naturally difficult. Amoebic dysentery in particular is common in Iraq and, for example, in a survey at Tarmiya, a district situated centrally some forty miles from Baghdad, Bailey, 1957, reported a "relatively high" incidence of 54.8 percent. In this connection it may be said that the texts are not silent concerning the passage of blood per rectum (e.g., AMT 107,2 4; KAR 191 ii 16; BAM II 152 iii 8), but clearly any of several conditions might be similarly described. It is thus only the complication of liver abscess which convincingly suggests the ancient occurrence of amoebiasis, even though this is still not the only cause of it. Liver abscess was treated by minor surgery, and both the diagnosis and the first proper presentation of the relevant text are owed to Labat, 1954. The account is not well preserved, and the word "abscess" does not in fact occur on the extant portion of the text, but in that an incision was to be made between the "third" and "fourth" ribs (Labat suggests that these would to-day be called the eighth and ninth, and that the Babylonians counted the ribs in the reverse order to that of modern practice) it is difficult to think of another condition which might have been so treated.

As to the bladder, this was very largely a matter of stone (*abnu* or *aban muštinni*), stricture (*hiniqtu*) and *mūṣû*. The latter condition may have been serious and it will be worth while to assess the evidence carefully.

The disorder in question was taken by Thompson, 1926, following an original suggestion by Jensen, to be gonorrhoea. He was followed in this opinion by Labat, TDP xxviii, and albeit cautiously, by Sigerist, 1951, and the idea seems also to have inspired Köcher to see veneral disease in his BAM II, Nos. 114 and 117. But to judge, for instance, from the relevant chapter in Bloomfield, 1958, gonorrhoea, as one looks backward in time from the present day, is lost sight of within a few centuries. Thus the original focus, or foci, of infection could be virtually anywhere, and so especially for a communicable disease—as opposed, say, to a nutritional, or known endemic, disease—an assertion that gonorrhoea was prevalent some three thousand years earlier in Mesopotamia would require the very strongest of support. The fact that the Babylonians themselves did not associate *mūṣu* with intercourse does nothing to encourage the idea.

As for the essential symptoms, these are that the patient has pain on micturition and passes water which, in its simplest description, is "wine-coloured" (haematuria), and is so defined as to suggest the additional complication of a purulent discharge; and what we would suggest for *mūṣû* in place of gonorrhoea is urinary schistosomiasis—or bilharziasis, as previously called—with septic infection of the bladder. This condition is geographically right: indeed, *Schistosoma haematobium* is one of the present day scourges of the country with incidence of the uncomplicated disease up to 100 percent in some villages, and the Iraq medical journals treat almost endlessly of the problem. But what makes the identification especially attractive is that associated with the *mūṣû*-disease are *mūṣû*-stones. Thompson considered that the stones cured the disease; but since by the definition he himself quotes these stones "come from within the urinary passage" of the patient himself, they must, in fact, be calculi. Such stones often form in the disease with the ova of the species as their nuclei, and indeed it is these eggs which make for the whole mischief of the condition. Their sharp terminal spines pierce the wall of the bladder and thus cause the urine to become blood-stained, while, accumulating in enormous numbers, many in time form a calcified lining to the bladder making it difficult for calculi to escape, and causing pain and stricture. It need only be mentioned that gonorrhoea is usually included in the differential diagnosis of schistosomiasis to suggest that, in the septic condition, the two diseases present a similar clinical picture.

It remains to add that not only have the characteristic ova of *S. haematobium* been discovered in Egyptian mummies as is described on p. 177, but also, in an important historical study for Iraq, Zakaria, 1959, reports the finding of shells of the schistosoma host snail *Bulinus truncatus* in the mud-brick walls of Babylon, in the sides of excavation cuts at Tel 'Aqeir, and in the ziggurat at 'Aqar Quf. Both observations lend all support to the conclusion that schistosomiasis is old in the Near East.

HYDATID DISEASE

Before leaving the field of parasitology, one other worm, *Taenia echinococcus*, may be mentioned here, not because it is an intestinal parasite of man—it in fact infests mainly the dog, wolf, jackal, dingo and fox —but, because through association with the dog, man may become infected with larval forms of the parasite in the shape of hydatid disease. For the modern picture in Iraq, together with some historical allusions to "water bladders" as apparently described in the Talmud (no reference given) and Hippocrates (*Aphorisms*, VII. 55), see Kelly and Izzi, 1959.

The cycle of the disease is complicated, and in what will prove to be a journey of arguments, we turn first to the livers of sheep, noting here that, whatever the anatomical failings elsewhere, the divining priest's knowledge of this organ was second to none The reason for this interest is to be sought in the massive part which the science of hepatoscopy played in Babylonian times, and indeed the technical language of the science and of the liver itself is complex and detailed in the extreme. Now, of these many words, one, *erištu*, bids fair to mean those round, grey cysts which may be made on the sheep's liver by the *Taenia echinococcus* embryos as they emigrate thereto through the portal vein radicles from the stomach or upper part of the small intestine. We may say this both because of the descriptions (thus the *erištu* may be like a "mustard-seed," a "chick-pea," a "bud," "flattened," or "deeply incised," etc.), and also because it is recorded as being found on several different parts of the liver. What follows in the suggested cycle of hydatid disease is that dogs ingest these cysts by eating the liver and offal of sheep, and the ova, passed in the faeces and adhering to their coats or skin as the result of sundry associations with the dust, is passed on to man, and particularly to children who so often put hand to mouth, by direct contact. One cannot say that this happened: but one may suspect in the nature of things that it happened since, in Mesopotamian antiquity, sheep were abundantly eaten on the frequent festival days throughout the year, and their offal must often have been available to the prowling dogs of the times.

CUTANEOUS LEISHMANIASIS

Our concern in this section is with a condition of the skin which every visitor to Baghdad will have seen on his first day, namely the so-called Oriental Sore or "Baghdad boil." There are two things to say about it so far as Mesopotamian antiquity is concerned. For the first one cannot do better than quote what Thompson, 1906, wrote almost sixty years ago. "An instance of the Baghdad boil occurs in an Assyrian astrological report to the king from a priest. 'Concerning this evil of the skin, the king, my lord, hath not spoken from his heart. The sickness lasts a year; people that are sick (therefrom) all recover.' Now the boil is popularly supposed to last a year."

To this statement one need only add that the record is to be trusted, the observation that the affection, in general, continues for about a year is perfectly true, while the king's ignorance on the subject is perhaps to be explained on the grounds that he was an Assyrian, not a Babylonian.

As to the second point, our understanding of the old skin nomenclature is badly frustrated by a lack of descriptional detail, but the word *ibāru* may very reasonably be taken as referring to the permanent scar which the sore leaves on the skin. The evidence comes from the physiognomical omina, and in one text the parts which may be so affected are stated as being the right or left cheek, the right or left side of the throat, and the right arm. At this point the text breaks off, but the document is starting well for what may be taken to be a list of those exposed parts of the skin which may be attacked. A further text reads, 'If on his face, on the right side, there is an *ibāru*-scar," the cheek in fact being a favorite site for the sore.

One may add that, in an appealing study, Lamborn, 1955, suggests that, perhaps even more than the dog, the horse and ass are main reservoirs for the infection in modern Iraq, this being carried to man via the flies which endlessly hover around the affected eyes of these animals. Somewhat curiously, we have direct evidence in more than one text for eye disease in the wild ass of antiquity, while economic texts from all periods provide many references to the making and supply of horse- and donkey-blinkers, the latter being no more than an essential veterinary requirement against the all-pervading and ubiquitous fly.

There is nothing yet to say concerning the corresponding disease of Kala-azar (visceral leishmaniasis). This is of recent description in Iraq as elsewhere, but the disease might be old since the argument for an Iraq focus of Kala-azar, not thought to be the result of communication between Iraq and a neighbouring endemic focus, is advanced by Pringle, 1956.

EYES AND EARS

What diseases flies and dust brought to the eyes of ancient Babylonians we may never fully know, but a question asked in one of the eye incantations (AMT 10,1 iii 11), "why should they sting you, the sand of the river, the flowerdust of the datepalm, the pollen of the fig, the straw of the winnower?" gives an immediately oriental setting to the whole subject of Mesopotamian ophthalmology.

It is in fact probable that what the Babylonians simply called "eye disease," and which heads the conditions for treatment given in the therapeutic texts, was either the common and expected condition of ophthalmia or else speaks for the scourge of trachoma. Difficulties of this kind, where so much depends upon the meaning of a single term, are seldom easy to resolve; but cases where the eyelashes were said to be growing into the eyes (AMT 16,1 17), and whether the diagnosis be trichiasis or entropion, provide at least some secondary evidence for trachoma since both of these conditions most commonly occur in old cases of the affection.

Alike serious was the famous case referred to in Hammurabi's Code which concerns a physician who "has cut into the *nakkaptu* of a man with a bronze lancet (?) and (by so doing) has destroyed one of his eyes" (Law 218). Neither the old idea that this situation speaks for cataract, a disorder of advancing years and one which primitive Babylonian skills are not likely to have controlled, or that it involved scarification, (Oppenheim, 1962), personally commend themselves. If one must pronounce judgment it is preferable to accept the conclusion of von Soden, 1949, that *nakkaptu* means "der untere Teil der Stirn, der Augenbrauenbogen"—although extending "superciliary ridge" in this instance to include the inner canthus area—and then to follow Singer *apud* Driver and Miles, 1955, in seeing the operation as one for abscess of the lacrymal sac. This condition is common in the East, as Singer states; but the main attraction of the diagnosis is that it makes an understandable legal case. In acute dacrocystitis pain may be so intense that something must be done, yet the patient still has the use of his eyes. Probed or cut into in some way the pain may be relieved, but in the absence of strict antiseptic measures sight may certainly be lost through the action of the purulent discharge. More than anything else it is necessary to recognize that, in the first instance, the problem belongs to the history of law.

One might, with the greater confidence, think that eyes which were "full of flesh" described the serious condition of tumours if a difficult word *alikam* was more understandable in this connection. But in that an *ašītu* might turn itself into a *ṣillu* or "shadow" (AMT 9,1 32)—the latter being rightly seen by the Chicago dictionary to mean an opaque spot or discoloration in the eye—more certainty attends the subject of ulcers. Indeed this observation provides an essential clue for the meaning of the etymologically related term *ašû* when applied to other parts of the body, while the fact that "shadows" could occur in the *lamassat īni*, approximately "guardian of the eye," suggests that this term denotes the cornea. Probably of no great consequence was the *še'u*, literally "barley-corn," which has been previously thought to mean a "stye," but more realistically will have been a very small cyst. A whole Tablet in one of the physiognomical omen series is devoted to the meaning of *še'u*-cysts when found on different parts of the skin, and for the eye these are said to arise in one of the "corners" (inner or outer canthus), in the surface epithelium (*quliptu*, from a verb *qalāpu*, "to peel," and which may itself suggest that shredding of the epithelium was a common observation), or on the eye borders. It is not likely to have been a subject for treatment. Amongst other conditions squint was variously described, exophthalmos was expressed by a verb *zaqāpu*, "to stand out" or "protrude," while a unique case of short-sightedness in which "a man can recognize another (when approaching) only from a distance of between sixty and thirty *gar*," —the *gar* being a linear unit of twelve cubits, or approximately twenty feet—features in a text published by Köcher and Oppenheim, 1957.

"Ear disease" appears to have been largely a matter of tinnitus and varying degrees of deafness, but for a probable description of suppurative otitis media see Thompson, 1931; Sigerist, 1951.

PNEUMONIA

It is a misfortune that lung diseases in general are very difficult to isolate in the Babylonian record. This is not for want of evidence (*cf.* Thompson, 1934) but more because of ancient failings in the way of differentiation, and our inability to do anything much with mere references to cough, chest pain, expectoration or constriction. Fully conscious of this disappointment, Sigerist, 1951, nevertheless sought to see bronchitis in the syndrome "If the patient suffers from hissing(?) cough, if his wind-pipe is full of murmurs, if he coughs, if he has coughing fits, if he has phlegm. . . ." But the probability is that this is not a syndrome at all, and that the little word "or" must be mentally supplied four times before each of the "ifs" following the first. One then has a kind of block treatment for five separate disorders, a procedure for which there are other parallels in Babylonian medicine.

The first of all lung diseases, however, must have been pneumonia, for despite its sub-tropical rating the winter nights in Babylonia are bitterly cold, the local inhabitants may not have been any more circumspect in their choice of clothing than they are to-day, and pneumonia has always dogged the steps of sufferers from deficiency or parasitic disorders. For such reasons what the Babylonians simply called "lung disease" (*muruṣ hašê*) must surely have included pneumonia whatever else it may sometimes have meant. The suggestion at least appears to suit all occurrences of the term, and it may be particularly significant that in AMT 55,1 6, "lung disease" is cited as a possible sequel to *bu'šānu*, or scurvy, since, in the latter, there is a predisposition to bacterial infections and especially to pneumonia.

Perhaps the disease one would most like to know more about is *šu'ūlu limnu*, or the "Evil Cough." It was evidently serious—hence the capital letters may be considered legitimate—and certainly one context is against its meaning pneumonic plague. Whether it could turn out to be pulmonary tuberculosis must depend upon further evidence, but this is perhaps the place to say that all other approaches to this vast subject appear to be presently frustrated. The following section suffers accordingly.

MENINGITIS

A "disease" called anciently *qāt etemmi* or *ṣibit etemmi*—the two terms appear to be virtually synonymous—may be suggested here as being history's first pronouncement on the subject of meningitis. By the latter term we mean meningitis in all or many of its varieties and associations so that by modern standards the idea is something of a disease complex. The word *etemmu* means a "ghost" and it is not known why the condition should have been thought of as a "ghostly attack." It is well possible that the Babylonians did not know either.

The symptoms of the disease are not given all together in one place. This might have happened if the treatise *Sakīkū*, so constantly referred to in these pages by its editor's abbreviation of TDP, had been a book on diseases, and its purpose to describe them. In fact its symptoms and "fragmented syndromes" are largely classified according to unknown principles, and often a number of references needs to be collected in order that the full picture of a disease may be pieced together. When this is done for *qāt etemmi* the following details emerge.

The ghost diseases were classified in the therapeutic texts under disorders involving the neck, and painful neck was a symptom (AMT 47,3 iii 20; 97,4 18; TDP 34 15), also pain in the back (TDP 112 16). The neck specifically might be "drawn in" or "retracted" (*karû, kitarrû*, TDP 76 62; 84 32), a word one may use in preference to Labat's "contracté." Equally characteristic was the symptom usually expressed by the ideogram SAG.KI.DIB.BA or "(severe) frontal headache"—whence Köcher's "migräne" (BAM I XIV) is at least the right general idea—and no less than ten successive entries in *Sakīkū* (TDP 32 10ff.) begin with this symptom in the descriptions of *qāt etemmi*. This headache might persist without interruption "from sunrise to sunset" (TDP 34 13). At the same time the ears might "roar" (*šagāmu*), and this symptom is the first to be given in an explanatory text published by Thompson, 1924. The patient could also become "hot and cold" (TDP 34 17), that is, feverish—the same phrase is still used even to-day in Arab countries to describe a fever—and AMT 88,4 + 96,8 expands on this theme by saying that the patient "is hot and cold, his (that is, the ghost's) terror is so close to him that he cannot rest by day or night, and his voice does [not] appear to be(?) like his own voice." In this passage there is probably enough that is certain to suggest a condition of delirium.

Against this background it is unfortunate that the observation of TDP 86 53: "If his neck is retracted [properly 'fallen (back)'], and if his legs are drawn up and he lies sleepless," should be assigned a different diagnosis to *qāt etemmi*, for this is the typical position of the patient with meningitis who, lying on his side with legs drawn up in flexion at hip and knees, thus tries to lessen the tension on the inflamed nerve roots of his back. But the next line which is concerned with neck retraction and neck pain does indeed state the disease as being *qāt etemmi* so that we may be encouraged to think that the two entries are in some sense parallel to each other. Not unequivocal is the insistence of TDP 34 18, as also entries in KAR 182, that vertigo (*sīdānu* and certain associations of the parent verb *ṣâdu*) might be found in a given case, this being an identification promoted by the Chicago Assyrian Dictionary and one which seems to be sound both medically and philologically. A yet further development that might occur is implied by the fact that *qāt eṭemmi* had some close relationship to epilepsy. It is found in several texts in a position immediately, or closely, following the known words for epilepsy, a placing which could suggest either apoplexy and meningeal haemorrhage or else a convulsive state of other origin. Eye pain (TDP 34 16), epistaxis (*ibid.* 14) and most interestingly "grinding of the teeth" (*ibid.* 60 42) are further symptoms of a disease complex for which the general line of interpretation does not stand in doubt.

Finally two long syndromes assist to unite together several of the features already mentioned while at the same time adding a few more. The first is LKA 88 1-8 where the causing agency is given as an *eṭemmu murtappidu* or approximately a "wandering ghost." It is not known what this means but it is likely to have indicated a variety of the basic concept of ghost disease, and up to a point the given symptoms would appear to uphold the suggestion. These include headache, noises in the ears, neck retraction, chest pains, cold sweats, some paralysis of limbs and also mental symptoms of which *ašuštu* is certainly depression and *hūṣ hepi libbi* probably an hysterical reaction.

The symptoms of the second text have been presented by Thompson, 1929, and Ebeling, 1931. The scene is that the sick man "has a frontal headache, his ears 'roar,' his eyes are brightly shining, his neck hurts, one of his sides is paralyzed, his kidneys are affected, his mind is confused, and his legs 'contain *rimûtu*' (here perhaps indicating a flaccid condition of the limbs)."

In this syndrome significant details are the brightly shining eyes, indicative of contracting pupils; an element of delirium(?) suggested by the disturbed mental state; the hemiplegia, a virtually certain interpretation since both noun and verb are in the singular; and lastly the unmistakable signs of meningeal invasion. In that the cause of the illness was stated as being due to yet another kind of "ghost" (*eṭem rīdâti*) there is additional support for the prominence of the meningeal symptoms. At least as a platform for argument it may be suggested that the case was one of leptomeningitis.

EPILEPSY

It will be unnecessary to elaborate unduly in this section on the symptom complex of epilepsy which has been recorded from practically every country where medical records have been kept. But it may nevertheless be interesting to catch some glimpses of it through eyes as yet not accustomed to understand what they saw.

The two words most concerned are *antašubbû* and *bēl ūri* which commonly occur together and which are not doubtful as indicating respectively the major and minor epileptic attack. The latter is defined in the explanatory text of Thompson, 1924, as a condition in which "the patient's eyes, right and left, are inverted (or, turn inwards)." That they might, in fact, also roll upwards is suggested by the name itself, for the demon *bēl ūri* was the "lord of the roof-top [or, roof-beam(s)]," and nothing will suit better than to think that, to the observer, it was as if some influence at this height was temporarily drawing the attention of the patient. By comparison far more was written about *antašubbû*. The opening lines of *Sakīkū*, Tablet X (TDP 80 1ff.) describe the head of the fallen patient being turned to the left or right, *amšā* of hands and feet (that is, clenched fists and similar inversion of feet with bending of toes), deviated eyes (*nabalkutā*), saliva exuding from the mouth, and the condition *iharrur*, probably an onomatopoeic word and descriptive of the stertorous breathing which follows upon the period of apnoea. As I have attempted to show (1957), differentiation appears to have been made between tonic and clonic forms, for a new text relates that an epileptic might fall, "his eyes remaining wide open . . ., but without involuntarily jerking hands and feet [*sc.*, as in the clonic phase of the major attack]."

The rest of the story is told in Tablet XXVI of the *Sakīkū* treatise, a Tablet which is entirely given over to the symptomatology of fits, epileptic and other. This Tablet is represented by STT I 91, and an as yet unpublished tablet, BM 47753. Interesting here is a first reference to auras for Obv. 24 states that a diverse "pricking" sensation may be felt in trunk or limbs in advance of the attack. For such paraesthesiae the verb *zuqqutu* which was used is altogether appropriate, for it indicates a plural aspect of *zaqātu*, "to sting or prick." A different kind of aura, not fully understood at present (Rev. 10), is of interest because of the subsequent conclusion that "a major attack (*antašubbû*) will overtake him in the *nīdūtu* [that is, in open country not built up with houses], or in the corner [of a house]." In this pronouncement the logic is not perfect, but the places mentioned are typical sites for the epileptic to seek in his concern to avoid publicity for the ensuing event. A further entry (Rev. 18) informs that a case where the patient "cries u_s-*a-a-i*" (thus giving the actual transliteration of the signs), "or [?] cries like an animal, saliva flows from his mouth and his neck is pressed down to the left, is *antašubbû*." That the word "left" is immaterial may easily be forgiven: but it must surely be that in u_s-*a-a-i* we hear the first epileptic cry in medical history.

It should be mentioned that, within this setting, a word *bennu* has caused some trouble. It is commonly found on slave contracts—the previous owner guaranteeing the slave against *bennu* for a stated period of time—and although its original identification with epilepsy was not based on any more sufficient evidence than contextual suitability and the parallels of Graeco-Egyptian slave contracts, the suggestion has since gained support from a lexical text which explains the Sumerian, and original, word *an-ta-šub-ba* by *miqtu* (approximately, "the falling disease") and *bennu*. Despite this equation, the symptoms now given in STT I 89, iv, 192-5, point in another direction. The relevant passage presents some difficulty but appears to say, "if his [the patient's] chest often hurts him, and however [?] much it hurts or however[?] long it retains the pain . . . [text not understood] and he becomes nervous and fearful, a *bennu*-attack has seized him. For such a man an attack may occur at a gateway, at an animal enclosure or at a river." This is not epilepsy but probably a heart condition, excited respectively by a slope(?) frights or sudden noise(?), and cold. Since the duration of the attack appears not to have been diagnostically important, one could perhaps think that the text speaks for angina pectoris and coronary thrombosis considered as one condition.

It is of interest that a therapeutical text published in STT I 57, mentions, in line

31ff., the procedure to be followed for a case in which *antašubbû* seizes a baby, but in default of supporting details the diagnosis can hardly be pressed beyond that of infantile convulsions.

CHILDBIRTH AND CONGENITAL DISORDERS

Findings under this heading will contain little that stands in much need of commentary. In pregnancy, miscarriage caused by some event in which a second party was involved was the concern of the law; but from his clinical experience the ancient priest recorded recurring fits of vomiting on the part of the future mother (TDP 208 81), an observation which suggests toxaemia of pregnancy, and he noted also bilateral oedema of the feet and ankles (TDP 206 71) which was doubtless either the same condition or else may be explained as caused by pressure on the veins by the enlarging uterus. A yellowness of the face (*ibid.* 200 1) would probably have had more to do with anaemia than anything else: it certainly did not mean, as the record was pleased to think, that the child would be a boy.

In normal circumstances childbirth is likely to have been performed by means of the birth stool of which clay models have been found, and a midwife (*šabsūtu*) will often have been in attendance in urban communities. The main concern was undoubtedly that of difficult parturition, described in the texts as "If a woman has difficulty in giving birth" (*e.g.*, AMT 67,1 iv 6ff.). Treatment by means of Solanum berries—a virtually certain identification —tells something of what this meant, for the *Solanaceae*, which are poisonous, are well known to have antispasmodic properties. When the worst happened and the child died in the womb there is sufficient evidence from two literary texts (the Epic of Creation, VII 43, and a new text of the Legend of Etana, Tablet I) to indicate that extraction (*nasāhu*) of the foetus was attempted using a *kakku*, or some kind of surgical instrument, probably in much the same way as Celsus so beautifully describes in his *De medicina*, VII. 29. That this was done is not remarkable in view of the urgency of the situation; but more ambitious was the Caesarean section which might be performed on a mother who had died in labour and where it was suspected that the child was still alive. The evidence for this statement, which is quite certain, comes from a legal text of the Old Babylonian period published by Oppenheim, 1960. The writer has since informed me that other such texts are now known, and in that the mothers were all slave-girls the practice appears to have been prompted more by economic than humanitarian reasons.

Passing to the birth itself, it is of interest that there is good evidence for conjoined twins in Mesopotamian antiquity. The source of such information is an omen series known as *šumma izbu*, literally "If a foetus . . .," which is concerned with observations on human, as also animal, offspring and whether or not delivered at full term. A new edition of this work is promised from Chicago, but in the meantime one may quote CT 27,4 16 with 6 12: "If a woman gives birth to twin male children and their *libbu* [approximately 'internal system'] is common to both;" or again (same texts, lines 20 and 16), "If a woman gives birth to twin male children and they are joined at their backbone"; or again (*ibid.* 23 and 20), "If a woman has given birth to twins and they are attached by their ribs." The first of these conditions is not quite certain since *libbu* is a somewhat indefinite word; the second suggests union at the sacrum; the third may be xiphopagus. In the latter condition junction is specifically at the lower sternum, but this would in fact mean that a majority of the ribs would be joined also. What the same series has to say concerning monstrosities has not yet been scientifically analyzed.

Congenital disorders are often difficult to recognize as such, but congenital blindness is mentioned in the *šumma izbu* omina, and second millennium economic texts from Chagar Bazar in northern Mesopotamia name certain blind girls or women who, as makers of reed mats, were the historical predecessors of the basket-maker. For technical reasons the word *hašikku* has been established as meaning "deaf-mute," and the words *dunnamû*, *lillu* and *ulālu* belong in the description of aments and congenital idiocy. Among speech disorders *hasû* and *šassā'u* qualify to denote persons with a tendency to lisp or sibilate, both words being possibly onomatopoeic in origin. Also interesting is the entry of TDP 136 63, "If his testicles are atrophied and he has a continuous stammer and bites his lips," for one could think that even the ancient writer was prepared to relate the stammering and sexual inadequacy. In this sentence the word translated "atrophied" literally means "rolled up tight" (thus "reduced in size"), and the condition could also mean that the patient "might have intercourse with the high-priestess of his [city-]god" (TDP 136 62). This is quaint language, but in that the high-priestess was not permitted to have children, the obvious inference is that the phrase, not to be taken literally, represented the Babylonian way of describing sterility. Nothing can yet be said from the medical point of view on the subject of dwarfs.

TYPHOID AND DIPHTHERIA

The story of typhoid will take us into the world of folk medicine. An Old Babylonian Lamastu text edited by von Soden, 1954, describes this demoness, already encountered in connection with ascariasis, in the following curious way:

She is small of hand but long of fingers, Her finger-nails are long and her knuckles[?] . . . . She has entered the house, skirting the door socket, She has slipped in by the door socket to kill the child. Seven times on his belly has she struck him.

The text continues with a direct command to the demon to withdraw her finger-nails from the child. . . .

This may not at first sight look like anything very specific, let alone typhoid: but it should first be mentioned that Lamastu was above all responsible for diseases of the young and this would be one of the proper age groups in which to seek typhoid should there be any suggestion that it was an ancient disease in Babylonia. Understandably many text-books present the disease largely from the aspect of an unprotected person entering a typhoid zone in adult life, but the picture for native communities living in areas of permanent risk is rather different. Among such an apparent immunity to typhoid diseases in later life has been noted, thought to arise solely as the result of widespread early infection in childhood. In any event the typhoid group occupies a first place amongst infectious diseases of modern Iraq, and the "Bulletin of Health and Vital Statistics" published annually by the Iraq Government shows typhoid as a known cause of infant deaths in the country, even during the first year.

As to the diagnosis, all turns upon the "fingernails" pressing into the abdomen of the child, for in typhoid the diagnosis is never certain until there first appears in this region a sparse rash of "rose-spots." These are 2 to 3 mm in diameter, and the text-book number for such initial lesions is six to ten. The passage mentions "seven," but if a single number had to be chosen seven would be quite appropriate since this has always been a somewhat vague numerical indication in Semitic thought. In terms of the diagnosis both this number and the site of the spots serve to distinguish it alike from paratyphoid and from typhus.

Diphtheria is again very largely a disease of children and early life, and it is well known in Iraq. It is, in fact, far more of a "tropical" disease than is often supposed, having a known antiquity in the East and being familiar to Aretaeus under the name of "Egyptian" or "Syrian ulcers."

One would have liked some reference to tonsillar membranes before suggesting that Mesopotamia lay within the ancient boundaries of the disease, but a tentative diagnosis is not quite impossible without it. What is provided has to do with a disease(-complex?) known as *qāt Gula*, described in the Fortieth and last Tablet of *Sakīkū* which is given over to the symptoms and diagnoses of infantile disorders.

In this disease (see TDP 228 90ff.), the understandable symptoms are that the baby shows signs of suffocation (*unappaq*), its "larynx is constricted" (*urus-su haniq*), it will not feed at the breast, it is burning with heat and it is seized with *bu'šānu*. The latter condition, properly "the stinking disease," has been already discussed under scurvy. But diphtheria is a "stinking disease" also, and it is particularly the combination of angina, laryngeal constriction, high temperature and an associated foetid smell which makes the case for the diagnosis. That the evidence thus presented speaks more for laryngeal diphtheria than for other forms can be no more than chance. Since in TDP 84 29 and 34 the phrase *urus-su iharrur* will mean "laryngeal stridor"—*iharrur* being the same word as was used to describe the harsh, laboured breathing of the epileptic —the translation "larynx" would appear certain. The most that is known about tonsils is that they are perhaps represented by the word *nipkū* (TDP 84 n. 163).

POISONS

This short section need initially do no more than call to mind the menace of snake-bite and scorpion sting. Sundry texts refer to both. So far as snakes are concerned the picture for Iraq has been clearly presented by Landsberger, 1934, the authorities there cited, and Corkill, 1932 and 1939. Omitting the hoodless cobra, *Naja morgani*, which is recorded only East of the Tigris, the two poisonous snakes of the country are the Levantine viper, *Vipera lebetina*, and the horned viper, *Cerastes cornutus*. That the first of these may have been indicated by a word *kurṣindu* or *kurṣiddu* is more probable than certain, but no doubt attaches to the identification of *ṣēru qarnu*, "the horned snake" with the Cerastes. It is small in size with a prominent triangular head, and on the Lamastu amulet reproduced earlier in this chapter it is represented by the "triangle and tail" shown to the right of the field. For the scorpions of Iraq the latest statement is that of Pringle, 1960. In this study the writer shows that ten, or possibly eleven, species inhabit the country, and that of these only two produce serious stings. In antiquity, as to-day, the greatest risk will have been to children.

Outside of animal poisons it is likely

that the *kamūnu,* "mushroom" and an as yet unidentified fungus, *katarru,* caused trouble in their time, for both words are included in the "list of diseases" of CT 19, 3. But from the medical point of view greater interest attaches to the symptoms of a toxic state of alcoholic origin (text in Küchler, 1904). The case concerns a man who "has been drinking *šikaru*-beer, his head is affected, he has some amnesia, is incoherent in talking, his attention cannot be held, and his eyes remain motionless [ophthalmoplegia]."

In such a condition there was obviously some kind of alcoholic poisoning at work, and if one were to think that it arose after but a single night's *divertissement* one might incriminate some by-product of the brewing for the ocular palsy involved. But, in that the priest or physician would hardly have bothered to elaborate on ordinary symptoms of drunkenness which he would well have known and which do not persist, it is preferable to think that the ophthalmoplegia belongs rather to Wernicke's encephalopathy, and that the attendant amnesia and disorientation are features of Korsakow's psychosis. So interpreted, the record stands at a far distance from 1881 when the condition was first described, but it is not one which will have known many boundaries either in space or time.

LEPROSY

The problem of leprosy in different parts of the ancient world can easily become something of a philologist's nightmare if one would wrestle too much with the possible meanings of mere words. More than with any other disease a simplicity of approach is almost a first requirement.

For Babylonia the essential argument concerns a word *saharšubbû* and a known synonym *garābu* which, when it was first seen from the curse formulas of certain Kassite boundary stones to be indicative of any unyielding skin disease which might mean that the unfortunate sufferer had to roam outside the city walls like the wild ass, was translated "lepra" as early as 1877 by Oppert, one of the first decypherers of Akkadian. Belser, in 1894, followed him by seeing the same condition as "eine aussatzartige Krankheit," and so far as translation is concerned Assyriological opinion has not changed since. To-day numerous boundary stones have been unearthed and few details need to be supplied in our understanding of *saharšubbû* as a disease which "covered" or "clothed" the skin and which often required the excommunication of the sufferer (*cf.* Nougayrol, 1948.) That a more lenient view of the matter was taken in the Neo-Babylonian period where one first meets a word *garbānūtu,* translated by the Chicago Assyrian Dictionary as "office of the person in charge of lepers," is a possibility that needs to be considered.

Thus far there is nothing to say about symptoms, the only additional clue provided by the boundary stones (as also one other source) being that the condition had an association with "dropsy" (*aganutillû*). This has nothing to do with leprosy which does not involve the lymphatics, but it is probably not against the general argument in that, forced to scrape a meagre living from the refuse dumps and by begging, many lepers—if such there were—must often have fallen a victim to the additional evil of hunger oedema.

In personal belief only one observation may convincingly elicit support for the opinion that leprosy was an ancient disease in Mesopotamia. This is a statement of the Old Babylonian omen text published by Köcher and Oppenheim, 1957, which, literally translated, says that "If the skin of a man exhibits 'white *pūṣu*-areas,' or is 'dotted with *nuqdu*-dots,' such a man has been rejected by his god and is to be rejected by mankind."

In this passage it seems well possible that the two main types of leprosy are being contrasted, the neural type in which the *pūṣu*-lesions (from a verb meaning "to be white") will represent the characteristic whitish or depigmented macular areas, and the nodular (or lepromatous) type where the nodules will have been the *nuqdū*. In that an economic text mentions "one fattened . . . *gumāhu*-bull, whose limbs are perfect, whose skin has no *pūṣu*(-spot)," it will be seen that *pūṣu's* were not only white (or whitish) but also of a certain size and not raised in the skin. They would seem at least to be as good a claimant for maculae as "skin-dots" are for nodules. The further notice of excommunication—the rejection of man following the rejection of god—provides good support for the proposal.

While the Babylonians knew nothing of tissue resistance no one will think it remarkable that such careful observers, and at so early a period, appear already to have classified the disease in terms of two of its major features. Differential p r o b l e m s there might have been, but it is precisely the simple contrast of types which may be thought to point the diagnosis. One must travel a long way through ancient writings before meeting it again.

ABBREVIATIONS

AMT THOMPSON, R. C.: *Assyrian Medical Texts from the Originals in the British Museum*. Oxford, Humphrey Milford, 1923.
BAM KÖCHER, F.: *Die babylonisch-assyrische Medizin in Texten und Untersuchungen*, I and II. Berlin, de Gruyter, 1963.
CT Various authors and dates. *Cuneiform Texts from Babylonian Tablets in the British Museum*. London, British Museum.
KAR EBELING, E.: *Keilschrifttexte aus Assur religiösen Inhalts*. Leipzig, Hinrichs. 1919.
LKA EBELING, E.: *Literarische Keilschrifttexte aus Assur*. Berlin, Akademie-Verlag. 1953.
STT GURNEY, O. R., and FINKELSTEIN, J. J.: *The Sultantepe Tablets*, Vol. I. London, The British Institute of Archaeology at Ankara, 1957.
TDP LABAT, R.: *Traité akkadien de Diagnostics et Pronostics médicaux*. Leiden, Brill, 1951.

REFERENCES

ABDULNABI, M., et al., 1961: The nutritional value of some Iraqi foodstuffs. *J. Fac. Med. Baghdad*, NS 3: 10.
BAILEY, V. M., 1955: Notes on the incidence of human parasites in Samawa, Iraq. *Bull. End. Dis. (Iraq)*, 1:250.
———, 1957: An intestinal parasite survey in a rural district of Baghdad, *Bull. End. Dis. (Iraq)*, 2:148.
BAQIR, H., 1960: Prevalence survey of Bilharziasis, Ascariasis and hookworm infection in Mussayeb Project. *Bull. End. Dis. (Iraq)*, 3:24.
BLOOMFIELD, A. L., 1958: *A Bibliography of Internal Medicine: Communicable Disease*. Chicago, University Press.
CORKILL, N. L., 1932: *Snakes and Snake Bite in Iraq*. London, Baillière, Tindall and Cox.
———, 1939: Snake specialists in Iraq. *Iraq*, 6:45.
DRIVER, G. R., and MILES, J. C., 1955: *The Babylonian Laws*, Vol. II. Oxford University Press.
EBELING, E., 1931: *Tod und Leben nach den Vorstellungen der Babylonier*, 85. Berlin and Leipzig, de Gruyter.
KELLY, T. D., and IZZI, N., 1959: Pulmonary Hydatid disease in Iraq, with a review of history and life history of the parasite. *J. Fac. Med. Baghdad*, NS 1: 115.
KINNIER WILSON, J. V., 1956: Two medican texts from Nimrud. *Iraq*, 18:142.
———, 1957: Two medical texts from Nimrud—continued. *Iraq*, 19:44.
———, 1962: The Nimrud catalogue of medical and physiognomical omina. *Iraq*, 24:61.
KÖCHER, F., and OPPENHEIM, A. L., 1957: The Old-Babylonian omen text VAT 7525. *Arch. f. Orientforschung*, 18:62.
KRAUS, F. R., 1936: Ein Sittenkanon in Omenform. *Z. Assyriologie*, NF 9:113.
KRAUSE, A. C., 1934: Assyro-Babylonian ophthalmology. *Ann. Med. Hist.*, NS 6:42.
KÜCHLER, F., 1904: *Beiträge zur Kenntnis der Assyrisch-Babylonischen Medizin*, 32. Leipzig, Hinrichs.
LABAT, R., 1954: A propos de la chirurgie babylonienne. *J. Asiat.*, 242:207.
LAMBORN, W. A., 1955: The haematophagous fly as a possible vector of Leishmania. *Bull. End. Dis. (Iraq)*, 1:239.

LANDSBERGER, B., 1934: *Die Fauna des alten Mesopotamien.* Leipzig, Hirzel.

MUAZZAM, M. G., et al. 1960: Hepatic ascariasis. *J. Trop. Med. Hyg.,* 63:95.

NOUGAYROL, J., 1947: Textes et documents figurés. *Rev. J. Assyriologie,* 41:41.

———, 1948: Sirrimu . . . *J. Cuneiform Stud.,* 2:206.

OPPENHEIM, A. L., 1960: A caesarian section in the second millennium B.C. *J. Hist. Med. Allied Sci.,* 15: 292.

———, 1962: Mesopotamian medicine. *Bull. Hist. Med.* 36:97.

PRINGLE, G., 1956: Kala Azar in Iraq: Preliminary epidemiological considerations. *Bull. End. Dis. (Iraq),* 1:275.

———, 1960: Notes on the scorpions of Iraq. *Bull. End. Dis. (Iraq),* 3:73.

SIGERIST, H. E., 1951: *A History of Medicine,* Vol. I *Primitive and Archaic Medicine.* New York, Oxford University Press.

THOMPSON, R. C., 1906: The Folklore of Mossoul. *Proc. Soc. Bib. Arch.,* 28:78.

———, 1924: A Babylonian explanatory text. *J. Roy. As. Soc.,* 1924:452.

———, A 1924: Assyrian medical texts. *Proc. Roy. Soc. Med. Hist.,* 17/1:23.

———, 1926: Assyrian Medical Texts. *Proc. Roy. Soc. Med. Hist.,* 19/3:38, 40.

———, 1929: Assyrian prescriptions for the "Hand of a Ghost." *J. Roy. As. Soc.,* 1929:819.

———, 1931: Assyrian prescriptions for diseases of the ears. *J. Roy. As. Soc.,* 1931:8.

———, 1934: Assyrian prescriptions for diseases of the urine, etc. *Babyloniaca,* 14:108.

———, 1934: Assyrian prescriptions for diseases of the chest and lungs. *Rev. d'Assyriologie,* 31:1.

VON SODEN, W., 1949: Kleine Beiträge zum Verständnis der Gesetze Hammurabis. *Arch Orientální,* 17/II: 365.

———, 1954: Eine altbabylonische Beschwörung gegen die Dämonin Lamaštum. *Orientalia,* NS 23:337.

ZAKARIA, H., 1959: Historical study of *Schistosoma haematobium* and its immediate host, *Bulinus truncatus,* in central Iraq. *J. Fac. Med. Baghdad,* NS 1:2.

Chapter 16

Diseases in the Bible and the Talmud

MAX SUSSMAN

INTRODUCTION

THE Bible, the Apocrypha, the Pseudepigrapha[2] and the Talmud provide a vast amount of material of medical interest. That this was so was recognised long ago and has given rise to an extensive literature, the standards of which have varied considerably. A great deal of this literature is occupied with simple problems of diagnosis and possibly the best example of this is that dealing with the nature of the biblical leprosy.[3] Very little of the literature deals with questions of methodology. It will be the purpose of this review to provide a brief methodological introduction on the basis of which previous work may be assessed and also to give some account of the extent of the material at hand. To do justice to a critical review of the literature would, however, require far more detailed treatment than space here will allow. The review will, in the main, be limited to a discussion of the actual descriptions of disease and some of the many case histories presented in the Bible and the Talmud. Therapeutics, for example, will not be dealt with. A further limitation will be that in the main only human disease will be considered; discussion of a great deal of material of veterinary interest must be left for another occasion.

The nature and structure of the Biblical and Apocryphal literature will be well enough known not to require discussion here and only a brief account of the talmudic literature will be provided. The Talmud may be divided into two general parts, the Mishnah[4] and the Gemara.[5] The Mishnah was edited towards the end of the second century C.E. and is a collection of legal rulings and opinions together with some ethical teachings. The Gemara is a far more extensive work than the Mishnah and is the edited collection of the discussions relative to the Mishnah by the post-mishnaic teachers. There are two Talmuds, the Palestinian and the Babylonian, of which the latter is the more extensive and important. The Palestinian Talmud was probably edited at the beginning of the fifth century C.E. and the Babylonian Talmud somewhat later in the fifth century C.E. The talmudic texts contain the body of fundamental post-biblical Jewish law, together with an admixture of folklore, history, mythology, mysticism, a good deal of anatomy, physiology, medicine, surgery, therapeutics and much else.[6]

METHODOLOGY

The combined consideration of biblical and talmudic disease may be justified in that the Bible and the Talmud, with the exception of the New Testament and parts of the Apocrypha and Pseudepigrapha, represent a continuous tradition but it must also be recognised as quite artificial for many reasons. In the first place the source material was composed over a very long period of time and it is perilous to impute any uniformity of purpose or descriptive criteria to the authors or editors. Though this lack of uniformity may be obvious as between the Bible and the Talmud, it also exists in the separate components (e.g., Old Testament and New Testament) and books of the Bible and the strata of the Talmud. Secondly, complications are provided by the purpose for which the texts were composed. Thus, whereas some parts of the Bible are clearly historiographic in purpose, others are homiletic[7] and yet others are legal[8] or for guidance in ritual.[9] A similar classification would apply to the talmudic literature. The crux of the matter is that few disease descriptions in either Bible or Talmud are intended to be for the purposes of diagnosis and few are, therefore, diagnostic.

Disease descriptions in the Bible and Talmud may be divided into two general types. The first of these may be called the *biographical description* and is a case history in the true sense of the word but with reservations to be noted below. The second may be called the *formal description*. It is a generalised account of a pathological condition and may vary from a brief mention to a lengthy discussion of a disease. Both these types of description have their own inherent advantages and disadvantages. Thus, whereas the case historical account of a disease may appear easier to diagnose than a diffuse formal account, care must be exercised because of the danger of regarding as diagnostic legendary material which may be misleading. This leads to the important point that the historical validity of case disease descriptions must be assessed before a scientific diagnosis may be made. Part of this assessment should consist of an examination of the motive for the writing of the account. A case in point is that of the final illness and cause of death of Titus as given in the Talmud,[10] which relates how a gnat or mosquito[11] crawled up his nose and stung into his head for seven years, suggesting a chronic condition as the cause of death, if nothing more. The brief mention by Suetonius[12] would suggest that the cause of death was an acute fever. Titus' entry into the Holy of Holies in Jerusalem would account for the desire of the authors of the talmudic account to burden him with a terrible death and thus account for the discrepancy. Indeed, Suetonius records[12] that Titus had a sin on his conscience when he died. The significance of the gnat probably lies in the field of mythology. The Talmud also relates[10] that at autopsy something like a sparrow or dove was found in Titus' head and that its mouth was made of copper and its nails of iron. Preuss[13] regards this as possible evidence of a tumour, the copper mouth possibly representing old haemorrhages and the iron nails representing areas of calcification. Suetonius' account seems much more plausible than that of the Talmud, especially if motive is taken into account. The cause of death of Titus remains unknown but is unlikely to have been cerebral tumour.

Formal descriptions are, on the other hand, frequently heterogeneous, that is different conditions having similar presenting signs and symptoms are considered to be one and the same disease. The confusion in the Talmud between pain of gastric origin and that of cardiac origin and *vice-versa*—a confusion still encountered among the lay public—is a simple example. A similar heterogeneity in the Hippocratic writings is well recognised[14] and certainly exists throughout ancient and mediaeval medical literature. In addition, the lack of a technical terminology makes the identification of some formal descriptions difficult in modern terms and in those places where the ancients made use of a technical terminology we have only philological methods of deciphering it, unless a parallel usage can be found elsewhere.

Caution must be exercised in attributing disease conditions to the particular time which may be stated in the text, when the text itself may have been composed many years after the time to which it refers. Thus the several New Testament accounts of cures obtained by Jesus of Nazareth were composed after his death and one cannot be certain that the signs and symptoms described by the writers of the gospels were those observed by Jesus. It is not impossible that fragmentary legends available to the Evangelists were modified to conform with disease conditions known in their own day and in their own experience, but different from those known to Jesus. Moreover, it must be noted that embellishment of a legend gives rise to a heterogeneous description. An important step in any assessment of the sources should, therefore, be an examination of the historical background of the text.

Since the Old Testament descriptions of disease are intended for legal and ritual use, certain curious circumstances arise which have been noted by Preuss[15] in connection with the biblical "leprosy."[16] It is very likely that there was no problem about the diagnosis of the fully developed condition of "leprosy" and that the biblical text is concerned with the initial stages of the disease when diagnosis might be difficult. If, for the sake of argument, we accept such a hypothesis it would be misleading to identify a fully developed syndrome from a description concerned only with an incipient disease. Furthermore, it must be noted that the scientific identification of disease, which is our purpose here, is almost certainly not identical with that of the biblical priest, whose concern was with ritual impurity, a condition with no known patho-physiological basis.

The talmudic sources are somewhat easier to assess than the biblical ones, mainly because their whole purpose is to be explanatory to the Old Testament. Thus, descriptions tend to be more extensive and frequently give sufficient information on which a diagnosis may be based. In using talmudic texts to clarify an Old Testament source it is necessary to remember that a talmudic suggestion may be incorrect. The Talmud almost certainly contains much restrospectively conceived legislation. That is, laws and traditions which were current in talmudic times, were connected to a biblical text, often by devious means, and the idea was allowed to grow up that these laws and traditions originated in biblical times. A similar retrospective attribution of a talmudic description to a biblical disease could be very misleading. However, the commentators of the Middle Ages may be useful in weighing up some talmudic texts since it is believed that some of them were in possession of old traditions.[17]

A considerable portion of that part of the Talmud which is of medical interest is concerned with veterinary pathology and, in particular, with the problems of prognosis in veterinary disease. This is because according to Jewish ritual law an animal is considered unfit for food (trepha) if the disease, injury or deformity from which it is suffering is one from which it is unlikely to survive for twelve months.[18] Apart from this there are certain conditions which make an animal trepha no matter how long it survives. It is very likely that animal pathology was particularly well understood in talmudic times because limited autopsy after slaughter was obligatory and full autopsies must frequently have been carried out. Very strict criteria of health were also applied to the fitness of priests[19] for the tabernacle and temple services. In connection with the rules of priestly fitness a large number of orthopaedic and ophthalmic diseases are recorded.

It is notable that most of the conditions discussed in the Bible and the Talmud are those that may be identified by external observations of the body. This is no doubt due to religious objection to human autopsy that existed in antiquity.[20]

Some of the best talmudic descriptions and discussions of disease in both men and animals can be theoretical and this demands care on the part of the medical historian. The Talmud[21] considers the cases of animals with a wide selection of congenital and acquired abnormalities. Strong independent evidence is required before it can reasonably be asserted that the conditions discussed there were ever seen. It is, unfortunately, sometimes difficult to separate the theoretical descriptions from the ones due to practical experience.[22] Occasionally a statement is clearly in error such as, that an animal which has no kidneys is viable.[23] This is probably due to a complete lack of knowledge of the importance of the kidneys.

In dealing with methodology in general, one final point must be made with reference to the literature. Some authors have made determined efforts to extract the maximum of detail from the texts they have examined and it is somewhat disappointing to realize that some of their expositions owe more to ingenuity than a clear assessment of how much modern scientific superstructure the ancient texts will support.

In the remaining part of this review diseases and malformations will be discussed under convenient systematic headings.

THE ALIMENTARY TRACT

Man has probably always been very conscious of the alimentary canal and our sources mention a number of diseases of interest. A condition of the mouth *ṣaphdina*[24] is described in which anything brought into contact with the molars leads to bleeding of the gums. *Ṣaphdina* was regarded as serious and progressive leading eventually to involvement of the intestines. The disease is regarded as being scurvy.[25] Whether *ṣaphdina* is related to σηπεδών (= putrefaction) is uncertain[26] but such a relationship would support pyorrhoea or secondary infection of the congested spongy gums of scurvy.

There is a reference to toothache and the risk of "white matter" appearing on the gums in the absence of treatment.[27] This is probably pyorrhoea. Dental caries was known from the earliest times and in Proverbs[28] it is regarded as the punishment of the traitor, though this is probably not why Jeremiah complained of the condition.[27] Dental extraction was practiced in talmudic times[30] and the edentulous priest was regarded as unfit for the temple service.[31]

Congenital oesophageal atresia was clearly recognised[32] and known to be incompatible with life but a reference[33] to a congenital perforation of the oesophagus which is compatible with life is not easily identifiable unless it is congenital tracheo-oesophageal fistula. It is possible, however, that we have here an inaccurate description. The muscular and mucous coats of the oesophagus were recognised in cattle[34] and perforation of both these coats was not considered compatible with indefinite survival whereas perforation of either one or other did not have a fatal prognosis. It seems reasonable to assume that the same was thought to be true of the human infant. Since the condition was thought compatible with survival it could never have been seen post mortem in an infant and suggests that the congenital perforation of the oesophagus of the Talmud was a theoretical entity.[35]

A series of conditions referred to in the Talmud as heart diseases are almost certainly really gastric in origin. Thus ke'ev lev[36] is literally "heart pain" but if cardiac pain were really meant it is curious that the dyspnoea of angina and the relation to effort were not recognised. Since the disease did not have a sufficient mortality to be noted it is unlikely to have been myocardial infarction. The simple reference to the severity[37] of ke'ev lev suggests pain of gastric origin as a reasonable diagnosis. Possibly a related condition is ḥulsha de'liba or "faintness of the heart"[38] which is a general faintness associated with lack of food.

The reference to sickness in Deut. 7: 15 is taken by one talmudic teacher[39] to refer to diseases of the gall-bladder. Preuss[40] points out that this interpretation probably depends on the similarity of the Hebrew ḥoli and the Greek $\chi o\lambda\acute{\eta}$. There are no texts which would suggest any recognition of specific biliary disease. Another condition known as ḥoli me'ayim is probably enteritis of varied aetiology and was considered to be a serious condition,[41] sometimes associated with distension.[42] The association of dysentery with fever is noted in the New Testament[43] and the cases in which sudden death occurred[44] are almost certainly Asiatic cholera. It is interesting to note that dysentery was regarded as endemic amongst the priests at Jerusalem.[45]

Infestation with intestinal helminths was known in talmudic times[46] but there is little useful information to make diagnosis possible.

Imperforate anus was recognized[47] and the following ingenious treatment was suggested, for simple cases. Oil was to be applied to the anal cleft and at the point at which a translucency appeared a crosswise tear was made. There is also reference[48] to a condition which is difficult to identify and is termed pik'a which means cleft or fissure. The treatment described provides no clue as to the nature of the disease. Hananel[49] considers it a condition of the anal cleft but Rashi[50] refers it to the anus. It could be fissure-in-ano or pruritus ani and it is not possible to be certain.

Haemorrhoids were well known and are frequently mentioned.[51] The certainty of the diagnosis depends on the reference to straining at stool as an aetiological factor.[52] Rashi[53] suggests that suspension on the buttocks leading to excessive passive opening of the anus is meant. The use of dates in the treatment of haemorrhoids,[54] presumably for their laxative effect, supports the identification.

THE RESPIRATORY SYSTEM

There is very little discussion of diseases of the human respiratory system in the Bible or the Talmud. This is the more remarkable in view of the large body of knowledge that accumulated about diseases of the respiratory system in animals. Diseases of the throat were known[55] and some were considered serious[56] but it is impossible to make a diagnosis.

An epidemic disease known as askara causes many deaths in an outbreak[57] and mostly affects children,[58] the attacks frequently coming on at night.[59] The condition was regarded as beginning in the viscera and finally attacking the mouth.[60] Death was due to asphyxia.[61] Preuss's suggestion[62] that the disease is diphtheria is probably correct.

Blood issuing from the mouth may originate either in the lungs or in the liver.[63] This may include the haemoptysis of pulmonary tuberculosis but what was meant by blood originating in the liver is quite uncertain. It seems most unlikely that bleeding oesophageal varices were known to be connected with disease of the liver.

Difficulty is presented by a mishnaic text[64] which reads *ruaḥ kezarit ba'ah alaw* literally "a short spirit came upon him." H. Danby[65] translates this as lockjaw and in this seems to follow J. Cohn[66] who follows Bertinoro,[67] in the commentary by the latter. The identification as lockjaw must be a mistake and Epstein's translation[68] as asthma seems much more likely to be correct. Epilepsy seems reasonably excluded since it is clearly mentioned immediately before our phrase.

GENITO-URINARY SYSTEM

Fertility was of great importance in antiquity, giving rise to fertility rites in the ancient heathen religions of the Near East. It is not unlikely that a considerable amount of biblical legislation, which was later subjected to careful definition by the Talmud, finds its origin in the importance of fertility in a community of tribal origin.[69] This importance of fertility is the source of a great interest in the genitalia and their abnormalities. Apart from this few diseases of the genito-urinary system can be identified.

A man who had his penis cut off[70] was not allowed to marry an Israelite woman. Traumatic avulsion of the penis may have been common in tribal wars and the references to cases in whom only parts of the glans remained,[71] suggest that accidental partial amputation of the glans, possibly during circumcision, may have occurred in the talmudic era. Another rule, that a man with an abnormal opening in his penis was not allowed to marry into the community,[72] probably refers to epispadias and hypospadias. The talmudic reference to "one who passes water in two places"[73] has been identified as either epispadias or hypospadias[74] but could also have been watering-can perineum, possibly following stricture due to gonococcal urethritis. Either might explain the discussion of the different grades of the condition.[75]

The disease of Lev. 15:1ff. is most probably gonorrhoea. It was a urethral discharge and the indication that it would occasionally decrease[76] shows that a transition from acute to chronic urethritis, possibly with stricture, was recognized. Any doubt that we are dealing with a genital discharge is set aside by the position of the text of this disease description between that of menstrual uncleanliness[77] and seminal pollution (? spermatorrhoea).[78] The biblical gonorrhoea was limited to men which may not be surprising in view of the somewhat greater difficulty in recognizing gonococcal urethritis in the female. A clear differentiation between a urethral discharge and semen was made in mishnaic times[79] in that erection is said to be required only for the production of the latter.

Unilateral and bilateral cryptorchidism was recognized[80] and either condition disqualified a priest from temple duty. The description "if he has no stones"[80] could, of course, cover the case of castration. It is not possible to ascertain whether it was known that in human cryptorchids the testes were undescended and atrophic and not in fact missing, though in the case of animals the anatomy of the scrotum and testes was well understood.[81] Hydrocoele is probably meant[82] when describing one with abnormally large testes reaching down to his knees, but almost certainly includes cases of hernia into the scrotum.

The treatment of stone in the urinary bladder is discussed in the Talmud[83] and Rabbi Judah the Prince, the editor of the Mishnah, is said to have been a sufferer.[84] The suggestion that the patient should urinate on the doorstep and watch for the stone is ingenious.[83] The Midrash[85] refers to prostatic obstruction when stating that old men have to strain to pass water and that as a result they occasionally defaecate before they can urinate. Retention of urine was a great fear and people were advised to pass urine at the first urge since any delay might lead to retention and swelling of the abdomen.[86]

Two forms of intersex were recognized. The first was the *androgenes*[87] of whom it was not possible to tell whether they were men or women. Most *androgenes* were believed to ejaculate semen or to menstruate,[88] though it was believed that some were able to do both.[89] This is clearly a description of the human hermaphrodite though one cannot take the details too seriously. The second condition was that of the *tumtom*, a person whose sex was unknown until he was "cut open."[90] Some of the cases under this heading were cryptorchids[91] but others[92] were clearly some form of intersex.

THE EARS AND NOSE

Abnormalities of the ear disqualified the priest from duty and one can recognize several deformities of the pinna, e.g., the cauliflower ear,[93] the abnormally small ear,[93] atresia of the external auditory meatus.[94] Deafness was, of course, recognized in biblical times[95] and it is instructive to note that the talmudic teachers knew that mutism is frequently due to deafness.[96]

Certain abnormalities of the nose also disqualify the priest, such as absence of the bridge of the nose.[97] Polyps are mentioned in the Mishnah[98] and Preuss[99] considers that nasal polyps are meant.[100] Anosmia may have been known and is suggested by the test applied to a woman who claims to be unable to smell.[101] Epistaxis is mentioned in the Talmud[102] together with the suggestion, amongst others, that the sufferer should have cold water poured over his head.

THE EYE

Of all diseases, blindness was much feared in olden times and it provided an example of suffering which was used to teach people charity. Congenital blindness,[103] blindness in one eye[104] and blindness in both eyes were all known. The causes of blindness cannot be identified but measures designed to prevent it are described. Thus one was to wash one's hands after meals for fear of getting salt[105] in the eyes, since this would cause blindness.[106] Blindness due to injury was known; interesting examples are blindness due to a fall[107] and after a blow on the head.[108] Another text[109] suggests that it was known that caustic substance could cause blindness. A possible example is the disease in Tobit.[110]

Several ophthalmic diseases are mentioned in the Bible[111] of which *dak* is some form of corneal opacity.[112] Conjunctivitis is probably intended when the Talmud[113] speaks of discharging, pricking, congested, watering eyes with inflammation. The *ra'atan*,[114] a disease in which there was lachrymation, a thick mucous discharge from the eyes and nasal discharge, may have been trachoma. Rabbi Johanan warned of the danger of the flies which feed upon the discharge—remarkably enlightened advice.

THE BLOOD

Diseases of the blood were not recognized as such in biblical or talmudic times and are considered together here for the sake of convenience.

There is considerable doubt about a condition in the Bible[115] called *yerakon*. Preuss[116] ingeniously, and probably correctly, identifies the disease as some form of anaemia. The green neonate[117] may have been a case of icterus neonatorum and it is interesting to note that circumcision was delayed in these infants for fear of fatal haemorrhage. In connection with circumcision we also have a clear recognition of haemophilia.[118] If two boys of one mother died as a result of circumcision then the third was not to be operated upon.

THE NERVOUS SYSTEM

The nervous system is, perhaps, the least recognized system where the ancients are concerned. The Bible and Talmud are no exception, and there is not a single case in either of a disease in man specifically referred to the brain, spinal cord or nerves.[119]

Headache was a frequent complaint and was much feared;[120] it was not unknown in biblical days.[121] An interesting idea is recorded in the name of Rabbi Judah of Siknin, that in twin sisters (? identical) the one feels the headache of the other.[122] We may note here that the post-alcoholic hangover is not a modern disability.[123] The demon *palga* of the Talmud[124] was regarded as migraine in the Middle Ages.[125]

Cerebral haemorrhage is clearly indicated in the apocryphal description of the death of Alkimos.[126] The disease of Philopater[127] is more uncertain but suggestive of Parkinson's disease though his survival and continued activity speak for a brief seizure possibly of grand mal. The many cases of paralysis in the New Testament[128] are quite impossible to diagnose but the rapid cures suggest that some at least were hysterical in origin, others[129] may have been simulated paralysis for the purpose of begging.

A frequently noted condition is epilepsy and only a few cases can be discussed here. Bileam[130] described himself as "fallen down, yet with open eyes," a condition which Preuss[131] suggests is consistent with epilepsy. The New Testament clearly describes several cases of epilepsy[132] though the cures obtained are difficult to understand if the diagnosis is correct. The Talmud also knows of epilepsy.[133]

An interesting case of seizures is that of Paul of Tarsus.[134] Paul's condition has been widely accepted as epilepsy but this diagnosis is fiercely contested by Fenner[135] who claims that some of the signs and symptoms of genuine epilepsy are missing. In particular, the absence of intellectual degeneration and lack of reference to sleep after the seizure are noted. Though it must be admitted that the New Testament does not paint a picture of Paul as a classical epileptic, it has to be remembered that all we have is Paul's own description of his suffering. When this is taken into account, a diagnosis of epilepsy again seems reasonable and so far as Fenner's above noted criticisms are concerned we have to note that epilepsy can probably be associated with great intellectual distinction[136] and that the amnesia after a fit would make Paul's account quite unreliable. In a difficult case of this sort it is wisest to reserve judgment and note that Paul's blindness may have been hysterical.

THE SKIN AND HAIR

Much ink has been spilt on the subject of the biblical "leprosy" and, paradoxically, the only thing that is reasonably certain is that it is not a uniform description of clinical leprosy as known today. Examination of the patient by the priest was carried out every seven days[137] in which time leprosy would not have changed sufficiently for the change to be recognizable. In addition, the description of the disease[138] does not include any reference to the severe mutilation due to leprosy. The quarantine to which sufferers were subject[139] has led to the belief that the text in Leviticus is concerned with public health measures in true leprosy. It has to be noted, however, that if the disease covered the patient completely, he was allowed to remain in the camp[140] suggesting that the quarantine was not, in fact, a public health measure.

A closer examination of the text suggests that biblical leprosy was a heterogeneous condition and that, perhaps, all the then known skin diseases were divided into two groups, those that made the patient unclean and those which did not. Possibly only chronic skin diseases were included in the classification. The whiteness of the skin[141] is suggestive of vitiligo or leucoderma. The condition in which patients had a snowy white condition of the skin[142] from which they recovered after a time is possibly not leprosy but psoriasis. The condition called Shehin[143] which is also the name of Job's skin disease may be eczema.[144]

Preuss has reviewed the literature up to 1923[145] and provided an analysis of the whole complex of diseases mentioned in Lev. 13.[146] He is of the opinion that some of the conditions in Lev. 13 are true leprosy and the reader is referred to Preuss[13] for an exhaustive discussion.

A condition called enabta is mentioned in the Talmud[147] and is regarded as a "messenger of the angel of death." The name and treatment suggests that it was a berry-like tumour which could be white or black. It is tempting to regard the black form as malignant melanoma though Jastrow[148] tentatively suggests that it was a carbuncle. Preuss[149] connects enabta with enab a disorder of the eye[150] and therefore suggests that enabta is also a disorder of the eye. Suitable support is adduced from Galen's use of the term $\sigma\tau\alpha\varphi\dot{v}\lambda\omega\mu\alpha$ for a condition in which the pupil is subject to tumor and dolor, is swollen like a grape and turns white. Clearly, it is impossible to be certain what sort of disease enabta really was, but it illustrates some of the difficulties which may be encountered.

Loss of hair was thought to be brought about occasionally by shock.[151] Depilatories were known and that their application could give rise to inflammation and its sequelae.[152] Total alopecia in priests was distinguished from partial alopecia in which a fringe of hair remained reaching from ear to ear,[153] but only because the former is unsightly. A case of alopecia, probably physiological, is mentioned in the Bible[154] and also a case of generalized hypertrichosis.[155] The hairy mole was also known.[156]

MENTAL DISORDERS

In antiquity disorders of behaviour were commonly believed to be due to possession by evil spirits. A similar general idea can be traced in the talmudic attitude to mental disease[157] but it is not very strong and on the whole an enlightened view of mental disease prevailed. The identification of the mental disease of the Bible and the Talmud in terms of modern terminology is, however, far from simple.

The Old Testament reports the cases of the madness of Nebuchadnezzar[158] and that of Saul[159] but neither case is identifiable in terms of a specific mental condition. There are suggestions of legend in Nebuchadnezzar's sudden cure after seven years of severe psychosis. The problem with Saul is that apart from the depression he suffered, which in view of the prevailing political conditions was almost certainly not endogenous, there are according to Preuss[160] indications of status epilepticus. He is probably right in thinking that Saul's depression was not pathological at all.

Matters are not any simpler with the New Testament mental disease descriptions. They are so brief and so set on magnifying the miracle of their cure that we can say little about them. It is notable, however, that possession by devils and evil spirits is given as a cause of the disease, an idea foreign to the Old Testament.[161]

The Mishnah deals with mental disease in order to define legal capacity but this gives little in the way of information about types of disease prevalent in mishnaic times. It is notable, however, that it was known that periodic changes can occur in mental disease so that the patient is apparently healthy at times[162] and that disturbances of consciousness and judgement may accompany acute disease.[163] The Talmud also notes that the state of the body influences the mind.[164]

Male homosexuality is, of course, mentioned in the Old Testament[165] and is discussed from the moral and legal point of view,[166] in the Talmud. We have evidence of extensive homosexual practice from various classical sources[167] which probably accounts for the talmudic discussion, while the biblical prohibition is almost certainly connected with prohibition of pagan rites. A similar reason for the prohibition of bestiality[168] may be assumed. Several talmudic souces suggest that bestiality was not uncommon.[169] Tribadism, though not noted in the Old Testament, is mentioned in the New Testament.[170] The talmudic sources suggest that the practice may have been common.[171]

DEFORMITIES, DISABILITIES AND TRAUMA

Under this heading we may note a few curious conditions which are recorded in the talmudic literature.[172]

Levi bar Sisi developed a limp after demonstrating how he could touch his toes with his finger tips while keeping his knees extended.[173] It is probably justifiable to consider this as prolapse of an intervertebral disc brought about by exercise; it is reported that the lesion was permanent.[174]

Traumatic avulsion of the hand in a child brought about by a roller is reported in the Jerusalem Talmud[175] and surgical amputation of the hand as a life-saving measure after injury was known.[176]

Acromegaly is probably indicated by the phrase ". . . whatsoever man he be that hath a blemish . . . hath anything maimed or anything too long."[177] Atrophy and contralateral hypertrophy may be meant by the Talmud[178] when speaking of one with one very strong and one very weak arm. A fear is expressed[179] lest the children of two dwarf parents be midgets.

Gynaecomastia disqualified the priest[180] and lactation in a man is reported.[181] Since the extraordinary nature of this phenomenon is noted it may be justifiable to consider it as real and not legendary.

Among the disqualifying disabilities of the priests there is mentioned umbilical hernia.[180]

SOME CASE HISTORIES

Jacob[182]

When Jacob fought the angel and the latter saw that he could not subdue Jacob, he touched him on the thigh. Jacob limped thereafter but was not totally incapacitated. The prohibition of the sciatic nerve for food which follows in this account suggests that Jacob suffered a lesion of the sciatic nerve. It is not impossible, however, that we have here the vestiges of a legend of how the touch of an angel paralyses, or even the first case of slipped disc!

The Ten Spies[183]

The Talmud[184] reports how the tongues of the ten spies, who gave the bad reports about Canaan, lengthened until they reached the umbilicus and that worms passed from the tongue to the umbilicus and *vice versa*. This is clearly a legendary disease which even now does not seem inappropriate for those who say the wrong thing at the wrong time.[185]

Jehoram[186]

Jehoram was punished by an incurable intestinal disease.[187] Two years later his "bowels fell away" and he died in pain. This is suggestive of a chronic condition of the large bowel leading to a massive rectal prolapse and death from infection and shock.

Antiochus[188]

The text reports how Antiochus was afflicted with sudden abdominal pain. He was then in a manic condition and fell off his wagon, sustaining severe injuries. Later worms grew from his body and there was a painful falling away of the flesh and he emitted a foul odour and died. It seems clear that the episode of pain and the final cause of death were separate entities. The cause of the pain is clearly unidentifiable but the worms, the falling away of the flesh and the odour are suggestive of a foul, suppurating wound.

Rabbi Eleazar and Rabbi Ishmael[189]

These two talmudic teachers were so fat that if they stood opposite one another a yoke of oxen could be driven under their protruding abdomens. Rabbi Eleazar was given a sleeping draught, taken into a marble hall where his abdomen was opened and many baskets of fat were removed from him. This requires no interpretation. It is a report of two cases of extreme obesity with apronectomy in one case.

ACKNOWLEDGMENTS

I wish to thank Professor Scott Thomson for many valuable suggestions. The help of Dr. K. M. Laurence and Mr. R. Shields, F.R.C.S.E., is also gratefully acknowledged.

NOTES AND REFERENCES

1. The following abbreviations are employed:- C.E.= Of the Common Era. J. D. = M. JASTROW, *A Dictionary of the Targumim, the Talmud Babli and Yerushalmi and the Midrashic Literature*, Pardes, N.Y. 1950. M = Mishnah. Other abbreviations, all of which are standard, will be found in the Index Volume of *The Babylonian Talmud*, Ed. I. EPSTEIN, Soncino, London 1935-52 pp. 1-6. Texts employed are as follows: For the Old Testament the translation by the Jewish Publication Society of America, Philadelphia, 1917. For references to the Apocrypha see *The Apocrypha and Pseudepigrapha of the Old Testament in English*, Ed. R. H. CHARLES, Oxford 1963. References to the Babylonian Talmud are to the Vilna Edition published by Rom, an English translation by Epstein is available (see above). For Mishnah references, see DANBY's translation n. 65 below. For the Jerusalem Talmud, see *Talmud Yerushalmi*, Manes, Krotoschin 1866. Midrashic references are to *Midrash Rabbah* Ed. H. FREEDMAN and M. SIMON, Soncino, London 1939. Gen. R., Ex. R. etc., in the notes refer to the Midrash on the appropriate pentateuchal text. Tosefta references are to the version appended to the Code of Alfasi in the Vilna Edition published by Rom. Aboth d'Rabbi Natan will be found appended to Tractate 'Abodah Zarah in the Vilna Edition of the Babylonian Talmud.
2. See, for example, SORSBY, A., (1958): Noah—An Albino. *Brit. Med. J.* ii, 1587.
3. Lev. 13:1ff. see below Section X.
4. *s.v. Mishnah* in J. D. p. 857.
5. *s.v. gemara* in J. D. p. 255.
6. For further detail of the talmudic and midrashic literature see H. L. STRACK, *Introduction to the Talmud and Midrash*, Meridian, N.Y. 1959.
7. Lev. 26:14ff.
8. Ex. 21:1ff.
9. Lev. 1:1ff.
10. Git. 56b.
11. *s.v. yattush* in J.D. p. 603.
12. Suetonius, *The Twelve Caesars*, trans. R. Graves, Penguin, 1957.
13. J. PREUSS, *Biblisch-Talmudische Medizin*, Karger,

Berlin, 1923 p. 236-7. This excellent treatise on biblical and talmudic medicine in all its aspects has not been superseded but is now out of print. It will be referred to below as BTM.

14. See NEUBERGER, M., 1910: *History of Medicine.* Oxford, Vol. I, p. 154.

15. BTM p. 370.

16. See below Section X.

17. See M. LIBER, *Rashi,* Jewish Publication Society of America, Philadelphia, 1938, for information on Rabbi Solomon ben Isaac a leading mediaeval commentator.

18. Ḥul. 42a.

19. Lev. 21:16.

20. See. I. JACOBOVITS, *Jewish Medical Ethics,* Philosophical Library N. Y. 1959. p. 132ff.

21. In Tractate Ḥullin.

22. It is occasionally interesting to follow the lines of development of the legal-cum-ritual aspects of pathology from the Talmud through the legal codes. Thus the Talmud discusses wounds or lesions of the heart and regards any such lesion which penetrates a chamber of the heart as leading to a fatal prognosis. A ruling in Karo's code (*Shulḥan 'Aruk, Yoreh Deah* 40.6), which mentions traumatic avulsion and congenital absence of the heart, may be traced back to the text in Ḥul. 42a. The historian of disease has to learn from this how easy it is to pass from fact to the "hypothetical case" by route of talmudic dialectics.

23. M. Ḥul. iii. 2.

24. A. Z. 28a.

25. So. BTM p. 196; also J. D. s.v. *ṣaphdina* p. 1295.

26. BTM. p. 196.

27. Git. 69a.

28. 25:19.

29. Lam. 3:16.

30. Pes. 113a.

31. M. Bek. vii, 4; Bek. 44a.

32. Nid. 23b.

33. *loc. cit.*

34. Ḥul. 43a.

35. In a difficult case such as this there may be hidden evidence of the occasional autopsy in talmudic times. If this is right, congenital tracheo-oesophageal fistula may have been seen at such an autopsy in an adult or in an infant dying of some other obvious cause.

36. Shab. 11a; Ber. 55a.

37. Shab. 11a.

38. Shab. 10a; Ta'an. 7a.

39. B. M. 107b.

40. BTM p. 215 n.7.

41. Soṭ. 42b.

42. A.R.N. 41:1. The word used is *she'tapuaḥ,* which means "was blown up" (= distended) s.v. *tapaḥ* I in J. D. p. 1685. J. GOLDIN mistranslates this as "was afflicted" in his *The Fathers According to Rabbi Nathan,* Yale Judaica Series No. 10, Yale University Press 1955.

43. Acts 28:8.

44. 'Erub. 41b.

45. J. T. Shek. v. 2.

46. B. M. 107b.

47. Shab. 134a.

48. A.Z. 28b.

49. *ad loc.* A.Z. 28b see margin in Vilna ed. of the Talmud publ. by Rom.

50. *ad loc.* A.Z. 28b; see n. 49.

51. e.g., Ket. 111a; Ber. 55a.

52. Ber. 55a: Shab. 81a.

53. *ad loc.* Ber. 55a.; see n. 49.

54. Ket. 10b.

55. Tosef. Shab. xiii, 8; M. Yom. viii. 6.

56. M. Yom. *loc. cit.*

57. Yeb. 62b.

58. Ta'an. 27b.

59. Soṭ. 35a.

60. Shab. 33a.

61. Lev. R. xviii. 4.

62. BTM p. 179.

63. Git. 69a.

64. M. Bek. vii. 5.

65. H. DANBY in *The Mishnah,* Oxford 1933. p. 538.

66. J. COHN in his translation and commentary to *Mishnah Bekhoroth* in the Itzkowski-Kanel Edition of the Mishnah, Berlin and Leipzig 1887-1933.

67. See commentary by Obadiah of Bertinoro in *Mishnayoth Tifereth Yisrael* Vol. V. Rom. Vilna 1891. *ad loc.* M. Bek. vii. 5.

68. See n. 1. *ad loc.* Bek. 44b.

69. According to Lev. 15:19ff. a woman is considered unclean for seven days after the end of menstruation. If an average normal period lasts five days and another seven days are added, there is a total of twelve "unclean" days. In a twenty-eight day cycle the first possible post menstrual intercourse would then take place about the time of ovulation. By this method a high probability of conception can be achieved. The belief in antiquity that ovulation occurred at the time of menstruation notwithstanding, this whole practice could be an ingenious fertility system.

70. Deut. 23:2.

71. M. Yeb. viii. 2; Yeb. 75b.

72. Tosef. Yeb. x.4.

73. Yeb. 76a.

74. BTM p. 251.

75. Yeb. 76a.

76. Lev. 15:2, ". . or his flesh be stopped from his issue. . . ."

77. Lev. 14.

78. Lev. 15:16ff.

79. Tosef. Zab. ii. 2.

80. M. Bek. vii. 5.

81. M. Bek. vi. 6.

82. Bek. 44b.

83. Git. 69b.

84. B.M. 85a.

85. Lev. R. xviii. 1.

86. Bek. 44b.
87. Tosef. Bik. ii. 2.
88. Nid. 28a; Tosef. Zab. ii.2.
89. loc. cit.
90. Yeb. 83b.
91. Ḥag. 4a.
92. Bek. 42b.
93. M. Bek. vii.4.
94. Tosef. Bek. iv. 5.
95. Ps. 38:14.
96. M. Ter. i. 2: J. T. Ḥag. i. 75.
97. Bek. 43b.
98. M. Ket. vii. 10.
99. BTM p. 340.
100. Following Shab. 109a.
101. B.B. 146a.
102. Git. 69a.
103. M. Meg. iv. 6.
104. Ḥag. 2a.
105. The salt of Sodom was regarded as very caustic.
106. 'Erub. 17b. See also Shab. 108b. and Ned. 81a.
107. Lev. R. xxxi. 4.
108. Tosef. B.K. ix.9.
109. Num. R. vii. 1.
110. Tobit 2:10ff.
111. Lev. 21:20.
112. BTM p. 301.
113. A.Z. 28b.
114. Ket. 77b.
115. Deut. 28:22.
116. BTM pp. 187-190.
117. Shab. 134a.
118. Yeb. 64b.
119. But see Ḥul. 51a for the case of a sheep with paraplegia due to a lesion of the spinal cord. cf. LEIBOWITZ, J. O., 1960: Traumatic (?) paraplegia as reported in the Talmud. Med. Hist., 4:350.
120. Shab. 11a; Tem. 16a; J.T. Ber. ii. 4 and freq.
121. I Chron. 4:10.
122. Pesiḳta d'Rab Kahana, Ed. Buber, Lyck 1868 p. 47a. cited by Preuss, BTM p. 348, n. 11.
123. Ned. 49b.
124. Pes. 111b.
125. s.v. garad in 'Aruch.
126. I Macc. 9:55.
127. III Macc. 2:22.
128. e.g., Matt. 4:24; Matt. 8:6; Acts 9:33.
129. e.g., Acts 3:1ff.
130. Num. 24:4.
131. BTM pp. 341-2.
132. Mark 9:17; Luke 9:39.
133. Bek 44b.
134. II Cor. 12:7ff.; Galat. 4:13ff.; Acts 9:3; Acts 22:6 and 26:12.
135. F. FENNER, 1930: Die Krankheit im Neuen Testament. Leipzig.
136. See for example R. J. Z. WERBLOWSKY, 1962: *Joseph Karo: Lawyer and Mystic.* Scripta Judaica IV, Oxford, p. 284.
137. Lev. 13:4.
138. Lev. 13:1ff.
139. e.g., Num. 12:14.
140. Lev. 13:12.
141. Lev. 13:13.
142. Ex. 4:6; Num. 12:10; II Kings 5:27.
143. Lev. 13:18.
144. BTM p. 390ff.
145. BTM pp. 369-74.
146. BTM pp. 374-90.
147. A.Z. 28a.
148. s.v. enabta in J.D. p. 1091.
149. BTM p. 304.
150. See M. Bek. vi. 2, also s.v. enab in J.D.
151. Ex. R. xxiv. 4.
152. B.M. 86a.
153. M. Bek. vii. 2.
154. II Kings 2:23.
155. Gen. 25:25 and Gen. 27:11.
156. Tosef. Bek. v.2; Ket. 75a.
157. R.H. 28a.
158. Dan. 4:29ff.
159. See I and II Sam.
160. BTM p. 357 and see I Sam. 19:24.
161. Mark 1:21; Luke 4:31; Luke 5:1; Matt. 8:28; Luke 8:26 and freq.
162. Tosef. Ter. i.3.
163. M. Nid. ii. 1.
164. B.B. 16a.
165. Lev. 18:22 and 20:13.
166. San. 73a; M. Kid. iv. 13 and 15.
167. See for example Josephus *Bell. Jud.* IV. ix. 10.
168. Ex. 22:18; Lev. 18:23; Lev. 20:15-16.
169. Kid. 81b; A.Z. 22a-22b; J.T. San vi. 6.
170. Romans 1:26.
171. J.T. Git. viii. 10: Shab. 65a.
172. See BTM p. 266ff. for an exhaustive treatment of lameness.
173. Ta'an. 25a.
174. J.T. Ber. i. 8.
175. J.T. Mak. ii. 1.
176. J.T. Naz. ix. 3.
177. Lev. 21:18.
178. Bek. 3a.
179. Bek. 45b.
180. Mishnah at top of Bek. 44b.
181. Shab. 53b.
182. Gen. 32:36ff.
183. Num. 14:37.
184. Soṭ. 35a.
185. But see BTM p. 211.
186. II Chron. 21:14ff.
187. II Chron. 21:17.
188. II Macc. ch. 9.
189. B.M. 83b.

Chapter 17

Records of Diseases in Ancient China

LU GWEI-DJEN AND JOSEPH NEEDHAM

It is a very long time since any review of this subject has appeared in a Western language, but we are now able to take advantage of a great movement which has been going on during the past fifty years in China for the advancement of the history of medicine in that civilisation. This movement has been closely allied with a revaluation of the practice of traditional Chinese medicine by those who have taken a special training in it. Many valuable works have been written in Chinese on the history of Chinese medical art and science. So far, however, all this material has remained practically unassimilated by sinologists and other Western students of Chinese culture. Thus, for example, most of the dictionary definitions in common use are quite out of date. Among the works which we have used in preparing the present contribution is the brilliant monograph of Yü Yün-Hsiu on ancient nosology, or what might be called pathognostics, the recognition and classification of individual disease entities. Western historians of medicine should be aware that the treatise of Wu Lien-Tê and Wang Chi-Min (K. C. Wong and Wu Lien-Teh) on Chinese medicine (nearly always the only one they know) may be described as the very small exposed piece of an iceberg, 90 per cent of which is "beneath the water," i.e., in the Chinese language and therefore inaccessible to most historians of medicine. During the past fifteen years the study of Chinese medicine has redoubled in activity; a great number of rare medical books from ancient and mediaeval times have been republished in photographic form, and some ancient texts have been reproduced in the modern colloquial (*pai-hua*) style, "translated"as it were from the ancient (*ku-wên*) style, either abridged or complete. We feel therefore that we need offer no apologies for differing from former translations and identifications. Limitations of space in the present work will preclude us from giving any justification of our statements but these will be found in our more extended publication, Volume 6 of *Science and Civilisation in China*.

The sources from which information concerning diseases current in ancient Chinese civilisation during the one and a half millennia before the beginning of the present era are (a) the oracle-bone writings of the second half of the second millennium B.C.; (b) epigraphical (especially sphragistic) evidence in the form of seals and other objects found in tombs during the first millennium B.C., and (c) the texts of the various classical writings ranging from the *Shu Ching* and the *Shih Ching* not long after 1000 B.C. to the first of the great dynastic histories, the *Shih Chi*, completed in 90 B.C., and the great medical classic, the *Nei Ching*, which took its present form probably about the first century B.C. This material provides, all told, a quite astonishing wealth of technical terminology. Although an analysis of it is not yet fully complete, it gives already a firm basis for conclusions as to what diseases were known. Perhaps the greatest difficulty is the imprecision of definition of some of the terms, but in fact they are much clearer than one might expect before one investigated the subject. Moreover the great continuity of Chinese civilisation is not to be overlooked here. Almost unique among the cultures, China possesses continuous traditions of interpretation in this field, directly linking the "sorcerer-physicians" of the second millennium B.C. with the profoundly learned and enlightened medical exponents of the Ming dynasty (16th cent. A.D.).

It would be possible to marshal our material in several ways. For example, purely chronologically, listing texts and their content, or purely nosologically, listing diseases and the terminology relating to them, but both these would be extremely dull and we shall therefore adopt a mixture of approaches. Moreover, we can only give a limited number of examples. We propose to bring the story down to the end of the first century B.C., but in so doing we intend to utilize the *Nei Ching* only in part; we cannot mention all the diseases which are described in that fundamental medical classic. It will be convenient also to consider diseases in the light of the macrocosm-microcosm theories current in early Chinese medicine. The physicians of the Chou period, which lasted most of the first millennium B.C., were extremely conscious of the relation of diseases to geography, to the prevailing climate, and to the seasonal changes of the year. They shared very markedly therefore the Hippocratic conception of "airs, waters and places."

The oldest form of Chinese writing is that which is found on the scapulae and tortoise-shells used for divination in the Shang Kingdom (15th to 11th cents. B.C.). From this was derived the scripts found on bronze vessels recovered from tombs of the Chou period (first eight centuries of the first millennium B.C.). Chinese writing was then stylised into approximately its modern form after the first unification of the empire under the Chhin dynasty in the third century B.C.

The radical *ni*, under which the great majority of diseases were later classified, is revealed by the oracle-bones to have been the pictogram of a bed (Fig. 1). Of the twenty or more medical terms which are found on bronze inscriptions, some four of these are clearly recognisable already on the oracle-bones. For example *chi*, which invariably meant subsequently epidemic disease in general, shows a man alone or lying on a bed with the arrow of the disease shooting into him (Fig. 2). The word *chieh*, in great use afterwards to designate an "itching scabies-like epidemic," i.e., infectious fever preceded by rash, shows again a man lying on a bed, but the spots are actually indicated (Fig. 3). *Li*[1] also means an epidemic fever, and in this case the oracle-bone form seems to show a scorpion (for that is the meaning of the phonetic in this case) occupying the bed alone with little remaining of the patient—or perhaps the maggot-like object is the patient and the scorpion is represented by the little l (Fig. 4). Another term for epidemic disease, *i*, combines the disease radical with a phonetic which is a pictogram of a hand holding a stick (Fig. 5). This however has been found only on bronzes. The last of these oracle-bone terms is the word *nio,* which combines the disease radical with pictograms for tiger and hand; the drawing in Figure 6 is com-

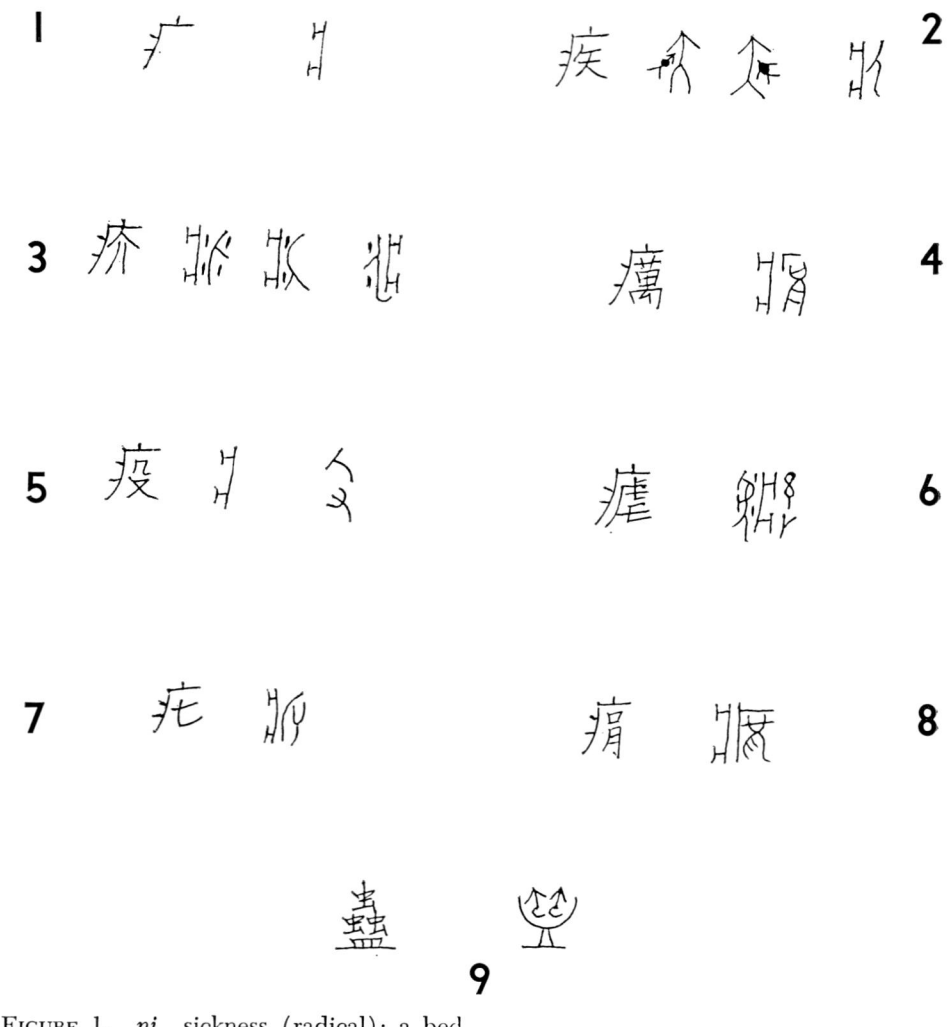

FIGURE 1. *ni* sickness (radical); a bed
FIGURE 2. *chi* epidemic disease; a man with arrows of disease attacking him, or lying on a bed
FIGURE 3. *chieh* "itching scabies-like epidemic"; a man with spots of rash lying on a bed
FIGURE 4. *li*[1] epidemic fever; man, bed and scorpion
FIGURE 5. *l* epidemic fever; bed, and hand holding stick; the patient belaboured by the disease
FIGURE 6. *nio* fever (later more specifically malarial); man, bed and spots, with other pictographic components of unknown significance
FIGURE 7. *pi* thin scabs or lesions on the head; bed, and unknown pictographic components
FIGURE 8. *yuan* arthritic pains; bed, and other pictographic components of unclear significance
FIGURE 9. *ku* poison or disease; insects or worms within a vessel

plex and the significance of it is not clear. This word came in later ages to be confined to fevers of malarial type but in ancient times it was used for all kinds of fevers.

Among the bronze forms we find the word *pi* (Fig. 7), which means thin scabs or lesions on the head, suggesting eczema and lichen or alopecia or psoriasis, for which there were other words used later on. On the bronzes we also find *yuan* (Fig. 8), which signified arthritic pains in the joints.

The medical content of the oracle-bones is, of course, far from exhausted by mere consideration of the few technical terms which had at that early time been developed. Many inscriptions show that divination was made about illnesses without using technical terms. From these we know that there were diseases of the special sense-organs, eyes, ears, etc., dental troubles, speech defects, abdominal diseases, dysuria, diseases of the extremities, including beri-beri-like syndromes, pregnancy abnormalities, and diseases of women and children. We also know of epidemic diseases coming at a particular time of year and causing death. All these they mention without recourse to a technical phraseology. There is one other oracle-bone term of great interest, however, and that is the poison or disease *ku* (Fig. 9). This pictogram indicates insects or worms within a vessel, and while we know that in later ages *ku* did indicate particular poisons prepared artificially by man, there is also reason to think that it referred to a particular disease. This has been identified by Fan Hsing-Chun and others as schistosomiasis, partly because the term *ku* so often occurs in combination with the term *chang* (ku^1 *chang*, ku^2 *chang*) and hence indicates without any doubt oedematous conditions of various kinds, in particular ascites. And the *Nei Ching* describes similar syndromes; certainly in schistosomiasis the liver and spleen become enlarged and ascites occurs when the disease is chronic.

Another valuable source of information about the diseases of the late Chou period (the Warring States period) i.e., the fifth to third centuries B.C., consists of seals which have been discovered in excavated tombs, and these have the further interest that they demonstrate an early development of specialisation in medical practice. Thus, for example, we have the seal of physician Wang, who specialised in speech defects (*yin*); physician Chang, who claimed to be able to cure external lesions (*yang*); physician Kao, who specialised in the care of ulcers (*yung*); physician Kuo, who dealt with oedematous conditions (*tso*), very likely beri-beri; physician Thu, who specialised in removing nasal polypi (*hsi*), and physician Chao, who was expert in psychological diseases ($yü^1$). These are only a few examples taken from those which Chhen Chih has collected. From the Warring States period also we have a large number of records written with ink on strips of wood or bamboo, but these have not so far yielded much of medical interest. Medical material on this medium indeed there is, but coming largely from the army records of the Later Han dynasty (1st and 2nd cents. A.D.) which is later than the period we are discussing here.

We come now to the relation of diseases to the seasons. The *Yüeh Ling* (Monthly Ordinances) is admittedly an ancient text, but opinions differ as to its date; some would put it as late as the third century B.C., when it was incorporated in the *Lü Shih Chhun Chhiu*, as also later in the *Li Chi*, but internal astronomical evidence tends to put it back further (seventh to fifth centuries B.C.). In the course of its description of the activities proper to the different seasons, information is given about what is likely to occur if the weather is entirely unseasonable. Thus if autumn or summer weather comes in the spring, or autumn weather in the summer, or spring weather in the winter, there will be great epidemics (*ta i, ping i, chi i, yang yü i*). In one of these cases the word *li*[1] is used and it is here to be taken as standing for another word *li*[2] and not for *lai*, which it often could do afterwards, for *lai* specifically refers to leprosy. As we shall see later, the first indication of leprosy occurred just about the sixth century B.C. (see Editorial Addendum). Now *chieh* is referred to as one of the evils of spring weather coming in winter, and although this meant scabies from early times, it must here be translated as an "itching scabies-like epidemic." Under any winter conditions exanthematic typhus is perhaps to be suspected, but the descriptions of *chieh* also sometimes include convulsions, with arched back, and speechless "lock-jaw," so that the word may sometimes have been used for tetanus. We shall suggest immediately below a more satisfactory meaning for it. The *Yüeh Ling* text has other interesting features. It says, for example, that if cool spring weather comes in what would normally be a very hot summer, there will be much *fêng kho*, i.e., tonsillitis, bronchitis, pneumonia, etc. It also says that if hot summer weather comes in autumn, there will be many cases of fever (*nio chi*). This is the word later appropriated to malarial fevers, but in the ancient times of which we are now speaking it was simply associated with rapid alterations of shivering cold and hot fever. The text also says that if the hot rainy season continues into the autumn there will be many cases of *chhou chih*, i.e., diseases involving sneezing, such as colds and catarrhs with some fever. The last part of the text says that if spring weather occurs in the last month of winter there will be many troubles of pregnancy, especially miscarriages and stillbirths (*thai yao to shang*). A possible explanation for this might be shocks to the body caused by going out without sufficient clothing. Another feature of this particular kind of unseasonableness was described as a high incidence of *ku*[3] *ping*. *Ku*[3] *ping*, literally "obstinate diseases" or "enfeebling diseases" (*fei*) might be described as those in which the patient is enfeebled and cannot easily help himself; such "handicapped" people were not considered fit to take part in social affairs. The *Kuliang Chuan*, one of the three great commentaries on the *Chhun Chhiu* (Spring and Autumn Annals) of the State of Lu (722-481 B.C.) defines four forms of handicap which prevented social competency. These were *thu*, some kind of skin disease of the scalp, *miao*, some kind of eye defect, possibly ankyloblepharon or Horner's syndrome but more probably trachoma, thirdly *po*, lameness, often no doubt congenital, and fourth *lu*, a term which means a hunchback or a person with arthritic limbs, the descriptions also covering rickets in advanced form, and osteomalacia. We shall mention this again. Texts of the centuries just subsequent to the *Yüeh Ling* (if we may regard it as of the late seventh century B.C.) begin to differentiate clearly between tertian and quartan malaria, the former being generally termed *hsien* or *tien* and the latter *kai* or *chiai*. There is considerable reason, however, for believing that at some of these periods *chiai nio* was a joint expression used for a disease of slow development ending in haemoptysis, which we can identify as tuberculosis.

Another interesting description of the seasonal incidence of disease occurs in the *Chou Li* (Record of Institutions of the

Chou Dynasty). Although much of the material in this book may well date from the Chou period, its compilation must undoubtedly be considered a work of the Early Han (2nd cent. B.C.). It gives a detailed account of what the people of that time considered the ideal democratic organization of the State. Here in Chapter 2 we read as follows:

Each of the four seasons has its characteristic epidemics (li^1 *chi*). In spring there come feverish aches and headaches (*hsiao shou chi*); in summer there are "itching scabies-like epidemics" (*yang chieh chi*); in autumn there are malarial and other fevers (*nio han chi*); in winter there are respiratory diseases (*sou shang chhi chi*).

How is one to interpret these technical terms? No doubt the feverish aches and headaches of the spring refer to influenza, catarrhs, etc., but the "itching scabies-like epidemics" of the summer were certainly far more serious. In the light of the passage which we have just studied in the *Yüeh Ling*, it would seem that cerebrospinal fever (meningococcal meningitis, spotted fever) may have been one of the important components of these epidemics, for the course of the disease links together both severe rash, fever and convulsions. Here epidemic encephalitis is less likely, though it certainly occurred widely in North China down to our own times, and one must also leave a place for scarlet fever and other less important infectious diseases. In the autumn, apart from malaria, one would naturally also think of dysentery of both kinds and gastroenteritis (enteric fever caused by *Salmonella*, etc.) as composing the content of the words *nio han chi*, i.e., epidemics caused by a cold, internal or external. The winter picture almost certainly involved pneumonia, acute and chronic bronchitis, and similar pulmonary affections. This is obviously indicated by the words used, which suggest the rising of the *pneuma* into the region of the lungs, with coughing and difficulty in breathing. Among the epidemics of summer and autumn one would obviously also want to leave place for typhoid-type diseases and perhaps staphylococcal bacteremia, though tuberculosis would hardly have been classified as an epidemic. The word later universal for diarrhoea and dysentery, $li,^3$ does not seem to occur much in texts of this date earlier than the *Nei Ching* itself.

Now let us see what diseases we can find in the texts of the "airs, waters and places" type. In the *Lü Shih Chhun Chhiu*, chapter 12, we read as follows:

In places where there is too much "light" (*chhing*, clear) water, disease of the scalp (*thu*), (alopecia, ringworm, psoriasis, etc.) and goitre (*ying*) are commonly found. In places where there is too much "heavy" (*chung*, turbid) water, people suffering from swellings and oedematous ulcers of the lower leg ($thung^1$) are commonly found and there are many seriously affected who are unable to walk at all (*pi*). Where sweet (*kan*) water abounds, men and women will be healthy and handsome. Where acrid (*hsin*) water abounds there will be many skin lesions, such as abscesses (*chü*) and smaller boils (*tso*); where bitter (*khu*) water abounds there will be many people with bent bones (*wang yü*).

These technical terms are of much interest. The scalp diseases (*thu*) we have already met with, but this is the first time that we encounter goitre, for which the term *ying* is characteristic and indubitable. In the next sentence the term $thung^1$ (more correctly written in medical usage $thung^2$) associaed with *pi*, which means lame in both feet, and bedridden, strongly suggests beri-beri, indeed the wet form. This term occurs again in a much older text, in one of the poems in the *Shih Ching* (approximately 8th cent. B.C.) where it is associated with another word, *wei*, both meaning ulceration of the lower leg. The commentators of the Book of Odes describe it as a disease of swampy places, where, no doubt, the vitamin in the stored grain was destroyed by moulds. The word *wang* resembles that for oedema in general (*chung*), which is to be distinguished from the terms for ulcers—*yung* if oedemtous and unbroken, *chü* open, much worse and generally fatal. The probable identification of beri-beri in the *Shih Ching* as well as in the *Lü Shih Chhun Chhiu* is accepted by Hu Hou-Hsüan and Chhen Pang-Hsien, who indeed find evidence of it as far back as the oracle-bones themselves, but there only with reference to disease of the feet. It is pleasant to hear of one place at least where people were healthy and handsome, but immediately afterwards we learn of places where *chü* were plentiful; *chü* means carbuncles, furuncles and perhaps also cancer, while *tso* refers to smaller skin lesions such as acne. Rickets and osteomalacia are certainly to be recognized in the last sentence. The bronze script form of *wang* is a pictogram of a person with a crooked back, and many famous people of antiquity are said to have been deformed in this way, even the great Duke of Chou himself. $Yü^2$ undoubtedly means hunchback; it occurs in the expression $yü^2$ *lu*, which we find in Chapter 7 of the *Huai Nan Tzu* book (approximately 120 B.C.). There Tzu-Chhiu at fifty-four years of age "had an illness which left his body deformed. He was so bent that his coccyx was higher than his head and his sternum was so lowered that his chin was bent below the level of his spleen." There can be no doubt that rickets and osteomalacia were widespread in ancient China. There are a number of other valuable texts of the "airs, waters and places" type, such as, for instance, those found in the *Huai Nan Tzu* book (Chapter 4) and in the *Nei Ching, Su Wên* itself (Chapter 12), where the endemic diseases are related to the different regions of the Chinese oikoumene, but we have not space to quote them or analyse them in the present paper.

Nosological data in the *Shih Ching* (Book of Odes, c. 8th cent. B.C.) have been analyzed in detail by Yü Yün-Hsiu but there is a special difficulty here because these ancient folk-songs naturally took advantage of poetic licence and it is not always easy to be sure that the disease terms are being used in their proper medical sense; some of them may have been used for malaise or depression in general. Nevertheless *shou chi* (feverish headaches), *shu* (enlarged neck glands, perhaps goitre, tuberculosis, or Hodgkin's Disease) and *mêng, sou* (various forms of blindness) are all of interest. Nosological data derived from the *Tso Chuan*, the greatest of the three commentaries on the *Chhun Chhiu* already mentioned, are more reliable and also much more abundant. More than forty-five consultations or descriptions of diseases occur in these celebrated annals. Perhaps the most important is the consultation dated 540 B.C. which the Prince of Chin had with an eminent physician, Ho, who had been sent to him by the Prince of Chhin. Physician Ho, as part of his bedside discourse, included a short lecture on the fundamental principles of medicine which enables us now to gain great insight into the earliest beginnings of the science in China. Especially important is his division of all disease into six classes derived from excess of one or other of six fundamental, almost meteorological, *pneumata* (*chhi*). Excess of Yin, he says, causes *han chi*, excess of Yang, *jê chi*, excess of wind, *mo chi*, excess of rain, *fu chi,* excess of twilight influence causes *huo chi,* and excess of the brightness of day causes *hsin chi.* The first four of these are subsumed in the later *Nei Ching* classification under *jê ping,* diseases involving fever; the fifth implies psychological disease and the sixth cardiac disease. This classification into six is of extreme importance, because it shows how ancient Chinese medical science was independent of the theories of the Naturalists which classified all natural phenomena into five groups associated with the Five Elements. Chinese medicine never lost entirely its sixfold classification; but that is a long story which cannot be told here. Physician Ho diagnosed the illness of the Prince of Chin as *ku*[1], by which he did not mean the artificial poison nor, so far as we can see, schistosomiasis, but rather some kind of physical exhaustion and melancholia arising from excessive commerce with the women of his inner apartments.

There is interest in every one of the medical passages in the *Tso Chuan.* For example, in 638 B.C. a deformed (*wang*) sorceress, doubtless suffering from rickets or osteomalacia, was to be burnt as a remedy for drought, but a sceptical statesman, Tsang Wên-Chung, intervened and said that other means would be much more efficient, so this method was not used. Two years later we hear of Chhung Erh, the son of Prince Hsien of Chin State, who suffered from *phien hsieh,* i.e., his ribs were so distorted and deformed as almost to meet in front of the sternum. Moved by scientific curiosity perhaps, the Prince of Tshao succeeded in getting a view of him while in the bathhouse. In an episode of 584 B.C., a certain country was described as dangerous for giving people a disease named *o.* Although in this particular case the disease endemic there seems to have been beri-beri, because there is talk of oedematous leg swellings and waterlogged feet (*chui*); we encounter the same term again (*o ping*) in the *Lun Yü,* the discourses of Confucius, dating from about a century later. One of his disciples, Po Niu, suffered from *o ping,* and the universal interpretation of all the commentators since that time has been that this disease was leprosy. We do not find the term *lai* (cf. p. 226) at such an early date, but there seems no reason to reject so old and continuous a tradition that this was the first mention of leprosy in Chinese literature.

Another case relating to 569 B.C. was death by heart disease (*hsin chi, hsin ping*); it happened to a general, Tzu-Chung (Kung tzu Ying-Chhi of Chhu), who was greatly distressed after a military failure, and we may regard it as angina pectoris brought on by anxiety. Soon afterwards the term *shan* came into use to denote this disease, the symptoms and psychosomatic nature of which are so characteristic. We find this word in the *Nei Ching,* used instead of the term just mentioned for the parallel *hsin thung* of the *Shan Hai Ching,* that ancient geographical text which belongs to the middle of the Chou period. In B.C. 565 the *Tso Chuan* notes another case of *fei chi,* some kind of chronic disablement which prevented the normal life of a minister's son. Hydrophobia is also fairly clearly indicated in an entry connected with B.C. 555, where a mad dog (*chi kou* or *hsia kou*) entered into the house of Hua Chhen, a minister of Sung State. The word *khuang* was used indiscriminately for the mad dog itself and for the disease which it caused. Towards the end of the *Tso Chuan* we have a story relating to 497 B.C., in the latter part of the life of Confucius, which includes the famous remark that "only he who has thrice broken his arm can make a good leech."

We have already mentioned the *Shan Hai Ching*. This is a strange book full of legendary material which reached its present form probably about the second century B.C., but which contains much far older material. Many legendary and mythological elements pervade its descriptions of the mountains and forests of the Chinese culture-area, the spirits proper to be worshipped by travellers in any particular region, and also the peculiar plants and animals and their virtues. More than thirty herbs, beasts and stones are recommended to ward off various diseases, and this is where the nosological interest comes in. Many terms we have already met with, e.g., epidemic fevers (i, li^1), epidemics with rash (*chieh*), oedematous swellings (*chung*), goitre (*ying*), rodent ulcers (*chü*) and eye defects, probably trachoma (*mi*). Ku^1 disease is also mentioned. Yu^1 we have not encountered previously; it means both swellings in the neck and also torticollis or palsy. If equivalent to yu^2 or *chan*, the commentators interpret it as paralysis agitans or senile tremor, but it may also stand for an affection called yu^3 *chui*. This consisted not of large swollen lymph glands or the parotitis of mumps but small wart-like tumors on the head, neck and extremities which recall verruca, the multiple warts produced by a rickettsia. Another of the disease terms met with in the *Shan Hai Ching* is *chia*, which undoubtedly refers to a massive infestation with intestinal worms (ascariasis or oxyuriasis). This brings us to the great period of Han case-histories and so to the work of Shunyü I.

During the Warring States, Chhin and early Han periods there were two great schools of medicine; the earliest grew up in the western state of Chhin, the other was located in the eastern seaboard state of Chhi. From Chhin came the physician Huan, whose attendance on the Prince of Chin in 580 B.C. long remained famous; and the physician Ho, already mentioned as examining another Prince of Chin forty years later, also came from there. More celebrated than either was Pien Chhio, about whom there is much to be said, but as the records concerning him do not give us very much in the field of disease nomenclature, we must pass him over here. Shunyü I is a different matter. Born in 216 B.C. in Chhi, he studied under Kungsun Kuang and Yang Chhing, practising medicine successfully from about 180 B.C. onwards. In 167 B.C. he was accused of some crime and taken to court but acquitted after the supplication of his youngest daughter. As he had been attending on the prince and lords of Chhi, he was summoned to answer an enquiry from the imperial court some time between 164 and 154 B.C., then released again and continued in practice until his death about 145 B.C. It is owing to this perquisition by the imperial authority that we possess today the records of some twenty-five detailed case-histories which Shunyü I reported. For every one we have the name of the patient, the circumstances in which the disease was contracted, the details of the attendance of Shunyü I, the treatment which he prescribed, the explanations which he gave of his diagnostic reasoning, in which the pulse played a very prominent part, and finally the ultimate result. We also have the answers which Shunyü I gave to eight general questions, answers which throw a flood of light upon the general conditions of medical education and practice in the second century B.C. Bridgman, who has given us a pioneer study of Shunyü and his times, concludes that the general level of Chinese medicine thus revealed was in no way inferior to that of the contemporary Greeks, and in this judgement we concur. For the present purpose the point is that the clinical descriptions are so detailed that we can see exactly what Shunyü I meant, at any rate by his own technical terminology.

Let us first look at some of the less severe cases which Shunyü I was able to cure, or at any rate relieve for a time. In a child, *chhi ko ping* was clearly difficulty in breathing, probably influenza or catarrh, perhaps acute laryngitis; some fever is implicit in the explanation. In a palace superintendent, *yung shan* was evidently vesical schistosomiasis, accompanied by haematuria, urinary retention, vesicular calculi, perhaps prostatorrhoea. Other similar cases, however, were too far gone to recover, for example, a police chief who seems to have had bladder cancer accompanied by intestinal obstruction due to heavy ascaris infestation (*chia*). The Chief Eunuch of the Palace of Chhi fell into a river and got very cold and wet, so his *jê ping* due to *han* was surely bronchitis or pneumonia; Shunyü I gave antipyretic drugs and pulled him through. Then the Queen Mother of Chhi had *fêng tan*, which is clearly interpretable as acute cystitis, probably connected with nephritis. She had haematuria, but she got better under Shunyü's treatment. An old nurse of the princely family had *jê chüeh*, with hot and swollen feet—this may have been gout accompanied by chronic alcoholism, or possibly simply a traumatic infection of the extremities. *Chhiu chih* was clearly dental caries and one of the Grand Prefects of Chhi had it. One of the concubines of the Prince of Tzu-Chhuan had a difficult childbirth; Shunyü I gave nitrate and obtained the rejection of post-partum blood clots. A young courtier had *shen pi*, traumatic lumbago or muscular strain caused by trying to lift a heavy stone, together with dysuria, perhaps caused by compression of the hypogastric plexus; he also got better. By means of a vermifuge prepared from the gingko tree, a girl was cured of an intense *Enterobius* infestation (oxyuriasis). Here the description is particularly precise because this was termed *jao chia*, and already by this time there were several other terms (*hui, chiao, pa*, etc.) for other types of intestinal parasites. Another case of *pi* was that of a young prince who had acute lobar pneumonia but recovered.

One of the more striking features of Shunyü I's practice was the way in which he was able to give a long-term diagnosis. For example, on one occasion he was asked to give a general health check-up of the serving-maids of the Prince of Northern Chhi, and among these he found one, named Shu, who was certainly not ill but in his opinion was going to be. She was, he said, suffering from *shang phi*, and this must have been tuberculosis because it ended in a sudden and fatal haemoptysis some six months later. No one would believe that Shunyü I was right in saying that she was ill, but events confirmed his opinion. On another occasion he was alarmed by the appearance of a slave of a client of the Prime Minister of Chhi, who again, in his view, had a *shang phi chhi*, although the man himself did not feel particularly ill. Shunyü I said that he would not last through the following spring, and he did not. Here the clinical description suggests hepatic cirrhosis, almost certainly of parasitic origin, caused by liver flukes (hepatic distomiasis); jaundice was apparent, and the case might also have been one of acute yellow atrophy of the liver. One of the most extraordinary cases reported by Shunyü I was that of another royal physician, by name Sui. He must have been interested in Taoist arts, for he had himself prepared elixirs from the "five mineral subsances," and when Shunyü I saw him he was suffering from *chung jê*, apparently in this case a pulmonary abscess, presumably brought on by arsenical or mercury poisoning. Shunyü I warned him that it would be hard to avoid a fatal result, and in fact some months later the abscess burst through under the clavicle and Sui died. Another man had what Shunyü I described as *ping khu tho fêng*, i.e., some progressive paralysis, possibly disseminated sclerosis, possibly progressive muscular dystrophy.

More rapidly fatal in termination were other cases. A palace chamberlain had a peritoneal abscess, perhaps a perforating ulcer (*chü*, leading to *chung jê*); perhaps the perforation was due to heavy ascarid infestation. Another man died of *fei hsiao tan* with delirious fever (*han jê*). This would have been acute hepatic cirrhosis, probably caused by liver and blood flukes. In this case the Royal Physician of Chhi had diagnosed and treated quite wrongly. It is curious that down to this time we have not found the characteristic term for cholera (*ho luan*), but it seems that Shunyü I may well have had a case of it among his records, for a minister of the Lord of Yang-hsü died of "penetrating pneuma" (*tung fêng*), the description of which suggests total failure of digestion, intense diarrhoea, possibly due to enteric fever, perhaps to cholera. The word *shan* appears again in a combination *mu shan*, where it clearly refers to an aortic aneurism which caused the death of a general. The last case we shall mention was that of a Court Gentleman of Chhi who had a fall from his horse on to stones; the resulting traumatic abdominal contusion followed by intestinal perforation of a gut probably already weakened by parasitic infestations of one sort or another was termed *fei shang*, i.e., injury not to the lung but to the tract (*ching*) of the lung. This brings us to the last subject which we can touch upon here, namely the medical system of the *Nei Ching*.

The *Nei Ching* was, we think, approximately already in its present form by the first century B.C. The full title under which it is commonly known is the *Huang Ti Nei Ching* (The Yellow Emperor's Manual of Esoteric Medicine), consisting of two parts, the *Su Wên* (Pure Questions and Answers) and the *Ling Shu* (Spiritual Pivot). This was the recension which came from the editorship of Wang Ping in the Thang dynasty, but it is probable that this was not the recension which the Han people had. Another one, known as the *Huang Ti Nei Ching, Thai Su*, which was edited a hundred years or so earlier than Wang Ping, by Yang Shang-Shan in the Sui period, and which has only in very recent times come to light, may be considered nearer the original text of the Han. The *Nei Ching* system of diagnosis classified disease symptoms into six groups in accordance with their relation to the six (*n.b.* not five) tracts (*ching*)which were pursued by the pneuma (*chhi*) as it coursed through and around the body. Three of these tracts were allotted to Yang (Thai-Yang, Yang-Ming, Shao-Yang) and three to Yin (Thai-Yin, Shao-Yin and Chüeh-Yin). Each of them was considered to preside over a "day," one of six "days," actually stages, following the first appearance of the feverish illness. In this way differential diagnosis was effected and appropriate treatment decided upon. These tracts were essentially similar to the tracts of the acupuncture specialists, though the acupuncture tracts were composed of two sixfold systems, one relating to the hands and the other to the feet, and crossing each other

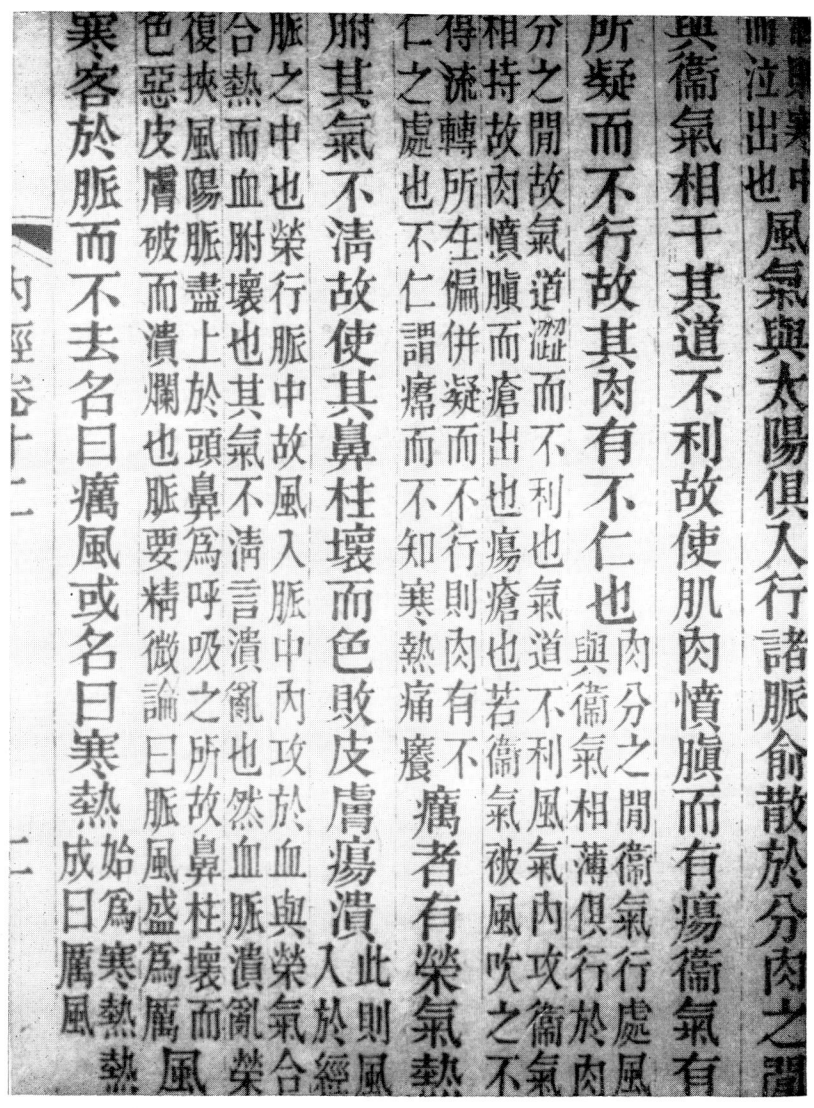

FIGURE 10. One of the pages of a chapter in the *Nei Ching*, which includes the description of leprosy as leading to the erosion of the nose and breakdown of the skin (fifth and sixth columns from the right).

like the cardinal (*ching*) and decumane (*lo*) streets of a city laid out in rectangular grid arrangement. Moreover by the time of the *Nei Ching* the physicians had achieved full recognition of the fact that diseases could arise from purely internal as well as from purely external causes; the ancient "meteorological" system explained by physician Ho had therefore been developed into a more sophisticated sixfold series, namely *fêng, shu, shih, han, sao, huo*. As external factors, they could be translated as wind, humid heat, damp, cold, aridity and dry heat; but as internal causes we could name them blast (*cf.* van Helmont's *blas*), fotive *chhi*, humid *chhi*, algid *chhi*, exsiccant *chhi* and exustive *chhi*. It is interesting to notice the partial parallelism with the Aristotelian-Galenic qualities, which were part of a quite different, fourfold, system.

In the brief remaining space of this contribution it would be impossible for us even to sketch the aetiological and diagnostic system of the *Nei Ching*, but it is fair to say that it provided an elaborate classificatory framework into which the results of keen clinical observation could be fitted. A rather comprehensive theory of medicine, both diagnostic and therapeutic, was now available. Interpreting a whole millennium of clinical tradition, the physicians of the Former Han dynasty were able to combine into one science the influences of external factors on health, the abnormal functioning of internal organs whether by excess or defect, and the manifestation and inter-relationship of symptoms; using the concepts of Yin Yang (the two fundamental forces in the universe), Wu Hsing (the five elements), Pa Kang (the eight diagnostic principles), and Ching Lo (the circulatory system of the *chhi*). The five elements had not been part of the most ancient Chinese medical speculations; they derived from another school, that of the Naturalists (Yin-Yang chia) whose greatest exponent and systematiser had been Tsou Yen (c. 350 - 270 B.C.) Five-element theory (a lengthy discussion of which will be found in *Science and Civilisation in China*, Volume 2) was so influential and so widespread in all the non-medical sciences and proto-sciences of ancient (and mediaeval) China that the physicians could not remain unaffected by it, but in taking it into their theoretical disciplines they added a sixth unit or entity to conform with their sixfold categories. Thus there were five Yin viscera (liver, heart, spleen, lungs and urino-genital organs) and five Yang viscera (gallbladder, stomach, large and small intestines and bladder) recognised by all schools. To these the physicians added a further entity in each category, the *hsin pao lo* (pericardial function) and the *san chiao* (three coctive regions); and the particular interest of this lies in the fact that these additions represented physiological operations rather than morphologically identifiable structures. The six "viscera" could thus correspond readily with the six *chhi*, the six tracts, and so on. It must not be supposed that the state of Chinese medicine at the time of the *Nei Ching* synthesis was destined to remain unchanged through the following nearly two millennia of autochthonous practice, on the contrary there were great developments, many elaborations, and a proliferation of diverging schools, but if we are to think of any presentation of Chinese medicine as classical, this is what deserves the name.

The ancient Chinese physicians were extremely conscious of the temperature regulating and perceiving systems of the human body, so that although they had no means of measuring temperature accurately, the observation of subjective chill or fever, together with algophobia or algophilia, was extremely important for them. By this time also the study of the pulse and its modifications had advanced to a highly developed state. All we can do here is to illustrate some of the disease syndromes which they recognised, for among them there are not a few which can be identified rather clearly in modern terms.

All fevers were placed in the category of Shang Han diseases and termed diseases of temperature (*jê ping*). Every sign which is still examined today, pain, perspiration, nausea, etc., short of the results of modern physico-chemical tests, was studied by them and meant something to them. For example *fu man,* or abdominal fullness, was an important sign. This could mean oedema (*chung*). The *Nei Ching* actually says that "fluid passing into the skin and tissues by overflow from above and below the diaphragm forms oedema." It could also mean ascites occurring in liver cirrhosis, heart failure, and especially schistosomiasis, undoubtedly so common in ancient China. *Fu man* was also accompanied by the excretion of watery faeces with undigested food (*shih i*) found in gastro-enteritis, cholera, and the like. *Fu man* was also called *fu chang* and *tien*. This latter word is a good example of a word which can be pronounced in two ways; if one said *tien* it meant abdominal distension, but if one pronounced it *chen*, then it meant various forms of madness and in the binome *chen hsien,* epilepsy. It is clear from the clinical description that from Han times onwards the terms *lao fêng* and *lao chung* referred to tuberculosis. The term *fêng* by itself always had the connotation of convulsion or paralysis; it might be regarded as a violent *pneuma*, in distinction from the mild *pneuma* (*chhi*), which was part of the physiology of the normal body. Other forms of *chung fêng,* therefore, were hemiplegia (*phien khu*) and cerebral haemorrhage giving full apoplexy (*fei*). Among the fevers (*wên ping*) we now find fairly clear descriptions of diphtheria, as *shê pên lan* (lesions at the root of the tongue), doubtless complicated by streptococcal infections. Diphtheria is also clearly denoted by *mêng chü,* "fierce ulcer" (of the throat). Hepatic cirrhosis caused by liver and blood flukes was now called *kan jê ping,* tuberculosis *phi jê ping,* pneumonia *fei jê ping.*

It does not always follow that the organs referred to in descriptions (in the three preceding cases, liver, spleen and lungs respectively) were those to which we might refer the diseases today. Rather these were the organs concerned with the six tracts already spoken of, each one of which was connected with an organ. Of the malarial types of fever (*chiai nio*) we have already spoken. The terminology now continued with little change, but one disease, *tan nio,* may be identified with relapsing fever caused by *Borrelia* spirochaetes as Sung Ta-Jen has suggested.

One last word on diabetes. Polyuria was recognised as the sign of a special disease in the *Nei Ching,* where it is called *fei hsiao.* Han ideas about this illustrate the principle of successive involvement or shifting (*i*) when some pathological influence spreads from organ to organ in the body. Thus in *fei hsiao* the cold *chhi* in the heart passes over into the lungs and the patient excretes twice as much as what he drinks. Though the characteristic name for diabetes (*hsiao kho, hsiao chung*) had not been developed by the end of this period, there can be little doubt that diabetes was here in question. The sweetness of the urine was discovered a good deal later, indeed in the seventh century A.D. We have discussed in another place the Chinese knowledge of diabetes, and the theories which were held about it.

The fact that mummification was not practised in ancient Chinese civilisation has no doubt militated against the acquisition of a mass of concrete evidence concerning many of the diseases from which people suffered in those times, such as has been developed for ancient Egypt. As far as we know, almost nothing has been done on the pathological anatomy of the skeletons which have been excavated from ancient tombs in China, whether in the Neolithic or in the Chou, Chhin and Han periods. Since there must be a mass of skeletal material in the Chinese museums, it may be that this task could still be accomplished with valuable results by Chinese archaeological pathologists. However, the study of the written records of ancient China from the middle of the first millennium B.C. down to the beginning of our era, shows that they have preserved a veritable mass of information concerning the diseases prevalent in those times, and although the study of human remains themselves may bring precious confirmation of what the writings reveal, it may well be that on balance the written records when fully analyzed will present a broader picture than the study of the skeletal remains themselves alone could ever give us.

NOTE

The system of romanisation adopted in this paper follows that of Wade-Giles with the substitution of an *h* for the aspirate apostrophe. Detailed references and all the Chinese characters for the personal names and technical terms will be provided in Volume 6 of *Science and Civilisation in China*. Since the Chinese language contains many homophones, several different technical terms may be indistinguishable when in romanised form; in such cases we indicate differences by superscript figures.

REFERENCES

BRIDGMAN, R. F., 1955: La Médecine dans la Chine Antique," *Mélanges Chinois et Bouddhiques*, 10:1.
CHHEN CHIH, 1958: "Hsi Yin Mu Chien chung Fa-hsien-ti Ku-Tai I-Hsüeh Shih-Liao (Ancient Chinese Medicine as recorded in Seals and on Wooden Tablets)", *Kho-Hsüeh Shih Chi-Khan*, 1:68.
CHHEN PANG-HSIEN, 1937: "Chung-Kuo I-Hsüeh Shih (History of Chinese Medicine)", Commercial Press, Shanghai, 2nd. ed. 1957.
FAN HSING-CHUN, 1953: "Chung-Kuo Yü Fang I-Hsüeh Ssu-Hsiang Shih (History of the Conceptions of Hygiene and Preventive Medicine in China)", Jen-Min Wei-Sêng, Peking, 1954.
HU HOU-HSÜAN, 1943: "Yin Jen Ping Khao (A Study of the Diseases of the Shang (Yin) People (as recorded on the oracle-bones)", *Hsüeh Ssu* (no. 3,) 73, (no. 4), 83.
NEEDHAM, J., WANG LING, HO PING-YÜ, LU GWEI-DJEN et al., 1954—: *Science and Civilisation in China*," 7 vols. in about 10 parts, University Press, Cambridge.
NEEDHAM, J. and LU GWEI—DJEN: "Proto-Endocrinology in Mediaeval China", Stefan Milcu Festskrift, Bucarest (in the press).
SUNG TA-JEN, 1948: "Chung-Kuo Ku-Tai Jen Thi Chi-Shêng-Chhung Ping Shih (On the History of Parasitic Diseases in Ancient China)" *I-Hsüeh Tsa Chih*, 2 (no. 3/4): 44.
WANG CHI-MIN and WU-LIEN TÊ (K. C. WONG and WU LIEN-TEH) 1936: "History of Chinese Medicine." Nat. Quarantine Service, Shanghai (1st. ed., 1932).
YÜ YÜN-HSIU, 1956: "Ku-Tai Chi-Ping Ming Hou Su I (Explanations of the Nomenclature of Diseases in Ancient Times)", Jen-Min Wei-Sêng, Shanghai, 1953. Cf. Nguyen Tran-Huan, *Rev. Hist Sci.* 9:275.

EDITORIAL ADDENDUM

During the course of editorial work, we asked Drs. Lu and Needham whether they could satisfactorily state that leprosy occured in China by the sixth century B.C. Their reply, given below, seems worthy of inclusion here.

"We should rather like to stand by our assertion that the Chinese did diagnose leprosy fairly specifically in the first millennium B.C. In order to explain why we feel this, we should perhaps give our translation of the celebrated text concerning the disciple of Confucius (*Lun Yü*, VI. 8):

Po-Niu was suffering from leprosy. When Confucius went to visit him he would only touch his hand through the window (for the discase was a disfiguring one). The Master said "How fortunate to find him still alive! What a dreadful fate! That such a (sensitive) man should suffer such an illness! That such a (sensitive) man should suffer such an illness!"

The last phrase of Confucius repeats itself. This version is probably different in several particulars from other translations which you will find in the books, but we have built it upon the consensus of opinion of commentators throughout the ages. For example, Pao Hsien, who was writing in the Later Han period, sometime between 25 and 55 A.D., was quite certain that the reason for the shaking of hands through the window was because the disease was a disfiguring one. On the other hand, as I think we mentioned in our paper, the general opinion of commentators all down through the ages was that the disease (*chi*) from which Po-Niu was suffering was specifically *o chi*. Now although these words *o chi* translated literally mean 'evil disease,' this particular phrase has always had the connotation of leprosy. To prove this you only have to go to the great dictionary compiled by Hsü Shen in 100 A.D., the *Shuo Wên Chieh Tzu*, where he defines leprosy (*lai*) as *o chi*. Then the link with the later term *ta fêng*, still usual today, was given by the great pathological writers of the seventh century A.D., Sun Ssu-Mo in his *Chhien Chin Yao Fang* of 650 and Chhao Yuan-Fang in his *Chu Ping Yuan Hou Lun* of 610. They all define *o chi* as being equivalent to *ta fêng*. In this way we have a continuous record throughout many centuries. Moreover the term *ta fêng* is already present in the great medical classic, the *Nei Ching Su Wên*, which we date as approximately of the second century B.C. Here it is equated again with *lai*. I think you will agree that the descriptions are fairly specific, because at various places the text says that in this disease the skin of the patient is whitish with much eruption and decay, leading for example to the destruction of the pillar of the nose. Besides these patients have stiffness of the joints and lose all their external hair, including bread and eyebrows. Lastly, I might refer to the possibility that in some of the descriptions of seasonal diseases which we quoted in our paper, as in the *Yuëh Ling* and the *Lü Shih Chhun Chhiu*, texts of the later half of the first millennium B.C., there is a word which can be pronounced in two ways, *li* and *lai*. We had generally interpreted this as to be pronounced *li* and to mean epidemics in general, but if it should be taken in its sense of *lai*, then it would refer to leprosy specifically".

Chapter 18

Disease in Antiquity: Ancient Greece and Rome

ADAM PATRICK

ONE of the earliest and best-known of the epidemics of classical times was the Plague of Athens, which occurred in 430 B.C. The Peloponnesian war, which marked the decline of Athenian greatness, had begun in 431, with Athens and Sparta the leading city states on the opposing sides. Langer (1940) says that the basis of the war was the existence in Greece of two great rival systems of alliances. The strategy of the Athenians was to avoid a land battle, in which they would almost certainly be defeated, remain within their walls, and let their country be ravaged by the enemy. They could support themselves by their control of the sea, and hoped to wear down the Peloponnesians by constant raids.

In 430 a great plague broke out in Athens. The nature of this plague has never been proved, and the problem is perhaps insoluble. The history of the war, down to 411, was written by Thucydides with what H. J. Rose (1934) calls "minute and scientific a c c u r a c y." Thucydides fought in the war, and himself had an attack of the plague, and this accounts for a medical interlude in a military history, an unusual intrusion. He describes the plague, and gives a detailed account of the symptoms. "I shall narrate its actual course," he says, "and describe the symptoms, a study of which should enable a diagnosis to be made." His expectation has not been fulfilled.

Thucydides says (II, 47-52): "At the very beginning of the summer [430 B.C.], the Peloponnesians and their allies invaded Attica, and proceeded to ravage the countryside. Their invasion had not lasted many days when the plague showed itself for the first time among the Athenians. . . . No scourge so destructive of human life is anywhere on record. The physicians had to treat it without knowing its nature, and it was among them that the greatest mortality occurred. It is said that the malady had its origin in Ethiopia, whence it descended into Egypt and Libya. . . . Then it suddenly fell upon the inhabitants of Athens, by way of Piraeus, and rose also to the upper city. . . . I had an attack myself, and saw others who suffered from it."

He goes on to describe the symptoms: "Suddenly, in the midst of good health, a man would first be seized by an intense heat of the head, and redness and inflammation of the eyes; the pharynx and the tongue became red, and the breath foetid. Sneezing and hoarseness followed, and then the chest became affected, and coughing was severe. Then came vomiting, of every kind of bile, and retching. The skin was livid red, and came out in a raised eruption and sores. It was not so very warm to the hand, but the patients felt it burning hot, and could not endure the lightest covering. They wanted to be left naked, and would have liked most of all to throw themselves into cold water; and, indeed, many who were not closely watched jumped into the water-tanks. They were tormented by an unquenchable thirst, and were restless and sleepless. The body, at the time of the illness, was not wasted, and when patients died, as many of them did on the seventh or the ninth day, they might still have some strength left. If they got over the crisis, the disease went down into the bowels, producing severe ulceration and diarrhoea. Most of the deaths in this later stage were due to weakness. The malady spread through the whole body, and if a man got over the acute stage, it might attack the extremities, affecting the genitals, the fingers, and the toes. Many recovered with the loss of these, and some lost their eyes as well."

Thucydides adds that the Athenians suffered additional hardship through the crowding into the city of the people from the country districts. Since no houses were available these had to live in stifling huts. They perished in wild disorder. Bodies of dying men lay one upon another, and half-dead people rolled about in the streets.

Could Thucydides have foreseen how posterity would fail to agree about the nature of the plague, he would have been surprised as well as disappointed. The Roman poet Lucretius (99-55 B.C.) ends his long poem *De rerum natura* with a free translation of Thucydides' description of the plague of Athens. H. A. J. Munro (1860) published his edition of Lucretius about a hundred years ago, and pokes fun at the medical writers for their diversity of opinion on the nature of this epidemic.

He says in his commentary (II: 39): "I have looked into many professional accounts of this famous plague, and writers, almost without exception, praise Thucydides' accuracy and precision, and yet differ most strongly in the conclusions they draw from the words. Physicians—English, French, German—after examining the symptoms, have decided that it was each of the following: typhus, scarlet, putrid, yellow, camp, hospital, jail, fever; scarlatina maligna; the Black Death; erysipelas; smallpox; the oriental plague; some wholly extinct form of disease. Each succeeding writer at least throws doubt on his predecessor's diagnosis."

The impression I have gained from reading various comments is that those commentators who have had most experience of the pestilences commonly make a diagnosis of typhus fever or of smallpox. It seems to me that any opinion on the nature of this epidemic must be based on clinical grounds. Considerations such as that this disease or that was not known to exist at the time, are not valid, and should be left out of account: this might be the only instance of it. Some of the symptoms described by Thucydides are of a general character; others, of which the rash is one, are more specific. Of the two words used to describe it, the first (*phluktaina*) means an eruption raised above the level of the skin; the second (*helkos*) seems to point to some breach of continuity in the skin surface. The rash of smallpox would be better described by these terms than that of any of the other epidemic diseases.

In 396 B.C., thirty-five years after the plague of Athens, the Carthaginians besieging Syracuse were stricken by a plague. Here are some of the points in the description given by Diodorus Siculus (II: 14, 70; *floruit* 44 B.C.): "After the Carthaginians had seized the suburb, and plundered the temples of Demeter and Athene, the plague struck the army. . . . It began with a catarrh, then a swelling on the throat; gradually burning sensations ensued, pain in the sinews of the back, a heavy feeling in the limbs. Then dysentery supervened, and pustules (*phluktainai*) upon the whole surface of the body. In most cases this was the course of the disease, but some lost their reason and their memory. . . . Death came on the fifth day, or the sixth at the latest, amidst such terrible tortures that all looked upon those who had fallen in the war as blessed." Zinsser (1942) comments: "The disease as described by Diodorus—again like the epidemic in Athens—corresponds about as closely as can be expected of ancient descriptions, to the severe confluent type of smallpox, in which death on the fifth or the sixth day is not exceptional."

The books which are known as the *Hippocratic Collection* are regarded as differing much in the periods in which they were written, but at one time they, or some of them, were attributed to Hippocrates, whose date was put in the fifth century B.C. Many short case-records are included in the *Epidemics,* in some of which a diagnosis can be hazarded. In none, however, can it be made so conclusively as in the epidemic described near the beginning. Parts of the book read like extracts from the report of a medical officer of health. It begins: "In Thasos during the autumn, from the equinox to near the setting of the Pleiades [Sept. 21-Nov. 2], there were frequent rains, gentle and persistent, with southerly winds. . . . Early in the spring, however, a few patients went down with ardent fevers, but these were mild, causing haemorrhage but seldom, and no deaths. Many people had swellings beside one or both ears, in most cases unaccompanied by fever, so that confinement to bed was unnecessary. In some cases there was slight fever, but the swellings always subsided without complication; none suppurated. They were flabby, large and spreading, without inflammation, and cleared up without leaving a sign. . . . Few women suffered. Either early, or after a little time, painful inflammation might occur in one or both testes, sometimes with fever, and the pain severe." The description of an outbreak of mumps is easily recognizable, and the complication of orchitis is characteristic.

A feature of the case summaries is the numerous instances of puerperal sepsis or of septic abortion. The first case record is that of Philiskos, who took to bed with fever and sweating. On the third day he passed black urine, and the urine continued to be black until he died on the sixth day. Had this been an isolated case with mention of black urine, one might have thought it to be blackwater fever, but in several others the same adjective (*melas*) is used to describe the urine colour. This is the usual adjective for black, but Liddell and Scott say that sometimes it may signify a shade no deeper than dark.

Kriton had pain in the great toe, which became red and swollen. He died on the second day, probably from streptococcal septicaemia.

A man dined well and drank too much. Vomiting, fever, and pain in the right hypochondrium followed, then fever and chills, and on the eleventh day he died. This may have been a case of acute cholecystitis.

Python probably had lobar pneumonia. His illness began with trembling, at first in the hands, then fever and depression. He was worse on the fourth day—constipation is noted; and on the sixth the sputum was reddish; on the tenth day there was sweating and crisis.

There are several references to what is probably relapsing fever. One is that of Chairion, who had fever and pain in the head after drinking. On the third day there were a rigor, convulsions, and delirium; an exacerbation on the fifth day; and on the seventh, sweating and crisis. There was a relapse on the ninth day, fever was acute on the fourteenth, and on the fifteenth there was vomiting. Crisis occurred on the seventeenth, and recovery was complete by the twentieth.

Erysipelas occurred in a severe form, associated sometimes with gangrene. Many suffered from it all over the body when the exciting cause was a trivial accident, or a very small wound. Many, even when undergoing treatment, suffered from acute inflammation, and erysipelas would quickly spread widely in every direction. There was a great deal of suppuration and sloughing of tissue. The discharge which formed was not like pus, but was foetid and associated with gangrene. (This description is reminiscent of what we are told of the erysipelas and gangrene which were rife in hospitals in the pre-antiseptic days of the mid nineteenth century.) Fever was sometimes present and sometimes not. When the symptoms resulted in suppuration, the patients usually recovered, but, if there was no suppuration, deaths were numerous. The course of the disease was the same, in whichever part of the body it occurred. Many lost the arm and the forearm. If the side of the trunk was attacked, gangrene spread to the front and to the back. Sometimes the bones of the thigh, or the leg, or the whole foot, were bared. The worst case of all was when the pubes and the genital organs were involved.

The most severe and troublesome disease, as well as the most often fatal, was consumption. Many cases began in winter. In early spring, of those who had taken to bed, most died; none of the others lost their cough. In the autumn they all took to bed, and there were many deaths. Most of those who died had been ill for a long time. These, for the most part, when they became worse, showed the following symptoms: frequent shivering; often continuous and acute fever; unseasonable and copious cold sweats throughout; the bowels sometimes constipated and sometimes loose, but always bad diarrhoea towards the end; great wasting. There was frequent coughing throughout, and great distaste for food. Dropsy tended to develop, and there were shivering fits and delirium when death was imminent.

Here is the description of Pott's curvature, from the book on *Joints*: When the spinal vertebrae are drawn into a hump by disease, most patients are incurable, especially when the curvature has formed above the attachment of the diaphragm. ... When hunchback comes on in children, before the body has completed its growth, the legs and the arms grow to full length, but the spine does not grow correspondingly, and these parts are defective. When the hump is above the diaphragm the ribs grow, not laterally, but forward, and the chest becomes edged in front. The patient becomes short of breath and hoarse, because the cavities for the breath are small. These patients, as a rule, have hard and immature tubercles in the lungs, for the curvature arises, in most cases, from such gatherings. Cases in which the curvature is below the diaphragm, are sometimes complicated by affections of the kidney, and parts about the bladder. There may be abscesses in the lumbar region, and about the groins."

There are references to tetanus. Two aphorisms are: "Spasm intervening upon a wound is deadly;" and "Those who are attacked by tetanus either die in four days, or recover."

One of the commonest causes of fever must have been malaria, and many short fevers, which were followed by sweating, were probably of that type. One of the *Aphorisms* says: "It is in autumn that diseases are most acute, and, in general, most deadly; spring is healthiest, and its diseases the least dangerous." Benign tertian malaria, which has lain dormant during the winter, has a special tendency to relapse in the spring, and primary infections begin to show themselves. The seriousness of the autumn malaria is due to its being malignant subtertian, the seasonally-named *aestivo-autumnal*. Again: "Fevers that, without intermitting, grow worse on alternate days, are dangerous"—probably subtertian. Also: "An exact tertian reaches a crisis in seven periods at most." I might mention a personal experience of the seasonal difference between the two types of malaria referred to. In 1916-18, when the British armies in Macedonia were stricken with malaria, the proportion of benign tertian relative to malignant tertian cases in patients who came to base hospitals in Malta was 50:1; in the autumn the figures were reversed, 1:50.

One other short passage might be quoted about malaria, from the *Treatise on Airs, Waters, Places*: "I wish now to treat of waters, those that bring disease or good health. Such as are marshy, standing, and stagnant, must in summer be hot, thick, and smelly; in winter they are frosty, cold, and turbid, and so very conducive to phlegm and sore throats. Those who drink such waters always have large firm spleens, and hard, thin, hot stomachs; and their shoulders, clavicles, and faces are emaciated. The fact is, the flesh dissolves to feed the spleen, and so they are very thin."

He here describes the debilitated subjects of chronic subtertian malaria.

As in Greece, so in Rome: the best-known systematic account is an early one, and it is also that from which most information can be obtained about the occurrence of diseases. Celsus (*floruit* A.D. 14-37) is the most famous of the Roman writers on medicine; yet he remains a shadowy figure, and almost nothing is known of his life. His medical writings, *De Medicina,* in eight books, form three compact volumes in the Loeb library. The treatise on medicine was part of an encyclopaedia which dealt also with other subjects, but of these only a few fragments remain.

The disease which he describes in greatest detail is that which occupies the same position in the *Hippocratic Collection*, the malarial fevers. Though there may have been an interval of four hundred years between the writing of the two accounts, the descriptions are substantially alike. We might go further and say that they show no radical difference from those written today. The fevers which have regularly recurring paroxysms, are characteristic, those which show some irregularity are not so well defined. A possible reason is that he included in the group some fevers which were not malarial. He says: "Now there follows the treatment of fevers, a class of disease which is exceedingly common. One of the fevers is quotidian, one tertian, and a third quartan. At times certain fevers occur in even longer cycles, but they are infrequent. Quartan fevers have the simplest characters. Nearly always they begin with a shivering, then sweat breaks out, and the fever ends. There are two classes of tertian fevers: one begins and ends in the same way as quartan, but is distinguished from it by having a single day free, and a recurrence on the third day.

The other is far more dangerous. It does indeed recur on the third day, but out of the forty-eight hours, about thirty-six are occupied by the paroxysm; nor does the fever entirely subside in the interval, it merely becomes less intense. This class most practitioners term 'hemitritaion.'" He describes here the subtertian illness. The cycle of the parasite occupies about forty-eight hours, but the circulating parasites are of all ages, and are not collected into one, or two, generations as in benign tertian. Hence the fever tends to be continuous, though perhaps irregularly so. The danger in the malignant type arises from the multiple capillary thrombosis which tends to occur in the brain and other organs of the body.

Celsus goes on: "Quotidian fevers vary, and have many forms. Some begin with a feeling of heat, others with a chill, others with a shiver. (I call it a chill when the extremities become cold, and a shiver when the whole body shakes.) Some attacks cease and are followed by a period of complete freedom; in others, the fever lessens, yet some remnants of it persist till the next paroxysm. Others may coalesce, so that there is no remission; instead, a continuous fever. In some the hot stage is acute; in others, bearable. In some the height of the fever is equal on successive days, in others, unequal: slighter one day, higher the next." This last group probably includes the double tertians, in which there are two generations of parasites in the blood, equally, or almost equally, spaced in time. This double character is common in benign tertian malaria; one of the generations may drop out and leave a simple tertian. When malarial injection came to be used in the treatment of general paralysis, it was found that the initial fever which was generated by the injected parasites, tended to be of the double tertian type. One generation might give more severe fever than the other.

A constitutional disease with which Celsus is familiar is diabetes. He says: "When the amount of urine passed is greater than that of the liquid taken, even though it is passed painlessly, it gives rise to wasting and the danger of consumption."

In a discussion of ulcers in the mouth, Celsus says: "By far the most dangerous ulcers are those the Greeks call *aphthai*. This is certainly so in children, in whom they often cause death, though in adults they are not dangerous. These ulcers begin in the gums, then invade the palate and the whole of the mouth; and thence they pass to the uvula and the throat. In that case the child has little chance of recovery." These aphthae are not the lesions of our aphthous (or vesicular) stomatitis. Liddell and Scott define the Greek original as "an erysipelatous eruption in the mouth;" but they are our cancrum oris or noma, a spreading gangrene, which may start on the inner surface of the angle of the mouth, and erode away part of the cheek. I never saw a recovery. I should doubt if they are seen now, but at one time they might occur in debilitated children after a severe attack of measles or whooping-cough. The Loeb editor, Spencer, refers here to an epigram (xi: 91) of the Latin writer Martial (A.D. c.40—c.104), which is really an epitaph: "Canace, the daughter of Aeolis, lies buried in this tomb; little Canace, whose seventh winter was her last. Alas for the guilt and the crime of it! Thou, passer-by, who art quick to weep, may lament here, not the shortness of life, but something sadder than death, the way death came. A dreadful canker wasted her face, and settled on her tender mouth, and consumed her very lips before they were surrendered to the smoky pyre. If it had to come with so ill-timed a flight, fate should have come by another path. But death hastened to close the channel of her charming speech, lest her tongue might have power to bend the stern goddesses."

I should like to refer to two diseases in Celsus' surgical sections: abscess of the liver, and stone in the bladder. He says of the former: "If the liver suffers from an abscess, the same general treatment is to be carried out as in any other internal suppuration. Some even make an incision over the liver with a scalpel, and bore through into the abscess with a cautery." For the diagnosis of an abscess in the liver it must be a large one; and if the treatment described is to be carried out with confidence, there must be a visible swelling. This makes it likely that the abscess was amoebic and not pyogenic in nature; and the development of this bold treatment suggests that such abscesses were common. The passage is presumptive evidence of the prevalence of amoebic dysentery.

The operation for stone in the bladder is described in great detail. What we find surprising is that the imagined patient is a boy of nine or ten. Spencer says in a footnote: "Stone was evidently c o m m o n among schoolboys then, as it continued to be in this country up to within living memory." Hirsch supports these impressions: "As early as the Greek period we find precise references to the frequency of stone in childhood; for example, in the *Hippocratic Collection;* in Aretaeus; in Galen, who speaks of it as a malady proper to boys;" and he adds a large group of modern figures.

There are several references to affections of the skin. Sir Erasmus Wilson was able to identify a number of them, and in 1863 published a paper on the dermo-pathology of Celsus. Here are some of his remarks: "Celsus enumerates forty to fifty cutaneous affections in Books 3, 5, 6, and 7. He arranged his descriptions remedially—according to their amenability to hygiene and diet, to drugs, and to surgical measures; and, secondly, topographically, according to the parts of the body affected. In the fifth book *erysipelas* is treated as a complication of wounds. *Carbuncle* is to be treated by the cautery. *Therioma* is an ulcerous condition, which would seem to be either a syphilitic ulcer of the constitutional type, or it belongs to an allied affection, lupus. [The term *therioma* is no longer in use.] *Ignis sacer* is an ill-conditioned ulcer, either lupus or syphilis. The term 'pustule,' as used by the ancients, signified any eruption of the skin raised above the surface, not necessarily a prominence containing pus. *Pernio,* the chilblain, is described. *Struma* is a tumour of the skin in which, beneath a crust of pus and blood, things like small glands arise (moist granulations). They occur chiefly in the neck, but also in the axillae, groins, and side of the trunk. *Kerion* [Greek *kērion* = Latin *favus*] is a suppurative inflammation of the hair follicles. In the treatment of *scabies* (meaning *eczema*) Celsus used a cerate containing sulphur, wax, tar and oil." But this might be the itch.

The question of the history and prevalence of leprosy has always been a difficulty, as various writers have noted. Castiglioni remarks that it began to appear in Europe in the early Middle Ages (sixth and seventh centuries). As in biblical times, other diseases were confused with it, especially psoriasis and probably syphilis. This confusion with other diseases affecting the skin makes the assessment of its real prevalence difficult; but in addition there is a great deal of verbal confusion. In a footnote at the beginning of his historical account, Hirsch, gives eight lines of synonyms which have been applied to it.

In 1841-42 J. Y. Simpson published three notable papers on leprosy. He says: "Leprosy has had at different times and by different authors a great variety of appellations. In the writings of Aretaeus, Aëtius, and later Greek physicians, the disease is described as *elephantiasis* . . *Lepra* was a word which the Greek physicians usually applied to different forms of scaly eruption, but not to any tubercular (that is, nodose) d i s e a s e. Hence *elephantiasis Graecorum* and *lepra Arabum* are leprosy."

Celsus' description of elephantiasis is as follows:

Elephantiasis is almost unknown in Italy, although very common in other parts of the world. The entire body is attacked, down even to the bones. The surface of the body is thickly studded with macules and tubercles, red at first, then turning brown: The skin is remarkably irregular, in some places hard, in others unnaturally thin; in some places roughened, with scales. The trunk swells, the face [*os*], calves and legs swell. When the disease is of long standing, the fingers and the toes are sunk under the swelling. Fever supervenes, and this may easily destroy a patient, overwhelmed by such a multitude of troubles.

Wilson comments: "Nothing can be clearer or more perfectly illustrative of that terrible disease, *elephantiasis Graecorum*."

Writing about later times, Castiglioni says: "There were terrible epidemics which destroyed entire cities, sometimes accompanied by inundations and earthquakes, which were frequent in Italy in the first centuries of our era." The specific nature of these epidemics is not clearly indicated by the writers of the period. As Galen himself has stated, the Greek term *loimos* indicates any severe disease with a high mortality which attacks a large number of people at the same time. The Latin words *pestis* and *pestilentia* have the same meaning, and were often used to indicate any sort of misfortune. It is therefore difficult to identify the plagues of antiquity. Even though bubonic plague was without doubt one of the most frequent, it is probable that other serious contagious diseases were often manifest in these centuries.

Castiglioni singles out five great periods of pestilence: (1) that which followed the eruption of Vesuvius in A.D. 79, a terrible epidemic which spread through the Campagna; (2) the plague of Orosius, which broke out in 125, and killed thirty thousand Roman soldiers who had been sent to defend the colony; (3) the plague of Antoninus, or of Galen, which lasted from 164 to 180—it began in the eastern frontiers of the empire, and was carried to Rome by the army in 166—Castiglioni says this was probably typhus, but might have been bubonic plague—the diagnosis remains in doubt; (4) the plague of Cyprian, which lasted from 251 to 266: probably smallpox in view of its highly contagious nature, and the frequent affection of the eyes; (5) in 312, another severe epidemic of smallpox.

There is some uncertainty about any early appearance of plague, most formidable of epidemic diseases. Hirsch quotes, indirectly, from Rufus of Ephesus to the effect that the contemporaries of Dionysios make mention of a certain disease as "pestilential buboes, acute and very dangerous to life, which are to be seen mainly in Libya and Egypt and Syria." Rufus gives a description of this disease according to observations made of it in Libya by two physicians, Dioscorides and Poseidonius, who lived in Alexandria early in the Christian era. It leaves no doubt as to what the disease was. The passage runs: "They give a full account of it in their book on the plague, which was prevalent in Libya in their time. The symptoms were acute fever, pains, general bodily disturbance, and delirium; the appearance of large hard buboes, which did not go on to suppuration, not only in the usual places, but also behind the knees and at the elbows." Hirsch comments: "There are no other certain references to the plague in the writings of earlier or contemporary physicians, or in those of the later period of antiquity."

In the sixth century there occurred the great plague of Justinian. Justinian was emperor from 527 to 565. Chambers's Encyclopaedia says of him: "His reign . . is the most brilliant in the history of the late Empire. He restored the Roman empire, at least in outward appearance, to its ancient limits, and re-united the East and the West under a single rule." But in the course of the article there is no reference to the plague. This epidemic is said to have spread mainly from the Egyptian port of Pelusium, and gradually pervaded the known world. Richard Mead said it began in A.D. 543, spread the infection over all the earth, and lasted fifty-two years. An account is given by Procopius, the Byzantine historian, who was a contemporary. At the height of the epidemic, he said, more than ten thousand people died each day, and it became impossible to bury all the bodies. The emperor himself seems to have suffered from the disease. According to Procopius, half the inhabitants of the Byzantine empire had died by 565. Gibbon (1788) discusses these numbers. Procopius said that a myriad of myriads of myriads died, an impossible number; but even if there were a myriad of myriads, that would furnish a hundred millions, "a number," says Gibbon in a footnote, "not wholly inadmissible." L. F. Hirst (1953) comments: "Doubtless this calamity played an important part in giving rise to the Dark Ages."

REFERENCES

CASTIGLIONI, A., 1947: *A History of Medicine.* trans. E. B. Krumbhaar. London, Routledge and Kegan Paul.

CELSUS.: *De Medicina.* trans. W. G. Spencer. 1948: London, Heinemann (Loeb Library).

DIODORUS SICULUS.: *History* trans. C. H. Oldfather. 1933: London, Heinemann (Loeb Library).

GIBBON, E.: 1909: *Decline and Fall of the Roman Empire.* London, Methuen.

HIPPOCRATES.: *Works.* trans. W. H. S. Jones. 1923: London, Heinemann (Loeb Library).

HIRSCH, A., 1883: *Handbook of Geographical and Historical Pathology.* trans. C. Creighton. London, New Sydenham Society.

HIRST, L. F., 1953: *The Conquest of Plague.* Oxford, University Press.

LANGER, A., 1940: *Encyclopaedia of World History.* London, Harrap.

LIDDELL, H. G., and SCOTT, R., 1864: *Greek-English Lexicon.* Oxford, Clarendon Press.

MARTIAL.: *Epigrams.* trans. W. C. Ker. 1920: London, Heinemann (Loeb Library).

MUNRO, H. A. J., 1860: *Commentary on Lucretius,* 4th Ed. Cambridge, Deighton Bell.

PATRICK, A., 1925: Spring relapses in benign tertian malaria. *Ann. Trop. Med. Parasit. 19:*387.

ROSE, H. J., 1934: *Handbook of Greek Literature.* London, Methuen.

THUCYDIDES.: *History of the Peloponnesian War.* trans. C. F. Smith. 1920: London, Heinemann (Loeb Library).

WILSON, E., 1863: On the dermopathology of Celsus. *Brit. Med. J.,* 2:446 and 465.

ZINSER, H., 1942: *Rats, Lice and History.* London, Routledge.

SECTION IV

SOMATIC DISEASES (INDIVIDUAL)

Chapter 19

Tuberculosis*

DAN MORSE

INTRODUCTION

IN discussing the antiquity of tuberculosis this chapter will be concerned with two problems. First, did tuberculosis exist in America in prehistoric times—and by prehistoric is meant before Columbus discovered a new continent in 1492; secondly, how far back can we trace the disease tuberculosis in the Old World? Sources of evidence of the existence or non-existence of tuberculosis in ancient times can come not only from early literature but also from art forms, and human remains.** It would also seem important to know what would be the expected appearance of an art representation whose live model had tuberculosis, and what picture tuberculosis pathology left in specimens of dried bone.

Pathology of Bone Tuberculosis

Luck (1950), in his book *Bone and Joint Diseases*, makes the following statement: "The clinical diagnosis of skeletal tuberculosis is not a simple matter, and according to published statistics, *a substantial percentage of error occurs even in the best of our orthopedic clinics.*"

If one is confronted solely with a dried bone specimen, this error will certainly be multiplied many times. Furthermore, if one attempted to make a diagnosis of tuberculosis from a prehistoric bone specimen, the only chance one would have of making even a good guess would be if the spine was involved. Tuberculosis in other locations would be indistinguishable from too many other diseases (Girdlestone, 1940; Steindler, 1952).

Tuberculosis involving the spine can cause different pathological results depending upon virulence of the organism and resistance of the victim. The principal characteristics of spinal tuberculosis may be enumerated as follows. Tuberculosis of the spine usually involves only a few vertebrae. Pronounced bone destruction occurs—with little or no regeneration. The pathology is confined almost exclusively to the vertebral bodies with little or no involvement of the neural arches, transverse and spinous processes. As the disease advances, the bone in the vertebrae becomes eroded and decalcified so that the pressure of the body weight causes forward collapse of the vertebral bodies to give the characteristic deformity—angular kyphosis

---

* Some of the material and photographs in this chapter have appeared in previous publications. See Morse, 1961; and Morse, Brothwell, and Ucko, 1964.

** Spinal disease has been found in pre-human animals such as dinosaurs and the cave bear. All of these as far as can be determined can be classified as hypertrophic arthritis and bear no resemblance to spinal tuberculosis as we know it exists in man, (Moodie, 1923; Pales, 1930).

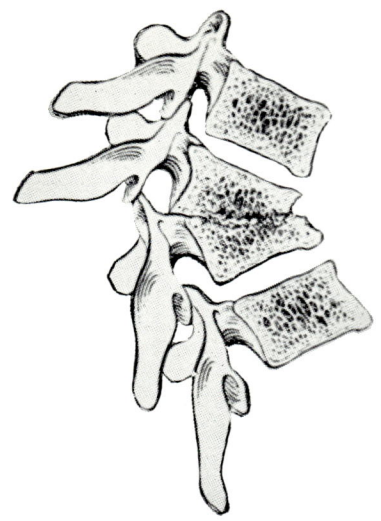

FIGURE 1. Destruction of Vertebral Bodies Resulting in the Characteristic Angular Deformity of Spinal Tuberculosis (Girdlestone 1940)

(Fig. 1). Many other conditions could be confused with tuberculosis especially if only dried specimens were available (Table I).

In the past there was a widespread misconception that kyphosis was synonymous with spinal tuberculosis. The frequent references in literature that prehistoric hunchback art invariably means tuberculosis are not consistent with clinical facts. Besides the various diseases mentioned in the differential diagnosis table, developmental, nutritional, and endocrine disturbances are much more frequent causes of all sorts of spinal deformities, including kyphosis, than tuberculosis (Morse, 1961) (Fig. 2).

PREHISTORIC TUBERCULOSIS IN AMERICA*

Since there was no recognized written language in prehistoric America, evidence of the existence of pre-Columbian tuberculosis can come from only two sources: art forms and skeletal remains.

Art Forms in America

Webb (1936), in his book on the history of tuberculosis, makes the following statement which has been repeated frequently in the literature:

In pre-Columbian bronzes from Peru, in effigy water bottles in Indian burial grounds in many of our states, and from the pictographs of the cliff-dwellers, all pre-Columbian, the unmistakable picture of Pott's disease is portrayed. This picture is so characteristic that orthopedic surgeons are certain that Tuberculosis alone can be the cause.

TABLE I
DIFFERENTIAL DIAGNOSIS:
TUBERCULOSIS OF SPINE

Chronic pyogenic osteomyelitis
Traumatic arthritis
Crush fractures
Malignancy
Typhoid spine
Sarcoidosis
Actinomycosis
Blastomycosis
Coccidioidomycosis
Rheumatoid arthritis
Osteitis deformans (Paget's disease)
Osteochondritis (Calve's—Scheuermann's)
Neuroarthropathies

* It is generally agreed that America was originally populated by Asiatics who came over by way of the Bering Straits thousands of years ago. This did not occur by one group of Asiatics crossing the straits at one specific date but probably came about by many migrations over a span of hundreds of years. It is possible that some of these migrants may have brought tuberculosis with them—but it is quite unlikely that contagious tuberculosis could have survived very long because of the almost total lack of opportunity for spread in a primitive society. Another possible source of introduction of tuberculosis into America would be the Vikings (1000 A.D.) Brøndsted (1960). Actually, discovery of America by the Vikings has never been definitely proven. Recent *unpublished* excavations suggest that the Vikings settled in Nova Scotia for approximately one and one-half years. C[14] dates this occupation at about 1000 A.D. Whether this Viking occupation meant sufficient contact with the native Indians to transmit diseases has not yet been determined.

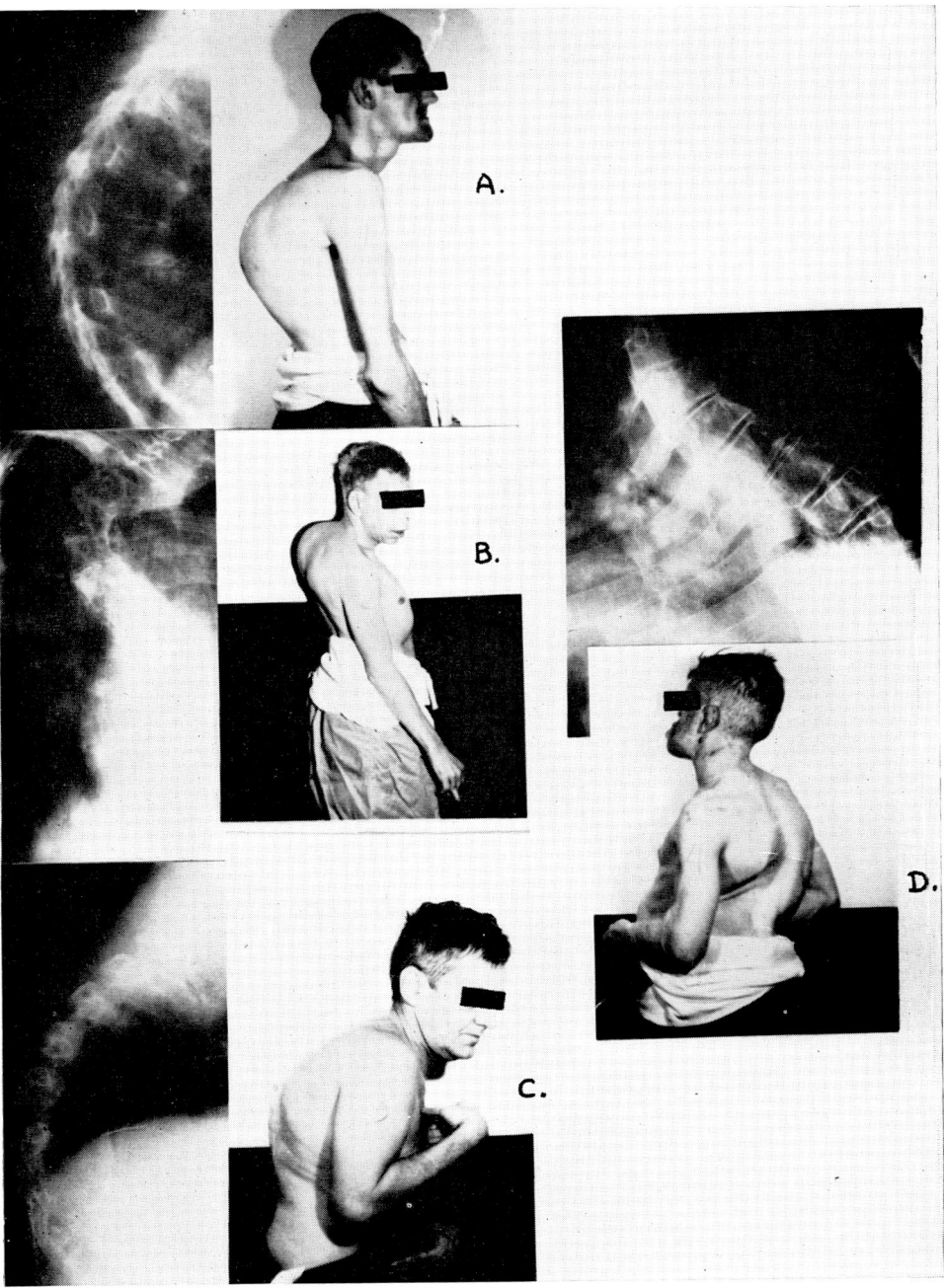

FIGURE 2. Spinal deformities not caused by tuberculosis. Four kyphotic deformities in residents of Lincoln State School, Lincoln, Illinois, an institution for the mentally retarded. Roentgenographic interpretation by Dr. Edward Wood, Consulting Radiologist. A. Wedging of vertebral bodies and marginal spurring; B. Compression deformity of thoracic vertebrae with no actual bone lesion; C. Fusion of several of vertebral bodies suggesting a congenital lack of segmentation; D. No significant bone changes.

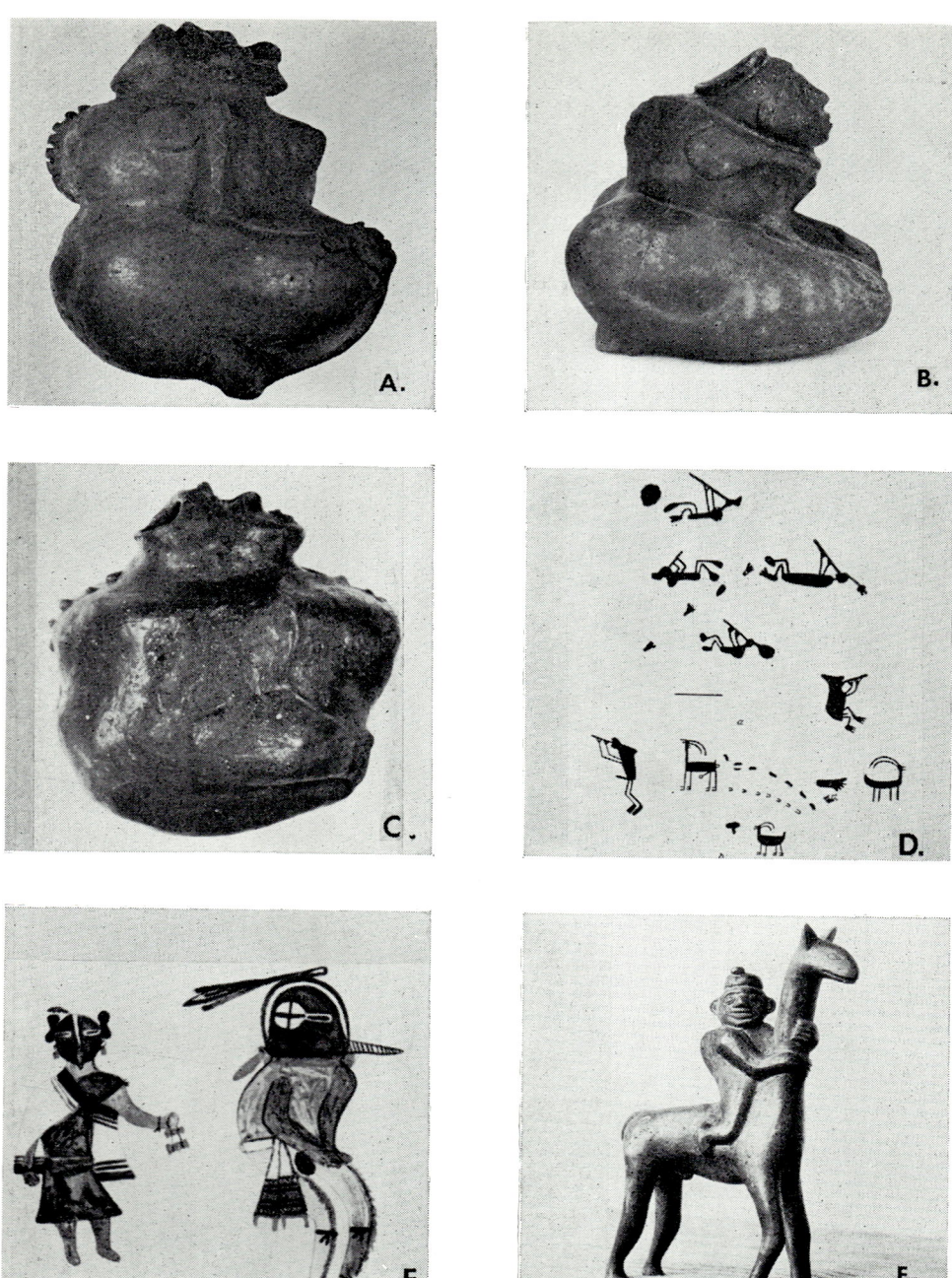

FIGURE 3. Humpback art in America.

A, B, C, Hunchback water bottles from the Memphis area. Mississippi culture. (*c.* 1400-1600 A.D.) A. Obtained by Gregory Perino for the Gilcrease Foundation from the Banks Site, Crittenden County, Arkansas. It was found with Burial 70 and represents an old "crone" (toothless old woman); B. Found February, 1959, at Bradley Site, Crittenden County, Arkansas, by T. W. Gitchell of West Memphis, Arkansas. The vessel is 6½ inches high. Note the smooth curve of the hump (Collec-

Pre-Columbian Bronzes from Peru

The Inca of ancient Peru manufactured many metal and ceramic effigies and figurines depicting various diseases and deformities. The exact number of authentic humpback figures from this area is unknown, but it is certain that they are very rare. Two are located in the U. S. National Museum. The figure of a man on a llama (Fig. 3. F) is 4½ inches tall and is made of silver. It came from Cuzco, Peru, and is pre-Spanish in time. The culture is Inca. Webb (1936) states that several of these bronzes are in the possession of the Army Medical Museum, and illustrates one in his book. However, after correspondence with the Curator of the Museum and a personal visit, only one bronze figure was found to be on exhibition, and no information as to its authenticity could be obtained. The figure pictured in Webb's book could not be located.

Hunchback art can be found sparsely scattered throughout Mexico and Central and South America (Linne, 1943). Instances of figurines, vases, and pottery vessels depicting spinal deformities have been located in Venezuela, Guatemala and Costa Rica. Many of these may be historic. On some, it is doubtful whether the intention was to portray deformity, but may represent a natural position of bending over, or the desire on the part of the artist to create symmetry. Others strongly suggest a pack on the back and not a deformed spine.

Effigy Water Bottles

In a small area of the central United States, including the contiguous corners of Arkansas, Tennessee, Mississippi, and Missouri, dozens of pottery bottles representing severely deformed hunchbacked human beings have been found by archaeologists (Fig. 3, A, B, C). These water bottles have been identified as having been manufactured by people of a prehistoric Indian culture (late Mississippian). However, some recent dating strongly suggests that at least some of these may have been post-Columbian in time. The Banks Site, Crittenden County, Arkansas, where many of these vessels were found, gave a carbon date of 1535 A.D. Artifacts found elsewhere in this "Memphis" area suggest that other sites may also have a late date. It might be assumed that the original potters used human beings as their models for these vessels. The literature contains many references that these hunchback bottles are evidence of the existence of tuberculosis in the American Indian before contact with the white man. The writer personally has seen many of these bottles and viewed photographs of many others (see: Holmes, 1882-3, 1898-9). The majority do not show

---

tion of Dr. Dan Morse); C. Obtained by Gregory Perino at the Banks Site. It accompanies Burial 244 with three other vessels. It represents a male with bulging chest. Note the five spinous processes. Height 6 inches (Gilcrease Collection); D. Pictograph from Arizona showing humpbacked flute players. From Bulletin 65 B. A. E., Smithsonian Institution (see text); E. Female and male "Kokopolo" of the Hopi Katcina cult. The male is a humpback (from drawing in twenty-first annual report of the Bureau of Ethnology, Smithsonian Institution); F. Catalog No. 210366 (Division of Archaeology) from Cuzco, Peru. Silver, 4½ inches tall. The culture is Inca and is pre-Spanish. Received by the U. S. National Museum in 1901. (Photograph— courtesy of the Smithsonian Institution).

the typical angular deformity of tuberculosis, but have smooth rounded curves with the prominent individual spines evenly spaced throughout the curvatures.

Pictographs in the Southwest

In Figure 3 D, a pictograph group is shown which was found near ruin 5 in Canyon Hagoe (Kidder and Guernsey, 1919). Webb (1936) quoted these as representing spinal deformities due to tuberculosis with involvement of the spinal cord. Some of the group are pictured lying down, which is supposed to portray paralysis of the legs and the playing on the flute is suggestive of occupational therapy. The figures which show priapism are supposed to be the result of spinal irritation. On the pictographs throughout the Southwest there have been many instances showing these hunchback flute players. From a medical standpoint, it is not likely that individuals with tuberculosis of the spine, in such an advanced stage as suggested in the pictures, would lie on their backs playing flutes—they would be dying. Elsie Parsons (1938) states that the humpback flute player displayed on the rock walls throughout the Southwest is not a human being, but represents an insect. This is the antecedent of the historic Hopi Katcina character called Kokopolo. The function of male Kokopolo (Fig. 3, E) is to chase women. Once he has seduced them he gives them a few presents which he carries in his hump. According to Roberts (1932), in a study of the historic Zuni, it is believed that the humpbacked flute player may represent a rain priest, and that the frequently associated figures of horned toads and insects are symbols of magic. Furthermore, it is very likely that all of the pictographs in the Southwest made by prehistoric Indians were not meant to be accurate reproductions of certain objects, but have a legendary or mythologic meaning.

Skeletal Remains in Prehistoric America

The following fifteen cases of possible spinal tuberculosis are worthy of description.

Case 1 (Putnam, and Whitney, 1886). The first suggestion of tuberculosis in a prehistoric American Indian was from a published report by William F. Whitney (1886), of Harvard Medical School. Apparently, the Curator of the Peabody Museum, F. W. Putnam, asked Whitney to make a study of their osteological collection. One skeleton (catalogue No. 17223) from a stone grave mound near Nashville, Tennessee, showed an extreme case of anterior angular spinal curvature. "The disease had destroyed almost the whole of the bodies of the lower cervical or upper dorsal vertebrae and they had then become united in a firm mass."

Case 2 (Moore, 1913; Hrdlička, 1913). Skeleton 227,730 is now located at the Department of Physical Anthropology, Smithsonian Institution, Washington. It was found by Clarence Moore in 1913 in a mound at Sorrel Bayou, Iberville Parish, Louisiana. The specimen was first examined and reported by Ales Hrdlička. Hrdlička stated that: "The bones of skeleton show considerable disease. They represent what is either tuberculosis or a very pronounced form of arthritis at the lower dorsal and especially the upper lumbar vertebrae, with moderate curvature forward of the spine." This specimen was seen and photographed by the present writer. All of the available vertebrae showed a moderate amount of osteoporosis and there was lipping of all bodies. Lumbar vertebrae 1, 2, 3, and 4 had a moth-eaten appearance on the surfaces of the bodies with numerous tiny bony nodulations. The pathologic appearance of this specimen has no resemblance to tuberculosis.

Case 3 (Hooton, 1930). This case was reported on by E. A. Hooton (1930), and the specimen was from the large amount of skeletal material excavated under the direction of A. L. Kidder from 1915 to 1924. It is now located in the Peabody Muesum of Harvard. The specimen (catalogue No. 60,-280) was a middle-aged female, from Pecos Pueblo, New Mexico. It is thought to belong to Glaze IV time period which would date it to about 1600 A. D. This would not only be post-Columbian but also post-Spanish; according to Hooton's monograph, the Spanish first visited the Pecos Village in 1540 under the leadership of Francisco Coronado. The specimen had a "degenerative arthritis deformans of the right shoulder joint and Pott's disease (?) or spondylitis deformans."

The present writer has not seen this specimen but from the photograph pictured in Hooton's book it appears that there is involvement of six vertebrae firmly fused together with only a slight deformity. It is more likely that this was a severe arthritis than tuberculosis.

Case 4 (Valcarcel, and Garcia-Frias, 1940). In 1940, J. E. Garcia-Frias, Medical Director of the Clavegoya Sanitarium of Jauja, Peru, reported a case of what was thought to be tuberculosis in a Peruvian "mummy." Mummies found in certain arid areas of Peru are completely dehydrated so that flesh, skin, hair and organs remain in a dried but well preserved state. By process of rehydration the tissues can be brought back to some resemblance of their original histology. This particular mummy was named *Jorobado* (meaning hunchback) and was supplied to Garcia-Frias by the National Museum of Archeology of Lima, Peru. The roentgenogram showed marked spinal kyphosis and compensating lordosis. From the ninth dorsal vertebra to the first lumbar, there was destruction of the bodies and fusion. Histological studies of the hydrated lungs showed a large amount of fibrous tissue in the right apex. Garcia-Frias concluded that the combination of spine and lung disease showed that tuberculosis is the most likely diagnosis, and the present writer agrees, although other conditions are not completely ruled out.

The mummy *Jorobado* was pronounced to be prehistoric by Dr. Valcarcel, Department of Anthropology, University of San Marcos, Lima, Peru, and his co-workers from Cuzco. However, a pre-Columbian date is difficult to establish in certain instances because there is no distinct dividing line between pre- and post-Columbian. For a considerable time after the discovery of America, many Peruvian Indians had the same customs, used the same materials, and lived the same as their ancestors had for years before America was discovered by the white man.

Case 5 (Requena, 1945; Dupouy; and Cruxent and Rouse, 1958). "The first unmistakable evidence of pre-Columbian tuberculosis in America" is a statement of Antonio Requena when describing a diseased human vertebra found in January, 1945, by himself, Dupouy and Cruxent. A personal communication from Walter Dupouy revealed that this specimen was found at the Al Palito archaeological site in the state of Carobobo, Venezuela, and that any material excavated at El Palito would certainly be dated long before Columbus discovered Venezuela in 1498.

From the description and the photographs in the report (Requena, 1945) there are two features atypical of spinal tuberculosis: (1) failure of the collapse of the diseased vertebrae to create an angular deformity, and (2) the fistulous tract, described and pictured in the photograph, opening posteriorly, would occur very rarely in tuberculosis involving the lower thoracic vertebrae.

Cases 6 (Rochester Museum Catalogue No. A. P. 526), *7* (Cat. No. A. P. 529) and *8* (Cat. A. P. 530). In 1952, three cases of suspected spinal tuberculosis from New York State were reported on by William Ritchie (1952). Each case came from a separate Indian culture. All three are in the possession of the Rochester

Museum of Arts and Sciences. According to Ritchie, there is no doubt of the pre-Columbian origin of all three. The actual specimens or the photographs and the roentgenograms were submitted for opinions as to diagnosis to seventeen different medical specialists including pathologists, orthopedic surgeons, radiologists, and several physicians interested in prehistoric pathology.

A. P. 526 belonged to the prehistoric Iroquois occupation (Ritchie, 1944) and would date to about 1200 A. D.

A. P. 529 belonged to the Owasco culture—dating somewhere between 500 and 1200 A. D.

A. P. 530 was Middle Point Peninsula and would date to about 2000 years ago.

The medical specialists, in evaluating these three cases, were not unanimous in their opinions. The majority favored the diagnosis of tuberculosis as likely for all three but, as expected, there was considerable disagreement. The difficulties of diagnosing tuberculosis from the dried specimen alone are obvious. The one atypical finding common to all three of Ritchie's specimens is the extensive involvement (in A. P. 526, 12 vertebrae; in A. P. 529, possibly 17; and in A. P. 530, at least 7).

Case 9 (O'Bannon, 1957; Lichtor, 1957). This specimen was discovered by L. G. O'Bannon during a series of excavations undertaken by the Tennessee Archaeological Society in 1957. The skeleton came from Montgomery County, Tennessee, Site M. T. 17. It is from a stone box burial and belongs to the Mississippi culture which existed in this area between 1000 A. D. to 1600 A. D., according to Professor Thomas M. N. Lewis. The specimen is an adult, but the sex could not definitely be determined. In this specimen the involvement of seven vertebrae and the excessive bony regeneration are certainly not typical of the pathology caused by tuberculosis.

Case 10 (Judd, 1954). This is a child burial from Pueblo Bonito excavated by Neil Judd in 1947. The date is 828-1130 A. D. The specimen is now on display at the Armed Forces Medical Museum. There is partial destruction of bodies of T-12 and L-2, with almost complete destruction of L-1. There is fusion of the bodies of these three vertebrae. In addition there is a 45° angular deformity.

This case is quite typical of tuberculosis with destruction of vertebral bodies, angular deformity, and very little bone regeneration.

Case 11 (Stewart). This specimen was excavated in 1955 at Nanjemoy Creek, Maryland, by Dr. Dale Stewart. The culture is Algonquin; the date is prior to 1608. No European contact material was found. All the lumbar vertebrae were involved and there are angulation deformities. Regeneration of bone is a prominent feature with complete fusion of L-1, 2, and 3. The specimen is on display at the Armed Forces Medical Museum, Washington D. C.

This case could be tuberculosis, but the amount of regeneration of bone is atypical. It also is probably not pre-Columbian in date.

Case 12 (Nash). This case is located in the burial mound at Chucalissa, a State Park near Memphis, Tennessee. The burial is one of many excavated by the Parks Archaeologist, Charles Nash, and is permanently on exhibition. It is a male, age about thirty, and belongs to the Mississippi culture. Five radio carbon dates have been obtained on this site, ranging from 1027 to 1617 A. D. This burial is in the upper part of the mound, which suggests that it probably belongs closer to the 1600 date. Although this is pre-white contact it is probably not pre-Columbian. Disease newly introduced could travel faster than the actual settlement of the territory by intrusive Old World peoples.

This specimen comes very close to being typical tuberculosis. Two vertebrae are involved. The bodies of T-7 and 8 have apparently been completely destroyed and are not present. T-7 and 8 are fused at their inferior and superior articular processes, giving an acute angular deformity (about 45°). There is very little, if any, evidence of new bone formation.

Case 13 (McGuirr and Dickson). In the early 1930's a farmer was digging for relics at the Crable Indian Site (Morse, 1960), Kerton Township, Fulton County, Illinois. Since it was known that Dickson was interested in bone pathology, anything which the farmer found that appeared abnormal was brought to Dickson Mound Park. One of the specimens was a conglomeration of ten vertebrae, T-5 through to L-1, all firmly fused together. The angle of deformity was about 160°. The vertebrae were firmly fused through the bodies. Crable is a Late Mississippi site—carbon dates are *circa* 1400 A. D. (Crane and Griffin, 1959).

The most prominent feature of this specimen is the extreme deformity and tremendous regeneration of bone. Involvement of ten vertebrae and massive bone regeneration would not favor a diagnosis of tuberculosis. Many burials have been excavated by individual archaeologists at Crable (Smith, 1951; Morse, 1960), but no other similar specimens have been found.

Case 14 (Morse, Morse and Emmons, 1961). This is burial 6, Emmons Cemetery, Fulton County, Illinois. It is a male of about thirty-six years of age. The normal height would have been about 5 feet 6 inches, but, due to the deformity and telescoping of the spine, the individual was probably about 6 to 8 inches shorter. The culture is Middle Mississippi, with definite Old Village influence (Morse, Morse, and Emmons, 1961). No carbon dates have so far been reported, but from the associated artifacts, the approximate date would be 1200 to 1300 A. D. All five lumbar and the lower three thoracic vertebrae are involved in a twisted compressed mass of bone destruction and massive regeneration. The fifth lumbar is pathologically fused to the sacrum. There are many holes in the anterior surfaces of bodies of the involved vertebrae suggesting multiple fistulae. There is a 45° angular deformity which probably was compensated by positioning. The most prominent feature, besides the bony regeneration, is the telescoping of the vertebrae. Although the neural arches are fused as are the transverse processes, the spinal canal is patent.

The features which are somewhat contrary to a diagnosis of tuberculosis are the extensive involvement (8 vertebrae and sacrum) and the large amount of bony regeneration.

Case 15 (Hiles and Dickson). This specimen found near the surface in a Hopewell Mound, was brought in to Dickson Mounds State Park by a retired construction worker and given to Don Dickson. It was the skeleton of a young male of about twenty-five years of age, probably with a height of 5 feet 3½ inches (possibly 2 inches less because of deformity). There was a severe destructive osteomyelitis involving the bodies of thoracic vertebrae 5 to 10. Almost all of the body of T-8 was completely destroyed, with partial destruction of T-7 and about three quarters of T-9. Two small sinus openings were seen on the anterior surface of T-10. The neural canal was patent. Small nodules were seen on the surface of the bodies of T-5 to 11, indicating a moderate amount of bony regeneration with bone destruction, while there was a fusion of the bodies of T-8 and 9, involving a small part of the transverse processes. A 45° angular kyphosis was caused by the collapse of bodies of T-8 and 9.

With such an extensive destructive lesion, death must have occurred rather rapidly. Bony regeneration as seen by the nodules and fusion of two of the vertebrae must have occurred early. This would suggest that this was an acute severe type of osteomyelitis and of short duration. However, it certainly could have been caused by tuberculosis. Since it would seem that this individual was either late Hopewell or post-Hopewell Woodland, the probable date would fall between 200 A. D. and 1000 A. D.

Summary of the American Cases

A series of fifteen cases of spinal diseases occurring in the American Indian is presented. Of the fifteen only four (Cases 4, 10, 12, and 15) could be considered typical enough to give a strong presumption of spinal tuberculosis. In one of these cases (Nash), a pre-Columbian date is not certain. Even in these four, other disease could have caused the same picture. In most of the remaining eleven cases, tuberculosis could have been the cause, but certain features are present which would throw some doubt on such a diagnosis.

Early Tuberculosis in the New World: A Final Comment

A contagious disease when introduced for the first time in any population is likely to be more severe because of a general *lack of population resistance*. This was true when many contagious diseases, such as measles, scarlet fever and smallpox, were first introduced to the aborigines of America after the discovery of this continent by Columbus. For example, smallpox, according to Stearn and Stearn (1945), had a mortality rate in the colonial population of 12 to 14 per cent while, in the Indians, the mortality rate was 55 to 90 per cent. Untreated tuberculosis always seemed to be a more serious disease in the Indian than in the intrusive alien. Many medical authorities, however, who have had experience in tuberculosis control among the Indians believe that this apparent severity as measured by excessive mortality and morbidity rates is not due to any racial lack of resistance, but is caused wholly by unfavorable environmental influences.

Those who claim that tuberculosis did exist in prehistoric America offer evidence from two sources: prehistoric art and reports of spinal disease found in skeletal remains. The much quoted early American art forms, when carefully analyzed, add nothing to these claims. The fifteen cases of spinal disease reported here are not too impressive, since some were of doubtful pre-Columbian origin and only four could be considered typical enough to represent presumed cases of tuberculosis.

Undoubtedly there have been other specimens of suspicious spinal tuberculosis found that have not come to our attention or have never been reported. If tuberculosis did exist in certain prehistoric American populations, there should be *many* cases of typical spinal tuberculosis found among the large amount of excavated skeletal material (see Morse, 1961, p. 491), and if only a rare case of tuberculosis-like spinal disease is found, the possibility that tuberculosis is the cause becomes extremely unlikely.

ANCIENT TUBERCULOSIS IN THE OLD WORLD

In the Old World evidence of tuberculosis should be looked for in the early literature, art forms and human remains.

Early Literature

The antiquity of tuberculosis has been firmly established by numerous descriptions of the disease found in early literature. The fact that it is mentioned in some of the earliest known literature indicates that it must have been present even before this for many generations.

Hindu Writings

In the religious hymns of the Hindu, called Vedas, which were written in Sanskrit, are found many references to tuberculosis. Apparently it is impossible to determine the exact date of the origin of the Vedas. It is believed that before they were actually put into writing they were transmitted from generation to generation orally. The oldest of the Vedas, called the *Rig-Veda*, probably started around 2,000 to 1500 B.C. (Sarma, 1939; Sigerist, 1961). It consists of some 1028 hymns all of which probably did not originate at the same time. In the *Rig-Veda*, tuberculosis is mentioned under the name of *Yaksma* (Wise, 1845) and the cure of *Yaksma* is the subject of an entire hymn (Meinecke, 1927).

The *Athawa-Veda* consisting of 731 hymns, prayers, incantations and charms is thought to have been composed around 1200 B.C. (Sigerist, 1961). Tuberculosis is called *Balasa*. *Balasa* can be cured by a "charm" and scrofula is described.

The *Yajur-Vedas* were written later. Consumption is described in detail and the causes, cures and prognoses are discussed (Castiglioni, 1941). The following quotation is from *Yajur-Veda*:

A consumptive who has good digestion, is not emaciated and is at the beginning of the disease, the physician can cure.

A consumptive who eats little, who is failing, who has diarrhoea, dyspnea, tumefaction of the scrotum and the abdomen, that one, the physician anxious to be a man of renown, will abandon.

The physician who wants great fame cures a man attacked by consumption.

Early Chinese Evidence

In the ancient literature of China there is considerable amount of confusion both as to the time of the actual origin of the literature and the distinct possibility that changes have been made through editing from time to time. This is especially true of any articles and books that could be considered medical. The first mention of tuberculosis in China was supposed to date back to the legendary period of the five rulers,* particularly to emperor Shen Nung who ruled around 2700 B.C. He was the second of the five rulers and has been called the founder of agriculture and the father of medicine. The Chinese gave him credit for the invention of the plough and the discovery of the curative value of plants (Lawall, 1927). In the literature which is supposed to have originated with emperor Shen Nung are many remedies for the cure of consumption. These included the testicles of the dog, the lungs of a hog and the flesh of the crow. For an expectorant the extract of the human excrement was advised; for haemoptysis the warm blood of a stag or the urine of a woman or child.

Babylonia and Assyria

Numerous accounts referring to tuberculosis are found in Babylonian and Assyrian literature, but as with India and China the chronology is somewhat uncertain. Somewhere around 1700 B.C. Hammurabi, king of Babylonia, compiled his famous Code of Laws (Johns, 1903). Tuberculosis is not specifically mentioned in the Code of Hammurabi, but in the following excerpt the term "fever" may include tuberculosis. "When a seignior married a woman and a fever has then seized her, if he has made up his mind to marry another, he may marry (her) without divorcing his wife whom the fever seized; she shall live in the house which he built and he shall continue to support her as long as she lives."

Castiglioni (1941) quotes the following from Mesopotamian medical texts dating to about 675 B.C. "The sick one coughs frequently, his sputum is thick and sometimes contains blood, his respirations give a sound like a flute, his skin is cold but his feet are hot, he sweats greatly and his heart is much disturbed. When his disease is extremely grave, his intestines are frequently opened."

Jastrow (1917), states that all diseases

---

* The legendary era of Chinese history (Gowen and Hall, 1926) was from circa B.C. 3000 to B.C. 2000. Actually there were Ten Emperors; five of these were called the *Five Rulers,* the first, Fu Hsi, being the greatest (Walker, 1955).

in early Babylonia and Assyria were thought to be caused by demons. The demon of wasting disease was called "Ashakku." To effect a cure the demon must be driven from the body. Some of the methods of getting rid of the culprit were: the water treatment, that is by pouring water over the patient; burning various items in the fire (such as an image of the demon, or things like onions, dates, palm clusters or bits of sheep hides); and by drugs, which had to be nasty smelling and tasting in order to convince the demon to leave.

There has been some difficulty in getting accurate translations of Mesopotamian medical texts, and it has been suggested that closer cooperation should occur between the physician and the Assyriologist (Oppenheim, 1962).

Egypt

At the present time, Ancient Egyptian literature gives us little evidence regarding tuberculosis. However, it is certain that tuberculosis must have been quite prevalent throughout Egyptian history, and it is therefore rather strange that Egyptian literature has so little to say that could be interpreted as referring to this disease (Jonckheere, 1948).

Greek and Roman Literature

An excellent and complete review of consumption in classical antiquity has already been written by Meinecke (1927). Quoting from this article, he says; "The many positive references to consumption found in Greek and Roman writers, both medical and non-medical establish beyond a doubt their full acquaintance with this malady. They manifest a keen observation of the symptoms, causes, development and dreaded results of the disease . . ."

The earliest reference in Greek literature occurs in Homer (800 B.C.). Later many others wrote about "phthisis." Among them were Sophocles, Euryphon, Herodotus, Euripides, Plato, Galen and the most famous of all, Hippocrates (460-357 B.C.).

Hippocrates' description of the disease around 400 B.C. would be considered accurate today if one would wish to describe a case of far advanced untreated pulmonary tuberculosis.

Roman writers who refer to tuberculosis include Plautus (184 B.C.) Cicero (43 B.C.), Horace (8 B.C.), Seneca (65 A.D.), Celsus (50 A.D.) and many others.

Other Literature

According to both Webb (1936) and Meinecke (1927), the Old Testament of the Bible contains a few passages that may be interpreted as referring to tuberculosis.

Alexander of Tralles (c. A.D. 600) a Byzantine physician, recommended for treatment of consumption, change of air, sea voyages and asses' milk. Tuberculosis appeared in Arabian medical literature, as seen in that of Rhazes (A.D. 900) and Avicenna (1000 A.D.).

Throughout the Middle Ages (400 A.D.—1400 A.D.) which brings us up to about the time of the discovery of America, much evidence is found in the literature that "The Great White Plague" existed even in epidemic form. It was during this time that touching by royal personages was supposed to cure the "King's Evil" (tuberculosis).

Art Forms in the Old World

There are many artistic representations from Ancient Egypt which have, at some time or another, been claimed to show various types of deformity, including hunchback. For almost all these cases there are two possible interpretations; firstly that they do indeed represent actual physical deformity, and secondly that the apparent deformity is due not to actual physical causes but to stylistic convention.

Three of the earliest figures which may well show deformity have been published as presumably Predynastic (before 3000 B.C.) in date.

The first (Schrumpf-Pierron, 1933) is made of clay and was reported to have been found by Bedouins in the Assuan desert. It represents an emaciated human with angular kyphosis of the thoracic spine,

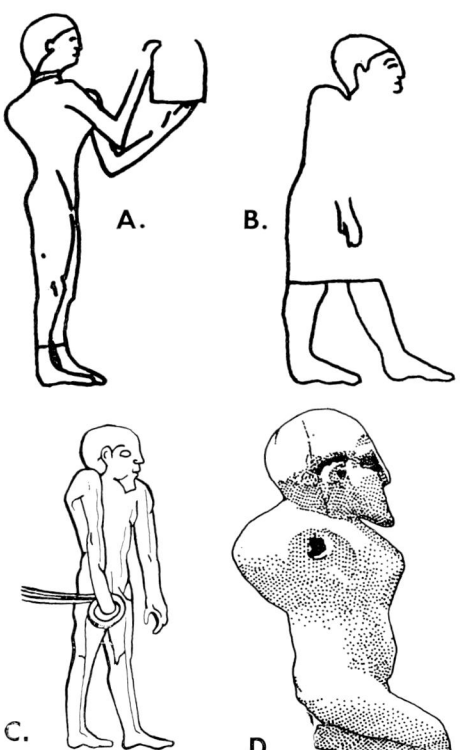

FIGURE 4. A. Drawing of a bas-relief representing a *serving girl* from Gizeh-Tomb No. 45, IVth Dynasty (Old Kingdom). B. Found in tomb of Beni Hassan XIIth Dynasty (Middle Kingdom *c*. 2000 B.C. C. Deformed servant of Ti, leading dogs, Old Kingdom (?). The deformity appears to be more a scoliosis than a kyphosis. D. Predynastic (?) wood statue (10 cm. in height) portraying a bearded male with a hunchback deformity, now located in the Brussels Museum.

and is crouched in a clay vessel (Fig. 5, A, B). The general Predynastic date assigned to it in the original publication has been generally accepted but may well be suspect.

The second possible Predynastic representation with spinal deformity (Wellcome Historical Museum) is a small standing ivory likeness of a human with arms down the sides of the body and bent forwards at the elbows (Capart, 1905). The head is carefully modelled with facial features indicated. The figure is shown with a protrusion of the back and on the chest (now extensively broken). It has been variously claimed to be of Predynastic, Semainean, and Archaic date, but recent re-analysis (Ucko, 1962) of this and the large group of similar ivory figurines (all of unknown provenance) suggests that it may be post-predynastic in date.

The last example of possible spinal deformity usually claimed to be of Predynastic date is a wooden statue in the Brussels Museum (Jonckheere, 1948). This is described as a bearded male with facial features indicated, kneeling and possibly phallic. Many of the details can no longer be made out for the arms and much of the legs are destroyed. The figure is shown with a large modelled and rounded hunchback and angular projection of the sternum. (Fig. 4 D). As with the other examples which have been claimed to be of Predynastic date, this figure has no known provenance (it was originally in the Amherst Collection) and has been dated to the Predynastic period on stylistic grounds only (Gilbert, 1948).

Turning to historic Egyptian representations which may represent tuberculosis deformity, there are several examples already in the Old Kingdom which may be interpreted as artistic ineptitude, stylistic convention or representation of actual de-

FIGURE 5. A. B. Egyptian clay (red) statuette of an emaciated man with an angular kyphosis of the thoracic spine reported to be dated *circa* 4000 B.C. (see text) and found inside a clay bowl. Height of statue 30 cm. (from Schrumpf-Pierron). C. Photograph of a painting furnished by Henry G. Fischer, Associate Curator, Department of Egyptian Art, Metropolitan Museum of Art, New York. The original painting is still on the tomb wall of *Ipuy* near Thebes. It is dated XIXth Dynasty (*c.* 1300 B.C. The subject is a deformed (?) gardener engaged in raising water with a shaduf (water elevator).

formity (Slomann, 1927; Smith, 1946; and Jonckheere, 1948). One of the most suggestive of these representations is that of a bas-relief serving-girl (Fig. 4 A) from a IVth Dynasty Gizeh Tomb No. 45 (Lepsius, 1904). This representation could well be thought to show localized angular kyphosis. More problematic are the slightly later representations of a servant of *Ti* leading dogs (Jonckheere, 1948; Smith, 1946; Steindorff, 1913), shown in Figure 4 C, and the priest *Ankh Oudges* (Koefoed-Petersen, 1956; Madsen, 1906; Slomann, 1927; Smith, 1946) which may more probably be results of profile representation than representation of deformity.

From the Middle Kingdom also there are examples of possible deformity. In a tomb at Beni Hasan (Newberry, 1894) is a representation which may show pronounced localized angular deformity of the cervical-thoracic spine (Fig. 4 B).

In the New Kingdom and later, figures with deformity probably caused by tuberculosis continue to be portrayed (Jonckheere, 1948). For the present purposes it suffices to illustrate just one example (Fig. 5 C) from the tomb of *Ipuy* (Davis, 1927). In this painting can be seen a "deformed" gardener engaged in raising water with a shaduf (water elevator).

This short review shows that throughout the history of Egypt occur representations which to modern eyes resemble tuberculous involvement of the spine. Assuming that not all the cases mentioned above are due to conventions of style or artistic ineptitude, it should be noted that the type of deformity portrayed angular kyphosis) is what would be expected to occur in recovered cases of spinal tuberculosis untreated by either surgery or splinting. Most of the prehistoric hunchback American art forms do not have angular deformities but smooth rounded curves, the cause of which, would more likely be non-tuberculous conditions (Morse, 1961).

Human Remains

Neolithic

The oft-quoted earliest European case of human tuberculosis ever found consisted of some upper dorsal vertebrae occurring in a young male adult and described first by Paul Bartels (1907). The specimen was excavated in 1907 from a Neolithic cemetery near Heidelberg. From the excellent photographs and radiographs appearing in several publications there is clearly partial destruction of the bodies of the 4th and 5th thoracic vertebrae with complete fusion. The adjoining 3rd and 6th vertebrae appear to be unaffected. An angular kyphotic deformity is present and apparently there is minimal involvement of the neural arches and spinous processes.

This gives a fairly typical picture of what could be the result of healed spinal tuberculosis; however, most of the other conditions in the differential diagnosis table could also be the cause.

Brothwell (1961) reported on two possible cases of spinal tuberculosis in early British skeletal material belonging to the Saxon period.

Earlier Work on Possible Egyptian Tuberculosis

In 1891 Grebart discovered near Thebes forty-four well-preserved mummies, identified as the priests and priestesses of Amûn from XXIst Dynasty. These were examined at the Medical School in Cairo, and one of them showed unmistakable evidence of having tuberculosis. This mummy, Nesperehân, has been referred to many times in the literature, being described in detail by Elliot Smith and Ruffer (1910), Cave (1939) and others. This was an adult male showing partial destruction of the lower thoracic and upper lumbar vertebrae creating an angular kyphotic deformity. In addition there was a huge abscess occupying the area of the right psoas muscle.

Following this announcement, other reports of tuberculosis is Egyptian material came to light. Douglas E. Derry (1938) provided a valuable summary of the cases of Egyptian spinal tuberculosis excavated up to that time (see Table II). Of special interest were four of the cases (4, 5, 6 and 7) in two graves close to each other. One grave contained two adults, a man and a woman with spinal tuberculosis. In the other grave, two of the three occupants had tuberculous-like spinal disease.

Waterman (1960) commented on an additional case of suspected tuberculosis from Gizeh (Abu-Bakr, 1933). The atypical feature of this case was massive involvement of the neural arches.

Regarding tuberculosis of other tissues of the body, it is to be expected that in view of the more uncommon occurrence of well preserved mummies, the evidence would be more slender than for spinal tuberculosis. In fact only one likely case has so far been described, in a Byzantine mummy of a Nubian woman from the island of Hesa (Elliot Smith and Wood Jones, 1908). In this mummy the left lung was collapsed and shrunken and was firmly bound to the chest wall by a series of adhesions. Such lesions could well be the result of pulmonary tuberculosis.

TABLE II
SUMMARY OF EGYPTIAN SPINAL TUBERCULOSIS EXCAVATED PRIOR TO 1938.
(Derry, 1938).

Sex, Date and Provenance	Age	Vertebrae Involved	Remarks
1 Woman—Nubia, 2000 B.C.	21 years	1st, 2nd and 3rd lumbar	2nd entirely destroyed, 3rd centrum much eaten away
2 Man—Nubia, 3000 B.C.	Adult (young)	11th and 12th thoracic	11th almost completely gone. 12th central part of body excavated
3 Man—Nubia, 3000 B.C.	Adult	1st, 2nd, 3rd and 4th lumbar	2nd almost completely destroyed and much of 1st and 3rd; 2nd, 3rd and 4th fused together
In same grave—4 Man—Nubia, 3000 B.C.	Adult	8th, 9th and 10th thoracic	9th completely eroded; all three fused together and enclosing a large abscess cavity
5 Woman Nubia, 3000 B.C.	Adult	1st and 2nd sacral	Abscess cavity involving 1st and 2nd sacral segments; surface of bone near cavity much damaged by suppuration
In same grave—6 Man—Nubia, 3000 B.C.	Adult	10th and 11th thoracic	Body of 10th almost entirely destroyed
7 Boy—Nubia, 3000 B.C.	9 years	10th, 11th and 12th thoracic, 1st and 2nd lumbar	All severely damaged and distorted and fused together, forming one irregular mass
8 Woman—Dier el Bahry, 1500 B.C.	Old	8th—12th thoracic	11th thoracic completely gone and most of 10th; 1st and 3rd lumbar commencing ulceration of body; right sacro-iliac joint disorganised
9 Woman Sakkara, 3300 B.C.	Aged	8th—12th thoracic, 4th and 5th lumbar, 1st sacral	11th thoracic completely gone and most of 10th; 5th lumbar body gone; new bone growth from 4th lumbar to base of sacrum

TABLE III
NUBIAN SPECIMENS

Catalogue No.	Location	Pathology
182 C (Fig. 5)	Sacrum. All lumbars and T-12	Destruction—Fusion—Kyphosis—narrowed neural canal
182 B (Fig. 5)	First four lumbars and T-12	Destruction—Fusion—Kyphosis—Probable T. B. plus arthritis
20 A (Fig. 5)	T 1, 2, and 3	Kyphosis—Destruction—Arches not involved
182 E	One lumbar (?)	Abscess in anterior surface of body. Is Derry's Case 1 (See Table II)
182 E	Five upper thoracic (from a different individual)	Destruction of bodies. Kyphosis—minimal arch involvement
182 E	Two lower thoracic (represents another individual)	Destruction of bodies—very little regeneration

The Nubian Specimens

A considerable number of pathological specimens from Nubia were until recently part of the collection of the Museum of the Royal College of Surgeons in London. During World War II, these specimens were moved to a basement room, but did not escape Nazi bombing, which destroyed much of the priceless collection of the Museum. They have now been transferred to the British Museum (Natural History).

Rowling (1960) reviewed some of this Nubian pathological material and more recently Morse, Brothwell and Ucko (1964) have examined the four specimens described by Rowling as well as two additional specimens which have apparently not been described or published previously (Table III, Fig. 6).

The most probable date of all of these specimens is Middle Kingdom. Except for one (Catalogue No. 182E) vertebra, none of the specimens resemble any of the cases reported by Derry (Table II) and we can presume that they represent additional examples of Egyptian spinal tuberculosis.

All these show what could be typical tuberculosis involvement although there may have been a co-existing hypertrophic arthritis present.

Another possible case of tuberculosis was presented to the British Museum (Natural History) by University College, London, through the kind recommendation of Dr. A. J. Arkell of the Egyptology Department. The material came from Gurob (Brunton and Engelbach, 1927, with a skeletal report by Elliot Smith). The skeletal material of interest to this study was from a male, and only a portion of the skeleton was present, all bones were in very poor condition. These scattered bones are thought to be from a royal prince, Pa-ra^c Messu, who was the son of Seti I, of the New Kingdom. Apparently in an effort to preserve these bones they had been covered with paraffin wax, but this has offered little protection and the bones have started to crumble and portions are detached. The bone specimens included several vertebrae, for the most part badly damaged. Two vertebrae

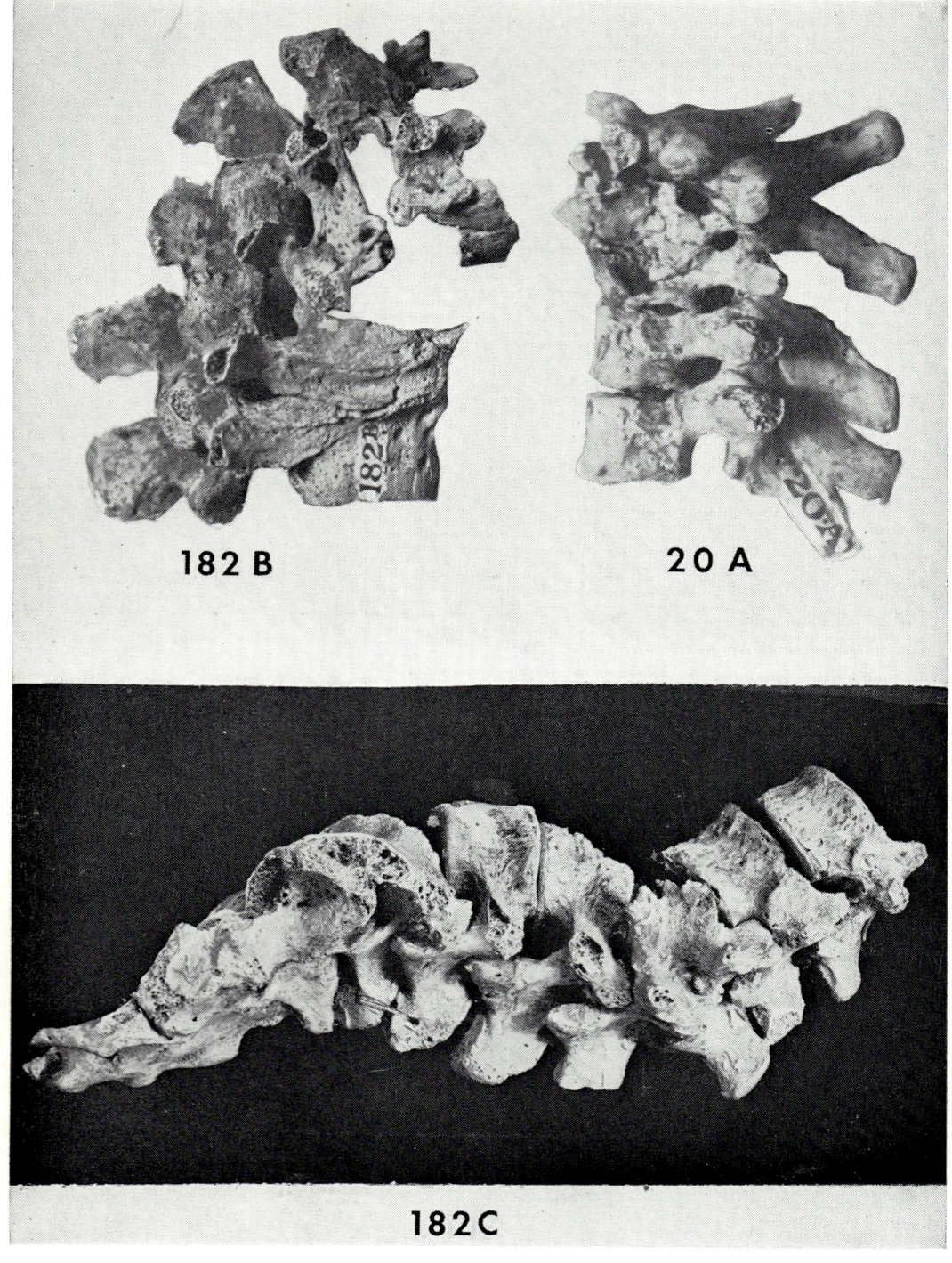

FIGURE 6. Nubian specimens, Middle Kingdom, from Royal College of Surgeons, London. Now located in the British Museum (Natural History). See text.

from the lumbar region showed definite fusion of the bodies, but only at the margins of the centra—a change typical of advanced osteoarthritis and spondylitis. There was also some involvement of the neural arches but only at the articular facets. There is perhaps a 20° angular kyphosis present in this part of the column, although exactly how much, is obscured by breakage.

The Nagada Specimens

In 1895 Flinders Petrie and J. E. Quibell (1896) uncovered a large cemetery near Nagada in Upper Egypt. Over two thousand burials were excavated. Because of the flexed position of the bodies and the accompanying grave goods, including many varieties of pottery unlike anything previously found in Egypt, Petrie announced that he had found a "New Race." It is now known that this was in fact part of the Predynastic population of Egypt.*

As a result of a request by Karl Pearson in 1894, Petrie sent back to University College, London, remains of 400 skeletons. A number of anthropologists, including Herbert Thompson, Ernest Warren (1897) and C. D. Fawcett (1901) measured the skulls from this collection, certain postcranial remains also being studied by Warren (1897). Strangely, no one seemed to pay much attention to any pathology exhibited by these Nagada skeletons, or at least, if anyone did, the results were never published.

In 1962, the Department of Pathology of Cambridge University placed on loan to Don Brothwell, of the British Museum (Natural History), a considerable amount of pathological Egyptian skeletal material, which had apparently been in the department since the turn of the century. Catalogue information on this material was no longer available, although it was certainly known to be Egyptian, and numbers were still evident on most of the specimens. At least one of the specimens, numbered T. 52, (Petrie, 1920), came from Petrie's excavations at Nagada, and indeed it is highly likely on available evidence that all these specimens, which are now more fully described (Morse, Brothwell and Ucko, 1964), are from this site. The numbering seems to have been done by one individual, and corresponds to Nagada numbering. It is certain that the Nagada skeletons became dispersed to some extent after their arrival in England, and the Biometric Laboratory in London appear to have retained only those bones of biometric interest to them. The transfer of the pathological material to Cambridge is therefore not an unreasonable assumption, although some mixture with pathological material from other Egyptian sites cannot at this late date be entirely excluded.

If all this material is accepted as being from the site of Nagada, it is of course still far from easy to date each of the specimens by means of the numbers. Many of the tombs, whose numbers are given on the bone specimens, were not described, while others contained no distinctive material from a dating point of view. All that can be said of the dating is that in terms of probability, some of the vertebral specimens are very likely to be Predynastic but that some may be of Old, Middle or early New Kingdom date, but cannot be of a more recent date than *circa* 1400 B.C. It would seem unnecessary to describe

---

* In 1953, five specimens consisting of human hair and skin from Petrie's Nagada material were submitted to W. F. Libby at the Institute for Nuclear Studies, University of Chicago, by the Department of Archaeology and Ethnology, Cambridge University, England. The carbon dates ranged from 3791 ± 300 B.C. to 2767 ± 300 B.C. None of the cases of spinal pathology quoted in this paper have been carbon dated.

TABLE IV
NAGADA

Catalogue No.	Location	Pathology	Comment
B 107	One lumbar	Abscess and bone destruction	Specimen in Petrie's "Prehistoric Egypt" has same No.
T 52 (Fig. 6)	Five lower thoracic	Destruction—Fusion— Kyphosis	Listed in Petrie's "Prehistoric Egypt"
T 7 (Fig. 6)	Ten vertebrae	Destruction and fusion	Probable T. B. plus arthritis
586	Five lower thoracic	Destruction of bodies—severe deformity	90% angular kyphosis 45% rotation to right
T 53 (Fig. 6)	Three lumbar	Destruction of bodies	Typical of T. B.
853	Two lower thoracic	Destruction of bodies. Fusion	Very likely T. B.
60	Thoracics 11 and 12, Lumbar 1.	Destruction of body of T 12 with little bone regeneration	Very likely T. B.

again in detail, each of these specimens from Nagada.

Thirteen cases have sufficiently typical pathological change to be considered as tuberculosis and seven of these can be regarded as quite typical. The case of involvement and pathology in these latter cases is summarized in Table IV (see also Fig. 7).

It is, alas, only too likely that the possible tuberculous cases reviewed in this paper represent only a part of those which have actually been found at Egyptian sites. Indeed, it is significant that the probable Nagada cases should have been quietly lingering for seventy years in a museum without description! These were at least well preserved, but how many in a fragmentary condition have been discarded as so much valuable skeletal material has been in the past? From the evidence it seems highly likely that tuberculosis was established by predynastic times, and was certainly in evidence by very early dynastic times. Further evidence, however, especially from well dated Predynastic sites is still very much needed.

SUMMARY

It is quite impossible to say whether tuberculosis has always affected the hominid line throughout its evolution. Clearly some form of tuberculosis has been with man since Neolithic times.

In the New World tuberculosis was probably introduced during post-Columbian times, but the Vikings could (just conceivably) have brought it over a few hundred year earlier.

ACKNOWLEDGMENTS

The author wishes to acknowledge the help given by Dr. Peter J. Ucko, University College, London, in the preparation of the portions of this chapter pertaining to Egyptian Archaeology.

REFERENCES

BARTELS, PAUL, 1907: Tuberkulose in der jüngeren Steinzeit. *Arch. Anthop.*, 6:243.
BRØNDSTED, JOHANNES, 1960: *The Vikings.* London, Cox and Wyman.
BROTHWELL, DON, 1961: The Palaeopathology of Early British Man. *J. Roy. Anthrop. Inst.*, 91:318.
BRUNTON, G., and ENGELBACH, R., 1927: The Tomb of Pa-ra^c Messu. Gurob, London, British School of Archaelogy in Egypt, Mem. XLI, pp. 19-26.
CAVE, E. J. E., 1939: The evidence for the incidence of tuberculosis in Ancient Egypt. *Brit. J. Tuberc.*, 33: 142.

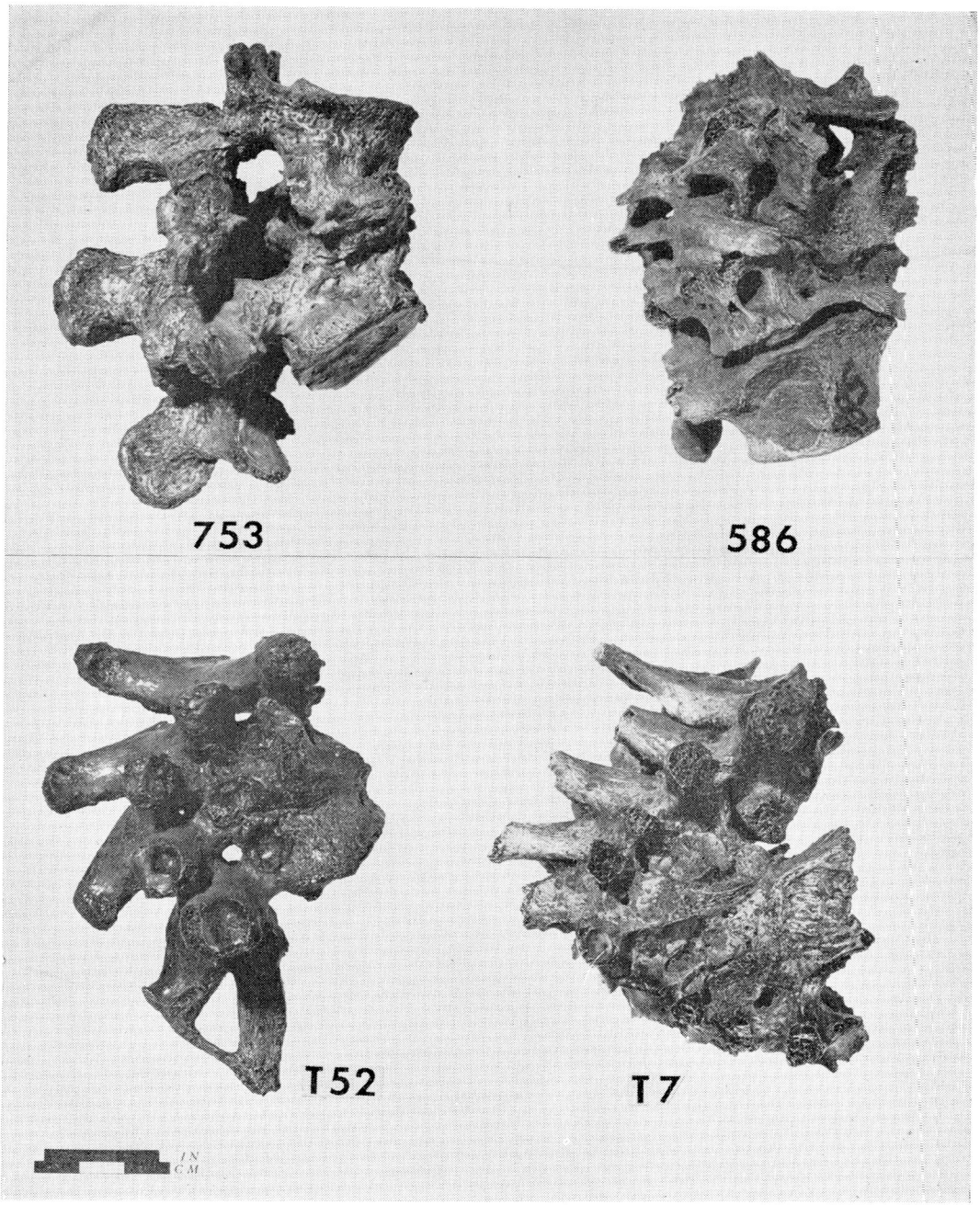

FIGURE 7. Pathological vertebrae described in the text under "Nagada Specimens" from the Department of Pathology, Cambridge University. These specimens from Ancient Egypt closely resemble tuberculosis. They were probably excavated by Sir Flinders Petrie in 1895. All are thought to date before 1400 B.C. Some are probably Predynastic (before 3000 B.C.).

CAPART, J., 1905: *Primitive Art in Egypt*. Philadelphia, Lippincott.

CASTIGLIONI, ARTURO, 1941: *A History of Medicine*. New York, Alfred A. Knopf.

CRANE, H. R., and GRIFFIN, J. B., 1959: University of Michigan, Radiocarbon Dates IV, *Amer. J. Science*, Radiocarbon Supplement, *1*: pp. 181 and 186.

CRUXENT, J. M., and ROUSE, I., 1958: *An Archaeological Chronology of Venezuela*. Washington, D. C., Social Science Monographs VI, Pan American Union, Vols. 1 and 2.

DAVIES, N., DE G., 1927: Two Ramesside Tombs at Thebes. *Robb De Peyster Tytus Memorial Series. Vol. V*, London, Pl. XXIX.

DERRY, D. E., 1908: Anatomical Report. *Archaeol. Survey of Nubia. Bull. No. 3.* Cairo, National Printing Dept.—Egypt, pp. 31-32.

DERRY, D. E., 1909: Field Notes. *Archaeol. Survey of Nubia Bull. No. 4.* Cairo, National Printing Dept.—Egypt 22-28.

DERRY, D. E., 1938: Pott's Disease in Ancient Egypt. *Med. Press*, 197:1.

FAIRSERVIS, W. A. JR., 1959: *The Origins of Oriental Civilization*. New York, New American Library (Mentor Books).

FAWCETT, C. D., 1901: A second study of the variation and correlation of the human skull with special reference to the Naqada crania. *Biometrika*, 1:408.

GARCIA-FRIAS, J. E., 1940: La tuberculosis en los antiquos Peruanos. *Actualidad Med. Peruana*, 5:274.

GILBERT, P., 1948: La Date de la figure de Bossu. *Chronique d'Egypte*, No. 45, p. 35.

GIRDLESTONE, G. R., 1940: *Tuberculosis of Bone and Joint*, New York, Oxford University Press.

GOWEN, H. H., and HALL, J. W., 1926: *An Outline of the History of China*. New York, Appleton.

HOLMES, W. H., 1882-83: Ancient Pottery of the Mississippi Valley. Washington, D. C., Bureau of American Ethnology, Fourth Annual Report, pp. 425-426.

HOLMES, W. H., 1898-1899: *Aboriginal Pottery of the Eastern United States*. Washington, D. C., Bureau of American Ethnology, Twentieth Annual Report, p. 95.

HOOTON, E. A., 1930: *Indians of Pecos Pueblo*. New Haven, Yale University Press.

HRDLIČKA, A., 1913: A report on a collection of crania and bones from Sorrel Bayou, Iberville Parish, Louisiana, *J. Acad. Nat. Sci. Philadelphia*, 16:97.

JASTROW, MORRIS., 1917: Babylonian-Assyrian Medicine. *Ann. Med. Hist.*, 1:232.

JOHNS, G. H. W., 1903: *The Oldest Code of Laws in the World*. Edinburgh.

JONCKHEERE, F., 1958: Le bossu des musees royaux d'art et d'historie de Bruxelles. *Chronique d'Egypte*, No. 45:25.

JUDD, N. M., 1954: *The Material Culture of Pueblo Bonito*. Washington, D. C., Smithsonian Miscellaneous Collections, 124.

KIDDER, A. V., and GUERNSEY, S. J., 1919: *Archaeological Explorations in Northeastern Arizona*. Bureau of American Ethnology Bulletin 65:195.

KOEFOED-PETERSEN, O., 1956: *Catalogue des Bas-Reliefs et Peintures Egyptiens*. Copenhagen, Glyptothéque ny Carlsberg.

LAWALL, C. H., 1927: *Four Thousand Years of Pharmacy*. London, J. B. Lippincott.

LEPSIUS, C. R., 1904: *Denkmaler Aegyptens V. Aethiopens Vol. II*. Leipzig.

LICHTOR, J., and LICHTOR, A., 1957: Paleopathological evidence suggesting pre-Columbian tuberculosis of the spine. *J. Bone Joint Surg.*, 39A:1398.

LINNE, S., 1943: Humpbacks in Ancient America. *Ethnos*, 8:161.

LUCK, J. V., 1950: *Bone and Joint Diseases*. Springfield, Thomas.

MADSEN, H., 1906: Ein Künstlererisches Experiment im alten Reiche. *Z. Aeg. Sprach.* 42:00.

MEINECKE, BRUNO, 1927: Consumption in Classical Antiquity, *Ann. Med. Hist.*, 9:379.

MOODIE, R. L., 1923: *Paleopathology*. Urbana, University of Illinois Press.

MOORE, C. B., 1913: Some aboriginal sites in Louisiana and in Arkansas. *J. Acad. Nat. Sci. Philadelphia*, 16: 13.

MORSE, DAN, 1961: Prehistoric Tuberculosis in America. *Amer. Rev. Resp. Dis.*, 85:489.

MORSE, DAN, BROTHWELL, DON, and UCKO, PETER J., 1964: Tuberculosis in Ancient Egypt. *Amer. Rev. Resp. Dis.* 90:524.

MORSE, D. F., 1960: The Crable Site. *Central States Archaeological Journal*, 7: No. 4.

MORSE, D., MORSE, PHILLIS, and EMMONS, M. 1961: The Southern Cult. The Emmons Site, Fulton County, Illinois. *Central States Archaeological Journal*, 8: No. 4.

NEWBERRY, P. E., 1894: *Beni Hasan. Part II*. London,

O'BANNON, L. G., 1957: Evidence of tuberculosis of the spine from a Mississippi box burial. *Tennessee Archaeologist*, 13:00.

OPPENHEIM, A. LEO, 1962: Mesopotamian Medicine. *Bull. Hist. Med.*, 26:107.

PALES, D. L., 1930: *Paleopathologie (Tuberculose Prehistorique.)* Paris, Masson.

PARSONS, E. C., 1938: The humpbacked flute player of the Southwest. *Amer. Anthrop.*, 40:337.

PETRIE, W. M. F., 1920: *Prehistoric Egypt*. London, Bernard Quaritch.

PETRIE, W. M. F., and QUIBELL, J. E., 1896: *Nagada and Ballas*. London, Bernard Quaritch.

REQUENA, A., 1945: Evidencia de Tuberculosis en la America pre-Colombia. *Acta Venezolana*, 1:00.

RITCHIE, W. A., 1944: *The Pre-Iroquoian Occupations of New York State*. Rochester, Rochester Museum of Arts and Sciences.

RITCHIE, W. A., 1952: Pathological evidence suggesting pre-Columbian tuberculosis in New York State. *Amer. J. Phys. Anthrop.*, 10:305.

ROBERTS, F. H. H., 1932: *The Village of the Great Kivas.* Bureau of American Ethnology Bulletin III, p. 150.

ROWLING, J. T., 1960: *M.D. Thesis.* University of Cambridge, Unpublished.

SARMA, P. J., 1939: The Art of Healing in the Rigveda. *Ann. Med. Hist.*, 1:541.

SCHRUMPF-PIERRON, B., 1933: Le Mal de Pott en Egypte 4,000 ans avant notre ère. *Aesculape*, 23:295.

SIGERIST, H. E., 1961: History of Medicine Vol. II. *Early Greek, Hindu and Persian Medicine*, New York, Oxford Un. Press.

SLOMANN, H. C., 1927: Contribution de la Paleo-Pathologie Egyptienne. *Bull. Mem. Soc. Anth. Paris,* VIIe Sér., t. VIII: 62.

ELLIOT SMITH, G., and DERRY, D. E., 1910: Anatomical Report. *Archaeological Survey of Nubia No. 5.* Cairo, National Printing Dept.—Egypt, pp. 21-22.

ELLIOT SMITH, G., and WOOD JONES, F., 1908: The Anatomical Report. *Archaeological Survey of Nubia, Report of 1907-1908.* Cairo, National Printing Dept. Egypt.

ELLIOT SMITH, G., and RUFFER, M.A., 1910: Pottsche Krankheit an einer ägyptischen Mumie aus der Zeit der 21 Dynastie (um 1000 V. Chr.). in Karl Sudhoff's *Zur historischen Biologie der Krankheitserreger*, Leipzig, pp. 9-16.

SMITH, H., 1951: *Crable Site.* Ann Arbor, University of Michigan Press, Anthropological Papers No. 7.

SMITH, W. S., 1946: *A History of Egyptian Sculpture and Painting in the Old Kingdom.* London, Figs. 171, 176, 157.

STEARN, E. W. and STEARN, A. E. 1945: *The Effect of Smallpox on the Destiny of the Amerindian.* Boston, Bruce Humphries.

STEINDLER, A., 1952: Postgraduate lectures on orthopedic diagnosis and indications. *Tuberculosis of the Skeletal System*, Section A, Vol. III, Springfield, Thomas.

STEINDORFF, G., 1913: *Das Grab des Ti II.* Leipzig.

UCKO, PETER J., 1962: *Prehistoric Anthropomorphic Figurines of the Ancient Near East and Aegean* (Unpublished Ph.D. thesis, London University).

WALKER, K., 1955: *The Story of Medicine*, New York, Oxford Un. Press.

WARREN, E., 1897: An Investigation of the Variability of the Human Skeleton with especial reference to the Nagada Race. *Phil. Trans.*, 189B:135.

WATERMANN, R., 1960: Palaeopathologische Beobachtungen an altägyptischen skeletten und Mumien. *Homo, 11*:167.

WEBB, G. B., 1936: *Tuberculosis; Clio Medica.* New York, Paul B. Hoeber.

WHITNEY, W. F., 1886: Notes on the anomalies, injuries and diseases of the bones of the native races of North America. *Annual Reports of the Peabody Museum of American Archeology & Ethnology*, 3: 433.

WISE, T. A., 1845: *Hindu System of Medicine*, London.

Chapter 20

Paraplegia

J. THOMPSON ROWLING

PARAPLEGIA is one of the most striking and dramatic catastrophes which affect mankind. To the observer the picture is clear-cut, the signs beyond dispute and the clinical condition of the patient not easily forgotten. To the patient it is an unredeemed disaster. It is not surprising that a clear description of the condition should have been given as early as the Old Kingdom in Egypt.

Paraplegia is due to a lesion of the spinal cord, causing paralysis and anaesthesia of the body below the level of the affected segment of the cord. The commonest cause is a fracture of the spine with mechanical division of the cord. Less common causes are pressure on the cord due to tumours, or to pressure from a tuberculous abscess associated with spinal tuberculosis. The onset of symptoms in the latter cases is gradual and the paraplegia may take weeks or months to develop. The clinical picture varies with the site of the lesion. In the upper cervical region the muscles of respiration are paralyzed (together with all four limbs) and death takes place within a few minutes. This is the mode of death in judicial hanging. At the other extreme a lesion of the lowest part of the cord may produce bladder disturbance and perineal anaesthesia only. The commonest site of traumatic paraplegia is at the dorso-lumbar region producing paralysis from the waist downwards. If the cord is completely divided, as it usually is in traumatic cases, no recovery is possible, as the unmyelinated nerve fibres of the cord have no power of regeneration. However, if the paraplegia is due to external pressure, particularly that associated with tuberculosis, it may be incomplete and some degree of recovery is possible, although perhaps unlikely.

In traumatic paraplegia the injury is followed by a period of spinal shock, lasting a few days, associated with retention of urine and complete flaccidity of the muscles. After this passes off reflex micturition is established whereby the bladder empties itself spontaneously when full, without the patient's conscious knowledge or control. The muscles regain their tone and muscular spasms may develop. The complete lack of voluntary control and of sensation persists.

In traumatic cases the initial mortality from the injury to the spine is low, but death during the ensuing months used to be very common. As recently as 1948 the mortality in the first two years was as high as 56 per cent (Nichols, 1965). This has been greatly reduced at the present time. Death is due primarily to urinary and respiratory infections, atrophic ulceration and bedsores.

Paraplegia in Egypt can be studied from two aspects. Firstly, there are the contemporary references to the condition in the papyri, and secondly there is the evidence derived from mummies and bony specimens.

The most important references to paraplegia occur in the Edwin Smith papyrus. This is a formal exposition of injuries and their treatment and deals with surgical conditions only. It is notable for its rational and logical approach and for the clinical acumen and observation of its author, contrasting with the somewhat confused and less rational outlook of the other medical papyri. The work consists of a series of case histories, each consisting of a clinical description of the injury, a diagnosis and instructions concerning treatment. The injuries commence with the head and descend through the neck and trunk, but at the forty-eighth case—an injury to the dorsal spine—the text ceases, the papyrus being incomplete. Glosses have been added to each case description, possibly about the end of the Old Kingdom, serving to interpret some of the archaic phraseology of the original text. The papyrus itself dates from the New Empire, probably the XVIIIth Dynasty, but it seems certain that it is a copy of an Old Kingdom text (Breasted, 1930).

Cases thirty-one and thirty-three merit quotation in full:

Thirty-one

Instructions concerning a dislocation in a vertebra of his neck.

If thou examinest a man having a dislocation of a vertebra in his neck, shouldst thou find him unconscious of his two arms and two legs on account of it, while his phallus is erected on account of it and urine drops from his member without his knowing it; his flesh has received wind, his two eyes are bloodshot; it is a dislocation of a vertebra of his neck extending to his backbone which causes him to be unconscious of his two arms and his two legs. If, however, the middle vertebra of his neck is dislocated, it is an *emissio seminis* which befalls his phallus.

Thou shouldst say concerning him, "One having a dislocation of a vertebra of his neck, while he is unconscious of his two legs and two arms . . . and his urine dribbles. An ailment not to be treated." Gloss A. As for "A dislocation (wnh) in a vertebra of his neck" he is speaking of a separation of one vertebra of his neck from another, the flesh which is over it being uninjured; as one says, "It is wnh," concerning things which had been joined together, when one has been severed from another."

The phrase "his flesh has received wind," refers to distension of the bowel with gas which is a common but transient feature in paraplegia.

Thirty-three

Instructions concerning a crushed cervical vertebra in his neck.

If thou examinest a man having a crushed cervical vertebra in his neck and thou findest that one vertebra has fallen into the next one, while he is voiceless and cannot speak; his falling head downwards has caused that one vertebra crush into the next one; and shouldst thou find that he is unconscious of his two arms and his two legs because of it, thou shouldst say concerning him; "One having a crushed vertebra in his neck; he is unconscious of his two arms and his two legs and he is speechless. An ailment not to be treated."

Gloss A. As for "A crushed vertebra in his neck" he is speaking of the fact that one vertebra of his neck has fallen into the next, one penetrating into the other, there being no movement to and fro. Gloss B. As for "His falling head downwards has caused that one vertebra crush into the next" it means that he has fallen head downwards upon his head, driving one vertebra of his neck into the next."

It is noted that in both cases the verdict is given as "a case not to be treated." With quadriplegia the prognosis must have been extremely bad and it seems likely that death took place within a few hours or days of the injury.

One other interesting observation may be made. In the thirty-first case it is specifically stated that "the flesh which is over it being uninjured . . ." In the second case, the impaction of the upper vertebra into the lower one at the site of the fracture dislocation is specified, "one penetrating into the other, there being no movement to and fro." Neither of these observations could have been made without post mortem dissection. It has been suggested that the surgical pathology might have been observed during the process of embalmment of the dead man. However, it is likely that the cases which are described occurred among the working population rather than the nobility. If we accept the papyrus as being a copy of an Old King-

FIGURE 1a. Tuberculosis of the lumbar spine.

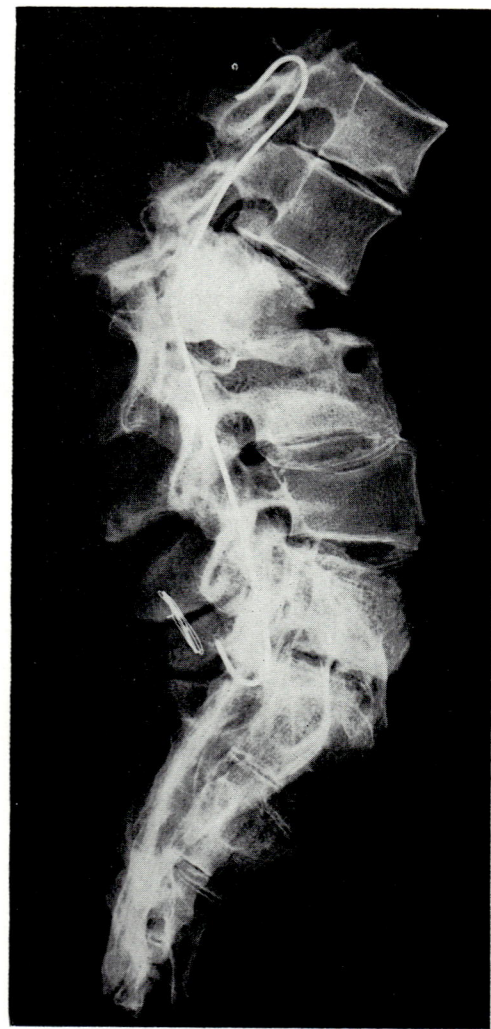

FIGURE 1b. Radiographic appearance of the vertebrae shown in Figure 1a.

dom document, and on the present evidence this seems reasonable, the working population of the time did not enjoy the benefits of mummification which became so universal after the New Empire period. We must, therefore, credit the author with sufficient authority and interest in surgical pathology to perform deliberate post mortem examinations on his patients and with sufficient knowledge of anatomy to both recognise the lesion and give an extremely accurate description of it.

The evidence of paraplegia derived from the study of mummies and bony specimens is more equivocal. Many specimens of spinal fractures are known, one of the largest collections being in the Nubian specimens of the Royal College of Surgeons of England. These are all bony specimens, however, without any surviving soft tissue and the occurrence of cord damage with bony injury is so variable that it is impossible to say with certainty whether paraplegia was present in a particular case or not. Certainly some of these specimens are compatible with traumatic paraplegia. Similarly, tuberculosis of the spine was common in specimens from all periods of Egyptian history, and it is reasonable to suppose that paraplegia associated with it was not unknown. Although the evidence of spinal tuberculous lesions is unequivocal, we have only presumptive evidence of secondary soft tissue changes in the cord. This presumptive evidence is extremely strong. Figure 1 shows a typical tubercular lesion of the lumbar spine, and similar lesions of the dorsal spine, where the occurrence of paraplegia is more common, are frequent.

The Nubian Collection contains two femora, two radii and ulnae, the pelvis, sacrum and associated lumbar vertebrae all numbered 178 A and, therefore, presumably from the same body, which is given as of Predynastic date. In many ways these bones represent the most interesting specimens in the collection.

The femora present a remarkable appearance (Fig. 2, 3). The bones are light and fragile. Great sheets of new bone extend from the linea aspera and from the sites of most other muscle attachments. Separate sheets representing the adductor magnus, the adductor brevis and the adductor longus can be made out. The changes are most marked on the right, where there is a great quantity of addi-

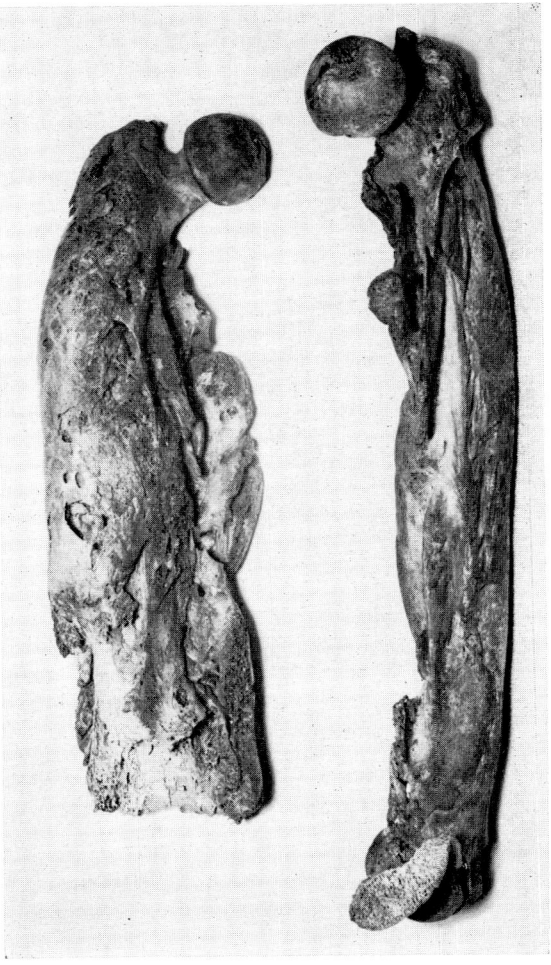

FIGURE 2. Ectopic new bone. Right and left femora.

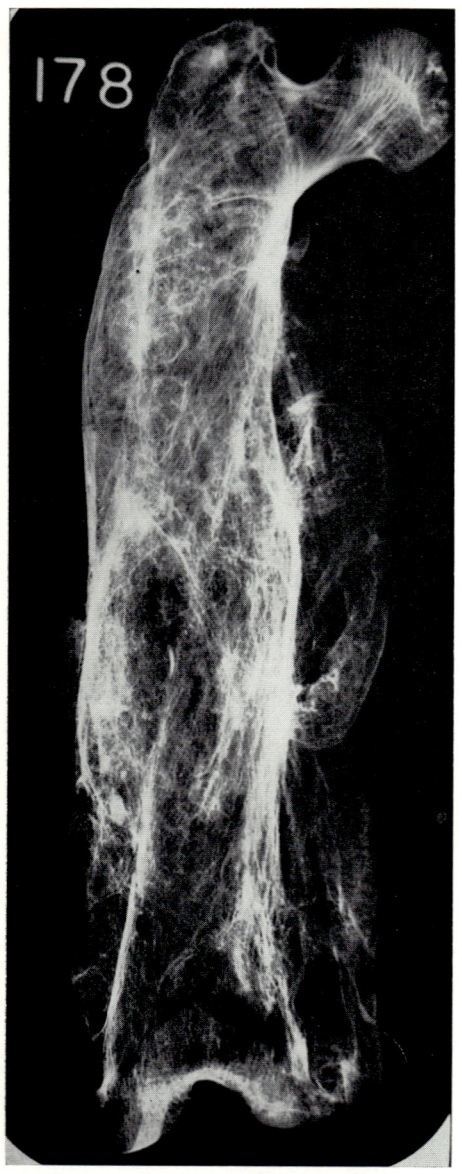

FIGURE 3. X-ray appearances of the right femur shown in Figure 2.

tional spongy new bone on the anterior aspect of the shaft. The edges of these bony sheets are smooth, regular and quite clear cut. Deep to the spongy mass, the cortex of the femur appears. It does not look in any way abnormal. The right femur is broken in its lower third—post mortem—and no signs of a sequestrum or cloacae can be seen in it, a finding confirmed in the X-ray studies (Fig. 3).

The pelvis shows similar, but less exuberant new bone formation over the dorsum ilii and labrum of the acetabulum.

Both radii and ulnae are short, hard and dense. They are extremely bowed with the convexity anteriorly, the bones being fixed in three quarter supination by ossification of the interosseous membrane, and the general deformity of the bones. No trace of movement can have taken place at the radio-ulnar joints (Fig. 4). The elbow and wrist joint surfaces appear to have been functional on both sides. The humeri are relatively normal in length and appearance.

The sacrum and lumbar vertebrae are hard and dense. A posterior defect in the neural arch is seen in the upper two sacral segments (Fig. 5).

These findings present a fascinating diagnostic riddle. At first sight the appearances of the femora suggest chronic osteomyelitis, but the lightness of the bones and the entire absence of sequestra, together with the regular deposition of the new bone in sites anatomically conforming to muscle attachments, refute this diagnosis.

The suggestion of simple widespread myositis ossificans is not supported by the relatively thin forearm bones or the normal vertebrae.

The question arises as to whether the changes could be associated with osteogenesis imperfecta, particularly that of the cystic type of the disease. The bones are from an adult, and there are no traces of epiphyseal lines on X-ray plates of the specimens. The medulla is not actually expanded in either femur, and the anatomy of the trabeculae, particularly the *calcar femorale*, is almost normal, without

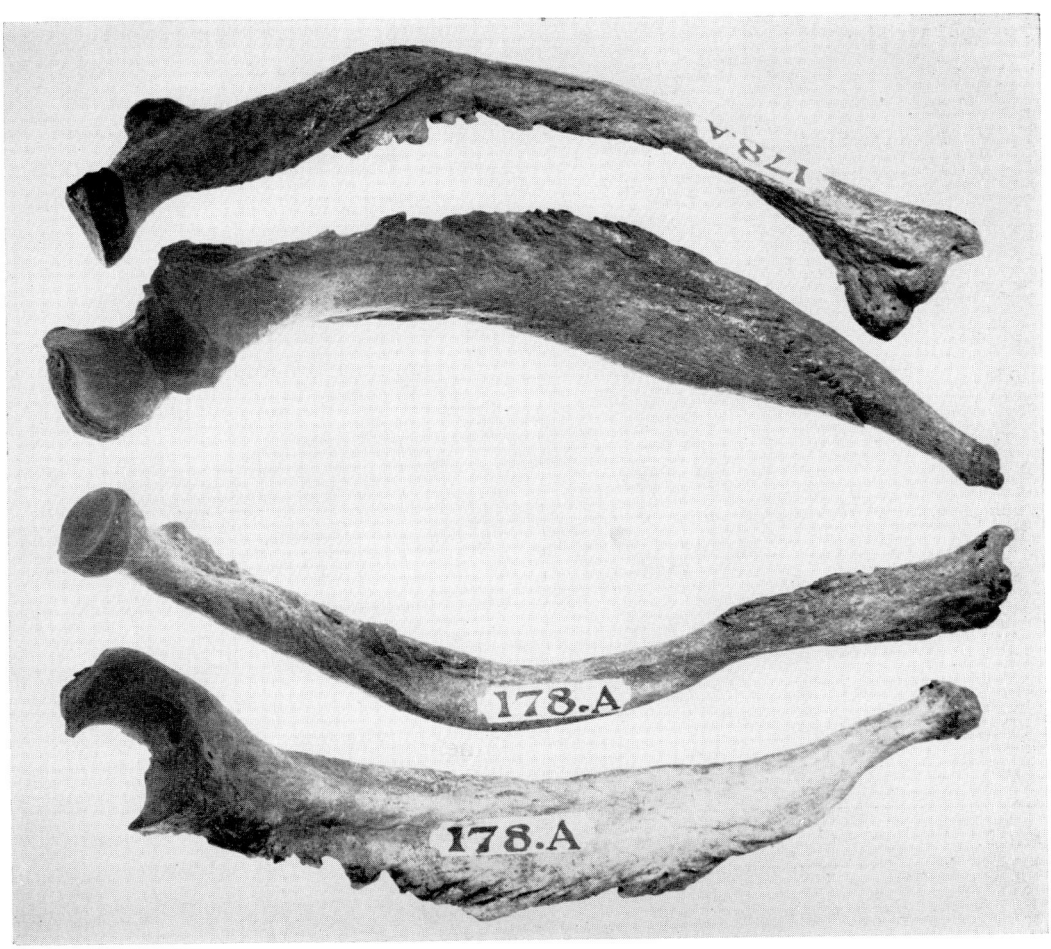

Figure 4. Radii and ulnae from the same specimen.

the cystic expansion so characteristic of dysplasia of bone. The cortex of the bone, especially that of the left femur, is almost normal in diameter, the new bone being essentially extra-cortical and corresponding too closely to the muscle attachments to suggest a medullary dysplasia. Finally there is none of the fusiform distribution of new bone which one would associate with the fractures of osteogenesis imperfecta, and the condition of the articular surfaces suggests an age at which the disease would be exceptional.

It seems likely that the changes were the result of paraplegia, possibly incomplete, and probably associated with the spina bifida. This suggestion provides a solution compatible with all the facts at our disposal. It has been known for many years that there is a tendency for ectopic new bone to form in paraplegic patients. These bony deposits affect the pelvis and thigh only. They are never found higher, nor do they affect the leg below the knee. They have a tendency to follow the planes of muscles and fascia, and characteristically consist of regular, cancellous bone with large lacunae at intervals. The time of onset is variable, but they may start as early as twenty days after the onset of

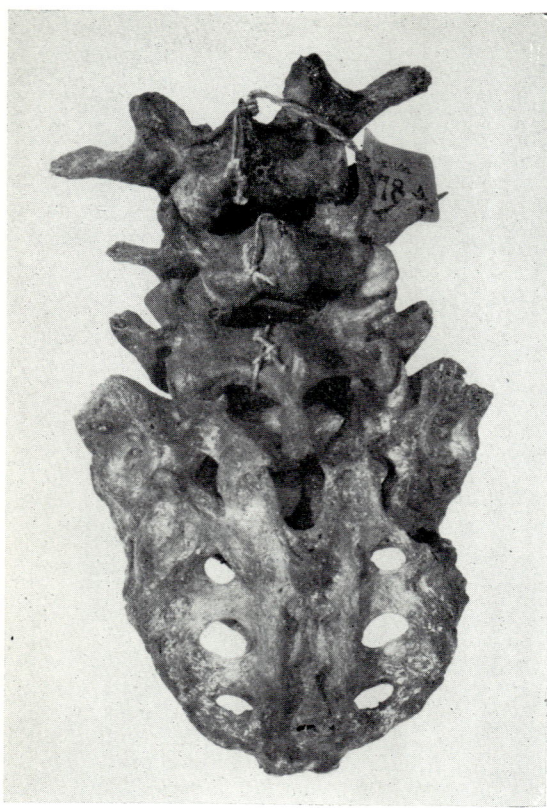

FIGURE 5. Sacrum and lower lumbar vertebrae from the same specimen.

paraplegia. The joints are characteristically unaffected. The changes are irreversible, but the process may undergo spontaneous arrest at any time. A full description of the condition is given by Lodge (1955, 1956). The present specimens fulfill all these criteria and a diagnosis of ectopic ossification, secondary to paraplegia is by far the most satisfactory explanation of the femoral changes.

If we accept this supposition, the sacral defect suggests the cause of the paraplegia. The spina bifida is, of course, at the level of the cauda equina, but it may well have been associated with a cord defect above.

On this hypothesis, the changes in the forearm bones might be due to the fact that the patient was forced to progress, using his hands to support his body weight. This alone would not account for the deformity, but if it were accompanied by a dietary deficiency, more particularly rickets, the deformity would be explained. It must be admitted, however, that rickets appears to have been extremely rare in Egypt.

It might be argued that a Predynastic Egyptian would have had little chance of surviving with paraplegia. We know, however, from the extent of the changes in ankylosing spondylitis, that invalids were often well cared for in Predynastic times, and the fact that the patients survived long enough to develop these gross changes, whatever the cause, indicates that he was cared for remarkably well.

REFERENCES

BREASTED, J. H., 1930: *The Edwin Smith Surgical Papyrus.* Univ. of Chicago Press.
LODGE, T., 1955: *Recent Advances in Radiology,* 3rd Ed. London, Churchill.
LODGE, T., 1956: *Acta Radiol.,* 46:435.
NICHOLS, A., 1965: Personal Communication.

Chapter 21

Syphilis

C. W. GOFF

The interpretation of the evidence for disease in ancient bones is much like the problem confronting linguistic experts interpreting an ancient scroll. There are many approaches. Champollion encountered many difficulties, not unlike those of a palaeopathologist, yet he broke through the barriers of early Egyptian writing. Investigators over the past twenty years have become much more cautious in expressing opinions as our fund of knowledge has increased concerning the *Treponema*. Our lack of knowledge of specific features prevents positive identifications.

Since the concept of a general treponematosis, with at least four subdivisions or varieties was accepted, pathologists and epidemiologists have become far more cautious. The genetic study of bacteria has shown the importance of mutations and of the environmental influence on prevalence and distribution of diseases which these microorganisms may cause. Such factors are particularly evident in the treponemal disorders of pinta, yaws, endemic syphilis (*bejel*) and venereal syphilis.

Hamlin (1939), in summarizing the evolutionary affinity of the treponematoses —yaws and syphilis—believes that the evidence points to the greater antiquity of these disorders in the New World. His suggestion that syphilis is the more recent morphologic type fits these theories. Hackett (1963), writes "the more likely it is that the four treponemas are a consecutive series starting with pinta."

Hudson (1963), believes that medical history provides many illustrations of the conversion of yaws to endemic syphilis and later to venereal syphilis, and Willcox (1960) agrees with this suggestion. The process, according to Hudson, can take the form of a sudden mutation or can be a progressive but gradual change. Incidentally, he believes that the process can be reversed and syphilis can become yaws. The Treponemata may diverge but never lose their ability to regain their former status of infectibility. If Hudson is correct, the differential diagnosis of treponemal bony lesions of antiquity becomes less important. To him venereal syphilis is a disease of advanced urbanization and endemic syphilis and yaws a disease of villages and the unsophisticated. They occur for practical purposes only in man. On the other hand, the possible infectibility of other animals poses problems. Originally some animal form of the treponemal infection may have existed and been transmitted to man.

HISTORICAL

Venereal syphilis seems to have suddenly increased its virulence about 1495-1496. An epidemic took place in the Mediterranean area, which spread throughout

Europe. It probably was caused by the introduction of either a new strain or a mutant of powerful infectability. Likewise, in areas of the New World in particular, an immunity had been developed. The exact situation may never be known, but, according to many reliable investigators, yaws was endemic in the New World about that time. Other treponemas more than likely were also present, circumstances altering infectibility. Climate, culture, soil, and possibly an animal vector not yet recognized, are aspects influencing the variety or degree of intensity of pathological processes. These must be taken into account, when describing characteristic bone lesions.

Shattuck, in 1938, believed the evidence to indicate that syphilis was an ancient disease of Yucatan and Guatemala. Certainly yaws was imported from Africa. He states that "nearly all the more deadly epidemic diseases known in the New World since its discovery by Columbus have been imported from the Old World within historic times." He believed that the Indians of Mexico and Central America today show (1) a low incidence of syphilis; (2) a very high proportion of clinical latency, and (3) almost a complete absence of syphilis of the internal organs. In discussing yaws and syphilis, he inclines to the view that syphilis was quite prevalent and severe among certain tribes of North American Indians in distant times. Just what this means is not clear. Of course, he is largely writing about the clinical picture with which he was familiar. However, he used Wassermann serology to determine a positive degree of syphilis. We know now that false positives are quite common. All of his quotations, which are many, seem to indicate that syphilis was diagnosed clinically amongst the American Indian rather haphazardly and with the greatest of ease. This indicates to me that many are not reliable. In his summary, he states "evidence derived from pre-Columbian bones points to the existence of syphilis among Indians before contact with Whites, but Hrdlička seems to hold the contrary opinion."

PALAEOPATHOLOGY

Anyone who has shown a dried bone to a present day pathologist for an opinion relative to any abnormality that the bone might show, knows well the diversity of opinions which may be obtained. There is always a tendency to think of the dried bone in terms of a fresh human specimen. Thus, an active soft tissue lesion over a frontal bone of the cranium, if extensive and foul, carries with it the impression for the pathologist, that the bone itself must be equally involved, foul and necrotic. This is particularly true with our modern day approach to syphilis and yaws, and it is most unfortunate. Because of this, the study of ancient dried specimens has not been particularly rewarding.

In a recent personal communication, Hackett suggested that the first proposition to overcome this dilemma should consist of a statement of conditions necessary in order to establish an opinion of the pathologic typology. This approach is reasonable, logical and although most difficult will be adopted.

TYPICAL LONG BONE VENEREAL SYPHILIS

An American Indian skeleton of recent date from Red Rock Valley, Arizona, was given in 1931 to the late Professor E. A. Hooton, by Earl Morris (P.M. N 1267. Carnegie Institute Expedition). The skeletal remains, of a man over forty years were partially mummified; and displayed gross defects of frontal and nasal bones, eroded

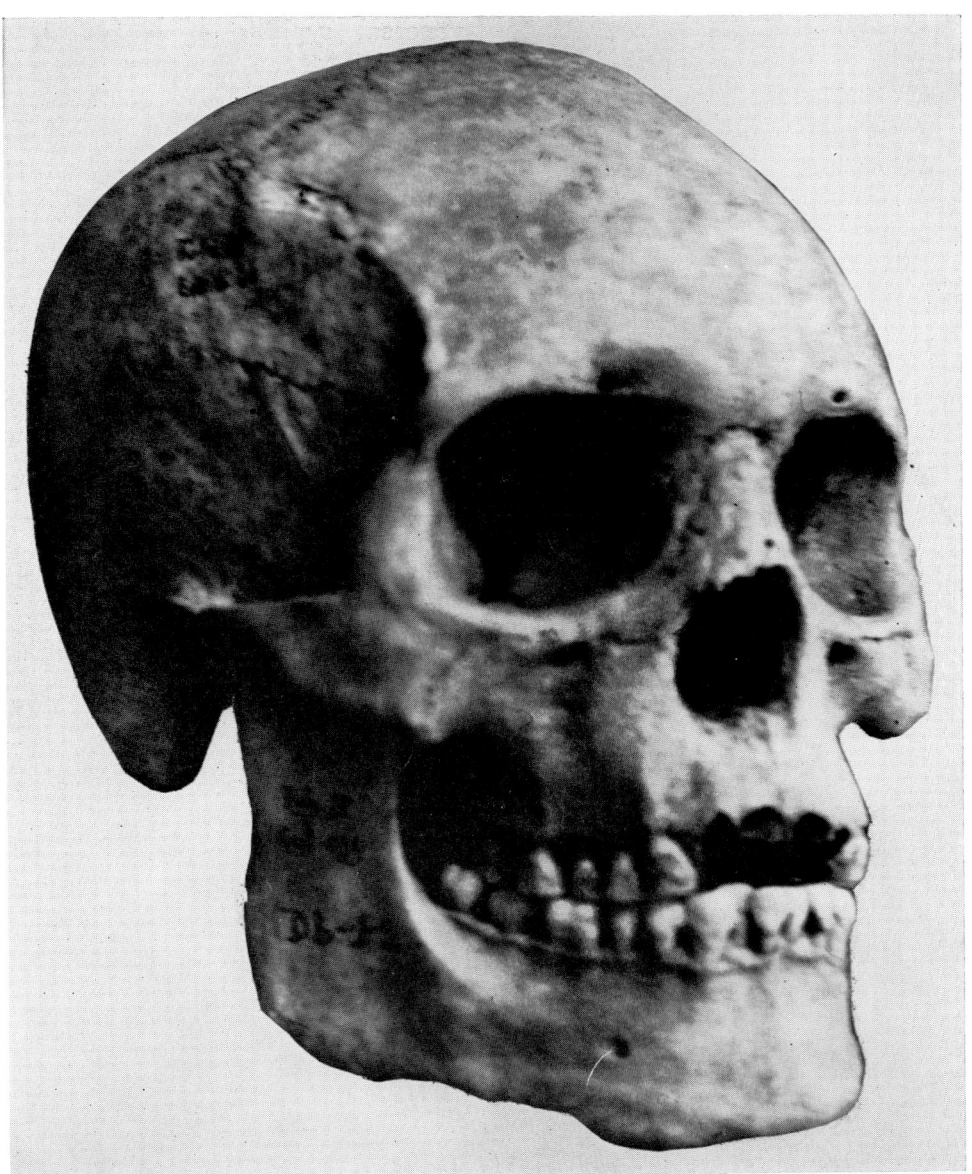

FIGURE 1. Osteitis of the frontal, facial, and zygomatic margins with thickening. Productive osteitis about nares with some chronic necrosis and partial healing. Loss of four upper incisor teeth. Diagnosis—Leprosy? (Mexico, Cueva de la Candelaria. Probably pre-Columbian).

by disease prior to death. One tibia was given to me as representative of bone syphilis. It is reproduced in Figures 13, 14, and 15 to illustrate typical long bone syphilis.

This tibia shows a chronic inflammatory, irregular layer of less dense bone, located under what must have been an equally irregular periosteum overlaying the dense bony cortex. There were no sinuses as might be expected if the disease was tuberculosis or a pyogenic osteitis. The medullary or endosteal bone was granular, coarse, lattice-like in structure, almost filling the usually sparsely occupied bony cavity of the shaft. This reactive endosteal bone is typical. The lack of any treatment accounts for the marked osteitis, which apparently ran its course. This to me is an example of long bone venereal syphilitic osteomyelitis of classical significance.

Many published references of the past two hundred years, with the diagnosis of venereal syphilis, must always remain doubtful. They indicate, however, the rarity of cranial bony involvement. The tibia was a more common site of reaction, with enlargements, irregularities and gummata. Archaeologists have not unearthed many preserved skulls retaining stigmata of venereal syphilis.

After extensive experience of dried bone specimens in many museum collections in Europe and the New World, I have been able to arrive at some tentative but nevertheless "working" conclusions as to what

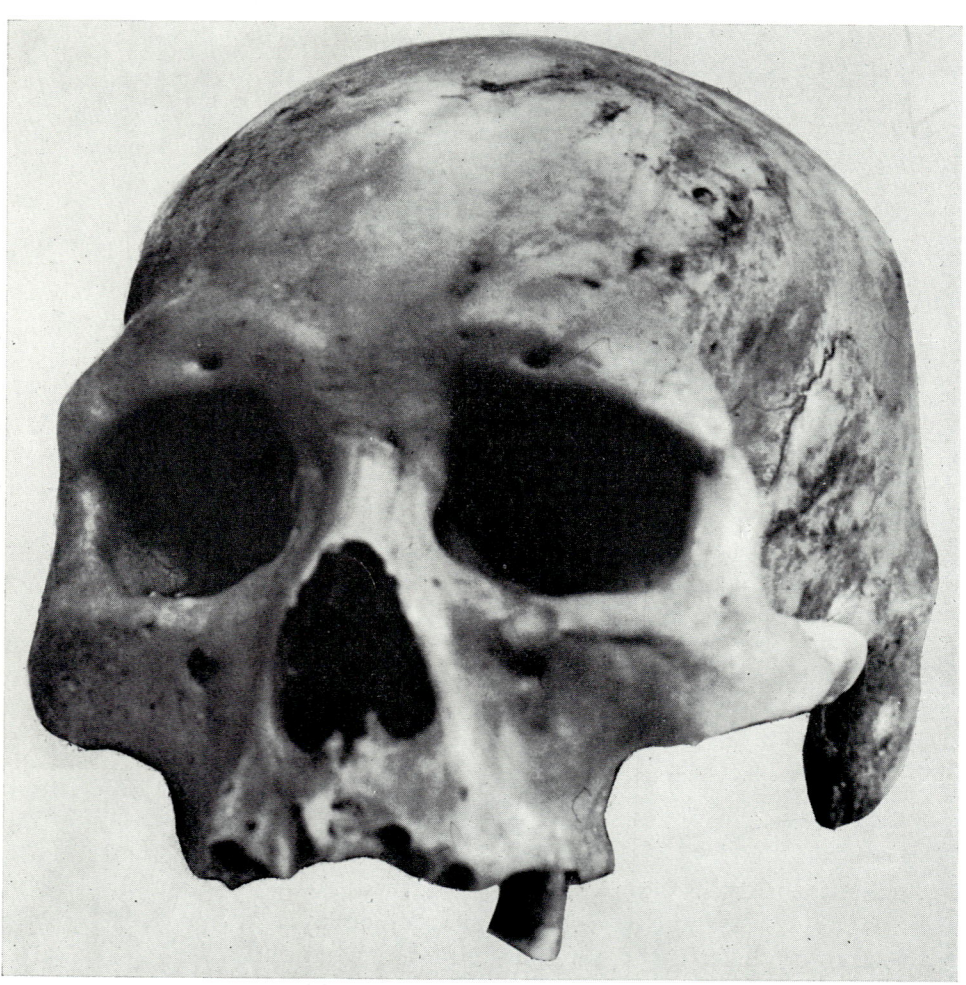

FIGURE 2, Osteitis, plastic variety, a smooth thickening as if paint had been applied. Diagnosis—Yaws? (Mexico, Cueva de la Candelaria).

changes reasonably constitute venereal bony syphilis. Having applied numerous laboratory methods and tests, to have cut, sectioned, polished, stained, macerated and ground bones, the stark limitations of our capacity to prove clearly the presence of syphilis as a causal agent has been frustratingly emphasized to me. Until new methods are developed, I see nothing for it but to accept gross descriptive traits in the absence of other cranial evidence.

What about cranial bone syphilis? How does syphilitic cranial osteomyelitis differ from yaws and other destructive bone inflammation of the calvarium, facial and mandibular regions? In the first place, widespread cranial (venereal) syphilis is preceded by the formation of gummatous areas in the scalp, periosteum and even within the cancellous bone of the skull. These are necrotic, rounded, discrete, lumpy swellings that involve all layers, and destroy bone. Multiple sequestra are formed. Sometimes these are totally liquified, leaving large bony defects. In the dried bone specimens, varieties are observed proceeding from a lumpy, general periostitis, to great, irregular, defects of both cranial tables (see Fig. 3). Virtually no healing has taken place, whereas in Figures 4 and 5 healing has been reasonably satisfactory, although a bony defect remains with its margins closing toward the center.

OLD DESCRIPTIONS OF BONY SYPHILIS

Many writers during the past 100 years, have described the skull, tibia and fingers as the most commonly involved bony areas affected by venereal syphilis. Two conditions were formerly described. An inflammatory process, beginning in the periosteum, was soon separated from the underlying bone, which became necrotic.

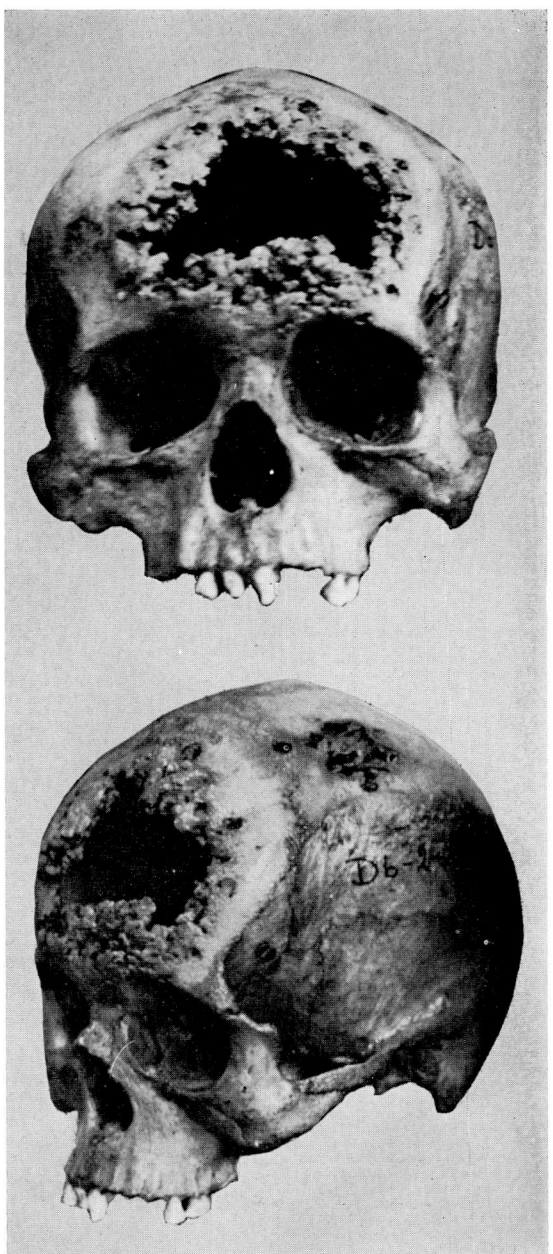

FIGURE 3. Irregular, gummatous osteitis with sequestrations, multiple, small and large. Diagnosis—Venereal syphilis (Mexico, Cueva de la Candelaria).

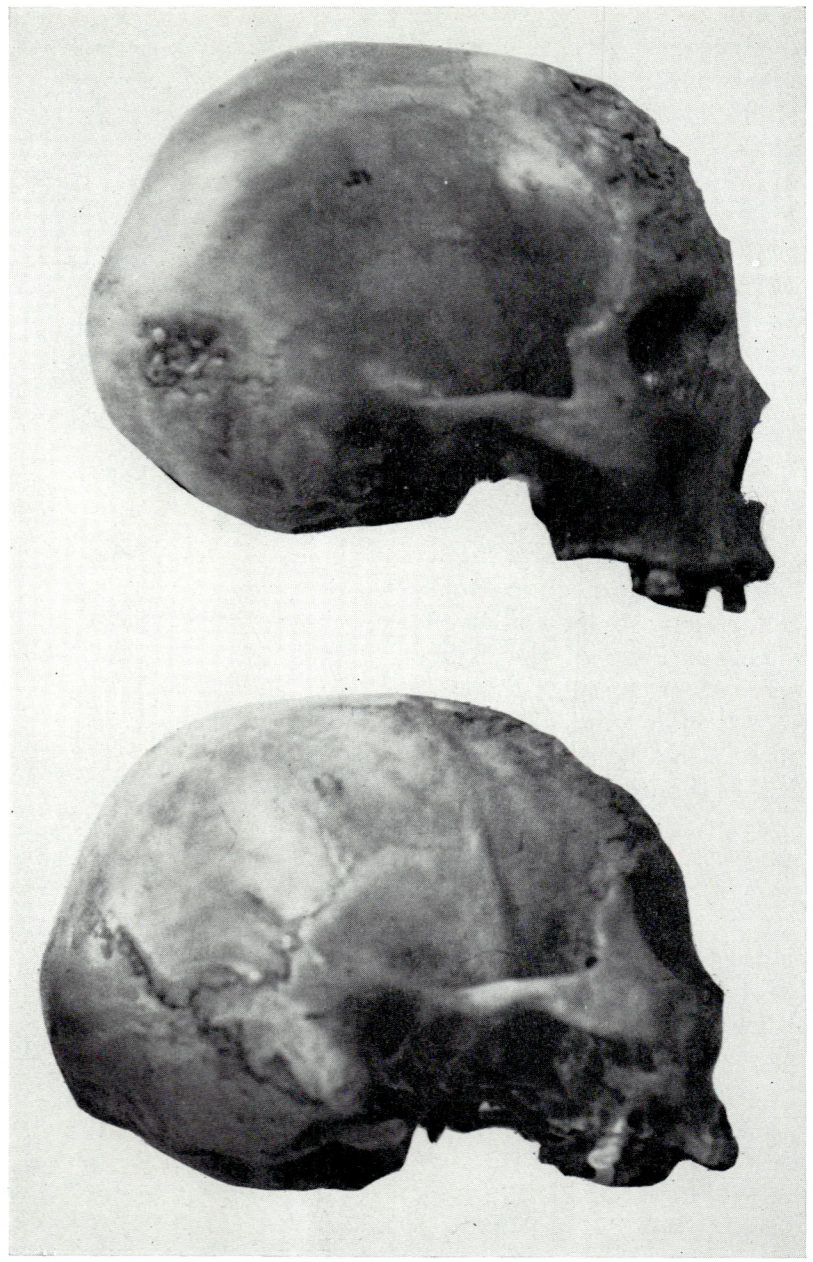

Figures 4-5. Two crania with similar erosions, and with frontal, parietal, and occipital osteitis with sequestration of upper specimen. Diagnosis—Venereal syphilis (Mexico, Cueva de la Candelaria).

FIGURE 6. Side and top views of an irregular, periostitis, general over the entire cranial surface. Diagnosis—Variant of venereal syphilitic periostitis (Mexico, Cueva de la Candelaria).

The skin over these necrotic areas frequently formed ulcers and sloughed away. A sequestrum formed within the necrotic bone, and discharged through the broken down areas. Sometimes, however, instead of becoming necrotic, a new growth of spongy bone formed on the surface of the old cortex, appearing rather like a coating of pumice stone when the bone dried. The formation of this sub-periosteal new bone produced irregular surfaces beneath the skin. Thickening followed, affecting the shafts of long bones but involving any part of the cranium.

Instead of forming on the original and normal surfaces of bones, new bone was sometimes produced along the Haversian canals so that the bone became ivory-like, weighing much more than normally. This process has been described as occuring also in clavicles, sternum and the phalanges of the hands. Cranial bone syphilis is more likely to develop defects and sequestra, because of a less satisfactory circulation.

Gummata have been described singly or in numbers involving the whole space of the marrow cavity. This in turn causes an expansion of the bony cortex.

DIAGNOSIS OF LATE CONGENITAL SYPHILIS

Numerous older descriptions of hereditary syphilis point to the presence of Hutchinson's teeth as the most characteristic trait of this disorder. The upper incisors of the second (permanent) set are affected. They consist of peg shaped teeth with vertical notching about their lower margins.

Jonathan Hutchinson considered these features to be pathognomic of inherited syphilis and described them as follows: "The central incisors are short and narrow with a broad vertical notch on their edges and their corners rounded off. Horizontal notches or furrows are often seen but as a rule have nothing to do with syphilis." Such teeth are described as disappearing by the age of twenty to twenty-five years. By that time the irregular edges have worn off. Thus, in adults we would not expect to find Hutchinson's teeth. I have never observed any in museum specimens or archaeological burial material.

Differential Diagnosis

Multiple myeloma, appear as small, not especially necrotic, translucent defects in the cranium. They rarely coalesce. Of course, other bones are involved, notably ribs and a few long bones.

A giant cell tumour is fairly unmistakable, as is a fibro-sarcoma, osteogenic sarcoma and carcinomatous metastases. Malignant bone tumours are usually so destructive that there is no evidence of any healing process taking place about the marginal defects. A roentgen examination of dried bones easily confirms the diagnosis.

Tuberculosis of bone causes softening, collapse and sinus formation. Garre's osteomyelitis is quite local and discrete. A syphilitic gumma is at times so slow in forming that evidence of healing, taking place simultaneously with the inflammatory destructive process, is commonly observed. Paget's disease of bone, probably of great antiquity, is an unmistakable disorder. Roentgen examination indicates great density and specific mosaic arrangement of bone cells.

GEOGRAPHICAL DISTRIBUTION OF VENEREAL SYPHILIS

The evidence for the distribution of venereal syphilis and yaws is most confusing, especially for tropical regions. Hackett (1963) believes that venereal syphilis could have been present around the Mediterranean, but was absent elsewhere in Europe. The treponemal diseases were probably brought to the Americas sometime during the past 10,000 to 40,000 years, at the time of a migratory wave from Asia or by sea from the East Indies. They then adapted to the New World environments. Contact with other parts of the world, such as the Pacific Islands and Northwest Africa were possible, as we know, but have not been proven conclusively.

Hudson postulated that each stage in human development would have had the kind of Treponemal infection appropriate to its development whether in the Old or New World. Epidemiologists have long regarded Haiti as a focal point concerning yaws in the New World. It was this island which was the first point of contact under village living conditions during the early conquistadores times. It may have given rise to the "romantic" concept of the introduction of venereal syphilis to Europe by the crews of Cristobal Colon in 1493, and it is of course equally true that yaws may well have been introduced to Haiti from Africa. Syphilis may have developed as a mutation or was separately introduced. Why do the pre-Columbian Indian skeletons of the West Indies show so few stigmata of venereal syphilis? My study of a small collection in Santo Domingo fails to reveal a single lesion that resembles cranial bone syphilis or even yaws. Other West Indian collections show an occasional specimen that I accept as characteristic.

EXAMINATIONS FROM PRE-COLUMBIAN BURIALS

The reader must now realize that treponematous ancient bone lesions have been difficult if not impossible to identify scientifically in the light of our present knowledge. Certain pathologic concepts, especially of soft tissues, taught to most investigators, have been the background on which their interpretations of bone diseases are based. It has been difficult for most investigators to conceive the difference between venereal and non-venereal syphilis as related to bone lesions. Nevertheless, since the modern classification includes all four varieties, such a differentiation should be possible. Hudson states that "bones which show these multiple changes have been found all over the earth." Why not Santo Domingo?

Hackett (1963), believes "Reports of examinations of human bones of all periods, especially before 1500, from all parts of the world, in museums and elsewhere, should be reviewed for descriptions of lesions likely to have been caused by treponematoses, and bones not so reported should be carefully examined."

Williams (1932) began this study and stressed the difficulties of the diagnosis of venereal syphilis in ancient bones and also the difficulty of dating them. Williams concludes that there is no certain evidence of venereal syphilis in "pre-Columbian" bones in Egypt or Europe, but he accepts some of the osteological evidence of venereal syphilis in American Indians before 1492.

World War II epidemiology has developed many new and interesting facets with which data may be historically correlated. For example, the false positive serological reactions may be caused by a number of infectious diseases, even if there is no clinical evidence of syphilis. Later individuals may develop systemic lupus erythematosis, rheumatoid arthritis, yaws, pinta, leprosy and a haemolytic anaemia. Any or all may give a false positive serological reaction. These indicate that an-

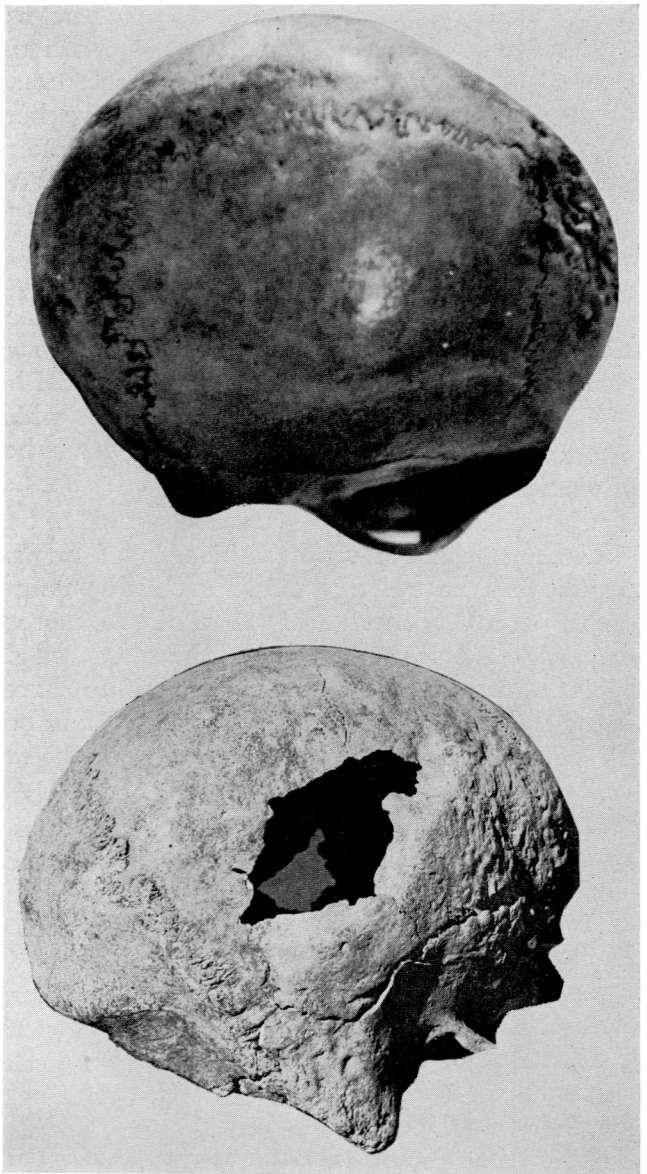

FIGURE 7. Venereal syphilis, gumma of parietal with massive necrosis and sequestration, surrounded by irregular syphilitic periostitis. Zaculeu (900-1000 A.D.), Guatemala Highlands.

cient bony material cannot be tested satisfactorily by any of the laboratory procedures in use at present. The gross examinations of today continue to be the methods on which a presumptive opinion is rendered.

NEW MATERIAL

In the highlands of Guatemala, at Zaculeu, two crania were excavated with a certain date of 900 - 1000 A.D. Presumably the individuals were Mam-speaking Indians. These I described in 1953. An ex-

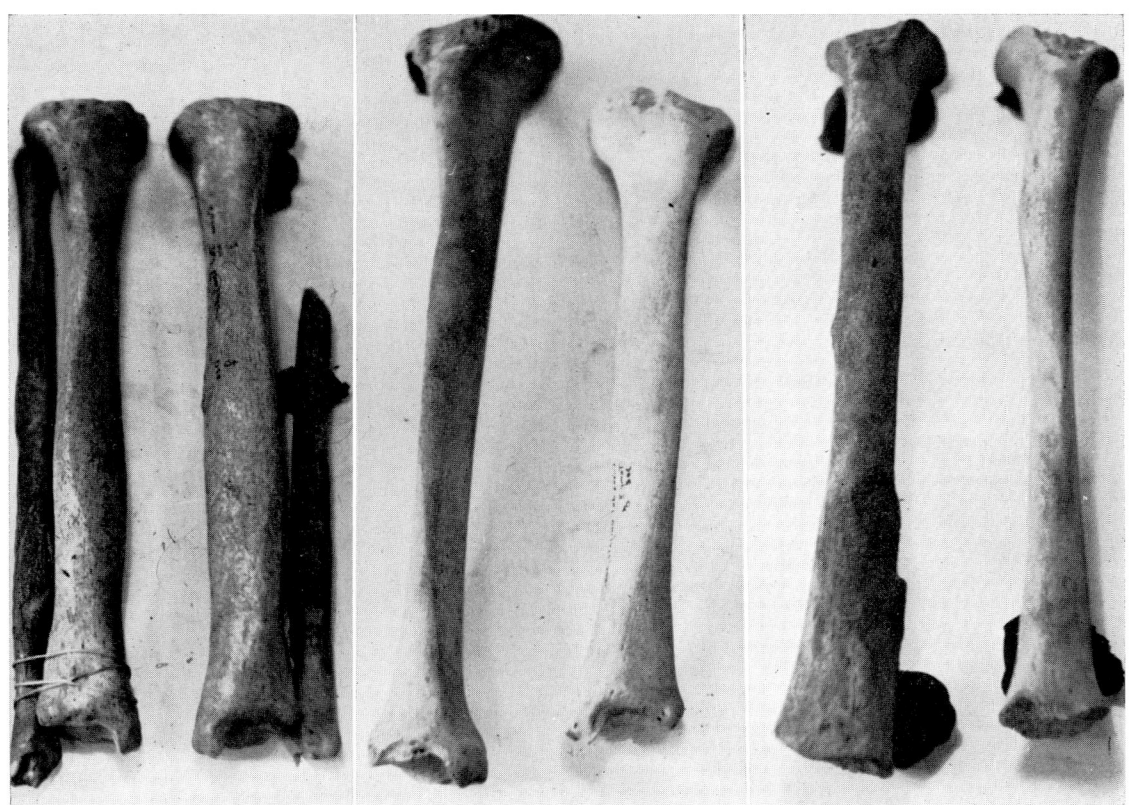

FIGURE 9. Productive osteitis, fusiform shafts of tibia, bilateral. Diagnosis—Venereal syphilis; congenital, (Mexico, Cueva de la Candelaria).

FIGURE 10. Same as 9 but less severe. (Mexico, Cueva de la Candelaria).

FIGURE 11. Same. Productive periostitis and overgrowth of an adolescent pair of tibiae, same individual. Diagnosis—Venereal syphilis, sabre shins, (Mexico, Cueva de la Candelaria).

haustive examination, tests of many kinds including serological, microscopic, electron microscopic and chemical, failed to produce a conclusive diagnosis. The gross examination presented evidence which I accepted, as that of cranial (venereal) syphilis. (Fig. 8).

New Specimens, Mexico

Over the past few years, the Coahuila Caves of Mexico in the Candellaria Mountains have yielded treasures of cultural material. To give them a date has been difficult. Some artifacts are quite unsophisticated and the date of 500 A.D. has been established by Mexican archaeologists; other archaeologists reserve a later date, perhaps even as late as the sixteenth century for some of the material. Thus an elusive timing element of this enormous collection of osseous material faces investigators and confounds conclusions. More accurate determinations will soon be published by the Museo Nacional de Antropologia, Mexico, D.F.

More than twenty skulls and nearly a hundred long bones, showing pathological changes, are now in the National Museum.

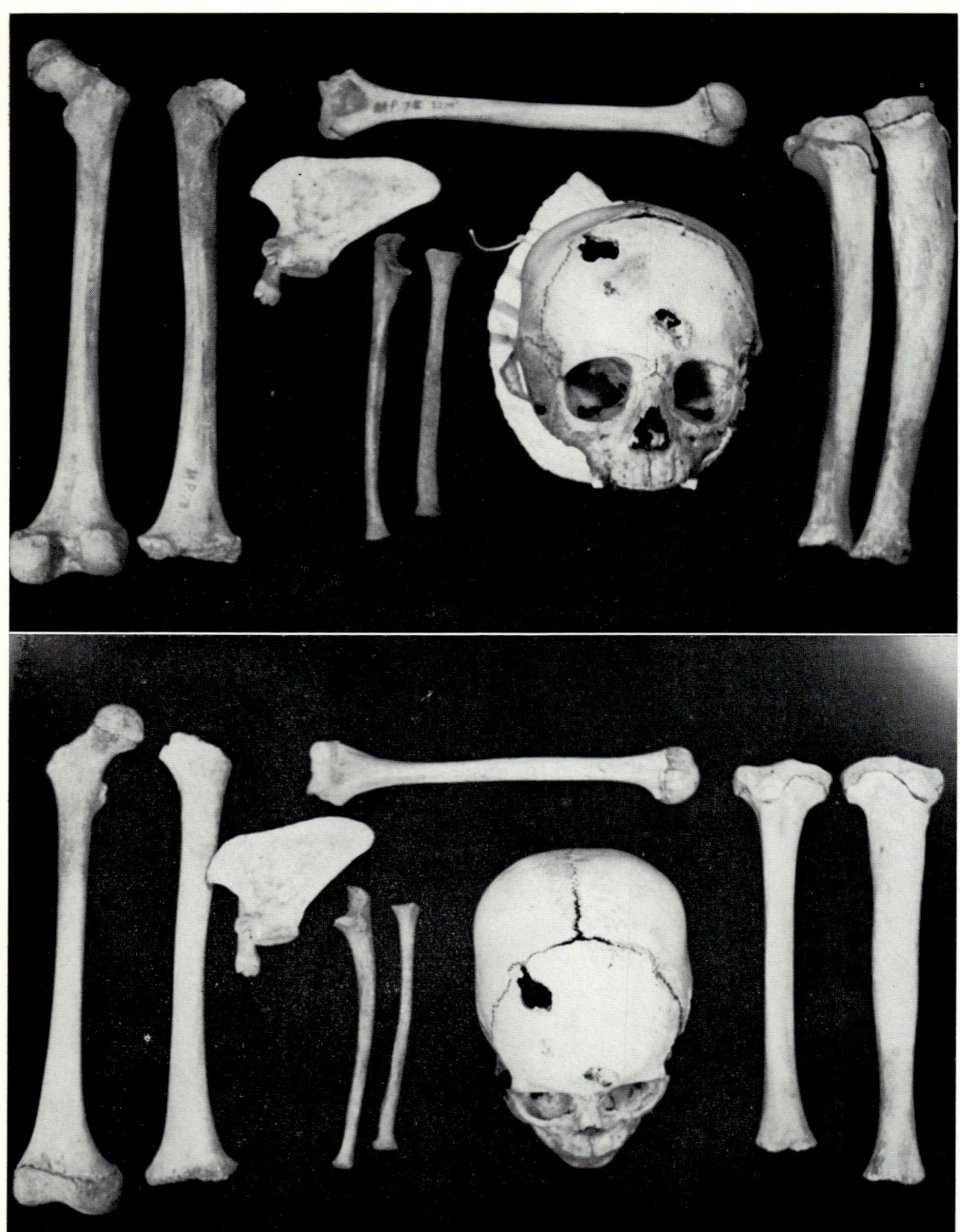

FIGURE 12. Burial material, bones of the same individual, a child of eight years. Yale expedition to Machu Picchu, Peru, (*c.* 1100-1200 A.D.). Congenital syphilis. Peabody Museum, New Haven, (No. 51-9210). Note typical sabre shin with overgrowth and perforating osteitis of cranium.

My inspection and comparative study of these indicate and confirm earlier investigators opinions. I believe several show evidence of a treponematosis of bone, presumptive yaws, evidence of venereal syphilis, and at least one skull with characteristics of leprosy. Full descriptions must await publication by Eusebio Davalos Hurtado.

Pablo Martinez Del Rio has portrayed three skulls from this amazing collection (Texas Archaeological Society Bulletin, 1953). They have the osseous traits that I believe are characteristic of gross cranial (venereal) bone syphilis. I have reproduced other skulls from the collection that are variants of venereal osseous syphilis of the cranium. A few long bones are included. One skull (Fig. 1), for comparative purposes, is reproduced from the same mortuary cave which I believe might represent cranial leprosy as described by Møller-Christensen (1961). In leprosy the nasal cartilages are eroded while in syphilis there is an osteomyelitis of the bony nares, with a depressed appearance. In leprosy a combined effect is frequently achieved, with loss of the upper incisor teeth plus palate erosion. The two conditions, leprosy and venereal syphilis of cranial bones, may be confusing when the cranium is not intact as a specimen.

There were several skulls with lesions not unlike those reported by Stewart and Spoehr (1952), who believed their child's cranium represented yaws. A significantly different bony marking, seen in Figure 2, shows plastic osteitis in the parietal and zygomatic regions. It could not have been a pyogenic inflammatory osteitis, because of the smooth continuity of bony surfaces. A variant of treponematosis is most likely, but which one was not clear. Tuberculosis would have shown sinuses, which were absent. Endemic syphilis seemed the most likely diagnosis, but this is not final.

Other descriptions of syphilis in bony remains from Mexico are few. Stewart describes a female tibia from Xochicalco as showing "productive, irregular surface, bowing and probably syphilis," although he does not positively identify the osteitis.

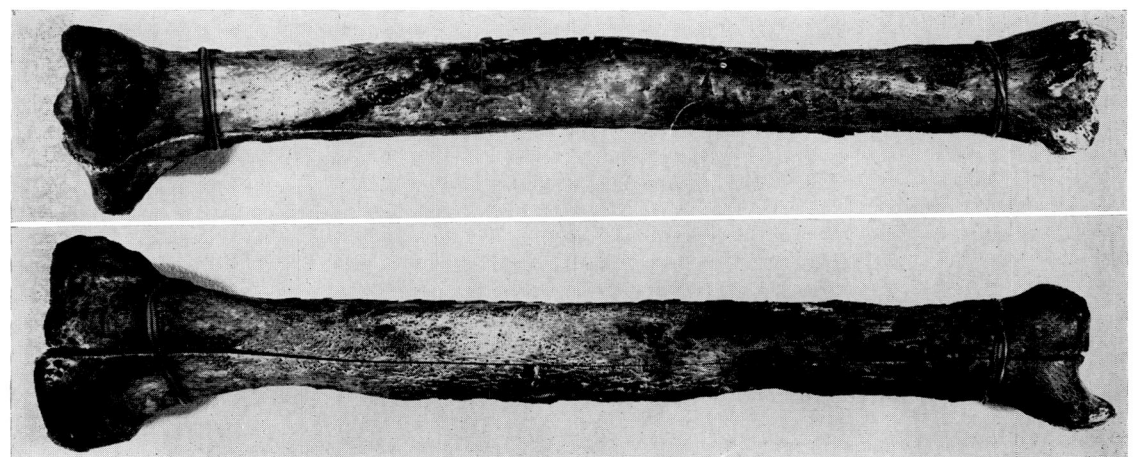

FIGURE 13. Modern Indian tibia (1850?). Venereal syphilis. Peabody Museum Gift (Professor E. A. Hooton). Severe osteitis with classical details of venereal bony syphilis. Has been cut longitudinally. A small section was cut for microscopic examination from one edge.

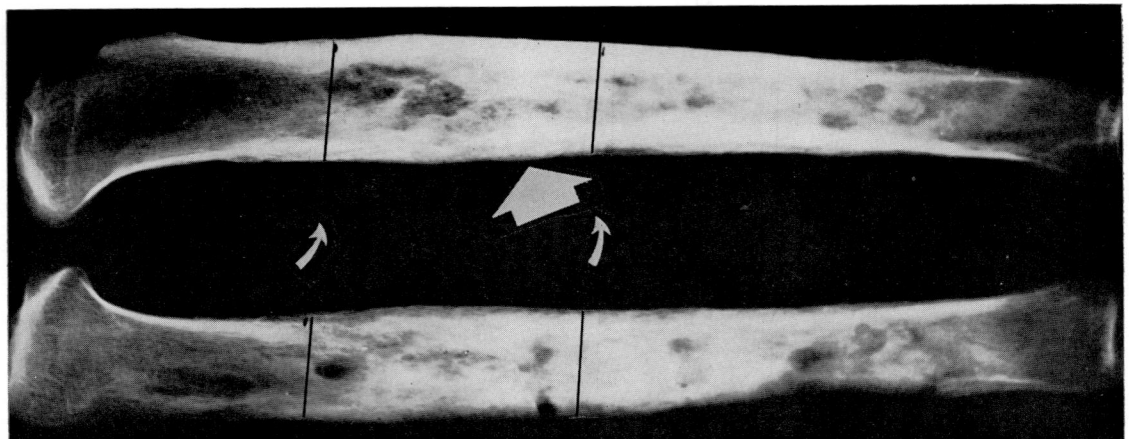

FIGURE 14. Marked osteitis plus total medullary involvement. No sequestration seen on roentgen plates. Diagnosis—Venereal syphilis, untreated. (Modern North American Indian). Same as Figures 13 and 15.

FIGURE 15. Longitudinal section of syphilitic tibia (as in Fig. 14; section between black lines). KKK shows double and triple overgrowth of cortex. (Section was printed as a mirror image of Roentgen view, large white arrow). Represents an acceptable example of bone syphilis.

From his photograph, it does not seem to be tuberculosis, yaws nor pyogenic osteitis. It probably fits syphilis best. It is from a pre-Columbian site, dated 900 - 1000 A.D.

Peru

I have reproduced a few of the long bones and the cranium of a child from Eaton's collection at Yale, obtained on the Bingham expedition to Machu Picchu in 1913. These appear to be typical of congenital syphilis and were so considered by Eaton.

The long bones of Figures 9, 10, 11 are from the Candelaria Cave. These are identifiable as varieties of congenital syphilis of childhood, while one is an adult osteomyelitis of venereal syphilis of the tibia.

SUMMARY OF PATHOLOGIC CRITERIA OF BONE SYPHILIS

As I have indicated, difficulties in distinguishing a probable cause of bone lesions observed in ancient material are great. Tuberculosis, diseases of cancellous bones of several varieties, tumors of bone, and other destructive disorders are especially significant to historical epidemiologists. Archaeologists may have difficulty in dating with sufficient accuracy the burial material in question. Modern pathologists have not improved diagnostic techniques of dried bone, with the possible exception of Frost (1963) and Barker (1964). The field of true differentiation, if this is ever going to be possible, will no doubt come from similar biochemical approaches. On the other hand, this may be wishful thinking.

REFERENCES

ANON., 1962: Diseases of Ancient Man. *Brit. Med. J., 1:* 852.
ASHMEAD, A. S., 1895: *Pre-Columbian Leprosy*. Chicago. A. Med. Assoc. Press.
BARKER, P., 1964: Personal Communication.
BARTLETT, R. C., 1960: Present status of serological tests for syphilis. *Hartford Hospital Bull., 15:*24.
BINGHAM, HIRAM., 1930: *Machu Picchu*. New Haven, Conn., Yale Univ. Press, for the National Geographic Soc.
BLOCH, IWAN, 1904: *Das erste Auftreten der Syphilis, (Lustseuche) in der europäischen Kulturwelt*. Jena, Gustav Fischer.
BROTHWELL, D. R., 1961: The palaeopathology of early British Man. An essay on the problem of diagnosis and analysis. *J. Roy. Anthrop. Inst., 91:*318.
BROTHWELL, D. R., 1963: *Digging up Bones*. London. British Museum (Natural History).
COOPER, SAMUEL, with notes by STEVENS, A. H., 1835: Philadelphia. Griggs and Elliot. 4th Edition (American).
FROST, H. M., 1963: *Bone Remodelling Dynamics*. Springfield, Thomas.
GOLDSTEIN, M. S., 1957: Skeletal pathology of early Indians in Texas. *Amer. J. Phys. Anthrop., 15:*299.
GOLDSTEIN, M. S., 1963: Human Paleopathology. *J.A.M.A., 55:*100.
HACKETT, C. J., 1963: On the Origin of the human treponematoses. *Bull. W.H.O., 29:*7.
HACKETT, C. J., 1951: *Bone Lesions of Yaws in Uganda*. Oxford, Blackwell.
HAMLIN, H., 1939: The Geography of Treponematosis. *Yale J. Biol. Med., 12:*29.
HUDSON, E. H., 1963: Treponematosis and Anthropology. *Ann. Intern. Med., 58:*1037.
LANCEREAUS, E., 1868: *A Treatise on Syphilis. Historical and Practical*. London. The New Sydenham Society. Vol. 1.
LUIS AVELEYRA ARROYO DE ANDA MANUEL MALDONADO-KOERDELL and PABLO MARTINEZ DEL RIO with the collaboration of IGNACIO BERNAL and FREDERICO ELIZONDO SAUCEDO, 1956: Cueva de la Candelaria. *Instituto Nacional de Antropologia e Historia*. Vol. I. Secretaria de Educacion Publica, Mexico.
MARTINEZ DEL RIO, PABLO, 1953: A Preliminary Report on the Mortuary Cave of Candelaria, Coahuila, Mexico. *Texas Archeological Soc., 24:*208.
MAY, J. M., 1950: Medical Geography: Its Methods and Objectives. *Geographic Review, 60:*9.
MØLLER-CHRISTENSEN, VILH., 1961: *Bone Changes in Leprosy*. Munksgaard, Copenhagen.
MOODIE, ROY L., 1923: *Paleopathology*. Urbana, Univ. of Ill. Press.
MOORE, JR., M. B., 1963: The Epidemiology of Syphilis. *J.A.M.A., 186:*71.
SHATTUCK, G. C., 1938: *A Medical Survey of the Republic of Guatemala*. Washington, D. C., Carnegie Institution of Washington.
STEWARD, J. H., 1950: *Handbook of South American Indians*. Vol. 6. U. S. Government Printing Office, Washington, D. C.

Stewart, T. D., and Spoehr. O., 1952: Evidence on the Paleopathology of Yaws. *Bull. Hist. Med.*, 26:538.

Stewart, T. D., 1956: *Skeletal Remains from Xochicalco, Morelos. Estudios Antropologicos publicados en homenaje al doctor Manuel Gamio,* Mexico, D. F. 132-155.

Sudhoff, Karl, 1912: *Graphische und Typographische Erstlinge der Syphilisliteratur aus den Jahren 1495 und 1496.* München, Carl Kuhn.

Willcox, R. R., 1960: Personal communication.

Williams, H. V., 1932: The origin and antiquity of syphilis. The evidence from diseased bones. *Arch. Path.,* 13:779.

Woodbury, R. B., and Trik, A. S., 1953: *The Ruins of Zaculeu, Guatemala.* Special Articles: New Evidence of Pre-Columbian Bone Syphilis in Guatemala, by C. W. Goff, pp. 312-319. Richmond, Wm. Byrd Press.

Chapter 22

Evidence of Leprosy in Earlier Peoples

VILHELM MØLLER-CHRISTENSEN

Little is known about the early history of leprosy, and although it is generally believed that the disease is mentioned in some early literature, including the Bible, and that the highest incidence of the disease in Europe was in the Middle Ages, there is very little reliable evidence on which medical historians can base an opinion. The only way of obtaining definite evidence of the occurrence of early leprosy is to make use of the fact that the lepromatous type of leprosy causes characteristic and permanent changes in the bones; one important way of obtaining such evidence is the difficult and laborious task of locating mediaeval graveyards and examining their contents. With this in mind, I have carried out such a study in Denmark and examined four mediaeval leprosy hospital cemeteries (Naestved, Svendborg, Spejlsby and Bornholm) which comprise altogether nearly 1000 graves (Møller-Christensen, 1953, 1961).

In this present paper, however, I do not wish to limit myself to this Danish evidence alone but to consider world-wide evidence for early leprosy. Although some mention may be made of the literature, it seems more important here to discuss as fully as possible the direct skeletal evidence of this disease. First, it would seem advisable to review the variety of bone changes which may occur in leprosy.

BONE CHANGES IN LEPROSY
Leprous Changes in the Skull

A number of well defined bony changes have now been noted in the skull, particularly by Møller-Christensen, Bakke, Melsom and Waaler (1952), and Møller-Christensen (1953, 1961). In order to be as precise as possible in the description of these changes, it has been thought advisable to discuss them according to the area involved, (and see Fig. 1 a,b).

The Pyriform (Nasal) Aperture. Normally, the margins of the nasal aperture are fairly sharp, especially in European peoples. However, in a number of pathological conditions (including leprosy, yaws, syphilis and chronic facial lupus) there may be considerable destruction of bone in this region. But of these diseases, leprosy appears to be the most selective and specific with regard to its effect upon the face.

Selective recession in this region is evident for example in some of the medieval Naestved (Danish) material, and in Britain it was first noted in a mediaeval skull from Scarborough (Brothwell, 1958.)

The Anterior Nasal Spine. Møller-Christensen et al. (1952) have demonstrated atrophy of the anterior nasal spine in leprosy, both on skeletal material and in living patients. It was probably Glück (1897), however, who first described this

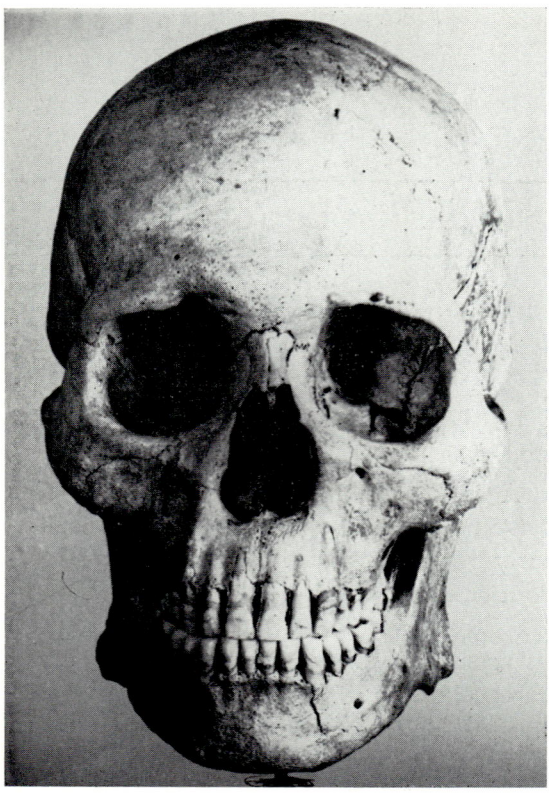 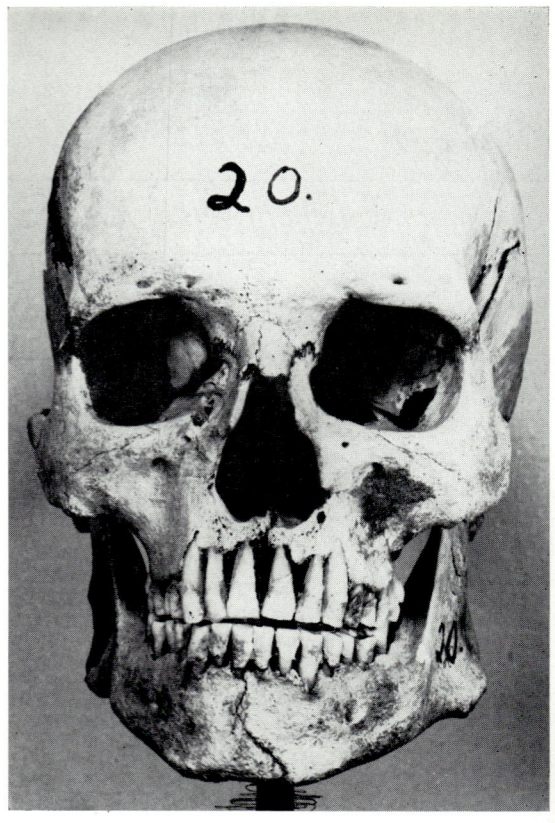

FIGURE 1. a: Skull of an adult male with *facies leprosa*: second degree atrophy of the maxillary alveolar process (m.a.p.), with loosening of the frontal incisors (1—1) and third degree atrophy of the anterior nasal spine (a.n.s.) and pronounced *inflammatory changes* of the superior surface of the hard palate (s.s.h.p.)

b: Skull of an adult male with *facies leprosa*: third degree atrophy of the m.a.p. with pronouced loosening of the frontal incisors (2, 1 1, 2) and third degree atrophy of the a.n.s. and pronounced inflammatory changes of the s.s.h.p.

change in the skull of a modern leprosy patient. In his specimen and in many of the early Danish leprosy skeletons which I have excavated, the anterior nasal spine is replaced by an osteoporotic area. The term *facies leprosa* was applied by me to the condition characterised by:

1. Atrophy of the anterior nasal spine.
2. Atrophy and recession of the maxillary alveolar margin, confined to the incisor region, beginning centrally (at prosthion), resulting in loosening and possibly loss of the incisors.
3. Inflammatory changes on the superior surface of the hard palate.
   The inflammatory changes of the hard palate constitute the basic pathology of this condition. They must always be present for a diagnosis of leprosy to be made. The other conditions (1) and (2) may not always be present.

THE NASAL CAVITY. There may also be bony reactions within the nose. Glück (1897) notes in his specimens: swelling, deformity and osteoporosis of the vomer,

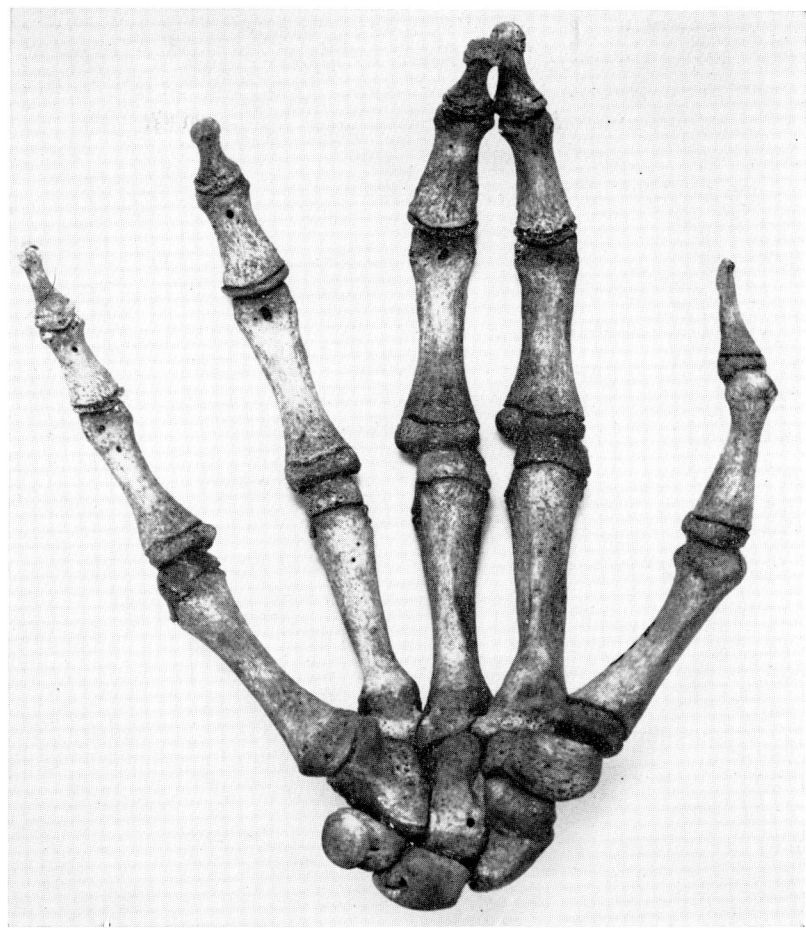

c: Volar aspect of the right hand of a six year old child showing "enlarged nutrient foramina." Danish Mediaeval specimens.

osteoporosis of the turbinate bones, and some evidence of pathological erosion in both cases. The inferior conchae of Danish specimens also showed changes suggestive of a chronic inflammatory process, and the majority of skulls had changes in the *facies nasalis* of the palate (Møller-Christensen, 1953, 1961).

THE MAXILLARY ALVEOLAR PROCESS. Perhaps the most reliable bony change which characterises leprosy is the localized recession of the upper alveolar margins at the incisors. The area of bone destruction is far too limited to be mistaken for periodontal disease, and to my knowledge, syphilis, yaws, chronic lupus, or cancrum oris, are most unlikely to produce these features. I have described in detail this "pathomorphological" character in numerous Danish skeletons from Naestved and other sites, and on the evidence provided by these I have suggested a specific term for the lesion: *paradentosis leprosa*.

THE PALATE. As early as 1888, Zwillinger and Läufer noted the occurrence of a perforated palate in some cases of leprosy, and Glück (1897) described two skulls showing perforation, thinning and osteoporosis of the palate. In the Mediaeval Danish material, more than 90 per cent of the cases examined showed inflammatory changes of the *facies nasalis*. A number had perforations of the hard palate and inflammatory changes of the *facies oralis*.

Changes in the Long Bones

Several authors have commented on the occurrence of periostitis of the long bones, but only as an occasional phenomenon. Hirschberg and Biehler (1909) demonstrated surface irregularities of the tibia and ulna. Chamberlain, Wayson and Garland (1931), in a study of Hawaiian leprosy patients, observed periosteal changes of the long bones at the ankle and wrist. Murdock and Hutter (1932) also note in one of their cases "periostitis and deformity of the fibula," and Paterson (1956) also says that other bones, such as the tibia, may be affected. However, it was not until the study of excavated skeletal material (Møller-Christensen, 1953) that the extent and types of changes were more fully appreciated (Fig. 2). In this report I stated that there are: "some characteristic and often severe changes in the tibia and fibula—in the tibia often appearing in the form of *vascular grooves on the lateral surface* that always appeared to be accompanied by pathological changes in the feet. The fibulae are rarely attacked in the same degree. . . . There appears to be a causal connection between the presence of leprosy and the occurrence of these changes as among thirty complete skeletons vascular grooves were found in twenty-five cases (i.e., in 83.3 per cent) where characteristic trophoneurotic changes in the hands or feet or the presence of *facies leprosa* make the diagnosis of leprosy highly possible."

Perhaps a less misleading term than "vascular grooves" would be transverse striations, which does not imply so specifically an association with overlying vessels. The term "striations" has already been used by Greig (1931) in discussing mild periostitis.

Osseous Changes of the Hand and Foot

Numerous authors have discussed the bony changes of the hands and feet in leprosy patients, usually from radiographic evidence, and it would be out of place to discuss the literature in detail here.

Hand changes are found principally in the phalanges. The metacarpals are rarely involved. The joints may show all degrees of arthritis with ankylosis, subluxation and dislocation. Barnetson (1950) summarises leprous reactions in the phalanges thus: "The principal change is atrophy and disappearance of bone without productive changes unless secondary infection has occurred. Atrophy begins at distal margins of terminal phalages but destruction may become greater in more proximal bones. Lesions are bilateral and are frequently stated to be symmetrical. Degree of bone change varies with duration of leprosy, extent of nerve involvement, age of patient, trauma, secondary infection and other factors." Sometimes, as some of the cases of Oberdoerffer and Collier (1940) show, the distal phalanges may be lost without much apparent involvement of the more proximal ones. Lastly, the nutrient foramina of the phalanges (which are usually tiny apertures in the normal person) may be considerably enlarged (Fig. 1c). Murdock and Hutter (1932) concluded that the enlarged nutrient foramina "are atropic responses to pathologic leprous vascular supply," having found about a 30 per cent frequency in their leprosy patients.

The form of bone changes in the foot

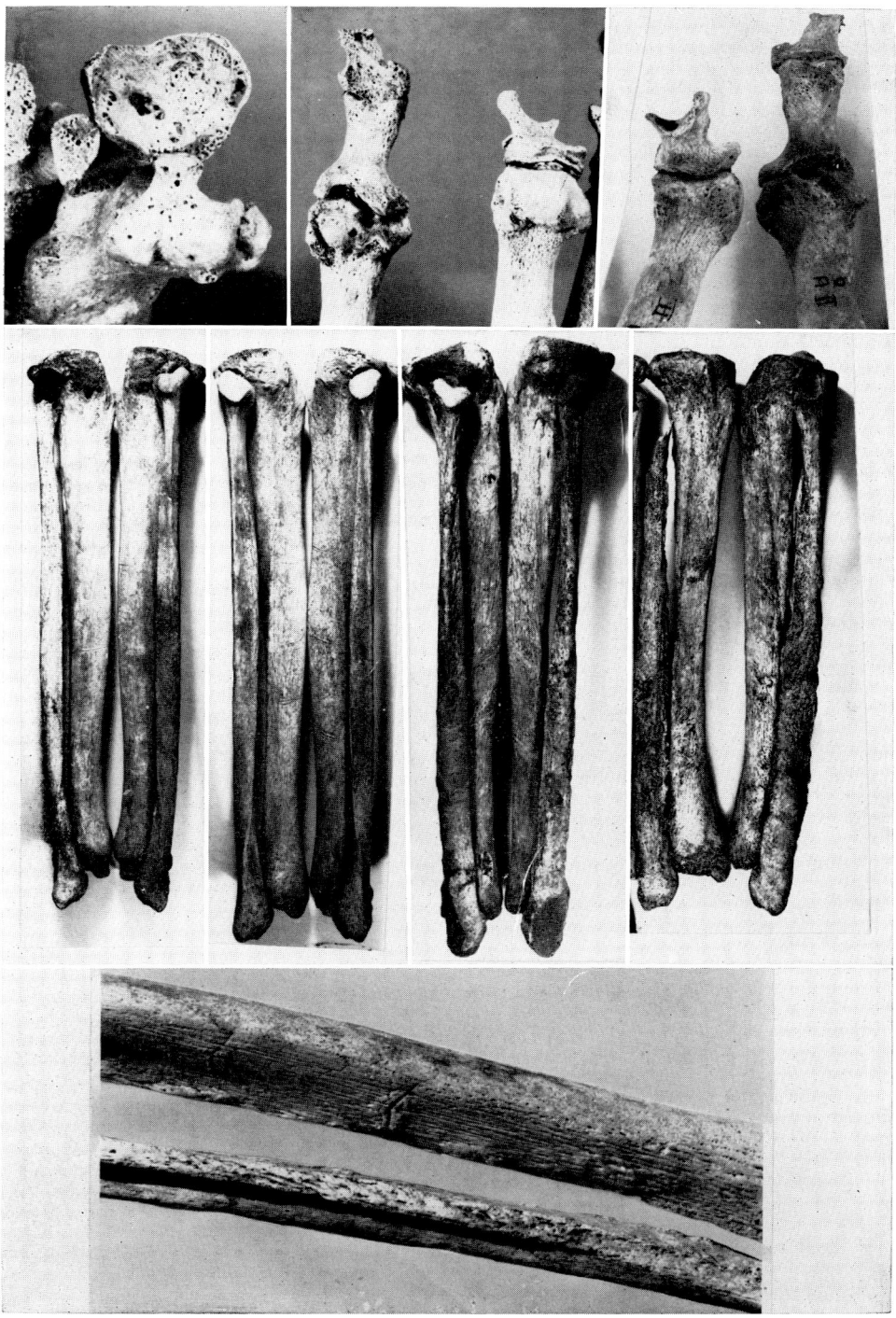

FIGURE 2. a. b and c: Destructive changes in the basic joint of the big toe, and the phalanges, characteristic of leprosy. d, e, f, g and h: Vascular (?) grooves on tibiae surrounded by areas with periostitis, and fibulae with sub-periosteal reactions. Danish Mediaeval specimens.

in leprosy (Fig. 3) and the sequence are different from the changes found in the hand in the following respects:

1. The tarsal bones are frequently involved,
2. The metatarsal bones are frequently involved,
3. The phalangeal changes frequently start at the metatarsophalangeal joints.

THE MATERIAL EVIDENCE FOR EARLY LEPROSY

It is an unfortunate fact that about 80 to 90 per cent of the archaeological skeletal material in museum and university collections consists entirely of skulls, complete or fragmentary, and one is therefore often required to diagnose tentatively diseases such as leprosy and syphilis from the characteristic changes found in parts of the skull alone.

In the case of leprosy, pathological changes in parts of the skull other than the maxillary bone (*facies leprosa*) have never been observed, except for the orbital anomaly known as *usura orbitae* (Møller-Christensen and Sandison, 1963). Contrasting with these specific changes found in leprosy, in syphilis the skull vault is the predominant part of the skull to show characteristic changes (Møller-Christensen and Inkster, 1965).

During 1962, I spent six months as a Carlsberg Research Fellow travelling in England, Scotland and France examining the extensive archaeo-osteological collections for signs of leprosy, syphilis and tuberculosis in earlier peoples. Alone or in teamwork with six British and French anthropologists, leprologists and pathologists I examined in all about 18,000 skulls, skeletons and mummies. The degree of certainty of a diagnosis depended of course on the nature of the human remains, and how complete they were.

If a cranium displaying *facies leprosa* was found, it was considered as a *possible* case of leprosy. If the tibiae and fibulae showed no pathological changes, or had not been preserved, the case was not regarded as being a sufficiently proven one of leprosy.

Only when a cranium with *facies leprosa* was accompanied by tibiae and fibulae showing typical pathological changes, bilaterally and symmetrically, was a more firm diagnosis of the lepromatous type of leprosy made. Fairly certain diagnosis was only possible when marked changes also occurred in preserved hand and foot bones. In practice, this severely confined the number of certain cases, but when all the very doubtful cases were omitted, the collaborators usually agreed as to the other diagnoses. But the possibility of less advanced cases of leprosy being overlooked also cannot be excluded, and the figures given here for leprosy must therefore be regarded as an absolute minimum.

The results of this investigation of a total sample of about 18,000 specimens are given in Table I.

Egypt

From Egypt there is only one certain case of leprosy, in the now famous mummy from Aswan (El Bigha) in Nubia, first described and depicted by Elliot Smith and Derry (1909). As they state: "We now put on record photographs of the hands and feet of a man of the early Christian period (about the sixth century) in a condition strongly suggestive of leprosy." When I visited England in 1962, it was impossible to find the right hand and foot (Fig. 3), there being only the left hand and foot remaining (in the care of The Royal College of Surgeons, London.)

TABLE I

THE OCCURRENCE OF LEPROSY IN VARIOUS EARLY POPULATIONS, BASED ON A STUDY OF 18576 SPECIMENS (SKULLS OR SKELETONS, COMPLETE OR FRAGMENTARY). NUMBERS IN PARENTHESES INDICATE THE NUMBER OF CASES TENTATIVELY IDENTIFIED

Region and Period	Number Studied	Number of Tentative cases	Questionable cases of leprosy	Date
Egypt				
(6000 B.C.-600 A.D.)	1,844	(2)		550 A.D.
Palestine (Lachish)				
(700-600 B.C.)	695			
France				
Neolithic (4000 B.C.)	192			
Gallo-Romain	237			
Merovingian (481-750 A.D.)	233	(1)		550 A.D.
Others (750-1600 A.D.)	1,191			
Parisian (1200-1700 A.D.)	5,300			
Marville (1200-1400 A.D.)	735			
British Isles				
Neolithic (3000-1700 B.C.)	178			
Bronze Age (1700-500 B.C.)	224			
Iron Age (500 B.C.-43 A.D.)	54			
Romano-British (43-425 A.D.)	544	(5)		550 A.D.
Saxon (425-850 A.D.)	656	(4)		1200 A.D.
Mediaeval (1066-1550 A.D.)	274			
Hythe (1200-1600 A.D.)	607			
Spitalfields (1200-1600)	640			
Moorfields (c. 1665)	112			
18th Century	743			
19th Century	775			
Denmark and Sweden				
1200-1500 A.D.	1,500	(310)		1200 A.D.
1500-1650 A.D.	500	(5)		1600 A.D.
Other Continentals (19th cent.)	260			
Africa (19th century)	122		(1)	1800 A.D.
Asia (19th century)	346			
Australia (19th century)	251		(1)	1800 A.D.
Pacific Isles (19th century)	239			
U.S.A. (19th century)	124	(1)		1800 A.D.
Total	18,576			

Further mummy material, this time at the British Museum (Natural History), revealed in the radiographs of one head,* changes which could well be attributed to *facies leprosa*. Yet further evidence of early Egyptian leprosy came to light in Cambridge, at the Duckworth Laboratory of Physical Anthropology, where a female skull from the same cemetery as the

---

* It is highly probable, that this mummy head was from the same mummy body as the leprous hands and feet (Warren Dawson, personal communication).

FIGURE 3. a: Dorsal aspect of the feet of an adult female with typical leprosy mutilation. Danish Mediaeval.

mummy material also displayed *facies leprosa*.

These two cases (considering the mummy hands, feet and head as being from one person), one with definite signs and one with possible signs of leprosy are dated to about 500 A.D. Elliot Smith and Wood Jones (1908); Elliot Smith and Derry (1909); Elliot Smith and Dawson (1924) and are hitherto the only cases of leprosy found in all the thousands of carefully examined mummies and skeletons from ancient Egypt.

Palestine

I examined the remains of about 600 individuals from the burial site of Lachish near Jericho (Risdon, 1939) but without finding any suggestion of leprosy or syphilis.

The British Isles

From England we now have evidence of nine cases of leprosy, the oldest dating back to about 600 A.D. (Brothwell 1958, 1961; Møller-Christensen and Hughes,

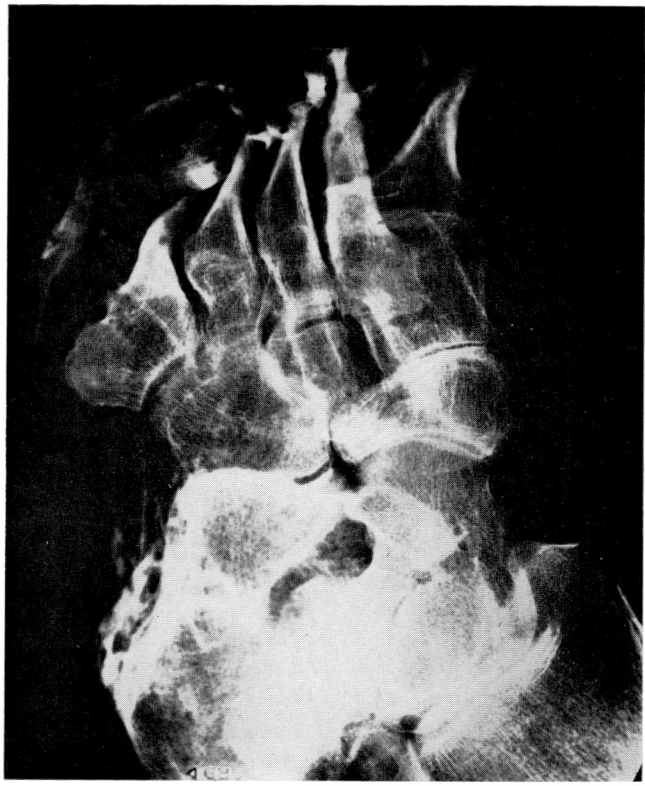

b: The left foot (X-ray) of the early Christian mummy from Nubia (El Bigha). See Elliot Smith and Derry, 1910.

1962; Møller-Christensen and Inkster, 1964; Wells, 1962).

France

Among 700 skeletons from France dating from the Neolithic period until about 1000 A.D., I found only one case of possible leprosy (dated to about 500 A.D.). From the more recent remains in the Catacombs of Paris, where we examined 5300 skulls (Møller-Christensen and Jopling, 1963) we found no cases displaying *facies leprosa*, although four calvaria were syphilitic. An exact dating of these cases is at present impossible, but they are probably more recent than the year 1500.

Denmark and Sweden

In Denmark and Sweden leprosy occurs very commonly in the mediaeval leprosy hospital cemeteries, but although we have 310 cases of probable leprosy, no case is older than 1250 A.D. (Møller-Christensen, 1953, 1961).

* * *

Although far more skeletons need to be examined before any statement can be made as to the precise antiquity of the occurrence of leprous changes in earlier populations, the analysis of skeletal material so far does strongly suggest that the disease was little in evidence in early Western cultures before the first few centuries A.D. Its appearance in Egypt and northern Europe, on present evidence, would seem to have been at a similar time. Rather than deriving the spread of leprosy from a small focal point in Africa or Europe, it seems far more likely that the disease spread westwards from Asia. Early records strongly suggest leprosy was present in the East, certainly in China (see also Lu and Needham's contribution here), prior to its earliest occurrence in the West, and it would seem most important that early Asiatic skeletal material be closely examined for confirmatory evidence of the disease.

LEPROSY IN BIBLICAL AND OLD MEDICAL TEXTS

The absence of material evidence of leprosy in Bible lands prior to about the middle of the first millenium A.D. may seem to contradict actual statements in the Bible. A more critical look at Biblical translation does, however, provide an alternative explanation. Cochrane (1961) and Gramberg, Swellenbrebel, Nida and Wallington (1961) have pointed out that the Hebrew word *tsara'at* in the Old Testament, at present translated as leprosy, is probably not identical with the disease we now know by the word *leprosy*.

Early Egyptian literature is noticeably silent as regards leprosy. Warren R. Dawson, in a personal communication to me, writes "In the ancient medical papyri no case of leprosy or its treatment is mentioned, unless to be concealed in the many names of diseases which have not yet been identified with certainty. The last case in the Ebers Papyrus (109.18-110.9) deals with the treatment of what appears to be an eruptive condition of the skin. This disease is called in the text by a vernacular name which affords no clue to its identity —*The carnage of Khons (the Moon-god)*.

Dr. Ebbell considers this disease to be *lepra mutilans* (*Acta Orientalia*, VII. 42; *The Papyrus Ebers*, Copenhagen, 1937, p. 127). I think this identification very doubtful and founded upon insufficient evidence.

"Although the medical papyri afford us no information as to leprosy, there is another class of papyri, of which all known examples, twenty-one in number, have recently been published: (I. E. Edwards, Hieratic Papyri in the British Museum, 4th Series, *Oracular Amuletic Decrees of the late New Kingdom*, London, 1959.) In thirteen of these twenty-one papyri, which date from XXI to XXIII Dynasties, mention is made of a disease called *sbh*, a word that has survived in Coptic as ⲋⲱⲑⲣ. This latter word occurs frequently in the Coptic Biblical texts where its equivalent in the Septuagint and in the Greek New Testament is λέπρα (e.g., Lev. 13:2; Matt. 26:6; Luke 5:12, etc). Whatever interpretation modern medical opinion places upon λέπρα, will apply therefore to the Coptic word and to its hieroglyphic prototype.

"The twenty-one texts mentioned above are written in hieratic script on narrow strips of papyrus which were rolled up and carried on the person as protective amulets. They purport to be oracular decrees given by the gods, whereby the deities promise to keep the wearer safe from all kinds of diseases, from accidents and misfortunes, from the assaults of demons and hostile gods, as well as from acts of malice and oppression by ill-disposed human beings. They also promise to keep every part of the body healthy and sound. The organs and parts of the body are enumerated from the top of the head to the soles of the feet, as well as internal organs—heart, liver, spleen, kidneys, etc. The formulae are similar but not identical in all cases. In one of those formulae, which occurs in thirteen of the twenty-one papyri, we read: 'we shall keep him/her safe from leprosy (*sbh*), from blindness . . . throughout his/her entire lifetime.' These papyri, though magical texts, are a valuable supplement to the medical papyri on account of the many anatomical and medical terms that occur in them."

As regards my own opinion in this matter, I am still doubtful whether the disease *sbh* is identical with leprosy as we know it, and even Hippocrates did not identify it correctly. Under the term leprosy, Hippocrates described a scaly affection which occurred frequently in the spring, itched violently and was easy to heal. When the Old Testament in the days of the Septuagint (about 150 B.C.) was translated into Greek, the word *tsara'at* was translated as λέπρα, and this disease has never been the modern disease leprosy. The lepromatous type of leprosy is identical with the disease *elephantiasis graecorum*, which is first mentioned and described by the Alexandrian medical school. Possibly the *sbh* disease is identical with the Hippocratic *lepros*, but not with the *elephantiasis graecorum* (lepromatous leprosy).

ACKNOWLEDGMENTS

I wish to acknowledge the tenure of a Carlsberg Foundation Research Fellowship which enabled me to make this investigation in Britain and France over a period of six months in 1962.

For permission to examine the material in their care, I am grateful to Dr. Kenneth P. Oakley, British Museum (Natural History); Professor J. C. Cunningham and Miss Jessie Dobson, both of the Royal College of Surgeons, London; Dr. E. Ashworth Underwood, Director of the Wellcome Historical Medical Museum, London; Dr. J. C. Trevor, Director of the Duckworth Laboratory, University of Cambridge; Professor G. J. Romanes, Department of Anatomy, University of Edinburgh; Professor G. M. Wyburn, Department of Anatomy, University of Glasgow; Mr. Warren R. Dawson, Bletchley, Bucks.; Mlle A. M. Vidal-Hall, The French Embassy, London; Mlle A. de Pitteurs, Ministère des Affaires Étrangéres, Paris; and Mme. Gessain, Musée de L'Homme, Paris.

It has been a great pleasure to work in collaboration with the following: Don Brothwell, David Hughes, R. G. Inkster, W. H. Jopling, A. T. Sandison, John Walter and Calvin Wells.

REFERENCES

BARNESTON, J., 1950: Osseous changes in neural leprosy. Radiological findings. *Acta Radiol.* 34:47.
BROTHWELL, D. R., 1958: Evidence of leprosy in British archaeological material. *Med. Hist.* 11:287.
BROTHWELL, D. R., 1961: The palaeopathology of early British man. *J. Roy. Anthrop. Inst.* 91:318.
BROTHWELL, D. R., and MØLLER-CHRISTENSEN, VILH. 1963: Medico-historical aspects of a very early case of mutilation. *Danish Med. Bull.* 10:21.
CHAMBERLAIN, W. E., WAYSON, N. E., and GARLAND, N. H. 1931: Bone and joint changes of leprosy: roentgenologic study. *Radiology.* 17:930.
COCHRANE, R. G., 1961: Biblical leprosy. *Christian Medical Fellowship Quarterly* (I.V.F.) 24
ELLIOT, SMITH, G., and WOOD-JONES, F., 1908: Report of the human remains. *Archaeological Survey of Nubia. Report of 1907-1908.* Cairo
ELLIOT SMITH, G., and DERRY, DOUGLAS E., 1910: Anatomical report. Bulletin No. 6. *Archaeological Survey of Nubia.* Cairo.
ELLIOT SMITH, G., and DAWSON, W. R., 1924: *Egyptian Mummies.* London. Allen and Unwin.
GLÜCK, L., 1897: Die Lepra der oberen Athmungs—und Verdauungswege. *Conf. int. Lèpre* (Berlin) 1:18.
GRAMBERG, K. P. G. A., SWELLENBREBEL, J. L., NIDA, E. A., and WALLINGTON, D. H., 1961: Leprosy and the Bible. *The Bible Translator.* Amsterdam.
GREIG, D. M., 1931: *Clinical Observations on the Surgical Pathology of Bone.* Edinburgh, Oliver and Boyd.
HIRSCHBERG, M., and BIEHLER, R., 1909: Lepra der Knochen. *Derm. Z.* 16:415, 490.
MAC ARTHUR, W., 1953. "Leprosy" in the British Isles. *Leprosy Review,* London. 24:00.
MOODIE, R. L., 1923: *Paleopathology.* Illinois.

Morse, Dan, 1961: Prehistoric tuberculosis in America. Amer. Rev. Resp. Dis. 83:489.

Møller-Christensen, V. 1953: *Ten Lepers from Naestved in Denmark.* Copenhagen, Danish Science Press.

Møller-Christensen, V., 1961: *Bone Changes in Leprosy.* Munksgaard, Copenhagen.

Møller-Christensen, V., Bakke, S. N., Melsom, R. S., and Waaler, E. 1952: Changes in the anterior nasal spine and the alveolar process of the maxillary bone in leprosy. *Int. J. Leprosy.* 20:335.

Møller-Christensen, V., and Hughes, O. R., 1962: Two early cases of Leprosy in Great Britain. *Man,* 62:177.

Møller-Christensen, V., and Inkster, R. G., 1965: Cases of leprosy and syphilis in the osteological collection of the Department of Anatomy, University of Edinburgh. In press, *Danish Med. Bull.* 12:11.

Møller-Christensen, V. and Jopling, W. H. 1964: An examination of the skulls in the Catacombs of Paris. *Med. Hist.* 8:187.

Møller-Christensen, V. and Sandison, A. T., 1963: Usura orbitae (cribra orbitalia) in the collection of crania in the Anatomy Department of the University of Glasgow. *Path. Microbiol.* 26:175.

Murdock, J. R., and Hutter, H. J., 1932: Leprosy; a roentgenological survey. *Amer. J. Roentgenol,* 28: 598.

Oberdoerffer, M. J. and Collier, D. R., 1940: Roentgenological observations in leprosy. *Amer. J. Roengenol.,* 44:386.

Pales, L., 1930: *Paleopathologie et Pathologie Comparative.* Paris.

Paterson, D. E., 1956: Bone changes in leprosy. *Leprosy in India,* 28:128.

Risdon, D. L., 1939: A study of the cranial and other human remains from Palestine excavated at Tell Duweir (Lachich) by the Wellcome-Marston archaeological research expedition. *Biometrika, 31*:00

Sigerist, H. E., 1951: *A History of Medicine, I.* New York, Oxford U.P.

Wells, C., 1962: A possible case of leprosy from a Saxon cemetery at Beckford. *Med. Hist.* 6:383.

Chapter 23

Evidence on the Palaeopathology of Yaws*

T. D. STEWART AND ALEXANDER SPOEHR

INTRODUCTION

Most of our knowledge of yaws has been acquired relatively recently and in the course of establishing its differentiation from syphilis. The coexistence during historic times of these treponemal diseases in the limited and largely tropical regions where yaws is endemic not only has confused the clinical and epidemiological picture of yaws but has obscured the historic relationship of the two diseases. As a result, there are those (as for example, Butler, 1936) who believe "that yaws, so-called, is simply the 'native interpretation' of [syphilis]" (p. 85); and on the other hand those (for example, Turner and Chesney, 1934), who claim "that at the present time syphilis and yaws are different diseases, and that yaws is not just syphilis modified in some mysterious fashion by circumstances which operate in the tropics" (p. 184).

The weight of medical opinion at present favors a clear distinction between syphilis and yaws, and it is on this basis that we make the same distinction in the field of palaeopathology. This view is well stated by Turner (1937) as one of the conclusions of the Yaws Commission of the Rockefeller Foundation: "Whether [the Treponema group] of organisms were derived from a common stem, and, if so, at what period differentiation occurred can only be surmised. There is no convincing evidence that differentiation first occurred within historic times, for certainly yaws and syphilis have presented much the same clinical and epidemiological features over the entire period for which adequate descriptions of these diseases are available" (p. 503).

In reviewing the historical relationship of these two diseases, it is important to point out that syphilis made itself known in Europe, suddenly and in epidemic proportions, just after Columbus returned from his first voyage to America; in fact to that part of America—the West Indies—where yaws is now endemic. Also it appears (Stewart, 1940) that shortly after the discovery of America a disease, which is interpreted as syphilis, spread rapidly among the American Indians. As compared to the evidence for its presence in pre-Columbian Indian remains, the spread of this disease in America in post-Columbian times appears again to have been almost epidemic in proportions.

One plausible explanation of these phenomena, assuming them to be essentially true, is that Europe and America each had its own variety of syphilis and therefore may have exchanged treponemata for which the recipient populations lacked immunities. In support of this explanation

* From *Bull. Hist. Med.*, 26: 538, 1952.

may be cited the claims of various medical historians for the existence of syphilis on both sides of the Atlantic prior to Columbus' first voyage (cf Holcomb, 1941; Williams et al., 1927). If the evidence for syphilis in Europe prior to 1493 is rejected on the grounds of inadequate osteological evidence or dubious literary interpretations, then it must be conceded that the disease came from the New World. However, even the evidence for syphilis in America before 1492, as already remarked (Stewart, 1940), is not impressive.

Another explanation is that the treponema for syphilis may have developed or differentiated locally in the West Indies from the treponema for yaws, and that the Spaniards unwittingly carried the new disease first to Europe, then to the rest of the Americas, and eventually to other parts of the world. This explanation assumes both that syphilis is of recent origin and that yaws was present in America in pre-Columbian times. The idea that syphilis may have differentiated recently from yaws seems to rest on an observation attributed to Manteufel. According to Turner and Chesney (1934, p. 184, footnote 2), he "is said to have observed a gradual change in the character of experimental yaws lesions in rabbits as a result of continued passage of the yaws virus through these animals. The experimental yaws lesions changed in such a way that they took on the characteristics of syphilitic lesions and eventually . . . could scarcely be distinguished from such lesions." Turner and Chesney's own work did not support these observations.

As for the existence of yaws in America before the visit of Columbus, there is as yet no osteological proof. Indeed, Ashburn (1947) is inclined to the belief that yaws was brought in by Negro slaves just after the discovery.

Only Europe and America have been taken into consideration in these historical reconstructions for the good reason that only here are the medico-historical records fairly reliable. Because we know almost nothing about the prehistoric existence of yaws and syphilis, it is important to add to our knowledge anything in this line wherever it turns up. Therefore we are presenting what we believe to be a case of prehistoric yaws discovered on the other side of the world; on the Island of Tinian in the Mariana Island group of the Western Pacific. So far as we are aware this is the earliest evidence of this disease thus far reported.

SOURCE OF MATERIAL

The Mariana Islands, of which Guam, Saipan, and Tinian gained prominence during World War II, are in the area where yaws currently exists (Butler, 1936, pp. 69 ff.). These islands were discovered by Magellan in 1521, but were, not colonized by the Spaniards until 1668. Even then the Marianas remained fairly isolated until taken over by the United States and Germany at the time of the Spanish-American War. By the latter date the population was found to be heavily afflicted with yaws and possibly also syphilis. Subsequently all the islands but Guam passed into the hands of the Japanese.

The record of discovery in the Pacific tells nothing about the diseases in which we are interested. Therefore it has been assumed that yaws, but not syphilis, was present in the aboriginal Chamorro population of these islands, as well as in the rest of Micronesia at the time of discovery. However, since the rigors of Spanish rule resulted in near depopulation of the Marianas in early historic times, this disease could have been introduced or re-introduced there in the course of repopulation by Filipinos and other island natives. Judging by the history of other native communities, syphilis was carried to the Marianas by visiting European ships, but whether or not it became established in the native population is masked by its similarity to yaws. This timing is consistent with the belief, expressed above, that syphilis did not exist in this part of the world prior to the coming of the Spanish. On the basis of this reasoning any evidence of such a disease from pre-historic village sites would pertain to yaws.

The human skeletal material which we are here describing was excavated by the second author in 1950 during the Chicago Natural History Museum's 1949-50 Expedition to the Mariana Islands. A major objective of this expedition was to push forward the understanding of man's early history in that section of the Pacific called Micronesia, in which the Marianas lie. Most of Micronesia is administered today by the United States, as a United Nations trust territory, the islands having been captured from Japan in World War II.

The characteristic feature of most of the surviving prehistoric sites in the Marianas is groups of stone columns or pillars. Locally these are called *latte*. Many of the Chamorros believe that these *latte* mark ancient graveyards that still remain the homes of ghosts of the dead of olden times. The *latte* sites are indeed burial places. They are for the simple reason that the stone pillars, or *latte*, were the foundation posts for houses, and the ancient dwellers of these houses had the custom of burying their dead under the house, or immediately to the seaward side of the house. The stone pillars are not grave markers, but merely the most solid type of house post available to the original builders.

One of the expedition's objectives was to find a *latte* village site as undisturbed as possible. Work was begun on Saipan, but no large village site was found intact and undisturbed. Moving on to Tinian, the Museum party was fortunate in locating a relatively intact village, which was called the Blue Site, as it lay directly back of a beach designated by invading American Marines during the war as Blue Beach.

The Blue Site consisted of a dozen large, stone-pillared houses strung end-to-end back of, and paralleling the beach. Back of the houses, the earth was found to contain areas densely filled with charcoal and broken utilitarian pottery, indicating that this was the area where the cooking was done. Burials were found as usual under the houses and toward the seaward side. In every case the burials were secondary; that is, they consisted of an assortment of disarticulated bones, sometimes representing more than one individual and never including all the bones of one individual. In the old days the Chamorro preserved the skulls and a few of the other bones of the dead. What they did with the rest of the skeleton in the case of the Blue Site burials is unknown.

Burial no. 2 (Fig. 1) which contained some of the remains of at least two individuals and most of the pathological bones to be described, was found beneath the floor of a large *latte* house. The soil here was very shallow and overlay limestone bedrock. Surface depth to the top of the burial was only 1.05 feet, although the burial lay directly on the limestone bedrock, in a shallow depression. The type of culture with which the burial was associated existed up to the time of contact with Europeans, though no material of foreign origin was found at the site. How far back in time this culture goes is uncertain. Nevertheless, it is likely that the Blue Site is pre-historic, although not very ancient;

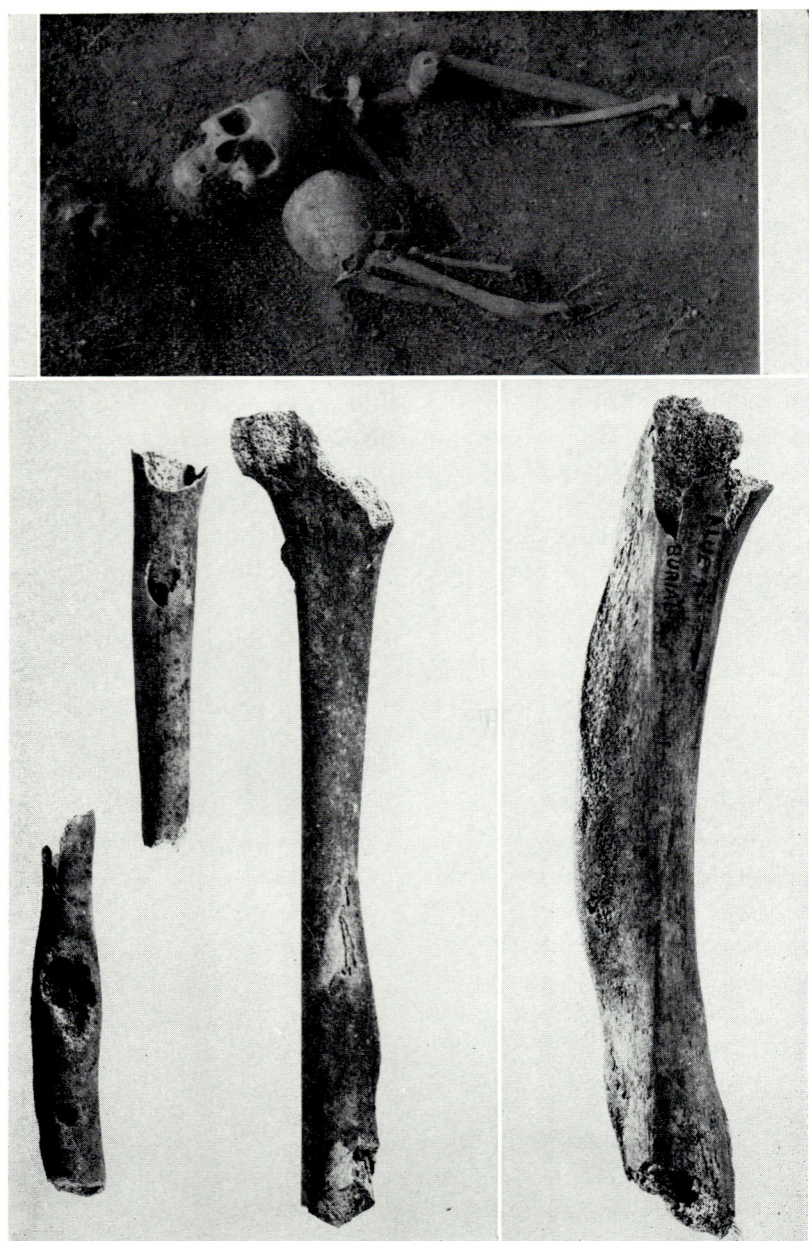

FIGURE 1. *Upper*: Burial no. 2 at Blue Site, Island of Tinian, in the Marianas group. The skull facing down is the one shown in Fig. 2. *Lower*: The three pathological bones on the left (radius?, humerus, femur) are from burial no. 2; the one on the right (tibia) is from burial no. 1. (½ natural size)

probably it does not go back before 1200 A.D.[1]

DESCRIPTION

After the collections of the expedition arrived in Chicago, the first author visited the Museum and was shown the pathological skull from Blue Site burial no. 2 (Figs. 2 and 3). He was impressed by the fact that the crater-like lesion on the frontal was unlike anything he had seen in American Indian crania, and for this reason he asked to be allowed to study the specimen. This request was kindly granted and in due time the skull arrived at the National Museum in Washington. Further examination showed that this individual had died during the age interval between the eruption of the second and third molars. The age at death was around thirteen to fourteen years, judging by the developmental state of the still deeply embedded M3's. Since yaws, unlike syphilis, is acquired most frequently in childhood, the age of this specimen is significant.

Yaws affects other bones besides the skull, much as does syphilis, hence it is important to know whether any of the associated bones belonged to this individual, and if so, whether they too show lesions. Inquiry produced the three incomplete pathological bones shown in Figure 1. Although these bones appear to be in the same subadult stage as the skull, loss of the epiphyses makes it impossible to narrow down the age determination. However, the lesions are compatible with those on the skull. Subsequently Dr. Spoehr discovered part of a pathological subadult tibia among the remains from Blue Site burial no. 1. There is reason to doubt, because of the size of this bone (Fig. 1), that it belonged with the other pathological bones, but here again the occurrence of such lesions in the tibia of a subadult is highly suggestive of yaws.

The following description of the gross bone lesions can be checked by the reader in the photographs already referred to and in the x-rays (Figs. 4-7).

Skull

The mid-frontal crater-like lesion (Figs. 3 and 4) is irregularly shaped and longer in the transverse direction (maximum diameter 28 mm) than in the sagittal (maximum diameter 18 mm). Within the crater the bone surface is rough and porous; the deepest parts are 3 mm below the rim. Immediately surrounding the crater and largely confined to the elevated bone forming its outer slope, is an irregular zone of porous bone. Beyond this is a zone of fine striations. Together, these zones, probably representing increased vascularity, extend about 18 mm at their greatest width (superiorly), and only a few millimeters elsewhere (left side, inferiorly). In spite of being porous and striated, the bone around the crater is glossy and thus is indicative of a chronic rather than an acute inflammatory reaction.

The frontal bone also bears a second and smaller lesion on the left side just above the temporal line (Figs. 2 and 5). The shape of this one is circular (7 mm in diameter) and the depth is only 1 mm. Although the edges are sharp, only the anterior edge is slightly raised. Except for this small area anteriorly there is little evidence of reaction.

---

1. Since this was written, a tridacna shell from a small refuse dump associated with the house where burial no. 2 was found has yielded a Carbon-14 date of A.D. 854 ± 145 (personal communication from W. F. Libby, Institute for Nuclear Studies, University of Chicago). Since the evidence indicates that the house and refuse dump are the result of a single and probably not a very long occupation, it can be inferred that the date of the shell and the date of the skeleton are within very narrow limits the same. Certainly there is no doubt that the skeleton is pre-contact by a very comfortable margin.

Figure 2. Top, front and left side views of the pathological skull from Blue Site burial no. 2. Tinian, Marianas Islands (slightly less than ½ natural size).

Other lesions are to be seen on the parietals (Figs. 2 and 5). Between the midpart of the sagittal suture and the right parietal eminence or boss the outer table and the diploë have been destroyed in several places. Surrounding these main lesions are areas where the outer table is porous, suggesting beginning destruction of the underlying diploë. Otherwise there is little surface change. The total area of involvement is roughly triangular and extends from near the midline to a point near

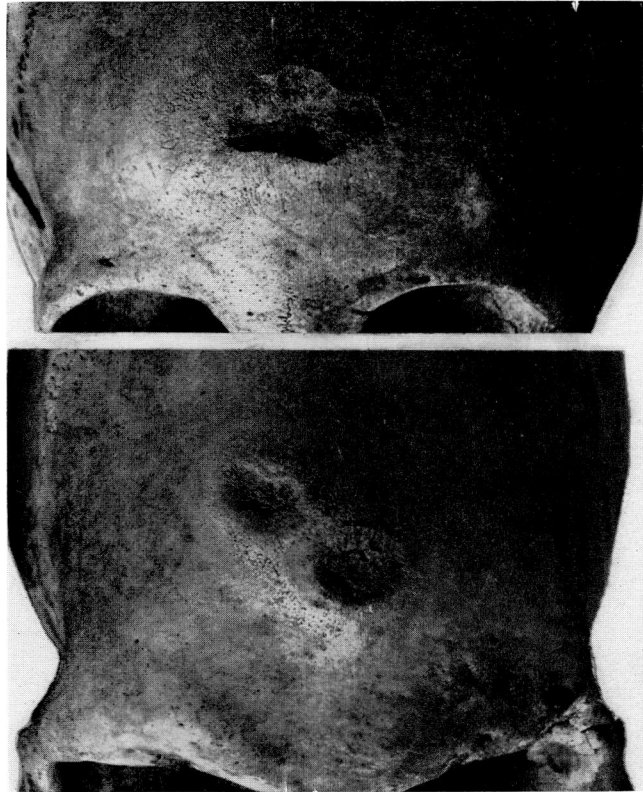

FIGURE 3. *Upper*: Close-up view of crater-like lesion on forehead of skull from Blue Site burial no. 2. *Lower*: Close-up view of similar lesion on forehead of Australian skull from Arnhem Land (USNM 380, 453).

the boss, a distance of 65 mm. Unlike the lesions on the frontal, these have edges that are undercut.

The right parietal, in addition, shows a small oval lesion near the sphenoidal angle just below the temporal line. The cavity of the lesion has a maximum diameter of 8 mm, which is greater than the opening in the external table. A perforation of the inner table may have occurred post mortem. In appearance this is a lesion expanding out of the diploë.

On the left parietal, just below the apex of the boss, is a single large area of destruction having the appearance of partly confluent multiple lesions (Figs. 2 and 5). This area is almost circular (nearly 40 mm in diameter) with sharp, undercut edges inferiorly, and with less distinct edges superiorly merging into a porous outer table. In two places at the deepest parts of the destroyed area the inner table is gone. It is impossible to say now whether these holes are original or were produced in the course of cleaning the specimen. Except where these holes occur, the lesions are only 2 mm deep. Also, there is little evidence of reaction in the surrounding outer table.

Turning now from the skull vault to the face, there is only one other gross lesion to be described. This appears on the left malar bone just below the orbital border (Fig. 2). Unlike the others, it has rounded and sometimes indistinct borders, yet the

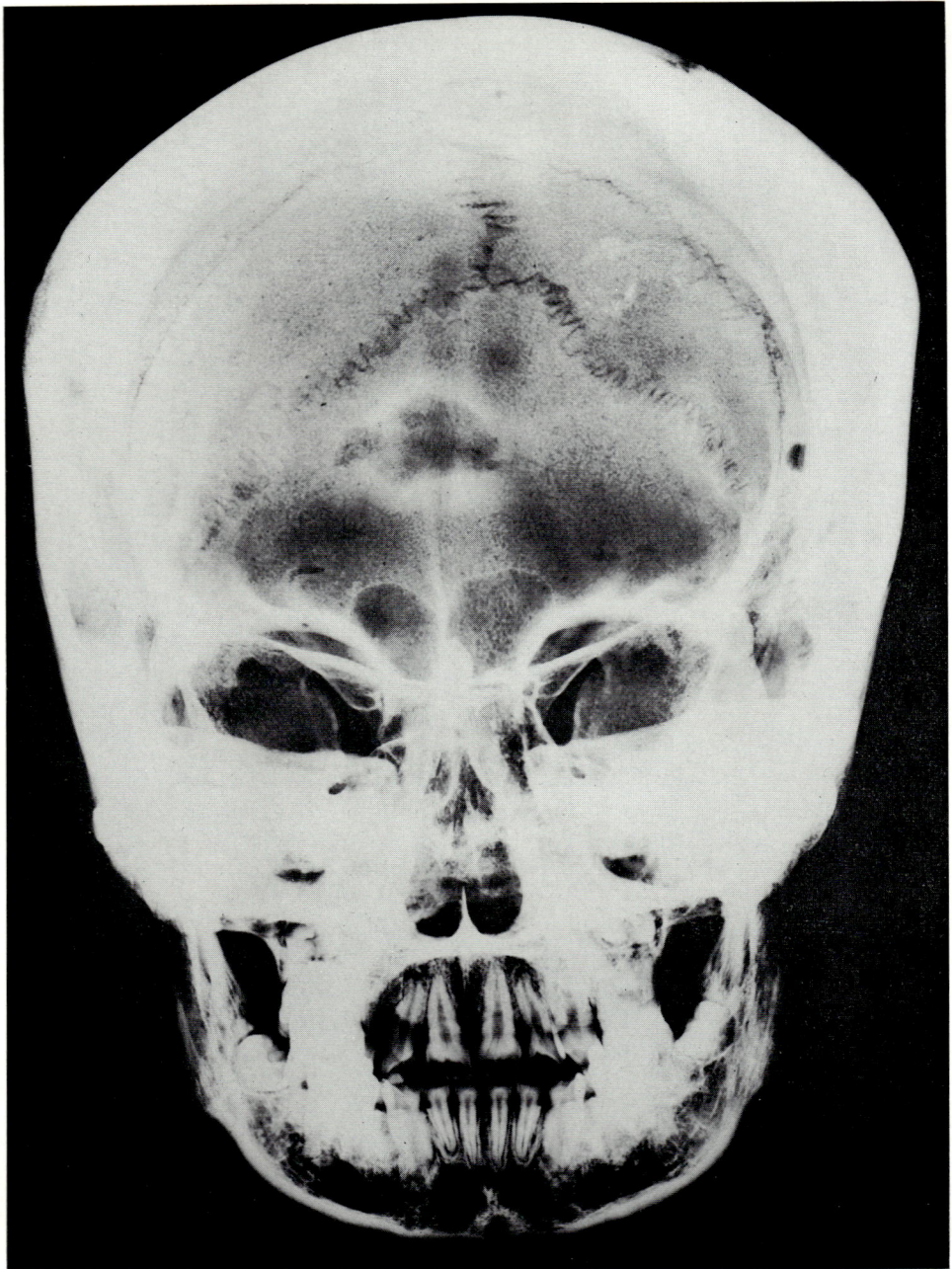

FIGURE 4. Antero-posterior x-ray view of skull shown in Figure 2.

general effect is a small circular pit about 6 mm in diameter and 2 mm in depth. The irregularity of the bone surrounding this pit is taken to indicate a long-standing inflammatory reaction.

Long Bones

The pathological subadult bones accompanying the skull (Fig. 1) were parts of a femur, a humerus and a radius(?). The femur (left) lacks the distal one-fourth

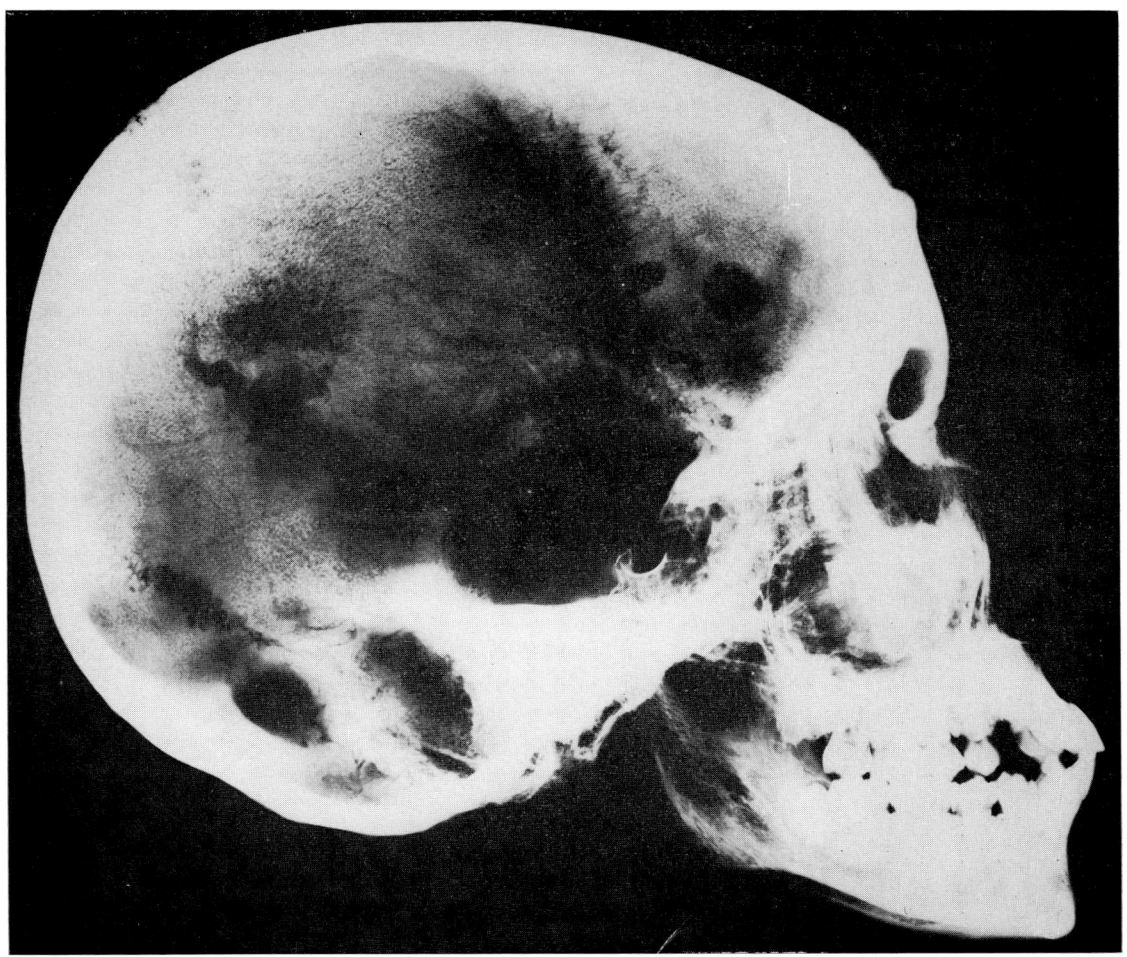

FIGURE 5. Lateral x-ray view of skull shown in Figure 2.

and shows the proximal epiphyses ununited. The humerus (left) is the proximal half without the head. The radius (right?), if correctly identified, is the proximal half without the proximal end; it is so distorted as to make positive identification impossible.

Grossly the fragment of humerus is the least abnormal. However, at the juncture of its upper and middle thirds is an opening through a thickened cortex. This opening expands irregularly (from 6 to 13 mm) as it approaches the outer surface, and its edges are somewhat rounded and glazed. Surrounding the opening is a zone of inflammatory reaction. Judging from what can be seen through this opening and by examination of the x-ray (Fig. 6), it appears that a large cavity within the cortex communicates with the opening. The cortex covering this cavity is very thin. Probably the opening was a sinus draining the cavity.

Whereas the cortex of the humerus shows localized internal thickening, that of the femur and radius (?) shows localized external thickening. In the case of the femur, the process is limited to the anterior

FIGURE 6. X-ray views of three bones (radius?, humerus, femur) shown in Figure 1.

and lateral aspects of the distal half of the shaft. Here apparently the new bone has been laid down in layers on the outer surface. These layers are not fully united either with each other or with the original diaphyseal surface and are now tending to separate. Probably this indicates an early acute stage of periostitis.

The radius (?) gives evidence of this process at its broken distal end. But elsewhere in this fragment the process seems to have gone on to fuller cortical consolidation and cavitation. At one point a large (14 × 22 mm), irregular-shaped pit has formed, as if a number of small cavities had coalesced, causing the surface to break down. The x-ray (Fig. 6) reveals other smaller cavities.

The subadult tibia (left; both ends missing) from burial no. 1 (Fig. 1) represents a stage of involvement comparable to that in the radius (?). Thickening of the cortex has progressed so far that the usually distinct anterior border is now swollen and distorted, giving the appearance of anterior bowing (Fig. 7). Also, several, more or less confluent pits have formed. From the appearance of these pits, especially their physical shape and undercut edges, it would seem that they had begun as cavities within the thickened cortex.

DISCUSSION

The bone lesions here described, except the crater-like kind on the frontal, are indistinguishable from those seen in the bones of many adult American Indians from outside the tropics. The skeletal remains of subadult American Indians from the same areas are not thus involved. It is doubtful whether a further distinction could be made from the skeletal "sites of election," even if the skeleton from Tinian were complete, because the individual pattern of bone involvement probably is

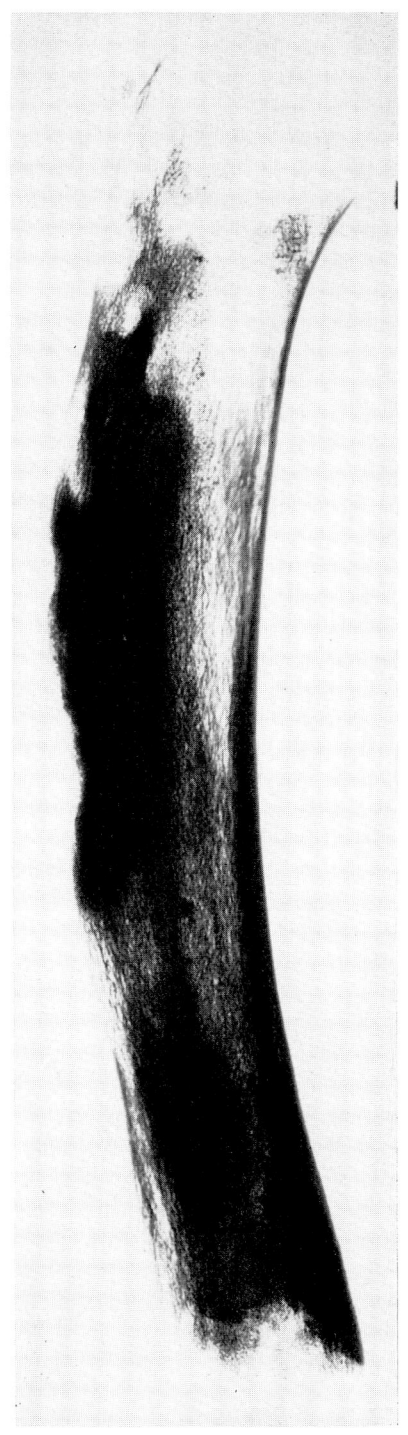

FIGURE 7. X-ray view of tibia shown in Figure 1.

quite variable. Age, then, is the main distinguishing feature between these manifestations of disease in geographically widely separated places.

It is generally agreed that the disease affecting the bones of nontropical, adult, American Indians, in a manner like that here described, must have been syphilis. Logically, therefore, the disease in the present case should be yaws, both because of the geographical area from which the bones come and because of their youthful age. As already mentioned, yaws, unlike syphilis, is acquired mainly during childhood, before the period of heterosexual activity. Also, and again as already mentioned, the antiquity of the present case would rule out syphilis.

That the bones in a case of yaws should show lesions similar to those of syphilis is to be expected. This is the conclusion to which Williams (1935) came after reviewing all the evidence: "As to the involvement of bone, it seems to me that it must be admitted that roentgenograms show that the bones are often affected in yaws and that the pictures are often like those seen in cases of syphilis" (p. 627).

The emphasis which we have placed on the crater-like lesions of the frontal bone in our presumptive case of yaws may be unwarranted; we cannot be sure that this is a typical lesion of yaws. However, there are skulls of Australian aborigines in the National Collections with similar but less sharply defined lesions in the same location (Fig. 3). Australia is another place where yaws is endemic. Indian skulls, on the other hand, sometimes show small pitted scars on the frontals; very rarely do Indian skulls apear 'worm eaten" as in the classical portraits of syphilitic skulls.

Williams (1935) detected a difference between syphilis and yaws in the involvement of the skull. He says:

One point of difference from syphilis may be mentioned here. The involvement of the outer surface of the skull, especially the frontal and parietal bones, in a gummatous periostitis has long been regarded as frequent in severe cases of tertiary syphilis, but it is rarely seen today. The worm-eaten appearance of such a skull is well known. There is practically no evidence that yaws produces this type of skull. Sir Arthur Keith informed me of a skull of that type from Society Islands, said to be that of a patient with yaws, but without other proof. Georg described a doubtful case from the Dominican Republic, which was probably a case of syphilis. (p. 613)

Probably some confusion enters into the picture here due to different stages of the disease having been studied. Quite likely clinical records on yaws, and especially roentgenograms, relate mostly to adults who have long been infected. In the Chamorro child from Tinian here presented the disease probably had not reached the chronic stage. Indeed, some of the lesions appear to be early. This reminds us that many observers of yaws claim that bone lesions begin the secondary stage, rather than, as in syphilis, in the tertiary stage (cf. Hackett, 1946).

At this point perhaps we should return to our original historical perspective. Our specimen is evidence that yaws has been present among the peoples of the Western Pacific since prehistoric times. If yaws is still more ancient in that area, and if a population develops any resistance to this disease after many generations of infection, it would be natural to find a bone reaction that is modified as compared to that produced by syphilis in a relatively unprepared population. Actually, herein is the significance of the Tinian specimen: That it gives evidence not only of the presence, but of the character, of yaws in the Western Pacific before the introduction of syphilis.

The authors are grateful to Sister Charles Regina and the X-ray Department of Georgetown University Medical School for making the x-rays of the specimens here described; they acknowledge also the kind assistance of Drs. William J. Tobin and Aubrey O. Hampton of Washington, D.C.

REFERENCES

ASHBURN, PERCY MOREAU, 1947: *Ranks of Death; a Medical History of the Conquest of America*, Ed. by Frank D. Ashburn. New York, xix + 298 pp.

BUTLER, CHARLES S., 1936: *Syphilis sive Morbus Humanus. A Rationalization of Yaws So-Called.* Brooklyn, xii + 137 pp.

HACKETT, C. J., 1946: Review of references to bone lesions on yaws. *Trop. Dis. Bull.*, vol. 43:1091.

HOLCOMB, R. C., 1941: The antiquity of congenital syphilis. *Bull. Hist. Med.*, vol. 10, no. 2, pp. 148-177.

STEWART, T. D., 1940: Some historical implications of physical anthropology in North America. *Smithsonian Misc. Coll.*, 100:15.

TURNER, T. B., 1937: Studies on the relationship between yaws and syphilis. *Amer. J. Hyg.*, vol. 25:477.

TURNER, T. B. and A. M. CHESNEY, 1934: Experimental yaws: Comparison of the infection with experimental syphilis. *Bull. Hopkins Hosp.*, 54:174.

WILLIAMS, H. U., 1935: Pathology of yaws, especially the relation of yaws to syphilis. *Arch. Path.* 20:569-630.

WILLIAMS, H. U., JOHN P. RICE and JOSEPH RENATO LACAYO, 1927: The American origin of syphilis. *Arch. Dermat. Syph.* 16:683.

Chapter 24

The Evidence for Neoplasms

DON BROTHWELL

INTRODUCTION

THE identification of neoplastic growths is perhaps one of the most interesting and at the same time difficult aspects of palaeopathological studies. Most of the evidence to date is in the form of bone changes, and these range from an additional bony mass representing the participation of bone-forming cells in the growth, to the elimination of bone substance by osteolysis producing what might be called a "negative" or "mould" of the original tumour surfaces. Because the material evidence is more revealing than the art and literary data, it will be discussed first.

It has been stated by a number of previous workers that tumours are rare in skeletal collections. Certainly such pathology was uncommon, for as some neoplasms are clearly correlated with age—and as in most prehistoric and early historic populations few lived into middle or old age—it is to be expected that more would die from the diseases and traumas of infancy to early adulthood than anything.

Some, however, did live into old-age and of course individuals in the lower part of an age pyramid are not completely immune to neoplastic growths. Indeed, one of the points I wish to emphasize in this survey is the likelihood that the scarcity of tumours has been *overemphasized* in the past—a fact which in itself may have depressed some detailed searching. Except perhaps for a few skeletal series from Egypt it is also surprising how few skeletal collections have ever been examined with the thoroughness necessary to detect some disease processes—and all too often in the case of nineteenth century excavated material, only the skull now remains, with the rest or most other bones having been reburied or thrown away as of no great anthropological value! Thus, we can by no means claim as yet that large samples of early skeletons from various regions have been studied sufficiently exhaustively for tumour evidence to be counted as absent. It might be mentioned as an example that in my own "reworking" of various early skeletal series in various parts of Europe and North Africa, I have come across pathological changes not reported or identified by those who had previously examined the remains, and this includes cases of tumors. There is also evidence that earlier workers have not recognised the significance of some changes, and it is inevitable that doubtful conditions will not receive the same degree of description or attention as obvious diseases. It seems worth expanding on this point further as it may be especially significant from the point of view of neoplastic disease.

In his detailed anatomical work on early Nubian skeletons, Derry (1909) has this to say of one Middle Nubian (C-group) burial; ". . . . as regards signs of other disease, only one of any interest was encountered. This was found in the skeleton of a woman who had suffered from some necrotic process in various bones. The left humerus had broken across in the grave at a weak spot caused by some ulceration of the bone which has resulted in a cavity in the medulla communicating with the surface by a round hole. Around this hole there has been periosteal inflammation. A similar pathological process had attacked the left clavicle, which had also broken at a corresponding weak spot. The sternum had a hole of the same character, while the entire bone was much inflamed. The left scapula had a large hole through the supraspinous fossa, surrounded by spicules of inflammatory bone. The same disease may probably explain the condition of the vertebrae (particularly in the cervical region), which were eaten away by a form of rarefying osteitis" (p. 45). As yet I have been unable to ascertain the present whereabouts of this skeleton, but it is evident that this morbid change demands critical re-examination. Were the holes in fact of neoplastic origin, and was the "inflammation" surrounding these areas in fact marginal reactions following tumour development? Such a possibility cannot be ruled out on the present limited evidence.

A second instance where a neoplastic origin could well have been considered but was not, is in the skeleton of a female of New Kingdom date from Nubia (grave No. 204/15) described by Batrawi (1935). He writes: "This is an interesting case of unexplained erosion of the base of the skull. . . . . . . The process largely affected the left half of the sphenoid bone, thus resulting in its rarifaction. The condition was acute, since there is no thickening of bone, and it very probably affected the left eyeball" (p. 186). In Plate XV of Batrawi's study, there is not only a view of the sphenoid lesion but also a view of the top of the vault where there appears to be further localised reactions, including a frontal perforation. The specimen clearly deserves careful re-examination with a view to its possible neoplastic involvement.

Pathological versus Post-mortem Change

The problem of pseudo-pathology has already been considered in this book by Calvin Wells. It would be easy to expand on this subject as regards pseudo-tumours; from "carcinomatous" destruction of bone in the temporal area of a Winchester Saxon and frontal of an Iron Age individual from the Bernese Oberland to "myelomatosis" changes in a medieval youth from Scarborough—all in fact post-mortem changes. Similarly insect destruction, as for example in the femur from a burial site in the Island of Socotra (Brothwell, 1963, Plate 7A), can simulate neoplastic destruction of bone.

What is far more important to emphasize here is the fact that not only might post-mortem changes produce "pseudo-tumours" but they may also *obscure to varying degrees changes resulting from the presence of true tumours*. This fact has not been stressed sufficiently in the past, especially as perhaps the majority of early skeletons being excavated or already available for study are defective to some extent —through post-mortem changes. Those who have worked on early skeletal series will appreciate the range of these changes. Moreover, where breakage and soil erosion take place, it seems particularly liable at, say, the seat of a thin-walled enchondroma, a giant-cell tumour of bone, an erosive extra-osseous fibrosarcoma, or even a fragile-boned osteosarcoma. In the case of some tumours, considerable osteolytic destruction of bone may take place with but little if any marginal reaction, with the result that even slight surface changes taking place after burial might obscure such slight reactions or 'blur' the sharpness of the original margins of the tumour. To take first a modern case as an example, Greig (1931) shows the changes to the skull of a man who had carcinomata extending from the left medial canthus. This resulted in the destruction of the "left and most of the right nasal bones, the ethmoidal labyrinth and most of the left half of the pars orbitalis of the frontal bone" (p. 215). This area of the skull is frequently broken and eroded in early skeletons, and one might well reflect that such a case would in all probability be missed in some early series.

This problem was presented forcibly to me recently while studying early Swiss skeletons. One skull (D. 10) in the Kantonsmuseum, Liestal, dated to the "Frühmittelalter" period, displayed much destruction of part of the frontal region. The right upper orbital margin—and extending past the nasal bones to the beginning of the left upper orbital margin—displays much destruction of bone. This looks at first sight like post-mortem erosion and has obviously been taken as being so by Hug (1959) in his survey of early skeletal material from Baselland. However, the irregularity and pitting is quite unlike soil "erosion" present in other parts of the skull or in Swiss skeletons generally. What makes some form of pathological process a certainty is the clear reaction in the right orbital area of the frontal. This is a subperiosteal change taking the form of fine pitting, a few larger perforations and possibly slight thickening of the bone (quite distinct, it must be emphasized, from *usura orbitae*). The greater part of the eroded frontal margin is irregular with the reaction being limited in extent, extending onto the outer surface of the bone for about 15 mm at the most, and on to the inner surface for a little over 10 mm. This reaction is unlike that in the orbit in that it has a more "etched" appearance with some large depressions and irregularities. Although it is impossible to be certain, I think it by no means improbable that this is a case of a destructive tumour in the region of the frontal sinuses, with a very limited marginal reaction (the extent of the neoplasm and bone destruction could have been similar to the case described by Greig, 1931, p. 214, Fig. 210).

Differential Diagnosis

The separation of tumours into varieties, when one has only the dry bones to deal with, is no easy task. However, I have endeavoured in the following review of early tumour evidence to indicate whenever possible the most likely neoplasm to have caused the bone changes in each case. It must therefore be kept in mind that although a division of the specimens into more specific tumour categories has been attempted, it is only extremely tentative for some of the cases, and I do not wish to be considered dogmatic in this matter. For the most part, my terminology is that defined by Jaffe (1958) and Coley (1960).

PRIMARY BENIGN TUMOURS

Of all the major classes of tumour which can leave evidence of their occurrence in the ancient dead, this is by far the most commonly occurring category. Also, by the varied forms of these tumours, a number of different types have been identified —albeit tentatively.

Osteoma

IVORY OSTEOMATA. A not uncommon occurrence on the skull vaults of early skeletal series are small circular mounds of dense osseous tissue. A particularly large example is given in Fig. 6a, in an Egyptian of Roman date. In a recent survey of their occurrence in earlier British populations (Brothwell, 1961), seventeen cases were noted, their distribution on the skull being as follows:

tosis on the right femur. This is on the lateral aspect and takes the form of a prominent mound which is internally composed of cancellous bone (Fig. 7e). This may well be a large osteoma with post-mortem erosion—exposing the inner spongy bone—but alternatively an osteochondroma should be considered. From

Population	Frontal	Parietal	Other Regions	Total
British (Neolithic-Anglo Saxon)	7	6	4	17

Hooton (1930) has also shown their prevalence in early Pecos Pueblo Indians, and gives the following frequencies for "button osteomata":

the New World, Goldstein (1957) notes a large rounded tumour on the humerus of an early Texas Indian which could be of this type (Fig. 7f). Moodie (1926) has

SKULL OSTEOMATA, PECOS PUEBLO

Total Examined	Number with Tumours			Total Affected
	Young Adult	Middle-aged	Old	
581	2	7	4	13

Incidentally, it may be noted here that the mandibular, auditory, palatal and maxillary tori are not uncommon in some populations, and display a similar structure. However, it is very debatable whether they should be considered as pathological and they are now recognised more as discontinuous variants of anthropological interest (Brothwell, 1959).

OSTEOCARTILAGINOUS EXOSTOSIS. These do not appear to be so common as the ivory osteoma. This may be because some observers do not distinguish between the two, with the result that all have been mistakenly considered to be ivory osteoma. At the National Museum, Denmark, a Roman skeleton of a woman from Juellinge is on display which shows a large hyperos-

described a pre-Columbian skull from Ancon, Peru, which appears to have this form of osteoma in the left orbit.

A cancellous exostosis has recently been located radiographically on the left tibia of a desiccated Predynastic body from Egypt (P. K. Gray, 1964, personal communication).

Osteochondroma

A Vth Dynasty Egyptian femur has previously been considered to be an osteosarcoma, but Rowling (1961) considers it far more likely to be an osteochondroma. A consideration of the macroscopic pathology (Fig. 1) certainly makes the latter diagnosis far more likely. Similarly, the Juellinge case may be an osteochondroma rather than a cancellous exostosis.

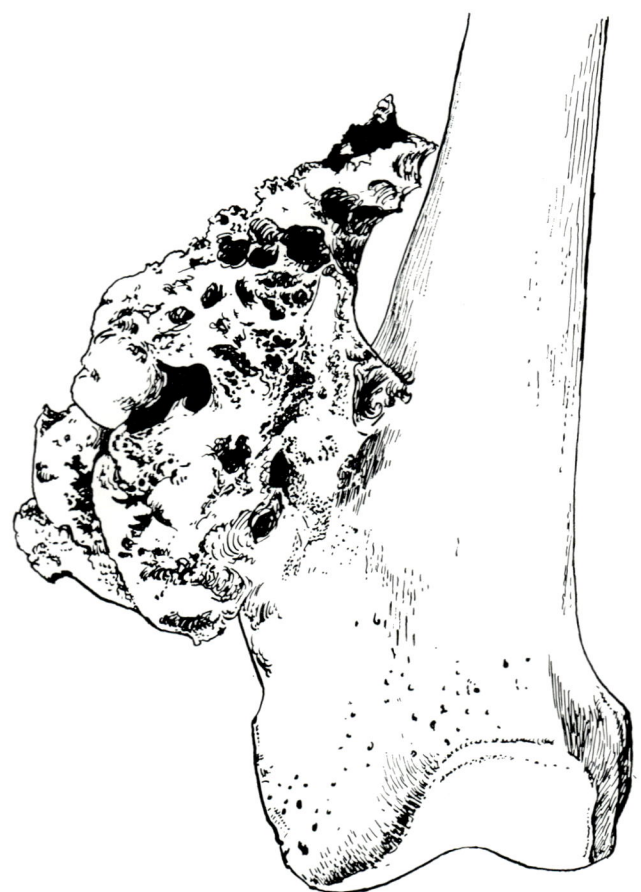

Figure 1. Lower end of a Vth Dynasty femur, showing a large irregular bone mass, probably the result of an osteochondroma.

Chondroblastic Tumour / Giant Cell Tumour

One early Nubian humerus (187A) now part of the British Museum (Natural History) collection displays a massive enlargement of the proximal end. The head region is approximately twice its original size, although its shape is very far from normal, the joint being severely disorganized. Most of this globular swelling (Fig. 7d) is intact and is covered with a thin continuous layer of bone. Although the surface is slightly irregular in parts, it nevertheless contrasts noticeably with the macroscopic dry-bone appearences in an osteosarcoma and osteochondroma. Radiographically, the inner cancellous tissue does not appear to be generally very dense. Considering these various factors together, it is possible that this specimen represents an old chondroblastic tumour. Certainly the proximal end of the humerus is a common site for this neoplasm in modern populations, perhaps a further point in favour of this diagnosis. Alternatively, the site of this lesion is a relatively common one for a giant cell tumour of bone, and considering also the macroscopic details, this must be considered as a secondary diagnosis.

Osteoid Osteoma

Other than typical ivory osteomata it is by no means an easy task to ascertain the reason for smooth low swellings on bones. A Saxon tibia from Sedgeford, East Anglia, presented such a problem (Brothwell, 1961). In the central part of the shaft, at the anterior border, is an elliptical smooth swelling about 100 mm in length. Post-mortem breakage shows the swelling to be mainly compact bony tissue. Certain diagnosis seems at present impossible and it seems equally likely to be an osteoid osteoma or alternatively the result of a low-grade chronic abscess. As mentioned elsewhere, the Saxon femur described by Wells (1964 a), might also be considered as a further possible case.

Giant Cell Tumour / Aneurysmal Bone Cyst

From the Saxon site of Finglesham in Kent, the left femur of adult skeleton 44 shows marked pathological change (Fig. 7c). On the medial side of the shaft just above the distal condyles is a pronounced swelling some 120 mm. long. Owing to the post-mortem bone erosion, the internal structure of the swelling is uncertain, but appears likely to have been mainly a thinly cancellated inner zone which is surrounded by a bony shell varying in thickness. Although interpretation in such a case must be regarded with much caution, it is worth questioning whether this may have been an old giant cell tumour, or perhaps an aneurysmal bone cyst. Again, it might be noted that this femoral position is one of the commonest (if not most common) sites of the giant cell-tumor today.

CONDITIONS SIMULATING PRIMARY BENIGN TUMOURS

Post-traumatic Bone Swellings

It would seem worth while considering under this heading various swellings which can be described as myositis ossificans or ossifying haematoma. As they can be confused with true tumours, their inclusion here is worthwhile. Although not common, they have a long history, in that the first discovered femur of the Middle Pleistocene variety *Homo erectus* (*Pithecanthropus*) displays a large prominent irregular bone protuberance on the medial

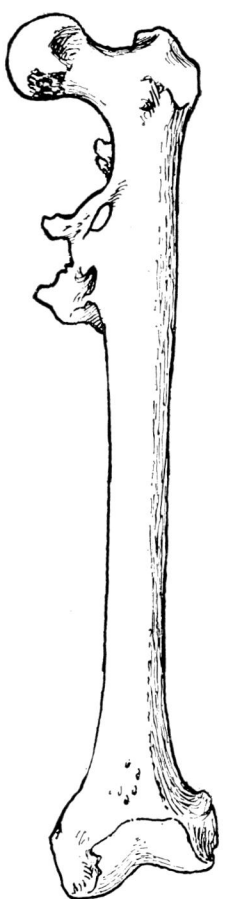

FIGURE 2. The first discovered *Homo erectus* (*Pithecanthropus*) femur, showing the medial exostosis.

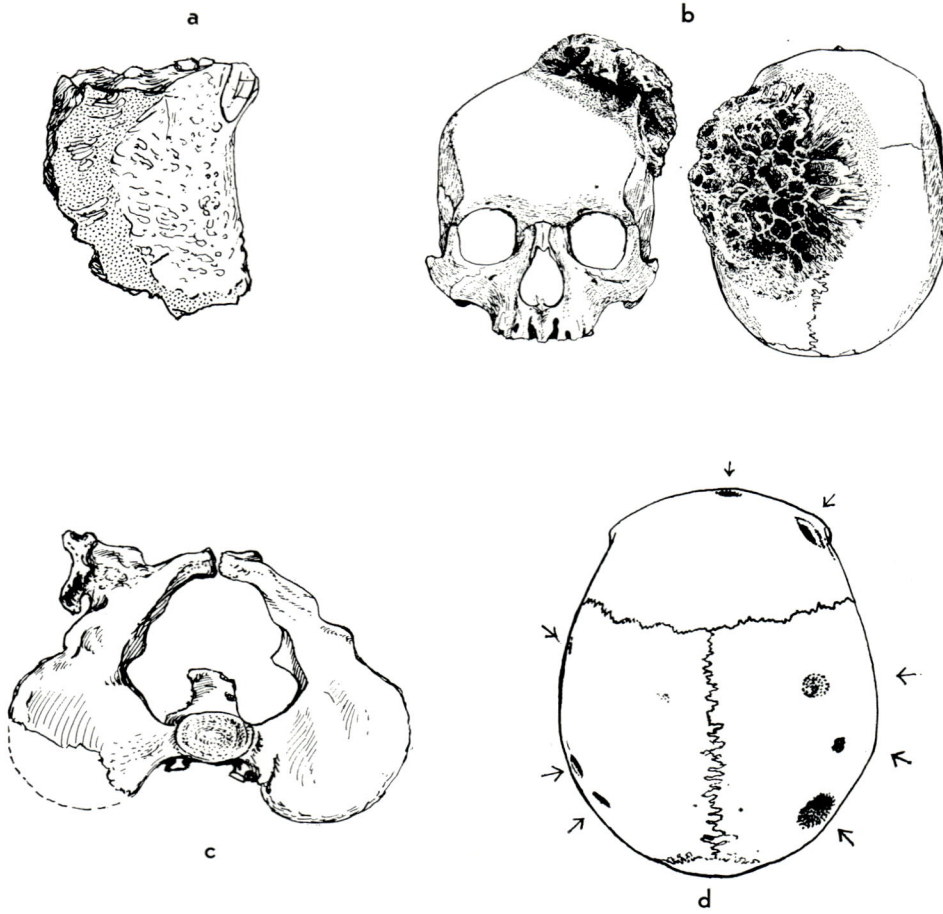

FIGURE 3. (a) Section through the right first incisor region of the Kanam mandible, showing the pathological mass of new bone on the lingual surface.
(b) Facial and top views of an early Peruvian skull, showing a possible osteosarcoma or reaction to a meningioma on the left side. After MacCurdy, 1923.
(c) The abnormal pelvis from Boarhills, Fife.
(d) Prehistoric skull from Pyrennean cave of Joan d'os, showing multiple perforations in the skull vault (arrows point to the seven visible) After Fusté, 1955.

aspect of the proximal half of the femoral shaft (Fig. 2). Following this, the next earliest cases are from the Neolithic; for example, the long barrow femur from Wiltshire, where the irregular projection is shallower but extends 116 mm. down the posterior aspect of the shaft (Brothwell, 1961). From a Bronze Age cist near Boarhills, Fife, Waterston (1926) described a large extra-pelvic bone mass near the left acetabulum which he considered the result of a severe injury resulting in periosteal tearing, an effusion of blood into the area, and finally ossification of the haematoma (Fig. 3c). Ruffer (1921) has also noted a traumatic osteoma in the pelvis of a Coptic mummy. A Yorkshire Bronze Age skull also displays bony swelling which seems best explained as the result of an ossifying haematoma. The thickened cranial vault in the region of the coronal suture is associated with extensive fracturing in the fronto-parietal region, which shows healing and might be contemporary (Brothwell, 1960). A further possible calcified haematoma has been described in a Saxon femur from Norfolk (Wells, 1964; who prefers to consider it as evidence of scurvy), although the diagnosis of an osteoid osteoma must be considered a feasible alternative.* As trauma was not an uncommon experience of earlier peoples, as evidenced by the numbers of fractures and cranial injuries, such traumatic conditions are to be expected.

Condylar Hyperplasia / Ossifying Chondroma

As the following two cases of tumour-like swelling are probably not true neoplasms but the result of a unilateral condylar hyperplasia, they also demand mention here. An alternative explanation, but a very doubtful one, is that the swellings are the result of an ossifying chondroma (such as described by Coley, 1960). In both of these early specimens there was, as a result of the condylar enlargement, marked unilateral elongation of the face and deflection of the chin to the side opposite the swelling, and clearly there would also have been promalocclusion. A large condylar expansion of this type is seen in a Danish mandible of the Roman period from Varpeler, now in the Anthropological Laboratory of the University of Copenhagen (Fig. 6b). A further case, of Saxon date, is in the Duckworth Laboratory of Physical Anthropology collection at Cambridge, from a Cemetery at Melbourn, Cambridgeshire. There is no evidence that either of these deformities is associated with fracturing.

SECONDARY INVOLVEMENT OF BONE BY BENIGN NEOPLASMS

Meningioma / Angioma

Lambert Rogers (1949) describes two Egyptian skulls displaying cranial hyperostosis which may well indicate a meningioma. A Ist Dynasty skull from Helouan has a fairly rounded swelling on the external aspect of the right parietal, with some change in the same region endocranially. He concludes that these changes are the result of a parasagittal meningioma. (The large vascular channels to be seen on the side of the hyperostosis and leading to it support this suggestion.) From Meydum, a XXth Dynasty skull shows a "diffuse honeycomb type of hyperostosis" mainly situated on the right parietal but extending over the sagittal suture onto the left parietal and also over the coronal suture onto the frontal. He suggests that this is a reaction to an angioblastic or sarcomatous meningioma. A Roman skull from Radley, Berks., has been previously described by me as showing vault changes which may have been produced by a sarcoma (Brothwell, 1961). However, in modern cases of sarcoma involving the cranial vault, spicular calcification may only be seen when very "soft" radiation is employed (Shanks and Kerley, 1950), but heavy calcification as in this Roman specimen is not the rule. On further consideration, therefore, taking into account the nature and extent of bone growth on the external table (Fig. 6c) and the slight but definite changes endocranially, it now seems more reasonable to interpret the changes as caused by a meningioma or angioma.

---

* Since completing my MS with this opinion, Wells has withdrawn his original diagnosis in favour of an osteoid osteoma. See *Brit. J. Radiol.*, 1965, 38:393.

In New World material, Moodie (1926) has described a skull (No. 158) from Chavina, Peru, with considerable external hyperostosis in the frontal area "due to the presence of a meningioma during life."

Intradiploic Epidermoid Cyst

On the skull of a Late Dynastic Egyptian (Cambridge, E. 270) there is a large crater. This is situated just anterior to the vertex of the vault and has its long axis a little to the right of mid-line. In length, it is approximately 80 mm to the margins of the crater; in breadth about 70 mm. The depression has the general shape of a low-angled trephine hole, with the perforation through the inner table much smaller than the external margin (Fig. 8a, b). The concavity is, however, noticeably roughened and the external margins are raised above the level of the outer table. On the non-crater side of this raised margin the bone has an "etched" and highly vascular appearance. It must be emphasised that close examination reveals no evidence of endocranial change. Two healed cranial injuries—one a large sword cut—are situated within 30 mm of this crater, both on the left side.

Jelsma (1959) has discussed in detail the erosive properties of epidermoid cysts, which may involve both the inner and outer skull tables. Although some are thought to be congenital, others may well be associated with trauma. In position, size and form there is some similarity between modern cases and the Egyptian case (the case noted by Shanks and Kerley, 1950, described as a "sub-pericranial dermoid tumour," indicated that the marginal area may be ridged). This interpretation of the Dynastic Egyptian anomaly thus seems worth considering, although it cannot be considered the only possibility by

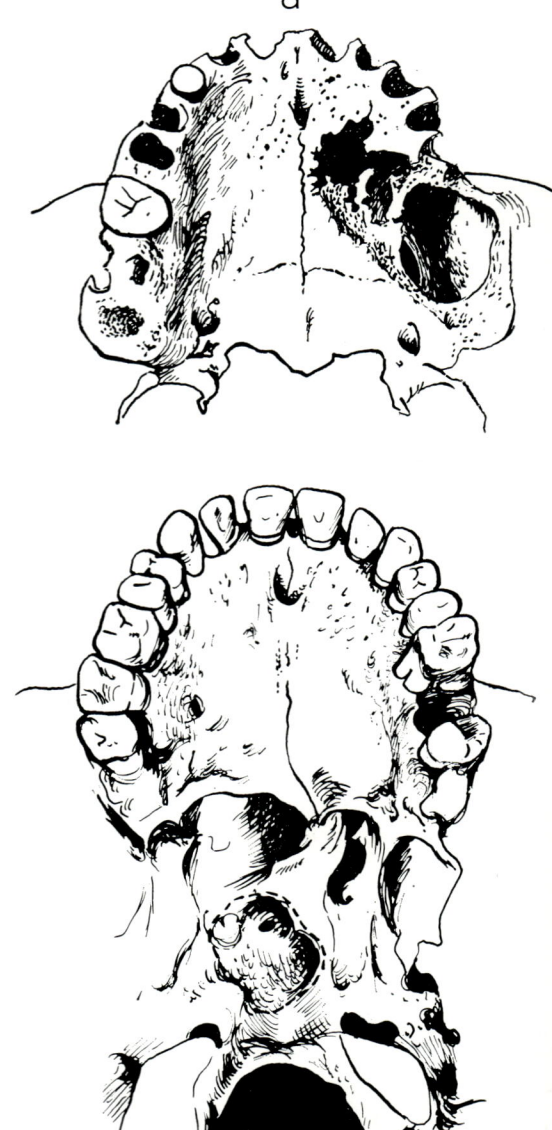

FIGURE 4. (a) The palate of an early Peruvian showing considerable destruction in the left maxillo-alveolar region.

(b) Perforated sphenoidal sinus in a Yorkshire Saxon skull.

any means. Angioma, relatively common in the skull as shown by modern surveys, must be considered a likely alternative diagnosis.

Pituitary Adenoma

Grana, Rocca, and Grana (1954) have suggested that bone changes in the region of sella turcica in a Pre-Hispanic Peruvian skull may be the result of adenomatous expansion.

Neurofibroma

A debatable case of neuro-fibromata has been reported in a Saxon from Dunstable, Bedfordshire (Brothwell, 1961). The second and third lumbar vertebrae display fairly smooth and mainly rounded cavities within the cancellar tissue of the vertebral bodies, the remaining bone being as strong as the other normal vertebrae. The vertical external surfaces of these two bodies show some slight roughening and pitting, but no other changes were noted. The nature of the cavities and strength of surrounding bone cannot be considered typical of spinal tuberculosis—although the disease was certainly in evidence in Saxon England. A reasonable alternative is that the vertebral erosions are the result of neuro-fibromata extending into the bodies, but the diagnosis must remain extremely tentative.

Sebaceous Cyst (Wen)

In a report by Duckworth (1906) on Saxon skeletons from Mitcham, Surrey, he describes a male skeleton which displays an extensive depression on the left parietal (the specimen is presumably still at Cambridge, but I have not yet located it). The true nature of such minor deformities are particularly difficult to ascertain, and Duckworth could only conclude that it was either the "sequel to chronic inflammation consequent on the presence and growth of a wen or sebaceous tumour," or the result of an old healed wound.

Benign Nasopharyngeal Tumours

Restricted destruction of bone through the gradual compression and extension of benign soft tissue tumours—and of the kind which leave rounded or at least smooth margins to the cavity produced—may be particularly difficult to identify. This is well exemplified by a Saxon skull (S26 Hull Museum) from Yorkshire, excavated by Mortimer (1905), but not recognised as diseased until my own examination of the specimen a few years ago (Brothwell, 1961). The vomer was deflected to the left side, and pathological destruction of bone in the roof of the nasal cavity had caused the exposure of the sphenoid sinus with further penetration through to the endocranial surface in front of the sella turcica (Fig. 4b). There was also a large perforation in the region of the right spheno-palatine foramen, the right side of the palate was noticeably roughened, and both upper and lower teeth on the right side were almost covered with calculus. From this evidence, one can tentatively "reconstruct" the presence of a slowly growing soft tumour, situated rather to the right and at the back of the nasal cavity and extending into the sphenoidal sinus. A fibroma or extensive polypi seem particularly likely to have produced these changes, although other possibilities can not be excluded.

In the Anthropology Laboratory of the University of Copenhagen there is a skull from Slagslunde, Denmark, which is an undoubted case of a soft tumour of the facial region, in this case with massive bone destruction. Pollen analysis of associated soil matrix suggests a Neolithic date. The skull is from a male individual, probably a young adult on the evidence of dental attrition. The tumour appeared to have its original centre in the nasal cavity, but to have extended asymmetrically from there. As seen in Figure 9 (a and b) the vomer and conchae are missing and the left side of the nasal aperture has been destroyed. Medially, the growth had continued upwards with the consequent loss of most of both nasal bones, and to the left with the loss of the inner margin of the left orbit (there is some deflection of bone here). It has penetrated as far as the left zygomaxillary suture and into the left maxillary sinus. To a much smaller degree, it appears to have penetrated through into the right maxillary sinus. The lower margin of the pyriform aperture is lowered and there is much rounding towards the external alveolar surface—suggesting the extension of the tumour labially. Downwards and backwards growth has also resulted in the loss of much of the palate, especially in the posterior region (Fig 9a). Although the bone loss is considerable, it is important to note that the bone margins which indicate the degree of tumour spread are smooth and rounded, suggesting a clearly defined and fairly rounded neoplastic growth, indicative of a benign process.

In New World material, Moodie (1926) has described the extensive destruction of bone by an "unknown type" of tumour. The main tumour mass appears to have been situated towards the back of the mouth and mainly within the upper jaw (Fig 4a), but it extends anteriorly and perforates the facial surface of the alveolus. It seems more likely that these changes were the result of a benign process, although malignancy can not be completely ruled out.

MALIGNANT PRIMARY TUMOURS OF BONE

Osteogenic Sarcoma

Perhaps the earliest example of a malignant tumour in man is to be seen in the Kanam mandibular fragment from East Africa (Lawrence, 1935), probably of Lower/Middle Pleistocene date. Although some details of the tumour (Fig. 3a) are obscured by heavy fossilization, it is clearly seen to be fairly extensive, extending onto the lingual and labial surfaces of the jaw in the region of the symphysis, and is somewhat irregular in form. The extent and thickness of this extra bone eliminates the possibility of fracturing with subsequent heavy callus development and probably also of low grade inflammation with much sub-periosteal new bone. Of the tumours, an osteogenic sarcoma "fits" the macroscopic picture well, although there is some difference of opinion on this diagnosis.

From Egypt, a Vth Dynasty femur has, since its initial description (Elliot Smith and Dawson, 1924), been referred to many times as an osteogenic sarcoma. On the medial aspect of the distal half of the femoral shaft is a large irregular bone mass, but there is no evidence of periosteal reaction round the tumour base and the growth does not show spiculation (Fig. 1). In view of these latter points the original diagnosis is by no means satisfactory and, as mentioned previously, Rowling (1961) has suggested that the lesion is more likely to be a benign osteochondroma. Elliot Smith and Dawson (1924) also refer to two cases of "sarcoma" in the proximal region of two humeri from Vth Dynasty Gizeh graves, but no illustrations or extensive descriptions are given. In view of the probably mistaken diagnosis of their femoral case, it would therefore seem best to accept these two further instances of

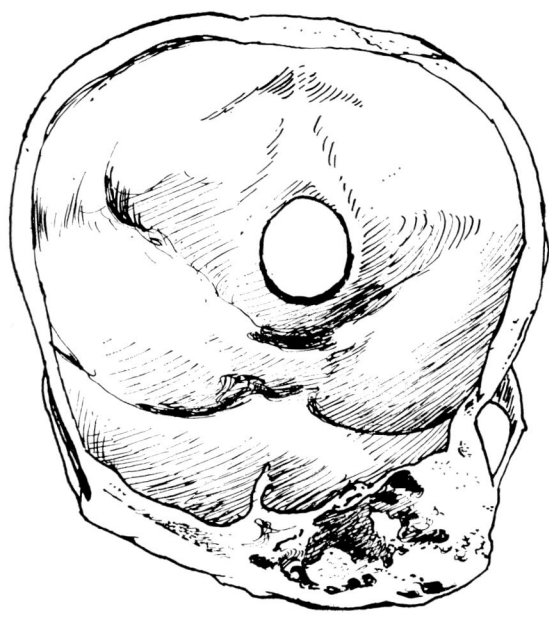

FIGURE 5. Male Peruvian skull showing massive growth of bone over the left frontal, possibly the result of a tumour. After Grana, Rocca, and Grana, 1954.

sarcoma with reserve, until a more detailed description is forthcoming.

From the catacombs of Kom el-Shougafa in Alexandria, a pelvic tumour of a Roman date was described by Ruffer and Willmore (1914). The tumour is represented by compact bony tissue with numerous cavities interspersed and which do not have communication with the exterior. Although it was considered to be an osteosarcoma, in the form and structure of the osseous mass it cannot be considered in any sense typical, and one might equally interpret this abnormality as being a long standing chondromatous process.

One of the most typical cases of osteogenic sarcoma has been described in an Iron Age skeleton from Münsingen, Switzerland (Hug, 1956). The tumour involves the left humerus head and extends down about a third of the shaft, the thickness of the osseous mass increasing proximally (Fig. 7b). Although post-mortem erosion appears to have caused some irregularity at the tumour, the macroscopic details are still very clear. In particular the "sunburst" structure so typical of the osteogenic sarcoma is clearly seen, with the axis of the spicules at about right angles to the axis of the shaft. Even in such apparently clear cut instances, however, caution is necessary; Jaffe (1958) has pointed out that other tumours are known to simulate the macroscopic appearance of an osteosarcoma, and gives an instance of "sunburst" structure which was in fact the result of an osteoplastic metastasis from a carcinoma of the prostate.

A remarkable British case of an osteosarcoma is seen in Saxon grave 2 from Standlake, Oxon (Fig. 6d). D. F. Roberts (unpublished report) writes: "A young male, aged twenty to thirty, of light build. An advanced case of osteogenic sarcoma, the primary tumour being situated at the distal end of the left femur, where its dimensions were approximately 10 inches (vertically) X 11 inches (horizontally); a fracture had clearly occurred at an early stage, the tibia being displaced laterally 4 to 5 inches. Numerous secondary lesions." This is certainly the largest case so far recorded for an earlier population, and it is unfortunate that removal of this spicular mass from the grave was not possible in one piece. However, the osteosarcoma (now in the British Museum of Natural History collection) is still in a number of large pieces, and clearly shows its structure (Fig. 11b).

From the New World, MacCurdy (1923) has attributed the extensive tumour on the skull of a pre-Columbian Peruvian to the osteogenic sarcoma variety. Situated mainly on the left side of the vault, it extends across the sutures from the parietal region onto the frontal

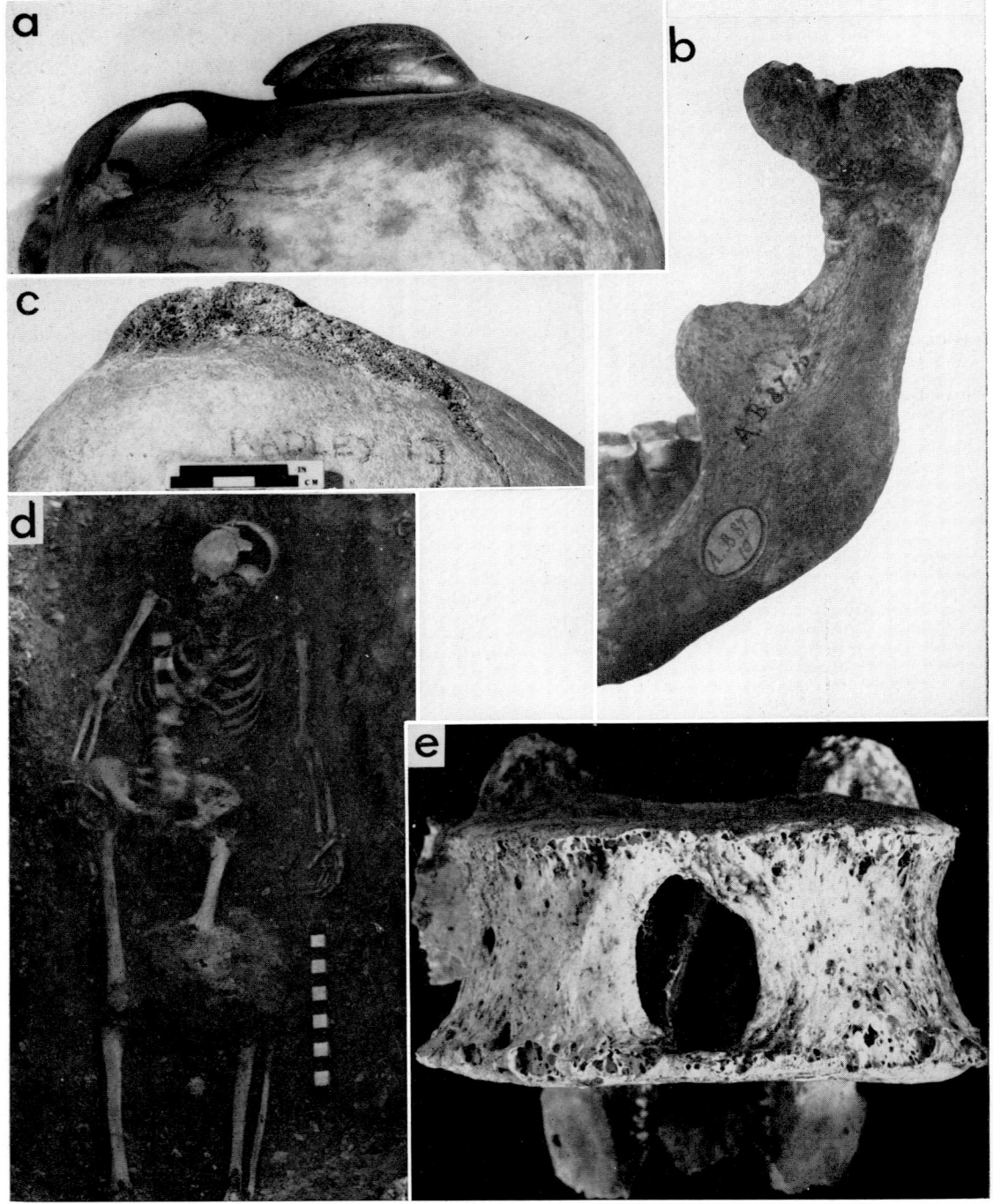

FIGURE 6. (a) Right side of the Roman Egyptian skull.
(b) Detail of left ramus of Roman from Varpeler (Denmark).
(c) Vertex of the Radley skull vault.
(d) The Standlake Saxon with massive tumour at left knee.
(e) Nubian vertebra with perforation through into the body.

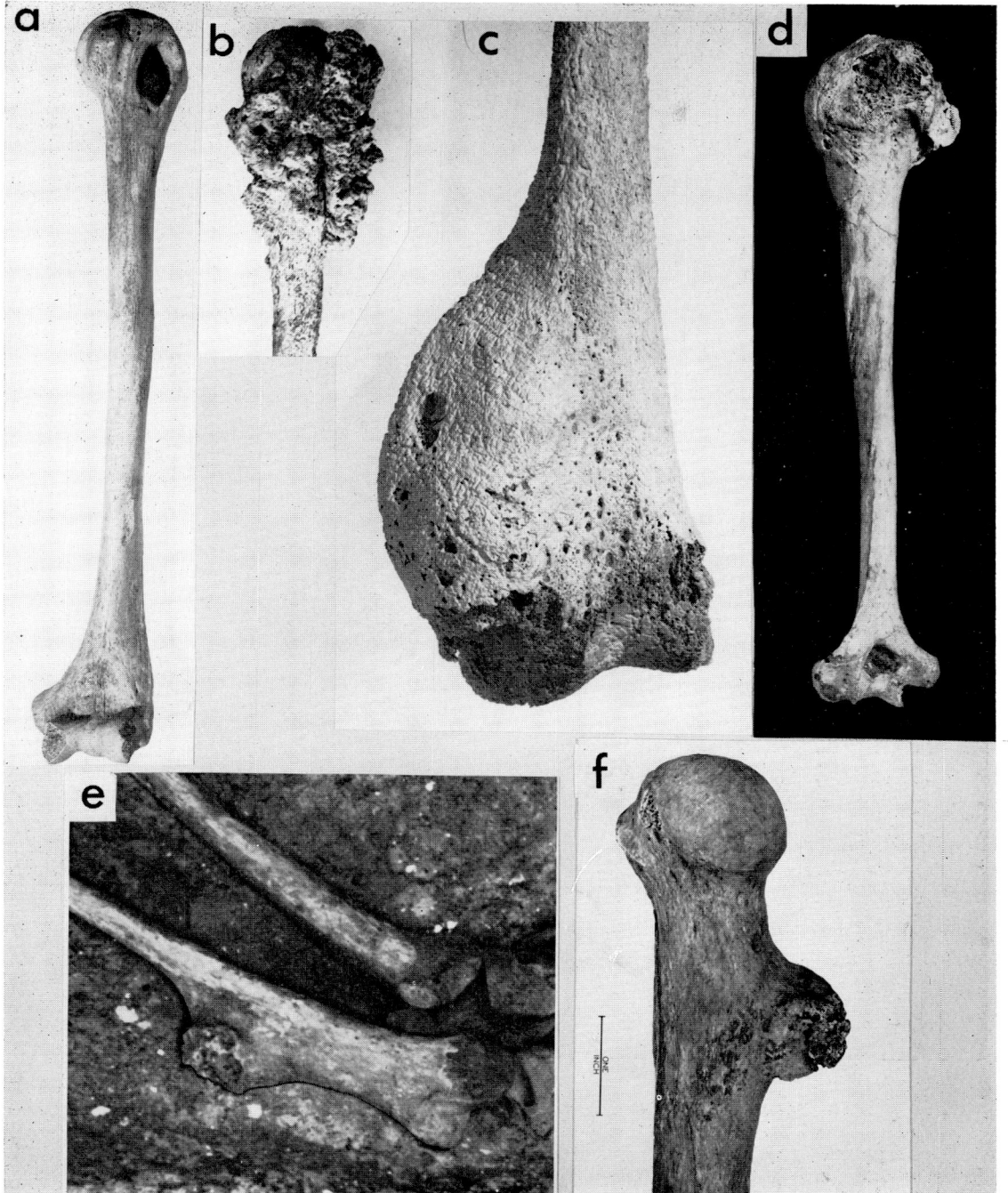

FIGURE 7. (a) The humerus of Neolithic date from West Kennet.
(b) Proximal end of an Iron Age humerus from Münsingen.
(c) Distal end of the Finglesham Saxon femur.
(d) A Nubian humerus showing massive swelling at the head.
(e) Distal parts of both femora of the Roman Juellinge body.
(f) Proximal end of an early Texas Indian humerus.

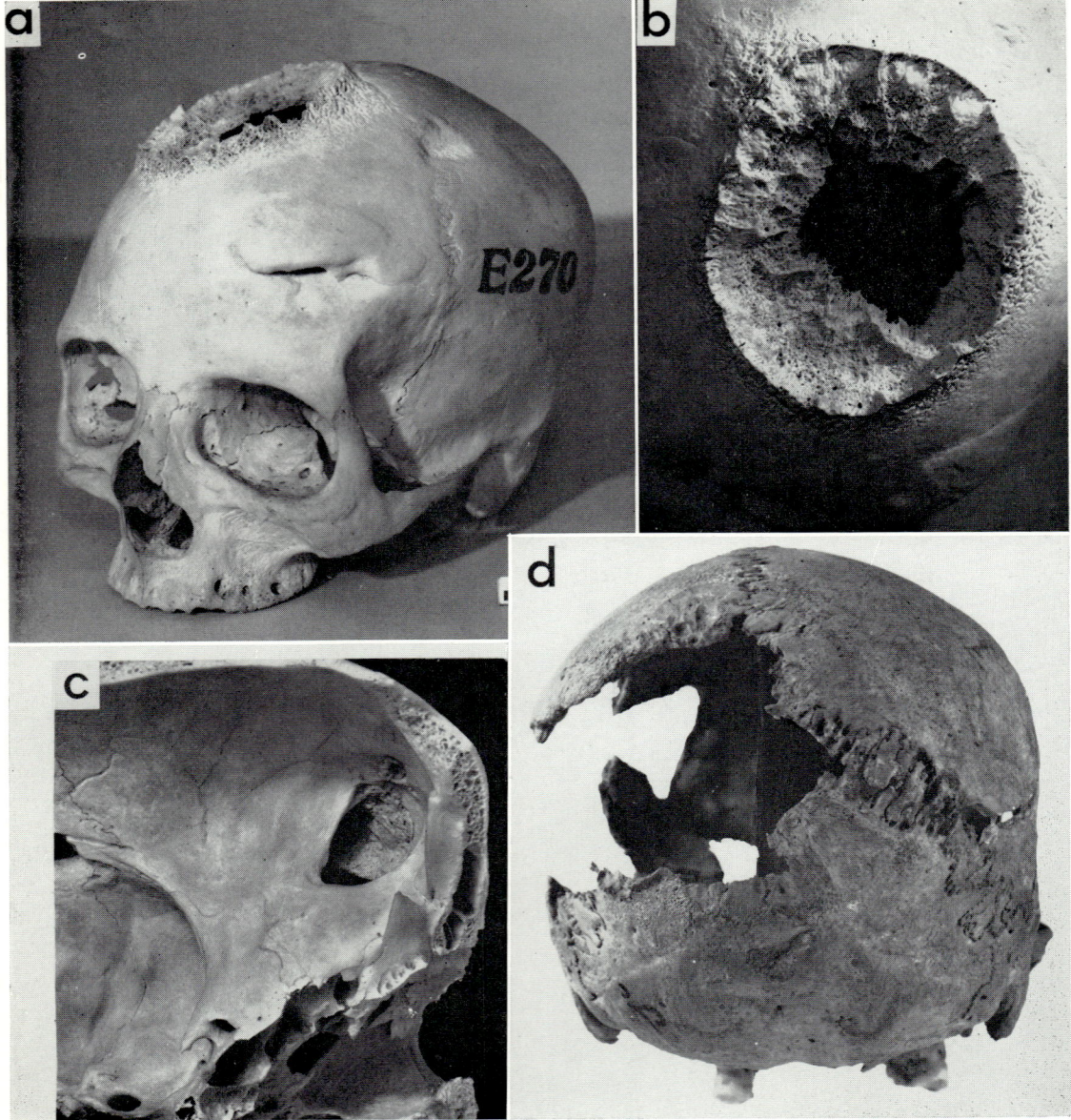

Figure 8. (a) And (b) general view and close-up of the neoplastic lesion in the late Dynastic Egyptian skull E 270.

(c) Endocranial view of the Nubian orbital lesion.

(d) Posterior view of the Nørregard Neolithic skull.

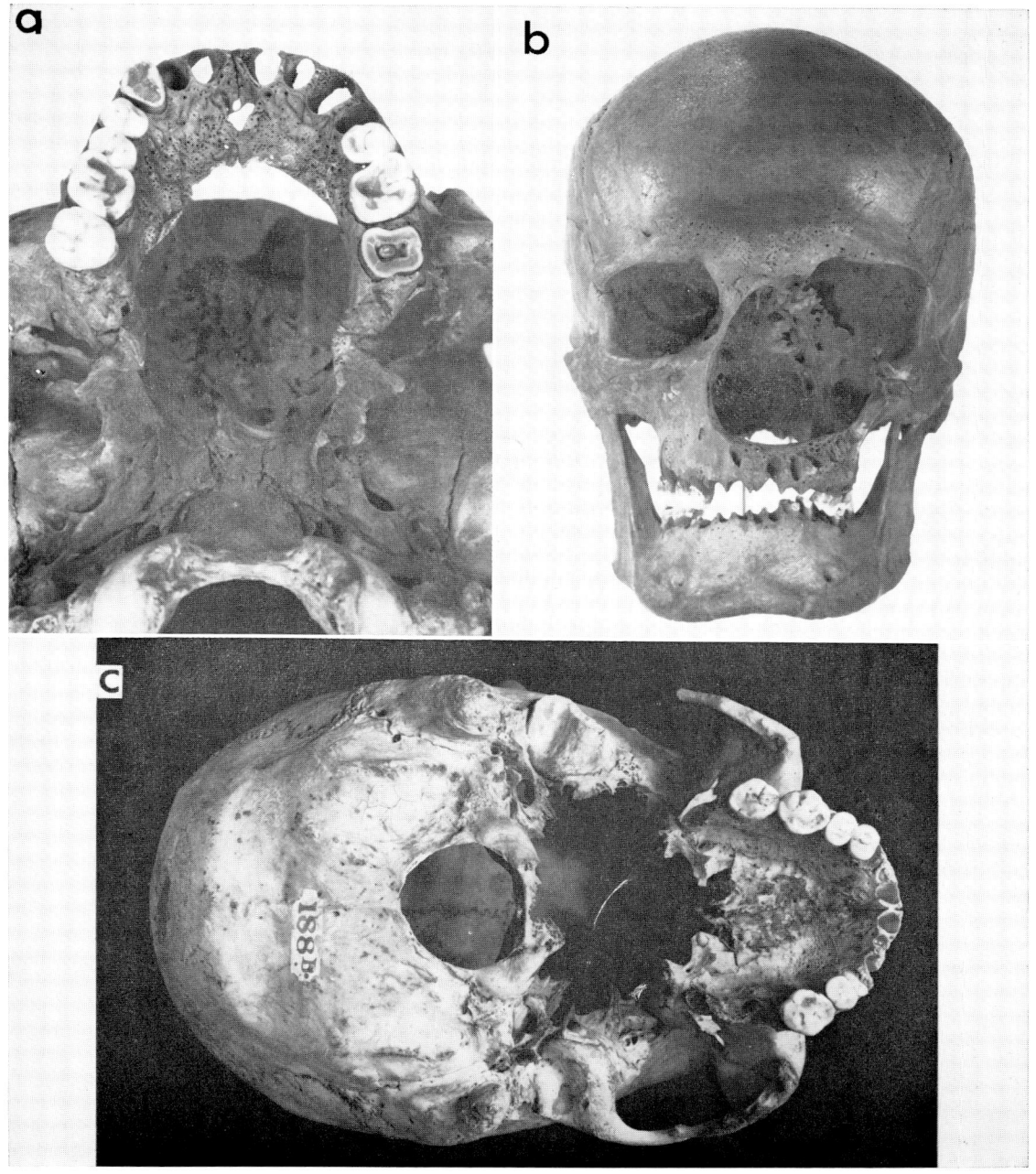

FIGURE 9. (a) And (b) close-up of palatal region and general view of the Danish skull from Slagslunde.
(c) Basal view of a pre-Christian Nubian skull, showing massive ante-mortem destruction of bone.

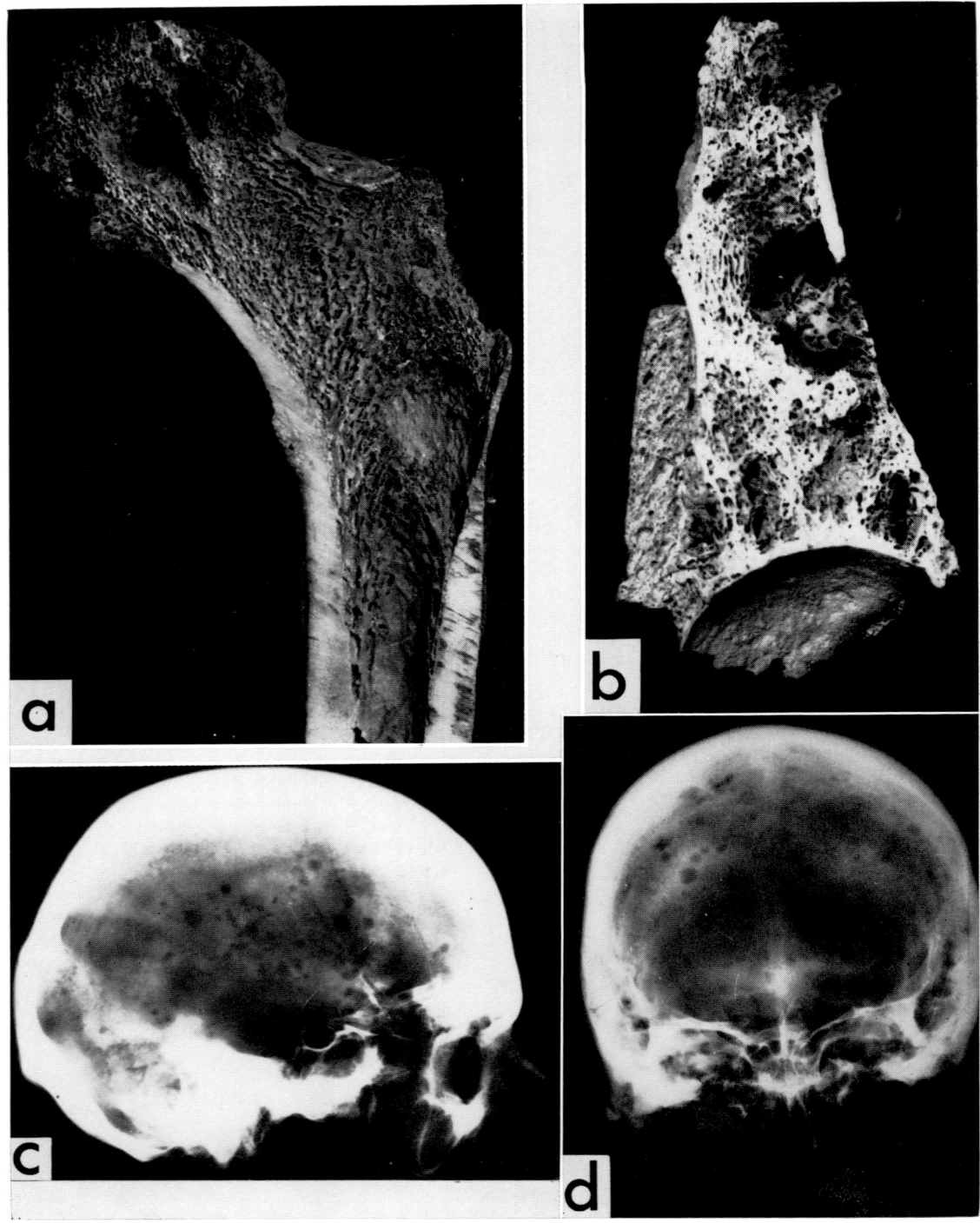

FIGURE 10. Parts of the skeleton from grave 38, Kérpuszta, Hungary: (a) femur, (b) pelvis, (c) and (d) radiographs of lateral and facial aspects of the skull.

Fig. 3b). The tumour mass is fairly circular and macroscopically displays a large honeycomb appearance. Radiographically, there is marked spiculation. This could be an osteogenic sarcoma, but as in the Radley Roman skull, seems worth reconsideration as a possible meningioma or angioma.

Myelomatosis

If we accept the early cases noted in the literature, and consider the chance nature of these discoveries in a limited skeletal series, one might well conclude that this malignant multiple tumour was by no means uncommon in earlier populations.

From the Pyrenees, Fusté (1955) has recorded the presence of "punched out" lesions in a late Neolithic skull (Fig. 3d). The holes were fairly rounded, but "etched" and with somewhat irregular margins, and with a restricted "areola" of surface reaction. Such changes are not inconsistent with myelomatosis but nor are they with secondary metastases. Another claim for Neolithic myelomatosis has been made by Ackerknecht (1953), but without further description.

From the Hungarian tenth to eleventh century site of Kérpuszta, grave 38 contained a skeleton with multiple osteolytic lesions (Nemeskéri and Harsányi, 1959). In view of the pattern of tumour spread in the post-cranial skeleton, as well as in the skull (Fig. 10), the diagnosis was given as myelomatosis, although again the alternative of metastases can not be ruled out.

Wells (1964) has described two British mediaeval skulls, from Kent and Suffolk, which display multiple cranial lesions. Although hesitating as to a final diagnosis, he considers the evidence "may indicate that myelomatosis is rather more probable than carcinoma."

From the New World, two pre-Columbian cases have been described; an adult male (Ritchie and Warren, 1932) with extensive deposits throughout the skeleton, and a child of about ten years (Williams, Ritchie, and Titterington, 1941). Regarding this second specimen, myeloma is of course extremely unlikely in a child of that age. It would seem more likely to have been an example of secondary neuroblastoma or some such malignancy of childhood.

Thus, at least six archaeological instances of multiple malignant lesions have been considered as possible evidence of myelomatosis. The number is not large, but surprising in that it *exceeds* that for cases of metastatic lesions. At present, there is a deficiency of literature as to possible differences in the osseous pathology of these two malignancies, and I have been unable to locate satisfactory data on the variable expression (in terms of lesion size, shape etc.) to be found in these conditions. In terms of statistical probability, a consideration of modern frequencies is revealing. Myelomatosis represents under 20 per cent of occurring bone malignancy (Coley, 1960), and of course the total number of primary bone malignancies is overshadowed by the numerically more important metastatic group of bone lesions. Even allowing for differences in the form of the age pyramid in human populations in space and time, one would still expect metastic lesions to be the "common" finding, with myelomatosis a rare occurrence. It could be that myelomatosis was far more prevalent in some earlier communities, but this seems most unlikely, and I for one would prefer to see all these multiple lesion cases considered as metastases until it can be proved otherwise.

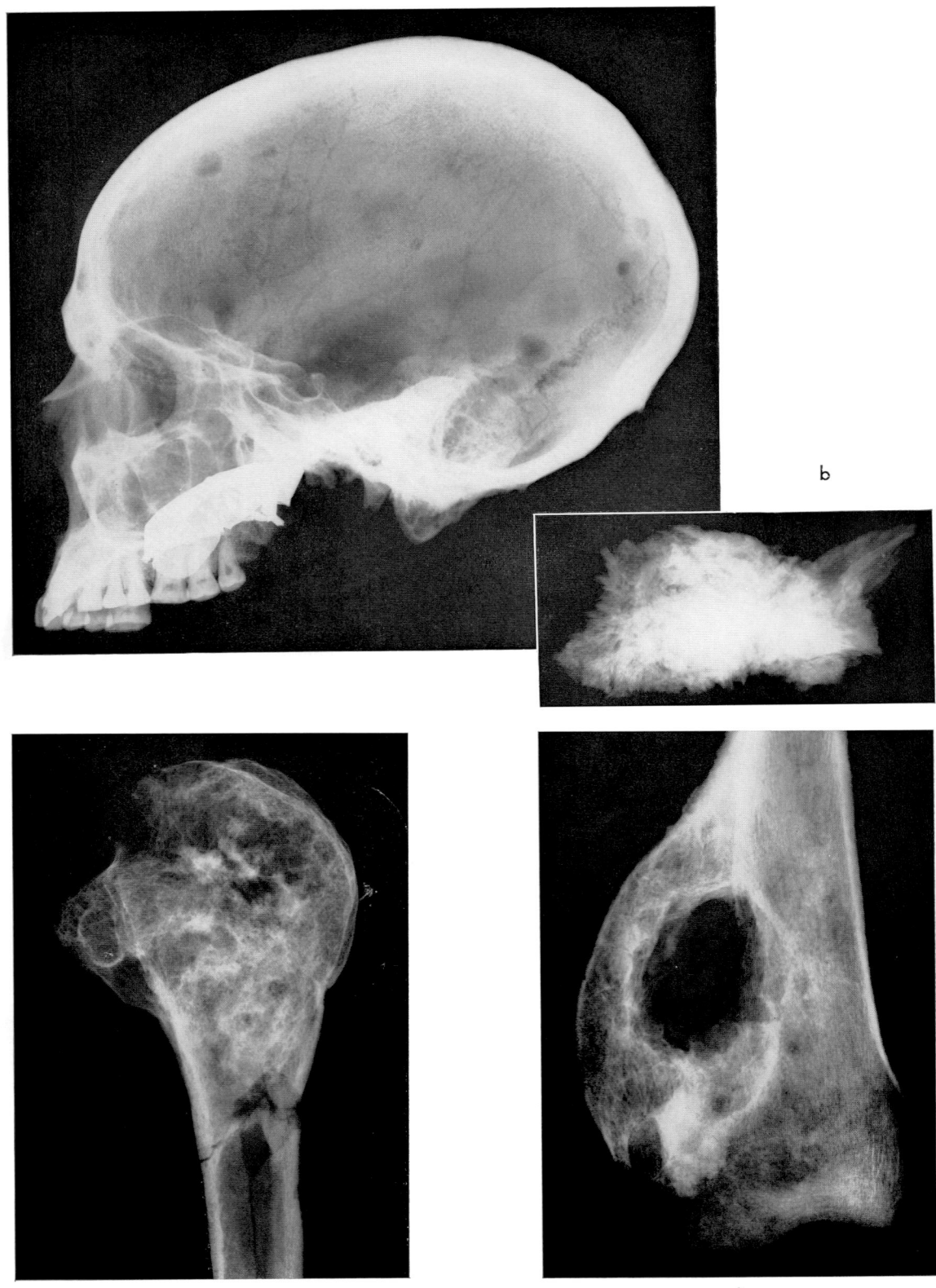

PRIMARY CARCINOMA AND SECONDARY INVOLVEMENT OF BONES BY MALIGNANT DISEASE

The identification of malignant tumours of this type has already been discussed to some extent in relation to the problem of pathology and post-mortem changes. Also it has been suggested that previous workers may have sometimes failed to realise the significance of such malignancy changes in earlier skeletons. Nevertheless, some cases are available, and these at least show that malignancy occurred in antiquity and is present in various populations.

Evidence of Primary Deposits

Derry (1909) described in detail a destructive process occurring at the cranial base of a Pre-Christian Nubian (c.4th-6th cent. A.D.). The entire bony structure from the cribriform plate of the ethmoid back nearly to the foramen magnum on the basi-occipital, and with nearly all the sphenoid, is missing (Fig. 9c). The nasal turbinals and vomer had also disappeared and the posterior margins of the palate were also probably involved. The margins of the bone surrounding this area are very irregular and "rarified" in texture, strong evidence that the neoplasm was osteolytic and malignant. Elliot Smith in a comment on Derry's findings suggests that this is indeed a malignant process "originating in the nasal mucous membrane or sphenoidal sinus," and it could well have been the seat of a primary carcinoma.

A III-Vth Dynasty skull from Egypt provides further evidence of a primary carcinoma (Wells, 1963). By surface inspection, transillumination, and radiography, it was demonstrated that there was not only a primary lesion, but also about twenty-six secondary deposits scattered over the skull (Fig. 11a). The primary destruction involved the left maxillo-alveolar region, part of the palate, medial and lateral pterygoid laminae, and part of the inferior concha. Surrounding much of the margin of this primary lesion is a zone of osteitic reaction.

A further possible early case is from the Iranian site of Tepe Hissar (3,500-3,000 B.C.). Destruction of the left maxillary-alveolus, palate and antral wall (Krogman, 1940) could well be the result of carcinoma.

No further mention need be made of the Swiss skull D.10., except to re-affirm that the bone destruction may have been initially due to a neoplasm and this could well have been a carcinoma.

Finally, Elliot Smith and Derry (1910) note a Middle Nubian male skeleton "in which the sacrum had suffered extreme erosion, perhaps due to rectal cancer" p. 29; perhaps the result of a chordoma.

Metastases

The problem of distinguishing neoplastic metastases from myelomatosis has already been discussed, and it need only be said that very probably some of the

---

FIGURE 11. (a) Radiograph of a III-Vth Dynasty Egyptian skull with the region of the maxillo-alveolar tumour delimited by lead foil; 'shadows' of some of the secondary vault tumours being also visible.

(b) Part of the Standlake Saxon tumour mass, showing "sunburst" appearance at the margins.

(c) Radiograph of humerus in Figure 7 (d).

(d) Radiograph of femur in Figure 7 (c).

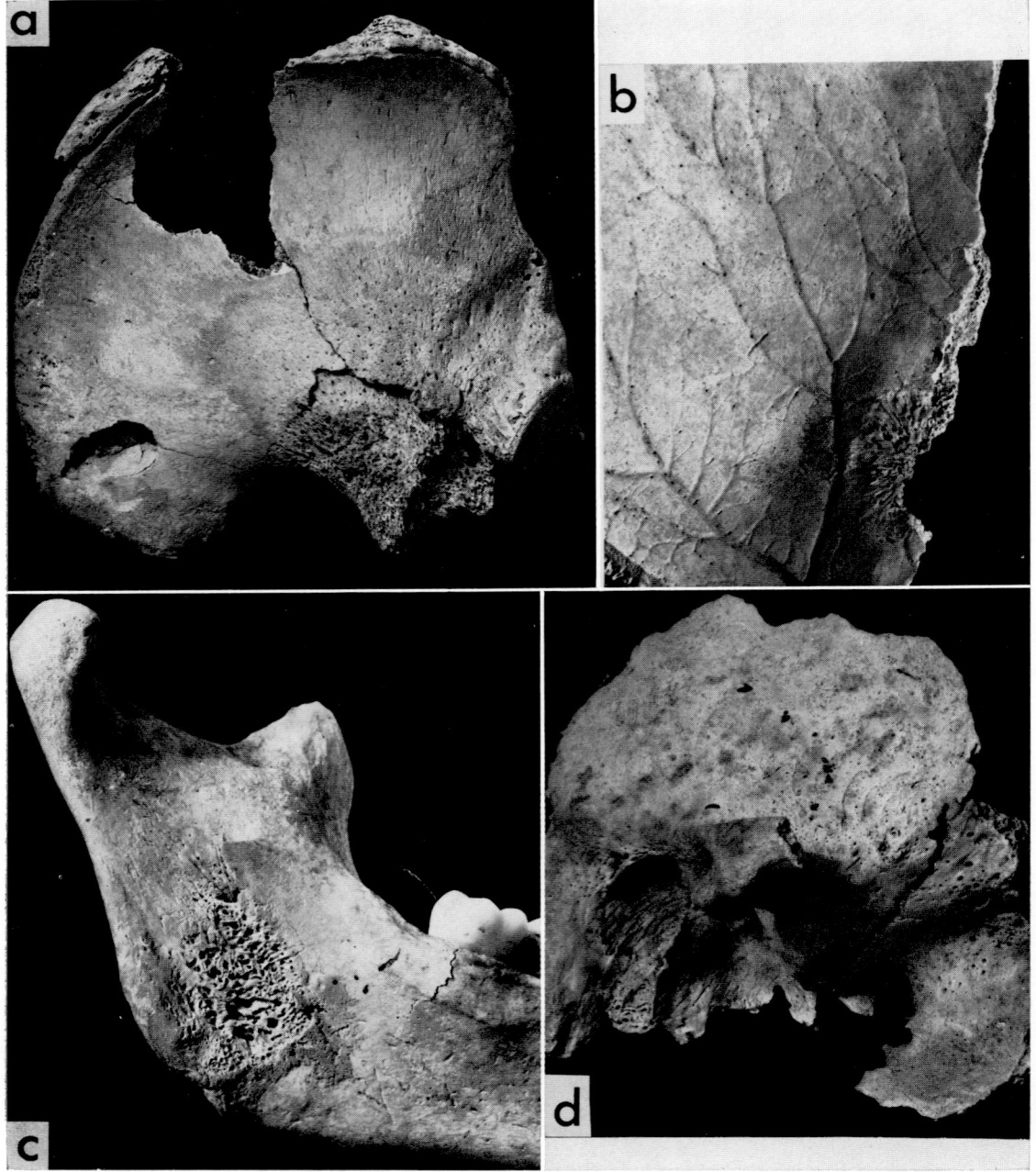

FIGURE 12. Parts of skeleton from grave 30, Winchester.
(a) Pelvis, with large surface reaction above and in front of the greater sciatic notch, and a small "punched-out" area half way along the anterior border.
(b) Endocranial surface of a parietal fragment, with "etched" appearance at lower right corner.
(c) Inner aspect of the left ramus.
(d) Part of the right temporal and sphenoid, with restricted bone reaction on the sphenoid wing.

cases considered as the latter may in fact be the former. It is to the credit of Møller and Møller-Christensen (1952) that they considered their medieval case from Aebelholt Abbey, Denmark, as most probably showing metastases and not multiple myeloma (although this alternative could not be ruled out). The specimen consisted only of the skull of a female. Five lesions were present in the parietals and frontal and are "surrounded by crenated or spongy osseous matter." They conclude that the primary growth elsewhere may have been a carcinoma, or a sarcoma. Moreover, as a result of considering recent data on the tendency of different varieties of carcinoma to form bony metastases they state that "it seems possible—with decreasing likelihood—that the young woman from Aebelholt Abbey was suffering from primary cancer of the breast, the thyroid, the uterus, the lungs, the ovaries, the skin" (p. 341). Speculation of this type, following a consideration of modern frequencies, is perhaps justified to some extent—especially when it is a matter of choice between a rare or relatively common tumour—but it must be remembered that world populations vary as regards tumour incidence, and must equally have done so in the past.

From Egypt, the metastases associated with a primary carcinoma in a III-Vth Dynasty skull have already been described. Also the case reported by Derry (1909) as a "form of rarefying osteitis" which I suggest needs re-examining (if it can be found) as a possible case of metastases.

From Winchester a recently excavated Saxon skeleton has produced further evidence of changes which seem best explained as metastases. The skeleton (No. 30, now in our collection) is fragmentary, but sufficiently complete to demonstrate that lesions occurred in various parts of the skeleton. In the skull, the lesions vary from irregular osteolytic destruction of bone with or without marginal reactions (Fig. 12b,c,d) to rounded punched-out lesions. In the pelvis, there is an extensive involvement of the innominate fragment (partly destructive but also with new bone formation); also on the same piece is a rounded punched-out area (Fig. 12a). In the vertebral column, there is clear evidence that osteolytic destruction of some vertebral bodies had commenced but was not advanced.

Finally, from the New World, Hooton (1930) has recorded an interesting case (No. 59802) from an early level at Pecos Pueblo. However, although he notes: "Metastatic tumor involving right lower dorsal vertebrae, right and left radius, right ulna" (p 320), he does not describe the lesions in detail, and no illustrations are given. In the extensive study of Peruvian trephination by Grana et al. (1954) they include the Pre-Hispanic skull 144 which shows a number of fairly circular perforations which are not of surgical origin. This specimen must also be considered as perhaps showing metastatic lesions.

PROBLEMATICA

Because of the sparsity of description and doubts as to diagnosis, or because the pathology does not permit a more definite decision as to the neoplastic nature of the lesion, a few speciments must be considered in this non-specific category.

In 1893 a femur with pathological changes was obtained from an Etruscan tomb in Tarquinii (Haddow, 1936). The tumour was considered to be a "periosteal sarcoma" of the distal femoral area. However, a search through more recent literature on Etruscan skeletal material has failed to reveal further mention of this specimen, and it therefore seems unwise to accept this earlier diagnosis until a modern confirmation of the disease is obtained.

Elsewhere (Brothwell, 1961) I have referred to a large and deep perforation in the region of the surgical neck of a Neolithic humerus from West Kennet, Wiltshire (Fig. 7a). Wells (1962) originally suggested the lesion to be indicative of an abscess as a result of a penetrating wound. It has puzzled me, however, that the cavity is so rounded and smooth, with so little evidence of surrounding inflammatory reaction considering the extent of the "trauma" (the restricted area of smooth new bone below the cavity may be caused by the stripping away of the periosteum for other reasons). Further consideration and discussion suggests that this lesion should be considered—if only to be rejected in time—as possibly the result of a neoplasm. Whether a simple cystic process, a form of osteoclastoma, or a chondroma is a decision best left for the future.

In a Pre-Hispanic skull from Peru, Grana et al. (1954) have noted a large frontal tumour centered mainly over the left orbit. The separation of the skull "cap" by a horizontal cut above the supraorbital ridges, displays an extra bone mass which appears to be predominantly spongy in nature (judging from the photograph) and certainly this swelling is not the result of an ivory osteoma within the frontal sinuses (Fig. 5).

From a Middle Kingdom Nubian site Elliot Smith and Dawson (1924) illustrated an upper lumbar vertebra with a well defined cavity within the cancellous tissue of its body; situated in mid-line and opening onto the anterior surface (Fig. 6e). They considered the condition to be probably indicative of tuberculosis, but Rowling (1960) is critical of this and suggests that the lesion might be neoplastic. The strength of the surrounding bone—with no evidence of inflammatory changes—supports this latter view. The radiographic appearance of this vertebral anomaly is suggestive of a haemangioma or even erosion by an aortic aneurysm, and clearly this specimen should be considered as possibly further evidence of an early tumour.

Elliot Smith and Wood Jones (1910) have called attention to a puzzling lesion in the left upper orbital region of one of their early Nubian cases. The frontal table of the orbit displays a well defined perforation through from the orbit into the interior of the skull, the endocranial opening being larger than the orbital one. Although the margins on the endocranial side are generally rounded and smooth, there are some patches of irregular bone (Fig. 8c). This lesion clearly demands consideration as possibly the result of a benign orbital tumour (? dermoid cyst). On the other hand, as Ingalls (1953) points out, chronic inflammatory masses (i.e., granulomas) can produce such perforations, and the bone changes can resemble those of true tumours.

A final case worth mentioning is a Danish Neolithic skull (Nørregard I.PMD 271) in the University of Copenhagen. At first sight the bone destruction might be mistaken for post-mortem erosion or a suppurative condition following trephination, but closer inspection revealed that the posterior part of the vault more likely suffered osteolytic and neoplastic destruction. Also, at the margins of this diseased area was a restricted meningioma-like reaction (Fig. 8d), breakage unfortunately obscuring the full extent of this. Clearly, this must be considered further evidence of early tumour development, the nature of the bone changes suggesting perhaps a malignant condition.

EVIDENCE FROM SOFT TISSUES

In view of the rarity of naturally dried bodies and well preserved (and unwrapped) mummies, it is to be expected that direct soft tissue evidence of tumours is rare indeed.

Granville (1825) noted in the pelvis of an Egyptian mummy he dissected, various changes which strongly suggest the presence of an ovarian tumour possibly a cystadenoma. He writes: "The ovarium and broad ligament of the right side were enveloped in a mass of diseased structure, while the Fallopian tube of the same side was sound; but the uterus itself was larger than normal, while the remains of a sac were found connected with the left ovarium; all of which, connected with the appearance of the abdominal integuments, leave no doubt of ovarian dropsy."

From Nubia, Elliot Smith and Wood Jones (1908) have reported a case of true cervical polyp; although of course, cervical polyps are usually not truly neoplastic. Elsewhere in this volume, Sandison illustrates a simple squamous papilloma of skin (see his Fig. 5).

LITERARY AND ART EVIDENCE

These media are far from satisfactory as indicators of tumour varieties in earlier peoples. As regards Egyptian literature, the tumour evidence has been recently reviewed by Ghalioungui (1963). Both in the Ebers and Hearst papyri tumours and swellings are considered to some extent. From these descriptions, it is possible to identify tentatively a ganglionic mass, a polypoid tumour, and a sebaceous cyst. There is no good indication of malignancy, except perhaps for the tumours of Chons, which were considered the most terrible. Haddow (1936) comments on more recent historic interest in tumours and points out that Galen, Paul of Aegina, and Celsus, for example, were aware of the difference between benign and malignant growths. Long (1928) recalls the story that Democedes of Crotona (520 B.C.) healed Darius' wife Atossa of a cancer of the breast. Although in fact the disease really appears to have been inflammatory mastitis, the case is of interest as an early claimed record of tumour pathology (Sandison, 1959).

Zammit and Singer (1924) describe two Neolithic human representations from Malta which have tumour-like swellings, the art style being such that the interpretation must remain extremely tentative. Both are female figures; in one case from Mnaidro (No. 29) an "immense tumour . . . projects from the abdomen," while the other case (No. 49) displays a large swelling below the region of the left groin. A Neolithic human representation from Sesklo, Greece, and another from Vinča, in Jugoslavia, also show tumour-like swellings. In these two instances, the anomaly is in the region of the throat (Peter Ucko, personal communication).

Aldred and Sandison (1962) have argued that the monuments suggest that the Pharaoh Akhenaten suffered from an endocrinopathy which had been caused by a pituitary adenoma.

CONCLUSIONS

Although I have attempted to classify, as well as review, the evidence for ancient tumours, it must be emphasized that any such break-down is for the most part very tentative. It is my opinion, however, that such classifications are useful and give a better indication of the range of tumours which might be identified and may have occurred in earlier populations. Skeletal evidence seems more likely than any other to give fairly precise details of this class of diseases. Finally, in studying such ancient pathology full consideration should be given to the differences between post-mortem and disease processes, especially as it is likely that previous workers have overlooked evidence of some neoplasms for this reason.

ACKNOWLEDGMENTS

It is a pleasure to record my thanks to various colleagues with whom I have discussed a number of the specimens described in this paper, and who offered helpful criticism. These are: Drs. H. A. Sissons, A. M. Barrett, A. T. Sandison and Donald Teare; Professors E. Uehlinger, A. E. W. Miles and H. Hamperl; and Mr. J. Thompson Rowling. Dr. Peter Gray kindly supplied me with unpublished radiographic details of an Egyptian case, and Drs. Marcus Goldstein and J. Nemeskéri permitted me to reproduce—and provided prints of—early tumour cases which they had reported. Special thanks for museum facilities are given to Dr. Balslev Jørgensen (University of Copenhagen) and the Directors of the National Museum of Copenhagen, Natural History Museum (Bern), Baselland Museum (Liestal) and Duckworth Laboratory research collection (Cambridge). Dr. D. F. Roberts kindly permitted me to quote from his unpublished report on the Standlake Saxon case, and Dr. H. W. Catling of the Ashmolean Museum, Oxford, provided photographs of this specimen for my use. Most of the drawings are the work of Miss Rosemary Powers (B.M.N.H.)

REFERENCES

ACKERKNECHT, E. H., 1953: Palaeopathology: a Survey. In: *Anthropology Today*, pp. 120-127, Chicago University Press.

ALDRED, C. and SANDISON, A. T., 1962, The Pharaoh Akhenaten: a problem in Egyptology and pathology *Bull. Hist. Med.*, 36:293.

BATRAWI, A. M., 1935: Report on the human remains. *Mission Archéologique de Nubie 1929-1934*, Cairo, Government Press.

BROTHWELL, D. R., 1959: The use of non-metrical characters of the skull in differentiating populations. In *Ber. 6 Tag. dtsch. Ges. Anthrop. Kiel*, pp. 103-109. Göttingen, Musterschmidt-Verlag.

BROTHWELL, D. R., 1960: The Bronze Age people of Yorkshire: a general survey. *Advancement of Science, 16*:311.

BROTHWELL, D. R., 1961: The palaeopathology of early British man: an essay on the problems of diagnosis and analysis. *J. Roy. Anthrop. Inst., 91*:318.

BROTHWELL, D. R., 1963: *Digging up Bones*. London, British Museum (Natural History).

COLEY, B. L., 1960: *Neoplasms of Bone and Related Conditions. Etiology, Pathogenesis, Diagnosis, and Treatment.* New York, Hoeber.

DERRY, D. E., 1909: Anatomical report. *Archaeological Survey of Nubia*, Bull.3:29.

DUCKWORTH, W. L. H., 1906: Notes on crania and bones. In: Excavations in an Anglo-Saxon burial ground at Mitcham, Surrey. *Archaeologia*, 60:61.

ELLIOT SMITH, G., and DAWSON, W. R., 1924: *Egyptian Mummies*. London, Allen and Unwin.

ELLIOT SMITH, G., and DERRY, D. E., 1910: Anatomical report. *Archaeological Survey of Nubia*, Bull.5:11.

ELLIOT SMITH, G., and WOOD JONES, F., 1908: Anatomical report. *Archaeological Survey of Nubia*, Bull.2:29.

ELLIOT SMITH, G., and WOOD JONES, F., 1910: Report on the human remains. *Archaeological Survey of Nubia. Report of 1907-1908*, 11:375. Cairo, Ministry of Finance.

FUSTÉ, M., 1955: Antropología de las poblaciones pirenáicas durante el período neo-eneolitico. *Trab. Inst. 'Bernardino de Sahagún' Antrop. y Etnologia, 14*:109.

GHALIOUNGUI, P., 1963: *Magic and Medical Science in Ancient Egypt*. London, Hodder and Stoughton.

GOLDSTEIN, M. S., 1957: Skeletal pathology of early Indians in Texas. *Amer. J. Phys. Anthrop., 15*:299.

GRANA, F., ROCCA, E. D., and GRANA L., 1954: *Las Trepanaciones Craneanas en el Perú en la Época pre-Hispánica*. Lima, Maria.

GRANVILLE, A. B., 1825: An essay on Egyptian mummies, with observations on the art of embalming among the Ancient Egyptians. *Philos. Trans. Roy. Soc. 0*: 269.

GRAY, P. K., 1964: Personal communication.

GREIG, D. M., 1931: *Clinical observations on the surgical pathology of bone*. Edinburgh, Oliver and Boyd.

HADDOW, A., 1936: Historical notes on cancer from the MSS. of Louis Westenra Sambon. *Proc. Roy. Soc. Med., 29*:1015.

HOOTON, E. A., 1930: *The Indians of Pecos Pueblo. A study of their skeletal remains*. New Haven.

HUG, E., 1956: *Die Anthropologische Sammlung im Naturhistorischen Museum Bern.* Bern, Natural History Museum.

HUG, E., 1959: *Die Anthropologische Sammlung im Kantonsmuseum Baselland.* Liestal, Kantonsmuseum Baselland.

INGALLS, R. G., 1953: *Tumors of the Orbit and Allied Pesudo Tumors.* Springfield, Thomas.

JAFFE, H. L., 1958: *Tumors and Tumorous Conditions of the Bones and Joints* London, Kimpton.

JELSMA, F., 1959: *Primary Tumors of the Calvaria.* Springfield. Thomas.

KROGMAN, W. M., 1940: The skeletal and dental pathology of an early Iranian site. *Bull. Hist. Med.*, 8:28.

LAWRENCE, J. E. P., 1935: Appendix A (p. 139). In: L. S. B. Leakey, *Stone Age Races of Kenya,* London, Oxford University Press.

LUCK, J. V., 1950: *Bone and Joint Disease.* Springfield, Thomas.

MACCURDY, G. G., 1923: Human skeletal remains from the highlands of Peru. *Amer. J. Phys. Anthrop.*, 6: 217.

MØLLER, P. and MØLLER-CHRISTENSEN, V., 1952: A mediaeval female skull showing evidence of metastases from a malignant growth. *Acta Pathol. Microbiol. Scand.* 30:336.

MOODIE, R. L., 1926: Studies in paleopathology. XVIII. Tumors of the head among Pre-Columbian Peruvians. *Ann. Med. Hist.*, 8:394.

MORTIMER, J. R., 1905: *Forty Years' Researches in British and Saxon Burial Mounds of East Yorkshire.* London, Brown.

NEMESKÉRI, J. and HARSÁNYI, L., 1959: Die Bedeutung paläopathologischer Untersuchungen für die historische Anthropologie. *Homo*, 10:203.

RITCHIE, W. A., and WARREN, S. L., 1932: The occurrence of multiple bony lesions suggesting myeloma in the skeleton of a pre-Columbian Indian. *Amer. J. Roentgenol.*, 28:622.

ROGERS, L., 1949: Meningiomas in Pharaoh's people. Hyperostosis in Ancient Egyptian skulls. *Brit. J. Surg.*, 36:423.

ROWLING, J. T., 1960: Unpublished M.D. Thesis. University of Cambridge.

ROWLING, J. T., 1961: Pathological changes in mummies. *Proc. Roy. Soc. Med.*, 54:409.

RUFFER, M. A., 1921: *Studies in Paleopathology of Egypt.* Roy L. Moodie (Ed). University of Chicago Press.

RUFFER, M. A., and WILLMORE, J. G., 1914: A tumour of the pelvis dating from Roman times (250 A.D.) and found in Egypt. *J. Path. Bact.*, 18:480.

SANDISON, A. T., 1959: The first recorded case of inflammatory mastitis: Queen Atossa of Persia and the physician Democedes. *Med. Hist.*, 3:317.

SHANKS, S. C., and KERLEY, P., (Eds.), 1950: *A Text-Book of X-ray Diagnosis.* London, Lewis.

WATERSON, D., 1926: A stone cist and its contents found at Piekie farm, near Boarhills, Fife. *Proc. Soc. Antiq. Scot.*, 61:30.

WELLS, C., 1963: Ancient Egyptian pathology, *J. Laryng. Otol.*, 77:261.

WELLS, C., 1964a: The radiography of ancient bones. *X-ray Focus*, 5:2.

WELLS, C., 1964b: Two Mediaeval cases of malignant disease. *Brit. Med. J.*, 1:1611.

WELLS, L. H., 1962: Report on the inhumation burials from the West Kennet Barrow. Appendix I. in: S. Piggott. *The West Kennet Long Barrow—Excavations 1955-56.* London, H.M.S.O.

WILLIAMS, G. D., RITCHIE, W. A., and TITTERINGTON, P. F., 1941: Multiple bony lesions suggesting myeloma in precolumbian Indian aged ten years. *Amer. J. Roentgenol.*, 46:351.

ZAMMIT, T., and SINGER, C., 1924: Neolithic representations of the human form from the Islands of Malta and Gozo. *J. Roy. Anthrop. Inst.*, 54:67.

Chapter 25

An Eruption Resembling That of Variola in the Skin of a Mummy of the Twentieth Dynasty (1200-1100 B.C.)*

M. ARMAND RUFFER, AND A. R. FERGUSON

THE body from which the skin was taken was that of a tall man of middle age. It was brought to the notice of one of us by Professor G. Elliot Smith during his investigations into the process of mummification as illustrated in the royal mummies in the Cairo Museum of Antiquities. The body was the seat of a peculiar vesicular or bullous eruption which in form and general distribution bore a striking resemblance to that of small-pox. The portion of skin which we were permitted to remove, and which forms the subject of the present note, was taken from the abductor surface of the right thigh. The eruption on the inner surface of the thigh was, as the drawing shows (see Fig. 1), a closely set vesicular one, and it was in this situation that the general resemblance to small-pox was most noticeable.

Small portions of skin were treated by the following method:[1]—(1) The tissue was softened in a solution of sodium carbonate mixed with alcohol (alcohol, 100 parts; water 15 parts; 5 per cent. solution sodium carbonate, 60 parts); (2) this solution was replaced by 30 per cent. alcohol, and the tissue gradually brought thereafter into absolute alcohol, and embedded in paraffin.

A reference to Figure 2 (a low-power drawing of a microscopical section) shows that the superficial epithelial covering is very much disintegrated, all traces of Malpighian layer and its papillae having disappeared. No nuclear staining is discernible in any of the sections, a considerable number of which were stained and examined. The skin is everywhere broken up into a series of deeply staining lamellae or blocks. The dermis shows a more definite structure, and its wavy fibrillae and bundles are easily discernible. No distinct

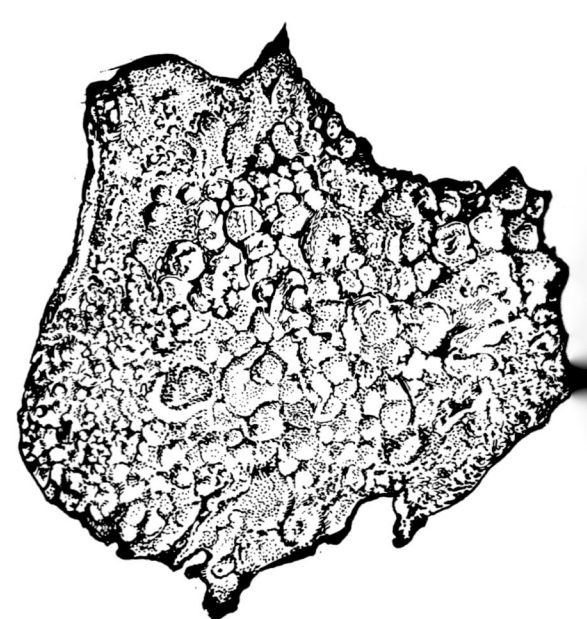

FIGURE 1. Naked-eye view of skin.

* From J. Path. Bact., 15:1, 1911.
1. See note by DR. RUFFER, Brit. Med. Journ., London, 1909, vol. i. p. 1005.

FIGURE 2. Microscopic section of skin under low power.

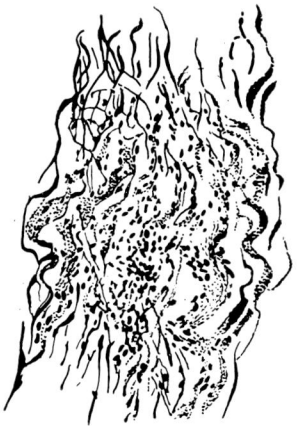

FIGURE 3. Section through dermis; Gram and eosin; Zeiss, DD, compens. ocular 6.

vessels, however, can be made out. On looking at the skin layer with a planatic magnifier, the presence of the dome-shaped vesicles is clearly demonstrated. They must have originated and developed in the middle of the prickle layer, *i.e.*, in the situation in which the small-pox eruption is first seen.

In the fully matured state of the vesicles, as they are present in the skin under consideration, their bases are formed by the deepest (Malpighian) layer, whilst the elevated superficial layers of the epidermis form their roofs. In one or two of the sections examined there are traces of the vertical septa and curtains which subdivide the developing vesicle in small-pox.

The structure of the dermis has been much less interfered with, and wavy or curling hyaline fibrillae of the fibro-areolar tissue are as distinct as in many similar sections from freshly fixed tissues. There are no traces of cellular infiltration beneath the vesicles.

Sections stained by Gram's method reveal very large numbers of bacteria, the large majority of which are strongly Gram-positive. By far the largest proportion of these occur in the connective tissue of the dermis, where they are met with either in dense clusters or diffusely s p r i n k l e d throughout the tissue, following the lines of separation of the fibrillar bundles. Occasionally, however, they are seen to follow the track of what may have been a small vessel, the direction of which is more or less oblique to the surface (see Fig. 3). Owing to the tenacity with which the epithelial layer retains the Gram's stain, the presence of bacteria amongst the epithelial remains is impossible to establish. Careful search, however, in sections

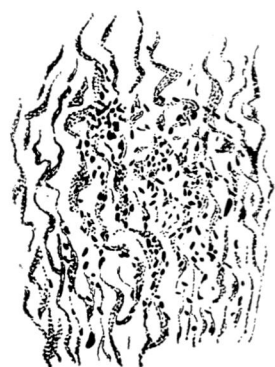

FIGURE 4. Section through dermis; methylene-blue; Zeiss, DD, compens. ocular 6.

stained with methylene-blue, leaves no doubt as to their presence here also. They appear to be more numerous in the neighbourhood of the vesicles than elsewhere. The organisms present in the largest numbers are short plump bacilli, often swollen at one end, so as in many instances to resemble one of the drum-stick bacilli. Others are distinctly beaded in form or have a torpedo shape. A cluster of bacilli with such characters bears a superficial resemblance to a group of diphtheria-like organisms. A few micrococci also occur; these are more apparent in sections stained with methylene-blue (see Fig. 4).

It is certainly unusual to find the sub-epithelial tissue so invaded by bacteria in small-pox as in the skin under consideration. Nor do we wish to maintain that these organisms played any part during the progress of the malady (supposing it to have been small-pox); but, after careful examination of a large number of sections, we are of opinion that these bacteria were present in the body at the time of death, though they have probably multiplied enormously after death. It may be fairly surmised that bacteria already present in the tissues might in some cases greatly multiply locally between the time of death and the mummification proper.

The specimen which we have described thus provides several features of quite exceptional interest, among which may be mentioned:

1. The probable existence of small-pox as evidenced by as characteristic an eruption as the conditions of preservation of such ancient material permits.

2. The conservation of the form of minute organisms such as bacteria after such a phenomenal period.

3. The demonstrability of bacteria in mummified tissues by modern staining methods.

Chapter 26

Evidence of Endemic Calculi in an Early Community

DON BROTHWELL

In a pathological report by Wood Jones (1908) on early Nubian skeletons, he writes "It is somewhat remarkable that, considering the great numbers of well-preserved bodies that have been examined, no calculi have as yet been found. It is not likely that they have been overlooked in these cemeteries, and it is therefore legitimate to conclude that, as compared with their present-day frequency in civilized races, calculi were very rare among the ancient Egyptians" (p. 56). Moreover, in a footnote to this report Elliot Smith writes, "The first ancient Egyptian body that I ever saw *in situ* ... was a prehistoric youth with a vesical calculus. Although I have been constantly on the look-out for other examples ever since then, I have never seen another case, although close upon ten thousand bodies must have been examined either by Dr. Wood Jones or myself in Nubia and Egypt. I have seen two cases of renal calculi, both in Ancient Empire graves in Egypt and one case of gall-stones (in a mummy of the New Empire)." Ruffer (1910) describes three further urinary calculi found in a predynastic skeleton excavated by Sir Flinders Petrie. The length and breadth dimensions for two of these were given as 45 x 30 mm and 40 x 25 mm, the respective weights of the three being 30 gm, 24 gm and 11.7 gm. The chemical analysis of the smallest stone was given as: water 6.5 per cent; organic matter 34.8 per cent; P_2O_5 37.6 per cent; MgO 19.7 per cent; CaO .8 per cent. Eggs of parasites were not recognisable.

Considering the particularly favourable burial conditions generally found throughout Egypt it is reasonable to assume that calculi would certainly have been identified in most cases had they occurred. However, the same cannot be said of burials in some other regions of the world. In Britain for example, such calculi would not be easily identified in some grave soils. Mortimer (1905) noted calculi in a Bronze Age burial in Yorkshire, but this is exceptional. Indeed, it is more than likely that the majority of archaeologists finding such calculi would, at present, fail to realise what they were, and would not preserve them. This point was demonstrated rather forcibly to me during my recent participation in a Dark Age cemetery excavation in Somerset. The day after describing to some of the excavators the nature of calculi and the slender chances (but nevertheless worth considering) of finding such anomalies, one of the group did recognise calculi within a pelvis and carefully removed and preserved them.

Because of the paucity of cases of urinary calculi in earlier cemeteries, their relatively common occurrence at Jebel Moya would demand special note. The site of Jebel Moya, in the Sudan, was excavated over a number of seasons by Sir

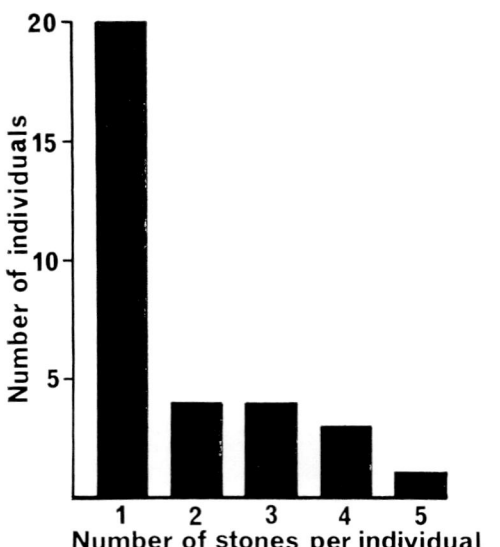

FIGURE 1. Distribution of the number of calculi per individual in a sample of thirty-two Jebel Moyans.

Henry Wellcome (Addison, 1949). A sample of the skeletal material recovered was studied by Mukherjee, Rao and Trevor (1955) although this represented the study of only a small part of the 2883 located graves, dated to the first millennium B.C. Further, this monograph did not include a report on the palaeopathology. Unfortunately, it is difficult to determine with exactness the sample number from which the calculi cases were derived. Not all the located graves were cleared, and many which could be did not, in fact, contain well preserved and complete skeletons. Although the estimate must be regarded as tentative, it seems likely that only 1500 graves were of value anthropologically, and of these, thirty-two (2.1 per cent) contained urinary calculi. Although the frequency of urinary calculi is very variable in world populations (Halstead, 1961), even the Jebel Moyan 2.1 per cent contrasts so markedly with the findings in early Egypt generally as to suggest its mild endemicity in this population.

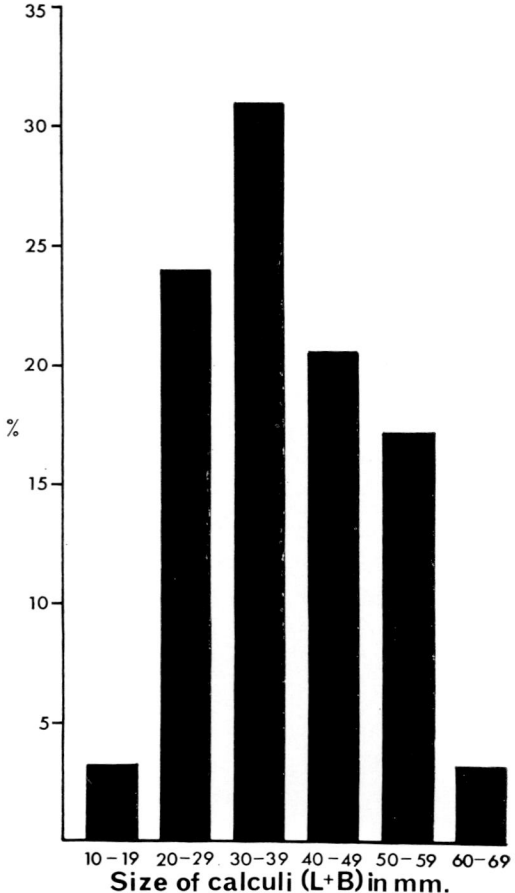

FIGURE 2. Percentage frequencies of calculi sizes in the Jebel Moya series (twenty-nine were measurable).

As shown in Figure 1, in most of the Jebel Moyan cases, only one stone was present (or at least was identified and retained) in each individual. Most of the calculi were spherical or egg-shaped, and in size (taken here as being the maximum diameter *plus* the greatest diameter at right angles to it) the calculi ranged from 18 mm to 61 mm. The distribution of these sizes given in Figure 2, shows that the greatest number of calculi were in the size range 30-39 mm. In twenty-eight cases, an age separation into adults and children was possible, there being in this sample twenty-seven adults and only one child (about seven years old). Owing to the unreliability of the sexing on the original field cards (now in the Museum of Archaeology and Ethnology, Cambridge), separation into males and females could not be attempted.

The Jebel Moyan calculi which were broken and revealed the interior structure, in no case revealed parasite eggs. Two (100/2075 and 100/2597) were analysed in the Department of Mineralogy, British Museum (Natural History), by the X-ray diffraction technique, and both proved to consist of calcite and apatite.

Thus, the Jebel Moyan cemetery shows clearly that, with careful excavation, some idea of the endemic nature of calculi in earlier populations can be hoped for. Also that frequency differences can be shown to occur in earlier times even in such a restricted area as Egypt and the Sudan.

REFERENCES

ADDISON, F., 1949: *Jebel Moya*. 2 vols. London, Oxford, University Press.

HALSTEAD, S. B., 1961: Bladder stone in Thailand. A review of the problem. *Amer J. Trop. Med. Hyg.*, 10:918.

MUKHERJEE, R., RAO, C. R., and TREVOR, J. C., 1955: *The Ancient Inhabitants of Jebel Moya* (Sudan). Cambridge, University Press.

MORTIMER, J. R., 1905: *Forty Years' Researches in British and Saxon Burial Mounds of East Yorkshire*. London, Brown.

RUFFER, M. A., 1910: Remarks on the histology and pathological anatomy of Egyptian mummies. *Cairo Scientific Journal*, IV:1.

WOOD JONES, F., 1908: Pathological report. *Archaeol. Survey Nubia*, Bull. 2:55.

Chapter 27

A Review of the Palaeopathology of the Arthritic Diseases

J. B. BOURKE

KELLGREN, Lawrence and Aitken-Swan (1953) investigated an industrial population in England and estimated the prevalence of rheumatic complaints at about 40 per cent. Most of this sample were then examined in detail and could be classified into four groups: osteoarthritis, rheumatoid arthritis, intervertebral disc disorders and "pains of undetermined nature." Similar findings have also been noted in several other countries (Expert Committee on Rheumatic Diseases of the World Health Organization 1954).

The physical anthropologist studies these diseases in order to record the extent of their history, and also contributes to the study of their natural history. The magnitude of the former problem is evident when it is remembered that the Neandertal skeleton from La Chapelle-aux-Saints shows many arthritic changes (Straus and Cave, 1957). The latter aspect may provide information about the changing expression of the disease or even its aetiology or it may provide forensic medicine with another aging method.

Epiphyseal union and dental eruption patterns have long been regarded as reliable criteria for the aging of skeletal remains. These methods only allow differentiation up to about twenty-four years. Auxiliary methods such as skull suture closure and the morphology of changes at the pubic symphysis have been developed.

McKern and Stewart (1957) studied a large series of the remains of American Korean War dead, and showed that skull suture closure has only a very general relationship with age and that as a guide to age determination it is of little use. Singer (1953), Genoves and Messmacher (1959) and Powers (1962) have reported similar findings. Todd (1920, 1921) described a series of changes at the pubic symphysis which are related to age; and McKern and Stewart (1957) have developed and refined this method. Arthritic changes can also be used as an aging criteria.

The rate of development of arthritic changes when considered against race, historical time and age may provide some information not only about an aging criteria, but also about the underlying pathology. Joint diseases are burdened with a multiplicity of names, many of which imply the same pathology. Before discussing examples of joint diseases the main pathological features as they effect skeletal remains will be briefly described.

PATHOLOGY

First, the normal anatomy of joints must be considered. They are classified into:
a). Synarthroses.

They are immovable unions between separate bones or between the shaft and the epiphysis of the same bone.

b). Diarthroses.

They are freely moveable joints. They are formed by the cartilage covered articular surfaces of two bones, the epiphyses and the synovial membrane. The articular cartilages are separated by a thin layer of synovial fluid. At rest, the articular cartilages are in direct contact.

c). Amphiarthroses.

They are slightly moveable joints. They are subdivided into:
  i). Syndesmoses—a ligamentous joint such as is found between the lower ends of the tibia and the fibula.
  ii). Symphyses—both bony surfaces are covered by hyaline cartilage, but they are separated by a mass of fibrocartilage and not by a synovial cavity. The pubic symphysis and the intervertebral joints are examples.

The diarthroses and the amphiarthroses are the joints that are affected by arthritic diseases.

Microscopically, the articular cartilages have a preponderance of intercellular substances over cells. Since the articular cartilage is almost avascular the nourishment of the articular surfaces is of great importance. The lateral articular areas have a delicate perichondrium which is continuous with the synovial membrane and contains many capillaries. The central area has no perichondrium and the surface is formed by an acellular matrix. The nourishment of the superficial layers of the central articular area is entirely dependent upon the synovial fluid. This low threshold of nourishment must be remembered when the function is impaired. The protein content of the synovial fluid is less than half that of lymph, and this is probably one of the reasons for the weak resistance of this part of the articular cartilage to disease and damage. It also accounts for the readiness with which it degenerates and the slowness or absence of repair.

Experimental destructive lesions of the lateral area of the cartilage are followed by rapid and excessive cartilage formation, whereas the central area shows almost none. The regenerative powers of the lateral area is reflected in the "lipping" found in chronic arthritis.

Diseases of joints can be classified thus:

a. Acute Arthritis—(i) suppurative. (ii) non-suppurative.

b. Tuberculous Arthritis.

c. Chronic Arthritis—(i) Rheumatoid Arthritis. (ii) Ankylosing Spondylitis. (iii) Psoriatic Arthritis. (iv) Osteoarthritis.

d. Gout.

e. Tumours.

f. Other Arthropathies.

Acute Arthritis

(i) SUPPURATIVE ARTHRITIS. The common infecting organisms are the staphylococci and streptococci, the knee and hip joints being most frequently involved (Heberling, 1941). Infections of the intervertebral joints are commonly caused by the tubercle bacillus, and the rare infection by other organisms will give a picture very similar to Pott's disease of the spine. The Brucella organisms and the typhoid bacillus rarely cause a spondylitis. Syphilis can cause changes in the vertebral column. Radiography may help in the differentiation of these infections.

In severe cases, the articular cartilages may be eroded and the underlying bone exposed. The ligaments are softened and give way so that the joint becomes disorganised and dislocated. Frequently, the infection is mild and the synovial membrane only is involved, and there is no joint destruction. Following tissue destruction there will be fibrous union. It may be cartilaginous or bony if the articular surfaces are destroyed.

(ii) NON-SUPPURATIVE ARTHRITIS. This group includes most of the acute arthritides. The inflammation is confined to the synovial membrane and there is no tissue destruction or permanent joint damage.

Tuberculous Arthritis

Further comments are unnecessary here, and the reader is referred to Morse's chapter on tuberculosis.

Chronic Arthritis

Chronic arthritis does not mean a chronic joint inflammation, but a slowly progressive disease.

(i) RHEUMATOID ARTHRITIS. This disease commonly occurs in women (in whom it is three times more frequent than in men) between the ages of twenty and forty. In rheumatoid arthritis, the small joints of the hands and feet are principally involved, but the larger joints may be affected later. The hip usually escapes. It is marked by remissions and exacerbations and at any time the the disease may become arrested, but the joint changes are irreversible. Typically the hands and feet are deformed. Ulnar deviation of the hand is a classical finding.

The disease may arrest at the stage of a synovitis or more usually, the articular cartilage may be involved. The synovial membrane grows over the articular cartilage from the side and forms a thick vascular covering (pannus), which adheres to the cartilage and erodes it. The cartilage is damaged by granulation tissue which is formed in the superficial layers of the epiphysis as part of the inflammatory reaction. The cartilage is destroyed and adhesions form between the two layers of the pannus covering the articular surfaces, and the joint cavity may be obliterated. Fibrous ankylosis of the joint develops and bony ankylosis may follow. In the rheumatoid process there is no development of osteophytes at the joint margins.

Radiological examination may show erosion of the bone at the insertion of the joint capsule. Osteoporosis occurs earlier in the disease than the above change but it is common in any condition which results in limb disuse, and the radiological density of archaeological material is very variable.

It must be stressed that rheumatoid arthritis is a slowly progressive and debilitating disease and that an accurate diagnosis can only be made in the more severe examples.

(ii) ANKYLOSING SPONDYLITIS. Ankylosing spondylitis is a disease which affects males more frequently than females in the ratio of about eight to one. Hersh, Stecher, Solomon, Wolpaw and Hauser (1950) studied fifty families and concluded that ankylosing spondylitis could be attributed to an autosomal dominant with variable penetrance. In most cases the age of onset is between fifteen and thirty-five (Hench, Slowcumb and Polley, 1947).

Ankylosing spondylitis is characterised by a polyarthritis which is followed by a bony fusion of the apophyseal, costo-vertebral and sacro-iliac joints. Ossification occurs in the spinal ligaments and in the outer fibres of the annulus fibrosus.

Radiography shows characteristic changes. Sacro-iliitis is the first demonstrable sign. The vertebrae appear squared. Calcification is frequently found in the anterior and posterior spinal ligaments. Ankylosis and bony fusion of the apophyseal joints also occurs. There is a slow loss of the lumbar curve and a gradual and continual thoraco-lumbar kyphosis. The disc spaces are usually unimpaired. Calcification and subsequent ossification may continue until the typical picture of a "bamboo spine" occurs.

(iii) PSORIATIC ARTHRITIS. This is a syndrome of psoriasis and rheumatoid arthritis. Psoriasis can only be diagnosed from soft tissues and so this diagnosis cannot be made from osteological material alone.

(iv) OSTEOARTHRITIS. Diarthroses and amphiarthroses show changes that are part of a degenerative process, but the production of these changes is different and distinctive even if the result is almost identical.

Osteoarthritis in diarthrodeal joints is a degenerative disease which is as common in men as it is in women and it is a disease of later life. The larger joints are more commonly involved than the smaller ones. One of the main aetiological factors is trauma. Degenerative arthritis is also associated with the aging of joint tissues.

Erosion of cartilage is commonest over the areas of contact where the greatest movement, strain and weight-bearing occur. As a result of the gradual loss of articular cartilage elasticity, the subchondral bone is no longer protected from the irregular effects of weight and pressure, and the changes of degenerative arthritis set in.

Osteoarthritis is a degeneration of articular cartilage and bone. The cells and matrix of the cartilage degenerate and the smooth surface becomes roughened. The cartilage cells swell, burst and disappear. The softened cartilage is gradually worn away until the underlying bone is exposed, and this also degenerates, but later undergoes condensation and hardening. Cartilaginous excrescences are formed at the articular cartilage margins and cause lipping of the joint edge. They increase the available articular surface and may be compensatory. They tend to ossify so that atrophied bone is surrounded by a ring of excrescences which restrict movement. These changes have typical radiological signs.

Cartilage loss is the primary lesion and it is followed by bone production. The process of osteoarthritis is a vicious circle of changing mechanical conditions and subsequent attempts at structural adaptations.

Osteoarthritis in Amphiarthrodeal Joints. The changes that occur at one of the amphiarthrodeal joints of the vertebral column will be discussed. The vertebral bodies are joined by an intervertebral disc and form a series of slightly moveable joints that have no synovial cavity or membrane.

The intervertebral disc is composed of an annulus fibrosus and a nucleus pulposus which contains about 80 percent water. It is almost incompressible and inelastic. The nucleus is bounded laterally by the strong elastic lamella of the annulus fibrosus, and vertically by the cartilaginous end plates of the intervertebral disc.

Disc degeneration is a biological degeneration, in which a change in the composition of the tissues occurs as well as water loss, which is the principal feature. Normally, the disc is an avascular structure but degeneration is accelerated if the disc becomes vascularised. Except where calcification has occurred, the only radiological sign of disc degeneration is the narrowing of the intervertebral space.

Escape of the nucleus pulposus may occur either after injury or spontaneously. It can occur as a prolapse or herniation of disc material through the annulus fibrosus or the cartilaginous plates.

Schmorl's nodes (Schmorl, 1932) are formed when disc material is herniated into the vertebral body. It cannot be detected radiologically until the disc material is outlined by a shell of reactive bone. Schmorl found this in 38 percent of spines at post-mortem examination.

With disc degeneration, narrowing of the intervertebral space results in a slight tilting of the vertebral bodies around the fixed axis of the apophyseal joints. The oblique compression of the disc causes a forward protrusion of the disc substance compressing the relaxed anterior fibres of the annulus. As the anterior margin of the disc slowly protrudes, it carries with it a tent-like elevation of the periosteum from the surfaces of the adjoining vertebral bodies. In these elevations new bone formation occurs and osteophytes develop which rarely undergo bony ankylosis.

Spinal osteophytes are rare in the mid-line anteriorly, but they commonly occur in the antero-lateral position at the side of the anterior longitudinal ligament. The attachment of this ligament is so secure that it is rarely displaced, and consequently periosteal elevation and subperiosteal bone growth occur on either side of it.

Osteophytes can grow to a great size at the margins of the vertebral bodies, but these are not osteoarthritic. They are the consequence of collapse of the intervertebral disc and protrusion of the annulus. Osteoarthritis affects the apophyseal joints of the vertebral column and true osteoarthritic osteophytes are found at these articulations.

Much confusion is caused by the indiscriminate use of the terms osteoarthritis and osteophytosis. Since they imply a distinct pathogenesis in different groups of joints it is desirable that the correct terminology be used.

Gout

In acute gout, urate crystals are laid down in the superficial layers of the cartilage as a chalky white deposit. The crystals are frequently surrounded by an area of necrosis. The principle results of chronic gout are:
(I) Disintegration of the joint space by urates.
(II) Degeneration of the cartilage and development of osteoarthritis.
(III) Obliteration of the joint space by urates.
(IV) Chronic synovitis.

In osteological remains it will be impossible to separate gout from osteoarthritis. A firm diagnosis should be possible if the soft tissues are present.

Tumours

Reference may be made to Brothwell's chapter on tumours (pp. 320-345).

Miscellaneous Arthropathies

(1) CHARCOT'S ARTHROPATHY. It principally occurs in tabes dorsalis, syringomyelia and diabetes mellitus. It usually affects the large joints of the lower limbs. It is a progressive disorganisation of an insensitive joint when exposed to continuing trauma. It is an example of the osteoarthritic process in excess, and has a typical radiological appearance.

(II) LOOSE BODIES. They are frequent incidental findings. They may be formed from pieces of synovial membrane which have become calcified and they cause osteoarthritis. This can only be diagnosed in mummies.

It should be remembered that other diseases such as acromegaly, rickets, haemophilia, ochronosis and osteochondromatosis may have an associated arthritis.

This is not intended to be a complete account of the pathology of the arthritic diseases, but rather a resumé of what the palaeopathologist should look for in osteological remains. It should be possible to make a fuller diagnosis if soft tissues are present.

The following diagnoses can be contemplated in osteological remains:

1. Chronic suppurative arthritis. A suggestion as to the aetiological factor may be possible in certain remains.
2. Tuberculous arthritis.
3. Rheumatoid arthritis.
4. Ankylosing spondylitis.
5. Osteoarthritis.
6. Osteophytosis.
7. Charcot's arthropathy.

We can now consider palaeopathological examples.

PALAEOPATHOLOGY

The Palaeopathology of Chronic Suppurative Arthritis and Charcot's Arthropathy

It is difficult to distinguish between these conditions unless there are other signs of an osteitis or osteomyelitis with joint involvement.

Wells (1962) reported a possible example of suppurative arthritis in the head of the left humerus of a young woman from an Early Saxon burial ground at Caister-on-Sea, near Great Yarmouth. The rest of the humerus was normal, but the scapula and clavicle were missing. He commented that the slight eburnation of the affected surface suggested continued movement at the joint after the damage occured. From the appearance of the bone alone, Wells was unable to exclude Charcot's arthropathy.

Brothwell (1963) reported a similar case in the head of the left humerus from a Mediaeval individual from Scarborough. The proximal end of the bone was deformed and the length reduced. Further examination of the other remains (right radius, pelvis and some damaged lower limb bones) showed no evidence of osteitis or osteomyelitis. Charcot's arthropathy cannot be excluded, but the deformity was probably due to suppurative arthritis.

Brothwell (1961, 1963) described a tibia from a Saxon Cemetery at Thurgarton, Nottinghamshire which has a sinus opening at its proximal end. This bone was one from a disturbed burial area and it was not associated with any other remains. The proximal shaft showed changes typical of osteomyelitis. The tibial condyles were sclerotic. There was derangement of the intercondylar area. On this evidence, it was suggested that the knee joint was infected and that a chronic suppurative arthritis had supervened.

The skeletal collection of the Natural History Museum, London, contains the remains of two individuals from the same grave at Abydos, being dated to the Ist Dynasty of which one consisted of a femur, sacrum and both humeri. The remains were of a female, and both humeri showed almost symmetrical changes of gross osteoarthritis with loss of the humeral heads (see Fig. 1). The other bones showed no pathological changes. The symmetry of these changes suggested a neuropathic rather than an infective process. The most probable diagnosis was Charcot's arthropathy.

Brothwell (1965) reported a tentative case of joint changes from Brucellosis in a Bronze Age individual from Jericho.

The difficulty of distinguishing between chronic suppurative arthritis and Charcot's arthropathy is evident from the examples above. Careful preservation and examination of remains should help in making this distinction.

The Palaeopathology of Rheumatoid Arthritis

The documentation of rheumatoid arthritis will remain obscure until archaeologists appreciate that to make this diagnosis the small bones of the hands and feet must be carefully preserved and examined. The incidence of the classical picture of the rheumatoid hand is low. Many of the definitive changes that precede this deformity can only be detected by radiography.

Brothwell and Møller-Christensen (1963) have reported a seventh century case of mutilation of a male of about thirty-five from Tean in the Scilly Isles. They published a radiograph of the right hand which showed punched out areas at two interphalangeal joints and one distal interphalangeal joint. These changes are commonly found in patients with rheumatoid arthritis and this may be an example of rheumatoid arthritis.

The Palaeopathology of Ankylosing Spondylitis

Zorab (1961) in a review of the historical background of ankylosing spondylitis described an X-ray investigation of eight mummies. He found no proven case of ankylosing spondylitis.

At Cambridge, the author recently came across two fused areas of vertebrae which had hitherto been unconnected. On examination it was found that both pieces had the same identification marks and fitted together. Subsequently, it was found that the specimen belonged to the collection of Egyptian vertebral columns which will be described below. Petrie (1901) recorded that a grave in the B cemetery (sequence date 66) at Hou contained a skeleton whose "spine was completely ankylosed from the top to the sacrum." The identification marks also corresponded.

This specimen for which no other remains could be found comprised all the vertebrae except the atlas. The sacrum was absent. It was fused into one unit

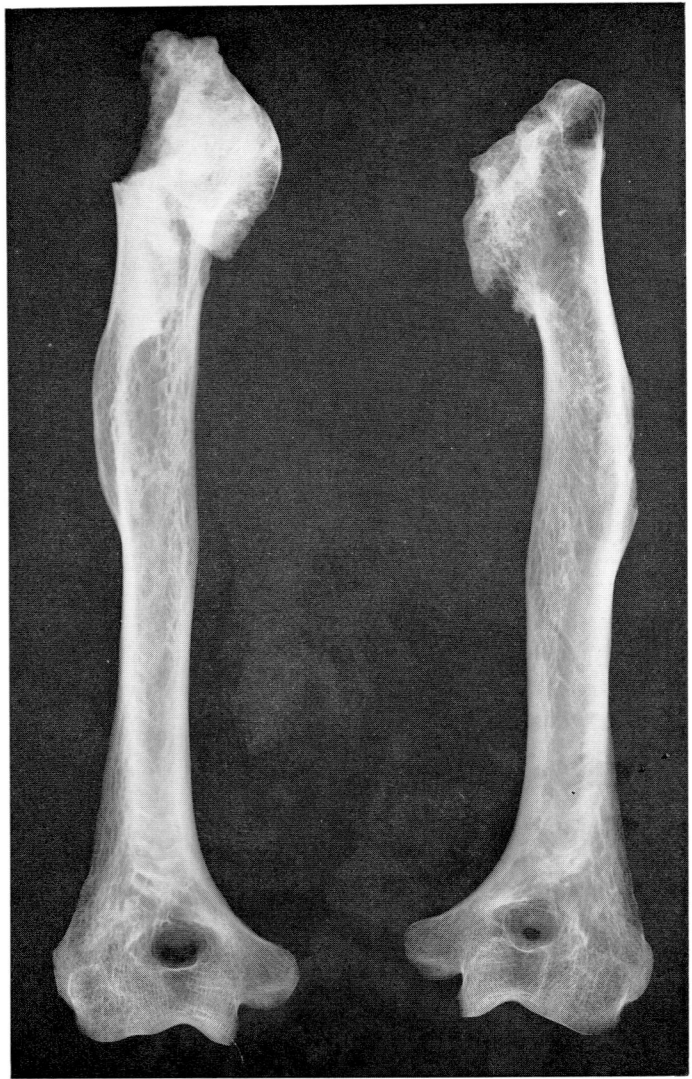

Figure 1. Radiograph of both humeri from a First Dynastic female from Abydos showing gross osteoarthritic changes with loss of both humeral heads.

with an increased thoracic lordosis and a decreased lumbar lordosis. The radiograph showed squaring of the vertebral bodies and calcification of the spinal ligaments. The apophyseal joints were heavily calcified (see Fig. 2). The features of this specimen support a diagnosis of ankylosing spondylitis.

At the same time another specimen of six fused mid-thoracic vertebrae was found, being also probably from Hou, and dates from the XIIth Dynasty (cemetery Y). This specimen showed fusion of the apophyseal joints. The radiograph (see Fig. 3) showed a calcified mass between two vertebral bodies which could well be calcified disc material. The vertebral bodies were squared and calcification had oc-

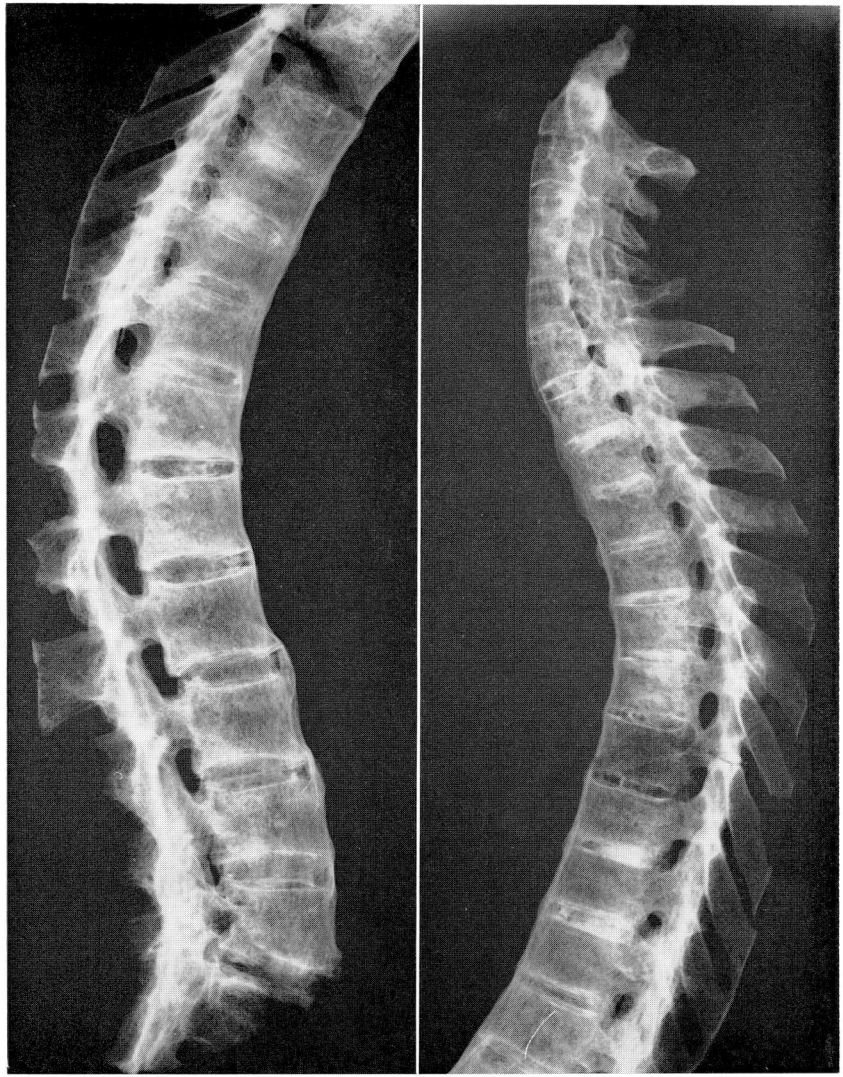

FIGURE 2. Two radiographs of the spine of an individual from Hou. They show an increased thoracic lordosis and a decreased lumbar lordosis, and squaring of the vertebral bodies with calcification of the spinal ligaments and apophyseal joints.

curred in parts of the spinal ligament. This is another clear example of ankylosing spondylitis.

The skeletal collection of the Natural History Museum, London, contains a specimen from the New Race cemetery at Naqada of five fused lumbar vertebrae and part of the sacrum. The bony elements were fused together at the apophyseal joints and at the margins of the vertebral bodies. The radiograph (see Fig. 4) showed the typical changes of ankylosing spondylitis.

From these examples it should not be assumed that ankylosing spondylitis was a common disease in Ancient Egypt. We

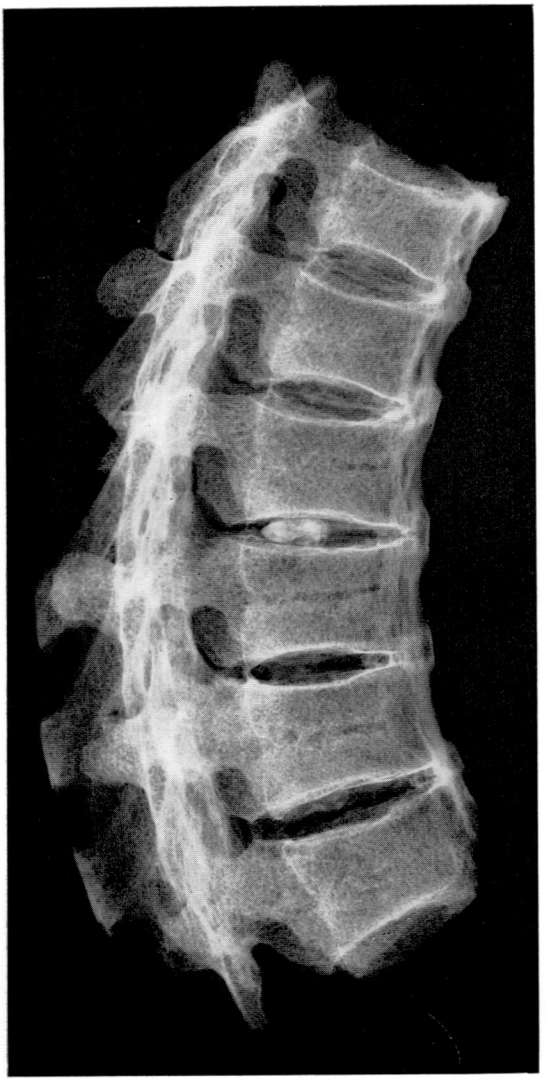 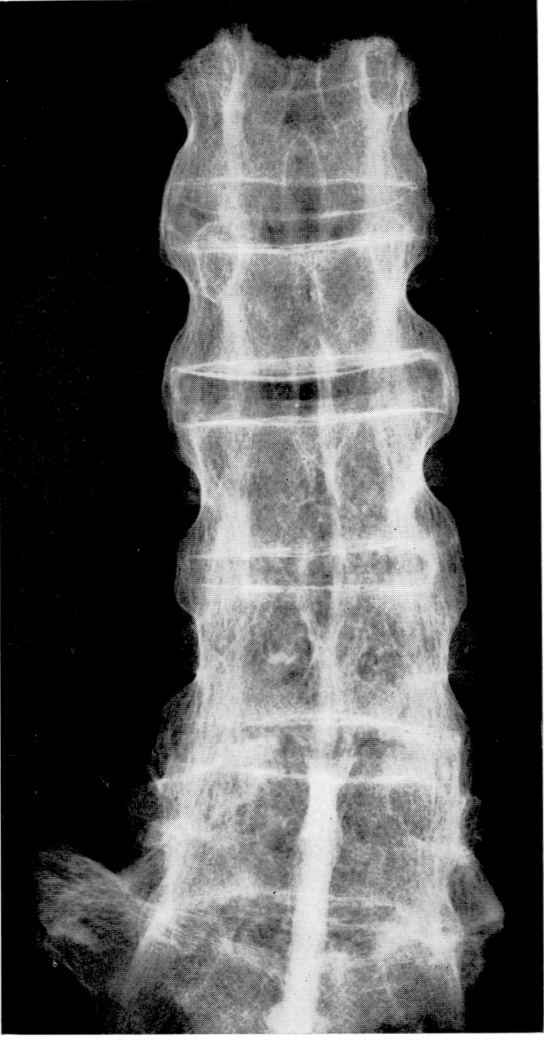

FIGURE 3. Radiograph of six fused mid-thoracic vertebrae probably from an individual from Hou. It shows squaring of the vertebral bodies and calcification of the spinal ligaments and apophyseal joints. There is also a calcified mass between two of the vertebral bodies.

FIGURE 4. Radiograph of five fused lumbar vertebrae and part of the sacrum of a New Race individual from Naqada. It shows calcification of the ligaments at margins of the vertebral bodies and squaring of the vertebral bodies.

know that skeletal abnormalities tended to be saved by excavators and at the same time we do not have reliable estimates of the population size from which the specimens were derived. It is certain that ankylosing spondylitis as we now recognise it was an uncommon disease in Ancient Egypt. Former claims for its great frequency were the result of terminological confusion with osteophytosis.

The Palaeopathology of Osteoarthritis

This disease is perhaps the best documented in palaeopathology. Straus and Cave (1957) and others have described gross osteoarthritis in the fossil man from La Chapelle-aux-Saints. Ackernecht (1953) has described it in Cro-Magnon man, and Vallois (1949) commented that chronic osteoarthritis was a specific disease of Paleolithic man. Ruffer and Rietti (1912) noted that the majority of lesions found in Egyptians, coming from a period of more than three thousand years were typical of chronic osteoarthritis. Elliot Smith and Dawson (1924) considered it to be the main disease of Ancient Egyptians and Nubians. Numerous reports of the disease in more recent material is available (Shore, 1935a; Inglemark, Møller-Christensen and Brinch, 1959; Brothwell, 1961).

In spite of the great frequency of osteoarthritis few studies have been made of the rate of development of changes at various sites except in the vertebral column. Osteoarthritis occurs at two sites in the vertebral column: the anterior atlanto-axial and apophyseal joints.

At the anterior atlanto-axial joint Shore (1935a) found changes in 32.2 per cent of an unsexed and unaged series collected at the University of the Witwatersrand, South Africa and from the Anatomy School, Cambridge. The incidence of osteoarthritic changes in a British series of sixty-two males and fifty-five females was 60 per cent and 45 per cent respectively (Bourke, 1964).

Shore (1935a) stressed the important point that apophyseal joint osteoarthritis was different from "osteoarthritis of the spine" which corresponds to osteophytosis. He described his findings on the same material as above on the basis of a presence or absence score. The incidence of osteoarthritis was highest in the thoraco-lumbar region (65 per cent). The incidence in the cervico-thoracic and cervical areas was 30 per cent and 23 per cent respectively. These figures referred to the areas of maximum involvement.

Roche (1957) studied a series of four hundred and nineteen vertebral columns of recent American Whites and Negroes of both sexes ranging from twenty to ninty-nine years of age. The incidence of osteoarthritis at the apophyseal joints was greater in the Whites than the Negroes and in the older skeletons rather than the younger ones and in females more than males. By sixty years of age the male incidence was equal to or greater than the female. Unfortunately, he did not mention the relation of changes to the regions of the vertebral column.

Inglemark et al. (1959) have made an extensive study of the vertebral column pathology in two hundred and fifteen skeletons from the mediaeval cemetery at Aebelholt in Denmark. Inglemark discussed his results on the basis of presence or absence of osteoarthritis. There were peaks of incidence in the mid portion of the cervical region, and at the transition between the upper and middle thirds of the thoracic region. These changes were progressively more common with age. The degree of osteoarthritis was greatest in the central areas of the cervical and thoracic regions and also in the lower part of the lumbar region. The average degree of involvement did not alter significantly in the various regions. Inglemark also showed that the frequency and degree of these lesions were independent of one another.

Shore and Inglemark did not agree completely on the distribution of osteoarthritis throughout the column. This difference could be due to the composition of Shore's series or to cultural and racial factors.

Osteoarthritic changes at the other joints have been infrequently studied. Brothwell (1961) reported on some British material from Neolithic to Saxon times. Heine (1926) studied a series of a thousand and two autopsies for osteoarthritic changes. Collins (1949) quoted and discussed these results and showed the rate of development and pattern of distribution of osteoarthritis. Grade III (Collins' scale) of osteoarthritic changes was not found until after the age of thirty at the knee joint and the age of forty at the other joints. The rate of development of osteoarthritis was highly variable at the joints of the human body. With further work, it may be possible to devise another ageing method for skeletal remains between the ages of thirty and seventy.

Many specimens have been described which show the joints that can be involved and the range of changes that occurs. Ruffer and Rietti (1912) and Ruffer (1918) gave many illustrations. Brothwell (1963) showed an excellent example of severe osteoarthritis of the hip in a Romano-Briton which has a very "lipped" acetabulum and deformed femoral head (see Fig. 5). In the same monograph he illustrated a knee joint with gross osteoarthritic changes at the femoral and tibial condyles (see Fig. 6).

Many palaeopathological specimens will be diagnosed as osteoarthritis of a particular joint, and no fuller diagnosis made because of lack of information from the study of the rest of the skeleton.

The Palaeopathology of Osteophytosis

The bony eburnations on the margins of the vertebral bodies have been given many names, but the authors were using different terms to describe osteophytosis. It has

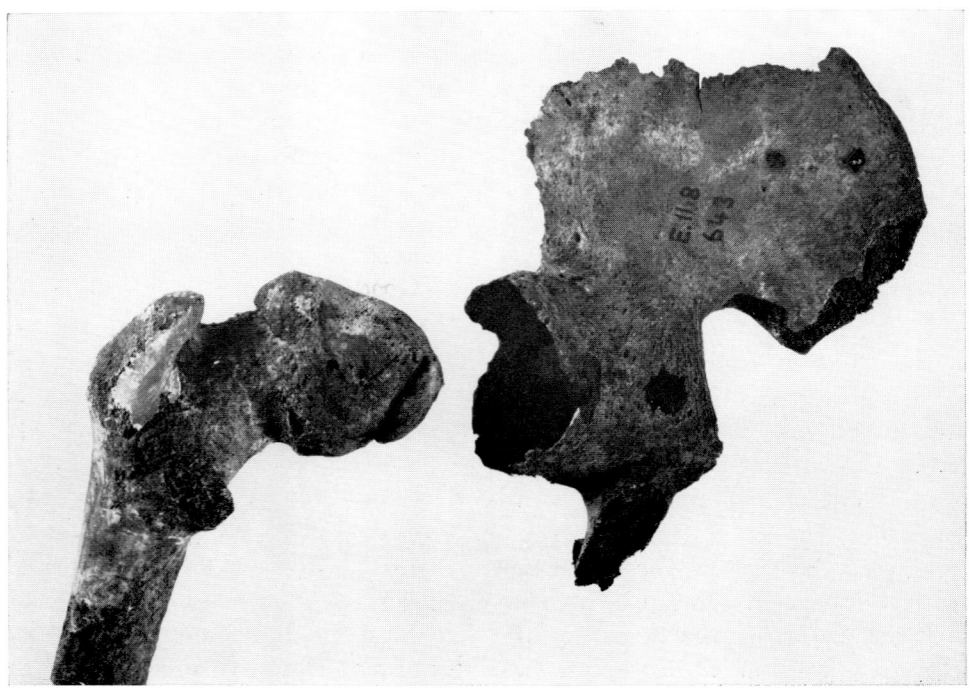

FIGURE 5. Photograph showing severe osteoarthritis of the hip in the remains of a Romano-Briton. (By courtesy of the British Museum. [Nat. Hist])

FIGURE 6. Photograph of the femoral and tibial condyles showing gross osteoarthritic changes in an individual from the mediaeval cemetry at Scarborough. (By courtesy of The British Museum [Nat. Hist.]).

been found in Neandertal man (Straus and Cave, 1957), Ancient Egyptians (Ruffer, 1918; Ruffer and Rietti, 1912), Eskimos, Whites and Pueblo Indians (Stewart, 1947), Danes (Inglemark et al., 1959) and Britons (Brothwell, 1960).

There is no absolute method of measuring the degree of severity of osteophytosis, but subjective grading has been used in several studies and appears to be gaining acceptance. Results are given below on four series of vertebral columns which were examined for osteophytosis. In assessing the degree of osteophytosis, each vertebra was divided into four quadrants and each quadrant assessed on an arbitrary scale from 0 to 3. The score for each vertebra was obtained so as to give an index of involvement and they were collected into three groups and expressed as a percentage of the total number of vertebrae present. An index of from nil to three denoted little or no involvement, while four to seven indicated moderate involvement and eight to twelve, severe involvement. All the readings were made by one observer and any column that showed evidence of other pathology was excluded.

The following series were studied, all being adult:

1. An Egyptian series (see Fig. 7) of one hundred and forty-three columns from Hou of a date prior to 1,500 B.C. It was impossible to sex or age this series.

2. An Iron Age series (see Fig. 8) of thirty-one males and twenty-three females from Maiden Castle, Dorset (Goodman and Morant, 1940).

3. An Anglo-Saxon series (see Fig. 9.) from Great Chesterford, Cambridgeshire of twenty-six males, thirty-four females and nineteen unsexed individuals.

4. A modern series of fifty-two males and forty-eight females from St. Bride's Church, Fleet Street, London (see Figs. 7, 10 and 11). All were of known age. The results are summarized below:

For the Male and Female Series

All the male and female series showed an area of pronounced involvement in the

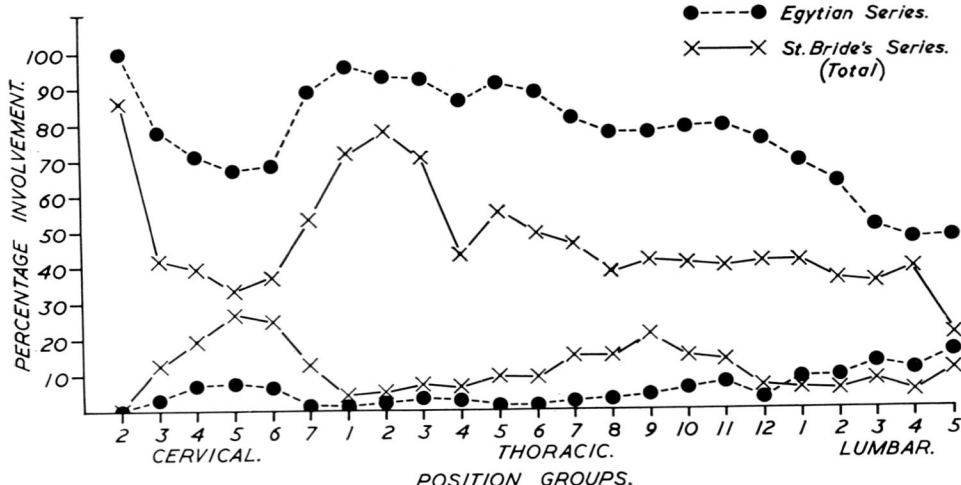

FIGURE 7. Graph for the Egyptian series and the total St. Bride's series. The top two lines are for slight osteophytic involvement and the bottom two lines are for severe involvement.

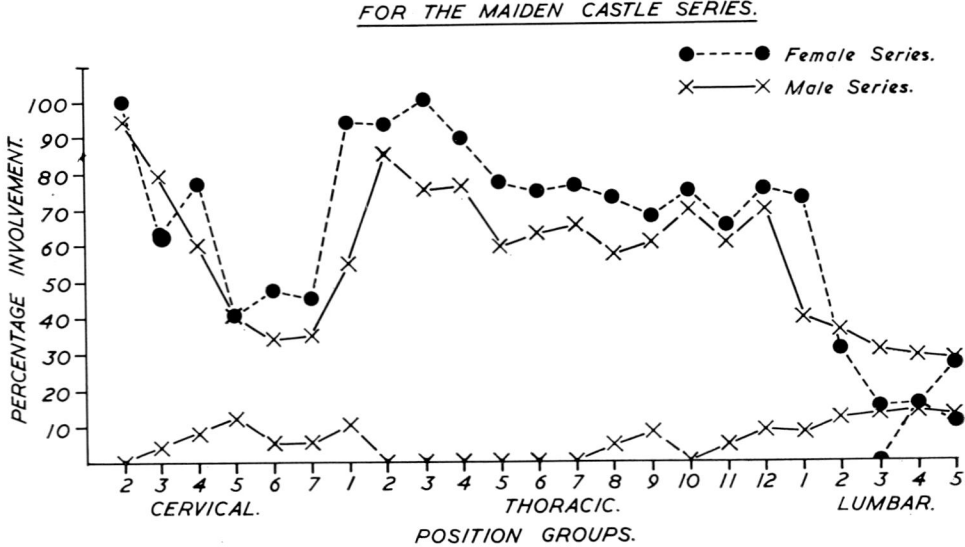

FIGURE 8. Graph for the Maiden Castle series. The top two lines are for slight involvement with osteophytic changes. The bottom line is for severe involvement in the male series. The female series has severe involvement only in the third, fourth and fifth lumbar vertebrae and this is also shown.

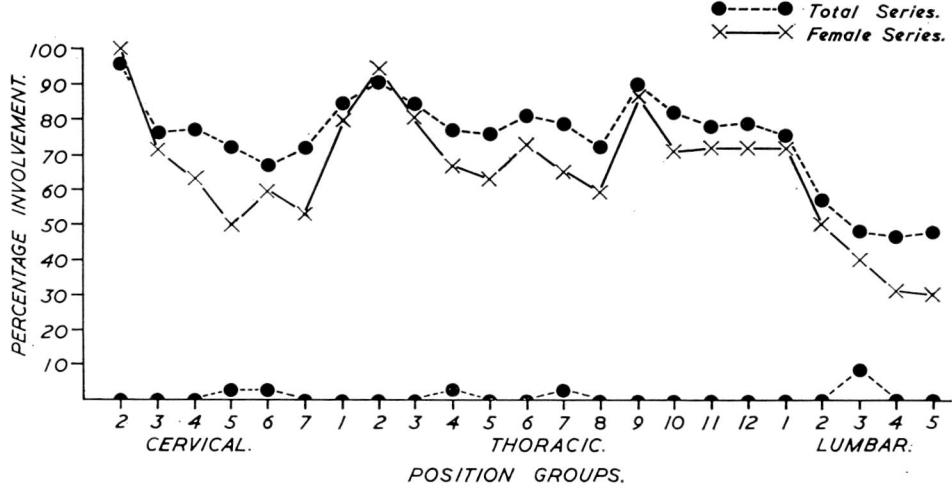

FIGURE 9. Graph for the female and the total Great Chesterford series. The top two lines are for slight involvement with osteophytic changes. The bottom line is for severe changes in the total series.

cervico-thoracic area with a peak incidence between the fourth and sixth cervical vertebra.

The male and female Maiden Castle and the male Great Chesterford series showed a steady increase in the severity of osteophytosis with descent of the column, and the involvement was greatest in the lower lumbar area. The female Great Chesterford series had a thoracic peak of involvement at the eighth thoracic vertebra and there was an area of less involvement between the ninth thoracic and the first lumbar vertebra. There was a steady increase in involvement with descent of the lumbar spine.

The male and female St. Bride's series showed a steady increase in involvement with descent of the thoracic and lumbar regions. The severity was greatest in the lower thoracic area in both sexes.

For the Total Series

All four series showed an area of cervico-thoracic involvement. It was distinct in the Egyptian, Maiden Castle and St. Bride's series with maximum involvement at the fifth or sixth cervical vertebra. There was an increase in involvement and severity with descent of the thoracic column in the Egyptian, Maiden Castle and St. Bride's series. The Great Chesterford series had an area of slight involvement in the mid-thoracic region. All four series showed marked involvement of the lumbar region.

For the Aged Series (St. Bride's Series Only)

The male and female series showed an increase in involvement up to the sixth decade in males and the fourth decade in females. There appeared to be an increase

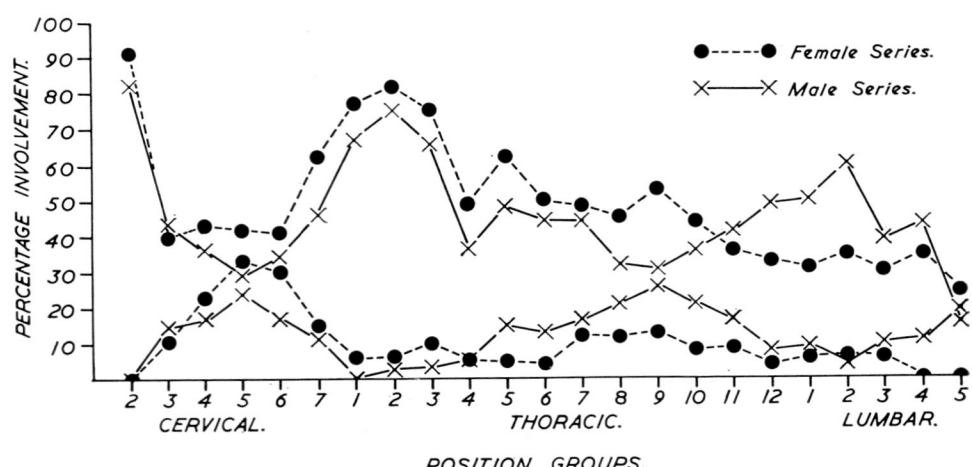

FIGURE 10. Graph for the male and female St. Bride's series. The top two lines are for slight involvement with osteophytic changes, and the bottom two lines are for severe involvement.

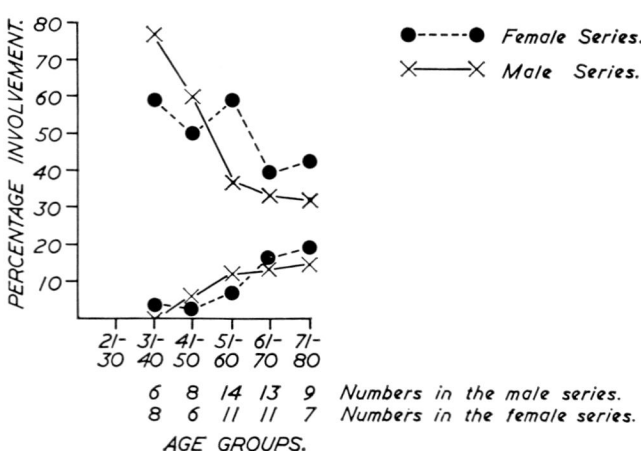

FIGURE 11. Graph for the male and female St. Bride's series. The top two lines are for slight involvement with osteophytic changes, and the bottom two lines are for severe involvement.

in severity with age in both sexes. There was a sudden increase in severity during the fifth decade in the female series.

The total series showed a steady increase in involvement and severity with age.

Stewart (1947) has studied the incidence of osteophytosis in Eskimos, Pueblo Indians and Whites. He combined both males and females into one series and excluded those under the age of forty. He found a general pattern of increasing amounts of lipping from above downwards in each of the three regions of the vertebral column. The maximum amount of lipping did not occur in the last vertebra of each region, and he postulated that they were transitional in character and have more lipping on their upper rather than lower borders. He noted that involvement of the articular facets bore little relation to the lipping of the vertebral bodies.

The Eskimos had little lipping anywhere except in the lumbar region, but they had many osteoarthritic facets especially in the mid-cervical and lower thoracic regions. The Pueblos showed more osteophytic lipping in each region, but had few articular facets involved and these were mainly in the thoracic region. The Whites had maximum lipping in the lower cervical region, and the lumbar and lower thoracic vertebrae followed in that order. In this group, the articular facet changes were minor and occurred mainly in the mid-cervical and upper thoracic regions.

Shore (1935b) used the term polyspondylitis marginalis osteophytica to describe osteophytosis, and has studied a series of one hundred and six vertebral columns from the collections of the anatomy schools at the Universities of Cambridge and Witwatersrand. He concluded that the liability of the vertebrae to marginal osteophytes increased with descent of the column. There were well defined "outcrops":

1. The cervical outcrop was marked off from the thoracic by a minimum zone at the junction of the first and second thoracic vertebrae.
2. The thoracic outcrop was greater than the cervical and was separated from the lumbar outcrop by diminution zone at the twelfth vertebra.
3. The lumbar outcrop was greater than the thoracic outcrop, but fell to a minimum at the lumbo-sacral junction.

Shore's first conclusion (1. above), was confirmed by the present study, and also it showed that the lumbar outcrop tended to be greater than the thoracic. Stewart's findings about diminution zones were confirmed by Shore, but not so in this investigation.

Brothwell (1960) examined thirty Bronze Age skeletons which were collected from many parts of Britain. The vertebrae were the most affected with arthritic lipping and the lumbar region was the most involved, the thoracic region to a lesser extent and the cervical region least. For comparison he examined fifty skeletons from Iron Age and Saxon periods and in them the thoracic and lumbar vertebrae were the most affected, each roughly to the same degree, but the cervical vertebrae were the least involved.

Inglemark et al. (1959) in their study of the remains from Aebelholt found a cervical, thoracic (eighth thoracic) and lumbar (second lumbar) peak of involvement. The frequency of osteophytosis increased with age. The severity was greatest in the cervical region and there was an almost constant level of severity in the rest of the column except for a slight increase in the lower area. The cervical region was the only area to show a positive correlation between the frequency and the degree of the changes.

Ingelmark found that generally the changes in the small joints were graver than those at the vertebral bodies. Separate vertebrae showed marked and sometimes significant differences between the frequencies of lesions on the bony surfaces and the small joints. This showed that the two processes were not closely related. Both became more frequent with age.

Hooton (1930) studied five hundred and three adult or sub-adult skeletons of Indians from Pecos Pueblo who lived during the ninth and twelfth centuries. "Spondylitis deformans" (another term for osteophytosis) was present in 13 per cent, but any positional variation was not mentioned.

Goldstein (1957) studied fifty-six skeletons of Indians from Mexico dating from 800 to 1,700 A.D.; 62.5 per cent showed lipping of the lumbar vertebrae. The incidence in the thoracic and cervical vertebrae was not mentioned.

McKern and Stewart (1957) in their study of American Korean War dead noted that the distribution of osteophytosis was erratic, but that the greatest changes were concentrated in the lower thoracic region.

The St. Bride's series (see Figs. 7, 10 and 11) was probably composed of people of a greater age than the other series. The severity of involvement in the cervical and lower thoracic vertebrae was increased relative to the other series and a generalised distribution of these changes throughout the whole column had not occurred. This is further evidence that continuing trauma is an important factor in the pathogenesis of osteophytosis.

There was great variability in the distribution of osteophytosis throughout the vertebral column and this was due to cultural and racial factors. Kellgren (1961) has stressed this point in relation to clinical medicine.

McKern and Stewart (1957) and Stewart (1957, 1958) both investigated the use of the incidence of osteophytosis as an aging method. As a preliminary, any difference between the sexes must be investigated for racial variability.

Willis (1924) and Stewart (1957, 1958) both studied osteophytic development with age. Willis studied the lumbar vertebrae of an aged series of six hundred and twenty-five spinal columns from the Todd collection at Western Reserve University. He found that up to twenty-five years of age there was practically no lipping. In the group twenty-five to twenty-nine years, there were five spines showing the first degree of hypertrophy. Between thirty-five and thirty-nine, there appeared to be a very marked degree of change and this was greatest in the heavy type of spine and least marked in the slender. He related his findings to the build of the individual and stated that changes rarely occurred before the thirty-fifth year of life and were progressive thereafter.

Both Stewart's papers described the same material. His series consisted of American War Dead from Korea and skeletons from the Terry collection at Washington University, St. Louis. Unfortunately, this series contained only seventeen females compared with eighty-seven males from the Terry collection and three hundred and sixty-seven American soldiers. By itself osteophytosis did not permit close aging of the skeleton, but Stewart added that in skeletons of the white race the absence of grade ++ vertebral lipping (Stewart's scale) usually indicated an age of under thirty, and conversely the presence of extensive lipping, including grades ++ and +++ at some vertebrae usually indicated an age of forty. The qualification about race is important for Roche (1957) who worked on the Terry collection found that the incidence of osteophytosis was slightly higher in Whites than Negroes.

The present work showed that for the male series, involvement and severity increased with age and that severe lipping was absent until after the age of forty. This corroborates the work of Willis and Stewart. For the female series, severe lipping was present in the third and fourth decades of life to a small degree, but there was a sudden increase in the fifth decade, and the level remained almost constant. In a survey of seven thousand men and women from Leigh in Lancashire, Kellgren, Lawrence and Aitken-Swan (1953) investigated the incidence of disc degeneration with age for the male and female populations. They found a similar pattern to those described above.

One definite fact that has emerged from these studies is that if an individual showed severe lipping his age was over forty if male and over thirty if female, and probably over forty. This only applies to the "White" races and further work is needed before it can be applied to other racial groups.

The Palaeopathology of Gout

Elliot Smith and Dawson (1924) described an Egyptian mummy from the Christian period with multiple joint tophi and ulceration of the skin over the tarsal bones. Around the joints there were massive deposits of a chalky material which was proved to be uric acid on analysis. This specimen also showed changes of gouty arthropathy.

CONCLUSIONS

The palaeopathology of the arthritic diseases is far from complete. Much basic description remains to be carried out. This can only be achieved by archaeologists preserving all human remains, and by palaeopathologists considering the whole skeleton and not isolated bones. All techniques should be used whenever possible.

Population studies of osteoarthritis may give a more detailed and exact aging method for the higher age groups. Further studies of osteophytosis could contribute to our knowledge of its aetiology.

ACKNOWLEDGMENTS

My thanks are due to Dr. J. C. Trevor of the Duckworth Laboratory of Physical Anthropology, Cambridge; The Rector of St. Bride's Church, London, and Dr. F. Steel, Curator of the human remains at St. Bride's; and Dr. K. P. Oakley of the Sub-Department of Anthropology at the British Museum (Natural History) for allowing me to study material in their care.

My thanks are also due to Mr. G. Walter for photographic assistance.

REFERENCES

ACKERNECHT, E. H., 1953: Palaeopathology: a survey. *Anthropology Today*. Chicago, University Press.

BOURKE, J. B., 1964: The reliability of measurements of the atlas as a criteria for the sexing of human remains. Unpublished study.

BROTHWELL, D. R., 1960: The Bronze Age people of Yorkshire; a general survey. *Advancement of Science.*, 16:311.

BROTHWELL, D. R., 1961: The palaeopathology of early British man: an essay on the problems of diagnosis and analysis. *J. Roy. Anthrop. Inst.*, 91:318.

BROTHWELL, D. R., 1963: *Digging up Bones*. London, British Museum (Natural History).

BROTHWELL, D. R., 1965: The palaeopothology of Early-Middle Bronze Age remains from Jericho. In: K. M. Kenyon, *Jericho II*. London, Palestine Exploration Society.

BROTHWELL, D. R., and MØLLER-CHRISTENSEN, V., 1963: Medico-historical aspects of a very early case of mutilation. *Danish Medical Bulletin.*, 10:21.

COLLINS, D. H., 1949: *The Pathology of Articular and Spinal Diseases*. London, Arnold.

ELLIOT SMITH, G., and DAWSON, W. R., 1924: *Egyptian Mummies*. London, Allen and Unwin.

EXPERT COMMITTEE ON RHEUMATIC DISEASES, 1954: First report. W.H.O. Technical Report Series No. 78.

GENOVES, SANTIAGO T., and MESSMACHER, M., 1959: Valor de los patrones tradicionales para la determinacion del edad por medio de las suturas en craneos mexicanos. *Cuadernas del Instituto de Historia.*, Serie No. 7:7.

GOLDSTEIN, M. S., 1957: Skeletal pathology of Early Indians in Texas. *Amer. J. Phys. Anthrop.*, (n.s.) 15: 299.

GOODMAN, C. N., and MORANT, G. M., 1940: The human remains of the Iron Age and other peoples at Maiden Castle, Dorset. *Biometrika.*, 31:295.

HEBERLING, J. A., 1941: A review of two hundred and and one cases of suppurative arthritis. *J. Bone Joint Surg.*, 23:917.

HEINE, J., 1926: Über die arthritis deformans. *Virchow Arch.*, 260:521.

HENCH, P. S., SLOCUMB, C. H., and POLLEY, H., 1947: Rheumatoid Spondylitis, questions and answers. *Medical Clinics of North America.*, 31:879.

HERSH, A. H., STECHER, R. M., SOLOMON, W. M., WOLPAW, R. and HAUSER, H., 1950: Heredity in ankylosing spondylitis. A study of fifty families. *Amer. J. Hum. Genet.*, 2:391.

HOOTON, E. A., 1930: *The Indians of Pecos Pueblo. A study of their skeletal remains*. New Haven, Yale University Press.

INGLEMARK, B. E., MØLLER-CHRISTENSEN, V., and BRINCH, O., 1959: Spinal joint changes and dental infections. *Acta Anatomica.* Supplement 36. Vol. 38.

KELLGREN, J. H., 1961: Osteoarthritis in patients and populations. *Brit. Med. J.* 2:1.

KELLGREN, J. H., LAWRENCE, J. S., and AITKEN-SWAN, J., 1953: Rheumatic complaints in urban populations. *Ann. Rheum. Dis.*, 12:5.

McKERN, T. W., and STEWART, T. D., 1957: *Skeletal Changes in Young American Males*. Technical report. Headquarters Quartermaster Research and Development Command. Natick., Mass.

PETRIE, W. M. F., 1901: *Diospolis Parva (The cemeteries of Abadiyeh and Hu)*. London, Special Extra Publication of the Egypt Exploration Fund.

POWERS, R., 1962: The disparity between known age and age as estimated by cranial suture closure. *Man.*, 62: 52.

ROCHE, M. B., 1957: Incidence of osteophytosis and osteoarthritis in four hundred and nineteen skeletonised vertebral columns. *Amer. J. Phys. Anthrop.* (n.s.) 15:433.

RUFFER, M. A., 1918: Arthritis deformans and spondylitis in Ancient Egypt. *J. Path Bact.*, 22:152.

RUFFER, M. A., and RIETTI, A., 1912: On osseous lesions in Ancient Egyptians. *J. Path. Bact.*, 16:439.

SCHMORL, G., 1932: Uber osteitis deformans Paget. *Virchow Arch.*, 283:694.

SHORE, L. R., 1935a: On osteoarthritis in the dorsal intervertebral joints. *Brit. J. Surg.*, 22:833.

SHORE, L. R., 1935b: Polyspondylitis marginalis osteophytica. *Brit. J. Surg.*, 22:850.

SINGER, R., 1953: Estimation of age from cranial suture closure. *J. Forensic Med.*, 1:52.

STEWART, T. D., 1947: Racial patterns in vertebral osteoarthritis. *Amer. J. Phys. Anthrop.*, (n.s.) 5:230.

STEWART, T. D., 1957: The rate of development of vertebral hypertrophic arthritis and its utility in age estimation. *Amer. J. Phys. Anthrop.*, (n.s.) 15:433.

STEWART, T. D., 1958: The rate of development of vertebral osteoarthritis in American Whites and its significance in skeletal age identification. *The Leech.*, 28:144.

STRAUS, W. L., JR., and CAVE, A. J. E., 1957: Pathology and the posture of Neanderthal man. *Quart. Rev. Biol.*, 32:348.

TODD, T. W., 1920: Age changes in the pubic bone. I. The male white pubis. *Amer. J. Phys. Anthrop.*, 3: 285.

TODD, T. W., 1921: Age changes in the pubic bone. II. Pubis of male negro-white hybrid. III. Pubis of white female. IV. Pubis of female negro-white hybrid. *Amer. J. Phys. Anthrop.*, 4:1.

VALLOIS, H. V., 1949: Paleopathologie et paleontologie humaine. *Homenaje a Don Luis de Hoyos Sainz.*, 1: 333.

WELLS, C., 1962: Joint pathology in ancient Anglo-Saxons. *J. Bone Joint Surg.*, 44B:948.

WILLIS, T. A., 1924: The age factor in hypertrophic arthritis. *J. Bone Joint Surg.*, 6:316.

ZORAB, P. A., 1961: The historical and prehistorical background to ankylosing spondylitis. *Proc. Roy. Soc. Med.*, 54:415.

Chapter 28

Osteitis Fibrosa in a Skeleton of a Prehistoric American Indian*

HENRI STEARNS DENNINGER

Editorial Comment

This paper is reprinted as an excellent example of fibro-osseous disease in a prehistoric skeleton. Some explanation of the title is indicated, however, due to the considerable advances in the understanding of fibro-osseous disease since Denninger noted this case.

This is, in fact, an instance of *polyostotic fibrous dysplasia;* this condition is well reviewed by Harris *et al.* (1962). The term *osteitis fibrosa* (formerly used to indicate all fibro-osseous lesions) is now limited to the condition of the bones in hyperparathyroidism. The changes in Denninger's case are predominantly unilateral; the left femur shows a typical shepherd's crook appearance; there is probably some increase in the amount of bone in the skull but there is also some normal bone. All of these features indicate fibrous dysplasia and not osteitis fibrosa cystica, which is a generalised state.

We are indebted to Dr. Mary E. Catto, Lecturer in Orthopaedic Pathology in the University of Glasgow, for advice in the interpretation of this case and other examples of bone disease mentioned elsewhere in this book.

Reference

Harris, W. H., Dudley, H. R., and Barry, R. J. 1962: The Natural History of Fibrous Dysplasia. An Orthopaedic, Pathological and Roentgenographic Study. *J. Bone Joint Surg.* 44A:207.

IN an attempt to understand prehistoric American Indian life one must consider not only the cultural aspects, such as artefacts and type of burial, but also the pathologic conditions encountered. Anomalous or pathologic conditions are in themselves sufficient to modify the existence of an individual. An unusual example of a pathologic condition was excavated by the University of Chicago Archaeologic Survey, from mound F14, Fulton County, Lewistown, Ill., during the summer session of 1930.

The skeleton is that of a man approximately thirty-five years old. He was a cripple with an affliction involving practically the entire left side of the body. During life the left leg was bent over the

---

*Reprinted from *Arch. Path.*, 11:939, 1931. Copyright by American Medical Association.

right at about the level of the knee. The left foot and toes were markedly extended and permanently fixed. The right leg curved toward the median line. Walking must have been difficult, if not almost impossible, as even the pelvic and body axes were obliquely alined. The left half of the face was askew. A more or less constant rheumatic type of pain must have played no small part in his existence. Yet with this man there were found many distinctive artefacts, such as bear's teeth, bone needles, a shell necklace, discoidals and an effigy pipe. All these factors indicate that he probably held some important position among his fellows, and it becomes more significant when one considers the paucity of material expression associated with other burials in the same mound.[1] He may even have been the traditional "medicine man."

The exact chronological position of this skeleton is necessarily a matter of estimation. The mound in which it was found belongs to the last of the three bluff cultures, which precedes the Columbian era by several centuries, and is subsequent to the Hopewell culture (about 2,000 years old). Thus, its age may safely be approximated at 1,000 years.

A study of the skeleton indicates that the man to whom it belonged was afflicted with osteitis fibrosa, as evidenced by the following observation:[2]

The left femur presents the most striking pathologic changes in the upper third of the bone. These changes consist of a marked medial bending and an extensive enlargement and overgrowth of bone, with evidence of cavity formation. The bending of the head and neck of the femur on its shaft is such that the axes of these form an angle of approximately 45 degrees downward, and is probably not due to previous fracture. This must have required the man to hold the leg in a marked position of coxa vara, crossing in front of the right leg at about the level of the knee. The enlarged portion of the femur begins slightly above the junction of the upper and middle third of the shaft. Here there is a bony overhanging crest. The anterior surface of this swelling is relatively smooth and presents few perforations or porosities. Those present are mainly at either end of the enlargement. The perforations appearing on the lateral surface seem to have been made during the work of excavation. Bony trabeculations and perforations make up the posterior surface of this region of the femur, and in the upper part bony cords radiate from the margin of the femoral head to the trochanter. The lower portion of the posterior surface presents many projecting osteophytic growths.

The interior of the enlarged portion forms a cyst or large cavity. In its upper portion the cavity is filled with soft, fine, irregular, sponge-like bone, derived from the pathologic changes in the bone marrow. The cortical bone about these

---

1. The apotheosis of cripples is by no means a new idea. Well known examples of this are the deification of achondroplastic dwarfs by the Egyptians as described by: RUFFER, SIR M. ARMAND: Studies in the Paleopathology of Egypt ed. by R. L. Moodie, Chicago, University of Chicago Press, 1921, pp. 35-49 DAWSON, W. R.: Dwarfs and Hunchbacks in Ancient Egypt, Ann. M. Hist., 9:315, 1927.

2. As here considered, osteitis fibrosa is a primary disease, in contrast with the secondary fibrous osteitis often definitely associated with inflammatory lesions. Primary osteitis fibrosa "is a distinct clinical entity, is very liable to be confounded with osteomalacia, osteitis

---

deformans, or with central bone tumors, and is very puzzling to the practitioner," according to R. L. KNAGGS (The Inflammatory and Toxic Diseases of Bone, New York, William Wood & Company, 1926, p. 1246), and I may add that our position as palaeopathologists, diagnosing only from the osseous remains, is at times no less dilemmatic.

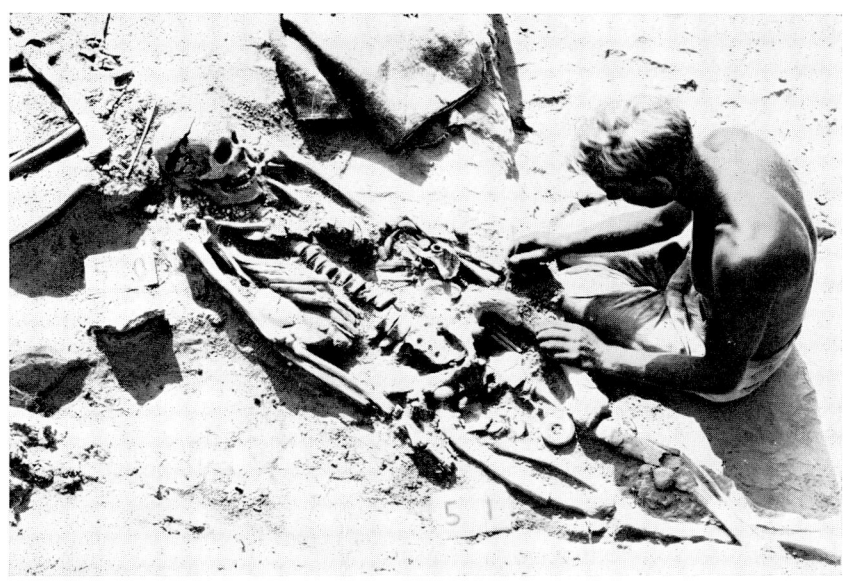

FIGURE 1. View in situ of skeleton (F 14-50) showing osteitis fibrosa, Mr. George Neumann removing the left femur (see fig. 2). The number 51 refers to a skeleton the skull of which lay directly beneath the fragile pelvis of this specimen and was responsible for considerable damage. Note coxa vara position of left leg.

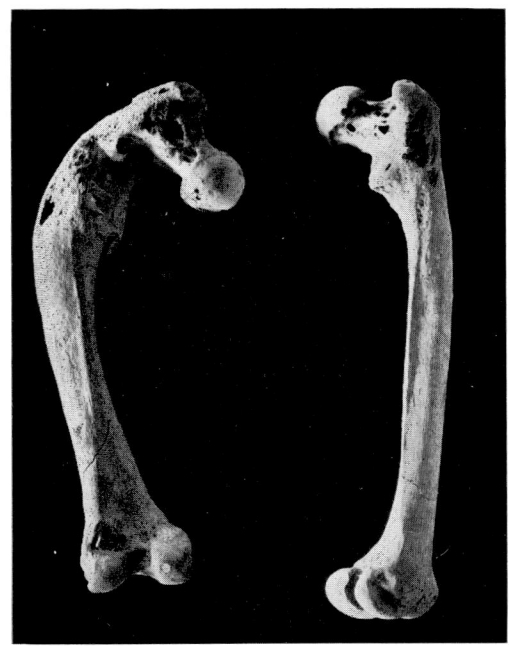

FIGURE 2. Posterior view of the left femur, and posterolateral view of the right.

cavities is very fragile and in places exceedingly thin. A lacelike trabeculation associated with spongy bone is found to extend through the entire marrow cavity on the lateral side. Just above the lateral condyle on the popliteal surface is an eroded, perforated, ulcer-like area, connecting with the peculiar bone formation described in the marrow cavity. This appears to be an early stage in the peripheral invasion of the disease.

In the right femur the pathologic changes are found in the proximal fourth of the bone, and are by no means so extensive. The normal contours are increased in size, with a nodular enlargement above the intertrochanteric crest. The surface here is perforated by hiati, from 4 to 8 mm in diameter, leading into a cavity, continuous with that of the bone marrow. Spongy bone is also found in this cavity. In the

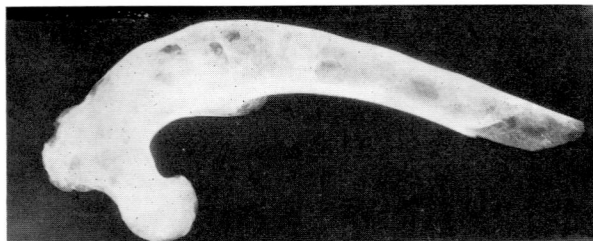

FIGURE 3. Roentgenogram of the left femur showing cyst formation, thinning of cortical bone, trabeculation and abnormal deposition of bone.

unaffected portions the cortex of both femurs is of normal thickness.

The right tibia exhibits a diffuse nodular swelling slightly above the middle of the shaft. On the medial surface there are two deeply eroded areas in the cortex, the smaller of which (2 by 4 cm) is surrounded by a slightly raised, eburnated ring of bone. Its base is composed of soft spongy bone and irregular bony trabeculae around numerous hiati which lead into the marrow cavity. The larger eroded area (3 by 7 cm) is similar in appearance, except that the margin is eburnated only in the back and here blends with a diffuse nodulation on the posterior surface. The left tibia was badly broken *in situ* by roots, and only the lower third of the shaft shows nodular enlargements.

The fibulae of both the right and the left legs present all the pathologic changes described. There are nodules associated with eroded areas, and cavities indicative of either cysts or fibrous areas, in various portions of the shaft. Such regions contain spongy bone and trabeculae and have a rarefied cortex. It may be noted here that in the long bones none of the articular surfaces are involved.

The bones of the right foot appear to be unaffected, while those of the left foot show a general increase in porosity of surface. Only the first and second metatarsals of the left foot show any significant pathologic changes. The first metatarsal contains spongy bone throughout and an eroded perforation on the distal end of the dorsal surface. The second metatarsal shows only an eroded area at the proximal end. As the bones of the left foot were uncovered *in situ*, they delineated the position of a marked talipes equinus.

In the right os coxae only the ischiopubic ramus and ischial tuberosity are involved by a single cystic swelling extending from the center of the tuberosity into the ramus. The erosion of the surface in this region extends upward to the posterior surface of the acetabular portion of the ilium. A small osteophyte is present at the attachment of the transverse ligament to the acetabular margin. The left os coxae had been crushed in the grave, but sufficient material remains to show the extensive pathologic changes that had taken place. The portion of the ilium above the acetabulum is ballooned out with a series of cavities to a width of 6.5 cm. The lateral and medial surfaces are much eroded, and many of the hiati show eburnated margins. The interiors of the cavities contain trabeculations and spongy bone. The sacral articular surface is missing.

Sacral involvement is primarily on the left side, as in the pelvis. The enlargement is directed laterally, so that the anterior sacral foramina become ellipses. These

Osteitis Fibrosa in Skeleton of a Prehistoric American Indian

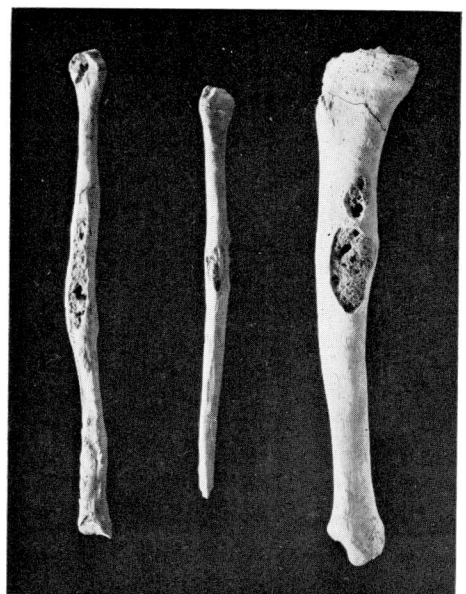

FIGURE 4. Medial views of right and left fibulae and of right tibia.

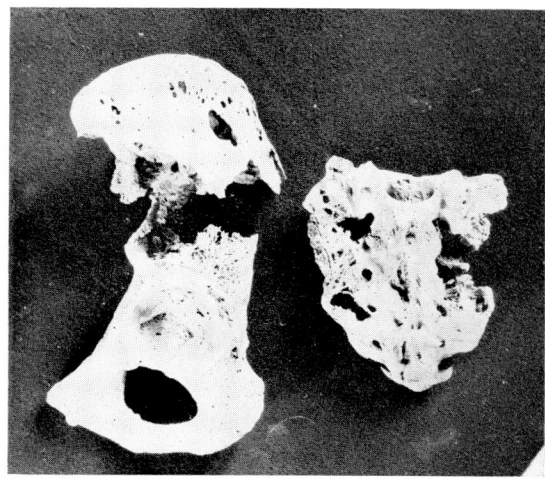

FIGURE 5. Lateral aspect of left os coxae and posterior view of sacrum. The loss of osseous material in the lower portion of the ileum and that in the right side of the sacrum are due to crushing in the grave, and do not represent a pathologic condition. However, the extensive involvement of the bones is evident. The first coccygeal segment has ankylosed with the sacrum.

open into a large cavity between the first and fifth sacral vertebrae, continuous with the posterior foraminae and with the sacral canal. The articular surface of the left side is much eroded, jagged and perforated, indicating that this joint formed no barrier to the extension of the disease process. A bending of the lower portion of the sacrum to the right may be due to long continued pathologic action.

Extensive changes are to be found in some of the ribs. The first left rib is about three times the size of its opposite in width and breadth, but its length is not increased. Its surface is somewhat eroded at the head, and the body is a mere cavity with thin cortical bony walls. The second, third, fourth and tenth ribs are unaffected. Of the remaining ribs of the left side it was possible to save only the seventh, ninth and eleventh. These are merely fragile laceworks of bone, with only the heads retaining their identity. As their condition is ineffable, one must rely on the accompanying illustrations. On the right side the ribs are unaffected, except for a small portion on the anterior surface at the angle of the fifth.

In the vertebrae the pathologic changes are similar to those described in the other bones. The seventh cervical vertebra has cavitous transverse processes and spine, with characteristic erosion and perforations of the surface. Like changes are also found on the left tranverse processes of the fifth to the tenth thoracic vertebrae, and the ninth, tenth and eleventh show involvement of the right transverse processes. The body of the eighth thoracic and that of the second lumbar vertebrae have been perforated by the pathologic process. An incipient perforation is seen in the body of the sixth thoracic vertebra.

The involvement of the skull by the disease is limited to contiguous portions of

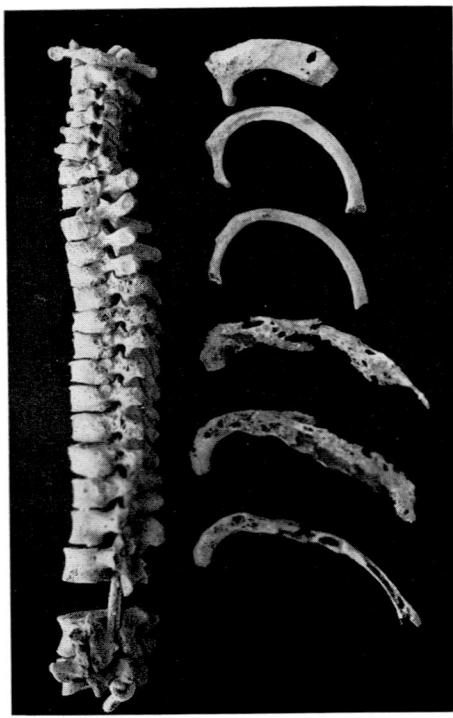

FIGURE 6. Vertebral column viewed from the left side; the third lumbar vertebra was broken, a space representing its location. At the right are the first, second, third, seventh, ninth and eleventh ribs of the left side, of which the second and third are normal.

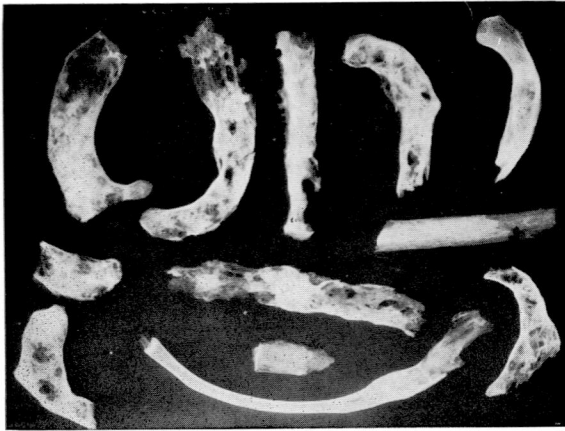

FIGURE 7. Roentgenogram of ribs, showing extensive cyst formation and rarefaction.

the left maxilla, palatine bone and pterygoid process of the sphenoid. The anterior portion of the palate is also affected, causing a dissymmetry of the nasal aperture on the left side. The palate here is 1.5 cm thick, but is of normal size posteriorly. It bulges somewhat below the normal patalal plate and presents an eroded surface. The pterygoid region on the left side is composed of a hard, porous enlargement of bone with only narrow ridges suggesting the pterygoid laminae. The base of the normal right pterygoid process measures 1.2 cm across, while the base of the affected side measure 2.5 cm, the pterygoid fossa being essentially eliminated. The palatine canal and other foraminae in this region are patent.

The foregoing material appears to me to indicate clearly that here for the first time is a case of osteitis fibrosa in a prehistoric American skeleton. Osteitis deformans has been encountered, but differs from osteitis fibrosa in important features, both in its effect on the living person and in the condition of the bones after death.[3]

Osteitis fibrosa is a multiple disease of bone characterized by resorption of bone and its replacement by a fibrous connective tissue, associated with formation of cysts. The disease very seldom appears after the age of forty and may occur before puberty. In recent years many cases have been found in association with parathyroid adenoma. The pathologic changes in the bones show: (1) extensive resorption of bone, which I have termed, macroscopically, erosion (2) irregular, malformed

---

3. There is a recent tendency to consider osteitis fibrosa, osteitis deformans and osteomalacia as a clinical group under the term of osteodystrophia fibrosa. However, from a palaeopathologic point of view it seems advisable to retain them as entities so that any remains of a doubtful pathologic nature may be more critically approached, and thus rescued from the oblivion imposed by an all inclusive term.

deposition of bone throughout the fibrous growth;[4] (3) formation of cavities associated with swelling beyond the normal contours of the bone; (4) rarefaction of cortical bone in the involved regions only, the cortical bone elsewhere being of normal thickness and consistency, and (5) a characteristic nodular appearance, as the disease possesses no uniformity of distribution or progress.

Roentgenograms of the entire skeleton have been made, and those of the bones described in detail substantiate the pathologic changes enumerated in foregoing paragraphs.[5] The left femur and ribs present the most characteristic features of the disease (see figs. 3 and 7). Involvement of the skull in the specimen is limited to the left side of the base as described, the roentgenograms showing nothing typical of the condition, the calvarium being normal in thickness and exhibiting no rarefaction.

Osteitis deformans, on the other hand, differs in the following characteristics: 1. As the bone marrow is transformed into fibrous tissue and ultimately into bone, it encroaches on the inner surface of the cortex increasing its thickness. 2. A similar change takes place on the outside through the subperiosteal deposition of layers of bone. Thus in osteitis deformans the cortex may become 1 cm or more in thickness, which characteristically is uniform throughout the length of the bone, and is seldom associated with the formation of cysts. In typical cases the thickness of the calvarium is increased four or more times the normal. Further, osteitis deformans is a disease of middle age, rarely beginning before the age of forty. These characteristic obviously do not obtain in the specimen described.

It is noteworthy that this case may be classed among the rarities because the involvement is so extensive, although practically unilateral.[6]

REFERENCES

GESCHICKTER, C. F. and COPELAND, M. M.: Osteitis fibrosa and giant cell tumor. *Arch. Surg.*, 16:169, 1929.

PAGET, J.: Brit. M. J., Dec. 16, 1882; *Tr. Med. Chir. Soc.* 14:37, 1877.

SUTHERLAND, C. G.: The differentiation of osteitis deformans and osteoplastic metastatic carcinoma. *Radiology*, 10:150, 1928.

WILDER, R. M.: Hyperparathyroidism: Tumor of the parathyroid glands associated with osteitis fibrosa. *Endocrinology* 13:231, 1929.

WILLIAMS, H. U.: Human paleopathology. *Arch. Path.* 7:839, 1929.

——— An ancient syphilitic skull. *Ann. M. Hist.*, 2:523, 1930.

---

4. It is understood that the replacement of bone by fibrous tissue can be seen only in fresh specimens, and it is during this active stage of the disease when fibrous tissue predominates, that the bending occurs.

5. Some of the bones that appear normal grossly show in the roentgenograms small areas of rarefaction, which may indicate an incipient process of the disease, but which are indistinguishable from the condition known as spotted atrophy. Such areas are found in the right calcaneus, the right and left astragalus, several of the ribs on the right side and the bodies of the sixth, eighth and ninth thoracic and in the second lumbar vetebrae. The acetabula, the heads of the femurs and a few of the vertebral bodies show minute osteo-arthritic changes which are negligible.

6. Since the foregoing observations were submitted for publication two prehistoric skeletons showing pathologic changes have been reported, for which the diagnosis of osteitis fibrosa has been suggested. The first, showing a bilateral involvement of the humeri only, was described by LEON PALES in his recent book, Paléopathologie et pathologie comparative, published in Paris by Masson & Cie, 1930. The second is reported by E. A. HOOTON, as follows: "Catalogue no. 60061, sex doubtful, aged 35-39: Diagnosis, (1) Spondylitis deformans; (2) Periostitis of mid shaft of left tibia. H. U. Williams, 'probably osteitis fibrosa,'" *The Indians of the Pecos Pueblo: A Study of Their Skeletal Remains*, New Haven. Yale University Press, 1930.

Chapter 29

Porotic Hyperostosis or Osteoporosis Symmetrica

J. LAWRENCE ANGEL

This change in marrow space and outer lamina of parts of the skull vault, sphenoid, orbit roof, face, and long bones has been recorded often in ancient Egyptians (Moodie, 1931; Wood-Jones and others quoted by Hamperl and Weiss, 1955), at Early Neolithic Catal Hüyük (Anatolia) and Nea Nikomedeia (Macedonia) in 7—6th millenia B.C. with 12 mm thick vault, in Bronze Age and later Greeks and Cypriotes (Angel, 1955, 1964, 1966, in Neel, 1951), in Etruscans, Italians, and other Mediterranean populations (Welcker, 1888), in Hindus and Indonesians (Welcker, 1888; Müller, 1935) in Ainu and Japanese (Koganei, 1894) and Chinese, in Peruvians (Welcker, 1888; Hrdlička, 1914; Williams, 1929; Hamperl and Weiss, 1955), Maya (Williams, 1929; Hooton, 1930), Arkansas Indians (Neel, 1951), Florida Indians in the Southeast and Pueblo Indians of the Southwest United States (Williams; Hooton op. cit.). It is frequent also in Negro Africans (Welcker, 1888; Williams, 1929) but occurs very rarely indeed in North Europeans (Gejvall, 1960). This distribution fits quite well the major pattern of P. falciparum malaria and the Old World occurrence of the thalassemias (Chernoff, 1959; Bannerman, 1961) and of sicklemia (Singer, 1962). But in the New World thalassemia is rare and apparently Post-Columbian (Lisker, 1962; Arends 1963) and the malarias were absent (Dunn, 1965), though other anemias could have occurred. It is logical to ask, therefore, whether thalassemia is typically the cause of this type of symmetrical overgrowth of spongy bone in ancient populations (Neel, 1951; Angel, 1964) and whether P. falciparum malaria was a prehistoric human disease (Jones, 1909; Ackerknecht, 1945) which could act as a prime selective factor for thalassemia (Bruce-Chwatt, 1965).

These two questions are important for our understanding of the ecology and disease adaptation of man in relation to his evolution especially during and after the Farming Revolution (Neolithic) when populations were beginning to become too dense for the food supply. I shall deal with these questions step by step, against the background of ancient society so far as known.

Porotic hyperostosis was first fully described by Welcker (1888) under the name Cribra orbitalia. He noted correctly its occurrence most often as cribrous areas in orbit roofs as well as on parietal and occipital bones. He stressed that the diploë is thickened or even created in an area (orbit roof) normally lacking it, and that the round or oval holes perforating it join sinuous hollows in the spongy bone, often branching, and often having microscopic columellae of bone reaching out through them. He notes the vessel-grooves which are most obvious in the orbit roof location and observes that the cribrous area "machen vollkommen den Eindruck, als ob sie durch die Gefässe geprägt und modelliert seien," hoping soon to confirm this possible causal relation with blood vessels by dissection of soft parts.

Hrdlička (1914) and then Hooton (1930) and Williams (1929) used the term osteoporosis symmetrica to describe the bony porosity visible in the skull vault as well as orbits, and recently Hamperl and Weiss (1955) introduced the term spongy hyperostosis both to stress the fact that the key skull change is an increase in spongy bone and to avoid the unfortunate semantic confusion with old age osteoporosis. I am using the term porotic hyperostosis in order to combine these two descriptions and fit the observed bony swelling and porosity seen in the long bones as well in fully developed examples of this disease.

In prehistoric and ancient children's skulls from Greece and Cyprus (Angel, 1964) the thickening of diploic marrow space is greater around the original ossification centers where likewise the perforations are circular, relatively cylindrical, and usually perpendicular to the outer table of the skull. Further away from the parietal boss, for example, the perforations are more angular and are more oriented along radiating lines out from the center, as they would be if formed by vessels attached to the growing osteogenic periosteum (Fig. 1). In cross section almost all the diploë appears to be new bone usually without traces of the original outer table as claimed and illustrated by Hamperl and Weiss (1955) for their Peruvians. Some Indian crania from North America (*cf.* U.S.N.M. 327107 from New Mexico) show a maze-like winding pattern of blood-vessel grooves, like coral formation.

There is wide variation in the skull bones affected; often the maxilla, zygomatic bone, greater wing of the sphenoid (and sometimes the orbital plate of the frontal bone) show an almost unbelievable increase in bone marrow space (Fig. 1) with little porosity of the surface.

There is very wide variation in the degree of porotic hyperostosis. This ranges from the extreme in which the whole skull vault shows thickening with radiating "brush" or "crew-cut" trabeculae and thinned outer cortex (Letterer, 1949; Hamperl and Weiss, 1955; Hooton [and Williams] 1930; Wintrobe, 1951) via the porous thickening previously described to skulls with parietals which show some thickening and a cicatricial irregular rayed appearance of the cortex (Sjövall 1934); this latter I consider trace or slight *healed* "osteoporosis" and find in adults rather than children. The extreme and medium degrees of "hyperostosis" occur almost always in young children, except in some Neolithic pre-Greek adults with vaults up to 19 mm thick.

A cribrous rayed appearance on the *inside* of the frontal bone occurs in some newborn Middle Bronze Age Greeks from Lerna. In more osteoporotic form this is mentioned by Williams (1929) as occurring also in ancient peoples and in modern Germans and Japanese children. This is clearly a different response because of the quite different reaction to periosteum in a growing skull. I don't know if this is the usual first stage of porotic hyperostosis; Baker (1964) says that in thalassemia bony changes are visible in the first six months, but not marked.

With the delay in osteoclastic activity, inadequate calcification and osteoid formation of rickets (Weinmann and Sicher, 1947, pp. 259-270) a considerable diploic hyperplasia occurs. But this is not inter-

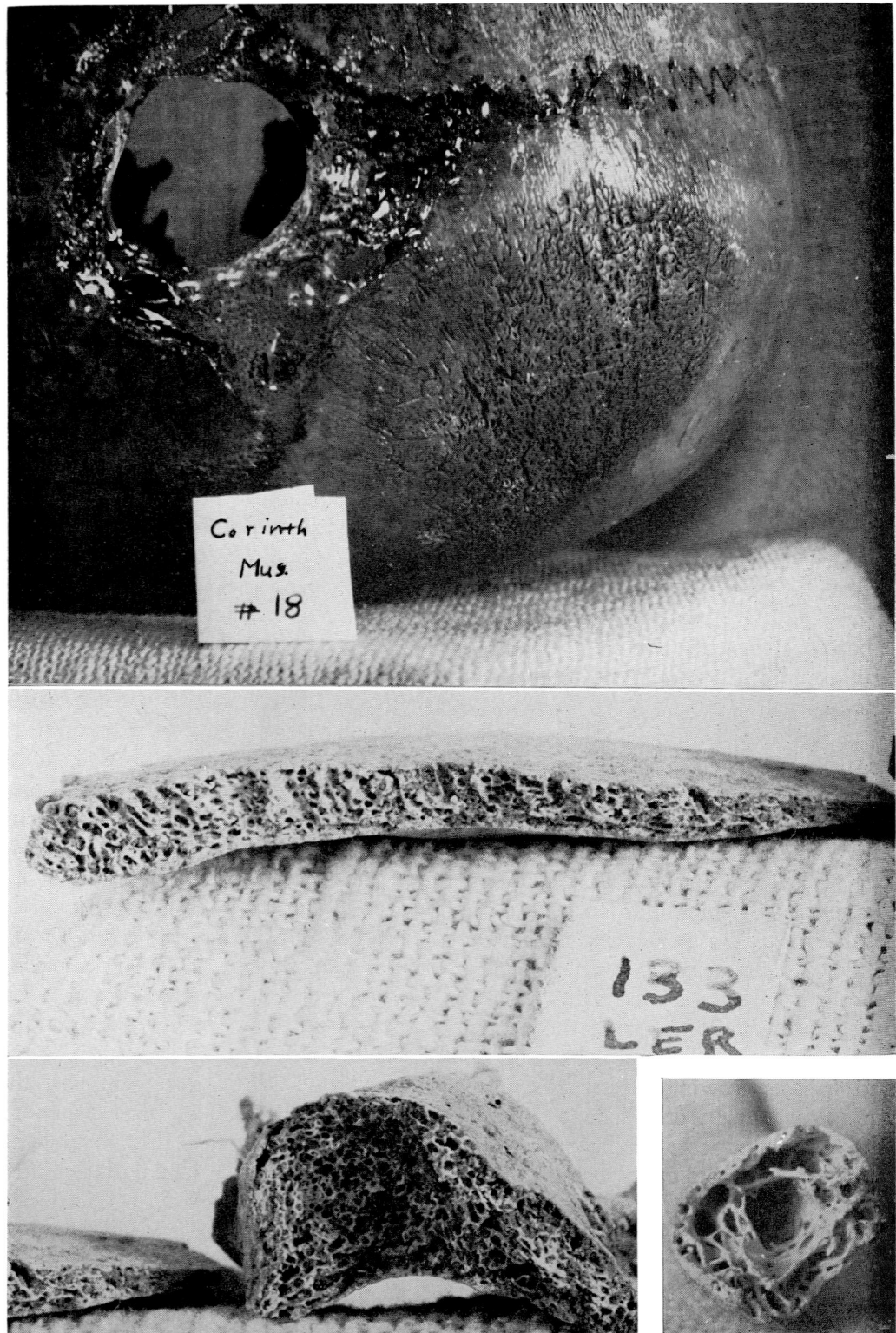

nally porotic; there is an excess rather than a thinning of bone in the skull marrow space; and the external appearance of added bone near the sutures likewise differentiates it from our spongy hyperostosis as typically seen (Letterer, 1949; Hamperl and Weiss, 1955), though in the healed state one might confuse the two conditions if only the skull vault is available (Hamperl and Weiss, 1955). It is in the long bones and ribs that the distinction is quite clear. Rachitic bones show peripheral enlargement of the growing areas with irregular periosteal bone, poorly calcified at first, lacking normal remodelling, and having inadequate epiphyseal length growth (Weinmann and Sicher, 1947). Long bones and ribs belonging to children with symmetrical porotic hyperostosis show a thin cortex, sometimes porous (cf. Hooton 1930, on Mayas), immensely enlarged marrow space with thin coarse trabeculae crossing it, or even a semicomplete inner shell representing the unremodelled bone at an earlier age as shown in Figure 2. This is clear in Bronze Age children from Lerna (Greece) and Bamboula (Cyprus), and of Early Neolithic Nea Nikomedeia (Angel, 1966).

This combined picture of enlarged marrow space and thinned or porous but active cortical bone growth fits exactly the X-ray and autopsy findings in severe childhood hemolytic anemia (Caffey, 1937; Wintrobe 1951, chap. 11; Letterer, 1949; Hamperl and Weiss, 1955; Dacie, 1960; Baker, 1964; Moseley 1965) in which the hyperactivity of bone marrow in producing red blood cells to compensate for the shortened life span of these cells forces a large increase in the medullary spaces and a matching increase in veins and arteries on the *active* periosteal areas (not the endocranial nor the mid-shaft bone areas, except in foetal life probably).

Of the possible varieties of anemia, thalassemia and sickle-cell anemia would seem to be the most likely choices because of the close coincidence in Old World geographic distribution (Chernoff, 1959; Livingstone, 1958; Singer, 1962) as well as the virtual identity in long bone and skull changes (Stein, Stein and Beller, 1955; Caffey, 1937; Hooton, 1930; Williams, 1929; Wintrobe, 1951; Baker, 1964; Moseley, 1965).

There are at least two apparent objections to this identification. First, there

---

FIGURE 1. *Upper half*: left parietal bone in top view of 128 C, a child of about two from Mediaeval Corinth (South Stoa West, 1938 excavation of C. Morgan and Doreen Canaday) to show marked degree of porotic hyperostosis.

Middle and lower left: sections through parietal bone and left zygoma of 133 Ler, a child just over two years old from Middle Bronze Age Lerna (Trench BE, Burial 27, 1956 excavation of J. L. Caskey and G. Bass) showing the great thickening of diploë and thin cortex in a vault bone and creation of thick spongy tissue in a face bone where it is normally completely absent.

Lower right: cross section through ulna of 40-3 CCB, a child of almost five years in Late Bronze Age Bamboula, at Episkopi in South Cyprus (Tomb 40, 1954 excavation of J. L. Benson) to show weblike trabeculae around unremodelled inner shell. All of these photographs illustrate effects of expansion of erythropoietic bone marrow.

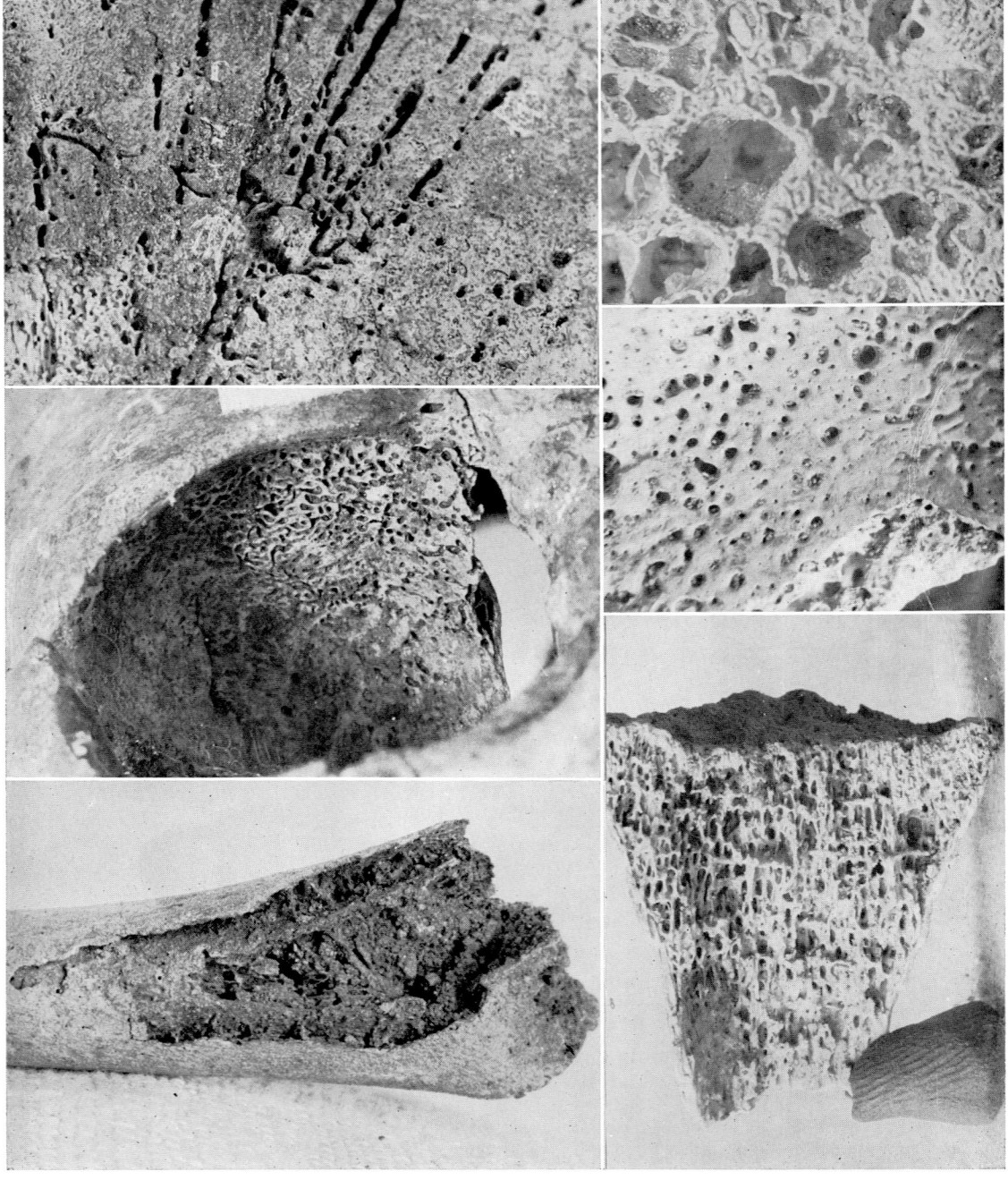

FIGURE 2. *Upper left*: left parietal bone in side view of 10 Ler, an infant less than one year old from Middle Bronze Age Lerna (Area A, Grave 11, 1952 excavation of J.L. Caskey and E. Caskey) showing the radial arrangement of porosities around the ossification center in the rapidly growing hyperostotic vault apparently for convenient rich connection with periosteal blood vessels.

are no autopsy findings or X-rays of heterozygotes for thalassemia or sicklemia. Presumably there are some bony changes. The frequency of spongy hyperostosis, "osteoporosis" or cribra orbitalia (Welcker, 1888; Koganei, 1894; Hrdlička, 1914; Hooton, 1930; Angel, 1955) of slight or trace degree in adults or older children ranges from 10-50 per cent or more and shows chronological changes (Table I) as would be expected with change in ecology and culture (*cf.* Livingstone, 1958) if this slight or "healed" cranial porosity does indeed identify thalassemia heterozygotes with their slight blood changes. Second, there are no ancient skulls or skeletons of young children which show the extreme "brush" or crew-cut skull pictured for modern thalassemia and sicklemia homozygotes.

The first objection demands autopsy checks of proven carriers of Ss or Tt genetic makeup; though it is quite proper to find a wide range of bone response to thalassemia since this term includes several different molecular changes in the α and β chains of the globin part of the hemoglobin molecule (Ingram, 1963; Bannermann, 1961), which affect rate of formation and hence lead to varying excesses of foetal hemoglobin. Hence the variation in thickening of marrow cavity and in cortical porosity, the much greater frequency of slight and trace than medium degrees in adults (Table I) and the limitation of marked changes to child skeletons all fit the hypothesis that spongy hyperostosis represents thalassemia or sicklemia. Yet Moseley (1965) shows that iron-deficiency anemia in late suckling children

---

Upper right: detail of perforations in outer table of parietal bone of U.S.N.M. 377454, a child of under two from Canaveral, Florida, to illustrate differences in size of perforations (for arteries and for veins?).

Middle left: right orbit roof of 103 Ler, a child of almost two and one half years from Middle Bronze Age Lerna (Trench BD, Burial 27, 1956 excavation of J.L. Caskey and E. Courtney) to show typical "cribra orbitalia" where blood flow through the orbit apparently exceeded that through the skull vault since the vault is thickened greatly but not porous.

Middle right: right orbit detail of 20a C (U.S.N.M. 349899b) a child from Early Bronze Age Corinth (Cheliotomylos shaft, 1930 excavation of T. L. Shear and F. O. Waage) to show spongy bone where it is normally absent and pattern of subperiosteal vessels.

Lower left: right humerus of 204 Ler, an infant under one year old from Middle Bronze Age Lerna (Area DE, Grave 64, 1956 excavation of J. L. Caskey and E. Caskey) to show original long bone shaft, untouched by osteoclasts, with new marrow cavity and spidery trabeculae formed outside it in order to increase marrow for red blood cell formation; the skull showed thickening and cribra orbitalia.

Lower right: coronal section through upper (knee) end of tibia of 40-3 CCB from Late Bronze Age South Cyprus, a child of almost five (Bamboula Tomb 40, 1954 excavation of J. L. Benson) to show lines of growth arrest indicating short periods of illness plausibly connected with erythroblastic anemia (thalassemia?) which probably was the cause of cribra orbitalia, greatly thickened and osteoporotic vault, and peculiar trabeculation of long bones as shown in Figure 1, lower right.

TABLE I

PERCENTAGE FREQUENCIES OF DIFFERENT DEGREES OF POROTIC HYPEROSTOSIS IN SKULLS OF ADULTS AND OF CHILDREN FOR SUCCESSIVE PERIODS IN GREECE IN ORDER TO SHOW CHRONOLOGICAL CHANGE POSSIBLY ASSOCIATED WITH MALARIA AND CHANGING ECOLOGY AND CULTURAL LEVEL

Period Date	Adults			Children			
	Trace	Slight-medium	N	Trace	Slight	Medium-extreme	N
Romantic 1800	33	4	199				0
Turkish 1400	35	10	51				3
Medieval 600	6	6	83			10??	10
Roman 120 A.D.	19	5	95		12		8
Hellenistic 300 B.C.	9	1	121		(20)		5
Classic 680	1	0	114	(12)			8
Early Iron Age 1150	4	0	92	10	5		21
Late Bronze Age 1450	7	1	210	2	2		44
Royal graves (Mycenae)	4	4	23				1
Middle Bronze Age 2000	7	6	124	2	4	4	49
L. Neolithic & Early Bronze Age 3500	14	10	52	17	17	17	12
Early Neolithic Macedonia	23	45	22	14	21	21	14

can duplicate the porotic hyperostosis of sicklemia. And there are no studies of possible marrow thickening in anemia produced by chronic malaria directly.

The second objection we can meet more easily. Neolithic or Bronze Age conditions of life for young children were severe: extremely anemic children would usually die before the disease reached the extreme stage possible with modern care and hospital treatment and transfusions. The heterozygotes too might have a shortened life span in some cases.

Ever since the identification of thalassemia by Cooley (Cooley and Lee, 1925) it has been clear that the disease, being inherited, varies in frequency between families in the areas where it occurs. Proof of the hypothesis that thalassemia depends on a recessive gene (Caminopetros, 1938; Chini and Valeri, 1949; Neel, 1951; Banton, 1951) meant that families producing one or more homozygous recessive anemic children must include heterozygous parents and a number of heterozygotes among relatives. But what maintained the frequency of such a harmful recessive gene? At the same time that Allison (1955), Lehmann (1959) and others were showing that sicklemia heterozygotes as such were for viscosity reasons somewhat more resistant to *P. falciparum* malaria than normal people, Cepellini (1959), Silvestroni and Bianco (1949, 1959), Neel (1951, 1953, 1959), Banton (1947), Fessas (1959) and others showed that in Italy, Sardinia, Cyprus and Greece the gene frequency for thalassemia was high in malarious areas and lower in nonmalarious areas. The lack of a thalassemia peak in the Rome area particularly disturbs this correlation. This may result from postmedieval migration into Rome or from inadequate sampling or most probably from some other protective gene. Thus, it is probable that thalassemia, like sicklemia, is a balanced polymorphic condition protective against malaria in the heterozygous state and that in such formerly malarious areas as the lower Po valley, the lowlands of Sardinia, the marshy south coast of Cyprus west of Limassol, or the occasionally swampy coast of the Bay of Argolis as well as Boeotia, the Macedonian plain, and other areas in Greece and the Konya plain in Anatolia, we might expect heterozygous parents to be slightly more fertile than normal people.

The selective effect must be strong, because the heterozygotes in thalassemia may be themselves somewhat anemic and less vigorous than normal (cf. Fraser *et al.*, 1964).

Can we expect to find evidence first that in Neolithic or Bronze Age settlements thalassemia (= porotic hyperostosis) showed an excess in certain families and, second, that these families were more fertile than others at the same site?

The clearest indication of such evidence comes from the Middle Bronze Age cemetery at Lerna, excavated by Jack Caskey from 1950-58 for the American School of Classical Studies at Athens. Here we have over 230 individual graves dug next to or under the houses occupied by various families. By combining adjacent grave-groupings I have arrived at a minimum number of thirteen "clans" or extended families. Several families plausibly contain immigrants, as indicated by actual pottery (not imitations) from Crete, Central Cyclades and Western Balkans, but there is no way to distinguish families of unmixed pre-Greeks (if they exist) since the intrusions of Greek-speakers had occurred a few generations earlier than the bulk of the Middle Bronze Age graves. But the population is clearly heterogeneous, with every probability that some came from nonmalarious areas and others (autochthonous, Anatolians, Macedonians and Cretans) from plausibly malarious areas.

We can divide the families up into four groups on the basis of porotic hyperostosis: (a) those with none (three families); (b) those with porotic hyperostosis in children and infants only (three families); (c) those in which it occurs in both children and adults (five families), and (d) those in which it is limited to adults. The major difficulty in demographic study of ancient populations is the estimate of fertility, which obviously cannot be safely inferred from the number or proportion of dead infants and children. It is possible to classify female pelves according to the roughenings, bone erosion, lipping and areal extension of the places of ligament attachments around the sacroiliac joints and especially the pubic symphysis; such changes are virtually limited to female pelves and presumably reflect the mechanical stress effects of pregnancy and childbirth, though clearly the individual response to such stress will be far from

TABLE II

DEATHS AND ESTIMATES OF FERTILITY AS RELATED TO OCCURRENCE OF POROTIC HYPEROSTOSIS (THALASSEMIA?) IN DIFFERENT CLASSES OF TOMB GROUPS ("FAMILIES?") IN THE MIDDLE BRONZE AGE CEMETERY AT LERNA, IN GREECE

	Porotic hyperosotosis occurring in:				
	None	Children only	Children & adults	Adults only	Total
	a	b	c	d	
Infants (0-1)	18	27(6)	32(11)	7+1	84+1(17)
Children (2-14)	7	19(2)	17(2)	5	48(4)
Male adults (15-x)	7	7	32(6)	9(2)	55(8)
Female adults	9+1	7+2	25+2(3)	6	47+5(3)
Births, estimated from changes in bony pelvis	48	51	149	18	266
Births per female	4.8	5.7	5.5	3.0	5.1
Survivors per female	2.3	.6	3.7	.8	2.6—
Number of tomb groups	3	3	5	2	13
Number of skeletons	4.1+1	60+2(8)	106+2(22)	27+1(2)	234+6(32)

NOTE: the number of instances of porotic hyperostosis in any category is shown in parentheses immediately after the number of skeletons in that category. Exact incidence is uncertain since 84 skeletons have fragmentary skulls or none, especially among the infants and children. Lack of enough females to account for dead juveniles in 3 family groups and too few dead infants in one group are compensated, e.g. " + 2."

constant. Yet these are the only physical indices we have for an estimate of fecundity.

Using this uncertain index as seen in Table II we note that the number of children born per woman is (a) five, (b) six, (c) five, and (d) three—in the four types of family, suggesting a decrease rather than increase in childbearing with adult spongy hyperostosis, logical if these represent the more anemic carriers of thalassemia (c, d); and infecundity from malaria (a, d). The estimated number per woman of children who reached adulthood (i.e., who were not buried) in the four groups is (a) two, (b) half, (c) four, and (d) one. This reflects the excess deaths of children with severe hyperostosis (b, and d ?), presumably homozygotes, and probably a greater viability of children lacking the full disease but born of heterozygous parents (c). This would provide the extra fertility needed to balance the loss of children in groups b and d. Presumably groups b, c and d all have heterozygous parents (though fewer or healthier in group b where none were identified in the cemetery). Together these groups produced 2.6 "live" children per woman, exceeding the average of the "nonthalassemic" group a. These data are too approximate to show if this excess fertility (.3 per woman) is statistically valid. Yet these data parallel those of Fraser et al. (1964), where parents are properly known.

These data from Lerna point to the kind of balance which would maintain (but not increase) the frequency of a balanced polymorphic gene such as that for thalassemia and to the kind of kinship segregation expected in a heterogeneous population. But there is no proof of active malarial selection for thalassemia if we accept the hypothesis of identity with hyperostosis. Table I suggests, indeed that hyperostosis and malaria had become stabilized in the preceding millennium, and that a decrease may have been starting, perhaps because of improved drainage and farming methods.

If in pre-Pleistocene times man's ancestors had developed tolerance for and symbiosis with *Plasmodia malariae* and *vivax* (Dunn, 1965), his intolerance for *Plasmodium falciparum* shows its mutational evolution to be quite recent, so that Gatto's (1960) and Zaino's (1964) Mediterranean theory of origin for the thalassemias may be accurate, even though Zaino's details of timing depend partly on the incorrect assumption that American Indians brought both malaria and thalassemia with them across the Bering Straits about 20,000 B.C. If the thalassemias, sicklemia and other abnormal hemoglobins and also favism (G6Pd deficiency) all depend on a new and almost lethal *falciparum* parasite then these protective mutations could have occurred and started to increase in anopheline foci around the Eastern Mediterranean by Upper Palaeolithic times: the area of greatest mutational variety is a logical place to look for the origin of these polymorphisms and their first balance against falciparum malaria. The tendency of the first farmers to settle in marshy areas (soft soil) like the Konya and Macedonian plains would explain the extremely high development of porotic hyperostosis (thalassemia and sicklemia presumably) in skeletons at Catal Hüyük and Nea Nikomedeia as contrasted with those at Khirokitia (dry valley), Kephala on Kea (rocky headland), Karatas-Semayük (Lycian mountain valley). Greek mythology (Jones, 1909) and archaeology record extensive marsh-drainage in the Bronze Age, not to mention improved irrigation methods. The Roman Empire increase in malaria goes with breakdown of really careful individual farming and likewise parallels the data of Table I.

There is no reason why thalassemia might not have developed in response to some other parasitic disease, specifically amebiasis. It is likewise most probable that where thalassemia does not occur in modern populations, as apparently in Central and South American highland areas, the development of hyperostosis may follow the need for extra red cell reserves in rapid growth during childhood or at puberty (personal communication, Paul Baker). We need further and more careful study of the various possible patterns of bone response to increased activity of blood-forming marrow in order to distinguish between possible causes. Hookworm is a possible cause of anemia in the New World.

REFERENCES

ACKERKNECHT, E. H., 1945: The History of Malaria. *Ciba Symposia*, 7:51.

ALLISON, A. C., 1955: Aspect of Polymorphism in Man. *Cold Spring Harbor Symposium Quant. Biol.*, 20: 239.

ANGEL, J. L., 1951: Population Size and Microevolution in Greece. *Cold Spring Harbor Symp. Quart. Biol.*, 15:343.

——— 1955: Human Biology, Health, and History in Greece. *Yearbook of Amer. Phil. Soc.*, 1954:168.

———1960: Physical and Psychological Factors in Culture Growth. *Fifth Internat. Congr. of Anthr. and Ethn. Sci.*, Philadelphia 1956; 666-670 E. F. C. Wallace ed., University of Pennsylvania Press.

——— 1964: Osteoporosis: Thalassemia ?, *Amer. J. Phys. Anthrop.* N. S. 21:369.

——— 1966: Porotic hyperostosis, anemias, malarias and marshes in the prehistoric Eastern Mediterranean. *Science*, 153:760.

ANGEL, J. L., in press: Late Bronze Age Cypriotes from Bamboula. In J. L. Benson, Bamboula. Philadelphia, University Museum.

ARENDS, T., 1963: Current Status of the Study of Abnormal Hemoglobins in Venezuela. *Sangre* (Barcelona) 8:1.

BAKER, D. H., 1964: Roentgen manifestations of Cooley's anemia. *Annals N.Y. Acad., Sci.* 119:641.

BANNERMAN, R. M., 1961: *Thalassemia*. N. Y., Grune and Stratton.

BANTON, A. H., 1951: A Genetic Study of Mediterranean Anemia in Cyprus, *Amer. J. Hum. Genet.* 3:47.

BLUMBERG, B. S., 1961: *Proceedings of the Conference on Genetic Polymorphisms and Geographic Variations in Disease*. New York, Grune and Stratton.

BRUCE-CHWATT, L. J., 1965: Paleogenesis and paleoepidemiology of primate malaria. *Bull. Wld. Hlth. Org.* 32:363.

CAFFEY, J., 1937: The Skeletal Changes in the Chronic Hemolytic Anemias (erythroblastic anemia, sickle cell anemia and chronic hemolytic icterus). *Amer. J. Roentgenol.* 37:293.

CAMINOPETROS, J., 1938: Recherches sur l'anémie erythroblastique infantile des peuples de la Mediteranée orientale. *Ann. Méd.,* 43:27.

CEPELLINI, R,. 1959: Blood Groups and Haematological Data as a Source of Ethnic Information. In G. E. W. Wolstenholme and C. M. O'Connor, eds. *Medical Biology and Etruscan Origins*. Ciba Symposium. Boston, Little Brown.

CHERNOFF, A., 1959: The Distribution of the Thalassemia Gene. *Blood* 14:899.

CHINI, V., and C. M. VALERI, 1949: Mediterranean Hemopathic Syndromes. *Blood,* 4:989.

COOLEY, T. B., and P. LEE, 1925. Series of Cases of Splenomegaly in Children with Anemia and Peculiar Bone Changes. *Tr. Am. Pediatr. Soc.,* 37:29.

DACIE, J. V., 1960: *The Hemolytic Anemias. Part I. The Congenital Anemias,* 2nd ed. New York, Grune and Stratton.

DUNN, F. L., 1965: On the antiquity of malaria in the western hemisphere. *Hum. Biol.* 37:385.

FESSAS, PH., 1959: The Hereditary Anemias in Greece. In. J. H. P. Jonxis and J. G. Delafresnaye, eds. *Abnormal Hemoglobins*. Oxford, Blackwell, pp. 260-266.

FRASER, G. R., G. STAMATOYANNOPOULOS, C. KATTAMIS, D. LOUKOPOULOS, B. DEFARANAS, C. KITSOS, L. ZANNOSMARIOLEA, C. CHOREMIS, P. FESSAS, A. G. MOTULSKY, 1964: Thalassemias, abnormal hemoglobins and glucose-6-phosphate dehydrogenase deficiency in the Arta area of Greece: diagnostic and general aspects of complete village studies. *Annals N.Y. Acad. Sci.* 119:415.

GABALDON, A., 1949: Malaria Incidence in the West Indies and South America. Ch. 31 in M. F. Boyd, *Malarialogy.,* 1:764. Philadelphia, Saunders.

GATTO, I., 1960: Origine della thalassemia. *Proc. of the 7th Int. Congr. Internat. Soc. of Hematology,* Rome II. Pensiero Scientifico 3:413.

GEJVALL, N-G., 1960: *Westerhus. Mediaeval Population and Church in the Light of Skeletal Remains*. Håkan Ohlssons Boktrykkeri, Lund.

HAMPERL, H., and P. WEISS, 1955: Über die spongiose Hyperostose an Schädeln aus Alt-Peru. *Virchow's Archiv* 327:629.

HARRIS, H. A., 1933: *Bone Growth in Health and Disease*. London, Oxford University Press.

HOOTON, E. A., 1930: *Indians of Pecos*. New Haven, Yale University Press. xxvii and 391 pp.

HRDLIČKA, A., 1914: Anthropological Work in Peru in 1913, with Notes on the Pathology of the Ancient Peruvians. *Smithson. Misc. Coll.* 61 (18). Smithsonian Institute, Washington.

INGRAM, V. M., 1963: *The Hemoglobins in Genetics and Evolution*. Columbia University Press, New York.

JONES, W. H. S., 1909: *Malaria and Greek History*. University of Manchester Historical Series No. VIII. University Press, Manchester.

KOGANEI, Y., 1894: Beiträge zur physischen Anthropologie der Aino I. Untersuchungen am Skelett. *Mitt. Med. Facultät Kaiserl.-Japanischen Univ.* 2. Tokio. 250 pp.

LEHMANN, H., 1959: The Maintenance of the Haemoglobinopathies at High Frequency. in J. H. P. Jonxis and J. F. Defafresnaye, eds., *Abnormal Hemoglobins*, Oxford, Blackwell, pp. 307-321.

LETTERER, E., 1949: Über den "Bürstenschädel" und seine Bedeutung. *Zentralblatt Allgem. Pathol. Pathol. Anat.,* 85:244.

LISKER, R., 1962: Estudios sobre algunas caracteristicas genéticas hematológicas en la población mexicana. II: Frecuencia de hemaglobinas anormales en México. *Gaceta Médica Mexicana* 93:289.

LIVINGSTONE, F. B., 1958: Anthropological Implications of Sickle Cell Gene Distribution in West Africa. *Amer. Anthrop.* 60:533.

MELLANBY, SIR E., 1950: *A Story of Nutritional Research. The Effects of Some Dietary Factors on Bones and the Nervous System. The Abraham Flexner Lectures Series No. Nine*. Williams and Wilkins, Baltimore.

MOODIE, R. L., 1931. Roentgenological studies of Egyptian and Peruvian mummies. *Mem. Field Mus. Nat. Hist., Chicago,* 3:66 pp.

MOSELEY, J. E., 1965: The palaeopathologic riddle of "symmetrical osteoporosis." *Am. J. Roentgenol., Rad. Ther. and Nucl. Med.* 95:135.

MOTULSKY, A. G., 1960: Metabolic Polymorphisms and the Role of Infectious Diseases in Human Evolution. *Human Biol.* 32:28.

MÜLLER, H., 1935: Osteoporosis of the Cranium in Javanese. *Amer. J. Phys. Anthrop.,* 20:493.

NEEL, J. V., 1951: The Population Genetics of Two Inherited Blood Dyscrasias in Man. *Cold Spring Harbor Symp. on Quant. Biol.,* 15:141.

——— 1953: Haemopoietic System. Ch. 24 in A. Sorsby, ed., *Clinical Genetics,* pp. 446-475. London, Butterworth.

——— 1959: Genetic Aspects of Abnormal Hemoglobins. in J. H. P. Jonxis and J. F. Delafresnaye, eds., *Abnormal Hemoglobins,* pp. 158-180. Oxford, Blackwell.

SILVESTRONI, E., and I. BIANCO, 1949: Microcytemia, Constitutional Microcytic Anemia, and Cooley's Anemia. *Amer. J. Hum. Genet.,* 1:83.

SILVESTRONI, E., and I. BIANCO, 1959: The Distribution of Microcythaemias (or Thalassemias) in Italy. in J. H. P. Jonxis and J. F. Delafresnaye, eds., *Abnormal Hemoglobins.* Oxford, Blackwell.

SINGER, R., 1962: The Significance of the Sickle Cell in Africa. *The Leech,* 32:152.

SJÖVALL, E., 1934: Ein Vergleich zwischen der symmetrischen Osteoporose des Schädel und dem Reparationsstadium des Kephalhämatoms. *Kungl. Fysiogr. Sällskapets i Lund,* Forhandlingar V, No. 6.

STEIN, I., STEIN, R. O. and M. L. BELLER, 1955: *Living Bone in Health and Disease.* Philadelphia, Lippincott.

WEINMANN, J. P., and H. SICHER, 1947: *Bone and Bones. Fundamentals of Bone Biology.* St. Louis, Mosby.

WELCKER, H., 1888: Cribra Orbitalia, Ein ethnologisch-diagnostisches Merkmal am Schädel mehrerer Menschenrassen. *Archiv Anthrop.,* 17:1.

WILLIAMS, H. U., 1929: Human Palaeopathology, with Some Original Observations on Symmetrical Osteoporosis of the Skull. *Archiv. Pathol.* 7:839.

WINTROBE, M. W., 1951: *Clinical Hematology.* 3rd ed. Philadelphia, Lea and Febiger.

ZAINO, E. C., 1964: Paleontologic thalassemia. *Annals N.Y. Acad. Sci.* 119:402.

Chapter 30

A New Approach to Palaeopathology: Harris's Lines

CALVIN WELLS

THE comparative morphology of early historic and prehistoric Man has long been a major concern of the physical anthropologist. Its basic technique is well known: the available skeletons are submitted to the tedium of description, measurement and statistical analysis in order to estimate the peculiarities of the group, their range of variation and the relationship, if any, between them. It is a technique which has undoubtedly borne abundant and valuable fruit, though whether the yield has been commensurate with the vast labours devoted to it seems more open to doubt. This traditional approach to the study of early populations has always suffered from the static nature of the information it provided. Today, most physical anthropologists feel that a more dynamic understanding of mankind is needed, hence the increasing attention which is being given to comparative physiology, problems of maturation, aging and growth studies. This is the attitude towards modern populations, at least: with early communities such an approach is more difficult. Even when dealing with mummies the physiology of a group remains almost wholly inaccessible. We may sum up the situation by saying that we can see an early individual at only one moment of his existence—the moment of death. It is a snap-shot, not a moving picture. It is true that from the structure and pathology present in the skeleton at that moment much can be inferred about what led up to the final appearance, but that is quite different from being granted a diachronic or perspective view of a person at several stages of his life history.

It was in an attempt to break through this barrier and to gain a more dynamic view of some early populations that a new application of radiography was devised (Wells, 1961).

MECHANISM OF BONE GROWTH

During childhood and adolescence a typical long bone grows in a way which, stripped to its essentials, is as follows. At each end of the main shaft of the bone is a detached cap—the epiphysis—separated from the shaft by a zone of cartilage. Elongation occurs as a result of this "growth cartilage" depositing new bone on to the shaft more or less as a stalagmite is built up. But eventually the epiphyses fuse with the rest of the bone after which no further increase in length can take place. If during the growing period a child is attacked by some illness this orderly process of development may be interrupted for an indefinite time. Growth may be slowed or even stop altogether. With recovery, it starts again but when it does so a transverse line of calcification is left in the bone (Fig. 6) adjacent to the growth cartilage, as a record of what has

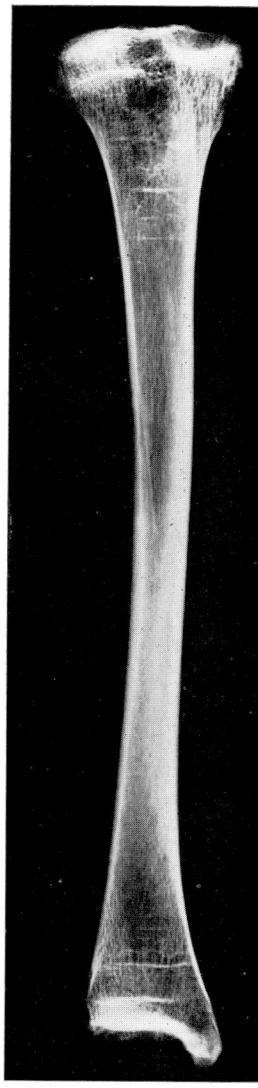 

FIGURE 6. An early Saxon tibia showing a strong Harris's line about an inch away from the ankle joint. Radiograph by Ilford, Ltd.

FIGURE 7. A Late Saxon tibia. About seven Harris's lines are present at the distal end. Proximally they have been lost owing to post-inhumation damage to the bone. Radiograph by Ilford, Ltd.

happened (Park and Richter, 1953). These lines resemble, therefore, the well known "hunger bars" found on the primaries and tail feathers of hawks and other captive birds. These "bone scars" were first intensively studied by H. A. Harris (1933) after whom they are named, although they had been recognized macroscopically even before the discovery of X-rays (Wegner, 1874). If repeated illnesses—or, apparently, episodes of starvation—attack the child a series of Harris's lines will be laid down and will record with some precision the amount of morbidity suffered during the growing period (Fig. 7).

TABLE I

Site	Description	Average Number of Lines Per Bone (Index of Morbidity)	n
Crichel and Shrewton, (Dorset)	Bronze Age Round Barrow inhumations. Probably pastoralists	0.8	17
Shouldham, (Norfolk)	12th-15th century Gilbertine Priory	1.6	23
Red Castle, Thetford, (Norfolk)	Late Saxon—possibly of Frisian or Flemish origin	1.8	42
Eriswell, (Norfolk)	Early Saxon	2.1	34
St. Catherine, Thorpe, Norwich, (Norfolk)	Late Saxon	2.2	45
Burgh Castle, (Suffolk)	Middle Saxon	2.6	226
Thornham, (Norfolk)	Early Saxon	3.4	36
Caister-by-Yarmouth, (Norfolk)	Middle Saxon	5.1	139

RESULTS: COMPARATIVE MORBIDITY

Here, then, is a method not only of estimating the sequence and incidence of disease processes in one individual but also of comparing morbidity rates between different populations. This can be done by calculating the average number of lines occurring in a selected bone from a sample of each population. Table I gives the results for eight groups based on radiographs of tibiae.

These findings are striking and unexpected because all the radiographed bones were normal to the naked eye. The differences call for some explanation in terms of genetic background, environment, diet, habitation, mode of life, etc. This variation in childhood morbidity can be further emphasized by noting the percentage of individuals in each group who show no lines at all.

It is interesting to see that the figures in Table II are almost inversely related to the

TABLE II
ABSENCE OF LINES

Site	% Individuals Without Any Harris's Lines
Shouldham	60.0
Crichel and Shrewton	50.0
Thetford	42.8
Thorpe	33.3
Eriswell	32.4
Burgh Castle	26.7
Thornham	22.5
Caister-on-Sea	10.7

Index of Morbidity in Table I, and that in the Caister group, with the highest number of lines, only one person in ten got through his childhood without having an illness severe enough to arrest growth.

It is important to recognize that, although the presence of a Harris's line reveals that some pathological process has occurred, it gives no clue as to its nature.

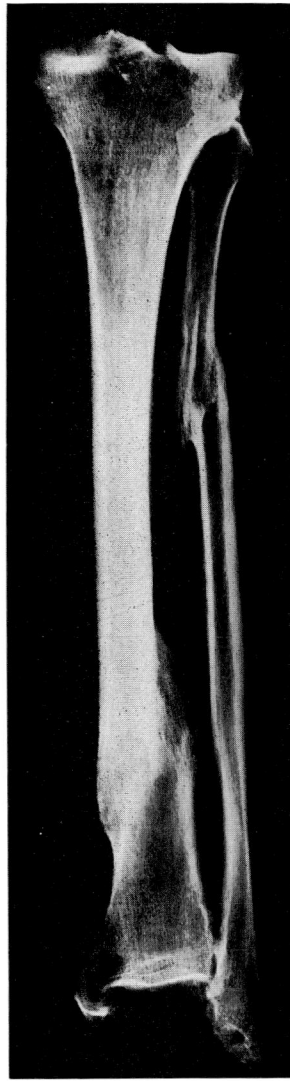

FIGURE 8. The left tibia and fibula from the same individual as Figure 7. Both bones have sustained a severe fracture and as a result almost all their Harris's lines have been removed. One faint remnant of a line can still be seen in the proximal end of the tibia. Radiograph by Ilford, Ltd.

At present it does not even seem possible to distinguish between a line due to a short febrile illness such as measles and one that results from a three month famine. Possibly lines produced by scurvy are an exception to this rule. Moreover, the use of this radiological technique to count Harris's lines assumes that once a bone scar has been formed it remains permanently in the bone and can be detected at any subsequent time. In general this is true but there are a few exceptions. If a limb has been immobilized for some months by a fracture or from poliomyelitis the resulting osteoporosis may leach out some or all of the lines on the affected side and when, with return of function, normal density is restored they will not be redeposited (Fig. 8).

The distance of any Harris's line from the end of a bone indicates the diaphyseal length of the bone at the time the line was produced and from this it is possible to estimate the age of the child at that time. Thus we have, registered in the bone, a record of the major illnesses which have attacked the child in the first fifteen or sixteen years of life, together with the age at which they occurred. This gives a sequential or diachronic view of an individual's health which does not seem to be readily obtainable in any other way. It can also be used to distinguish one group from another even if their Index of Morbidity should happen to be identical. After counting the total number of Harris's lines laid down in each year of life from birth to fifteen it is a simple matter to calculate what percentage of all lines occurs at any age. In Figure 1 the distribution for Burgh Castle and Caister is shown.

This shows a remarkable difference between the two populations. A higher rate of morbidity is found at Caister for each year until the age of ten, when it is surpassed by Burgh Castle, the rate from which reaches its maximum at eleven years, whereas the peak for Caister is at nine years. At eight years the Burgh Castle children register only 6 per cent

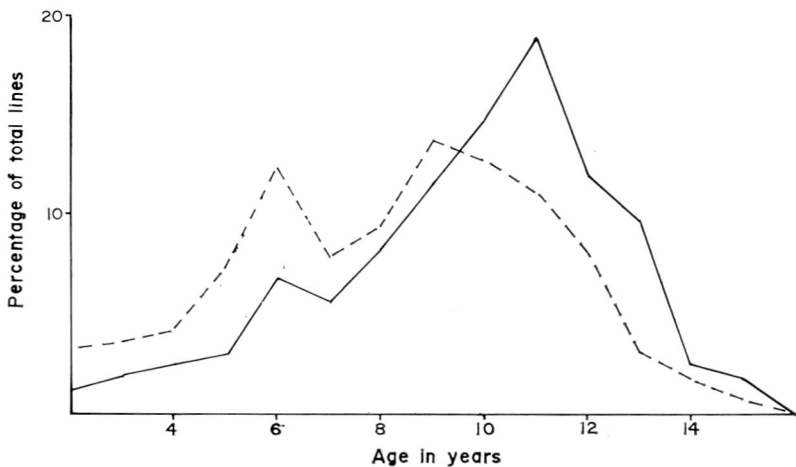

FIGURE 1: BURGH CASTLE AND CAISTER-ON-SEA.
Percentage of total lines occurring at each year of age
——— = Burgh Castle   ---- = Caister.

	2	3	4	5	6	7	8	9	10	11	12	13	14	15	16	
Burgh =	5	6	8	10	24	20	29	41	52	67	43	35	9	6	0	= Lines
Caister =	10	11	13	23	38	24	29	42	39	35	26	10	6	3	0	= Lines

of their total morbidity in contrast to 12 per cent for Caister children of that age. By thirteen years these differences are reversed: at Burgh Castle 10 per cent of their morbidity is affecting them whilst only 3 per cent falls on the Caister group.

Although these percentages are expressed on a yearly basis it is probably unduly optimistic to fix with such precision the age at which any line was formed: there are practical difficulties and theoretical uncertainties here and a less vulnerable way of presenting the two communities is shown in Table III where the percentages are gathered into three lustra: 2 — 6 + years, 7 — 11+ and 12 — 16.

The results are significant. In addition to the higher morbidity of the Caister people (as is already known from the lines-per-bone count) they are found to have twice the Burgh Castle percentage of their morbidity in the first lustrum, rather less in the second and only about a half in the third. In other words the Caister group gets a much higher proportion of its ailments before the age of seven, and much less after the age of twelve, than the Burgh Castle people. If each population is divided into no more than two phases, early and late childhood, it is found that Caister had 61.5 per cent of its total morbidity in the 2 — 9 + period and only 38.5 per cent thereafter. For Burgh Castle the equivalent figures are 40.3 per cent and 59.7 per cent.

The result of taking the sexes separately

TABLE III
LINES PER LUSTRUM

Site	Percentage of Total Lines Occurring at Age:		
	2—6+	7—11+	12—16+
Burgh Castle	16.1	61.7	22.2
Caister	32.0	55.4	12.5

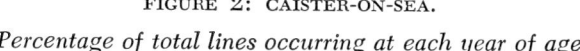

FIGURE 2: CAISTER-ON-SEA.

Percentage of total lines occurring at each year of age

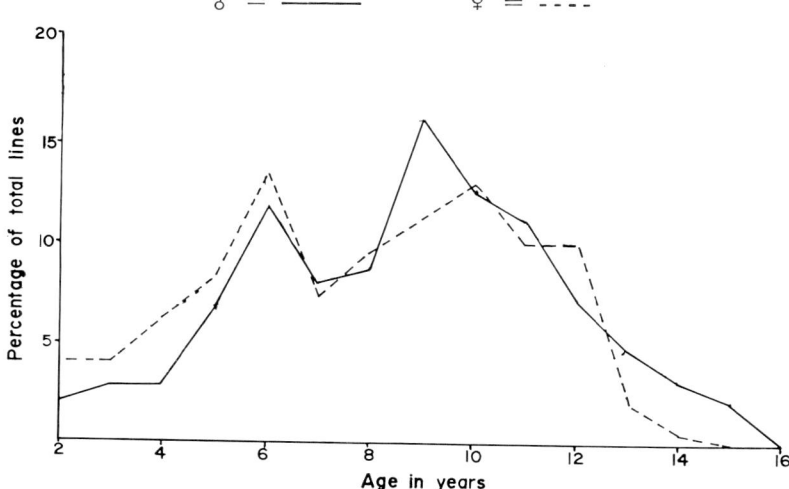

for one of these communities is shown in Figure 2 where it is seen that the girls have a higher percentage of their morbidity occurring in the early years of life.

With the males and females split the reduction in numbers of lines occurring at each year of age becomes statistically weaker and offers a strong incentive to group them by lustra as was done for the total group. The result of doing so is shown in Table IV.

TABLE IV
CAISTER: HARRIS'S LINES AND AGE

Sex	Percentage of Lines Occurring at Age:			Numbers of Inhumations
	2–6+	7–11+	12–16+	
Male	26.5	58.3	15.2	27
Female	39.2	50.0	10.8	28

No doubt this sex difference admits of various interpretations. Almost all anthropologists agree that an outstanding functional difference in ancient populations is the earlier age at death of the females as compared with the males. This is commonly assumed to be due to the hazards of parturition in the absence of skilled obstetric intervention but this explanation is not a wholly satisfactory one and there is evidence that other factors play a part. Specifically, for these Saxon groups I would suggest that, if they are considered in terms of social structure, they emerge, to the best of our knowledge, as strongly patrilineal, patrilocal and patriarchal; devoted to an ethos of male superiority and dominance; and pursuing a culture in which the highest status and prestige values were accorded to the role of the warrior, the sea-rover and the man of brawn. With this as the cultural ideal it is but a short step to recognize that boys would be more highly prized than girls and that from infancy onward they may well have been more favoured, probably getting the lion's share of the food (at least when supplies were short) whilst their sisters, from an early age, had to accept a lower status, to content themselves with the leftovers of the repast and accustom themselves to being the undernourished drudges of their menfolk. Some such organization within the family group may explain the apparent fact that the girls start getting their illnesses at an earlier age than the boys, despite the inherent advantage conferred by their XX chromosome pattern.

DENTAL CARIES

Because the whole point of this technique is to recover information not available in any other way it is, therefore, not easy to correlate it with other aspects of a group's pathology. Nevertheless, some attempt can be made to do so. An obvious starting point is the dentition, with special reference to the caries rate. It might reasonably be assumed that a high average number of lines per bone in a community, reflecting as it does a high incidence of morbidity in childhood would be associated with a high rate of dental caries.

Table V shows the only evidence so far available.

TABLE V
LINES PER BONE AND CARIES RATE

Site	Surviving Teeth	Number Carious	% <2.0	Lines Per bone <2
Crichel and Shrewton	85	0	0.0	0.8
Thetford, Red Castle	638	10	1.5	1.9
			>2.0	>2
Thorpe, St. Catherine	710	28	3.9	2.2
Thornham	563	18	3.2	3.4
Thetford, St. Mary	879	19	2.1	3.6
Caister	1614	47	2.9	5.1
Shouldham	215	16	7.4	1.6

From this it is clear that there is no point by point correlation between the line frequency and the caries rate of these populations but if they are split into two broad categories, those with fewer and those with more than two lines per bone, we find that six of the seven groups are similarly divided into those with less and those with more than 2.0 per cent of dental caries. For the remaining group, Shouldham, the correlation fails completely. It has one of the lowest rates of lines per bone but the highest of all the caries rates. So far not enough populations have been examined to know how to interpret these results. It may be that the hint given by the other six groups would prove sterile on further investigation. On the other hand there may be a broad correlation between line frequency and dental decay in early populations but for some reason the Shouldham group is an exception to the rule. Perhaps their conventual life had some bearing on what is found, either as a result of peculiarities of diet or of some other unknown factor. It may be relevant to remember that this is the latest of the communities discussed here and they probably lived at a time of worse climatic conditions than any of their predecessors.

Clearly, we can arrive at no firm conclusion from the evidence of Table V. The facts should make us alert to the possibility of a positive correlation between dental disease and the general morbidity of childhood but they do not carry proof. At least it can be said that the evidence, as far as it goes, does no violence to common sense.

Some further clues can, however, be quarried from this material. In Table I the Index of Morbidity is seen to be 5.1 for the Caister people. This can be split for the sexes, the values being 5.3 for males and 4.9 for females—a difference which is not significant. In view of this it might possibly be expected that closely similar dental decay rates would exist for both the men and women. In fact the incidence of caries is 1.9 per cent in males, 5.0 per cent in females.

The explanation of this is probably to be found in Figure 2 and Table IV where it was seen that despite a similar total line frequency for both sexes there is a considerable difference in the distribution of lines by age. The boys have only a quarter of all their lines before the age of seven whilst the girls have 40 per cent or half as much again. This hints that, at least within any one community, it is not so much the total number of illnesses which affects an individual's teeth as the time at which they occur. And at Caister the heavy incidence of disease in early childhood amongst the girls occurs precisely during those years when the formation of dental tissue is most active and, therefore, when the teeth are at their most vulnerable. Whereas the boys have a better chance to consolidate the formation of their teeth in the first half of childhood before their peak of morbidity hits them at a later and less sensitive stage in their dental evolution.

PARADONTAL ABSCESS

Some slight further support for this suggestion may be adduced if we examine another aspect of dental pathology. One of the commonest of all abnormalities is paradontal abscess which, in these early English populations, is commonly seen in adults, often in association with well marked attrition of the teeth. In children it occurs far less frequently. A high proportion of these abscesses probably results from husks of grain or spicules of bone becoming lodged between the neck of the tooth and the bony rim of the socket. For our present purposes we need only note that if their eventual appearance is in any way connected with events occurring during the phase of tooth formation then, as with dental caries, we might expect a significantly higher incidence of them amongst the women. If (as in fact there is every reason to believe) their occurrence is mainly the result of accidental mishaps affecting the jaw well after the teeth have completed their development then there should be little difference in the sexual incidence.

Table VI shows the results for the Caister-on-Sea population.

The difference between males and females shown here is not significant and the table confirms that there is no correlation between the Harris's lines laid down in childhood and the subsequent development of paradontal abscess.

TABLE VI
CAISTER-ON-SEA PARADONTAL ABSCESSES

	Males	Females
Number of identifiable tooth positions	1456	1181
Number of paradontal abscesses	74	75
Percentage of positions with abscesses	5.1	6.3
Total number of individuals	48	46
Number of individuals with abscesses	27	23
Percentage of individuals with abscesses	56.2	50.0

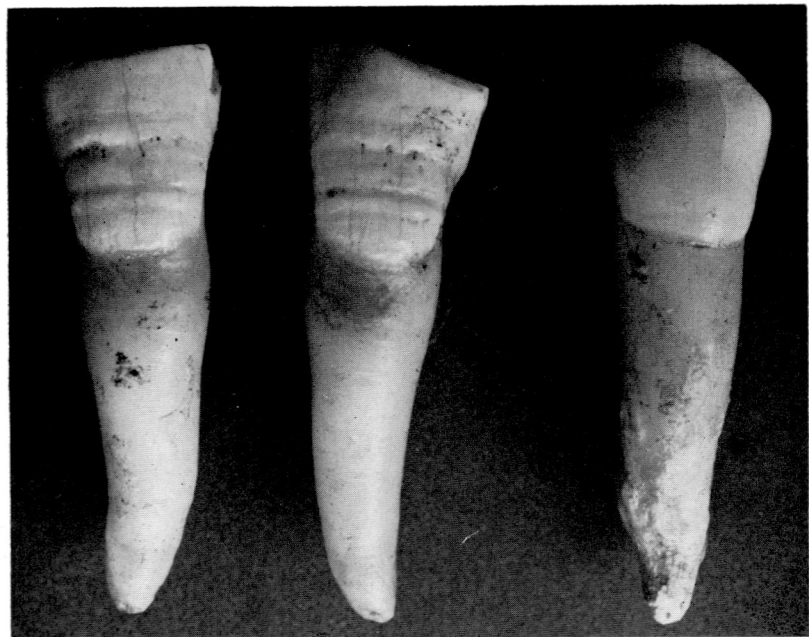

Figure 9. A pair of Anglo-Saxon incisors and a Mediaeval canine. All three teeth show hypoplastic defects. Photograph Hallam Ashley.

DENTAL HYPOPLASIA

Another morbid condition of the teeth which might be expected to have some positive correlation with the number of Harris's lines is enamel hypoplasia. Its pathology is still not fully understood but it is generally accepted as being due to systemic disease affecting the teeth during the period of their development. It appears as irregular ridges and pits in the enamel of the tooth and as bands of discolouration which may take various forms such as opaque chalky lines or zones of darker pigmentation, etc. (Fig. 9) On any one tooth several of these hypoplastic zones may occur, often merging into one another. It is impossible to determine the exact age at which any zone was produced but their overall frequency and size are usually thought of as being in some degree a measure of general illness during childhood. Rather than count the hypoplastic elements on all the teeth, which is difficult to interpret because of the inevitable duplication that is recorded, the investigation is better limited to two teeth only: the canine and the third molar.

The canine is a long tooth of simple shape which often displays hypoplastic defects with conspicuous clarity. Its calcification in the jaw starts as early as the fourth or fifth month of life, its crown is completed at about six or seven years of age, and it erupts at about eleven. It registers, therefore, morbid processes affecting it during the first half of childhood. The third molar is an unsatisfactory tooth from many points of view. It often fails to erupt, it may be misplaced and partly obscured by the M2, it is commonly reduced in size and malformed. These facts make it the least suitable of teeth for studies of hypoplasia. It has, however, one outstanding advantage: its calcification begins at about eight to ten years, very much later than any other tooth, and its crown is not completed before thirteen or even sixteen years of age—the end of the period of skeletal growth. This means that it gives (and is the only tooth to give) a record of the morbidity resulting from events in the second half of childhood.

The little evidence at present available does hint at a slight positive correlation between the Index of Morbidity and hypoplasia, but it is certainly very weak. This is, perhaps, due to the small amount of work that has yet been done on the problem but if it should eventually be found that there is little or no positive correlation here it may suggest that there is some basic difference between the morbid processes which produce Harris's lines and those responsible for hypoplasia. It might, for instance, be possible to show that, as between bones and teeth, there is a differential in the response to either infectious diseases or inadequacy of diet.

AGE AT DEATH

It is not uncommon to find individuals with a dozen or even a score of lines at each end of a long bone and this presumably represents some sort of fairly severe illness every six months or so throughout childhood. Nowadays, in modern paediatric practice it is common to find that a child who is much prone to illness in early life will, with careful treatment and nurturing, eventually turn into a healthy, well-built teenager apparently none the worse for his early maladies. However, it is far from certain that a similar resilience would be shown by archaic populations who lacked the resources of modern scientific therapy. In the absence of skilled medical supervision we might expect that a series of juvenile illnesses, if at all severe, would burden the child with a degree of ill-health which he was never able to overcome.

With this possibility in mind it becomes of interest to examine the relationship between the average number of Harris's lines in a group and its mean age at death. Owing to the loss of neo-natal and infant burials the true mean age at death can never be known. The estimate used here is the mean age at death of all those individuals who survived to the age of eighteen and beyond. The results obtained by doing this will be much higher than the true rate for the group but they will be a perfectly good comparison of the adult part of the communities. Table VII shows the relationship between bone scar frequency and age at death from four different sites.

The evidence given by this table is remarkably interesting. It shows that although Caister has far fewer line-free in-

TABLE VII
HARRIS'S LINES AND AGE AT DEATH

Site	% individuals with no Lines	Index of Morbidity	Mean Adult Age at Death		
			Males	Females	Combined
Shouldham	60.0	1.6	33.8(10)	41.7(11)	39.2(21)
Thetford, Red Castle	42.8	1.8	37.9(16)	30.5(17)	33.9(33)
Thorpe St. Catherine	33.3	2.2	36.4(27)	29.8(15)	34.1(42)
Caister	10.7	5.1	37.5(38)	32.0(39)	34.5(77)

dividuals and a much higher Index of Morbidity than the other three sites yet its average age at death is substantially the same as at least two of them. Only the mediaeval priory group from Shouldham seems to live longer and here we have the uncertainty of dealing with the smallest numbers. (The somewhat older average of these people may partly explain their higher dental caries rate which has already been noted).

This parity of age between Caister, Thetford and Thorpe is most revealing now that we know, what we had no means of detecting before, that their childhood illness rates were very different. It suggests that, whatever may have been the nature of the many morbid episodes afflicting the Caister children, the group as a whole possessed enough biological resilience, enough reserve of reparative and vital energy, and a nutritional level which, despite occasional shortages, was sufficient to bring them up to the same constitutional standard as the other communities by the time adult life was reached. This surprising conclusion makes us wish to know more about the nature of the processes producing the Harris's lines. But as already noted, there is no way of doing this, and any conclusions drawn can only be inferred on a basis of probability.

There is much evidence that the diseases which in mediaeval and later times became the major epidemics were unknown or rare in the Saxon period. Bubonic plague, cholera, typhus, typhoid and smallpox probably do not concern us. Even our own common exanthemata—measles, rubella and scarlatina may not have been present. It is probable that repeated episodes of winter bronchitis and summer dysenteries may have been as likely a cause as any for a substantial majority of the lines found in these populations, although malaria may also have been common. In addition to this it is possible that the chief cause of all may have been recurrent periods of famine or malnutrition rather than any infection or other illness (Stewart and Platt, 1958). In a series of tibiae of Kalahari Bushmen, chosen on account of their rigorous environment, the Index of Morbidity was the extremely low one of only 0.3 lines per bone.

But the relationship between the age at death and the number of lines can be refined a little further. Instead of contrasting different populations Figure 3 shows the result of plotting line frequency against individual age for a single group of people.

This contradicts to some extent the evidence of Table VII which showed that, as between one community and another, the average number of lines has little affect on the mean age at death. Figure 3 reveals that among the Caister people there is a moderate correlation between the number of lines and the individual age at death. But if we remove from the scattergram the nine individuals in the top right corner (that is the few people who died young after having had unusually high morbidity in childhood) the correlation almost vanishes and we come very close to random scatter. Is it too much to infer that in removing these individuals we have picked on the very ones who did in fact suffer from precisely those sporadically occurring diseases—rheumatic valvular heart lesions, chronic nephritis, bronchiectasis, etc.—that would quite certainly lead to a permanently impaired constitution and a diminished expectation of life?

STATURE

Bone not only grows, it exhibits another activity—it matures. Modern studies have shown that growth and maturation are in-

A New Approach to Palaeopathology: Harris's Lines

FIGURE 3: CAISTER-ON-SEA: Index of Morbidity and Age at Death
x = Males; o = Females.

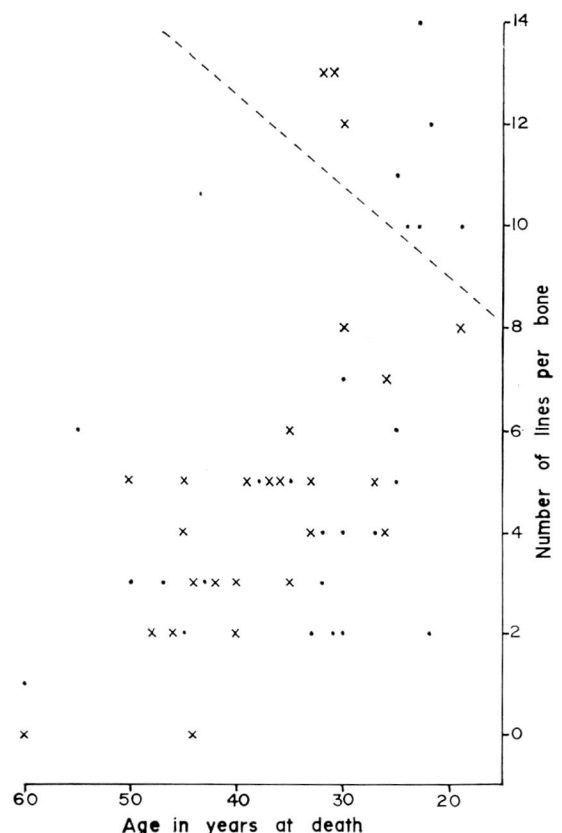

FIGURE 4: CAISTER-ON-SEA.
Lines and stature
x = ♂
o = ♀

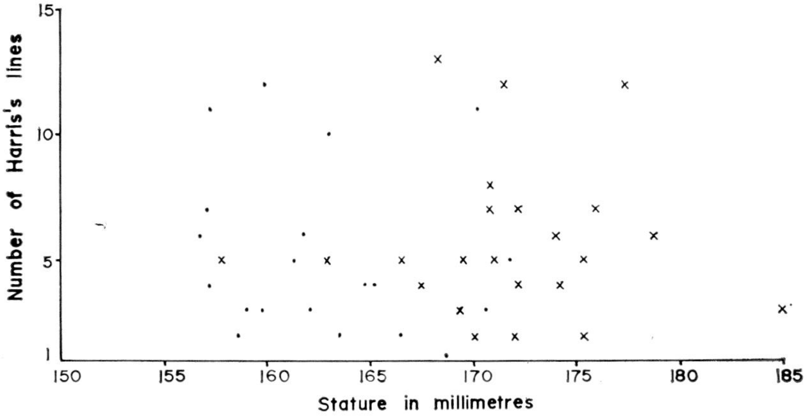

dependent processes in which maturation can and usually does continue even when growth is arrested. The bone scars are the result of cessation of growth for some indeterminate period. If a person has many such episodes of arrested growth it is reasonable to assume that the final stature will be reduced unless maturation of the bones is also interrupted, which does not seem to happen. We may expect, therefore, that the individuals of shortest stature will be those with the highest number of lines. Figure 4 is a scattergram of the Caister population plotting stature against the Index of Morbidity.

It shows no correlation between the two but it is extremely unlikely that the Caister people differed from modern groups in being subject to arrested maturation coincidentally with arrested growth. Another explanation of the random scatter must be found and it appears to lie quite simply in the inadequacy of the sample. There is a very wide range of variation about the mean for the stature of this group and with the small population at our disposal this wide dispersal seems effectively to mask a correlation that would almost certainly be revealed in a really large sample.

One interesting relationship does emerge, however. Wells (1960) has drawn attention to work on domestic animals by Hammond (1958), which suggests that because of the relatively higher rate of post-natal development in the caudal as compared with the cephalic half of the body, failure of nutrition or the onset of some morbid process will arrest caudal more than cephalic growth. Wells thinks that this might be found to apply to Man and to support his contention he adduces evidence from discrepancies in stature reconstruction formulae. In the material presented here the individuals with the highest post-natal morbidity can be identified by their Harris's line count. We can also use arm length as a measure of cephalic growth and leg length as a measure of caudal growth. If caudal growth is arrested by disease or malnutrition as has been suggested the relative diminution of leg length would give a higher intermembral index in those individuals so affected. (The intermembral index is the length of the humerus-plus-radius expressed as a percentage of the length of the femur-plus-tibia.) Figure 5 shows the results of plotting the number of Harris's lines against this index.

It shows that there is a positive relationship between the two and it is of interest that the shortening of the legs by illness (it is unlikely to be a lengthening of the arms) can be detected in this population

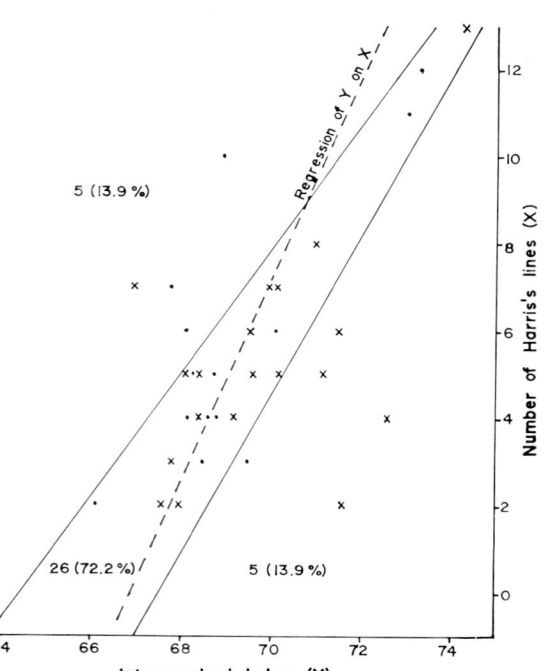

FIGURE 5: CAISTER-ON-SEA.

Lines and intermembral index

x = ♂
o = ♀

through the intermembral index even though diminution of stature as such eluded scrutiny. This correlation is far from perfect but is quite good in view of the smallness of the sample.

TECHNIQUE

First, the choice of bone: Harris's lines may be formed in any long bone but the greater its length the more widely spaced they are likely to be, the easier it will be to decide when they were deposited, and the less likely will any be missed through overcrowding. The obvious choice, therefore, is the femur. Against this is the fact that the length and shape of the thigh bone make it relatively expensive to X-ray, whilst the complexity of the multiple epiphyses at its proximal end creates difficulties and ambiguities of interpretation. The fibula often shows the lines very clearly and many fibulae can be X-rayed on a single film but its fragility and poor preservation in most burial grounds are an insuperable barrier to its routine use. In practice the tibia is found to have more advantages and fewer drawbacks than any other bone and all figures quoted in this study are derived from tibial X-rays. Probably the combined evidence from two or more different bones would be found to give even better results, especially in groups with a high proportion of damaged tibiae, but no such procedure has been used here. Both left and right tibiae should be examined because it is occasionally found that some of the fainter lines fail to appear in one or other side. When this happens the side with the fewer lines is discarded. There is no point in averaging the pair when it is the maximum number of morbid episodes which is being sought. The bones ought to be as perfect as possible. Some damage to the epiphyseal ends can often be ignored but the full diaphyseal length must be present. If the outer cortex of the bone is broken, sand or silt may get into the medullary cavity and mask the radiographic appearance. Only bones in which the epiphyses have fused should be used when a complete record of lines is needed. All figures quoted in this article are based on the use of adult bones only.

Ideally all individuals in the group should be examined but it is rare that every tibia is intact, and usually a sample has to suffice. To estimate the age of deposition of any line its distance from the proximal or distal end of the bone is measured. It has been arbitrarily assumed that the birth length of all tibiae is 90 mm (a deliberately high figure). Post-natal growth is, therefore, $x - 90$ mm (where x is the final length) and this commonly ranges from $230 - 340$ mm or even beyond these limits. It has also been assumed that of this post-natal growth three-fifths is from the proximal, two-fifths from the distal end. This is uncertain and no doubt even to look for a constant of this kind is an over-simplification. Owing to variations in the rate of proximal or distal growth at different ages the proportion of growth from each is almost certainly a fluctuating one throughout childhood. Any error introduced here is similarly repeated in all the populations that have been examined and, provided these fluctuations of growth rates are broadly similar in most human groups, their final comparison should be valid.

From known average lengths of tibiae at each year of growth until the age of sixteen a correction has been made according to the final length of the bone, short of or in excess of the mean. If this were not done a line 120 mm from the proximal end of a long tibia might represent an age of four years, whilst the same distance in a short bone would fall in the pre-natal section. For the practical task of recording the lines from the X-ray films it is convenient to match each bone according to its length against a prepared scale of similar length graduated in "years of age" proximally and distally. The lengths of these scales increase by increments of 10 mm and each bone is measured against the scale it most closely matches. It is worth noting here that a transverse line may be seen as a rare anomaly in a part of the bone which corresponds to its pre-natal length. The interpretation of a line in this position is uncertain. It is very doubtful whether they are true Harris's lines and they are best ignored. Owing to the medullary absorption and subperiosteal deposition which accompany linear growth it is impossible to recover from adult bones a Harris's line which has been deposited in the first couple of years of life. They are, however, obvious enough if the bones of children are being radiographed and it is interesting to find that the trauma of birth seems usually to produce one.

The best exposure can only be found empirically but owing to the high radiographic contrast in the texture of dried bones a non-screen technique usually gives better results than cassettes with intensifying screens and an average adult exposure using a film such as "Ilfex" would then be 70Kv. 20 m.a. 2 secs. at 3' 6". Scanography enables the length of the bones and the distance between lines to be measured with precision from the film but it is a technique which sometimes results in non-recognition of very faint lines.

It will be seen that inconsistencies appear in the results presented here. There is no doubt that a lot more information is needed before accurate understanding of these data will be possible and future refinements of technique and interpretation must be sought in the light of further knowledge about the variability of bone growth under different circumstances.*

ACKNOWLEDGMENTS

The many radiographs on which this study is based were all taken and processed by Ilford Ltd., to whom my grateful thanks are due. Miss E. B. Green, Keeper of Archaeology, Norwich City Museums, has kindly allowed me access to material in her charge.

---

*Addendum. Since the above was written, work by W. A. Marshall has shown that absorption of Harris's lines during childhood is more common than was formerly believed. This must add further difficulties to the interpretation of these structures.

REFERENCES

Hammond, J., 1958: 'Darwin and animal breeding.' in *A century of Darwin*, edited by S. A. Barnet, London: Heinemann.

Harris, H. A., 1933: *Bone Growth in Health and Disease.* Oxford Univ. Pres.

Park, E. A. and Richter, C. P., 1953: Transverse lines in bone: mechanism of their development. *Bull. Johns Hopkins Hosp.*, 93:234.

Stewart, S. J. and Platt, B. S., 1958: Arrested growth in the bones of pigs on low-protein diets. *Proc. Nutrit. Soc.* 17:1.

Wegner, G., 1874: Ueber das normale und pathologische Wachstum der Roehrenknochen. *Arch. Path. Anat.*, 61:44.

Wells, Calvin., 1961: A new approach to ancient disease. *Discovery*, 22(12):526.

Wells, L. H., 1960: Differences in limb proportions between modern American and earlier British skeletal material. *Man*, 60:181.

Chapter 31

Thinning of the Parietal Bones in Early Egyptian Populations and Its Aetiology in the Light of Modern Observations

THOMAS LODGE

INTRODUCTION

THE cranial vault or dome of the skull varies in thickness in different individuals and in the same individuals at different ages. Studies in skull thickness (Oldberg, 1945; Getz, 1960) on the whole suggest "a clear tendency to greater thickness with increasing age, both of the frontal and parietal bones" (Getz, 1960) though there may be a marked senile atrophy (Young, 1959) after the age of sixty. When the calvarium is excessively thickened the condition is known as *hyperostosis cranii* which may be local (usually in the frontal bone and known as *hyperostosis frontalis interna*) or general. Occasionally, a thin cranial vault is encountered and though it is of no intrinsic pathological significance it may facilitate the production of a fracture in the event of injury to the head. Very rarely a skull may be seen in which the thinning is not only extreme but is localised. In nearly every such case the localised thinning is situated in the posterior part of the parietal bone and is bilateral and symmetrical. The symmetrical nature of the condition which goes under the name of bilateral thinning of the parietal bones has suggested to some people an association with parietal foramina, a congenital and familial condition of small holes one on either side of the mid-line sagittal suture which is also found in the posterior parts of the parietal bones. The two conditions, however, are probably distinct and separate processes. Moore (1949) thought enlarged parietal foramina to be no more than incomplete examples of symmetrical parietal thinness but Johnstone (1955) and more recent observers are convinced that there is no connection between the two.

HISTORY

In 1858 Sir George Humphrey called attention to a skull in a Cambridge Museum which showed unilateral parietal atrophy and this was followed by further descriptions; for example, in 1865 by Professor Sir William Turner (an adult female with bilateral changes) whilst Wrany (1866) listed four cases and amongst three further examples published by Broca (1875) was a skull presented to the Val-de-Grace Museum in Paris by Baron Larrey, Surgeon-in-chief to Napoleon. Greig published his first case in 1892 and in another paper in 1917 added a second which, by chance, proved to be the brother of the first. Radiography was not well developed at that time and Greig's published illustration of the calvarium of the second case is not as clear as a modern radiograph would be. In the present writer's opinion the illustration is not typical of the condition as we see it to-day and as it appeared in ancient Egyptian skulls but is much like enlarged parietal foramina (O'Rahilly and Twohig, 1952). However, it is sometimes accepted as bilateral parietal thinning. Paterson (1900) added three cases and since that time cases have been added sporadically to the literature on the subject. The paper to which all writers refer is that of Greig in 1926 (see below).

Thinning of the parietal bones of the type described has been reliably recorded mostly in the elderly. Steinbach and Obata (1957), however, claim an example in a twenty-eight-year-old man with a diagnosis of eunuchoidism and Camp and Nash's (1944) series include ten cases under the age of thirty. Steinbach and Obata suggest that there may be two distinct but similar clinical entities; one a progressive disease of the elderly and the other a congenital lesion remaining unchanged throughout life. The incidence of this cranial state is hard to estimate as, even when found at necropsy, it is not always recorded unless it is pronounced: suffice it to say that, even with modern diagnostic methods, it is an uncommon finding. Both sexes may be affected in about equal proportions. Unproductive of symptoms, it rarely leads the affected individual to the doctor and most of the modern examples have resulted from deliberate surveys or appeared as a chance finding at necropsy or when the skull has been radiographed following injury. Similarly, its geographical distribution is poorly charted. Subirana and Wackenheim (1965), discovering a case recently in Strasbourg, considers it extremely rare in modern France, despite earlier reports by Broca (1875), Sauvage (1869) and Carrière (1874). Jackson (1957) has reported its occurrence in Australia and I have seen it in a skull in the City Museum, Suva, Fiji. It is uncommon but not rare in Great Britain; I have personally collected fourteen cases.

Camp and Nash in America were able to find by radiological methods 119 cases, the largest in any one series, whilst Carrière noted the condition in 0.4 per cent of 1,000 dry skulls. The classic description is that of Greig (1926) who referred to two types: one a shallow flat depression of quadrangular or triangular shape and the other a groove or sulcus: they were situated in the parietal bones, the quadrangular form usually having its long axis in the sagittal (antero-posterior) plane and the grooved one in the coronal (transverse) plane. Rarely there are double grooves and rarely also the condition may be unilateral as in a case described by Nashold and Netsky (1959). In all true examples the loss of bone is of the outer table of the skull and of the diploic marrow cavity as if the cranial vault were being eroded from without inwards. The depressions can be readily appreciated by palpation but surprisingly seem not to have been considered worthy of comment by the persons affected. Combing the hair has sometimes called attention to them. Greig stated that the inner table was normal but it can be extremely thin as in the case of two elderly sisters seen by the present writer; one of whose skulls was almost paper thin in the affected areas (Fig. 1). Because of this, the condition is obviously of some medico-legal importance as a blow with a sharp instrument would penetrate such a weakened area with considerable ease.

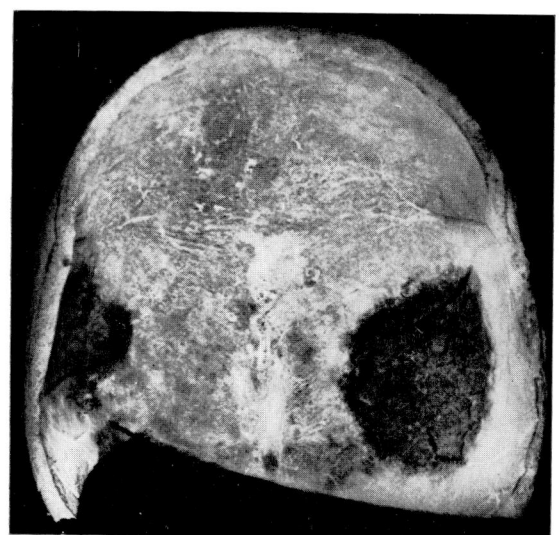

FIGURE 1. Parietal thinning. Parchment thin areas in skull of woman of 86.

AETIOLOGY

The cause remains unknown but many associated factors have been cited. The commonest factor is old age. All the fourteen cases seen by the present author were above sixty-seven years of age. Camp and Nash recorded the condition in people under thirty but this must be accounted rare. Epstein (1953) postulated a relationship between parietal thinning and osteoporosis of either postmenopausal, senile or idiopathic types, a theory partly supported by the finding (Gershon-Cohen, Schraer, and Blumberg, 1953) of moderate skeletal osteoporosis in five cases (3.3 per cent) of 130 normal elderly people, two-thirds of whom had general nutritional osteoporosis. If osteoporosis is a factor in causation it is probably not the only one. In Greig's view parietal thinning was a congenital dysplasia. Gershon-Cohen and colleagues suggest that as the parietal bones are the last to develop and are more liable to injury and to congenital lacunae, focal diploic atrophy may occur as a normal variation, merely becoming more noticeable in the elderly. Certainly the parietal bone is more often the site of the congenital defect known as *cranio-lacuna* than any other bone of the vault but if his supposition were correct minor and early forms of parietal atrophy would surely be apparent on radiographs of skulls in younger age groups. Jonathan Hutchinson thought that syphilis might be a factor and Humphrey suggested a stretching of the occipito-frontalis tendon. Elsewhere it has been ascribed to the fact that there are no muscle attachments in this area of the parietal bones over which the *galea aponeurotica* moves freely and hence these are the likeliest areas to reflect the effects of decreased osteoblastic activity (Steinbach and Obata, 1957). Elliot Smith (1907, 1927) always strongly maintained that the condition must be due to some such factor as pressure on the outer table causing its absorption. It certainly seems true (Lodge 1963) that bone generally has decreased power of resistance to lateral pressure. It is not without interest that Paget knew of this condition which he considered (as did Virchow and Rokitansky) to be in the nature of senile atrophy because in a recent French article (Reboul *et al.* 1962) a defect similar to parietal thinning is described as an atypical feature of Paget's *osteitis deformans* though it is probably a coincidental occurrence of the two conditions in one patient. *Osteoporosis circumscripta* is a quite different condition which occurs anywhere in the vault and is clearly a form of focal Paget's disease of the skull (Collins and Winn, 1955).

PARIETAL THINNING IN ANCIENT EGYPTIAN SKULLS

The condition is seen in the mummies of Meritamon and Tuthmosis III in the New Empire period (Elliot Smith, 1912 a and b). In both, the overlying scalp is

SAUVAGE, H. E., 1869: *Recherches sur l'état senile du crâne.* Thèse de Paris. No. 315.

STEINBACH, H. L., and OBATA, W. G., 1957: The significance of thinning of the parietal bones. *Amer. J. Roentgenol.* 78:39.

STEWART, T. D., 1964: Personal communication.

SUBIRANA, M. and WACKENHIEM, A., 1965. Osteolyse de la Table Externe et du Diploé des Parietaux. *Ann. de Radiologie.* 8:645.

TURNER, W., 1865: On some congenital deformities of the human cranium. *Edin. Med. J., 11*:133.

VIRCHOW, R., 1854: Ueber die Involutionskrankheit der platten Knochen *Verh. phys-med. Ges. Wurzburg.* 4: 354.

WILSON, A. K., 1944: Roentgenological findings in bilateral symmetrical thinness of the parietal bones (senile atrophy). *Amer. J. Roentgenol., 51*:685.

WILSON, A. K., 1947: Thinness of Parietal Bones. *Amer. J. Roentgenol., 58*:724.

WRANY, A., 1886: Abnorme Veite der Foramina Parietalia. *Vjschr. prakt. Heilk.* 89-90:108.

YOUNG, R. W., 1959: Age changes in the thickness of the scalp in white males. *Hum. Biol., 31*:74.

Chapter 32

Biparietal Thinning in Early Britain

DON BROTHWELL

Although T. W. Lodge in the previous paper has discussed the pathology and aetiology of this anomaly, and has indicated that it has been identified in earlier remains (especially Egyptian), it would seem opportune to emphasize three further important points by reference to British archaeological specimens. These points may be best summarized as follows:

1. It cannot be assumed that skeletal material from sites excavated some time ago, and now reported on, did not have cases of this biparietal thinning. Specimens 1 and 2 demonstrate clearly that it was not always identified and described by earlier workers.
2. It is likely that the thinning may be far from symmetrical, and in such instances may need carefully separating from depressions resulting from depressed cranial injuries and inflammatory processes. See case 2.
3. Slight degrees of biparietal thinning demand very careful examination of the cranial vault, with radiographic evidence or vault thickness data ("contouring" the region involved) as confirmatory evidence. See case 3.

1. Yorkshire, Saxon

Near the northern margin of Barrow C.34 near Garton, Yorkshire, J. R. Mortimer (1905) excavated a Saxon cemetery. Of sixty individuals, one adult (No. 47) and one child (No. 2) were buried with the head to the east. The skull of this adult female (No. 47) is now in the Mortimer collection of Hull Museum (recorded as S.25) and I was interested to note this (then first) British case of biparietal thinning—not previously recorded by those working on Yorkshire material.

Considering the degree of dental attrition, the woman was at least middle-aged. On either side of the sagittal suture are two nearly circular depressions (Fig. 1a). The left one was about 40 mm in maximum length and 41 mm in breadth, being some 25 mm from the sagittal suture. The right depression was slightly larger (length 55 mm, breadth 50 mm) and was only about 21 mm from the sagittal suture. Both are smooth depressions which go down to, but do not perforate, the inner table. There is no evidence of inflammatory changes surrounding these concavities. The endocranial surface is not affected.

2. Roman, Berks

During work on our own collections at the British Museum (Natural History), I came across another case. Again the individual is female, and although the dentition is far from complete, tooth loss and attrition suggest that she was at least mid-

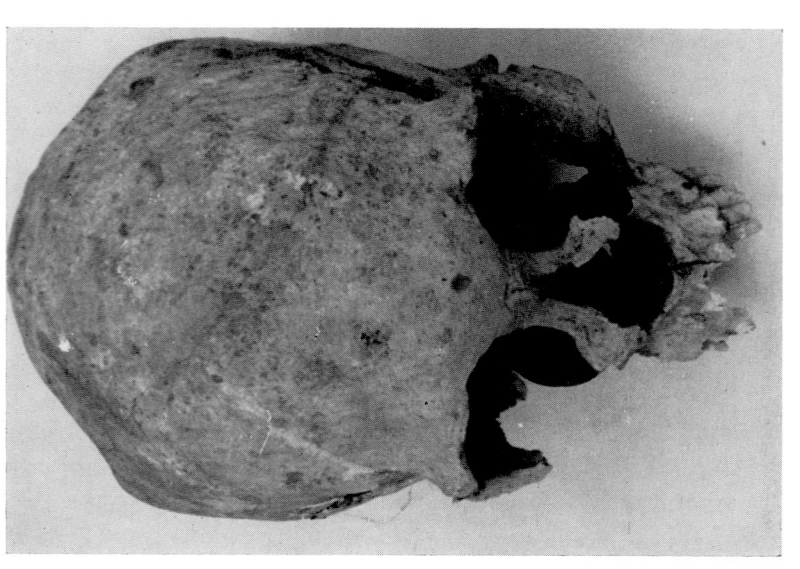

FIGURE 1. a) Frontal view of the Hull Museum Saxon specimen, with biparietal thinning well marked.

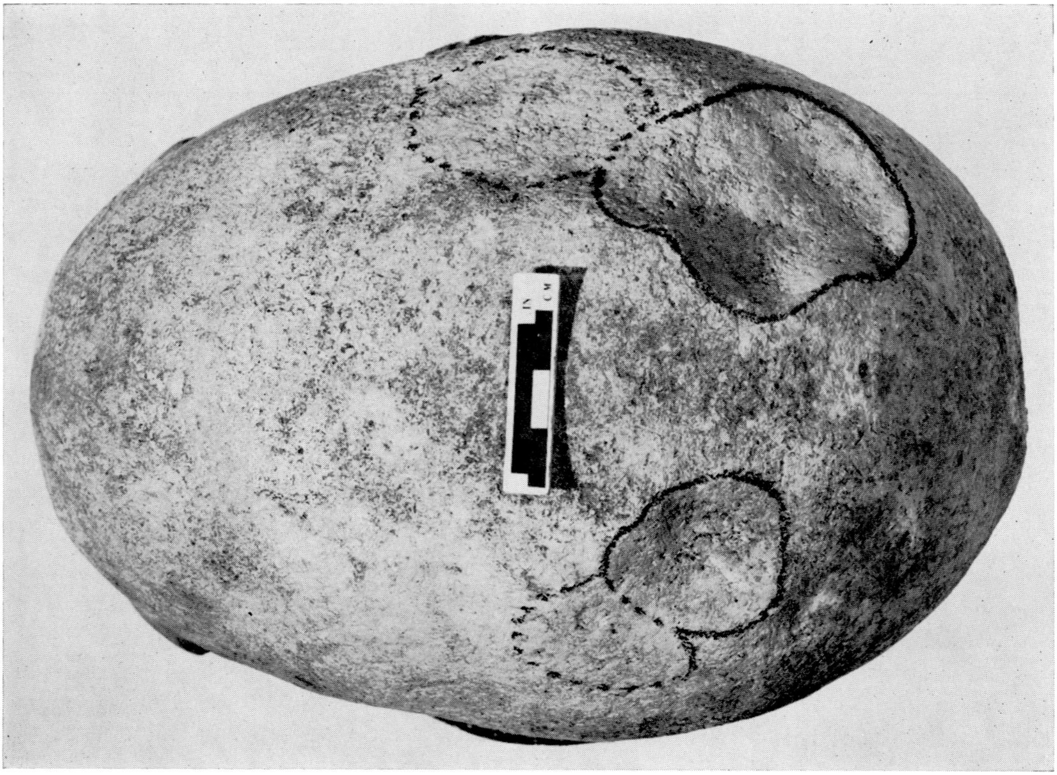

b) The Roman case, showing irregular areas of thinning.

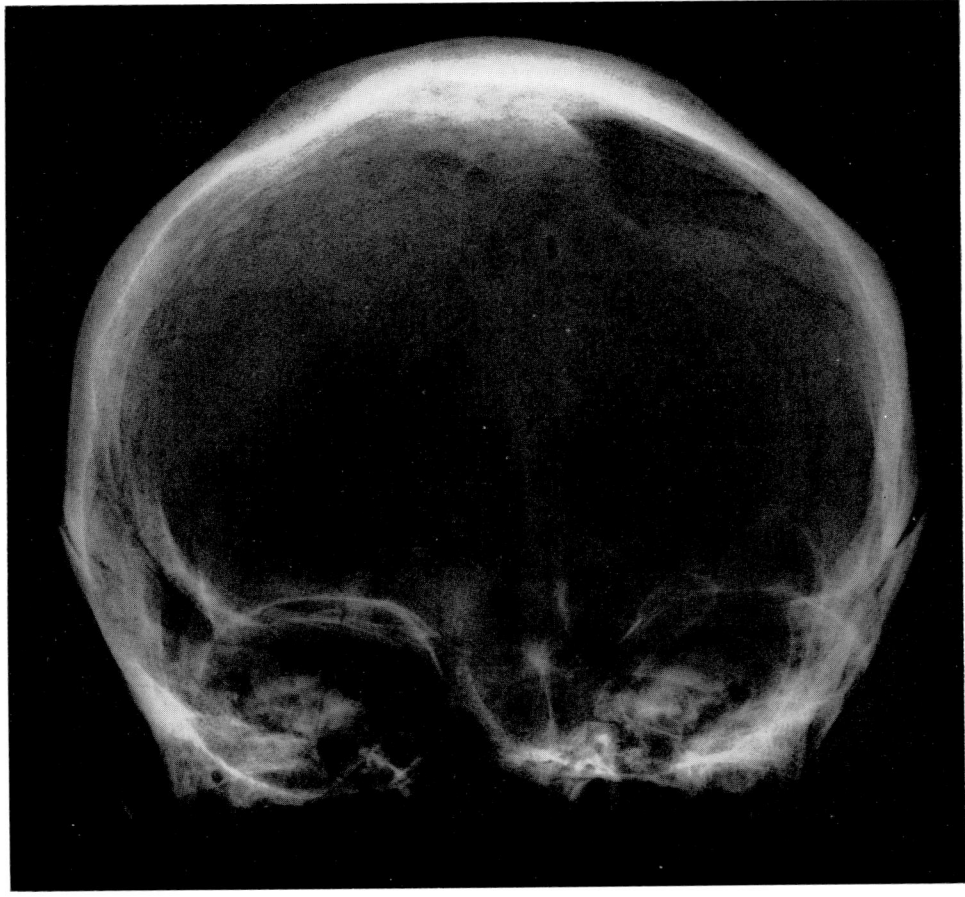

FIGURE 2. Frontal radiograph of the Great Chesterford skull.

dle-aged. She is part of a series of Romano-British skeletons from Frilford, Berks., (B.M.N.H. E. 11.8.388) and again the nature of the parietal thinning seems to have been overlooked, although excavated over fifty years ago.

Unlike the Yorkshire Saxon case, the depressions are more asymmetrical, and less regular in contour (Fig. 1b). The exact limits of these anomalies are not so easy to define owing to the low-angled grading-off of the depressions. On both sides there appears to be a more posterior deeper zone followed by a shallow area. Nevertheless, this does not preclude it from being bilateral parietal thinning, and indeed, as far as alternatives are concerned, there is no evidence to suggest that these are partial trephinations or the result of a suppurative scalp infection in two areas. There is of course, no reason why bilateral thinning should be perfectly symmetrical as regards size and shape, any more than "symmetrical osteoporosis" of the skull is.

3. Saxon, Cambridgeshire

A final case which has come to my notice is from a Saxon cemetery at Great Chesterford, excavated recently by Miss

V. Evison, and as yet unpublished. Skull 116 is from a female individual who, judging by the degree of dental attrition is likely to have been well over forty years of age. In both parietals, and in positions approximately 50 mm from the sagittal suture, there is an elliptical area of thinning (Fig. 2). The significance of this specimen is that it must represent a very early stage of biparietal thinning for there is hardly any depression. Indeed, the thinning is mainly the result of the flattening of the rounded external parietal contour, and clearly the next stage would have been the formation of concavities. Although the thinning was quickly confirmed by X-ray (and in the absence of contradictory evidence such as inflammatory changes), it does emphasize the need for very careful examination of the skull vault if more certain evidence of its frequency in earlier populations is to be obtained.

REFERENCES

MORTIMER, J. R., 1905: *Forty Years' Researches in British and Saxon Burial Mounds of East Yorkshire.* London, Brown.

Chapter 33

Historical Notes on Some Vitamin Deficiency Diseases in China*

T'AO LEE

The discovery of the vitamins is a very important event in the medical history of the early twentieth century.

In 1907 E. V. McCallum began, by using the rat, to study why animals fail to develop on diets composed of purified food substances.[1] Several years later he discovered that the diet he gave the rats was not pure, and that his early successes were due to the vitamins in certain constituents of the previous rations on which he had fed them. In 1913 Osborne and Mendel reported their independent discovery of a substance in butterfat, which had a stimulating effect on growth, and afterwards found to be contained in cod liver oil. This substance is now known as Vitamin A. In 1897 Eijkman showed that a paralytic disease, polyneuritis, could be produced by a diet composed largely of rice and that it could be cured by feeding on rice polishings. Later it was demonstrated by his assistant, that the rice polishings furnished a hitherto unrecognized dietary essential, now known as Vitamin B. The cause and cure of scurvy has been known for centuries, but not until 1919 was it fully established that scurvy results from want of Vitamin C. Vitamin D is the dietary factor that prevents and cures rickets in children. Although the disease was known as early as the second century of the Christian era, the specific antirachitic vitamin, designated as Vitamin D, was not established until 1922.

Although the discovery of vitamins is very recent, the diseases due to vitamin deficiency appeared very early. Night-blindness in the fifth century B.C.,[2] rickets in the second century,[3] scurvy in the sixteenth century, beriberi[4] in the seventeenth century, have been reported in Europe. According to Dr. Wu Hsien's research[5] on "Chinese diet in North China," the total calories of Chinese diet is very high, but the protein calory is not good enough; the contents of Vitamins B and C are sufficient, but Vitamins A and D are usually scant. Therefore the diseases due to vitamin deficiency in China such as xerophthalmia, nyctalopia, beriberi, and rickets are very prevalent from olden times until the present day.

Chinese writers show no acquaintance with scurvy. Fruits and vegetables grow abundantly in China and this disease must have been of rare occurrence. Greek, Roman and Arabian writers also have no knowledge of it. However, scurvy probably existed in the northern parts of Europe and Asia, ever since they were occupied by man. The lack of records is probably due to the low educational status of the people. After the discovery of America, long voyages became more common, and scurvy became a familiar disease. Another disease, pellagra, is very seldom

* From *Chinese Med. J.*, 58:314, 1940.

seen in China. It is difficult to find a detailed description of these two diseases, scurvy and pellagra, in Chinese references. Therefore discussion in this paper is limited to three diseases, nyctalopia, beriberi and rickets.

NYCTALOPIA

In China, nyctalopia was first recognized by Ch'ao-yuan-fang[6] early in the seventh century. In his book, *Ch'ao's General Treatise on the Etiology and Symptoms of Diseases* we read as follows:

A man who can see in daytime and cannot see at twilight is called sparrow eyed which means that his vision is very much like that of a sparrow which cannot see anything at night.

Sun-szu-mo (581-673 A.D.) in his book *Thousand Golden Remedies*[7] called it sparrow blindness. It is also known as hen blindness by the common people. These two terms are meant to connote "blind at night."

The cause of the disease has been considered to be due to the liver disturbance in the *Essentials of Ophthalmology* which was said to be written by Sun-szu-mo.[8] Yü-t'uan[9] in 1531 in his book The *Principal Record of Medicine* stated:

Why is a man afflicted with sparrow eye unable to see anything at twilight and his vision is restored next morning? It is due to exhaustion of the liver. The blood of the eye comes from the liver. When the liver is exhausted, the vision becomes obscure.

Even though the ancient Chinese did not know the real cause of nyctalopia, they knew that it was caused by nutritional disturbance. Wang-k'en-tang[10] in the early seventeenth century (1607) attributed this disease to insufficiency of sex hormone which he called *Yuan-yang*. He also knew

that if the food suited the patient this disease could be cured without treatment.

As early as the seventh century, Sun-szu-mo[11] in his book *A Supplement to Thousand Golden Remedies* stated that night-blindness could be cured by pig's liver. *The Essentials of Ophthalmology*, the guide for native practitioners on diseases of the eye, contains descriptions of eighty-one eye conditions of which two undoubtedly refer to night-blindness. In this book the animal livers (pig and sheep) have been considered as basic remedies. In *Sheng-hui-fang*[11] written in 980 A.D. by the imperial medical officials of Sung-t'ai-tsung, a decoction of *Atractylis ovata* and pig's liver, had been used for treating night-blindness. *Lung-mu Medical Treatise on Ophthalmology* written in the thirteenth century[12] stated that *Blantago major* and *Asarum sieboldi*, Mig. were commonly used for this disease.

P. G. Mar and B. E. Read[12] in 1936 examined twenty animal products and 65 vegetable substances listed as remedies for night-blindness in the Chinese Herbals, and found many of them contain vitamin A or provitamin A up to 10 or 20 times the average value for commercial cod liver oils. In their figures, the livers showed the highest values in the animal group, and *Atractylis ovata* showed the highest values in the vegetable group. From this we may conclude that the Chinese in the seventh century were not only able to recognize this disease, they also had a reasonable conception about its effective treatment.

BERIBERI

The people who lived in North China up to the third century B.C. had as their common food, wheat, beans, corn, millet, vegetables, all of which are rich in Vitamin B. Although ancient Chinese writings have recorded a disease with a similar symptom to beriberi, yet none of them have given a detailed account of it. In the *Book of Poetry* edited by Confucius (551-479 B.C.) there was mentioned a disease named *Wei-chung* which means "foot swelling."[14] Tso-ch'iu-ming in his book *Tso-chuan* in the sixth century B.C. mentioned a disease named *Chung-t'ui* or leg swelling.[15] Both, *Wei-chung* and *Chung-t'ui*, have been considered as beri-beri, but there is not sufficient evidence to confirm this opinion.

In the *Nei-ching* there are numerous references to beriberi. It is designated by the name *Wei-pi* and *Chueh*. Chinese ancient writers declared that it was one and the same disease. However, this is still an undecided question in medical history.[16]

Coming down to the Ch'in Dynasty (240-207 B.C.) when the Chinese empire grew larger, the southern areas were also dominated by Chinese. Polished rice is the chief diet of southern Chinese, and Vitamin B deficiency diseases are the natural outcome. Szu-ma-ch'ien (145 B.C.-?) in his book *Historical Records*[17] stated that:

The inhabitants of the Yang-tze River basin and along the South Sea Coast make salt from water. Rice and fish are their chief food. As agricultural produce is plentiful, all of the people have an easy life. Therefore they become more lazy day by day.

From the second century B.C. the Chinese began to suffer from a disease, which was called *Huan-feng* "snail pace," or *Shih-pi* "wet and paralysis." In the Chin dynasty this disease is known by such terms as *Chiao-chung* "foot injury," or *Chiao-juo* "weak foot." A letter of Emperor Liang-wu-ti (464-459) to his minister, called it *Chiao-ch'i* a term which has been selected for use in modern literature.[17] It has been supposed that the miasma, arising from the earth, came first in contact with the feet and so caused the trouble.

In the early seventh century, the first classical description of this disease appeared in *Ch'ao's General Treatise on the Etiology and Symptoms of Diseases*, written by Ch'ao-yüan-fang. in this book the author classified beriberi into three types, oedemic, non-oedemic and cardiac types. This classification is very much like the present forms which are known as wet, dry, and cardiac types.[18]

The history of beriberi in China has been mentioned in *The Thousand Golden Remedies*, written by Sun-szu-mo[7] which says:

Beriberi is a disease commonly mentioned in medical books. However very few of the ancients suffered from it. Since the fifth year of the Emperor Chin-huai-ti (311 A. D.) the Chinese left the North for South China. Many officials were attacked by this disease. Two celebrated physicians, Chih-fa-tsun and Yang-tao-jen, earned good reputations in South China, as beriberi specialists. Many officials were cured by them. Shen-shih, a Buddhist priest, in the Sung and Chih dynasties in the fifth century, has composed a book named *Shen-shih's Prescriptions* based on Chih-fa-tsun's and Yang-tao-jen's formularies, in which more than one hundred recipes for beriberi are included. During Wei and Chow dynasties in the fifth to the sixth century it seems no such disease was discovered in North China. The well-known medical books composed at this period by Yao-seng-yuan and Hsu-wang in North China, do not mention this disease in detail. It is probably due to the great difference in customs and in the climates of North and South China. Therefore the people of Shansi and Hopei provinces did not recognize it.

Recently some Chinese officials and scholars, who have never been in the Yang-tze River valley have also suffered from this disease. This is due to the change of customs and to the mixture of Northern and Southern races.

From this short history we may conclude that Chinese physicians began to recognize beriberi in the Yang-tze River Basin in the fourth century and that no such disease was found in North China until the sixth century. As it was a new disease, the Chinese still could not differentiate it from other diseases. Medical literature[19] used to record nutritional oedema as beriberi. In 529 A.D. General Hou-ching besieged Tai-cheng city in which there were a hundred thousand inhabitants, and twenty thousand soldiers. After a prolonged siege, a large number of people in the city suffered from a disease, which was characterized by oedema of the body and shortness of breath. Eight or nine patients out of ten died from it. It has been said that this disease was beriberi. But we are perhaps justified in saying that it was probably nutritional oedema rather than beriberi. In 605 General Liu-fang of Sui dynasty attacked Lin-i, the northern part of Indo-China. Four or five soldiers out of ten got oedema of the feet and died. This was also confused with nutritional oedema.

Many theories relating to the etiology of beriberi had been advanced in China. As this disease was prevalent in the south where humidity was very high, therefore miasma or a moisture from the earth was believed to be the cause of this disease.[6] But as it also occurred in high altitude and dry places, the moisture theory was untenable, then the exhaustion of the kidney or sexual indulgence took its place.[20]

Another cause attributed by Imamura in his book *The Essentials of the Treatment of Beriberi* in 1861 was water toxin. However, his explanation is not different from the miasma theory.[19]

As milk and cream were not commonly used by the Southern Chinese, it was supposed that the beriberi of Mongolia and Manchuria was due to toxin in milk or its products. Therefore they forbade patients to eat milk and meat, both of which contain significant quantities of Vitamin B.[21]

Though the ancient Chinese did not know the real cause of beriberi, it seems, however, that from very early days they had an efficient treatment. In the fifth century, Shen-shih already had collected more than eighty prescriptions in his book. A century later, Sun-szu-mo[7] stated that he could cure this disease in ten days if the patient believed in him. Many patients died from delay of treatment by not taking his advice. He strongly recommended two prescription, *Chu-li-tang* and *Pa-feng-tang*, as specific prescriptions for beriberi. In these two recipes figured many drugs such as *Silver divaricatum*, *Apricot kernel* and Chinese pepper which contained vitamin B. Not only was he skilful in curing this disease, he also knew how to prevent it. He stated that decoction of millet bran boiled into congee had such an effect.[11]

In the *Private Prescriptions of an Official*[21] written by Wang-t'ao in 753, A.D. many valuable writings and recipes have been collected. The drugs commonly used in his book are such as soybean, areca nut, orange peel, all of which contain a significant amount of Vitamin B.

Tung-chi who lived in the eleventh century and was the author of *Treatment of Beriberi*, suffered from beriberi himself. Therefore he studied the disease for the rest of his life. His book mentions many kinds of seeds which are recognized to be the parts of plants that contain a large amount of this vitamin.[22]

In this connection, it may be mentioned that Yang and Read (1940) who have made an investigation on Vitamin B content of Chinese Beriberi Remedies of plant origin, come to[23] the following conclusion:

Most of the seeds, especially that of plantain, contain significant quantities of Vitamin B. The values of mulberry leaf, loquat leaf and carpenter weed are also high. The Vitamin B content of barks and stems is low, roots contain a moderate amount.

RICKETS

The history of rickets in China dated from very remote times. The earliest indication of anything suggesting rickets has been found in the "Book of Poetry" edited by Confucius.[15] He did not give any detailed description of the disease, but mentioned two deformities, *chü-ch'u,* pigeon breast and *ch'i-shih,* hunchback, which seem to point definitely to rickets. A more unequivocal reference to the disease is found in *Lülan* written by Lü-pu-wei[24] in the third century, B.C. He recorded that:

If bitter water had been used for drinking in certain places, the inhabitants will suffer from a disease producing crooked legs and hunchback.

This makes it almost certain that rickets is referred to, as there is no other likely disease which causes such deformities and at the same time attacks such a large proportion of the population.

During the early seventh century we begin to recognize it in a medical reference[6] *Ch'ao's General Treatise on the Etiology and Symptoms of Diseases* written by Ch'ao-yuan-fang. In this book we read:

The teeth are the end of the bone and the marrow develops into teeth. If the constitution of an infant is very weak, the marrow of the maxillary bone will not be sufficient to develop into teeth and thus dentition is delayed.

After a child's birth, its blood vessels, bones and joints grow stronger day by day. When his kneecap fully develops he is able to walk. We know that the bone is developed from marrow. The delicate baby has a very scanty marrow, therefore the bone is also underdeveloped. This cause the child's inability to walk at the proper age.

This shows that the disease was recognized as two separable clinical entities, delayed dentition and walking. In the same century another important clinical entity of rickets, the patency of anterior fontanelle, was observed by Sun-szu-mo.[7] He stated:

Here is a child who has a big head. Anterior fontanelle patent. His body is wasting and his color is yellow. He is unable to walk at three or four years of age.

From the above quotations it would seem to indicate quite definitely that the disease was present during the seventh century of the T'ang dynasty. Coming down to the eighth century a perfect account of the clinical appearance of the disease was given in *The Private Prescriptions of an Official* by Wang-t'ao. He recorded the enlargement of the head, the smallness of the legs and arms, the delaying closure of the anterior fontanelle, pigeon breast and anaemia. Chien-i, the father of Chinese pediatricians in the tenth century[25] and Wang-k'en-t'ang in the seventeenth century record many typical cases of rickets in their books.

Since then nothing of importance has been added to the above description on the symptomatology of the disease. It is a notable fact that the Chinese physicians did not recognize the whole picture of the disease, and they thought that each symptom was a separate disease.

Regarding the etiology of rickets, it had been considered to be due to the exhaustion of the kidney and the liver. They also thought of common cold to be the cause. It may be said that the supposed causes of rickets in the Chinese literature are mere speculations and so do not deserve a detailed discussion.

As the onset of this disease is insidious, the ancient Chinese physicians did not recognize it at an early stage. They did not know how to prevent it, although a few ridiculous methods of prophylaxis were mentioned.

The treatment of rickets in China is not so well advanced as the treatment of night-blindness and beriberi. Tiger bones, antelope horn, and rhinoceros horn are commonly used in the prescriptions. Whether drugs of this kind contain Vitamin D is a matter of question, but may have hormones to which their reputed good effects on rickets may be due. Stalactites which contain chiefly calcite, are also commonly used.

SUMMARY

1. The history of vitamin deficiency diseases in China dated from very remote times. Most of the diseases, such as night-blindness, beriberi and rickets, were early recognized in the history of China.

2. Nyctalopia was first recognized in the seventh century and a reasonable and efficient treatment was also formed.

3. Beriberi was first recognized in the fourth century in the Yang-tze River Basin and it was not found in North China until the sixth cenutry. Some efficient drugs for its treatment have been recorded in the fifth century.

4. Rickets was first recognized in the third century B.C. The symptoms of the disease were regarded as many separate clinical entities in Chinese medical books. The methods of treatment did not conform to our modern idea as noted in the treatment for night-blindness and beriberi.

REFERENCES

1. SURE BARNETT: *Vitamins in Health and Disease*, Balitmore: Williams & Wilkins, 1963.
2. BURTON CHANCE, Ophthalmology, *Clio Medica*, 1939.
3. FINDLAY, LEONARD: *Rickets, A Short History of Some Common Diseases*, 1934.
4. GARRISON, F. H.: *History of Medicine*, London: Saunders, 1924.
5. WU, HSIEN: General Treatise on Nutrition, 1929.
6. CH'AO YÜAN-FANG: Ch'ao's General Treatise on the Etiology and Symptoms of Diseases. 18, 28, 31, 48, 50.
7. SUN SSU-MAO: The Thousand-golden Remedies 5, 6, 7.
8. SUN SSU-MAO: The Essentials of Ophthalmology.
9. YU T'UAN: The Principal Records of Medicine.
10. WANG K'EN-T'ANG: *Standard Methods for Diagnosis and Treatment*. 4, 7.
11. SUN SSU-MAO: *Formulary, A Supplementary to Thousand-golden Remedies*. 11.
12. PAO KUANG TAO JEN: The Lung-mu Medical Treatise on Ophthalmology.
13. MAR, PETER G. and READ, B. E.: Chemical examination of Chinese remedies for night-blindness, *Chin. J. Physiol.* 10: No. 2, 1936.
14. CONFUCIUS: Book of Poetry.
15. TSO CH'IU-MING: Tso-chuan.
16. WONG and WU: History of Chinese Medicine, *Chin. Med. Ass.* Shanghai, 1932.
17. SZU MA-CHIEN: Historical Records.
18. KAO, C. L.: Infantile beriberi in Shanghai, *Chin. M. J.* 50:324, 1936.
19. IMAMURA: The Essentials of the Treatment of Beriberi.
20. WANG TAO: The Private Prescriptions of An Official. 18, 19, 21, 36.
21. CH'EN M'ING-LEI and OTHERS: A Compilation of Ancient and Modern Books; Section on Medicine. 189-192.
22. TUNG CHI: General Treatment of Beriberi.
23. YANG, E. F. and READ, B. E.: Vitamin B Content of Chinese Plant Beriberi Remedies, Vol. 15, No. 1, 1940.
24. LÜ PU-WEI: Lü-lan.
25. CH'IEN I: Ch'ien's Direct Methods for Treating Children's Diseases.
26. WANG K'EN-T'ANG: Standards for the Diagnosis and Treatment of Children's Diseases.

Chapter 34

Major Congenital Anomalies of the Skeleton: Evidence from Earlier Populations

DON BROTHWELL

During the routine examination of skeletal series excavated from historic and prehistoric sites, it is usual to note evidence of minor congenital anomalies. Extra mental foramina, Inca bones, intra-malar or intra-parietal sutures, sacral "spina bifida," and such anomalies have varying incidences which might be classed from "very rare" to "not uncommon." These appear to have no effect on the physical performance of the individual, and a full survey of them would certainly be out of place here. For further literature on these minor variations, one is referred to Le Double (1903, 1912) and Brothwell (1963).

Because of the paucity of examples of major congenital deformity it is not possible to enter into a detailed survey of the various categories of such disorders, giving relevant archaeological examples. It would thus seem sufficient to describe the few specimens available rather as a series of case histories, although some subdivision of the specimens into categories of deformity can be attempted. The anomalies cover a wide range of deformity, and there is now sufficient evidence to indicate the survival of such defective people in earlier—even prehistoric—societies. Survival at one place and in one period does not of course imply survival at all times and places, and, as in modern primitive populations, one would expect reactions on the part of early communities to physical deformity of this nature.

Because congenital anomalies of the jaws have been fully dealt with elsewhere in this book by Alexandersen, further comment is not necessary. Also, inherited blood variants, which may cause identifiable bone changes, have also been considered elsewhere by Angel, and do not therefore warrant further consideration.

Talipes Equinovarus

This is by far the most common congenital deformity of the foot. The fact that it occurs once in every 1,000 births (Shands, 1951), and could well occur in such a frequency in earlier populations, makes it a likely deformity to be discovered in ancient human remains.

From Egypt, both art and mummy evidence have been given in support of its occurrence. Certainly the drawings of marked bilateral clubfoot (Fig. 1) are convincing, and seem hardly likely to be due to style or artistic inefficiency. On the other hand, the mummy cases, which have been readily accepted in the past, I think deserve critical re-examination.

FIGURE 1. Possible Egyptian art evidence of club foot; a. and b. Beni Hassan (XI-XIIth Dyns.), c. and d. El Amarna. (After M. A. Ruffer)

a. The Priest Khnumu-Nekht (XIIth Dynasty)

Cameron (1910), who undertook a detailed study of this mummy found no marked abnormality in the tibiae, except for marked platycnemia, but of the left foot he writes: "The left foot of the priest exhibits a well-marked degree of club-foot (talipes varus), the thickening of the skin and superficial tissues over the region of the cuboid being still quite evident" (p. 42-3).

Through the kindness of T. Burton-Brown of the Manchester Museum, I have recently been able to examine the foot, and take radiographs of it, in detail. Although the bones of the foot are in good condition, the soft tissues are poorly preserved and somewhat moth-eaten in appearance. Contrary to Cameron's observation, I could find no convincing evidence of tissue thickening, but there is certainly post-mortem deformity of the soft tissue. On the left side, there is some indication of compression due to tight binding. Three of the toes display large flattened areas (in two cases with mummy cloth adhering) which can only have been caused by binding, and demonstrate the degree of post-mortem deformity which can take place. The breadth of the foot in the region of the cuboid/medial cuneiform area is about 70 mm, but the breadth across the distal articular region of the metatarsals is only about 57 mm, there being noticeable post-mortem squashing together of the metatarsals. As the left foot is no longer in articulation with the tibia (and talus) and can be handled separately, it is possible to study the articular facets on the calcaneum. The anterior, medial and posterior facets for the talus are within the normal range of shape variation and the sulcus calcanei is well marked and normal in appearance.

The radiographs (Fig. 5a, b, c.) confirm the normal appearance and structure of the arch of the foot, and show the degree to which there is anomalous (post-mortem) squashing together of the metatarsals. The internal architecture of the tarsal bones, especially the calcaneus, is of normal pattern, and in view of the fact that the internal organization of these bones in equinovarus deformity is unusual (Raybuck and Manter, 1959), the normality in this case is further proof of its incorrect diagnosis in the past.

b. Pharoah Siptah (XIXth Dynasty)

As in the previous case, the left foot only is involved. Previously the concensus of opinion has been in favour of equinovarus deformity, although an alternative diagnosis of poliomyelitis has not been ruled out. Again, I should like to see a critical restudy of this specimen—it has certainly not received the detailed analysis it deserves.

The external form of the foot certainly does not rule out at least some modification at a post-mortem date, and it may be significant, as Rowling (1961) has pointed out, that no such foot deformity is reproduced in contemporary portraits.

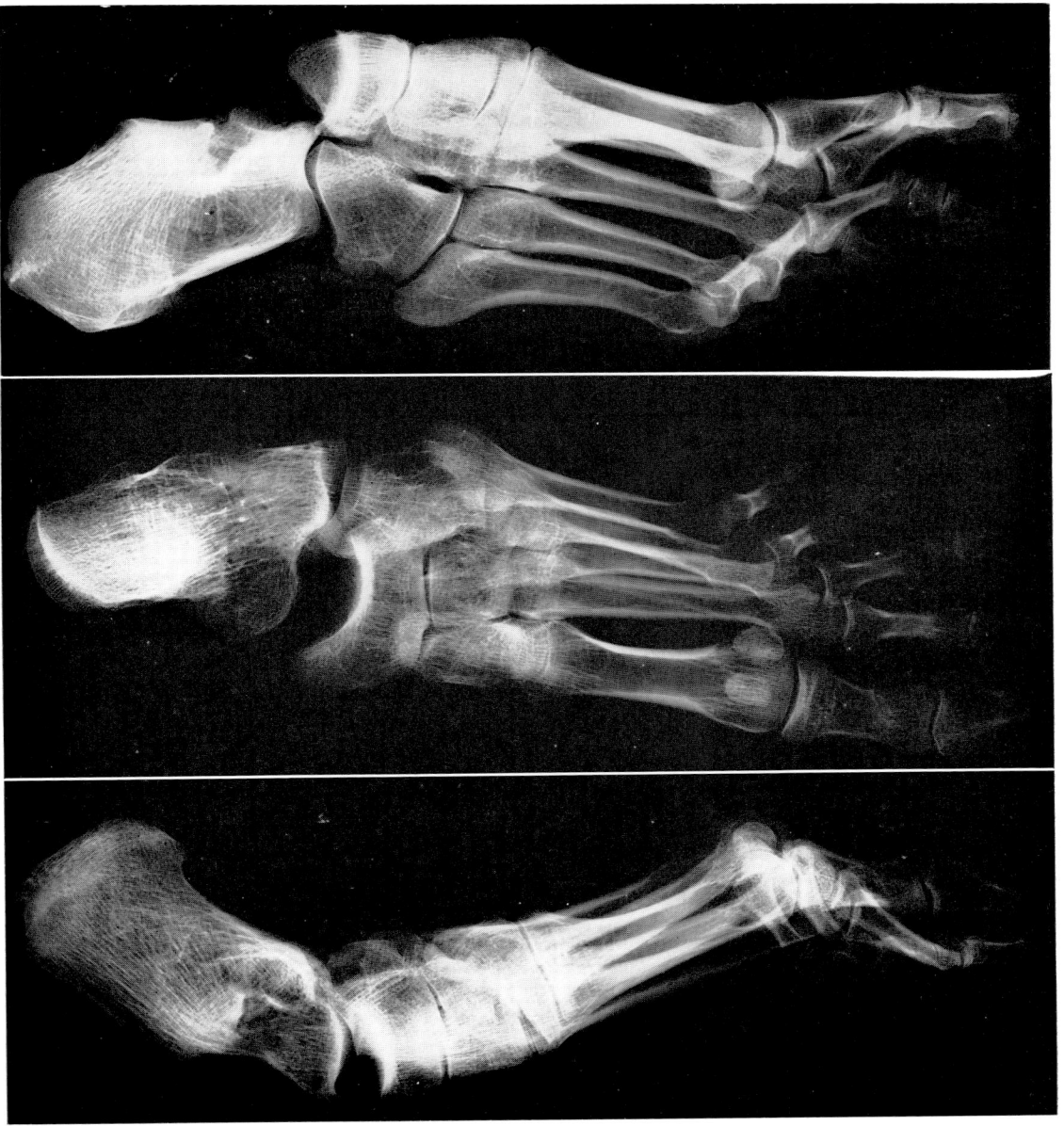

FIGURE 5. a. b. and c. Radiographs of three views of the foot of Khnumu-Nekht. Courtesy of T. Burton-Brown, Manchester Museum.

c. Nether Swell (British Neolithic)

Between the years 1867-74, a long barrow was excavated at Nether Swell, Gloucestershire, revealing three skeletons and parts of five others (Greenwell, 1877). From the fragmentary remains of one disturbed individual (Swell. ii. 3) has been reconstructed an undoubted case of talipes equinovarus—certainly the most firmly established case of all (Fig. 6a, b, c). This anomaly did not come to light until a visitor (Professor V. Møller-Christensen) working on early British skeletons in our (B.M.N.H.) collection suspected abnormality. Later, on closer study, and with the reconstruction of this left foot as far as the fragmentary remains permitted, it became clear to me that this was a severe equinovarus deformity. Working downwards and distally, the tibia may be considered first. Although there is some damage to the bone, the maximum length was probably 351 mm. The proximal articular surfaces appear to be normal in conformation, but the main shaft is abnormally rounded, with no well-defined anterior border (dimensions at the nutrient foramen being $TiD_1 = 26$ mm $TiD_2 = 23.7$ mm). The distal articular area is also noticeably anomalous, particularly at the

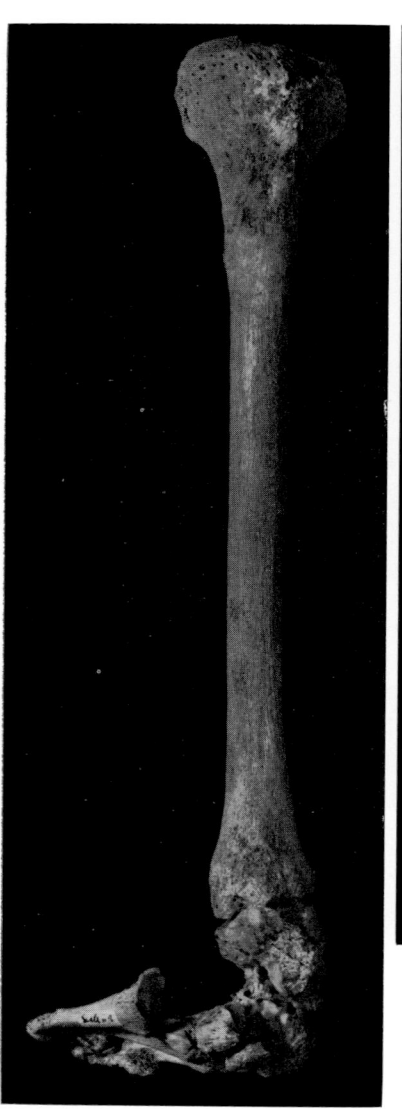

Figure 6a.

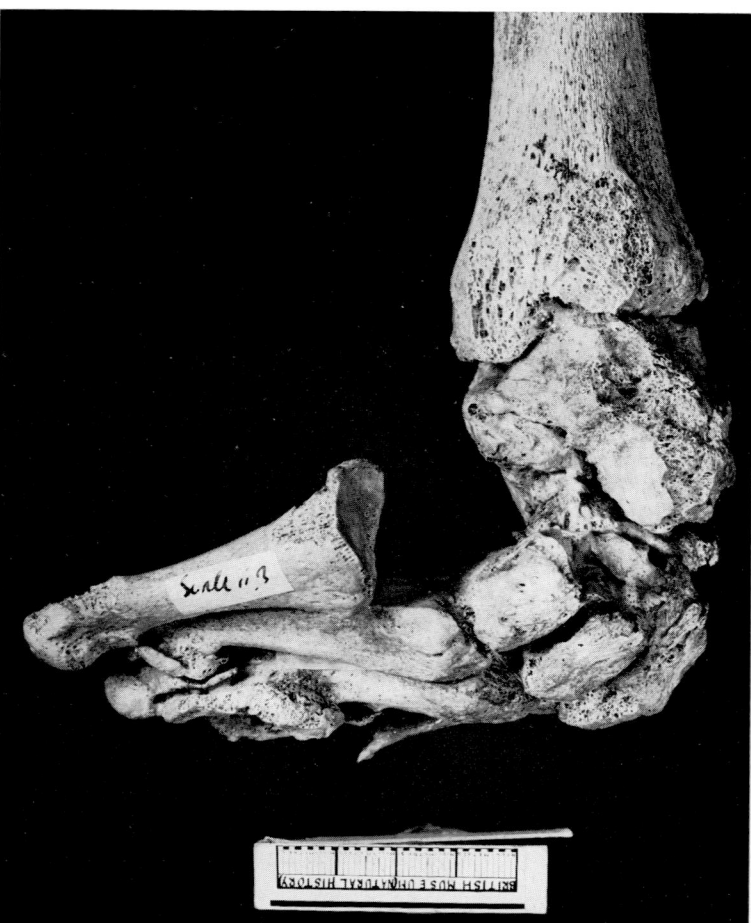

Figure 6b.

medial malleolus (where there is probably additional arthritic deformity superimposed).

The talus and the calcaneum (which is damaged) are very considerably deformed, but what seems to be the "correct" articulation for these bones, and in relation to the tibia, has been attained in the reconstruction. As in the case described by Raybuck and Manter (1959) the calcaneus shows the greatest modification of the tarsal bones. Although the navicular is missing, the cuboid—which is far more irregular than normal—can be articulated with the calcaneus. Because of the extreme inversion of the more distal bones of the foot, and the anomalous nature of the tarsal articulations, the cuboid would appear to form a small pseudo-heel to the foot.

All five metatarsals are present, although in one the distal half is lacking. They are of normal size, but at the articular ends and in parts of the shafts, there is some shape modification. Of special interest are the extra, somewhat flattened areas of bone formed on at least three of these metatarsals (Fig. 6c). Owing to the extreme inversion and posteriorly-directed twisting over of the foot, what should have been the superior aspects of the metatarsals now in fact form the under surface. Thus, in walking the individual would have applied weight to the more tender "upper" surface of the foot (shown clearly in Fig. 6a, b.) Although infection may have been a factor in the development of these flattened areas of extra bone, it would seem more satisfactory to regard them as part of the process of callosity formation,

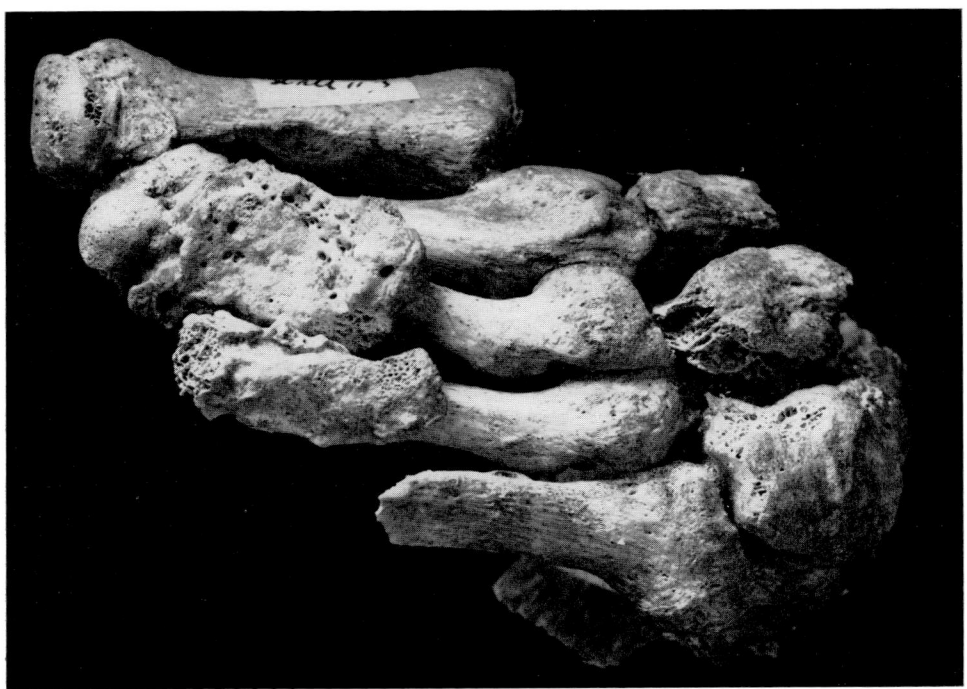

Figure 6c.

FIGURE 6. a. b. and c. Neolithic case of clubfoot from Nether Swell. *a*, general view; *b*, detail of the front view; *c*, detail of the walking (normally the superior) surface of the foot. B.M.(N.H.) collection.

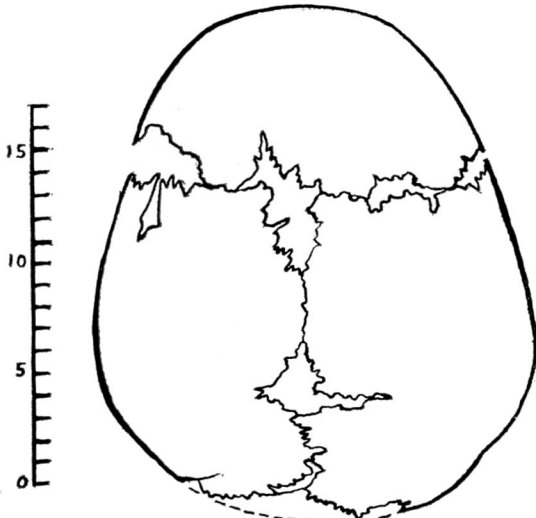

FIGURE 2. Hydrocephalic skull of a child from a Neolithic site at Seeburg, Germany. (After H. Grimm)

as a result of the continual application of this surface of the foot to the ground when walking.

Hydrocephaly

Although only about a quarter of the cases of hydrocephaly may be regarded as congenital (Laurence, 1958), it is convenient to consider this disease here. There is, of course, no way of determining from ancient skulls showing hydrocephaly whether it was initiated during prenatal development, or was caused by perinatal trauma or anoxia, or post-natally through infection or tumour growth.

a. Seeburg, Neolithic

By far the earliest case yet recorded is from Seeburg in Germany (Grimm and Plathner, 1952), dated to the Neolithic period. The individual was a child of about five or six years. Although the skull vault was somewhat broken, there is sufficient to demonstrate that the maximum length of the skull must have been about 230 mm (Fig. 2).

b. Norton, Romano-British

From a Yorkshire site has been excavated a second undoubted case. The individual was a young adult, probably of male sex (Trevor, 1950). In maximum length, the skull was 216 mm, and in breadth 176 mm, the cranial capacity being estimated at 2,600 cc (some 1,000 cc above the Romano-British average).

c. Egypt, Roman Period

Again the individual was a young adult and probably of the male sex, the detailed study of this interesting skeleton being undertaken by Derry (1913). Although the face is within the normal size range, the skull vault is expanded into an enormous dome (length 230 mm, and breadth 184 mm) whose cranial capacity was in excess of 2,900 cc (Fig. 7). Associated with this enormous cranial enlargement were changes in the left side of the post-cranial skeleton, clearly resulting from hemiplegia following the cerebral disease. In the case of the humeri, ulnae, femora and tibiae, the right bone is noticeably more robust than the left. There is also a marked difference in the type and degree of femoral torsion; also asymmetry of the sacrum, and noticeable growth retardation in the left side of the pelvis.

d. Other Possible Cases

A number of other examples of hydrocephaly have been recorded in the literature—although they are all of a far more questionable nature. One of slight degree was recorded in a Ist Dynasty skeleton from Sakkara (Batrawi and Morant, 1947), and a similar slight (and thus very questionable) instance in the Iron Age series from Lachish (Risdon, 1939). From Egypt, it has been suggested that the alleged Smenkh-ka-re was mildly affected

Figure 7 a. Normal Nubian skull (*left*), compared with the Roman period hydrocephalic (*right*). B.M.(N.H.) collection.

by this deformity (Elliot Smith, 1912), although this opinion has not gone unchallenged. Rowling (1961) has mentioned that one of the daughters of Akhenaten is represented by a statue which displays a head size which might be regarded as indicative of mild hydrocephaly.

Although about 25 per cent of modern hydrocephalies show some degree of mental defect, there is no way of telling from the skull alone the mental abilities of the individual. Thus, it was considered more reasonable to exclude these examples from the section on the evidence for mental defect.

Scoliosis

Although there is evidence that, at least in peoples of European origin, scoliosis may occur in nearly 2 per cent of the group (Shands and Eisberg, 1955), the number of vertebral columns with marked bone changes is very small indeed. Thus, although a few cases of single mild vertebral wedging (of a scoliotic not kyphotic type) have been described, there is, to my knowledge, only one case so far where there is marked congenital deformity and the spine is sufficiently complete for the column to be articulated and examined in detail. This skeleton, belonging to the Beaker (Bronze Age) period was excavated by Stuart Piggott (1940) from a barrow on Crichel Down, Dorset. Although this skeleton of a man, became well known for the large trephined hole in the skull (Fig. 7), it was not until recently that the post-cranial skeleton was exam-

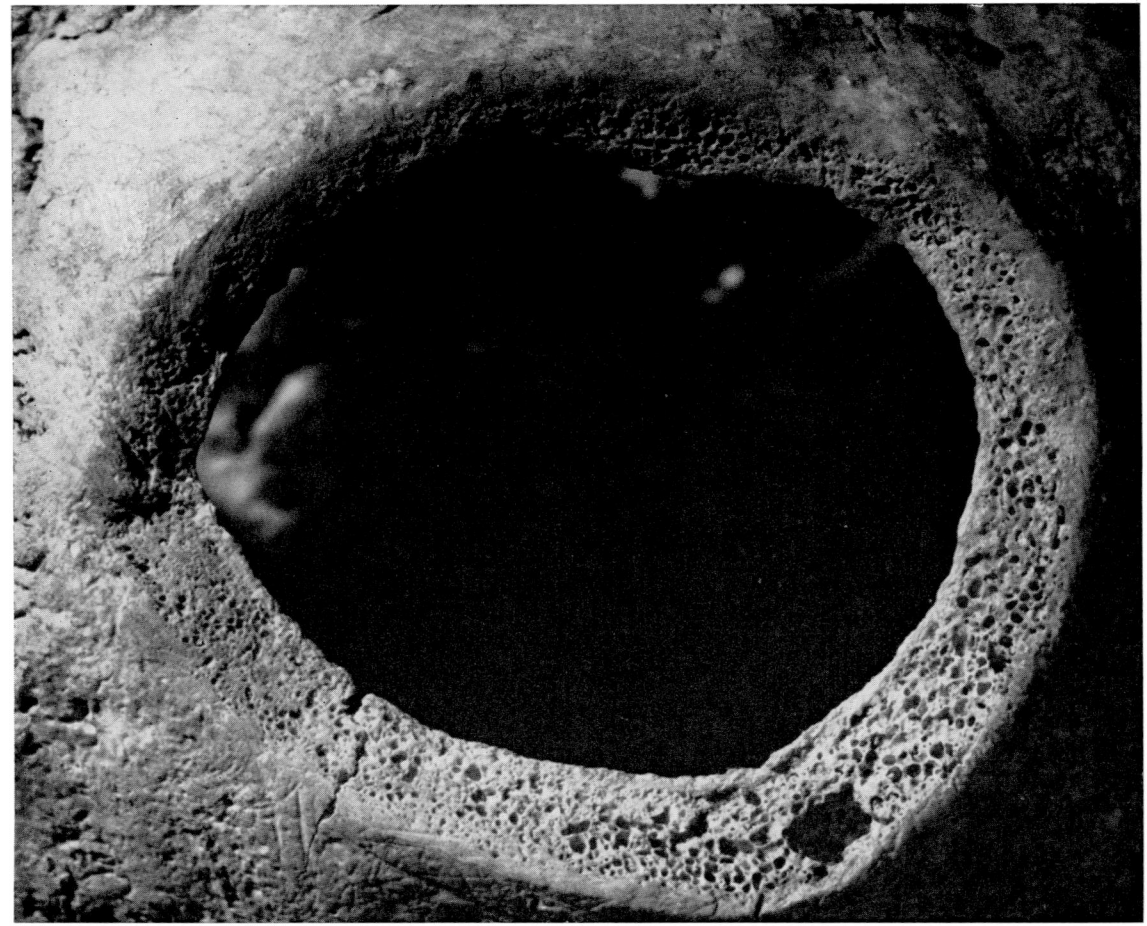

Figure 7 b. Close-up of the Crichel Down trephine hole. Courtesy of the Duckworth Laboratory of Physical Anthropology. Cambridge (Also 7c).

ined in detail and revealed the scoliotic deformity (Brothwell, 1961), both the fourth and fifth thoracic vertebrae being markedly abnormal in shape (Fig. 7c). The reason for this would appear to be that two additional centres of ossification had been present and had given rise to two wedge-shaped "hemivertebrae," each fusing on to one of the whole vertebrae. As these hemivertebrae are placed at opposite sides of the spine, the deformity at one vertebra must have been partly cancelled-out by the deformity on the other—although the column was clearly still scoliotic to some extent. Irregularity on the upper and lower surfaces of both these vertebrae suggests that intervertebral disc trouble had developed, and, in view also of the medium degree of osteo-arthritic change on eleven other vertebrae, pain may well have occurred at times. The association of this disorder with a trephine hole (albeit unhealed) makes the skeleton of special interest, and one might speculate as to whether the skull operation was performed to alleviate the back disorder.

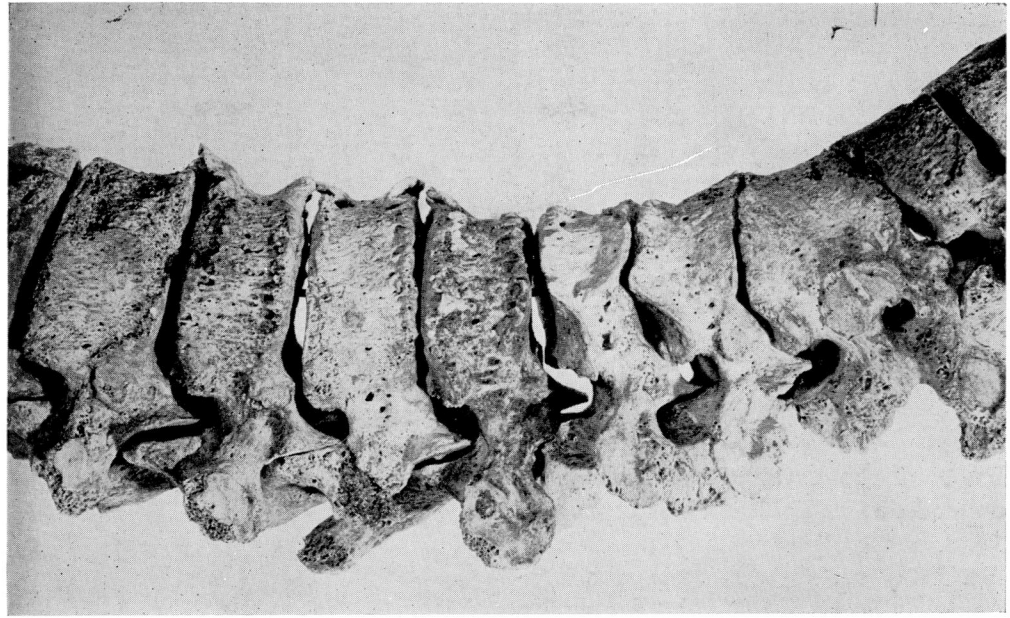

FIGURE 7 c. Part of the Crichel Down vertebral column, showing one of the fused hemivertebrae (third from right).

Klippel-Feil Syndrome

Barclay-Smith (1911) reported in detail on an incomplete skeleton from Sakkara. It was inadequately dated, but certainly Ptolemaic or earlier. From various features of the skull, the individual was considered to be a young adult female. All regions of the vertebral column show conspicuous anomalies, summarized by Barclay-Smith as follows (pp. 145-6).

"*Cervical Region.*—(a) The presence of eight vertebrae to which the term cervical may be applied. (b) The atlas is synostosed to the occipital bone. (c) The axis and 3rd cervical vertebra are fused together and form a composite mass. (d) The 7th cervical vertebra has an interrupted neural arch. The interruption is dorsi-median, the two laminae being mutually independent and the spinous process subdivided. (e) The 8th cervical vertebra is provided with a small unilateral cervical rib; otherwise it presents features characteristic of a cervical vertebra.

Thoracic Region.—(a) The joints between the 11th and 12th thoracic vertebrae are peculiar in that the opposed articular processes are lumbar in type on the one side and thoracic in type on the other.

Lumbar Region.—(a) The lumbar vertebrae are all more or less distorted, suggesting a lateral curvature in this region. (b) In the 3rd, 4th and 5th lumbar vertebrae the neural arches are interrupted. (c) In the 5th lumbar vertebra there are two interruptions, a part of the neural arch being an independent ossicle.

Sacral Region.—(a) The sacrum as a whole is distorted, presenting a lateral curvature. (b) The 1st sacral vertebra is partially lumbarised. Its neural arch presents two interruptions, and in consequence is in part an independent ossicle."

Unfortunately the conclusions to his detailed study—"that this spine evinces marked reaction to prolonged and excessive usage, that this excessive usage commenced at a fairly early stage of life, and that the possessor was a contortionist"—must be considered highly dubious. The type and extent of the vertebral anomalies is clearly indicative of a congenital disorder, and Wells (1964) has suggested that this may be an example of Klippel-Feil syndrome.

Achondroplasia

Because of the characteristic features of the upper face, cranial base, and especially the long bones, it is usually possible to make a fairly certain diagnosis of achondroplastic dwarfism from skeletal remains. Not the same degree of certainty can be hoped for from art representations, although in some cases artistic likeness is striking.

a. Old World Evidence

By far the most cases come from Egypt, either in the form of tomb illustrations and statues or as human remains. There seems little doubt that the dwarf could and often did occupy exalted positions in court. Ruffer (1911) called attention to the Vth Dynasty statuette of Chnoum-hotep from Saqqarah (Fig. 3a), a Predynastic drawing of the "dwarf of Zer" from Abydos, and a Vth Dynasty dwarf drawing from the tomb of Deshasheh—all probably achondroplastics. A VIth Dynasty carving of Seneb at Gizeh displays this dwarf clearly scaled against a wife of normal stature. The gods Ptah (Fig. 3b) and Bes also assumed dwarf size.

Of more value is the skeletal evidence, which establishes achondroplasia in Egypt as early as Predynastic times. Jones (1932) described an interesting but fragmentary

FIGURE 3. a. The court official Chnoum Hotep, depicted as an achondroplastic dwarf;

FIGURE 3. b. Egyptian amulet of the God Ptah, also shown as an achondroplastic. Drawings by Rosemary Powers.

Badarian skeleton with, surprisingly, a quite normal skull both in size and shape. In contrast to this, however, the radii and ulnae are short and robust bones typical of achondroplasia. The humeri are also shorter, the left humerus being particularly so—and more robust—than the right. A second case which may be Predynastic (certainly pre-1400 B.C.) is now in the Duckworth Laboratory of Physical Anthropology at Cambridge. This consists of both femora and tibiae, showing the typical short shafts but relatively large articular ends.

In the tomb of King Zer (1st Dynasty), an achondroplastic humerus has been reported by Bleyer (1940). From the early

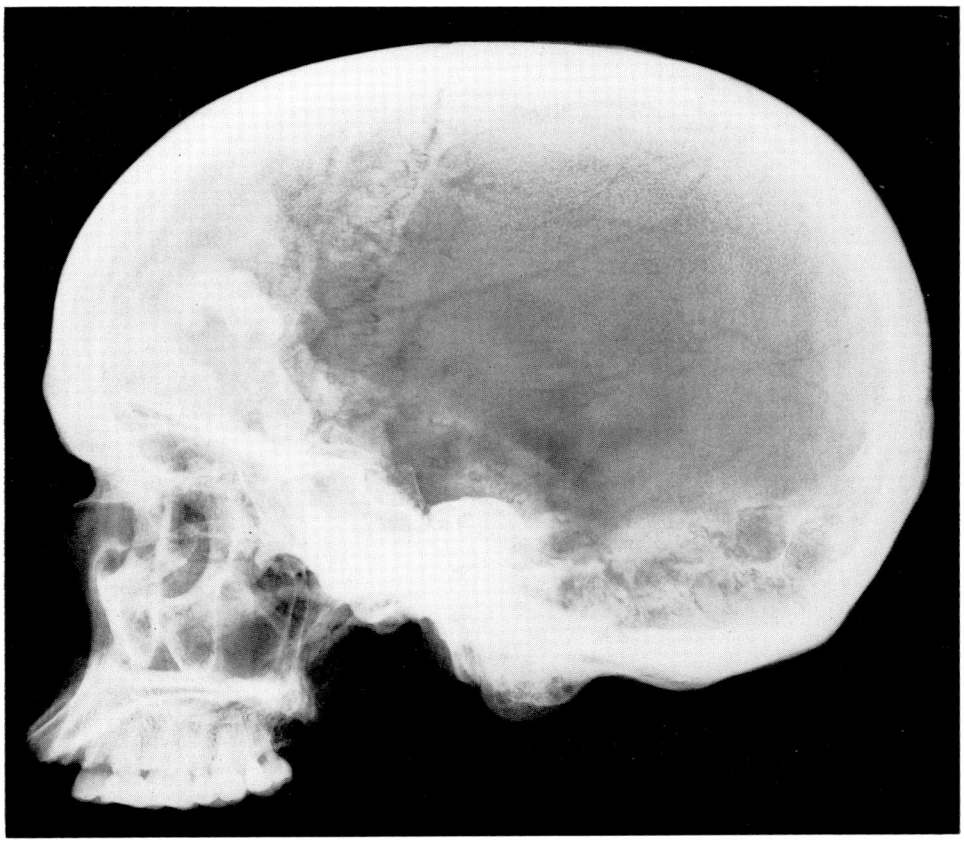

FIGURE 8 a. Radiograph of the achondroplastic skull from the tomb of King Mersekha, showing a typical high forehead and retarded nasal area.

Dynastic tomb of King Mersekha (MacIver, 1901) two further instances are known (specimens now in B.M.(N.H.) collection). One achondroplastic is represented by a typical skull (Fig. 8) and short squat long bones; the second by long bones only. An XVIIIth Dynasty skull recovered from the temple of Tuthmosis IV at Thebes shows very characteristic changes to the face and cranial vault which clearly indicate achondroplasia. Seligmann (1912) had incorrectly described this specimen as cretinous, but Keith (1913) later criticized this diagnosis.

Prehistoric achondroplasia has not yet been described in human remains from Europe, but there is one definite Frankish case from Coxyde in Belgium (kindly shown to me by Professor F. Twiesselmann).

b. New World Evidence

The valuable review of American Indian dwarf skeletons by Snow (1943) demonstrates clearly that they were not rare (considering the numbers of skeletons excavated). As early as 1875, Jefferies Wyman described human remains from a shell mound on Huntoon Island, Florida. Although achondroplasia is not certain in this case, the bones would certainly appear to have been of an adult dwarf. Fowke

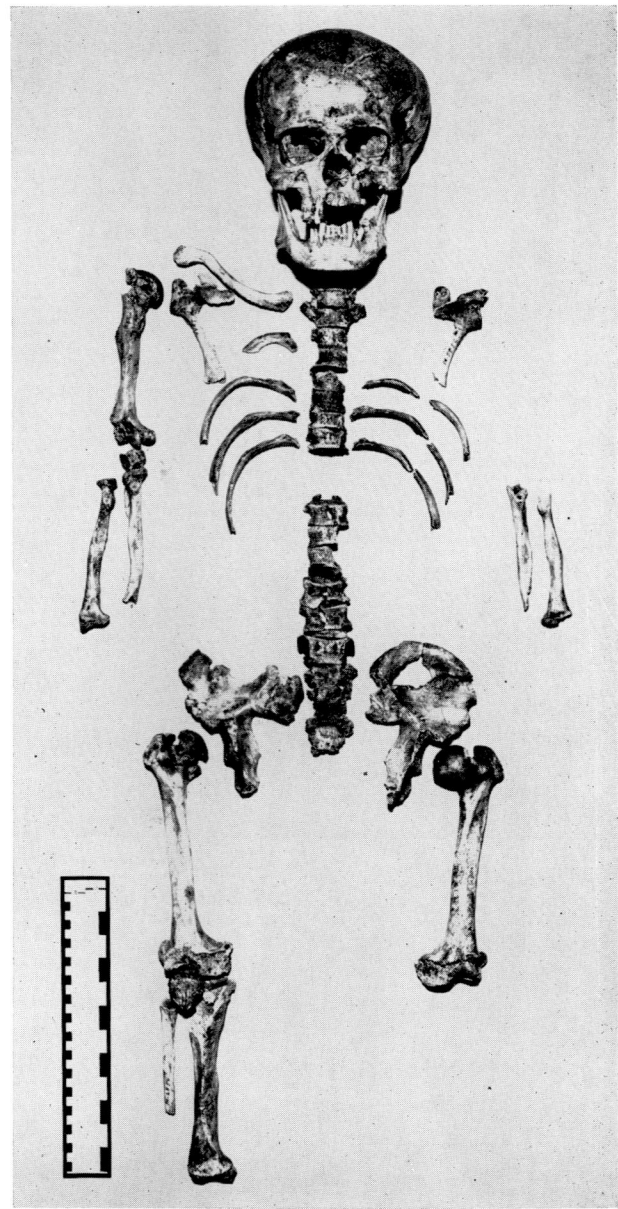

Figure 8 b. Male dwarf skeleton from Moundville, Alabama. Courtesy of Professor C. Snow and Alabama Museum (Also 8c.).

(1902), on the other hand, described dwarf bones from one of the Ohio Adena mounds near Waverly, with features of the skull and long bones which definitely suggested achondroplasia. A further skeleton from Florida, excavated at Belle Glade, also had dwarf proportions—but, alas, has since been lost as in the case of the other two.

Of special interest are the two certain

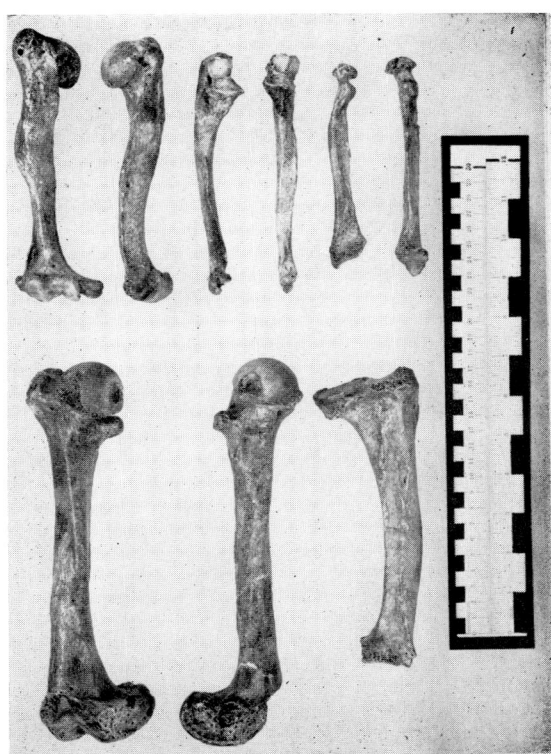

Figure 8 c. Long bones of the female achondroplastic dwarf from Moundville, Alabama.

achondroplastic dwarf skeletons from Moundville, Alabama, which have been described in detail by Snow (1943). Both of these prehistoric individuals were adult, one being male and the other female. Much of each skeleton survives, and both show the same abnormally short robust long bones (Fig. 8b and c) and retarded appearance of the nasal region of the face.

Partial Aplasia

Evidence of primary aplasia of bones is rare. Two early instances have been reported, both from British cemeteries.

a. Fibulae, Dunstable

There is to my knowledge only one instance in archaeological series of the partial suppression of a long bone, in this case the fibula. This is in the skeleton N.S.91 of the Saxon/Mediaeval series from one of the Five Knolls barrows, Dunstable Downs (Dunning and Wheeler, 1932). Because this case is very little known, and because the skeletons seem no longer to be available for study, it is worthwhile to give the published findings in some detail. Doris Dingwall (1932) who undertook the analysis of this skeletal series writes:

"The right tibia has no facet on its posterolateral aspect for the fibula. The posterior aspect of the tibia has a heavy bony ridge bounded laterally by a deep rough groove. The ridge is an exaggeration of the line which normally divides the origin of the tibialis posticus from that of the flexor longus digitorum. It is suggested that the ridge is due to the crowding of the origin of tibialis posticus and soleus muscles on the tibia in the absence of the bony fibula. The shaft of the bone is rotated so that the internal malleolus is directed antero-medially. The left tibia has a small fibular facet, circular in shape, and 4 mm by 3 mm in diameter. The diminutive facet indicates that the head of the left fibula was not fully developed. The shaft of the bone shows the same excessive development of the ridge on the posterior surface as the right fibula. Inferiorly the area on the tibia for articulation with the fibula is normal but for the presence of a small cartilaginous facet of crescentic shape, separated by a sharp ridge from the articular facet for the talus. It is suggested that this facet articulated with the lower end of the fibula, so that the external malleolus of the fibula articulated not only with the lateral aspect of the talus but also with the crescentic facet on the infero-lateral aspect of the tibia. This rare condition is bilateral. The general configuration of the bones of this subject indicate that the fibulae were suppressed to a varying degree as regards the shaft, the right fibula also being suppressed as regards the head and superior anticulation with the tibia. Both tibiae were outwardly rotated on the shaft and the feet were carried in the position of valgus" (p. 217).

b. Basi-occipital, York

From a Romano-British burial ground at York, a series of well preserved skeletons were excavated during the previous century, the skulls now being in the collection of the British Museum (Natural History). In one skull, probably from an adult male, there is extreme platybasia and complete fusion of the atlas vertebra with the occipital at the posterior half of the foramen magnum (Brothwell, 1958). Of special note, however, is the complete absence of basi-occipital, although the body of the sphenoid has not taken part in this aplasia. There seems little doubt that very noticeable head-posture deformities would have resulted from these abnormalities.

Dysplasia of the Hip Joint and Sequelae

The frequency of congenitally determined hip deformities, leading to dislocation, varies considerably between and within human racial groups. It appears to be relatively common in Italy, Germany, Japan and Czechoslovakia, but rare in China and among Negroes. Among the Lapps and Norwegians, who have received much attention as regards hip dislocation, frequencies of 5.0 and 4.2 per cent have been estimated on skeletal series (Wessel, 1918; Getz, 1955). However, considering modern cases as a percentage of the regional Norwegian populations from which they were derived, the incidence of congenital hip dislocation varies from 0.05 to 0.24 per cent (Getz, 1955), some of the differences in the seven samples available being statistically significant.

In view of this evidence that hip dysplasia and dislocation is by no means uncommon in some populations (and indeed the fact that this pelvic region is more affected by congenital deformity than any other part of the body), it is not surprising that a number of cases are known in early skeletal series. It may be significant that evidence of inadequate development of the hip joint and dislocation has as yet only appeared in archaeological remains from Europe and Africa. With

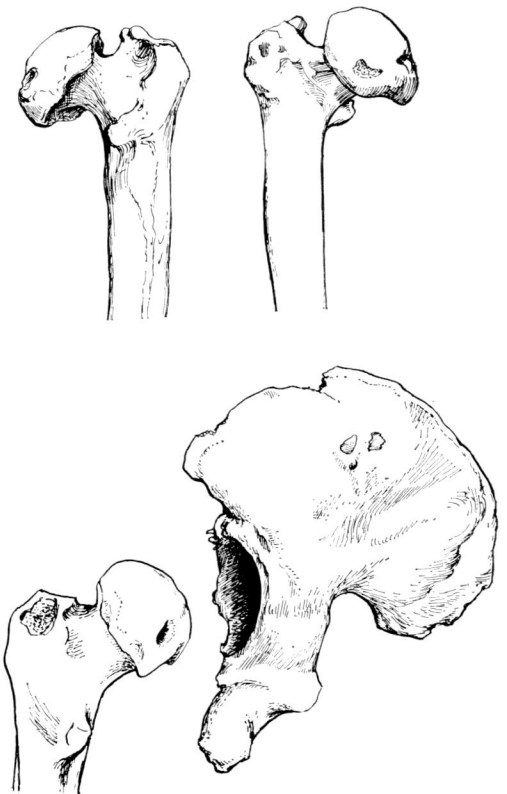

FIGURE 4. a. Deformed Neolithic femur head, probably indicative of congenital hip dysplasia. After L. Pales.

FIGURE 4. b. Shallow acetabulum and deformed femoral head in a Saxon from Surrey. B.M.(N.H.) collection. Drawing by Rosemary Powers.

regard to early medical texts, the treatise *On The Articulations* by *Hippocrates* discusses in some detail dislocation and the varying degrees of luxation and subluxation. Clearly hip disorders of this nature were well known in some parts of the early Mediterranean world by the end of the first millenium B.C.

a. France, Neolithic (Fig. 4a)

Pales (1930) gives an account of a femur showing changes in the proximal region indicative of hip dysplasia. He records "tête-fémorale en 'tampon de wagon,' col court, trapu; mais sans hyperproductions osseuses. Le grand trochanter était bas mais il convient de tenir compte sur le dessin, d'une fracture posthume ancienne de la partie supérieure de cette saillie. Le petit trochanter est volumineux."

b. Britain, Saxon

I have discussed these instances in some detail elsewhere (Brothwell, 1961). In the case of the Saxon (or Dark Age) skeleton from Lincolnshire, the female displayed a shallow left acetabulum, and a free-riding and somewhat deformed left femoral head in articulation with a false joint surface near the lesser trochanter (Fig. 9a). The left femur, patella and tibia are noticeably more slender than the corresponding bones on the right.

From the Saxon cemetery of Guildown, Surrey, one individual displayed a shallow inadequate acetabulum and flattened femoral head (Fig. 4b).

A further British case came to my notice recently in Hull Museum. The femur is not well dated, but is certainly Saxon or pre-Saxon. The femur head is markedly deformed (Fig. 9b, c), there being much flattening, irregularity at the margins, and a remarkable antero-posterior broadening.

c. Greece, Early Iron Age

Angel (1946) notes one case of hip dislocation, but does not give a detailed description.

d. Nubia, Christian period

Elliot Smith and Wood Jones (1910) have described a remarkable group of hip joint deformities which are all very similar. All five women in question were from one small district, three being from the same cemetery. This evidence of the close proximity of the cases, and the modern findings on the inheritance of hip joint dysplasia (Hart, 1952), strongly suggests that these five women were related. Certainly they all belonged to the Christian colony which settled around the temple of Philae, and as Elliot Smith and Wood Jones point out, they only noted two other cases in nearly six thousand bodies which they examined. The first case they describe will suffice to demonstrate the nature of the abnormality.

"The right femur shows the effects of long-standing separation of the epiphysis of the head. The head of the bone, of normal size and shape, was found in a perfectly normal acetabulum, but it had no skeletal connexion whatever with the rest of the femur.

The condition of the femur is remarkable, and there is no trace of any pathological process other than the separation of the epiphysis. The bone has no neck, for an oblique surface stretching from the great trochanter downwards to the base of the lesser trochanter takes its place and completes the upper end of the bone.

The articulation appears to have been formed between this oblique surface and the free head of the bone lying within the acetabulum, the great trochanter having apparently been displaced directly inwards owing to the entire absence of the neck.

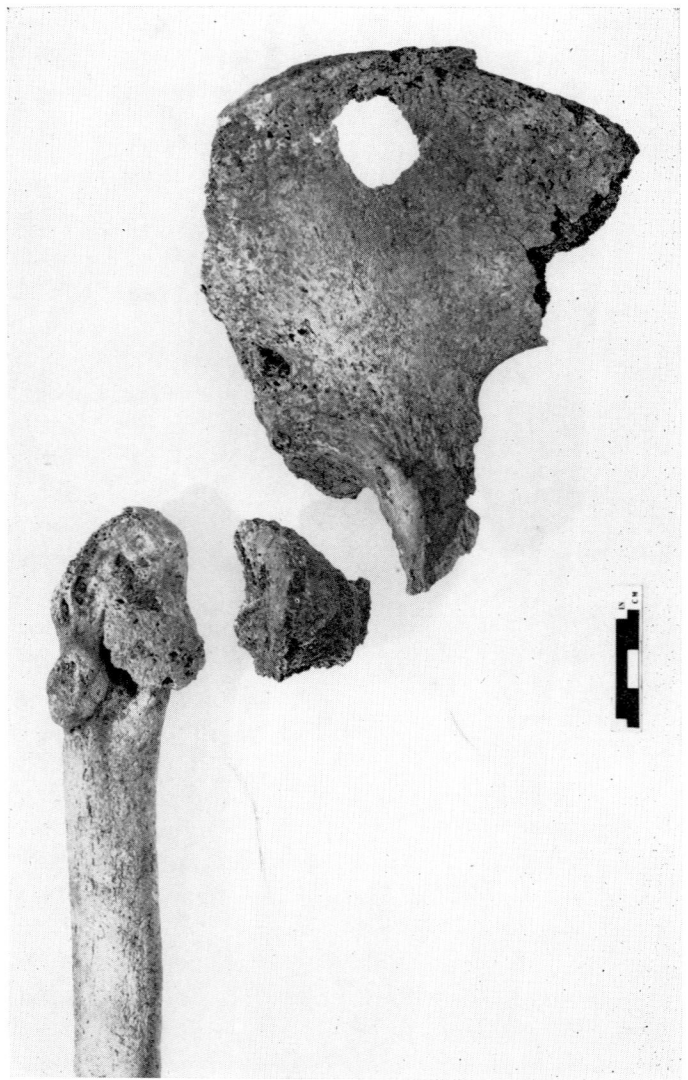

FIGURE 9 a. Anomalous hip joint in a skeleton from Lincolnshire. Courtesy of the Duckworth Laboratory, Cambridge.

The femur does not seem to have been much displaced in an upward direction, and was probably held in place by an abnormally-developed capsule. There were no signs of arthritis about the joint, and the very curious condition appears to have produced remarkably little ill result" (p. 324).

It is obvious that the primary anomaly was at the femoral head epiphysis—but it is not clear from their description what mild dysplastic features might have been present at the acetabula.

It may perhaps be mentioned here that the controversial figure of the Queen of Punt from the temple of Hatshepsut (XVIIIth Dynasty) has been considered by Slomann (1927) and some more recent writers as a case of bilateral hip deformity. This must remain dubious, however.

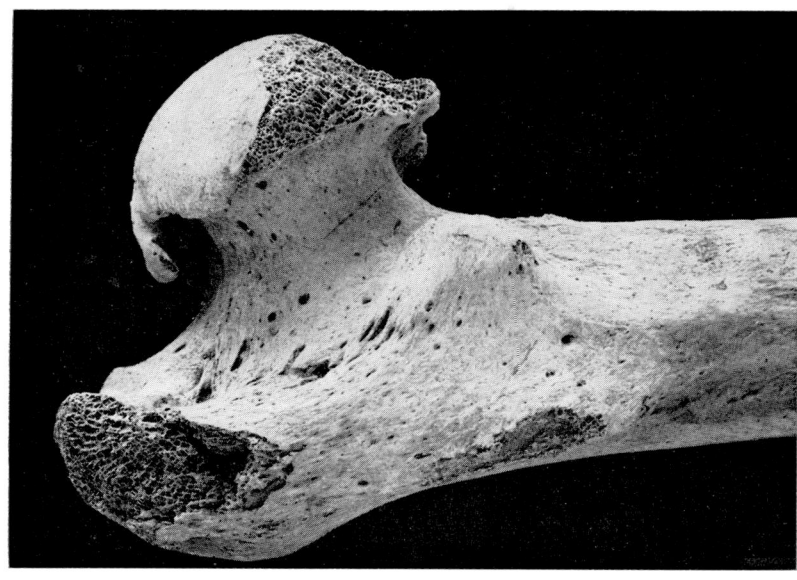

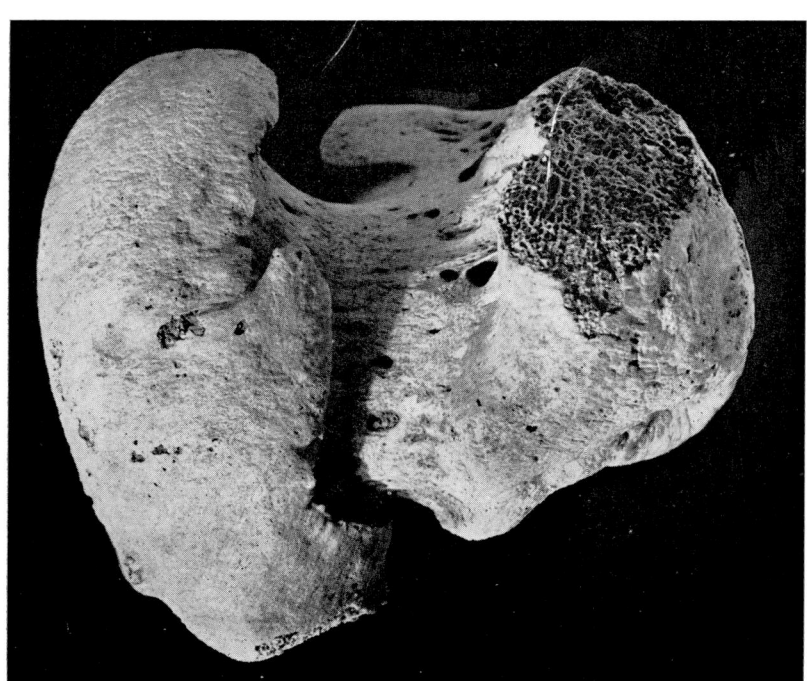

FIGURE 9 b. and c. Posterior and superior views of the deformed Saxon (?) femur head from Yorkshire. Courtesy of Hull Museum.

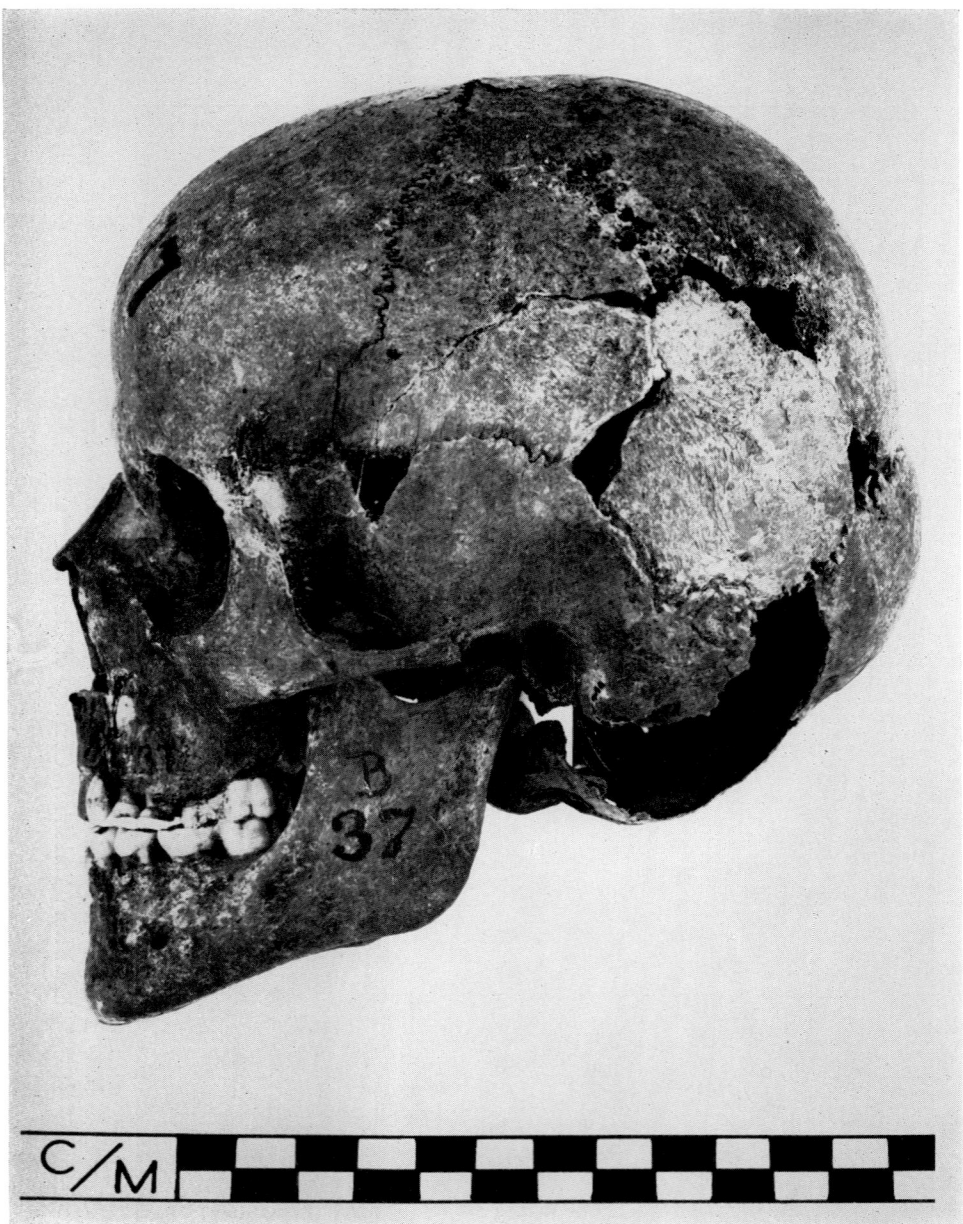

Figure 9 d. Skull of a mongolian idiot child from Breedon-on-the-Hill, Leicestershire.

e. Early Peruvian Evidence

Moodie (1923) notes briefly a pelvis of a Peruvian showing "effects of luxation of femur and formation of new acetabulum." The specimen is not well illustrated, but the rim of the true acetabulum does appear to be defective and somewhat rounded. This would appear to be the only case so far reported from ancient sites in the New World.

Anencephaly

Lortet and Gaillard (1907) and Pales (1930) have noted the presence of a mummified human case of anencephaly amongst a series of monkey mummies excavated at Hermopolis, Egypt. A detailed description does not, however, appear to have been published. Whether the specimen is human also appears to be in doubt.

Congenital Mental Defect

Excluding deformities such as hydrocephaly and acrocephalosyndactyly, where there may sometimes be mental retardation, definite examples of mental defect occur in the form of mongolism and microcephaly.

a. Mongolism

As yet there is only one highly probable case of mongolism in early skeletal remains (Fig. 9d). This specimen (the skull only) was recovered by me during part of a series of rescue excavations on Breedon-on-the-Hill, in Leicestershire (Brothwell, 1960), and is probably late Saxon in date. The most notable features of this skull are: the skull is microcephalic, the cranial capacity being about 300 cc smaller than the smallest comparative skull of the same developmental age; hyperbrachycephaly, a most unusual feature for an Anglo-Saxon, is associated with a very globular vault; the sphenoid body is small and the basi-occipital has a noticeably high angle; all the vault bones are thin; and finally, maxillary smallness contrasts with a fairly normal sized mandible. As all these features are common in mongolism it seems reasonable to consider the Breedon case as an example of this chromosomally determined abnormality, namely trisomy of chromosome 21.

b. Microcephaly

Early cases of true microcephaly are as rare as the mongol, and in view of the much greater frequency of mongolism-- relative to microcephaly—it is in fact less likely that the latter condition will appear in early cemeteries. Although a claim of late prehistoric microcephaly has been made, the earliest likely case is from Donnybrook, Ireland (Frazer, 1879), probably of tenth century A.D. date (Martin, 1935). A more recent case, no earlier than the twelfth century, occurs in the ossuary at Hythe, Kent.

Apparent evidence of microcephaly can be exemplified from prehistoric and protohistoric art representations, but it is uncertain whether such instances can be taken as intentional.

Osteogenesis Imperfecta

Rowling has described in the section on paraplegia, a remarkable case of Nubian pathological change, showing gross deformity in both forearms and in the legs. Although he has given reasons why he would prefer paraplegia rather than osteogenesis imperfecta as the primary cause of these skeletal abnormalities, it must nevertheless be remembered that this alternative cannot be eliminated altogether.

Acrocephaly (?-syndactyly)

There is still much confusion regarding the use of the terms acrocephaly and oxycephaly, and whether acrocephaly is quite distinct from acrocephalo-syndactyly or is purely a milder expression of the same syndrome. For myself, I would prefer the use of the term oxycephaly for skulls showing only premature obliteration of the coronal suture—with resulting "tower skull." Oxycephaly is not uncommon and the minor nature of the deformity excludes it from further discussion here. Genetic studies by Blank (1957) suggest that deformity resulting from widespread and premature vault suture obliteration should be kept distinct from skull changes associated with syndactylous deformity.

The case to be discussed must be considered as something more than simple oxycephaly. On the other hand, as the post-cranial skeleton is missing, it is not possible to decide whether it is acrocephaly or acrocephalosyndactyly. The skull is part of the Nubian collection given to the Royal College of Surgeons in 1908, and now preserved in the B.M.(N.H.) It is Early Christian in date (being from Grave 13B, Cemetery 35). Only the vault of the skull remains and would appear to be from a child of about two or three years of age (in view of the fact that it may be a small skull for its age, size is not a very reliable indicator of age in this case). The most notable abnormality is the complete obliteration of the coronal and sagittal sutures. Much of the lambdoid suture is also obliterated and the squamous portion of the left temporal is united with the parietal. The shape of the vault, with a disproportionately broad basal area, suggests that obliteration took place only a short time before death; certainly there is no evidence of compensatory growth in the upper vault region as yet, although cranial thinning may have commenced.

Clearly had the child lived, considerable vault changes would have taken place, similar to those described in detail recently for Apert's syndrome (Glass, 1958).

Diaphyseal Aclasis

Singer (1961) has described a female negroid skeleton from Que Que, Rhodesia, showing anomalous bone development attributable to diaphyseal aclasis. Multiple spongy exostoses are present and the right tibia and fibula are united near the distal and proximal ends.

GENERAL CONCLUSIONS

Although the whole range of congenital developmental errors known to occur in modern populations has not been noted as yet in earlier peoples, there is sufficient evidence to demonstrate that a variety did occur. The discovery and study of such deformities is of special value in ascertaining the likelihood and frequency of their survival in prehistoric and proto-historic communities, and helps to mirror the attitudes of earlier communities towards such obvious deformities.

REFERENCES

ANGEL, J. L., 1946: Skeletal change in Ancient Greece. *Amer. J. Phys. Anthrop.*, 4:69.

BARCLAY-SMITH, E., 1911: Multiple anomaly in a vertebral column. *J. Anat. Lond.*, 45:144.

BATRAWI, A., and MORANT, G. M., 1947: A Study of the First Dynasty series of Egyptian skulls from Sakkara and of an Eleventh Dynasty series from Thebes. *Biometrika*, 34:18.

BLANK, E., 1957: Acrocephalosyndactyly of Apert's type; Genetic Study. Unpublished thesis, University of London.

BLEYER, A., 1940: The antiquity of achondroplasia. *Ann. Med. Hist.*, 2:306.

BROTHWELL, D. R., 1958: Congenital absence of the basioccipital in a Romano-Briton. *Man*, 58:73.

BROTHWELL, D. R., 1960: A possible case of mongolism in a Saxon population. *Ann. Hum. Genet.*, 24:141.

BROTHWELL, D. R., 1961: The palaeopathology of early British man: an essay on the problems of diagnosis and analysis. *J. Roy. Anthrop. Inst.*, 91:318.

BROTHWELL, D. R., 1963: *Digging up Bones*. London, British Museum (Natural History).

CAMERON, J., 1910: The anatomy of the mummies. In: M. A. Murray, *The Tomb of Two Brothers*, pp. 33-47. Manchester, Sherrat & Hughes.

DERRY, D. E., 1913: A case of hydrocephalus in an Egyptian of the Roman period. *J. Anat. Lond.*, 48:436.

DINGWALL, D., 1932: A barrow at Dunstable, Bedfordshire. Part II. The skeletal material. *Archaeol. J.*, 88:210.

DUNNING, G. C and WHEELER, R. E. M., 1932: A barrow at Dunstable, Bedfordshire. Part I. The archaeological evidence. *Archaeol. J.*, 88:193.

ELLIOT SMITH, G., 1912: *The Royal Mummies*. Catalogue général des Antiquités égyptiennes du Musée du Caire, LIX.

ELLIOT SMITH, G., and WOOD-JONES, F. 1910: Report on the human remains. *Archaeological Survey of Nubia. Report of 1907-1908*, 11. Cairo, Egyptian Survey Dept.

FOWKE, G., 1902: *Archaeological History of Ohio* Columbus (Quoted by Snow, 1943).

FRAZER, W., 1879: Description of a sepulchral mound at Donnybrook containing human and other remains referable to the tenth or eleventh century. *Proc. Roy. Irish. Acad.* 2:29, 116.

GETZ, B., 1955: The hip joint in Lapps and its bearing on the problem of congenital dislocation. *Acta Orthop. Scand.*, Supplement XVIII (18)

GLASS, D. F., 1958: Acrocephalosyndactyly. *Trans. Europ. Orthodont. Soc.* (Reprinted 1-23.)

GREENWELL, W., 1877: *British Barrows. A Record of the Examination of Sepulchral Mounds in Various Parts of England*. Oxford, Clarendon.

GRIMM, H. and PLATHNER, C. H., 1952: Über ein jungensteinzeitlichen Hydrocephalus von Seeburg in Mansfelder Seekris und sein Gebiss. *Deutsch. Zahn, Mund Keiferheilk.*, 15, 1.

HART, V. L., 1952: *Congenital Dysplasia of the Hip Joint and Sequelae*. Springfeld, Thomas.

HIPPOCRATES (Trans. F. ADAMS) *The Genuine Works of Hippocrates*. 1939: Baltimore, Williams and Wilkins.

JONES, E. W. A. H., 1932: Studies in achondroplasis. *J. Anat. Lond.*, 66:565.

KEITH, A., 1913: Abnormal crania—achondroplastic and acrocephalic. *J. Anat. Lond.*, 47:189.

LAURENCE, K. M., 1958: The natural history of hydrocephalus. *Lancet*, 1152.

LE DOUBLE, A. F., 1903: *Traité des variations des os du crane de l'homme*. Paris, Vigot.

LE DOUBLE, A. F., 1912: *Traité des variations des os de la colonne vertébrale*. Paris, Vigot.

LORTET, L. C., and GAILLARD, C., 1907: La faune momifiée de l'ancienne Egypte et recherches anthropologiques. 3ᵉ ser., Extr. Arch. Mus. Hist. Nat. de Lyon. (Quoted by Pales, 1930.)

MacIVER, R., 1901: *The Earliest Inhabitants of Abydos*, Oxford, Clarendon.

MARTIN, C. P., 1935: *Prehistoric Man in Ireland*. London, MacMillan.

MOODIE, R. L., 1923: *Paleopathology: an Introduction to the Study of Ancient Evidences of Diseases*. Illinois, University Press.

PALES, L., 1930: *Paléopathologie et Pathologie Comparative*. Paris, Masson.

PIGGOT, S., 1940: A trepanned skull of the Beaker period from Dorset and the practice of trepanning in prehistoric Europe. *Proc. Prehist. Soc.*, 6:112.

RAYBUCK, H. E., and MANTER, J. E., 1959: An anatomical study of a case of clubfoot. *Anat. Rec.*, 133:677.

RISDON, D. L., 1939: A study of the cranial and other human remains from Palestine excavated at Tell Duweir (Lachish) by the Wellcome-Marston Archaeological Research Expedition. *Biometrika, 31:* 99.

ROWLING, J. T., 1961: Pathological changes in mummies. *Proc. Roy. Soc. Med.* 54, 409.

RUFFER, M. A., 1911: On dwarfs and other deformed persons. *Bull. Soc. Archéol Alexandrie*, 13:1.

SELIGMANN, C. G., 1912: A cretinous skull of the Eighteenth Dynasty. *Man*, 12:17.

SHANDS, A. R., 1951: *Handbook of Orthopedic Surgery*. St. Louis, Mosby.

SHANDS, A. R., and EISBERG H. B., 1955: The incidence of scoliosis in the state of Delaware. *J. Bone Joint. Surg.*, 37A:1243.

SINGER, R., 1961: A skeleton with diaphysial aclasis. *S. Afr. Archaeol Bull.*, 17:14.

SLOMANN, H. C., 1927: Contribution à la paléo-pathologie égyptienne. *Bull. Mém. Soc. Anthrop. Paris*, 8: 62.

SNOW, C. E., 1943: Two prehistoric Indian dwarf skeletons from Moundville. *Alabama Museum Paper* 21. University, Alabama.

TREVOR, J. C., 1950: Notes on the remains of Romano-British date from Norton, Yorks. In: *The Roman Pottery at Norton, East Yorkshire*, pp. 39-40. Leeds.

WELLS, C., 1964: *Bones, Bodies and Disease*. London, Thames & Hudson.

WESSEL, A. B., 1918: Laaghalte slefter i Finmarken. *T. Norsk. Laegeforen.*, 38:337.

WYMAN, J., 1875: Fresh water shell mounds of the St. John's River, Florida. *Peabody Acad. Sci.* (Quoted by Snow, 1943.)

Chapter 35

Hernia in Egypt

J. THOMPSON ROWLING

The presence of a hernia is a striking abnormality, usually only too apparent to both the patient and his medical attendant at the present time;—and no doubt equally so in Ancient Egypt.

Abdominal herniae are of two types, the majority being congenital in origin, and a minority being acquired. In congenital inguinal herniae, however, a potential sac may be present throughout childhood, only manifesting itself as a frank hernia during adult life. The basic anatomy of all herniae consists of a defect in the muscular abdominal wall through which a sac of peritoneum protrudes into the subcutaneous tissues. This sac contains bowel or other viscera which are perpetually threatened by the risk of strangulation, whereby their blood supply is cut off by the constriction of the neck of the sac. Should this occur the affected viscera within the sac become gangrenous and death is almost inevitable without surgical intervention. The common sites for hernia are at the umbilicus and in the groin, where their presence is usually marked by an obvious swelling, and by associated pain and discomfort. The occurrence of strangulation is to some extent a matter of chance. A small hernia, particularly a femoral hernia, may cause fatal strangulation at a relatively early stage. Conversely a large hernia may be present for many years without causing more than minor discomfort.

Hernia is relatively common in animals, the pig and the bear being examples. In the former the majority are congenital inguinal herniae.

The evidence available to us suggests that hernia was neither more nor less common in Egypt than it is to-day. The condition was certainly recognized by the Egyptians, and a clear description is given in the Papyrus Ebers. Some tomb reliefs also show unequivocal evidence of herniae.

Egyptian Representations of Hernia

Egyptian portraiture, particularly the "portrait statue" buried with the dead, tended to avoid the depiction of any bodily defect. An example of this is seen in the case of the Pharaoh Siptah, whose mummy shows a gross equinus deformity of the left foot, which is not reproduced in his contemporary portraits. A reason for this may be found in the sixty ninth chapter of the Book of the Dead: "I have become a spirit, I have been judged, I have come and I have avenged my own body. I have taken up my seat by the divine birth-chamber of Osiris and I have destroyed the sickness and suffering which were there. I have become mighty. . . ." The passage suggests that the disease from which the deceased died, and, by inference any other disease, was not reproduced in the Sahu or spiritual body of the hereafter. The care taken by the embalmers to restore the mummy to perfection, ilustrated by the covering of bedsores by gazelle skin in the mummy of Nesi-Tet-Nab-Taris in the XXIst Dynasty, also demonstrates the point. For evidence of herniae in Egyptian art we must therefore turn from the formal portrait to the informal illustration of the scenes of everyday life.

Within the conventions of Egyptian art a certain latitude was allowed whereby individual characteristics could be depicted, much as the mediaeval stonemason was allowed to portray his friends and enemies in his carvings after the main sculpture was completed. Undoubted cases of umbilical herniae are seen in some of the figures shown on the walls of the chapel of the tomb of Ptah-Hetep, dating from the Old Kingdom, at Sakkara. One of a group of papyrus carriers is shown with a small but unmistakable umbilical swelling. There is also a less clearly defined swelling in the umbilical area of one of the figures depicted in the boating scene. In the tomb of Mehou, also in Sakkara, and dating from the VIth Dynasty, two further examples of umbilical herniae are seen. There is also some suggestion of scrotal swellings in some of the figures, but the pathology here must remain obscure, and there is insufficient evidence to postulate the presence of inguinal herniae. A description of the scenes is given by Ghalioungui (1963).

The Medical Papyri

There is only one clear reference to inguinal hernia in the eight great medical papyri. This occurs in the hundred and sixth column of the Papyrus Ebers, and reads:

"Instructions concerning a swelling of the covering on his belly's horns:

If thou examinest a swelling of the covering on his belly's horns, above his pudenda, then shalt thou place thy finger on it and examine his belly, and knock (?) on thy fingers; if thou examinest his ? that has come out and hast arisen by his cough, then shalt thou say concerning it; it is a swelling of the covering of his belly; it is a disease which I will treat. It is heat on the bladder in front in his belly which causes it to fall downwards, and the return is likewise. Thou shalt heat it in order to shut it up in his belly; thou shalt treat it as the Priest of the goddess Sekhmet treats." Ebbell (1937).

This translation must be accepted with some reserve, as an accurate rendering is a matter of the greatest difficulty. The "belly's horns" refers to the inguinal region above the inguinal ligament. Seen from in front, the pudenda and inguinal ligaments are aptly likened to an animal's head and horns, the terminology being as descriptive as the Graeco-Roman phraseology in use to-day. The demonstration of the hernia by coughing and its reduction by pressure are noteworthy. The treatment recommended is undoubtedly surgical, but whether the reference to heat is a mistranslation, as seems likely, or a description of cauterisation of the area, is obscure. It is interesting to note that no mention is made of trusses or similar restraints on the hernia.

The description is an important one, and it might be considered permissable to paraphrase it using reasonable deductions as to the author's meaning:

"Instructions concerning a subcutaneous swelling in the inguinal region:

If you examine a swelling beneath the skin in the inguinal region, above the pubic symphysis, you should place your finger on it, and examine it by pressure exerted by your fingers; if you examine his swelling which has appeared after he has coughed, then you should say concerning it: it is a subcutaneous abdominal swelling; it is a condition which I will treat. It is heat [distension?] of the bladder in the anterior part of the abdomen which causes the swelling to come down, and reduction of the hernia is similar. You should heat it [operate upon it?] in order to confine the hernia within the abdomen; you should treat it as the surgeon treats it."

We have no further description of the operation, nor have we any objective evidence from mummies of any such operation having been attempted.

It seems likely that by the time mummification became common, in the New Empire period, operative surgery was no more than a hallowed tradition of the Old Kingdom, included in the literary works on medicine, but never actually practised. This would account for the fairly frequent references to operative procedures in the papyri, and the complete absence of surgical scars in mummies.

It is, perhaps, surprising that there is no clear description of the onset of strangulation in herniae to be found in the papyri.

Hernia in Mummies

We have no absolute proof of hernia in mummies, although it seems likely that among the many thousands of mummies which have been examined indications of herniae have been missed. The embalmers would undoubtedly empty any hernial sac filled with viscera which they encountered, and this may account for the absence of objective evidence.

The scrotum of Ramesses V was very large and bulky, suggesting the presence of either a large inguinal hernia or else a hydrocoele or cyst of the epididymis. The mummy was relatively well preserved and it is surprising that Elliott Smith, who unwrapped it, makes no comment on the inguinal canal in his two published descriptions (Smith, 1912; Smith and Dawson, 1924).

The scrotum of the mummy of Merneptah was removed during the process of embalment, and it is possible that it contained a strangulated hernia with gangrene, which the embalmers considered detrimental. No description is given of the inguinal canal.

CONCLUSIONS

From the evidence available external hernia was probably common in Egypt. No reference is found in the papyri to the control of hernia by trusses or simple bandaging, nor is strangulation described. Nevertheless operative treatment is mentioned, although we have no evidence that it was undertaken, at least after the Old Kingdom period.

REFERENCES

EBBELL, B., 1937: *The Papyrus Ebers*. Levin and Munksgaard, Copenhagen.

ELLIOT SMITH, G., 1912: *The Royal Mummies*. Musée du Caire, Cairo.

ELLIOT SMITH, G., and DAWSON, W. R., 1924: *Egyptian Mummies*. Allen & Unwin, London.

GHALIOUNGUI, P., 1963: *Magic and Medical Science in Ancient Egypt*. Hodder and Stoughton, London.

SECTION V

SOMATIC DISEASE (REGIONAL AND SYSTEMIC)

Chapter 36

Diseases of the Skin

A. T. SANDISON

The skin being one of the more extensive tissues of the body is relatively susceptible to trauma and other influences. Sigerist (1951) points out that skin eruptions are not rare in undeveloped societies of the present day and are treated by rational methods. Further, the skin is readily accessible to simple examination and therefore it is likely that there would be evidence of skin disease in the ancient literature.

The precise diagnosis of diseases of the skin is, however, fraught with difficulty. Classification based on a secure pathology awaited Ferdinand Hebra (1816-1880) and later Paul Gerson Unna (1850-1929). It would, therefore, seem probable that descriptions of skin lesions in the ancient literature would not always lend themselves even to provisional diagnosis. To some extent this proves to be true; difficulties of interpretation also rise because of the nomenclatures adopted by the Greeks and Romans.

Saggs (1962) states that the ancient Babylonian literature contains reference to icterus of skin, skin pustules and ulcers of the mouth and lips. Krumbhaar (1937) indicates that the Library of Assurbanipal contained references to pediculosis, scabies and possibly leprosy.

The Holy Bible contains references to the botch, scab and itch in Chapter 28 of Deuteronomy. Bennett (1896) believed that the Sixth Plague of Egypt was an epidemic of carbuncles; carbuncles have also been thought to be Hezekiah's disease. Job's affliction has attracted much attention and has been variously attributed to staphylococcal infection, leprosy or cutaneous leishmaniasis (Bennett, 1896; Shapter, 1834). Guy (1955), however, regards it as some form of prurititic psychosomatic disease which might have been lichen planus, psoriasis, atopic eczema, chronic urticaria, dermatitis herpetiformis or generalised exfoliative dermatitis. Certainly, Job used a potsherd to scratch his skin.

Much confusion exists with regard to leprosy in the Bible (Leviticus, 13). True leprosy was probably rare and *Zaraath* often translated as leprosy probably included such various conditions as simple vitiligo, eczema, psoriasis, alopecia, boils and myiasis (Bennett, 1896; Brim, 1936; Preuss, 1911; Shapter, 1834). Meenan (1955) suggests that much Biblical leprosy was psoriasis and notes that Naaman, Captain of the Hosts of the King of Syria, was cured by washing in the River Jordan. Meenan belives that scabies may also have been included in the "leprosy" group of diseases.

Brim (1936) points out that Esau suffered from hypertrichosis. Grant (1951) has suggested that the *Mark of Cain* may have been a rosacea induced by guilt and shame. He also notes that Theocritus (3rd cent. B.C.) suggested that pimples on the nose (? rosacea) might be psychosomatic and result from telling a lie.

The Ancient Egyptian literature refers extensively to skin disease; this was treated by magico-religious and by cosmetic methods (Sigerist, 1951). The Papyrus Ebers (Ebbell, 1937) refers not only to skin diseases but contains recipes for the treatment of baldness, including alopecia areata, the preservation of the hair, the concealment of grizzling of the hair and eyebrows, the darkening of the white cicatrices which result from burns and the treatment of wrinkles of the skin. Icterus, pruritis of the anus and genitalia and foetid intertrigo are also considered.

Skin diseases proper, which have been identified with varying degrees of confidence, include "eating ulcers," phagadaena, suppurating inguinal adenitis, erysipelas, possible herpes zoster, weeping eczema, discharging exanthem of the scalp, scrofuloderma, lepra mutilans and sebaceous cysts (Ebbell, 1937; Sigerist, 1951; Rowling, 1961). It must be reiterated that such diagnoses can only be provisional. Sigerist (1951) points out that later Coptic manuscripts indicate extensive knowledge of numerous skin diseases.

The Ancient Persian *Videvdad* contains references to skin diseases which include scabies and possibly leprosy. Herodotus (Bk. 1:138) indicates that Persians suffering from leprosy and the scab were isolated and forbidden the city. Skin diseases in men and horses were cured by immersion in the River Tearus in Thrace (Bk. IV:90).

Chadwick and Mann (1950) indicate that in *Corpus Hippocraticorum* some fairly firm diagnoses may be made. Erysipelas is certainly identifiable. Among the commoner lesions mentioned are rashes, sometimes pustular in type, which accompany fevers. The skin in fevers was known to be dry and that of consumptives thought to be smooth. Generalised pruritis and scratch-like eruptions were known. Helminthic infections were said to be accompanied by diarrhoea and a red scaly skin. Simple vitiligo is mentioned and chilblains are well described. Pedunculated warts were known to occur in children (*Aphorisms* III:26) and possible skin cancer and rodent ulcer (*Aphorisms* VI:45 and *Traditions in Medicine* 19).

Meenan (1955) states that the Greeks divided skin lesions into three main types. *Psora* was a moist ulcerated and pustular condition; *lepra* was scaly and *leichen* was tuberculous. True leprosy was known as *elephantiasis graecorum*: *lepra* included cases of psoriasis. It is clear that *elephantiasis graecorum* and *lepra* conditions were sometimes confused and the term *lepra arabicum* was sometimes used to designate true leprosy.

In an important paper Grant (1951) states that the Greek physicians certainly recognised acne which they knew as $\mathring{\iota}ον\theta οί$ (ionthoi): Aristotle described this in sufficient detail for identification. That $\mathring{\iota}ον\theta οί$ was associated with puberty is implied in the meaning of this word in the singular—$\mathring{\iota}ον\theta ος$ (ionthos)—which means first growth of the beard. The Roman writers, e.g., Pliny and Celsus, used the word *varus* for acne but later Cassius (3rd cent. A.D.) states that the disease is called $\mathring{α}κμας$ (akmas) since it occurs at puberty $\mathring{α}κμή$ (akme). Aetius (5th cent. A.D.) used the word acne. Grant points out that the Ancient Egyptians used the word "aku-t" in the Papyrus Ebers and that this was transliterated by Budge as "boils, blains, sores, pustules and any inflamed swelling."

That the Romans were skilled in dermatology is unquestioned. Celsus (1st cent. A.D.) has much to say about skin diseases in Books 3, 5, 6 and 7 of *De Medicina*. He enumerates almost fifty cutaneous affections (Rosenthal, 1961). An early but still valuable commentary on Celsus as a dermatologist came from the pen of Erasmus Wilson (1863). Celsus indicates clearly that true leprosy was rare in Italy. Among the conditions which may be recognised with reasonable certainly in Celsus are chilblains (*pernio*), erysipelas, noma (*aphthai*), carbuncle (*carbunculus*), phagadoenic ulcers (*therioma, ignis sacer*) which Wilson thought must be syphilitic or tuberculous, cutaneous and axillary abscesses (*phyma, phygethlion*), acne (*varus*), boils (*furunculus*), purulent folliculitis (*kerion*), sycosis barbae (*sycosis*), pediculosis (*phtheriasis*), herpetic eruptions (*phlyctenae*), infantile pemphigus (*epinyctis*), acute and chronic eczema (*impetigo* and *scabies*), lichen (*papulae*) and pityriasis (*porrigo*).

Celsus also describes corns (*clavus*), epidermal cysts (*ganglia, meliceris, atheroma*), hordeolum (*hordeolum*), mollusca (*acrochordon*), warts and condyloma acuminatum (*thymion*), plantar warts (*myrmecia*) and skin cancer (*carcinoma*). Krumbhaar (1937) states that Celsus also noted soft chancre. It appears that the term *eczema* was in fact used by Dioscorides and Galen.

It is interesting to recall that the Roman gynaecologist Soranus was conversant with the skin pigmentation which occurs in pregnancy; these changes were also known to Susruta in Ancient India. Susruta also appears to have described leprosy. Smallpox was described by Eusebius (4th cent. A.D.) and by the Mediaeval Arabs, e.g., Rhazes. Haly Abbas described the lesions of measles and anthrax (Krumbhaar, 1937).

Thorwald (1963) points out that the Aztec Xipetotec, God of the Spring, sent sores, carbuncles and scabies. Wells (1964) points out that the *huacos* of South America provide fairly firm evidence of the lesions of *uta* (South American cutaneous leishmaniasis) and possibly of blastomycosis and Oroya fever.

PALAEOPATHOLOGICAL STUDIES

That the skin may be well preserved in mummified and dried bodies was shown by Ruffer in 1911. He demonstrated persistence of dermal nuclei in predynastic bodies and showed that in mummies the epidermis of the hands and feet was often well preserved. In some instances sweatglands, blood vessels, adipose tissue, deeper muscle and hair follicles were seen. In later papers Ruffer (1912, 1913) showed skin structure to be well preserved in bodies from the Persian period (525-332 B.C.) and also in Coptic bodies (400-500 A.D.).

Simandl (1928) also noted persistence of skin structures in Egyptian mummies: Wilson (1927) made sections of the skin from bodies of Cliff-dweller Indians. Sandison (1955) published a photograph of a section of mummy skin showing persistence of epidermal nuclei (see Figs. 1 and 2). Leeson (1959) published electron micrographs of the skin of an American Indian body of uncertain date.

In 1911 Ruffer and Ferguson described an XVIIIth Dynasty mummy from Deir-el-Bahri which bore a vesicular skin eruption which, in form and general distribution of the lesions, resembled smallpox. A sample of skin removed from the adductor surface of the right thigh showed, on histological examination, dome-shaped vesicles which must have originated in the middle of the prickle-cell layer. In more mature vesicles the base was formed by the deepest Malpighian layer and the roof by an elevated superficial layer of epidermis. There were

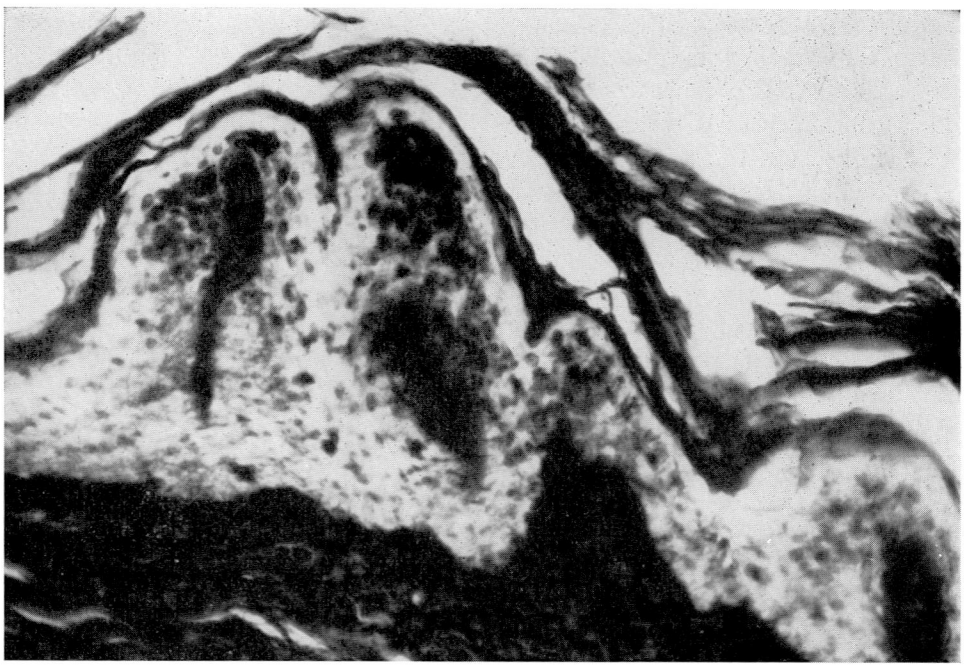

FIGURE 1. Skin from female Egyptian mummy stained by phosphotungstic acid-haematoxylin to show persistence of epidermal nuclei. x 400. (Courtesy Stain Technology)

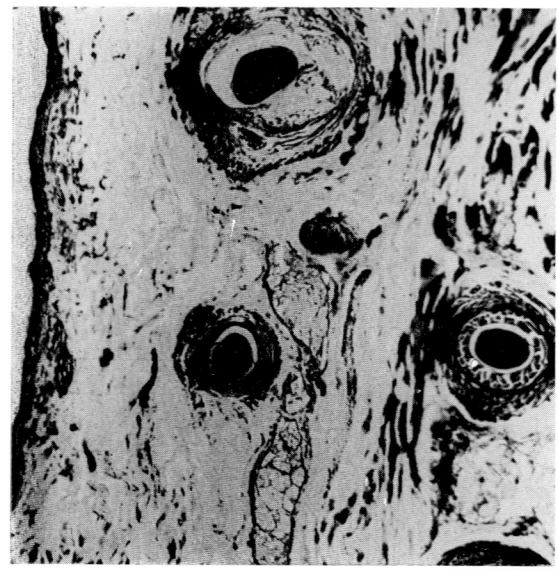

FIGURE 2. Scalp of male Egyptian mummy stained by Heidenhain's iron haematoxylin to show hair follicles and sebaceous glands. x 130.

vertical septa and curtains such as are seen in variola lesions. The dermis appeared normal and free of cellular infiltrate but bacteria were present; Ruffer believed the organism to have been present before death but to have multiplied extensively between the time of death and onset of mummification.

Elliot Smith (1912) described the mummy of Pharaoh Ramesses V. On the surface of the pudenda, lower abdominal wall and face there was a well-marked papular cutaneous eruption, the distribution of which was regarded by A. R. Ferguson as highly suggestive of smallpox (see Fig. 3). Ruffer (1914) recalled this case and countered criticisms by Unna of the case published by Ruffer and Ferguson (1911).

Elliot Smith (1912) had also described

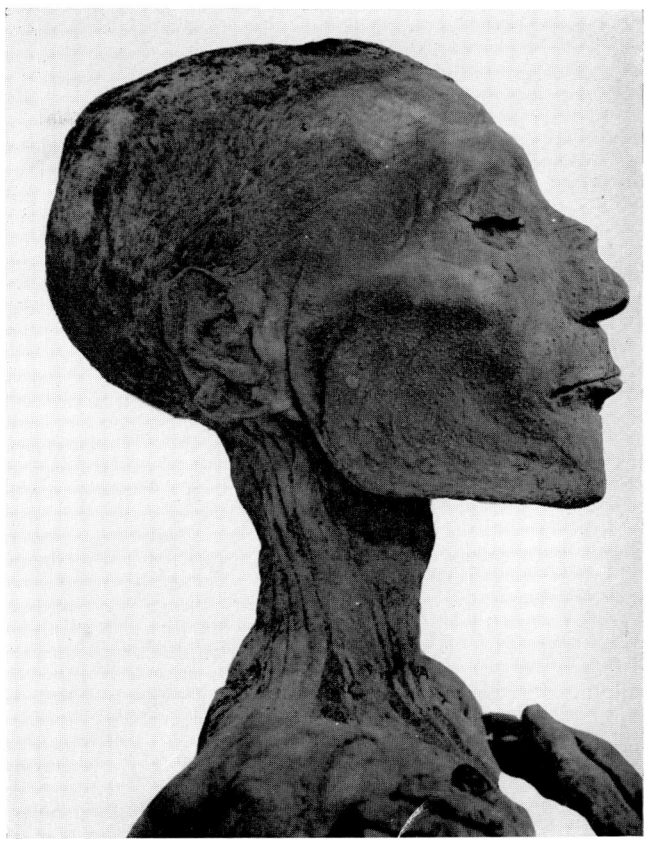

FIGURE 3. Head of Pharaoh Ramesses V to show eruption strongly suggestive of variola. (Courtesy Service of Antiquities)

comedones of the type seen in old men over the frontal region of Ramesses II. Ruffer (1914) confirmed this and stated that comedones were not uncommon in mummies. I have seen senile comedones on the face of an elderly Egyptian male mummy and have prepared sections which confirm the diagnosis of senile acne (see Fig. 4).

Ruffer (1914) also described possible ante-mortem ulcers with some reaction on the heel of the older women found in the tomb of Amenophis II. The body of Ramesses IV showed an appearance on the penis which may have been a post mortem artefact but might equally be an ulcer; his body showed also another 2.0 x 1.0 cm apparent ulcer with raised edges present on the back of the right scapula extending from the posterior lip of the glenoid fossa to its lower half.

One thigh of Queen Inhapy of the New Kingdom showed stigmata which Ruffer thought similar to those of *lichen*. The mummies of Tuthmosis II, Amenophis II (and to a lesser extent of Tuthmosis III) showed raised macules on the skin of the thorax, shoulders, arms, hands, buttocks, legs and feet. The head was unaffected. The lesions varied in diameter from pinpoints to about 1.0 cms. The nature of these lesions is obscure: possibly they represent some familial disease process but

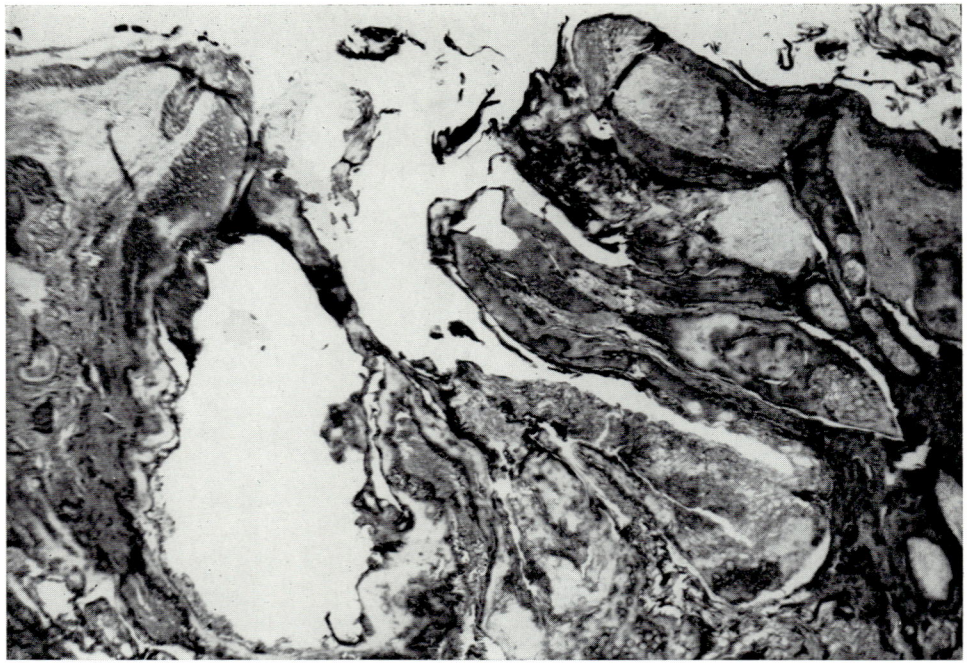

FIGURE 4. Senile acne in elderly male Egyptian mummy stained by phosphotungstic acid haematoxylin. x 200.

it has been suggested that they indicate an embalming artefact.

Wilder (1904) made tentative diagnoses of infantile eczema in a Utah Indian baby body and of infective dermatosis in a Peruvian dried body.

I have not been able to trace any report in the literature concerning neoplasms of the skin in mummies or dried bodies. I have, however, seen a small filiform squamous papilloma of the skin of the hand in a female Egyptian mummy of probable late dynastic date (see Fig. 5). Other sections of skin from the same specimen show marked hyperkeratosis and a rather thinned-out epidermis (see Fig. 6). These changes suggest that the skin in this individual was abnormal and that the appearences perhaps approximate to those of solar keratosis.

CONCLUSIONS

A study of the ancient literature reveals that diseases of the skin were common in antiquity. The works of Celsus provide ample evidence of a wide variety of cutaneous diseases.

The skin of some mummies and other dried bodies may be well preserved and further histological studies might reveal concrete evidence of the existence of some of these skin diseases.

REFERENCES

BENNETT, R., 1896: *The Diseases of the Bible.* London, Religious Tract Society.

BRIM, C. J., 1936: *Medicine in the Bible*: the Pentateuch. New York, Froben Press.

CHADWICK, J., and MANN, W. N., 1950: *The Medical Works of Hippocrates.* Oxford, Blackwell.

EBBELL, B., 1937: *The Papyrus Ebers: the Greatest Egyptian Medical Document.* Copenhagen, Levin and Munksgaard.

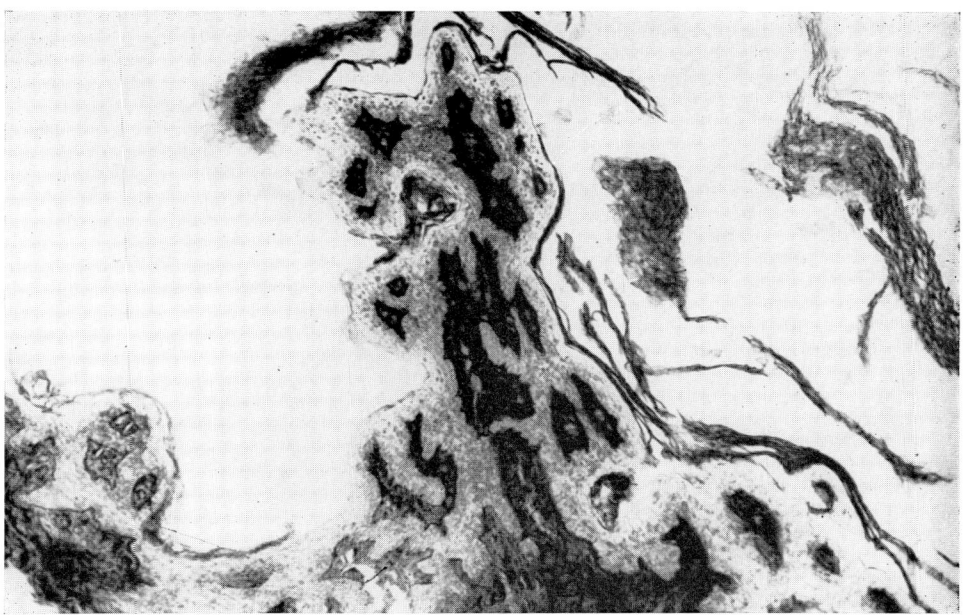

FIGURE 5. Small squamous papilloma of mummy hand stained by phosphotungstic acid haematoxylin. x 150. (Courtesy Scottish Society History of Medicine)

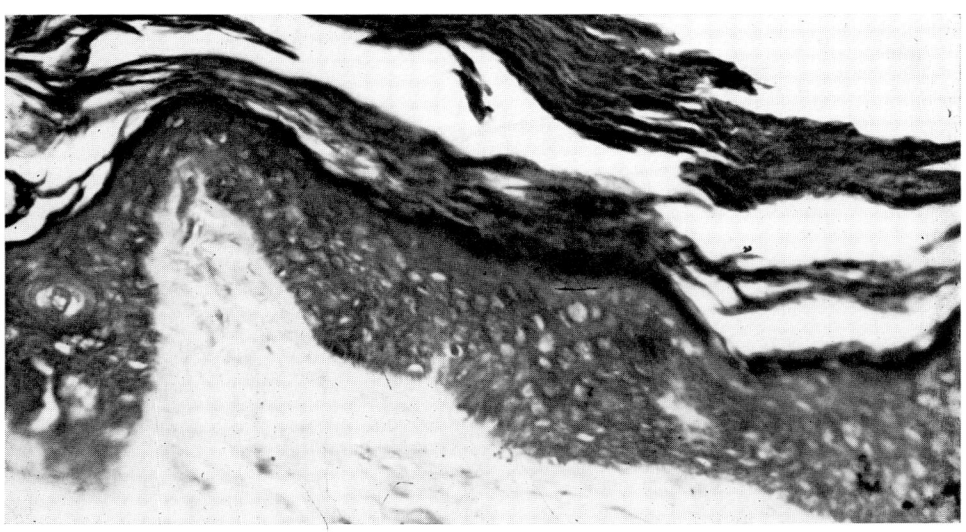

FIGURE 6. Skin from same specimen to show changes suggestive of a solar keratosis. Stained by Heidenhain's iron haematoxylin. x 300.

GRANT, R. N. R., 1951: The History of Acne. *Proc. Roy. Soc. Med.*, 44:647.

GUY, W. B., 1955: Psychosomatic Dermatology circa 400 B.C. *Arch. Derm. (Chicago)*, 71:354.

KRUMBHAAR, E. B., 1937: *Clio Medica: Pathology.* New York, Hoeber.

LEESON, T. S., 1959: Electron Microscopy of Mummified Material. *Stain Technol.*, 34:317.

MEENAN, F. O. C., 1955: A Note on the History of Psoriasis. *Irish J. Med. Sci.* No. 351:141.

PREUSS, J., 1911: *Biblisch-talmudische Medizin.* Berlin, Karger.

ROSENTHAL, T., 1961: Aurelius Cornelius Celsus: His Contributions to Dermatology. *Arch. Derm. (Chicago).*, 84:129.

ROWLING, J. T., 1961: *Disease in Ancient Egypt*: evidence from Pathological Lesions found in Mummies. M.D. Thesis, University of Cambridge.

RUFFER, M. A., 1911: Histological Studies on Egyptian Mummies. *Mem. Inst. Égypt.* Fasc. 3.

RUFFER, M. A., 1913: On Pathologic Lesions Found in Coptic Bodies (400-500 A.D.) *J. Path. Bact.*, 18:149.

RUFFER, M. A., 1914: Pathological Notes on the Royal Mummies of the Cairo Museum. *Mitt. Ges. Med. Naturw.*, 13:239.

RUFFER, M. A., and FERGUSON, A. R., 1911: An Eruption Resembling That of Variola in The Skin of a Mummy of the Twentieth Dynasty. *J. Path. Bact.*, 15:1.

RUFFER, M. A., and RIETTI, A., 1912: Notes on Two Egyptian Mummies Dating from the Persian Occupation of Egypt (525-332 B.C.). *Bull. Soc. Arch. Alex.* 14:1.

SAGGS, H. F. W., 1962: *The Greatness that was Babylon.* London, Sidgwick and Jackson.

SANDISON, A. T., 1955: The Histological Examination of Mummified Material *Stain Technol.*, 30:277.

SHAPTER, T., 1834: *Medica Sacra or Short Expositions of the More Important Diseases Mentioned in the Sacred Writings.* London, Longman Press.

SIGERIST, H. E., 1951: *A History of Medicine.* Vol. I. *Primitive and Archaic Medicine.* New York, Oxford University Press.

SIMANDL., I., 1928: A Contribution to the Histology of the Skin and of the Muscle of an Egyptian Mummy. *Anthropologie (Prague).*, 6:56.

SMITH, G. E., 1912: *The Royal Mummies.* Cairo, Musée du Caire.

THORWALD, J., 1963: *Science and Secrets of Early Medicine.* London, Thames and Hudson.

WELLS, C., 1964: *Bones, Bodies and Disease.* London, Thames and Hudson.

WILDER, H. M., 1904: The Restoration of Dried Tissues with Especial Reference to Human Remains. *Amer. Anth.* 6:1.

WILSON, E., 1863: On the Dermo-Pathology of Celsus. *Brit. Med. J.*, ii:446, 465.

WILSON, G. E., 1927: A Study in American Palaeohistology. *Amer. Nat.*, 61:555.

Chapter 37

Diseases of the Eyes

A. T. SANDISON

THE eyes and their appendages are readily accessible to simple clinical and even casual observation. We should, therefore, anticipate that information concerning ophthalmic disease might be available in the ancient medical literature. This supposition proves to be well founded and such information is more abundant and more readily interpreted than that concerning other organs or systems.

Arrington (1959) has pointed out that destruction of the eye was a severe judicial penalty in Ancient Babylon and among the Semitic nations. There are references in the code of Hammurabi to the treatment of eye diseases and to fees for success and punishment for failure of such treatment. There are references in Babylonian literature to abscesses about the eye, to *ophthalmia* (conjunctivitis), purulent conjunctivitis and possibly to trachoma. Night blindness was recognized in the Ancient Orient. Glaucoma and cataract do not appear to have been distinguished. Kinnier Wilson (page 198) suggests that dacryocystitis, corneal ulcers, exophthalmos and shortsightedness may also be identified. According to Chance (1939) verdigris, copper dust and yellow sulphide of arsenic were used therapeutically. Copper has a long and honoured place in ocular therapeutics.

Laulan (1962) suggests that among the Hebrews congenital and acquired blindness was not rare. He claims that corneal leucoma may be identified. There are numerous references to blindness and to behaviour towards blind persons in the Holy Bible. Blindness is often regarded as a divine retribution. For example, blindness was inflicted on the men of Sodom (Gen. 19:11) and on the Syrian Army (2nd Kings 6:18). The blindness of Saul of Tarsus may be interpreted as possibly hysterical and transient. Samson was deliberately blinded by the Philistines (Judges 16:21). There are numerous references in the Gospels to the healing of the blind by Jesus Christ.

During the Neolithic period in Egypt women used malachite (copper carbonate) as an eyepaint. In later Predynastic times galena was added to the female ornamentarium. In Ancient Egyptian literature there are references to royal eye specialists. Examples of identifiable eye diseases have been culled from the Ebers, Edwin Smith, Hearst and Coptic medical papyri (Arrington, 1959). These include blepharitis, trichiasis, ectropion, chalazion, pterygium, pinguecula, chemosis, inflammations and granulations, leucoma, ophthalmoplegia, iritis, staphyloma, possible tumours and, of course, blindness. Blindness is commonly portrayed in Ancient Egyptian art: musicians appear to have sometimes been blind (Wells, 1964). Feigenbaum (1958) believes that trachoma was probably included among cases of running eye and purulent conjunctivitis. Laulan (1962) sees evidence for night blindness, cataract and possible menstrual iritis in the female. The relationship of strabismus to fractured skull was recognized.

Feigenbaum (1957, 1958) derived evidence of *seasonal ophthalmia* in Ancient Egypt from inscriptions on kohl-boxes of both Middle and New Kingdom date. Pliny suggested in his writings that ophthalmia was seasonal in Egypt and Feigenbaum's scholarly approach adds considerable weight to this. Feigenbaum (1958) also pointed out that substances recommended in Papyrus Ebers for the treatment of ophthalmic conditions included antimony sulphide, copper compounds (verdigris and bluestone), lead sulphate, zinc carbonate, red iron ore, aluminates, malachite and castor oil leaves. Several of these could be therapeutically efficacious. Chance (1939) states that the Ancient Egyptians epilated for trichiasis.

The Vedic books of Ancient Hindu origin include references to couching for cataract and to excision of pterygium. Pandit (1957) suggests that occurrence of trachoma can be deduced from the literature of Vedic Era and from the pre-Susruta, Susruta and post-Susruta periods. The evidence here appears to be slender. There is, however, little doubt that conjunctival inflammation was extremely common and may have included follicular conjunctivitis. Copper was used therapeutically. Blindness was common in Ancient India and in Ancient Persia (Arrington, 1959; Sigerist, 1961). Chinese medicine of the T'ang period referred to entropion but not to cataract (Chance, 1939).

If we turn now to Ancient Greece we find evidence of ophthalmic disease in the earlier period in the form of votives to Asclepius, testifying to the alleviation of eye disease. The *Corpus Hippocraticorum* includes mention of conditions identifiable as blepharitis, hordeolum, ectropion, trichiasis, conjunctivitis, pterygium and keratitic ulcers. Some authorities believe that trachoma may be identified in *Corpus Hippocraticorum* and in *Plutus* (Chance 1939). Trachoma is believed to have been common during the Peleponesian War and was treated by blepharoxysis. Copper peroxide was also used therapeutically. According to Arrington (1959) glaucoma and cataract were not distinguished clearly. Feigenbaum (1959) has discussed at length the problem of the recognition of glaucoma in the classical period. Chance (1939) states that operative treatment was available for tarsal anomalies and for hypopyon.

Ophthalmia is mentioned frequently in *Epidemics, Airs, Waters and Places, Aphorisms,* and *Coan Prognosis* (Chadwick and Mann, 1950): at this time it implied inflammation of the eye but Galen later restricted the term to conjunctival inflammation. Feigenbaum (1956) believes that *Behcet's Syndrome* (aphthous stomatitis, ulcers around the genitals and recurring uveitis) may be identified in the third Book of *Endemic Disease*. If this attribution is acceptable (and Feigenbaum puts forward a good case) we would appear to have an example of an endemic disease which over the centuries has become sporadic and uncommon.

Corpus Hippocraticorum also attached great importance to changes in the eyelids and to ocular signs in general medicine. These signs included photophobia, lachrymation, strabismus, pupillary inequality, ptosis, enophthalmos, exophthalmos, hemianopia, nystagmus, nyctalopia, amaurosis and amblyopia.

Aristotle clarified concepts of myopia,

hypermetropia and presbyopia and during the Hellenistic period advances were made in ophthalmic anatomy and physiology following the observations of Herophilus and Alcmaeon. Galen's knowledge of ocular anatomy was little more advanced but from *De re Medicina* (Bk. 6:VI and Bk. 7:VII) we obtain useful evidence of eye disease. *Ophthalmia* is discussed at length and distinction made between moist and dry ophthalmia and xerophthalmia. Chance (1939) believes that trachoma may be identified not only in *De re Medicina* but also in the writings of Pliny and Pedanius Dioscorides. Chance believes that at this period trachoma was widespread in the Mediterranean littoral. Other conditions identified in the works of Celsus include proptosis, carbuncles of the eyelids, stye, sebaceous cysts, chalazion, trichiasis, distichiasis, ectropion, lagophthalmos, pterygium, dacryocystitis, staphyloma and cataract (Chance, 1939; Dollfus, 1958; Arrington, 1959).

Rufus of Ephesus is said to have distinguished glaucoma and suffusio (Arrington 1959). Much of Galen's writing on the eye has been lost but he appears to have had a fair knowledge of anatomy and physiology. The essential concept of monocular and binocular vision were known. Galen described photophobia, snow blindness and eclipse blindness. Cataract, a trachoma-like disease and hypopyon were treated surgically; sophisticated instruments and collyria were used by Roman ophthalmic surgeons (Dollfus, 1958). Galen's concept of glaucoma remained imprecise.

Antyllus is said to have devised a plastic operation for coloboma. Aetius of Amidea (6th cent. A.D.) listed sixty-one affections of the eye and other compilations were made by Alexander of Tralles (6th cent. A.D.) and Paulus Aegineta (7th cent. A.D.). The mediaeval Arabs made great advances in optics and founded ophthalmic hospitals in Baghdad, Cairo and Damascus. Trachoma was still confused with other diseases but interesting conditions described include pediculosis of the lids, ocular animal parasites, cataract, lacrymal fistula, pannus, symblepharon and visual errors (Chance, 1939).

PALAEOHISTOLOGICAL STUDIES

Unfortunately, ocular diseases leave no traces on skeletal remains. It is possible, however, that usura orbitae (cribra orbitalia)—the presence of bilateral symmetrical cribriform change in the orbital roof—might be due to some change in the lacrymal glands (Møller-Christensen and Sandison 1963). Wells (1964), however, suggests that it might be due to some unidentified deficiency disease or to a panophthalmia which lead to blindness. Usura orbitae is common in skulls, especially of children, from Ancient Egypt but relatively uncommon, for example, in Anglo-Saxon skulls.

I have given elsewhere (Sandison, 1957) an account of the eye in Egyptian mummies. The Ancient Egyptians made no attempt to preserve the globe of the eye which receded into the back of the orbit during mummification. Sometimes the eyes were simulated by the insertion of artefacts; these included pads of linen on which the iris might be depicted. On one occasion, in the case of Ramesses IV, small onions were used: model eyes made of white stone into which a black portion was inlaid to represent the iris and pupil were also used (see Fig. 1). Dried bodies from the Torres Straits and Polynesia have shown very effective artificial eyes made from shells. A scholarly and valuable account of artificial eyes in Egyptian statues is given by Lucas (1948).

Ruffer (1911) noted on gross examina-

460

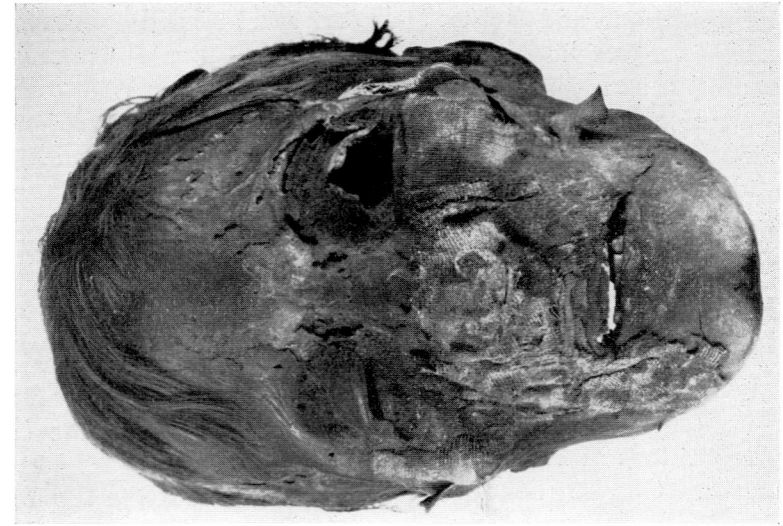

FIGURE 2 Head of a male Egyptian mummy to show apparently empty orbits.

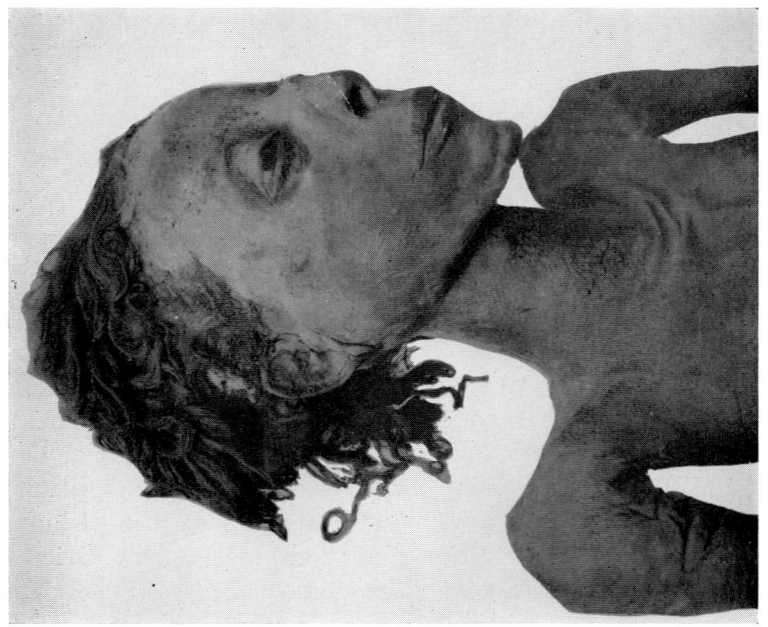

FIGURE 1. Head of mummy of Princess Nes-ta-neb-ashru to show artificial eyes. (Courtesy Service of Antiquities)

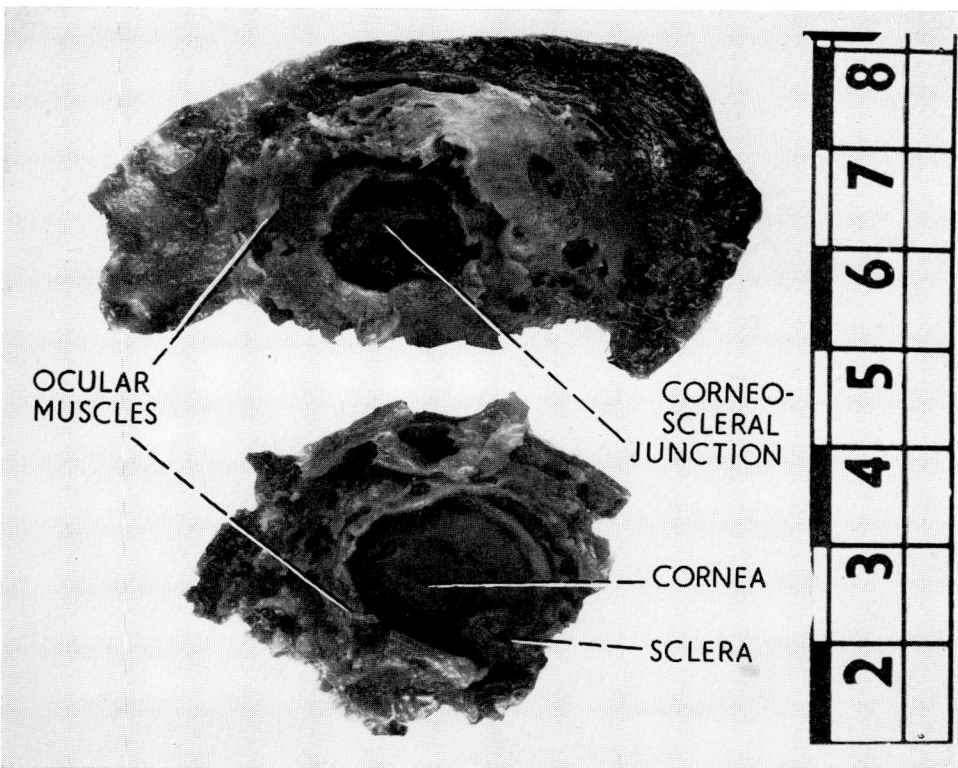

FIGURE 3 Orbital contents of another mummy head seen from behind after rehydration and removal.

tion that ocular muscles remained in an Egyptian mummy and that the shrunken eyeballs were present in two others of the XXVIIth to XXIXth dynasties. He made histological sections of the ocular muscles and of the optic nerves from sand burial bodies of the Ptolemaic period. Williams (1927) noted ocular remnants in a died Peruvian body of date about 700 A.D. but microscopy revealed no features of note.

This paucity of studies is not surprising as the appearances of mummy orbits are distinctly unpromising (see Fig. 2). I have shown, however, (Sandison, 1957) that rehydration of mummy heads may reveal remnants of the globes of the eyes which slowly unfold and come to approach the partially opened palpebral fissure (see Fig. 3). These remains comprised sclerotic, cornea, irido-corneal angle and parts of the pigmented coats in which melanin may be convincingly demonstrated (see Fig. 4).

In my sections micro-organisms of putrefactive type were seen as well as an intrusive larva of a *Piophila* fly (see Fig. 5). It was tempting to regard this as possible evidence of ocular myiasis. Ocular myiasis in Egypt is, however, a disease of shepherds, persons not likely to have been mummified. This phenomenon is interpreted as the result of post-mortem deposition of eggs which developed partially on the cadaver before checked by the mummification process.

A later study by Rohen (1959) de-

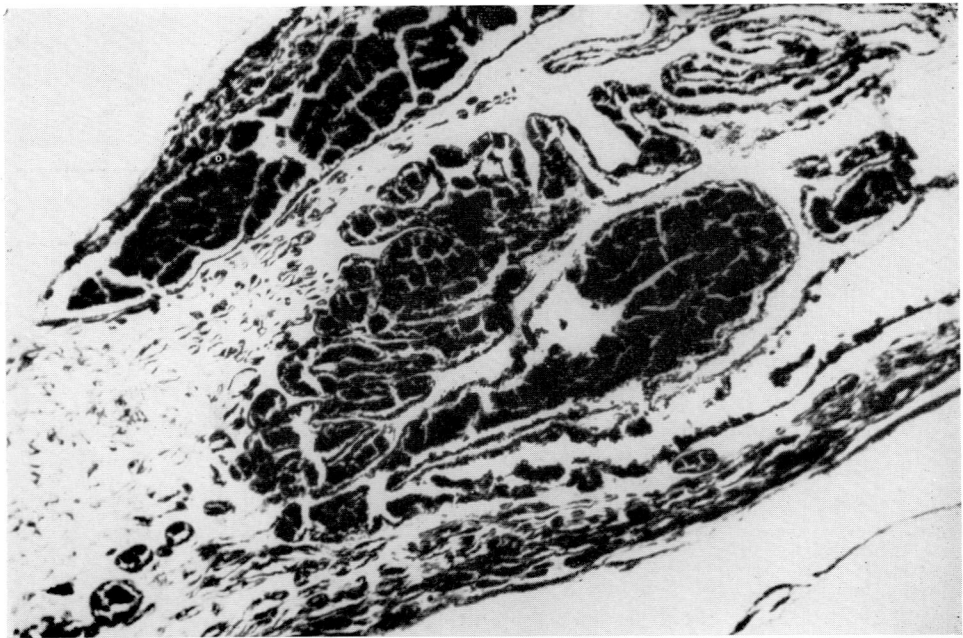

FIGURE 4 Internal coats of mummy eye at region of the angle to show persistence of melanin. Eosin x 100.

scribed the histological structure of the eyes of a series of dried bodies from the Canary Islands (Teneriffe and Grand Canary). The iris was shown to be pigmented and the uvea, ciliary body and sclera were demonstrated. No pathological change was noted.

These rather sparse studies suggest that further examinations of the eyes of mummies or dried bodies might reveal evidence of gross ocular disease.

CONCLUSIONS

The literature of Ancient Societies bears conclusive evidence that ophthalmic diseases were common. Some attributions may be doubtful but many of these diseases may be identified with considerable accuracy. That these diseases often resulted in blindness is evidenced by the frequency with which the condition is mentioned in the literature and portrayed artistically.

Occasional studies of the eyes of mummies and dried bodies have shown reasonable structural preservation and further investigations might give concrete evidence of ophthalmic disease of the more gross types.

REFERENCES

ARRINGTON, G. E., 1959: *A History of Ophthalmology.* New York, M.D. Publications.

CHADWICK, J., and MANN, W. N., 1950: *The Medical Works of Hippocrates.* Oxford, Blackwell.

CHANCE, B., 1939: *Clio Medica. Ophthalmology.* New York, Hoeber.

DOLLFUS, M. A., 1958: Les Instruments d'ophtalmologie chez les gallo-romains. *Arch. Ophtal. (Paris)*, 18: 633.

FEIGENBAUM, A., 1956: Description of Behcet's syndrome in the Hippocratic Third Book of endemic diseases. *Brit. J. Ophthal.*, 40:355.

FEIGENBAUM, A., 1957: Archaeological evidence of the occurrence of regular seasonal ophthalmias in Ancient Egypt. *Janus.*, 46:165.

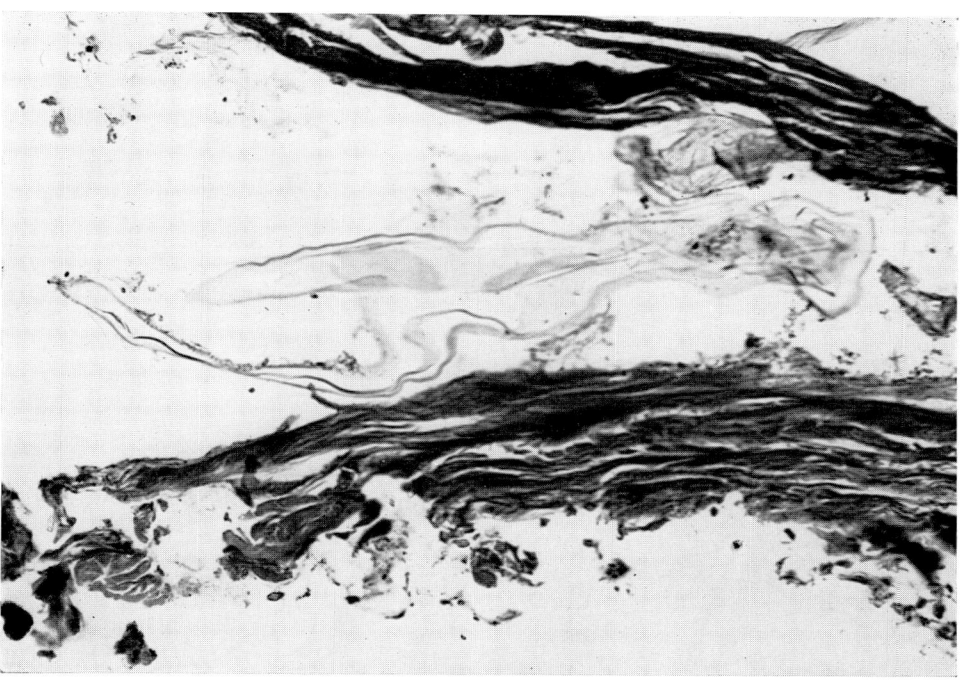

FIGURE 5 Intrusive larva of probable Piophila fly. This is almost certainly a postmortem invader and *not* an example of myiasis. Phosphotungstic acid-haematoxylin stain x 300.

FEIGENBAUM, A., 1958: History of ophthalmia (including trachoma) in Egypt: evidence for its seasonal occurrence in antiquity. *Acta. Med. Orient.*, 17:130.

FEIGENBAUM, A., 1959: Le terme "glaucome," son histoire et son premier usage populaire connu dans l'Antiquité pour designer une certaine espèce de cécité. *Le Scalpel*, 112:396.

LAULAN, R., 1962: Les maladies des yeux dans l'ancienne Egypte et chez le peuple d'Israel. *Presse. Med.*, 70:2477.

LUCAS, A., 1948: *Ancient Egyptian Materials and Industries*, 3rd Ed. London, Arnold.

MØLLER-CHRISTENSEN V., and SANDISON, A. T., 1963: Usura orbitae (cribra orbitalia) in the collection of crania in the anatomy department of the University of Glasgow. *Path. Microbiol.*, 26:175.

PANDIT, Y. K. C., 1957: History of trachoma in Ancient India. *Proc. All.-India Ophthal. Soc.*, 17:173.

ROHEN, J., 1959: Histologischen Untersuchungen an Augen altkanarischer Mumien. *Homo.*, 10:35.

RUFFER, M. A., 1911: Histological studies on Egyptian Mummies. *Mém. Inst. Égypt.*, Tome 6, Fasc. 3.

SANDISON, A. T., 1957: The eye in the Egyptian mummy. *Med. Hist.*, 1:336.

SIGERIST, H. E., 1961: *A History of Medicine*. Vol. II. *Early Greek, Hindu and Persian Medicine*. Oxford, University Press.

WELLS, C., 1964: *Bones, Bodies and Disease*. London, Thames and Hudson.

WILLIAMS, H. U., 1927: Gross and microscopic anatomy of two Peruvian mummies. *Arch. Path.*, 4:26.

Chapter 38

Disease in the Ear Region

WILLIAM McKENZIE AND DON BROTHWELL

THE interpretation of bone disease in an early skull has many difficulties. In particular, the reaction of bone to infection may vary from one generation to another, and therefore the type of disease cannot be *confidently* recognised by the bony change. Before examining the evidence for temporal bone disease in earlier groups it may help to distinguish the three main types of change in ear disease which were seen twenty years ago before the use of antibiotics, and are occasionally seen now. The first type is a virulent infection, which can pass through the middle ear and mastoid and kill in forty-eight hours. The whole petrous temporal bone may be softened if this happens, but it is doubtful whether anything could be recognised after the soft tissues had decayed after death. The second type is an abscess formation with destruction of the mastoid cells. This is the most conspicuous type of ear infection, and is called to mind by most physicians when they think of mastoiditis. It represents a considerable amount of resistance to infection, and if it appears, the interval between the onset of infection and the appearance of an abscess may be quite long, perhaps four or six weeks. The abscess finally enlarges under the skin, it "points" to use an old fashioned term, and its size and shape depends on the place where the pus has passed through the mastoid process. The common situation in a child lies above and behind the mastoid antrum; here it can easily be seen and felt, for it lies immediately under the skin. Another point of emergence lies medial to the tip of the mastoid process. This is a *Bezold's abscess,* and it nearly always occurs late in the disease and leads to a large ill-defined swelling of the neck, because the abscess is confined by the fascia of the sternomastoid muscle. The last type is simply a continuous discharge from the ear due to infection of the middle ear and mastoid without abscess formation.

THE MEDICAL LIMITATIONS OF ANCIENT MAN

It would seem reasonable to ask, at the beginning of this review, what evidence there is of the recognition and treatment of ear disorders by earlier peoples. In particular we might ask whether the more dramatic abscess formation would be thought to be associated with ear infection in the past and what surgical treatment of mastoid disease might have been undertaken. The answer is of course made more difficult because few surgeons at the present time have seen how these mastoid abscesses behave. They became scarce twenty years ago when antibiotics came into general use, and there have been few, if any, records of late disease, such as *Bezold's abscess,* since World War II. In the nineteenth century, a mastoid abscess behind the ear was incised by Sir William Wilde of Dublin as a regular form of treatment, and this incision came to be known as Wilde's incision. It is certainly doubtful whether anyone before his day did more than an occasional incision of the abscess, and no-one would recognise that a more successful form of treatment lay in the removal of the mastoid cells. In less advanced parts of the world today individuals can still be found suffering from mastoid abscesses which have burst through the skin to produce a discharging sinus. The patient does not always die from the discharging sinus and, of course, the infection may clear up if antibiotics are available. It is unlikely, however, that any otologist in recent times and in advanced societies has seen a mastoid abscess burst through the skin and heal spontaneously, without treatment, but this must have happened in the past. It is difficult therefore to imagine what change must have occurred in the mastoid area when recovery took place, and to know whether the mastoid cells were simply absorbed to form a cavity, or whether their shape was to some extent restored as the patient recovered. Today, if a surgeon removes the mastoid cells, and if later the mastoid process is explored surgically, the mastoid cells are found to be absent, although the denser outer bony cover may grow to obliterate the surgical opening in the mastoid itself.

There is one other type of mastoid infection which has been commonly noted throughout surgical records, and this is an inflammatory disorder occurring in the middle ear and a mastoid antrum which is devoid of mastoid air cells. The mastoid process is, in fact, acellular, and the disease is limited to the antrum and middle ear. This infection is often chronic and associated with the formation of a *cholesteatoma*, an expanding cyst of epithelium. It is still seen in hospital clinics and it may well have been the cause of ear disease in early man. If it produces an abscess, the swelling may be seen behind the ear, although the thick bone of the mastoid process usually prevents spread in any direction except towards the brain.

On reflection, it is clear that there were only two possible courses for a mastoid abscess before modern surgery; it could have burst through the skin or into the cranial cavity, or it could cause a swelling under the skin which might have been cut by a knife. It must be remembered that there was no efficient anaesthesia. In view of the antiquity and wide distribution of alcoholic drinks in ancient cultures, the sick may sometimes have been reduced to a satisfactory state of stupor before a knife was used, (and some minor surgery was no doubt performed without preparation), but even so, there would be no time to search for the mastoid, and one quick movement to open the abscess would probably be all that could be tolerated.

Enough has been said of the difficulties of interpreting past ear disease by reference to recent cases, and of the very unlikely occurrence of surgical intervention. But this does not mean that the Ancients were not aware of ear disease. The Ebers Papyrus not only discusses minor damage and inflammation of the external ear, but also gives prescriptions for: "If the canal is painful. . . . If its opening discharges. . . . If meanwhile it grows fatty . . . ." Similarly the Berlin Papyrus gives prescriptions for "pressure in the ears" or a discharge (Ghalioungui, 1963) Mesopotamian literature also has something to say about ear disease. As Thompson (1931) and Sigerist (1951) note otitis media was described: "fire extends into the interior of his ear and it dulls the hearing". The very

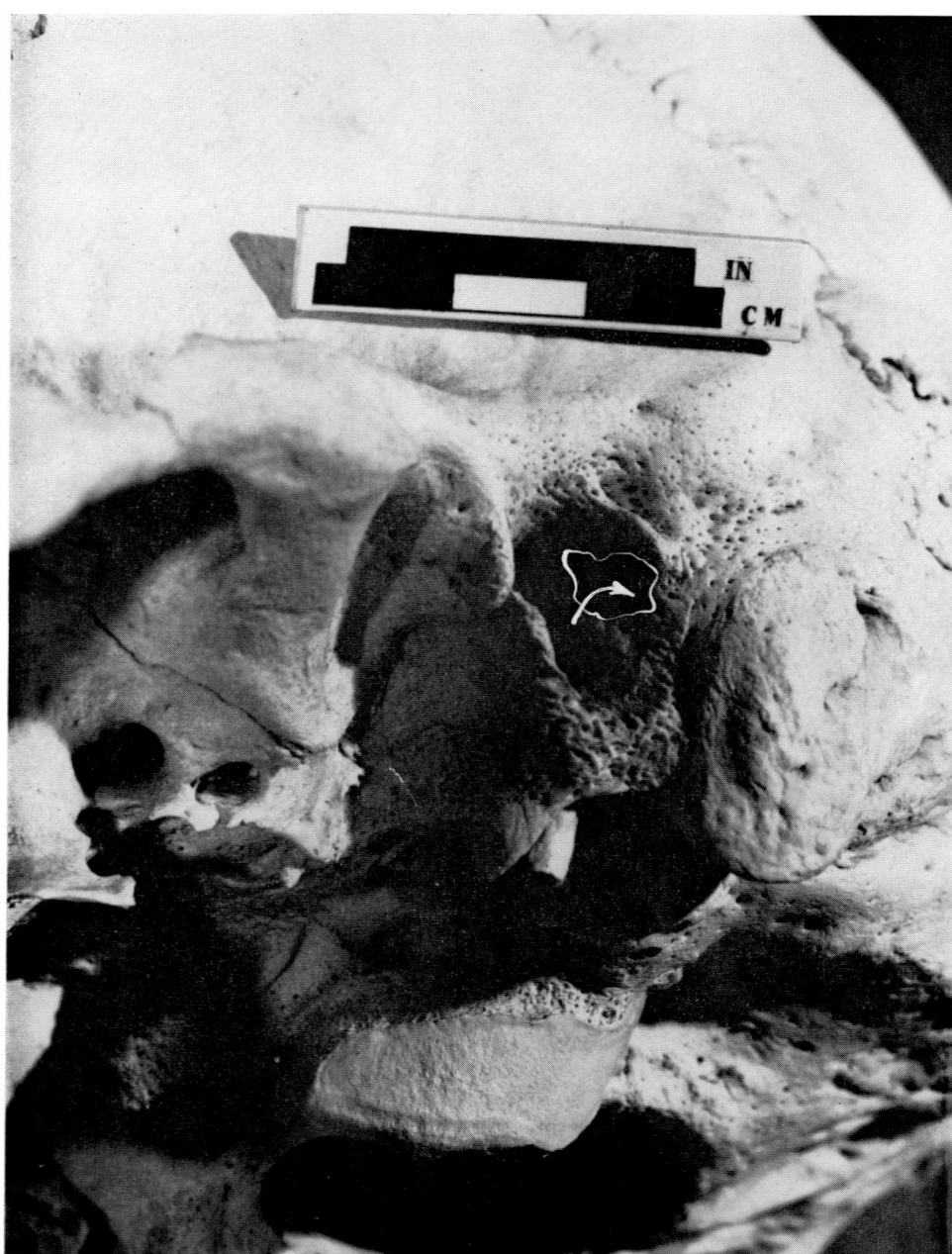

FIGURE 6. Ear region of the Early Dynastic Tarkhan skull. A white line and arrow show the perforation in the meatal wall. Courtesy of the Duckworth Laboratory of Cambridge.

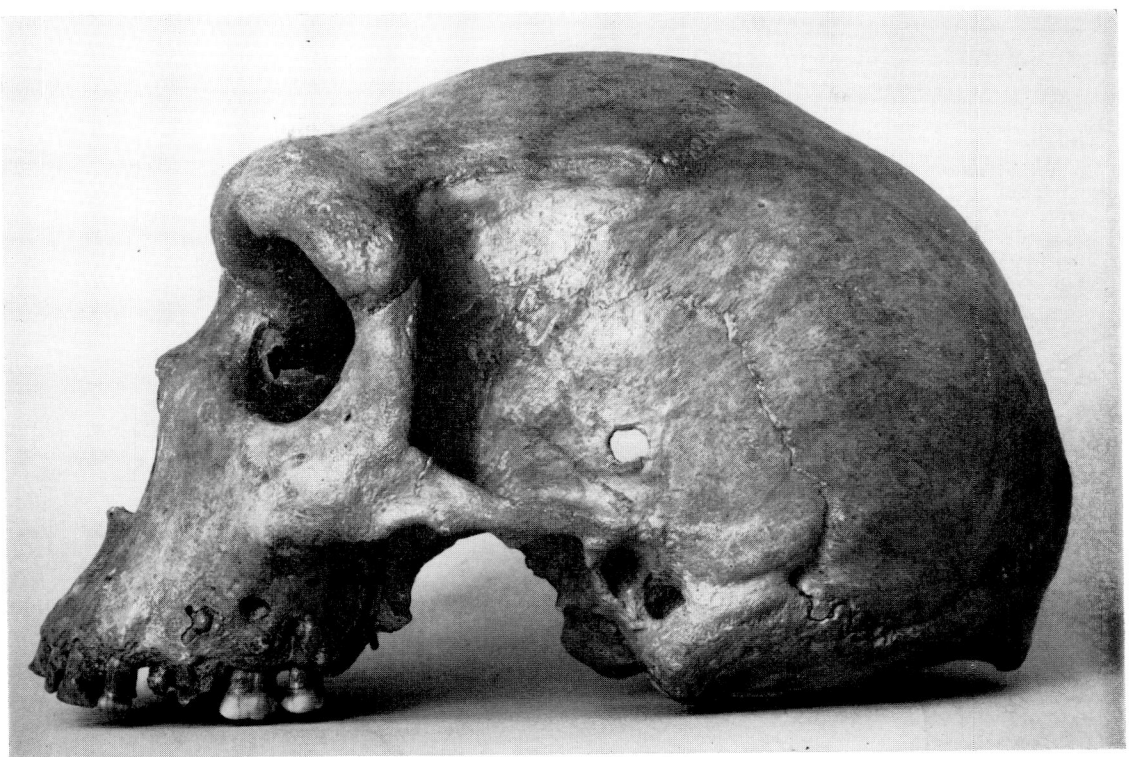

FIGURE 7a General lateral view of the Rhodesian skull, displaying both oral and mastoid disease.

painful, interior swelling, with exuding pus and offensive smell, which is described, was treated by the application of various drugs and by fumigations.

Although this early literary evidence suggests that surgical procedures were not considered, an early Dynastic Egyptian skull from Tarkhan (Fig. 6) makes it necessary to qualify this statement. This skull, described in more detail by one of us elsewhere (Oakley, Brooke, Akester and Brothwell, 1959), shows a well defined breach in the meatal wall (Fig. 6) which might well be the sinus of a mastoid abscess. Although the sinus usually occurs behind the spine of Henle, and not in front as in this case, it is nevertheless suggestive of early mastoiditis. Of special interest is the fact that a fairly well healed trephine hole is situated some distance (but directly above it) on the parietal bone of the same side. Although this association could be accidental, it seems equally probable that trephination was resorted to as a curative procedure. If this is so it demonstrates that surgery could be considered by earlier man as a cure for ear trouble, but that it was not directly undertaken at the place of the ear disease, but was more in the way of a magico-medical practice.

OTHER EARLY CASES OF EAR DISEASE

Southern Africa

The two earliest examples of temporal bone pathology are both from Africa. Of these, the most famous is the Rhodesian Skull, the robust Upper Pleistocene man discovered in 1921 (Fig. 7a). Yearsley

(1928) has described the pathology of the left temporal bone in some detail. It is not our intention to discuss here the perforation in the squama, which he considered the result of a blow (a doubtful statement), or the "pus" track (also highly debatable). It is the mastoid lesion which seems most likely to be genuinely pathological (Fig. 1). Although Yearsley was not prepared to state that the bone abnormality was caused by "middle ear suppuration and mastoid abscess," he put the following points as "very suggestive evidence."

"1. It is unquestionable that this individual suffered from serious and extensive dental caries with alveolar abscesses, a condition which affords strong presumption of chronic oral sepsis, probably involving the tonsils, circumstances which predispose to chronic pharyngitis and middle ear trouble.

"2. The mastoid process is of the pneumatic variety, a type in which mastoid abscess is most prone to occur.

"3. The loss of the posterior part of the outer attic wall and the posterior tympanic spine suggests disease of the tympanum in the neighbourhood of the *aditus ad antrum*. This point cannot, however, be relied upon absolutely, as the condition of the tympanic ring may have been caused by the clearing away of matrix.

"4. The cavity A communicates with the tip cells, and these are exposed by a breach in the tip cortex. This latter may be accidental, but the edges of the cortex and the cell partition walls suggest a pathological perforation rather than one made accidentally after death.

"5. No connection between the cavity A and the mastoid antrum could be made out. This is against extension from the middle ear, but is not, in my opinion, of paramount importance, since the communication may have become shut off by crumbling bone. The existence of such a communication might be decided by laying open the *tegmen tympani* and *tegmen antri*.

"6. The perforation of an acute mastoid abscess through the tip of the process (the so-called *Bezold's mastoiditis*) is not uncommon in modern times. This may account for the opening in the tip in the Rhodesian skull. It is, further, possible that this may have been the cause of death, the pus tracking down under the neck muscles and deep cervical fascia into the thorax" (pp. 61-62).

Generally his argument is very reasonable, but we disagree with two of his statements. In the first place, chronic oral sepsis is not a cause of ear disease. Secondly, his suggestion that the middle ear and antrum have been separated by crumbling bone is unusual and makes the diagnosis of *Bezold's abscess* unlikely.

The second African specimen is probably of late Stone Age date, and is known as the Boskop skull (Fig. 2). It was found in 1913 on a farm near the village of

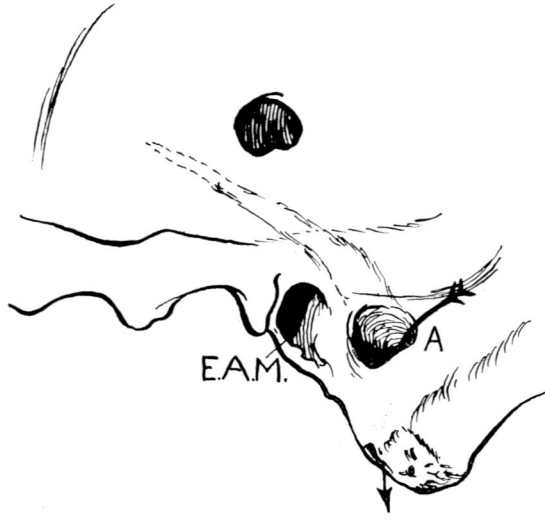

FIGURE 1. Diagram of the lesions in Rhodesian Man. (See Yearsley's description.)

FIGURE 2. Diagrammatic representations of the erosion in the petrous temporal of the Boskop skull. (Courtesy Professor Ronald Singer.)

Boskop, Transvaal. The temporal bone condition has recently been described in excellent detail by Singer (1961) and it would seem best to quote him. He states:

"In the postero-superior roof of the external auditory meatus is a smooth-edged circular hole (designated A), approximately 3.5 mm in greatest diameter. It communicates internally with a large cavity which, in turn, has perforated into the endocranial cavity in the region of the superior petrosquamous suture. This perforation (designated B) is almost rectangular, the long axis being in an antero-posterior plane. It measures approximately 4 mm in width. Where A and B communicate is the region of the epitympanic recess, and the posterior extension and erosion of B is in the region of the tympanic antrum. The endocranial aspect of B is smooth medially (i.e., the petrosal surface), while laterally, extending up from it on to the inner aspect of the squama, is an irregular large eroded area. Posteriorly B has perforated the thin bony plate between the tympanic antrum and the sigmoid (lateral) sinus. This orifice (designated C) has a thick-walled sickle-shaped anterior edge, which is decidedly of pathological nature, and not for the passage of a vein as proposed by Keith. Behind C the bony floor of the sigmoid sinus presents a small rounded, eroded area. Usually perforation of an abscess in the antrum into the sigmoid sinus results in the formation of a thrombosis in the sinus.

"This pathological picture has undoubtedly been produced by a cholesteatoma which, as a complication of chronic middle ear infection, has the peculiar physico-chemical property of eroding bone, and is frequently followed by intracranial complications" (p. 103-4).

We would agree with him that the evidence does suggest possibly a cholesteatoma (although the mastoid is not described as acellular). It must also be remembered, especially in view of the "smoothness" of the fractured bone edges, that post-mortem change can simulate pathological change, and the Boskop temporal bone is certainly not well preserved.

Egypt

Elliot Smith and Dawson (1924) have stated that mastoid disease was very common in Egypt and in Nubia. This appears to be an overstatement because only about six cases appear to have been noted, of a total of at least 10,000 Egyptians and Nubians. Of these, we have already mentioned the Tarkhan skull. We have also been able to examine a second specimen, originally figured by Elliot Smith and Dawson (1924, Fig. 68). It is Predynastic in date, and the oldest example of temporal bone disease so far from that region (Fig. 3). It has now been transferred from the Royal College of Surgeons to the British Museum (Natural History). Unfortunately, owing to bomb damage in the

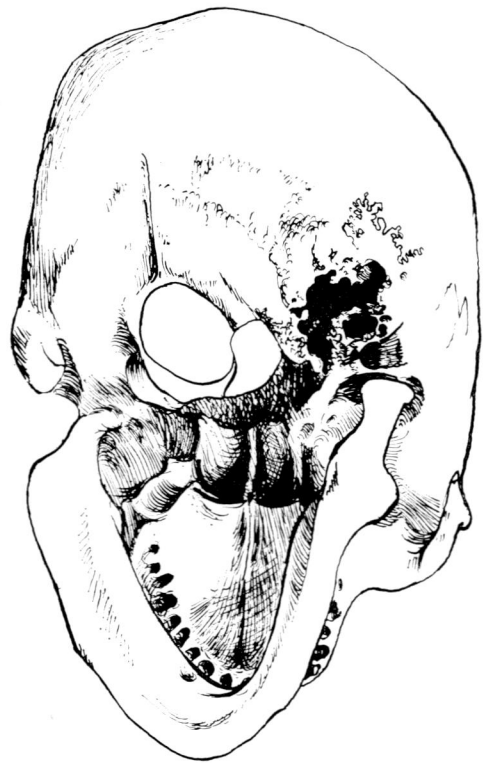

FIGURE 3. Predynastic Egyptian skull showing considerable destruction of the mastoid region. From a photograph by Elliot Smith and Dawson, 1924.

Second World War part of the specimen, including part of the diseased area, is no longer available. There is little doubt that the ante-mortem destruction of the right mastoid region resulted from acute mastoiditis with widespread erosion. The erosion reaches the lateral sinus, and may have caused a sinus thrombosis, but there is no evidence of intracranial extension apart from this, and the cause of death is uncertain. There is no sign of healing, and the individual probably died from the extensive inflammation in the ear region.

Also of Predynastic or Early Dynastic date is the case described by Derry (1909) of a man showing evidence of mastoid disease which had: "probably commenced in the middle ear, and destroyed the anterior wall of the left external auditory meatus and tympanic plate, and has involved the temporo-mandibular articulation, the condyle of the mandible being destroyed by the pathological process. It had spread backwards into the mastoid antrum and opened into the lateral sinus. The antrum communicates with the exterior by several apertures at the site of the suprameatal triangle" (p. 51). Although the description suggests mastoid disease in most respects, it is interesting to note apparent involvement of the temporo-mandibular joint, a feature not seen today.

A further case is described by Elliot Smith and Wood Jones (1910) in an adult female of the New Empire period (Fig. 4). They write: "The condition of the specimen showed it to be an example of septic disease of the antrum, which perforated outwards on the surface of the temporal bone, and had led to periostitis and necrosis of the surface of the bone around the site of perforation.

"Behind the right external auditory meatus there was a definite patch of the temporal bone in a condition of inflammation and necrosis; above the disease area

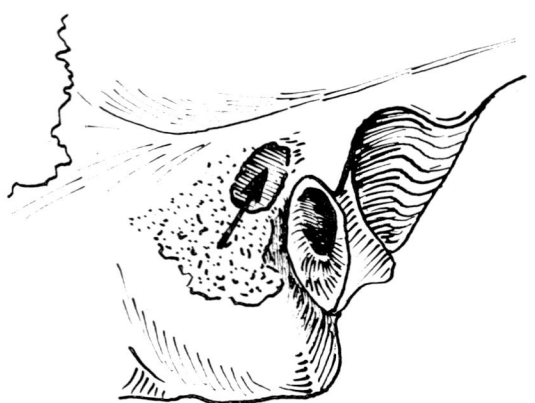

FIGURE 4. Perforation of the suprameatal triangle by a mastoid abscess. Modified from Elliot Smith and Wood Jones, 1910.

Disease in the Ear Region

an oval sinus communicated with the antrum of the mastoid. The sinus was situated in the suprameatal triangle, and at its bottom lay necrosed bone showing the opening of numerous mastoid cells. The infection of the bone below the sinus was for the most part superficial, and was the result of periostitis spreading downwards from the discharging mouth of the sinus" (p. 284).

Derry (1909) also notes a female of Coptic dates with: "suppuration in the walls of the right auditory meatus, causing a large round abscess cavity, destroying both anterior and posterior walls of the meatus and communicating with the middle ear" (p. 48). Also dated to the first millenium A.D., was another Nubian case of ear disease described by him. Batrawi (1935) has briefly described another obvious case of mastoid infection, in a man of Meroitic date (Cemetery 214, Body 161A. Plate XV, Fig. 4).

Britain and Ireland

Only one prehistoric case of questionable mastoid infection is known in this area, and it was reported in a male skull from the Bronze Age cemetery-cairn of Knockast, in Ireland (Fig. 7b). In the general excavation report by Hencken and Movius (1934), Professor E. A. Hooton writes:

"The region of the right glenoid fossa, auditory meatus, and mastoid process shows clear evidence of an extensive infection which was completely healed. . . . The hinder wall of the auditory meatus near the aperture shows a perforation opening upon the surface of the mastoid process, the larger portion of the exterior part of which has been sloughed away showing, however, scarification and repair. The auditory meatus is considerably enlarged and it seems probable that an abscess of the middle ear drained through the posterior wall and the mastoid, with-

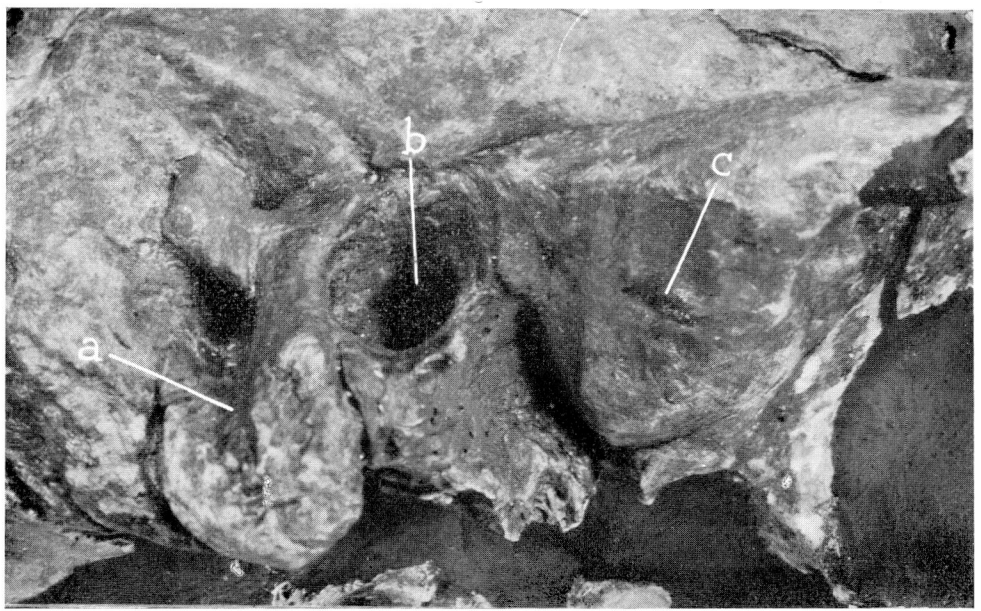

FIGURE 7b The right ear region of the Bronze Age Irish skull from Knockast. Mastoid process (a), and meatus (b) are indicated. Courtesy of the Proceedings of the Royal Irish Academy.

out, however, causing a general infection of the mastoid cells."

As in one of the Nubian specimens, it is puzzling that the temporo-mandibular joint is involved. Furthermore, the state of the remaining mastoid cells is not clear from the description. Thus it must remain a questionable case, and certainly calls for another examination.

Wells (1962) claims two more cases of mastoid infection, both from Norfolk, and both of Anglo-Saxon date. His first specimen (Fig. 5), is less convincing, and it would be useful to have a second opinion on this specimen. Post-mortem soil erosion can be highly selective, and in the mastoid area can simulate to a remarkable degree mastoid infection (we could list various such cases in the B.M.[N.H.] collections). In the Wells' Case I, there has clearly been some post-mortem erosion of the bone surface, and it is not clear whether this might have affected the mastoid cells. Further study would be desirable.

The second case, however, certainly shows mastoid disease (Fig 5) and appears to be the result of a cholesteatoma. A cholesteatoma occurs nearly always with a sclerotic and not a cellular mastoid, and it is thus a pity that comments on the radiographic appearance of the mastoid area were not given.

CONCLUSIONS

We have presented evidence of ear infection in Upper Pleistocene, late prehistoric and early historic populations, and have described the types of bone change noted. In some cases, it has been possible to study the original specimens. All our conclusions were however made after emphasizing the difficulties in interpreting the progress and characteristics of ancient ear disease from our knowledge of ear disease today.

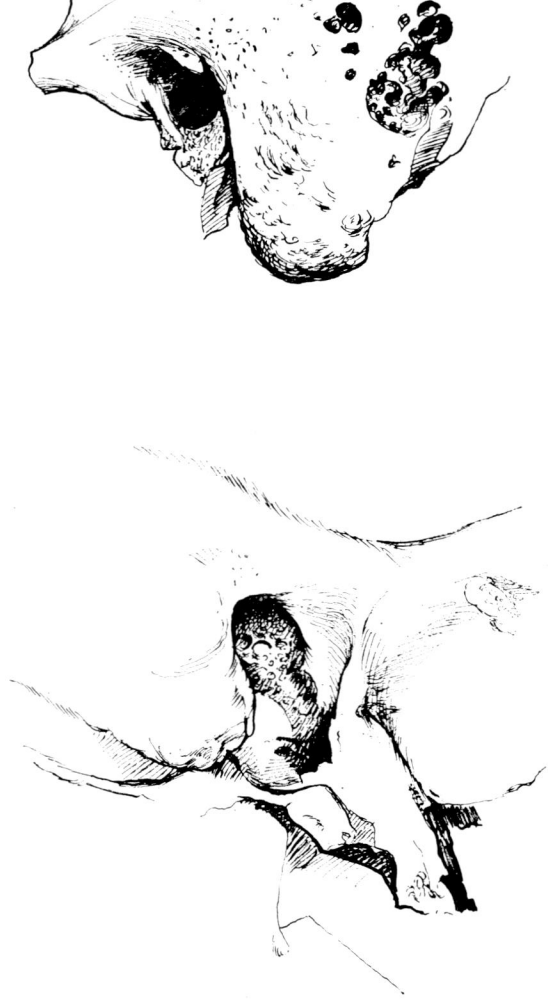

FIGURE 5. The Saxon ear specimens described by Wells (1962). Case 1 is *above*, and Case 2 *below*.

REFERENCES

BATRAWI, A. M. EL., 1935: Report on the human remains. *Mission Archeologique* de Nubie. 1929-1934. Government Press, Cairo.

DERRY, D. E., 1909: Anatomical report. *Archaeol. Survey Nub.*, Bull. 3., 29-52.

ELLIOT SMITH, G., and DAWSON, W. R., 1924: *Egyptian Mummies*. London, Allen and Unwin.

ELLIOT SMITH, G., and WOOD JONES, F., 1910: Report on the human remains. *Archaeol. Survey Nubia. Report 1907-1908.*, 11, Ministry of Finance, Cairo.

GHALIOUNGUI, P., 1963: *Magic and Medical Science in Ancient Egypt.* London, Hodder and Stoughton.

HENCKEN, H. O., and MOVIUS, H., 1934: The cemetery-cairn of Knockast, *Proc. Roy Irish Acad., 41*:232.

OAKLEY, K. P., BROOKE, W., AKESTER, A. R., and BROTHWELL, D. R., 1959: Contributions on trepanning or trephination in ancient and modern times. *Man, 60*:122.

SIGERIST, H. E., 1951: *A History of Medicine,* Vol. I. Oxford U. P.

SINGER, R., 1961: Pathology in the temporal bone of the Boskop skull. *S. A. Arch. Bull., 16*:103.

THOMPSON, R. C., 1931: Assyrian prescriptions for diseases of the ears. *J. Roy. Asiat. Soc.,* p. 1.

WELLS, C., 1962: Three cases of aural pathology of Anglo-Saxon date. *J. Laryng. Otol., 76*:931.

YEARSLEY, M., 1928: The pathology of the left temporal bone of the Rhodesian skull. In: *Rhodesian Man and Associated Remains.* British Museum (Natural History), London. 59-63.

Chapter 39

Degenerative Vascular Disease

A. T. SANDISON

It is generally agreed that at the present time degenerative vascular disease causes more deaths than any other combination of diseases. This is to some extent a reflection of longer expectation of life due to the conquest of the important infections; although in antiquity fewer persons survived to an age at which cardiovascular degeneration became probable we should at least expect to find evidence of such degeneration in more aged people.

Various writers have speculated upon the possible effects of vascular disease in altering the course of history. Wright (1952) has stated "One might well marshall strong evidence to prove that nations have fallen and peace conferences have failed because of cerebral arteriosclerosis in the leaders." Maclaurin (1930) suggests that among eminent historical personalities who suffered from degenerative vascular disease are included Henry VIII of England, "Bloody" Mary, Charles II of Britain, Philip II of Spain, the Holy Roman Emperor Charles V, and Frederick the Great of Prussia.

Before discussing evidence of vascular disease in antiquity it might be advisable to outline current views on the classification of such diseases. It is clear that medical historians are sometimes confused in these matters, e.g., Sigerist (1951) talks loosely of "arteriosclerosis" of the aorta of the Pharoah Merneptah and Thorwald (1963) of "arteriosclerosis" in mummies when clearly atheromatous degeneration was present.

ATHEROMA or endarteritis deformans is a common disease and virtually inevitable in old age; rarely it attains significant degree in young persons. It is a patchy intimal lesion associated with degeneration of the deeper lamellae and accumulation of lipid material. The medial coat is secondarily affected and the internal elastic lamina may stretch and fragment so that the process extends deeply: often in these circumstances there is atrophy of the muscle of the media. This process tends to be sectoral when the artery is studied in cross section. The process affects the aorta or smaller vessels or both. The lumen is gradually occluded and atrophy of the tissues served by the affected artery may occur. If the lumen of the vessels is blocked by supervening thrombosis there may be sudden severe ischaemia and infarction, e.g., of the myocardium. Atheroma may occur early and be severe in diabetics, leading to gangrene of the lower limbs.

ARTERIOSCLEROSIS is a diffuse condition affecting virtually the entire arterial tree and is the result of hypertension either idiopathic (essential) or secondary to renal disease. The larger arteries show medial hypertrophy followed by fibrosis. The internal elastic lamina is thickened and may reduplicate into the intima; later the elastic fibres may degenerate and fragment. Smaller arteries are hyalinised with intimal proliferation producing concentric thickening. In very old persons senile arteriosclerosis may develop when fibrosis occurs without preceding hypertonus. Severe arteriosclerosis may result in cardiac failure, secondary changes in the kidney leading to renal failure or haemorrhage from rupture of a cerebral artery (apoplexy).

Atheroma and arteriosclerosis may occur in the same person but the term "atherosclerosis" is unwarranted as the conditions are aetiologically quite distinct.

MEDIAL CALCIFICATION or Mönckeberg's sclerosis occurs in old persons: transverse rigid annuli of calcium form in the mid-zone of the medial coat after hyaline degeneration of muscle and connective tissue. The arteries of the lower limbs are most often affected and thrombosis and gangrene may result. *Other arterial lesions* are much less important in our consideration. *Endarteritis obliterans* is a reactive change with obliteration of the lumen and occurs in arteries close to chronic ulcers or infections and in the kidney in chronic glomerulonephritis. *Syphilis* affects the aortic media by extension from the adventitia along the vasa vasorum. Elastic tissue is destroyed and this mesarteritis may go to aneurysm. Sometimes the aortic valve is damaged and becomes incompetent; the process may involve the ostia of the coronary arteries causing myocardial ischaemia. Syphilis may also result in endarteritis of the smaller vessels in any tissues of the body. *Thromboangeitis* affects the entire neurovascular bundle of the limb in relatively young persons. Inflammatory change and thromboses are prominent and gangrene results. Smaller arteries may be affected in acute rheumatism, temporal arteritis and polyarteritis nodosa, etc., conditions in which there is an altered state of sensitivity.

ANEURYSMS are dilatations of arteries produced by several diseases, e.g., syphilitic mesarteritis, severe ulcerative atheroma, medial degeneration of Erdheim (commonly resulting in dissecting aneurysm), congenital deficiency of the muscular coat (as in cerebral artery aneurysm causing subarachnoid haemorrhage), trauma, myotic infection and from associated ulcerative processes close to the arterial wall.

MYOCARDIAL DISEASES. The commonest are myocardial fibrosis and infarction, the result respectively of slow or rapid occlusion of a coronary artery. Prolonged hypertension leads to marked ventricular hypertrophy and eventual cardiac failure. Less commonly the myocardium is damaged in acute rheumatism, malnutrition, toxaemia or by inflammatory processes of bacterial or viral type.

Endocardial Diseases. The most important cause in acute rheumatism; there may be acute inflammation or chronic change following acute episodes. The latter causes valvular disease of the heart predominantly affecting the mitral and aortic valves.

Subacute or acute bacterial endocarditis may complicate rheumatic valvular disease or congenital abnormality. Occasionally chronic valvular disease of insidious onset and non-rheumatic origin produces severe changes. The aortic valve may be damaged by syphilis. Occasionally in old people there are small atheroma-like degenerative lesions with some calcification.

It is therefore clear that cardiovascular disease may be of varied pattern and may affect not only heart and blood vessels but also produce important secondary changes in the kidneys, limbs, brain and other organs.

THE GROWTH OF KNOWLEDGE OF THE CIRCULATION

That Neolithic man was aware of the importance of the heart is witnessed by his depicting for magical purposes the heart on cave paintings of animals. His successes in trephining imply possible familiarity with meningeal vessels and venous sinuses. Sigerist (1951) suggests that phlebotomy may have been performed for the magical release of spirits based on the symptomatic relief of tension by menstruation or by epistaxis in certain fevers.

Mesopotamian medicine never countenanced human dissection and the heart was regarded as the seat of the intellect. The Papyrus Ebers appears to show that the Ancient Egyptians (c. 1500 B.C. and probably earlier) were aware of the movement of the heart and transmission of impulses along vessels to the limbs—"the heart speaks out of vessels of every limb" and the pulse appears to have been studied. The Ancient Chinese believed that blood moved in the arteries and was controlled by the heart. Study of the pulse was an important empirical factor in their medical practice.

Around 480 B.C. Empedocles considered the heart to be the centre of the vascular system distributing *pneuma* by the blood vessels, a view refuted by the Coan School. About this time also Diogenes of Alexandria made a detailed study of veins. In the book "On the Heart" of the *Corpus Hippocraticorum* some anatomical knowledge possibly based on animal dissection is evident. The auricles, valves, chordae tendineae and columnae carneae were recognized and according to Singer (1925) the concept of valve competence had been established. Aristotle (384-322 B.C.) recognized that arteries and veins usually ran together. It is believed that around 300 B.C. Herophilus of Chalcedon made the first dissection of the human body; he regarded pulsation as an intrinsic quality of arteries and studied the pulse, possibly timing it with the aid of a clepshydra.

Erasistratus of Chios (c. 290 B.C.) described the atria, chordae tendineae and valves and came near to appreciating that blood circulated; he knew that during life the arteries contained blood and that consequently arteries and veins must communicate. Rufus of Ephesus (1st cent. A.D.) claimed that in systole the cardiac apex strikes the chest wall, a fact which had to be rediscovered in mediaeval times.

Galen (130-200 A.D.) studied the pulse and noted variations—softness or hardness of the arterial wall. The authority of his views stifled research in mediaeval times. During the thirteenth century, however, in Cairo Ibn-al-Nafis concluded that blood circulated from the heart through the lungs.

Human dissection began in Bologna in the thirteenth century; Leonardo da Vinci (1452-1519) made admirable drawings of the heart and peripheral vessels and made models to demonstrate valve action. A new era opened with the publication of *De Fabrica Corporis Humani* by Andreas Vesalius in 1543 but the vascular system was one of the least satisfactory books. Realdus Columbus (1516-1559) showed that cardiac systole is synchronous with arterial expansion. Coiter (1534-1576) observed the hearts of living animals and noted that auricular preceded ventricular contraction. Fabricius ab Aquapendente (1537-1619) made excellent descriptions of the cardiac valves.

Assisted by this laboriously accumulated evidence William Harvey (1578-1657) speculated on the function of the vascular system, made some experiments and opened a new era in medicine with the publication of *De Motus Cordis* in 1628. Willius and Keys (1941) believe Harvey's work to be "in reality the fundamental contribution on which the modern concepts of anatomy and physiology of the heart and circulation are based." Years later the final piece of evidence was revealed by the microscope of Malpighi and it was confirmed that tiny capillary channels link the arteries and veins.

For an interesting account of the discovery of the circulation of the blood see Doby (1963); Singer (1925) may also be consulted.

THE GROWTH OF VASCULAR PATHOLOGY

Pathological knowledge was even more slowly acquired with isolated sparse observations. Krumbhaar (1937) states that references to paralytic stroke and heart disease occur on tablets in the Library of Ashurbanipal; a much better case can be put up for knowledge of respiratory disease (Sigerist, 1951). Mesopotamian medicine regarded symptoms as diseases and translators face considerable translational difficulties.

There is considerable evidence in Papyrus Ebers that the Ancient Egyptians were familiar with syncope, angina pectoris, oedema, ascites, haematuria, haemorrhoids, varicose veins and possibly arteriovenous aneurysms. Ebbell (1937) also believes that hemiplegia was recognized. Ghalioungui (1963) believes that tachycardia, extrasystoles and dyspnoea of effort in cardiac disease were noted. In Papyrus Edwin Smith there is a possible reference to septic cerebral thrombosis. The Egyptians attempted a pathology based on changes in *vessels* of different kinds arising in the liver but despite embalming evisceration no formal dissection was carried out and *vessels* included blood vessels, tendons, nerves and muscles. The Ancient Hindus believed that dropsy was sent by the God Vishna and jaundice caused by demons. In the Vedic Books retention of urine, strangury, dropsy, paralysis and heart diseases may be recognized (Sigerist, 1961). The Ancient Maya believed that gods sent disease changes which appear to have included oedema, jaundice, cachexia, haematemesis and sudden death (Goetz and Morley, 1951).

The Hippocratic School recognized angina pectoris, apoplexy (relating this to advancing age), possibly cerebral arteriosclerosis, gangrene of the extremities (possibly vascular in origin), oliguria and oedema. Clark (1963) has discussed in detail apoplexy in Hippocratic writing and concludes that they were dealing with a heterogeneous selection of cases, giving a fairly consistent clinical picture. Ascites was related to liver disease and association with haemorrhoids was known. Erasistratus also noted stone-like hardness of the liver (cirrhosis) in cases of dropsy. It is interesting to recall that Ruffer (1911a) claimed to have demonstrated fibrosis in the liver of an Egyptian mummy: his description, however, is not that of fully established cirrhosis. Ghalioungui (1963) believes that bilharzial cirrhosis may have existed in Ancient Egypt. Probably it also occurred in Ancient China. Galen believed that splenic affection and haemorrhoids were the cause of ascites: we now appreciate that these result from portal hypertension. Galen also misinterpreted pulse irregularity as the result and not the cause of cardiac dropsy. Celsus (1st cent. A.D.) described paracentesis abdominis for ascites and Soranus of Ephesus (2nd cent. A.D.) noted genital oedema as part of anasarca.

Rufus of Ephesus related other cases of dropsy to changes in the kidney as did Aetius (6th cent. A.D.) and Avicenna (980-1036 A.D.). This was re-established by William of Saliceto (1201-1280 A.D.) and finally consolidated by Bright in the nineteenth century. Aretaeus of Cappadocia (2nd-3rd cent. A.D.) related anasarca to heart disease, hemiplegic paresis to contralateral cerebral lesions and treated aneurysms by ligation. Galen also studied aneurysms and noted that their pulsation could be controlled by pressure. Aetius recorded production of aneurysms by surgical trauma and treated aneurysm by section between double ligatures. Paulus of Aegineta (7th cent. A.D.) adopted similar measures. Paulus also described an association between ascites and carcinomatosis. Avenzoar (1094-1160) described serous pericarditis. Thorwald (1963) cites a pre-Columbian South American statuette which has been interpreted as showing nephritic oedema. Sarsaparilla is said to have been used by the Aztecs as a diuretic for renal complaints and late Inca chronicles include references to kidney inflammation and dropsy.

Bauhinus (16th cent.) described cerebral haemorrhage and Baglivi (1668-1706) investigated the cerebral haemorrhage which killed his master Malpighi (Ficarra, 1942): this was associated with hypertrophy and dilatation of the left ventricle and unilateral renal atrophy and must be regarded as hypertensive. Wepfer (1620-1695) related clinical phenomena to cerebral haemorrhage: his own necropsy showed aortic calcification.

Isolated observations during the fifteenth to sixteenth centuries included acute pericarditis, malposition of the heart, the distinction between aneurysm and haematoma, aortic sclerosis and serous transudates in cardiac disease. During the seventeenth century tricuspid endocarditis, ruptured aortic aneurysm, acute and chronic valvular disease, cardiac syphilis and aortic aneurysm were recognized. In 1761 Morgagni published *De Sedibus* containing descriptions of structural changes in angina pectoris, acute endocarditis including the gonorrhoeal variety, arteriosclerotic cerebral haemorrhages, rupture of aortic aneurysm, etc. In 1797 Baillie described rheumatism of the heart and in 1806 Corvisart published the first systematised clinico-pathological correlations in vascular disease. Rokitansky in 1852 described the important arterial diseases.

Curiously the recognition of coronary artery disease as an important cause of cardiac disablity and death was long delayed. Angina pectoris is noted in *Coan Prognosis* of Corpus Hippocraticorum and is also said to have been noted by Coelius Aurelianus (c. 400 A.D.). The Earl of Clarendon gave an excellent description of the life and death in 1632 of his father who suffered from angina pectoris.

William Harvey briefly recounted the case of Sir Robert Darcy who suffered from nocturnal angina and died with dropsy. Autopsy showed rupture of the left ventricle and "apparent impediment to passage of blood from the left ventricle into the arteries." John Hunter suffered from angina pectoris for twenty years and at autopsy in 1793 was found to have diseased coronary arteries. In 1878 Hammer gave the first satisfactory account of coronary thrombotic occlusion but only in 1912 did Herrick clearly describe the clinical features of sudden coronary occlusion and cleared up much confusion.

It is interesting to note that although lathyrism of the nervous system was known to the Hippocratic School (*Epidemics II*) there is no evidence of affection of the vascular system. There are no acceptable accounts of gangrenous ergotism until 857 A.D. at Xanten in Germany: thereafter numerous epidemics are described in mediaeval France (Barger, 1931).

OTHER SOURCES

Other instances of probable vascular disease have been culled from the literature. Rowling (1961) suggests that the Papyrus Ebers may contain a description of coronary diseases, that the mortuary inscription of Weshptah may indicate his death by cerebral haemorrhage, and that an Old Kingdom tomb relief may show death by coronary or cerebrovascular accident. Bruetsch (1959) also illustrates the sudden death of Sesi (VIth Dynasty) and postulates that this may have been due to coronary artery disease. All of these are speculative. Rowling has also suggested that expectation of life in Ancient Egypt was probably less than forty years: if so, degenerative vascular disease cannot have been a statistically important cause of death. Needham and Lu (page 235) claim that cerebral hemorrhage, hemiplegia and aortic aneurysm may be recognised in the Ancient Chinese literature.

Many writers have searched the Holy Bible and Talmud for cases which may be diagnosed. Brim (1936) gives a long list which includes apoplexy, effort syndrome, angina pectoris, arteriosclerosis and heart block; all of these are extremely fanciful and scarcely worthy of consideration. Nor does Bennett's (1896) suggestion that Jeroboam suffered an embolism bear scrutinising: this may have been a hysterical paralysis. Krumbhaar (1937) suggests that Nabal (1st Sam. 25, 37) died from apoplexy and this is not unreasonable. In the opinion of the present writer the account of the life and death of Herod the Great given by Flavius Josephus in "The Jewish War" is entirely compatible with degenerative cardiorenal disease and it is hoped to elaborate this elsewhere.

Seneca, who died in 65 A.D., described his own ailment—"the attack is very short and like a storm; it usually ends within an hour . . . to have any other malady is only to be sick; to have this is to be dying"—in terms strongly suggesting coronary artery insufficiency (Marvin, 1957).

PALAEOPATHOLOGICAL EVIDENCE

We are fortunate in being able to derive direct evidence of vascular disease by studying mummified or desiccated bodies. These have been obtained in greatest number from Ancient Egypt but also from Peru, Canary Islands, North European peat bogs, etc. Fortunately in such bodies the blood vessels may be sometimes well preserved and available for histological study. It is also possible to obtain evidence of vascular degenerative disease when accompanied by calcification by radiography of unwrapped or wrapped bodies. Moodie (1931) described radiological evidence of tortuous, sclerosed artery over the scapula, ribs and interosseous region of the forearm in a Predynastic Egyptian body.

The first paper of which I am aware is that of Czermak (1852). Czermak, a distinguished Viennese laryngologist, teased out tissues from two Egyptian mummies, examined these microscopically and made surprisingly detailed observations. In one instance he observed aortic calcification. The next observation came from Shattock in 1909. Elliot Smith had given him a portion of the aorta of the Pharaoh Merneptah (died c. 1224 B.C.): this had been noted by Elliot Smith to show extreme calcareous degeneration with the formation of bone-like plaques. Shattock made frozen sections of these and examined these unstained and stained by Haemalum and Eosin and Carbol Thionin. He demonstrated long, parallel, wavy lamellae of elastic tissue with interlamellar calcium which dissolved in 10 per cent hydrochloric acid. There was also striation parallel to the elastic lamellae thought to be probably muscle. The intima was unaffected.

Elliot Smith had also noted tortuous calcified temporal arteries in the mummy of Pharaoh Ramesses II (died c. 1232 B.C.). This fact and a description of the aorta of Merneptah are recorded in his monumental work on the royal mummies (1912).

The publications of Sir Marc Armand Ruffer, collected posthumously in 1921, are our most valuable source of information of vascular disease in Egyptian mummies. Ruffer pointed out that mummy arteries are frequently well preserved and often much flattened—"looking as if they had been well ironed"—unless they have undergone marked fibrosis or calcification. Ruffer was priviledged to study a relatively large number of specimens, mainly from the New Kingdom but covering a wide range of time.

Among the phenomena he noted were a hard calcareous plate in the aorta of a XXIst Dynasty specimen and multiple calcareous change with maximal lesions just above the bifurcation of the aorta and calcific atheroma of the left subclavian artery in a mummy of the XXVIIIth to XXXth Dynasty period. This specimen showed also atheroma of the common carotids and calcific atheroma of the common iliacs; calcific change was also obvious in the femoral and profunda arteries. Ruffer concluded from the state of the costal cartilages that this mummy was not that of an old person.

The mummy of a man of the Greek period not over fifty years old at the time of death showed atheroma of the aorta and brachial arteries. In a XXVIIth Dynasty mummy calcareous changes were noted in the thoracic aorta. Similar patchy change was also noted along the length of the aorta of a Coptic desiccated body; this was most marked above the aortic bifurcation. Calcification of the posterior tibial artery was noted in the mummy of a woman of the XXIst Dynasty and in the posterior tibial, peroneal and ulnar arteries of further XXIst Dynasty specimens. In one female body of the same period with calcification of the peroneal artery bed sores were also present. Other tibial, peroneal and ulnar vessels from specimens of the same period and vessels of a mummy of the Persian period examined by Ruffer showed no evidence of disease.

The degenerative process noted by Ruffer in these specimens showed a predilection for sites of origin of smaller arteries so that Ruffer reasonably concluded them to be of atheromatous nature. Histological sections were prepared after rehydration, decalcification by Marchi's solution, embedding in paraffin and section in the usual way. Sections were stained by haematoxylin alone and by Van Gieson's fluid and proved to be of reasonable technical quality. It is much to be regretted that Ruffer did not use specific stains for elastic tissue. His findings are illustrated by coloured drawings not easily interpreted but there seems to be little doubt that these are atheromatous lesions. Ruffer was not successful in confirming fibrosis in vessels which he considered to be "fibroid,' i.e., arteriosclerotic (Ruffer 1910a, 1911a, 1911b).

In 1931 Long described a most interesting female mummy—that of the Lady Teye—aged about fifty years from the XXIst dynasty at Deir-el-Bahari. The heart showed calcification of one mitral cusp and thickening and calcification of the coronary arteries. The myocardium is said to have had areas of patchy fibrosis and the aorta a "nodular arteriosclerosis." The renal capsule was thickened, many of the glomeruli were fibrosed and the medium-sized renal arteries were sclerotic. This case appears to be of hypertensive arteriosclerotic type with associated atheromatous change.

It is interesting to note that although the heart was usually left *in situ* in Egyptian mummies and was on occasion examined by Ruffer no other example of cardiac pathology has been published. Ruffer noted that valves could sometimes be demonstrated and showed histologically that muscle striation might persist. Wilder (1904) also noted remains of the heart in a dried Basket-maker Indian body and good preservation of blood vessels in a dried Peruvian body. Wilson (1927) found no evidence of vascular disease in Basket-maker Indian bodies but Williams (1927) described arteriosclerosis with calcified thrombus in a dried Peruvian body dated around 700 A.D. Shaw (1938) utilising well-preserved Canopic material of XVIIIth Dynasty date described the appearances of pulmonary, hepatic and superior mesenteric arteries; the latter showed fibro-elastic thickening.

I felt that the descriptions of previous workers were difficult to interpret in view of terminological changes and the fact that they are accompanied usually by rather inadequate drawings and published the conclusions made following a detailed investigation of the stainability of vascular elastic fibres in mummified and dried human tissues (Sandison, 1963). A battery of modern staining techniques were utilised and showed that delineation of elastic laminae of blood vessels was a valuable aid in the recognition of vascular changes. These changes were also capable of being adequately photographed by both monochrome and colour methods. I had previously shown that it was possible to prepare frozen sections of mummy tissues and to demonstrate persistence of sudanophilic lipid (Sandison, 1959). This was obviously important in the investigation of atheromatous lesions.

On the basis of these staining methods, I investigated, using technical methods described elsewhere (Sandison, 1955, 1957), a limited amount of late dynastic Egyptian material available to me and

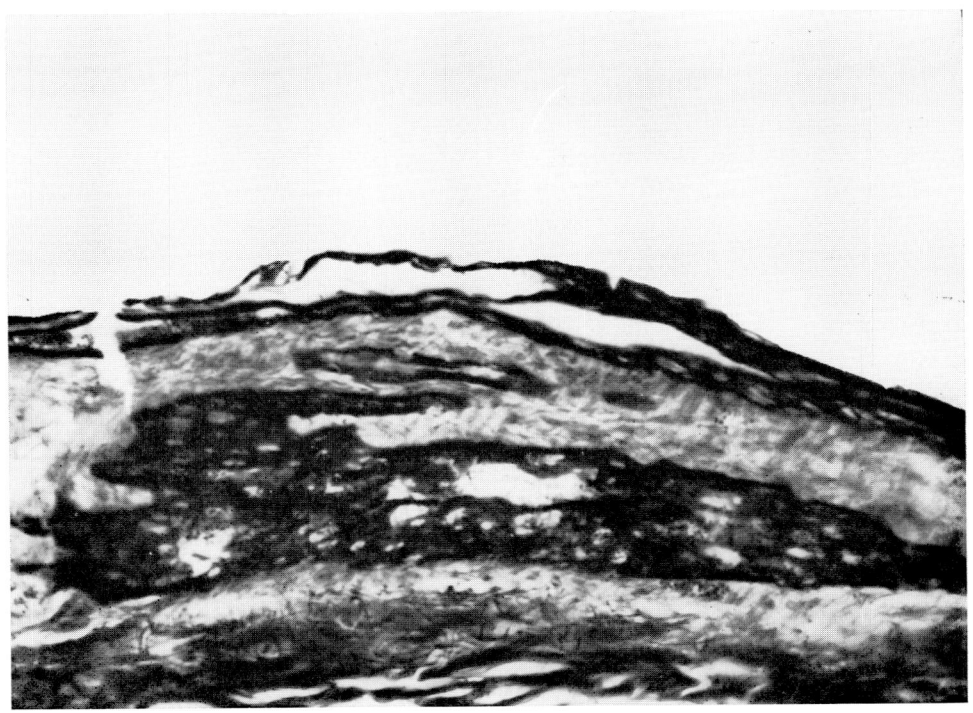

FIGURE 1. Calcification of the media of thyroid artery in a male mummy. Phosphotungstic acid haematoxylin x 630. (Courtesy of Medical History)

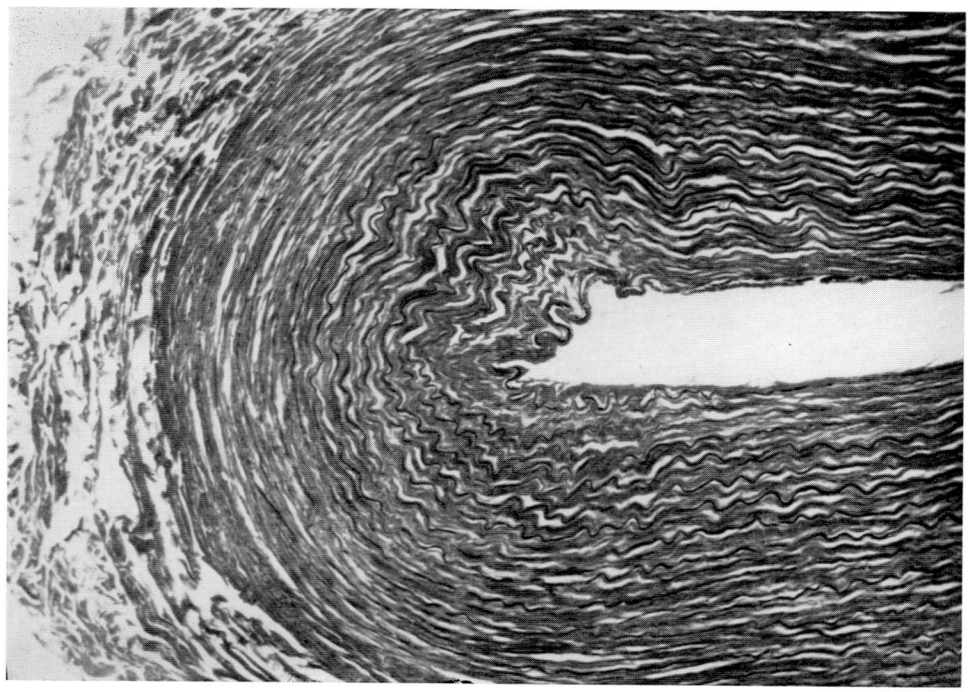

Figure 2. Carotid artery of male mummy; fibrous tissue alternates with the elastic laminae. Verhoef's elastica-van Gieson x 250. (Courtesy of Medical History)

showed that blood vessels were usually well-preserved (Sandison, 1962). Medial calcification was noted in the thyroid artery of a male mummy (see Fig. 1); it is interesting to note that Osborne (1963) has observed that the thyroid artery even in young persons may undergo calcification. It proved possible to demonstrate fibrosis, reduplication of the internal elastica lamina as well as intimal thickening of small arteries (see Figs. 2 and 3). These changes indicate arteriosclerosis and it will be recalled that Ruffer was unable to prove this histologically with the technical methods then available although he was convinced that by palpation some of the vessels he studied were "fibroid."

I also noted an artefact appearance produced by the flattening of mummified arteries to which Ruffer had drawn attention. This was a separation and splitting of the artery coats somewhat reminiscent of dissecting aneurysm. It was observed, however, that an amorphous material was present in these spaces and frozen sections show that this was sudanophilic lipid sometimes with calcium also present (see Fig. 4). It was concluded that the spaces represent shrinkage and splitting of atheromatous plaques. Sections taken at other levels confirmed the presence of sectoral atheromatous intimal thickening of atheromatous types (see Fig. 5); in one such thickening sudanophilic lipid was clearly demonstrated.

My studies using improved technical methods have therefore given clear histological confirmation of Ruffer's claims to have shown atheromatous, arteriosclerotic and calcific changes in mummy arteries. Sandison and Brothwell (unpublished work) have also shown that structure is

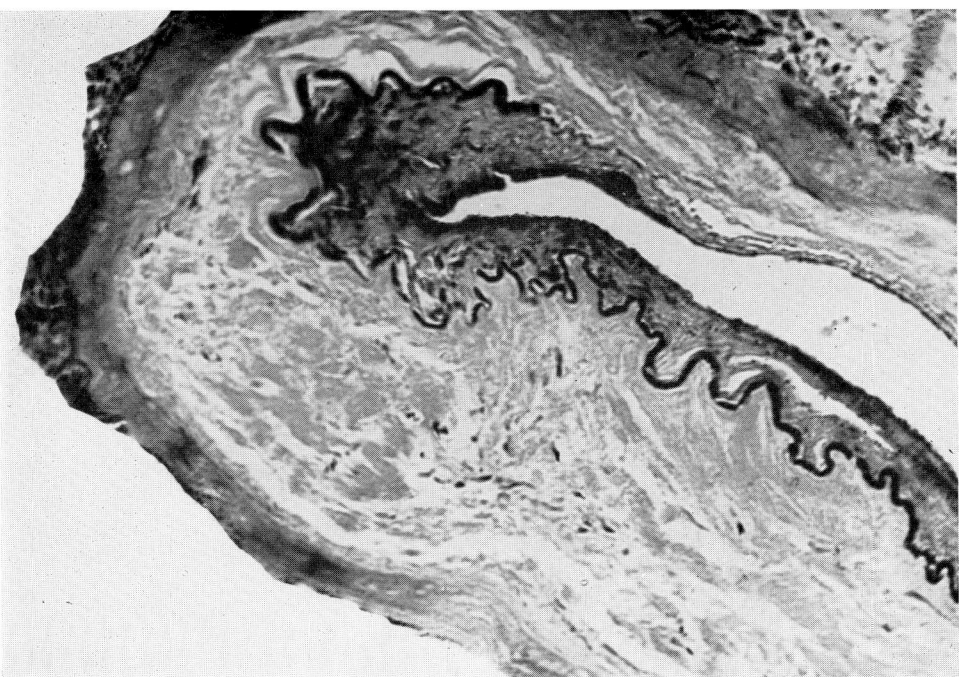

FIGURE 3. Reduplication of internal elastic lamina of tibial artery of elderly female mummy. Heidenhain's iron-haematoxylin x 400. (Courtesy of Medical History)

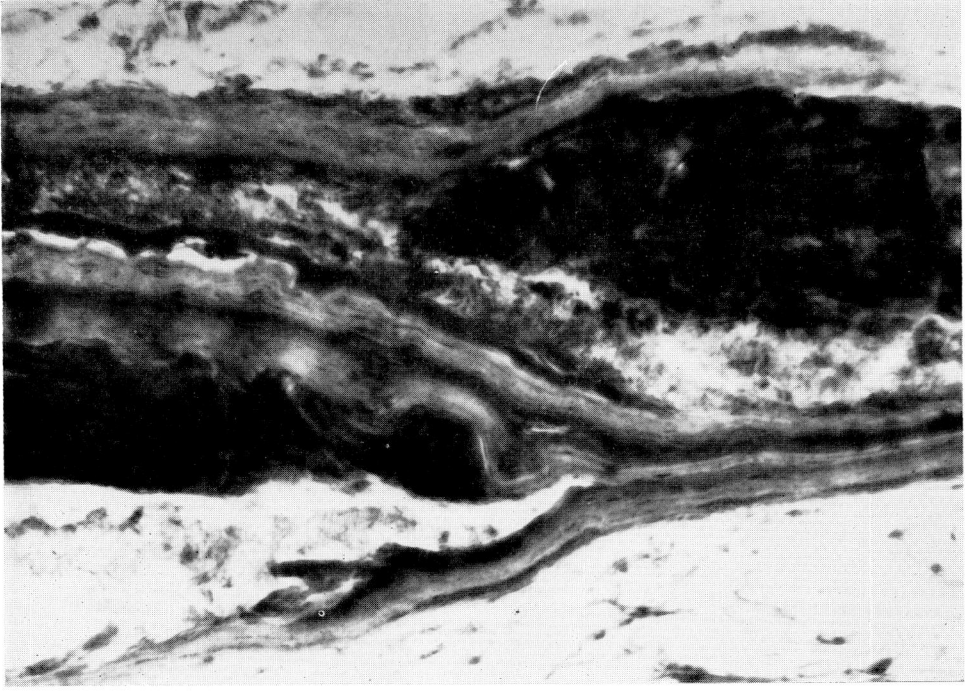

FIGURE 4. Frozen section to show lipid in atheromatous lesion of tibial artery of elderly female. Sudan Black x 400. (Courtesy of Medical History)

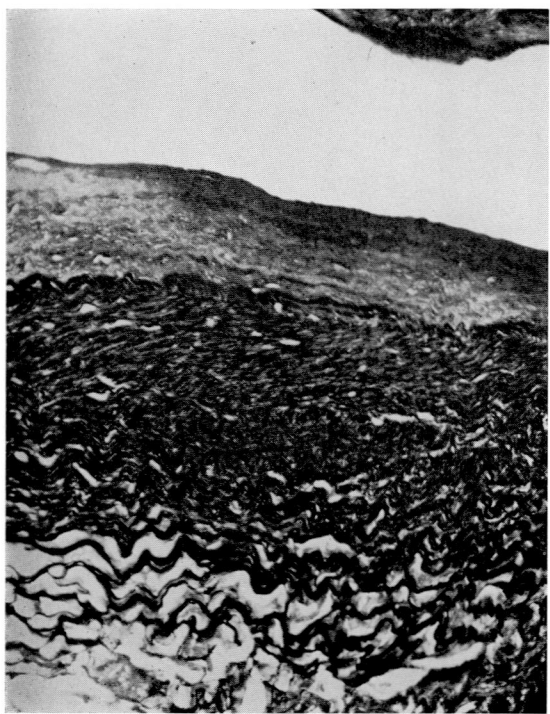

FIGURE 5. Intimal thickening of atheromatous external carotid of male mummy. Heidenhain's iron-haematoxylin x 250. (Courtesy of Medical History)

well preserved in the major pulmonary arteries of an anthracotic lung of a Guanche body (see Fig. 6). Modern histological techniques might therefore be applied with advantage to other desiccated human remains.

Among other interesting isolated observations may be included the demonstration by Aichell (1927) that blood vessels and fibrous tissues of the heart may be preserved in Moorleichen. This was to some extent confirmed by Schlabow et al. (1958). McGee (1894) reported that blood vessels were still recognisable in the body of Pizarro exhumed 350 years after his assassination.

Porter (1905) gave an interesting report on the exhumation of John Paul Jones (whose body had been coffined in spirit for 113 years). Jones had complained terminally of hydrothorax, ascites and jaundice; at the delayed necropsy the heart was still flexible and striation could be detected microscopically. The kidneys showed evidence of chronic renal disease, the vessels being sclerotic and the glomeruli fibrosed.

It will be recalled that Long (1931) described hypertensive nephrosclerotic changes in the mummy of the Lady Teye.

There are a few further reports of renal disease in antiquity. Ruffer (1910a) reported congenital unilateral left renal hypoplasia (sometimes a cause of hypertension) and possible renal abscesses and in another mummy of the XVIIIth to XXIst Dynasty period noted multiple renal abscesses which he considered to have been produced by *Bacillus coli*. Insufficient detail is given to evaluate this claim. A most important observation by Ruffer was his discovery in the kidneys of two mummies of the XXth Dynasty of numerous calcified ova of *Bilharzia haematobia* among the straight tubules (Ruffer, 1910a and b).

Lithiasis of the renal tract has been reported also in ancient material. It is clear that the Assyrians were familiar with lithiasis and differentiated between hard and soft stones. Shattock (1905) described and analysed renal calculi from a IInd Dynasty tomb; oxalates and conidia were noted. A vesical calculus discovered in the nostril of a XXIst Dynasty Priest of Amen contained uric acid covered by phosphates (Smith and Dawson, 1924). Ruffer (1910a) described three mixed phosphate-uric acid calculi from a predynastic skeleton and Williams (1926) a bladder stone from a Basket-maker Indian body.

CONCLUSIONS

It is obvious from both literary and palaeopathological sources that, despite

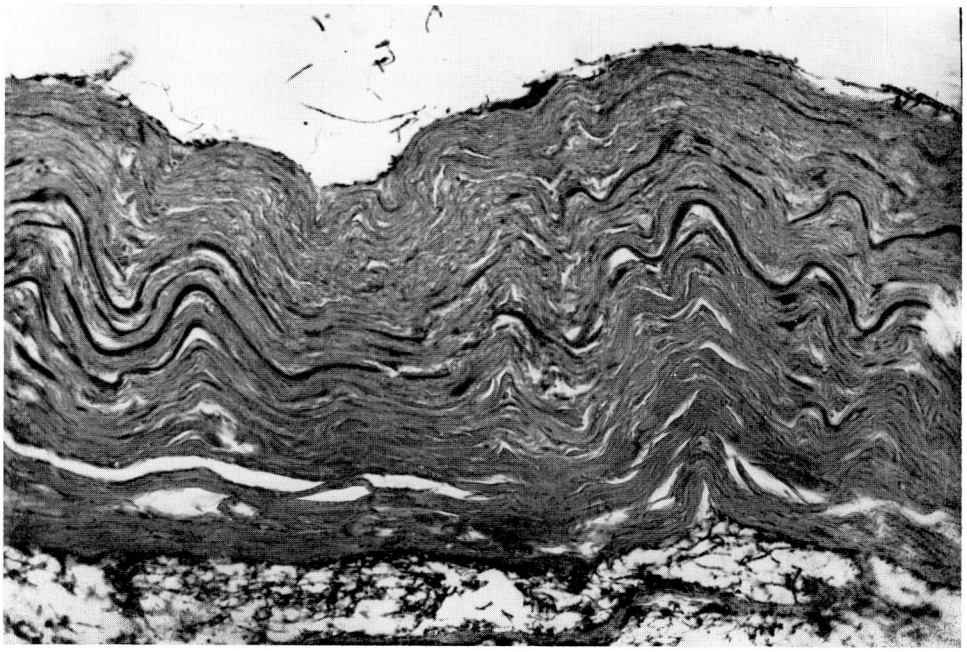

FIGURE 6. Pulmonary artery of male Guanche body. Phosphotungstic acid haematoxylin x 300.

the relatively short expectation of life in ancient societies, individual persons suffered from vascular degenerative disease. In some instances, for example, in the case of Pharaoh Ramesses IInd, this was probably related to senility but in others, as exemplified by some of Ruffer's mummies clearly not these of elderly persons, due, as happens at present, to the onset of vascular disease in relatively young persons.

It is equally clear that there is no indication of cardiovascular syphilis in ancient remains and no distinct references in the medical literature until the seventeenth century. While this is not very strong evidence it is further support for the view that syphilis was unknown in the Old World until the post-Columbian period. The only iatrogenic disease process noted in our investigation is the production of aneurysm by the surgeon or phlebotomist.

A fairly distinct picture emerges of the types of vascular disease which may be recognised in ancient societies. In Ancient Egypt we have evidence of vascular degenerative disease occurring over thousands of years from the Predynastic until the Christian Coptic period. The literature suggests that apoplexy, angina pectoris, aneurysm and cirrhosis of the liver with portal hypertension were observed by physicians. The actual mummies show clear evidence of atheroma with and without calcification, hypertensive arteriosclerosis, and medial calcification (Mönckeberg's sclerosis) of arteries. In one instance we have changes indicating hypertensive renal change, myocardial fibrosis and coronary artery disease. Other mummies have shown unilateral renal hypoplasia, probable renal abscess, lithiasis of the renal tract and renal bilharziasis. There is also a case of possible hepatic fibrosis.

Apoplexy and heart disease may have occurred in Ancient Mesopotamia and apoplexy and degenerative cardio-renal disease in biblical times. For these and the classical periods (apart from evidence of disease in Graeco-Egyptian mummies) we must rely on literary sources. There is, however, reasonable evidence of apoplexy and hemiplegia, vascular gangrene, aneurysms, angina pectoris and dropsy resulting from cardiac and renal disease and possibly from carcinomatosis. There is good reason to believe that portal hypertension evidenced by ascites, splenomegaly and haemorrhoids and resulting from hepatic cirrhosis also occurred.

From Peru around 700 A.D. we have palaeopathological evidence of degenerative arterial disease and from mediaeval Europe and Arabia literary indications of renal disease with dropsy, of pericarditis and of gangrenous ergotism. Renaissance writings following the introduction of the necropsy reveal evidence of cerebral haemorrhage, pericarditis, malpositions of the heart and aortic valve disease. With the advent of the seventeenth century, physicians were recognising endocarditis, rupture of aneurysms, valvular disease of the heart and cardiovascular syphilis. The nineteenth century saw the elucidation of the morbid anatomy of almost all the important diseases of the heart, blood vessels and kidneys.

Much has been written in recent years on the causes of degenerative vascular diseases but we are, in fact, still in doubt concerning the mechanisms which operate in their production. There seems little doubt that advanced age is the main factor in the development of medial calcification. Certainly arteriosclerosis results from prolonged hypertension; the latter may be essential or secondary. Secondary hypertension is clearly related to pre-existing chronic renal disease, e.g., glomerulonephritis; essential hypertension is probably, but not absolutely certainly, related to release of renal pressor substances. Hypertension may accompany gout or lead poisoning but in the vast majority of cases there is no such association. Probably there is a familial tendency in many cases; excessive eating, drinking and hard physical labour have all been suggested as aetiological factors without clear proof. Smoking does not appear to be important.

More recently attention has been focussed on atheroma, particularly of the coronary arteries where the development of myocardial fibrosis and infarction is a fairly sensitive indicator of significant arterial change. It is clear that the incidence of atheroma increases with age and that the local distribution of the lesions in the arterial tree is to an extent decided by local physical strains. Hypertension plays no part in the production of atheroma in the general circulation although it is important in the pulmonary circulation in producing pulmonary artery atheroma.

Attention has been focussed on the importance of cholesterol and fats of animal origin. It is clear that atheroma is less common in human societies living on subsistence diets than in others where the diet is liberal and rich in animal fats and produces changes in the plasma lipoprotein levels. Endocrine factors also influence liproprotein patterns and it is clear that the higher oestrogen levels of younger women play a part in preventing atheroma.

Smoking has a baneful influence in cases of severe occlusive vascular disease and of thromboangeitis but there is no evidence that it is an aetiological factor in the development of atheroma. Nor is there evidence that alcohol is an aetiological factor;

Ruffer (1911b) showed that atheroma occurs frequently in devout Moslems who are prohibited the use of alcoholic beverages.

It is now fairly clear that within circumscribed social classes these persons undertaking strenuous work or exercise are less liable to develop myocardial infarction than those leading sedentary lives. The group undergoing strenuous effort may, however, suffer from angina pectoris.

"Psychological" factors have also been implicated, but since the psychological make-up of any person may affect his choice of occupation, degree of exercise, diet, smoking and drinking habits etc., it is clear no concrete assistance is likely to be gained from such considerations.

The knowledge which has been derived from studies of vascular disease in ancient societies may be considered in the light of the above observations. We have the best concrete evidence of vascular disease from Ancient Egypt and it is fortunate that we are also well informed about their social life. We may certainly eliminate tobacco as a cause of their vascular diseases; that alcohol was enjoyed and occasionally abused is clear from tomb paintings (Sigerist, 1951). Beers and wines were drunk but fortified spirits were not then available and it is probable that alcoholism was not a social problem. The common man in Ancient Egypt probably had scant opportunity to enjoy meat, the diet being largely vegetable. It cannot be denied, however, that the well-to-do, whose bodies would automatically be embalmed, may have lived well and certainly were not infrequently rather obese.

It is also probable that members of those social classes whose bodies were regularly mummified did not undertake hard physical work; we know also that there was no cult of regular and strenuous muscular exercise such as obtained in Classical Greece. We have no reason to believe, however, that life was any less traumatic psychologically in Ancient Egypt; the "wear and tear" of modern life is probably a convenient myth.

While lack of physical exertion may have played a part in the development of vascular disease in Ancient Egypt we should be unwise to draw any firm conclusions on the aetiology of such diseases.

Cumston (1926) stated "Historical pathology gives descriptions of diseases in all ages that can be exactly applied to the diseases of today, thus showing that there are diseases of all times and of all places. These maladies form the common property of human pathology; they are the inevitable result of the conflict of man with his surroundings and his struggle with the general forces of nature . . . the fact is, of course, that the majority of the diseases which successively appear in medical history are not 'new diseases;' they are but diseases which with the progress of science have been recognised as 'types apart.'"

When Cumston penned these words the iatrogenic diseases were but a small cloud on the horizon; the cloud now looms larger but this apart we must agree with his conclusions. The degenerative vascular diseases amply support his thesis.

REFERENCES

Aichell, O., 1927: Über Moorleichen, nebst Mitteilung eines neuen Falles (2½ Jähriges Mädchen von Röst in Dithmarschen). *Anthrop. Anzeig.*, 4:57.

Barger, G., 1931: *Ergot and Ergotism*. London, Gurney and Jackson.

Bennett, R., 1896: *The Diseases of the Bible*. London, Religious Tract Society.

Brim, C. J., 1936: *Medicine in the Bible*. New York, Froben.

Bruetsch, W. L., 1959: The earliest record of sudden death possibly due to atherosclerotic coronary occlusion. *Circulation*, 20:438.

CLARK, E., 1963: Apoplexy in the Hippocratic Writings. *Bull. Hist. Med.,* 37:301.

CUMSTON, C. G., 1926: *An Introduction to the History of Medicine.* London, Kegan Paul, Trench, Trubner.

CZERMAK, J., 1852: Beschreibung und mikroskopische Untersuchung zweier ägyptischer Mumien. *S. B. Akad. Wiss., Wien.,* 9:427.

DOBY, T., 1963: *Discoverers of Blood Circulation. From Aristotle to the times of Da Vinci and Harvey.* London, Abelard-Schuman.

EBBELL, B., 1937: *The Papyrus Ebers: the Greatest Egyptian Medical Document.* Copenhagen, Levin and Munksgaard.

FICARRA, B. J., 1942: Eleven famous autopsies in history. *Ann. Med. Hist.,* 4:504.

GHALIOUNGUI, P., 1963: *Magic and Medical Science in Ancient Egypt.* London, Hodder and Stoughton.

GOETZ, D., and MORLEY, S. G., 1951: *Popol Vuh: the Sacred Book of the Ancient Quichè Maya.* London, Hodge.

KRUMBHAAR, E. B., 1937: *Clio Medica: Pathology.* New York, Hoeber.

LONG, A. R., 1931: Cardiovascular renal disease: report of a case of 3000 years ago. *Arch. Path. (Chicago).,* 12:92.

MCGEE, W. J., 1894: The remains of Don Francisco Pizarro. *Ann. Anthrop.,* 7:1.

MACLAURIN, C., 1930: *De Mortuis.* London, Cape.

MARVIN, M. M., 1957: In *You and Your Heart* by Marvin, H. M., Jones, T. D., Page, I. H., Wright, I. S. and McCarty M. New York, Signet Key Books.

MOODIE, R. L., 1931: *Roentgenologic Studies of Egyptian and Peruvian Mummies.* Chicago, Field Museum.

OSBORN, G. R., 1963: *The Incubation Period of Coronary Thrombosis.* London, Butterworths.

PORTER, H., 1905: The recovery of the body of John Paul Jones. *Century Magazine,* 70:927.

ROWLING, J. T., 1961: Disease in Ancient Egypt: evidence from pathological lesions found in mummies. M. D. Thesis, University of Cambridge.

RUFFER, M. A., 1910a: Remarks on the histology and pathological anatomy of Egyptian mummies. *Cairo Sci. J.,* 4:1.

RUFFER, M. A., 1910b: Note the presence of "bilharzia haematobia" in Egyptian mummies of the Twentieth Dynasty (1250-1000 B.C.). *Brit. Med. J.,* 1:16.

RUFFER, M. A., 1911a: Histological studies on Egyptian mummies. *Mém. Inst. Égypt.* 6(3).

RUFFER, M. A., 1911b: On arterial lesions found in Egyptian mummies (1580 B.C.-525 A.D.). *J. Path. Bact.,* 15:453.

RUFFER, M. A., 1921: *Studies in the Palaeopathology of Egypt.* Chicago, University of Chicago Press.

SANDISON, A. T., 1955: The histological examination of mummified material. *Stain Technol.,* 30:277.

SANDISON, A. T., 1957: Preparation of large histological sections of mummified tissues. *Nature,* 179:1309.

SANDISON, A. T., 1959: Persistence of sudanophilic lipid in sections of mummified tissue. *Nature,* 183:196.

SANDISON, A. T., 1962: Degenerative vascular disease in the Egyptian mummy. *Med. Hist.,* 6:77.

SANDISON, A. T., 1963: Staining of vascular elastic fibres in mummified and dried human tissues. *Nature,* 198:597.

SCHLABOW, K., HAAGE, W., SPATZ, H., KLENK, E., DIEZEL, P. B., SCHÜTRUMPF, R., SCHÄFER, U., and JANKUHN, H., 1958: Zwei Moorleichenfunde aus dem Domslandsmoor Gemarkung Windeby, Kreis Eckenförde. *Praehist. Z.* 36:118.

SHATTOCK, S. G., 1905: A prehistoric or predynastic Egyptian calculus. *Trans. Path. Soc Lond.,* 56:275.

SHATTOCK, S. G., 1909: A report upon the pathological condition of the aorta of King Merneptah. *Proc. Roy. Soc. Med.,* 2 (Pathological Section):122.

SHAW, A. F. B., 1938: "Histological study of the mummy of Har-Mosĕ, the singer of the Eighteenth Dynasty (circ. 1490 B.C.). *J. Path. Bact.,* 47:115.

SIGERIST, H. E., 1951: *A History of Medicine,* Vol I: *Primitive and Archaic Medicine.* New York, Oxford University Press.

SIGERIST, H. E., 1961: *A History of Medicine,* Vol. II. Early Greek, Hindu and Persian Medicine. New York, Oxford University Press.

SINGER, C., 1925: *The Evolution of Anatomy.* London, Kegan Paul, Trench, Trubner.

SMITH, G. ELLIOT, 1912: *The Royal Mummies.* Cairo, Musée du Caire.

SMITH, G. ELLIOT, and DAWSON, W. R., 1924: *Egyptian Mummies.* London, Allen and Unwin.

THORWALD, J., 1963: *Science and Secrets of Early Medicine.* London, Thames and Hudson.

WILDER, H. M., 1904: The restoration of dried tissues, with especial reference to human remains. *Amer. Anthrop.,* 6:1.

WILLIAMS, G. D., 1926: An ancient bladder stone. *JAMA,* 87:941.

WILLIAMS, H. U., 1927: Gross and microscopic anatomy of two Peruvian mummies. *Arch. Path. (Chicago).,* 4:26.

WILLIUS, F. A., and KEYS, T. E., 1941: *Cardiac Classics.* London, Kimpton.

WILSON, G. E., 1927: A Study in American palaeohistology. *Amer. Nat.,* 61:555.

WRIGHT, I. S., 1952: *Vascular Diseases in Clinical Practice.* Chicago, Year Book Publishers.

Chapter 40

Respiratory Disease in Egypt

J. THOMPSON ROWLING

OUR knowledge of respiratory disease in Egypt is very incomplete and in some cases fragmentary. It is derived from the study of mummies, from references in the medical papyri, and from certain circumstantial evidence which throws an indirect light on the probable nature and incidence of respiratory conditions.

THE EVIDENCE FROM MUMMIES

The lungs were removed during the process of embalment from a very early period,—certainly as early as the Old Kingdom. Until the XXIst Dynasty they were placed in either a Canopic jar or Canopic chest, usually preserved with either dry natron or a solution of natron. The lungs were under the care of one of the "Four Children of Horus," the god Hapi, who is represented as an ape-headed figure, and who was himself under the protection of the goddess Nephthys. The god himself originally represented one of the four pillars supporting the sky, and from this the northern region, but his association with the lungs probably dates from the predynastic period and remained fixed and undeviating throughout Egyptian history. His protection of the lungs applied to the living as well as the dead and an uncompromisingly magical papyrus of the XXth Dynasty reads: "Thou shalt not take thy stand in his liver, in his lung, in his heart, in his kidneys, in his spleen, in his intestines, in his rib, nor in any flesh of his body. Imsety, Hapi, Duamutef and Queb-Snewef, the gods who are in his body, are against thee" (British Museum No. 10687).

After the XXIst Dynasty a change in theological opinion occurred whereby the mummy itself was made to represent the form and abiding place of the spirit. Previously, the body had been preserved as being essential to the *Sahu*, or spiritual body, while the identity and individuality of the deceased was preserved in the portrait statue, buried with him, which the immaterial parts of the soul could either inhabit or animate. The mummy itself was now regarded as the sole abiding place of the spirit, preserving the identity of the deceased in its own right, and was therefore made as complete as possible. The viscera were accordingly replaced within it, while the portrait statue was omitted from the tomb as being no longer necessary. Preservation of the viscera was still essential, however, nor were they without the protection of the Children of Horus. Each group of viscera was washed and preserved, and made into small parcels, each containing a wax figure of the appropriate god, before being returned to the abdomen. The lungs are found as quite small parcels and were clearly dried and collapsed before they were wrapped. No attempt was made to replace them specifically in the chest.

With a few exceptions the lung tissue is found in a very poor state of preservation. In most Canopic jars the lungs are reduced to an amorphous powder, which is useless to the palaeopathologist. One brilliant exception to this occurred in the case of Har-Mosě, a musician of the XVIIIth Dynasty whose viscera were preserved in a Canopic chest (Shaw, 1938).

The lungs were removed from the body during the process of embalment through a left sided abdominal incision, corresponding either to a left paramedian approach or to a left Rutherford Morison incision. The abdominal viscera were first removed, after which an incision was made in the diaphragm from within the abdomen and the lungs removed without disturbing the heart or great vessels. This was no easy matter for the embalmers, and in some mummies the heart has clearly been inadvertently removed and replaced in its original position. The presence of pleural adhesions clearly makes removal of the lung by this method much more difficult, and many mummies show adherent fragments of lung tissue in the chest. We have, therefore, positive evidence of gross inflammatory changes in the lung and pleura, although it is not possible to differentiate between adhesions caused by empyema, pleurisy, pneumonia, or tuberculosis. It seems likely that all these conditions were relatively common, however.

PATHOLOGICAL CHANGES IN LUNG TISSUE

In bodies which have not been eviscerated, as in Predynastic sand burials, and Coptic and Byzantine burials, the lungs are found as small flattened structures, lying on the posterior chest wall, and having the appearance of cardboard. It is difficult to detect any pathological changes in such lungs. Smith and Dawson (1924), record a case from a Byzantine cemetery in Nubia, where the left lung was firmly adherent to the chest wall by a series of dense adhesions, strongly suggestive of empyema, but such evidence is rare.

In embalmed bodies it is usually impossible to detect pathological changes from the naked eye appearance of the lung. The preparation of histological specimens from mummified tissue has presented great difficulties. Most mummified tissue is hard and dry and impossible to cut with a microtome, although there are exceptions to this. The tissues are readily rehydrated with normal saline, but they then become soft and jelly-like, losing their structure. Ruffer (1909), overcame the difficulty by combining a fixative such as formol or alcohol, with the rehydrating fluid, so that the tissue was fixed and hardened during the process of rehydration. By this method fairly good histological sections of lung can be obtained. Cellular structure is largely lost, and it is exceptional to be able to identify cell nuclei. Nevertheless, the outlines of the tissue planes are clear, and gross microscopical changes, such as those of hepatisation, can be demonstrated. Curiously enough Egyptian bacteria are fairly easily stained and identified, although the picture is clouded by the presence of yeasts, moulds and putrefactive bacteria which have proliferated at the time of the embalment.

Ruffer has described two cases dating from the XXth Dynasty, one of which shows hepatisation of the lung, and the other histological changes compatable with pneumonia. He also describes a third case, dating from Greek times, showing pneumonic changes (1910).

In 1938 Shaw examined sections from the viscera of an XVIIIth Dynasty mummy from Thebes. This was the singer, Har-Mose, whose grave was found below that of Sen Mut, the architect of the mortuary temple of Queen Hatshepsut, suggesting that he was either a relative or servant of Sen Mut. The body was contained in an anthropoid coffin with a lute laid beside it. The body itself was in an extremely poor state of preservation. Examination showed an old impacted fracture of the neck of the right femur. The bones were those of an elderly man about 5 feet 2 inches tall. The viscera, however, were preserved in a Canopic chest, wrapped in linen, and lying on a layer of loose, rather dirty natron. (Natron is a naturally occurring mineral, much used by Egyptian embalmers, whose composition is approximately sodium carbonate 35 per cent, sodium bicarbonate 25 per cent, sodium chloride 14 per cent, sodium sulphate 3 per cent). The viscera were unusually soft and pliable and Shaw found it possible to cut paraffin sections without preliminary rehydration. The quality of these sections was excellent. Examination of the right lung showed a flattened, rather distorted structure although the three lobes could be defined. The hilum had been cut across with a knife. The main bronchus showed traces of surviving epithelium in which mucous glands could be identified. The hilar nodes were loaded with carbon, and anthracotic foci were scattered throughout the parenchyma. There was no evidence of silicosis, using crossed Nichol prisms. The lower lobe was more solid than the middle and upper lobes, and the alveoli were filled with what was probably inflammatory exudate, justifying a diagnosis of pneumonia, probably of the hypostatic type.

The finding of anthracosis in this mummy presents an unexpected problem. Ruffer (1910), has also recorded similar changes, and the writer has found carbon granules in the lung of a mummy of the XXVIth Dynasty. It is difficult to see how the Egyptians, particularly Har-Mose, who was a singer, should be exposed to the inhalation of large quantities of sooty air. Copper smelting was, of course, common in Egypt, but it was performed with open furnaces in the northern parts of the country or in Sinai or elsewhere outside Egypt itself. The rock tombs rarely show evidence of soot on the roofs, and it is difficult to believe that lamps and cooking fires would produce much smoke. Without further evidence it is difficult to draw any conclusions from the findings, although the evidence of the presence of anthracosis seems quite definite. It may also be noted that Long (1931) described changes which he attributed to pneumoconiosis in the lungs of a mummy of the XXIst Dynasty which he examined.

CIRCUMSTANTIAL EVIDENCE OF RESPIRATORY DISEASE

Pneumonia, particularly hypostatic pneumonia, is commonly the immediate cause of death in patients confined to bed with fractures, particularly of the spine and femur. There are very large numbers of fractures of both the spine and femora known in mummies, many of which are un-united and it is logical to assume that a terminal pneumonia was present in a high proportion of these cases.

"Ankylosing spondylitis" was particularly common in Egypt. Specimens are common in mummies throughout the whole course of Egyptian history, although the maximum incidence seems to have occurred about the Ptolemaic period. In some specimens the whole spine from the cervical vertebrae to the sacrum are fused into a single bar of bone. The costo-vertebral joints are restricted in their movements and respiration must have been very grossly embarrassed. It is reasonable to assume that acute respiratory infection must have been the cause of death in many of these cases.

Tuberculosis of bone, particularly spinal tuberculosis, was also relatively common in Egypt, and many specimens from the Nubian collection, lately in the Royal College of Surgeons, show clear evidence of Pott's disease. Bone tuberculosis is never a primary condition, but results from blood borne spread, usually from a lymphatic focus elsewhere. The three common origins for this are the cervical lymph nodes, the mesenteric lymph nodes and the hilar nodes associated with pulmonary tuberculosis. The latter is common in modern times and there is no reason to suppose that pulmonary tuberculosis was less common in Egypt.

EVIDENCE FROM THE PAPYRI

Of the eight great medical papyri known to us, only the Papyrus Ebers sheds much light on the Egyptian view of respiratory disease. The translation of the papyri presents many difficulties and our present versions should be accepted with some reserve.

A case from the 37th column of the Papyrus Ebers is very suggestive of a diagnosis of bronchiectasis. It reads: "If thou examinest a man with an obstacle, and he produces expectoration, and his disease under his breast sides is like a latrine cave, then shalt thou say thereof: It is due to accumulations in his breast sides. Thou shalt prepare for him strong remedies to drink . . ." (Ebbell, 1937).

The 38th column of the papyrus mentions shivering in association with chest conditions, and mentions "purulency," although the context is obscure.

The 39th column mentions a condition compatable with advanced phthisis: "If thou examinest a man for advanced illness in his chest, whose body shrinks, being altogether bewitched; if thou examinest him and dost not find any disease in his belly, but the 'hnwt' of his body is like 'pjt,' then shalt thou say to him: It is a decay of thy inside. Thou shalt prepare for him remedies against it . . ."

The 53rd column gives a list a twenty remedies "to expel cough," the number of the prescriptions implying that the symptom was very common. The 55th column gives a series of remedies which Ebbell translates as "to eradicate asthma." There is some doubt as to the translation here, however.

It is interesting to note that haemoptysis is not mentioned, nor is there any reference to the character of the sputum, with the exception of the reference to purulent sputum already quoted. The papyrus implies that examination was limited to inspection and palpation of the chest. There is no reference to direct auscultation. (The crepitus of fractured ribs, or possibly surgical emphysema is noted in the Edwin Smith papyrus.)

CONCLUSION

The evidence which we have at present suggests that respiratory disease was relatively common in Egypt, although it is difficult to be certain which conditions predominated. It seems certain that pneumonia was the cause of death in many cases, and that pulmonary tuberculosis was fairly common. It is likely that chronic bronchitis and carcinoma of the lung were both rare. The riddle of the finding of anthracosis in some specimens remains unsolved. A. T. Sandison and D. R. Brothwell (personal communication) have noted similar changes in the lung of a Guanche body from the Canary Islands.

REFERENCES

Ebell, B., 1937: *The Papyrus Ebers*. Levin and Munksgaard. Copenhagen.

Long, A. R., 1931: Cardiovascular renal disease; report of a case of three thousand years ago. *Arch. Path.*, *12*:92.

Ruffer, M. A., 1909: Preliminary note on the histology of Egyptian mummies. *Brit. Med. J.*, *1*:1005.

Ruffer, M. A., 1910: Remarks on the histology and pathological anatomy of Egyptian mummies. *Cairo Sci. J.*, *4*:3.

Shaw, A. F. B., 1938: A histological study of the mummy of Har-Mosĕ, the singer of the eighteenth dynasty (c. 1490 B.C.) *J. Path. Bact.*, 47:115.

Smith, G. E. and Dawson, W. R., 1924: *Egyptian mummies*. Allen and Unwin, London.

Chapter 41

Disease of the Alimentary System in Egypt

J. THOMPSON ROWLING

The alimentary canal is the structure least amenable to preservation by the embalmer, and our knowledge of disease in the stomach and bowel in Egypt is hampered by the almost total lack of material which can be investigated. At best, the bowel is a thin and delicate structure, ill suited to preservation, and when this is associated with rapid putrefaction by its own bacteria in a hot climate the poor results of embalming are to be expected. Paradoxically, it is in the sand burials of Predynastic times, and in the bodies of Coptic and Byzantine date where evisceration was not practised that we see the best preservation. In Predynastic bodies it is sometimes possible to analyse the contents of the bowel and Elliot Smith and Netolitzky have been able to identify barley, millet of the species *Panicum colonum* (now no longer cultivated), root tubers of *Cyperus esculentus* and fish of the species *Tilapia nilotica* (1923).

The references to gastro-intestinal disease in the papyri are also very difficult to interpret. They occur principally in the "medical" papyri and lack the clarity and logical thought which characterise the Edwin Smith surgical papyrus, and the surgical sections of the Papyrus Ebers. However, by combining our sources of information, a rather cloudy picture of Egyptian gastro-enterology can be obtained.

PEPTIC ULCERATION

Peptic ulceration is regarded, with some justification, as a disease of modern times associated with stress. It might therefore be wrong to expect to find it in so stable and feudal a society as that of Egypt in its great periods of civilisation, and certainly we have no pathological evidence of its occurrence. Nevertheless there is a passage in the Papyrus Ebers which is compatible with a description of haemorrhage from a gastric or duodenal ulcer. It is found in the 39th column and reads thus: "If thou examinest his obstacle in his cardia, and thou findest that he has been changed, and has turned deathly pale [Lit. has crossed the channel to the beyond], his mind goes away, and his cardia becomes dry, then shalt thou say of him: it is a blood nest which has not yet attached itself; thou shalt let it descend by means of remedies. . . . There comes in this case from his mouth or from his anus like pig's blood, after it is fried." There can be little doubt that the description refers to a brisk upper intestinal haemorrhage, the description of the subsequent melaena being particularly apt. It might be argued that the bleeding could be associated with a carcinoma of the stomach. However, these growths seldom give risk to sudden acute bleeding. Certainly the bleeding described originated in the upper digestive tract otherwise fresh, rather than altered blood would have been passed. On balance, the description best fits the bleeding seen after erosion of the gastro-duodenal artery by a posterior duodenal ulcer. It is interesting to note that there are no descriptions or remedies for chronic post-prandial pain in the papyri. It seems likely that the regularity of the pain associated with chronic peptic ulceration would have impressed itself on the Egyptian physician, but we do not find any reference to this.

CARCINOMA OF THE STOMACH

This appears to have been very rare in Egypt. It is a disease of older people, and the relatively low life expectancy in Egypt may have a bearing on this. We have no pathological evidence of its occurrence, and indeed there is no definite case of any bony metastatic deposits having been found in mummies. One or two descriptions of disease in the papyri are not incompatible with the diagnosis, but these are far too vague and inconclusive to confirm such a pathology.

GALLSTONES AND CHOLECYSTITIS

In view of the close relationship of the two conditions it is reasonable to suppose that the presence of stones indicates cholecystitis in a majority of cases.

Figure 1 shows the gall-bladder from the mummy of a priestess of Amen of the XXIst Dynasty. This contains many stones, all of approximately the same size, and probably infective in origin rather than metabolic. The wall of the gall bladder appears thin, but this is probably accounted for by the process of dehydration. The mummy itself was in an excellent state of preservation. It is doubtful if the gallstones contributed in any way towards death.

There are two descriptions in the Papyrus Ebers which are compatible with a diagnosis of either cholecystitis or appendicitis, the first possibly associated with peritonitis. Neither description is definite enough to draw any firm conclusion, however. They are found in the 40th column and read as follows: "If thou examinest a

FIGURE 1. Gall-bladder from the mummy of a priestess of Amen, showing multiple stones.

man with an obstacle in his cardia, and he vomits, being very ill, and he suffers from it, as from a wound, then shalt thou say: It is a seizure of purulency which has not yet attached itself. Thou shalt prepare for him. . . ." The suggestion of pain on movement—as from a wound—is interesting, as is the reference to purulency. The second description reads: "If thou examinest a man for an affection of his cardia, thou shalt lay thy hand upon him; if thou find that it has attached itself on the right side, then shalt thou say. . . . Thou shalt prepare for him a remedy against it . . . drunk for four days If thou examinest him, after this has been done, and thou findest this, his disease, remaining as formerly, then shalt thou prepare for him strong remedies . . . until it goes away from him, and he gets well." Both the suggestion of localization on the right side and the interval of four days for improvement are compatible with both inflammation of the gall bladder or appendix. (At times it is extremely difficult to differentiate between the two even now, and thus our earlier colleagues may perhaps be excused some lack of precision here.)

APPENDICITIS

The writer is not aware of any instance of appendicitis in either Predynastic or Pharaonic Egypt, but a case has been described by Elliot Smith and Wood Jones of a Byzantine body where the pathological changes were definite and conclusive. The case is recorded in the second bulletin of the Archaeological Survey of Nubia (1908). The body was that of a woman who had been buried virtually without evisceration. The appendix was lying on the pelvic brim and was drawn over to the left side. It was adherent to the pelvic wall by dense adhesions, indicating a severe attack of appendicitis, probably complicated by an appendix abscess.

VOLVULUS

The Papyrus Ebers contains a case description in the 25th column which merits consideration: "If thou examinest one who suffers from phlegm, with colicky pains, and whose belly is stiff through it, and who has pain in his cardia when his phlegm is in his belly and does not find a way to come out, nor is there any way by which it can come out, then it shall rot in his belly, and not being able to come out it grows into a twist in the bowel. If it will not grow into a twist in the bowel, so that it grows to . . . (*wnn-mt*), he shall evacuate it and get well immediately. If he does not evacuate it, for a twist in the bowel, thou shalt prepare for him remedies for evacuation, so that he can get well immediately." The three cardinal points of the syndrome are colicky pains, abdominal distension or rigidity ("stiffness"), and constipation. It is implied that bowel movement is an indication of cure. Although it is not possible to take the diagnosis very far, the picture is compatible with one of sigmoid volvulus, becoming gangrenous if "it shall rot in his belly," or spontaneously untwisting if "he shall get well immediately." The fact that the Egyptian diet was largely farinaceous lends some support to this diagnosis. The description also suggests that the post mortem appearances were familiar to the writer. Intestinal obstruction from other causes, or acute appendicitis with paralytic ileus from peritonitis are other possibilities. Although confused, the description suggests that the physician was familiar with the acute abdomen, but only imperfectly aware of the possible pathological causes. Exploration of the abdomen was outside his province.

DIARRHOEA AND CONSTIPATION

Remedies for both diarrhoea and constipation are given at some length in the Papyrus Ebers, the Hearst Papyrus, the Berlin Medical Papyrus, and the Chester Beatty Papyrus. Rectal bleeding is sometimes mentioned but it is nowhere possible to identify any specific pathology assigned to either symptom.

HAEMORRHOIDS AND RECTAL PROLAPSE

It is not always easy to differentiate between these two conditions, particularly in bodies which have become dry and dehydrated. Wood Jones has pointed out that shrinkage of the soft tissues causes the outer coats of the rectum to be pulled in towards the pelvic wall, leaving a relatively redundant mucosa which tends to give the appearances of a false rectal prolapse and which may also be mistaken for piles.

It seems likely, in spite of this, that haemorrhoids were common in Egypt. The Chester Beatty Papyrus is virtually a treatise on anal and rectal diseases, and large parts of the Ebers and Hearst papyri are concerned with anal conditions. The impression is given that proctitis and *pruritus ani* were common, as was *fistula-in-ano*. Rectal bleeding and diarrhoea were also common, but the suggestion is one of inflammatory rather than neoplastic origins for these symptoms.

CONCLUSION

There are innumerable pathological changes of the alimentary system about which we have no knowledge and insufficient information on which to base even speculation. Achalasia of the cardia, carcinoma of the stomach, regional ileitis, ulcerative colitis and even carcinoma of the colon and diverticulitis are beyond the scope of study until we obtain further information. Few of the conclusions drawn in this paper would stand up to an investigation demanding positive proof of pathology and they must remain as probabilities only.

The best hope of clarifying the position in the future probably lies in the careful examination of Predynastic bodies, preserved by sand burial without evisceration.

REFERENCES

SMITH, G. E., 1923: *The Ancient Egyptians*. London.
SMITH, G. E. and JONES F. W., 1908: The anatomical report, *Archaeological Survey of Nubia*. Cairo, 1:32.

Chapter 42

Diseases of the Reproductive System

A. T. SANDISON AND CALVIN WELLS

FEMALE GENITAL SYSTEM

From earliest times man has desired fertility in his women, his domestic animals and his crops: there is abundant evidence of magic rites directed to this end. From the Palaeolithic period there are many figurines of the *Venus of Willendorf* type which emphasise the sexual organs and secondary sexual characteristics. There are also cave drawings of pregnant women, and men pursuing women. Later cultures also produced figurines and representations of fertility mother-goddesses with prominent vulvae and well-developed or accentuated breasts (see Fig. 1): we shall return later to the importance of the breast in this context. There are also many early amulets of a sexual nature (Huyghe, 1962: Sigerist, 1951).

Symbols of the Sumerian Goddess Nintu have been identified as formalized representations of the uterus and so has the *ankh* symbol of ancient Egypt: this appears on amulets of Ta-urt, a goddess of childbirth, and elsewhere. The head of Hathor is sometimes portrayed within a uterine symbol and there are similar appearances in Etruscan terra-cottas. Apuleius in *The Golden Ass* describes emblems carried in the progress of Isis, applying terms used to designate parts of the uterus. Certain votives from Cos probably represent the female pudendum (Plaut, 1959).

In Sumerian times the moon was believed to control the menses and there were probably early taboos against coitus during menstruation: the ancient Hebrews practised a strictly ritualized timetable for coitus which was forbidden in the puerperium and during and immediately after menstruation. This empirical programme, in fact, ensures coitus about the time of ovulation, when chances of conception are maximal. In ancient China menstrual coitus was thought to be wrong (Gulik, 1961). Coitus in pregnancy was permitted by the ancient Hebrews but was avoided in Egypt. It was also forbidden by the Greeks on ritual occasions. Frazer (1933) notes that Attic matrons strewed *Agnus castus* under their beds as an anaphrodisiac at the festival of the Thesmophoria.

Population control among many preliterate peoples was no problem where numbers were reduced by famine, war, disease and high infant mortality. Sexual taboos and prolonged lactation probably limited conception. Infanticide was widely practised. Similar conditions obtained in early civilizations: the emphasis was on promotion rather than prevention of conception. From ancient Egypt have come down innumerable amulets to promote fertility and probably none that can be recognized as of contraceptive intent (see Fig. 2). Aphrodisiac charms, incantations and drugs—such as the mandragora fruits found in Tut-ankh-amun's tomb—were

FIGURE 1. Chalk figurine. Grime's Graves, Weeting, Norfolk. Neolithic. This crudely carved fertility figure is an interesting variant of the more familiar "Aurignacian Venuses." London, British Museum. Photograph British Museum.

widely used. Rarely, for social reasons and among prostitutes, deliberate abortion might be desired. There are, therefore, recipes both to produce and prevent miscarriage. The ancient Hebrews and Persians regarded abortion as sinful: Flavius Josephus refers to women who induced it. There were strict laws against abortion in Assyria where it was a capital offence (Saggs, 1962). The Hippocratic Oath (probably Pythagorean in origin) proscribes abortion but other works (probably Cnidian) in the *Corpus Hippocraticorum* show it was considered legitimate. The philosphers Plato and Aristotle believed it was justifiable. Similar views were held in Rome where it was practised by prostitutes and other women. The Incas certainly used quinine and cathartic drugs, perhaps with abortifacient intent.

Janssens (1963) has described an exceptionally interesting case of a second century A.D. Gallo-Roman woman excavated at Tongeren. In her pelvis was a stiletto-like instrument made of animal bone. He suggests that it was used by the woman in an attempt to make herself abort and that its presence was not suspected by whoever buried her because she had died suddenly from embolism whilst she was alone.

Contraception was probably also used sometimes for personal and sometimes for medical reasons. In ancient Egypt prolonged lactation probably played a part but there were also magical recipes and

Talmud mentions anal coitus and the use of the vaginal sponge. Athenaeus refers to female eunuchs being used for sexual traffic in ancient Lydia, and Thorwald (1963) states that in Babylonia some temple priestesses (*mustarrestu*) were sterilized: the latter is based on slender evidence (Saggs 1962). In ancient India there is evidence of magical recipes, intravaginal applications and possibly some form of coitus reservatus: the latter however, may have had sexual rather than contraceptive significance.

From the Graeco-Roman period we have clear evidence of the condonation of abortion and infanticide. The excavation of a single Athenian well has yielded the remains of 175 newborn babies (Angel 1945) and at Hadrumetum, Tunisia, from 126 funerary urns 118 newborn cremations were recovered (Soleil et al., 1958). Contraceptive practices were also used, e.g., coitus interruptus, postcoital removal of semen, the doubtful use of goat bladder condoms in Imperial Rome, vaginal insertions, anointing of the genitalia, oral potions and magic. From the Mochican culture of ancient South America there is pottery evidence of anal coitus and fellatio (Posnansky 1925) and of bestiality (see Fig. 3), but these are probably variant practices rather than of contraceptive significance (Himes, 1936). Linné (1943) refers to many examples of Amerindian infanticide if children were deformed.

In the early Classical period foetal anatomy was studied by Alcmaeon (5th cent. B.C.), Diocles of Carystus (c. 400 B.C.), and others. Herophilus (*fl.* 300 B.C.) described the ovaries and Aristotle was familiar with the effects of female castration in animals. He also described the human uterus but not until Soranus of Ephesus (A.D. 98-138) was a reasonably accurate description forthcoming (Plaut,

FIGURE 2 Concubine Figure. Egypt, probably from Deir-el-Medineh. XVIIIth Dynasty. Such figures have fertility significance. Coll. A. T. Sandison.

more rational vulvar and vaginal insertions to prevent conception. The story of Onan indicates that coitus interruptus was known to the Hebrews while the early

FIGURE 3 Pottery vase. Argentina, Diaguita, pre-Inca period. This shows a man in the act of intercourse with a llama but it is, perhaps, chiefly remarkable for the splendid facial expression of the performer. London, British Museum. Photograph British Museum.

1959). Rufus of Ephesus (*fl.* A.D. 100-117) regarded the uterus and vagina as separate organs and knew that the tubes were inserted into the uterus.

In most ancient societies parturition was carried out in a squatting position, often supported by bricks or on a birth-chair. Many early figurines show labour in progress. In North America the pre-Columbian "Mound Builders" made earthenware pots which vividly depict the birth of a child (Wakefield and Dellinger, 1937). A Mexican statuette shows the child's head fully out of the vagina (Hamy, 1906): a Cypriot limestone group from Golgoi shows the woman seated in a birth-chair (Cesnola, 1885): and in a Peruvian pottery group she sits in the lap of a helper while the midwife delivers the child (Engelmann, 1884). A Luristan bronze pin of the first millenium B.C. shows the baby's head emerging in an occipito-posterior, face to pubes, position: so, too, does a composite Nigerian wood carving (Witkowski, 1894).

There are, as might be expected, references to monstrous births in ancient societies, e.g., in Babylon. Kinnier Wilson (page 203) notes that conjoined twins are described in the Ancient Mesopotamian literature and that embryotomy was sometimes practised. The Chaldeans associated monstrous births with astrological occurrences as later did Ptolemaeus Claudius of

Alexandria. Empedocles of Agrigentum (5th cent. B.C.) believed that monstrous births resulted from seminal abnormality but Aristotle thought that female elements were also implicated. Galen, however, adhered to the view that seminal influences were paramount. Soranus of Ephesus believed that maternal influences, e.g., sights or sounds, affected the development of the foetus.

Plutarch implies that Thales believed that some monsters resulted from intercourse with animals. Other authors who mention monsters include Pliny, Livy and St. Augustine. Hippocrates, Anaxagoras, Alcmaeon and Menander were interested in avian teratology.

The Spartans are said to have destroyed monsters for eugenic reasons. An anencephalic monster was, however, carefully preserved in the necropolis of Hermopolis (Glenister, 1964).

There are references to plural pregnancy, e.g., in the Bible (Genesis 38:28) and *Corpus Hippocraticorum*. A Mexican manuscript (Cod. Vatic. 3773) shows a Maya goddess who has just given birth to twins. The *Popol Vuh* describes how the maiden Xquic—whose name means "blood of a woman," perhaps with reference to menstruation—magically conceived twins. After they were born they were placed by their enemies on an ant-hill in an unsuccessful attempt to destroy them. This probably reflects a common method of infanticide among the Quiché (Goetz and Morley, 1950). In the Westcar papyrus triplet birth is mentioned: and a mediaeval painting depicts the, obviously unexpected, delivery of quadruplets (Schlesinger 1962). Dawson (1927) describes the mummies of two Roman period children who were almost certainly twins (see Fig. 4.). In *Corpus Hippocraticorum* and also in Aristotle (*Hist. Anim.* 7:4) superfoetation is discussed.

Spontaneous and traumatic accidental abortion is recorded from Assyrian and Hittite times (Sigerist 1951: Saggs 1962) and in the Bible (Exodus 21:22, Numbers 12:12). There are recipes in the *Atharveda* of ancient India designed to prevent miscarriage. Spontaneous abortion was clearly a major medical problem in Graeco-Roman times since this subject is considered at length in *Corpus Hippocraticorum* and the works of Soranus. Herodotus (3:32) tells us that Cambyses injured his sister-wife so that she miscarried and died.

The literature of ancient Egypt, Greece and Rome details empirical pregnancy tests: a further sophistication in Mesopotamia, Greece and Rome was the attempt to determine foetal sex by observation of the patient. In ancient China the sex of the foetus was divined by study of the maternal pulse and in ancient Egypt by carefully detailed experiments on germinating cereals (Ghalioungui et al., 1963). These authors repeated the tests and found that pregnancy urine caused significantly less inhibition of germination than non-pregnancy or male urine: the sex of the foetus, however, did not affect the rate of cereal growth. Manger (1933) performed similar tests.

As early as 1000 B.C. the risk of puerperal infection seems to have been appreciated in ancient India since the *Ayur-Veda* advises careful trimming of the midwives' nails and prescribes post-partum vaginal fumigation. At this time also the whims of pregnant women were invariably satisfied in an attempt to prevent malformation of the foetus and Susruta says that deformed babies may be produced by premature "bearing down" on the part of the mother. Post-puerperal infection is also

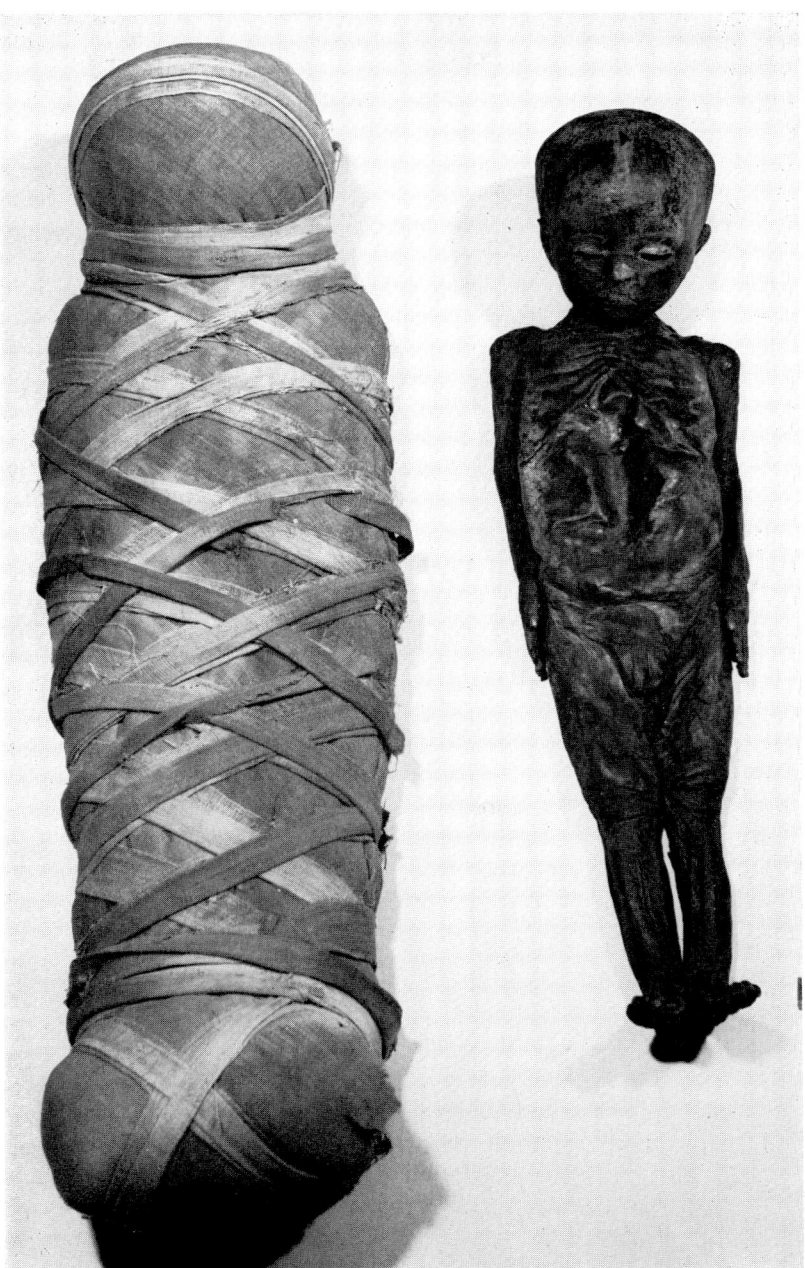

FIGURE 4 Two mummies. Egypt, Thebes. Roman period. These two mummies were found in a specially designed double size sarcophagus which is unique. They are both aged about three years and appear to be twins. Edinburgh, Royal Scottish Museum. Photograph Tom Scott by courtesy of Mr. Cyril Aldred.

clearly described in *Corpus Hippocraticorum* and recognized as a serious condition (e.g., *Epidemics I*, case XI).

Cephalic version may have been practiced in pre-Hippocratic times (Jameson 1936). In Ancient Egypt and Classical Greece vertex presentation was preferred but in Greece breech presentations were sometimes left (Sigerist, 1961). In *Diseases of Women* of *Corpus Hippocraticorum* transverse, arm and foot presentations are described: hand presentation is noted in the Bible (Genesis 38:28). In difficult cases, e.g., transverse lie, in Classical times embryotomy was employed. Soranus considered abnormal presentations, e.g., transverse and foot, in detail and found that podalic version sometimes made embryotomy unnecessary. In the *Susruta Samhita* eight abnormal presentations are described including breech, transverse, and single and double arm presentation. Abnormal positions of the placenta are specifically mentioned and it seems to recognize pseudocyesis, contraction ring, embolism and rupture of the uterus. Fatal signs are a "deranged sense perception of the mother, convulsions, displacement or contraction of the reproductive organs (yoni), a peculiar pain like the after-pain of childbirth, cough, difficult respiration, or vertigo" (Bhishagratna 1911). It is not difficult to fit a potentially fatal condition to each of these clinical signs. Difficult labour is described in Ancient China, Mesopotamia, and in the Bible (Genesis 35:17) and also occurred in ancient Egypt since there are charms to avert this, and actual remains showing death in childbirth have been recovered. Abnormal presentations also occurred in ancient China (Cianfrani, 1960).

The *Lex Regia* of Numa Pompilius (700 B.C.) required removal of the child before burial of pregnant women, and occasionally a live child may have been obtained. Immediate post-mortem Caesarean section is advised in the *Susruta Samhitá* if the mother dies in labour from convulsions or other causes.

Scholars have pointed out the tragedy of barrenness in the eyes of the Hebrews and the jubilation resulting when the elderly nulliparae, Sarah and Hannah, at last conceived. In those elderly primiparae and others (such as the Shulamite woman, Elisabeth who bore John, and Manoah's wife who bore Samson) birth was not difficult (Green-Armytage 1945). Green-Armytage has culled examples of traumatic abortion, premature labour, rapid delivery, precipitate labour with syncope, ruptured perineum and possible hydramnios. These interpretations seem to be reasonable. He also mentions the paucity of references hinting at puerperal sepsis but later suggests that in Numbers 5:21 we have evidence of peritonitis.

In *Corpus Hippocraticorum* we find a good grasp of the basic physiology and pathology of pregnancy, labour and the puerperium. Attention was paid, for example, to pigmentary changes in the breast and skin during pregnancy. Possibly pseudocyesis was recognised. There is also a fairly rational discussion of sterility within the limits of knowledge then available, e.g., obesity was known to hinder conception. Kleegman (1961) says that the thought of Aristotle, Socrates and Plato shows awareness of the possibility of male sterility; usually, however, in antiquity this was only considered when impotence was present. Sterility was obviously not rare; nor were abortions in the first trimester and failure to go to term. Abortion was noted to complicate enteritis and following venesection. Abnormalities of the placenta were known and the concept of foetal death *in-utero* was established. Elliot Smith (1930) suggests that the ancient Egyptians were deeply influenced by a complicated pattern of beliefs about the placenta. Indeed, a placenta with umbilical cord appears as a ceremonial standard on the famous palette of Narmer. It was known that molar change occurred in some pregnancies; hydatidiform mole was recognised in *Corpus Hippocraticorum* and by Diocles of Corystus and named by Aetius of Amidea (A.D. 502-575) (Ober and Fass, 1961). Pregnancy hyperemesis was probably known, also hydropic foetus and hydramnion. Infection by erysipelas, genital bleeding in pregnancy and watery discharge before labour were known to be serious matters. Retention of placental membranes occurred as well as acute uterine inversion. Labour was sometimes difficult and protracted; attention was paid to prevention of perineal lacerations. Puerperal sepsis was known to be of ominous import. Puerperal insanity is almost certainly to be recognised in *Coan Prognosis* and a further case where drowsiness, headaches and convulsions occurred in pregnancy may be interpreted as evidence of pregnancy toxaemia. The Weisman collection contains a pre-Columbian Mexican statuette which has been interpreted as showing Caesarean section for pre-eclampsia. Kinnier Wilson (page 203) gives evidence for toxaemia and oedema in pregnancy in Ancient Mesopotamia. An important concept in Greek and Indian obstetric theory was the idea that maternal thoughts and impressions affected the appearance of the child. Soranus, for example, thought that women who looked at monkeys near the time of their conception would produce ape-like children. This notion of the influence of maternal impressions became extremely popular in Europe after the Renaissance and is widely held even today. Recent work has shown that the foetus may, indeed, be affected by factors influencing the mother's mind.

In Soranus' *Gynaecology* we find a sophistication of Hippocratic concepts so that sterility is described in relation to somatic types. Abortion, molar pregnancy and metritis were commonly seen. Hyperemesis occurred, as did haemorrhage following miscarriage or labour.

Infantile mortality was high in all ancient societies; mummies of foetuses, of infants and of children are not rare from ancient Egypt. The Archaeological Survey of Nubia revealed great numbers of bodies of children of all ages up to puberty (Smith and Wood-Jones, 1908).

When we turn from diseases of women in the obstetric sense to those now in the province of the gynaecologist, we have little accurate information about the less literate ancient societies. From India there is a wealth of literature preoccupied with sexual virility, fecundity and coital pleasure: in this atmosphere gynaecology must have been regarded as important but not much definite information is available to the non-specialist in the language. The *Susruta Samhita* contains references to conception, pregnancy and women's diseases. Cianfrani (1960) states that in the Brahmanistic period dysmenorrhoea, coital injuries, retained placenta, utero-vaginal tumours and possibly ovarian tumours were described. Similarly, we have little information from ancient Persia which was largely dependent on foreign physicians from Egypt and Greece. A Peruvian *huaco* appears to show a gynaecological examination in progress (Goldman and Sawyer, 1958).

Green Armytage (1945) notes the absence of Biblical references to strictly gynaecological complaints in contrast to obstetric matters, but he suggests that Mark 5: 25 may refer to metropathia haemorrhagica. It might equally describe bleeding from a mucous or fibroid polyp, or even from piles. In Egypt, however, gynaecological matters are considered in the Kahun (Griffith, 1898), Ebers and London Medical Papyri. Much of this is para-obstetrical and concerned with sterility, fecundity and divination of pregnancy or foetal sex. There is clear evidence of uterine prolapse and the uterus was believed to be capable of wandering about the body, causing various symptoms of the "hysteria" of later periods. There is also discussion of amenorrhoea, menorrhagia, leucorrhoea, vaginal discharges and other ailments, pruritus ani, urinary affections and possibly of phlebitis in women. Ghalioungui (1963) suggests that metritis and patulous cervix also occurred. Some scholars, relying on an alleged characteristic smell, believe there is good evidence of gynaecological cancer: it probably did occur but no proof is available and in this view we are supported by Sigerist (1951) who quotes Sudhoff to this effect.

There is a wealth of gynaecological material in *Corpus Hippocraticorum* especially, but not only, in *Diseases of Women*. Unfortunately, this last is available only in the original and in a good French translation by Littré (1853). Amenorrhoea and menorrhagia were frequent and believed to reflect uterine disease. Multiparae were less liable to amenorrhoea and hypomenorrhoea. Dysmenorrhoea and dyspareunia also occurred. Vaginal and peripudendal discharges and ulcers, genital pain, genital atrophy and affection of the pudendum by erysipelas are all noted. Abnormalities of the os and cervix were known, including indurations which may have been cancer. Tumours of the uterus, again possibly cancerous, are described here and by other writers e.g., Archigenes of Apameia (*fl.* A.D. 110) (Long, 1928). Many votive offerings in the form of a uterus have been found on Graeco-Roman sites (Rouse, 1902). Some, as illustrated by Wells (1964a) have small appendages which may perhaps be intended to represent fibroids. Similar uterine *ex votos*, sometimes made of silver, are known from Ceylon (Hildburgh, 1908).

Much attention was given to uterine displacement, to prolapse and procidentia. True prolapse appears to have been not rare; but in the explanation of psychosomatic "hysterical" conditions in women uterine displacement was probably often falsely implicated. Displacement was also invoked to explain sterility in some cases.

Soranus of Ephesus wrote a surprisingly rational and sophisticated treatise on gynaecology which may still be read with interest and profit in a good English translation by Temkin (1956). Soranus discarded the belief that uterine wanderings were the cause of disease but made an intensive study of uterine prolapse. He found that prolapse occurred in the senile woman as well as after labour and classified it into partial or complete. It is interesting to note that Paulus Aegineta (A.D. 625-690) suggested vaginal hysterectomy in the treatment of procidentia.

Soranus also distinguished amenorrhoea from menstrual retention and in the aetiology of amenorrhoea noted masculine habitus, athleticism, anaemia, cachexia and obesity. He observed premenstrual tension and noted that dysmenorrhoea was often cured by pregnancy. He also saw cases of nymphomania. Paulus Aegineta recommended clitoridectomy in nymphomania (Cianfrani, 1960).

Soranus was familiar with vaginal atresia, and disease affecting the os and cervix uteri. Genital abscesses, ulcers, growths and fistula occurred in his practice. He saw polyps, warts and fissures of the vulva and described clitoral enlargement. It is interesting to note that in ancient China a large clitoris was regarded with suspicion since most Mongoloid women appear to have a small clitoris (Gulik, 1961). Soranus further discussed changes in the uterus and classified these as soft, scirrhous and sclerotic. Oribasius may have seen calcified fibroids (Cianfrani, 1960). Aetius of Amidea also distinguished ulcerated from non-ulcerated uterine tumours and probably saw carcinoma of the cervix. He is also said to have recognised parametritis as a distinct process. Celsus described imperforate hymen (Cianfrani, 1960).

PALAEOPATHOLOGICAL MATERIAL

The female Royal Mummies of the Cairo Museum, in the instances where they have been unwrapped, provide an oportunity to inspect the genitalia of Ancient Egyptian women (Smith 1912). In some cases no published descriptions are available and in others, e.g., the woman in the coffin lid of Setnakhte, the perineum was covered by a linen pad. A similar pad, elaborately fixed through the embalming wound, covered the perineum of Queen Henut-tawy. The pudenda of Queens Nofretari and Sitkamose were covered by resinous paste. The rima pudendi in Princess Merytamūn is described as normal, the pudendum of Queen Nedjmet as untreated while the external genitalia of Queen Inhapy are described in considerable detail, the vagina and uterus apparently being also present. The labia majora of the Lady Rēi are still in close apposition and pubic hair is present; usually body hair is lacking in mummies. Henut-empet's vagina is packed with linen and the woman from the tomb of Amenōphis II shows an open rima which has been stuffed with linen from within the pelvis. Nes-ta-neb-ashru's labia majora were pulled together to conceal the rima. It is interesting to recall that in the moulded carapaces of mummies from the Archaic period the genitalia as well as the breasts were delineated with care, this obviously being considered to be important. Wilder (1904) described a parous os uteri in the dried body of a woman of the Basketmaker Indians. Granville (1825) had also diagnosed child-bearing in his Ptolemaic mummy. In some Peruvian dried bodies the pudenda are well preserved, as in a female body examined by one of us (A.T.S.) where traces of pubic hair remained.

There have been many reports, not all reliable, of death in labour with the foetus impacted in the pelvis. Regöly-Mérei (1961) illustrates one and amongst fifty female skeletons of reproductive age at Aebelholt Møller-Christensen (1958) found five with foetuses. One of these was a breech presentation. Another, an abortion of about the fifth month, he thinks was probably a coffin-birth. As a rule it is impossible to decide whether death occurred during labour or in late pregnancy.

With regard to abnormalities, Williams (1929) cites a personal communication from D. E. Derry stating that Princess Hehenit of the XIth Dynasty of Egypt had a narrow pelvis and presumably died while still a young woman soon after a difficult labour, having developed a vesico-vaginal fistula. The Archaeological Survey of Nubia unearthed the remains of a deformed Coptic negress who died in childbirth because of severe pelvic insufficiency (Derry 1908). The body was lying on its back with abducted thighs and slightly flexed knees. The pubic portion of the pelvis was missing permitting full view of a large foetal head with the occiput engaged at the lower border of the ischial tuberosities;

it was fixed in the right occipito-anterior position with the chin against the left sacro-iliac joint. There was over-riding of the parietal bones, and, to a lesser extent, of the frontals. After removal of the foetal head it was seen that the right sacro-iliac joint was absent, the bones being fused. That this was probably a congenital Naegele's deformity was evidenced by the malformation of the rest of the pelvis: its right side was small and contracted producing an oblique elongation of the cavity. Obstetrically this may be a catastrophe. Greenhill (1955) says it is more common than is usually believed and Wells (1957) records a (presumably South African) example of uncertain provenance.

This is only one of many skeletal diseases which may deform bones—including the pelvis. Rickets, though rare in antiquity, is occasionally found. It has even been diagnosed in pre-human contexts (Moodie, 1930) and in non-human material (Schmerling, 1883; Abel, 1912). So, too, has osteomalacia (Walther, 1925). In human burials it appears with increasing frequency in Europe throughout the Middle Ages (Nemeskéri and Harsanyi, 1959) and, by producing flat pelves, must have made a substantial contribution to maternal and foetal mortality. Ruffer (1914) specifically notes a case in which the sacrum was affected. A pelvic tumour in an ancient Egyptian has been described by Ruffer and Willmore (1914) and a few others have been recognised. Spondylolisthesis has been found in a woman of the Neolithic period (Messeri, 1959) and it seems to have been common in some early American groups (Congdon, 1932). The forward displacement of the lumbar vertebrae in this condition must have increased the chance of abnormal foetal presentations. Congenital dislocation of the hip (see Fig. 5) has often been found

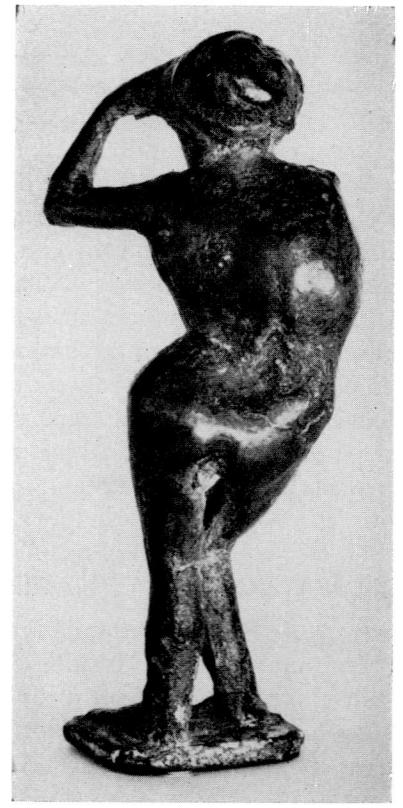

FIGURE 5 Bronze figurine. Hellenistic. H. 7.5 cm. An apparently unique representation of congenital dislocation of the hip. Private collection. Photograph Hallam Ashley.

in ancient cemeteries (e.g., Pales, 1929; Wells, 1963) and the pelvic asymmetry with marked scoliosis which may accompany it must also, at times, have led to malpresentations. It is portrayed with great sensitivity in an apparently unique Hellenistic figurine.

One of us (C.W.) has recovered from a Late Saxon burial ground at Thetford the skeleton of a woman in which the sacrum shows evidence of recent septic infection with osteitis of the pelvic surface and sinuses entering the bone (see Fig. 6). It is just possible that this may have been due to extension from a pelvic abscess associated with a puerperal infection.

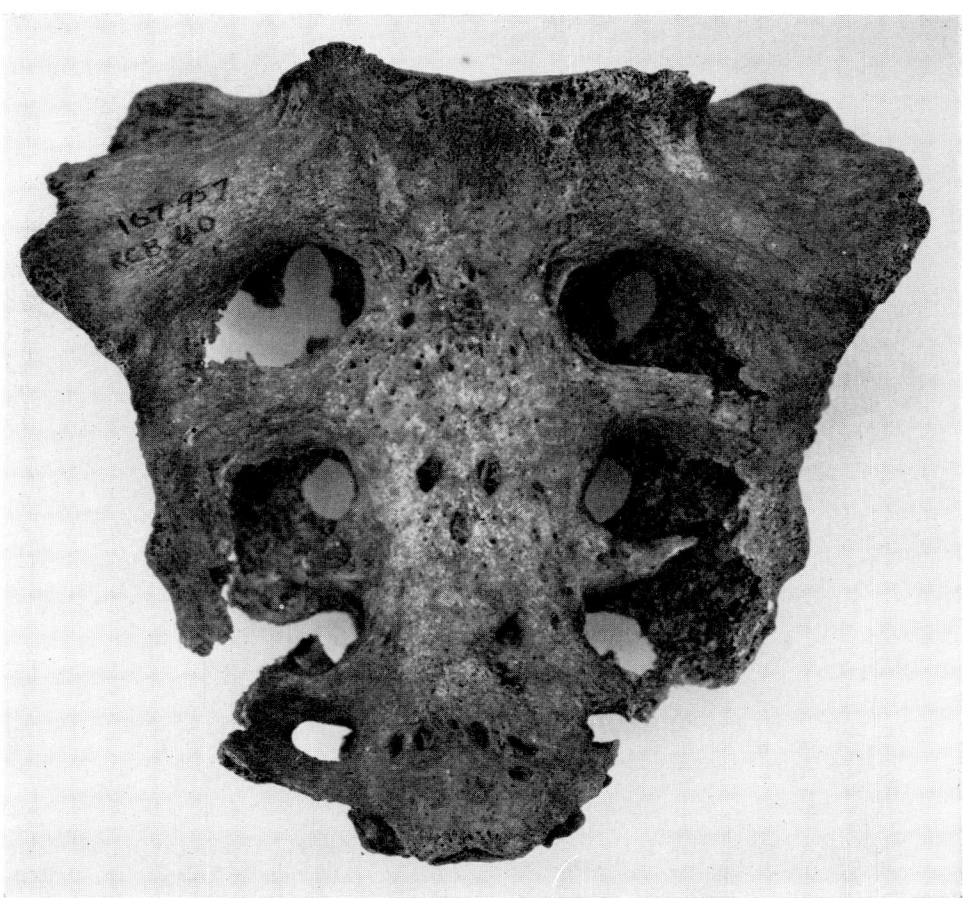

FIGURE 6 Sacrum. Thetford, Norfolk. Late Saxon period. Extensive osteitis of the anterior surface with sinuses penetrating the bone. Norwich, Castle Museum. Photograph Ministry of Public Building and Works.

Foetal, as well as pelvic deformities, no doubt accounted for many maternal deaths, the most important condition being hydrocephalus (see Fig. 7). This has been described in Neolithics (Grimm and Plathner, 1952), Ancient Egyptians (Derry, 1913; Batrawi and Morant, 1947), Romano-Britons (Trevor, 1950), Ancient Patagonians (Verneau, 1903; Moodie, 1928) Merovingians (Pfeiffer, 1900) and others. No doubt most of the surviving examples did not begin to develop until after birth but we may infer that ante-partum cases were not rare. Hydrocephalus has also been recognised in Tanagra terracottas (Regnault, 1894) and Soranus describes its operative treatment in labour.

Some slight evidence survives to supplement textual and skeletal evidence of difficult labour. Baglioni (1937) describes a second to third century A.D. marble bas-relief of a birth scene. One of the central figures appears to be a surgeon who is holding up an instrument which very closely resembles an obstetric forceps (Graham, 1950). This seems not improbable in view of the finding of elaborate vaginal specula of the Bossi type at Pom-

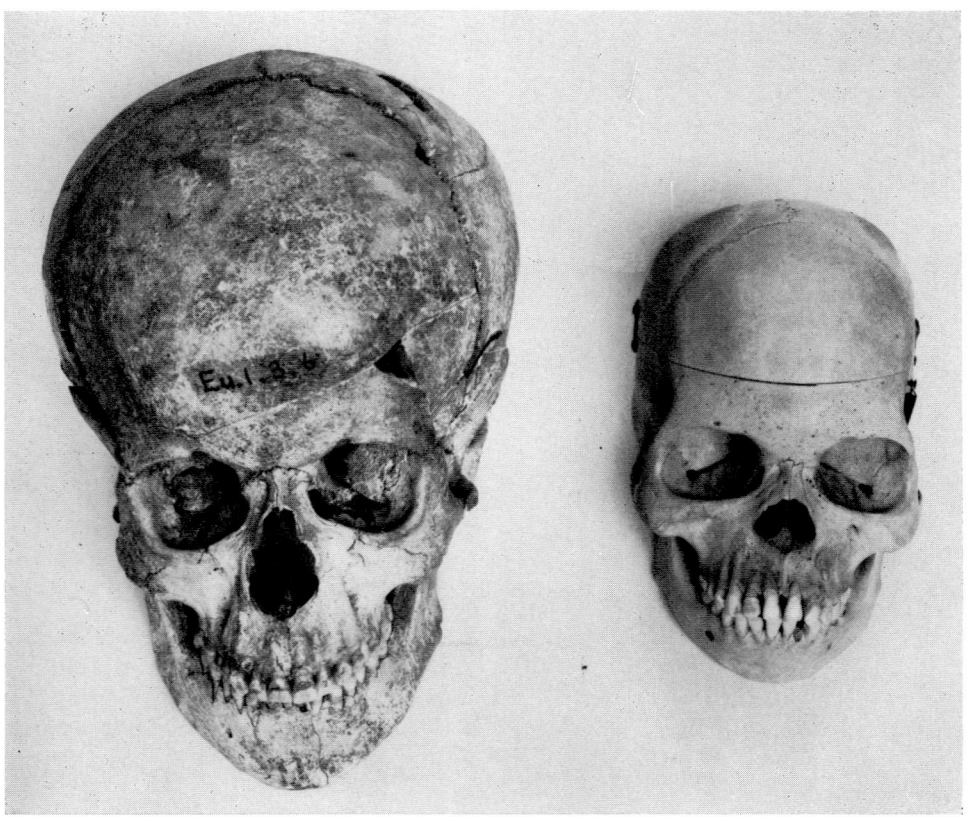

Figure 7 Hydrocephalus. Norton, Yorkshire. Roman period. The intracranial volume of this skull is estimated to be 2600 cc. Cambridge, Duckworth Laboratory. Photograph Hallam Ashley by courtesy of Dr. J. C. Trevor.

peii. Further finds of early instruments include those in a grave of La Tène Iron Age date from Obermenzing, Bavaria (Navarro, 1955) and others of the same period from Kiskoszeg, Hungary (Sudhoff, 1913). The first contained a retractor that could have been used for vaginal inspection, in the second were two hooks which might serve as obstetric crochets.

It is possible in theory, though difficult in practice, to recognise the effects on the child of trauma occurring during birth. Fay (1959) suggests that the famous classical portrait head of Menander portrays very precisely the effects of a left parietal lesion acquired as a birth injury. Many examples of apparent hemiplegia with facial involvement are found in figurines and masks throughout the world and are probably not all to be explained simply by artistic distortion. Some may be representations of traumatic births. Wells (1961) records a decapitated Late Iron Age skull showing hemiatrophy, probably from this cause. He also notes (1964b) another decapitated Romano-Briton with a left obstetric shoulder injury of the fractured epiphyseal type as described by Scaglietti (1938). In this case it was associated with atrophy of the left femur, tibia and fibula—perhaps a congenital hemiplegia.

Smith and Dawson (1924) described the unembalmed skeleton of a sixteen-year-old ancient Egyptian girl whose abdominal cavity contained the remains of a six-month foetus. The skeleton had multiple fractures of both forearms and skull. The authors postulated that she had been murdered by outraged male relatives when her condition of illegitimate pregnancy became evident, the forearm fractures having been sustained while she attempted to ward off blows to her head. Klausner (1961) points out that even in modern times in Oman an unmarried expectant mother would, on discovery, be murdered by her father or brothers. Cyril Aldred in a personal communication, however, cautions us that this girl's death might equally well have resulted from armed robbery.

Granville (1825) noted in his report to the Royal Society on the dissection of the Ptolemaic mummy (suffering from "Porrigo decalvans") which now bears his name, that the abdominal skin was wrinkled and that the right ovary and broad ligament were enveloped in mass of diseased structure: further, the uterus was enlarged and the remains of a sac were attached to the left ovary. Granville was not in doubt that this was a case of "ovarian dropsy" and Rowling (1961) reasonably concludes that it may have been bilateral ovarian cystadenocarcinoma.

Wood Jones (1908) described two cases of prolapse from the Archaeological Survey of Nubia—one of gross rectal prolapse in a young girl and the other of vaginal prolapse caused apparently by a stalked polypus, possibly cervical in origin—and stated that Elliot Smith noted prolapse of rectum and uterus in predynastic women.

BREAST

From earliest times the female breast has been selected as a symbol of fertility; it is often shown in exaggerated form in the so called Venuses of the Palaeolithic period in association with prominent buttocks. Mammary prominence is also seen in fertility figures from many parts of the world. Sometimes the breasts were inlaid with precious metals and associated with the portrayal of the female escutcheon or vulva; the latter may be large and prominent as in a Cycladic fertility Goddess of the third millenium B.C. (Huyghe, 1962). Sometimes the breasts were supported by the hands or were suckling a child. In Japanese art the suckling of old women is often shown.

For a study of the human breast in art forms reference may be made to Siemens (1952) and to Levy (1962). In some cultures the ideal breast aesthetically was small and virginal, e.g., in Classical Greece, while in others, e.g., Ancient India, it was large and full. Ancient Egyptian art often faithfully portrays the breast from the small, firm conical organ of the nubile girl to the pendulous structure of old age. There are poetical allusions to breasts in most literatures: in Ancient Egyptian (Erman, 1927), Babylonian (Rowton, 1962), Biblical Song of Solomon, Classical, e.g., Greek Anthology and Ovid (Licht, 1932; Levy, 1962).

The Code of Hammurabi refers to amputation of the breasts of wet nurses as punishment for the substitution of a second child for one who had died under their care and the same fate was suffered by some adultresses in Ancient India; later, Christian saints also suffered this mutilation for their faith (Schechter and Swan, 1962).

In Ancient Mesopotamia and Egypt lactation by mother or wet nurse usually continued for three years. In the London Medical and Ebers Papyri there are recipes for the promotion of milk flow; human milk was also used as a therapeutic agent in Assyria and Ancient Egypt. The Ancient Egyptians were also aware of the changes which occurred in the breast when conception supervened. There are references to breast diseases in Babylonia (Saggs, 1962). The Ebers papyrus mentions treatment of diseases of the breast by local inunction and the Edwin Smith surgical papyrus refers to probable acute mammary abscess (Case 39). Case 45 has been interpreted as a breast tumour but might better be regarded as superficial presentation of caseous tuberculous osteitis. A Coptic manuscript fragment from the Monastery of Deir-el-Abiad deals with the treatment of breast disease (Bouriant 1888; Sigerist 1951).

In our opinion there is no convincing evidence of breast carcinoma in Ancient Egyptian Literature, a view supported by Warren R. Dawson (Sandison, 1962). Nor has breast cancer been detected in Egyptian mummies although Ruffer (1911) examined mummy breast tissue. On theoretical grounds it seems probable that a scirrhous carcinoma would produce collagenisation recognisable in microscopic sections.

As early as the IInd Dynasty, external wrappings of poorly preserved bodies were moulded so as to delineate breasts, nipples and pudenda (Emery, 1961). We have noted realistic conical false breasts modelled from linen on a New Kingdom mummy. This was even done with the genitalia of monkeys (Lortet and Gaillard, 1907). Smith and Dawson (1924) reported a female mummy of the Roman period, the hardened carapace of which recalled to them representations of Aphrodite; there were skilfully moulded breasts with nipples of copper discs. A similar disc was placed at the navel. Even during the XXIst Dynasty when embalming attained its most sophisticated level no attempt was made to pack the female breasts although face, trunk and limbs were so treated. The breasts in mummies are usually flattened, wrinkled structures as in the bodies of Queens Inhapy, Sitkamose, Nedjmet, Neskhons, Princess Nes-ta-neb-ashru, the Lady Rēi etc. (Smith, 1912). Queen Nofretari's breasts are apparently atrophic and senile, as are those of the elder woman in the tomb of Amenophis II and the unknown woman in the coffin lid of Setnakhte. Those of Queen Nedjmet are markedly pendulous and senile but the breasts of Queens Henut-tawy and Neskhons are the large full breasts of plump young women. Neskhons had loose abdominal skin and large prominent nipples and was therefore parous.

Queen Makerē presents an interesting picture of enormously enlarged, almost certainly lactating, breasts which, pulled away from the front, reach the inferior thoracic margin in the mid-axillary line. The bust was moulded on a foundation of cloth for the final wrapping. The abdomial skin was loose and wrinkled and the body of an infant was also present so that recent pregnancy seems likely.

Wilder (1904) described parous breasts and cervical os in a Basket-maker Indian woman. In a female Peruvian body examined by one of us (A.T.S.) the breasts were small and flattened as in Egyptian mummies.

From Classical Greece we have considerable evidence of breast disease. Many votives of breasts have been recovered from temple sites and Long (1928) and La Roe (1947) illustrate one from the Meyer Steinegg collection which was thought to represent an ulcerating tumour. This diagnosis is almost certainly ill-founded. Ancient statues showing polymastia and polythelia in goddesses such as

Diana may reflect knowledge of these abnormalities. Amazons are described without comment in *Airs, Waters and Places* of *Corpus Hippocraticorum*. Schechter (1962) discusses in detail the ancient references to Amazons and commentaries by mediaeval scholars. He concludes that disputes about the methods of self mutilation by the Amazons may have stimulated thought on procedures in breast surgery.

We know that Greek and Roman women supported the breasts by a strapless tightly worn strophion as shown in the Ten Girls Mosaic Pavement of the Piazza Amorina, Sicily. In the Hellenistic period it seems to have been fashionable to compress the breasts and to appear very willowy but Soranus of Ephesus advocated loosening the bands in later pregnancy to accommodate the enlarging breasts. In *Corpus Hippocraticorum* the female breast tissues are described as spongy and loose. It was believed that if a woman conceived a male child the nipples were upturned, if a female child downturned. In *On Glands* and also in *Diseases of Women* inflammation, abscess and corruption of milk are described. Suppuration of the breast is also mentioned in *Coan Prognosis*. Recipes to promote lactation indicate that this was sometimes a problem in Classical times. It was known that extended lactation may prolong amenorrhoea and that sudden regression of the breasts in pregnant women may herald miscarriage or mole. Other phenomena described include pain in and bleeding from the nipples. A reference in *Aphorisms* to production of "milk" in non-pregnant, non-parous women may imply involutional nipple discharge due to mammary duct ectasia.

A common error in medical histories is to see Queen Atossa of Persia (Herodotus 3:133) as an early example of breast cancer. Sandison (1959) has discussed this at length and concluded that it was probably a case of superficial inflammatory mastitis. Long (1928) says that Archigenes of Apameia (*c.* 100 B.C.) noted occasional tumours of the male breast.

In Roman times Soranus paid particular attention to the female breast in gynaecology. He noted the simultaneous enlargement of breasts and uterus at puberty and their atrophy after the menopause. He refers to the enlargement, slight pain and prominent vessels of the breast in pregnancy. He knew that damage to the foetus led to decrease in breast size and that this presaged miscarriage. He thought maternal milk (and that of wet nurses) was unsuitable for infants until twenty days after labour. He forbade the massage of engorged breasts and particularly handling of the nipples because of the risk of abscess formation. He knew that suppression of lactation might also lead to breast abscess: in Ancient Rome young animals were used to empty human breasts. Soranus also details the testing of milk and prescribes for its poor flow. He gives exhaustive instructions on the selection of wet nurses paying attention not only to breast and nipple configuration but also to general health and moral character.

Breast cancer was well-recognised in Ancient Rome and therapeutic indications are given by Celsus (25 B.C.-A.D. 50), Galen (A.D. 130-200) and Leonidas (1st cent. A.D.). Paulus of Aegineta (7th cent. A.D.) commented on the frequency of cancer in the female breast and its poor prognosis (Lewison, 1955).

MALE GENITAL SYSTEM

Sigerist (1951) states that Neolithic man castrated animals and knew the effects of this procedure. The Hittites and Ancient Greeks had myths with a castration content. According to Morse (1934) human castration was performed in Ancient China as early as the Chou dynasty (*c.* 1000 B.C.) by a stroke of the knife or strangulation of the parts. Gulik (1961) says that during the Ming period (A.D. 1368-1644) there were many eunuchs at the Imperial Chinese Court at Peking. Castration appears to have been long practised in many societies, e.g., Babylonia (Saggs, 1962) and Assyria (Mack, 1964). The Ancient Egyptians knew that it resulted in sterility. Eunuchs could not be priests nor members of the congregation in Israel. The Israelites swore oaths upon the testes (Mack, 1964). There are references to eunuchs in *Corpus Hippocraticorum,* in the Apocryphal *"Wisdom of Solomon"* in St. Matthew 19: 12 and to spadones in Ancient Rome. Wells (1964) illustrates a beautiful Romano-British castration clamp, dredged from the river Thames, which was used in the rites of Cybele. A similar but less ornate one was found near Basle. Early reports (Purcell 1893) which hinted at the practice of female castration per vaginam in Australian aborigines seem to have been based on a misunderstanding of an operation of the infibulation type (Basedow 1927).

In Ancient Mesopotamia boys were not circumcised and it now seems fairly clear that the Hebrews acquired this custom from the Egyptians. From Tepe Gawra (a preliterate period) a stone model phallus shows circumcision. This is probably due to early Semitic influence and indicates that circumcision antedates Abraham and Moses. Baudouin (1910; 1914) reviews the prehistoric evidence for circumcision, tracing it back to the Neolithic period. He illustrates stone tools which he thinks may have been used for the purpose and also a bronze knife of curious shape from Condren. He links the operation with a solar ritual and sees it as an expiatory sacrifice to a divine creator. In Mesopotamia penis and testes received the attention of physicians. The Assyrians severely punished those who injured a man's genitals or caused traumatic miscarriage in women, the penalty depending on the social status of the persons concerned. The Ashurbanipal tablets refer to spermatorrhoea, emissio seminis, a probable purulent urethral discharge with may have been gonorrhoeal, discharge of blood from the penis, impotence, urinary retention and hard and soft renal calculi (Sigerist, 1955; Saggs, 1962; Thorwald, 1963).

We also know a good deal about diseases of the male genitalia in Ancient Egypt. Further, genitalia are often well preserved in otherwise poor mummies and it is possible that much importance was attached to such preservation. Nevertheless, care must be taken in interpreting these appearances. Cameron (1910) thought that Nekht-Ankh, of the Middle Kingdom, had been incised but suggestions of this kind must be viewed with caution: as we shall see, mummy genitalia present curious appearances probably of artefact nature.

Nevertheless, it is clear that male circumcision was widely practised and was performed about the age of puberty. From the Royal Mummies we know that Prince Wab-khy-senu was still not circumcised when he died at about the age of eleven years. The state of the genitals of Prince Sipair aged about six years is uncertain but the general opinion is that he was, in fact, not circumcised. Hebrew circumcision was performed in infancy.

There is clear evidence of circumcision in the adult mummies of Pharaohs Amenophis II, Tuthmosis IV, Ramesses IV, and many others. The question of female circumcision in ancient Egypt, referred to by Strabo, has been raised. In a personal communication Cyril Aldred informs us that there is no good evidence of the practice in Ancient Egypt.

Penile diseases are mentioned in the London Medical Papyrus and in the Edwin Smith Surgical papyrus there is a clear reference to priapism in Case 31, one of spinal injury (Breasted, 1930). Papyrus Ebers also discusses the male genital system, including penile itching, impotence, priapism, and refers to treating efflux of the male member and to purulency which have very tentatively been equated with gonorrhoea (Ebbell, 1937). Diseases constricting the male and female genital organs and the post-operative phase in circumcision are discussed. Other identifications include probable orchitis, hydrocoele and, in the Hearst papyrus, perineal abscess (Leake, 1952). Scenes from the wall of the Tomb of Mehou, VIth Dynasty, show men with scrotal swelling, phallus hypertrophy and possible gynaecomastia. This might be the result of bilharziasis (Ghalioungui, 1963). Herodotus (2:181) tells of the Pharaoh Amasis who was impotent with his Greek wife Ladice, but not with others. This was presumably psychological.

Studies of the Royal Mummies (Smith, 1912) are frustrated by the fact that tomb-robbers sometimes broke off the phallus, as in the cases of Ramesses II and Sethes II, by the apparent removal of the pudenda during embalming in the cases of Tuthmosis III and the unknown persons C and E, by pressure of the pudendum into a leaflike structure against the perineum as in Tuthmosis I and II and probable removal of scrotum and subsequent damage to the penis in the case of Merneptah.

We can state, however, that Amosis I had a long phallus. Two Pharaohs—Ramesses IV and V—had puzzling lesions. Ramesses IV had a probable ulcer at the junction of shaft and glans penis: Smith (1912) was unable, however, to exclude post-mortem origin. Ramesses V appears to have had a bubo in the right groin and changes in the pudendal skin which resemble those produced by smallpox.

From Ancient Greece we find phallic votives at Corinthian Asclepaiedae suggesting that genital diseases occurred which were of such a nature as to demand attempted relief. Philolaus of Tarentum (c. 450 B.C.) regarded the genitals as one of four vital organs, the others being brain, heart and navel. Both men and women were believed to produce semen which, if mixed after coitus, remained in the womb and initiated pregnancy (Sigerist, 1961). As early as 280 B.C. Erasistratus of Chios named the prostate (Mettler, 1947).

The *Corpus Hippocraticorum* refers to cutting the prepuce, to urethral and perineal abscesses and to other testicular swellings, including an excellent account of mumps orchitis. Evidence of hydrocoele is less certain. There are references to delayed puberty, to excessive sexual indulgence as a cause of disease and to priapism, and also to sexual asthenia in Scythians, to eunuchs and to effeminacy.

Gangrene of the genitalia was seen: retraction of the penis and testes was regarded as an ominous sign. Renal and vesical calculi in adults and children are mentioned: with the passage of small renal stones, strangury, and genital pain and bladder diseases. Celsus describes an uncertain, sometimes fatal, genital disease in which excessive seminal discharge occurs apart from coitus or nocturnal dreams. Gonorrhoea seems to have been common in ancient China during the pre-T'ang (Sui A.D. 589-618) and T'ang period (A.D. 619-907). Genital disease with urethral stricture and arthritis is described. It remained common in the Ming period and protective sheaths may have been used. Syphilis is said to have occurred and to have been relieved by mercury (Gulik 1961). The Mixtec Gods Xochipilli and Xolotl-Nanautzin caused sickness of the sexual organs (Thorwald 1963).

Finally, we come to the concept of the hermaphrodite. Aphrodite, wife of Hephaetus, slept with Hermes and produced Hermaphroditos, a double-sexed being who was a minor Greek deity. This story was embellished and complicated by Ovid. As a normal youth Hermaphroditos was seen by the nymph Salamis who conceived a passion for him. When he resisted her advance she cried out to the Gods asking that she should never be parted from him nor he from her. Their bodies fused and became as one (Guirand, 1961; Graves, 1955). According to Young (1937) the first written reference to Hermaphroditus was made by Theophrastus, who says that his statues were garlanded. Aristotle, Hippocrates and Galen are said to have recognized hermaphrodites: in *On Epidemics* of *Corpus Hippocratiocorum* there are references to masculinisation of women. Herodotus mentions the women-like men or Enarees of Scythia. Young also says that in the early Roman Republic hermaphrodites were forbidden to marry, cast into the Tiber and were also killed in Umbria and Etruria. By the time of Pliny, however, they were permitted to marry. The Talmud contains concise rules for the behaviour of hermaphrodites, who are also mentioned in Ancient Chinese literature (Gulik, 1961).

They were often shown in Graeco-Roman art, and this subject is considered in detail by Delcourt (1961); Young (1937) illustrates the bronze statuette from Mirecourt now in the Epinal museum and points out that the erect hypospadic penis suggests that the statuette was modelled from the life.

Dieke (1956) discusses the subject of hermaphrodites at some length and illustrates many examples. Statues showing the condition vary greatly in explicitness. A Roman copy in the Berlin State Museum and a statue from Pergamon in Istanbul Museum have representations of fully developed female breasts in conjunction with complete male genitalia, just as in the Epinal specimen. The *"Goldener Fingerring"* Hermaphroditos from Sicily, also in Berlin, is similarly detailed. So, too, are a terracotta in Palermo and a Pompeiian wall painting now in Naples Musuem. A Vatican statue of Dionysos is less explicit genitally but has clear gynaecomastia. Dieke also illustrates an androgynous votive offering from Amnisos, Crete. The differences in degree to which male and female characters are portrayed in these works may indicate clinical knowledge of differing degrees of hermaphroditism.

The many examples of these figures certainly show the interest aroused by the condition and it is worth noting that even in the classical world the concept of bisexuality was extended beyond the idea of the minor god Hermaphroditos. Such portrayals are found beyond the influence of Graeco-Roman art, however. Dieke illustrates a bisexual ancestor figure from the Sudan and sees roots of this concept in Siva Ardhanarisvara in a rock temple at Elephanta as well as in a wall painting of a nature god in the temple of the Pharaoh Sahure. Further afield, African figurines show the condition not uncommonly although it is less certain here what the underlying psychology of the representation may be. Some of these figures may be subconscious rather than deliberate presentations of the condition.

Margetts (1951) discusses the problem

of Queen Hatshepsut of Egypt. She appears in male attire, as a god, wearing the ceremonial false beard. In murals she is sometimes shown without breasts and, as a child, is portrayed with male genitalia. Little is known of her sex life but the extent of this artificial change from female to male may possibly represent a real intersex state in the queen.

CONCLUSIONS

With regard to diseases and abnormalities of the female genital system we cannot do better than quote the words of Littré (1853) "Au reste, ce tableau des affections utérines qui affligaient les femmes grecques, il y a plus de deux mille ans, est tout à fait semblable à celui que nous avons présentement sous les yeux: et il est évident que rien, dans leur existence, ne les mettait, plus que ces femmes, à l'abri de ces maladies si fréquentes et si pénible."

This view is, in our opinion, amply substantiated by the material presented in this paper. Perhaps this is not a surprising conclusion. Conception, pregnancy and labour are biological phenomena which have probably changed very little over the millenia and we should expect that the train of pathological consequences which they bring should remain relatively unchanged.

As at present when 10 per cent of marriages are sterile, barrenness was a problem in antiquity. The desire was for children since many perished before puberty was attained: contraception was therefore probably little in demand.

Spontaneous abortion occurred and all pregnancies did not go to term. Malpresentations presented problems as they do today. Puerperal sepsis was a serious disease. Gynaecological complaints—amenorrhoea, leucorrhoea, menorrhagia, prolapse of the uterus, diseases of the cervix, body of the uterus and of the vulva—were probably as frequent as they are in present day practice.

Similarly, we are not surprised to learn that inflammatory mastitis was common in antiquity. Much acute inflammation of the breast is puerperal: it is, however, possible to state with some confidence that, contray to present experience, the *Staphylococcus aureus* is unlikely to have been the commonest aetiological agent. The current predominance of staphylococcus over streptococcus appears to be related to the indiscriminate use of antibiotics. That breast cancer occurred in antiquity is also clear but it is unlikely to have been as prevalent as at present since there is little doubt that multiple pregnancy and breast feeding protects women against its supervention.

With regard to the male genital system it is obvious that urethritis was fairly common in ancient societies. It seems likely that at least a proportion of this was gonorrhoeal in original. Prostitution was common in most ancient societies and would be likely to encourage spread of venereal urethritis: there is, however, no evidence of syphilis in the Ancient Old World. Orchitis, including that complicating mumps, appears also to have occurred: hydrocoele was probably also known. Hermaphrodites occurred in several ancient societies and gave rise to a rich mythology.

REFERENCES

ABEL, O., 1912: *Grundzüge der Paläobiologie der Wirbeltiere.* Stuttgart.

ANGEL, J. L., 1945: Skeletal material from Attica. *Hesperia, 14*:279.

BAGLIONI, S., 1937: Conoscevano gli antichi l'uso del forcipe ostetrico? *Fisiol. e Med., 8*:169.

BASEDOW, H., 1927: Subincision and kindred rites of the Australian aboriginal. *J. Roy. Anthrop. Inst.* 57:154.

BATRAWI, A., and MORANT, G. M., 1947: A study of the First Dynasty series of Egyptian skulls from Sakkara

and of an Eleventh Dynasty series from Thebes. *Biometrika.*, 34:18.
BAUDOUIN, M., 1910: La préhistoire de la circoncision. *Arch. Prov. Chir. Paris.*, 19:100.
BAUDOUIN, M., 1914: Les opérations chirurgicales culturelles: l'origine prehistorique de la circoncision. *Arch. Prov. Chir. Paris.*, 23:41.
BHISHAGRATNA, K. K. L., 1911: *An English translation of the Sushruta Samhita.* Vol 2.
BOURIANT, M. U., 1888: 'Fragment d'un livre de médicine en copte thébain'. *C. R. Acad. Insc. Belles Lettres.*, 15:374.
BREASTED, J. H., 1930: *The Edwin Smith Surgical Papyrus.* Chicago, University of Chicago Press.
CAMERON, J., 1910: In Murray, M. A. *Tomb of two Brothers*, Manchester, Museum Handbooks.
CESNOLA, L. P. DI, 1885: *A descriptive Atlas of the Cesnola collection of Cypriote antiquities in the Metropolitan Museum of Art, New York.* Berlin.
CHADWICK, J., and MANN, W. N., 1950: *The Medical Works of Hippocrates.* Oxford, Blackwell.
CIANFRANI, T., 1960: *A Short History of Obstetrics and Gynecology.* Springfield, Thomas.
CONGDON, R. T., 1932: Spondylolisthesis and vertebral anomalies in skeletons of American aborigines, with clinical notes on spondylolisthesis. *J. Bone Joint Surg.* 14:511.
DAWSON, W. R., 1927: On two Egyptian mummies preserved in the museums of Edinburgh. *Proc. Soc. Antiq. Scotland.*, 61:290.
DELCOURT, MARIE, 1961: *Hermaphrodite* trans. Jennifer Nicholson, London, Studio.
DERRY, D. E., 1908: Anatomical Report B. *Archaeological Survey of Nubia.* Bulletin 3.
DERRY, D. E., 1913: A case of hydrocephalus in an Egyptian of the Roman period. *J. Anat. Phys.*, 47:436.
DIEKE, W., 1956: Die antiken Hermaphroditen. Eine paramedizinische Studie. *Zbl. Gynäk.*, 78:889.
EBBELL, B., 1937: *The Papyrus Ebers, the Greatest Egyptian Medical Document.* Copenhagen, Levin and Munksgaard.
EMERY, W. B., 1961: *Archaic Egypt.* Harmondsworth, Penguin.
ENGELMANN, G. J., 1884: *Die Geburt bei den Urvölkern.* Vienna.
ERMAN, A., 1927: *The Literature of the Ancient Egyptians.* London, Methuen.
FAY, TEMPLE, 1959: "The Head." A neurosurgeon's analysis of a great stone portrait. *Expedition.* 1(4):12.
FRAZER, J. G., 1933: Spirits of the corn and of the wild. Vol. 1. (Vol. 7 of 3rd ed. of *The Golden Bough*), London, Macmillan, p. 116 n.2.
GHALIOUNGUI, P., 1963: *Magic and Medical Science in Ancient Egypt.* London, Hodder Stoughton.
GHALIOUNGUI, P., KHALIL, SH., and AMMAR, A. R., 1963: On an Ancient Egyptian method of diagnosing pregnancy and determining foetal sex. *Med. Hist.*, 7:241.
GLENISTER, T. W., 1964: Fantasies, facts and foetuses. *Med. Hist.*, 8:15.
GOETZ, DELIA, and MORLEY, S. G., 1950: *Popol Vuh: the sacred book of the Ancient Quiche Maya.* Univ. Oklahoma Press.
GOLDMANN, L., and SAWYER, A. R., 1958: Ancient Peruvian medicine. *J. Hist. Med.*, 13:10.
GRAHAM, HARVEY (Pseudonym), 1950: *Eternal Eve* London, Heinemann.
GRANVILLE, A. B., 1825: An essay on Egyptian mummies; with observations on the art of embalming among the ancient Egyptians. *Phil. Trans. Roy. Soc.* Pt. 1, 269
GRAVES, R., 1955: *The Greek Myths*, Vol. 1. Harmondsworth, Penguin.
GREEN-ARMYTAGE, V. B., 1945: *Some Obiter Dicta in Obstetrics and Gynaecology.* London, Devonport Press.
GREENHILL, J. P., 1955: *Obstetrics.* 11th ed. Philadelphia, Saunders.
GRIFFITH, F. LL., 1898: *Hieratic Papyri from Kahun and Gurob*, London, Quaritch.
GRIMM, H., and PLATHNER, C. H., 1952: Über ein jung steinzeitlichen Hydrocephalus von Seeburg im Mansfelder Seekreis und sein Gebiss. *Deutsch Zahn, Mund, Kieferheilk.* 15:456.
GUIRAND, F., 1961: In *Larousse Encylopaedia of Mythology.* London, Hamlyn.
GULIK, R. H. VAN, 1961: *Sexual Life in Ancient China.* Leiden, Brill.
HAMY, E. T., 1906: Note sur une statuette mexicaine en wernérite representant la deesse Xcuina. *J. Soc. Américanistes Paris*, 3:1.
HILDBURGH, W. L., 1908: *Notes on Sinhalese magic.* J. Roy. Anthrop Inst, 38:148
HIMES, N. E., 1936: *Medical History of Contraception.* Baltimore, Williams and Wilkins.
HUYGHE, R., 1962: *Larousse Encyclopaedia of Prehistoric and Ancient Art.* London, Hamlyn.
JAMESON, E. W., 1936: *Clio Medica: Gynaecology and Obstetrics.* New York, Hoeber.
JANSSENS, P. A., 1963: Eeen benen stilet ouit een Galloromeins graf (Tongeren). Communication 22 Sept. Geschiedenis en Geneeskunde et Antwerpen's Geneeskundige Dagen.
KLAUSNER, S. Z., 1961: Sex Life in Islam. In *The Encyclopaedia of Sexual Behaviour.* Ellis, A., and Abarbanel, A., (eds.) London, Heinemann.
KLEEGMAN, SOPHIA J, 1961: Infertility in Women. In *The Encyclopaedia of Sexual Behaviour.* Ellis, A. and Abarbanel, A., (eds). London, Heinemann.
LA ROE, ELSE K,, 1947: *Care of the Breast.* New York, Froben Press.
LEAKE, C. D., 1952: *The Old Egyptian Medical Papyri.* Kansas, University of Kansas Press.

LEVY, M., 1962: *The Moons of Paradise: Some reflections on the appearance of the female breast in art.* London, Barker.

LEWISON, E. F., 1955: *Breast Cancer.* London, Bailliere, Tindall and Cox.

LICHT, H., 1932: *Sexual Life in Ancient Greece.* London, Routledge.

LINNE, S., 1943: Humpbacks in ancient America. *Ethnos.* 8:161.

LITTRÉ, E., 1853: *Oeuvres Complètes d'Hippocrate.* Tome Huitieme, Paris, Bailliere.

LONG, E. R., 1928: *A History of Pathology.* Baltimore, Williams and Wilkins.

LORTET, L. C., and GAILLARD, C., 1907: La faune momifiee de l'ancienne Egypte. *Arch. Mus. Nat. Hist. Lyon.,* 9:1.

MACK, W. S., 1964: Ruminations on the Testis. *Proc. Roy. Soc. Med.,* 57:47.

MANGER, J., 1933: Untersuchungen zum Problem der Geschlechts diagnose aus Schwangerenharn. *Deutch. Med. Wschr.,* 59:885.

MARGETTS, E. L., 1951: The masculine character of Hatshepsut, Queen of Egypt. *Bull. Hist. Med.,* 25:559.

MESSERI, P., 1959: Un caso di spondilolistesi in epoca neolitica. *Arch. Antrop. Etnol.,* Firenze, 89:267.

METTLER, C., 1947: *History of Medicine.* Philadelphia, Blakiston.

MØLLER-CHRISTENSEN, V., 1958: *Bogen om Aebelholt kloster.* Copenhagen, D.V.F.

MOODIE, R. L., 1928: The paleopathology of Patagonia. *Ann. Med. Hist.,* 10:314.

MOODIE, R. L., 1930: Suggestion of rickets in the Pleistocene. *Amer. Surg.,* 10:162.

MORSE, W. R., 1934: *Clio Medica: Chinese Medicine.* New York, Hoeber.

NAVARRO, J. M. DE, 1955: A doctor's grave of the Middle La Tene period from Bavaria. *Proc. Prehist. Soc.,* 21:231.

NEMESKERI, J., and HARSANYI, L., 1959: Die Bedeutung paläopathologisches Untersuchungen für die historische Anthropologie. *Homo.,* 10:203.

OBER, W. B., and FASS, R. O., 1961: The Early History of Choriocarcinoma. *J. Hist. Med.,* 16:49.

PALES, L., 1929: Arthropathie coxo-femoral bilaterale chez un homme prehistorique. *Presse Medic.,* 53:872.

PFEIFFER, L., 1900: Cor-Bl. allg. Aerzt. Thüringen. 29:426.

PLAUT, A., 1959: Historical and Cultural Aspects of the Uterus. *Ann. N. Y. Acad. Sci.,* 75:389

POSNANSKY, A., 1925: Die erotischen Keramiken der Mochicas und deren Beziehungen zu occipital deformierten Schädeln. *Abhandl. Anthrop. Ethnol. Urg.,* 2:67.

PURCELL, B. H., 1893: Rites and customs of Australian aborigines. *Z. Ethnol. Berlin,* 25:286.

REGNAULT, F., 1894: Deformations craniennes dans l'art antique. *Bull. Soc. Anthrop. Paris, 4 ser.,* 5:691.

REGÖLY-MEREI, G., 1961: Beiträge zur Geschichte der Krankheiten. (Über einige interessante pälaopathologische Fälle). *Therap. Hungarica.,* 9:3.

ROUSE, W. H. D., 1902: *Greek votive offerings. An essay in the history of Greek religion.* Cambridge Univ. Press.

ROWLING, J. T., 1961: *Disease in Ancient Egypt: evidence from pathological lesions found in Mummies.* M.D. thesis, Univ. of Cambridge.

ROWTON, M. B., 1962: The use of the permansive in classic Babylonian. *J. Near East Stud.* 21:233.

RUFFER, M. A., 1911: Histological studies on Egyptian mummies. *Mém. Inst. Égypt.,* 6:(3).

RUFFER, M. A., 1914: On the diseases of the Sudan and Nubia in ancient times. *Mitt. Gesch. Med. Naturwissensch.* No. 58, 13:453.

RUFFER, M. A., and WILLMORE, J. G., 1914: A tumour of the pelvis dating from Roman times (250 A.D.) and found in Egypt. *J. Path. Bact.,* 18:480.

SAGGS, M. W. F., 1962: *The greatness that was Babylon.* London, Sidgwick and Jackson.

SANDISON, A. T., 1959: The first recorded case of inflammatory mastitis—Queen Atossa of Persia and the physician Democedes. *Med. Hist.,* 3:317.

SANDISON, A. T., 1962: *An Autopsy Study of the Adult Human Breast.* Bethesda, National Cancer Institute Monographs.

SCAGLIETTI, O., 1938: The obstetrical shoulder trauma. *Surg. Gynec. Obstet.,* 66:868.

SCHECHTER, D. C., 1962: Breast mutilation in the Amazons. *Surgery,* 51:554.

SCHECHTER, D. C. and SWAN, H. 1962: Of saints, surgical instruments, and breast amputation. *Surgery,* 52:693.

SCHLESINGER, B. E., 1962: Paediatrics in classical art. *Brit. Med. J.* 2:1671.

SCHMERLING, P. C. 1883: *Recherches sur les ossements fossiles.* Liege.

SIEMENS, H. W., 1952: Über die Form der Weiblichen Brust, insonderheit den Descensus mammae. *Virchows. Arch. path. Anat.* 332:101.

SIGERIST, H. E., 1951: *A History of Medicine.* Vol. I. *Primitive and Archaic Medicine.* New York, Oxford University Press.

SIGERIST, H. E., 1961: *A History of Medicine.* Vol. II. *Early Greek, Hindu and Persian Medicine.* New York, Oxford University Press.

SMITH, G. ELLIOT, 1912: *The Royal Mummies.* Cairo, Musee du Caire.

SMITH, G. ELLIOT, 1930: *Human History.* London, Cape.

SMITH, G. ELLIOT, and DAWSON, W. R., 1924: *Egyptian Mummies.* London, Allen and Unwin.

SMITH, G. ELLIOT, and WOOD JONES, F. 1908: *The Anatomical Report.* Archaeological Survey of Nubia, Bulletin No. 2.

SOLEIL, J., MULLER, P., and RICHARD, J., 1958: Contribution a la détermination de l'âge des enfants sacrifiés à Carthage (Étude des chapeaux de dentine calcines). *Ann. Med. Lég.* 38:17.

SUDHOFF, K., 1913: Chirurgische Instrumente aus Ungarn. Prähist. Zeitschr. 5:595.

TEMKIN, O., 1956: trans. *Soranus' Gynaecology.* Baltimore, Johns Hopkins Press.

THORWALD, J., 1963: *Science and Secrets of Early Medicine.* London, Thames and Hudson.

TREVOR, J. C., 1950: Note on the human remains of Romano-British date from Norton, Yorks. In Hayes, R. H. and Whitley, E. *The Roman Pottery at Norton, East Yorkshire.* (Roman Malton and District Report No. 7), Leeds.

VERNEAU, R., 1903: *Les anciens Patagons.* Monaco.

WAKEFIELD, E. G. and DELLINGER, S. C., 1937: Artefacts found among the remains of the "Mound Builders." *Bull. Hist. Med.* 5:452.

WALTHER, P. F. VON, 1825: Ueber das Alterthum der Knochenkrankheiten. *J. Chir. Augenh.* Bonn, 8:16.

WELLS, CALVIN, 1961: A human skull from Runham, Norfolk. Pt. 2. The Skull *Norfolk Archaeol.* 32:312.

WELLS, CALVIN, 1963: Hip disease in ancient Man. Report of three cases. *J. Bone Joint Surg.* 45B:790.

WELLS, CALVIN, 1964a: *Bones, Bodies and Disease.* London, Thames and Hudson.

WELLS, CALVIN, 1964b: An early case of birth injury. *Dev. Med. Child. Neurol.* 6:397.

WELLS L. H., 1957: Sacro-iliac synostosis and pelvic deformity. *S. African Med. J., 31:*1067.

WILDER, H. H., 1904: The Restoration of dried tissues with especial reference to human remains. *Amer. Anthrop.* 6:1.

WILLIAMS, H. U. 1929: Human palaeopathology with some original observations on symmetrical osteoporosis of the skull. *Arch. Path.* (Chicago), 7:839.

WITOWSKI, G. J., 1894: *Les accouchements dans les beaux-arts.* Paris.

WOOD JONES, F., 1908: The Pathological Report. *Archaeological Survey of Nubia,* Bulletin No. 2.

YOUNG, H. H., 1937: *Genital Abnormalities, Hermaphroditism and Related Adrenal Diseases.* London, Bailliere, Tindall and Cox.

Chapter 43

Endocrine Diseases

A. T. SANDISON AND CALVIN WELLS

THE ancients had, of course, no idea of the function of the endocrine organs: indeed, this is a quite modern concept. There is, nevertheless, some evidence of endocrine disease in ancient societies.

We shall consider first the thyroid gland, about which we have reliable information. A relief from Dendera in Egypt representing Cleopatra shows a fullness of the neck which might perhaps be goitrous. Cyril Aldred, however, has indicated in a personal communication that fullness of the neck was an artistic convention in Ptolemaic times. Elsewhere Needham and Lu state that goitre occurred in Ancient China. Galen recognized the thyroid gland as a distinct structure. Endemic goitre in the Alps is mentioned by Juvenal and Pliny the Elder. Rolleston (1934), however, points out that probably in Classical times cervical tuberculous lymphadenitis was sometimes confused with early goitre. Soranus of Ephesus noted swelling in the neck after labour which he considered to be some sort of bronchial tumour: this may, however, have been thyroid enlargement of pregnancy. Celsus suggested operative treatment for goitre but removal was probably not performed until the time of Albucasis (1013-1106). In *Coan Prognosis* there are references to exophthalmos.

Haddow (1936) illustrates a statuette with marked swelling of the neck. He says of it ". . . its source was thought to be Veii. The advanced age of the patient, the shape and size of the growths—rounded and large as oranges—their apparent fixity, together with the expression of pain on the face, so well rendered by contracted muscles, closed eyes and slightly opened mouth, suggest malignant disease of the thyroid rather than simple adenomatous goitre." No certain conclusion can be reached but it is an interesting piece. Duran (1963) illustrates a Mayan statuette from Kaminal Juyu which shows bilateral exophthalmos which he thinks may be thyrotoxic. Wells (1964) illustrates a Peruvian jug of the Mochica period and suggests the same diagnosis (see Fig. 1).

Adult hypothyroidism is less certainly recognizable in early records but Parrot (1878) and Regnault (1902) refer to the possibility that the Egyptian god Bes, usually thought of as an achondroplastic (Regnault, 1910), might be intended to represent myxoedema.

Seligmann (1912) described a Theban skull of the XVIIIth Dynasty which showed hypoplasia of bones laid down in cartilage, with arrest of growth in the floor of the posterior fossa. The forehead was prominent and there was arrested development of the nasal bones and nasal processes of the superior maxillae. Seligmann believed that the skull was cretinous and that the condition of the nasal bones permitted exclusion of achondroplasia. Keith

Figure 1. Pottery vase. Peru. Mochica period. H.28 cm. Possibly intended to portray thyrotoxicosis. London, British Museum. Photograph British Museum.

(1913) however thought it was achondroplastic; Brothwell (1963) also believes that the typical depressed nasal region is diagnostic of achondroplasia. Tietze-Conrat (1957) refers to a statement of Martial implying that cretinous dwarfs were sought after in Rome. Ruffer (1911) notes that Athenaeus amusingly said that the poet Arisastus was so small that no one could see him.

Ghalioungui (1963) suggests that the Ancient Egyptians were familiar with polyuria and diabetes mellitus. In the Kahun papyrus there is a recipe "Treatment for a woman thirsty" which breaks off at this point (Griffith, 1898). Certainly in Classical times diabetes mellitus was recognized as a disease entity and Celsus wrote of a disease characterised by polyuria and wasting and Aretaeus (2nd cent. A.D.) used the word *diabetes* meaning a syphon. Interestingly enough, in the Ayur-Veda and Susruta Samhita of Ancient India there is a reference to sweet urine in association with a syndrome of burning sensation in the hands and feet. Needham and Lu mention elsewhere that diabetes occurred in Ancient China although the sweet nature of the urine was not recognized till the seventh century A.D. Many examples of gangrene of the toes and feet have been recovered from ancient burials and some of these might have been due to diabetes. Unfortunately, we are never able to exclude the numerous other possible causes of this condition.

The pineal was known to the Greeks and the pituitary was well-known to Galen, although not so named till the time of Andreas Vesalius (1514-64). The ancient authors believed that the pituitary secreted mucus: the first inkling of its endocrine nature awaited the researches of Pierre Marie in the late nineteenth century. The first mention in the medical literature of gigantism, possibly acromegalic, did not appear till 1567. Herodotus described in Book 7:117 of his History a man Artachaees of the Achmaenid family, who was about 8 feet 2 inches (four fingers' breadth short of five royal cubits) high and had the loudest voice in the world; also in Book 9:83 he mentions among the Persian dead at Plataea the skeleton of a man 7½ feet tall (5 cubits). Schlaginhaufen (1925) diagnosed acromegaly in a Swiss neolithic skeleton but his evidence is unduly slender. Ruffer (1911) considered that some Ancient Egyptian scenes depict men of considerably greater stature than their fellows. If this were not a racial trait or artistic convention it might indicate pituitary overactivity. In this connection it may be recalled that Smith and Wood Jones (1908) discovered two skeletons of considerable stature in a Ptolemaic cemetery at Meris. The femora measured 532 and 523 mm in oblique length. No evidence of pathological cause was described, as might be expected.

Wells (1964) notes an even taller body from an Anglo-Saxon cemetery at Burgh Castle, Suffolk. The femora in this case measure 574 and 575 mm in oblique length (see Fig. 2). In this example it might be argued that it is not a pathological gigantism but merely an individual at the upper limit of the curve of normal variation of human stature. Some cases of pituitary gigantism are associated with undue fragility of the bones with a tendency to multiple fractures and the fact that the Burgh Castle giant had severe, though well healed, fractures of his right humerus, both ulnae, right tibia, and right talus and calcaneus might give slight additional support for the diagnosis of a true pituitary dysfunction. L. H. Wells (1963) mentions a probable case of Romano-British gigantism described originally by Cheselden. Wells (1964) notes that the coins of Maximinus Thrax (A.D. 235-238), who is known to have been a man of great stature, sometimes hint at an appearance of acromegaly.

Brothwell (1963) describes a probable acromegalic from Ancient Egypt preserved in the British Museum (Natural History). Keith (1931) describes in great detail a twelfth century skull found in the cemetery of the cathedral of Gardar, at the southern tip of Greenland (see Fig. 3). He diagnoses it as acromegalic "produced by normal . . . action" in primitive man and sees analogies between it and the Rhodesian and Neandertalian skulls. This is an exceptionally interesting cranium because it is almost certainly a descendant of the early Norse settlers of Greenland, a people in whom there is some evidence

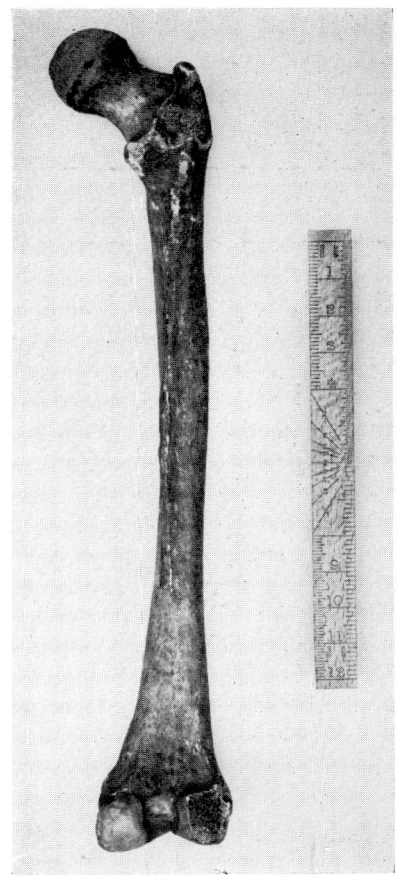

FIGURE 2. R. femur. Burgh Castle, Suffolk. Middle Saxon period *c.* 650 A.D. The oblique (physiological) length of this bone is 57.5 cm. Norwich, Castle Museum. Photograph Hallam Ashley.

to suggest widespread pituitary dysfunction, including gigantism. The Egla, Gretla, Burnt Njal and other sagas contain many detailed references to men of immense stature and unusual appearance. Often it is possible through the genealogies to trace the inheritance of these characters, as in the line of Ulf, Skallagrim, Egil Skallagrimson and their collaterals. This gigantism seems to have become endemic in the Fyrdafjord district of Sogn (Norway) and was commonly associated with personality changes. Sometimes the nick-

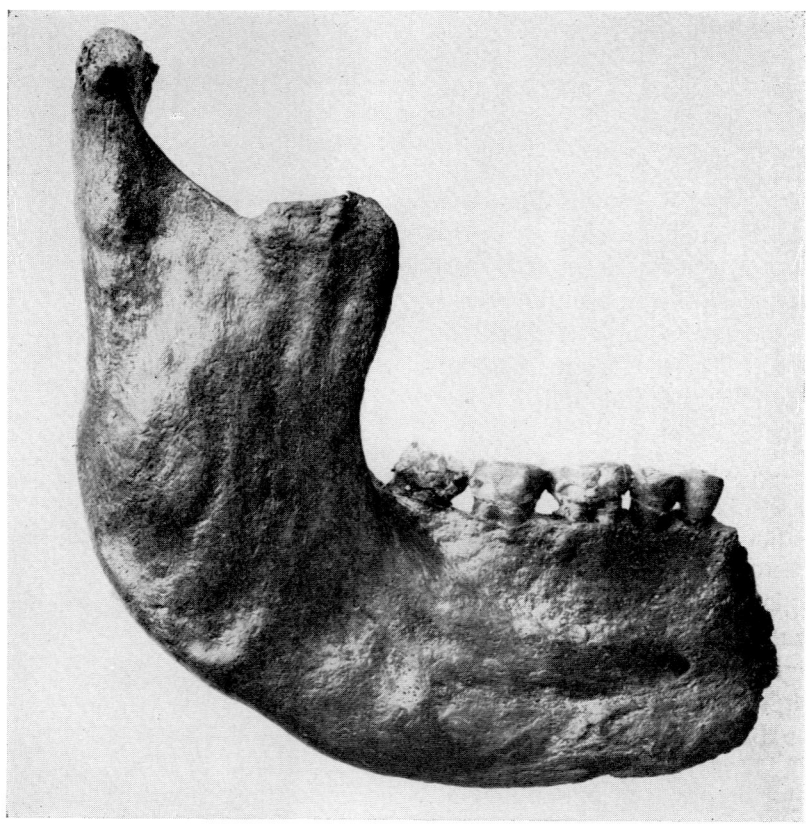

FIGURE 3. The damaged mandible of the Gardar skull. From the superior part of the condyle to the most anterior point on the inferior margin of the body is 15.3 cm. It must originally have been considerably longer. Photograph by courtesy of Professor Folke Henschen, Stockholm, Karolimska Sjukhuset.

names of these individuals betray features of pituitary disease. Ofeig "Clumsy Foot" and the famous berserker Thorir "Long Chin" are examples. Both Egil and other members of his family seem to have combined a mental and physical precocity with a tendency to premature senility which may possibly indicate a correlated adrenal dysfunction. Grettir the Strong, who was a collateral descendant of the Skallagrim line seems to have been sluggish in behaviour and to have had some genital hypoplasia. It may be that his gigantism was associated with elements of pituitary hypofunction, a condition that may also be detectable in Egil's granddaughter Thorbjorg the Fat. After the emigration to Iceland three of these giant families can be recognised by the same features appearing in their descendants. The Gardar skull may be an extreme example of this stock but similar, less pronounced, specimens have been recovered in the same cemetery (Perkins, 1931).

A parallel may be found at Tanum in South Sweden. Here rock-tracings have been found showing sea-raiders of immense size, many being prognathous with a suggestion of the acromegalic chin, whilst some are remarkably fat.

It is worth noting that Larger (1916) has discussed the concept of "racial gigantism" as a cause of extinction of fossil species. The giant Saurians have been seen as victims of this process and Moore (1955) asks "Did the tall Cro-Magnon people finally succumb to gigantism and are the Shillucks of to-day fated to follow them?" Mahoudeau (1916) rejected Larger's hypothesis. Todd (1914) refers to the *Anakim*—legendary giants of Hebrew mythology—and examines the factual basis for the belief. This concept of racial gigantism is closely related to that of pachyostosis in early animal groups (Volz, 1902; Kaiser, 1960): a concept which has recent been invoked to explain modern cranial hyperostosis (Casati, 1959). We feel, however, that these hypotheses are extremely dubious. Soranus of Ephesus described pigmentary disturbances of the facial skin in pregnancy and areolar pigmentation is noted in the Susruta Samhità. This we know now to be related to the melanocyte stimulating hormone of the pituitary and it may be significant that the Norse sagas also comment on skin pigmentation in some of the berserkers (Gretla Saga XIV). Lentigo is described in *Corpus Hippocraticorum*.

The ancients were certainly not aware of the existence of the parathyroids and adrenals despite some claims that the latter were known to the Hebrews (Schumacher, 1963). Nevertheless, gross distortion of bones, indicating a phase of demineralization has been several times recognised from ancient burial grounds. Denninger (see page 371) suggests osteitis fibrosa in the case of a pre-Columbian Indian with multiple cavitation in the skeleton and in whom the left femur was bent like a shepherd's crook, but elsewhere the editors of this volume indicate that this is probably an instance of fibrous dysplasia. Wells (in press) has diagnosed as osteogenesis imperfecta a closely similar femur, bent like a hair-pin, from an Anglo-Saxon cemetery.

Soranus of Ephesus described clitoral enlargement which is sometimes related to adrenal abnormality. Some Bible scholars have seen in the description of Esau a possible example of hyperadrenal hypertrichosis and have also suggested that Samson and Goliath were perhaps pituitary giants and that Judith and Jezebel were hyperthyroid subjects, but all of these attributions rest on very slender evidence. Thompson (1954) illustrates a Boeotian terracotta of a grossly fat woman which might be a representation of Cushing's syndrome, due to hyperfunction of the adrenals (see Fig. 4).

There are some interesting references to phenomena which we now realise are probably related to sex hormone activity; for example, in *Aphorisms* of *Corpus Hippocraticorum* it is stated that eunuchs do not suffer from gout or become bald: that only post-menopausal women develop gout and that gout in males does not occur before puberty. Further, in *Diseases of Women* of Corpus Hippocraticorum obesity is related to sterility and in such cases there may be endocrinopathy. Soranus made similar observations and also noted that amenorrhoea tends to occur in women of masculine habitus: such women may also have abnormal sex hormone levels.

Aldred and Sandison (1962) have discussed in detail the case of the Pharaoh Akhenaten (see Fig. 5) and conclude from study of the monuments that he presented an acromegaloid facies and a eunuchoid obesity. It is suggested that Akhenaten was an endocrinopath who may have had a pituitary adenoma which at first showed some alpha-cell activity but later led to pressure hypofunction of the pituitary pa-

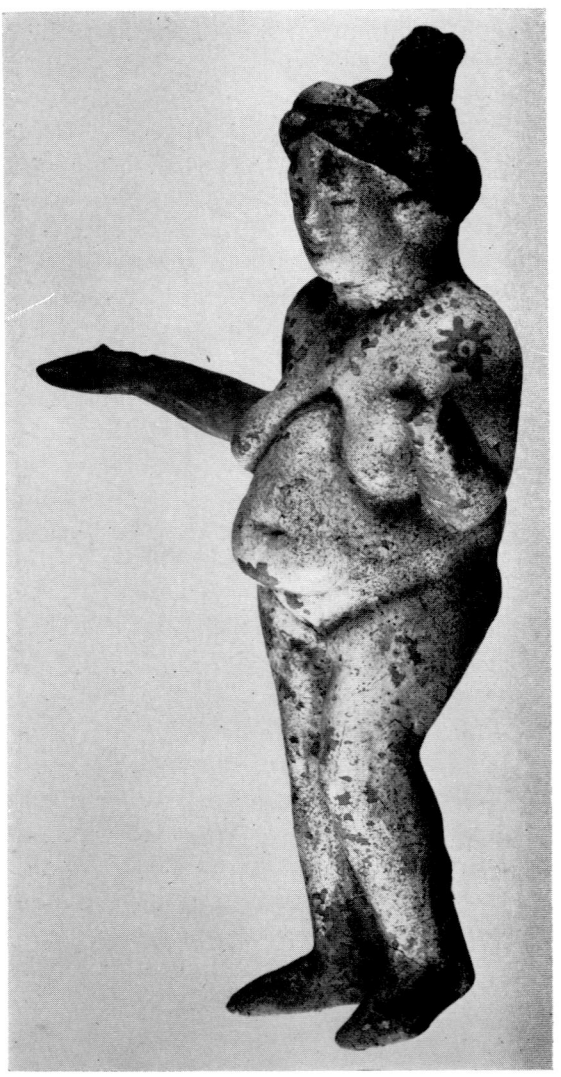

FIGURE 4. Terracotta figurine. Boeotia, fourth century B.C. H.19.2 cm. Perhaps mere middle age obesity but possibly intended to represent Cushing's disease or some other endocrine disorder. London, British Museum. Photograph British Museum.

FIGURE 5. Limestone portrait head of Akhenaten. Egypt. XVIIIth Dynasty, c. 1365 B.C. Berlin, Ägyptisches Museum 14512. Photograph Staatliche Museen zu Berlin.

renchyma. Further, remains believed to be those of Akhenaten are in Cairo Museum but to date have not been adequately published (Elliot Smith (1912)). Nevertheless, the data at present available are not inconsistent with this view. Fuller publication is awaited with interest. Ghalioungui (1963) has suggested that the breast fullness may have been due to liver disease caused by bilharziasis.

The subject of dwarfism may also be considered here. Many dwarfs are of the achondroplastic variety, i.e., growth failure is due to a defect in endochondral ossification (see Fig. 6); others of the Lorain-Levi type are due to hypo-pituitarism (Sandison 1958). Wells (1942) and Hrdlicka (1939; 1943) mention skulls of Peruvian midgets and Smith (1912) refers to small crania from IIIrd Dynasty in Ancient Egypt. The latter may have been from pygmy rather than pathological individuals. The remains of an achondroplastic dwarf from the Badarian period of predynastic Egypt have been described by Hughes Jones (1932) and Bleyer (1940). According to Tietze-Conrat (1957) the Pharaoh Pepi I had a dwarf from Punt. Ruffer (1911) published an interesting paper on dwarfs and other deformed persons in Ancient Egypt. Ruffer took into account various art forms, e.g., tomb paintings and reliefs, statuettes, etc. and concluded that dwarfs were probably not rare in Ancient Egypt. Many, he thought, could conclusively be shown to be achondroplastic, e.g., Chnoum-hotep, whose statuette, discovered at Sakkara, showed typical disproportion between trunk and limbs and a large head. Porak (1890) and other writers commented that the Egyptian god Bes is represented as an achondroplastic. Parrot (1878) also refers to the god Ptah as an example of this condition but Regnault (1902) and Vassal (1956) think he was derived from a foetus and enshrined foetal elements in his worship. Bes, too, has foetal connections: statuettes of this god are known in which foetal bones have been placed in a receptacle hollowed out of the back of the figure. The concept of racial gigantism has been noted above. Similar notions of doubtful validity have been invoked to explain modern and prehistoric pygmy races, such as those described by Dor (1903). Poncet and Leriche (1903) refer to ethnic achondroplasia and Smith, H.D. (1912) to ethnic dwarfism.

Many other typical anchondroplastics are found in works of art. They are shown on a Greek vase (Pottier, 1906; Singer, 1928), in a Hellenistic bronze (Huyghe, 1962), in a second century A.D. relief from

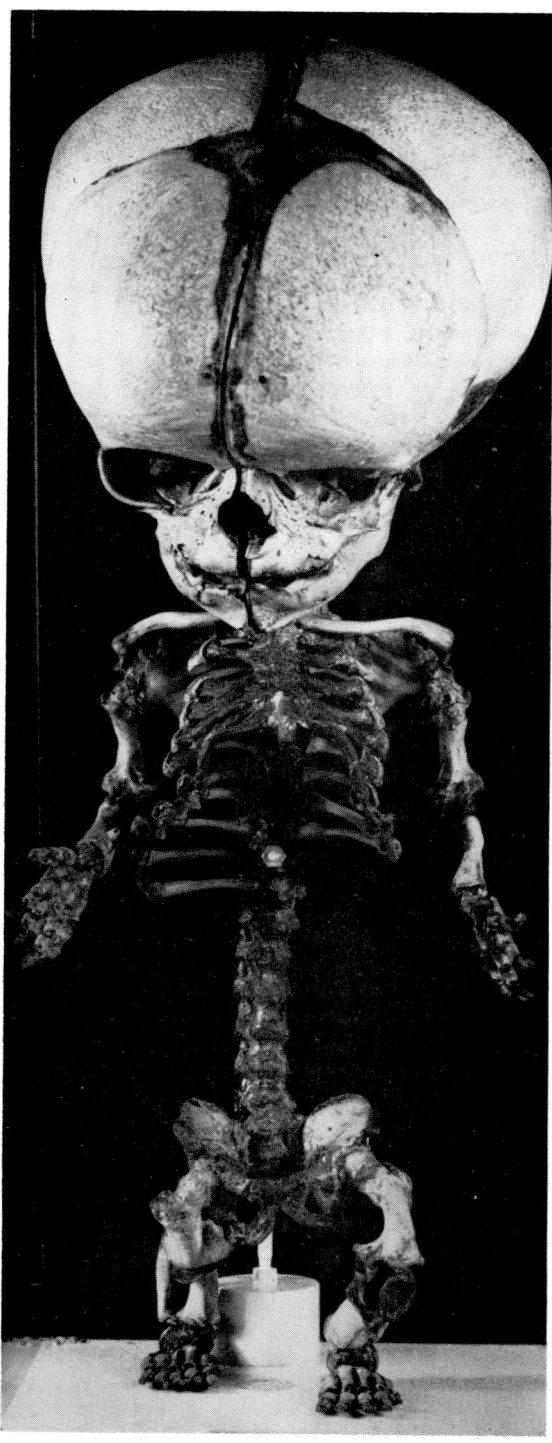

FIGURE 6. Achondroplastic dwarf. London, Royal College of Surgeons, S.59.2. Photograph R.C.S. This specimen was presented to the College in 1842 by Robert Liston. It is only 29 cm high (compare this with the femur in Fig. 2).

India, in sixteenth century Benin bronzes (Wells, 1964) and elsewhere. It is possible that Turold's groom, as shown in the Bayeux tapestry is intended to be a dwarf of this kind. They are extremely common in Renaissance painting.

Ruffer considered that other dwarfs might be cretinous but adduces no evidence for this although he thought that evidence might be obtained from Roman studies. Tietze-Conrat (1957) cites evidence for dwarfs in Ancient Rome from Quintillian, Plutarch, Martial and Suetonius and indicates that there was probably a morbid taste for monstrosities in Imperial times. Geikie-Cobb (1947) names two dwarfs from the Graeco-Roman period —Pheletas of Cos, tutor to Ptolemy in 300 B.C. and Coropas, dwarf of Julia, niece of Augustus.

Ruffer also points out that the Hebrews forebade access to the Temple by dwarfed persons. Tietze-Conrat refers to a statuette of a dwarf from the T'ang period and states that from the earliest times in Ancient China dwarfs acted as jesters. He illustrates a Mayan sixth century A.D. pottery figure of a hunch-back and states that hunch-backs and dwarfs are illustrated from the Inca Empire.

Ruffer also analyses the famous representation from Deir-el-Bahri of the Queen of Punt and rejects both the conclusion that she was steatopygous and the suggestion of achondroplasia. He concluded the lordosis to be important but was unable to come to a definite diagnosis. It is, however, interesting to recall that over a wide area of Europe from Spain to Russia and the Near East there are Palaeolithic *Venuses* showing pronounced curvature of the spine and sacrolumbar adiposity. Zammit and Singer (1924) discuss Neolithic representations of female obesity.

Obesity was not uncommon in antiquity and, as at present, it was probably often related to overeating, as in the case of Jeshurun (Deuteronomy 32: 15). However, it may sometimes have been related to endocrinopathy; as already mentioned, in Graeco-Roman times obesity was related to female sterility. Brim (1936) alludes to a Greek terracotta alleged to represent adiposis dolorosa. Cianfrani (1960) suggests that Aurignacian statuettes give evidence of glandular dystrophy. A Mochican pot from the Chicama valley illustrated by Wells (1964) shows morbid obesity, perhaps related to Fröhlich's syndrome (see Fig. 7).

There is an excellent bas-relief of an obese musician of the Middle Kingdom from Ancient Egypt in the Rijksmuseum van Oudheden Leiden. If we turn to the Royal Mummies (Smith (1912) we find examples of very stout women, e.g., Queens Inhapy and Henut-tawy and of slim or emaciated women, e.g., the Lady Rei, Queens Nofretari and Nedjmet, and of corpulent men, e.g., the High Priest Masaherta and of very emaciated men, e.g., Pharaoh Tuthmosis IV.

Some authorities claim that baldness may be related to endocrine changes: while there is considerable doubt about this, in general it may be useful to indicate here that in Ancient Egypt charms, recipes and amulets abound for the treatment of baldness and grey hair. Baldness (complete or partial) was noted by Elliot Smith (1912) during the examination of the mummies of Pharaohs Tuthmosis I, II and III, Amenōphis II and III, Sethōs I, Ramesses II, Merneptah and Ramesses III, IV and VI. Premature baldness is specifically referred to in the Icelandic sagas as affecting Skallagrim, Egil and other sufferers from the pituitary dyscrasias which we noted above. Long hair was noted in Tuthmōsis IV, Siptah, Ramesses V and

FIGURE 7. Pottery jug. Peru, Chicama valley. Mochica period. H.26.2 cm. Cambridge, University Museum of Archaeology and Ethnology 24.189 Morbid obesity. Photograph by courtesy of Dr. G. H. S. Bushnell.

Djed-ptah-ef-ànkh. These latter were, at least in the cases of Siptah and Ramesses V, young men. It is interesting that considerable alopecia was also noted in Queens Nofretari, Nedjimet and Henut-tjemhu and the unknown woman B. For a discussion of depilation of body hair in antiquity reference may be made to Sandison (1963). Domitian is said to have used tweezers to pluck out the genital hairs of his many mistresses (Cianfrani, 1960). It is clear from excavations in Nubia (Smith and

Wood Jones, 1910) that the ancient Egyptians and Nubians were relatively glabrous with scanty hair on body and face except for the chin, while Syrian bodies of the seventh century A.D. showed hairy limbs and bodies.

CONCLUSIONS

We have some evidence of endocrine disease; diabetes mellitus and goitre appear to have been well known and possibly exophthalmos. There is some evidence in favour of the existence of gigantism and a possible example of pituitary endocrinopathy in the case of Pharaoh Akhenaten. References in the classical writings give some weight to the possibility that endocrine related sterility may have occurred. Obesity and baldness which have rather doubtful endocrine relevance appear to have been common but the former was probably often due to overeating and the latter to genetic influences.

REFERENCES

ALDRED, C., and SANDISON, A. T., 1963: The Pharaoh Akhenaten: A problem in Egyptology and pathology. *Bull. Hist. Med.*, 36:293.

BLEYER, A., 1940: The antiquity of achondroplasia. *Ann. Med. Hist.*, 2:306.

BRIM, C. H., 1936: *Medicine in the Bible.* New York, Froben.

BROTHWELL, D. R., 1963: *Digging up Bones* London, British Museum (Nat. Hist.)

CASATI, A., 1959: Le iperostosi intertabulari del cranio come fatto di variabilità normale. *Arch. Antrop. Etnol. Firenze.*, 89:127.

CIANFRANI, T., 1960: *A Short History of Obstetrics and Gynecology.* Springfield, Thomas.

DOR, H., 1903: Les pygmées néolithiques en Suisse. *Bull. Mém. Soc. Anthrop. Lyon.* 22:171.

DURAN, C. M., 1963: Surgery of the Mayas. *Abbottempo*, 1(4):14.

GEIKIE-COBB, I., 1947: *The Glands of Destiny.* London, Heinemann.

GHALLIOUNGUI, P., 1963: *Magic and Medical Science in Ancient Egypt.* London, Hodder and Stoughton.

GRIFFITH, F. LL. 1898: *Hieratic Papyri from Kahun and Gurob.* London, Quaritch.

HADDOW, A., 1936: Historical notes on cancer from the MSS. of Louis Westenra Sambon. *Proc. Roy. Soc. Med.*, 29:1015.

HRDLIČKA, A., 1939: Normal micro-and macrocephaly in America. *Amer. J. Phys. Anthrop.*, 25:1.

HRDLIČKA, A., 1943: Skull of a midget from Peru. *Amer. J. Phys. Anthrop.*, n.s. 1:77.

HUYGHE, R., 1962: *Larousse Encyclopaedia of Prehistoric and Ancient Art.* London, Hamlyn.

JONES, E. W. A. H., 1932: Studies in achondroplasia. *J. Anat. Phys.*, 66:565.

KAISER, H. E., 1960: Untersuchungen zur Vergleichenden Osteologie der fossilen und rezenten Pachyostosen. *Paläeontographica.*, 114(A):113.

KEITH, A., 1913: Abnormal crania. Achondroplastic and acrocephalic. *J. Anat. Phys.* 47:189.

KEITH, A., 1931: *New discoveries relating to the antiquity of Man.* London, Williams and Norgate.

LARGER, R., 1916: L'acromégalie-gigantisme, cause naturelle de la dégénérescence et, partant, de l'extinction des groupes animaux actuel et fossiles. *Bull. Mém. Soc. Anthrop. Paris* 6ᵉ sér., 7:22.

MAHOUDEAU, 1916: Comments on Larger (Q. V.) *Bull. Mém. Soc. Anthrop. Paris*, 6ᵉ sér., 7:22.

MOORE, SHERWOOD, 1955: *Hyperostosis cranii.* Springfield, Thomas.

PARROT, J., 1878: Sur la malformation achondroplastique et le dieu Phtah. *Bull. Soc. Anthrop. Paris*, 3ᵉ sér. 1: 296.

PERKINS, M., 1931: Acromegaly in the Far North. *Nature*, 128:491.

PONCET, A., and LERICHE, R., 1903: Nains d'aujourdhui et nains d'autrefois; nainisme ancestral; achondroplasie ethnique. *Bull. Soc. Anthrop., Lyon*, 22:178.

PORAK, 1890: De l'achondroplasie. *Nouv. Arch. Obstet. Gynéc.*

POTTIER, 1906: Une clinique grecque au Vᵉ siècle. Monuments et mémoires de la fondation Piot.

REGNAULT, F., 1902: (Deux squelettes atteints d'achondroplasie). *Bull. Mém. Soc. Anthrop. Paris*, 5ᵉ sér., 3:175.

REGNAULT, F., 1910: *Divinités pathologiques.* Paris.

ROLLESTON, H., 1934: Endocrine Disorders. In *A Short History of Some Common Diseases.* 87. Bett, W. R., Ed. London, Oxford University Press.

RUFFER, M. A., 1911: On Dwarfs and Other Deformed Persons. *Bull. Soc. Arch. Alex.* 13:1.

SANDISON, A. T., 1958: Autopsy Findings in a Case of Pituitary Dwarfism. *Arch. Dis. Child.*, 33:469.

SANDISON, A. T., 1963: The use of natron in mummification in Ancient Egypt. *J. Near East. Stud.*, 22:259.

SCHLAGINHAUFEN, O., 1925: Die menschlichen Skeletreste aus der Steinzeit des Wauwilersees (Luzern) und ihre Stellung zu anderen anthropologischen Funden aus der Steinzeit. Zurich, Rentsch.

SCHUMACHER, H. B., 1936: The Early History of the Adrenal Glands. *Bull. Inst. Hist. Med.*, 4:39.

SELIGMANN, C. G., 1912: A cretinous skull of the 18th Dynasty. *Man*, 12:8 p. 17.

SINGER, C., 1928: *A Short History of Medicine*. Oxford, Clarendon Press.

SMITH, G. ELLIOT, 1912: *The Royal Mummies*. Cairo, Musée du Caire.

SMITH, G. ELLIOT, and WOOD JONES, F., 1910: The Human Remains. *Bull. Archaeol. Surv. Nubia*. Vol. 2.

SMITH, H. DOROTHY, 1912: A study of pygmy crania, based on skulls found in Egypt. *Biometrika*, 8:262.

THOMPSON, DOROTHY B., 1954: Three centuries of Hellenistic terracottas. *Hesperia* 23:72.

TIETZE-CONRAT, E., 1957: *Dwarfs and Jesters in Art*. London, Phaidon.

TODD, T. W., 1914: Palaeolithic giants and dwarfs. *Cleve. Med. J.* 13(8):533.

VASSAL, P. A., 1956: La physio-pathologie dans le panthéon Égyptien: les dieux Bès et Phtah, le nain et l'embryon. *Bull. Mém. Soc. Anthrop. Paris*, 10ᵉ sér., 7:168.

VOLZ, W. 1902: Proneusticosaurus, eine neue Sauropterygier-Gattung aus dem unteren Muschelkalk Oberschliesens. *Palaeontographica*, 49:121.

WELLS, CALVIN, 1964: *Bones, Bodies and Disease*. London, Thames and Hudson.

WELLS, J. R., 1942: A diminutive skull from Peru. *Amer. J. Phys. Anthrop.*, 29:425.

WELLS, L. H., 1963: Stature in earlier races of mankind in Science in Archaeology, D. Brothwell and E. Higgs (eds.) London, Thames and Hudson.

ZAMMIT, T., and SINGER, C., 1924: Neolithic representations of the human form from the islands of Malta and Gozo. *J. Roy. Anthrop Inst.*, 54:67.

Chapter 44

Urology in Egypt

J. THOMPSON ROWLING

Age plays an extremely important part in the incidence of urological disease. Wilm's tumour of the kidney (nephroblastoma) is as unknown in the adult as the evils of prostatic hypertrophy are in childhood and, in general, urological conditions select their peaks of incidence throughout the different decades of life. In considering the probable type of disease present in Egypt the average expectancy of life therefore becomes of some importance. In general this may be said to have been low. Egyptian cemeteries have a high proportion of children and young adults buried therein, and portrait statues and tomb reliefs tend to portray younger rather than older subjects. This evidence is no more than circumstantial. We have, however, an almost unbroken series of royal mummies from Seknenre, the last king of the XVIIth Dynasty, to the XXIst Dynasty. These include the mummies of the reigning Pharaohs and their immediate families. Of the forty-eight mummies in the series described by Elliott Smith (1912), forty-four can be ascribed to a specific age group with reasonable certainty. The mummy of Tut-ankh-amun can also be added to this series. The criteria for this assessment rest on the presence or absence of hair, its colour, both on the scalp and beard, the degree of wear of the teeth, the appearance of the skull sutures, the ossification of the costal cartilages and the fusion of the epiphyses. Clearly the assessment of age can only be approximate, but on these criteria the series can be divided into four groups. The first include children up to puberty, the second from puberty to the age of thirty, the third from thirty to sixty and the fourth over sixty. The results are given in Table I.

These figures, taken as they are from the royal mummies, represent the best cared for section of the population, at a period when the country was at the apex of its material prosperity. Deaths from violence are excluded from the series, the only known one being that of Seknenre. Premature babies, such as those found buried in the tomb of Tut-ankh-amun are also excluded. The general expectation of life was therefore short, and, in the series reviewed, averaged 39.8 years. Petrie (1923) also considered Egyptian life expectancy to be short, estimating that 25 per cent of the population died before puberty.

There are, however, many instances of longevity. Pepy II was probably six years old when he began to reign, and reigned a total of ninety-four years. The inscriptions of Bekenkhonsu of the New Empire period give the dates of the principal periods of his life and, if these are accurate, they indicate that he was at least one hundred and eighteen at his death (1320-1202 B.C.). The Abydos stele of Ramesses IV

TABLE I
ROYAL MUMMIES OF XVIII DYNASTY TO XXI DYNASTY

	Children Birth-Puberty	Young Adults Puberty—30	Middle Age 30—60	Old Age Over 60	Total
Number of Mummies	4	16	16	9	45
Percentage	8.8%	35.6%	35.6%	20%	100%

mentions that Ramesses II reigned sixty-seven years.

The traditional figure for the limit of life in Egypt was a hundred and ten years, corresponding to the biblical "threescore years and ten." To exceed this was considered improper, but in the case of sorcerers their skill was said to increase until the figure of one hundred and ten was attained. It seems likely that they remained at this age for a considerable time.

It is, therefore, to be expected that the incidence of benign prostatic hypertrophy would be low, that malignant disease of the kidney and prostate would be low, and that stone and infective diseases would be more common. It is known that schistosomiasis was not uncommon in ancient Egypt and the incidence of carcinoma of the bladder occurring as a complication of the infestation may, therefore, be assumed to be high.

There is some evidence to suggest that true operative surgery was only practised in the Old Kingdom and reached its peak at this period. The Edwin Smith Surgical Papyrus dates from this era (although the papyrus itself is a copy of the original work which was made in the New Empire period), and speaks so surely and authoritatively on surgical matters that it is impossible not to believe that the author was experienced in surgery. The "medical" papyri, on the other hand, speak of surgery as of a lost art, giving the traditional instructions for surgical operations as "thou shalt treat it as the Priest of the Goddess Sekhmet treats it . . ." but never elaborating on the method or its details. The impression is given that surgical treatment was a hallowed tradition only, described but never practised in the New Empire period and that surgery had by then become a matter for wonder rather than practice. Certainly we have no evidence of surgical scars on mummies and the whole tradition of Egyptian thought tended towards preservation of the body from injury, both in the living and the dead. It was probably this trend which led to the preservation of useless and painful carious teeth, rather than their extraction.

One curious point about Egyptian urology was absence of all references to catheters. Our first proper description of catheters comes from Celsus in the first century, yet the Egyptians were quite able to make small tubes. Copper in all forms was common from Predynastic times and silver soldering was known from the time of Hetepheres in the IVth Dynasty (Lucas, 1948). One of the stelae from the New Empire period shows a Syrian settler in Egypt drinking wine from a cask by using a small copper tube of a size quite suitable for catheters. Yet we have no indication that catheters were ever made or used. Acute retention was known for there are many remedies in the medical papyri to enable a man to pass water. Retention of urine after spinal injury is also described in the Edwin Smith Papyrus as "his urine dribbles from him . . ." but the distinction between retention with overflow and frequency does not seem to have been appreciated. The fiftieth column of the Papyrus Ebers gives remedies "to remove the urine which runs too often . . ." and "to remove constant running of the urine. . . ." These may refer to either frequency of micturition or to retention with overflow. Nowhere in the papyri are these conditions differentiated.

During the process of mummification the kidneys were left *in situ* in the abdomen. The Egyptians were aware of the importance of the kidneys and held them as sacred. The thirtieth chapter of the *Book of the Dead* refers to the heart and kidneys alone amongst the viscera: "May naught be against me in the presence of the great god, the lord of Amentet. Homage to thee, O my Heart! Homage to you, O my Reins!" (Papyrus of Nu.). The role of the kidneys in the production of urine was probably not appreciated. The *Secret Book of the Physician* from the Papyrus Ebers and the Berlin Medical Papyrus states: "There are two vessels to the bladder; it is they which give the urine." This may, of course, refer to the ureters, but it is more likely that they are the iliac arteries.

INFECTIVE LESIONS

Cystitis

We have no objective proof that cystitis was common in Egypt, but it is a common condition in all parts of the world and the frequent references to "burning of the water," "heat in the bladder" and "frequency" in the papyri are suggestive.

Schistosomiasis

The bilharzia worm, *Schistosoma haematobium*, occurs throughout Africa, but is particularly common in the Nile Valley. It is estimated that about 50 per cent of the population of Egypt suffer from the disease at the present time, and it is likely that this figure has not altered much since Pharaonic times.

Among the many dark and sombre modes of existence favoured by parasites, that of the Schistosoma must be one of the strangest. The worm is a small, sexually differentiated Trematode, about 1 cm in length, whose normal habitat is in the plexus of veins round the bladder in man. The female lays her eggs in these veins. They eventually ulcerate through the bladder mucosa and are passed in the urine. If this finds its way into fresh water the membrane of the ovum ruptures and the embryo or miracidium attacks its intermediate host, the snail (*Bulinus contortus*), entering the snail's liver. Here the miracidium liberates large numbers of sporocysts which mature to become small, free-swimming cercariae, which pass out into the water in which the snail lives. The cercariae are able to penetrate human skin whence they pass to the human liver, where the male and female worms mature and mate. The worms pass down the portal blood stream to the pelvis and by a process which is not fully understood they migrate from the portal system to the perivesical plexus. The whole cycle takes about six weeks.

The parasites produce a severe cystitis, with haematuria, increased frequency of micturition and dysuria. In these cases the bladder is peculiarly liable to malignant changes and carcinoma of the bladder is not uncommon even as early as the third decade of life. Vesical calculi are often formed and sometimes ova are found within them. Occasionally the ova have been found in association with the appendix or large bowel.

Haematuria is a striking symptom and the Papyrus Ebers contains twenty prescriptions for its cure. In Egypt, the two predominant causes of haematuria are bilharzia and neoplasia of the renal tract. In view of the relatively low life expectancy in Egypt, and the fact that primary neoplasia is essentially a disease of later life, it seems reasonable to suppose that most of the references in the papyri refer to bilharzia. Steuer and Saunders (1959) relate the 'aaa' disease, so frequently mentioned in the papyri, with bilharzia and associate it with the parasite $hrrw.t$, which they believe was recognised by the Egyptians as the causative agent. In this, however, they may be going rather further than is justifiable on the evidence available. Hoeppli (1959) reviews the literary evidence for relating 'aaa' disease with bilharzia.

Objective proof of the presence of the parasite in Egypt was obtained by Ruffer in 1910 when he was able to demonstrate the calcified ova of *Schistosoma haematobium* in the kidneys of two mummies of the XXth Dynasty. Shattock had specifically searched for the ova in vesical calculi some years previously, but had failed to find any. The writer has observed an object somewhat resembling the ovum of the parasite in the submucous layer of the large bowel of a Ptolemaic mummy.

The evidence for the existence of bilharzia in Egypt is therefore conclusive. There is, however, very much more doubt as to exactly how much the Egyptians knew of the parasite and whether they did, in fact, relate it to the haematuria from which they so frequently suffered.

Renal Abscesses

In 1910, Ruffer gave an account of a mummy of the New Empire period in which both kidneys contain multiple abscesses in which gram negative bacilli were found. These may have been pyaemic in origin. There was no suggestion of pyelo-nephritic changes (Ruffer, 1921).

Renal Tuberculosis

We have no objective evidence of renal tuberculosis, but in view of its known origin by blood stream infection and the prevalence of osseous tubercle—also disseminated by the blood stream—it seems logical to deduce that renal tuberculosis was not uncommon in Egypt.

Calculi

It is said that the remains of about 30,000 mummies have been examined, but out of this large number only a few examples of urinary calculi have been found (excepting the series from Jebel Moya, discussed by Brothwell elsewhere in this book). These stones would, of course, be easily missed, but their rarity in a hot country is surprising. No mention of stone is made in the medical papyri.

One of the earliest examples of stone was found in Predynastic body in which only the bones have survived. Three stones were found by Petrie and analysed by Ruffer. On analysis they were found to consist of mixed phosphates and uric acid. They were probably in the bladder.

A large vesical calculus was found in the nostril of a priest of Amen of the XXIst Dynasty, where it had been placed by the embalmer. This was found to consist of a uric acid nucleus, surrounded by phosphates. It measured 6.5 cms in diameter.

Four renal stones have been found in the body from a IInd Dynasty tomb at Maga-el-Dier. The soft tissues had disappeared, but the stones lay opposite the second lumbar vertebra and were therefore probably renal. The largest, 1.6 cms in diameter, was found to be an oxalate stone.

Tumours

We have no positive proof of urinary tumours in Egypt, but it seems at least very probable that malignant disease of the bladder, arising as a complication of bilharzial cystitis, was common at all periods. Primary malignant disease of the kidney and bladder, unrelated to infection, was probably rare.

Circumcision

Circumcision was practised almost universally in Egypt and the custom probably dated from Predynastic times. Certainly it is found in mummies of the Old Kingdom. In the more plentiful mummies of the New Empire and later periods it is practically always seen. Herodotus (II.37) states that "they practise circumcision,

FIGURE 1. Circumcision. From the tomb of Ankh ma Hor, Sakkara.

while men of other nations—except those who have learnt from Egypt—leave their private parts as nature made them." The practice is finally seen to disappear in Coptic times, when the salt-preserved bodies of the time show no evidence of the operation.

Circumcision was almost certainly performed just before puberty. The tomb of Amenōphis II contained the mummy of a boy of about eleven years old, showing the common "Horus lock" of hair, and clearly uncircumcised.

A relief in the tomb of Ankh-ma-hor, at Sakkara, dating from the VIth Dynasty, shows the operation being performed (Fig. 1). This was probably done by a priest, although as the Egyptian surgeon was often referred to as a priest of the goddess Sekhmet the distinction may be superfluous. The suggestion that the operation is being performed at puberty is given support by the Horus lock of the patients, which can be made out fairly clearly in the left hand figure, as contrasted with the adult hair style of the surgeons, although the height of all the figures is similar. The patient on the left is having his hands restrained by an assistant, while the right hand patient supports himself on the surgeon's head. Both surgeons operate seated. The details of the operation cannot be made out owing to erosion of the stone, but there is no doubt that a knife is being used in the right hand scene.

A remedy for bleeding following circumcision is mentioned in the Papyrus Ebers. "Remedy for a prepuce (?) which is cut off, (circumcised) and from which blood comes out: 'dsrt,' honey, cuttlebone, sycamore, fruit of 'dsjs,' are mixed together and applied thereto."

A rather exceptional finding is that the body of Prince Sipaari from the series of Royal Mummies of the New Empire, shows some evidence of circumcision, although he was only five or six years old. Elliott Smith, who unwrapped the mummy, describes it as "probably circumcised," and comments on the unusually early age for this to be found.

CONCLUSION

Our knowledge of Egyptian urology is regrettably scanty and to improve it a great deal more work on both the macroscopical and microscopical appearances of the urogenital system in both embalmed and sand-buried bodies is required. Urology was studied as a specialty in Egypt and the Egyptians can lay undisputed claim to be the fathers of a subject which is only now coming into its own.

REFERENCES

Hoeppli, R., 1959: *Parasites and Parasitic Infections in Early Medicine and Science*. Singapore, University of Malaya Press.

Lucas, A., 1948: *Ancient Egyptian Materials and Industries*. London, Arnold.

Petrie, W. M. F., 1923: *Social Life in Ancient Egypt*. London, Constable.

Ruffer, M. A., 1921: *Studies in the Palaeopathology of Egypt*. Chicago, University of Chicago Press (1910 pages reprinted).

Smith, G. E., 1912: *The Royal Mummies*. Cairo, Musée du Caire.

Steuer, R. O. and Saunders, J. B., 1959: *Ancient Egyptian and Cnidian Medicine*. Berkeley, University of California Press.

Chapter 45

Palaeostomatology

H. BRABANT

INTRODUCTION

THE study of palaeostomatology is of more than historical interest. Once we allot palaeostomatology its proper place in the general context of the study of maxillary evolution, we find it shedding new light on dental disorders of our own day, some of which are as yet little known, not to say unknown.

The present study is based on personal research on more than 3,200 intact or fragmentary skulls and on about 50,000 ancient teeth. It derives also from a thorough examination of the casts of most of the classic fossilized remains, and from a critical study of several hundred publications. In the limited space at my disposal, I cannot aim to give here a complete list of all the publications studied, but a detailed enumeration can be found at the end of my publications mentioned in the bibliography of this study. These publications also provide a detailed exposition of the methods used in the study of ancient teeth, and I would refer the interested reader to them. The findings set out in this article are principally drawn from subjects of European stock and from the white-skinned peoples of the United States. The research we are at present pursuing has in fact shown that what is known of the dental palaeopathology of African, Asian and South American peoples must be completed in both breadth and depth before a real synthesis can be offered; a beginning has been made, but it is still too soon to set out the results.

In the first part of this study, I propose to deal with abnormalities in the teeth and with malocclusions; in the second part, with their diseases and lesions. Before proceeding further, it is worth asking oneself whether abnormalities in the number, shape and position of the teeth (or at least some specified abnormalities, whose incidence has varied since prehistoric times) could have their origin in the progressive diminution of size and in the changes in the shape of the teeth and of the masticatory part of the mandible.

In a recent review of this field (Brabant and Twiesselmann, 1964a) we have shown that from the australopithecines to our own time, passing through the *Homo erectus*, Neandertal Man and *Homo sapiens,* the teeth have diminished progressively in size, but not uniformly. The reduction in size has been most marked in the premolars, then for the third and second molars; next comes the first molar, and then the canine. The incisors are those which by contrast have shown least decrease in size. The vestibulolingual diameter of the teeth has in general decreased in more or less the same proportions as the proximodistal diameter. The shape of the dental arch has changed radically in the course of time: in our day the teeth are implanted

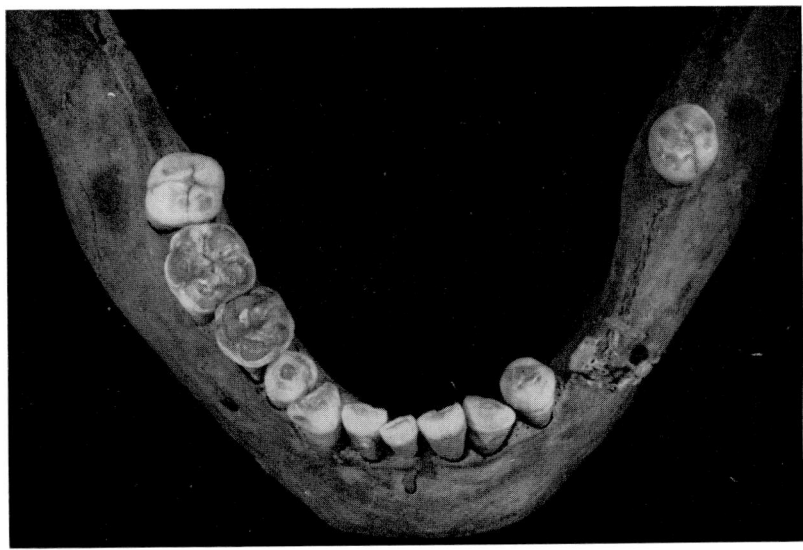

FIGURE 1a. Mandible of a Mediaeval adult male from Coxyde, Belgium. There is hypodontia of 5_1.

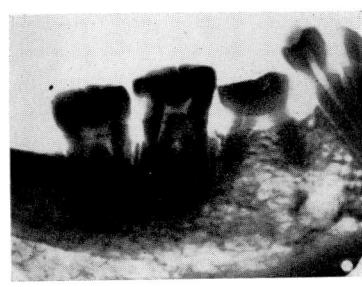

FIGURE 1b. Neolithic mandible (Belgium) with evidence of hypodontia, and retention of the milk molar.

along an arc showing a much wider curve, and the bi-canine line has shifted slightly to the rear. The bony portion corresponding to the molars has diminished in length by 32 per cent, while the sum of the proximo-distal diameters of the molars has shown a similar proportional decrease, 34 per cent. It does not appear therefore that the molars of present-day man need necessarily be cramped when they erupt by the smaller extent of bone that is now reserved for them. The study of the *corpus mandibularis* in *norma lateralis* shows that all the horseshoe portion that corresponds to the seat of the tongue's musculature and of the sub-hyoidien diaphragm *appears unchanged*. We have come to the conclusion that the emergence of the chin has come about passively; those parts of the mandible whose size has decreased are those which tend normally to disappear when, with the loss of teeth, the structure becomes a bony arc, bare of alveolae, and whose upper branch becomes thinner.

ABNORMALITIES IN THE TEETH

Hypodontia

Abnormalities in the number of teeth in man have been observable as far back as the Palaeolithic stage, but it is not easy to determine exactly how frequently they then occurred. Examples of such congenital absence are given in Figure 1, a and b. As early as the Pleistocene period (Twiesselmann and Brabant) hypodontia was concerned, as it is today, with the lateral upper incisors, the premolars (particularly the second), and the third molars, both upper and lower. One may put the incidence of agenesis for all of these teeth at about 0 to 27 per cent (depending on the series) of individuals studied. The third molars by themselves account for between about 10 and 26 per cent; the upper lateral incisors are those effected in 0.1 to 0.8 per cent of cases, and the premolars in about 0.5 to 1 per cent. In mediaeval times, the incidence shows a slight increase (Brabant and Twiesselmann, 1964) and, at the present day, for the three groups of teeth, it can be put at between about 10 and 40 per cent. In the past as today, it was the last tooth in each group that was most likely to be missing, except for the central lower incisor which is more often absent than is the lateral.

Naturally, the frequency of dental agenesis has varied in different populations, and it seems always to have been greater in female than in male subjects, (although we cannot state this with absolute certainty.)

Agenesis of the wisdom tooth, which appears to have been very rare during the Palaeolithic period, has increased in frequency through the ages, and Ruffer (1920), like ourselves, has observed that its absence could sometimes occur even in a wide arch.

Hyperodontia

Hyperodontia, or the development of extra teeth, has been observed in man as early as the Pleistocene period (Twiesselmann, Brabant and Kovacs, 1962) but remains rare during the Neolithic (0.01 to 0.05 per cent of individuals). Its incidence increases slightly in the Gallo-Roman period and the Middle Ages to reach, at the present day, estimated frequencies ranging from 0.15 to 3 per cent according others (Brabant, Klees and Werelds, 1958). An early deciduous example is given in Figure 1, d. Finally it seems that in the past, as now, supernumerary teeth have occurred more often in men, whereas agenesis has occurred more frequently in women. Here again however, it is impossible to be dogmatic.

To sum up, hypodontia and hyperodontia have shown a slow and progressive increase since Palaeolithic times, but the increase in the incidence of hyperodontia is quite clearly less than that of hypodontia. It must also be emphasized that there are cases where one can observe in one and the same dental arch both supernumerary teeth and agenesis (Brabant *et al.*, 1958).

ABNORMALITIES IN SIZE AND SHAPE

Abnormalities of size and shape of teeth are observable from Palaeolithic times onward. Let us consider the chief among them.

Microdontia

Microdontia of certain teeth has been observed by the present writer as far back as the Neolithic period; that of the lateral incisor (peg-shaped tooth) being the least rare (0.1 to 0.6 per cent individuals). In the Middle Ages, among the ancient peoples of Western Europe, we have found similar evidence of this abnormality (Fig. 1c). At the present time, the rate of incidence can be set at between about 0.5 and 1 per cent. In certain teeth, therefore, microdontia frequencies are increasing slowly.

Macrodontia

Macrodontia, by twinning or fusion of teeth, has not been observed by the present writer, nor by others to my knowledge, before the beginning of our own era (which does not mean that one will never encounter it in prehistoric and proto-historic periods). In the course of our studies

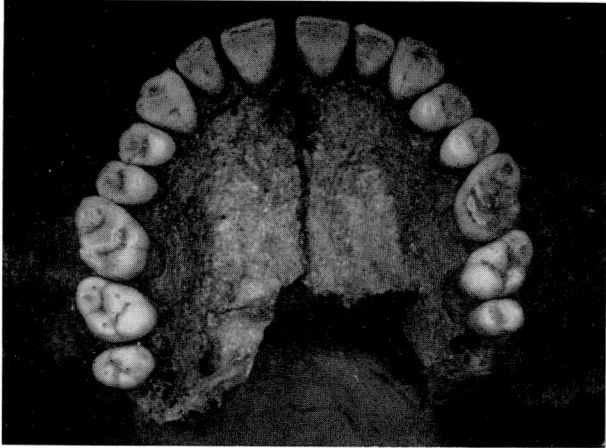

FIGURE 1c. Microdontia of the third molars. Mediaeval Coxyde palate.

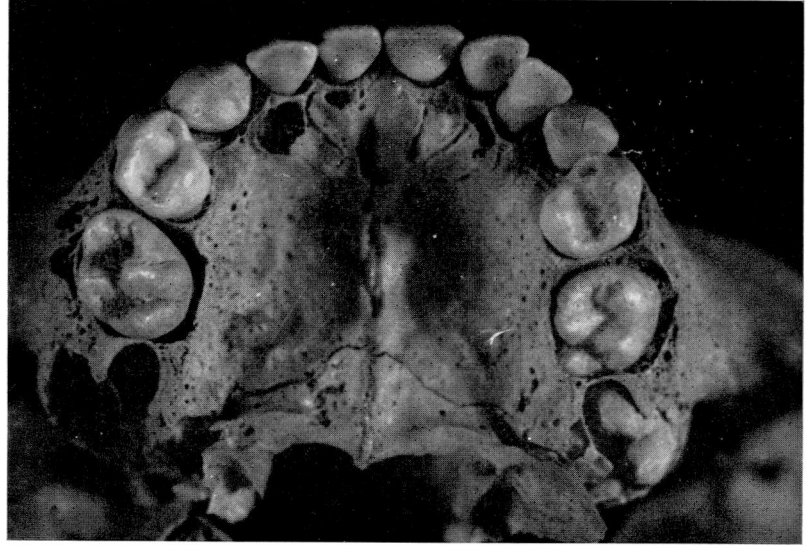

FIGURE 1d. Hyperodontia in a child's palate, from the Humic cemetery of Mös (Hungary).

some cases have been found which are up to 2,000 years old. At the present time, the incidence remains small (less than 1 per cent).

Shovel-shape Incisor

The "shovel-shaped" incisor has been observed in the australopithecines, the *Homo erectus* group and in Neandertalers, with a high rate of incidence. It is found also in Europe during Mesolithic times, and then during the Neolithic (Fig. 2a). The frequency varies between 16 and 53 per cent of individuals in the series of skulls from this latter period. In certain series, the percentages are much lower: 1 to 7 per cent. This tooth-formation is not at all exceptional in the Middle Ages, and

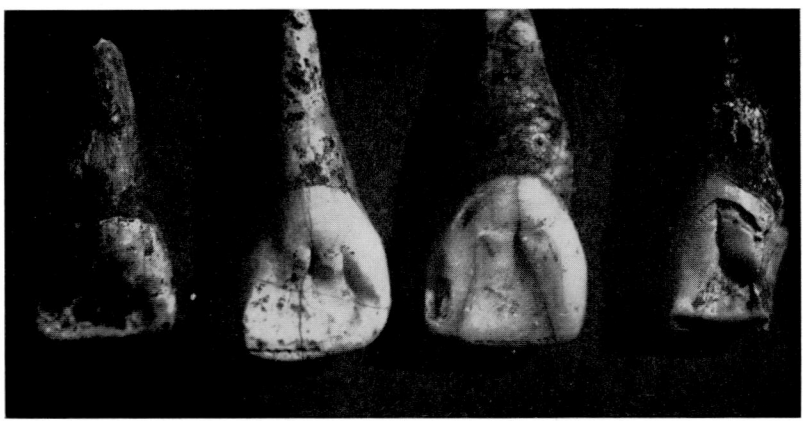

FIGURE 2a. Shovel shaped incisors from the Neolithic crematorium of Les Matelles, France.

even in our own day. We have observed cases in France and in Belgium. It is however certain that its frequency has progressively declined in Europe since the prehistoric period. It would be interesting to gather data on its present incidence in European populations in order to establish precise comparisons, which we have not yet had an opportunity to do..

The association of a lingual tubercle in the incisors, and the "shovel-shaped tooth," has several times been observed in Palaeolithic and Neolithic series (Carbonnel, *in* Brothwell, 1963), and I have also encountered cases of it. This association is rare in modern European man. It is not yet possible to give precise percentages.

Paracingular Invagination

Paracingular invagination of the incisors is fairly common in certain of our series of prehistoric mouths. It is found about 31 per cent of incisors in the Neolithic population of Matelles (Montpellier), and particularly in the "shovel-shaped teeth" generally. In modern Europeans, this invagination seems to have become more common, as it is said to be found in approximately 1.8 to 14 per cent of incisors. But the latter figures are taken from various authors, and are not the results of personal research.

Enamel Variations

INTERRADICULAR EXTENSION OF THE ENAMEL in molars can, it seems, be observed right from Palaeolithic times in Europe. It is found again in the Middle Ages (Fig. 2b) and occasionally at the present day in France, Belgium and Switzerland (Moeschler, 1964). It is still difficult to determine rates of incidence, and new research on the subject is needed (Brabant and Twiesselmann, 1964b).

ENAMEL PEARLS or "enamelone" are often found in association with this interradicular extension of the enamel. Its frequency varies greatly with the various series, but is generally low in Europe. However, it is certain that so-called "mongoloid" dental characteristics (shovel-shaped teeth, taurodontia, interradicular extension of enamel, etc.) can be found in Europe during prehistoric and historic periods, and can still sometimes be observed in our own day, but less frequently (Brabant and Twiesselmann, 1964).

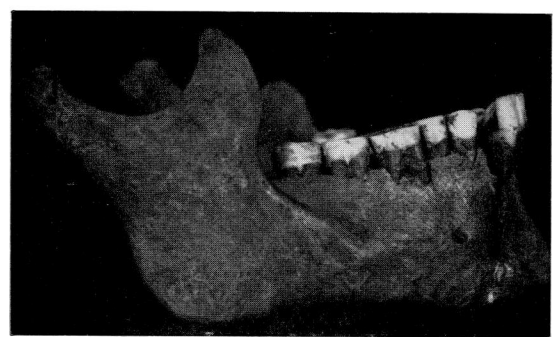

FIGURE 2b. Interradicular extensions of enamel. Mediaeval Coxyde mandible.

Premolar Crowns

Variations in the shape and size of premolar crowns are, in modern man, different from those observed in Palaeolithic man. In the latter, the first premolar was generally a little larger than was the second. In modern man, it is usually the reverse. The second lower premolar showed then, and still shows, a tendency to molarisation, with two secondary cusps forming at the expense of the lingual cusp. In all the series of jaws available to me for study, both prehistoric and later, I have several times found cases of second premolars with not only three well-developed cusps but also a disto-lingual portion which is clearly molarised, with several cusplets. From Neolithic times onward, there is a fairly wide variety in the forms of the triturant aspect of the premolars. We do not have space here to go more fully into this question. Variations in the shape and size of molar crowns in ancient specimens are often difficult to study, due to the wearing down of these teeth.

The progressive reduction in size of the hypocone and the metacone is very noticable in certain modern groups but it was already observable in certain groups of fossil man. Examples of it have also been found from the Neolithic. The reduction of cusps from this time onwards seems to the present writer, less clear and less rapid than other authors have taken it to be (see Brabant and Twiesselmann, 1964). The first lower molar shows most regularity in its formation from Palaeolithic times on; the second lower molar, during the same period, has developed into its present form most quickly and unmistakably; while the lower third molar seems in general to have undergone less simplification than the second.

Carabelli Tubercle and Hollows

The Carabelli tubercle, furrow and pit have for a long time puzzled odontologists as to their significance. We will not discuss this problem here, but will merely point out that we have found the tubercle and the Carabelli hollow in certain teeth in all the series of dentitions from ancient times we have examined. Extensive dental attrition often makes it difficult to determine the existence of the tubercle. Its incidence varies greatly according to the series of teeth examined: for the first upper molar it ranges from about 1.5 per cent to sometimes as much as 25 per cent. It is difficult to draw from such differing frequencies any phylogenetic conclusions. According to De Jonge (1963), the Carabelli tubercle can also be observed at the present time on the lower molars; it is too rare for it to be possible to establish percentages. In ancient teeth we have found three abnormal cases of the tubercle which could, according to De Jonge, be grouped with Carabelli tubercles of the lower molars.

On the second molar, only rare instances of this tubercle have been found. It is on the other hand a little more frequent on the third molar: 0.4-0.6 per cent. For the sake of comparison, it may be recalled that, at the present time, the tubercle and the Carabelli furrow and pit are found in about 11-70 per cent of white subjects on the first molars, in 0.2-23 per cent of individuals for the second molars, and in 0.4-10 per cent of individuals for the third molars (see Brabant and Twiesselmann, 1964).

THE PARAMOLAR TUBERCLE is found in the human dentition from Palaeolithic times on, but much less frequently than the Carabelli tubercle. It can be found on the third molars, but it is mainly situated on the second molar. In our series of teeth from prehistoric or Mediaeval man, we have noticed several cases of it. It is nevertheless difficult to establish any reliable percentage in the present state of our knowledge. It seems, however, that one might estimate the incidence, for Mediaeval groups, at between 0.3 and 1.2 of the specimens examined.

Roots

Variations in the shape and number of roots are fairly common among ancient teeth. Sometimes, abnormalities in the shape of the dental root appear, independent of abnormalities in the crown—but this is not a hard and fast rule. One could cite as an example the *radix praemolarica*, the so-called *Carabelli root* accompanying certain peculiarities in the crown.

The coming-together of the roots in molars seems to be on the increase since the end of the Palaeolithic period. It has been claimed that this closing-up was related to the shortening of the jaws, but the reverse might well be equally true (Brabant and Twiesselmann, 1964). One might ask whether this closing-up is a phenomenon of the same order as the increasing frequency, from Neolithic times onward, of malformations of the roots, mainly in the last two molars.

The roots of the incisors show, exceptionally, a tendency to bifidity. I have found just one case of this among ancient teeth, except for a very slight proportion of lateral upper incisors with a distinct but very shallow syndesmo-radicular furrow. The upper canines exhibit root bifidity in only very exceptional cases; whereas the lower canines, both ancient and modern, reveal from 1 to 10 per cent bifidity. On this point our findings confirm the conclusions of Alexandersen (1963).

From about the Neolithic period onwards, the first upper premolar has usually two roots (in about 66 per cent of subjects): one root is less frequent (30 per cent of subjects), and three even more unusual (4 per cent). The second premolar most frequently has one root (80-85 per cent of cases), but sometimes two (12-15 per cent) or three (1-2 per cent). It is nevertheless impossible at present to establish any very certain statistical comparisons between the various prehistoric and recent populations.

The first lower premolar has for a very long time shown a single root in the majority of cases (approx. 96 per cent), and two complete or vestigial roots in 3 to 4 per cent of cases. The second lower molar shows a single root in 97 to 98 per cent of cases, two roots in the rest. Three-root forms with their roots more or less individualized can also be found, not only among Modern teeth, but also among Gallo-Roman and Mediaeval teeth. In the case of the Anthropoids and australopithecines, two roots are the general rule. When the lower premolars do have a double root, one of them is generally vestibular, the other lingual. But, in very rare cases, one may be mesial, the other distal (Brabant *et al.*, 1958).

The upper molars also exhibit root variations and abnormalities, in prehistoric dentitions as much as among those from historic or modern times. Thus, the first molar, which in 80 to 90 per cent of cases has three independent roots, shows partial root fusions in 10 to 19 per cent of cases and, very rarely, a total fusion (less than 1 per cent according to our own findings, but up to 5 or 6 per cent for other authors: (see Brabant and Twiesselmann, 1964).

For the second upper molar, the three roots are independent in only about 45 per cent of cases—the remainder being accounted for by fusions, usually partial, between the palatal root and the mesio-vestibular root (26 per cent), or between the palatal and one of the vestibular roots (25 per cent). Complete fusion with occasionally a single canal accounts for from 3 to 4.5 per cent.

Finally, in the case of the third upper molar, the total fusion of roots shows an incidence of from 15 to 22 per cent, fully independent roots occurring in 21 to 22 per cent of cases, and the rest exhibiting partial fusions. These proportions seem to hold good from at least the beginning of our own era.

The lower molars can also present examples of extra roots. Cases of this have been found in Mesolithic and Neolithic specimens. It is difficult to give firm percentages at the present state of our knowledge, the results varying greatly between the different series examined. The question of the role of genetic factors calls, in this connection, for further study.

Taurodontism

The taurodont molar (Fig. 3) should not, in our opinion, be considered as a peculiarity of the Neandertaler, although it is very frequently found at that group. Taurodontism is found among certain white peoples of the present day, but its frequency is not yet easy to determine exactly (Brabant and Kovacs, 1961).

The molars of both jaws can exhibit this particular shape, but it is principally found in the upper jaw and especially in the second molar. The "mesotaurodont" form is that found most often; hyperotaurodontism is rarer, and hypotaurodontism is rather difficult to define and indeed open to doubt, as we hope to show in a later study.

Anisodontia

The inequality of shape and size of analogous teeth in one and the same dental arch, i.e., anisodontia, is not rare among Neolithic, Mediaeval or Modern teeth. It is found particularly among the two last molars, particularly the upper ones.

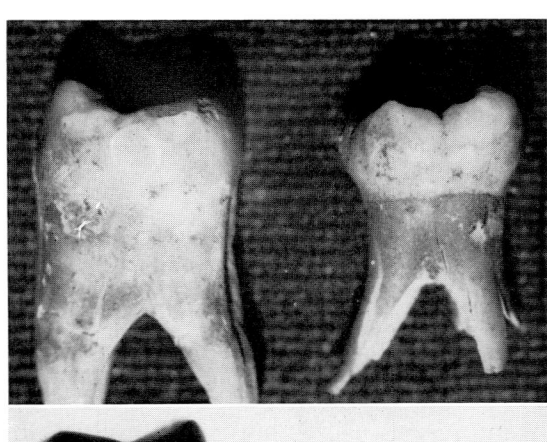

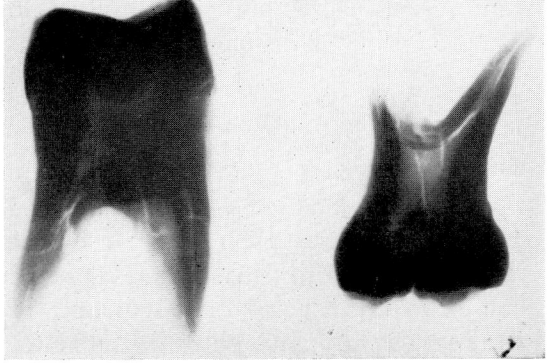

FIGURE 3. Neolithic examples (external appearance and X-rays) of taurodonty. Deciduous and permanent molars from Les Matelles, France.

Impaction

Abnormalities in the eruption and position of teeth, as well as abnormalities of articulation, have perceptibly increased in frequency since Palaeolithic times, when they are rarely found. From an incidence of less than 1 per cent at this period, they reach a level of from 5 to 15 per cent in the series of Mediaeval teeth we have examined, and they seem to have doubled or trebled in frequency since then. The most typical situation is that of dental impaction, especially of the permanent canine and of the wisdom tooth. Examples can be found from the end of the Palaeolithic period onwards, but they are more frequent from the beginning of our own era. For example, at least one third molar was embedded in 1.5 to 5 per cent of the individuals in our series of Mediaeval jaws (at the present time, the frequency varies from 7 to 40% of individuals, according to various authors).

Malocclusion

Malocclusions have likewise shown an increase from Neolithic times on, ranging from a very few per cent to sometimes more than 50 per cent in the last two centuries. New research in this field is necessary. Investigations often run into difficulties owing to the poor preservation of the facial bones in some skeletal series.

DENTAL DISEASES

Dental Caries

Caries has, whatever one may say, an extremely long history in man. If it is true that we find erroneous interpretations given in works tracing the evolution of dental caries through prehistoric and later periods, it is because many observers have confused "true" with "false" caries, the latter being damage produced postmortem. These false caries have recently been well studied by several authors and in particular by Werelds (1961, 1962). On this point we share the opinion of several authors and in particular Krogman (1938), who writes: "Caries certainly antedates civilization." The disease was undoubtedly less frequently encountered then than at present.

During the Palaeolithic, it was already in existence, despite statements to the contrary by various writers. It is however still impossible to form any exact idea of how common it was at that period. In the Mesolithic period, there are grounds for estimating its incidence at between 2 and 10 per cent of all teeth. For the Neolithic and metal ages, the incidence is at about the same level, contrary to the beliefs of various authors. In the Gallo-Roman period and in the Middle Ages (Fig. 4a), the level increases slightly to about 5 to 14 per cent. In our day, it occurs in about 50 to 90 per cent of all teeth (of course, with some variations between different populations, and according to standards of dental care, etc.)

The average number of teeth affected by caries in the adult individual shows a progressive increase from the beginning of our era. In prehistoric times, the mean was generally from 1 to 3 cases of caries in each individual. Exceptions can however occasionally be found. *Polycaries* is therefore not a wholly recent pathological phenomenon, despite the contrary view one often hears expressed. Cases can be found as far back in hominid antiquity as we are able to go, but it was then incontestably less common than it is nowadays.

I referred earlier to the average number of caries *per adult* individual because dental caries has in fact been for a long time a mainly adult disease, appearing generally between twenty to thirty years

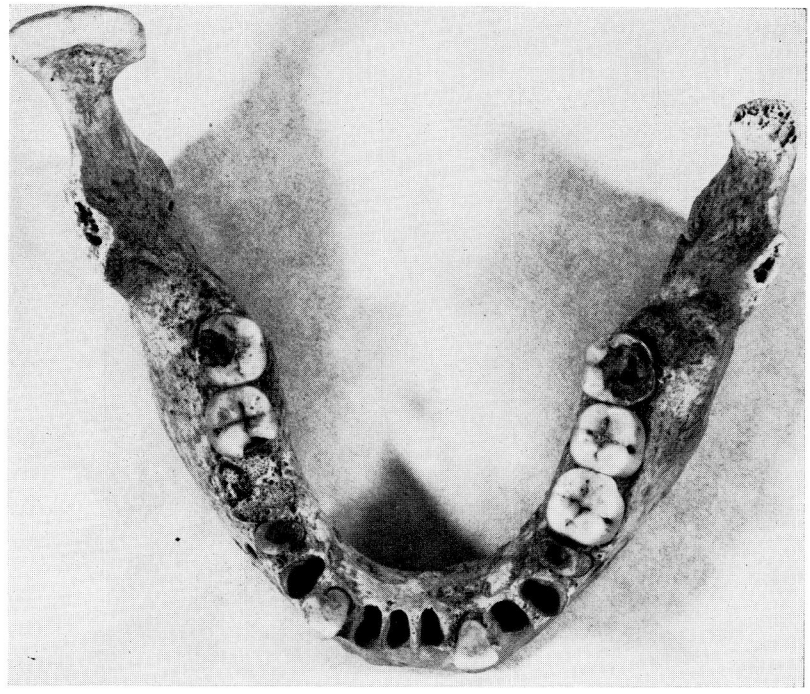

FIGURE 4a. Polycaries. Mediaeval mandible from Nivelles (Belgium).

of age and often later. Since the average expectation of life was not high, before our own era and up to the Renaissance, we can easily understand how caries may have seemed very rare, on superficial examination of the evidence. But, even at the beginning of our era, and particularly in the Mediaeval period, we find a lowering of the age of the first onset of caries; and it is known that at the present day the first molar begins to be affected with caries in many children one or two years after its eruption, that is to say between seven and ten years.

In the Neolithic and metal ages, on the other hand, between 10 and 35 per cent of adults seem to have had dental caries. At the Gallo-Roman period, according to our own findings and the study of the literature, 30 to 40 per cent of adults were affected. At the end of the Middle Ages, the proportion increases to 50 per cent or sometimes higher. At the present time the percentage reaches 90 to 100 per cent of adult whites, though there are some exceptions. Differences are naturally found in the various populations studied, for which reason we give average figures only.

There is however no certain evidence, despite some statements to the contrary, that in other ages caries attacked different parts of the teeth, or preferred other sites in the teeth, from those affected nowadays. Nor is there anything to indicate that caries developed less quickly within the tooth than it does nowadays, although this at first sight attractive idea has sometimes been propounded. It is a possible hypothesis but has not been proved. Nothing in any of my own investigating has lent any support to it.

Was there formerly any relation between the shape of the cranium, the constitutional type, and caries, as some authors have maintained? I do not think so, and various recent authors share this opinion (Pierleoni, 1961). But there is no doubt that some races exhibit a greater tendency than others to be affected by caries (Proell, 1934).

Finally, my own observations and those of other writers give us grounds for maintaining that from Mesolithic times onward, the teeth affected earliest and most markedly by caries have remained the same up to our own day. The teeth most often affected by caries are the molars, and of these particularly the first permanent molar. It is the lower first permanent molar that is most often affected; next, the upper first molar. Those least often affected are the canines and the lower incisors. It is clear therefore that there has existed for a very long time, if not always, a characteristic degree of susceptibility to caries corresponding to each type of tooth. Modern theories of dental caries do not yet afford any satisfactory explanation of this phenomenon. The first permanent molar has exhibited through many millenia a sort of "pathological supremacy" in this respect, which, if not absolute, is at least the general rule, particularly among white-skinned peoples.

To sum up: as we look at the history of caries, it appears that its evolution through the ages has taken five main directions:

—extension of the caries to an ever-increasing number of individuals;

—extension of its range, i.e., affecting a larger number of teeth in each individual;

—progressive extension of the caries to the anterior teeth (in early times almost untouched by it);

—gradual lowering of the age at which the permanent teeth become susceptible to it;

—Encroachment finally upon the deciduous teeth, which were generally free from it in primitive times.

These observations underline the necessity of taking into account the age of the subjects examined when one is establishing statistics for the incidence of caries in ancient populations. This many authors have unfortunately omitted to do.

It seems that the deciduous teeth may be in the process of following, though somewhat tardily, the same evolution as has already been carried through in the permanent dentition. If however one compares these two evolutionary patterns, one observes that, with regard to abnormalities and to caries, the temporary dentition has until very recent times shown much more evolutionary stability, and a greater resistance to caries, than have the permanent teeth (in spite of the contrary opinion sometimes stated).

Although the influence of present-day feeding habits (especially in the excessive consumption of carbohydrates) on the progressive deterioration of the teeth does not appear to be negligible, it is certain that we cannot draw any strict parallel between this deterioration which has been in progress since prehistoric times, and the evolution of human feeding habits. The influence of heredity on predisposition to caries is estimated variously by different authors, who ascribe to it from 10 to 50 per cent of the responsibility as a cariogenic factor.

Nevertheless, if the hypothesis that we have proposed in an earlier paper, that there exists not one carious dental disease, but two or perhaps more (Klees, 1961; Klees and Brabant, 1962), be accepted,

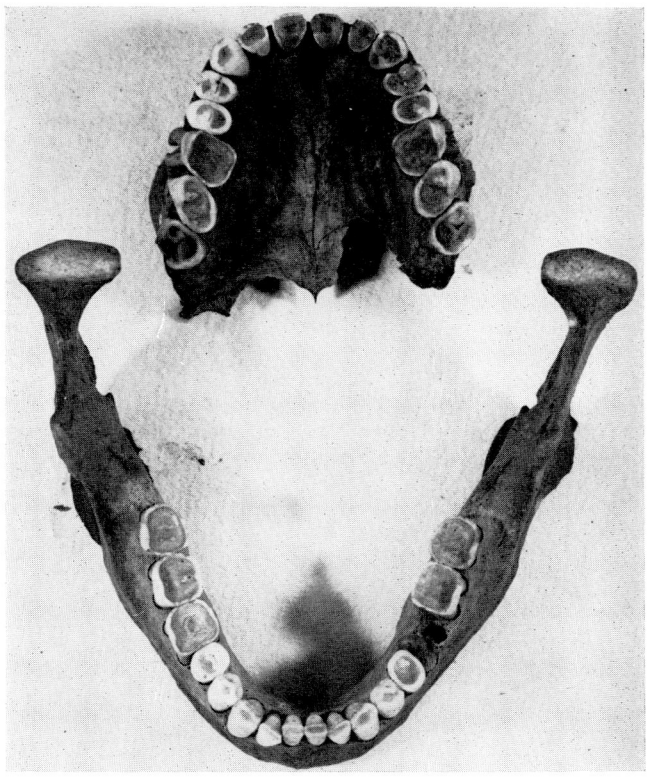

FIGURE 4b. Marked dental attrition. Mediaeval Coxyde.

two things become easier to understand. On the one hand, there is the slow and progressive extension, through the workings of heredity, of one of the most ancient forms of the disease, corresponding with the gradual involution of the maxillodental complex; on the other, there is the sudden blossoming-forth, two or three centuries ago, of that type of carious disease which is provoked by the abuse of carbohydrates.

Wear

In prehistoric times, wear was much more marked than that observed nowadays; but there were then different degrees of wear corresponding to the different categories of tooth, to individuals, and to populations. On average it was after the age of thirty that this wear became very appreciable in the adult. Although the pulp would react to this wear by an accelerated production of "secondary" dentine, this formation was not always sufficient to prevent the denuding of the pulp and its inevitable consequence: necrosis of the pulp and alveolar osteitis. This latter lesion is found under carious teeth fairly frequently from Neolithic times on.

In populations of the Gallo-Roman or Mediaeval periods (Fig. 4b), dental wear remains in general more pronounced than in populations of our own day, but, considered as a whole, a little less frequent than in prehistoric times. But naturally, here too, there are discernible differences in populations and individuals. The type of wear most frequently found in molars of European populations has been, and still is, in our own day, that called "helicoidal." "Ad palatum" wear is not so very rare, despite some statements that it is.

The cause of this heavy wear in ancient teeth might well lie in slow mastication, in the type of labidodontal occlusion attained by the adult, and, in some cases, in a diet offering more resistance to chewing. This wear is not, in the opinion of the present writer, due as has sometimes been maintained, to the eating of earth or stone dust in food, or to repeated "stress," to which the individual reacts by grinding his teeth continually, thereby wearing them down excessively. One might recall, for the sake of comparison, that at present dental wear is, apart from fairly rare exceptions, relatively slight. This is not, however, true in individuals who have lost a large number of teeth (which gives those remaining more work to do), or in those individuals who possess very powerful jaws and strong teeth.

Parodontoses

This disorder is dealt with in another chapter (see Alexandersen, p. 558), and I therefore do not propose to discuss it here.

REFERENCES

ALEXANDERSEN, V., 1963: Double-rooted human lower canine teeth. In D. R. BROTHWELL (ed.) *Dental Anthropology*. Pergamon Press, London.

BRABANT, H., KLEES, L. and WERELDS, R., 1958: *Anomalies, mutilations et tumeurs des dents humaines*. 1 vol. Ed. J. Prélat.

BRABANT, H. and KOVACS, I., 1961: Contribution à l'étude de la persistance du taurodontisme dans les races modernes et de sa parenté possible avec la racine pyramidale des molaires. *Bull. Group. Int. Rech. Sci. Stomat.*, 4:232.

BRABANT, H. and TWIESSELMANN, F., 1964: Observations sur l'évolution de la denture permanente humaine en Europe occidentale. *Bull. Group. Int. Rech. Sci. Stomat.*, 7:11.

BRABANT, H., 1965: Observation sur l'évolution de la denture temporaire humaine en Europe. *Bull. Group. Int. Rech. Sci. Stomat.*, 8:235.

BROTHWELL, D. R. (Ed.), 1963: *Dental Anthropology*, Pergamon Press, London.

DE JONGE, T. E. 1963: Le tubercule de Carabelli dans les molaires de la mâchoire inférieure. *Bull. Group. Int. Rech. Sci. Stomat.*, 6:147.

HARDWICK, J. L., 1960: The incidence and distribution of caries throughout the ages in relation to the Englishman's diet. *Brit. Dent. J.*, 108:9.

KLEES, L., 1961: Quelques conclusions tirées de l'étude histologique des rapports existant entre les formations organiques de l'émail et le dévelopement des caries dentaires. *Bull. Soc. Sc. Méd. Grand-Duché de Luxembourg*, 98:477.

KLEES, L. and BRABANT, H., 1962: Contribution radiographique et histologique à l'étude des caries de l'émail des faces proximales des dents. *Bull. Group. Int. Rech. Sci. Stomat.* 5:119.

KOVACS, I., 1964: Personal communication.

KROGMAN, W. M., 1938: Quoted according to BRABANT and TWIESSELMANN (1964a).

MARTIN, R. and SALLER, K. 1959. *Lehrbuch der Anthropologie*, G. Fischer, Stuttgart.

MOESCHLER, P., 1964: Personal communication.

PIERLEONI, P., 1961: La carie dentale nei tipi costitozionali studiata con metodologica statistica. *Riv. Ital. Stomat.*, 16:1420.

PROELL, F. W., 1934: *Klima und Zivilisation in ehrer Auswirkung auf Körper Zähne*. Kommission der Berlinischen Verlaganstalt, Berlin.

RUFFER, A., 1920: Study of abnormalities and pathology of ancient Egyptian teeth. *Amer. J. Phys. Anthrop.*, 3:335.

TWIESSELMANN, F., BRABANT, H. and KOVACS, I. 1962: Sur une anomalie dentaire rare et méconnue de l'Homme néanderthalien de Spy. *Bull. Int. Rech. Sci. Stomat.*, 3:452.

WERELDS, R., 1961: Observations macroscopiques et microscopiques sur certaines altérations *post mortem* des dents. *Bull. Group. Int. Rech. Sci. Stomat.*, 4:7.

WERELDS, R., 1962: Nouvelles observations sur les dégradations *post mortem* du cément et de la dentine des dents inhumées. *Bull. Group. Int. Rech. Sci. Stomat.*, 5:554.

Chapter 46

The Pathology of the Jaws and the Temporomandibular Joint

V. ALEXANDERSEN

In the preceding chapter, Professor Brabant has discussed the subject of disease and its variability in the dental tissues of earlier populations. Evidence for early diseases and anomalies in the jaws extends, however, far beyond that of the teeth themselves, and in the following study, it is my intention to review as broadly as possible all such extra data relating to oral health in the past. Traumatic injury to the jaws will be considered separately, in a later section of this book.

DISTURBANCES IN NORMAL GROWTH AND DEVELOPMENT

Congenital Malformations

CLEFT palate and clefts in the alveolar process of the maxilla are the congenital malformations most often met with in the jaws. The alveolar deformities appear in skulls as defects in the anterior part of the maxillary alveolar process. Cleft palate involves a larger or smaller part of the bony palate. Sometimes only the posterior part of the bony palate shows a defect which is situated in a position corresponding to the median palatine suture. Clefts in the palate may be associated with defects in the alveolar process. The soft tissues which are more or less involved in the defects will not be considered here. It should be noted, however, that in some cases minor defects in the bony palate are covered by mucosa and thus cause the individual no trouble at all.

In a number of recent European and American populations the incidence of cleft lip and cleft palate is 1 to 2 per cent. In skeletal material only a part of these defects can be observed as the soft tissues no longer exist. Greene (1963) mentioned that studies reporting on differences among ethnic groups agree that Negroes less frequently have cleft lip and cleft palate deformities than Whites. Ivy (1962) reached the tentative conclusion that these deformities occur about one fifth as frequently in Negro babies as in white children born in Pennsylvania. On the other hand mixed races (according to the review by Greene ((1963) on the epidemiology of cleft lip and cleft palate) have a somewhat higher rate than Whites.

Mortality is high among infants born with cleft deformities. Visible defects in the face have possibly resulted in the children not being cared for sufficiently if they were born into a so-called primitive or preliterate community (Thompson Brooks and Hohenthal, 1963). Consequently, when examining excavated series of skeletons, we might expect to find the more serious types of congenital malformations among infants. In spite of the difficulties in food intake and the increased disposition to infection, some infants with cleft palate manage to reach adult age without any medical and surgical treatment. Where prehistoric and early historic periods are concerned the number of well preserved skulls of infants with fairly intact palates and alveolar processes is limited and to my knowledge cleft deformities have not been reported from such material. Judging from skulls of older individuals cleft palate occurred in the Old as well as in the New World.

Elliot Smith and Wood Jones (1910) mentioned an Egyptian skull found by Derry which showed marked cleft palate. This was possibly the same case as that described in detail by Derry himself (1938). Brothwell (1961) described the skull of a child dating from the Saxon period that showed cleft palate. Besides this case from Great Britain there are a number of cases on record from pre-Columbian South and North America (MacCurdy, 1923; Thoma, quoted from Moodie, 1923; Thompson Brooks and Hohenthal, 1963). Two cases of cleft palate and one case of alveolar cleft have been found among some 700 ancient as well as more recent Eskimo skulls preserved in Danish collections (Pedersen, 1941; Balslev Jørgensen, 1953).

Most of the skulls with congenital cleft malformations have been thoroughly described and illustrated by the authors mentioned. The dorsal part of the palate showed a median cleft in the Saxon child, in the two Eskimo skulls and likewise in the skull found among 422 Peruvian skulls examined by MacCurdy. In cases with isolated cleft palate, the palate is sometimes remarkedly broad in proportion to its length. This is also seen in modern patients and Dahl (1960) mentions that it is not known whether the broad, short palate represents a developmental tendency or is a result of the operation to close the cleft in the palate. The skulls are thus of interest in this connection. Aplasia of one or more teeth may occur as seen in the Eskimo skull in Fig. 1. Alveolar cleft visible as a notch in the region of central incisors associated with aplasia of these teeth were described in an Eskimo skull by Balslev Jørgensen (1953) and in a California Indian by Thompson Brooks and Hohenthal (1963). More extensive were the malformations in a middle-aged woman from the XXVth Dynasty mentioned by Derry (1938). The pre-maxillary part of the maxilla and the horizontal plates of the palatine bones were not present in the skull. This caused a marked reduction of the size of the hard palate. Several teeth were not developed. This applies to the incisors, the left canine and the second and third molars. The mandible was normal.

The skulls of California Indians described by Thompson Brooks and Hohenthal (1963) occupy an exceptional position, being dissimilar (in regard to the defects) from the skulls already mentioned and recent skulls with congenital defects depicted by Veau (1931), Derry (1938) and Bøhn (1964). The Indian skulls show large defects in the palates and the alveolar processes. The authors draw attention to multiple exostoses and the osteoporitic appearance of some bones in the skulls which, according to them, are also signs of a developmental anomaly. The defective palates in two skulls, and the alveolar defect in one—associated unilaterally with lack of incisors, canine and both premolars—cannot with confidence be

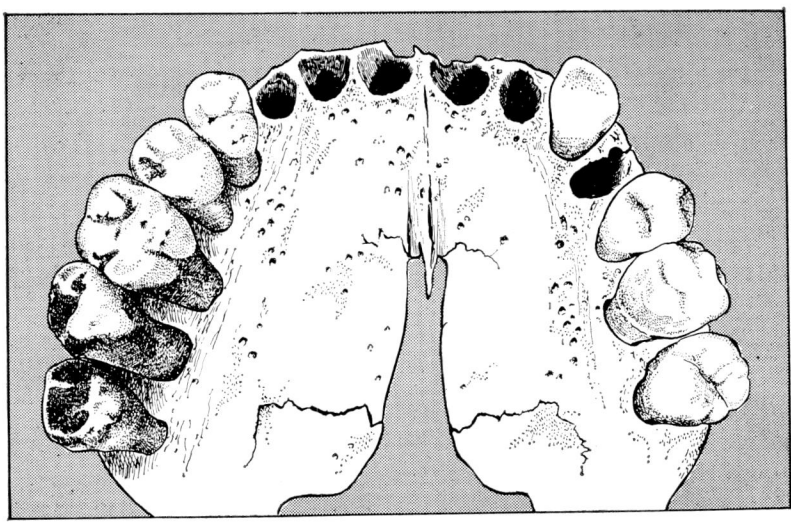

FIGURE 1 Cleft palate in an adult female Eskimo (Collection of the Laboratory of Physical Anthropology, Copenhagen. No. 506/B 1923). *Foramen incisivum* is not present. In the asymmetric dental arch there is aplasia of the left first molar. Calculus has been formed on the occlusal surfaces of molars in the right side of the jaw. By courtesy of Professor P. O. Pedersen.

considered congenital malformations. Further investigations on the original material may reveal pathological conditions of a different nature.

Acquired Abnormalities and Diseases of Bone

Disturbances in normal growth and development of the jaws are manifested in various ways. One or both jaws may be diminished or enlarged in size, the form and structure of the bones are sometimes changed and malocclusions are often found. A multitude of causes may be responsible for the clinical variations observed. In the case of skulls it is impossible in many cases to give any more than a very tentative explanation of the aetiology and the pathogenesis.

Mandibular Micrognathia

Mandibular micrognathia is often a result of delay or arrest of growth in the mandibular condylar region brought about by infections or injuries to the temporomandibular joint in childhood (Greer Walker, 1956). Evidence of infections located at the temporo-mandibular joint in skulls from prehistoric and early historic periods is scant. Evidence of infectious diseases near the temporomandibular joint is likewise exceedingly rare compared with the common infections of the periapical alveolar bone. It is also remarkable that evidence of suppurative infections in the middle ear and the mastoid process is reported only rarely in literature dealing with ancient skulls. Thus it is not surprising that an adult Nubian skull (Collection of the Laboratory of Physical Anthropology of Copenhagen University, No. 25-119) with profound changes in the dimensions of the mandible caused by arrest of growth of the left condylar process shows evidence more likely to denote injury to the jaw than infection in the

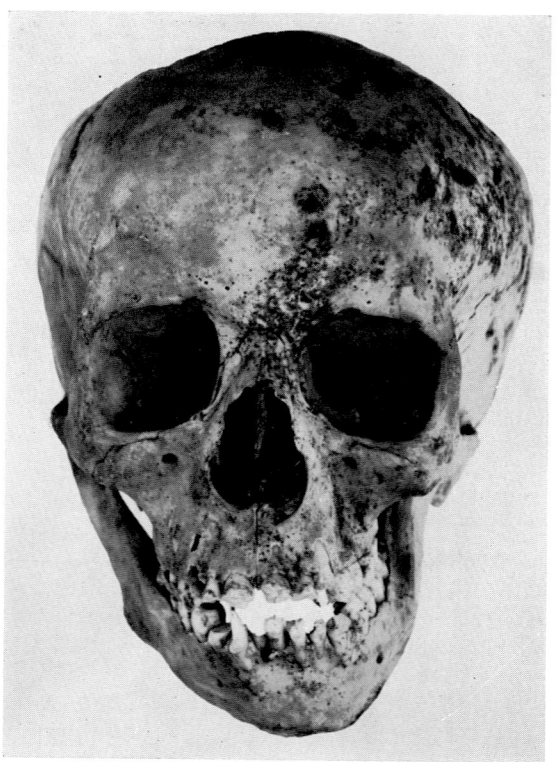

FIGURE 2. Nubian skull with asymmetry of the face brought about by arrest of growth in the left condylar process after injury. Fractures of the anterior teeth are post-mortem. (Collection of the Laboratory of Physical Anthropology, Copenhagen. No. 25-119, dating from *c.* 200-600 A.D.

temporo-mandibular region. The skull, which dates from *circa* 200-600 A.D. is shown in Figure 2 and 3. The asymmetry of the face developed gradually after an injury to the individual when about seven years old. The injury caused fracture of the left condylar process, damaged the anterior part of the mandible and resulted in fractures of the first molars, exposing the dentine but not the dental pulps. After the injury the left temporo-mandibular joint became disorganized and the result was ankylosis. The jaws have failed to develop fully on this side of the skull. Normal chewing—as can be seen from the lack of attrition—was made difficult.

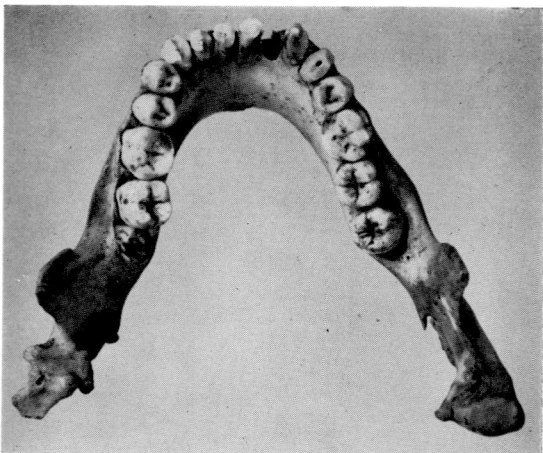

FIGURE 3 The mandible belonging to the skull shown in Figure 2. The change of form of the mandibular left condyle is interpreted as having been due to fracture of the condylar process when the individual was about seven years old. Fractures of the first lower molars also occurred on this occasion.

All teeth apart from the left third molar have erupted and attained good occlusion.

One Neolithic German skull, discussed by Mohaupt (1939) and Zuhrt (1962), also shows sequelae of an injury to the temporomandibular joint in childhood. To my knowledge no cases of micrognathia of the mandible, to such an extent that "bird-face" developed, have been recorded. Such cases of bilateral under-development of the mandible due to growth arrest have usually been initiated early in life.

Mandibular Macrognathia

Mandibular macrognathia becomes evident during youth and early adulthood. There should be every possibility of finding skulls with abnormally enlarged mandibles in view of their conspicuousness.

Acromegalic changes of the skull also involve the jaws, of which the mandible shows the most conspicuous alterations. Brothwell (1963b) depicted an Egyptian female skull with an extremely elongated face and enlarged mandible. This probably early Egyptian female is a more reasonable example of hyperpituitarism than the fragments of the skull and the lower jaw of an individual from Gardar, the Norse colony in Greenland during the late Middle Ages. These fragments were ascribed to an acromegalic person by Keith (1931). From the very thorough investigation of the remains of the jaw and the teeth by Pedersen (1944), it is concluded that it may be acromegaly, but definite evidence could not be provided either for or against the views advanced by Keith. In Figure 4 a skiagram of the lower jaw fragment is depicted showing the enlarged corpus and the elongated neck of the condyle. No spreading of the teeth was found, nor any atypical attrition on the occlusal surfaces. Hypercementosis was found on the molar roots.

Isager (1936) mentioned a skeleton excavated from the graveyard belonging to a late Mediaeval monastery, Øm Kloster, in Denmark, which showed evidence of acromegalic alterations. The mandible belonging to the skull was not found, but most of the other bones belonging to the skeleton were.

A disproportionate increase in the size of the mandible may lead to mesio-occlusion and anterior open bite. The Habsburg lower jaw is an example of this hereditary anomaly (Böhne, 1955). In some lines belonging to the House of Habsburg this anomaly has been manifested (sometimes associated with macrocheilia) from late in the fourteenth century.

Maxillary Micrognathia and Malocclusions

Maxillary micrognathia may occur in achondroplastic dwarfs. Keith (1913) discussed a female Egyptian skull showing this condition. The anterior part of the upper jaw is tilted upward. The upper face is low. The teeth are normal in number and shape. This skull was originally described by Seligman (1912) as a cretinous skull. Mongolian idiots have conspicuously small upper jaws associated with mandible of normal size. A possible case of mongolism in a Saxon population has been described by Brothwell (1960). Skulls may occasionally be found with disproportionate maxillae, without it being possible to suggest any cause for this (Zuhrt, 1956 and my observation of a Dutch skull from Tiel—Figures 8 and 9—dating from the beginning of the nineteenth century).

Among the objective symptoms of rickets are deformity of the jaws, malocclusions and hypoplastic defects visible on the tooth surfaces. Rihan (1930) made the diagnosis on a skull of a one and a half year old child from Pecos, the development of whose alveolar processes were

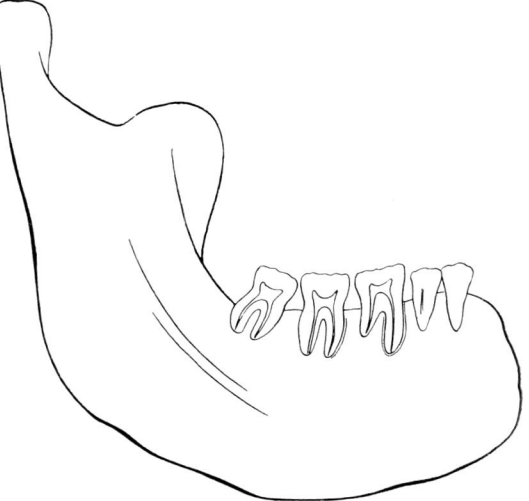

FIGURE 4 Skiagram of the mandible fragment belonging to a presumed acromegalic individual. The evidence is not conclusive, however, as outlined in the text. (Collection of the Laboratory of Physical Anthropology, Copenhagen, Gardar VII, dating from the late Middle Ages).

retarded with malalignment of the deciduous teeth. None of these symptoms is a pathognomonic sign, however, of a deficiency of vitamin D in the diet. Cases of rickets have been diagnosed by Nielsen (1911), from Danish Neolithic material in six instances, and from skeletal remains dating from the Danish Iron Age in three. No typical rachitic lesions in the skulls or in the long bones of skeletons dating from the Late Middle Ages were found in the extensive Danish collections from the monasteries of Øm and Æbelholt (Isager, 1936 and Møller-Christensen, 1958). In nine skulls from Æbelholt the teeth were marked by hypoplasia of the enamel, which might suggest a disturbance in the development of the teeth caused by D avitaminosis. Hypoplasia of the enamel, however, can be caused by many things, as any textbook on the pathology of the hard tissues of teeth shows.

According to Gejvall (1960) the effects of rickets were very rare in Swedish skeletal material from the Neolithic period to the Late Middle Ages. The results obtained in the Scandinavian countries have not been contradicted by the results obtained in other parts of the world where sunlight was scant during some parts of the year (Sigerist, 1951; Brothwell, 1963b). A natural explanation for the favourable condition is that infants were breast-fed for a longer period than nowadays, mother's milk being adequate with regard to its content of vitamin D, calcium and phosphorus (Poulsen, 1909; Møller-Christensen, 1958).

Malformation of the shape of the skull was intentionally brought about in some primitive societies. The effect on the development of the jaws has, to my knowledge only been touched on very slightly. Leigh (1937) noted (in artificially deformed pre-Columbian Peruvian skulls) a certain influence of the fronto-occipital deformity on the development of the palate. This, however, seldom led to malocclusions even though the palate became widened and asymmetrical. A book by Hooton (1930) on the Pecos Indians includes an appendix by Rihan on their dentitions. In deformed skulls Rihan noted a forward position of the glenoid fossa on the side of the skull which was deformed in the occipital region. However, judging from several photographs in the book, malocclusions had not developed.

Facial hemiatrophy is occasionally found in skulls. Gejvall (1960) described a Mediaeval skull from Westerhus in Sweden with atrophy of bones in the right side of the body and the left side of the skull including a thin left zygomatic arch, a small left mastoid process and the left half of the *corpus mandibulae* reduced in size in proportion to the right side. It was suggested by the author that poliomyelitis or a brain injury may have caused the pathological changes, these being most pronounced in the long bones. Derry (1913) described a case of hydrocephaly from Egypt. The left side of the body showed evidence of hemiplegia presumably caused by the cerebral disease. However, the facial skeleton does not seem to be involved in the deformity as it shows no asymmetry.

Unilateral hyperplasia of the mandibular condyle is not considered here but in the section on tumours.

Several abnormal skull types have been reported among ancient skeletal material according to pertinent literature reviewed by Brothwell (1963b). The influence of abnormal development of the base of skull on the development of the jaws and their relationship to each other has not been considered.

In otherwise normally developed skulls a number of cases are found in which the dental arches are not harmoniously developed; this may lead to malocclusions. Displacement of individual teeth is known from Neandertal jaws and from the dentitions of some Upper Palaeolithic skeletal remains. In the lower jaw from La Naulette a second premolar is rotated. The left canine in the juvenile Le Moustier skull is impacted. Premolars are rotated in the skull designated Skhūl IV (McCown and Keith, 1939). Conversely, the dentition of a child from Grimaldi shows displacement of teeth post-mortem and this is difficult to distinguish from certain developmental disturbances (Legoux 1962). Several examples of crowding of teeth, especially of the lower incisors, can be found in Upper Paleolithic jaws.

Abnormal jaw relations in sagittal, vertical or lateral directions appear (though the incidence is low) in Neolithic European populations, of which we have a sufficient number of skulls on which to base a reasonable judgement. The number of dentitions with malocclusions gradually increase until an almost contemporary high rate of incidence of some malocclusions is reached in late Mediaeval European groups (Brabant, 1960-63; Andrik, 1962); Lundström and Lysell, 1953; Lysell, 1958; Zuhrt, 1956).

The causes for the increase in frequency (especially of the more severe malocclusions in successive populations leading to modern civilized nations) have been discussed from a theoretical point of view by Andrik (1962), Begg (1954), Hunt (1961) and Mills (1963), to all of whom the reader is referred.

Diseases of Bone

A number of diseases of bone causing dysplastic growth of the jaws are described in textbooks of oral pathology. These diseases of bone are not frequent in modern populations and most of them have not as yet been found in skeletal remains from ancient times. To recognize pathologically changed bones is not the main problem; the latter is more that of distinguishing the possibilities with the few diagnostic methods available. Publications that appeared during the first decades of this century discussed the macroscopic appearance in skulls showing some of the bone affections known in those day. Hultkrantz (1908), thus discussed *dysostosis cleidocranialis,* rickets, cretinism, achondroplasia and *osteogenesis imperfecta.* X-ray examination is of great value in diminishing the gap between diagnosis from the macerated skull and in the living body, whose diseases are most accurately characterized histologically and clinically.

Cases of osteo-dystrophically deformed jaws are exceedingly rare. Pales (1930) called attention to a remarkable case of *leontiasis ossea* which had deformed the skull completely. The skull, of fairly recent date, was described and illustrated in 1755, and according to Boule and Vallois (1952) it was considered in the eighteenth century to be that of a fossilized man. Gullberg and Burkitt (1924) discussed the appearance of what was presumably also a rather recent skull, this time from New Guinea. The mandible belonging to the skull showed an enormous enlargement of the *corpus mandibulae,* the cause of which could not be determined by the authors. In Figure 5 the mandible of a Nubian child is shown. The mandible is the only preserved part of the skeleton of this child, who was about six months old. Neither the macroscopic nor the X-ray examination have given conclusive evidence of the cause of the enlarged rami.

A great deal of interest is centered upon rare anomalies or diseases of bones. According to a classification of hereditary

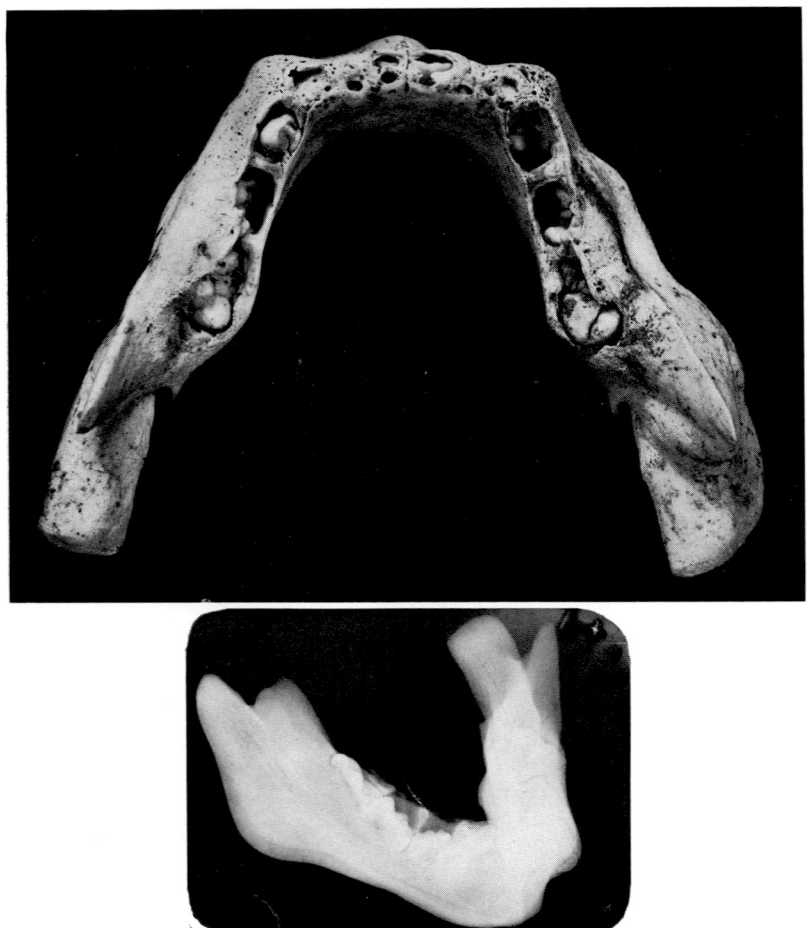

Figure 5 a, b. Mandible of a Nubian child about six months old. The enlargement of the mandibular rami gives a swollen appearance of the posterior part of the mandible. Minor morphological anomaly of the occlusal surface of M_1 dext. (b) X-ray of the jaw. (Collection of the Laboratory of Physical Anthropology, Copenhagen. No. 25-302, dating from about 200-600 A.D.).

oral diseases by Witkop (1958), some of them are more or less hereditarily determined. Demonstration of such disturbances in the development of the jaws in ethnic groups would not only tell us what diseases the people suffered from but also give some information to evaluate the state of endogamy in the group. On a basis of genetical consideration, Ferembach et al. (1962), interpreted congenital malformations of the sacrum and a high incidence of wormian bones as probable evidence of isolates having been formed in Epipalaeolithic North Africa. Demonstration of family relationships using such evidence is even more obvious.

PERIODONTAL DISEASE

The frequency of periodontal disease as far as this can be evaluated from skulls has been reported in a number of publications dealing with collections from pre-

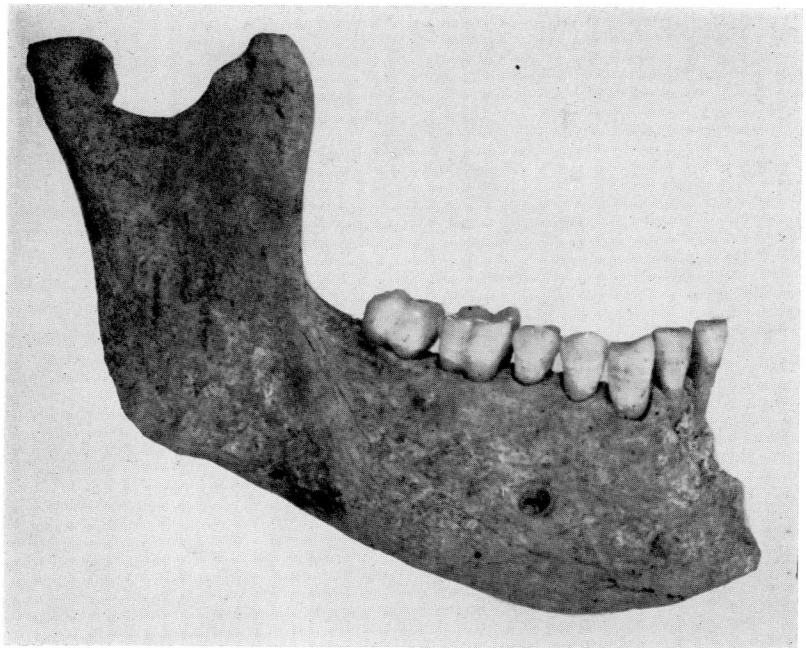

FIGURE 6 Mandibular fragment of a juvenile individual showing normal level and normal structure of the alveolar crest. Bronze Age, Syria *circa* 1900-1700 B.C. (Collection of the Laboratory of Physical Anthropology, Copenhagen. Tell Soukas, skull 14).

historic, early historic and mediaeval times in Europe. The same sort of changes in the alveolar bone were found in all the populations studied. However, the prevalence of different osseous changes varied to some extent. The presence of calculus and pronounced attrition was common in the dentitions surveyed. Disagreement prevails among authors with regard to the interpretation of the osseous changes observed, that is, whether due to physiological or pathological causes. It seems, therefore, reasonable to consider first of all some of the fundamental problems in the study of periodontal disease, before presenting the results obtained by different authors working on skulls.

A classification of periodontal diseases in living persons can be grouped according to the fundamental types of pathological processes, such as the inflammatory, degenerative and neoplastic ones, which differ in characteristics, origin and course (WHO report 1963). In skulls, only a part of the periodontium—the osseous part of it—can be studied. Since the soft tissues and clinical symptoms are invariably missing, it would be inaccurate to describe the changes observed in the bony alveolar process in terms of clinical disease entities.

One feature usually considered indicative of an inflammatory reaction in the alveolar bone is when the surface of the alveolar crest appears porous at the interproximal septa and crenated along the facial and lingual *limbi alveolares*. Ruffer (1921), Zuhrt (1956), (Brabant (1960-63) and Henkel (1961) have described the macroscopic change from a smooth alveolar surface to a surface which looks like a pumice stone in the marginal part of the tooth sockets and on the outer

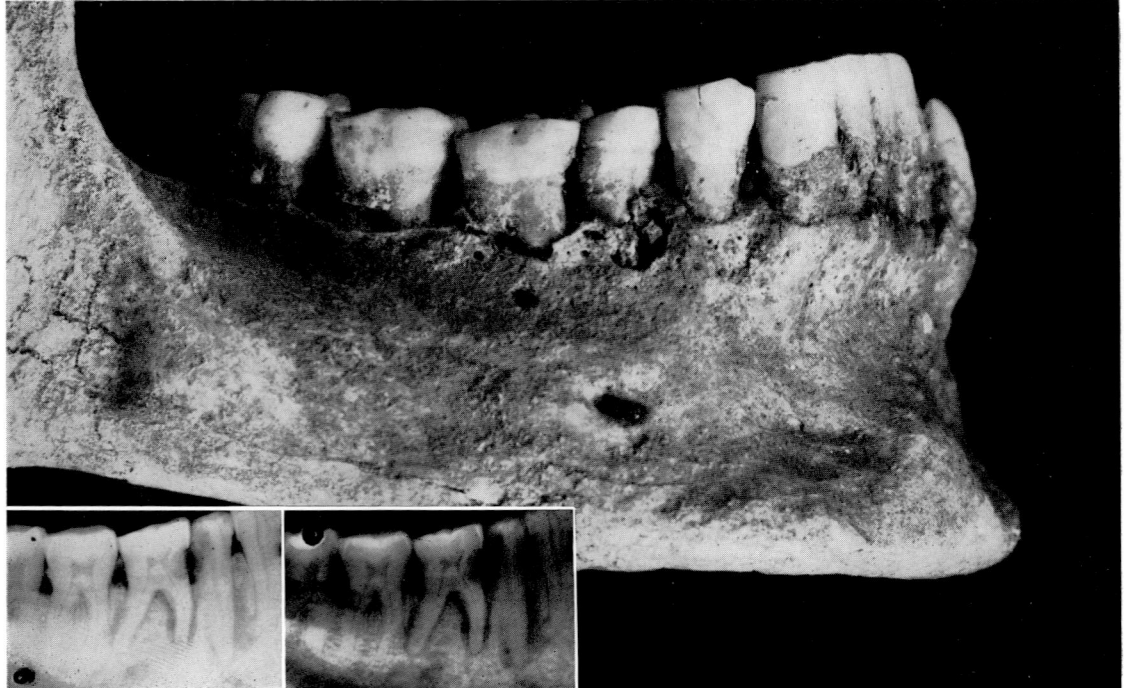

FIGURE 7. Mandibular fragment of an adult individual showing subgingival calculus on the roots of some of the teeth, increased distance between the cemento-enamel junction and the alveolar crest compared to the fragment shown in Figure 6. The alveolar crest is porous in the regions of the second premolar and the first molar. X-ray photograph of premolars and molars showing caries in P_2 and M_1 and osteitis around the roots of these teeth. Bronze Age, Syria, circa 1900-1700 B.C. (Collection of the Laboratory of Physical Anthropology, Copenhagen. Tell Soukas, skull 24)

alveolar surface. X-ray photographs of the porous alveolar crest reveal the absence of the *lamina dura* at the interproximal septa (Figs. 6-9). In many cases it is likely that macroscopic alveolar porosity is a result of inflammatory periodontal disease in the dentition of the skull examined, especially if indirect evidence such as maturity of age, calculus and loss of alveolar bone is present. In living persons inflammatory changes in the alveolar bone are very often associated with advanced age and various local predisposing factors (Løe, 1963).

Degenerative changes can occur in the periodontal tissues. When considering the skull alone, it is still impossible to decide whether extensive destruction of alveolar bone is due to degeneration caused by inherent or acquired changes in the bone, the fibres or the cementum of the periodontium, or due to inflammatory changes responding to determining local factors in a periodontium which is weakened by a low resistance in the individual.

One might expect to find atrophic changes in the periodontal tissues as well as in other tissues in the case of old people. Histological investigations of the alveolar bone in skulls have not yet been concentrated on the diagnosis of atrophy. Hammarlund-Essler (1956) found it difficult to study spongy bone in Mediaeval

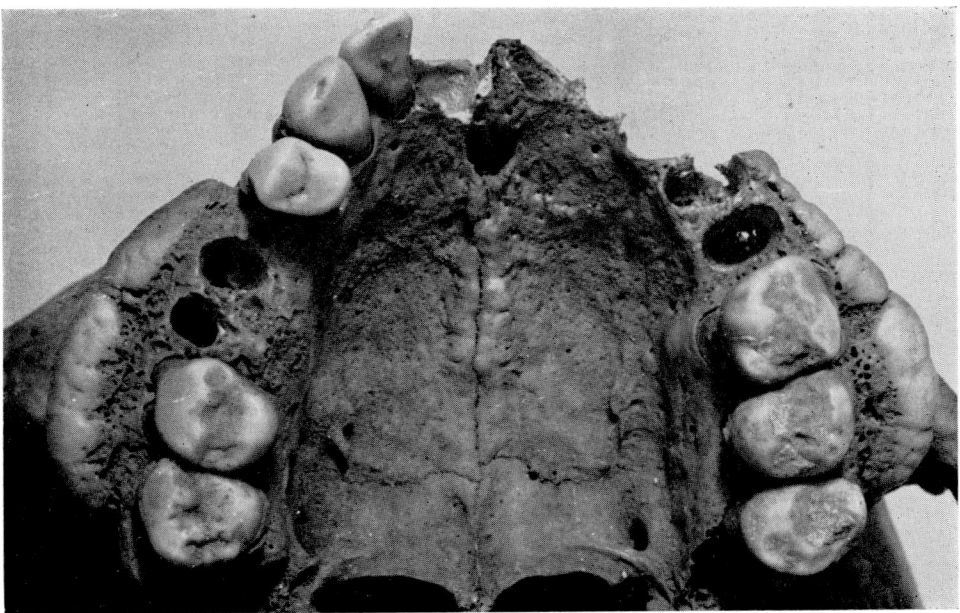

FIGURE 8. Occlusal view of an upper jaw with large *tori alveolares* and a small *torus palatinus*. Calculus is present on the occlusal surfaces of the molars in the left side of the jaw. The palatal root of the right M¹ had been completely exposed after movement in a lingual direction as a response to the atypical occlusion. Collection of the Anthropological Institute of the University of Utrecht. Early nineteenth century. Courtesy of Dr. J. Huizinga.

skulls. Jerusalem (1955) studied different forms of bone resorption, using compact bone from skeletal remains as a suitable medium. Further studies are urgently needed to distinguish between pathological forms of resorption and senile atrophy.

Neoplasms in the region of the jaws will not be discussed here, being referred to in the chapter on tumours. The most important part of the description of the bony periodontium in skeletons is still the registration and analysis of the marginal loss of alveolar bone. Epidemiological studies of the loss of alveolar bone in collections of skulls have been carried out by a number of investigators (Mellquist and Sandberg, 1938; Holmer and Maunsbach, 1956; Brabant, 1960-63, partly in collaboration with colleagues; Leigh, 1925-1937; and several others).

Mellquist and Sandberg (1938) as well as Brabant (1960) and later papers, recorded loss of alveolar bone of the horizontal type which exposes large or minor dentition. One value was given to each dentition according to the system of grading shown in Table I. It is regretted that other authors reporting on alveolar bone loss have not stated their method of recording loss of alveolar bone in detail. At present it is impossible to evaluate from the results alone whether the somewhat subjective way of estimating the amount of bone loss in a dentition shows a systematic tendency to underrate or overrate the real loss of alveolar bone. Thus the Swedish authors, Mellquist and Sandberg gave each dentition a value according to the tooth which had the root or roots most exposed, while Brabant estimated a mean

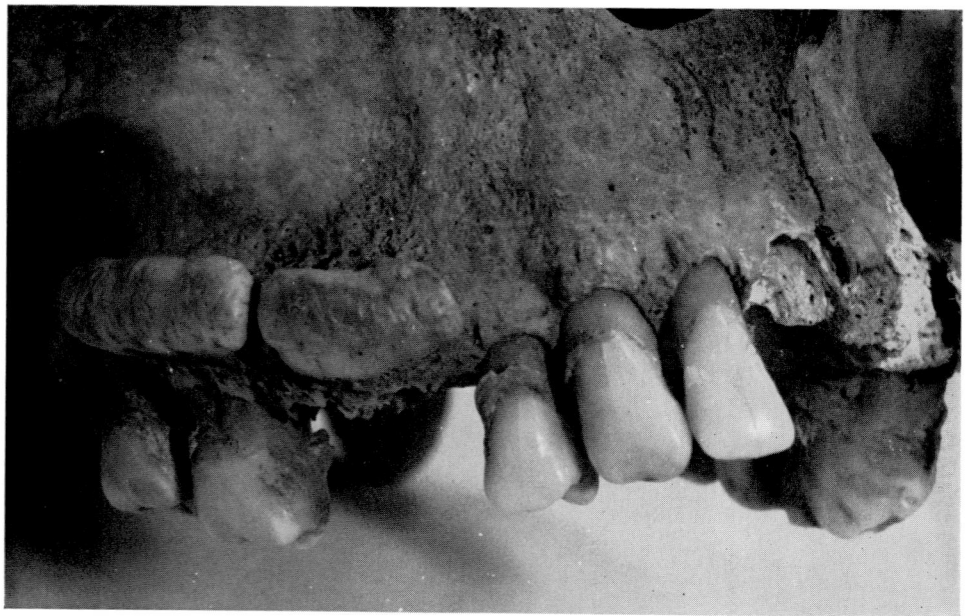

FIGURE 9. The right side of the upper jaw belonging to the Dutch skull shown in Figure 8. The exostoses are readily seen as well as the porosity of the alveolar crest. Supragingival calculus is present. The porosity above the exostoses might partly be due to the presence of the apices of the roots in this region. Atypical attrition of the facial surfaces of the molars and the premolars.

TABLE I
RECORDING OF ALVEOLAR BONE LOSS OF HORIZONTAL TYPE

Mellquist and Sandberg	Brabant
0 no atrophy	no atrophy
1 atrophy for about 1-3 mm	atrophy of less than a third of the length of the root
2 atrophy from 3 mm to half the length of the root	atrophy exposing half the length or more of the root
3 atrophy of more than half the length of the root	atrophy exposing almost the whole length of the root

from several teeth showing perhaps different amounts of alveolar bone loss. To Trachtenberg (1960), more than 4 mm of the root length exposed indicated atrophy of alveolar bone, while Leigh (1925-37) registered pathological alveoloclasia only when the roots were exposed appreciably more than could be explained by a slow gradual recession of the alveolar crest with age and physiological eruption of the teeth, for both of which he laid down norms. Resorption of septal and intra-alveolar bone was, according to Leigh, pathological.

In view of the lack of any uniform standards of recording alveolar loss of bone, the results obtained so far by epidemiological surveys by different authors are therefore not directly comparable. In order to decide whether loss of alveolar bone has occurred, what has to be studied is the normal level of the alveolar crest in rela-

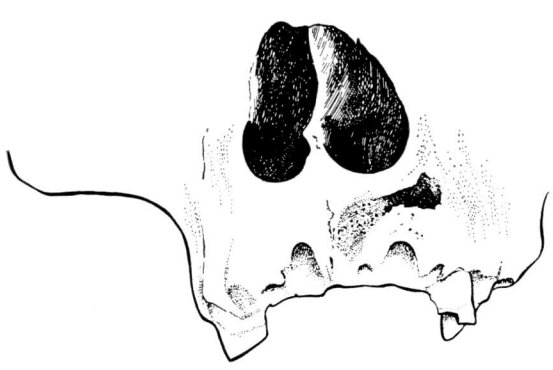

FIGURE 10. Groove on the facial surface of the maxillary alveolar process caused by a fistula originating from periapical osteitis round the apices of the canine and the first premolar. (Collection of the Laboratory of Physical Anthropology, Copenhagen. Nubia, site 25, dating from about 200-600 A.D.).

tion to the cemento-enamel junction. In populations from the Middle Ages or earlier historic and prehistoric times, attrition of the occlusal surfaces of the teeth is pronounced. This hinders the establishment of reference points on the cusps to which the alveolar bone height can be compared.

In children with erupted permanent teeth, and in juvenile skulls, the vertical distance between the cemento-enamel junction and the alveolar crest varies between 1 and 2 mm. This distance may be larger in the anterior part of the dentition around the incisors and the canines, or elsewhere in the dentition if the roots of the teeth are protruding (cf. Strahan, 1963 and Coppens, 1963). In adult and mature skulls the distance between the cemento-enamel junction and the alveolar bone tends to be larger. This is very well demonstrated by Holmer and Maunsbach (1956), who studied Neolithic Swedish skulls. They measured the distances (facially and lingually to the teeth) between the cemento-enamel junction and the most coronal point of the alveolar crest in about one hundred dentitions. The relationship between the same structures in the proximal septa was measured in material consisting of some 1500 X-ray photographs. In the upper jaws of adult and mature skulls the distance lingually to the teeth was uniformly about 3 mm, while the facial distances varied from about 2 mm in the anterior part of the dentition to a maximum of 4 mm at the second upper molars. Measurements of the septal heights from radiographs were shortest in the premolar regions (1.5 mm), longer in the medial incisor region and longest in the molar regions (approx. 2.6 mm). In the lower jaws the distances between the cemento-enamel junction and the limbus were uniform in the anterior and the premolar regions, namely about 3 mm lingually and about 4 mm facially. In the molar regions the distances were not larger but the second and third molars had distances facially and lingually of the same size or longer at the lingual surfaces compared to the other teeth in the lower dental arch. From measurements of the distances in the septal regions (using radiographs) the variation in the lower jaw was as little as 0.5 mm; the range from 2.0 mm to 2.5 mm.

It is important to note that the Swedish skulls examined—as is usually the case in skulls of prehistoric or early historic collections—showed a great deal of attrition. A certain positive correlation was found by Holmer and Maunsbach between the amount of attrition and the distances measured between cemento-enamel junctions and alveolar crests. Mellquist and Sandberg (1938), who studied Swedish skulls from the later part of the Middle Ages, also reached the same conclusion, namely that attrition and what they considered horizontal loss of marginal bone were positively correlated.

Philippas (1952) studied skulls from early historic and prehistoric times in Greece and showed that the distance between the occlusal surface and the alveolar crest of the lower first molar remained relatively unchanged throughout life. This distance, corresponding to the clinical height of the crown, was usually 7.5 mm. Picton (1957) confirmed this in rather limited material and added that in teeth other than the first lower molars the clinical crown also remained constant throughout life. In the event of pronounced attrition, elongation of the worn teeth is possible, thereby increasing the distance between the cemento-enamel junction and the alveolar crest. However, advanced or extreme attrition will not be completely compensated by elongation, as was believed by Parma (1948). Using different techniques on living persons as well as on skulls, Henkel (1961), Tallgren (1957), Murphy (1959) and others showed that advanced attrition may be accompanied by loss of facial height as well as loss of tooth height. This means that the clinical crown height sometimes decreases, or that if it remains constant in length over a considerable span of years—as supposed by Philippas and Picton—the alveolar crest is gradually lowered at the same time as the size of the tooth crown is reduced by attrition. In some skulls with evidence of advanced attrition (or perhaps antemortem loss of teeth) elongation of a few teeth can be demonstrated from the abrupt shift of the cemento-enamel junction in an occlusal direction and from local changes in the dentition. Taylor (1963) has recently shown this in collections of Moriori and Maori skulls. It is more difficult to determine whether all teeth in a dentition are elongated to some extent. From these considerations it can be deduced that recording of horizontal loss of alveolar bone according to amount of root surfaces exposed means recording of the amount of elongation plus the amount of loss of bone.

The apparently slow horizontal loss of alveolar bone which undoubtedly takes place is often considered as atrophy. This atrophy is not accompanied by conspicious changes in the structure of the bone regarded macroscopically, and Holmer and Maunsbach consider it a physiological involution of the alveolar bone. This is likewise the opinion of Christophersen (1939-1941), who examined Danish skulls dating from the Neolithic period to that of the Vikings. It was considered as being senile atrophy by Rygge (1913). Held (1938) considered this loss of bone as being genetically conditioned. However, further studies are necessary in order to clarify the pathogenesis of this resorption of the alveolar process.

The existence of pathological resorptive processes localized in the dentition of a skull can easily be compared with similar conditions in living persons. To recognize such pathological processes in skulls, postmortem changes should be considered and excluded. X-rays are of value in relating the conditions seen in skulls with conditions known from living persons, but radiographs definitely have disadvantages that render them, in some respects, less reliable than direct inspection of the dentition belonging to a skull (Teilade, 1960). Epidemiological surveys of periodontal disease in recent populations have demonstrated the importance of age and oral hygiene in the aetiology of periodontal disease. Differences observed among races and populations of different socio-economic positions reflect in a manner not yet fully understood the nutritional differences, variations in food consistency, general health, etc. (Løe, 1963). In order to interpret accurately the results obtained

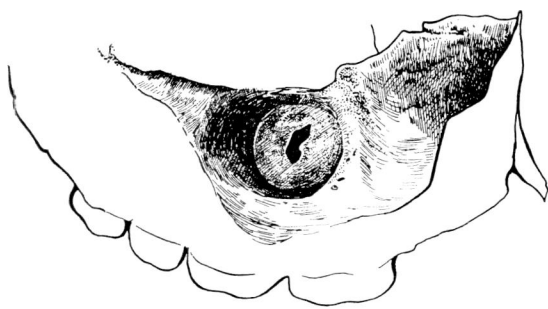

FIGURE 11. Radicular cyst extending into the maxillary sinus. The cyst originates from the apex of the palatal root of the carious first molar. (Collection of the Laboratory of Physical Antropology, Copenhagen. Århus, dating from the seventeeth century).

from analysis of the alveolar bone in dentitions of skulls the general conditions of the whole dentition and the whole individual should be taken into consideration. Sometimes tentative clinical diagnoses can be applied with reasonable probability to dentitions of skulls if this method is used. An accumulation of details in support of the diagnosis is brought forward from the dentition: for example, the presence of calculus, pattern of attrition, loss of teeth *ante mortem*, and tooth alignment. Further necessary information stems from an estimation of the age and sex of the individual and from a knowledge of the life of

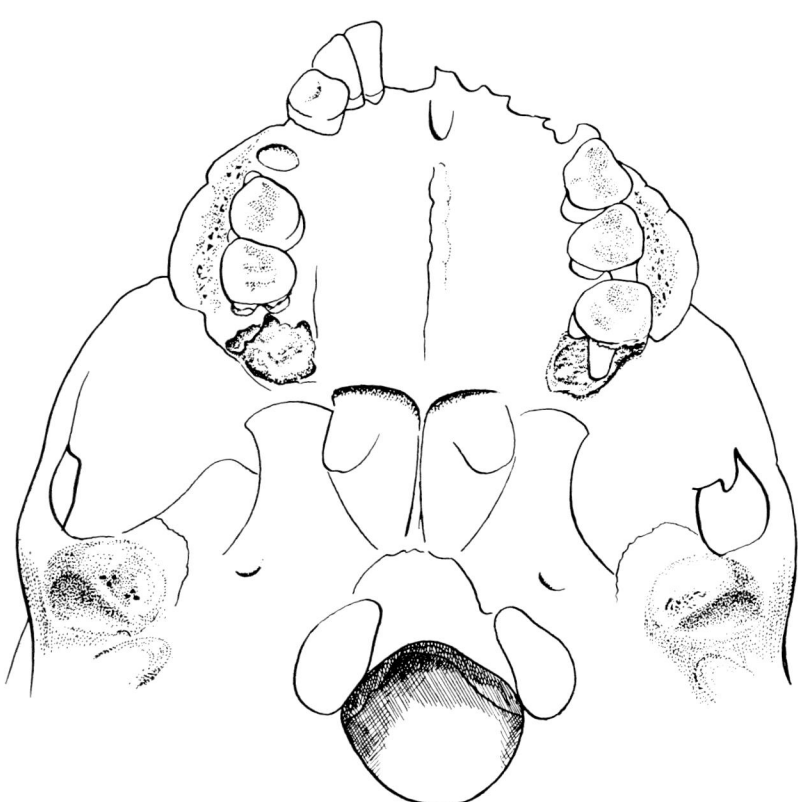

FIGURE 12. Bony cavities in the maxillary alveolar process behind the third molars. The origin of the cavities is discussed in the text. Evidence of pathological changes of the articular eminence is present in form of porosity. (Collection of the Anthropobiological Institute, Utrecht. Tiel (884 C), dating from the beginning of the nineteenth century).

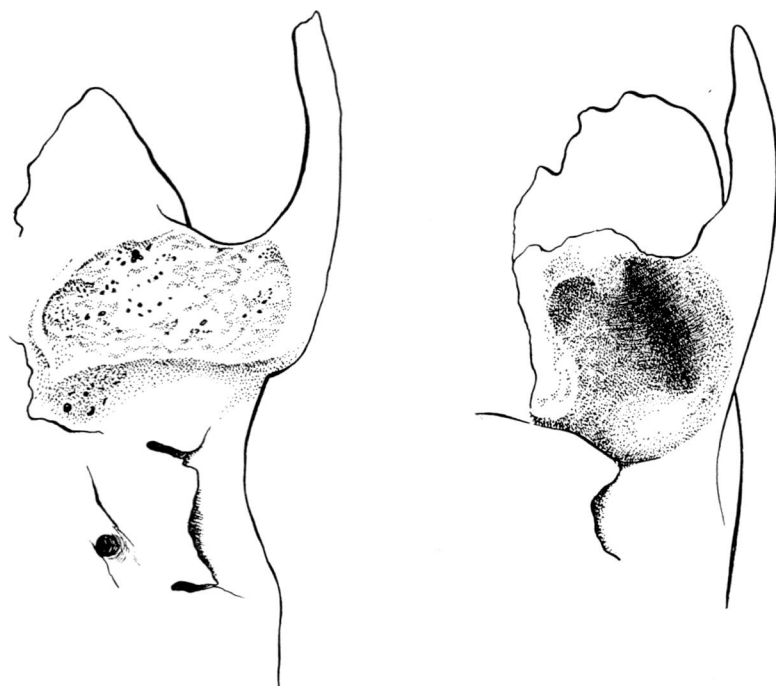

FIGURE 13. Alterations at the articular surface of the glenoid fossa and the articular eminence caused osteo-arthritis and a remodelling of the articular surfaces. Nubia 97-27 (about 2300-2150 B.C.) and Kællingbjerg Revle, (Danish Early Iron Age.) In Figure 14 is the corresponding mandibular condyle of the Danish skull. Collection of the Laboratory of Physical Anthropology, Copenhagen).

the community to which this individual belonged.

The Middle Ages

Surveys of the condition of the bony periodontium have been undertaken on a number of European populations dating from the Middle Ages. The results of some of these investigations will be discussed. Mellquist and Sandberg (1938) studied Scandinavian populations from about 1000 to 1700 A.D. A survey of a comprehensive Norwegian collection of late Mediaeval skulls was performed by Rygge (1913). Brabant and his collaborators (1960-1963) had at their disposal Belgian and French populations from the early and late Middle Ages comprising about 2000 individuals. Further information on the condition of the osseous periodontium in Mediaeval populations is found in German collections of skulls described by Zuhrt (1956) and Henkel (1961 and 1962). A collection from Geneva was examined by Held (1938). Trachtenberg (1960) and Schranz (1962) surveyed Polish and Hungarian materials. The reader is referred to Brabant and Schranz for references to papers not mentioned by the present author.

The results obtained from the examination of such European skeletal remains show some remarkable traits. Juvenile dentitions rarely show evidence of loss of alveolar bone. Zuhrt had at his disposal twenty-four skulls, of which one showed localized loss of alveolar bone. Brabant

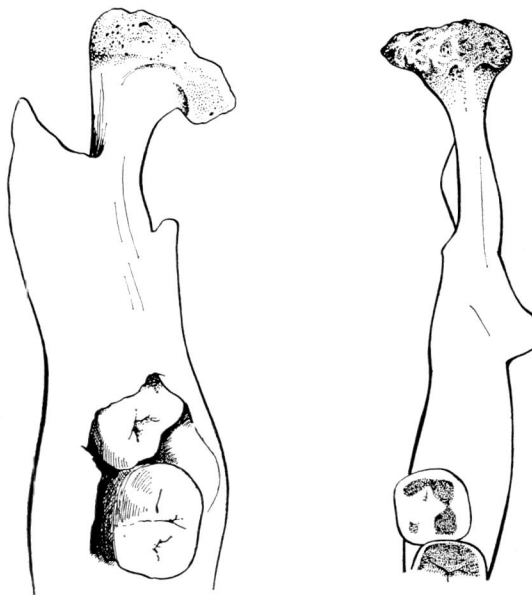

FIGURE 14. Osteo-arthritic changes of mandibular condyles. Nubia 185-466, (about 1580-1090 B.C.) and Kællingbjerg Revle, (Danish Early Iron Age). (Collection of the Laboratory of Physical Anthropology, Copenhagen).

reported five cases of incipient horizontal loss of bone among thirty-one dentitions from Belgian and French late Mediaeval series.

At adult, "mature" adult and even at senile age grades, cases without appreciable loss of alveolar bone can be found. Zuhrt examined fifteen dentitions belonging to individuals estimated to have been over fifty years old. Eight of these showed neither localized destructive processes nor generalized horizontal loss of alveolar bone. Brabant found several dentitions of old individuals in his Belgian material with alveolar loss of bone at stage 1 (cf. Table 1 and Fig. 15). In the age group thirty to fifty years, 7.4 per cent of 404 dentitions showed no alveolar loss of bone at all (see Fig. 15 and 16).

Table II gives a survey of the number of dentitions with alveolar loss of bone in various populations from the late Middle Ages. Some discrepancy between the results obtained by the authors might be due to different ways of judging the dentitions as a whole (cf. p. 562 and Table I). Mellquist and Sandberg studied only adult and mature skulls. To make the data given by Brabant (1963) in his paper comparable to the Scandinavian data, the present author has calculated the numbers in Table II omitting from Brabant's data the age groups twelve to eighteen and over fifty years of age. The method employed by Zuhrt to register alveolar loss of bone differed from those of the other authors mentioned. Zuhrt recorded individuals with evidence of periodontal disease using as many criteria as possible to decide whether alveolar loss of bone caused by a pathological process was involved. Differences in the proportion of older individuals in relation to younger ones in the populations studied may to some extent have influenced the results given in Table II.

TABLE II
LOSS OF ALVEOLAR BONE IN LATE MEDIAEVAL POPULATIONS

Group	No. of dentitions	% with bone loss	Author
Halland and Scania (Sweden)	1179	66	Mellquist & Sandberg (38)
Belgium and France	374	74	Brabant (1963)
Germany	122	56	Zuhrt (1956)
Poland 11th to 12th cent.	56	71	Trachtenberg (1960)
Poland 12th to 16th cent.	160	44	Trachtenberg (1960)

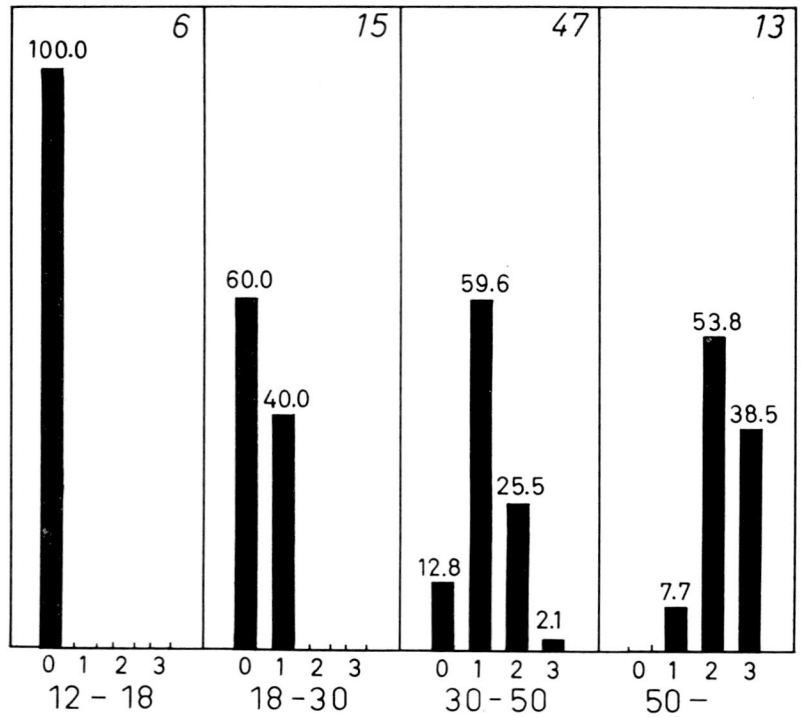

FIGURE 15. Loss of alveolar bone in Belgian late Mediaeval skulls registered according to Table 1. Calculations based on data given by Brabant (1963). The number of dentitions examined is shown (*upper corner*) as well as their distribution in each age-group according to the severity of alveolar bone loss.

In skulls, alveolar loss of bone is more often generalized in the dentition than localized as resorptions of a vertical type. The gradual exposure of the roots becomes evident in adult and mature age groups (see Fig. 15). In dentitions with a lowered alveolar crest it is usually possible to find here and there a porous surface of the alveolar bone upon the interproximal septa. In most cases, however, the structural changes are inconspicuous.

Vertical loss of bone becomes pronounced after the commencement of the fourth decade. It is thus most often seen in dentitions with much wear of teeth. The influence of attrition on the alveolar bone resorptions is a complicated problem. Both tend to increase as the individual grows older. A relation between attrition of an advanced degree and localized resorptions around such teeth can sometimes be pointed out in Mediaeval dentitions. (See also p. 563).

In old age the most serious results of horizontal and vertical types of alveolar resorptions are seen in combination with loss of teeth undoubtedly caused by periodontal disease. In mediaeval collections of skulls, the teeth around which the maximum loss of bone occurs are the molars. The anterior teeth are less involved by resorptive processes.

Characterizing single dentitions, Brabant (1960) found a juvenile with serious alveolar resorptions in the material from Coxyde. When the periodontal condition in a dentition diverges from the pattern usually seen in the group to which the dentition belongs, it is hard to provide an explanation. A systemic disease or nutritional disturbances may have played a part in this case. Mediaeval Norsemen from Greenland were examined by Mellquist and Sandberg (1938), and compared with contemporary Scandinavian skulls. They showed more advanced destruction of the alveolar bone and more teeth lost antemortem than the skulls from Scandinavia. Caries was not found in a single tooth from Greenland, nor was attrition more severe in this country. The major difference in diet between Greenland and that part of Scandinavia where the comparative material came from, was that vegetable foodstuffs were uncommon in Greenland. As a tentative explanation of the poor periodontal conditions in many Norsemen, Mellquist and Sandberg suggest scurvy.

The Mediaeval European populations surveyed reveal conditions which show greater similarity to recent communities at a so-called primitive cultural stage than to conditions seen in civilized European populations. The study by Hilming and Pedersen (1940) on periodontal diseases among recent Greenland Eskimos reveals conditions comparable to those described in Mediaeval European groups. The conclusions drawn by Hilming and Pedersen with regard to the Eskimos might be extended to include the medieval European groups.

The development of a resistant periodontium is promoted when the consistency of the food requires vigorous and prolonged chewing. Occlusal and interproximal attrition is favourable until advanced degrees of attrition are attained. The periodontal tissues are not destroyed until attrition has abolished the proximal contacts and made self-cleaning difficult, or carious cavities have developed in some of the teeth. With age, some loss of alveolar bone occurs in most individuals. If the alveolar loss of bone is pronounced and associated with vertical extensions along root surfaces, local causes can usually be recognized. Advanced periodontal destruction is found in older individuals, and in such cases the molars are very often involved.

The Neolithic and Mesolithic Periods

Brabant and Sahly (1962) published a review of their investigations on Belgian and French skulls dating mostly from the Neolithic and Mesolithic periods. The condition of the bony periodontium found by these authors in the Neolithic material did not show any essential differences from the conditions in Belgain and French Mediaeval collections of skulls. After the commencement of the third decade structural changes of the surface of the alveolar bone appeared in several individuals; this is considered as evidence of gingivitis. Loss of alveolar bone also became evident and in older individuals, that is in "mature" adults, advanced destruction of the supporting bone could occasionally be observed. Fewer dentitions with serious periodontal destruction were f o u n d by Sahly, Brabant and Bouyssou (1962) and Brabant and Brabant (1962) in Neolithic collections than in the Mediaeval Coxyde population (Twiesselmann and Brabant, 1960). This is shown in Table III and Figure 16. The proportion of older individuals was the same and the prevalences of degree 3 (cf. Table I) were 7 per cent and 17 per cent in the Neolithic and the early Mediaeval populations respectively. It was remarkable to find that teeth which were exposed with half or more of their roots were still firmly fixed in the jaws. The strongly built alveolar bone and fa-

TABLE III

THE DISTRIBUTION OF ALVEOLAR LOSS OF BONE IN NEOLITHIC AND MEDIAEVAL POPULATIONS

Loss of bone according to Table I.	Neolithic period Belgium No. %		Neolithic period Matelles No. %		Early Middle Ages Coxyde No. %		Late Middle Ages Belgium + Paris No. %	
0	21	26%	7	16%	95	25%	97	26%
1	36	44%	21	48%	90	23%	138	37%
2	19	23%	13	29%	133	35%	117	31%
3	6	7%	3	7%	67	17%	22	6%

Calculated from data in Brabant and Brabant (1962), Sahly, Brabant and Bouyssou (1962), Twiesselmann and Brabant (1960), Brabant (1960a, b, 1963) and Brabant and Twiesselmann (1960). No individuals whose age is open to doubt are included in the calculations which are based on the well-preserved skeletal remains of individuals over twelve years old.

vourable loading of the teeth during occlusion and articulation might be partly responsible for this.

Common to reports on Neolithic collections of skulls from France, Germany, Denmark and Sweden is the low incidence of serious loss of alveolar bone resulting in loss of teeth due to periodontal disease (Bouvet, 1922; Euler, Mohaupt, Christophersen in Euler, 1939a; Heuser and Pantke, 1959; Holmer and Maunsbach, 1956).

Brothwell (1959) studied British collections dating from Neolithic to Saxon times and registered recession of the alveolar bone in about 74 per cent of the 130 skulls examined. No criteria for estimating the amount of loss of bone were given. Brothwell considered extreme attrition, food impaction and accumulation of calculus as contributing factors of great importance to the alveolar recession. Such high figures for the prevalence of recession of the alveolar crest are only found when the horizontal loss of bone is registered. Trachtenberg (1960) found 64 and 71 per cent of individuals with alveolar atrophy in Polish Neolithic and early Mediaeval materials.

Sahly, Brabant and Bouyssou (1962) examined a small group of Mesolithic skeletal remains from France (Rouffignac). The oldest individual was the only one showing loss of alveolar bone to a considerable extent. Poitrat-Targowla (1962) studied the Epipalaeolithic population from Taforalt in North Africa and found loss of alveolar bone in thirty-two of the sixty-six jaws examined. This high prevalence in a population consisting mostly of young individuals rarely reaching much more than adulthood is due to the method of recording alveolar loss of bone and to the excellent possibilities of distinguishing between ante-mortem and post-mortem exposure of roots in the material examined. Resorptions of horizontal type were recognized and in some cases 3 to 5 mm of the roots were exposed. The exposed surfaces had taken on a browner colour than the root surfaces covered by bone, thus making it easy to detect post mortem alveolar loss of bone.

In the fifteen skulls from Teviec described by Pequart, Boule and Vallois (1937), half showed traces of an osseous reaction along the alveolar crest. This reaction (which is not precisely described) was supposed to be due to gingivitis caused by chewing hard things such as shells. Another possibility suggested by

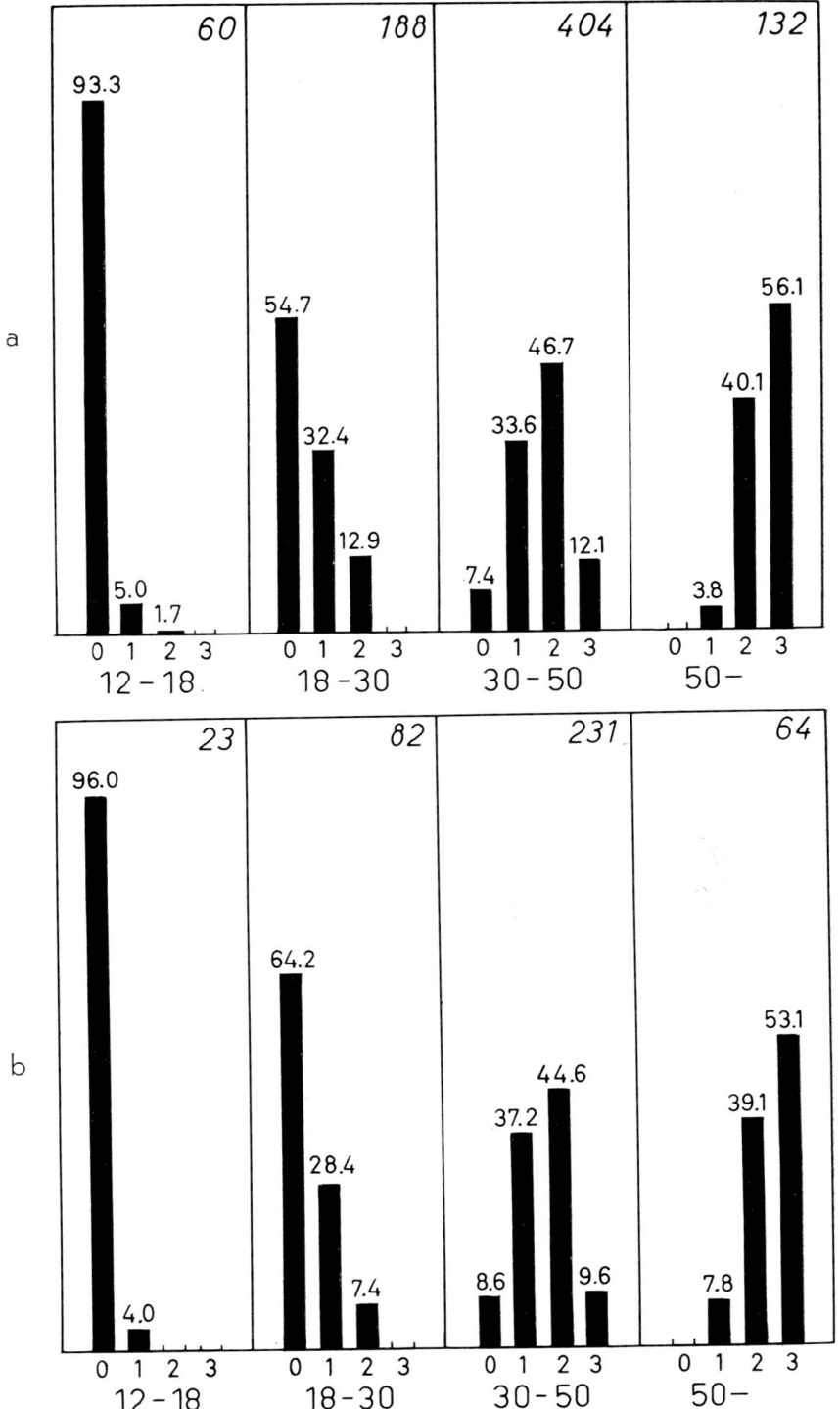

FIGURE 16. Loss of alveolar bone in Belgian early (a) Mediaeval and (b) Neolithic skulls registered according to Table 1. (Twiesselmann and Brabant, 1960; Brabant and Brabant, 1962). The distribution of dentitions examined is shown according to the severity of the alveolar bone loss in each age-group.

the author for the structural changes was that they were brought about by "rheumatism."

The Palaeolithic

Although very few of the numerous finds of Upper Palaeolithic man and Neandertal man involve mature and senile individuals, some of the skulls show changes in the alveolar bone that are likely to have been caused by periodontal disease. The extreme attrition accompanied by infection of the dental pulps after exposure and subsequent periapical osteitis had a pronounced effect on the course of pathological processes in the marginal periodontium. It seems more realistic when evaluating the more or less damaged, and now restored specimens of human Palaeolithic jaws not to distinguish too sharply between the pathological processes going on in the marginal and the apical periodontium. It is futile to discuss the initial cause of ante-mortem loss of teeth in these specimens. Even experts can differ, as in the recent controversy about the Tepexpan man from Mexico (Fastlicht, 1949 *contra* Moss, 1960). Another controversial case is that of the famous Neandertal man from La Chapelle-aux-Saints, which was claimed to be the earliest case of periodontal disease leading to extensive loss of teeth *intra vitam*. The original description by Choquet (1909) has not met with the approval of all experts. Thus Ruffer (1921) and Praeger (1925) both considered the post-mortem changes too extensive and the remaining two teeth too few to determine whether or not periodontal disease has been present. If the condition of the man from La Chapelle-aux-Saints is regarded as being an aggravated stage of that of the dentition of one of the skulls from La Ferrassie (Boule and Vallois, 1952, Figs. 160-162), there can be no doubt that resorptions must have been taking place here and there along the alveolar crest in the La Chapelle skull. A number of predisposing factors are present in both dentitions. Pulp exposure by extreme attrition may have led to periodically painful conditions with reduced function in parts of the dentition and accumulation of food debris. Loss of teeth with periapical osteitis by exfoliation led to movement of some of the remaining teeth. Food impaction occurred and all conditions favoured the contraction of a periodontal disease.

The mandible belonging to an adult individual from Ehringsdorf shows loss of alveolar bone due to periodontal disease (Virchow, 1920). Kötzschke (1960), after a study of the original mandible and X-rays of the jaw and teeth, was opposed to this view. From a cast of the jaw examined by the present author it is evident that the alveolar crest in the molar regions is located more apically on the roots on the lingual side of the teeth than on the facial side. Half of the roots of the molars are exposed. Post-mortem changes make it impossible, however, to determine accurately the general level of the alveolar crest in most regions of the tooth-row.

From Krapina, the only mandible of nine adult specimens with evidence of periodontal disease is the jaw called Krapina J. Besides a generalized lowered alveolar crest a more profound destruction of alveolar bone is present around the first right molar. Here the bony surface is porous and new formation of bone as limbal noduli are present lingually as depicted by Schranz (1962). In the description of the mandible, Kallay (1955) stresses that a great deal of calculus is found on the teeth, which are only slightly worn. According to Kallay, even a fragment of a tooth—perhaps of the missing first right premolar—is fixed in the supragingival calculus.

It has been suggested by some authors dealing with pathological changes in European Neandertal man that periodontal disease in some of the specimens was due to or associated causally with arthritis in the temporo-mandibular joints. This idea apparently goes back to Choquet (1909) and Boule (1911-1913). According to Baudouin (1913) and Kötzschke (1960), Choquet and Boule found pathological alterations in the dentition belonging to the La Chapelle-aux-Saints skeleton and morphologically changed articular surfaces of joints at several places in the skeleton, including the temporo-mandibular joints. The alveolar changes were called *poly arthritis alveolar-dentalis*, which was then the usual term for periodontitis. As far as published photographs and casts of the skull and the mandible can be used for accessing any pathological changes of the temporo-mandibular joints, these were of an osteo-arthritic nature.

In the present state of our knowledge there is little evidence of a causal relation between periodontitis and the temporo-mandibular joint. It is not permissible to deduce from evidence of periodontal disease that arthritis existed in the temporo-mandibular joint and explain the alveolar changes in the mandible of the adult Ehringsdorf specimen as the result of a common disease of the periodontium and the joints (Virchow, 1920). It is likely that Virchow misinterpreted the term used by Choquet, namely *polyarthritis alveolar-dentalis*. In a recent publication, Kötzschke (1960) draws attention to the coincidence of alveolar and joint alterations in Neanderthals. Besides the skull from La Chapelle-aux-Saints, the Krapina mandible J is brought forward as evidence of the theory that periodontal disease succeeded a joint disorder, for instance arthritis. In the Krapina J specimen, a fistula is believed to have been identified under one of the flattened mandibular condyles, both deformed by osteo-arthritis. It is likely that the lateral tubercle for ligamentous attachment which is sometimes found on the mandibular condyles of Neanderthal man is present in this Krapina mandible, and that no fistula exists (*cf.* Patte, 1956 and McCown and Keith, 1939).

It is well known that painful conditions in the joints influence their normal function, and Poitrat-Targowla (1962) mentioned the frequent coincidence in the Epipalaeolithic Taforalt population of osteo-arthritic changes in the temporo-mandibular joints and periodontal affections of a localized, circumscript type.

Among the Upper Palaeolithic men said to have suffered from periodontal disease, the "old" Cro Magnon man is not a good example since all his teeth are missing (either ante- or post-mortem) and the post-mortem changes of the alveolar bone are serious. This assertion was made by MacCurdy (according to Williams, 1929). Another doubtful case is one of the Obercassel skulls; the central incisors in the mandible being missing. Kötzschke (1960) is of the opinion that this is due to injury rather than to periodontal disease.

Bearing in mind that determination of bone resorptions along the alveolar crest can only with certainty be made from inspection of the original jaws, I found evidence of horizontal recession of the alveolar crest in casts of the mandible from Chancelade and in a fragment of the mandible belonging to Brno II. There is no doubt that the same alveolar changes occurred in the Upper Palaeolithic as in the dentitions from the much later Neolithic. In adult specimens with advanced attrition and ante-mortem loss of teeth, loss of alveolar bone is visible as a generalized recession of the alveolar crest (Rhodesian Man according to Brothwell, 1963; Skhūl V according to McCown and Keith, 1939). New formation of bone along the alveolar crest is found not only in the Krapina J mandible, but also in the upper jaw belonging to the Rhodesian Man—buccally and lingually opposite the first and second molars.

It is possible to find parallels in the Pithecanthropus group. From the original fragments of Pithecanthropus examined by me in Utrecht and from published illustrations and casts of "Sinanthropus" it is obvious that, generally speaking, the distance vertically between the cemento-enamel junction and the alveolar crest tends to be larger on the lingual side of the teeth than on the facial. This is most clearly seen in mandibles: in the mandibles from Ternifine (Arambourg and Hoffstetter 1963) the distance increases with advanced attrition and possibly also with age. Localized loss of alveolar bone has been recorded in the incisor region in the Heidelberg jaw (Schranz, 1962). Bouvet (1922) had already noticed the exposed lower incisors but was doubtful whether this was caused by a pathological resorption of the bone. This is not likely since Weinert (1938), in a description of a reconstruction of the jaw mentioned that the facial alveolar bone opposite to the incisors was never found. The jaw, when recovered, was broken at the symphysis. In the mandible belonging to *Pithecanthropus modjokertensis* a depression interproximally between the second and third molars is present but still filled with matrix. In a cast of a "Sinanthropus" upper jaw fragment, loss of alveolar bone has occurred around the posterior molars. The resorptions are most pronounced in the proximal area between the third and the second molars.

Schranz (1962) mentioned the jaw fragment belonging to *Meganthropus palaeojavanicus* (v. Koenigswald) as an example of atrophy of alveolar bone even among the oldest Hominidae. This example is not well chosen because post-mortem changes "caused" the atrophy which was clearly evident when I examined the original jaw fragment in Utrecht.

Evaluation of Palaeopathological Literature of Oral Hygiene and Socio-economic Conditions in the Aetiology of Periodontal Disease

Most investigators of the jaws of early skeletal remains discuss the presence of calculus on the teeth. As a sort of indicator of the state of oral hygiene at the time of death, the prevalence and distribution of calculus in the dentition is worth knowing.

In early skeletal remains, calculus is present surprisingly often on most or all teeth in the dentition. Calculus is most often of the supragingival variety. The courseness of the food eaten and the vigorous chewing required (as witnessed by the pronounced attrition) has not hindered calculus deposition in most ancient jaws. However, large amounts of calculus in a dentition have only rarely been found. The proportion of dentitions with large amounts of calculus in early and late Mediaeval populations was 8 per cent in skeletal remains from Coxyde (Twiesselmann and Brabant, 1960), 5 per cent in Picton's small material of Jutes from the sixth century (Picton, 1957), 7 per cent in the collection of jaws from Geneva studied by Held (1938), and between 3 and 25 per cent in the late Mediaeval populations from Scandinavia and Greenland examined by Mellquist and Sandberg (1938). It is likely that large amounts of calculus were formed in mouths where self-cleansing was difficult or impossible. Reduced masticatory function thus accompanied painful conditions in the oral cavity or in the temporo-mandibular joints. Loss of teeth and, in rare cases, disturbances in muscles such as paresis, could influence the useful process of self-cleansing by vigorous chewing.

The prevalence of calculus deposition in relation to individual age is difficult to establish with certainty. A number of skulls may have lost calculus by manipulations carried out after or during the excavation and cleaning of the skull. Zuhrt (1956) found that the number of dentitions with calculus increased with age until the fourth decade, after which it declined. This decline is mostly due to loss of the molars as a result of caries, extreme attrition and their sequelae: osteitis. Even in the fourth decade, only 34 per cent of the skulls showed calculus. From tables prepared by Brabant and collaborators (1960-1963), it is evident that some older individuals in the French and Belgian collections had no calculus. Henkel (1961) also shares the opinion that the amount of calculus in the dentition does not increase steadily with age. To all appearances, an equilibrium was formed in adult age between the amount of calculus able to be deposited on the tooth surfaces and the ability to keep the teeth clear of food-debris by chewing, possibly helped sometimes in some peoples by the use of toothpicks or some form of "toothbrush."

With increased age the apical limit of the calculus moved in an apical direction. From its location on the apical third of the crown in children and juvenile individuals it is (according to Henkel, 1961) located apically to the cemento-enamel junction in most individuals who have passed the middle of their third decade. The apical limit is not necessarily located identically on all teeth in a dentition. In the skulls studied by Henkel (1961) the distance between calculus and the alveolar crest was almost constant throughout life unless pronounced pathological changes of the alveolar bone were evident.

Very valuable studies by Leigh (1925-1937) on North American Indian and pre-Columbian Peruvians supplement and increase the information about these populations given by MacCurdy (1923), Moodie (1928) and Goldstein (1957), and are confirmed by later publications by Rabkin (1942-1943). Leigh studied the pathology of the teeth and jaws in relation to varied environmental and food conditions. Without going into the possible meaning of physical inheritance in relation to the differences found among various Indian tribes, in regard to susceptability to peridontal disease, Leigh pointed out the differences in food conditions. These varied according to the tribes concerned from meat-eaters (Sioux) and semi-sedentary tribes (Kentucky Algonquins and Arikara) whose diet consisted of meat but also of maize and beans, to the Zunis, with (as Leigh wrote in 1925) a mode of life and food closely related to our own civilization.

In the oldest group, the Kentucky Algonquins from the Indian Knoll burial site, no cases of "generalized periodontoclasia" (a term chosen by Leigh) were found. This tallies with the results obtained by Rabkin (1943) who studied the same tribe. The Sioux had excellent periodontal conditions too, although 13 per cent showed localized destruction of alveolar bone caused by local factors. In 36 per cent of the Arikara Indians alveolar bone destructions were found. By contrast, the

Zunis, with moderate attrition and more caries, calculus and stains on the teeth than the other tribes studied, had alveoloclasia involving all teeth in the dentitions in 56 per cent of the cases. The development of the jaws also seemed inferior to that of the other tribes.

The reasonable relationship between periodontal disease and the progress of civilization involving changes in eating customs, food preparation and choice of food, is well illustrated in the enormous collections of skulls from ancient Egypt and Nubia. This relationship is discussed in general terms by Kötzschke (1960). Detailed information on the actual alveolar alterations found in some Egyptian and Nubian populations was given by Ruffer (1921). To study the influence of domestication and civilization on the prevalence of dental and periodontal diseases, Grimm (1960) surveyed most of the available literature on dentitions of prehistoric and early historic skeletal remains from Mesopotamia. The materials already studied were so few and uncomprehensive that no conclusions could be drawn with regard to a certain relationship. It is therefore encouraging to see that papers dealing with dentitions of skeletal remains from the Mediterranean and Near Eastern countries have been published recently (Senyürek, 1952, Schaeuble, 1958, Dahlberg, 1960, Carr, 1960, Brothwell and Carr, 1962 and Brothwell 1965). It is to be hoped that sufficient data will, in the near future, be forthcoming from this important area to permit reasonable deductions concerning the relationship between developing civilization and the incidence of periodontal disease.

OSTEITIS

Osteitis in the jaws, originating from the teeth, is a common pathological process in ancient skeletal remains and in fact the most frequent inflammatory bony reaction observed in such remains.

Periapical Osteitis and Radicular Cysts

Periapical osteitis is the reaction of the periapical tissues to infections from an inflamed or—as a rule—a necrotic pulp. In skulls it can safely be deduced that a pulp has been involved in a pathological process if pulp exposure is visible by macroscopic inspection of a tooth.

Exposure of the dental pulp by attrition is a rather usual occurrence in most early human communities. The attrition advanced so fast that sufficient quantities of secondary dentine (to avoid exposure) never had time to form. Holmer and Maunsbach (1956) suggested that a pattern of wear which hollows out the dentine on the occlusal surfaces is more harmful to the dental pulp than a pattern where the enamel rim and the dentine core are on the same plane. Angel (1952) supposed that individual hereditary differences in the ability of the pulp to form secondary dentine could explain the differences observed in the frequency of pulp exposure (due to attrition) in different communities. For a discussion of this point of view, see Sahly, Brabant and Bouyssou (1962).

Occasionally periapical osteitis or radicular cysts can be found in skulls beneath intact teeth, that is, in teeth without caries or evidence of fractures and with little wear (Heuser and Pantke (1959), Ruffer (1921), Brabant (1960a). When anterior teeth are involved, injuries may be a possible cause of pulp degeneration and successive periapical osteitis. Periapical rarefactions can also be observed beneath intact molars, and the most natural explanation for this is that the pulps could not stand either the burden of mastication repeated day after day or some sort of abuse of the teeth practiced by the persons concerned. The forces necessary for chewing meat in primitive Eskimo communities have led to the development of chewing muscles half as strong again as those of civilized Europeans (Klatsky and Fischer, 1953). According to Wood Jones (1910) the food eaten in early agricultural communities likewise required great jaw strength—judging from contemporary peoples at the same technological stage. The great mechanical forces at work during mastication in prehistoric and early historic populations sometimes caused lateral displacement of teeth to such an extent that one or two roots became free of the alveolar process (Fig. 8). Molars with ante-mortem vertical fractures of crowns and roots have been reported by Brabant (1963) in Mediaeval specimens from Gutschoven in Belgium, by Pedersen (1944) in Mediaeval Norsemen in Greenland, and in skeletal remains of North American Indians (Leigh, 1925 and Hoyme and Bass, 1962).

Apart from attrition, caries is the most common cause of exposure of the dental pulp. Brabant (1960-1963) surveyed with his collaborators large collections of skulls dating from the Mesolithic Age to the end of the Middle Ages. Among the decayed teeth only 12 to 36 per cent showed superficial attacks of caries. The rest of the teeth with caries showed more profound lesions indicating that the speed with which caries progresses has not changed essentially from that day to this, although the abundance of secondary dentine often found (Zuhrt, 1956), may have delayed the time of exposure.

It is reasonable to consider that pulp exposure should lead to periapical osteitis. This theory was put forward by Mathis and by Clementschitsch (1939) and confirmed Holmer and Maunsbach (1956) working on Swedish Neolithic material. Using X-rays as a supplement to macroscopic inspection they disclosed that forty-five of the forty-six teeth with pulp exposures had periapical rarefactions.

Periapical osteitis in the dentitions of skulls is directly visible if the spreading pathological process has destroyed the outer surface of the jaw. Only periapical infections thus diagnosed are recorded—as a rule—in statistical reports on the rate of incidence of periapical osteitis in skeletal remains. The number of teeth with periapical osteitis obtained in this way is not necessarily the real number of diseased teeth. An X-ray analysis of the jaws with careful study of the periapical regions of the teeth is thus necessary in order to obtain results comparable with epidemiologic surveys of groups of living persons.

A distinction between different forms of periapical osteitis is sometimes undertaken in skulls, based on the criteria obtained from X-ray analysis of recent clinical material.

1. Periapical granuloma is a localized rarefaction of periapical bone which in a skull is limited by a well-defined border of porous, spongy bone.
2. Diffuse periapical osteitis forms an irregular bony cavity with osteoporitic walls.
3. Radicular cysts form smooth-walled cavities which sometimes have a parchment-like bony, and more or less intact covering that bulges from the bony surface.

(1. and 2. are sometimes called abscesses in pertinent literature.)

The classification given above mainly serves descriptive purposes. Even in live persons, the distinction between osteitis and radicular cysts can only be made with certainty on the basis of microscopic evidence (Shafer, Hine and Levy (1963)). Whether histological investigations on dry bones applied to this problem might help in pronouncing a more concise diagnosis of periapical infections remains to be seen. So far very little research has been done in this field.

Evidence of periapical osteitis is met with in skeletal remains of Neandertal and Upper Palaeolithic Man.

Brothwell (1963a) found pulp exposures caused by attrition in 11 to 40 per cent of skulls of European Neandertal and Upper Palaeolithic Man as well as in some North African and European Mesolithic skulls. In the groups mentioned, 1.4 to 5.0 per cent of the teeth examined showed pulp exposures. An instructive example of the exposure osteitis relationship is seen in one of the Neandertal dentitions from La Ferrassie, in whom pulp exposures caused by extreme attrition occurred in several teeth, resulting in visible periapical foci round the apices of the last premolar and the first molar in the left side of the mandible (see illustration by Boule and Vallois, 1952). It is reasonable to assume that caries was very rare in Palaeolithic groups, but even Mesolithic groups may show decayed teeth with a frequency of 7.7 per cent of the teeth examined (according to Brothwell, 1963a). An exceptional position is occupied by the Rhodesian skull, which has eleven carious teeth beneath which at least four periapical osteitic foci are found.

It is thus obvious that the most common causes of damage to dental pulps were active from at least the Mesolithic Age. This appears also from the pathology of the teeth and jaws of the Epipalaeolithic population from Taforalt (Poitrat-Targowla, 1962) and Mesolithic jaw fragments recovered from France and examined by Sahly, Brabant and Bouyssou (1962).

During the Neolithic period, carious teeth in Germany, Denmark and Sweden were few, and attrition, which was excessive in several populations studied, caused most of the periapical osteitic foci found (Euler and collaborators, 1939, Heuser and Pantke, 1959, Christophersen, 1939, Holmer and Maunsbach, 1956). In the French skulls dating from Neolithic to Roman times and studied by Hartweg (1945), periapical osteitis was, according to this author, a rather common occurrence. However, in the tables he supplied the number of teeth with periapical signs of infection was less than 1 per cent. Brothwell (1961) studied the incidence of periapical osteitis in the dentitions of six British groups dating from the Neolithic period to the seventeenth century. The incidence of periapical osteitis varied between 1.9 per cent and 5.2 per cent of the total number of sockets examined, but the incidence of caries fluctuated much more in the six groups from successive periods, (from 2 per cent to 20 per cent of the teeth examined in the respective groups). The pattern of ante-mortem tooth loss, through time, varied in the same way as the incidence of caries in all the groups except the Post-mediaeval London material. The increase in incidence of teeth showing signs of being affected with periapical osteitis was only slight and, this finding, despite the fact that the prevalence of caries and the ante-mortem loss of teeth seemed to be associated, can partly be explained by the equally relevant point that teeth affected with caries and successive periapical osteitis will ultimately be exfoliated. Exfoliation of teeth is a process which, in early historic times, was both common and quite rapid in its course. It is furthermore known that not all carious teeth are associated with periapical changes in the bone visible on the outer surface of the jaws. Brabant (1960b) mentioned in his discussion of a Belgian collection from Nivelles that about 37 per cent of the carious teeth had periapical osteitis. In the early Mediaeval Coxyde collection likewise examined by Brabant (Twiesselmann and Brabant, 1960) 56 per cent of the carious teeth had visible signs of periapical infections.

At this point it is of interest to draw attention to the publications by Brothwell and collaborators (1962-1964). In Egyptian populations dating from Predynastic to early Christian times, Brothwell, Wood Robinson and Carr found a close association between the incidence of teeth with pulp exposure due to attrition and that of teeth with periapical osteitic foci. An exception was the collection dating from the XXVIth to the XXXth Dynasty (about 500 B.C.) where the incidence of periapical infections did not decrease correspondingly with a decrease in the incidence of attritional pulp exposures. On the contrary, a slight increase of periapical osteitis foci to a level of about 9 per cent was found. This was most likely caused by other aetiological factors than attrition, for example, caries. In the Bronze Age remains from Jericho (Brothwell, 1964) as well as in the Predynastic Egyptians, a low incidence of carious teeth (2.3 per cent and 3.0 per cent respectively) was associated with a high incidence of periapical osteitic foci, (7 per cent of sockets examined). Comparative data were compiled by Brothwell and Carr (1962) on the prevalence of caries and periapical osteitis in approximately contemporaneous peoples, namely the Etruscans, Iron Age Britons, Ancient Greeks and Egyptians. The authors found only a slight difference in the incidence of periapical osteitic foci (3.2 per cent to 4.2 per cent) when comparing the British group with South European populations. The incidence of caries in the same populations varied between 4.0 per cent (the Etruscans) and 10.4 per cent (the Britons). The Egyptian material revealed an incidence of 8.7 per cent of carious teeth and 9.1 per cent of teeth with evidence of chronical apical infection.

In order to estimate the extent to which various aetiological factors influence the prevalence of periapical osteitis, the composition of the group in regard to individual age must be considered, and likewise the prevalence of teeth with pulp exposures should be based on dentitions with little post-mortem loss of teeth. X-rays are a necessary aid to the recognition of periapical osteitic foci.

Thorough investigations on occurences of periapical osteitic foci and their macroscopic variability have been made by Zuhrt (1956 and 1960) on late Mediaeval German material from Reckahn in Brandenburg. While the number of teeth affected by periapical chronic infection amounted to 4.45 per cent of the total number of teeth examined, the rate in the second decade was only 0.4 per cent; in the fourth, however, 5.5 per cent. In individuals of more than fifty years of age 11.2 per cent of their teeth were affected by periapical osteitic foci.

Numerous periapical osteitic foci were found in the skulls of different groups of Alemanns dating from the early Middle Ages and studied by Schwerz (1916) and Gröschel (1937). Periapical osteitis was most often found in mature and senile skulls. Several teeth in the same dentition sometimes showed evidence of periapical osteitis. Their distribution in the dentitions is shown in Figure 17, which also includes data from Zuhrt (1956) and Ruffer (1921). The Predynastic peoples described by Ruffer showed more anterior teeth affected by periapical infections than Etruscans (Brothwell and Carr, 1962) and French Neolithic groups (Hart-

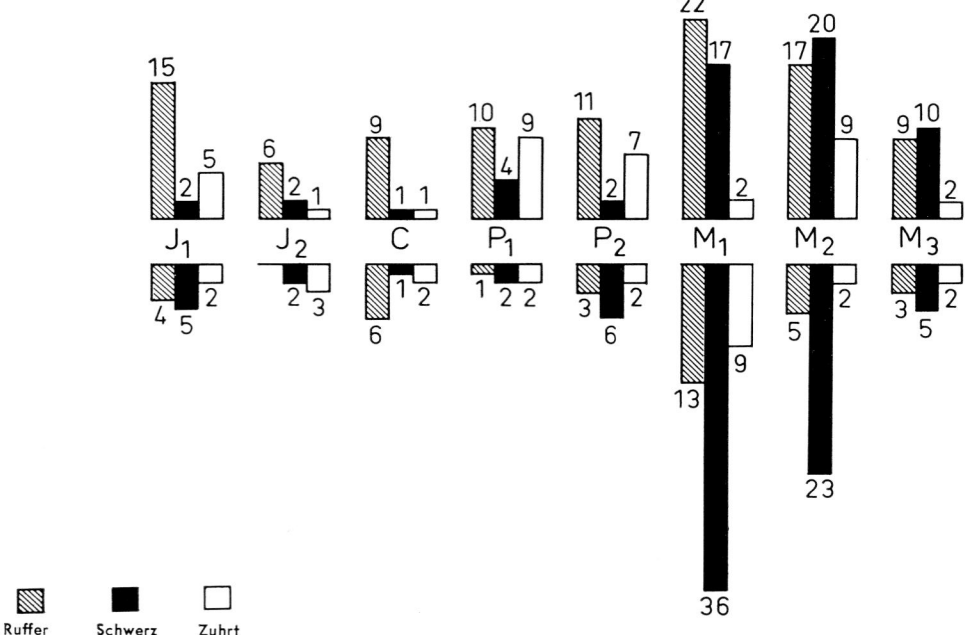

Figure 17. Susceptibility to periapical osteitis of teeth in predynastic Egyptians (Ruffer, 1921), in Alemanns dating from the fifth to the tenth Century (Schwerz, 1916), and in a late Mediaeval group from Reckahn in Germany (Zuhrt, 1956).

weg, 1945) which had no anterior lower teeth and only a few anterior upper teeth affected by osteitis. All the authors mentioned found the second premolars and the first and second molars often attacked by caries and chronic apical infection. In old age, attrition and periodontal disease may lead to osteitis throughout the dentition. As far as can be judged, injuries account for a very small number of periapical osteitic foci.

The resorption of bone caused by periapical osteitis will, after an elapse of time of varying length, remove the bony cover of the roots, thereby making the pathological process visible. This perforation of the facial and/or lingual or palatal bone surface is frequently large in the dentitions of skulls (see similar findings in recent material of human cadavers by Regan and Mittchell, 1963). The funnel-shaped canals described and depicted by Gorjanovic-Kramberger, (1908) as occurring in the Krapina J mandible are quite unlike the normal appearance. While the normal course is that the pathological process perforates near the root apices, the fistulae of the Neanderthal mandible are placed far from the apices of the teeth (see also Kallay, 1955, and Kötzschke, 1960).

A controversial case of a bony fistula or "boring" is that of an Egyptian mandible dating from the Old Empire (2500-2200 B.C.) (and described by Leek in more detail on page 702 of this book). Thoma (1917) and Hooton (1917) were both convinced that two holes had been bored through intact bone to an alveolar osteitic process around the apices of a first molar. Wingate Todd (1921) held the opinion that a fistula and a supernumerary mental foramen were present, and Weinberger (1948) concluded that there is not the slightest doubt that an operation had been performed. Bony fistulae sometimes occur in the jaws leading from the periapical focus to the alveolar crest inside the tooth socket, or on the outer surface of the bone visible as a groove (Fig. 10).

Determination of the size of periapical osteitic foci is a rather individual matter. Schwerz (1916) and Gröschel (1937), studying comparable groups of Alemanns, were not in agreement on the proportion of large and small foci in their respective materials. It is not known whether a real difference existed. The proportion of periapical cysts in relation to chronic abscesses have been evaluated by some authors using the criteria outlined on p. 577. The results of some investigations are presented in Table IV.

Radicular cysts are usually considered a result of a long-standing low-grade infection. They are comparatively rare and in the collection surveyed by Zuhrt (1956) they were most frequently found in the fourth decade or later in life. According to Zuhrt (1960) small circumscribed osseous cavities as evidence of a chronic inflammatory condition are often found under teeth with wide exposure of the dental pulp caused by attrition. Caries, on the other hand, lead more often to a more acute periapical osteitis, which, like a foreign-body reaction, tends to exfoliate the tooth. This applies to Neolithic and Mediaeval European collections studied by Zuhrt. It is necessary at this point to recall the warning given by Heuser and Pantke (1956) against a detailed distinction between acute and chronic forms of periapical osteitis in skeletal remains. By arranging series of teeth showing successive stages (from a superficial carious lesion to exfoliation of the root or roots) Zuhrt was able to deduce the usual course of the pathological process. A diffuse periapical osteitis was formed in most cases, which led to extensive destruction of bone along the sides of the root. Often supplemented by a certain loss of bone originating from the alveolar crest, the root was extruded and exfoliated—seemingly within a rather short period of time. In multi-rooted teeth, caries separated the roots in the course of its advancing destruction of the tooth. In several instances the root nearest to the site of pulp exposure was exfoliated first and most rapidly. Under the other root (or roots) more smooth-walled cavities were formed as evidence of a more chronic form of infection which

TABLE IV
PROPORTION OF RADICULAR CYSTS TO PERIAPICAL OSTEITIC FOCI

Author	Period	Geographic Region	Radicular Cysts	Periapical Osteitic Foci
Brabant and Sahly (1962)	Neolithic	Belgium	3	15
Brabant, Sahly and Bouyssou (1961)	Neolithic	France	1	7
Holmer and Maunsbach (1956)	Neolithic	Sweden	4	24
Twiesselmann and Brabant (1960)	Middle Ages	Coxyde (Belgium)	3	214
Brabant (1963)	Middle Ages	Spy (Belgium)	4	19
Brabant (1963)	Middle Ages	Gutschoven (Belg.)	3	50
Brabant and Twiesselmann (1960)	Middle Ages	Renaix (Belgium)	2	20
Zuhrt (1956)	Middle Ages	Germany	4	56

slowly led to loss of these roots. The bony defects in the jaws refilled with bone after loss of the teeth.

The chronic form of osteitis (sclerosing osteitis) was recorded by Bentzen (1929) in the dentition of a North American Indian child. Although such conditions are rare, more extensive use of X-ray examination would no doubt disclose more cases in skeletal remains.

Destruction of large parts of the alveolar bone involving several teeth and extending sometimes in the upper jaw to the nasal cavity or to the maxillary sinus system are occasionally found in ancient dentitions. Such wide-spread destruction cannot always be satisfactorily interpreted with regard to aetiology and pathogenesis, and a diagnosis is often omitted by authors describing these lesions.

In the lower jaw belonging to the adult Ehringsdorf specimen, a minor cavity in the alveolar bone was found round the apices of the double-rooted left canine. This cavity has been recognized by Virchow (1920) and Kötzschke (1960) as evidence of a periapical osteitis. A larger defect in the jaw located in the region of the right incisor is still filled with matrix. Whether this defect was caused by osteitis (presumably odontogenic as suggested by Virchow), or by trauma (as supposed by Kötzschke) or by yet another cause cannot be determined.

Gröschel (1937) described a mandible belonging to an Alemannic woman with a pathological process which had destroyed medullary bone in the alveolar process from the third molar to the second premolar. The destruction is considered as having been due to an odontogenic osteomyelitis. Wood Jones (1910) presented similar cases from Nubian series. Ruffer (1921) mentioned a skeleton from the IIIrd Dynasty cemetery at Tourah in which loss of bone occurred in the mandible, in the upper jaw, and also in the pelvis. It is evident that the perspectives opened up by such a case make it necessary to proceed with some caution when interpreting. Brabant (1960) considered odontogenic osteitis in the jaws to be present only when the pathological process was located around teeth with extensive caries, attrition, or with evidence of trauma or around teeth loosened by periodontal disease.

Non-odontogenic Infections and Specific Infections

The possibility exists of infections from surrounding tissues spreading to the jaws and causing osteomyelitis. Haematogenous spread of infections from foci in distant parts of the body is likewise a cause to be considered when interpreting osteomyelitis in the jaws. In addition there are the specific infections which are sometimes located in the jaws.

The maxillary sinus is occasionally involved in cases of periapical osteitis round the apices of upper premolars and molars (Fig. 11). Leigh (1937) found among pre-Columbian Peruvians that 10 per cent of the individuals of forty years of age or older had fistulae draining maxillary antra. Perforation of the hard palate was found in approximately 1 per cent of 348 Texan Indians studied by Goldstein (1957). The perforations were usually caused by periapical osteitis and observed in mature and senile skulls. Spread of infection originating from the sinus or the bottom of the nasal cavity to the alveolar part of maxilla or via the palate to the oral cavity is rare in ancient skeletal remains. Wood Jones (1910) mentioned an Egyptian skull with a perforated palate presumably caused by rhinitis. According to Pales (1930) evidence of sinusitis of non-dental origin was found by Moodie in pre-Columbian Peruvians. When a broad communication between the nasal and the oral cavities is present, as for instance in the cases illustrated by Risdon (1939), the primary cause is very difficult to establish.

As a rule the diagnosis of specific infections located in the jaws cannot be made from macroscopic inspection alone. Brabant (1963) published a photograph of a mandible from late Mediaeval Paris with structural changes of the bone in one of the ascending rami. Comparable changes were observed in the roof of the right orbit, and actinomycosis was suggested as a probable diagnosis.

If there are any characteristic lesions of a specific infection visible in other parts of the bone system than the jaws these will often be of great help in making a correct diagnosis of pathologic changes in the jaws. This is the case with leprosy. Thanks to the investigations of Møller-Christensen (1952, 1953 and 1961) of skeletons from mediaeval leprosaria in Denmark, it is now possible to diagnose leprous changes in the facial bones and jaws of skulls. In skeletons with characteristic leprous atrophy of bone in the hands and feet, some remarkable alterations in the alveolar process of maxilla were recognized. Loss of alveolar bone occurred in the region of the upper incisors—the central incisors being more involved than the lateral incisors and the canines. The rest of the dentition did not show localized loss of alveolar bone. Gingiva was not lost as fast as the bone because incisors were recovered belonging to skulls with no sockets left for these incisors. They had only been attached to soft tissues. The alveolar changes seen in skulls have been observed later in living patients (Michmann and Sagher, 1957). In some patients suffering from leprosy, a chronic inflammation in the nose may lead to perforations of the palate, a condition seen in several cases in Danish skull material from leprosaria. Resorption of the anterior nasal spine, and the occasionally seen evidence of "saddle nose," may contribute further to verification of a diagnosis of leprosy. Brothwell (1958 and 1961) noted leprosy in cases from Great Britain which hitherto had only been described from Danish collections, and other specimens have now been reported.

To my knowledge tuberculous affections of the jaws in skeletal remains have only been mentioned by Heuser and Pantke (1959). These authors suggested that some of the periapical osteitic foci found in Neolithic German skulls had been caused by tuberculosis or other specific infections such as actinomycosis. However, positive evidence was not brought forward for these diagnoses.

A diagnosis of syphilis is rather difficult to pronounce accurately in skeletal remains and information about syphilis located in the jaws in a skull dating from the early Middle Ages (from the Alemannic material described by Sticker 1935) has met with some scepticism. Sticker tells us frankly that pathologists consulted by him did not confirm his diagnosis of syphilis in the osseous changes in the left mandibular ramus and in the skull cap. Rabkin (1942) depicted the upper jaw of a North American Indian with perforation of the palate and loss of the anterior part of the alveolar process. Other parts of the skeleton also show pathological changes which Rabkin suggests could be due to syphilis—although he does not forget to mention yaws, which is endemic in some geographic regions in the American continents. The difficulties in distinguishing evidence of syphilis and yaws have been mentioned by Brothwell (1963). Hooton

(1930) studied the Indians from Pecos Pueblo, among whom he found some interesting specimens with inflammatory alterations in the nasal cavity, extending to the oral cavity through perforated hard palates or via the alveolar process. Other lesions in the vault of the skulls in these prehistoric skeletons looked like syphilitic changes. However sound diagnosis as to the nature of these changes could not be obtained, although Hooton consulted J. Ewing and H. U. Williams, both of them most competent pathologists.

Of great interest is the late Mediaeval Æbelholt material from Denmark, where Møller-Christensen (1958) found a clear case of syphilis treated with mercury. This caused a very osteoporitic surface along the marginal alveolar crest. Traces of mercury were present in the clavicle, which was examined by spectrographic analysis.

Cysts in the Jaws

Besides the periapical cysts already mentioned, a number of other cystic cavities may occur in the jaws.

Among developmental cysts, those forming from the incisor canal seem to be the most common. Stafne, Austin and Gardner (1936) found these cysts in about 1 per cent of the individuals studied in a recent population. Incisor canal cysts have occasionally been found in skeletal remains (Zuhrt, 1956, Brabant, 1963). As far as the Danish collections are concerned, incisor canal cysts have been recognized by me in a Neolithic skull (Holbæk Ladegard I) and in a Mediaeval jaw belonging to one of two named persons—Skjalm Hvide or Toke (Pedersen, 1955). The form and size of the anterior palatine foramen shows great variation. Taylor (1962) studied the variation in recent Moriori and Maori skulls and found the largest foramen as large as the socket of an upper central incisor. Goodman and Morant (1940) reported on an anterior palatine foramen of similar size in one of the skulls from Maiden Castle without mentioning any cystic cavity in the incisive canal.

In the case of cysts in the periapical regions with the roots of one or more teeth protruding into the cystic cavity, there exist on record cases of cysts beneath intact teeth, i.e., teeth showing little wear and no evidence of caries or trauma (Brabant, 1960a; Gröschel, 1937; Sticker, 1935). It is most likely that irritation from injured pulps initiated the development of the cysts, although this cannot be demonstrated with certainty.

Osseous destructions in the region of the third molars are occasionally found and sometimes diagnosed as cysts. Poitrat-Targowla (1962) diagnosed a dentigerous cyst round a third lower molar in one of the skulls from the Epipalaeolithic Taforalt population. Brabant (1963) discussed the diagnostic problems presented by a cyst-like cavity as large as a cherry stone; whether possible evidence of periodontal disease, a cyst or a tumour. In the skull shown in Figure 12 the same problems were present. In this Dutch skull cavities with porous walls were present distally to the third upper molars. The location of the cystic cavities distal to the tooth row is suggestive of fourth molar crypts or to primordial cysts.

Very large cystic cavities in the jaws or cysts not in close contact with teeth usually defy any precise diagnosis. Salama and Hilmy (1951) described a multilocular cyst in an Egyptian mandible dating from the Vth Dynasty. The cyst involved the mandible body from the region of the canine to the ramus in the same side into which the cyst extended. All the teeth in this side of the jaw were lost ante-mortem. The interior surface of the cavity was multilocular and rough. In a skull from the same period, Salama and Hilmy found a large cyst in the maxillary alveolar process with expansion almost to the midline of the palate. Risdon (1939) in a study of human remains from Lachish in Palestine and Moodie (1928a) in pre-Columbian Peruvians found cysts located in the upper jaw involving the hard palate, sinus and the nasal cavity. A residual cyst was recorded by Brabant (1962). Macroscopic inspection and X-ray examination of a skull cannot replace histologic evidence for a pathological process and the nature of large osseous defects in the jaws therefore remain unknown.

Focal Infection

Chronic infections in the periodontal tissues are considered important foci of infection which sometimes occasion effects in other parts of the body. In palaeopathological literature the theory is sometimes advanced that a relationship exists between oral foci and certain joint diseases.

Ruffer (1921) drew attention to the frequent coincidence of dental disease and *spondylitis deformans* and chronic osteoarthritis in ancient Egyptian and Nubian skeletons. Ruffer hesitated however, to conclude that a direct cause-and-effect relationship existed because recent material of Egyptians and Europeans in Egypt, according to him, did not confirm such a conclusion as oral foci were quite usual but chronic "spondylitis" rare.

While Moodie (1928b) showed an association between arthritis and dental diseases, Keith (1925) found some skeletons with teeth and joints affected by pathological changes, others with potential oral foci as well but with healthy joints, and still others which did not appear to have oral foci but nevertheless had pathological changes of joints. Hoyme and Bass (1962) examined skeletons of Indians from two sites in Virginia. Differences in arthritic changes in the minor joints as well as in the larger ones between the two sites were not paralleled by changes in the incidence of dental decay.

A possible relationship between osteoarthritic changes in the temporo-mandibular joints and dental disease has been suggested by Straus and Cave (1959) Poitrat-Targowla (1962) and Kötzschke (1960). Taylor (1963) studied collections of Moriori and Maori skulls in which, according to him, oral sepsis was prevalent as a result of excessive attrition with pulp exposure and periapical osteitic foci as sequelae. In skeletons of the Moriori Duckworth (1900) was struck by the frequency of osteo-arthritic changes. Taylor interprets this as lending support to the hypothesis that the two conditions mentioned are causally related.

With the object of investigating the possible correlation between oral foci in the form of periapical osteitis and spinal joint changes, Møller-Christensen and Brinch (1948) and later Ingelmark, Møller-Christensen and Brinch (1958) examined the late Mediaeval Æbelholt material from Denmark. The investigations concentrated on skeletons belonging to mature and senile individuals. Periodontal disease was not considered a potential infectious focus in these investigations. According to the authors its prevalence was low in the material surveyed. Teeth lost ante-mortem were to some extent considered as having been associated with periapical osteitis. The pathological changes recorded in the spinal joints were of a degenerative nature. Møller-Christensen and Brinch found that 17 per cent of the individuals had joint changes in the cervical part of the vertebral column without any oral foci as determined by the authors in dentitions without ante- or post-mortem loss of teeth. In the group of skeletons with oral foci, the incidence of individuals likewise having joint changes was 93 per cent. This coincidence led the authors to conclude that a causal relationship existed in these Mediaeval skeletons between degenerative changes in the cervical part of the spinal column and certain foci of dental infection.

Regarding the investigations on pathological changes in the whole spinal column, Inglemark was able to show significant differences in the incidence of pathological alterations of large and small joints in two groups of adult and mature skeletons, namely those having dental infective foci and those which had not. It is important to note, however, that a pathogenetic explanation of the apparent causal relationship between dental foci of infection and degenerative changes in the spinal column —both of which diseases are as old as mankind—has not yet been discovered. It is impossible to record in skeletons all potential foci of infection that existed in the individuals while alive, and it is also impossible to decide whether a potential focus had any distant effect. Nevertheless, the conclusions that could be drawn from the Mediaeval Æbelholt collection are interesting spurs to further research.

THE TEMPORO-MANDIBULAR JOINT

In literature concerned with palaeopathology, references to the pathological changes in the temporo-mandibular joint are few and scattered. Only the most obvious alterations of the bony articular surfaces have as a rule been mentioned. These conspicuous changes are usually considered as signs of osteo-arthritis.

Osteo-arthritis is a disease associated with the ageing process. There is good evidence to show that degenerative osteo-arthritic changes in the temporo-mandibular joint are dependent on the wear and tear suffered by the joint. In recent individuals the limit of tolerance is surpassed when the individual is over forty years old and changes are first initiated in the articular covering and later in the underlying bone. It is reasonable to conceive of marked alterations in the normal form of the articular surfaces as a final result of a gradual change. It is thus a somewhat subjective matter where to place the dividing line between macroscopic changes to the bony surfaces in the macerated joint as are within the (normal) limit of remodelling (to a changed functional load of the joint) and such changes as are pathological. This dilemma was stressed by Blackwood (1963) even with regard to minor histological changes noted in the mandibular condyle and its articular covering. Some authors go very far in permitting morphological changes of the articular surfaces in skeletal remains to be considered as functional—not pathological—modifications (Périer, 1948 and Schwarz, 1922). It is important in any interpretation of the articular surfaces of the temporo-mandibular joint in skeletal remains (as Périer, 1953, warned) that post-mortem changes should be recognized as such and excluded. This applies especially to porosity of the articular surfaces in skeletal material.

A good foundation for understanding what osteo-arthritic changes look like in skeletal remains is the study made by Bauer (1932). Macroscopic and microscopic findings in the joints of human cadavers, suffering in several instances from osteo-arthritis, were compared with the macerated bones from skeletal remains. The results obtained show general agreement with those of a number of other students on osteo-arthritic changes in skulls or human cadavers (see Bauer, 1932, and Blackwood, 1963, for literature).

On the condyle, the glenoid fossa and the articular eminence the following macroscopic bone changes are seen when the joint suffers from osteo-arthritis:

The articular surfaces are porous and rough (Figs. 12, 13 and 14). This is seen on the mandibular condyle and the articular eminence. The porosity which is not pathognomonic for osteo-arthritis is most often found on the upper surface of the condyle and on the anterior and lateral part of the articular eminence. Occasionally the porosity is located at the bottom of a large depression or groove in the upper or posterior surface of the condyle. Hrdlička (1941) suggested that this information of a median, rather deep fossa, revealed an incipient double condyle. When only a part of the articular surface shows porosity this tends to be a cribriform pitting. When the total articular surface is involved it is irregular and rough.

Osteophytic lipping is sometimes found where the joint capsule is attached to bone and wear and tear causes growth in form of exostoses from the periosteum. In rare cases enormous amounts of bone have been formed. The skulls from New Caledonia and the Loyalty Islands are classical examples (Sarasin, 1922; Vischer, 1921). The glenoid fossa in these skulls is filled with a bony plate in continuation of the flattened articular eminence. The corresponding condyles have a flattened upper surface and an irregular periphery. In such cases osteophytic lipping may also be seen along the lateral border of the joint beneath the root of the zygomatic arch.

Eburnation of sub-articular bone is sometimes found (Vischer, 1921; Schwarz, 1922; and Pales, 1930). In about 400 skulls from various geographic regions and various historical periods kept in Danish collections and examined by me with the object of finding pathological changes in the temporo-mandibular joint, only one case (Æbelholt Museum) showed a polished upper surface of the condyle due to wear. The corresponding part of the joint in the temporal bone was missing. In some cases, resorptions dominate over neo-formation of bone. The glenoid fossa becomes much enlarged while the articular eminence almost disappears. The size of the condyle is visibly reduced.

Ankylosis or hypomobility of the temporo-mandibular joint is most frequently caused by injuries or infections in or near the joint. When there is evidence of acute injury to the joints pronounced changes of the articular surfaces may occur. Similar malformations of the condyle in skeletal remains of juvenile and adult individuals may be the result of an inflammatory reaction in the joint. Infectious arthritis may be systemic or local in origin. It may be caused by spread of infection from the middle ear, the mastoid process, the parotid gland or the ramus of the mandible of non-specific or specific infections (Sarnat and Laskin, 1962). Infectious arthritis often leads to ankylosis—either fibrous or bony. This condition may also occur as a result of rheumatoid arthritis.

Complete ankylosis is a rare condition in skulls dating from ancient times. In a Mesolithic skull from Afalou (No. 14) which I examined at the Institut de Paléontologie Humaine in Paris, a part of the right mandibular condyle was attached to the glenoid fossa. The mandible was broken off beneath the condylar head. In the opposite side of the skull the temporomandibular joint was normal in every respect. Attrition of the teeth was more pronounced on teeth in the left side—the attrition being *ad palatum*. Extensive restoration of the upper jaw precludes any possibility of determining the movements of articulation. In a Mediaeval skull from Paris (Pipiniere 23251), examined at Musée de l'Homme in Paris, the mandibular condyle on the right side of the skull is attached firmly to the skull, while the ramus and the rest of the mandible is missing. On the left side the articular eminence is eroded and a new articular fossa developed as a depression in front of the glenoid fossa. The skull is characterized by large hyperostoses along the lateral margins of the maxillary alveolar bone and, on account of these, was described and illustrated by Brabant (1963).

In most cases of serious pathological changes of the temporo-mandibular joint it is impossible to decided what initiated the pathological process. From an examination of other joints of the skeleton it has been possible in a few instances to show that rheumatoid arthritis may have been responsible for the alterations observed in the temporo-mandibular joint. This applies to an Egyptian mummy from the Vth Dynasty described by May (1897, quoted by Karsh and McCarthy, 1960). An elderly Christian from Philae is considered a perfect example of true gout (Elliot Smith and Dawson, 1924). Among the joints affected are the left temporo-mandibular joint. Here osteo-arthritic changes were found and no white concretions as in many other joints. The concretions yielded the typical reactions to uric acid. In skeletal remains the conspicuous changes of the articular surfaces in the temporo-mandibular joint are usually referred to as osteo-arthritic, thus avoiding speculations as to their pathogenesis. Such speculations are tempting in cases with only one mandibular joint involved. It is known that acute arthritis may be the forerunner of osteo-arthritic alterations of the articular surfaces. Most often the left-sided mandibular joint is involved in unilateral cases, and Schwarz (1922), in a number of such cases, found loss of molars in the corresponding side of the jaws.

Evidence of osteo-arthritis is found in the Neandertal mandible Krapina J (Gorjanovic-Kramberger, 1908). The mandibular condyles are flattened and enlarged with a rough and porous upper surface and peripheral formation of new bone. Similar deformation of mandibular condyles has been found in the collection of skulls from New Caledonia, described and illustrated by Sarasin (1922). A funnel-shaped canal in the lateral exostosis on the left condyle was originally considered a fistula (Gorjanovic-Kramberger, 1908). This led to the hypothesis that an infectious arthritis had been present in the joint (Williams, 1929; Schranz, 1960). Kallay (1955), who redescribed the mandible, did not attach much importance to the small canal. It is known that a tubercle or an exostosis on the lateral aspect of the neck of the condyle is present in some Neandertal mandibles (McCown and Keith, 1939 and Patte, 1955). It is possible that the attachment of a ligament to the condyle of the Krapina J mandible resembled a fistula. Straus and Cave (1957) discussed the evidence of osteo-arthritis in the temporo-mandibular joints of the skull from La Chapelle-aux-Saints, the adult specimen from La Quina, and in one on the skulls from La Ferrassie. In the skull from La Chapelle-aux-Saints the enlarged and flattened glenoid fossa and the flat upper surface of the well-preserved right condyle as well as the rough articular surfaces make this diagnosis very likely. Postmortem alterations to the mandibles from La Quina and La Ferrassie make some caution necessary when interpreting the very deformed mandibular condyles as evidence of pathological distortion caused by osteo-arthritis as done by Straus and Cave. In a cast of the skull and mandible of Dolni Vestonice III there is evidence of osteo-arthritis especially in the left temporo-mandibular joint (Klima, 1950). In skeletal remains of Skhūl V and Skhūl VII the right and the left condyles respectively have been deformed by arthritis (McCown and Keith, 1939). In the corresponding glenoid fossae, pathological changes were also found.

Among the Epipalaeolithic skulls from Taforalt in North Africa, Dastague (1962) found one case of an enlarged, flattened glenoid fossa and eroded articular eminence whose articular surfaces were irregular and rough. Since the lesion was unilateral and the mandible belonging to the skull was missing, the aetiology of the alteration is not clear. In an excellent discussion by Dastague, the pros and cons of the case showing sequela of traumatic arthritis, infectious arthritis or osteo-arthritis are outlined. No definite conclusions can be drawn. Dastague also mentioned two mandibles from the Taforalt population showing evidence of osteo-arthritis.

From the European Neolithic Period, the subsequent Metal Ages and the Middle Ages, few cases of temporo-mandibular diseases with conspicuous changes of the articular surfaces have been reported. From Neolithic French, German and British collections, cases have been described by Le Baron (quoted by Moodie 1923), Kraus (1939), Henkel (1962), Cameron (1934) and Cave (1938). Gejvall (1960) mentioned one case among 139 Mediaeval Swedish skulls. From Denmark, Nielsen (1911) reported on a Neolithic case; and another, of marked remodelling of the articular surfaces and rough condylar articular surface (Figs. 13-14), was found by me in a skull dating from the Iron Age.

Møller-Christensen has called attention to three cases from the late Mediaeval Æbelholt collection, the three cases being found among some 800 skulls. To get an impression of the incidence of minor alterations of the articular surfaces, I have examined about one hundred and sixty well-preserved mandibles dating from the Neolithic Period and the Iron Age (Laboratory of Physical Anthropology, Copenhagen). All the skulls were those of adults. In some 10 per cent of the skulls examined small areas of rough or porous surface were present on the articular eminence and/or the mandibular condyles. In some cases marked remodelling of the articular surfaces was present.

Outside Europe the Nubian material described by Wood Jones (1910) showed examples of completely disorganized temporo-mandibular joints. Brothwell (1965) surveyed the pathology of human remains from Jericho dating from the E. B.-M. B. and Middle Bronze Age. Among twenty-six individuals, seventeen were arthritic but only one instance among sixteen skulls showed osteo-arthritis in the temporomandibular joint.

According to Pales (1930), the North American Indians showed evidence of temporo-mandibular arthritis more often than any other ancient population. Leigh (1929) found a general tendency among pre-Columbian California Indians towards a lengthening of the mandibular condyle in an anterior-posterior direction and a decrease in the depth of the glenoid fossa in addition to resorption of the articular eminence with age. Leigh (1925) moreover described an Arikara Indian with extensive and severe arthritis of the right temporomandibular joint. Rabkin (1942) reported on osteo-arthritis of the mandibular joint among Alabama Indians and so did Hoyme and Bass (1962) in two cases among thirty individuals from Clarksville, Virginia, (800 and 1630 A.D.). Further references are given by Pales (1930) on older pertinent literature.

Pathological changes are more often found in other joints in the skeleton rather than in the temporo-mandibular joints. This applies also to populations often affected by osteo-arthritis in the temporo-mandibular joints, for example, the Peruvian materials studied by Moodie and MacCurdy (see Pales, 1930). An exception to this rule is seen in the skeletons from New Caledonia and the Loyalty Islands in which—according to Vischer (1921)—the temporo-mandibular joints were attacked more often than other joints.

In a large collection of Indian skulls from Texas (c. 800 to 1700 A.D.) Goldstein (1957) found lesions of the glenoid fossa in 3.6 per cent of the adult skulls, in 2.9 per cent of the mature and in 8.6 per cent of the senile skulls. The lesions were presumed to be osteo-arthritic. Hooton (1930) reported from a collection of Pecos Indians that 3.5 per cent of the males and 9 per cent of the females showed osteo-arthritic changes of the glenoid fossa.

Pales (1930) very ably discussed various aetiological factors which could possibly influence the prevalence of osteo-arthritis in an ethnic group. He concluded that the occurrence of osteo-arthritis in the temporo-mandibular joint is very dependent on wear and tear of the joints, the anatomy of the articular surfaces being to a high degree dependent on the physiology of the joints.

ACKNOWLEDGMENTS

In preparing this review I have as far as possible compared information from the literature with a personal examination of original specimens or casts of early Hominids, Neandertal and Upper Palaeolithic man. From the period dating from the Mesolithic Age to the end of the Middle Ages acquaintance with the dentitions of skulls preserved in Danish collections formed a useful basis for evaluation of data obtained in other parts of the world.

I am very grateful to Professor G. H. R. v. Koenigswald (Geologisch Instituut, Utrecht) for his permission to study original material of early Hominids in his possession. My sincere thanks are due to Dr. R. Hartweg (Musée de l'Homme, Paris) and Professor H. V. Vallois (Institut de Paléontologie humaine, Paris) for permission to study skulls and jaw fragments. I also wish to express my gratitude to Dr. J. Balslev Jørgensen (The Laboratory of Physical Anthropology, Copenhagen) and Professor V. Møller-Christensen for kind permission to examine materials in their departments.

Professor P. O. Pedersen, Dean of the Royal Dental College, has given me constant support and advice for which I am grateful. Likewise I would like to thank Dr. J. Huizinga (Instituut voor Anthropobiologie, Utrecht) for kind permission to use the facilities of the Institute during my stay in Utrecht.

The photographic work was done by Mrs. I. Kragballe and Mr. J. V. Holm, the drawings were made by Mr. E. Leenders and the translation into English by Mr. D. Hohnen. I thank all of them for their valuable co-operation.

REFERENCES

ACKERMANN, F., 1953: *Le Mécanisme des Mâchoires*. Paris.

ANDRIK, P., 1962: Beitrag zur Anthropologie des Bissanomalien. *Z. Morph. Anthrop.*, 52:129.

ANGEL, J. L., 1952: The human skeletal remains from Hotu Cave, Iran. *Proc. Amer. Phil. Soc.*, 96:258.

ARAMBOURG, C., M. BOULE, H. VALLOIS, and R. VERNEAU, 1934: Les Grottes paléolithiques des Beni-Segoual (Algérie). *Arch. Inst. Paleontologie humaine.*, Mem. 13.

ARAMBOURG, C. and R. HOFFSTETTER, 1963: Le Gisement de Ternifine. *Arch. Inst. Paleontologie humaine.* Mem. 32.

BALSLEV JØRGENSEN, J., 1953: The Eskimo Skeleton. Contributions to the physical anthropology of the aboriginal Greenlanders. *Medd. om Grønland.* 146: no. 2.

BAUDOUIN, M., 1913: La polyarthrite alvéolaire depuis le quaternaire jusqu'à l'époque romaine. *Gaz. Med. France,* 17:397.

BAUDOUIN, M., 1919: Fracture mandibulaire guérie spontanément et datant de l'époque de la Pierre polie. *Bull. Acad. Nat. Méd. Paris.,* 81:52.

BAUER, W., 1932: Anatomische und mikroskopische Untersuchungen über das Kiefergelenk mit besonderer Berücksichtigung der Veränderungen bei Osteo-Arthritis deformans. *Z. Stomat.,* 30:1136, 1279 and 1334.

BECKER, C. J., 1952: Skeletfundet fra Porsmose ved Næstved. *Fra Nationalmuseets Arbejdsmark* 1952; p. 25.

BEGG, P. R., 1954: Stone Age Man's Dentition. *Amer. J. Orthodont.,* 40:298, 373, 462, 517.

BENTZEN, R. C., 1929: Dental conditions among the Mimbres People of Southwestern United States previous to the year 600 A.D. *Dent. Cosmos,* 71:1068.

BLACKWOOD, H. J. J., 1963: Arthritis of the mandibular joint. *Brit. Dent. J.,* 115:8.

BØHN, A., 1964: The course of the premaxillary and maxillary vessels and nerves in cleft jaw. *Acta Odont. Scand.* 64:463.

BÖHNE, C., 1955: Die Habsburger Unterlippe. *Zahnärztl. Prax.,* 6: Juni.

BOULE, M., 1911-13: L'homme fossile de La Chapelle-aux-Saints. *Ann. Paléont.,* 6:109; 7:21, 85; 8:1.

BOULE, M. and H. V. VALLOIS, 1952: *Les Hommes Fossiles.* 4th Ed. Paris, Masson.

BOUVET, P., 1922: *Les lésions dentaires des hommes préhistoriques.* Thèse de med. de Paris. Edit. A. Legrand, Paris. 105 pp.

BRABANT, H., 1960a: Etúde de la denture d'une communauté religieuse médiévale soumise à un régime non cariogène. *J. Dent. Belg.,* 50:651.

BRABANT, H., 1960b: Observations odontologiques et anthropologiques sur les ossements provenant des fouilles exécutées dans la cathédrale Sainte-Gertrude à Nivelles, Belgique. *Acta Stomat. Belg.,* 57:17.

BRABANT, H., and TWIESSELMANN, F., 1960: Etude de la denture de 159 squelettes provenant d'un cimetière du XIe siècle à Renaix, Belgique. *Rev. Belg. Sci. Dent.,* 15:561.

BRABANT, H., SAHLY, A., and M. BOUYSSOU, 1961: Etude des dents préhistoriques provenant d'un four crématoire néolithique situé aux Matelles, départment de l'Hérault, France. *Bull. Group. Int. Rech. Sci. Stomat.,* 4:382.

BRABANT, H., 1962 Contribution a l'étude de la paléopathologie des dents et des maxillaires. La denture en Belgique a l'époque néolithique. *Bull. Inst. Roy. Sci. Nat. Belg.,* 38:1.

BRABANT, H., and A. SAHLY, 1962: La paléostomatologie en Belgique et en France. *Acta Stomat. Belg.,* 59: 285.

BRABANT, H., 1963: Observations sur la denture humaine en France et en Belgique a l'époque Gallo-romaine et au moyen Age. *Bull. Group. Int. Rech. Sci. Stomat.,* 6:169.

BRABANT, H., and F. TWIESSELMANN, 1964: Observations sur l'évolution de la denture permanente humaine en Europe Occidentale. *Bull. Group. Int. Rech. Sci. Stomat.,* 7:11.

BREITINGER, E., 1939: Gutgeheilter Unterkieferbruch aus der Frühbronzezeit. *Sudhoffs Arch. Gesch. Med.,* 32: 103.

BROTHWELL, D. R., 1958: Evidence of leprosy in British archaeological material. *Med. Hist.,* London. 2:287.

BROTHWELL, D. R., 1959: Teeth in earlier human populations. *Proc. Nutr. Soc., London.* 18:59.

BROTHWELL, D. R., 1960: A possible case of mongolism in a Saxon population. *Ann. Hum. Genet.,* 24:141.

BROTHWELL, D. R., 1961: The palaeopathology of early British man: an essay on the problems of diagnosis, and analysis. *J. Roy. Anthrop. Inst.,* 91:318.

BROTHWELL, D. R., and H. G. CARR, 1962: The dental health of the Etruscans. *Brit. Dent. J.,* 113:207.

BROTHWELL, D. R., 1963a: *Dental Anthropology.* Oxford, Pergamon Press.

BROTHWELL, D. R., 1963b: *Digging up Bones.* London, Brit. Mus. (Nat. Hist).

BROTHWELL, D. R., 1965: The palaeopathology of the E. B.-M. B. and Middle Bronze Age remains from Jericho (1957-8 Excavations). Paper in Press.

CAMERON, J., 1934: *The Skeleton of British Neolithic Man.* London, Williams and Norgate.

CARR, H. G., 1960: Some dental characteristics of the Middle Minoans. *Man.,* 60:157.

CAVE, A. J. E., 1938: Remarks on certain Neolithic Skulls. *Proc. Roy. Soc. Med.,* 31:1373.

CHOQUET, J., 1909: Examen de l'appareil dentaire du crâne de l'homme préhistorique de la Chapelle-aux-Saints. *Comptes rendus du Ve congrès dentaire international.,* I. Berlin.

CHRISTOPHERSEN, K.-M., 1939: Odontologiske undersøgelser af Danmarks forhistoriske befolkning II. Om tændernes tilstand i Danmarks yngre Stenalder og Bronzealder. *Tandlægebladet,* 43:142.

CHRISTOPHERSEN, K.-M., 1940: Über die Zahnverhältnisse bei einer Volksgruppe der Wikingerzeit. *Acta Odont. Scand.,* 2:87.

CHRISTOPHERSEN, K.-M., 1941: Odontologiske undersøgelser af Danmarks forhistoriske befolkning IV. Om tændernes tilstand i Jernalderen. *Tandlægebladet,* 45:304. (English Summary.)

COMAS, J., 1948: *Bibliografia morfologica humana de america del sud.* Primera parte: texto. 208 pp. Mexico, D. F., Ediciones del Instituto Indigenista Interamericano.

COPPENS, L., 1963: An explanation for abrupt buccal and palatal retraction of the gingiva. *Internat. A. R. P. A.* Athens (reprint).

COURVILLE, C. B., 1950: Cranial injuries in prehistoric man with particular references to the Neanderthals. *Yearbook of Physical Anthropology, 1950*:185-205.

DAHL, E., 1960: Læbe- og ganespalte (English Summary) *Tandlægebladet, 64*:469.

DAHLBERG, A. A., 1960: The dentition of the first agriculturists. (Jarmo, Iraq). *Amer. J. Phys. Anthrop., 18*:243.

DASTAGUE, J., 1962: In: Ferembach, D., Dastague, J. and M.-J. Poitrat-Targowla. *La Nécropole Épipaléopithique de Taforalt Casablanca (Maroc oriental)*.

DAWSON, W. R., 1929: *Magician and Leech*. London, Methuen.

DERRY, D. E., 1913: A case of hydrocephalus in an Egyptian of the Roman Period. *J. Anat. Physiol., 47*: 436.

DERRY, D. E., 1938: Two skulls with absence of the premaxilla. *J. Anat., 72*:295.

DUCKWORTH, W. L. H., 1900: On a collection of Crania of the Moriori. *J. Roy. Anthrop. Inst., 30*:141. Quoted by Taylor (1963).

ELLIOT SMITH, G., and F. WOOD JONES, 1910: The archaeological survey of Nubia. Report on the human remains. Report for 1907-1908. Vol. II.

ELLIOT SMITH, G., and W. R. DAWSON, 1924: Egyptian mummies. London, Allen and Unwin.

ELLIOT SMITH, G., 1926: *Cambridge University Med. Soc. Mag., 4*:37 (quoted from *Dawson*, 1929).

EULER, H., 1939a: *Die Zahnkaries im Lichte vorgeschichtlicher und geschichtlicher Studien*. München.

EULER, H., 1939b: Über Lehren der Altertumsforschung für die marginalen Erkrankungen des Zahnhalteapparates. *Tandlægebladet., 43*:114.

EULER, H., 1940: Über die Entwicklung der Paradentose als Domestikationserscheinung und die dabei entstandene Kurve. *Paradentium., 12*:no. 1.

FASTLICHT, 1949: In: Terra, H. de, Romero, J., and T. Dale Stewart. *Tepexpan man*. New York, Viking Fund.

FEREMBACH, D., DASTAGUE, J., and M. J. POITRAT-TARGOWLA, 1962: La nécropole épipaléolithique de Taforalt (Maroc oriental). Edita-Casablanca Rabat, pp. 175.

GEJVALL, N.-G., 1960: *Westerhus. Medieval Population and Church in the Light of Skeletal Remains*. Lund, Boktryckeri.

GIESELER, W., 1951: Die süddeutschen Kopfbestattungen (Ofnet, Kaufertsberg, Hohlestein) und ihre zeitliche Einreihung. *Aus der Heimat., 59*:291. (quoted from Saller, 1962).

GOLDSTEIN, M. S., 1948: Dentition of Indian Crania from Texas. *Amer. J. Phys. Anthrop., 6*:63.

GOLDSTEIN, M. S., 1957: Skeletal pathology of early Indians in Texas. *Amer. J. Phys. Anthrop., 15*:299.

GOODMAN, C. N., and G. M. MORANT, 1940: The human remains of the Iron Age and other periods from Maiden Castle, Dorset. *Biometrika., 31*:295.

GORJANOVIC-KRAMBERGER, 1908: Anomalien und Krankhafte Erscheinungen am Skelette des Urmenschen von Krapina. *Die Umschau, 12*:623.

GREENE, J. C., 1963: Epidemiology of congenital clefts of the lip and palate. *Public Health Rep., 78*:589.

GRIMM, H., 1960: Aufbrauchs- und Verfallserscheinungen am Gebiss bei vor- und frühgeschichtlichen Menschenresten aus Mesopotamien. *Z. Alternsforsch.* 14:252.

GRIMM, H., and R. ZUHRT, 1960: Hallstadt-und La Tène-zeitliche Gesichtsschädelreste. *Deutsch. Zahn Mund Kieferheilk., 32*:265.

GRÖSCHEL, W., 1937: Pathologische Erscheinungen an den Zähnen und Kiefern der Alemannen aus den Begräbnisstätten der Merovingerzeit des Bezirkes Dillingen an der Donau. *Deutsch. Zahn. Mund. Kieferheilk., 4*:370.

GULLBERG, J. E., and A. M. BURKITT, 1924: An abnormal skull from New Guinea with remarks on the structure of the mandible. *J. Anat., 59*:41.

HAMMARLUND-ESSLER, E., 1952: Histologisk undersökning av tänder och käkben från medeltidsskelett. *Svensk tandläk. T., 45*:275:

HARTWEG, R,. 1945: Remarques sur la denture et statistiques sur la carie en France aux époques préhistorique et proto-historique *Bull. Soc. Anthrop.*, Paris 6—IX serie:71.

HEBERER, G., 1938: Die mitteldeutschen Schnurkeramiker. *Veröff. Landesanstalt Volkheitskd. Z. Halle.*, 10:00.

HEDEGARD, B., 1956: Parodontala Aldersförändringar. *Skand. Tandl. forenings 31 kongres in Stockholm*.

HELD, A. J., 1938: Considérations sur quelques documents anthropologiques. *Rev. Mens. Suisse Odont.*, 48:1168.

HENKEL, G., 1961: Zur Frage der Parodontopathien bei frühgeschichtlichen Schädelmaterial. *Deutsch. Stomat., 11*:195.

HENKEL, G., 1962: Zahn- und Kieferverhältnisse von Skeletten des Mittelalters (Michaeliskirsche zu Jena). *Deutsch. Zahn Mund Kieferhlk., 37*:330.

HENSCHEN, F., 1962: *Sjukdomarnas Historia och Geografi*. Bonniers Boktr.

HEUSER, H., and H. PANTKE, 1959: Untersuchungen an jungsteinzeitlichen Gebissen. *Stoma., 12*:148.

HILMING, F., and P. O. PEDERSEN, 1940: Über die Paradentalverhältnisse und die Abrasion bei rezenten ostgrönländischen Eskimos. *Paradentium., 12*:69.

HOLMER, U. and MAUNSBACH, A. B., 1956: Odontologische Untersuchungen von Zähnen und Kiefern des Menschen aus der Steinzeit in Schweden. *Odont. T., 64*:437.

HOOTON, E. A., 1917: *Harvard University African Studies. I. Oral Surgery in Egypt during the Old Empire.* Cambridge, Harvard University Press. Quoted by *Weinberger* (1948).

HOOTON, E. A., 1930: *The Indians of Pecos Pueblo. A Study of their Skeletal Remains.* Papers of the Southwestern Expedition 4. Yale University Press.

HOYME, L. E., and W. M. BASS 1962: Human skeletal remains from the Tollifero (Ha 6) and Clarksville (Mo 14) sites, John H. Kerr Reservoir basin Virginia. *Bur. Amer. Ethnol. Bull., 182*:329.

HRDLIČKA, A., 1941: Lower Jaw: Double Condyles. *Amer. J. Phys. Anthrop., 23*:75.

HUELKE, D. F., BURDI, A. R., and C. E. EYMAN, 1962: Association between mandibular fractures and site of trauma, dentition and age. *J. Oral Surg., 20*:478.

HULTKRANTZ, J. W., 1908: Über Dysostosis cleidocranialis. *Z. Morph. Anthrop., 11*:385.

HUNT, E. E., 1961: Malocclusion and civilisation. *Amer. J. Orthodont., 3*:406.

INGELMARK, B. E., 1939: The Skeletons; p. 149-209 In: Thordeman, B. Armour from the Battle of Wisby 1361.

INGELMARK, B. E., MØLLER-CHRISTENSEN, V., and O. BRINCH, 1959: Spinal joint changes and dental infections. *Acta Anat.*, Suppl. 36=1 ad Vol. 38.

ISAGER, K., 1936: *Skeletfundene ved Øm Kloster.* Copenhagen.

IVY, R. H., 1962: The influence of race on the incidence of certain congenital anomalies, notably cleft lip-cleft palate. *Plast. Reconstr. Surg., 30*:581.

JERUSALEM, C., 1955: Über die histologische Diagnose postmortal und intravital entstandener Knochendefekte. *Z. Morph. Anthrop., 47*:67.

KALLAY, J., 1955: Krapinaunterkiefer I. und Parodontose im Pleistozän Öst. *Z. Stomat., 49*:239.

KARSH, R. S., and J. D. MCCARTHY, 1960: Archaeology and arthritis. *Arch. Intern. Med.* (Chicago), *105*:640.

KEITH, A., 1913: Abnormal Crania—Achondroplastic and Acrocephalic. *J. Anat. Physiol., 47*:189.

KEITH, A., 1925: *The Antiquity of Man,* 2 vols. London, Williams and Norgate.

KEITH, A., 1931: *New Discoveries Relating to the Antiquity of Man.* London, Williams and Norgate.

KLASTSKY, M., and R. L. FISCHER, 1953: The *Human Masticatory Apparatus.* London, Kimpton.

KLIMA, B., 1950: Hrog ženy lovce mamutu v Dolních Věstonicích. *Archeologické Rozhledy, 2*:32.

KLINDT-JENSEN, O., 1960: Var det en yndig tid? *Skalk, 2*: no. 3.

KÖTZSCHKE, G., 1960: Parodontopathien im Paläolithikum. *Forum Parodontologicum, Vierjährliche Beilage der Deutschen Zahnärztlichen Z., 15*:1173.

KOZUBKIEWICZ, Z., and B. TRACHTENBERG, 1960: Stomatological investigations of the excavated human remains from Kaldus (XI-XII Cent.), Stary Brzesc (XII-XVI Cent.), Tum (XIII-XVII Cent.) and Brzesc Kujawski (XVI-XVIII Cent.). *Czas. Stomat., 13*:29.

KRAUS, E., 1939: In: EULER, H. *Die Zahnkaries im Lichte vorgeschichtlicher und geschichtlicher Studien.* München.

KRØMER, H., 1961: Kjevefrakturer og krigsodontologisk behandling. In: *Nord. Klinisk Odontologi, 4*:20.

LE BARON, J., 1881: *Lesions osseuses de l'homme préhistorique en France et en Algérie.* Thèse pour le Doctorat en Médicine. No. 262. Paris.

LEGOUX, P., 1962: Nouvelle étude anthropologique des "Negroïdes des Grimaldi." *C. R. Acad. Sci. 255*: 2276.

LEIGH, R. W., 1925: Dental pathology of Indian tribes of varied environmental and food conditions. *Amer. J. Phys. Anthrop., 8*:179.

LEIGH, R. W., 1929: Dental morphology and pathology of prehistoric Guam. *Mem. of the Bernice P. Bishop Museum, 11.*

LEIGH, R. W., 1929: Dental pathology of aboriginal California. *Dent. Cosmos., 71*:756, 878.

LEIGH, R. W., 1937: Dental morphology and pathology of pre-Spanish Peru. *Amer. J. Phys. Anthrop., 22*: 267.

LUNDSTRØM, A., and L. LYSELL, 1953: An anthropological examination of a group of medieval Danish skulls, with particular regard to the jaws and occlusal conditions. *Acta Odont. Scand., 11*:111.

LYSELL, L., 1958: A biometric study of occlusion and dental arches in a series of medieval skulls from northern Sweden. *Acta Odont. Scand., 16*:177.

LÖE, H., 1963: Epidemiology of periodontal disease. *Odont. T. 71*:479.

MACCURDY, G. G., 1923: Human Skeletal remains from the Highlands of Peru. *Amer. J. Phys. Anthrop., 6*: 217.

MATHIS, H., and F. CLEMENTSCHITSCH, 1939: Bericht über eine Untersuchung an Zähnen u. Kiefern der prähistorischen, historischen und gegenwärtigen Bevölkerung im Gebiete des Gaus Niederdonau. *Z. Stomat., 37*:1418.

MCCOWN, T. D., and A. KEITH, 1939: *The Stone Age of Mount Carmel.* Vol. 2: Oxford U.P.

MELLQUIST, C., and T. SANDBERG, 1939: Odontological studies of about 1400 medieval skulls from Halland and Scania in Sweden and from the Norse Colony in Greenland and a contribution to the knowledge of their Anthropology. *Odont. T.* Supp. No. 3 B.

MICHMAN, J., and F. SAGHER, 1957: Changes in the anterior nasal spine and the alveolar process of the maxillary bone in leprosy. *Int. J. Leprosy. 25*:217.

MILLS, J. R. E., 1963: Occlusion and malocclusion of the teeth of Primates. In: Brothwell, D. R. (Ed.) *Dental Anthropology.* Oxford. Pergamon.

MOHAUPT, F., 1939: Untersuchungen über die Karieshäufigkeit sowie das Auftreten anderer Gebisserkrankungen in der jungeren Steinzeit im engeren mitteldeutschen Gebiet. *Deutsche Zahnärztl. Wschr.* 42:222.

Mollison, Th. 1936: Zeichen gewaltsamer Verletzungen an den Ofnet-Schädeln. *Anthrop. Anz., 13*:79.

Moodie, R. L., 1923: *Paleopathology. An introduction to the study of ancient evidences of disease.* Illinois. University Press.

Moodie, R. L., 1928a: Studies in paleodontology. *J. Amer. Dent. Ass., 15*:1826.

Moodie, R. L., 1928b: *Pacific Dent. Gazette* 36: 669-674. Quoted by *Stewart* (1950).

Møller-Christensen, V., and O. Brinch, 1948: Tooth infection and spinal joint diseases in medieval Denmark. *Paradentologie,* 2:12.

Møller-Christensen, V., 1952: Case of leprosy from the Middle Ages of Denmark. *Acta Med. Scand.,* Suppl. CCLXVI: 101.

Møller-Christensen, V., 1953: *Ten lepers from Næstved in Denmark.* Copenhagen, Danish Science Press.

Møller-Christensen, V., 1958: Bogen om Æbelholt Kloster. (With English Summary) Copenhagen, Danish Science Press.

Møller-Christensen, V., 1961: *Bone changes in leprosy.* Copenhagen, Munksgård.

Moss, M. L., 1960: A reevaluation of the dental status and chronological age of the Tepexpan remains. *Amer. J. Phys. Anthrop., 18*:71.

Murphy, T., 1959: Compensatory mechanisms in facial height adjustment to functional tooth attrition. *Aust. Dent. J.,* 4:312.

Nielsen, H. A., 1911: Yderligere Bidrag til Stenalderfolkets Anthropologi. *Aarb. nord. Oldk. Hist.* p. 181.

Pales, L., 1930: *Palaeopathologie et Pathologie Comparative.* Paris, Masson.

Parma, C., 1948: Research on prehistoric jaws. *Paradentologie,* 2:123.

Patte, E., 1955: *Les Néanderthaliens; Anatomie, Physiologie et Comparaisons.* Paris. Masson.

Pedersen, P. O., 1941: Interview in *Dens Sapiens* 1:192.

Pedersen, P. O., 1941: Über angeborene Missbildungen im Bereiche der Mundhöhle bei grönländischen Eskimos und westgrönländischen Mischlingen. *Z. Rassenkunde.,* 12:20.

Pedersen, P. O., 1944: Dental Notes and a Chapter on the Dentition. In: Brøste, K., K. Fischer-Møller and P. O. Pedersen. The Mediaeval Norsemen at Gardar. *Medd. om Grønland.* 89: no. 3.

Pedersen, P. O., 1955: Tænder og tandsygdomme hos nogle historiske personer. *Tandlægebladet,* 59:197. (English summary).

Péquart, M. et S-J., Boule, M., and H. Vallois, 1937: Teviec. Stationnecropole mesolithique du Morbihan. *Arch. Inst. Paleontologie Humaine.,* Mem. 18.

Périer, A.-L., 1948: Introduction critique a la paleopathologie des organes maxillo-dentaires. *Paradentologie.,* 2:1.

Périer, A.-L., 1953: In Ackermann, F. Le méchanisme des mâchoires.

Periodontal Disease. Report of an Expert Committee on Dental Health. *W. H. O. techn. Res. Ser.* 207. 1961.

Philippas, G. G., 1952: Effects of function on healthy teeth: the evidence of ancient Athenian remains. *J. Amer. Dent. Ass., 45*:443.

Picton, D. A. C., 1957: Calculus, wear and alveolar bone loss in the jaws of sixth-century Jutes. *Dent. Pract.* 7:301.

Poitrat-Targowla, M.-J., 1962: In: Ferembach, D., Dastague, J. and M.-J. Poitrat-Targowla. *La nécropole épipaléolithique de Taforalt (Maroc oriental).*

Poulsen, K., 1909: Contributions to the Anthropology and Nosology of the East-Greenlanders. *Medd. om Grønland.,* 28: no. 4.

Präger, W., 1925: Das Gebiss des Menschen in der Altsteinzeit und die Anfänge der Zahnkaries. *Deutsche zahnärztl. Wschr.,* 28:88; 122.

Rabkin, S., 1942: Dental conditions among prehistoric Indians of Northern Alabama. *J. Dent. Res., 21*:211.

Rabkin, S., 1943: Dental conditions among prehistoric Indians of Kentucky. The Indian Knoll collection. *J. Dent. Res., 22*:255.

Regan, J. E. and Mitchell, D. F. 1963: Evaluation of periapical radiolucencies found in cadavers. *J. Amer. Dent. Ass., 66*:529.

Rihan, H. J., 1930: In: Hooton, E. A. The Indians of Pecos Pueblo.

Risdon, D. L., 1939: A study of the cranial and other human remains from Palestine excavated at Tell Duweir (Lachish) by the Wellcome-Marston Archaeological Research Expedition. *Biometrika, 31*:99.

Ruffer, M. A., 1921: *Studies in the Palaeopathology of Egypt.* Chicago, University of Chicago Press.

Rygge, J., 1913: Tandsygdommene i Norges middelalder samt kostholdet i dette tidsrum sammenlignet med nutidens. *Norske Tannlæge-foren. Tid., 22*:335.

Salama, N., and A. Hilmy, 1951: An ancient Egyptian skull and a mandible showing cysts. *Brit. Dent. J.,* 90:17.

Saller, K., 1962: Die Ofnet-Funde in neuer Zusammensetzung. Ihre Stellung in der Rassengeschichte Europas. *Z. Morph. Anthrop., 52*:1.

Sahly, A., H. Brabant, and M. Bouyssou, 1962: Observation sur les dents et les maxillaires du Mésolithique et de l'Age du Fer trouvés dans la grotte de Rouffignac, département de la Dordogne en France. *Bull. Group. int. Rech. Sci. Stom., 5*:252.

Sarasin, F., 1922: *Anthropologie des Néo-Calédoniens et des Insulaires des Loyalty "Nova Caledonia."* C. W. Kreidel (ed.), Wiesbaden.

Sarnat, B. G., and D. M. Laskin, 1962: *Diagnosis and Surgical Management of Diseases of the Temporomandibular Joint.* Illinois.

Schaeuble, J., 1958: Die hethitischen Grabfunde von Osmankayasi. *Wiss. Veröff. deutsch. Orient-Gesell.,* 71. Berlin, 85 p.

SCHRANZ, D., 1962: Zahnbetterkrankungen der längstvergangenen Zeit. *Z. Morph. Anthrop.*, 52:347.

SCHWARZ, R., 1922: Veränderungen im Kiefergelenk der Neu-Caledonier und Loyalty-Insulaner und ihre Bedeutung für die zahnärztliche Prothese und Orthodontie. *Schweiz. Mschr. Zahnheilk.*, 32: no. 8:373.

SCHWERZ, F., 1916: Pathologische Erscheinungen an Alemannenzähnen aus dem 5.-10. Jahrhundert. *Schweiz. Vierteljahrschr. Zahnheilk.*, 26:1.

SELIGMANN, C. G., 1912: A cretinous skull of the Eighteenth Dynasty. *Man*, 12:17.

SHAFER, W. G., M. K. HINE, and B. M. LEVY, 1963: *A Textbook of Oral Pathology*, Second ed. Philadelphia and London.

SIGERIST, H. E., 1951: *A History of Medicine*, I. Primitive and Archaic Medicine. New York.

STAFNE, E.-C., AUSTIN, L. T., and B. GARDNER, 1936: Median anterior maxillary cysts. *J. Amer. Dent. Ass.*, 23:801.

STEWARD, J. H., (ed.), 1950: Handbook of South American Indians. Vol. 6. *Smiths. Inst. Bureau of Amer. Ethnol. Bull.* 143.

STICKER, 1935: Zur Pathologie der Alemannen. *Verh. Ges. Phys. Anthrop.*, 7:28.

STRAHAN, J. D., 1963: The relation of the mucogingival junction to the alveolar bone margin. *Dent. Pract.*, 14:72.

STRAUS, W. L., and A. J. E. CAVE, 1957: Pathology and the posture of Neanderthal man. *Quart. Rev. Biol.*, 32:348.

TALLGREN, A., 1957: Changes in Adult Face Height. *Acta Odont. Scand.*, 15: suppl. 24.

TAYLOR, R. M. S., 1962: Non-Metrical studies in the human palate and dentitions in Moriori and Maori skulls. *J. Polynesian Soc.*, 71:83; 167.

TAYLOR, R. M. S., 1963: Cause and effort of wear of teeth. Further non-metrical studies of the Teeth and Palate in Moriori and Maori skulls. *Acta Anat.*, 53:97.

TEILADE, J., 1960: Usikkerheden ved røntgenologisk bedømmelse af marginalt knoglesvind. *Tandlægebladet*, 64:351.

TERRA, H. DE, J. ROMERO, and T. D. STEWART, 1949: *Tepexpan man*. New York, Viking Fund Public. in Anthrop. 11.

THOMA, K. H., 1917: Oral diseases in ancient nations and tribes. *J. Allied Dent. Soc. New York*, 12:327. Quoted by Moodie (1923).

THOMPSON BROOKS, S., and W. D. HOHENTHAL, 1963: Archaeological Defective Palate Crania from California. *Amer. J. Phys. Anthrop.*, 21:25.

TRACHTENBERG, B., 1960: Parodontosis in excavated human skulls from the Neolithic Age, Middle Ages and the XVI-XVIII Centuries. *Czas. Stomat.*, 13:889.

THORDEMAN, B., 1939: Armour from the Battle of Wisby 1361. *Kungl. Vitt. Hist. Antikv. Akad.*, 1:480 p.

TWIESSELMANN, F., and H. BRABANT, 1950: Observations sur les dents et les maxillaires d'une population ancienne d'age franc de Coxyde, Belgique. *Bull. Group. Int. Rech. Sci. Stomat.*, 3:99; 355.

VEAU, V., 1931: *Division palatine; anatomie, chirurgie, phonetique*. Paris.

VIRCHOW, H., 1920: *Die menschlichen Skelettreste aus dem Kämpfeschen Bruch von Ehringsdorf bei Weimar*. Jena.

VISCHER, A., 1921: Sur une modification très frequente de l'articulation temporo-mandibulaire dans les crânes de Néo-Calédoniens et d'Insulaires des Loyalty. *Schweiz. Med. Wschr.*, no. 30.

WALKER, D. GREER, 1956: The mandibular condyle. *Dent. Pract.*, 7:160.

WEINBERGER, B. W., 1948: *An Introduction to the History of Dentistry, Vol. 1*. St. Louis.

WEINERT, H., 1938: Dem Unterkiefer von Mauer zur 30 jährigen Wiederkehr seiner Entdeckung. *Z. Morph. Anthrop.*, 37:102.

WILLIAMS, H. U., 1929: Human paleopathology. *Arch. Path.*, 7:839.

WINGATE TODD, T., 1921: Egyptian Medicine: A critical study of recent claims. *Amer. Anthrop.*, 23:460.

WITKOP, C. J., 1958: Genetics and Dentistry. *Eugen. Quart.*, 5:15.

WOOD JONES, F., 1910: In: Elliot Smith, G., and F. Wood Jones.

ZUHRT, R., 1956: Stomatologische Untersuchungen an spätmittelalterlichen Funden von Reckahn (12.-14. Jh.). I and II. *Deutsche Zahn. Mund. Kieferheilk.*, 25:1.

ZUHRT, R., 1960: Stomatologische Untersuchungen an spätmittelalterlichen Funde. III. *Deutsche Zahn. Mund. Kieferheilk.*, 32:51.

ZUHRT, R., 1962: Hergang und Folgen eines doppelseitigen Unterkieferbruchs in schnurkeramischer Zeit. *Ausgrabungen und Funde.*, 7:6.

SECTION VI
ACCIDENTAL TRAUMA AND SURGICAL INTERVENTION

Chapter 47

Trauma and Disease of the Post-Cranial Skeleton in Ancient Egypt

PHILIP SALIB

NIL *novi sub sole.* This truism is well exemplified by the study of trauma and disease in Ancient Egypt. Even during the Archaic and Old Kingdom periods cultural levels were amazingly advanced so that it is not surprising that we are able to derive knowledge of disease and injury from sources other than actual remains.

For example, such conditions as deformities are depicted on temple walls and among thousands of statuettes recovered from tombs are some which represent Pott's disease and achondroplasia. The Edwin Smith Surgical Papyrus has also thrown light on the considerable knowledge of the Ancient Egyptians of traumatic surgery. We are impressed by the business-like way in which they presented their cases and decided on prognosis and treatment.

Increasing interest in Egyptology during the early years of this century led to the discovery of many tombs which have provided much useful information. Thousands of skeletons and many mummies have been examined; these palaeopathological studies give us a fairly comprehensive knowledge of disease and trauma of the post-cranial skeleton in these ancient times.

Congenital anomalies and deformities are known; for example, in the Egyptian Museum at Cairo the mummy of Pharaoh Siptah (XIXth Dynasty, c. 1215 B.C.) shows a left club foot (Salib 1962).

Most dwarfs were cases of achondroplasia (Fig. 1). They existed from the predynastic times throughout all the dynasties (Dawson, 1938). Achondroplastic dwarfs were respected by Ancient Egyptians: "Laugh not at a blind man nor tease a dwarf." They were employed in important jobs like key-holders, masters of wardrobes, treasurers or personal attendants to the royal family. Their skeletons were found in the royal necropolis (Dawson, 1938) denoting how privileged they had been. On the other hand, Pygmies, who are an abnormally small race, were imported from central Africa and used to perform sacred dances and less important duties. A period of five thousand years has not changed the clinical and pathological characteristics of this condition (Ruffer 1921).

Two skeletons presenting multiple deformities attributed to dyschondroplasia were found in one tomb. They lie now in the Anatomy Museum of the Kasr-el-Ainy Faculty of Medicine, Cairo (Salib, 1962). Genu recurvatum was also represented in drawings on the walls of the tombs of Ptah-Hotep and Edout (VIth Dynasty, 2460-2270 B.C.) at Sakkara near Cairo.

FIGURE 1. The dwarf Seneb and his family (Vth Dynasty, 2700 B.C.) discovered near the Pyramids of Giza by the expedition of the Vienna Academy of Science in 1927. This painted limestone group is displayed at the Egyptian Museum, Cairo. Note the sculptor's tactful disposition of the children to maintain the symmetry of the group without compromising the respect of the Head of the family. (Courtesy *J. Bone Joint Surg.*)

Deformities presumably due to infantile paralysis have been found in a few drawings. The stele of the XVIIIth Dynasty (1580 B.C.) at the New Carlsberg Glyptotek in Copenhagen shows a severely atrophied and shortened right leg with talipes equinus which might be the result of poliomyelitis (Salib, 1962; Sigerist, 1951).

By far the most common affections of the post-cranial skeleton were spondylitis and osteo-arthritis (arthritis deformans). No ethnic group in Egypt was immune and almost every adult was affected to a greater or lesser degree. It may possibly be true that during this period osteo-arthritis was as common in other countries as in Egypt and Nubia (Ruffer and Rietti, 1912; Zorab, 1961). The lumbar spine and sacro-iliac joints were very early involved. The weight-bearing larger joints were more commonly affected than those of the upper extremities and peripheral articulations. At times the intensity was so severe that it resulted in ankyloses, and the afflicted were left considerably crippled and incapacitated. Nevertheless, some patients lived for a long time in spite of their infirmity (El-Batrawi, 1935; Elliot Smith, 1910; Riad, 1955; Ruffer, 1921).

Syphilis was unknown in Ancient Egypt. Skeletal tuberculosis, on the other hand, was encountered in the form of Pott's disease, the lumbar region being most involved (Salib, 1962). Statuettes, drawings, skeletons and mummies depicted the disease in many instances. Vertebral body destruction occurred with evidence of abscess formation in acute and subacute cases, and of fusion with gibbus in chronic forms (Derry, 1938; Blankoff, 1958).

Spondylolisthesis was met with in a few skeletons. Others showed typical gout, and concretions revealed uric acid crystals on examination (Elliot Smith and Dawson, 1924).

Rickets probably existed but it seems to have been rare. Many bones, however, exhibit distortions which are otherwise difficult to explain (Elliot Smith and Wood-Jones, 1910).

Sepsis of the post-cranial skeleton was seldom encountered. This may perhaps be accounted for by a remarkable resistance to infection or, on the contrary, by a resistance so low that the victims died before appreciable evidence of disease was imprinted in the bones. The latter view is the more probable one, and is supported by the fact that pyorrhoea alveolaris was an extremely prevalent disease (Salib, 1962).

Bone neoplasms such as osteoma, exostosis and osteosarcoma occurred but they were uncommon.

Traumatic lesions were fairly common. Sprains were described in the papyri. Dislocations are, in general, rarely found. This is not surprising because, unlike fractures, reduced dislocations leave no permanent evidence of their occurrence (Salib, 1962). Many cases of fractures were, however, encountered, some united and others still under treatment.

Six thousand skeletons of all ages and historical periods have been examined. They date from the Predynastic period (before 3400 B.C.) to the early Christian era. One hundred and sixty cases of fractures were seen (Elliot Smith and Wood Jones, 1910). The commonest fractures of the post-cranial skeleton were those of the forearm bones (31 per cent); the left forearm was injured more often than the right because in defence, whether in games or in war, blows were averted by raising the left arm (Salib, 1962). A popular game is still practised among Egyptian villagers, called "El-Tahteeb," where two men use long canes both for attack and defence. Other bones also sustained fractures, namely the clavicle (14 per cent), femur (12 per cent), leg bones (10 per cent), humerus (7 per cent), ribs (6 per cent) and pelvis (4 per cent). Rarely fractures occurred in the hand and foot bones, scapula and sternum. In contrast to present-day experience there was a striking absence of fracture-dislocation of the ankle. Six cases of femoral neck fractures and five of separation of the proximal femoral epiphysis were found in one cemetery of the early Christian period. This is attributed to the fact that Christians used to wear

FIGURE 2. Fractures of the left radius and ulna with three pieces of splints made of bark (Museum of Anatomy, Faculty of Medicine, Cairo University). (Courtesy *J. Bone Joint Surg*).

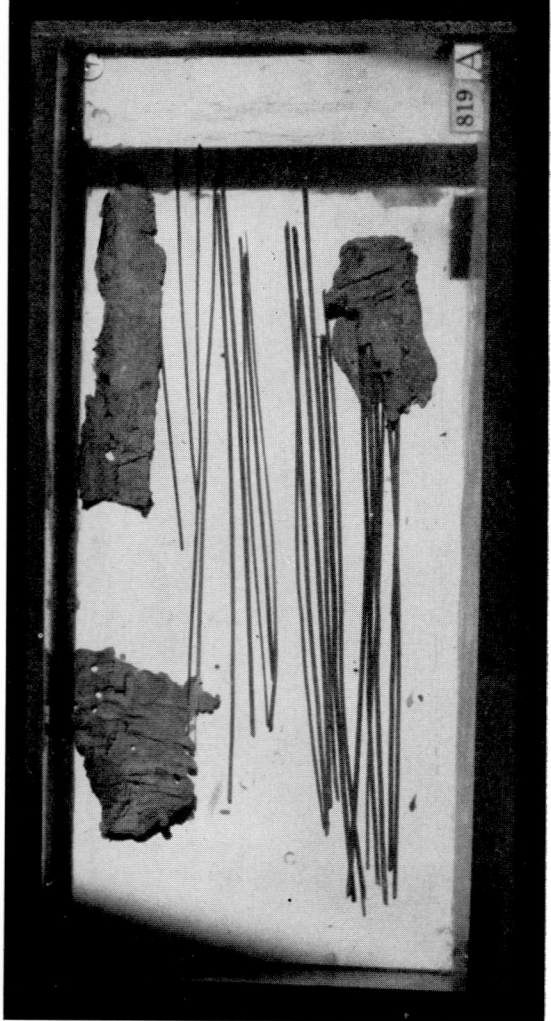

FIGURE 3. Ancient splints made from pieces of bark and straws (Museum of Anatomy, Faculty of Medicine, Cairo University). (Courtesy *J. Bone Joint Surg*).

boots and that they lived in rocky granite areas (Elliot Smith and Wood Jones, 1910).

Splint treatment was known as early as the Vth Dynasty of the Old Kingdom (2730-2625 B.C.), (Elliot Smith, 1908). The same kind of splint as well as the method of immobilization were in use over the centuries; splints found from the early Christian era are in all particulars identical to those used as early as the time of the Vth Dynasty. Several varieties of splints were employed; the mid-rib of the date-palm leaf, spongy strips of wood, bark of the Acacia tree or bundles of straw from coarse grass (Figs. 2 and 3). In contrast to modern practice more than two splints might be used to form a protective casing. Many turns of cotton bandage, with an average width of two inches, were made as a preliminary padding over which several splints were applied. Finally, more bandages were wound over the splints and sometimes a cord was tied over-all in several places to obtain more secure fixation. The fingers were not included in the splint but they were firmly bound with a separate linen bandage, evidently resulting in a considerable limitation of movement. In splint fixation, preliminary padding was always used and several different splints might be employed in combination to fix one fracture.

In spite of the many fractures encountered, very few splints were found; possibly they were removed before burial of non-mummified bodies (Elliot Smith and Wood Jones, 1910). On the other hand, some post mortem fractures were splinted. An example of this is the mummy of Pharaoh Siptah (XIXth Dynasty, 1215 B.C.), which had been maltreated by grave robbers and sustained a postmortem fracture of the arm, and was restored thereafter with splints (Salib, 1962).

Although the results of fracture treatment were strikingly good, cases of malunion and non-union have been found. Undoubtedly not all fractures received treatment. No healed fracture of the spine was encountered whereas sternal and rib fractures showed good union. All pelvic fractures healed, sometimes with malunion. Fractures of the bones of the upper limb gave variable results; those of a single bone of the forearm were fairly good. Fracture of the ulna was more common. Synostosis and deformities sometimes followed fractures of both bones of the forearm.

Fractures of the lower extremity involved the femoral neck with consequent malunion, coxa vara and excessive osteophytic formation. Subtrochanteric and femoral shaft fractures resulted in shortening with solid union, the best results being in the proximal third. Traumatic separation of the proximal femoral epiphysis was found in five cases. For some unexplained reason the early dynastic cemetery in the Shem Nishai and Dehmit area yielded the great majority of the sixteen cases of fracture of both tibia and fibula. Thirteen occurred in males and three in females. The results of fracture of a single bone of the leg were uniformly good, whereas fractures of both bones led to extremely bad results (Elliot Smith and Wood Jones, 1910).

Valuable information on Ancient Egyptian Medicine has been obtained from the papyri:

Name	Date	Location
Kahun	1900 B.C.	London
Edwin Smith	1700 B.C.	New York
Ebers	1550 B.C.	Leipzig
Hearst	1550 B.C.	Berkeley
Erman	1550 B.C.	Berlin
London	1350 B.C.	London
Berlin	1350 B.C.	Berlin (Leake, 1952)
Chester Beatty	1200 B.C.	London

Papyri were written in *hieratic* which was the rapid script written by hand, from right to left: *hieroglyphic*, however, was used on stone monuments and was written either way (Breasted, 1930). Of the eight chief medical papyri the most methodical is that of Edwin Smith, which is the only surgical text. Edwin Smith of Connecticut, U.S.A., bought his papyrus at Luxor, Egypt, in 1862. Until his death nothing was known of its contents. In 1906 his daughter presented it to the New York Historical Society which entrusted its study and translation to Professor J. H. Breasted in 1920. It took him nine years to complete this great task.

It dates back to 1700 B.C. and is probably a copy of an earlier manuscript which had been written at least a thousand years before, at some time in the Pyramid Age (3000-2500 B.C.) (Salib, 1962). The author of the original treatise is said to have been 'Imhotep,' the universal Father of Medicine.

This papyrus is the most important surgical treatise. Seventeen columns (three hundred and seventy seven lines) on the *recto* described forty-eight surgical cases, systematically arranged, beginning with head injuries and working downwards through the body. Each case is presented in an orderly way: the title, clinical examination, diagnosis, prognosis and treatment (Breasted, 1930). In every region the author starts by describing the simpler and more superficial lesions and goes on to the more complicated and dangerous ones. This papyrus revealed considerable power of clinical observation, based on knowledge and experience.

Sprain was described, accurately enough, as "rending of the two members although each is still in its place." Open fractures had a bad prognosis. In an open fracture of the humerus it says "If, however, thou findest that wound which is over the break, with blood issuing from it, and piercing through to the interior of his injury, thou shouldst say concerning him 'One having a break in his upper arm, over which a wound has been inflicted piercing through. An ailment not to be treated' " (Case 37). On the other hand, if the overlying wound is not communicating with the fracture site it is "An ailment with which I will contend."

Nothing could be more illuminating than to quote some of the observations, diagnoses and treatment in cases of injury to the post-cranial skeleton described in Edwin Smith papyrus.

Case 30 A sprain of cervical spine.

Case 31 Dislocation of a cervical vertebra. "If thou examinest a man having a dislocation in a vertebra of his neck, shouldst thou find him unconscious of his two arms and his two legs on account of it, and urine drops from his member without his knowing it; his flesh has received wind; his two eyes are blood shot, it is a dislocation of a vertebra of his neck . . . . an ailment not to be treated."

Case 32 Displacement of a cervical vertebra. "Look at thy breast and thy two shoulders; he is unable to turn his face that he may look at his breast and his two shoulders . . . . a sinking of a vertebra of his neck to the interior of his neck, as foot settles into cultivated ground . . . . an ailment which I will treat."

Case 33 A crushed cervical vertebra. "One vertebra has fallen into the next one, while he is voiceless and cannot speak; his falling head downward has caused that one vertebra crush into the next one; shouldst thou find that he is unconscious of his two arms and his two legs because of it . . . . an ailment not to be treated."

Case 34 Dislocation of the two clavicles. "shouldst thou find his two shoulders turned over and the heads of his two collarbones turned toward his face."

Case 35 A fracture of the clavicle. "place him prostrate on his back, with something folded between his two shoulders in order to stretch apart his collar-bone until that break falls into its place. Thou shouldst make for him two splints of linen, and thou shouldst apply one of them on the inside of his upper arm."

Case 36 A fracture of the humerus. The arm is fixed in splints after reduction.

Case 37 A fracture of the humerus with rupture of overlying soft tissue. "an ailment not to be treated."

Case 38 A split in the humerus.

Case 42 A sprain of the sternocostal articulations.

Case 43 A dislocation in the sternocostal articulations.

Case 44 Fractured ribs.

Case 48 A sprain in a spinal vertebra (Uncompleted). "thou shouldst say to him 'Extend now thy two legs and contract them both again.' When he extends them both he contracts them both immediately because of the pain he causes in the vertebra of his spinal column in which he suffers .... an ailment which I will treat." (Hussein n. d.; Lefebvre, 1956).

REFERENCES

BLANKOFF, B., 1958: *Un peu d'histoire de la tuberculose osteoarticulaire—Mal de Pott.* Brussels, Drukkerij.

BREASTED, J. H., 1930: *The Edwin Smith Surgical Papyrus.* Chicago, University of Chicago Press.

DAWSON, W. R., 1938: Pygmies and Dwarfs in Ancient Egypt. *J. Egypt. Archaeol.,* 24:185.

DERRY, D. E., 1938: Pott's Disease in Ancient Egypt. *Med. Press.,* 197:No. 5183.

EL-BATRAWI, A. M., 1935: Report on the Human Remains. *Mission. Archéol. Nubie. 1929-1934.* Cairo, Government Press.

ELLIOT SMITH, G., 1908: The Most Ancient Splints. *Brit. Med. J.,* 1:732.

ELLIOT SMITH, G., and DAWSON, W. R., 1924: *Egyptian Mummies.* London, Allen and Unwin.

ELLIOT SMITH, G., and WOOD JONES, F., 1910: Report on the Human Remains. *The Archaeological Survey of Nubia. 1907-1908.* Cairo, National Printing Department.

HUSSEIN, M. K., no date: *The Edwin Smith Papyrus.* Cairo, Mondiale Press.

LEAKE, C. D., 1952: *The Old Egyptian Medical Papyri.* Kansas, University of Kansas Press.

LEFEBVRE, G., 1956: *Essai sur la Médecine Égyptienne de l'Epoque Pharaonique.* Paris, Presse Universitaire de France.

RIAD, N., 1955: *La Médecine au Temps des Pharaons.* Paris, Librairie Maloine.

RUFFER, M. A., and RIETTI, A., 1912: On Osseous Lesions in Ancient Egyptians. *J. Path. Bact.,* 16:439.

RUFFER, M. A., 1921: *Studies in the Palaeopathology of Egypt.* Chicago, University of Chicago Press.

SALIB, P., 1962: Orthopaedic and Traumatic Skeletal Lesions in Ancient Egyptians. *J. Bone Jt. Surg.,* 44B:944.

SIGERIST, H. E., 1951: *A History of Medicine.* Vol. I. *Primitive and Archaic Medicine.* New York, Oxford University Press.

ZORAB, P. A., 1961: The Historical and Prehistorical Background of Ankylosing Spondylitis. *Proc. Roy. Soc. Med.,* 54:415.

Chapter 48

Cranial Injuries in Prehistoric Man

CYRIL B. COURVILLE

A SEARCH through the literature on Anthropology has brought to light considerable evidence that cranial injuries were not uncommon during the prehistoric period of human development. It seems quite clear that man's inhumanity to man, as manifested by wounds of the head, is a heritage of his past, and it seems worth while to review the various facts which have come to light regarding cranial injuries in ancient times.

It is often possible to examine excavated skulls showing injury and determine not only how the injury was inflicted, its immediate and ultimate effects, but perhaps even the purpose for which the injury was produced (if it was not actually accidental). So the student of anthropology is transformed into a coroner, and by careful examination of the remains is called upon to reconstruct the scene of violence of some millennia ago.

The study of extant crania showing marks suggestive of physical violence is also related to the beginning of the history of surgery. In prehistoric populations, probably little or no effort was made to treat cranial injuries. Even after trepanation was introduced, its application was more likely part and parcel of the magic arts practiced by the medicine man than a *bona fide* effort to treat a specific wound. It is true, of course, that any given ancient skull *per se* does not give any information as to contemporary symptoms of intracranial pressure (headaches), psychic phenomena, convulsive seizures (suggesting brain changes) for which this form of operation has ultimately come to be used. In the protohistoric period, however, there is now abundant evidence of cranial injury and trepanation occurring together (Courville and Abbott, 1942), indicating the first attempts at treatment of brain wounds by surgical measures, now a commonplace practice.

CRANIAL INJURIES IN THE PITHECANTHROPINES OF ASIA

In a report on a series of skulls found in the Middle Pleistocene deposits at Choukoutien, near Peking in China, Weidenreich (1943) noted the occurrence of fracture lines and depressions whose characteristics indicated that physical violence had been expended on the head, injuries which in some instances at least, the individual had survived. Brief mention will be made of the location and nature of these injuries as found in a series of fourteen skulls represented by cranial fragments of various size. In four of these specimens (Skulls VI, X, XI, XII) what appear to be unequivocal antemortem fractures were found (see Figs. 2 and 3 for typical features).

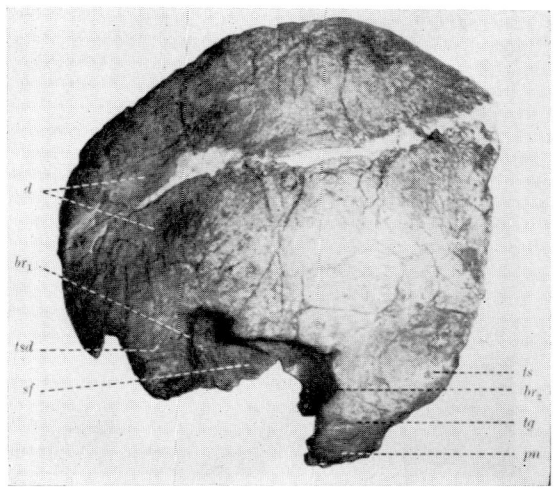

FIGURE 1 Depressed skull fracture of frontal bone in Ehringsdorf (Neandertal), skull, probably due to a blow with a blunt object. (after Weidenreich 1928) *d*. depression fracture. *or* pars nasal. *sf* opened frontal sinus *tg* glabella *ta*, ridge of eye brow, *fr* broken border of orbital cavity.

Skull VI, assumed to be that of a female of advanced age, is represented by four fragments—the right half of the frontal bone, a part of the left parietal bone, a smaller piece of the same parietal bone, and a fragment of the squamous portion of the right temporal bone. The left parietal bone presented "a typical depressed fracture outlined by sharp borders in the center of which the three fissures meet. There can be no doubt that the bone was hit in this place by a blow that depressed the bone and caused the fracturing."

Skull X, which is represented by most of the calvarium, was evidently that of a male of slightly advanced age. On the left side there was found a long deep depression running parallel to the sagittal suture Fig. 2b. "This depression begins near the vertex, crosses the coronal suture, and ends half way down the frontal squama on a broad fissure apparently of a later accidental provenance." A second injury was found on the right side. "This depression is circular, its diameter amounting to about 20 mm, and situated almost on the mid-line a little behind the vertex, mainly on the right side and affecting the sagittal margin of the left parietal bone . . . the inner table is broken around an oval line and projects into the cavity." The injury in this case was presumed by Weidenreich to be by an edged weapon.

In Skull XI, that of a female individual, which is also represented by the major portion of the calvarium, evidence indicative of a double injury was found. "The left [parietal] bone exhibits a typically semicircular, depressed fracture with fairly sharp borders situated just behind the vertex of the parietal tuberosity and at the edge of a large irregular defect which extends over the superior and posterior parts of the bone. This fracture is apparently the result of a heavy blow breaking the bone into smaller pieces along the transverse line that crosses the depressed area. The second injury occupies an almost identical place on the right side. The only difference is that the depressed area appears shallower, and no parts of the parietal bone were lost. They were recovered and used for the restoration. But it is evident that the depressed area represents a center from which three fissures start; one running backwards, one medially, and one laterally, the two later following the same straight line. In the first described injury the inner side is broken away; in the second it is in place but does show particular fracturing, except for a slight prominence of the broken pieces which border the transverse fissure." (See Fig. 2b, c, d). The injuries of the skull thus described are presumed to be due to blows from the impact of stones.

Skull XII also presented evidence of injury of a similar nature. This skull similar to the previous two specimens, was composed of the larger part of the calvarium. It was evidently from an adolescent male. "This injury, which has more the appearance of a cut, is found close to the sagittal margin of the parietal bone and extends from the suture just behind the vertex forward and laterally toward the coronal suture."

Figure 2a.

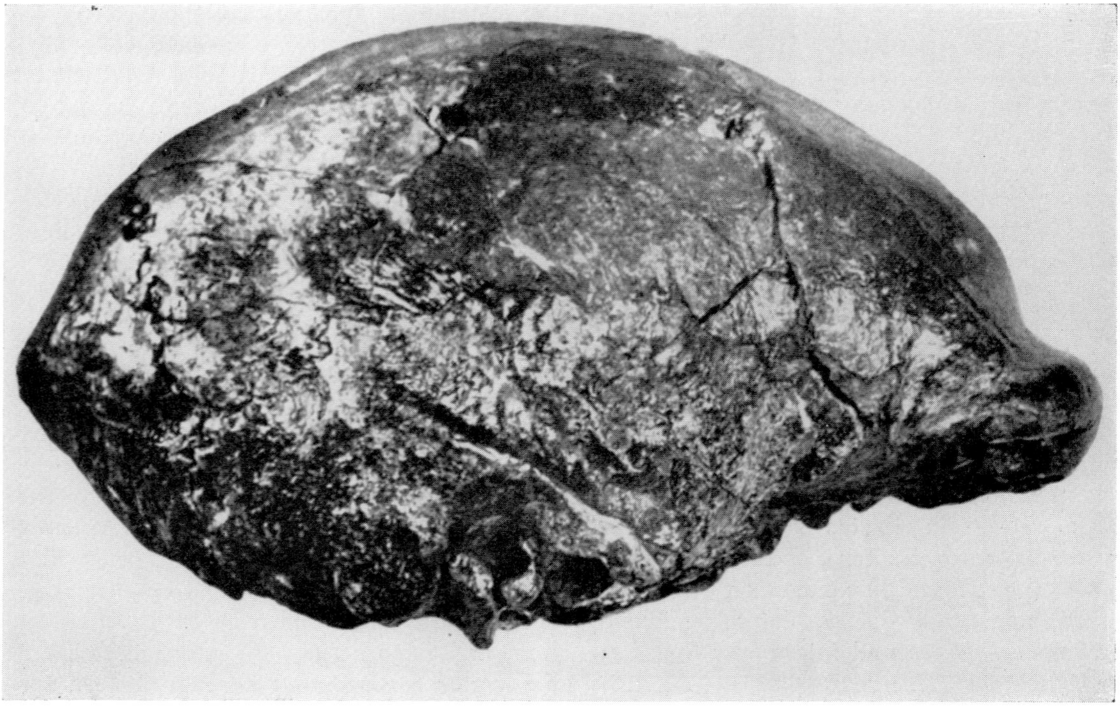

Figure 2b.

Figure 2c.

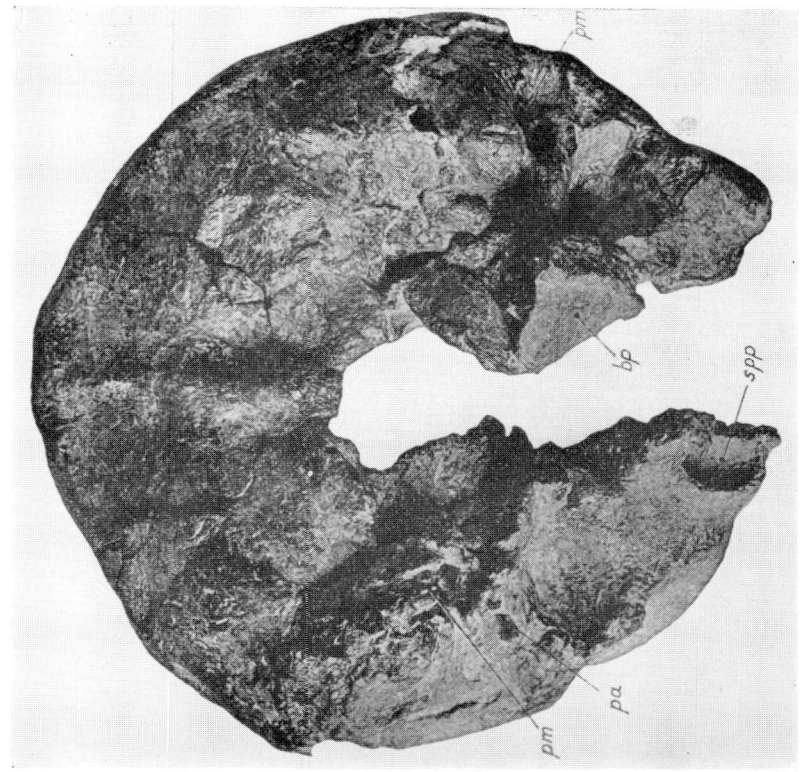

Figure 2d.

Figure 2. Cranial injury in *Homo erectus pekinensis*. *a*. Depressed fracture in left parasagittal region in skull X. *b*. Fracture lines (arrows) running through the left frontal bone and through diastases of frontal-parietal suture and into parietal suture line and from thence into the right parietal bone. Other minor fracture lines into both parietal bones. Lesions probably produced by heavy blows with a club to cranial vault. (Courtesy National Geologic Survey of China.) *c*. Severe linear fracture with depression of fragments (Skull XI). Injury probably fatal due to crushing blows with large rock. *d*. Basal view of same specimen.

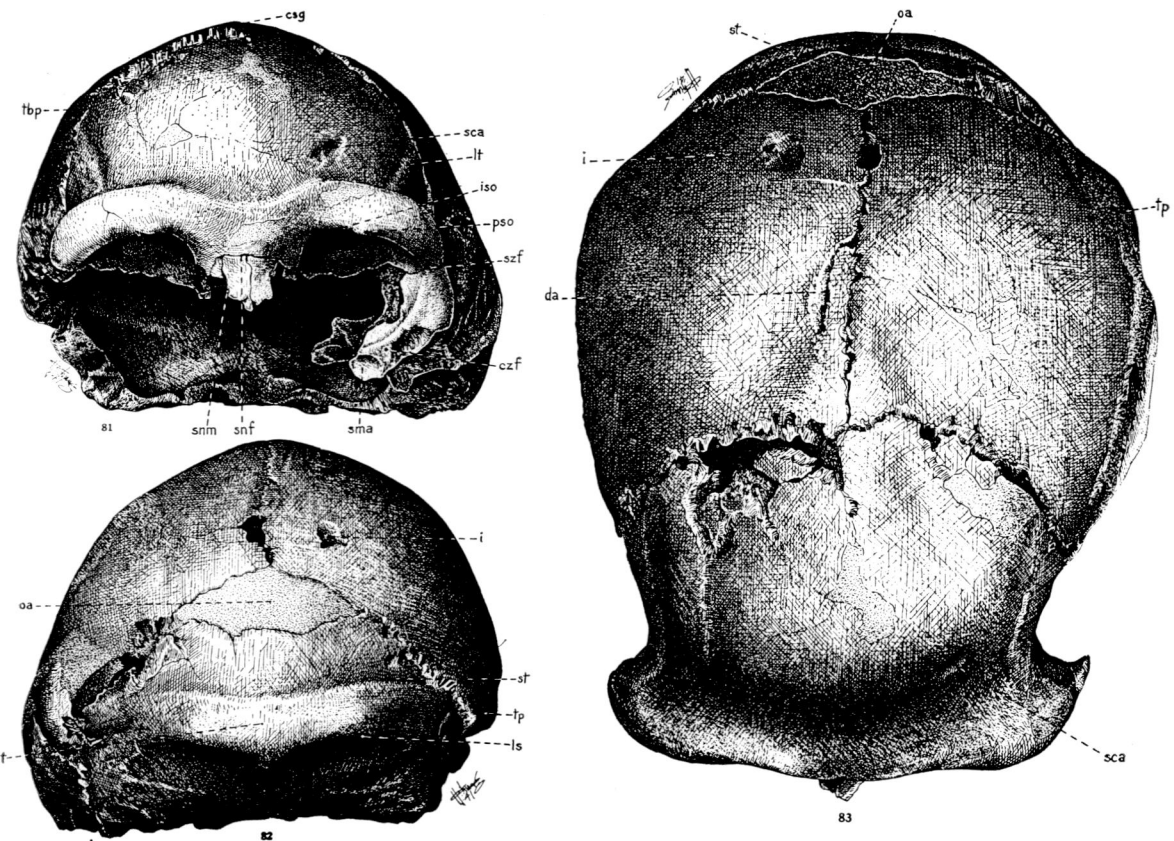

FIGURE 3 Cranial injuries (depressions in outer table) in *Homo erectus pekinensis*. *a.* norma frontalis, *sca*, scar; *b.* Normo occipitalis, *i* injury *c.* normo verticalis. *i*, injury; *sca*, scar.

These injuries are probably due to some type of stone axe. In addition to these larger and perhaps more characteristic injuries, a number of the specimens showed abrasions which were most abundant in Skull III. Weidenreich considered three possible causes for this type of mark: (1) falling stones in cave-ins; (2) cutting instruments in the hands of man, and (3) teeth marks of predatory animals. He concludes, however, that they were actually the effects of cutting instruments (Weidenreich, 1928).

The author summarizes his observations on the nature and causes of cranial injuries in this series of prehistoric skulls as follows:

If we recapitulate all the facts, the following is evident: (1) all *Sinanthropus* skulls lack the central parts of the base; (2) some of them show lesions or scars at the top of the skull the form of which indicates that they have been inflicted by axe- or knife-like implements; (3) some skulls exhibit, on the upper part of the vault, depressed fractures with radiating fissures indicating that the blow had hit the skull while it was impressible; (4) there are cranial bones and fragments of them with irregular indents at the injuries and scratches on the surface. My verdict is that the destruction of the base and the blows on top of the skull are the incidental work of man, although the possibility cannot be entirely excluded that at least those lesions which indicate that they were produced by pointed or blunt agents [also gross crushing injuries] may have been caused by stones falling from the roof of the cave on a living individual. Later on the skulls were broken as carrion by carnivores, probably hyaenas, which lived in the cave and cracked the bones as long as they were fresh.

... My early suggestion still stands, namely: that the strange selection of human bones [the calvaria] we are facing in Choukoutien has been made by *Sinanthropus* himself. He hunted his own kin as he hunted other animals and treated all his victims the same way. Whether he opened the human skulls for ritual or culinary purposes cannot be decided on the basis of the present evidence of his cultural life, but the breaking of the longbones of animals and man alike, apparently for the purpose of removing the marrow, indicates that the latter alternative is more likely. The remains of his meals became the prey of his predatory neighbors at the foothills of Choukoutien.

CRANIAL INJURIES IN THE PITHECANTHROPINES AND SOLO MAN OF JAVA

A report on another skull, now considered to be *Homo erectus* previously called (*Pithecanthropus robustus*), is also available for study (Weidenreich, 1945). This skull, represented only by the posterior part of the calvarium, was found by A. Von Koenigswald in the Trinil beds of Sangiran, Java. This calvarium resembled so closely that of Dubois, *Homo (Pithecanthropus) erectus* that its finder assumed it to be the same form (von Koenigswald, 1937). This specimen too showed evidence of a cranial injury.

The skull consisted of almost the entire occipital bone, including the major part of the occipital foramen and condyles, the outer portions of both temporal bones and approximately the posterior three-fourths of the two parietal bones. A portion of the maxilla, including almost the entire alveolar process, the palate, floor of the nasal cavities and maxillary sinus was also recovered. Of the marks of injury, the author writes:

After preparation and adjustment of those pieces which appeared to have been freshly broken it is obvious that the brain case had been crushed, apparently with great force. Cracks spread over the entire calotte, there are deep impressions where the occipital and temporal bones have telescoped into each other, and most of the fragments have been dislocated. A wide cleft also passes through both cap and base in an oblique direction from the right at the front to the left at the rear. Apparently, these injuries occurred not only before fossilization, but before the flesh had decomposed; otherwise mineralization could not have fixed the bones and bone fragments in such unnatural positions. An excellent example of this dislocation and subsequent fixation in the wrong place by fossilization may be observed on the fragment designated as bp. It is probably the basilar process of the occipital bone which was broken from its normal position by the blow cleaving the skull and has been turned to the left side of the outer surface of the base where it is still attached. The maxilla displays the same signs of violence. Several cracks extending over the plate have produced some dislocations; in particular, the left side of the alveolar process has been affected and moved inward.

In Java, there has also been recovered a group of Upper Pleistocene skull fragments which may be considered to be Neandertaloid or late advanced Pithecanthropines in type. Oppenoorth (1932) described fragments of eleven such skulls of *Homo soloensis* or Solo Man. These skulls consisted only of the brain pan, the facial parts and cranial base being missing, presumably having been purposely detached. In one of these specimens (Skull V), wounds of the skull were found in the occipital region. It was presumed that this individual had been killed by a blow on the back of the head (von Koenigswald, 1937).

CRANIAL INJURIES IN NEANDERTAL MAN

As good fortune would have it, an investigation of Neandertal crania indicates the fairly common occurrence of traumatic damage. It has been estimated that about 40 per cent of these skulls found in Europe, North Africa and Western Asia give unequivocal evidence of injury. Since Neandertal man apparently did not often survive beyond the age of fifty years, it is suspected that many of them died a violent death. The very fact that the brain pan alone is found (the basal or facial parts of the skull being missing) has suggested to some that this portion had been cut away to expose the brain as a tender titbit for the assailants. If this is the case, the Neandertalers were cannibals, and any associated fracture of the vault or base of the skull was but a mark of an intended lethal cranial injury.

Whatever the intelligence of Neandertal man and his contemporaries, it is clear that his world was one in which violence played an important role. And this violence was not limited to the bodies of man, for in the skulls of prehistoric animals as well the characteristic effects of trauma are to be found.

As yet, the only specimens implying injury to the head of prehistoric animals by Neandertalers or his human contemporaries are those of the skulls of the giant cave bear as found in the Drachenhole, Styria, South Austria. The skull of one of these animals still has a stone axe imbedded in it. This observation would imply that Neandertal man used hafted weapons, for such a blow could not have been delivered with an axe-head in the hand.

Cranial Injury in European Neandertalers

When workmen discovered the peculiarly shaped skull cap while blasting in the Neander Valley in Prussia in 1856, it was not realized for some time that this represented a distinct variety of fossil man. Although it had no evident signs of trauma (nor, for that matter, did the skulls or portions of skulls of similar type found at Gibraltar in 1848 and at Spy, Belgium in 1886), the discovery of a group of skulls at Krapina, Croatia in 1889 by Gorjanovic-Kramberger did suggest the occurrence of cranial injury as the cause of death. Here in a Palaeolithic (Mousterian) shelter were found portions of from ten to twelve skulls together with other bones which suggested that at least twenty individuals of both sexes and all ages were represented. The fragmentation of the skulls was so extensive as to suggest violence rather than accidental crushing. To strengthen this conclusion one specimen consisting of the supra-orbital portion of a frontal bone showed some cuts such as might have been produced by a flint knife or axe.

That not all representatives of Neandertal man in Western Europe met a violent end is well established by the discovery of fairly intact skulls and complete skeletons at La Chappelle-aux-Saints, Le Moustier and at La Ferrassie. While nothing has been written about traumatic lesions of the skull in this latter group, the published photographs of the male skull present faint linear markings that may indicate antemortem fracture lines and seem worthy of re-examination from this point of view. The skull recovered at La Quina

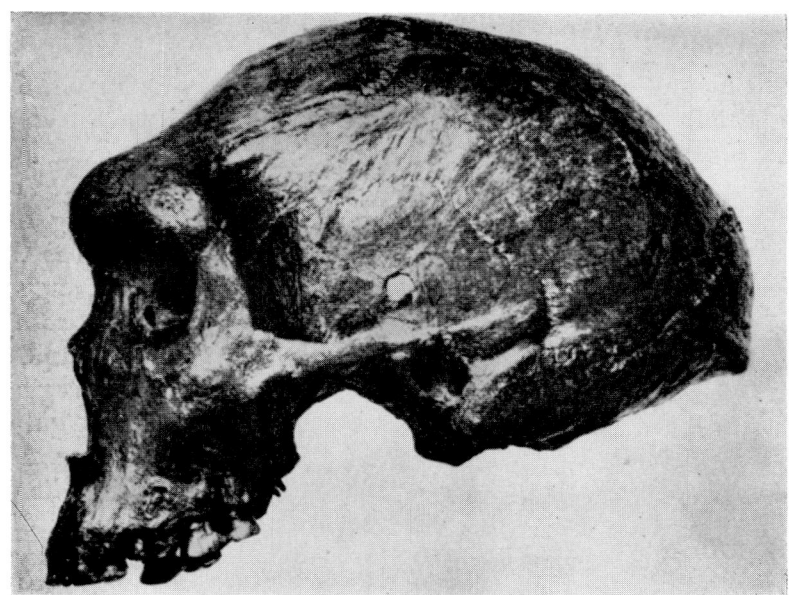

FIGURE 4a.

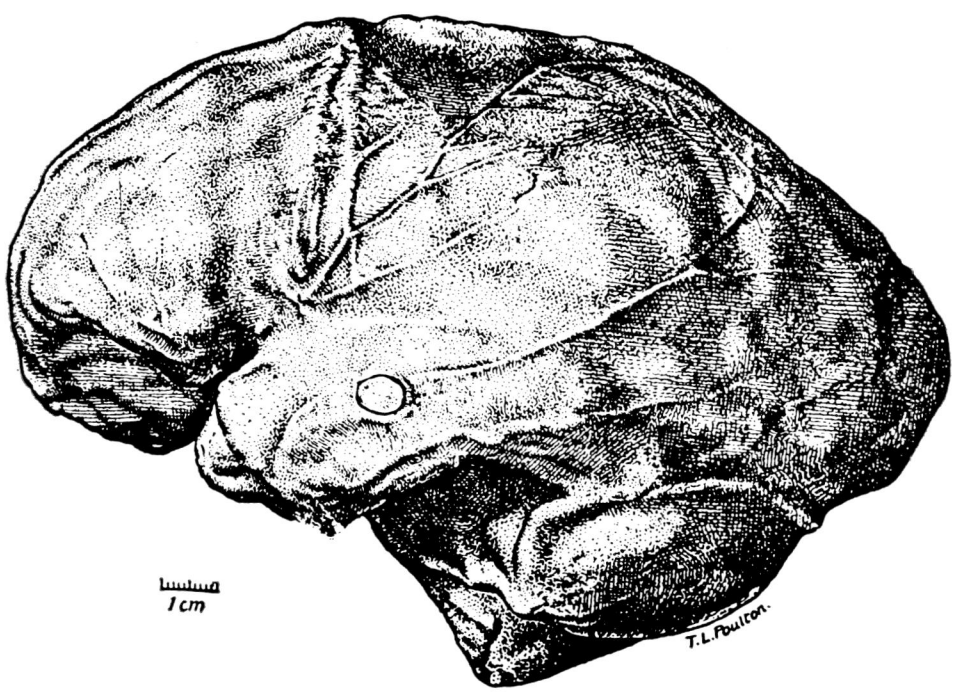

FIGURE 4b.

FIGURE 4. Cranial injury in Rhodesian Man. A. Rounded puncture wound of the left temporal region. B. Drawing of the endocranial cast showing a defect of the skull at a branch of the middle meningial artery. The location of this defect suggests that this individual may have died of an extradural haemorrhage secondary to the puncture wound.

(also considered as an accidental burial) was so broken that it would be difficult if not impossible to differentiate signs of trauma in life from postmortem crushing.

The subject of cranial injury in Neandertal man of Europe may be best illustrated by one of the discoveries at Ehringsdorf. This specimen was proved to be the skull of a young adult (estimated to be between eighteen to thirty years of age) possibly that of a female subject. The traumatic lesions (Fig. 1) are described by Weidenreich (1928) as "unmistakable dents on the frontal bone made partly by sharp, partly by dull stone implements, render it probable that the individual had been killed. The violence resulted also in breaks of the cranial bones and separation at sutures. The fact that the basal parts of the skull are missing, having broken away, lead to the conclusion that the skull was thus broken for the purpose of extraction of the brain [cannibalism]."

CRANIAL TRAUMA IN RHODESIAN MAN

Another Upper Pleistocene skull, that of the so-called Rhodesian man discovered in Northern Rhodesia at the Broken Hill Mine, may bear evidence of the effects of trauma. Details of the pathological changes were originally given by Yearsley (1928). The puncture hole (Fig. 4) in the squamous portion of the left temporal bone was described as follows:

It is strongly presumptive that the perforation B is not an instance of primitive "trepanning" but was due to a wound inflicted by some sharp instrument during life and was not the cause of death.

The perforation of which Yearsley speaks is a small round aperture with a slightly irregular margin. The photograph of the cranial interior suggests some chipping of bone around its margins and perhaps some fine radiating linear fracture marks. It is clearly not an effect of the hand-axe but rather that of a sharp pointed weapon, a rounded wooden or stone spear point or an antler prong of a primitive pick. Yearsley's opinion is that the sharp margins of the defect suggest its occurrence about the time of death (the mastoid defect, on the other hand, shows some rounding off of its margins, suggesting chronicity). It is also highly pertinent to know that the endocranial cast further shows this perforation to lie directly along the course of the inferior division of the left middle meningeal artery (Fig. 4). *It is entirely possible that Rhodesian man died of the effects of middle meningeal hemorrhage, either extra- or subdural, or both.*

INJURIES OF UPPER PALAEOLITHIC AND LATE STONE AGE CRANIA IN CHINA

A study of more recent generations of man, shown by their cranial morphology to be clearly *Homo sapiens sapiens,* sheds further light on the problem of ancient cranial injury. In a series of skulls recovered from the Upper Cave Choukoutien (considered to be Upper Palaeolithic) was found unmistakable evidence of violence. Such signs were found in four of seven crania sufficiently intact for critical study (Weidenreich, 1939).

Skull I. This skull displayed a typical round depressed fracture in the left temporoparietal region. If the skull here referred to is the same as that described by Weidenreich (1943) as the "Old Man" of the Upper Cave of Choukoutien, this depressed fracture occurred about midway along the coronal suture and appears to be less than an inch in diameter (Fig. 268. Plate LXXXIX of his monograph of "The Skull of Sinanthropus Pekinesis," 1943). The fracture line appeared to show no healing and, hence could only

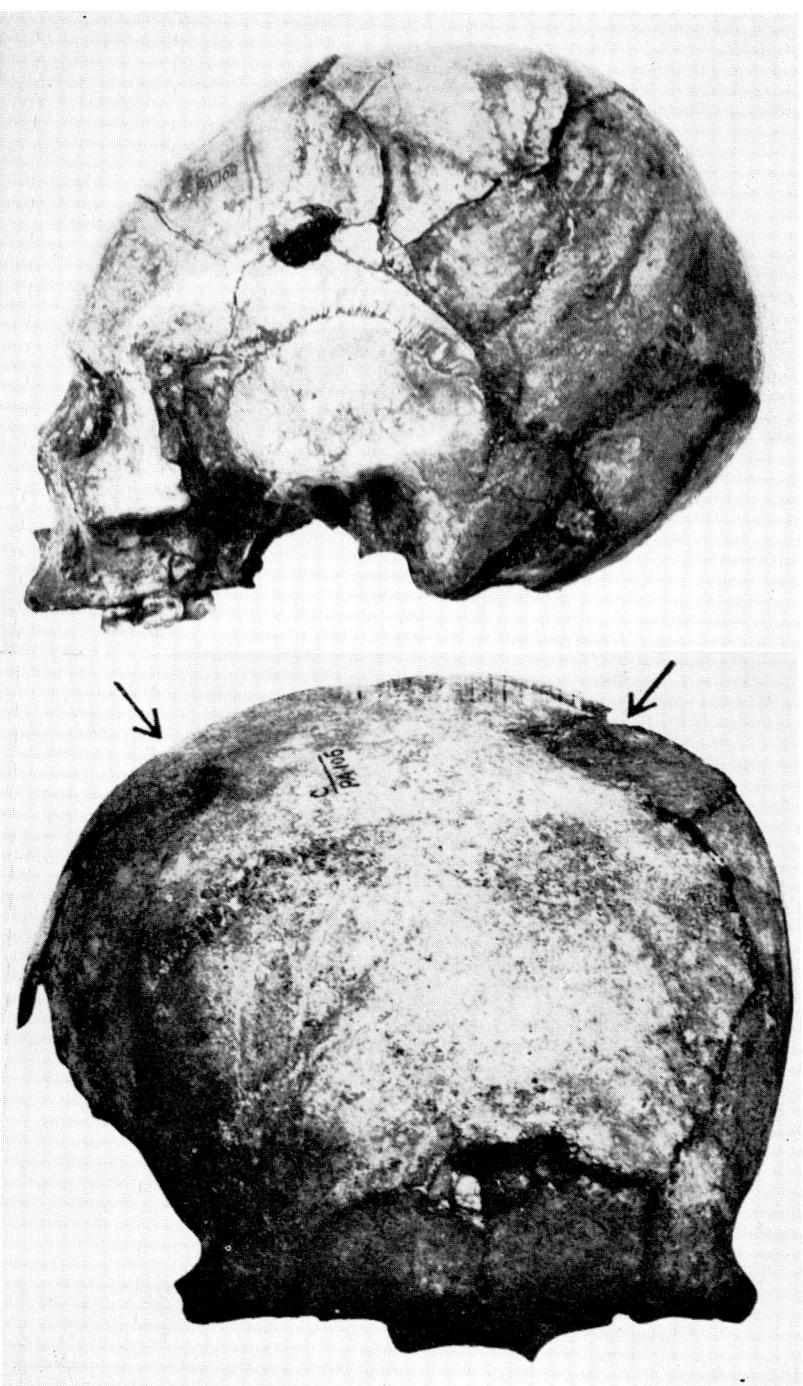

PLATE 5 Cranial injuries in Upper Palaeolithic specimens of Asia. *a.* Calotte of a man from the Upper Cave of Choukoutien, showing multiple linear and puncture wounds of the midfrontal and right and left parietal wounds (skull I). *b.* Another Upper Palaeolithic skull from this site, considered to be that of a woman (skull II) whose death was evidently due to an acute penetrating injury in the left parietal region, probably made by a pick-axe type of weapon. The penetrating defect was associated with a series of radiating linear fractures.

have been just antemortem, if not at or after the time of death (Fig. 51).

Skull II. This specimen is described as the skull of an adult female. The skull is shown in Fig. 267 and 268, Plate LXXXIX of the Weidenreich (1943) monograph, as well as in his 1939 article. It presents a rounded perforation about 20 mm in diameter just behind the lower portion of the left coronal suture, in the anterior and inferior part of the left parietal bone. Radiating fracture lines extend (1) anteriorly and slightly upward across the frontal bone; (2) upward, curving slightly forward to cross the anterior part of the left parietal bone; (3) backward and upward to the sagittal suture dividing the parietal bone into two unequal parts, and (4) backward and slightly downward to the mastoid region (Fig. 56). Weidenreich believed that this lesion was produced by "a spear-like weapon piercing the skull from above" (and laterally). Another picture of this skull (Weidenreich, 1943) shows multiple fractures of the right parietal region as well. Weidenreich believes that the radiating fracture lines (possibly those of the opposite side of the skull) were produced by crushing blows from club-like weapons. It is more likely that one weapon produced the penetrating wound and its radiating fractures, the fractures on the opposite side being produced by some other blunt object. The first was obviously a round, probably pointed weapon, such as a spear, with a head of sufficient weight to crack the skull as well as penetrate it.

Skull III. This cranium presented multiple fracture lines extending over the vault without displacement of the enclosed fragments. The impact of the major blow evidently fell on the left side of the frontal bone. If this is the same skull as portrayed in Figures 267 C and 268 C (Plate LXXXIX) of his 1943 monograph, the impact was sustained a little below the left frontal boss with fracture lines radiating (1) downward and backward to the left crossing the coronal suture just above the lateral orbital process; (2) upward and mesialward toward the vertex crossing the upper portion of the coronal suture to enter the parietal bone; (3) upward and mesialward crossing the frontal midline near the vertex to enter the right parietal bone, and (4) laterally to the right crossing the frontal bone quite horizontally to enter the right temporal region. Another fracture connects (3) and (4) bounding with them an equilateral triangular fragment with its apex almost in the midline. This would suggest multiple blows as Weidenreich states, perhaps with a club with a projecting point or knob.

Skull IV. Only part of the calvarium (the frontal and both parietal bones) of the skull is represented in this specimen. There was an irregular comminuted area about 20 by 30 mm in diameter with multiple indriven fragments found in the lower frontal bone in the midline. On the interior of the skull at this point was found a depression of the fragments of the inner table. Linear fracture lines extended horizontally from the lateral ends of this defect, that extending to the left meeting at right angles a second long linear fracture line extending to the left parietal region. Here a second depressed fracture with radiating lines was found. A third was found in an almost symmetric situation in the right parietal bone. The suggested selective location of these three depressed areas suggests the possibility of deliberate placement of the blows. The inner tables were fragmented in a characteristic fashion below the parietal depressions. The shape of the frontal lesion would suggest the possibility of its being made by a hand-axe, although the extensive radiating fractures would necessitate the blow being delivered by a helved weapon. Under such circumstances it would not be necessary to assume the action of both clubs and pointed weapons as Weidenreich has done.

Similar evidence of violence in Ancient China has been reported by Black (1925) who studied the skeletal remains of forty-five individuals from a Neolithic deposit in the caves of Sha Kuo T'un in Fengien, China. The skulls of these individuals of both sexes and all ages were badly broken, evidently by physical violence at the time of their death.

A study of the cranial lesions (in addition to supporting the thesis of Weidenreich and others that the calvaria had been broken into to get at the brain) indicate three possible mechanisms of cranial injury. Gross crushing injuries such as might be produced by large rocks projected with malignant purpose or resulting accidentally by rock falls in caves are characteristic. Only such an injury could readily account for the extensive fracture lines and gross displacement of fragments involving even the base of the skull. Gross comminution of the cranium may further imply the application of force by large clubs. The gutter-like depressions occasionally found also suggest that these clubs were not always large or massive but were of the type of billet club used throughout history and until recently in the South Seas. It is obvious that some type of pointed weapon was also used, in some instances perhaps a hand-axe but in others a more rounded point, such as the prong of a deer horn, hafted in some fashion to constitute a pick or halberd.

UPPER PALAEOLITHIC AND MESOLITHIC MAN IN EUROPE

Upper Palaeolithic peoples of Europe may be regarded as prototype of the modern European. Again we have good evidence of Stone Age violence. This is well shown in one of the original specimens of Cro-Magnon skulls described by Broca (1868, 1873). This incomplete specimen (the left parietal and the squamous portions of both temporal bones and the cranial basis are missing) is that of an adult woman who presents an incised wound about 50 mm long and 15 mm wide in the left frontal region. The wound lies obliquely and presents some eversion of its edges, from which radiating cracks may be seen. From its size, it is clear that the dura and leptomeninges were penetrated and the brain wounded (Fig. 6c). It was thought that the woman must have survived for some time, for the appearance of bone around the defect suggested to Broca that the wound had suppurated and healed. The nature of the wound suggests that it was produced by a stone axe, probably hafted in this case in order to deliver such a powerful blow. In the same group of bones, a femur of a male showed evidence of a healed fracture.

In a course of excavations at Laugerie-Basse on the bank of the Vézère opposite the site of Cro-Magnon, Massénat, Lalande and Cartailhac (1872) discovered a human skeleton, the remains of an individual who had apparently been crushed while asleep by the falling rock under which he lay. A large irregular defect was found in the left temporal fossa, which very likely represented a depressed skull fracture (Munro, 1912).

Another skeleton was found in 1888 by Teaux and Hardy near Perigeaux in the Commune of Chancelade Dordogne. As described by Testut (1889) it proved to be that of an Upper Palaeolithic male of about sixty years of age. In the left parietal region was an extensive irregular healed defect whose shape suggested an incised wound with radiating fractures inflicted with a narrow bladed blunt weapon. The victim evidently survived for some time for the wound was well cicatrized.

Another skull, also that of a female, was found by Lartet and Chaplain-Duparc among a number of bones recovered from a cavern at Sordes. E. T. Hamy, who studied these bones, found many evidences of trauma. In the case of the skull an old gaping defect was found in the right parietal region. It was assumed that heal-

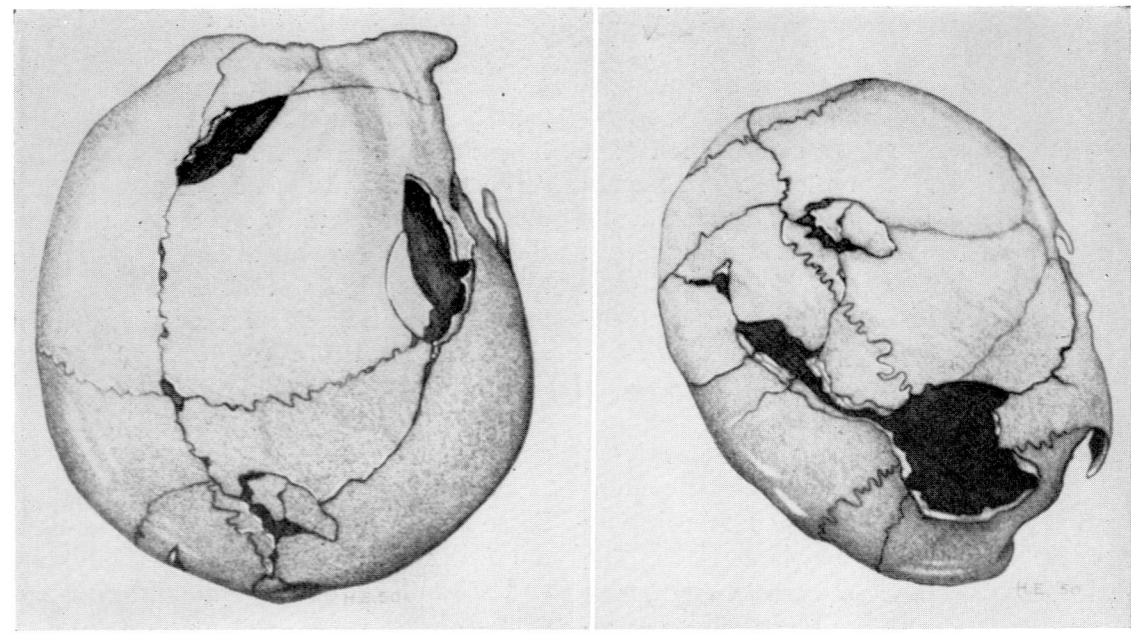

Figure 6a. Figure 6b.

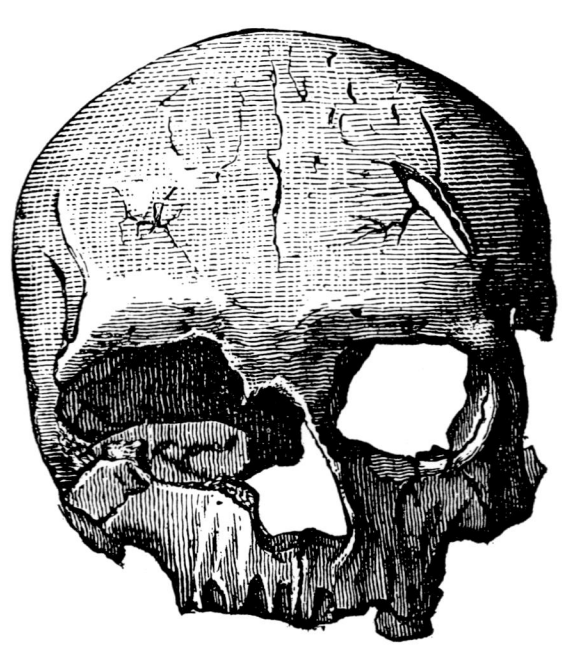

Figure 6c.

FIGURE 6d.

PLATE 6 Cranial Injuries in the late prehistoric and protohistoric periods. a. and b. Two traumatized skulls considered to be of Mesolithic culture found in "skull nest" at Ofnet Bavaria. The cranial injuries suggest blows with a stone-headed axe. c. Cranial injury in a Cro-Magnon female. Recent incised wound in left frontal region. d. Multiple linear fractures of the cranial vault presumed to be made by blow with a stone axe (after Holländer).

ing had taken place after fragments of bone had been removed, for the margins of the defect were well rounded off. The victim of the injury had undoubtedly survived her injury for many months if not years.

The skull of one of two skeletons of the late Palaeolithic (Magdalenian) period found at Obercassel, Germany, and studied by Verworn, Bonnet and Steinman (1914) disclosed evidence of an old injury of the right parietal bone. In this case, also, the victim must have survived for some time.

A different type of injury is portrayed by the specimen studied by De Boye. In this skull recovered from a cavern at Villevevard, France, were found embedded three tranchard transversal arrow points.

The skull described by Mollison (1936) as being found at Ofnet (skull 21) and considered to be that of an adult male of Mesolithic date, is one of very unusual interest. A series of penetrating wounds, evidently due to a sharp edged weapon such as an axe are found in the right frontal, the right temporal, the right parieto-occipital (2) regions, and at the vertex. The fracture lines about these penetrating defects are sharp, indicating that the wounds were recent, presumably the cause of the individual's death.

There is some question whether all the fragmented skulls in the "skull-nests" at Ofnet, Bavaria, represent the effects of injury (see Fig. 6a, b). One of these nests contained twenty-seven human skulls. The heads represented by these skulls were obviously removed for purposive placement in these nests. The presence of the skulls alone, particularly those of children imply burial by relatives.

INJURIES TO THE HEAD IN THE LATE PREHISTORIC AND PROTOHISTORIC PERIOD IN EUROPE

In cultures which are contemporary with early historic peoples, there is increasingly abundant evidence of injuries to crania. A number of such crania from the Neolithic to Iron Ages have been discovered in various parts of Europe. From Great Britain, for example, Rolleston (in Greenwell, 1877) described a skull of a man aged about thirty to thirty-five years found at Langton Wold, Yorkshire. A description of this Bronze Age skull included a comment on a shallow gutter wound. A more recent study of antemortem injury in early British crania by Brothwell (1961) gives frequencies shown in the following Table I.

TABLE I
DISTRIBUTION OF ANTEMORTEM INJURY TO THE SKULL (BRITAIN)

Period	Frontal	Parietal	Other Parts of the Vault	Face (general)	Nose	Total
Neolithic-Bronze Age	7	6	1	1	3	18
Iron Age—Anglo-Saxon	13	11	0	4	2	30
Total	20	17	1	5	5	48

Numerous other cases of cranial injury in European specimens have been described, but only two selected examples need be given here.

Höllander (1928) described a portion of a protohistoric calotte found at Reinbeck, Germany, which consisted essentially of a frontal bone marked by numerous fracture lines (Fig. 6d). He concluded that the fracture was the result of a blow with a stone axe in the left parietal region which shattered the cranial vault. There was evident healing in the fracture lines.

Another skull, described by Du Chaillu (1890), found in a graveyard of Roman date at Varpelev, Zeeland, Denmark, was evidently that of an adult male. This specimen presented evidence of trauma in the form of a triangular defect in the left frontoparietal region, made by some sharp-edged weapon (presumed to be a sword, but more likely a narrow-bitted axe). Short radiating fractures extended from the two lower angles of the defect; the upper angle was in line with the coronal suture.

CRANIOCEREBRAL INJURIES IN PROTOHISTORIC PEOPLES OF AFRICA, ASIA, THE SOUTH SEAS AND THE NEW WORLD

With the development of the protohistoric cultures which actually belong to Egypt, the Near East, Greece and Rome, injuries to the skull and brain became increasingly common (Courville, 1949). Similar studies on cranial injuries among the early Peruvians (Courville and Abbott, 1942), among the islands of Oceania (Courville, 1951) and the Indians of North America (Courville, 1948) and of California (Courville, 1952), from prehistoric as well as protohistoric and historic periods, have provided evidence of the gradual evolution of medical knowledge and treatment of cranial injuries. There is, however, still a need for further studies of craniocerebral injury and in particular with accurate and detailed statistics such as given in Hooton's (1930) classic analysis of the Indians of Pecos Pueblo. Not only does his data permit the study of injury through the dimension of time, but also according to age-grouping (Table II).

SUMMARY AND CONCLUSIONS

The examples given above, especially of late prehistoric and protohistoric specimens could probably be duplicated many times over if the crania in the numerous museum collections of the world could be examined specifically for injury.

But even this concise survey is sufficient to point out that before written history, man suffered a variety of head injuries. Since his industries were simple, industrial accidents were few, and most of the cranial wounds and fractures noted in ancient specimens are the result of blows delivered with malicious intent. Gross crushing injuries resulting from large stones or clubs are betrayed by multiple, extensive fractures, at times with deformation of the head. Simple linear fractures suggest blows with smaller clubs or blunt weapons. Penetrating wounds imply pointed and hafted striking weapons of stone or metal. Incised wounds with sharp edges bespeak weapons of better make and suggest that the flint axes were being polished or that bronze axes were in use.

REFERENCES

BLACK, D., 1925: The human skeletal remains from the Sha Kuo T'un cave deposits in comparison with those from Yang Shao Tsung and with North China skeletal material. *Palaeont. Sinica Ser. D.*, Vol. 1, fasc. 3.

BROCA, P., 1868: Sur les crânes et ossements des Eyzies. *Bull. Soc. Anthrop.* 2nd ser., 3:350.

———, 1873: The troglodytes, or cave-dwellers of the valley of the Vezere. *Ann. Report Smithsonian Inst.* p. 310.

BROTHWELL, D. R., 1961: The palaeopathology of early British man; an essay on the problems of diagnosis, and analysis. *J. Roy. Anthrop. Inst.*, 91:304.

COURVILLE, C. B., and ABBOTT, K. H., 1942: Cranial injuries of the Pre-Columbian Incas, with comments on their mechanism, effects, and lethality. *Bull. Los Angeles Neurol. Soc.*, 14:107.

TABLE II
SKULL FRACTURES CONSIDERED IN RELATION TO AGE-GROUP IN PECOS PUEBLO INDIANS. (DATA FROM HOOTON, 1930)

Age Group	Sub-Adult	Young Adult	Middle Aged	Old	Total Affected
Percentage of 581 individuals	13.04	8.70	52.17	26.09	3.96

COURVILLE, C. B., 1948: Cranial injuries among the Indians of North America. A preliminary report. *Bull. Los Angeles Neurol. Soc., 13*:181.

———, 1949: Injuries to the skull and brain in ancient Egypt. Some notes on the mechanism, nature, and effects of cranial injuries from predynastic times to the end of the Ptolemaic period. *Bull. Los Angeles Neurol. Soc., 14*:53.

———, 1951: Injuries to the skull and brain in Oceania, with reference to the mechanism and nature of such injuries, the measures used in protection against them, and their treatment particularly among the Melanesians. *Bull. Los Angeles Neurol. Soc., 16*:17.

———, 1952: Cranial injuries among the early Indians of California. *Bull. Los Angeles Neurol. Soc., 17*: 137.

DU CHAILLU, P. B., 1890: *The Viking Age: The early history, manners and customs of the ancestors of the English-speaking nations.* New York, Scribner.

GORJANOVIC-KRAMBERGER, K., 1889: Der Päläolithische Mensch und seine Zeitgenossen aus dem Diluvium von Krapina in Kroatien. *Mitteil. anthrop. Ges. Wien* (Sitzungsbericht) 29.

———, 1906: *Der diluviale Mensch von Krapina in Kroatien.* Studien über Entwicklungsmechanik der Primatenskelettes. Fasc. 2. Wiesbaden.

GREENWELL, W., 1877: *British Barrows. A Record of the Examination of Sepulchral Mounds in various parts of England.* London, MacMillan.

HÖLLANDER, E., 1928: *Eine Kulture-Sittengeschichte in Spiegel des Ärztes.* Berlin, Äskulap und Venus, p. 121-123.

HOOTON, E. A., 1930: *The Indians of Pecos Pueblo. A Study of Their Skeletal Remains.* Yale University Press, New Haven.

LEBARRON, J., 1881: *Lésions osseuses de l'homme prehistoric en France et en Algérie.* Paris, p. 47.

MASSÉNAT, E., LALANDE, P., and CARTAILHAC, E., 1872: *Un squelette humain de l'age du Renne a Laugerie-Basse.* Materiaux VII (Cited by Munro, R., 1912. pages 132 and 133)

MOLLISON, T., 1936: Zeichen gewaltsamer Verletzungen an den Ofnet-Schadeln. *Anthrop. Anzeig., 13*:79.

MUNRO, R., 1912: *Palaeolithic man and terramara settlements in Europe.* New York, Macmillan.

OPPENOORTH, W. F. F., 1932: *Homo (Javanthropus) soloensis. Een plistociene mensch van Java.* Dienst Mijnbouw Nederlandsch-Indie, Wetensch. Meded. No. 20, p. 49.

TESTUT, L., 1889: Recherches anthropologiques sur le squelette quarternaire de Chancelade. (Cited by Munro, 1912, p. 134).

VON KOENIGSWALD, G. H. R., 1937: Ein Unterkieferfragments des *Pithecanthropus* aus den Trinilschichten Mitteljavas. Amsterdam. *Prov. K. Akad. Wetensch,* 40:883.

VERNWORN, M., BONNET, R., and STEINMAN C., 1914: Der diluviale Menschbefund von Obercassel bei Bonn, *Die naturwissenschaften, 27*:625. (Also book of same title, Wiesbaden, 1919.)

WEIDENREICH, F., 1928: Der Schadelfund von Weimar Ehringsdorf, Jena, Fischer.

———, 1939: The duration of life of fossil man in China and the pathological lesions found in his skeleton. *Chin. Med. J., 55*:34.

———, 1943: The Skull of *Sinanthropus Pekinesis*. A comparative study on primitive hominid skull. *Palaeont. Sinica.,* n. s. D. No. 10, Lancaster Pa.

———, 1945: Giant early man from Java and South China. *Anthrop. Papers,* Am. Mus. Nat. Hist., New York. vol. 40, part 1.

WILLIAMS, H. U., 1929: Human Paleopathology. *Arch. Path., 7*:839.

YEARSLEY, M., 1928: The pathology of the left temporal bone of the Rhodesian skull. In *Rhodesian Man* by W. P. Pycraft. British Museum (Natural History), London.

Chapter 49

The Evidence for Injuries to the Jaws

V. ALEXANDERSEN

TRAUMATIC lesions of bones have been observed in skeletal remains of man from all periods, from Palaeolithic times to Middle Ages. Injuries to the skull were more frequent than injuries to any individual bones in the skeleton. Fractures of the jaws were nevertheless rare. This is made clear by Pales (1930), who surveyed the Palaeopathological literature available and reached the conclusions quoted above. Even to-day, fractures of the jaws form a small percentage of all fractured bones in injured persons brought to hospitals for treatment; according to Krømer (1961) the incidence has been estimated at about 2 per cent.

Brothwell (1963b) drew up a classified list of types of injuries seen in skulls; the type of injury is dependent on the sort of trauma. His classification can, moreover, be applied to jaw injuries.

1. *Heavy crushing* by large stones or clubs sometimes involve the facial skeleton. According to MacCurdy (1923) and Moodie (1927), several examples are known of crushing of facial bones in pre-Columbian finds made in Peru. Fractures of the upper jaws, the nasal bones and the malar bones were found in women as well as in men. In some of the skulls evidence of healing was found. In a report on human remains found in Nubia (Wood Jones, 1910) a number of cases of unhealed fractures of the facial bones are mentioned. Extensive crushing involving fractures of the zygomatic arch, maxilla and the mandible were reported. It is likely that mace-heads made of hard diorite were responsible for these fatal lesions in the cases dating from the Predynastic and Early Dynastic periods.

2. *Less extensive fracturing*, caused by blunt instruments, is a group which includes several of the published cases of healed fractures of the jaws. However, accidents such as falls may have been responsible for the injury in some of these cases.

3. *Thrust-wounds* by spears, lances, daggers, javelins and arrows sometimes involve the jaws. In a Neolithic Danish skull the end of an arrow which hit the middle of the face and perforated the palate is still preserved in the jaw (Fig. 1, Becker, 1952). In another Danish skull, dating from the Iron Age, an arrow-head was found in the maxilla (easily recognizable in an X-ray photograph taken of the jaw) (Klindt-Jensen, 1960). Twiesselmann and Brabant (1960) mentioned a mandible found in Coxyde in Belgium with the end of a harpoon still attached to it.

4. *Cut-wounds* by swords and axes can be diagnosed with greatest reliability in skulls found in soldiers' graves dating from historic periods when these weapons were known to be in use. Some cases are known from the Nubian excavations reported on

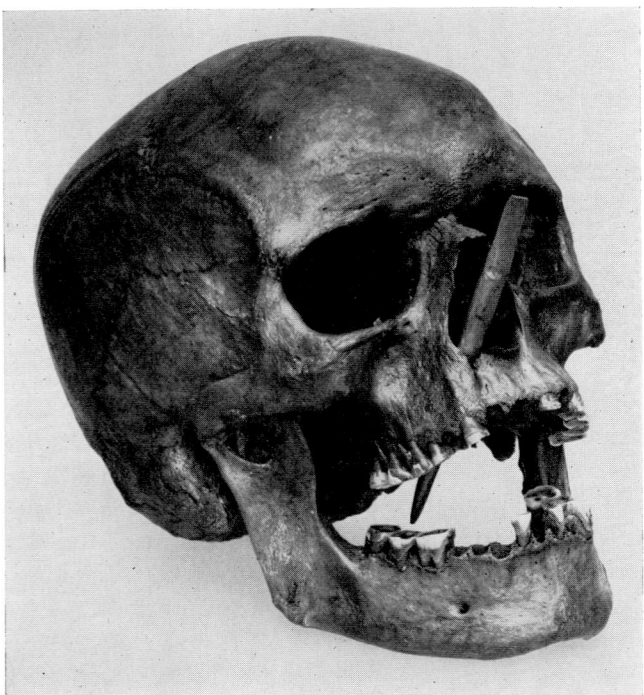

FIGURE 1 Danish Neolithic skull (Porsmose) with the end of an arrow fixed in the palate. Forty-three mm of the end of the arrow (total length 106 mm) has penetrated the palate. A fracture line is visible from the pyriform aperture to the socket of the right lateral incisor. Nationalmuseet, Copyright.

by Wood Jones (1910). Some of the cuts had not healed, but a skull dating from the Byzantine period, presumed to have belonged to a woman, had a healed fracture with no evidence of inflammation along the long fracture line stretching from the *processus frontalis* of the right maxilla (also involving the medial bones in the right orbit) to the frontal bone high on the vault. Gröschel (1937) described an Alemannic skull apparently damaged by a sword-cut separating the left half of the maxilla from the face. Among the skeletons excavated from the late Mediaeval Danish monastery of Æbelholt, that of an adult man revealed a fracture of the zygomatic arch on the left side of the head. The cut had opened the temporo-mandibular joint and cut a slice off the mandibular condyle. The injury was not fatal and the skull revealed that the left zygomatic arch had been drawn down beneath the *porus acusticus externus* without maintaining contact with the temporal bone. Deformity of the remaining part of the injured mandibular condyle had also occurred (Møller-Christensen, 1958). From the same monastery Møller-Christensen (1958) excavated the skeleton of an adult man with twenty-one sword-cuts. In the attempts made to kill him by separating his head from his body three dreadful cuts were delivered to the left side of the head and neck. The temporo-mandibular joint was hit twice, and the *corpus mandibulae* was fractured with multiple fracture lines.

At present fractures of individual facial bones are most frequent in the mandible with less involvement as one approaches the nose, the malar bones and the maxilla. In ancient skulls, very few healed fractures of the jaws have been reported, and thus insufficient data are available to make a comparison. The incidence of fractures cannot be deduced with any certainty on basis of healed fractures alone, as a number of unhealed fractures of the jaws undoubtedly existed that were associated with fatal injuries or at least with very serious lesions which caused death soon after the injury. In order to recognize unhealed fractures of the jaws it is necessary to be familiar with appearance of injuries caused by different kinds of weapons used in military actions and executions at different periods. It is well to remember that some fractures of the jaws may have been caused by blows or cuts delivered intentionally to the skulls post-mortem. Crushing of the Skhūl I, Pleistocene skull, with damage to the jaws, occurred not long after interment according to McCown and Keith (1939). In the case of some Mesolithic skulls recovered from caves in Southern Germany, Mollison (1936), Gieseler (1951) and Saller (1962) have reported instances of skulls having been buried without the skeletons. Such skulls show evidence of having been mutilated postmortem, including fractures of the jaws.

In order to form an impression of the incidence of fractures of the jaws the present author has surveyed much of the literature dealing with the palaeopathological evidence from Western and Northern European regions.

A healed mandibular fracture is known from the Teviec population dating from the Mesolithic period (Pequart, Boule and Vallois, 1937). The skeleton (T 16) is that of a male. It is preserved in the *Institut de Paleontologie Humaine* in Paris, where I had an opportunity of examining the skull. The fracture line passed through the socket of the right lateral incisor, which has been lost. It is not unreasonable to think that fractures of the left upper central incisor and the left lower lateral incisor occurred on the same occasion. Periapical abscess formation around the apex of the right central incisor in the maxilla, with root resorption, likewise suggest sequela of an injury to the tooth. Although some dislocation of the two halves of the mandible occurred before healing, normal ability to masticate was completely restored, judging from the occlusion of the dental arches and the attrition pattern. Viewed from below the *basis mandibulae* form an angle at the point of fracture instead of being curved in the frontal part of the mandible as is normally the case. Evidence of wide-spread infection in relation to even serious traumatic lesions is rare in early populations according to Wood Jones (1910), Williams (1929) and Pales (1930). This also applies to the mandible in the present instance, which has a little callus—rough and pitted—near the *basis mandibulae* while the alveolar process after healing has a smooth surface showing some porosity. Boule and Vallois think that the fragments could easily have been immobilized by the patient himself without bandages. This presupposes little dislocation of the fragments.

It is not possible to decide whether traumatic lesions occurred more often in people living in a hunting and gathering economy than in agricultural and more sedentary peoples. This is also borne out by references to the North American Indians where fractures of the jaws are not more frequent in one tribe than another, even though there are marked differences in economy. It is the opinion of Courville (1950) that as far as prehistoric skulls are concerned most of the cranial wounds and facial injuries were given. In thirty-eight cases the injuries to the skulls were located in the vault of skulls and in five other cases the noses were broken; five more showed other injuries located in the face. Brothwell recorded only healed fractures, or fractures showing some evidence of healing. Data of comparative value are available from ancient Nubian skulls described by Wood Jones (1910). Among the material examined by this author were five cases of skull vault injury, twelve of fractures of the face and three cases of fractured mandibles. The high incidence of fractured facial bones in the Nubian material will only be explained in the light of further information about the British and Nubian samples for example with regard to weapons most often used.

The Danish Neolithic collections, surveyed with regard to pathological evidence by Nielsen (1911), only showed two skulls with healed fractures of the bony nose and no fractures of the jaws.

Judging from published reports on healed fractures, the Middle Ages in Northern Europe did not result in any increase in the incidence of fractured jaws.

The large Belgian collections surveyed by Brabant and his colleagues (1960-1962), showed one case of injury to the upper jaw resulting in the loss of three incisors and two cases of fracture of the mandible (apart from the mandible mentioned above with the end of a harpoon fixed to the jaw). The fractures of the mandibles were located in the *angulus mandibulae* and the sub-condylar regions respectively. Both were healed with restoration of function, leading—in the case of condyle fracture—to asymmetric attrition. The Belgian material dates from the Gallo-Roman to the late Middle Ages.

An important contribution to the knowledge of traumatic injuries to bones is the publication by Ingelmark (1939) on skeletal remains of 1185 warriors from the Battle of Wisby on the island of Gotland in 1361. The injuries resulting from different weapons used in this battle are discussed, likewise the major difficulties in each case of distinguishing which weapon caused the injury and the differentiation of ante-mortem injuries from post-mortem damage. Six per cent of all blows delivered to the skulls involved injury to the mandible. All these fifteen injuries were due to cuts. Even fewer cuts involved the upper face. Four cuts involved the upper part of the maxilla and the zygomatic arch, and four noses were fractured. The number of maxillae examined was less than the total number of skeletons because some of the latter were very fragmentary. No injuries from arrows, lances or spiked maces were located in the jaws. The injuries to the mandibles were severe. Three horizontal cuts had removed parts of the alveolar process. In ten cases the cuts were delivered in a vertico-horizontal direction from above, seven from the right and eight from the left side; all were powerful blows.

No healed fractures of the jaws were found among some one thousand Danish skulls dating from the late Middle Ages and excavated from burial grounds belonging to monasteries, where treatment of wounded persons was effected (Isager, 1936 and Møller-Christensen, 1958). A skull revealing loss of the incisors originally ascribed by Isager to an injury, is now considered to have been caused by leprosy. (Møller-Christensen, 1958).

In the Mediaeval burial ground of Westerhus, on an island named Frösö in the Northern part of Sweden, Gejvall (1960) recorded healed ante-mortem lesions as well as unhealed injuries among 154 juvenile and older individuals. Only one of the injuries was located in the jaws.

It is not necessary to extend this enumeration further by using other geographical regions as starting points, for it should be evident that future studies are needed to account for differences in the incidence of injured facial bones in different populations. It is futile to discuss whether any treatment of jaw fractures was ever attempted. That infections have not seriously hindered the healing of such fractures is rather remarkable.

REFERENCES

BAUDOUIN, M., 1919: Fracture mandibulaire guérie spontanément et datant de l'époque de la Pierre polie. *Bull. Acad. Nat. Méd. Paris.*, 81:52.

BECKER, C. J., 1952: Skeletfundent fra Porsmose ved Naestved. *Fra Nationalmuseets Arbejdsmark* 1952; p. 25.

BRABANT, H., and TWIESSELMANN, F., 1960: Étude de la denture de 159 squelettes provenant d'un cimetière du XIe siècle à Renaix, Belgique. *Rev. Belg. Sci. Dent.*, 15:561.

BRABANT, H., and A. SAHLY, 1962: La paléostomatologie en Belgique et en France. *Acta Stomat. Belg.*, 59: 285.

BREITINGER, E., 1939: Gutgeheilter Unterkieferbruch aus der Frühbronzezeit. *Sudhoffs Arch. Gesch. Med.*, 32: 103.

BROTHWELL, D. R., 1961: The palaeopathology of early British man: an essay on the problems of diagnosis, and analysis. *J. Roy. Anthrop. Inst.*, 91:318.

BROTHWELL, D. R., 1963a: *Dental Anthropology*. Oxford, Pergamon Press.

BROTHWELL, D. R., 1963b: *Digging up Bones*. London, Brit. Mus. (Nat. Hist).

CAMERON, J., 1934: *The Skeleton of British Neolithic Man*. London, Williams and Norgate.

COURVILLE, C. B., 1950: Cranial injuries in prehistoric man with particular references to the Neanderthals. *Yearbook of Physical Anthropology;* 1950: 185-205.

GEJVALL, N.-G., 1960: *Westerhus. Medieval Population and Church in the Light of Skeletal Remains*. Lund.

GIESELER, W., 1951: Die süddeutschen Kopfbestattungen (Ofnet, Kaufertsberg, Hohlestein) und ihre zeitliche Einreihung. *Aus. der Heimat.*, 59:291. (quoted from Saller, 1962)

GRIMM, H., and R. ZUHRT, 1960: Hallstadt-und Tènezeitliche Gesichtsschädelreste. *Deutsch. Zahn. Mund. Kieferheilk.*, 32:265.

GRÖSCHEL, W., 1937: Pathologische Erscheinungen an den Zähnen und Kiefern der Alemannen aus den Begräbnisstätten der Merovingerzeit des Bezirkes Dillingen an der Donau. *Deutsch. Zahn. Mund. Kieferheilk.*, 4:370.

HEBERER, G., 1938: Die mitteldeutschen Schnurkeramiker. *Veröff. Landesanstalt Volkheitskd. Z. Halle.*, Bd. 10.

HUELKE, D. F., BURDI, A. R., and C. E. EYMAN, 1962: Association between mandibular fractures and site of trauma, dentition and age. *J. Oral Surg.*, 20:478.

INGELMARK, B. E., 1939: The Skeletons; p. 149-209 In: Thordeman, B. (1939) *Armour from the Battle of Wisby 1361*.

ISAGER, K., 1936: *Skeletfundene ved Øm Kloster*. Copenhagen.

KLINDT-JENSEN, O., 1960: Var det en yndig tid? *Skalk.*, 2: no. 3.

KRAUS, E., 1939: In: Euler, H. *Die Zahnkaries im Lichte vorgeschichtlicher und geschichtlicher Studien*. München.

KRØMER, H., 1961: Kjevefrakturer og Krigsodontologisk behandling. In: *Nord. Klinisk Odontologi*. vol: 4, 20.

LE BARON, J., 1881: *Lesions osseuses de l'homme préhistorique en France et en Algerie*. Thése pour le Doctorat en Médicine. No. 262. Paris.

MACCURDY, G. G., 1923: Human Skeletal remains from the Highlands of Peru. *Amer. J. Phys. Anthrop.*, 6: 217.

McCOWN, T. D., and A. KEITH 1939: *The Stone Age of Mount Carmel*. Vol. 2: Oxford, U.P.

MARTIN, R., 1928: *Lehrbuch der Anthropologie*. Munich, Fischer.

MOHAUPT, F., 1939: Untersuchungen über die Karieshäufigkeit sowie das Auftreten anderer Gebisserkrankungen in der jungeren Steinzeit im engeren mitteldeutschen Gebiet. *Deutsche Zahnaerztl. Wschr.*, 42:222.

MØLLER-CHRISTENSEN, V., 1958: *Bogen om Æbelholt Kloster*. (With English Summary) Copenhagen, Danish Science Press.

MOLLISON, TH., 1936: Zeichen gewaltsamer Verletzungen an den Ofnet-Schädeln. *Anthrop. Anz.*, 13:79.

MOODIE, R. L., 1927: Injuries to the head among the Pre-Columbian Peruvians. *Ann. Med. Hist.*, 9:277.

NIELSEN, H. A., 1911: Yderligere Bidrag til Stenalderfolkets Anthropologi. *Aarb. nord. Oldk. Hist.* p. 181.

PALES, L., 1930: *Paléopathologie et Pathologie Comparative*. Paris, Masson.

PEQUART, M. and S-J., BOULE, M., and H. VALLOIS, 1937: Teviec. Station-necropole mesolithique du Morbihan. *Arch. Inst. Paleontologie Humaine.*, Mem. 18.

SALLER, K., 1962: Die Ofnet-Funde in neuer Zusammensetzung. Ihre Stellung in der Rassengeschichte Europas. *Z. Morph. Anthrop.*, 52:1.

TWIESSELMANN, F. and H. BRABANT, 1960: Observations sur les dents et les maxillaires d'une population ancienne d'age franc de Coxyde, Belgique. *Bull. Group. Int. Rech. Sci. Stomat.*, 3:99.

WILLIAMS, H. U., 1929: Human paleopathology. *Arch Path.*, 7:839.

WOOD JONES, F., 1910: In: Elliot Smith, G. and F. Wood Jones.

ZUHRT, R., 1962 Hergang und Folgen eines doppelseitigen Unterkieferbruchs in Schnurkeramischer Zeit. *Ausgrabungen und Funde.*, 7:6.

Chapter 50

The Osteological Consequences of Scalping

H. HAMPERL

SCALPING, in the anatomical sense, consists of incising the skin over the skull down to the galea and the periosteum with a sharp object in a circular manner. The soft tissue can then be quickly removed from the cranial vault which then lies open, uncovered by periosteum. The circular incision may be carried out above or below the ears and therefore includes a larger or smaller area of skin.

Scalping had already been described in antiquity by Herodotus and there is ample evidence for it in a large number of cultures. Its most extensive practice, however, appears to have been during post-Columbian times in North America. Generally, it was performed on the head of a killed victim, but Friederici (1906) found forty-eight cases in the literature where scalping had been carried out in living persons, and where the victim had recovered and the wound healed. The changes found in skulls may vary, according to whether scalping has been carried out in (1) dead, or (2) living individuals.

1. In severing the scalp, the cutting instrument, especially if not particularly sharp, has to be pressed down and drawn through with some vigor, in order to dissect the whole tissue down to the bone. In so doing, the cranial vault may be scratched from the outside as with a saw.

In order to examine the possible production of such traces, we experimentally cut through the scalps of modern cadavers down to the bone using various instruments. If a sharp knife is employed, such as is usually done in autopsies today, no bone changes originate. There are no grooves or scratches, unless the knife is blunt and notchy and operates like a fine saw. However, if arrow heads or stone knives are used (Fig. 1), such as were available to the Indians in the early times, they have to be pulled to and fro a number of times until soft tissues in the cut have been dissected. As such instruments do not possess a smooth blade but rather operate like a saw because of their rough edge, numerous more or less parallel incisions or scratches result on the bony surface, as can be seen from Figures 2 and 3 of a macerated cranial vault so treated. The similarity of the traces on the bone produced in this manner, with those described in the literature as signs of previous scalping, is indeed striking.

Neumann (1940) describes the skull of a grown man (No. 896-101) with numerous parallel incisions or grooves over the cranial vault, reaching all around the circumference (Fig. 4).

Hoyme and Bass (1962) found similar incisions on the bones of four skulls. They were the skulls of two men (USNM 380865 [Fig. 5] and 380850), a young woman (USNM 380856) and a girl (USNM 381356).

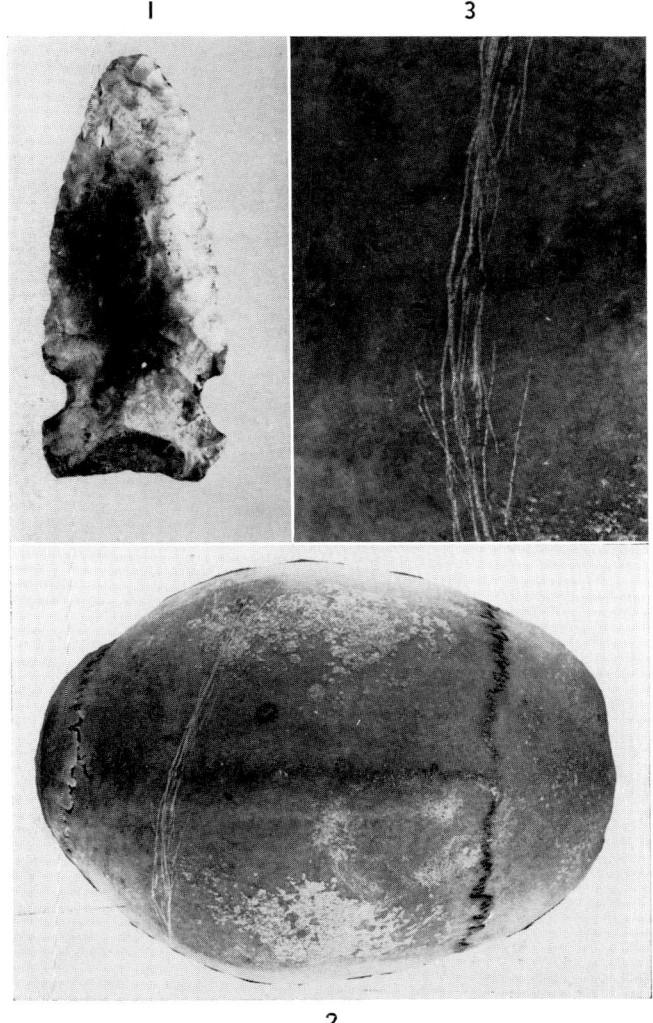

FIGURE 1. Indian spearhead as used for making the cuts in Figure 2. (Courtesy of Dr. Hans Reese).

FIGURE 2. Macerated skull of a man fifty-nine years old. The skin was cut, in the dead body, to the bone with an instrument as shown in Figure 1. The superficial grooves are clearly visible.

FIGURE 3. Same as Figure 2. Natural size.

Naturally, not every demonstrable incision on the cranial vault may be interpreted as the result of scalping, but only those which surround the crown of the head in a circular fashion. Other procedures were occasionally carried out on the skulls of corpses (see Friederici 1906).

2. If scalping was carried out on living persons and the individual expired soon thereafter, the findings naturally are identical with those encountered in scalping cadavers. However, if the victim survived for some time, certain changes may be expected in the skull. (see Fig. 6). The

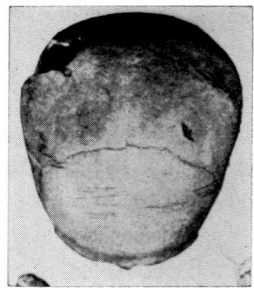

FIGURE 4. View of the skull F 896 showing scalping cuts. (After S. K. Neumann).

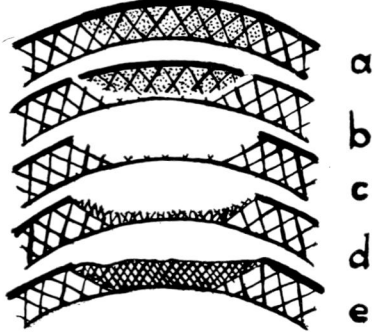

FIGURE 6. Schematic diagram of a sequence of changes, which may occur in the skull after scalping a living person who survived. (a) Immediately after scalping, the bone, deprived of its periosteum, becomes necrotic (black dotted area). (b) a granulation tissue has separated the dead bone (black dotted) from the living one. (c) the dead bone is shed; only the innermost part of the skull remains. (d) from this part and the edges begins the formation of new bone. (e) the new bone has a finer spongy appearance than the rest of the skull. The surface is covered by a thin continuous external layer, but remains depressed as seen in Figure 7. (After H. Hamperl and W. S. Laughlin).

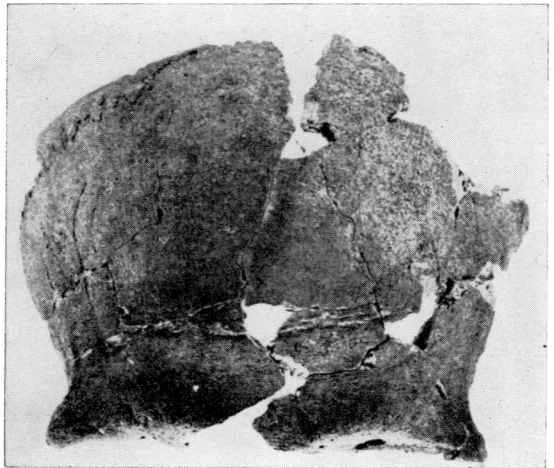

FIGURE 5. Skull USNM 380865 showing cut marks suggestive of scalping. (After L. E. Hoyme and W. M. Bass).

bone, deprived of its periosteum, will gradually exsiccate in its superficial portions and become necrotic (Fig. 6a), an inflammatory granulation tissue then separates it from deeper layers of living bone through demarcation (see Fig. 6b). A new spongy bone tissue (see Fig. 6d, e) is formed by the remaining inner bone layers, which is eventually covered by regenerating epidermis and the wound closes.

Primitive, and therefore especially credible reports (see, for instance, the information reported by Reese, 1940) prove that such an order of healing has indeed occurred. One would then expect to find in such a skull a shallow dent corresponding to the area of scalping. A completely similar finding was described by Hamperl and Laughlin (1959) (see Fig. 7) and interpreted as the result of an earlier scalping, which the victim survived for some time.

Snow (1941) has described changes on the skull in one case (No. M-2177) which he related to a possible earlier scalping. They include a groove, about a finger wide, leading around the cranial vault and corresponding more or less to a scalping incision (see Fig. 8). The edges of the groove are rounded, the base formed by compact bone being interpreted as a partial reconstruction of the *tabula externa* by the diploe. It was the opinion of the pathologist G. S. Graham, that in scalping the periosteum was only destroyed in the

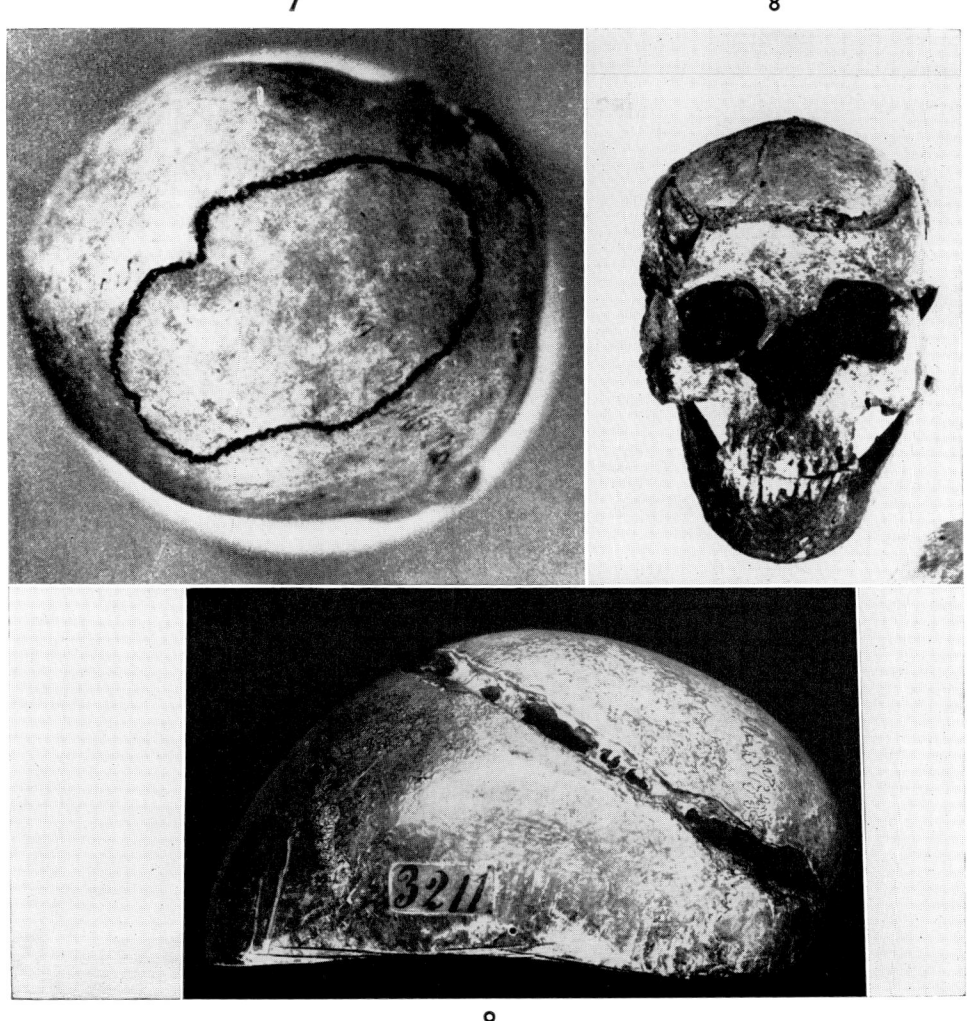

FIGURE 7. The extent of the depressed area of the skull outlined in black. (After H. Hamperl and W. S. Laughlin).

FIGURE 8. Skull No. M 2177 showing a pathological circular groove. (After C. E. Snow).

FIGURE 9. In the skull of a retarded eleven-year-old girl, one sees a circular groove, 3-7 mm broad, which has penetrated to the dura in both parietal bones and the occipital bone, with the exception of the tuber occipitale. It was caused by the compression of an elastic head band.

area of the groove, that is, in the area of the skin incision, while it remained intact in the center. However, that is hardly possible, as in typical scalping the periosteum is removed with the skin *in toto*.

The description leaves no doubt that the groove-like bone destruction is the result of an earlier inflammation. One would have to visualize a circular incision being carried out on the skull and leaving an infected wound, but without removing the skin with the periosteum as done in scalping. Possibly this could have been an attempted scalping or a circular incision in the skin for other—maybe therapeutic—reasons, which later involved the bone by inflammation.

To some degree, a similar picture is shown in the skull of a young mentally disturbed girl (see Fig. 9), which is in the collection of the Vienna Institute for Pathology (No. 3211). For years she had been wearing a tight fitting rubber band around her head, which at first destroyed the skin and later the cranial vault down to the dura through both the exerted pressure and the accompanying inflammation. Death had occurred as the result of meningitis.

REFERENCES

FRIEDERICI, G., 1906: Skalpieren und ähnliche Kriegsgebräuche in Amerika. *Inaug. Diss. Univ. Leipzig, Braunschweig.*

HAMPERL, H. and LAUGHLIN, W. S., 1959: Osteological consequences of scalping. *Hum. Biol. 31*:80.

HOYME, L. E. and BASS, M., 1962: Human skeletal remains from the Tollifero (Ha6) and Clarksville (Mc14) sites, John H. Kerr Reservoir Basin, Virginia. *Bureau of American Ethnology, Bull. 182*:329-400.

NEUMANN, G., 1940: Evidence for the antiquity of scalping from central Illinois. *Amer. Antiq., 5*:287.

REESE, H. H., 1940: The history of scalping and its clinical aspects. *Year Book of Neurology, Psych. and Endocrinology,* 3-19.

SNOW, Ch. E., 1941: Anthropological studies at Moundville Part 2 Possible evidence of scalping at Moundville. Paper 15. *Alabama Mus. Nat. Hist. Museum,* 55-57.

Chapter 51

Primitive Surgery*

ERWIN H. ACKERKNECHT

SPEAKING of "primitive surgery" is one of those arbitrary procedures which are to a certain extent unavoidable if we try to analyze primitive phenomena for a better understanding of our own cultural processes, and which are justifiable as long as we remain aware of their arbitrary character. Not before the second half of the Middle Ages were surgery and its practitioners regarded as different from or inferior to other methods or practitioners of the healing art. Then for 700 years they remained separated from the body of medicine. In modern society "surgery" has again become part of medicine—but as one of its "specialties." This again is without precedent. Though we find "specialists" also in primitive and archaic medicine (Egyptian, Peruvian, etc.), specialization there has other reasons and proceeds along dividing lines different from those observed in modern scientific medicine.[1] Surgery is, therefore, not a special field defined by the primitives themselves. We simply deal in the following with such procedures as would be mainly in the domain of the surgeon in our society. Practically, it means that we deal primarily with therapeutic measures which are of a definite technological interest.

WOUND TREATMENT

There is no tribe on record which does not in some way or other treat wounds. It would lead us too far to go into details here of the hundreds of different treatments recorded, the principles of which are very similar. Herbs or roots, often with astringent or disinfectant qualities, are applied to the wound in form of powders, infusions, or poultices. In rare cases animal substances, like powdered insects or cow dung, are used. Heat is rather widely employed to speed the healing process.[2] Wound treatment by primitives is in general regarded as "good."

In the treatment of wounds the Cherokee doctors exhibit a considerable degree of skill, but as far as any internal ailment is concerned the average farmer's wife is worth all the doctors in the whole tribe.[3]

Similar appreciative judgments can be easily collected from Oceania,[4] South

---

* Parts of this article were given as a paper at a Viking Dinner Conference on March 22, 1946. I wrote this article while working in the Institution of Human Morphology (Department of Anthropology, American Museum of Natural History, New York), a project financed by a grant from the Viking Fund. I am glad to express on this occasion my appreciation to the Fund and to its Director of Research, Dr. Paul Fejos. Originally published in *Amer. Anthrop., 49:25*, 1947.

1. ROSEN, 1944, p. 5 ff.: Ackerknecht, 1945, p. 37; Ackerknecht, 1946, 479 ff.

2. MARTIUS, 1844, p. 182; Harley, 1941, pp. 220, 222; Warner, p. 221.

3. MOONEY, 1891, p. 323.

4. LANDTMANN, 1937, p. 227.

America,[5] or Africa.[6] They are, of course, mainly based on the prompt healing of wounds, which can be explained equally well through the constitution of the patients or the absence of the highly "cultivated" germs of our cities and hospitals. To evaluate wound treatment in primitive society in such a general way is all the more difficult when we see in our own medicine rather different treatments applied with about similar results. Reports of deaths occurring during initiation rites in Melanesia and Africa[7] from infected scarification wounds should not be omitted from a general picture.

It is easier to appreciate certain isolated technical procedures in the course of wound treatment. For instance, the suturing of wounds (with sinews among North American Indians like the Carrier, Mescalero, Dakota, Winnebago;[8] with thorns among the Masai and Akamba,[9] and with the heads of termites among the Somali and Brazilian Indians[10]) is a very respectable accomplishment. Wound drainage too is reported from North American Indians.[11]

The stopping of blood vessels is a difficult problem for primitives. That they do not know the ligature[12] is not surprising, as it appeared in our own culture only with Celsus (1st century A.D.) and was rediscovered by Ambroïse Paré (1510-90). Primitives do use, more or less successfully, such diverse materials as powdered gum, charcoal, ashes, eagles' down, and bandages of bark or coconut fibre. Tourniquets are known in Africa, North America, and Oceania.[13] One of the best styptic methods is cauterization, practised in Africa,[14] and Oceania.[15] The only tribe known to suture *vessels* (with tendons) is the Masai.[16] We will encounter the Masai again as the primitive master surgeons. They are atypical, as their surgery is incomparably superior not only to that of all other primitive tribes reported, but also to the surgery of most civilized peoples up to the Renaissance. That even this highly developed surgery is not free from magic becomes obvious from such examples as the Masai, putting a dead fly into the wound, and binding one testicle around the left anterior leg of their cattle after castration so that the wound might close more quickly.[16a]

Occasionally, deep-seated arrows seem to be extracted skillfully and successfully.[17] In a limited area in East Africa, natives are *even able to suture intestines* opened by arrows or spears.[18] Intestinal wounds may heal sometimes,[19] but treatment in general seems inept.[20] Incarcerated hernias die. Umbilical hernias are, in many places, regarded as a sign of beauty. In North America hernias are occasionally

---

5. KOCH-GRÜNBERG, 1923, III, p. 274.
6. DRIBERG, 1923, p. 55.
7. LINTON, 1945, p. 301; HARLEY, 1941, p. 131.
8. MORICE, 1900-01, p. 22: STONE, 1932, p. 76.
9. LINDBLOM, 1933, p. 312; Merker, 1910, p. 181.
10. PARDAL, 1937, pp. 50, 161; Monfreid in Stephen-Chauvet, 1936, p. 76.
11. STONE, 1932, p. 76.
12. The ligature of an artery with copper wire among the Ba-Yaka mentioned by Tardy and Joyce, 1906, p. 50, and a ligature of North Carolina Indians are likely to be borrowed from the whites.

13. MALCOLM, 1934, p. 200.
14. BARTELS, 1893, pp. 282, 287.
15. HAGEN, 1899, p. 285.
16. MERKER, 1910, p. 181.
16a. *Id.*, p. 195.
17. E.g., Chartier in Stephen-Chauvet, 1936, p. 76; Routledge in Harley, 1941, p. 221: Grinnell, II, p. 147.
18. MERKER, 1910, p. 181; Monfreid and Maurice in Stephen-Chauvet, 1936, 76-80; Roscoe and Talbot in Harley, 1941, p. 222.
19. BARTELS, 1893, p. 284.
20. WEBB, 1933-34, p. 95: Catlin, 1876, I, p. 39.

bandaged.[21] The Déné bring a prolapsed uterus back into its original position and bandage.[22]

The oldest document in the German language, the so-called "Merseburger Zaubersprüche," is wound incantations. Homer's "Odyssey" (XIX, 457) mentions an incantation against bleeding. It is, therefore, not surprising that in addition to dressing, wounds are treated with magic songs among, e.g., the Apache,[23] Havasupai,[24] or Creek.[25] The Creek submit the wounded to the same kind of isolation as women after childbirth.[26] The wounded among the Banyankole can be nursed only by women without sexual relations.[27] The latex used in the treatment of wounds among the Mania plays a magic role.[28] The Maori exorcised arrowheads which they could not touch.[29] Heat used in the treatment of wounds, be it in the form of the "moxa" or not, has often a symbolical meaning and makes the disease spirit fly away.[30]

FRACTURES AND DISLOCATIONS

To base judgment concerning the quality of primitive surgery on mere excavation material has become impossible since Adolph Schultz has shown that well-healed fractures are numerous among wild gibbons and other primates which are not likely to enjoy treatment by professional bonesetters.[31]

As in the case of wounds, numerous reports emphasize again the good treatment of fractures, e.g., among the Creek, Winnebago,[32] the Barundi,[33] Bavenda,[34] Duke of York Islanders,[35] and Maori.[36] From these, and from the following tribes, the use of splints is reported: Chippewa,[37] Nez-Percé,[38] Hottentot,[38a] Tahitian,[39] Eskimo,[40] Original casts, made from leather, chicle, or clay, are used by the Shoshone,[41] the Lango,[42] the Jívaro,[43] and South Australian tribes.[44] That splints do not prevent bad healing was already emphasized by Martius.[45] Morice[46] sees the reason for such failures in the absence of proper setting of the fragments. As experience in Nias shows, even relocation does not always seem to prevent bad results.[47] The Cherokee[48] and the Dakota[49] are credited with

---

21. MORICE, 1900-01, p. 23; Opler, 1941, p. 217; Malcolm, 1934, p. 200.
22. MORICE, 1900-01, p. 24.
23. OPLER, 1941, p. 349.
24. SPIER, 1928, p. 284.
25. SWANTON, 1928, p. 617.
26. SWANTON, 1928, p. 625.
27. ROSCOE, 1923, p. 161.
28. VERGIAT, 1937, p. 171.
29. PARHAM, 1943, VI.
30. MORICE, 1900-01, p. 20; Sieroshewski, 1901, p. 105.

31. SCHULTZ, 1939, p. 571 ff.; id., 1944, p. 115 ff. Dr. Schultz' publications have very important implications on the problems of primitive pathology in general and natural selection, with which I have been unable to deal in this context. They should be perused by all interested in these problems. [The former paper is reprinted in this book. Eds].
32. BARTELS, 1893, p. 290.
33. MEYER, 1916, p. 142.
34. STAYT, 1931, p. 273.
35. BROWN, 1910, p. 185.
36 PARHAM, 1943, p. VI.
37. DENSMORE, 1928, p. 334.
38. SPINDEN, 1908, p. 257.
38a. SCHAPERA, 1930, p. 408.
39. ELLIS, 1853, III, p. 42.
40. WEYER, 1932, p. 329.
41. STONE, 1932, p. 82.
42. DRIBERG, 1923, p. 56.
43. STIRLING, 1938, p. 120.
44. BARTELS, 1893, p. 290.
45. MARTIUS, 1844, p. 182.
46. MORICE, 1900-01, p. 22.
47. KLEIWEG, 1913, p. 133.
48. OLBRECHTS, 1932, p. 71.
49. BARTELS, 1893, p. 290.

in New Guinea, Samoa, Tahiti, and Vaitupu.[87] The Masai operate even on abscesses of the liver and spleen.[88] In the same direction lies the lancing of a hydrocele in Vaitupu[89] or of an inflamed testis in Uvea. Fatal outcome of the incision of a hernia is reported by the same author.[90] Rare operations along similar lines are the opening of an empyema (Great Lakes);[91] pneumothorax by cautery in pleurisy and pneumonia in Uganda;[92] scarification of inflamed tonsils by the Masai;[93] multiple piercing of goitre,[94] and tenotomy.[95]

MEDICAL AMPUTATION AND EXCISION

With these operations we enter the field of the very rare. Observers of even the more surgery-minded tribes in America,[96] as well as in Africa[97] and Oceania,[98] emphasize the absence of amputation. Amputation seems most likely where nature, by freezing limbs, has already prepared the procedure. Thus we hear of the (very crude) amputation of frozen fingers among the Eskimo and Chippewa.[99] The Dama represent an isolated case of amputating crippled fingers and toes.[100] Whether the penis amputation photographed by Neuhauss[101] is of a medical nature is doubtful.

The Masai enucleate eyes and, for instance, amputate limbs with hopelessly complicated fractures with great skill.[102] They even have protheses.[103] They are thus apparently the only primitives to equal in this field the accomplishments of the ancient Peruvians[104] who, according to Roy Moodie, were far better surgeons than "any other primitive or ancient race of people."[105] We are not dealing further with the surgery of the ancient Peruvians because neither can they be regarded as "primitives," nor is it clear whether the numerous mutilations of nose, lips, and extremities seen on Peruvian pottery are spontaneous (a consequence of lepra, syphilis, or, most likely, leishmaniasis), or, if artificial, whether they are of a medical, religious or juridical nature.[106]

A remarkable local surgical accomplishment, reported by six different authors, is the excision of neck glands by native African doctors in the case of sleeping disease (trypanosomiasis).[107] Neck tumors are cauterized in Rhodesia.[108] The Galla and the Akamba remove the uvula.[109] The Tembu and Fingu operate on vaginal polyps.[110]

I have been able to find only one place outside of Africa where surgery equals similar levels, Vaitupu (Ellice Islands) in Polynesia. There subcutaneous lipomata, the elephantoid scrotum, tuberculous glands in the neck, old leprotic or yaw ulcers are removed successfully.[111] While

---

87. WHITING, 1941, p. 52; Turner, 1884, p. 141; Ellis, 1853, p. 44; Kennedy, 1931, p. 241.
88. MERKER, 1910, p. 183.
89. KENNEDY, 1931, p. 241.
90. ELLA, 1874, p. 50.
91. STONE, 1932, p. 84.
92. HARLEY, 1941, p. 222.
93. MERKER, 1910, p. 190.
94. MILNE, 1924, p. 251.
95. KAYSSER in Neuhauss, 1911, III, p. 77.
96. MORICE, 1900-01, p. 22; Grinnell, 1923, II, p. 147; Swanton, 1928, p. 625.
97. WEYER, 1932, p. 142; Van der Burgt, 1903, p. 363.
98. ELLA, 1874, p. 50.
99. WEYER, 1932, p. 328; Densmore, 1928, p. 333.
100. VEDDER, 1923, II, p. 90.
101. NEUHAUSS, 1911, I, p. 436.

102. JOHNSTON, 1893, II, p. 829; Merker, 1910, p. 193.
103. MERKER, 1910, p. 196.
104. PARDAL, 1937, p. 160.
105. MOODIE, 1927, p. 278.
106. PARDAL, 1937, Chapter VII, p. 217, 234.
107. BARTELS, 1893, p. 300; Harley, 1941, pp. 45, 219.
108. HARLEY, 1941, p. 222.
109. PAULITSCHKE, 1896, I, p. 184; Lindblom, 1933, p. 312.
110. LAUBSCHER, 1937, p. 11.
111. KENNEDY, 1931, p. 241 ff.

the Africans at least have iron knives at their disposal, these Polynesians operate exclusively with shark teeth.

I would regard with the greatest distrust the marvellous stories of the abdominal surgery of the Araucanians, based on old sources.[112] We know now that the shamanism of the Araucanians shows the closest resemblances to Siberian shamanism. The Araucanians practise the rare postmortem opening of the body, typical of Siberia,[113] as well as the old shamanistic trick of opening the body and cleaning the intestines. It is most likely that early observers have mistaken both customs for operations.

Equally spurious are reports on primitive "cataract operations."[115] The only clear-cut case in the literature is no less clearly of Arabic provenance.[116] If a case is described in sufficient detail, as by Morice,[117] it becomes obvious that we are not dealing with the operation of a cataract but a pterygium! Actually, I do not think that the one report on ovariotomy in Australia[117a] justifies counting this operation among the accomplishments of primitive surgery.

CAESAREAN SECTION

The Caesarean section is technically even more difficult than the operations mentioned above. It may be very old, but the earliest authentic reports of it stem from the 16th century, and only during the last sixty years has it become a routine procedure. One is thus inclined to dismiss report of primitive Caesareans as mere fable or misunderstanding. Neither self-inflicted rippings of the belly by desperate mothers, as have occurred even in our time,[118] nor the widely practised cutting out of the foetus when the mother has died can qualify as "surgery." The latter measure was prescribed by law in Rome as early as 715 B.C., and seems to be rather common in parts of Africa[119] and Oceania.[120] This performance, based mostly on magic representations, leads by no means automatically to the true Caesarian section. We might even disregard van der Burgt's repeated statements that the Kurundi do practise Caesarean section,[121] and dismiss primitive Caesarean section with a shrug of our shoulders if there did not exist a well-dated, most detailed, and most positive description of a Caesarean section coming from an observer whose reliability has, to our knowledge, never been challenged. The skill with which the operator acted leaves no doubt that he did not improvise, but followed a well-established procedure.

The observer was Robert Felkin and he saw the operation performed upon a 20-year-old woman in Kahura in Uganda in 1879.[122] Banana wine served as an anaesthetic and disinfectant. Haemorrhage was checked with a red-hot iron. The incision was made in the midline, between symphysis and umbilicus, and closed with iron nails. Temperature never rose above 101°F in the postoperative stage, and the wound was closed on the eleventh day.

Unfortunately, I feel unable to explain

---

112. E.g., CORLETT, 1935, p. 242.
113. ACKERKNECHT, 1948, p. 336.
114. CZAPLICKA, 1914, p. 233.
115. E.g., TEIT, 1900, p. 370.
116. HARLEY, 1941, p. 38.
117. MORICE, 1900-01, p. 27.
117a. MIKLUCHO-MACLAY, 1882, p. 26.

118. YOUNG, 1944, p. 12 ff.
119. BALIMA UNJORDO (Emin Bey, 1897, p. 393); Barundi (Meyer, 1916, p. 442); Kisiba (Rehse, 1910, p. 117); Thonga (Junod, 1927, II, p. 332).
120. MEAD, 1924, p. 133, 156; Mackenzie, 1927, p. 333; Ploss-Bartels, 1899, II, p. 310.
121. VAN DER BURGT, 1903, p. 363.
122. FELKIN, 1884, p. 928 ff.

why in 1879 there existed in Kahura in Uganda a black surgeon performing the Caesarean section safely and, in some respects, better than many of his contemporary white colleagues. It is suggested that, as we have already seen, East Africa shows generally a better surgery than any other region inhabited by primitives; that the widespread embryotomy[123] seems particularly frequent in East Africa;[124] but all this, of course, does not constitute a real answer to our question.

TREPANATION

The common surgical procedures described in the 1st to 4th section, and even the rare ones noted in the 5th and 6th, create a picture of primitive surgery which would be fairly consistent if we did not encounter among primitives, and not rarely but fairly frequently, an operation which up to the second half of the nineteenth century was regarded by modern surgeons as extremely dangerous and difficult: trephining of the skull. It is, of course, neither possible to survey here the extensive literature on primitive and prehistoric trephining, nor is it necessary for our purpose. We need not deal with technical details, and we can almost entirely omit European prehistoric trephining as not belonging properly to our subject. (In another context this material is of the highest importance as it is the sole existing tangible evidence of prehistoric medicine.) Peruvian trephining I have included to a limited extent. I do not think that Inca medicine can be called "primitive,"[125] but trephining in the Andean region preceded Inca civilization[126] and survived it into the twentieth century.[127]

The practice of trephining has been directly observed among the following tribes and nations: in the Balkans, in Daghestan,[128] and among the Berber; in Abyssinia,[129] Uganda,[130] and Nigeria.[131] It seems to be particularly frequent in Oceania (New Caledonia,[132] New Zealand,[133] New Guinea,[134] Uvea and the Loyalty Islands,[135] the Gazelle Peninsula and New Ireland,[136] New Britain,[137] the Solomons,[138] and Tahiti.[139] The centre of trephining in the Americas was undoubtedly the Peruvian highlands. But isolated evidences of trephining have been found all over North, Central and South America, from British Columbia down to Chile.[140] In the latter respect the situation is similar to the one encountered in European Neolithic trephining: besides numerous finds in France, we encounter isolated ones from Russia to Spain, from Sweden to North Africa. Although success in this complicated operation is explainable where it is a routine procedure, genesis and success of such

---

123. PLOSS-BARTELS, 1899, p. 306; Parham, 1943, p. VI.
124. PLOSS-BARTELS, 1899, p. 307; Roscoe, 1911, p. 54; id. 1924, p. 121.
125. See my article on the medical practices of the South American Indian in Vol. V of the forthcoming Handbook of the South American Indian, Smithsonian Institution.
126. MEANS, 1931, p. 446. My colleague, Junius Bird, was kind enough to inform me that the trephined skulls found at Paracas most likely come from the beginnings of the Christian era.
127. BANDELIER, 1904, p. 442.
128. GUIARD, 1930.
129. WÖLFEL, 1925, p. 1.
130. ROSCOE, 1921, p. 147; id., 1923, p. 161.
131. DRIBERG, 1929, p. 63.
132. GUIARD, 1930.
133. WÖLFEL, 1925, p. 14.
134. HAGEN, 1889, p. 257.
135. ELLA, 1874, p. 51; Ray, 1917, p. 273.
136. PARKINSON, 1907, p. 108; Crump, 1901, p. 167.
137. CRUMP, 1901, p. 167.
138. WÖLFEL, 1925, p. 10.
139. ELLIS, 1853, p. 43.
140. LEECHMAN, 1944; Shapiro, 1927; Wölfel, 1925, p. 19 ff.

isolated operations are hard to understand.

Prehistoric trephining was interpreted by Broca and many subsequent authors as largely magical (to liberate the disease spirit). Wölfel, on the other hand, regards trephining as a purely surgical measure in the case of skull fracture, caused mainly by two weapons—the slingshot and the club. The little we know of actual motivations of primitive trephining, however, does not allow a clear-cut decision.

In New Britain the operation obviously is used in the case of combat fractures of the skull—and yet it is also a means to obtain longevity.[141] In New Ireland, trepanation is applied in the case of skull fractures, as well as against "headaches" and "epilepsy" (both magically explained), and in children as a prophylactic against ill health in general.[142] In the Gazelle Peninsula the indication is surgical, but success depends entirely on magic remedies.[143] In Tahiti the indication seems purely surgical; in Uvea, purely magico-medical ("headache"). Almost all men are trephined there. The Uveans also trephine the tibia and ulna in cases of rheumatism.[144] In most Peruvian trephinings, the slingshot, club, or accident seems to have been the causative agent. Yet the form of some trephinings suggests "medical" indication, a fact which is admitted even by Wölfel.[145] Sometimes tumors or gummata seem to have been the reason for the operation.[146] It is remarkable that even such an apparently rational procedure as the Peruvian one was so carefully hidden by the natives for almost four hundred years that no chronicler or traveller recorded it before Bandelier (1904).

It has been shown repeatedly that diverse pathological conditions may produce lesions very similar to trephining.[147] That the bulk of our archaeological material consists nonetheless of true trephinings is suggested by the numerous unhealed trepanations which can be easily identified, and by the ethnographic record.

For Oceania, survival has been estimated by observers as high as 50,[148] or even 80 or 90 per cent.[149] For Peru, McGee concludes from his series of skulls a survival rate of 50 per cent, which is equal to European results in the second half of the nineteenth century.[150] The results of Tello and MacCurdy are even more favorable.[151]

RITUAL AND JUDICIARY MUTILATIONS

With trephinings we have reached the limits of surgery proper in primitive societies, but by no means the limits of activities which, objectively, are on the same level with amputations and incisions. On the contrary, in numerous places we see the same primitives who only rarely use a knife in the case of disease or traumatism fall into a veritable frenzy of cutting and chopping off when ritual or judiciary motives are involved.[152] Of this enormous field of ritual and judiciary mutilation I can here, of course, give only a very sketchy survey which, however, I hope

---

141. CRUMP, 1901, p. 168.
142. PARKINSON, 1907, p. 113; Crump, 1901, p. 168.
143. PARKINSON, 1907, p. 110.
144. ELLA, 1874, p. 51.
145. TELLO, 1913, p. 81; MacCurdy, 1923, p. 251; Wölfel, 1925, p. 32.
146. FREEMAN, 1924, p. 24; Wölfel, 1925, p. 14; Tello 1913, p. 79.

147. ALAJOUANINE and THURE, 1945, p. 71 ff.
148. ELLA, 1874, p. 51.
149. CRUMP, 1901, p. 168.
150. SHAPIRO, 1927, p. 266.
151. TELLO, 1913, p. 83; MacCurdy, p. 259.
152. In this respect trephining fits much better into the ritual than into the medical field, and this is one of the reasons why I would feel hesitant to discard Broca's old hypothesis as eagerly as Wölfel did.

will help us to a better understanding of primitive surgery.

I need scarcely mention such minor interventions as dental mutilations, ritual cicatrization, or head deformation. Amputation of the fingers for ritual reasons is well known to us from South and North American Indians.[153] The custom seems even more widespread in Africa and Oceania. In an excellent survey, Lagercrantz mentions no less than fourteen tribes in black Africa practising ritual finger mutilation.[154] Söderström gives almost the same number for Oceania.[155] Next to the fingers, the genitalia seem to offer a convenient target for religious zeal. No less than fourteen methods of operating upon the male genitalia are known.[156] There is no need to give details on circumcision and its variations.[157] Subincision, the opening of the male urethra, is, curiously enough, eminently unsuccessful when practised for purely medical reasons.[158] Its medical character in Fiji and Tonga is, to say the least, somewhat confused. Its magic character in Australia is clear. So far the most satisfactory explanation for its true nature has been brought forth by Dr. Ashley Montagu.[159] A surgically most remarkable genital mutilation, monorchy, the removal of one testicle, has been reported from the Hottentot, the Dama, from Abyssinia, the Loyalty Islands, the Carolines, Tonga, etc.[160] The female genitalia are submitted to clitoridectomy, infibulation, etc.[161]

For punishments, the fingers again become convenient objects[162] if the whole hand is not sacrificed.[163] The Seneca Indians performed a very neat amputation of half the foot upon their captives.[164] Offenders may be d e p r i v e d of their tongues,[165] or of their genitalia in case of adultery.[166] Compared to the rare incisions and even rarer amputations, ritual and judiciary mutilations are of an almost universal character in primitive societies.

THE SURGICAL PERSONNEL

Unfortunately most o b s e r v e r s have failed to inform us about the not unimportant item of who actually performs the operations described.

Only among the Masai[167] do we hear of a definite class of surgeons (treating humans and animals alike). In West Africa surgery is not in the hands of the witch doctors, but of the herbalists (mostly female), who operate with the mysterious assistance of a white fowl.[168] One of the religious societies of the Kiowa, the Buffalo Doctors, specialized in the treatment of wounds.[169] The Omaha Buffalo Society concentrated upon the magic treatment of accidents.[170]

It seems less surgery which is set apart as a specialty than bonesetting. We hear

---

153. DEMBO-IMBELLONI, 1938, p. 203; Morice, 1900-01, p. 23; Grinnell, 1923, II, p. 196; Karsten, 1926, p. 186; Preuss, 1890.
154. LAGERCRANTZ, 1935, p. 129 ff.
155. SÖDERSTRÖM, 1938, p. 24 ff.
156. MALCOLM, 1934, p. 200.
157. Hastings Encyclopedia of Religion and Ethics, III, p. 659 ff.
158. STEINEN, 1886, p. 129; Harley, 1941, p. 59.
159. MONTAGU, 1937, p. 193 ff.
160. LAGERCRANTZ, 1935, p. 199 ff.

161. Hastings Encyclopedia of Religion and Ethics, III, p. 659 ff.
162. Eleven African tribes mentioned by Lagercrantz, 1935, p. 129 ff.; for Oceania see Söderström, 1938.
163. UGANDA (Roscoe, 1921, p. 278); Bahr-el-Ghazal (Anderson, 1911).
164. PACKARD, 1901, p. 29.
165. MALCOLM, 1934, p. 200.
166. HRDLIČKA, 1908, p. 251; Lagercrantz, 1935, p. 132.
167. MERKER, 1910, p. 181.
168. KINGSLEY, 1899, p. 157.
169. MARIOTT, 1945, p. V.
170. FLETCHER-LA FLESCHE, 1911, p. 487 ff.

of Zuni bonesetters (mostly people struck by lightning);[171] of a clan of bonesetters among the Azande, using splints, massage, and mystic ointments;[172] of efficient, though much magic-using bonesetters among the Gio and Manos;[173] of bonesetters in Melanesia,[174] and at Ontong Java.[175]

Most frequently surgery seems to be done by the otherwise supernaturalistic medicine-man. It is the "wizard" who trephines in New Britain,[176] as well as in Bolivia.[177] The famous abscess opening among the Zuni is done by the "theurgist."[178] One of the first things a Creek medicine-man learns is how to treat gunshot wounds with "songs" and medicines.[179] Paviotso shamans cure wounds and injuries as well as illness.[180] Jívaro medicine-men set bones.[180a] In Kenya, surgery and bonesetting are in the hands of the medicine-man,[181] and in Uganda the medicine-man first chops the limbs off as an executioner, to treat the wounds later as a surgeon.[182]

It would lead too far to go here into the characteristics of medical specialization among primitives in general which differs considerably from medical specialization in our society. As far as the separation between medicine and surgery is concerned, it seems not to have advanced very far, if we are to judge from our scanty material. Our picture of the primitive surgeon as being probably more realistic and socially inferior is strongly influenced by the later Mediaeval and early modern situation in our own society. Such as expectation does not seem to be confirmed by the data available.[183]

DISCUSSION AND CONCLUSIONS

Primitive surgery has never reached the level of, for example, Alexandrian surgery as it is reflected in Celsus (1st cent., A.D.). Celsus speaks of the excision of tumors, of operation for aneurysm, hernia, and stone, of plastic surgery, of the amputation of the lower limbs and the resection of bones (the jaw included). The typical "surgery" of a Guarani tribe consists of cutting the umbilical cord, perforating the earlobe and the lower lip, and of opening abscesses.[184] Tribes with a very developed medicine, like the Liberian Manos, "are extremely conservative when it comes to surgery."[185] Their surgery is limited to bonesetting, blood-letting by small shallow incisions, circumcision and scarification of tribal marks.[186] There is no surgery in the Bhar-el-Ghazal;[187] neither surgery nor autopsies in Madagascar.[188] "Surgical cases are treated in the worst possible way, any intervention with a knife being looked on as absurd if not culpable. . . . Medical cases are generally treated more rationally than surgical cases."[189] There is no substantial difference between such isolated

---

171. SPIER, 1928, p. 285.
172. EVANS-PRITCHARD, 1937, p. 498.
173. HARLEY, 1941, pp. 17, 93.
174. CODRINGTON, 1891, p. 199.
175. HOGBIN, 1930-31, p. 165.
176. CRUMP, 1901, p. 167.
177. BANDELIER, 1904, p. 442 ff.
178. STEVENSON, 1904, p. 386.
179. SWANTON, 1928, p. 617 ff.
180. PARK, 1938, p. 59.
180a. STIRLING, 1938, p. 120.
181. BARTON, 1923, p. 74.
182. ROSCOE, 1911, p. 278.

---

183. See ROSEN, 1944, p. 5 ff.; Ackerknecht, 1946, p. 479 ff.
184. F. MÜLLER, 1928, p. 502.
185. HARLEY, 1941, p. 74.
186. *Id.*, p. 40.
187. ANDERSON, 1911, p. 264.
188. GRANDIDIER, 1908, IV, 428.
189. JUNOD, 1927, II, pp. 458-459.

judgments of observers and the results of our foregoing survey. *Primitive surgery is indeed poor in scope and quality.* Only in the more southern parts of East Africa and in certain Polynesian localities do we encounter a relatively well-developed surgery. The East African focus of good surgery is not limited to the Masai. Among their neighbours we have found such outstanding accomplishments as Caesarean section, intestinal suture, trepanation, excision of glands, etc. It is to be hoped that regional specialists will be able to throw more light on the reasons why we find these two local centres of a more highly developed surgery.

Logically there exist four possibilities why primitive surgery has not advanced further: that there was no need of surgery; that primitives lack technical skill; that they lack certain elements of knowledge; that other elements of their socio-mental makeup have been unfavorable to the development of surgery among them. All these possibiilties undoubtedly play a certain role, but their relative importance is by no means the same. It is a fact that one of the main objects of our surgical endeavours, cancer, is rare among primitives, whether for racial reasons, or simply because most of them never reach the cancer age. On the other hand, the ills to which the savages are not exempt are sufficiently numerous to have furnished enough incentive for a more developed surgery.

Surgery undoubtedly presupposes a considerable manual skill. But many primitives show such skill and yet, like the Eskimos, are very poor surgeons.

The great progress in modern surgery was made possible by an enormous increase in our knowledge concerning anatomy, anaesthesia and asepsis. *The anatomical knowledge of most primitives is notoriously bad.*[190] Yet explaining the lack of surgery by the lack of anatomy is only reformulating the problem in other terms. The lack of anatomy is undoubtedly due to certain objective limitations. Anatomical knowledge becomes valuable only when organized. The possibilities of organizing knowledge depend largely on the structure of society in general. Not accidentally have ideas concerning the social body so often colored anatomical ideas, and vice versa. And many primitive societies are rather amorphous. But this is only part of the truth. The knowledge of people is not only a question of what they are able to learn, but also one of what they want to learn—a question of interests and values. In this respect a comparison between the anatomical knowledge of primitives who make autopsies and those who do not has been very revealing to me. Surprisingly enough, both categories are equally ignorant of anatomy,[191] because even the dissectors are so strongly under the influence of supernaturalistic ideas that they overlook the anatomically obvious.[192] Anatomical ignorance of primitives seems mainly due to their supernaturalistic orientation. We must not forget that our own way of looking at the human body and bodily functions is rather unique compared to the attitude not only of primitives but also of most ancient civilizations.

190. E.g., Déné (Morice, 1900-01, p. 21); Omaha (Fletcher-La Flesche, 1911, p. 107); Chorti (Wisdom, 1940, p. 307); Brazilian Indians (Martius, 1844, p. 128); Pangwe (Tessmann, 1913, II, p. 128); Thonga (Junod, 1927, II, p. 332); Ba-Ila (Smith and Dale, 1920, I, p. 224); Kiwai Papuans (Landtman, 1927, p. 281); Sinaugolo (Seligman, 1902, p. 301).

191. Ackerknecht, 1948, p. 338.

192. Mere dissection did not improve Mediaeval anatomy either. More than 200 years of dissecting before Vesalius under scholastic auspices did not reveal the obvious errors of Galenic anatomy till the Renaissance shattered the authoritarian principle in science and replaced it by observation.

The absence of anaesthesia and asepsis in the modern sense is perhaps less important in the lag of primitive surgery than it appears at first sight. In spite of the lack of anaesthesia and a very rudimentary asepsis, the Masai have developed a quite creditable surgery. Many primitives have a considerable number of general and local anaesthetics at their disposal.[193] All observers agree upon the relative ease with which primitives overcome wound infection.[194] It is immaterial here whether this increased resistance is primarily constitutional or, as I am inclined to think, primarily due to the fact that no hospitals are at the disposal of primitives for the cultivation of particularly virulent strains of strepto- and staphylococci. The fact remains that primitives do, as far as wound infection is concerned, labour under less odds than did modern surgery in its beginning, and this might explain their comparatively excellent results in complicated operations like trephining whenever they did attempt them.

It seems, therefore, that the most satisfactory explanation for the particular character of primitive surgery lies in the direction of the limiting influence which supernaturalistic ideas among primitives exert upon the development of the operator's art. It seems the only way to explain the "mystery" of primitive surgery,[195] that is, the occurrence of major operations such as trephining and of wholesale ritual and judicial mutilation among people who otherwise possess but the rudiments of surgery.

We have mentioned already the presence of magico-religious elements in surgical practice in their proper place. We hope that it has become obvious that primitive surgery is, as little as primitive midwifery, the purely empirical half of primitive medicine, and different in principle from a much more supernaturalistic internal medicine (as to a certain extent it is in Egyptian medicine). "Superstition affected more or less all surgical operations."[196] Yet this active role of the magico-religious in primitive surgery is only a part, and perhaps the smaller one, of the influence of magico-religious representations on primitive surgery in general.

The negative influence of supernaturalistic ideas on surgery is very clear in all those cases where bodily mutilation in general is dreaded because of its detrimental influence in the future life of the ghost. "A Central African will not consent to an operation (not even tooth extraction), as it conflicts with the anticipation of his dismembered spirit.[197] For the same reason, the Tanala have no fear of death, but are very much afraid of mutilation.[198] To the Arab and Shawia, death is preferable to the loss of a limb.[199] It is obvious that amputation or other major surgery can hardly develop or be "diffused" under such circumstances. It is also clear that punishment by bodily mutilation in such tribes is damaging far beyond the physical disability it leaves. And it is understandable that in regions where mutilation is a customary form of punishment, people will dislike to undergo operations which externally would identify them with criminals.[200] Such attitudes of fear are not restricted to primitive societies. The Chinese, for instance, dislike for magic reasons

---

193. ELLIS, 1945. See also Stevenson, 1904, p. 386; Bandelier, 1904, p. 445; Freeman, 1924, p. 32: Harley, 1941, p. 70; Felkin, 1884, p. 928; Angus 1897-98, p. 324.

194. MALCOLM, 1934, p. 201; Bartels, 1893, p. 307.

195. BARTELS, 1893, p. 281.

196. MORICE, 1900-01, p. 22.
197. JOHNSTON, J., 1893, p. 335.
198. LINTON, 1933, p. 314.
199. HILTON-SIMPSON, 1913, p. 717.
200. E.g., ABYSSINIA, JANUS, 6: 289, 1902.

the spilling of blood to such an extent that they did not adopt blood-letting or a quite excellent surgery from the Hindus, though they did learn a great many other medical practices from them.[201]

Yet, as we have seen, this fear of mutilation is not general. In numerous tribes ritual mutilation is widely practised, and yet these tribes generally fail to develop medical amputation or other major surgery.[202] The same holds good for most of those who practise trephining. It is most likely that ritual mutilation is so far removed in their thoughts from practical considerations, their general orientation and their thinking about the human body and the most appropriate ways of treatment so different from ours, that it just never occurs to them that this mutilating technique might be useful or even life-saving when applied to infected complicated fractures, focuses of septicemia, tumors, etc.[203] The medical use of trephining or Caesarean section seems to be arrived at not as a result of a general approach, but on magic or empiric grounds in such an isolated way that it cannot influence the general status of surgery. The fact that relatively high technical accomplishments remain isolated without influencing the general level or orientation is rather frequent in primitive societies and by no means restricted to surgery. It is mysterious only as long as we suppose such technical accomplishments to be the result of more or less scientific thought or research as they would be in our society. In the case of the non-operating, ritual mutilators or trephiners, we deal, as in the case of the dissectors who were unable to learn anatomy, with a particular brand of "ignorance," an ignorance not of technical means, but existing in spite of technical means through different orientation, interests, and values. We must realize that such behavior is primarily dictated by magico-religious ideas and is not merely "irrational." (In some respects, it is even very logical.) Irrational behavior is a general psychological mechanism in humans like, for example, suggestibility, and modern empiricist surgeons can be subject to it as well as primitive trephiners,[204] while the supernaturalistic approach has been almost entirely eliminated from modern scientific surgery.

REFERENCES

1. *Abbreviations:* BAE—Bureau of American Ethnology; JAI—Journal of the Royal Anthropological Institute; AA—American Anthropologist; ZE—Zeitschrift für Ethnologie.

ACKERKNECHT, E. H., 1945: Primitive Medicine. *Trans. N. Y. Ac. Sc.* II, 8:26.

———, 1946: Natural Diseases and Rational Treatment in Primitive Medicine. *Bull. Hist. Med., 19:*467.

———, 1948: Primitive Autopsies and the History of Anatomy. *Bull. Hist. Med., 13:*334.

AITKEN, R. T., 1930: *Ethnology of Tubnai.* Honolulu.

ALAJOUANINE, TH., and R. THURE, 1945: Perte de substance cranienne consécutive à un traumatisme fermé. *Revue Neurologique,* 77:71.

ANDERSON, R. G., 1911: Some Tribal Customs in Their Relation to the Medicine and Morals of the Gour Peoples Inhabiting the Bahr-El-Ghazal. *4 Rep. Wellcome Trop. Res. Lab.,* Vol. B, pp. 239-278. Khartoum.

ANGUS, H. C., 1897-98: A Year in Azimba. *JAI,* 27: 324.

BANDELIER, A. L., 1904: Aboriginal Trephining in Bolivia. *AA,* 6, p. 440 ff.

---

201. RIVERS, 1924, p. 95.

202. I am indebted to Dr. David Bidney who drew my attention to the fact that lack of surgery in those instances which cannot be accounted for by supernaturalistic fear of mutilation can be explained on the basis of my hypothesis of a general supernaturalistic orientation of primitives in medicine (see Ackerknecht, 1946).

203. One of the many ritual mutilators reluctant to adopt medical amputation are the Cheyenne (Grinnell, 1923, II, p. 147). I am obliged to Dr. E. A. Hoebel for drawing my attention to the fact that, nevertheless, one case of medical amputation among the Cheyenne is on record (Llewellyn and Hoebel, 1941, pp. 122-123).

204. LERICHE, 1944.

BARTELS, M., 1893: *Die Medizin der Naturvölker.* Leipzig.
BARTON, J., 1923: Notes on the Kipsikis of Kenya. *JAI,* 53, pp. 42-78.
BEST, E., 1924: *The Maori.* 2 vols. Wellington.
BROWN, G., 1910: *Melanesians and Polynesians.* London.
CATLIN, G., 1876: *Illustrations of the Manners, etc. of the North American Indians.* London.
CODRINGTON, R. H., 1891: *The Melanesians.* Oxford.
CORLETT, W. TH., 1935: *The Medicine-Man of the American Indian.* Springfield.
CRUMP, T. A., 1901: Trephining in the South Seas. *JAI,* 21:167.
CULWICK, A. T. and G. M., 1935: *Ubena of the Rivers.* London.
CZAPLICKA, M. A., 1914: *Aboriginal Siberia.* Oxford.
DEMBO, A., and J. IMBELLONI, 1938: *Deformaciones intencionales del cuerpo humano de caracter étnico.* Buenos Aires.
DENSMORE, F., 1928: Uses of Plants by the Chippewa. *BAE Rep.,* 44:275, Washington.
DRIBERG, J. H., 1923: *The Lango.* Oxford.
———, 1929: *The Savage as He Really Is.* London.
———, 1930: *People of the Small Arrow.* London.
EHRENREICH, P., 1891: *Beiträge zur Völkerkunde Brasiliens.* Berlin.
ELLA, S., 1874, Native Medicine and Surgery in the South Sea Islands. *The Medical Times and Gazette,* 1:50. London.
ELLIS, E. S., 1945: *Primitive Anesthesia and Allied Conditions.* London.
ELLIS, W., 1853: *Polynesian Researches.* London.
EMIN BEY, 1897: Reise von Mruli nach der Hauptstadt Unjoros. *Petermann's Mitt.,* 25:179, 220, 388.
EVANS-PRITCHARD, E. E., 1937: *Witchcraft, Oracles and Magic among the Azande.* Oxford.
FELKIN, R. W., 1884: Caesarean Section in Uganda. *Edinb. Med. Jour.,* 29:928.
FLETCHER, A., and F. LA FLESCHE, 1911: The Omaha. *BAE, Rep.,* 27:15. Washington.
FREEMAN, L., 1924: Surgery of the Ancient Inhabitants of America. *Art and Archaeol.,* 18:21.
GRANDIDIER, A. and G., 1908: *Histoire physique etc. de Madagascar.* Paris.
GRINNELL, G. B., 1923: *The Cheyenne Indians.* New Haven.
GUIARD, E., 1930: *La trépanation cranienne chez les néolithiques et chez les primitifs modernes.* Paris.
HADDON, A. C., 1901: *Headhunters, Black, White and Brown.* London.
HAGEN, B., 1899: *Unter den Papuas.* Wiesbaden.
HARLEY, G. W., 1941: *Native African Medicine.* Cambridge, Mass.
HEGER, F., 1929: Neue Formen von Aderlassgeräten *Festchr. F. Schmidt,* p. 275 ff. Wien.
HILTON-SIMPSON, M. W., 1913: Some Arab and Shawia Remedies and Notes on Trephining in Algeria. *JAI,* 43:706.

HOGBIN, H. T., 1930-31: Spirits and the Healing of the Sick in Ontong Java. *Oceania,* 1:146.
HRDLIČKA, A., 1908: Physiological and Medical Observations among the Indians of the Southwestern United States and Northern Mexico. *BAE Bull.* 34.
JOHNSTON, JAMES, 1893: *Reality vs. Romance in South Central Africa.* New York.
JUNOD, H. A., 1927: *The Life of a South African Tribe.* 2 vols. London.
KARSTEN, R., 1926: *The Civilization of the South American Indians.* London.
KENNEDY, A. G., 1931: Field Note on Vaitupu, Ellice Island. *Mem. Poly. Soc.* 9. New Plymouth.
KINGSLEY, M. H., 1899: *West African Studies.* London.
KLEIWEG, DE ZWAAN, J. P., 1913: *Die Heilkunde der Niasser.* Haag.
KOCH-GRÜNBERG, TH., 1923: *Vom Roroima zum Orinoko.* Stuttgart.
LAGERCRANTZ, S., 1935: *Fingerverstümmelung in Africa.* ZE, p. 129 ff.
———, 1938: Zur Verbreitung der Monorchie. ZE, 70, p. 199 ff.
LANDTMAN, G., 1937: *The Kiwai Papuans.* London.
LAUBSCHER, B, J., 1937: *Sex, Custom and Psychopathology.* London.
LEECHMAN, D., 1944: Trephined Skulls from British Columbia. *Trans. Roy. Soc. Canada,* Sec. II, pp. 99-102.
LERICHE, R., 1944: *La Chirurgie à l'ordre de la vie.* Paris.
LILLICO, J., 1940: Primitive Blood-letting. *Ann. Med. Hist.,* Ser, 3, 2:133.
LINDBLOM, G., 1933: *The Akamba.* Upsala.
LINTON, R., 1933: *The Tanala.* Chicago.
———, 1945: *The Science of Man in the World Crisis.* New York.
LLEWELLYN, K. N., and E. A. HOEBEL, 1941: *The Cheyenne Way.* Norman, Oklahoma.
MACCURDY, G. G., 1923: Human Skeletal Remains from the Highlands of Peru. *Journ. Phys. Anth.,* 6:218.
MCKENZIE, DAN., 1927: *The Infancy of Medicine.* London.
MALCOLM, L. W. G., 1934: Prehistoric and Primitive Surgery. *Nature,* 133, p. 200 ff.
MARIOTT, A., 1945: *The Ten Grandmothers.* Norman.
MARTIUS, K. FR. PH. V., 1844: *Das Naturell, die Krankheiten, das Arztthum und die Heilmittel der Urbewohner Brasiliens.* München.
MEAD, M., 1924: *Coming of Age in Samoa.* New York.
MEANS, PH. A., 1931: *Ancient Civilizations of the Andes.* New York.
MERKER, M., 1910: *Die Masai.* Berlin.
MEYER, H., 1932: *Die Barundi.* Leipzig.
MIKLUCHO-MACLAY, N. V., 1882: Bericht über Operationen Austral Eingeborener. ZE. 12:526.
MILNE, E., 1924: *Home of an Eastern Clan.* Oxford.
MONTAGU, M. F. ASHLEY, 1937: The Origin of Subincision in Australia. *Oceania,* 8:193.

MOODIE, R. L., 1927: Injuries to the Head among the Pre-Columbian Peruvians. *Ann. Med. Hist.*, 9:277.
MOONEY, J., 1891: The Sacred Formulae of the Cherokees. *BAE, Rep.*, 7.
MORICE, A. G., 1900-01: Déné Surgery. *Trans. Canad. Inst.* pp. 15-28.
MÜLLER, F., 1928: Drogen und Medicamente der Guarani. *Publ. d'Hommage offerte au P. W. Schmidt.* Ed. W. Kopers. Wien.
NEUHAUSS, R, 1911: *Deutsch Neu Guinea*. 3 vols. Berlin.
OLBRECHTS, F. R., and J. MOONEY, 1932: The Swimmer Manuscript. *BAE Bull. 99*. Washington.
OPLER, M. E., 1941: *An Apache Life Way*. Chicago.
PACKARD, F., 1901: *History of Medicine in the United States*. Philadelphia.
PARDAL, R., 1937: *Medicina aborígen americana*. Buenos Aires.
PARHAM, H. B. R., 1943: Fiji Native Plants with Their Medicinal and Other Uses. *Polyn. Soc. Mem.* No. 16. Wellington.
PARK, W. F., 1938: *Shamanism in Western North America*. Evanston.
PARKINSON, R., 1907, *Dreissig Jahre in der Südsee*. Stuttgart.
PAULITSCHKE, P., 1896: *Ethnographie Nordost Africas*. Berlin.
PLOSS, H., and M. BARTELS, 1899: *Das Weib in der Natur und Völkerkunde*. Leipzig.
PREUSS, K. TH., 1890: Menschenopfer und Selbstverstümmlung in Amerika. *Festschrift für Adolf Bastian*. Berlin.
RAY, S. G., 1917: The People and Language of Lifu, Loyalty Islands. *JAI*, 47, p. 239 ff.
REHSE, H., 1910: *Kisiba Land und Leute*. Stuttgart.
RIVERS, W. H. R., 1924: *Medicine, Magic, and Religion*. London.
ROSCOE, J., 1911: *The Bazanda*. London.
———, 1921: *Twenty-five Years in East Africa*. Cambridge.
———, 1923: *The Banyankole*. Cambridge.
———, 1924: *The Bagesu*. Cambridge.
ROSEN, G., 1944: *The Specialization of Medicine with Particular Reference to Ophthalmology*. New York.
SCHAPERA, I., 1930: *The Khoisan Peoples of South Africa*. London.
SCHULTZ, A. H., 1939: Notes on Diseases and Healed Fracture of Wild Apes. *Bull. Hist. Med.*, 7, p. 571 ff.
———, 1944: Age Changes and Variability in Gibbons. *Am. J. Phys. Anthr.*, 2:1.
SELIGMAN, C. G., 1902: The Medicine, Surgery and Midwifery of the Sinaugolo. *JAI*, 32:297.
SHAPIRO, H. L., 1927: Primitive Surgery. *Nat. Hist.*, 27, 266.
SIEROSHEWSKI, M., 1901: The Yakuts. *JAI*, 31:61.
SMITH, E. W., and A. M. DALE, 1920: *The Ila-Speaking Tribes of North Rhodesia*. London.
SÖDERSTRÖM, J., 1938: Die rituellen Fingerverstümmelungen in der Südsee und in Australien. *ZE*, 70:24.
SPIER, L., 1928: *Havasupai Ethnography*. New York.
SPINDEN, H. J., 1908: *The Nez-Percé Indians*. Lancaster.
STAYT, H. A., 1931: *The Bavenda*. London.
STEINEN, K. V. D., 1886: *Durch Central Brasilien*. Leipzig.
STEPHEN-CHAUVET, 1936: *La médicine chez les peuples primitifs*. Paris.
STEVENSON, M. C., 1904: The Zuni Indians. *BAE*, 23:1. Washington.
STIRLING, M. W., 1938: Historical and Ethnographical Material on the Jivaro Indians. *BAE Bull. 117*. Washington.
STONE, E., 1932: *Medicine among the American Indians*. New York.
SWANTON, J. R., 1928: Religious Beliefs and Medical Practices of the Creek. *BAE Rep.*, 42:473. Washington.
SUMNER, W. G., and A. G. KELLER, 1927: *The Science of Society*. 4 vols. New Haven.
TEIT, J., 1900: *The Thompson Indians of British Columbia*. New York.
TELLO, J. C., 1913: Prehistoric Trephining among the Yauyos of Peru. *Int. Congr. Americanists, London, 1912*, p. 75 ff. London.
TESSMANN, G., 1934: *Die Bafia*. Stuttgart.
TORDAY, E., and T. A. JOYCE, 1906: Notes on the Ethnography of the Ba-Yala. *JAI*, 36:39.
TURNER, G., 1884: *Samoa a Hundred Years Ago*. London.
VAN DER BURGT, Y. M. M.: 1903: *Dictionnaire français Kirundi*. Bois le Duc.
VEDDER, H., 1923: *Die Bergdama*. Hamburg.
VERGIAT, A. M., 1937: *Moeurs et coutumes des Manias*. Paris.
WAFER, L., 1934: *A New Voyage and Description of the Isthmus of America (1680-88)*. Oxford.
WARNER, W. LLOYD, 1937: *A Black Civilization*. New York.
WEBB, T. T., 1933-34: Aboriginal Medical Practice in East Arnhem Land. *Oceania*, 4:91.
WEYER, E. M., 1932: *The Eskimos*. New Haven.
WHITING, J. W. M., 1941: *Becoming a Kwoma*. New Haven.
WISDOM, CHARLES, 1940: *The Chorti Indians of Guatemala*. Chicago.
WÖLFEL, D. J., 1925: Die Trepanation. *Anthropos*, 20:1.
YOUNG, J. H., 1944: *Caesarean Section*. London.

Chapter 52

Prehistoric and Early Historic Trepanation*

F. P. LISOWSKI

TREPANATION† of the human skull is the removal of a piece of calvarium without damage to the underlying blood-vessels, meninges and brain. It is possibly one of the earliest forms of surgical intervention on the head of which we have any authentic record and its practice is widely spread in space and time. In some parts of the world it is still practised in its early form by native medicine men. Trepanation as performed by man in prehistoric and early historic times shows an astonishing degree of technical skill. And certainly the number of survivals of this operation testify to the competence of the early surgeons.

For a long time medical science doubted the existence of healed prehistoric trepanations, since eighteenth and nineteenth century surgeons of the pre-antiseptic era rejected this procedure owing to the almost one hundred per cent mortality (Schröder, 1957). However, as evidence of trepanation appeared in the South Sea Islands and in North Africa, these doubts were gradually removed.

According to Rytel (1962) the first reference to trepanation dates from 1849 (Atlas de Morton, Cranea Americana). Bartucz (1964) claims that Dr. E. Kovaces of Hungary was the first to describe in 1853 an actual trepanation found at Vereb. Another of the earliest to be recognized as such was noted by E. G. Squier on a skull from Cuzco during his tour of 1863-65 through Peru (Stewart, 1958). He consulted Paul Broca, the noted French physical anthropologist of the time whose interest led to the recognition of Neolithic trepanations in France. And thus gradually more skulls came to light showing this early surgical interference. The realization that this practice has survived until the present day has greatly increased our knowledge of this operation.

The most important contributions on this subject are the study by Guiard (1930), the survey of European trepanations by Piggott (1940) and the more recent general review by Stewart (1958).

This particular account will deal with the prehistoric and early historic aspects of trepanation and note certain of its mediaeval features.

GEOGRAPHICAL DISTRIBUTION

A summary of the distribution of the various sites so far found in the world is given below together with the names of some of the main authors. The list is not fully comprehensive since the literature is

---

* From *Amer. Anthrop.*, 49:25, 1947.

† *Note: Trepanation*, or *trephination*, or *trephinement* is the removal of a disc of bone from the skull. And the instrument used is a *trepan* or *trephine*, which is derived from the Greek *trypanon*, a borer. The verb is to *trepan*, or to *trephine* or to *trepanize*.

rather scattered and new discoveries are frequent but it does indicate how wide spread the practice was. In each case the name of the country is followed by the number of sites found and the authors referred to. With regard to certain countries, although no actual case of trepanation has as yet come to light, there is strong presumptive evidence from the literature that it was practised.

Europe

Scandinavia—twenty sites (Fischer-Møller, 1936; Piggott; Karolyi, 1963). Iceland (Boev, 1959). England and Scotland—fourteen sites (Parry, 1921, 1936; McKenzie, 1936; Piggott; Brothwell, 1963). Ireland—five sites (Martin, 1935; Fleetwood, 1951; Rynne, 1962; Brothwell). France—seventy sites (Prunières, 1874; Broca, 1876a and b; Guiard; Giot, 1949; Karolyi). Germany—forty five sites (Brunn, 1936; Breitinger, 1938; Schröder, 1957; Karolyi). Switzerland—ten sites (Karolyi). Austria—two sites (Karolyi). Czechoslovakia—eighteen sites (Matiegka, 1928; Karolyi). Hungary—sixteen sites (Karolyi). Rumania—five sites (Russu and Bologa, 1961). Bulgaria—four sites (Boev). Yugoslavia (Giot and Desse, 1950; Boev). Albania (Giot and Desse; Boev). Greece (Littré, 1863). Italy—eleven sites (Maxia and Cossu, 1951; Karolyi). Iberian peninsula—sixteen sites (Piggott; Karolyi). Poland—two sites (Boev). U.S.S.R. (European and Asiatic parts)—Latvia (Boev), Ukraine (Rokhlin, 1964), Crimea (Bobin, 1964), Dagestan in the Caucasus (Terrier and Péraire, 1895; Guiard), Minusinsk in the upper Yenisei region (Boev), Vladivostok region (Montandon, 1926), Oglakti in southern Siberia (Tallgren, 1936).

Asia

China (Wong and Wu, 1936; Needham, 1954 and 1964; Boev). Japan—Hokkaido (Boev). India (Müller, 1959). Afghanistan (Giot and Desse; Roney, 1954). Pakistan—Hindu Kush region (Giot and Desse). Palestine (Parry and Starkey, 1936; Risdon, 1939; Giles, 1953; Oakley, Brooke, Akester and Brothwell, 1959).

Africa

Egypt—six sites (Elliot Smith and Jones, 1910; Ruffer, 1918; Batrawi, 1935 and 1957; Lisowski, 1954 and 1959). Kenya (Sood, 1960; Margetts, 1962). North Africa (Hilton-Simpson, 1913; Sudhoff, 1929; Oakley *et al*). Canary Islands (Sudhoff; Stewart, 1958; Karolyi).

North and Central America

Stewart (1958).

South America

In particular the region once occupied by the Inca empire (Bandelier, 1904; Moodie, 1923; Cabieses, 1957; Stewart; Oakley *et al*; Rytel).

Oceania

Mainly Melanesia (Crump, 1901; Ford, 1937).

DISTRIBUTION IN TIME

Here only the main outlines will be given together with certain additional information that has become available in recent years.

Moodie considers that the earliest European trepanations occurred some ten thousand years ago and Forgue (1938) too places the beginnings at the end of the Palaeolithic period. All these estimates, however, are hypothetical. It has been accepted by Broca, Lucas-Championnière (1878), Horsley (1888), Ruffer, Guiard,

Parry (1923), Piggott and Oakley *et al.* that this operation was practised during the Neolithic age. Piggott and Oakley *et al.* state that trepanation was performed occasionally by early Danubians (*c.* 3000 B.C.) and frequently by "battleaxe" people who constructed the chambered tombs in the Seine-Oise-Marne area of France (*c.* 2000 B.C.). So many skulls showing this trait were discovered in these tombs that it is probable that the operation had some ritual significance (Oakley *et al*). It seems that *circa,* 1900 to 1500 B.C., the south of France was a major centre for trepanation (Sudhoff, 1929; Stewart). Examples of this practice have also been reported from many regions of Neolithic Europe, and in particular, Denmark, England, Germany, Italy, USSR, the Balkans have revealed quite a large number of skulls.

In Europe trepanned skulls became rare after the Neolithic era, partly because in the later Bronze age and La Tène period the dead were mostly cremated (Regnault, 1936). Nevertheless a few examples are available from France (Guiard), Scandinavia (Piggott), Germany (Brunn, Breitinger), Czechoslovakia (Matiegka), Hungary (Bartucz), Rumania (Russu and Bologa), Bulgaria (Boev), USSR (Bobin) and other countries.

The Iron Age, early Historic, Greek, Roman and Mediaeval times all indicate that trepanation continued to be performed in Europe. This is known from actual specimens of those days, and from the relevant literature of the later period. Mediaeval evidence comes from England (McKenzie, Parry, Brothwell), Ireland (Martin, Fleetwood), France (Piggott), Germany (Brunn, Karolyi), Italy (Castiglioni, 1941), Czechoslovakia (Piggott), Hungary (Boev), Rumania (Russu and Bologa), Bulgaria (Boev) and other areas. According to Rytel, however, the frequency of this operation never reached the proportions nor the universality during Mediaeval times that it enjoyed in the Neolithic age. To this one has to add that skeletons of the former period unfortunately have aroused less interest than those of older burials and thus our knowledge in this respect is rather scanty.

In Greece, Hippocrates (*c.* 460 to 355 B.C.), advises trepanation for wounds of the head in one of his six treatises of his surgical classic (Littré). The account is detailed and meticulous and shows some experience, and although little or nothing is known of any actual specimens it is highly likely that this operation must have been performed fairly often.

Roman examples of this surgical interference are known from Gaul and also from Trier in Germany (Piggott). And from Rome we have the account by Celsus (*c.* 25 B.C. to 37 A.D.) of a method of operation which became standard in the surgical books of the Middle Ages. This apparently was adopted later by the Arabs. Celsus, whose method differed from that of prehistoric times, advises trepanation for head wounds and gives careful and precise instructions on methodology in his treatise *De medicina* (Spencer, 1948), and part of his *De artibus,* written between 25 and 35 A.D. Although Spencer believes that Celsus was a medical practitioner, Castiglioni (1941) considers that he was neither a physician nor a surgeon but a compiler from the works of others. Which ever he was, his text undoubtedly influenced the surgical world for many centuries.

Much later Rogerius Frugardi (*c.* 1170 to 1200 A.D.), otherwise known as Roger of Salerno, one of the greatest of the Salernitan surgeons, produced his text in which he deals at length with wounds of the head and brain, even giving the differential diagnosis of injuries of the skull and indications for trepanation (Castiglioni). However, his method was not dissimilar from that of Celsus.

From Ireland several interesting examples are available. A trepanned skull of a thirteen-year-old child, probably early Christian, was recovered from Collierstown in Co. Meath (Martin, 1935). Two further trepanations each of late Mediaeval date, one from Ballinlough (Co. Laois) and the other from Maganey Lower (Co. Kildare), were found during recent excavations. A fourth specimen was discovered in a stone-lined grave at the Abbey of Nendrum on Mahee Island in Strangford Lough (Martin). The abbey was destroyed in 974 A.D. by fire. It is highly likely that in those days "major surgery" was performed in monastic institutions (Fleetwood, 1951). Legend has it that Cennfaeladh, whose skull was fractured by a blow from a sword during the battle of Moyrath in Co. Down (637 A.D.), was operated upon by St. Bricin, the Abbot of Tuaim Drecain, an accomplished surgeon and scholar (Fleetwood).

In Yugoslavia, especially the south-western part, and also in northern Albania there is a long history of the practice of trepanation which persisted as late as the nineteenth century (Giot and Desse, Boev). The folklore of this region is rich in accounts of these operations (Leskin, 1919). Thus, the story is told of the physician who was secretly watched by his apprentice as he trepanned the forehead of the daughter of the Czar and extracted a beetle from her brain.

Outside Europe and apart from South America, the evidence of examples and historical accounts is more scant. This may be due to the fact that fewer exhumations have been carried out in these parts and further excavations may reveal more information.

In Asia the examples from Palestine are of some interest. There the oldest trepanation found so far comes from Jericho (Oakley et al.) and dates to the Bronze Age (c. 2000 B.C.). Risdon who excavated Tell Duweir (Lachish) discovered three Iron Age (c. 8th cent. B.C.) skulls, which were reported on by Parry and Starkey (1936). Giles (1953) reported a further case of trepanation belonging to that period.

From Roman Syria there is, for instance, indirect evidence by way of China (Needham, 1954). Ouyang Hsiu and Sung Chhi in their *Hsin Thang Shu* (New history of the T'ang dynasty) in 1061 A.D. state that the people of Ta-Chhin (Roman Syria) have physicians who can cure blindness by opening the brain and removing worms. Needham believes this to be "the solitary instance of any attention consciously paid in Chinese writing to early Western medical science."

Going further east one finds that both in ancient and recent times this type of skull surgery was performed in the region comprising present day eastern Afghanistan, northern Pakistan and Kashmir (Giot and Desse 1950; Roney, 1954). Although actual specimens of the earlier periods are rare, the lore of trepanation was very much current around c. 400 A.D. in that part of the world (Müller, 1959) and in the Tibetan region (Jungbauer, 1923). One of the accounts mentions that in ancient times students went to Takkasilā (Greek Taxila) in the north-western part of the Indian subcontinent to learn the arts and sciences. At that time a famous teacher Ātreya, king of the physicians, lived there, to whom prince Jīvaka went as an apprentice so as to learn the art of opening skulls. He watches his master extract a worm from the brain of a patient. And later when Jīvaka returns to his own country he trepans, using an "opening instrument" and thus removes centipedes. Legend has it that he later became the medical adviser to Buddha. Jīvaka is certainly famous in old Buddhist texts and in folklore. It is also known that these stories were carried eastward by Buddhist missions. However, in China these tales received certain additions, such as the use of acupuncture and feeling the pulse, which were not practised in ancient India. Thus the folklore became Chinese by adaptation. On the other hand the Tibetans took over the original stories and translated them without alteration. Müller feels that these folktales, and there are many, are based on fact and that trepanation must have been practised in this area.

So far the main part of China has not revealed any trepanations (Woo, 1964), though a few have been found in her peripheral provinces. Thus there is evidence from Tibet (Boev 1959), quite apart from the folklore traditions (Jungbauer 1923). Here medical knowledge probably came into the region through Kashmir. Tallgren (1936) reports a skull with an occipital opening from Oglakty in southern Siberia which was found in a cemetery dating to the Han dynasty (202 B.C. to 220 A.D.). And Montandon (1926) found another example from the Far East in a museum in Vladivostok; however, date and site are doubtful. These two regions at one time, of course, belonged to the northern part of the Chinese empire.

Although there are only these few skulls, classical Chinese literature does mention trepanation. In ancient times (c. 2700 to 1100 B.C.) there is supposed to have existed a physician Yü Fu who is alleged to have been able to expose the brain (Wong and Wu, 1936). However, he is a very ancient legendary character associated with surgery and, according to Needham, one of the interlocutors in the great classic of medicine *Huang Ti Nei Ching, Su Wên* (c. 2nd cent. B.C.). Wong and Wu also mention Hua Tho (c. 130 to 220 A.D.), the famous surgeon and discoverer of the use of anaesthetics. Legend has it that he offered to cure the headaches of the Wei emperor Tshao by opening the skull, an offer which was declined. Another account supposes that Hua Tho suggested trepanation to a famous warrior who thought the surgeon wanted to murder him and therefore had him beheaded. Reference already has been made to the Buddhist missions.

However, for the more detailed information on classical China given below, the author is indebted to Dr. Joseph Needham F. R. S. who kindly made this available. The *Pao Phu Tzu* book, by the famous physician and alchemist Ko Hung, written *circa* 300 A.D., is quoted by late scientific encyclopaedias as saying that Thai Tshang Kung (Shunyü I 205 to 150 B.C.) "used to cut open skulls of patients and arrange their brains in order." Here it is rather difficult to trace the original passage. Although there is some uncertainty, it is probable that the trepanation account of Hua Tho occurs in the Yuan period novel *San Kuo Chih Yen I* (The three kingdoms story) by Lo Kuan-Chung (c. 1364 A.D.). Nor is it known whether there are any more ancient accounts of this operation nearer the time of Hua Tho himself among the unofficial literature of the San Kuo or Liu Chhao periods.

In 1040 A.D. there was published a history of medicine entitled *Li Tai Ming I Meng Chhiu* (Brief lives of the famous physicians of all ages) by Chou Shou-Chung. He quoted two descriptions of trepanations from an earlier book, the *Yü Thang Hsien Hua* (Leisurely conversations of academicians) written between 960 and 1040 A.D., i.e., during the Sung dynasty.

This was also quoted in a later florilegium, the *Lei Shuo* by Wang Jen-Yü in 1136. The first account is of a metal worker who performed an operation for the extraction of a worm from a patient previously pronounced incurable. The other tells of a Taoist adept who was condemned to death for alleged arson, but who wished to demonstrate his skill at surgery before dying; a leper was therefore fetched, and the adept opened his skull and removed a cupfull of worms or parasites.

During this period too, the *Hsin Thang Shu*, quoted above, was published and this mentions the surgical practices in Roman Syria. All this seems to indicate that trepanation was probably practised during tenth and fourteenth centuries in China. Later in 1366 the *Cho Kêng Lu* (Talks while the plough is resting) by Thao Tsung-I appeared, in which it is said that Arabic physicians, of which there had been many in China since T'ang times, could open the skull and extract worms. Thus we have the classical literary references, but no actual skulls showing this operation. It is therefore conceivable that in time the skeletal evidence will be forthcoming.

The Ainu people in Hokkaido, northern Japan, are also supposed to have practised trepanation (Boev, 1959). Certainly Rytel (1962) mentions five Ainu skulls showing resection of the foramen magnum and of the alveolar process, and this evidence of surgical interference may indicate that other techniques, e.g., trepanation were not unknown.

Turning to Africa, one finds that so far in Egypt only six trepanations have been found, although more Egyptian skulls have been examined than of any other population. Chronologically the excavations have revealed the following: the oldest find is from Sesebi, Sudan (Lisowski, 1954) and belongs to the XVIIIth to XIXth Dynasty (*c.* 1200 B.C.), Batrawi (1935) reported one from Sakkara dating to the XXVth Dynasty (*c.* 600 B.C.); a bilateral trepanation was found in Sakkara (Lisowski) of possibly Ptolemaic date (*c.* 323 to 30 B.C.), a Meroitic skull (*c.* 50 to 200 A.D.) was reported earlier by Batrawi. Ruffer (1918) mentions one of circa 200 A.D. found near Alexandria, and Elliot Smith and Wood Jones (1910) described a Byzantine (395 to 638 A.D.) trepanation from Hesa near Aswan re-examined and also reported by Parry. In mentioning the above examples one has to consider the possible connexions between the practice of trepanation in Egypt and that practised in neighbouring regions. The Egyptian specimens roughly fall into three historic periods. To the first of these belong the Sesebi and the earlier Sakkara individuals who were more or less contemporary of the Palestinian trepanations. It is difficult, however, to relate these two centres since they are some eight hundred miles apart. The later Sakkara specimen, which dates to the Ptolemaic period, may be an example of the influence of Greek surgery. The other trepanations belonging to the period 50 to 600 A.D. are probably due to Roman influence.

From North Africa there is evidence that the technique this type of skull surgery has survived to the present day. Sudhoff (1929) gave reports from Libya, Hilton-Simpson (1913) and F o r g u e (1938) described the operative procedure from the Aures mountains in Algeria, and Oakley *et al.* gave a contemporary account of trepanation among the Tibu in Tebesti (Sahara). It might be argued that here one finds a degenerate form of Greek or classical surgery, though it is more likely that this practice dates to Islamic times. In Kenya too trepanation is still performed to this day (Sood, 1960; Margetts, 1962) where it may well have been introduced by the Arabs.

It is appropriate at this juncture when dealing with Asia and Africa to mention trepanation in the old Islamic world, a period during which their science and medicine flourished from the western borders of China right across Persia to Egypt, North Africa and Spain. It appears that the method of operation described by Celsus was adopted by the Arabs, for it is advocated in the treatises of the tenth to eleventh century surgeons Ali-abbas and Abdul-kassim (Piggott, 1940). In this connexion mention is made already that Thao Tsung-I wrote in 1366 of the skill of the Arabic physicians. In fact many of the latter had been in China since T'ang times.

With regard to South America the centre for trepanations was restricted largely to the central and southern parts of Peru and to the neighbouring part of Bolivia (Cabieses, 1957). It appears that more trepanned skulls have been found in this area than in all the rest of the world together (Stewart, 1958). According to Stewart the oldest specimens date to the period of c. fifth century B.C. to fifth century A.D., although Rytel (1962) considers that the oldest evidence of trepanation in Peru dates to circa 3000 B.C. It is also known that the Indians of this particular region continued these operations into post-Columbian times (Bandelier, 1904), and even today the practice of trepanation is not unknown (Oakley et al.).

MOTIVES FOR TREPANATION

Since science and magic are in their early stages indistinguishable (Needham, 1954), it is difficult to differentiate between ritual or magical and therapeutic motives underlying the practice of trepanation.

Dealing first with the performance of this operation in the living, one finds that various authors emphasise different aspects of the motivation. Broca (1876) decided that trepanations were performed for the relief of certain intracranial maladies and Horsley (1888) considered that all these surgical interferences were therapeutic. According to Lucas-Championnière (1912) the operation was done in order to cure a disease supposed to have its seat in the head or to remove splinters from a fractured skull—in the latter practising cerebral decompression. Ruffer believed that Neolithic people may have trepanned for injuries, but as most of the operated skulls show no signs of trauma, headache was probably the chief indication. Lucas-Championnière also mentioned that "according to the theory usually accepted, the operation was first performed from time immemorial on sheep for the relief of 'staggers,' and later man extended the application of the veterinary method to his species." However, Ruffer felt that this was pure speculation. Russu and Bologa (1961) also mention the practice of trepanation in connexion with staggers in sheep, and how the shepherds in Rumania thereby removed the larva of the *Multiceps multiceps* since ancient times. Thus it could be that the accounts in folklore of the extraction of beetles (Yugoslavia) and centipedes (Tibet) from the brain in man might be based on the trepanation of sheep, and may not be so far fetched as they might seem at first sight. Moodie (1923), Piggott (1940), and Russu and Bologa consider that the majority of operations were performed as a definite surgical treatment, either to repair a fracture of the skull or to alleviate headache. Guiard (1930), Regnault (1936), Forgue (1938) and Thompson (1938) believe that in prehistoric times various intracranial diseases were ascribed to evil spirits and therefore cure was obtained by letting these out of the skull. And since these operations were often followed by improvement in the patient's condition, the primitive surgeons persisted with this surgical intervention (Forgue). Castiglioni (1941) considers that trepanation owes its origin to a demonic or magic concept more than to the idea of therapy. In view of the Peruvian evidence, Stewart believes that trepanations were mainly performed in cases of skull fracture, though he does not exclude other motives for this operation. However, Guiard considered that since the procedure was such a frequent custom among the pre-Inca and Inca inhabitants of Peru and Bolivia, it bordered on a cult. Oakley et al. who described a skull from Tarkhan with a parietal craniotomy and which also shows signs of otitis media and mastoid inflammation, believed that in this case the operation must have been undertaken for clinical reasons. More recently Rytel stated that the surgical indications were both therapeutic and superstitious. Owing to the large number of trepanned skulls found in the chambered tombs in the Seine-Oise-Marne region of France Oakley et al. feel that here the operation had some ritual significance, though they consider that in other geographic regions the indications may have been of a clinical nature.

Our knowledge of prehistoric trepanation would be very poor indeed were it not for the fact that we know something about the practice of this operation in classical times and in the more recent past, and that it still is performed in widely separated parts of the world. In Hippocratic times medicine in Europe was no longer a branch of magic and religion but had begun to gain its own rightful place (Guthrie, 1958). Thus one finds that for wounds of the head Hippocrates advised early trepanation—within three days for contusion of bone, and secondary trepanation for infectious accidents—before the fourteenth day in winter and before the seventh day in summer (Littré). In his writings Celsus also notes that for cranial injuries trepanation is indicated (Spencer). And later in mediaeval times Rogerius Frugardi gives the same advice (Castiglioni). At the beginning of the last century Cornish miners insisted on having their skulls opened following injuries to the head (Lucas-Championnière). As late as the nineteenth century trepanations were performed in south-western Yugoslavia and northern Albania (Russu and Bologa) in cases of skull trauma and in nervous and mental diseases; and in the case of a blood feud where a person was marked for revenge the latter could escape by voluntary submission to this operation. From Algeria (Hilton-Simpson) it is known that in certain cases of head injury, usually a fracture resulting from blows from sticks or stones, trepanation is indicated, however, this surgical procedure is more often performed in cases of persistent headache. In present day Kenya where this operation is still practised, the most common indication is headache (Sood). Crump (1901) noted that trepanation was performed in Melanesia not only in cases of headache, epilepsy and insanity, but also as an aid to longevity. Ford (1937) studying this operation in the same group of Pacific islands, says that it was used in cranial injuries due to warfare, and for headaches and in some children of three to five years of age women cut openings into the foreheads to ward off future trouble from trauma—possibly an extension of surgical therapy to prophylaxis. And in Bolivia medicine-men still perform trepanation for head injuries (Oakley et al.). As a result of investigations among populations that still trepan Wölfel (1925) believes that injuries caused by blows and stone-slings are the main indications for trepanation. On the basis of Neolithic material in Denmark, Fischer-Møller (1936) comes to the same conclusions and goes on to observe that with the appearance of metallic weapons and helmets the neck proved to be a more vulnerable region for inflicting fatal injuries than the head. Russu and Bologa (1961) are of the opinion that the practice of trepanation may be related to the spread of the stone-sling and that only later on this surgical measure was used in other diseases in which the presenting sysmptoms were similar to those following cranial trauma.

The motive for posthumous trepanation was to obtain roundels of human skull bone (Dechelette, 1908). Apparently the object was to remove a piece of bone from the dead skull of one previously trepanned which included a bit of the healed rim from the earlier successful operation (Piggott). This type of trepanation was undertaken in prehistoric Europe and is practised in parts of Africa today (Oakley *et al.*). These roundels were usually of circular shape and often perforated and polished so as to be worn as a necklace. They had a superstitious significance and were used as charms, or amulets, or as a talisman to counter the demons (Broca, Regnault, Forgue). Even up to the Middle Ages it was supposed that powdered cranial bones possessed curative powers while roundels were worn as late as Gallic times (Ruffer).

Thus one may summarize the motives for trepanation in the living and the dead as follows.

In the living the indications can be considered under three headings:

1. Therapeutic, certainly in Hippocratic and later times: for head injuries such as fractures, especially depressed fractures, scalp wounds with or without an inflammatory process, concussion; and possibly in cases of lesions of a syphilitic nature in Peru (Rytel).
2. Magico-therapeutic, where in a sense the cause was considered to be evil spirits which had to be let out and the effect could be "therapeutic" at times: headaches, vertigo, neuralgia, coma, delirium, intracranial vascular catastrophies, meningitis, convulsions, epilepsy, intracranial tumours, mental diseases. And prophylactically to ward off trouble such as head injuries and to promote longevity in Melanesia (Crump).
3. Magico-ritual: e.g., as a ritual act in central France (Oakley *et al.*); in cases of feuds (Russu and Bologa).

The indications for post-mortem trepanation seem to have been in order to secure roundels for amulets.

SURGICAL PROCEDURE

In general craniotomies were performed on the left side (Guiard, Forgue, Piggott, Stewart). The reason for this (Russu and Bologa) was that traumatic lesions of the skull due to blows occurred in the majority of cases on this side since the adversary, usually right handed, was opposite the victim.

Most authorities (Lucas-Championnière, Ruffer, Moodie, Guiard, Forgue, Piggott) consider that in Europe the parietal bone was the most frequently trepanned skull element, followed by the frontal, occipital and rarely the temporal bones. Although Piggott points out that in a high proportion of Czechoslovak trepanations the frontal region was involved. Ruffer suggests that the high frequency of parietal selection was because this region was most easily accessible to the operator. The latter, squatting in front or behind the patient, held the head with his left arm or fixed it between his knees and operated with his right hand. It is of course well known that traumatic subdural haemorrhages can occur following blows in the parietal region. The first detailed side and site analysis was made by Stewart (1958) who, studying a series of 112 trepanations from Peru, found that 48.2 per cent had been operated on the left side, 29.5 per cent on the right, and 22.3 per cent in the median line. Of these, 53.6 per cent had been trepanned in the frontal region, 33.0 per cent in the parietal and 13.4 per cent in the occipital area. The frontal region also was the elected site for prophylactic trepanation in Melanesian infants (Ford).

Lucas-Championnière (1912) claimed that the sagittal suture was carefully avoided, implying that the primitive surgeons had some idea of the underlying anatomy of the superior sagittal sinus. Hilton-Simpson (1913) went so far as to state that sutures were never involved in trepanations. He based his views on his studies of the practice of craniotomy in the Aures mountains in Algeria, where the medicine-man observed two rules: that the opening must not involve the sutures and the dura mater must remain intact. That these views are not correct can be seen from observation of trepanned material in which the craniotomies have cut across the sagittal and other sutures and from the work of Stewart who has clearly shown that the sagittal, coronal and lambdoid sutures were quite often involved. Guiard and Maxia and Cossu also stated that the sutures were never respected.

For anaesthetic purposes the use of alcohol was not unknown in many parts of the world. Guiard states that the Serbians used grape wine and the people of Uganda palm wine, while the ancient Egyptians according to Parry and Sudhoff knew in addition the uses of opium. Similarly the Inca made use of alcohol as well as various preparations from the coca plant (Rytel). Oakley et al. report that for present day trepanations in Bolivia the medicine-men use *chicha*, a local drink, as an anaesthetic. However, it must not be forgotten that in cases of skull injury the patient often was unconscious thus facilitating sugical intervention, a fact that applied to most cases in Melanesia (Ford). On the other hand the Kabyles according to Hilton-Simpson never used anaesthetics when trepanning.

The earlier trepanations were performed most likely with the aid of instruments made of flaked stone, especially flint, of obsidian and of bone (Ruffer, Parry, Thompson, Stewart, Rytel). Probably other materials such as wood were employed also as aids. Later, instruments made of hardened copper were used which were fashioned with a rough edge and shaped like a wedge so as to prevent sudden penetration through the skull bone (Thompson). Russu and Bologa (1961) describe a saw of the La Tène period, discovered in Rumania, which they think was a trepanation instrument. This was found together with a cremation and various ceramics typical of a Celtic burial dating to *circa* second century B.C. The interment may have been that of a medicine-man. This well preserved saw is 11 cm long, has a half-moon-shaped blade and continues into a swan-neck-shaped stem which ends in a straight handle. The whole instrument is made out of one piece of iron, whose blade is thinner towards the serrated cutting edge and thicker towards the base. This wedge-shape would prevent the saw from cutting deeper than 5 mm to 7 mm into bone. Sudhoff and Ebert (1913) described a number of surgical instruments of the La Tène period which were found in Hungary. One of these is a bone saw while the others are retractors and elevators. The two authors believed that these were amputation instruments. However, Holländer (1915) who re-examined them considered that they were much too fine for such a brutal operation as amputation. He had the saw reconstructed and found that it could only be used on the skull since its blade was wedge-shaped thus limiting the depth of the saw cut. From this Holländer concluded that the instrument must have been employed in trepanning. The trepan was already in use at the time of Hippocrates and was held either between the palms and rotated by rubbing the hands together or rotated by a cross-piece and thong (Littré). Celsus described various trepans, a *meningophylax* for holding back the meninges when the border of the trepanned opening is manipulated and an instrument for removing the bone fragments following craniotomy (Spencer). Instruments such as these were excavated at Pompeii (Castiglioni). A more complete outline of the historic evolution of the various trepans from Hippocratic to recent times is given by Thompson (1938). In Algeria a variety of instruments such as scalpers, retractors, drills, saws, screws and elevators, were used (Hilton-Simpson) which will be men-

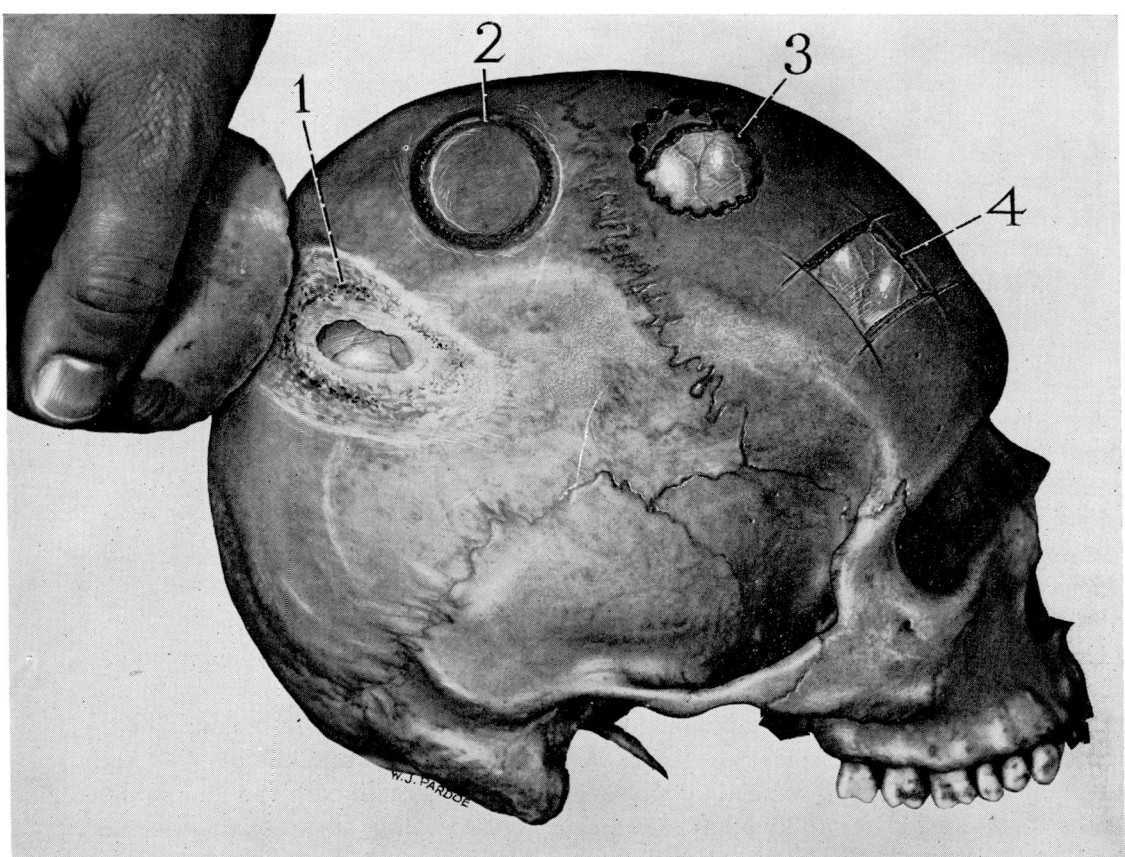

FIGURE 1. Different methods of trepanation: 1. scraping, 2. grooving, 3. boring-and-cutting, 4. rectangular intersecting incisions.

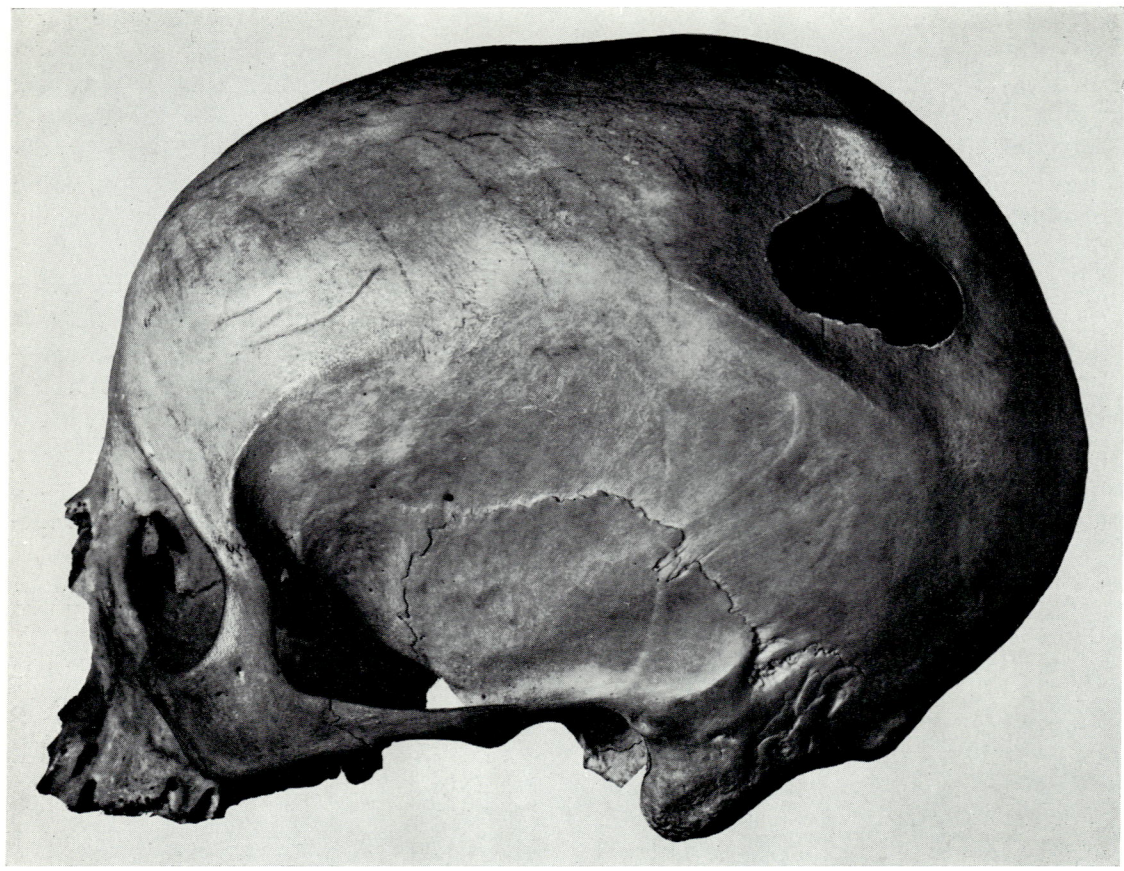

Figure 2. Skull from Sakkara, Egypt, showing trepanation on left side performed by scraping technique. The trepanation on the opposite side cannot be seen.

tioned when dealing with the methods of craniotomy. Sood has described to me the retractors, saws and elevators, mostly made of flattened nails, still used by medicine-men in Kenya. In Melanesia, the sharp edges of shells were utilized in addition to obsidian for making the skull openings (Crump). Ford further reported that shark's teeth as well as broken bottles and razors were used on these islands in more recent times. In South America during the classical Inca period a special T-shaped knife or "tumi" was employed for trepanation (Cabieses). This instrument has been adopted by the Peruvian Academy of Surgery as their emblem.

According to the literature (Littré, Broca, Lucas-Championnière, Ruffer, Parry, Guiard, Forgue, Piggott, Spencer, Stewart, Rytel) several methods of operation have been described, some of which are shown in Figure 1.

1. The scraping technique consists in removing the required area of bone by gradually scraping away, first the lamina externa and diploë, and then with considerable care the lamina interna to expose the dura mater. The resulting opening has of necessity widely bevelled edges and the removed part is in powder form Fig. 1, number 1).

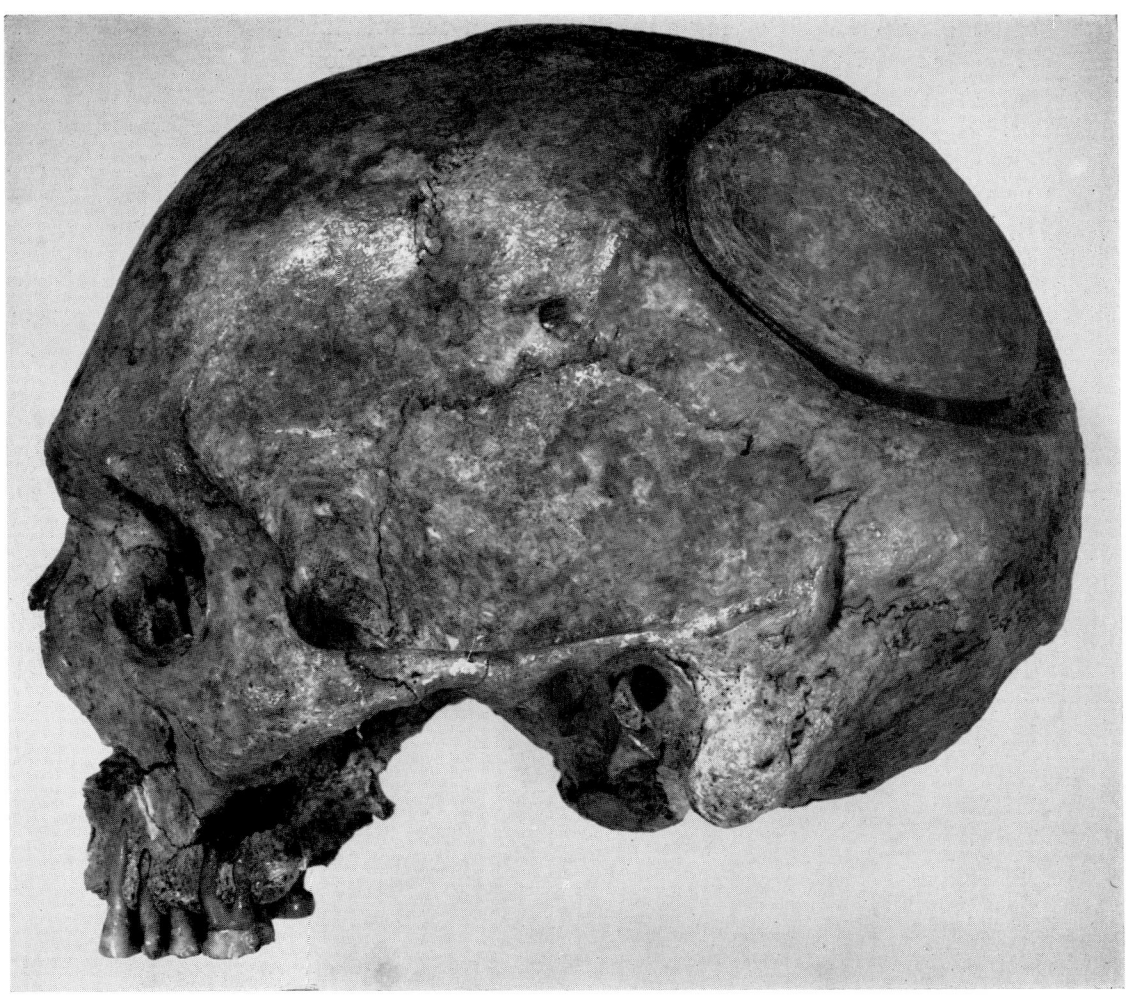

FIGURE 3. Crichel Down trepanation with roundel in place. Photograph by courtesy of the Duckworth Laboratory, Museum of Archaeology and of Ethnology, Cambridge.

This was probably one of the most common methods used, surviving even into the Italian Renaissance period (Piggott). In Rumania this procedure was practised only in mediaeval times (Russu and Bologa). For illustrative purposes an Egyptian trepanation is shown (Fig. 2).

2. The grooving method, in which a series of curved grooves are drawn and redrawn on the skull with a sharp instrument, until the bone between the grooves becomes loose and can be removed (Fig. 1, number 2). Probably this was the procedure by which roundels were obtained (Fig. 3).

This technique was also very frequently used in many parts of the world and is still performed at the present day in Kenya (Sood). Russu and Bologa state that this was the method of choice in prehistoric times in Rumania. In general the orifice in the external lamina is larger than the one in the internal lamina, thus giving a rather bevelled appearance. Ruffer believed that this was always the case, but according to Guiard certain trepanations whose diameter does not surpass 2 cms present practically perpendicular borders to the plane of the surface of the skull and are always circular.

3. In the boring-and-cutting technique the bone is perforated by a circle of closely adjoining perforations extending to the internal lamina, which are then connected by cuts with a sharp instrument, the latter more or less completely obliterating the serrated border. Finally the freed fragment is levered out (Fig. 1, number 3).

Lucas-Championnière considered that this type of operation was used in prehistoric times and based his supposition entirely on a single skull from Peru. On the other hand Stewart believes that this method probably was not practised outside Peru and there only occasionally. However, the boring-and-cutting procedure was described by Celsus in Roman times (Spencer). He advised in cases of more extensive cranial injuries that a hole is drilled with a trepan at the junction of the diseased and sound bone, close to this a second and a third, until the whole area is ringed by these perforations. Then a chisel is driven through from one hole to the next and so the intervening bone is removed. This surgical operation was later adopted by the Arabs and became standard in the Middle Ages. Thus Rogerius Frugardi recommended that in cases of depressed fractures a number of perforations are made around the affected area with a trepan and then the fractured bone is slowly raised taking care not to damage the underlying meninges. A variation or degeneration of the boring-and-cutting operation existed until recently in North Africa. Hilton-Simpson, studying the practice of trepanation among the Kabyles in Algeria, found that the procedure consisted in removing a circular portion of the scalp with a cylindrical iron punch that had been heated red-hot. Retractors were used to draw away the scalp, and next a small opening was cut in the skull by the confined use of a small drill which was spun between the palms of the hands. With a saw a small incision was made and care taken not to injure the dura mater. Each succeeding day the sawing process was repeated until the piece of bone to be removed was loose, this could take anything from fifteen to twenty days. Finally the part sawn away was lifted from the skull by an elevator.

4. The use of a trepan to remove a disc of bone from the skull (this method is not illustrated here).

This instrument seems to have been in current use in ancient Greece and at a later period in Rome. Hippocrates advised its use for a variety of head injuries, and stated that the craniotomy should be so performed that the trepan does not penetrate too quickly to the dura mater and that the cut fragment should be allowed to detach itself. During the operation the instrument was to be plunged frequently into cold water to avoid overheating the bone (Littré). According to Spencer, Celsus recommended the "crown trepan" or *modiolus* for treating smaller cranial injuries. A *modiolus* is a hollow cylindrical iron instrument whose lower edges are serrated and down its centre runs a fixed pin which is itself surrounded by an inner disc. As mentioned previously, for more extensive head trauma Celsus suggested the boring-and-cutting treatment.

5. A method in which four straight incisions are made, intersecting at right

angles and the in-between fragment is removed (Fig. 1, number 4).

This procedure was commonly adopted in Peru, though isolated finds of this type occur also in other parts of the world. Terrier and Péraire (1895) described such an example from Lizières in France dating to the Neolithic period. As already mentioned, from Palestine Parry and Starkey reported on two Iron Age skulls found by Risdon, and Giles described a further discovery from there showing an attempt at such an operation. Forgue notes that this type of procedure was also carried out by the Kabyles in Algeria; and according to Sood one skull from Kenya shows an attempt at such an intervention.

Rytel (1962) believes there has been what one might call an evolutionary trend in the methodology of trepanation. He considers that the earliest operative procedure began with rectilinear cranial incisions resulting in rectangular openings. This form of craniotomy then proceeded through the polygonal to the circular type of orifice. It was followed by a method of scraping away the bone with a rotatory technique thus making the hole lenticular, oval or circular in shape. And finally the practice evolved to the boring-and-cutting method described above and recommended by Celsus and adopted later by mediaeval surgeons.

Various attempts have been made to assess the time it took to perform a trepanation. Broca, repeating the procedure experimentally on adult post-mortem skulls, found that it took anything from thirty minutes to one hour to perform the operation. About the same length of time was also taken by present-day Peruvian surgeons when trepanning living heads with primitive implements under aseptic conditions (Cabieses). Lucas-Championnière found that the grooving technique took more than one hour to perform and therefore favoured the boring-and-cutting procedure as the method of choice in prehistoric times. At the other extreme is the method used by the Kabyles who, cutting little by little each day, could take up to twenty days to complete the operation.

The diameter of the trepanations varies from that of a small drill hole, a few millimeters across, to quite large openings of 82 mm x 62 mm (Regnault) or larger (Boev), though they are on the average between 30 mm to 45 mm across, one axis being longer than the other. Thus their shapes are frequently oval (in these cases the longer diameter tends to be anteroposterior) or triangular, the smaller openings tending to be round. Quadrangular orifices are produced by four right-angled intersecting incisions.

The majority of trepanned specimens show single openings. The skull with the highest number of trepanations so far discovered was reported by Oakley et al. and came from Cuzco in Peru. This has seven healed openings. Examples with two to three or even as many as five craniotomies have been found in various parts of the world.

Brief mention must now be made of the postoperative treatment of trepanations. Although Forgue (1938) inferred from analogy with present day practices that the early medicine-men used powdered charcoal, hot sand, cedar wood resin or even cinders from sacrifices for their dressings, this can only be conjecture. With regard to the actual trepanned opening Thompson (1938) considers that this was closed with a plate made from shell or other substances and that in some cases even a lead or other metal (Rytel) diaphragm was used, though Stewart believes that there is no good evidence of this practice. For haemostasis the Inca are reputed to have used extracts from the *Ratania* root and *Pumacbuca* shrub of the Andes which are rich in tannic acid (Rytel). Celsus recommended the use of vinegar to stop bleeding (Spencer). Although hardly anything is known of the ancient postoperative treatment of trepanations information on actual practices is available from Algeria and Melanesia. Hilton-Simpson (1913) reports that for dressings the Kabyles used daily applications of heated honey and butter and the stem of leaves belonging to the species of *labiatae*. This dressing was continued sometimes for as long as a month. Detailed accounts from Melanesia are given by Crump and Ford. The trepanned opening was washed with water of the unripe coconut, plugged with a piece of bark cloth and then covered with part of the inner bark or leaf of the banana palm which had been held over a fire. Then the skin flaps were replaced and stitched with a needle, the latter being made in some cases from the wing bone of a flying fox. And finally the head was bound with dried strips of banana stalks.

REPAIR AND SURVIVAL

Schröder (1957), dealing with the healing of trepanations, found that the endosteal callus produced by the diploë is small in amount and that the periosteal callus of the epicranium grows only very little. Thus the osseous regeneration is rather sparse. Apart from a few osteophytes the reaction at the margins of the opening only amounts to a very few millimeters. Similarly Pritchard (1946) has been able to show experimentally that in skull fractures in rats new bone formation is slight and confined to the fracture site when healing is uncomplicated by widespread haemorrhage or infection. In the latter event there is widespread subpericranial new bone formation with bone resorption.

The immediate bony area around the trepanation is radiologically more transparent from the periphery to the margin of the hole (Schröder). The same author warns that radiological differentiation between post-mortem and intra-vitam craniotomy is practically impossible and that Guiard went too far in his claims. The latter believed that he could differentiate radiologically whether an individual had survived the operation for several weeks, months or at least one year. The main diagnosis of healing at the margins is the macroscopic observation of the spongy diploë, the presence of occasional osteophytes and the character of the edges of the external and internal lamina. Healing is indicated by a closed or closing diploë and relatively smooth borders.

The region around some trepanations shows a circular area of osteitis surrounding the opening (Stewart, 1956). This takes the form of an osteoporotic pitting which can be seen as a halo around the craniotomy. The borders of the osteitis correspond to the edges of the orifice made in the scalp preparatory to trepanning. According to Stewart the halo indicates that the individual had lived after the operation and that some degree of infection had set in afterwards. Although it has been suggested that this is a chemical osteitis resulting from applications of medicaments to the wound, Stewart believes it is more likely a septic osteitis.

Thus when making a detailed study of this operation it is necessary to examine carefully the margins of the trepanation and the surrounding area of bone.

The survival rate following craniotomy was remarkably high as evidenced by skulls showing healed openings. Stewart, examining 214 trepanned skulls from Peru belonging to three collections in the United States, found that 55.6 per cent show complete healing, 16.4 per cent early stages of healing and 28 per cent no healing. Rytel too attests to the surgical skill of the Peruvians, of 400 trepanations 250 (62.5 per cent) showed healing. When examining the figures of Russu and Bologa one finds that of the Neolithic material two survived, one lived for a while and four died, whilst of the Mediaeval examples two survived and one died. The figures of Brunn show that thirteen survived and only three died, his material dates from the Neolithic to *circa* sixth century A.D. and comes from central Germany. And according to Crump the mortality rate was about 20 per cent in New Britain, in fact many deaths resulted from the original injuries rather than from the operation. The remarkable skull from Peru with seven healed trepanations and reported by Oakley *et al.* is proof of the skill of the early surgeons. In Kenya quite a number of individuals walk around having recovered from their second or even third trepanation (Sood).

The cause of death was very often the original injury. Complications from the actual operation such as haemorrhage, brain damage, severe shock, sepsis and meningitis further contributed to the mortality rate.

MISCELLANEOUS

Broca erroneously believed that the early surgeons trepanned mainly children and adolescents and he considered that this was due to the frequency of juvenile convulsions. However, the discoveries since have disproved this idea. Certainly children were trepanned and evidence for this is available from Peru (Stewart) and from Melanesia (Ford) where it was also practised prophylactically as cited above.

According to Guiard the practice of trepanation coincided with the presence of a dominant brachycephalic element but was absent from countries where a dolichocephalic population predominated. However, a careful examination of the material shows that this operation was also practised in those parts of the world where dolichocephalic and mesocephalic peoples were the dominent element.

In connexion with trepanation in man mention must also be made of the practice in sheep. Although Ruffer states that it was done for the relief of staggers, he felt that the theory of the veterinary origin of trepanation in man was purely hypothetical. In Rumania craniotomies were performed on sheep by shepherds since ancient times (Russu and Bologa) for the treatment of staggers. This disease which manifests itself by swaying and an uncertain gait is caused by *Coenurus cerebralis*, the larva of the *Taenia coenurus (Multiceps multiceps)* found in the brain of sheep, goats and other ruminants. According to the authors the skull is opened with a knife made of soft iron and the larvae are removed. However, only a few animals recover.

DIFFERENTIAL DIAGNOSIS

It is a fact that many unusual openings have been reported as true trepanations, although originating in devious ways. Thus the differential diagnosis is of some importance. Admirable attention has already been drawn to this in the exhaustive studies of Guiard and of Giot and Desse.

Openings in the skull may be produced by infective processes such as tuberculosis, syphilitic gummata, localized osteomyelitis, or mycoses or as a result of tumours like epidermoid and dermoid cysts, myelomas, secondary carcinomas and sarcomas. Traumatic conditions at birth or during early childhood may also appear as trepanned holes. Bircher (1908) has shown that some so-called trepanations in adults are the results of the use of certain weapons peculiar to the Middle Ages. Brothwell has also drawn attention to the action of beetles and porcupines or other rodents that can produce extensive destruction of bone. Similarly he points out that artefact openings may be due to a pick or other tool used during disinterment, or the cause may be a continual pressure of a sharp stone, or selective erosion of one region of the skull, all of which can give rise to a hole that may bear strong resemblances to a trepanation.

There are, however, three conditions that may give rise to mistakes in diagnosis since they occur in association with the parietal bone. These are enlarged parietal foramina, "Fenestrae parietalis symmetricae," and bilateral osteoporosis ("thinning") of the parietal bone.

Normally the parietal foramina are very small, but occasionally they may have a diameter or two or three centimetres (Broca, 1875; Spee, 1896; Le Double, 1903), this enlargement being due to a defect in development. According to Spee these sort of foramina are more frequent in males and on the right side. Apart from size, their site and number may vary too (Le Double), and the anomaly may also be h e r e d i t a r y (Weber and Schwarz, 1935). Cave (1928) has even reported two cases with bilaterally enlarged parietal foramina. These types can be easily mistaken for a trepanation. An example of this error is an Egyptian female skull of Roman date (c. 200 A.D.) found at Shurafa, Lower Egypt, with an enlarged right parietal foramen. This was reported by Derry (1914) as due to a dermoid cyst and considered as a trepanation by Horsley (1888).

With regard to "Fenestrae parietalis symmetricae" also known as "Catlin mark" (after an American family in whom this character occurs) quite a different situation obtains. Goldsmith (1941), who has also reviewed the literature, considers that this bilateral anomaly and the normal but highly variable paired parietal foramina are not one and the same thing, but have different origins though anatomically they may fuse. In fact he knows of skulls that show both the fenestrae and the enlarged parietal foramina. The fenestrae are oval to round in adults and have definite "healed" edges, and in the living head they are covered with skin and dura mater, but are clearly demonstrable radiologically. There seems no doubt that this condition may arise as a sport or a mutation. According to Goldsmith a survey of various collections of human skulls shows that the trait is not uncommon and that many supposed cases of trepanation really represent this hereditary anomaly.

The third condition which often gives rise to mistaken diagnosis is that of osteoporosis of the parietal bones. Although considered elsewhere in this book by Lodge (pp. 405) it seems worth considering the subject a little here since osteoporosis is found fairly frequently in ancient skeletons (Eve, 1889; Elliot Smith, 1906) and can so easily be confused with healed trephine openings. This manifests itself as a rarefaction of bone, due to diminished osteoblastic or increased osteoclastic activity, which results in a reduction in the amount of total bony substance without evidence of mineral deficiency (Schmidt, 1937; Grollman, 1963). Osteoporosis generally commences in those parts of the skeleton that are subjected to the greatest stress, e.g., the thoracic region of the vertebral column. However, ultimately it will involve other skeletal parts and even the skull. There is increased porosity and a decrease in the thickness of the cortex of the bone. Aetiologically the imbalance between osteoblastic activity and osteoclastic dissolution is the consequence of a variety of causes which will be mentioned briefly. (1) Nutritional deficiencies in vitamins, calcium and proteins will result in rarefaction of bone. Thus for example a lack of calcium may be demonstrated in individuals manifesting so called idiopathic, senile or postmenopausal osteoporosis (Grollman). Already long ago Lobstein (1834) and later Paget (1870) and Ferré (1876) suggested that osteoporosis of the parietal bones was due to senile atrophy. (2) Endocrine disturbances such as hyperthyroidism, hyperparathyroidism, acromegaly and Cushing's syndrome are often accompanied by osteoporosis (Grollman). (3) Congenital deficiency of osteoblastic function may also give rise to rarefaction of bone. This may also be hereditary as pointed out by Shepherd (1892) and Roger and Schachter (1941). (4) Reduced mechanical stress according to Grollman can also be a cause of osteoporosis.

The characteristic location of osteoporosis in the skull is in the parietal bones between the sagittal suture and the parietal eminence on both sides. In some cases the temporal, or frontal, or occipital bones may be involved bilaterally. The parietal osteoporosis forms a depression which is roughly triangular, quadrilateral, or trough-like in shape (Greig, 1926). There is no sharp line of demarcation between the depression and the adjacent normal external lamina and therefore the margins shelve gradually into the thinned area. Radiologically there is a general reduction in the density of the bone which becomes thin. Greig and Durward (1929) have drawn attention to the fact that the parietal foramina are not involved since the thinning process leaves a margin of bone about one centimetre in width. With regard to the incidence of osteoporosis of the parietal bones, Camp and Nash (1944) found that one in two hundred and seventeen heads showed this trait on radiological examination, with a mean age of over fifty years. And Humphry (1874) found that this abnormality is not confined to the human species only but is also present in the orang-utan. The error in diagnosing a trepanation is due to the fact that in parietal osteoporosis the normal bone shelves into a thinned area which is extremely fragile. The latter then breaks down post-mortem for one reason or another and leaves an opening with a more or less bevelled circumference and apparently healed. This type is exemplified by an Egyptian skull of the twelfth Dynasty (c. 1900 B.C.) and reported as a trepanation by Breasted (1930), its true condition was later diagnosed by Stewart (1952).

SINCIPITAL MUTILATION

This mutilation, practised in Neolithic times, is in the form of a T or L and consists of a series of cauterizations of the skin affecting the periosteum (Manouvrier, 1895). One line runs antroposterior following the sagittal suture and the other is at right angles and joins the two parietal eminences. In the majority of cases the operation was performed on women. As Stewart rightly points out, any damage to the scalp leading to a loss of blood supply to the bony vault may be followed by osteitis which can result in scarring of the bone. It is known from surviving medical records that in Mediaeval Europe thermal and chemical cauterization was applied to the head in cases of epilepsy and dementia (MacCurdy, 1905). Piggott remarks that there may have been some connexion between these mutilations and the practice of the tonsure. One might even ask the question whether this might not have been a form of baptism or branding. The examples of sincipital mutilation are few in comparison to the number of trepanations. According to Manouvrier (1895) this practice was restricted to a district north of present day Paris between the Seine and Oise. From Hungary Bartucz (1964) has reported one case and Zaborowski (1897) mentions that to the west of the Caspian Sea, the inhabitants of Dagestan practised a form of cauterization of the vertex of the head, similar to the sincipital operation, in order to prevent illness.

ACKNOWLEDGMENTS

My sincere thanks are due to Dr. Joseph Needham, F.R.S. and to his collaborators on the *Science and Civilisation in China* project for their kind help and for allowing me to use some of their valuable material. I wish to thank the Director, Dr. A. T. Lucas and Mr. A. B. O'Ríordáin, M. A. and Mr. E. Rynne, M. A. of the National Museum of Ireland, Dublin, for making the Irish trepanations available to me. I am also indebted to Dr. J. C. Trevor, Director of the Duckworth Laboratory, Cambridge, for allowing me to use the Crichel Down skull in Figure 3, to Dr. C. E. Oxnard for his criticism and to Mr. W. J. Pardoe for making the illustrations for Figures 1 and 2.

REFERENCES

BANDELIER, A. F., 1904: Aboriginal trephining in Bolivia. *Amer. Anthrop.*, 6:440.

BARTUCZ, L., 1964: Prähistorische Trepanation in Ungarn. Paper read at *7e Congr. Internat. Sc. Anthr. et Ethnol.*, Moscow.

BATRAWI, A., 1935: Report on the human remains. *Mission Archéologique de Nubie*, 1929-1934. Cairo.

BATRAWI, A., 1957: Personal communication.

BIRCHER, E., 1908: Schädelverletzungen durch mittelalterliche Nahkampfwaffen. *Arch. klin. Chir.*, 85:488.

BOBIN, V. V., 1964: Les recherches paléopathologiques des dernières années sur les ossements découverts en Crimée et dans le Caucase du nord. Paper read at *7e Congr. Internat. Sc. Anthr. Ethnol.*, Moscow.

BOEV, P., 1959: Trépanations d'intérêt historique (Trépanations historiques). *Bull. Inst. Morph. Acad. Sci. Bulg.*, 3:197.

BREASTED, J. H., 1930: *The Edwin Smith Surgical Papyrus.* Vol. 1. Chicago, University of Chicago Press.

BREITINGER, E., 1938: Zur Trepanation in der Frübronzezeit. *Anthrop. Anz.*, 15:73.

BROCA, P., 1875: Sur les trous parietaux et sur la perforation congenitale, double et symmetrique des parietaux. *Bull. Soc. Anthrop. Paris*, Sér. 2, 10:326.

BROCA, P., 1876a: Sur les trépanations préhistoriques. *Bull. Soc. Anthrop. Paris*, Sér. 2, 11:236 and 431.

BROCA, P., 1876b: Sur l'âge des sujets à la trépanation chirurgicale néolithique. *Bull. Soc. Anthrop. Paris*, Sér. 2, 11:572.

BROTHWELL, D. R., 1963: *Digging up bones: the excavation, treatment and study of human skeletal remains.* London, British Museum.

BRUNN, W. v., 1936: Über Trepanationen im sächsisch-thüringischen Kulturkreis. *Sudhoffs Arch. Gesch. Med.*, 29:203.

CABIESES, F., 1957: Personal communication.

CAMP, J. D. and NASH, L. A., 1944: Developmental thinness of the parietal bones. *Radiology*, 42:42.

CASTIGLIONI, A., 1941: *A history of medicine.* New York, Knopf.

CAVE, A. J. E., 1928: Two cases of congenitally enlarged parietal foramina. *J. Anat., Lond.*, 63:172.

CRUMP, J. A., 1901: Trephining in the South Seas. *J. Roy. Anthrop. Inst.* 31:167.

DECHELETTE, J., 1908: *Manuel d'archéologie préhistorique.* Paris, Picard.

DERRY, D. E., 1914: Parietal perforation accompanied with flattening of the skull in an ancient Egyptian. *J. Anat., Lond.*, 48:417.

DURWARD, A., 1929: Note on symmetrical thinning of parietal bones. *J. Anat., Lond.*, 63:356.

ELLIOT SMITH, G., 1906: Causation of symmetrical thinning of parietal bones in ancient Egyptians. *J. Anat., Lond.*, 41:232.

ELLIOT SMITH, G., and WOOD JONES, F., 1910: Report on the human remains. *The archaeological survey of Nubia, report for 1907-1908*. Vol. 2. Cairo, National Printing Department.

EVE, F. S., 1889: Bones of ancient Egyptians showing periostitis associated with osteoarthritis and symmetrical senile atrophy of the skull. *Trans. Path. Soc. Lond.*, 41:242.

FERRÉ, C., 1876: Sur l'atrophie sénile symmetrique des parietaux. *Bull. Soc. Anthrop. Paris*, Sér. 2, 11:423.

FISCHER-MØLLER, K., 1936: Trepanation og Kranielaesion i Stenalderen. Nyopdage de Tilfaelde fra Fyen. *Aarb. Nord. Oldkynd. og Hist.*, p. 109.

FLEETWOOD, J., 1951: *History of medicine in Ireland.* Dublin, Browne and Nolan.

Ford, E., 1937: Trephining in Melanesia. *Med. J. Aust.*, 2:471.

Forgue, E., 1938: Histoire de la chirurgie. In *Histoire générale de la médicine, de la pharmacie, de l'art dentaire et de l'art vétérinaire.* Ed. M. Laignel-Lavastine. Vol. 2, Paris, Albin Michel.

Giles, M., 1953: Crania from Tell El—Duweir. In *Lachish III: The iron age.* The Wellcome-Marston Archaeological Research Expedition to the Near East.

Giot, P. R., 1949: Les trépanations de la nécropole gauloise de Saint-Urnel en plomeur. Comparaison avec les autres trépanations préhistoriques de Bretagne. *Bull. Soc. Anthrop. Paris,* 10:59.

Giot, P. R., and Desse, G., 1950: Quelques documents sur les trépanations préhistoriques. *Pr. méd.,* 58:1283.

Goldsmith, W. M., 1941: Bilateral fenestrae in the parietal bones: more cases of the 'Catlin Mark.' *J. Hered.,* 32:301.

Greig, D. M., 1926: On symmetrical thinness of parietal bones. *Edinb. med. J.,* 33:645.

Grollman, A., 1963: *The functional pathology of disease: the physiologic basis of clinical medicine,* 2nd ed. New York, McGraw-Hill.

Guiard, E., 1930: *La trépanation crânienne chez les néolithiques et chez les primitifs modernes.* Paris, Masson.

Guthrie, D., 1958: *A history of medicine.* Rev. ed. London, Nelson.

Hilton-Simpson, M. W., 1913: Some Arab and Shawia remedies and notes on the trepanning of the skull in Algeria. *J. Roy. Anthrop. Inst.,* 43:715.

Holländer, E., 1915: Die chirurgische Säge. *Arch. klin. Chir.,* 106:316.

Horsley, V., 1888: Trephining in the neolithic period. *J. Roy. Anthrop. Inst.,* 17:100.

Humphry, G. M., 1874: Depressions in parietal bones of an orang and in man. *J. Anat., Lond.,* 8:136.

Jungbauer, G., 1923: *Märchen aus Turkestan und Tibet.* Jena, Diedrichs.

Karolyi, L. v., 1963: Daten über das europäische Vorkommen der vor-und frühgeschichtlichen Trepanation. *Homo,* 14:321.

Le Double, A. F., 1903: *Traité des variations des os du crâne de l'homme: et de leur signification au point de vue de l'anthropologie zoologique.* Paris, Vigot.

Leskin, 1919: *Balkanmärchen aus Albanien, Bulgarien, Serbien und Kroatien.* Jena, Diedrichs.

Lisowski, F. P., 1954: A report on the skulls from excavations at Sesebi (Anglo-Egyptian Sudan). *Actes 4e Congr. Internat. Sc. Anthr. et Enthnol., Vienne 1952,* 1:228.

Lisowski, F. P., 1959: Ägyptische Trepanationen. *Homo,* Suppl. 6:147.

Littré, E., 1863: *Hippocrates, Opera Omnia.* Vols. 7 and 9. Paris, Baillière.

Lobstein, J. F., 1834: *Lehrbuch der pathologischen Anatomie.* Vol. 1. Stuttgart, Brodhag.

Lucas-Championnière, J., 1878: *La trépanation.* Paris, Delahaye.

Lucas-Championnière, J., 1912: *Les origines de la trépanation décompressive.* Paris, Steinheil.

MacCurdy, G. G., 1905: Prehistoric surgery—a neolithic survival. *Amer. Anthrop.,* 7:17.

Manouvrier, L., 1895: Le T sincipital: curieuse mutilation crânienne néolithique. *Bull. Soc. Anthrop. Paris,* Sér. 4, 6:357.

Margetts, E. L., 1962: Personal communication.

Martin, C. P., 1935: *Prehistoric man in Ireland.* London, Macmillan.

Matiegka, J., 1928: La trépanation et des autres opérations sur la tête a l'époque préhistorique sur le territoire de la Tchécoslovaquie. *Anthropologie, Prague,* 6:41.

Maxia, C. and Cossu, D., 1951: Cranio dell'epoca nuragica con segni di trepanazione sincipitale in vita: Studio anatomo-radiografico. *Riv. antrop.,* 39:232.

McKenzie, D., 1936: Surgical perforation in a mediaeval skull with reference to neolithic holing. *Proc. Roy. Soc. Med.,* 29:895.

Montandon, G., 1926: Crâniologie paléosibérienne: néolithiques, mongoloïdes, tchouktchi, eskimo, aléoutes, kamtchadales, aïnou, ghiliak, negroïdes du nord. *Anthropologie, Paris,* 36:209.

Moodie, R. L., 1923: *Paleopathology. An introduction to the study of ancient evidences of disease.* Urbana, University of Illinois Press.

Müller, R. F. G., 1959: Schädeleröffnungen nach indischen Sagen. *Centaurus,* 6:68.

Needham, J., 1954: *Science and civilisation in China.* Vol. 1: Introductory orientations. Cambridge: University Press.

Needham, J., 1964: *Science and civilisation in China.* Unpublished.

Oakley, K. P., Brooke, W., Akester, A. R., and Brothwell, D. R., 1959: Contributions on trepanning or trephination in ancient and modern times. *Man,* 59:93.

Paget, J., 1870: *Lectures on surgical pathology.* 3rd ed. revised by W. Turner. London, Longmans.

Parry, T. W., 1921: The prehistoric trephined skulls of Great Britain, together with a detailed description of the operation probably performed in each case. *Proc. Roy. Soc. Med.,* 14:27.

Parry, T. W., 1923: Trephination of the living human skull in prehistoric times. *Brit. med. J.,* 1:457.

Parry, T. W., 1936: A case of primitive surgical holing in the cranium practised in Great Britain in mediaeval times, with a note on the introduction of trepanning instruments. *Proc. Roy. Soc. Med.,* 29:898.

Parry, T. W., and Starkey, J. L., 1936: Discovery of skulls with surgical holing at Tell Duweir, Palestine. *Man,* 36:233.

Piggott, S., 1940: A trepanned skull of the Beaker period from Dorset and the practice of trepanning in prehistoric Europe. *Proc. Prehist. Soc.,* n.s., 6:112.

PRITCHARD, J. J., 1946: Repair of fractures of the parietal bone in rats. *J. Anat., Lond.*, 80:55.

PRUNIÈRES, P. B., 1874: Sur les crânes artificiellement perforés et les rondelles crâniennes à l'époque des dolmens. *Bull. Soc. Anthrop. Paris*, Sér. 2, 9:185.

REGNAULT, F., 1936: La paléopathologie et la médicine dans la préhistoire. In *Histoire génerale de la médicine, de la pharmacie, de l'art dentaire et de l'art vétérinaire*. Ed. M. Leignel-Lavastine. Vol. 1. Paris, Albin Michel.

RISDON, D. L., 1939: A study of the cranial and other human remains from Palestine excavated at Tell Duweir (Lachish) by the Wellcome-Marston Archaeological Research Expedition. *Biometrika*, 31:99.

ROGER, H. and SCHACHTER, M., 1941: Los lagunas y los 'adelgazamientos' parietalis del cráneo de origen congénito. *Med. esp.*, 5:29.

ROKHLIN, D. G., 1964: Antiquity of pathological processes in fossil human bones in the USSR. Paper read at *7e Congr. Internat. Sc. Anthr. Ethnol.*, Moscow.

RONEY, J. G., 1954: Trephining among the Bakhtiari. *Bull. Hist. Med.*, 28:489.

RUFFER, M. A., 1918: Studies in palaeopathology: some recent researches on prehistoric trephining. *J. Path. Bact.*, 22:90.

RUSSU, I. G., and BOLOGA, V., 1961: Trepanationen im Gebiet des heutigen Rumänien. *Sudhoffs Arch. Gesch. Med.*, 45:34.

RYNNE, E., 1962: National Museum of Ireland: archaeological acquisitions in the year 1960. *J. R. Soc. Antiq. Ireland*, 92:139.

RYTEL, M. M., 1962: Trephinations in ancient Peru. *Bull. Pol. Med. Sci. Hist.*, 5:42.

SCHMIDT, M. B., 1937: Atrophie und Hypertrophie des Knochen einschliesslich der Osteosklerose. In *Handbuch der speziellen pathologischen Anatomie und Histologie*. Ed. O. Lubarsch and F. Henke. Vol. 9, pt. 3. Berlin, Springer.

SCHRÖDER, G, 1957: Radiologische Untersuchungen an trepanierten Schädeln (Neolithikum-Mittelalter). *Z. Morph. Anthrop.*, 48:298.

SHEPHERD, F. J., 1892: Symmetrical depressions on the exterior surface of the parietal bones. *J. Anat., Lond.*, 27:501.

SOOD, B. K., 1960: Personal communication.

SPEE, F. v., 1896: *Skeletlehre, Teil 2 Kopf*. In *Handbuch der Anatomie des Menschen*. Ed. K. v. Bardeleben. Jena, Fischer.

SPENCER, W. G., 1948: *Celsus: De medicina*. A translation. Vols 3. London, Heinemann.

STEWART, T. D., 1952: Personal communication.

STEWART, T. D., 1956: Significance of osteitis in ancient Peruvian trephining. *Bull. Hist. Med.*, 30:293.

STEWART, T. D., 1958: Stone age skull surgery: a general review, with emphasis on the New World. *Smithson. Inst. Rept.* (1957):469.

SUDHOFF, K., 1929: Trepanation. In *Reallexikon der Vorgeschichte*. Ed. M. Ebert. Vol. 13, p. 430.

SUDHOFF, K., and EBERT, M., 1913: Chirurgische Instrumente aus Ungarn. *Prähist. Z.*, 5:595.

TALLGREN, A. M., 1936: The South Siberian cemetery of Oglakty from the Han period. *Eurasia Sept. Ant.*, 11:69.

TERRIER, F., and PÉRAIRE, M., 1895: *L'opération du trépan*. Paris, Alcan.

THOMPSON, C. J. S., 1938: The evolution and development of surgical instruments, IV. The trepan. *Brit. J. Surg.*, 25:726.

WEBER, F. P., and SCHWARZ, E., 1935: Hereditary large parietal foramina. *Proc. Roy. Soc. Med.*, 29:122.

WÖLFEL, D. J., 1925: Die Trepanation. Studien über Ursprung, Zusammenhänge und kulturelle Zugehörigkeit der Trepanation. *Anthropos*, 20:1.

WONG, K. CHI-MIN, and WU LIEN-TEH, 1936: *History of Chinese medicine: being a chronicle of medical happenings in China from ancient times to the present period*. 2nd ed. Shanghai, National Quaratine Service.

WOO JU-KANG, 1964: Personal communication.

ZABOROWSKI, S., 1897: Le T sincipital. Multilation des crânes néolithiques, observée en Asie Centrale. *Bull. Soc. Anthrop. Paris*, Sér. 4, 8:501.

Chapter 53

Trepanation of the Skull by the Medicine-men of Primitive Cultures, with Particular Reference to Present-day Native East African Practice

EDWARD L. MARGETTS

INTRODUCTION

TREPANATION of the skull is the most fascinating surgical operation in the history of medicine. Archaeological finds indicate that it is also the oldest therapeutic procedure of which we have objective evidence. The bibliography of trephining is vast, there are thousands of references, and much controversy as to motive and technique. The extensive literature deals mostly with "prehistoric" trepanation, particularly Neolithic European or pre-Columbian South American. But in actual fact, the operation has been practised almost everywhere in the world and at all periods, from the New Stone Age certainly and perhaps even from an earlier period in the evolution of man.

The word "trepanation" is from the Greek *trypanon*, meaning "a borer," and dates from classical times. The more recent term "trephination" affords a variant, and is derived from the French. Essentially, both words mean the same, the making of a depression or perforation in the calvarium. Trepanation or trephination in the narrowest sense implies the boring of a hole through the intact skull of a living person. An extended meaning of this would include the making of a depression but not a hole in one or both tables of the cranium, or the removal of bone fragments already present from trauma or infection. The hole or depression is usually made by scraping (*raclage, grattage*), rasping, and cautery, but may also be accomplished by drilling, boring with a gimlet (*vrillage*) or knife-tip, cutting (*burinage*), and sawing (*sciage*).

In the previous chapter, Dr. Peter Lisowski has considered ancient trepanation, and thus the aim of this present review is not archaeological or historical in nature, but rather to survey trephining in recent times by traditional medicine-men in primitive cultures, with particular reference to the Kisii of Kenya and the Tende south of them in Tanganyika. For general reference to archaeological, anthropological and historical matters relating to trepanation, the reader is referred to Albu, 1889; Anonymous, 1887; Ballance, 1922; Belloni, 1965; Braune, 1875; Broca, 1877; Brothwell, 1963; Busacchi, 1935; Canu, 1908; Chipault and Daleine, 1893; Ciba Symposia, 1939 Ciba Zeitschrift, 1936; Courville, 1944; De Baye, 1880; Dechelette, 1908; De Celinski, 1838; De Nadaillac, 1892; Editorial, 1892; Editorial, 1916; England, 1962; Fletcher, 1882; Forgue, 1938; Genna, 1930; Guérin, 1951; Guiard, 1930; Guthrie, 1958; Höllander, 1927; Horne, 1894; L.A.S., 1879; Le Baron, 1881; Lucas-Championnière, 1878-1912; MacCurdy,

1905, 1924; Manouvrier, 1904; Marill, 1946; Mettler, C. C. and F. A., 1947; Mettler F. A., and C. C., 1945; Milne, 1907; Moodie, 1919-23; Munro, 1897; Oakley, Brooke, Akester and Brothwell, 1959; O'Connor and Walker, 1951; Parry, 1914-36; Piggott, 1940; Popham, 1954; Prunières De Marvejols, 1874; Rightmyer, 1965; Rogers, 1930; Ruffer, 1918; Scarpa, 1956; Smith, 1916; Sprengel, 1805; Stewart, 1958; Sudhoff, 1909 (translated 1926), Sudhoff in Ebert, 1924-32; Terrier and Péraire, 1895; Thompson 1938-42; Tillmanns, 1883; Vara López, 1949; Von Hovorka and Kronfeld, 1908-9; Walker, 1958; Williams, 1929; and Lisowski's contribution in this book.

TREPANATION IN THE PACIFIC ISLANDS

There are a few out of the way places in the world where cranial trepanation is still carried out by native traditional medical practitioners. One of the areas where the operation has recently been known is in Polynesia and Melanesia (Ackerknecht, 1946-7; Baudouin, 1913; Brodsky, 1936-8; Buschan, 1900-34; Courville, 1951; Crump, 1901; Ford, 1937; Guiard, 1930; Heyerdahl, 1952; Sanson, 1874; Stéphen-Chauvet, 1936; Von Hovorka and Kronfeld, 1908-9; Wölfel, 1925-36). Trepanation has not apparently been reported in Micronesia, the northernmost Pacific islands.

A number of reports have described the practice in Polynesia, the central south Pacific islands. As early as 1829, William Ellis, a missionary, reported "trepanation" in the Society Islands (Ellis, 1829). Actually this was removal of bone fragments after traumatic fracture, and not true trephining, the *making* of a depression or hole in the intact skull. He described how the natives repaired a bony deficiency in the skull by fitting in a piece of coconut shell. He also recounted the story that injured brain was removed and pig brain substituted—unsuccessfully, it is said, the person always becoming mad and dying! Coconut shell was undoubtedly chosen because of its hardness and its natural resemblance to the cranial vault—hard material indented with vein-like impressions, which infers a magical identification. Whether the shell was *inserted in* cranial deficits, or, as a dressing or protection placed *over* them, is not at all clear from the published reports (*vide infra*).

Trepanation was reported by others in Polynesia, particularly in the Society islands, which include Tahiti (Ellis, 1829; Henry, 1928; Topinard, 1875; Wölfel, 1925). Stewart visited the Marquesas in 1830, and mentioned that trepanation with a shark's tooth was carried out by the *tahunas* (*tauas*) or priests (Stewart, 1831). Natives still remember the operation there (Handy, 1923; Heyerdahl, 1941-52). Holed skulls from the Marquesas are in the Musée National d'Histoire Naturelle (Forgue, 1938) and the Musée de l'Homme (Heyerdahl, 1952) in Paris. Trephining was customary also in the Tuamotu Archipelago, between the Marquesas and Society Islands (Emory, 1942; Heyerdahl, 1952).

In Oceania, the practice of making holes in the cranium reached its height in Melanesia, the western islands south of the Equator (Heyerdahl, 1952; Parkinson, 1907-8; Wölfel, 1925). It has been reported from most of the larger islands, Fiji, Bismarck Archipelago, New Hebrides, New Caledonia and the Loyalty Islands. It is uncertain (Ford, 1937) whether the operation was indigenous to the Solomon Islands, though Wölfel (1925) claimed a skull from the island of Ysabel, and now in the Naturhistorisches Museum in Vienna, was trepanned.

In the Bismarck Archipelago, cranial

trephining was reported in New Britain, New Ireland, the Duke of York and also in some of the smaller islands. The instrument was an obsidian chip, a shell, or a shark's tooth. In New Britain (Neu Pommern) the operation was done only for fracture, usually from a sling-stone. The opening was scraped with a sharp stone, the fragments blown away, the scalp sutured, and elaborate dressings applied (Brodsky, 1936-8, Crump, 1901, Parkinson, 1907-8; Von Luschan, 1898).

Descriptions of holed skulls, photographs of surviving trepanned natives, and of the operative process are on record (Brodsky, 1936-8; Brown, 1910; Busacchi, 1929; Buschan, 1900-34; Ciba Zeitschrift, 1936; Crump-Horsley, 1901; Ford, 1937; Meagher, 1940; Parkinson, 1907-8; Pöch, 1907; Powell, 1883; Sanson, 1874; Seligmann, 1906; Underwood, 1951; Von Hovorka and Kronfeld, 1908-9; Von Luschan, 1898; Wölfel, 1925-36). Trepan instruments from the Bismarck Islands may be seen at the Department of Surgery in the University of Sydney (Brodsky, 1938) and skulls at the Naturhistorisches Museum, Vienna (Wölfel, 1925-36), and the Wellcome Historical Medical Museum in London. A New Ireland skull now at the latter museum was holed in eight different places (Brown, 1910, Ford, 1937, Underwood, 1951). The native operator (*tena-papait* or *tene a babait*) who did this job was surpassed only by the "civilized" seventeenth century European surgeon Hendrik Chadbourn, who is said to have bored holes twenty-seven times in the head of Philip, the Count of Nassau (Ballance, 1922; Guthrie, 1842; Solingen, 1684)! The collection of Bismarck Island trepanned skulls once in the Royal College of Surgeons Museum in London (Keith, 1925; Rogers, 1930; Seligmann, 1906) was destroyed by German bombing of the city during World War II.

In New Ireland (Neu Mecklenburg) the indications for trepanation were less specifically "surgical;" in addition to fracture it was recommended also for insanity considered to result from "pressure on the brain," and for "a beating or plucking sensation" (Crump, 1901; Brown, 1910). It was also done for epilepsy and headache. Trepannning was known in the Duke of York Islands (Neu Lauenburg) between New Britain and New Ireland, and in the easterly outlying islands, Gerrit Denys (Lihar) and Caens (Tanga) (Ford, 1937). A type of operation in New Ireland was interesting, in essence being the cutting of two or three vertical channels down the forehead (Parkinson, 1907; Crump, with Horsley, 1901). The operation became fashionable there and was promoted as an *amulet* (i.e., a protection) and as an aid to longevity (Crump, 1901). Ford stated that trepanning the skull has disappeared from New Britain and New Ireland within the last thirty years, as a result of government prohibition of slingstone fights rather than suppression of native medical practice (personal communication, Aug. 1959).

Trephining in the Loyalty Islands and New Caledonia (Nouvelle Calédonie) has been extensively reported (Baudouin, 1913; Ella, 1874; Ford, 1937; Hadfield, 1920; Martin, 1879; Nicolas, 1913; Ray, 1917; Riofrey, 1874; Sarasin, 1929; Turner, 1884; Waterston, 1908; Wölfel, 1925). True trepanation was reported by Ella (1874) in Uvea for headache and vertigo resulting from a blow on the head. A hole was made in the skull by scraping with a piece of glass or a shark's tooth. The defect was filled (or covered?) with a coconut shell. He noted that about half of those who underwent the operation died from it, but later writers have questioned this high figure (Ford, 1937). The point should be made that Ella's percentage might have been correct at the time of his residence in the Islands. Perhaps the mortality became less over the next few decades, as sling-stone fights decreased and therefore brain trauma and infection less frequent. Trepanning the cranium alone, without damaging the brain, may have quite a low mortality, as in East Africa today (*infra*). The inclusion or addition of a foreign substance such as coconut shell would probably increase the incidence of infection and raise the mortality rate. Sarasin (1929), reporting on the Loyalty Islands, observed a coconut shell plate *in situ* and he heard of a man who is said to have had five such plates in different parts of his head. Bark, banana leaf, and sea shell were used also for plugging or dressing the trepan hole, in the Loyalty as in the Bismarck Islands (Brodsky, 1936-8; Crump, 1901; Ford, 1937; Hadfield, 1920; Nicolas, 1913; Parkinson, 1907-8).

Eastern New Guinea and Papua would appear to be the westerly limit of the Pacific islands where trepanation has been practiced (Fritsch, 1907; Guiard, 1930; Hagen, 1899; Laloy, 1907; Wölfel, 1925). In New Guinea, cuts were made in the forehead to treat headache (Ford, 1937; Hagen, 1899; Pöch, 1907) and true perforation of the cranium has been reported (Fritsch, 1907; Fritsch, Pales, cited in Guiard, 1930; Laloy, 1907). It is not known for certain if the practice was known in the Warrior Reefs in the Torres Strait between New Guinea and the mainland of Australia (Ford, 1937) and in Australia itself. Most authors have questioned whether trepanation was known in New Zealand, though Wölfel (1925-36) was satisfied that a New Zealand skull in the Naturhistorisches Museum in Vienna was trepanned for fracture. In a country the size of New Zealand and with such a written history, it seems strange that more New Zealand trepanned skulls have not come to notice.

TREPANATION IN SOUTH AMERICA

While the trepanning operation might have reached Oceania from Peru (Heyerdahl, 1952) there are no adequate reports available that the custom prevailed in Peru during *historical* times, though current trepanning in Peru was mentioned in passing by Bandelier (1904), Freeman (1924), Hrdlička (1906) and Stewart (1950). Pre-Columbian trephination of the skull in Peru and elsewhere in South America is extensively documented (Balado, 1930; Bartels, 1913; Bazzocchi, 1947; Broca, 1867-77; Courville and Abbott, 1942; Courville, 1959; Daland, 1935; D'Harcourt, 1939; Fletcher, 1882; Freeman, 1918-24; Graña y Reyes, Rocca and Graña, 1954; Grimm, 1955; Lastres and Cabieses, 1960; MacCurdy, 1923; McGee, 1894; Mettler F. A. and C. C., 1945; Moodie, 1929; Muñiz and McGee, 1897; O'Connor and Walker, 1951; Quevedo and Sergio, 1943; Rogers, 1938; Rytel, 1956; Squier, 1877; Stewart, 1950-8; Tillmanns, 1883; Trelles, 1957-62; Walker, 1958; Weiss, 1949-58). Cranial trepanation was practiced by the Aymara Indians of Bolivia at the turn of the century (Bandelier, 1904-10; Chervin, 1907-8) and actively suggested there as a cure for headache as recently as 1950 (Brooke in Oakley *et al.* 1959). Freeman (1924) noted that: "In the mountains of Peru, Chili (*sic*) and Bolivia, trephining for fractures is still practiced (*sic*) occasionally by native medicine-men." Field studies to determine whether trepanation is currently undertaken in Peru and neighbouring countries are very much needed.

TREPANATION IN NORTH AMERICA

The operation was known in aboriginal North America before the white man came, and has been recorded from Alaska, British Columbia, the United States and Mexico (Anonymous, 1935; Cosgrove, 1929; Farquharson, 1881; Fletcher, 1882; Gillman, 1875-85; Goldsmith, 1922-45; Greenman, 1926; Hill-Tout, 1953; Hinsdale, 1924; Hinsdale and Greenman, 1936; Hinsdale and Cappannari, 1940; Holbrook, 1877; Hrdlička, 1906-39; Kidd, 1930-53; Leechman, 1944; Lumholtz and Hrdlička, 1897; Lumholtz, 1903; McGregor and Wadlow, 1951; Moodie, 1930; Shapiro, 1927; Smith and Hrdlička, 1924; Wakefield and Dellinger, 1939). There are no certain reports that it is still carried out by Indian tribes, but it is possible. Kidd (1946) notes the case of a British Columbian coastal Indian who came up to a missionary carrying a brace and bit, and begged him to bore a hole in his skull to let out the evil spirit causing his headaches! There are more than a dozen trepanned skulls from British Columbia sites, but they are archaeological and so fall outside the subject confines of this essay.

TREPANATION IN EUROPE

During the Middle Ages and right through to 1900, trepanation was an orthodox surgical procedure in Europe, particularly in fractures of the cranium, head injuries, and for various headaches, epilepsy, and mental trouble. With the spread of modern ideas about neurosurgery, the practice decreased in popularity before the turn of this century. The doctors of Cornwall until recently customarily trepanned miners who suffered head injuries, and perhaps not infrequently at the patient's own request for a "boring" (Fletcher, 1882; Lucas-Championnière, 1912; G. R., 1939; Hudson, 1877). With the growth of civilization and its technical improvements, trepanation declined in the heavily populated and advanced areas of the world. It did not always recede in the isolated backward locations where communication was limited; this was particularly true of mountainous places. Thus it was that this curious operation continued to be practised to the turn of this century (if not later) in the mountains of Albania, and of Serbia and Montenegro, (Boulongne, 1869; Durham, 1909; Editorial, 1892; Frilley and Wlahovitj, 1876; Trojanović, 1900; Védrènes, 1886).

TREPANATION IN WEST AND CENTRAL ASIA

The operations of cranial cautery and trepanation were widely practiced in the Caucasus Mountains of Daghestan, now in the U.S.S.R. (Krivyakin, 1887-8; Minkevich, 1897; Terrier and Péraire, 1895; Virsaladze, 1898; Zaborowski, 1897). It is not known if the custom is still extant there.

Incomplete trepanning is currently practiced in Persia. The Bakhtiari of the Zagros mountains in western Iran operate on the skull after head injury. The operation consists essentially of removing bone fragments or of scraping the skull to remove discoloration or bleeding. The inner table is not scraped through, i.e., a hole is not made (Roney, 1954). Bailey (1961) related an anecdote about a friend who was visiting the grave of Sir Victor Horsley in Iraq. Two Arabs came by and in conversation they revealed that they belonged to an ancient family which had included trepanners for eight or nine centuries.

Amongst the tribes inhabiting the mountain range of the Hindu Kush north of Kabul in north-east Afghanistan and in Dardistan, where India and modern Pakistan meet in the north, the practice of cranial cautery for disorders of the head was reported at the turn of the century (Zaborowski, 1897), and, as in Daghestan and elsewhere in primitive areas, particularly in Africa, this often is but little different in the operative intent as trepanning the outer table of the skull by cutting, scraping, boring and sawing.

TREPANATION IN AFRICA

Herodotus mentioned that the Libyans practised cauterization of the head on their children at the age of four years, "to prevent them from being plagued in their after lives by a flow of rheum from the head" (Bk. IV ch. 187). If the operation threw the children into convulsions, they were sprinkled with goat's urine. This was the earliest historical reference to anything resembling trepanation in an African people. The Libyans of Herodotus probably were Tuareg Berber nomads, the autochthonous inhabitants of North Africa. Prehistoric trephined skulls, dating long before Herodotus, were found by General Faidherbe in dolmens at Roknia, north of Constantine in Algeria. These skulls were examined by Broca and are apparently in the Musée Broca in Paris (Buckland, 1881; De Nadaillac, 1892; Fletcher, 1882; Guiard, 1930; Le Baron, 1881).

Considering the many thousands of skulls which have been excavated and examined, it is clearly significant that ancient trepanned skulls have not been found in abundance in Egypt; and it has not been reported that the true operation is practised there by traditional Arab or negro healers at the present time.

The operation of trephining to remove bone fragments after head injury is extant in the north-west corner of Africa. Trepanation is currently well known to Arab surgeons from Arabia itself, and has been since the early days of medical history. It is customarily practised by the Chaouïa (Shawia) Arabized Berbers of the Sahara Atlas, particularly in the Aurès mountains north of Biskra in Constantine, Algeria, near Timgad or ancient Thamugadi. It is not unknown on the plains north and south of the Atlas. Hilton-Simpson (1913-22) has written an account of the practice in the Aurès. The scalp over the site is branded away to bare the bone and to check haemorrhage, then, to let out the pus or blood, a hole is made in the skull by spinning a drill between the palms of the hands. Care is taken to protect the brain and its membranes, and to avoid the cranial sutures, which are believed to be the patient's destiny written by the hand of Allah. Using saws and elevators, "bad bone" is removed day after day. Finally, the wound is dressed with honey, butter and herbs, and allowed to granulate in. The operation is also known among the Kabyle Berbers closer to the coast, and westward to Morocco. Bertherand, as early as 1855, indicated that he had heard of "médecins maures" who practised trepanation for cranial fractures but he was not able to confirm the practice. It remained for Martin (1867) and Paris (1868) to document trepanation by the Kabyles. There has since been an extensive additional French literature (Lucas-Championnière, 1878-1912; Malbot and Verneau, 1897; Malbot, 1898; Védrènes 1885) which has been summarized (De Nadaillac, 1882; Duneau, 1886; Fletcher, 1882; Forgue, 1938; Guiard, 1930; Larrey, 1867; Lucas-Championnière, 1878; Terrier and Péraire, 1895). The native trepanners in Algeria

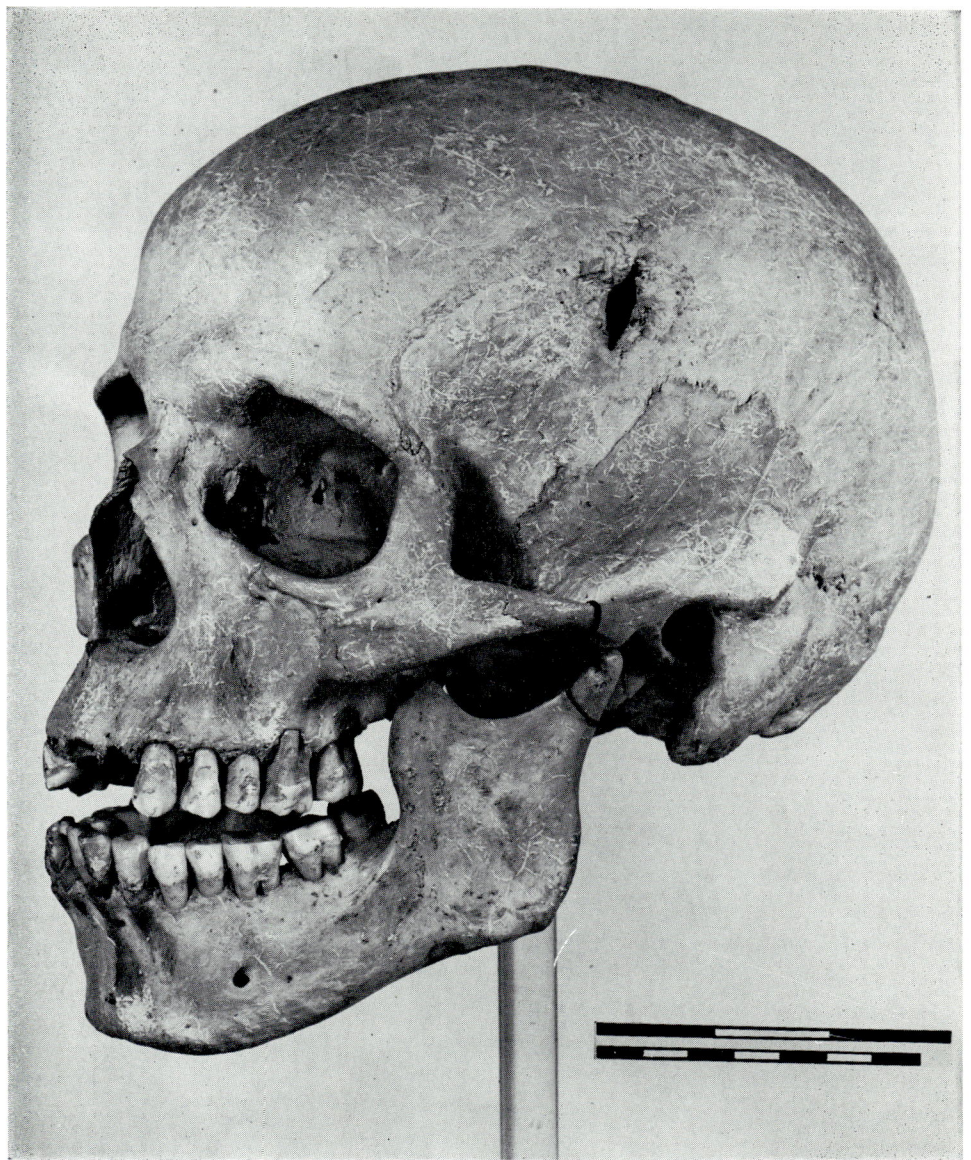

Figure 1. The Hilton-Simpson skull. Left parietal trepanation in a skull from Algeria. By courtesy of Dr. Audrey Butt and the Pitt-Rivers Museum, Oxford, England. Collected 1913.

were said to have taught their skill in "schools" according to Malbot and Verneau (1897), Malbot (1898) and Védrènes (1885) but Hilton-Simpson was unable to find out about this. Busacchi (1935) reproduced a photograph by E. Chantre "Kabyle trepané de Palestro" (Bertholon and Chantre, 1913). Bailey (1961) wrote that a skull found by a French surgeon in Morocco had been used to teach the trepanning procedure.

A skull (Fig. 1) collected by Hilton-

Simpson (1920) from El-Kantara in the Aurès may be examined in the Pitt-Rivers Museum at Oxford. It is probably trepanned but is "archaeological" rather than recent, having been dug up on the road and there is no ante-mortem history of the specimen. It may well be a Roman skull. The Hilton-Simpson collection of trepanning instruments from the Aurès and skull fragments from actual operations are in the same museum (Hilton Simpson, 1913-22; Butt, 1961).

Skulls and instruments from Algeria collected by French surgeons may be seen in the Musée National d'Histoire Naturelle (Jardin des Plantes) in Paris (Forgue, 1938; Guiard, 1930; Malbot and Verneau, 1897) and the Musée du Val-de-Grâce (Védrènes, 1885).

The Guanche of the Canary Islands, a now extinct aboriginal people, were probably a Berber group, who trepanned the skull (Beattie, 1930; Hooton, 1925; Meyer, 1896; Schmeltz, 1896; Tillmanns, 1883; Von Luschan, 1896-9) and cauterized or scraped the cranium to make cicatrizations like the Neolithic European sincipital-T (same references and also Bockenheimer, 1922; Chil y Naranjo, 1878; Lehmann-Nitsche 1903-5; MacCurdy, 1905). Scarred Guanche skulls may be seen in the Wellcome Historical Medical Museum (Underwood, 1951), and the Museum für Völkerkunde (Von Luschan, 1896-7) and trepanned Guanche skulls in the Museum of Santa Cruz in Tenerife (Hooton, 1925). The scarred and trepanned Guanche skulls described by Beattie (1930) and said by him to be in the Redpath Museum at McGill University are misplaced or lost, or at any rate the present curator has not been able to locate them.

The Teda (Tebou, Tibbu, Tebu, Ted, Toubou, Tubu, Tabu, or Tu) of isolated Tibesti in Tchad and southern Libya, are probably a Berber-Negro mixture. Dalloni (1935) noted information supplied by P. Nöel that the Teda surgeons of Kaouar explored cranial wounds and resected bone. Cline (1950) noted this also, but may have been paraphrasing Dalloni. The Cambridge expedition to Tibesti in 1957 documented the custom of trepanation by Teda native doctors (Akester, 1958-9; Steele, 1958). The one indication for the operation was headache. The technique was to scrape out fragments of skull with a sharp pointed instrument. One trepanned patient had a double operation and carried two pieces of bone about with him. The reports did not comment on the significance of the fragments, but they would certainly be a souvenir or symbol, and most likely an *amulet*.

There is practically no published information about trepanation in West and Central Africa. Talbot (1926) noted in Nigeria, "The only place where trepanning was carried out seems to have been in Bende District, where it is said that recourse was often had to it in case of serious illness. The patients almost invariably recovered, but not all doctors were capable of performing the operation." There is a skull in the Wellcome Museum said to be from Senegal and it does have an unhealed irregular engraved hole in the right frontal bone. There is insufficient history about this skull to report it in any detail as a possibly trepanned specimen.

No eye witness account of true trepanation in South Africa has been recorded. There is reasonable evidence from skulls that the operation was practised by the Bushmen (Drennan, 1937 and personal communication 1960). Drennan told me that he and Raymond Dart believed that ante-mortem tampering with the cranium was widespread among the Bushmen, and

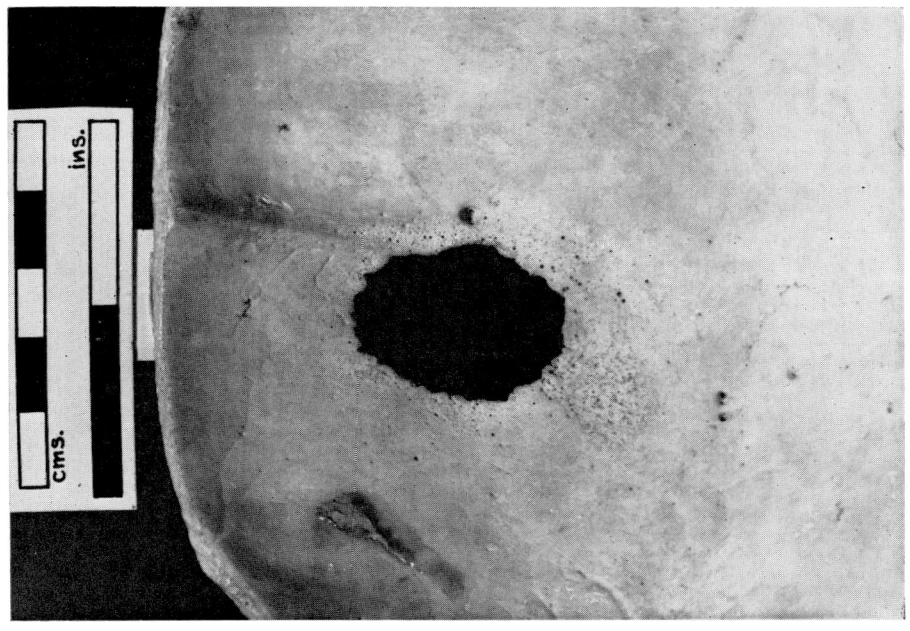

FIGURE 2b.

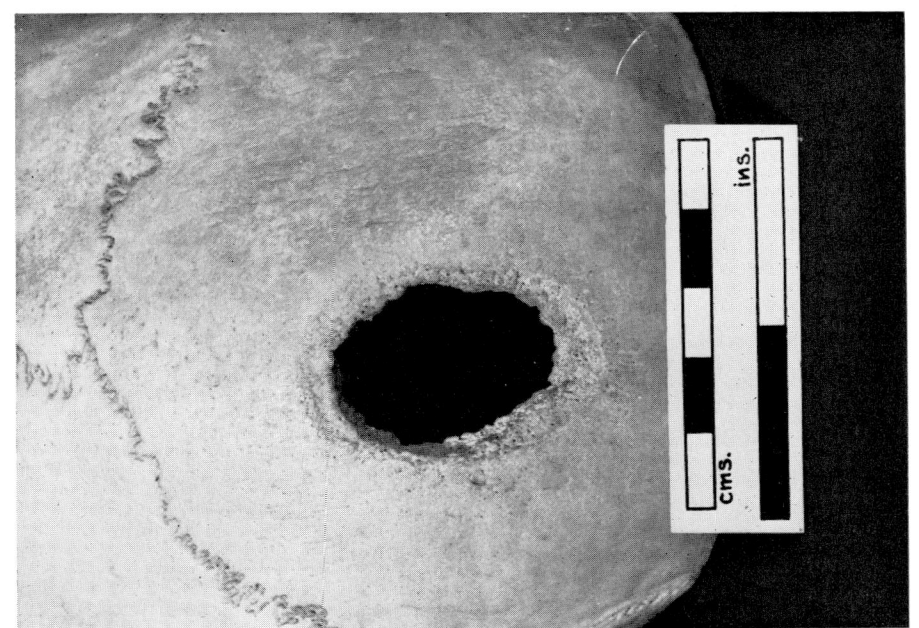

FIGURE 2a.

FIGURE 2. (a. and b). The Hailstone skull. Frontal trepanation in skull from Uganda. Note osteitis and healing at edges. Furnished by Dr. E. Ashworth Underwood and the Wellcome Historical Medical Museum, London, England, by courtesy of the Wellcome Trustees. Collected in the 1920's. Lugbara tribe, Uganda. 2(a) external surface. 2(b) endocranial surface.

was probably done with a cautery. Cautery of the scalp (and skull) is common in most primitive tribes of Africa.

There is much investigation yet to be done in South African natives so far as traditional medical practice is concerned, and perhaps with particular emphasis on the seeking out of trepanation procedures. Bloom (1962, 1964) has been on the verge of finding out, at first hand, information about this practice in the Zulu around Durban. He wrote (1962),

One *inyanga* suggested that when a patient's head is too hot, and aches severely, and throbs after having suffered a sharp blow, the cure is to make an opening on the area that was struck, and, when the bone is reached, to scrape it and then leave the wound to heal, by which time the headache will have subsided. This operation, similar to trepanning is known as *umhlahlo*, and the author regrets not having been able to discover the frequency with which it had been carried out by his informants, who were (perhaps understandably) reticent and evasive. The madness caused by a blow on the head is known as *ukuphambana*.

TREPANATION IN EAST CENTRAL AFRICA

The author's field experiences are confined mostly to East Central Africa (Kenya, Uganda, Tanganyika and Zanzibar) and particularly to Kenya, and it is proposed now to present more detail about trepan practices in that part of Africa.

A parietal bone showing a trephined hole was reported to have been found in Zanzibar and was said to be in the local museum (Ingrams, 1931). It could not be located by the curator in 1959. The Arab inhabitants and the aboriginals of the island in more recent times denied familiarity with true trepanation (Margetts, field observations 1955-9).

Trephining is practised by medicine-men in Somalia (Brotmacher, 1955; Drake-Brockman, 1912). A number of Somali surgeons questioned by me in 1958, near the Kenya-Somalia border in the vicinity of Wajir and Mandera, did not admit familiarity with the operation, but this does not mean the practice was not customary farther north, towards the horn of Africa. Trepanation is reported to be practised in Ethiopia though inhabitants at Moyale did not know about it.

The Ganda (Baganda, W a g a n d a), Nkole (Banyankole) and Soga (Basoga) of Uganda treated headache by cutting into the bone and then cauterizing (Fisher, 1911; Roscoe, 1921-3), and for fracture of the skull the Soga were said not to hesitate to operate and to probe, and remove pieces of bone (Roscoe, 1924). These tribes, quite acculturated, are inclined to be particularly reticent about divulging "tribal" informtaion, but so far as could be ascertained, they apparently do not trepan nowadays in the usual sense of perforating the intact skull (my field observations 1955-9).

The Lugbara (Lugbari, Lugwari, Laccara, Logbwari, Lugbwara, Louagonare, Lubare, Lugori, Lugwaret), in northwest Uganda, north of Lake Albert, trepanned the head "to let out the evil spirit which was causing an intractable headache." The operation was done with a knife having a four inch handle and a four inch blade with a double cutting edge and awl-shaped tip (Hailstone, 1961). The holed skull of a man who died after the operation was donated by Dr. John E. Hailstone to the Wellcome Museum, where it is displayed (Fig. 2a, b). In the Sudan, at the Kenya-Uganda border, West West North of Lake Rudolph, live the Topotha (Toposa, Taposa, Tabosa, Dabosa, Dabossa) and Driberg (1929) made passing mention that he had seen the operation of trepanning carried out by them "most efficiently."

Other doctors and anthropologists of experience in Uganda and the Sudan do not seem to be familiar with trepanation by the Lugbara and Topotha (correspondence).

TREPANATION BY THE KISII AND TENDE

Since the early days of British and German colonization, and no doubt before, traditional native healers in the hills east of Lake Victoria have operated on the cranium. The colonial service doctors, police and magistrates have documented the practice at length in government files. Trepanation in the narrowest sense, making a hole in the intact skull, is still carried out there by two Bantu tribes, the Kisii (Gisii or Gusii) of South Nyanza in Kenya, and, to a lesser extent, the Tende (Watende, Butende, Kulia, Kulya, Kuria, Kurya, Bukuria, Abakuria) farther south and into Tanganyika. How the custom of trepanation came to be so frequent in this area of East Africa is not understood. The practice was perhaps introduced by an itinerant patient or native medicine man, and was continued and promoted as a local custom which was encapsulated by the geography and limited migrations in the area. While government officials have always known about the practice of trepanning the skull by native medicine men in Kisii, the custom was apparently not recorded in other than government sources until 1958 (Coxon, 1962; Grounds, 1958; Margetts, 1958-61).*

---

*A moving picture film of the Kisii trepanation operation in a human subject is now available: "Maganga," distributed by Warner-Chilcott Laboratories, Morris Plains, New Jersey.

The Operation in the Kisii Highlands

Trephining in the Kisii highlands is done primarily for the complaint of headache (*ogwatigwa omotwe;** head, *omotwe;* ache or pain, *ogwatigwa*) after an injury to the head, with or without fracture of the skull.

Trepanation is not ordinarily done for headache without previous head injury, and the operation is *not* customary for psychosis, epilepsy, dizziness or spirit possession.

The operation is a simple but painstaking and sometimes long drawn out procedure, taking one to four hours. The operators are not usually specialized trepanners, but general medicine-men who do the operation as part of their total work. They are usually apprenticed until they are sufficiently skillful and responsible to do the operation on their own. They can learn from a non-relative or may be taught by their fathers. Women apparently do not practise the trephining art.

The "head surgeon" (singular *omobari omotwe*, plural *ababari emetwe;* surgeon, *omobari;* head, *omotwe*) may pray or go through other individualized magical procedures before the operation (*okobara*), but there is no set ritual. The patient's head may or may not be shaved and washed. He is then placed in a sitting or lying position and restrained. He is usually requsted to lie on a bed of leaves with a small log under his head. One operator rather preferred to have his patient lie on a small European style bed with his head over the edge, then to sandwich him by placing another bed upside down on top of him with a relative sitting at each corner of it! The scalp is incised in a linear or cruciate manner over the site of the

---

* Local Kisii dialect. There are no specific Kisii or Kiswahili terms for "trepanation"; the terminology is entirely descriptive.

headache and the flaps if need be are retracted by the fingers of assistants. As a rule nothing is added to the wound, but occasionally, a medicine (unidentified) is sprinkled in the site to assuage pain, and sometimes an agent like charcoal or local pressure is applied for haemostasis. Any fragments of bone, foreign bodies or clotted blood are removed, and any discolored bone or fracture line is removed by scraping the skull (*ekeore*) with a sharp scraping knife having an acutely curved tip, curved to avoid puncturing the dura and brain. The scraping is usually continued until the inner table is pierced and the brain membranes exposed. Less frequently, a saw is employed to make the hole (*enseke*) (Fig. 3a, b). Most operators are able to distinguish the cranial sutures from fracture lines, and seem to realize the danger of puncturing the dura, though in ignorance this is sometimes done in the case of subdural haematoma. Usually, both inner and outer tables of the skull are holed, but not always. After sufficient bone has been removed, the wound is washed with water. One *omobari* is said to have spewed water from his mouth onto the wound—no doubt an effective stream but not very aseptic.

Fat or butter may then be applied with a feather or other applicator. Sometimes herbal medicines are added to promote healing. The wound is usually allowed to heal by granulation; the scalp may rarely be sutured, in the common native fashion with figure-of-eight sutures over thorns. The operation is said to cause only very little pain, except initially as the soft tissues are cut and retracted. Anaesthesia is not employed. If any beer is drunk, it is probably taken by the operator, not the patient. "Pain killing" medicines are more likely to have a magical than a pharmacological effect. The *omobari omotwe* watches the patient carefully during the postoperative period, visiting him regularly until good prognosis is assured.

It is not uncommon for a patient to have multiple operations. Care must be taken to ensure the accuracy of claims made by primitives about the numbers of trepannings they have had. The native African is prone to exaggerate when giving his "medical history" (Margetts, 1958). One Kisii (*infra*) on different occasions varied the number of his borings from five to thirty. Multiple operations in Kisii are more likely to be enlargements of previous openings, rather than additional holes at different sites. The fee paid by the patient varies according to the demand of the *omobari,* to the circumstances, and perhaps least to the ability of the patient to pay—the amount charged, in money or in staple goods, has been known to vary from 40 to as much as 700 shillings in cash plus goods.

The mortality is low, perhaps 5 per cent. The bad results usually end up at government hospitals with local infections or meningitis, or are investigated by the police as deaths. A court case against an *omobari* usually starts out with a charge of murder, but almost invariably this is modified to one of manslaughter or practising medicine without a license. The courts generally recognize that the operator has no ill-intent in mind and perhaps is ignorant of the law, no excuse in a civilized and educated person, but often mitigating for an African illiterate.

THREE TREPANNED KISII NATIVES

It is now proposed to summarize briefly the case histories of three living Kisii tribesmen who had experienced trepanning operations. They were interviewed and photographed several times between 1958 and 1959.

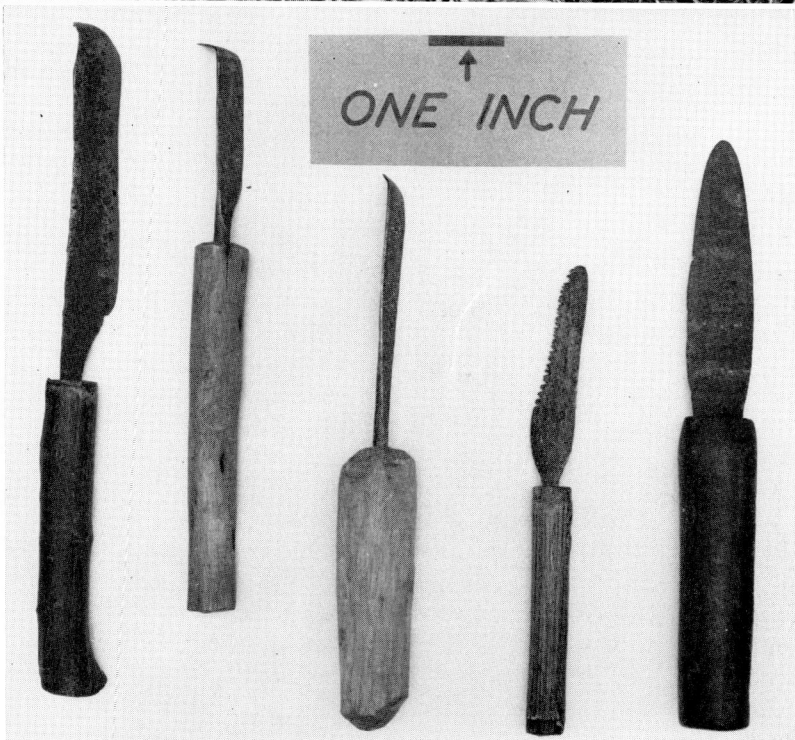

FIGURE 3 (a and b). Kenya skulls. (a) Two skulls, trepanned in life, with two knives, two scrapers with curved tips, and one saw. The skull on the left has been holed by scraping, that on the right by sawing. (b) Trepan instruments from Kenya. Collected by Dr. Brijendra Kumar Sood in 1956. Kisii tribe, Kenya.
Photographs by Margetts, 1958.

FIGURE 4. Case One. Right frontal trepanation in a surviving patient. Kisii tribe., Kenya. Photograph by Margetts, 1958.

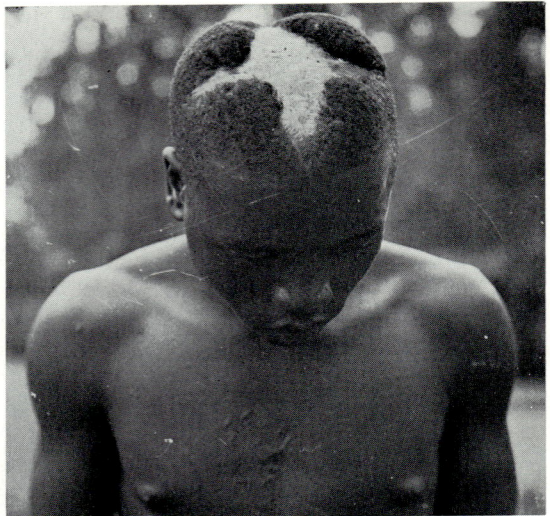

FIGURE 5. CASE TWO. 5 (a) Vertex of head in a surviving patient, showing extensive granulation within the area of a cruciate scalp incision. Note also, on the man's chest, "medicine cuts," with keloid formation. Medicine cuts are skin incisions in which protective or therapeutic magical or herbal substances have been rubbed. In this case they had nothing to do with the trepanation procedure. Kisii tribe, Kenya. Photograph by Margetts, 1958.

Case one (Fig. 4). This man was about thirty-five years old. He was a policeman on duty at the District Commissioner's office in Kisii. He suffered a blow to the forehead apparently without fracture. For persistent headaches several years after the injury he went to a native doctor who carried out a trepanation, probably by scraping with a curved knife. The patient was mentally normal, and at the time of examination he had no complaints of headache. Cured!

Case two (Fig. 5a, b, c). This man was serving six months for stock theft and four months for escape, in H.M. Prison, Nairobi. A tree fell on his head about 1 March 1958 and this produced a headache. He then had three trepan operations, about 15 April, 15 July and 15 November 1958. The reason given by him for *three* operations was "to finish the job." A vague answer like this was about all the information obtainable. His *omobari* was a famous doctor who had an enviable reputation that none of his patients would die. Before the operation, the patient was fed gruel containing a medicine "to stop the blood coming out too quickly." A powder was applied topically "to prevent pain." The patient claimed that only the cutting of the scalp was painful, scraping the bone did not hurt him. He said the operator used a curved scraping knife, and each session took about four hours. During this time he lay sandwiched between two European-style beds with his head sticking out, and helpful relatives sat one at each corner of the top bed, which was upside down over the patient. He said there was not much bleeding. After the procedure was completed, medicine was sprinkled over the wound and boiled fat was applied with a feather. A loose dressing was applied and the wound allowed to granulate.

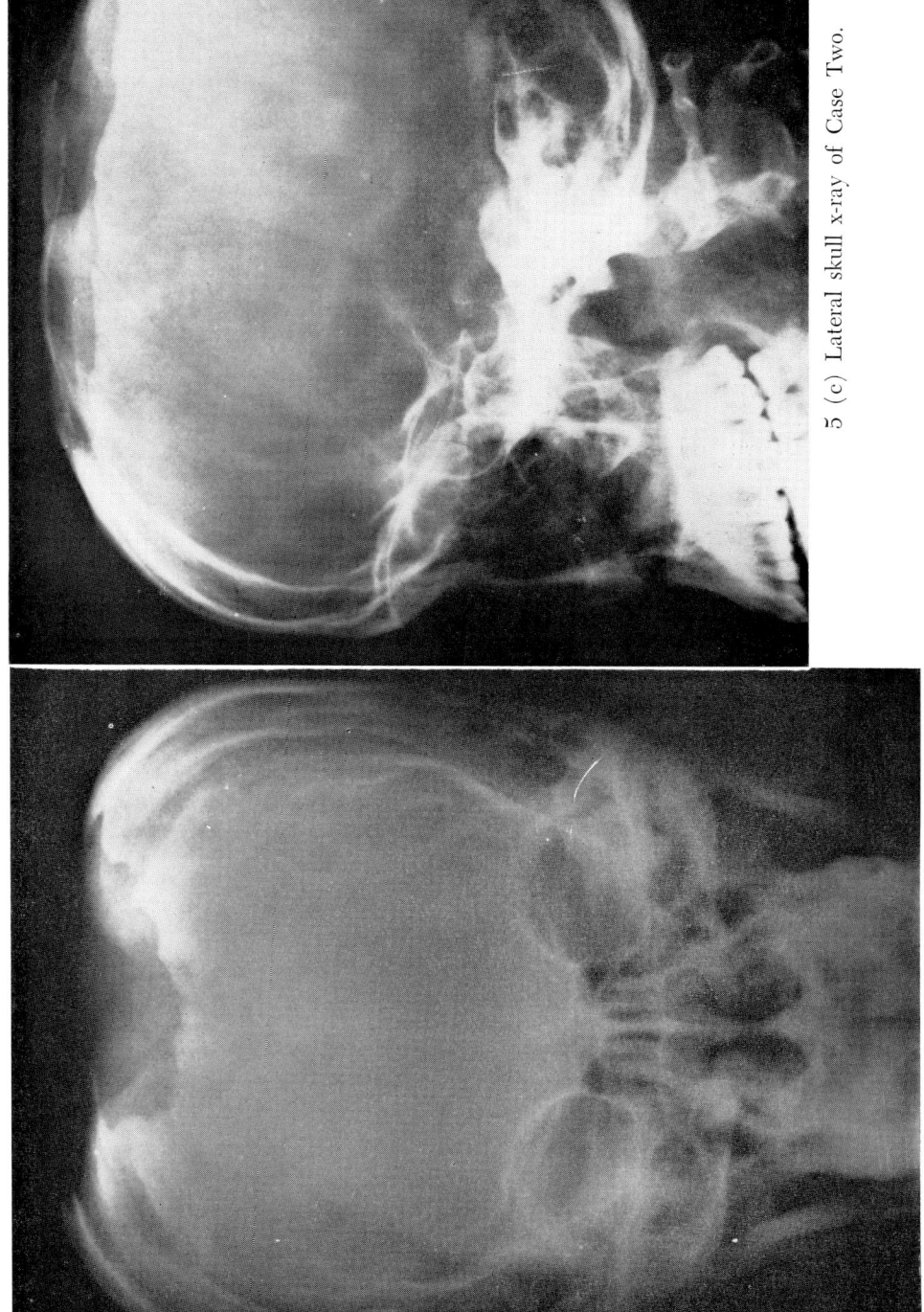

5 (b) Antero-posterior skull x-ray of Case Two.

5 (c) Lateral skull x-ray of Case Two.

Figures 5 (b and c): X-rays by Dr. Leslie R. Whittaker, King George VI (now Jomo Kenyatta National) Hospital, Nairobi; case of E. L. Margetts, 1958.

The operation was "to remove something that hurts the bone, he took the bad bone out."

The patient had to sell most of his property to pay the *omobari's* fee. He claimed he paid the doctor 700 shillings, one sheep, one goat, three chickens and three four-gallon drums of millet beer. This patient was somewhat withdrawn and depressed, but maybe this was because he was locked up in "King Georgi hoteli" (the prison at Nairobi). He did no work for fear it would "spoil" his head. He had no defineable physical illness and no diagnosable psychiatric one.

X-rays of the man's head revealed a single opening, both tables of the vertex of his skull being absent over nine square inches.

The man's sister was also trepanned twice by the same operator. Her husband had beat her on the head with a stick and caused her headaches.

Case three (Fig. 6a—f). The third example of trepanation was the most spectacu-

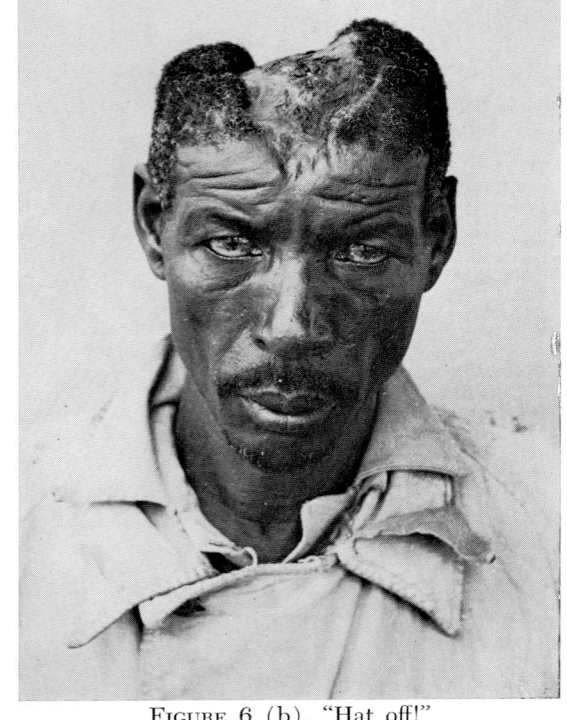

FIGURE 6 (b). "Hat off!"

FIGURE 6 (a). CASE THREE. "Hat on."

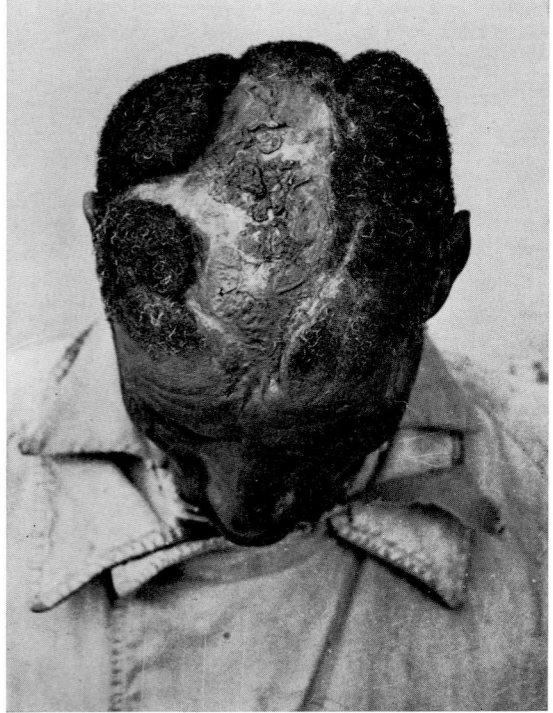

FIGURE 6 (c) "Hat off" again!
Photographs by Margetts, 1958.

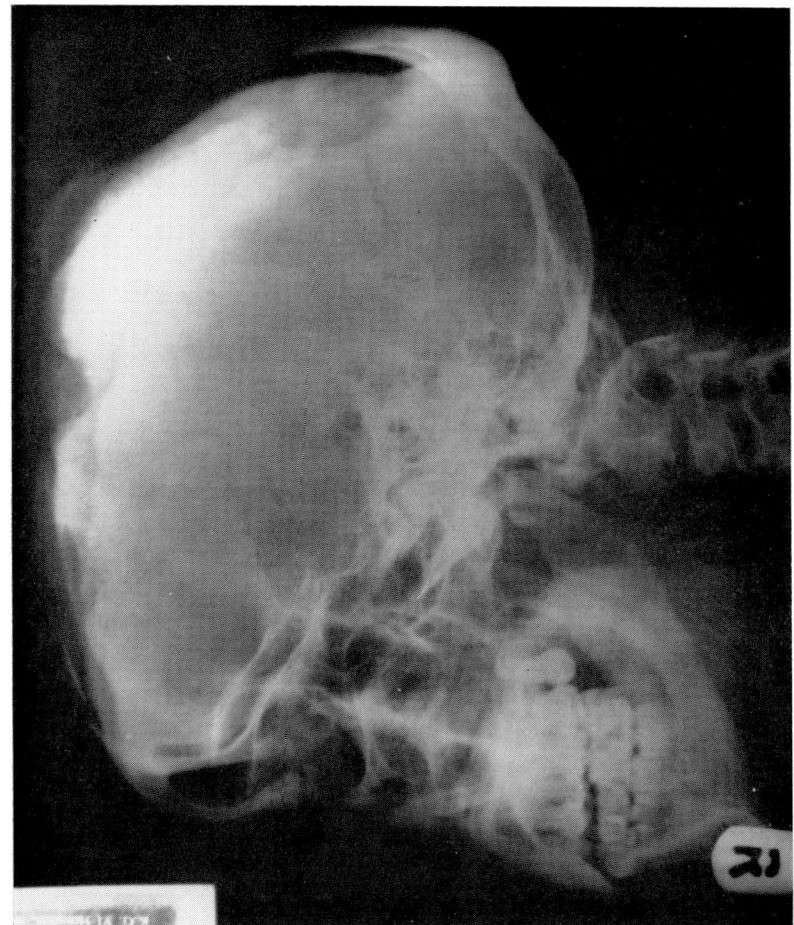

FIGURE 6 (e). Lateral skull x-ray of Case Three.

Figures 6 (d and e): x-rays by Dr. Leslie R. Whittaker, King George VI (now Jomo Kenyatta National) Hospital, Nairobi; case of E. L. Margetts, 1958.

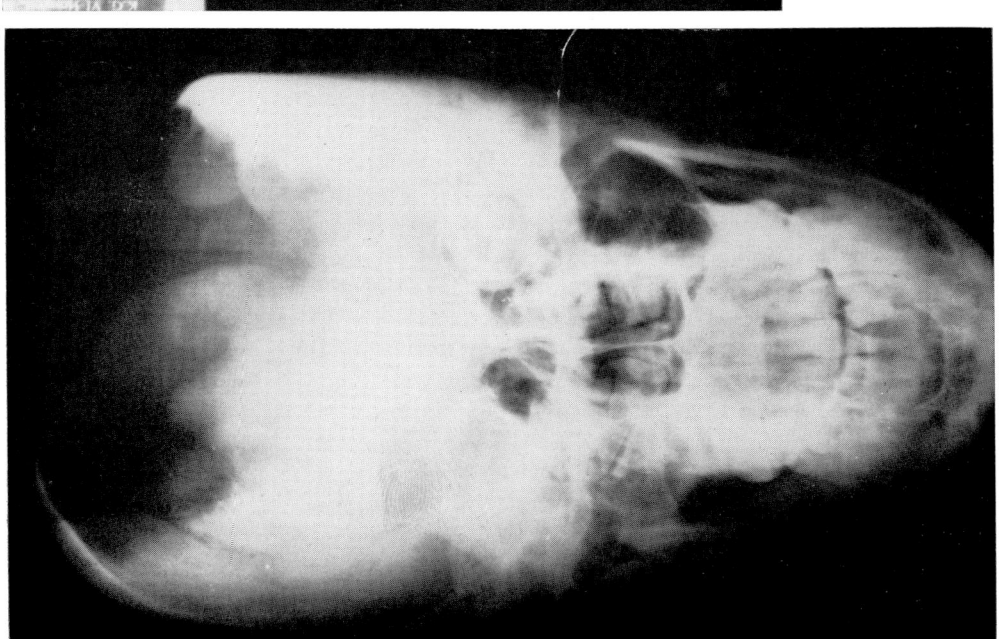

FIGURE 6 (d). Antero-posterior skull x-ray of Case Three.

Figure 6 (f). "Hat on" with his *omobari*. Note the Roman Catholic cross and beads around the neck of the *omobari*. He was a 'Christian sort of pagan.'

lar curiosity that one would ever hope to see. The author likes to describe him as "Hat on, hat off." When he had his battered old hat on, he looked unobtrusive and "normal" enough. But when he took his hat *off*, one was amazed to see the whole top of his head missing. X-ray photographs revealed an oval hole about 30 square inches in area in the vault of his skull!

This man was about fifty years old. He came to notice because he brought his son to see the European doctor. The son had been absent-minded, was neglecting his work in the father's shop, and was constantly reading the Bible and praying. He recovered without specific treatment. This was perhaps a schizophrenic withdrawal.

About 1940, "Hat on" was a tribal policeman. One day, while entering a hut, he hit his head on the door lintel. Subsequently, he developed vertex headaches. In 1945 he had a trepanation, and over the next seven or eight years he had several more operations. The exact number was not known. The fewest claimed by the patient was five—but his story varied and on one occasion he even claimed thirty. Enough of his skull was missing to make either the lower or the higher figure quite possible.

This man said his operations were very painful; nothing was given to assuage pain.

Because of the extensive deficit in "Hat on's" cranium, it was feared for his safety, so he was fitted with a plastic skull-cap to wear under his hat. At last report he was wearing this and quite proud of it.

The *omobari* who trepanned this patient was an interesting old fellow of about seventy or eighty years (Fig. 6 f). He had been doing the operation since he was a young man of twenty, and he could not recall how many patients he had trepanned, certainly well over a hundred. He was taught the technique by his father, and claimed he had never lost a patient. According to this operator, the only indication for trephining was headache following a blow. He used a curved scraper and disapproved of the saw. He expected to find a fracture at operation. He would know he was in the right place because the blood there was "black" (i.e., clotted). Many of the skulls opened probably never did show an actual fracture line, and the extensive amount of bone removed at some of the operations might have been the result of seeking for something that did not exist.

MOTIVES FOR TREPANNING THE SKULL IN PRIMITIVE SOCIETIES

The operation of trephining the cranium is of importance to archaeology, to anthropology and to medicine. There are few other subjects of study which integrate so well the interests of all these three specialties.

From the point of view of the medical sciences, trepanation is fascinating from the aspect of surgery, but perhaps even more so it is intriguing because of the psychology of the motives which lead to it.

To gain some understanding of these motives, it is necessary to examine the psychological and psychopathological makeup, firstly of the cultural milieu in which the operation is done, secondly of the patient and thirdly of the native surgeon.

While the custom of trephining seems a strange and particularly fascinating one to a stranger, it is commonplace and not considered to be a curiosity within the East African culture framework where it is practised. The *basic* mechanisms of thought, logic, emotional expression and behavior are the same in all cultures, and they are determined by genetic, instinctual and possibly by racial sources. These mechanisms vary in a fluid, *secondary* manner from culture to culture, depending on the advancement of the culture and on the opportunities and needs presented. These secondary mechanisms are determined by the differing impacts of the specific local culture, and by culture contact and exchange, *i.e.* the process of *including* within the social framework those values imposed or offered from outside societies and those requested or demanded by the specific inside society concerned, and of *excluding* by suppression, replacement or rejection some or all of the local values of the particular society under scrutiny.

The values of "civilization" in actual fact are very artificial, and consist mostly of added stresses and repressions which serve to mask the basic mechanisms common to all groups of mankind. The *basic* mechanisms of the mind would appear to be the same all the world over. Groups of men differ from their neighbors (which include the primitive and the more "evolved" or "involved" cultures) primarily because of the secondary factors, which can change within a chronological variable. This is the variable which is not taken into sufficient consideration in this day of world conflict. For people to change requires time.

From historical writings over the centuries, it is *known* that trepanation has been carried out for a variety of medical-surgical reasons. The indications included practically everything relating to the head—fractures, inflammations, localizations of pus and blood, epilepsy, madness, idiocy, moral degeneration, various head symptoms (headache, vertigo, deafness, etc.), the removal of foreign bodies from inside the head (real or faked, as in "stones of insanity" or *pierres de tête*) and the release of pressures, airs, vapors, humors, and demons and evil spirits.

When Neolithic European and pre-Columbian American trepanned skulls were studied, from the middle of the nineteenth century, the motives for the operation in prehistory had to be guessed at, since there were no writings available to explain why the procedure was done. It can be assumed that the motives in the prehistoric period were the same—multiple—as the ones that followed in historical times.

It is known also that the reasons offered for trephining in present day primitive cultures are not always the same. For instance, the Kisii motive is to relieve headache after a blow. Yet a few hundred miles away, the Lugbara motive is to let out an evil spirit.

If all the "reasons" for trepanation are put together, the common mechanism behind all the conditions in the list is *to remove something*. There is a *continuum*, from immaterial to material, from magic and thaumaturgy to science—evil spirits and demons, vapors, humors, air, hypothetical "pressures," actual pressures, real and fictional foreign bodies, "unknown substances," pus, blood, and finally, pieces of bone.

A given culture may utilize any one or several of these "reasons"—at the same time or at different chronological periods. The reasons chosen depend on the psychomental evolution of the individual and of his specific society at the particular time under observation. The reasons may vary whenever there is regression of the individual or his culture. Moreover, the custom may have "epidemic" significance, becoming a fad and a folly within the culture, depending on publicity and on the degree of regression.

There is only one motive for the operation which is not aimed at the "removal of something." This is the interesting modification of *trepanning as an amulet,* to protect the patient against developing a disorder which trepanning is known to "cure." Such was the "prophylactic" trepanning done in the Bismarck Islands (*supra*).

While not a motive to do the therapeutic operation, the obtaining of *cranial amulets or rondelles* has either magical, protective value or perhaps to a lesser extent a "curio" or "relic" significance. Amulets of trephined crania are not common in present day cultures where the operation is done.

Murder with *post-mortem* breaking of the skull in order to get brain matter and bone for witchcraft purposes and as ingredients of magical medicine is well known throughout Africa. Any regularly outlined opening, particularly a circular one, is more likely ante-mortem trepanning than post-mortem smashing of the cranium. If the edges of the opening are smooth, pending microscopic and X-ray analysis, and excluding disease and weathering, it is almost certain that the opening is the result of ante-mortem trepanning.

Intact skulls are used all over Africa for various magical purposes. Skulls may be holed post-mortem also for magico-religious purposes, as in West African ancestor cults (Stéphen-Chauvet, 1936).

The psychological makeup of the patient seeking a trepanning operation may now be summarized. The patient can demand relief because he believes that evil spirits are in his head ("spirit intrusion"), plaguing him as a result of his personal slight to a deceased relative, or to his transgression of a tribal prohibition or "taboo." He may believe that foreign bodies, such as stones, pieces of glass, or other material objects, have been projected into his head ("object intrusion") by an evil-working practitioner of black magic. This belief may exist in quite a *normal* person of a culture which customarily believes in this kind of magic, or it may be in a person who has paranoidal ideas relating to his head. His mental aberration may press him to seek and *demand* action by a medicine man.

Such patients are not necessarily masochistic at all. They have some sort of belief about the head relating to the presence there of something, and their magical explanation of it leads to the seeking of a magical cure. Even a natural illness causing headache, such as malaria, is explained on a magical basis of something *in* the head, which might be let out or removed through a hole.

There is a striking lack of overt anxiety and fear in the patient seeking a trephining. The reason would appear to be *security* generated by the forceful and positive suggestion and promise of cure by the native surgeon. The patient does not think in terms of possible mortality, he is secure in his hope or belief that the complaint will be relieved.

And now to review the possible motives of the trepanner himself. In modern primitive culture groups, he is usually a medicine man of reputation and high standing. His ministrations are meant to be for the good of his patient. On occasion, of course, he may be a deceitful and dishonest person, but this is unusual. He *believes* that he can cure his patient. The average medicine man and practitioner of white magic must be differentiated from the psychopathic minority who are cunning and dishonest, and who seek gain and personal gratification without regard to the patient's welfare.

The trepanner's beliefs are based partly on rational pathophysiology, partly on magic. He proceeds because he *knows* that in most cases the patient is relieved of his complaint, whatever it may be.

He may, on rational surgical grounds, remove bone fragments, bad blood, and the like. He may likewise erase a fracture line. A sick person complaining of feelings of *pressure* (variously described) may be cured by relieving this pressure. Thus any spheroidal organ, like the head, leads to the boring of a hole in it in order to afford an exit for the pressure-inducing agent. The other visible spheroidal organ, the eye, has for a similar reason been trephined within highly evolved cultures having developed a more elaborate surgical technology. Similarly, swellings resembling spheres may be punctured, as in case of the urinary bladder, ascitic abdomen, and the like.

In many primitive psychologies, the head is the *dwelling place* of the intellect and of the soul. The skull is a home, a house. The emotions and vital principle are located elsewhere, usually in the heart. It is reasonable then, to think of a hole in the head as an exit for the soul, or an inhabiting foreign spirit, as the door in a house. Mental activity associated with the head (thinking, perception) are located in the head and brain. Any abnormality of this mental activity may perhaps be treated by utilizing the natural orifices to put medicines into the interior of the cranium or to draw illness out of it. The primitive may believe that the auditory canals and the nostrils particularly lead directly to the brain. Even a better way in or out of the skull would be to make an artificial hole in it.

Not all culture groups who believe in the entrance of a maleficent spirit into the head, or the projection of foreign bodies, or semi-rational illnesses like "water in the head" or "maggots in the head" trepan the skull to seek relief, even though their basic theories of psychopathogenesis may be similar. This can only be explained by chance and by scattering as a result of migration.

As in the patient, the experienced native trepanner seems to be relatively free of anxiety about the outcome of his operation. Like the experienced and confident doctor anywhere, he is sure of his skill, and has developed a philosophy which protects him if things go wrong.

ACKNOWLEDGMENTS

Acknowledgment with thanks is made to a number of organizations which have supported the author's research over the years into the history of medicine, prehistoric archaeology, and primitive medicine—the Ciba Company of Canada (1950-1), McGill University (Faculty grants, 1952-4), the Humanities Research Council of Canada (1955), but most particularly to the government of the (then) Colony and Protectorate of Kenya (1955-9) and to the Director of Medical Services, Dr. T. Farnsworth Anderson, who encouraged the author's interests and study of present day traditional native healers.

REFERENCES

ACKERKNECHT, ERWIN, H., 1946: Contradictions in primitive surgery. *Bull. Hist. Med., 20:*184.

ACKERKNECHT, ERWIN H., 1947: Primitive surgery. *Amer. Anthrop.*, N.S., *49:*25.

AKESTER, A. ROGER, 1958: Tibesti—land of the Tebou. *Geograph. Mag. 31:*12.

AKESTER, A. ROGER, 1959: Trephining by a Tibu medicineman in Tibesti, 1957. *Vide Oakley et al.*, p. 94.

ALBU, ALBERT, 1889: *Die Geschichte der Trepanation und ihre Indikationen für die Jetztzeit.* Inaugural-Dissertation. Berlin, Emil Streisand.

ANONYMOUS, 1887: Prehistoric surgery. *Westminster Rev., 128:*538.

ANONYMOUS (re HRDLIČKA, ALEŠ), 1935: Alaska indians had brain surgeons 2,000 years ago. *Science News Letter. 28:*377.

BAILEY, PERCIVAL, 1961: Anecdotes from the history of trephining. *J. Int. Coll. Surgeons., 35:*382.

BALADO, MANUEL, 1930: Reseña descriptiva y bibliográfica sobre la trepanación prehistórica y en los pueblos salvajes. *La Semana Médica.*, Buenos Aires. 37:892.

BALLANCE, SIR CHARLES ALFRED, 1922: *The Thomas Vicary lecture. A glimpse into the history of the surgery of the brain.* London, Macmillan.

BANDELIER, ADOLPH FRANCIS, 1904: Aboriginal trephining in Bolivia. *Amer. Anthrop.*, N.S. 6:440.

BANDELIER, ADOLPH FRANCIS, 1906: Über Trepanieren unter den heutigen Indianern Bolivias. In: *Internationaler Amerikanisten-Kongress Vierzehnte Tagung Stuttgart 1904.* Stuttgart, W. Kohlhammer. Erste Hälfte. s. 81.

BANDELIER, ADOLPH FRANCIS, 1910: *The islands of Titicaca and Koati.* New York, Hispanic Society of America. *Vide* p. 172, 241.

BARTELS, MARTIN, 1913: Die Ausführung der geradlinigen präkolumbianischen Trepanation. *Deuts. Med. Wochenschr., 39:*2311.

BAUDOUIN, MARCEL, 1913: Trépanations préhistoriques de France comparées aux trépanations actuelles des Canaques. *Paris Chirurgical, 5:*47.

BAZZOCCHI, GIUSEPPE, 1947: Trapanazioni incaiche. *Chir. Ital.*, Belluno. *1:*434.

BEATTIE, JOHN, 1930: A note on two skulls from Tenerife. *Amer. J. Phys. Anthrop., 14:*447.

BELLONI, LUIGI, 1965: La trapanazione del cranio nella preistoria e nella storia. *Minerva Medica,* 56: n. 74; varia 1290; interventi 1305.

BERTHERAND, EMILE-LOUIS, 1855: *Médecine et hygiène des Arabes.* Paris, Germer Baillière. *Vide.* p. 40.

BERTHOLON, L. J. and CHANTRE, E., 1913: *Réchérches anthropologiques dans la Bérberie Orientale/Tripolitaine, Tunisie, Algérie.* Lyon, A. Rey. 2 vols. *Vide* vol. 1, p. 493.

BLOOM, LEONARD, 1962: *Some psychological concepts of urban Africans.* Thesis submitted in partial fulfillment of the requirements for the degree of M. Soc. Sc. in the department of psychology, University of Natal. Durban. Mimeographed and bound. *Vide* p. 49.

BLOOM, LEONARD, 1964: Some psychological concepts of urban Africans. *Ethnology,* 3:66.

BOCKENHEIMER, PH., 1922: Über Bregmanarben und ihre mutmassliche Entstehung nach Untersuchungen an Guanchenschädeln und nach Tierexperimenten. *Z. Ethnol.,* 54:130.

BOULONGNE, ALFRED, 1869: *Le Monténégro/le pays et ses habitants.* Paris, Victor Rozier. *Vide.* p. 45, 54.

BRAUNE, MAX, 1875: *Die Geschichte der Trepanation.* Inaugural-Dissertation. Berlin, Gustav Lange.

BROCA, PAUL, 1867: Cas singulier de trépanation chez les Incas. *Bull. Soc. Anthrop. Paris.* 2ᵉ sér. 2:403.

BROCA, PAUL, 1867: Trépanation chez les Incas. *Bull. Acad. Med.* Paris. 32:866.

BROCA, PAUL, 1877: *Sur la trépanation du crâne et les amulettes crâniennes à l'époque néolithique.* Paris, Ernest Leroux.

BRODSKY, ISADORE IRVINE, 1936: The operation of cerebral decompression as practised by the natives of New Britain fifty years ago. *Med. J. Aust.* 2:817.

BRODSKY, ISADORE IRVINE, 1938: The trephiners of Blanche Bay, New Britain, their instruments and methods. *Brit. J. Surg., 26:*1.

BROOKE, WINIFRED, 1959: Trephining by a medicineman in Bolivia, 1950. *Vide* Oakley, *et al.*, p. 93.

BROTHWELL, DON R., 1963: *Digging up bones/the excavation, treatment and study of human skeletal remains.* London, British Museum (Natural History).

BROTMACHER, LEON, 1955: Medical practice among the Somalis. *Bull. Hist. Med.;* 29:197.

BROWN, GEORGE, 1910: *Melanesians and Polynesians/their life-histories described and compared.* London, Macmillan. *Vide* p. 184.

BUCKLAND, A. W., 1881: Surgery and superstition in neolithic times. *J. Anthrop. Inst.,* 11:7.

BUSACCHI, VINCENZO, 1935: Le trapanazione del cranio nei popoli preistorici (neolitici e precolombiani) e nei primitivi moderni. *Atti e Memorie dell'Accademia di Storia dell'Arte Sanitaria.* Serie II, 1:64, 128.

BUSCHAN, GEORG, 1900: Trepanation. In: Matschie, Kustos P. (Herausgeber). *Handwörterbuch der Zoologie, Anthropologie und Ethnologie.* Breslau, Eduard Trewendt. Bd. 8, s. 97.

BUSCHAN, GEORG, 1934: Über primitive Trepanation. *Psychiatrisch-Neurologische Wochenschrift,* 36:308.

BUTT, AUDREY, (personal communications from 1961, and extracts from Pitt Rivers Museum subject catalogue, on African trepanation).

CANU, GEORGES, 1908: *Trépanation préhistorique et rondelles craniennes.* Paris thèse 328. Paris, Henri Jouve.

CHERVIN, ARTHUR, 1907-08: *Anthropologie bolivienne.* Paris, Imprimerie Nationale. Librairie H. Le Soudier. 3 vols.

CHIL Y NARANJO, 1878: Mémoire sur l'origine des Guanches ou habitants primitifs des îles Canaries. *Congrès International des sciences anthropologiques tenu à Paris du 16 au 21 août,* p. 167.

CHIPAULT, A. and DALEINE, E., 1893: Notes iconographiques sur l'histoire de la trépanation. *Nouvelle Iconographie de la Salpêtrière.* 6:292.

CIBA SYMPOSIA, 1939: (Various authors: Wakefield, E. G. and Dellinger, Samuel C.; Hrdlička, Aleš; Wehrli, G. A.; Robinson, Victor; Gerlitt, John; G. R.) Trepanation. *Ciba Sympos.* 1: Number 6; cover, 165-200.

CIBA ZEITSCHRIFT, 1936: (Various authors: Wölfel, Dominik J.; Wehrli, G. A.; Busch, J.; Gerlitt, John). Die Trepanation. *Ciba Z.,* 4: Nummer 39; cover, 1325-1355.

CLINE, WALTER BUCHANAN, 1950: *The Teda of Tibesti, Borku, and Kawar in the Eastern Sahara.* General series in anthropology/number 12. Menasha, Wisconsin, George Banta. *Vide* p. 31.

COSGROVE, C. B., 1929: A note on a trephined Indian skull from Georgia. *Amer. J. Phys. Anthrop.,* 13:353.

COURVILLE, CYRIL BRIAN, and ABBOTT, KENNETH H., 1942: Cranial injuries of the precolumbian Incas/with comments on their mechanism, effects and lethality. *Bull. Los Angeles Neurol. Soc.,* 7:107.

COURVILLE, CYRIL BRIAN, 1944: Some notes on the history of injury to the skull and brain. *Bull. Los Angeles Neurol. Soc.,* 9:1.

COURVILLE, CYRIL BRIAN, 1951: Injuries to the skull and brain in Oceania/with reference to the mechanism and nature of such injuries, the measures used in protection against them, and their treatment, particularly among the Melanesians. *Bull. Los Angeles Neurol. Soc.,* 16:14.

COURVILLE, CYRIL BRIAN, 1959: Cranioplasty in prehistoric times. *Bull. Los Angeles Neurol. Soc.,* 24:1.

COXON, ANN, 1962: The Kisii art of trephining. *Guy's Hosp. Gazette,* 76:263.

CRUMP, JOHN. A., 1901: Trephining in the South Seas. *J. Roy. Anthrop. Inst.,* 31:167. Discussion by Horsley, Victor.

DALAND, JUDSON, 1935: Depressed fracture and trephining of the skull by the Incas of Peru. *Ann. Med. Hist.,* N.S. 7:550.

DALLONI, M., 1935: Ethnologie. In: *Mission au Tibesti.* Mémoires de l'Académie des Sciences de l'Institut de France. Paris, Guathier-Villars, pp. 405-449. *Vide* tome 2, p. 433.

DE BAYE, BARON J., 1880: *L'archéologie préhistorique.* Paris, Ernest Leroux. *Vide* p. 213.

DE CELINSKI, M. F. T., 1838: *De Trepanatione Cranii.* Dissertatio Medico-Chirurgica. Berolini, Typis Nietackianis.

DECHELETTE, JOSEPH, 1908-1927: *Manuel d'archéologie préhistorique celtique et gallo-romaine.* Paris, Auguste Picard. 4 vols. *Vide* vol. 1 entitled *Archéologie préhistorique.* *Vide* p. 477.

DE NADAILLAC, MARQUIS, 1892: *Manners and monuments of prehistoric peoples.* Translated by Nancy Bell (N. D'Anvers). New York, G. P. Putnam's. *Vide* p. 257.

D'HARCOURT, RAOUL, 1939: *La médecine dans l'ancien Pérou.* Paris, Libraire Maloine. *Vide* p. 129, 196.

DRAKE-BROCKMAN, RALPH E., 1912: *British Somaliland.* London, Hurst and Blackett. *Vide* p. 160.

DRENNAN, MATTHEW ROBERTSON, 1937: Some evidence of a trepanation cult in the Bushman race. *S. Af. Med. J.,* 11:183.

DRENNAN, MATTHEW ROBERTSON, (from 1960). Personal communications.

DRIBERG, JACK HERBERT, 1929: *The savage as he really is.* London, George Routledge. *Vide* p. 63.

DUNEAU, O., 1886: Summary of: Védrènes, A. 1885. *Rev. d'Anthrop.,* Paris. 15:3 sér; tome 1, 536.

DURHAM, M. EDITH, 1909: Some Montenegrin manners and customs. *J. Roy. Anthrop. Inst.,* 39:85.

EDITORIAL, 1892: Prehistoric trepanning. *Cornhill Magazine.* 2nd ser. 19:512.

EDITORIAL, 1916: Ancient and primitive trephining. *Lancet,* 1:683.

ELLA, SAMUEL, 1874: Native medicine and surgery in the south sea islands. *Med. Times and Gazette,* 1:50.

ELLIS, WILLIAM, 1829: *Polynesian researches, during a residence of nearly six years in the south sea islands; including descriptions of the natural history and scenery of the islands—with remarks on the history, mythology, traditions, government, arts, manners, and customs of the inhabitants.* London, Fisher, Son, & Jackson. 2 vols. Vide vol. 2, p. 276.

EMORY, KENNETH P., 1942: Oceanian influence on American Indian culture. Nordenskiöld's view. J. Polynesian Soc., 51:126.

ENGLAND, IVOR A., 1962: Trephining through the ages. *Radiography,* 28:301.

FARQUHARSON, R. J., 1881: Amulets and post-mortem trepanation. Amer. Antiq. Oriental J., Chicago. 3:330.

FISHER, A. B., (née Ruth Hurditch). (1911): *Twilight tales of the black Baganda.* London, Marshall Brothers. Vide p. 61.

FLETCHER, ROBERT, 1882: On prehistoric trephining and cranial amulets. Contributions to North American Ethnology. U.S. Geographical and Geological Survey of the Rocky Mountain region. Washington, Government Printing Office. Vol. 5. pp. 1-32

FORD, EDWARD, 1937: Trephining in Melanesia. Med. J. Aust. 2:471.

FORGUE, ÉMILE, 1938: Histoire de la chirurgie. In: Laignel-Lavastine. *Histoire générale de la médecine, de la pharmacie, de l'art dentaire et de l'art vétérinaire.* Paris, Albin Michel. 1936, 1938, 1949. 3 vols. Vol. 2, pp. 350-450.

FREEMAN, LEONARD, 1918: Primitive surgery of the western hemisphere. JAMA, 70:443.

FREEMAN, LEONARD, 1924: Surgery of the ancient inhabitants of the Americas. Art and Archaeology. Washington, D.C., 18:21.

FRILLEY, G. and WLAHOVITJ, JOVAN, 1876: *Le Monténégro contemporain.* Paris, E. Plon. Vide p. 427.

FRITSCH, G., 1907: Über einen zweimal trepanierten Schädel. Z. Ethnol. 39:702.

GENNA, GIUSEPPE E., 1930: La trapanazione del cranio nei primitivi contributo alla sua conoscenza nella preistoria in Italia. Riv. di Antropol/Atti Della Società Romana di Antropologia. 29:139.

GILLMAN, HENRY, 1875: Certain characteristics pertaining to ancient man in Michigan. Annual Report of the Board of Regents of the Smithsonian Institution for 1875. Washington. pp. 234-245.

GILLMAN, HENRY, 1885: Further confirmation of the post-mortem character of the cranial perforations from Michigan mounds. Amer. Nat. 19:1127.

GOLDSMITH, WILLIAM M., 1922: "The Catlin mark" the inheritance of an unusual opening in the parietal bones. J. Hered., 13:69.

GOLDSMITH, WILLIAM M., 1945: Trepanation and the "Catlin mark." Amer. Antiq., 10:348.

"G. R." Vide ROSEN, GEORGE.

GRAÑA Y. REYES, FRANCISCO, ROCCA, ESTEBAN D. and GRAÑA R., LUIS, 1954: *Las trepanaciones craneanas en el Perú la época pre-Hispánica.* Lima, Imprenta Santa Maria.

GREENMAN, E. F., 1926: A report on Michigan archaeology. Amer. Anthrop. 28:310.

GRIMM, HANS, 1955: Vierfache Trepanation bei einer Inkamumie. Zbl. Neurochir., 15:212.

GROUNDS, JOHN G., 1958: Trephining of the skull amongst the Kisii. East African Med. J. 35:369.

GUÉRIN, DANIEL CHARLES JOSEPH, 1951: *Le problème des trépanations préhistoriques/interprétation actuelle.* Thèse pour le doctorat en médecine (diplome d'état). No. 980. Paris, D. P. Taïb.

GUIARD, ÉMILE, 1930: *La trépanation cranienne chez les néolithiques et chez les primitifs modernes.* Paris, Masson.

GUTHRIE, DOUGLAS, 2nd ed. 1958: *A history of medicine.* London, Thomas Nelson. Vide p. 8

GUTHRIE, GEORGE JAMES, 1842: *On injuries of the head affecting the brain.* London, J. Churchill. Vide p. 75.

HADFIELD, EMMA, 1920: *Among the natives of the Loyalty group.* London, Macmillan.

HAGEN, BERNHARD, 1899: *Unter den Papua's. Beobachtungen und Studien über Land und Leute, Thier-und Pflanzenwelt in Kaiser-Wilhelmsland.* Wiesbaden, C. W. Kreidel. Vide p. 257.

HAILSTONE, JOHN EDWARD, 1961: Personal communication.

HANDY, E. S. CRAIGHILL, 1923: *The native culture in the Marquesas.* Bernice P. Bishop Museum. Bulletin 9. Bayard Dominick expedition publication number 9. Honolulu. Vide p. 269.

HENRY, TEUIRA, 1928: *Ancient Tahiti/based on material recorded by* J. M. Orsmond. Bernice P. Bishop Museum. Bulletin 48. Honolulu. Vide p. 145.

HERODOTUS. The history of Herodotus. Translated by George Rawlinson. Everyman's Library nos. 405-6. London, J. M. Dent. 1910 (transl. 1858). 2 vols. Vide vol. 1, p. 360.

HEYERDAHL, THOR, 1941: Turning back time in the south seas. Nat. Geograph. Mag. 79:109.

HEYERDAHL, THOR, 1952: *American Indians in the Pacific/the theory behind the Kon-Tiki expedition.* London, George Allen & Unwin. Vide p. 655.

HILL-TOUT, CHARLES, 1953: The great Fraser midden. In: *The great Fraser midden.* Vancouver Art, Historical and Scientific Association. Vancouver, The Museum. Vide p. 8.

HILTON-SIMPSON, MELVILLE WILLIAM, 1913: Some Arab and Shawia remedies and notes on the trepanning of the skull in Algeria. J. Roy. Anthrop. Inst., 43:706.

HILTON-SIMPSON, MELVILLE WILLIAM, 1920: Shawia surgery. Proc. Roy Soc. Med., 13: Supplement; 47.

HILTON-SIMPSON, MELVILLE WILLIAM, 1921: *Among the hill-folk of Algeria/journeys among the Shawia of the Aurès mountains.* London, T. Fisher, Unwin. Vide p. 180.

HILTON-SIMPSON, MELVILLE WILLIAM, 1922: The Berbers of the Aures mountains, Algeria: a study of a primitive people. Scottish Geographical Magazine., 38:145.

HILTON SIMPSON, MELVILLE WILLIAM, 1922: *Arab medicine and surgery/a study of the healing art in Algeria.* London, Oxford University Press. *Vide* p. 30.

HINSDALE, WILBERT B., 1924: An unusual trephined skull from Michigan. *Papers of the Michigan Academy of Science, Arts and Letters.* 4:13.

HINSDALE, WILBERT B. and GREENMAN, EMERSON F., 1936: Perforated Indian crania in Michigan. *Occasional contributions from the Museum of Anthropology of the University of Michigan,* Ann. Arbor. No. 5. pp. 1-16.

HINSDALE, WILBERT B. and CAPPANNARI, STEPHEN C., 1940: Distribution of perforated human crania in the western hemisphere. *Papers of the Michigan Academy of Science, Arts and Letters,* 26:459.

HOLBROOK, W. C., 1877: Examinations of Indian mounds on Rock River, at Sterling, Illinois. *Amer. Nat.* 11:688.

HOLLÄNDER, EUGEN, 1927: *Äskulap und Venus / eine Kultur-und Sittengeschichte im Spiegel des Arztes.* Berlin, Propyläen. *Vide* p. 128.

HOOTON, ERNEST A. 1925: *The ancient inhabitants of the Canary Islands.* Harvard African Studies, Volume VII. Cambridge, Peabody Museum of Harvard University. *Vide* p. 153, 299.

HORNE, JOHN FLETCHER, 1894: *Trephining in its ancient and modern aspect.* London, John Bale.

HRDLIČKA, ALEŠ, 1906: Medicine and medicine-men. In: Hodge, Frederick Webb (Editor), *Handbook of American indians/North of Mexico.* Washington, Smithsonian Institution/Bureau of American Ethnology. Bulletin 30, Government Printing Office. Part 1. p. 836.

HRDLIČKA, ALEŠ, 1939: Trepanation among prehistoric people, especially in America. *Ciba Sympos.* 1:170.

HUDSON, ROBERT S., 1877: On the use of the trephine in depressed fractures of the skull. *Brit. Med. J.,* 2:75.

INGRAMS, WILLIAM HAROLD, 1931: *Zanzibar/its history and its people.* London, H. F. and G. Witherby. *Vide* p. 443.

KEITH, SIR ARTHUR, 1925: *The Antiquity of Man.* 2nd ed. London, Williams and Norgate, 2 vols. *Vide* vol. 1, p. 20.

KIDD, G. E., 1930: A case of primitive trephining. *Museum and Art Notes,* Vancouver. 5:85.

KIDD, G. E., 1946: Trepanation among the early Indians of British Columbia. *Canad. Med. Ass. J.,* 55:513.

KIDD, G. E., 1953: A case of primitive trephining. In: *The great Fraser midden.* Vancouver Art, Historical and Scientific Association. Vancouver. The Museum, pp. 19-21.

KRIVYAKIN, I., 1887: Trepanatsija bez trepana. *Voyenno-Meditsinskiǔ Zhurnal,* St. Petersburg. 160:145.

KRIVYAKIN, I., 1888: Trephining in Daghestan. *Lancet,* 1:138. (résumé in English from the Russian).

LALOY, L., 1907: La trépanation en Nouvelle-Guinée. *L'Anthropologie,* 18:718 (review of article by Fritsch 1907 *Vide*).

LARREY, FÉLIX HIPPOLYTE, LE BARON, 1867: (no title; note on trepanation of the skull in the Aures, referring to the reports of Amédée Paris and Th. Martin. *Bull. Acad. Méd.* Paris. 32:871.

L.A.S., 1879: (review of Lucas-Championnière, Just, 1878). *Amer. J. Med. Sc.* 78:489.

LASTRES, JUAN B. and CABIESES, FERNANDO, 1960: *La trepanacion del cráneo en el Antiguo Peru.* Imprenta de la Universidad Nacional Mayor de San Marcos, Lima.

LE BARON, JULES, 1881: *Lésions osseuses de l'homme préhistorique en France et en Algérie.* Thèse pour le doctorat en médecine présentée et soutenue le vendredi 1er juillet. Thèse no. 262. Paris. *Vide* p. 66.

LEECHMAN, DOUGLAS, 1944: Trephined skulls from British Columbia. *Trans. Roy. Soc. Canada.* Series III, 38, Sect. II: 99.

LEHMANN-NITSCHE, ROBERT, 1903: Notes sur des lésions de crânes des îles Canaries analogues à celles du crâne de Menouville et leur interprétation probable. *Bull. Mém. Soc. Anthrop.* Paris., 5 sér: 4:492.

LEHMANN-NITSCHE, ROBERT, 1905: Les lésions bregmatiques des crânes des îles Canaries et les mutilations analogues des crânes néolithiques français. *Bull. Mém. Soc. Anthrop.* Paris., 5 sér: 6:220.

LUCAS-CHAMPIONNIÈRE, JUST, 1878: *Étude historique et clinique sur la trépanation du crâne la trépanation guidée par les localisations cérébrales.* Paris, V. A. Delahaye. *Vide* L.A.S., 1879.

LUCAS-CHAMPIONNIÈRE, JUST, 1885: (Rapport. Sur un mémoire de M. Linon, médecin-major: *Sur cinq observations de plaie de tête;* sur une observation de M. Bélime, médecin-major: *Trépanation suivie de succès. Localisation cérébrale. Trépanation chez les Kabyles.*). *Bull. Soc. Chir.* Paris, n.s., 11:592.

LUCAS-CHAMPIONNIÈRE, JUST, 1912: *Trépanation néolithique, trépanation pré-Colombienne, trépanation des Kabyles, trépanation traditionelle.* Paris. G. Steinheil.

LUMHOLTZ, CARL and HRDLIČKA, ALEŠ, 1897: Trephining in Mexico. *Amer. Anthrop.,* o.s., 10:389.

LUMHOLTZ, CARL and HRDLIČKA, ALEŠ, 1897: A case of trepanning in North-Western Mexico. *Report of the sixty-seventh meeting of the British Association for the Advancement of Science,* held at Toronto in August 1897, 67:790.

LUMHOLTZ, CARL, 1903: Trephining among the ancient Tarahumares. *Lancet.,* 2:136.

MACCURDY, GEORGE GRANT, 1905: Prehistoric surgery—a neolithic survival. *Amer. Anthrop.,* 7:17.

MACCURDY, GEORGE GRANT, 1923: Human skeletal remains from the highlands of Peru. *Amer. J. Phys. Anthrop.,* 6:217.

MACCURDY, GEORGE GRANT, 1924: *Human origins/a manual of prehistory.* New York, D. Appleton. 2 vols. *Vide* vol. 2, p. 160.

MCGEE, W. J., 1894: Primitive trephining, illustrated by the Muñiz Peruvian collection. *Bull. Johns Hopkins Hosp.,* 5:1.

McGregor, J. C. and Wadlow, W. L., 1951: A trephined Indian skull from Illinois? *Amer. Anthrop.,* 53:148.

Malbot, Henri, and Verneau, R., 1897: Étude d'ethnographie algérienne les Chaouïas et la trépanation du crâne dans l'Aurès. *L'Anthropologie.,* 8:1, 174.

Malbot, Henri, 1898: La trépanation du crâne chez les Chaouïas. *Trav. Neurol. Chirurg.,* Paris. 3:10.

Manouvrier, L., 1904: Incisions, cautérisations et trépanations crâniennes de l'époque néolithique. *Bull. Mem. Soc. Anthrop. Paris.,* 5ᵉ sér. 5:67.

Margetts, Edward Lambert, 1958: Ethnopsychiatry in the field: an outline of the anthropological approach to the study of psychopathology and mental illness in African natives. CCTA/CSA/WFMH/WHO specialists meeting on mental health. Bukavu. 10-18 Mar. 1958. London, Document MH(58)14. Mimeographed.

Margetts, Edward Lambert, 1958: The psychiatric examination of native African patients. *Medical proceedings.* Johannesburg. Mental health number. 4: 679.

Margetts, Edward Lambert, 1962: Trepanation of the skull by primitive traditional medicine-men, with particular reference to East African practice. *Proceedings of the Third World Congress of Psychiatry.* Montreal, Canada. 4-10 June 1961. University of Toronto Press. McGill University Press. Volume II, p. 1298.

Marill, François-Georges, 1946: Trépanation crânienne préhistorique ou ostéomyélite de la voûte du crâne chez l'homme préhistorique. *Sem. Hop. Paris,* Supplement. 22: no. 27; 121 and no. 28; 140.

Martin, Jean, 1879: Deux crânes de Néo-Calédoniens, dont un trépané sur le front. *Bull. Soc. Anthrop. Paris.,* 3 sér: 2:719.

Martin, L. Th., 1867: La trépanation du crâne, telle qu'elle est pratiquée par les Kabyles de l'Aurès. *Le Montpellier Médecine,* 18:525.

Meagher, J. L., 1940: Trephining by natives of New Britain. *Brit. Med. J.,* 2:296.

Mettler, Cecilia C. and Mettler Fred A., 1947: *History of medicine/a correlative text, arranged according to subjects.* Philadelphia, Blakiston. Vide p. 793, 876, *etc.*

Mettler, Fred A. and Mettler Cecilia C., 1945: Historic development of knowledge relating to cranial trauma. Proceedings of the Association December 17 and 18, 1943 New York. Trauma of the central nervous system. Research publications. . . . Volume XXIV. Association for Research in Nervous and Mental Disease. New York, Williams & Wilkins. 24:1.

Meyer, Hans, 1896: Ueber die Urbewohner der Canarischen Inseln. In: *Festschrift für Adolf Bastien.* Berlin, Dietrich Reimer. ss.63-78.

Milne, John Stewart, 1907: *Surgical instruments in Greek and Roman times.* Oxford, Clarendon Press. Vide p. 126.

Minkevich, I. I., 1897: Trepanatsiya u Kavkezskikh gortsev i drugikh razlichnîkh narodov sravnitelnoye izsliedovaniye. *Meditsinskiŭ Sbornik.* Tiflis. No. 60; 1.

Montano, J., 1885: Rapport à M. le Ministre de l'Instruction Publique sur une mission aux îles Philippines et en Malaisie (1879-1881). *Archives des Missions Scientifiques et Littéraires.* 3ᵉ sér. 11:271.

Moodie, Roy L., 1919: Ancient skull lesions and the practice of trephining in prehistoric times. *Surgical Clinics,* Chicago., 3:481.

Moodie, Roy L., 1920: The use of the cautery among neolithic and later primitive peoples. *Surg. Clin.,* Chicago., 4:851.

Moodie, Roy L., 1923: *Paleopathology/an introduction to the study of ancient evidences of disease.* University of Illinois Press, Urbana. Vide p. 356.

Moodie, Roy L., 1929: Studies in paleopathology, XXIII/surgery in pre-Columbian Peru. *Ann. Med. Hist.* n.s. 1:698.

Moodie, Roy L., 1930: Studies in paleopathology, XXIV: prehistoric surgery in New Mexico. *Amer. J. Surg.* n.s. 8:905.

Muñiz, Manuel Antonio and McGee, W.J., 1897: Primitive trephining in Peru. In: *Sixteenth Annual Report of the Bureau of American Ethnology to the secretary of the Smithsonian Institution 1894-'95.* Washington, Government Printing office. Vide p. 1.

Munro, Robert, 1897: *Prehistoric problems/being a selection of essays on the evolution of man and other controverted problems in anthropology and archaeology.* Edinburgh, William Blackwood. Vide p. 191.

Nicolas, C., 1913: Note sur les interventions chirurgicales chez les Canaques. *Paris Chirurgical,* 5:19.

Oakley, Kenneth P., Brooke, W., Akester, R., and Brothwell D. R., 1959: Contributions on trepanning or trephination in ancient and modern times. *Man,* 59: no. 133; 93,

O'Conner, Desmond C. and Walker, A. Earl, 1951: Prologue. In: Walker, A. Earl (Editor) *A History of Neurological Surgery.* ch. 1, p. 1. Baltimore, Williams & Wilkins.

Paris, Amédée, 1868: Chirurgie arabe/de la trépanation céphalique/pratiquée par les médecins indigènes de l'Aouress (province de Constantine). *Gazette médicale d'Algérie.,* 13:25.

Parkington, Richard Heinrich Robert, 1907: *Dreissig Jahre in der Südsee/Land und Leute, Sitten und Gebräuche im Bismarckarchipel und auf den deutschen Salomoinseln.* Herausgegeben von Dr. B. Ankermann. Stuttgart, Strecker & Schröder. Vide p. 114.

Parkinson, Richard Heinrich Robert, 1908: Trepanaation bei den Südseeinsulanern. *Medicinische Blätter,* Wien., 31:290. Vide also same title in: *Z. Krankenpflege.,* 30:161, and *Deutsch Aerzte-Zeitung.* 18:415.

Parry, T. Wilson, 1914: Prehistoric man and his early efforts to combat disease. *Lancet. 1*:1699.

Parry, T. Wilson, 1916: The art of trephining among prehistoric and primitive peoples: their motives for its practice and their methods of procedure. *J. Brit. Archaeol. Ass.* n.s., 22:33.

Parry, T. Wilson, 1923: An address on trephination of the living skull in prehistoric times. *Brit. Med. J. 1:* 457.

Parry, T. Wilson, 1931: Neolithic man and penetration of living human skull. *Lancet. 2*:1388.

Parry, T. Wilson, 1936: A case of primitive surgical holing of the cranium practised in Great Britain in mediaeval times, with a note on the introduction of trepanning instruments. *Proc. Roy. Soc. Med.* (Sect. Hist. Med.), 29:898.

Piggott, Stuart, 1940: A trepanned skull of the beaker period from Dorset and the practice of trepanning in prehistoric Europe. *Proc. Prehist. Soc.* n.s., 6:112. Followed by Cave, A.J.E., The surgical aspects of the Crichel trepanation, p. 131.

Pöch, Rudolf, 1907: Reisen in Neu-Guinea in den Jahren 1904-1906. *Z. Ethnol., 39*:382.

Popham, Robert E., 1954: Trepanation as a rational procedure in primitive surgery. *U. Toronto Med. J., 31*:204.

Powell, Wilfred, 1883: *Wanderings in a wild country; or, three years amongst the cannibals of New Britain.* London. Sampson Low, Marston, Searle & Rivington.

Prunières de Marvejols, 1874: Sur les crânes artificiellement perforés à l'époque des dolmens. *Bull. Soc. d'Anthrop. Paris, 9*:185 (disc. par Broca 189, Hamy 202).

Quevedo, A., and Sergio, A., 1943: La trepanación Incana en la región del Cuzco. *Revista Universitaria,* Cuzco. No. 85.

Ray, Sidney H., 1917: The people and language of Lifu, Loyalty islands. *J. Roy. Anthrop. Inst., 47*:239.

Rightmyer, E. R., 1965: Prehistoric trephining. *Connecticut Med., 29*:239.

Riofrey, H. B., 1874: La migraine traitée par le trépan. *Tribune Médicale.* 6:202.

Rogers, Lambert Charles, 1930: The history of craniotomy/an account of the methods which have been practiced and the instruments used for opening the human skull during life. *Ann. Med. Hist.* n.s. 2:495.

Rogers, Spencer Lee, 1938: The healing of trephine wounds in skulls from pre-Columbian Peru. *Amer. J. Phys. Anthrop.* 23:321.

Roney, James G., Jr., 1954: The occurrence of trephining among the Bakhtiari. *Bull. Hist. Med.* 28:489.

Roscoe, John, 1921: *Twenty-five years in East Africa.* Cambridge University Press. Vide p. 147.

Roscoe, John, 1922: *The soul of central Africa/a general account of the Mackie ethnological expedition.* London, Cassell. Vide p. 300.

Roscoe, John, 1923: *The Banyankole/the second part of the report of the Mackie ethnological expedition to central Africa.* Cambridge University Press. Vide p. 161.

Roscoe, John, 1924: *The Bagesu and other tribes of the Uganda Protectorate/the third part of the report of the Mackie ethnological expedition to central Africa.* Cambridge University Press. Vide p. 119.

Rosen, George, 1939: Trepanation in Cornish miners. *Ciba Sympos. 1*:197.

Ruffer, Sir Marc Armand, 1918: Some recent researches on prehistoric trephining. *J. Path. Bact.,* 22:90. (Reprinted in Ruffer, *Studies in the palaeopathology of Egypt,* Chicago, University of Chicago Press. 1921. p. 194).

Rytel, Michael M., 1956: Trephinations in ancient Peru. *Quart. Bull. Northwestern Univ. Med. Sch., 30*:365.

Sanson, A., 1874: Sur les perforations artificielles du crâne chez les insulaires de la mer du sud. *Bull. Soc. d'Anthrop. Paris.* 2 sér: 9:494.

Sarasin, Fritz, 1929: *Ethnologie der Neu-Caledonier und Loyalty-Insulaner.* München, C. W. Kreidel, 2 Bde. Vide Bd 1, s.37.

Scarpa, Antonio, 1956: Le popolazioni attuali inculte praticano ancora la trapanazione del cranio? *Minerva Med., 47*:943.

Schmeltz, J. D. E., 1896: Trepanation bei den Ureinwohnern der Canaren. *Internat. Archiv Ethnog.* 9: 214. (redaction—re von Luschan, F.)

Seligmann, C. G., 1906: Note on a trephined skull from New Britain. *Man.* 6: no. 24; 37.

Shapiro, H. L., 1927: Primitive surgery/first evidence of trephining in the Southwest. *Natural History,* 27: 266.

Smith, Grafton Elliot, 1916: A note on the practice of trephining in early times and among primitive people, *J. Brit. Archaeol. Ass.,* n.s., 22:68.

Smith, Harlan I. (with a report by Hrdlička, Aleš), 1924: Trephined aboriginal skulls from British Columbia and Washington. *Amer. J. Phys. Anthrop.,* 7:447.

Solingen, Cornelis, 1684: *Manuale operatien der chirurgie* (etc). Amsterdam, J. Bouman.

Sprengel, Kurt, 1805: *Geschichte der Chirurgie.* Erster Theil. *Geschichte der wichtigsten Operationen,* Zweyter Theil. von 1805-1819. Wilhelm Sprengel. Karl August Kümmel. Halle. Vide Bd 1, s. 3.

Squier, E. George, 1877: *Peru/incidents of travel and exploration in the land of the Incas.* London, Macmillan.

Steele, P. R., 1958: Trephining in Tibesti. *St. George's Hosp. Gazette,* 43:92.

Stéphen-Chauvet, 1936: *La médecine chez les peuples primitifs (préhistoriques et contemporains). La médecine à travers le temps et l'espace.* Vol. 1. Paris. Librairie Maloine, Vide p. 71.

Stewart, Charles Samuel, 1831: *A visit to the south seas, in the U.S. ship Vincennes, during the years 1829 and 1830; with scenes in Brazil, Peru, Manilla,* the Cape of Good Hope, and St. Helena. New York. John P. Haven, 2 vols. *Vide* Vol. 1, p. 271.

STEWART, THOMAS DALE, 1950: Deformity, trephining and mutilation in South American indian skeletal remains. *In:* Steward, Julian H. (Editor), *Handbook of South American indians*. Washington, Smithsonian Institution Bureau of American Ethnology. Bulletin 143. Volume 6. p. 43.

STEWART, THOMAS DALE, 1950: Pathological changes in South American indian skeletal remains. *Ibid.* p. 49.

STEWART, THOMAS DALE, 1956: Significance of osteitis in ancient Peruvian trephining. *Bull. Hist. Med.,* 30: 293.

STEWART, THOMAS DALE, 1958: Stone age skull surgery: a general review, with emphasis on the New World. *Annual Report of the Board of Regents of the Smithsonian Institution/1957.* Publication 4314. Washington. United States Government Printing Office. p. 469.

SUDHOFF, KARL, 1926: Medicine in the stone age. *In:* Garrison, Fielding H. (Editor). *Essays in the history of medicine*. New York. Medical Life Press, p. 161. (translation by Riesman, David, of: Medizin in der Steinzeit, 1909).

SUDHOFF, KARL, 1924-32: Trepanation. *In:* Ebert, Max (Editor). *Reallexicon der Vorgeschichte unter Mitwirkung zahlreicher Fachgelehrter*. Berlin. W. deGruyter, *Vide* Bd. xiii, s. 430.

TALBOT, PERCY AMAURY, 1926: *The peoples of Southern Nigeria/a sketch of their history, ethnology and languages, with an abstract of the 1921 census*. London, Oxford University Press. 4 vols. *Vide* vol. 3, p. 943.

TERRIER, LOUIS-FÉLIX and PÉRAIRE, M., 1895: *L'opération du trépan*. Paris, Félix Alcan.

THOMPSON, C. J. S., 1938: The evolution and development of surgical instruments. iv. The trepan. *Brit. J. Surg.,* 25:726.

THOMPSON, C. J. S., 1942: *The history and evolution of surgical instruments*. New York, Schuman's. *Vide* p. 35, 56.

TILLMANNS, H., 1883: Ueber praehistorische Chirurgie. *Von Langenbeck's Archiv klinische Chirurgie,* 28: 775.

TOPINARD, PAUL, 1875: Des instruments de chirurgie de Taïti recueillis par M.A. Lesson. *Bull. Soc. Anthrop. Paris*. 2 sér. 10:619.

TRELLES, J. O., 1957: Las trepanaciones en la cirugía peruana prehistórica. *Neurocirugia*. Santiago. 15: 123.

TRELLES, J. O., 1962: Cranial trepanation in ancient Peru. *World Neurology*. 3:538.

TROJANOVIĆ, SIMA, 1900: Die Trepanation bei den Serben. *Correspondenz Blatt der deutschen Gesellschaft für Anthropologie Ethnologie und Urgeschichte*. München, 31:18.

TURNER, GEORGE, 1884: *Samoa a hundred years ago and long before. Together with notes on the cults and customs of twenty-three other islands in the Pacific*. London, Macmillan. *Vide* p. 339.

UNDERWOOD, E. ASHWORTH, 1951: Catalogue of an exhibition illustrating prehistoric man in health and sickness. *Publications of the Wellcome Historical Medical Museum/occasional papers series,* no. 3. London, Oxford University Press. *Vide* p. 20, 54.

UNDERWOOD, E. ASHWORTH, 1952: Catalogue of an exhibition illustrating the medicine of the aboriginal peoples in the British commonwealth. *Publications of the Wellcome Historical Medical Museum/occasional papers series,* no. 5. London, Oxford University Press. *Vide* p. 26.

VARA LÓPEZ, RAFAEL, 1949: La craniectomía a través de los siglos (discurso de apertura)/solemne apertura de curso/1949-1950. Universidad de Valladolid.

VÉDRÈNES, A., 1885: De la trépanation du crâne chez les indigènes de l'Aurès (Algérie) (province de Constantine). *Revue de Chirurgie,* Paris, 5:817, 907, 974.

VÉDRÈNES, A., 1886: Note sur la trépanation du crâne dans la principauté du Monténégro. *Revue Anthrop.,* Paris. 3 sér. 1:648.

VIRSALADZE, S. S., 1898: K voprosu o narodnoĭ meditsinĭe vōobshtshe i v chastnosti o trepanatsii cherepa u gortsev Dagestana. *Vestnik Obshtshestvennoĭ Higieni Sudebnoĭ i Prakticheskoĭ Meditsiny Izdavayemĭy Meditsinskim Departamentom*. St. Petersburg. No. 9. 2 sect.: 697.

VON HOVORKA, OSKAR and KRONFELD, ADOLF, 1908-9: *Vergleichende Volksmedizin/eine Darstellung volksmedizinischer Sitten und Gebräuche, Anschauungen und heilsattoren, des Aberglaubens und der Zaubermedizin*. Stuttgart, Strecker und Schröder. 2 Bde. *Vide* Bd. 2, s. 444.

VON LUSCHAN, F., 1896: *Vide* Schmeltz, J. D. E., 1896.

VON LUSCHAN, F., 1896: 1. Drei trepanirte Schädel von Tenerife. 2. Schädel mit Narben in der Bregma-Gegend. *Verh. Berliner Ges. Anthrop., Ethnol. Urgeschichte*. Jahrgang 1896:63, 65.

VON LUSCHAN, F., 1898: Trepanirte Schädel aus Neu-Britannien. *Z. Ethnol.,* 30:398.

VON LUSCHAN, F., 1899: Ueber Trepanation und verwandte Operationen bei den alten Bewohnern von Tenerife. *Comptes-rendus du XII Congrès International de Médecine*. Moscow, Yakovlev. Bd. II, s. 27.

WAKEFIELD, E. G. and DELLINGER, SAMUEL C., 1939: Possible reasons for trephining the skull in the past. *Ciba Sympos.,* 1:166.

WALKER, A. EARL, 1958: Primitive trepanation: the beginning of medical history. *Trans. Stud. Coll. Phys. Philadelphia*. 4 ser. 26:99.

WATERSTON, DAVID, 1908: Skulls from New Caledonia. *J. Roy. Anthrop. Inst.,* 38:36.

WEISS, PEDRO, 1949: *La cirugía del cráneo entre los antiguos peruanos*. Lima.

WEISS, PEDRO, 1958: Osteologia cultural/practicas cefalicas/l a. parte: cabeza trofeos—trepanaciones— cauterizaciones. *An. Fac. Med. Lima. Vide* p. 505.

WILLIAMS, HERBERT UPHAM, 1929: Human paleopathology/with some original observations on symmetrical osteoporosis of the skull. *Arch. Path.* 7:839.

WÖLFEL, DOMINIK J., 1925: Die Trepanation. Studien über Ursprung, Zusammenhänge und kulturelle Zugehörigkeit der Trepanation. Erste Studie. Die kulturellen Zusammenhänge und der einheitliche Ursprung der Trepanation in Melanesien und Amerika. *Anthropos.* 20:1.

WÖLFEL, DOMINIK J., 1936: Vom Sinn der Trepanation. *Ciba Z.,* 4: Nummer 39:1326.

WÖLFEL, DOMINIK J., 1936: Die Methoden der urgeschichtlichen und primitiven Trepanation. *Ciba Z.,* 4: Nummer 39:1331.

ZABOROWSKI, 1897: Le T sincipital.—mutilation des crânes néolithiques, observée en Asie centrale. *Bull. Soc. Anthrop. Paris.,* 4 sér; 8:501.

Chapter 54

Reputed Early Egyptian Dental Operation, An Appraisal

F. FILCE LEEK

For many years there has been much speculation as to the extent of medical, surgical and dental knowledge and practice amongst the ancient Egyptians. Contributions on early dentistry in particular have been made amongst others by Breasted (1930), Leake (1940), Junker (1927), Mariette (1845) Erman-Grapow (1921), Ruffer (1921), Ranke (1940), Ebbell (1937), Brown (1936), Lepsius (1849), Weinberger (1948), Hooton (1917) and more recently by Ghalioungui (1963).

The subject has been discussed from information acquired from hieroglyphics inscribed in tombs; from the three famous papyri dealing with medicine, namely the Ebers, Hearst and Edwin Smith; and from study of the skeletal remains which are preserved in various museums of the world.

In spite of some disagreement, especially by Warren R. Dawson (1948), the evidence has been generally accepted that surgical intervention took place in order to cure dental apical abscesses. This evidence is based mainly on the study of a IVth Dynasty mandible (Fig. 1) at present in the Peabody Museum, Harvard University, a photograph of which is shown in the books by Breasted (1930), Weinberger (1948), and others.

In this mandible all the teeth present show marked attrition, which led to the exposure of the pulp chamber of the lower right first molar and to the death of the pulp and abscess formation, as is shown by the rarified area of bone beneath the tooth in the radiograph (Fig. 2.). It also shows two holes penetrating the outer plate of the mandible on a level with the apex and in the direction of the anterior root of this tooth, one on its mesial and the other on its distal side.

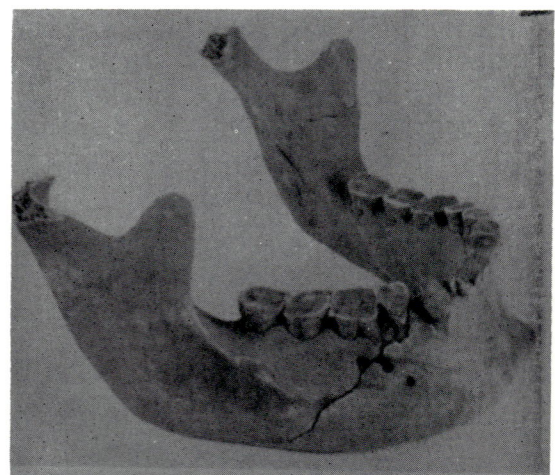

FIGURE 1. Fourth dynastic human mandible (c. 2900-2750 B.C.) The two "borings" above mental foramen were thought to be the work of an Egyptian dental physician or surgeon. From Weinberger: History of Dentistry, St. Louis, Missouri, 1948, The C. V. Mosby Company.

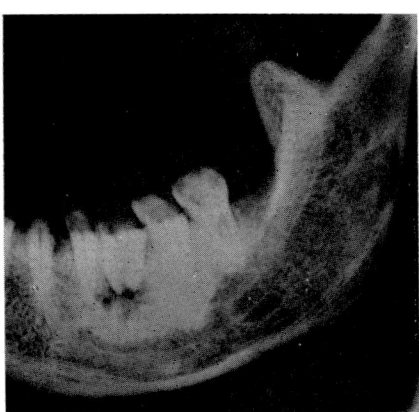

FIGURE 2. Roentgenogram of mandible shown in Figure 1. From Weinberger: History of Dentistry, St. Louis, Missouri, 1948, The C. V. Mosby Company.

In support of the surgical intervention hypothesis, it has been stated (Anon, 1919) that neither of these holes could have been the result of an abscess eroding its way naturally through the lateral wall of the alveolar process and so discharging because: (a) the direction of the hole is upward at an angle of 15° from the vertical and not horizontal, and (b) the abscess has not eroded its way through the alveolar process because the anterior opening goes through a thickness of 2.5mm of sound tissue before reaching the infected area, and the posterior hole passes through 3mm of sound tissue. And as a result of this reasoning the conclusion arrived at was, "The evidence of this specimen seems to establish beyond a reasonable doubt the existence of a rudimentary knowledge of oral surgery in the Old Empire."

Breasted (1930), in supporting this hypothesis, comments on the method employed as follows; "Of instruments, the treatise mentions only the 'fire drill,' employed when hot for cauterisation. A mandible disclosing a drill hole in the mental foramen for the purpose of draining an abscess under a molar tooth makes it evident that specialised surgical instruments of metal, presumably bronze, already existed in the age which produced our surgical treatise, but they are taken for granted by the ancient author."

Additional support to the theory that knowledge of drilling was known to the ancient Egyptians is given by reference to the tomb painting on the wall of the XVIIIth Dynasty tomb of the Rekhmara at Thebes, where there is a scene depicting a hand bow drill used in the construction of chairs (Fig. 3).

There is in the British Museum (Natural History) a mandible bearing on this subject. It was found by Sir Flinders Petrie at Abydos and dated by him to the Ist or IInd Dynasty. It was later presented by him to the museum (and is numbered AC 114/421).

As in the mandible in the Peabody Museum, the teeth have suffered considerably from attrition which exposed the pulp chamber of the lower left first molar with subsequent apical abscess formation (Fig. 4). There is a hole penetrating the outer plate, going downwards in the direction of the apex through sound tissue at an angle of approximately 10°, and the edges are circular and extremely cleanly cut (Fig. 5).

The conditions in both mandibles are so similar that it is only reasonable to con-

FIGURE 3. Showing the use of the hand bow drill as found upon the walls of the tomb of Rekhmara c. 1450 B.C. after D. A. Mackenzie

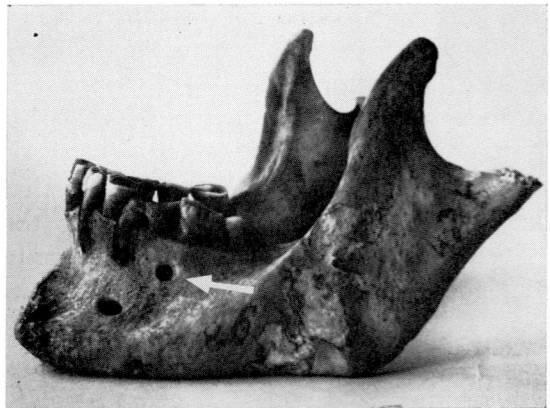

FIGURE 4. British Museum mandible (AC 114/421) showing two holes in the specimen. *Above* (*white arrow*), sinus or canal leading to apex of first molar tooth with clearly defined circular edges. *Below*, mental foramen through which passes a branch of mandibular nerve. This foramen is considerably larger than normal but has no pathological significance.

FIGURE 5. Enlargement of molar teeth shown in Figure 4, and the sinus (*white arrow*) piercing the alveolar plate (to show its backward direction.)

clude that the canals in both instances were made by the same agency.

On further examination of the mandible in the British Museum (Natural History) it is seen that the canal not only goes to the distal root of the molar tooth but that it penetrates at an obtuse angle of 140° from the border of the mandible—that is from behind forward (Fig. 5). It would be quite impossible for a hole to be made in this position with a straight drill.

In spite of Breasted's assertion that specialised instruments of bronze must have been known in the IVth Dynasty, it is quite inconceivable that there was knowledge of the right angled drill and there is indeed no record of the existence of a surgical drill of any kind, therefore these holes were not drilled by an operative procedure. The picture in the tomb of Rekhmara, with its presentation of a carpenter's drill, does not give support to his theory as it was painted during the XVIIIth Dynasty, approximately a thousand years later than the date ascribed to the mandible under discussion.

A clinical fact also supports this contention, since it is not necessary to drill holes in the bone to get relief from the pain caused by an apical abscess, but only to incise the gum and let the pus escape. The process of abscess expansion in itself initiates a sinus path through the outer alveolar plate, and it is important to note that such perforations are by no means an uncommon "end point" of apical infection in earlier populations. Finally, the point that had the canals been pathological ones they would have been horizontal, is not borne out by an examination of a large number of early Egyptian skulls where it is seen that various directions are taken. It seems therefore that the theory that surgical intervention took place in the cure of dental abscesses can no longer be regarded as tenable.

SUMMARY

The evidence supporting the assertion that the Ancient Egyptians bored through the bone to cure apical abscesses is presented, and in the light of further study the hypothesis is shown to be unacceptable.

ACKNOWLEDGMENT

Grateful thanks are due to Dr. K. P. Oakley, British Museum (Natural History) for permission to examine and photograph the collection of ancient Egyptian skulls.

REFERENCES

ANON., 1919: Oral Surgery in Egypt during the Old Empire. *Dent. Cosmos.*, 61:900.

BREASTED, J. H., 1930: *Edwin Smith Surgical Papyrus.* 2 vols. Chicago, Chicago Univ. Press.

BROWN, L. P., 1936: Appellations of the dental practitioner. *Dent. Cosmos*, 78:246.

DAWSON, WARREN R., 1948: Personal Communication to B. W. Weinberger. In: *History of Dentistry.* St. Louis, Mosby.

EBBELL, B., 1937: *The Papyrus Ebers.* Copenhagen, Levin and Munksgaard.

ERMAN-GRAPOW, H., 1921: *Aegyptisches Handworterbuch.* Leipzig.

GHALIOUNGUI, P., 1963: *Magic and Medical Science in Ancient Egypt.* London, Hodder & Stoughton.

HOOTON, E. A., 1917: Oral Surgery in Egypt during the Old Empire. *Harvard African Stud.*, 1:29.

JUNKER, HERMANN., 1927: Die Stele des Hofarztes 'Iry. *Z. Ägypt. Sprache und Altertumskunde. Leipzig*, 63: 53.

LEAKE, C. D., 1940: Ancient Egyptian therapeutics. *Ciba Sympos.*, 1:(10), 311.

LEPSIUS, C. R., 1849-59: *Denkmaler aus Aegypten und Aethiopien.* Berlin, pt. 11:25.

MARIETTE, A., 1845: *Les Mastaba de l'Ancienne Empire.* Vieweg, Paris. 203-205.

RANKE, HERMANN, 1940: Medicine and Surgery in Ancient Egypt. *Univ. Penn. Bi-Cent. Conf.*, pp 31-42.

RUFFER, SIR M. A. 1921: *Studies in the Palaeopathology of Egypt.* Chicago, Chicago Univ. Press.

THOMA, K. H., 1917: Oral diseases of ancient nations and tribes. *J. Allied Dent. Soc. New York.*, 12:327.

WEINBERGER, B. W., 1948: *History of Dentistry.* St. Louis, Mosby.

SECTION VII
MENTAL ABNORMALITY

Chapter 55

Mental Disorder in Antiquity

GERALD C. MOSS

The earliest connected accounts of the mind and its disorders are to be found in the writings of the Greeks. In the Old Testament it is usually taken for granted that madness and neurosis are god-sent,[1] and in the New Testament that they are due to evil spirits.[2]

Superstition, flourishing in our time, was universal in all primitive societies; with it and its congeners, magic and religion, went a belief in a number of supernatural powers; consequently mental disturbance was thought to be brought about by powers of this kind, by their interfering with or "possessing" its victim. There are many byways,[3] but such a concept is so well documented in Greek literature that it is convenient to take its appearance there as a point of departure. It would not be fruitful to pursue inquiry as far as the Middle Ages; for before they dawned, a resurgence of demonism and other forms of superstition, mainly of Eastern origin, had greatly impeded rational thinking; but in early Greek medical practice it is improbable that magic played much part, although doctors were influenced by religious beliefs (as they may be today) and incubations had an important role (as suggestion in various forms continues to have).[4, 5]

By convention, in Greek history, the Archaic Period (c. 800 or 750 to 500 B.C.) follows Homer; then come the Classical (the fifth and fourth centuries), the Hellenistic (after Alexander the Great), and the Roman Periods. The critical study of man followed the speculations of the sixth century philosophers, whose insight, though they lived at a time when no line

---

1. DEUTERONOMY 28:28.
2. JOHN 10:20.
3. FRAZER, J. G. 1925: *The Golden Bough* (abridged edition). London, MacMillan. This abridged edition is recommended.
4. EDELSTEIN, L., 1937: Greek Medicine in its Relation to Religion and Magic. *Bull. Hist. Med.*, 5:201. Incubation, the temple sleep and its accompanying dream, was an essential part of Asclepian rites; in an atmosphere of powerful suggestion it is not surprising that some sudden cures were obtained. Pausanias (2.27.1-3) described the temple of Asclepius at Epidaurus and its surroundings, and mentioned slabs—and some have been found—on which the names of men and women healed there were inscribed. For a farcical account of the proceedings at the shrine at Athens see Aristophanes, *Plutus* 659-742. The cult had its votaries far into the Christian era: one, e.g., was the neurotic sophist Aristides (c. A.D. 150). Walter Pater, in his *Marius the Epicurean*, Chap. 3, describes the experience of the young Marius in a Roman temple dedicated to the god.
5. DODDS, E. R., 1951: *The Greeks and the Irrational*. Berkeley and Los Angeles, University of California Press, p.115. Dodds pertinently remarks: In the [Epidaurian Temple] Record the cures are mostly represented as instantaneous . . . It is irrelevant to ask how long the improvement lasted: it is enough that the patient "departed cured" . . . Such cures need not have been numerous: as we see in the case of Lourdes, a healing shrine can maintain its reputation on a very low percentage of successes, provided a few of them are sensational.

was drawn between science and philosophy, should not be under-estimated; but even before them the simple moral reflections and aphorisms of poets such as Hesiod, Archilochus, Solon, and, a little later, Theognis contained wisdom and common sense.

It is proposed, beginning with the Greeks and relegating comment for the most part to notes, to call upon some ancient Greeks and Romans to speak for themselves; and though only a few will be able to hold the stage for a short time, it will be possible to observe a part, however small, of the development of language and outlook. For several reasons, more than passing notice of recondite psychology or philosophy would be out of place: Plato's psychology and philosophy, for example, important as they are, most fittingly find a place in the body of a short survey when they bear more upon clinical, as it were, than upon philosophical or moral aspects; further, much of their meaning may be lost or misinterpreted unless close attention is paid to their context and to the reservations that Plato often makes. Sexual deviation, which is discussed elsewhere, will not be considered.

The earliest professional descriptions are contained in the Hippocratic Collection (for the sake of convenience—there is no other justification—the name Hippocrates will be used for its compilers); later, those of Graeco-Romans and Romans often reach a high standard; and the works of laymen may also disclose an understanding of mental illness.

With regard to the intellectual development of the Greeks in general it has been well argued that the stages of the journey which saw a rational view of the nature of man establish itself are to be traced in the creations of epic and lyric poetry and in the plays;[6] with regard to medical psychology in particular the contributions of the historians Herodotus and the more discerning Thucydides, besides those of the philosophers, also deserve notice. Up to a point the stages can be traced in religion also: here it has to be remembered that, though the old mythology continued to permeate the literature, Xenophanes had attacked it in the sixth century, and it had dropped for the most part out of practical religion by the time of Pindar and Aeschylus and served chiefly as a vehicle for the literature.[7] But the tragic poets who used its themes had a purpose that went beyond entertainment: they knew that in them lay universal truths. And it is not only dramatists who are drawn to them today: to Freud and Jung they embody permanent but unacknowledged forces.[8] Finally, the masses could scarcely share the beliefs of the intellectuals: the gap between the two tended to become ever wider. Yet in all classical literature colloquial expressions may be found: it has words and phrases enough to correspond with our *blockhead, ninny, idiot, madman,* and the like.

* * *

The Greeks first had to "discover" the mind or intellect. The wide difference between Homer and the Classical Period is emphasized by the use and meanings of words in the two epochs. In Homer abstractions are undeveloped and his notions of psyche, *noös* (nous), and *thûmos* are quaint; moreover, he had no specific term

---

6. SNELL, B., 1953: *The Discovery of the Mind* (translated by T. G. Rosenmeyer). Oxford, Blackwell, *p.v.* and *passim.*

7. MURRAY, G., 1935: *Five Stages of Greek Religion.* London, Watts, p. 75.

8. HIGHET, G., 1957: *The Classical Tradition.* New York, Oxford University Press, pp. 523 ff.

for the living body;[9] it would seem, indeed, to a neurologist that he was thinking more of what is now called the body-schema than of something solid and compact. In the narrative portions of the Iliad and the Odyssey all action of consequence depends on anthropomorphic gods; but a personage, especially in the Odyssey, may attribute his own words and actions to the operation of a vague daemon or unspecified god or gods. The word *atê* often occurs: it means a temporary infatuation or mental aberration; never endogenous, *atê* is put into the subject's *thûmos* or its supposed seat, the *phrenes* (midriff or chest), by one of these supernatural beings. Later, it acquired the additional meaning of a punitive infatuation (judicial blindness) and even the ruin consequent on such folly; but attachment of the idea of punishment to it occurs only once in Homer. And in later Greek the word *mainomai* (be mad) was spoken and understood by all; in Homer it is used mostly of the frenzy of battle and the raging of weapons, fire etc.: thus Hector, girded for battle, went berserk (*maineto*)—and here autonomic accompaniments of his rage are described—and raged like Ares or like consuming fire in a forest.[10]

Turning from Homer to the archaic poet Theognis (*fl.c.* 520 B.C.), we find perhaps the earliest plea for human eugenics; the translation is by T. F. Higham:

> Ram, ass, and horse, my Kyrnos, we look over
> With care, and seek good stock for good to cover;
> And yet the best men make no argument,
> But wed, for money, runts of poor descent.
> So too a woman will demean her state
> And spurn the better for the richer mate.
> Money's the cry. Good stock to bad is wed
> And bad to good, till all the world's crossbred.
> No wonder if the country's breed declines,—
> Mixed metal, Kyrnos, that but dimly shines.[11]

In one of his Pythian Odes, the lyric poet Pindar (522-448? B.C.) tells of the birth, prowess, and death of Asclepius; the line ". . . and some he healed with soothing incantations" exemplifies one of the traditional rites.[12]

In the tragedies of Aeschylus (525-456 B.C.) the daemon has a more malignant form than it had in Homer and haunts and hounds their characters; it is no longer content to interfere with its victim, but wholly possesses him. A verb, *daimonân* (it is not found in Homer), has now established itself: it means to be possessed by and, often, be driven to madness by a daemon; and to this possession ruinous folly is attributed. In "*Seven against Thebes*" Ismene and Antigone lament the death of their brothers, slain by each other, and Ismene cries: "Alas, alas, O ye possessed of an evil spirit in your fatal folly."[13] In the Oresteian Trilogy the workings of a hereditary curse unfold; the second tragedy ends with a vivid picture

---

9. SNELL, B.: op. cit. pp. 1-22. *Thûmos* (θυμός) here and in ref. 17 and n. 38 is not a thŭmos (θύμος), thyme and, now, the thymus (gland): see Skinner, H. A. 1949: *The Origin of Medical Terms*. Baltimore, Williams and Wilkins, p. 346, and Liddell and Scott's Greek-English Lexicon. (abridged 1871)

10. HOMER, *Iliad* 15. 605-610: μαίνετο δ'ὡς ὅτ' Ἄρης . . . μαρναμένοιο Ἕκτορος. Cf. n. 16.

11. THEOGNIS 183-192, and in Higham, T. F. and Bowra, C. M. (Eds.), 1938: *The Oxford Book of Greek Verse in Translation*. Oxford, Clarendon, no. 188.

12. PINDAR, *Pythian* 3.51: . . . τοὺς μὲν μαλακαῖς ἐπαοιδαῖς ἀμφέπων,. Epaoidê, spell or incantation, occurs only once in Homer (Odyssey 19. 457).

13. AESCHYLUS, *Seven against Thebes* 1007: ἰὼ ἰὼ δαιμονῶντες ἄτᾳ.

of the madness of Orestes: he has hallucinations (*doxai*) of a terrifying form and is in a state of confusion (*taragmos es phrenas*).[14]

Ajax[15] and Heracles[16] are other legendary heroes who became insane; and Ajax, resolved on suicide after his fury has passed, utters: "Go quickly; no good doctor prescribes dirges and spells for an ill that needs the knife." The Medea of Euripides' (*c.* 480-406 B.C.) drama is driven by an unbearable conflict to kill her children. But it is noteworthy that Euripides makes her acknowledge that her compulsion arises in her innermost self or "heart," her *thûmos*; there is no suggestion that it was put there by a daemon.[17] In the *Bacchae* the results of the mass hysteria of the Maenads, women inspired to ecstatic frenzy by Dionysus, are revealed in their full horror.[18]

The characters in the plays of Aristophanes (448-380 B.C.), the chief representative of Attic Old Comedy, stigmatize each other freely: When the bird Euelpides calls the bird-seller Philocrates *melancholôn* he is merely calling him a fool.[19] *Elithios* (silly) and *mainomenos* (mad) have long since passed into popular speech.[20] In some lines in the "Frogs" Aristophanes employs a string of words that mean *stupid good-for-nothings, gaping idiots, mammy's darlings,* and *blockheads.*[21]

Much medical lore is to be found in Herodotus (*c.* 480- *c.* 425 B.C.), "The Father of History," and he is keenly interested in cases of psychosis (*maniê nousos*) and likes, after noting what others thought, to give his opinion upon the aetiology. He describes the madness of

---

14. AESCHYLUS, *The Libation Bearers* 1048-1062. Δόξαι (*doxai*), usually opinions, beliefs, judgments, sometimes means fancies, and those of Orestes were clearly visual hallucinations. ταραγμός (*taragmos*) is another form of τάραξις (*taraxis*); ἀταραξία (*ataraxia*), imperturbability or tranquillity of mind, is of course its opposite; hence the term *ataractic* (the adjective is found in Greek), now introduced into psychopharmacology. *Ataraxia* was one of the "static" or "negative" pleasures of Epicurus: for his use of the word see the Greek form in index in Bailey, C. 1926: *Epicurus, The Extant Remains* (with English translation and commentary). Oxford, Clarendon.

15. SOPHOCLES, *Ajax* 581-582:
πύκαζε θᾶσσον οὐ πρὸς ἰατροῦ σοφοῦ
θρηνεῖν ἐπῳδὰς πρὸς τομῶντι πήματι.
Tecmessa, who had seen Ajax slaughtering and mutilating sheep under the delusion that they were his enemies, reported (243-244) that he uttered words that no mortal taught him, but a daemon. It is a widespread belief among primitives that persons in abnormal mental states speak a special "divine" language (Dodds).

16. EURIPIDES, *Hercules Furens*. In this drama Hera sends Lyssa (madness, here personified) to seize on Heracles and cause him to slay his children. The word *lyssa* is found in Iliad 9. 239, 305 and 21. 542 and means the martial rage that possesses the hero; *cf.* ref. 10. In Iliad 9. 238-239 *mainetai* (rages) and *lyssa* appear in successive lines. In Iliad 8. 299 Teucer calls Hector a mad dog and an adjective of *lyssa* is used: the old name for rabies or hydrophobia thus has a very old origin.

17. EURIPIDES, *Medea* 1078-1080. Though there is no suggestion on Medea's part that she was possessed by a daemon, Jason at the end (1333-1335) can only attribute her action to an *alastôr*, the avenging daemon of unatoned guilt, here bloodguilt. This need cause no surprise: Euripides was an advanced thinker, but he was first a dramatist; *cf.* n. 37.

18. EURIPIDES, *Bacchae*. On Maenadism see E. R. Dodds, *op. cit.* pp. 270 ff.: on pp. 275-276 he refers to the snake-handling that is mentioned several times in the tragedy; *cf.* Sargant, W. 1959: *Battle for the Mind.* London, Pan Books.

19. ARISTOPHANES, *Birds* 14:
ὁ πινακοπώλης Φιλοκράτης μελαγχολῶν,.

20. ARISTOPHANES, *op. cit.* 523-524.

21. ARISTOPHANES, *Frogs* 989-991:
τέως δ' ἀβελτερώτατοι
κεχηνότες μαμμάκυθοι,
μελιτίδαι καθῆντο.
In *Knights* 1263 Aristophanes has a sly dig at the Athenians by referring to the city of the Athenians (*Athênaiôn*) as the city of Gapers (*Kechênaiôn*).

Cleomenes and the theories of its cause that were in vogue: always queer, he had no sooner returned to Sparta "than he was seized with madness."[22] The Spartans, Herodotus writes, asserted that Cleomenes had acquired the habit of taking his wine neat "and this was the cause of his madness;"[23] but his own opinion was that it was a punishment for what he did to Demaratus. The mad acts of Cambyses were popularly supposed to be due to a disease "called by some The Sacred Sickness," which Cambyses was said to have had from birth; Herodotus records his belief that there is no reason why a serious physical malady should not affect the brain.[24] We have more advanced thinking about mental illness here: Herodotus recognizes that it may have a physical cause even if he believed Cleomenes' condition to be a divine punishment; we see too that alcohol is already a supposed cause of insanity.[25]

Avoidance of continual nervous tension was a prescription well understand by Amasis, King of Egypt: When well-wishers remonstrated with him on his frequent revels, he replied that a bow kept strung breaks and is useless when needed; so it is with a man: if he did not relax "he would, before he knew it, be off his head or suffer a stroke."[26] The circumstances of the sudden blindness of Epizelus during the Battle of Marathon strongly suggest it was due to hysteria, though it would not have occurred to Herodotus to think so and he offers no explanation.[27]

The reflections of Thucydides (c. 460-c. 400 B.C.) on the Corcyrean revolution are a penetrating analysis of the cruelty and passions loosed by revolution and war alike:[28] his prediction that events such as he has described will recur as long as the nature of mankind remains the same has so far proved to be true; and many other parts of his work are proof of his knowledge of human motives and behaviour.

Plato (427-348 B.C.) applied himself to problems that are the daily concern of modern doctors, judges, and social workers. In the *Timaeus* the pathology and nurture of the soul are discussed: "Badness" is caused not only by physical constitutional defects, but also by education in bad social conditions; parents are also to blame.[29]

In the *Laws* it is laid down that if a son, believing his father to be suffering from mental derangement (*paranoia*), has qualms about laying a complaint he may consult the most senior guardians. They will advise him if he should lay it or not; if they think he should, they will provide legal aid; if the father is then found to be insane he shall be held to lack testamentary capacity and, though allowed to re-

---

22. HERODOTUS 6. 75: αὐτὸν αὐτίκα ὑπέλαβε μανίη νοῦσος. Herodotus, like Hippocrates and Aretaeus, wrote in Ionic Greek.
23. HERODOTUS 6. 84: καὶ ἐκ τούτου μανῆναι.
24. HERODOTUS 3.33.
25. CAMBYSES (Herodotus 3. 34) became enraged at hearing the Persians said excessive drinking had driven him mad. In Homer there are references to a fuddled state from wine in *Odyssey* 11. 61 and 21. 295 ff., but there are none to permanent mental impairment from it: in the first instance the shade of Elpenor attributed his undoing to an evil power as well as to wine: in the second, which is an exceptional case in Homer, wine alone is blamed for the infatuation (a verb from which *atê* is derived is used) of Eurytion the Centaur.

---

26. HERODOTUS 2. 173: λάθοι ἂν ἤτοι μανεὶς ἢ ὅ γε ἀπόπληκτος γενόμενος.
27. HERODOTUS 6. 117.
28. THUCYDIDES 3. 82-84.
29. PLATO *Timaeus* 86 B—90 D. And cf. n. 33.

main in his house, shall be treated as a minor for the rest of his life.[30]

In the *Republic* the base desires and instincts of low characters are compared to those that emerge in dreams; this passage has been regarded as a remarkable anticipation of Freud's views. The description of the tyrannical character that follows typifies the criminal psychopath;[31] he has been compared with Fielding's Jonathan Wild. In the *Republic* too valetudinarians and their symptoms, to which they cling in typical fashion, are considered: Asclepius would not treat them; for they were of no use to themselves or society; in the ideal state—and the word ideal is significant—judges and doctors will care for those of good physical and psychological constitution; but, it is asked: "Will they not leave the sickly in body to die, and put to death those who are incurably warped in mind?"[32]

Some of Plato's works contain long discourses upon the tripartite nature of the soul,[33] and upon madness. Madness is of two kinds, one produced by human disease, the other by an inspired departure from established usages. The latter kind is "bestowed by divine gift" and is subdivided into four parts.[34]

There are occasional references to mental illness and instability in the works of Plato's pupil Aristotle (384-322 B.C.).[35]

It is evident that belief in the supernatural origin of cerebral and mental disorders was losing its hold on the educated classes by the time it came under direct attack by Hippocrates. It is to be expected that, in a large number of medical treatises of different periods and by different hands, there would be much that is contradictory; but though a good deal of philosophy (and erroneous pathology, which it is not proposed to examine) is mixed with rational observation, there is little or no evidence of superstition.[36] In *"The Sacred Disease"* Hippocrates castigates those who hold that epilepsy and other neurological and psychological disorders are due to a particular god or power, and he affirms that the seat of epilepsy is in the brain.[37]

---

30. PLATO, *Laws* 929 D-E. παράνοια (paranoia), the word used here, clearly means nothing more than mental derangement; it occurs only three times in Plato, always with the same meaning, while μανία (mania, the general term for madness or insanity) occurs thirty times (Ast's *Lexicon Platonicum*). Hippocrates' παραφροσύναι (paraphrosunai, cf. n. 41) has the same meaning as *paranoia*; and the prefix "para" in these and similar words, of which there are a number, always indicates a wandering or deviating from the normal.

31. PLATO, *Republic* 571 A-576B. And *cf.* n. 33.

32. PLATO, *op. cit.* 406A-410A.

33. PLATO, *op. cit.* 434D-441C. The theme is taken up again at 543A-592B (on imperfect societies); there are good notes in H. P. D. Lee's translation (Penguin Classics, 1955) of the *Republic*. There is, *inter alia*, more about it in *Timaeus* (69A-90D): it is in this part (90A) that Timaeus makes the often quoted remark that man, whose divine power resides in the head, is like a tree with its root not in the earth, but in the sky. The analogy of the tripartite soul and the charioteer with a pair of winged steeds is in *Phaedrus* 246A ff. and is taken up again in 253C ff. The whole of the subject-matter is complex; but the general concept is, *mutatis mutandis*, not without some relation to modern psychopathological theory despite Plato's own reservations and his avowed concern with morals rather than psychology: *cf.* introduction above.

34. PLATO, *Phaedrus* 244A ff. and 265A ff. The "irony" of Socrates has to be taken into consideration: *cf.* introduction above.

35. HEIBERG, J. L., 1927: Geisteskrankheiten im klassischen Altertum. *Allg. Z. Psychiat.* 86:1. As well as citing Aristotle, Heiberg cites Xenophon, Aelian, and others.

36. But Hippocrates cannot be called irreligious; he is like any god-fearing person when he says (Jones 4. 422-423, *Regimen* 4. 87): Prayer indeed is good, but while calling on the gods a man should himself lend a hand.

37. HIPPOCRATES: JONES 2. 138, *The Sacred Disease*. It is necessary to study the treatise as a whole; on Powers mentioned and on some mentioned by Euripides see Dodds, E. R.: *op. cit.* pp. 77 ff. Though the common man may still have had such notions, the educated classes had mostly outgrown them; *cf.* n. 17.

He uses a variety of terms in the following aphorisms:

Fear or depression (dysthymia) that is prolonged means melancholia.³⁸
In spring occur melancholy, madness, epilepsy. . . ³⁹
Phrenitis supervening on pneumonia is bad.⁴⁰
Delirium with laughter is less dangerous, combined with seriousness it is more so.⁴¹

Hysteria (*husterika* or *ta husterika*) is often discussed: In *On the Nature of Women* the doctor is told what happens and what he should do if an attack of hysteria follows displacement of the womb towards the liver (Celsus and Aretaeus also accept this theory of the aetiology of hysteria without question): a widow is advised to become pregnant, a young woman to take a husband, other treatment also depends on age and marital status.⁴²

More in keeping with modern practice is the employment of digital pressure on the female patient in order to distinguish an hysterical from a convulsive attack: if she feels it, the attack is hysterical; if she does not, it is convulsive.⁴³

The Schools which flourished in Alexandria and other Greek cities during the Hellenistic Period fostered the arts and made noteworthy contributions, most of which are lost, to science and medicine; but in Rome indications of the detachment, long evident in Greece, of professional from folk and theurgic medicine were wanting. Yet, despite the heavy atmosphere of official ritual the Roman citizen could relax and enjoy the comedies of Plautus (254-184 B.C.) as the Athenian had enjoyed those of Aristophanes two hundred years earlier; and he too could laugh at allusions to weak-mindedness. In the *Menaechmi* a doctor asks: "What did you say he had? Say it again, old man. Is he bewitched or crazed? Tell me."⁴⁴

Talented Greeks, among them many physicians, were soon to flock to the capital of their masters. Not only did Greece, the captive, make the savage victor captive and bring the arts to rustic Latium;⁴⁵ she

---

38. HIPPOCRATES: JONES 4. 184-185, *Aphorisms* 6. 23: Ἢν φόβος ἢ δυσθυμίη πολὺν χρόνον διατελῇ, μελαγχολικὸν τὸ τοιοῦτον. The old word *thûmos* (in Plato it or a variant ranks second as a part of the tripartite soul) appears with the now familiar prefix "dys," so that we have dysthymia. Hippocrates elsewhere also has ἀθυμίη (athymia) and this is found in Aretaeus (see n. 65) and other writers: Thucydides (2. 51), for example, reported that the most terrible feature of the plague was the dejection (athymia) which ensued when anyone felt himself sickening. Just as we often make no distinction between *dysphasia* and *aphasia*, so did the Greeks make none between *dysthymia* and *athymia*.

39. HIPPOCRATES: JONES 4. 128-129, *Aphorisms* 3.20.

40. HIPPOCRATES: JONES 4. 194-195, *Aphorisms* 7. 12. Phrenitis (φρενῖτις), a very common term, means febrile delirium and it is best to preserve it in translation though delirium would serve; but see note on the next aphorism.

41. HIPPOCRATES: JONES 4. 190-191, *Aphorisms* 6. 53: Αἱ παραφροσύναι αἱ μὲν μετὰ γέλωτος γινόμεναι ἀσφαλέστεραι· αἱ δὲ μετὰ σπουδῆς ἐπισφαλέστεραι. The word *paraphrosunai* (the plural form is used) cannot be taken to mean delirium as we understand it today; cf. n. 30. It is probable that Hippocrates was contrasting our mania or hypomania with depression; cf. Celsus 3. 18. 20: Neque ignorare oportet leviorem esse morbum cum risu quam cum serio insanientium. The context of the *paranoias* in Jones 2. 148-149, *The Sacred Disease*, could conceivably justify their being translated by delirium though confusion would be better; but all that can certainly be inferred is, again, mental derangement. It is recommended that Jones's remarks (I. lix-lx) on words meaning delirium should be read with the points mentioned in mind.

42. HIPPOCRATES: Littré 7. 314-317, *On the Nature of Women*.

43. HIPPOCRATES: Littré 2. 522-523, *Regimen in Acute Diseases* (appendix).

44. PLAUTUS, *Menaechmi* 889-890:
   Quid esse illi morbi dixeras? narra, senex.
   Num larvatust aut cerritus? fac sciam.
Cerritus, one of the many terms used by Horace (see ref. 55), and *larvatus* were popular expressions. The derivation of *cerritus*, perhaps for *cerebritus*, is uncertain; see Palmer, A. 1955: *The Satires of Horace, Edited with Notes*. London, Macmillan, p. 307.

45. HORACE, *Epistles* 2. 1. 156-157:
   Graecia capta ferum victorem cepit et artis
   intulit agresti Latio.

also brought medical art of a type and quality new to Rome: Asclepiades of Bithynia, who died at an advanced age in Rome about 40 B.C. was, as far as is known, the first to make his mark there; only fragments of his works remain, but he is frequently mentioned by Celsus; Themison was his pupil. Archigenes practised at the time of Trajan (A.D. 98-117); Aretaeus of Cappadocia was greatly indebted to him, and Juvenal mentions him several times.[46]

The *De Medicina* of Aulus Cornelius Celsus is thought to be the work of a layman; it is known that he was living in the reign of Tiberius (A.D. 14-37), and his treatise is an exception to the general rule: it is written in Latin though necessarily based on Greek teaching; its sources were the Hippocratic Collection and the lost works of Asclepiades and others, and its *prooemium* is a concise and impartial summary of the history of medicine and its different Schools. A few extracts will demonstrate that Celsus' knowledge of some common mental disorders was not small: "I shall begin with insanity, and first that form of it which is both acute and found in fever. The Greeks call it *phrenêsis*" Passages which follow show he knew the prognosis of febrile delirium was that of the causative fever. He contrasts it with continuous mental derangement (*continua dementia*): when this is present, insanity is fully established. There are, he adds, several kinds of insanity: for some among insane persons (*ex phreneticis*) are sad, others hilarious.[47]

Discussion of the several kinds and their treatment occupies the whole of the eighteenth chapter of the third book; one of them is depression (*tristitia*), and of it he says: "There is another form of insanity, . . . It consists in depression, which seems caused by black bile."[48] The treatment advised is occasionally wise, often meddlesome though probably harmless, and sometimes brutal: some patients must be coerced by starvation, fetters and flogging; but fetters are to be used only when the patient is violent, for then he is not to be trusted. Celsus himself errs when he declares that madmen are sometimes duped not by their own mind but by phantoms (*imagines*, i.e., visual hallucinations), as the poets said Ajax and Orestes were; in others the mind itself is at fault: they simply lose their reason (*animo desipiunt*). In some cases deception on the part of the doctor may be justified, as when pretended legacies were from time to time announced to a wealthy patient who was in dread of starvation. The empty fears (*vani metus*) of this patient apparently dissolved though we are not told if their disappearance was permanent or not.

In a later chapter Celsus remarks that "one of the best known maladies is that which is called *comitialis*, or the greater;"[49] an accurate description of grand mal follows. In the fourth book he speaks of hysteria as a violent malady which arises from the womb of women; he gives its differential diagnosis from epilepsy, and the treatment he recommends includes the affusion of cold water.[50]

---

46. JUVENAL 6. 236; 13. 98; 14. 252.

47. CELSUS, *De Medicina* 3. 18. 1-3. On dementia, *cf.* n. 52.

48. CELSUS, *op. cit.* 3. 18. 17: Alterum insaniae genus est, . . . Consistit in tristitia, quam videtur bilis atra contrahere.

49. CELSUS, *op. cit.* 3. 23. 1. Celsus nowhere uses the name epilepsy, probably because it was held to be ill-omened; the name *comitialis morbus* was given to it because a meeting of the *comitia* was adjourned if anyone there was attacked by it; it was looked upon by most as a divine manifestation (*cf.* the name Sacred Disease).

50. CELSUS, *op. cit.* 4. 27. 1B.

The subject of mental disorder was handled in a variety of ways by men of letters: for instance, Cicero (106-43 B.C.) was seriously interested in it; Horace (65-8 B.C.) satirized the Stoic paradox that all unwise persons are mentally unsound, and told of the delusion of a man of Argos; Juvenal (born A.D. 50) aptly described dementia; and Martial (A.D. 40-104) ridiculed the selfish hypochondriac. In the year 87 B.C. the blind Stoic Diodotus was received into the house of Cicero's father, so that Cicero was familiar with Stoic tenets, which he was wont to discuss. He reviews topics at once of some terminological, nosological, and clinical importance in the *Tusculan Disputations,* in which his unknown interlocutor discourses upon conditions ranging from emotional lability to frank psychosis: he begins with certain irrational emotions called by the Greeks, that is, the philosophers, *pathè* (πάθη); he names them and thinks they are disorders (*perturbationes*) and not diseases (*morbi*). But the term unsoundness (*nomen insaniae*), he continues, means sickness and disease of the mind, that is, a condition of unhealthiness of soul which the philosophers have termed unsoundness.[51] After referring to amentia and dementia he confesses he cannot readily give the origin of the Greek term μανία (*mania*); Romans, he says, discriminate better between *insania* and *furor* (frenzy) than do the Greeks: the former term, from its association with *stultitia* (folly), is the more general one. He then complains that "what we call frenzy, they call *melancholia* (*quam nos furorem, μελαγχολίαν illi vocant*)," and he cannot agree that the mind is influenced by black bile: rather is it in many instances influenced by the stronger power of wrath, fear, or pain, in the sense in which one speaks of the frenzy of Athamas, Alcmaeon, Ajax and Orestes.[52]

In the *Academica* Varro speaks and refers to Zeno's idea of perception as a certain kind of external impression joined with the act of mental assent "which he called *phantasia* and we may call a presentation (*visum*)."[53] Later, he considers unreal presentations (*inania visa*) "whether depicted by the imagination, which we admit frequently to take place, or in slumber, or under the influence of wine or insanity." All presentations of this kind, he insists, are devoid of perspicuity.[54]

Horace's satire upon the Stoic paradox— and it is the longest of his satires—teems

---

51. Cicero, *Tusculan Disputations* 3. 7-8.

52. Cicero, *op. cit.* 3. 10-11. *Amentia* cannot have any meaning here but its modern one; but *dementia* need only mean, and often does only mean, mental derangement (*cf.* "*continua dementia*" of Celsus, ref. 47), in contrast to its obviously modern sense in the passage from Juvenal cited (ref. 58). The root of μανία [mania] comes in the related words μένος, μέμονα, μαίνομαι [mainomai], memini, mens, mind (King, J. E. 1927: Cic., Tusc. Loeb Classical Library, p. 236, n. 4). But the speaker was more worried by the lack of a Greek word to correspond with *furor* as he understood it, i.e., as a special type of insanity corresponding approximately to mania in its modern sense. It has been seen that, in Homer, *mainomai* indicated a state of frenzy; but it is true that *mania*, a noun not found in Homer, later became a general term for insanity; and even much later Caelius not only equated it with both *furor* and *insania*, but preferred the former to the latter term, which had been coined by philosophers (see refs. 69 and 70). The speaker's complaint seems justified; but in equating *furor* with *melancholia* he was quoting philosophers and not doctors, who would scarcely have agreed. To Celsus, who lived not very long after Cicero and who cited earlier authorities, melancholia certainly did not have the clinical features of furor in the sense of frenzy. We cannot wonder that, in such times, there was a groping after meaning; moreover, the idea of black bile seems to have bedevilled attempts to give melancholia its proper nosological position: it is often suggested that it is a systemic organic disturbance, as opposed to other functional mental disorders.

53. Cicero, *Academica* 1. 40.

54. Cicero, *op. cit.* 2. 51

with terms referring to mental states.⁵⁵ There can be no doubt that schizophrenia has existed from the earliest times: some of the patients described by Caelius must have had it;⁵⁶ and the man of Argos, whose history Horace relates, probably had it too: his delusion was that, though the theatre in which he sat and applauded happily was empty, he always thought he was listening to wonderful tragic actors; cured, it was thought, by the panacea hellebore, he fully recollected his former state and was upset at being deprived of "his dearest illusion (*mentis gratissimus error*)."⁵⁷

Juvenal depicts dementia as follows:

Worse by far than all bodily hurt is dementia: for he who has it no longer knows the names of his slaves or recognizes the friend with whom he has dined the night before, or those whom he has begotten and brought up. And by a cruel will he disinherits his own and makes over all his property to Phiale; well does she understand how to entice him with her *halitus oris*: not for nothing had she set herself up in a brothel years before.⁵⁸

Martial ridicules a hypochondriac, whom he calls Polycharmus, in these four lines:
Aegrotas uno decies aut saepius anno,
nec tibi sed nobis hoc, Polycharme, nocet:
nam quotiens surgis, soteria poscis amicos,
sit pudor: aegrota iam, Polycharme, semel.⁵⁹

Rufus of Ephesus practised during the reign of Trajan, but it is unlikely he went to Rome; he was the immediate predecessor of Galen and Aretaeus of Cappadocia. Writing upon the interrogation of the patient he stresses the importance of obtaining as complete a history as possible both from him and from his relatives. He makes it clear that, with the taking of the history, the examination has already begun: the patient's answers will not only indicate if he is mentally sound or unsound; deafness may be discovered, and his appearance may be an indication of physical strength or of weakness or disease. But it is mental disturbance—and Rufus gives suitable illustrations—which is most readily unmasked by the method of questioning.⁶⁰

Although Aretaeus the Cappadocian was Galen's contemporary in the second half of the second century, the one never referred to the other; whether this was due to professional jealousy or not is unknown. Familiar, as all the physicians were, with the clinical features of epilepsy, Aretaeus writes that the subject of an attack may have incontinence of urine, faeces, and semen.⁶¹ Hysterical suffocation, he observes, is after the form of epilepsy but convulsions are absent; he accepts the Hippocratic theory that hysteria is due to upward displacement of the uterus.⁶²

In the treatise upon the causes and symptoms of chronic diseases he says that, having described the epileptic attack, he wishes to draw attention to the condition of the patient suffering from the disease "when it is inveterate": he may become spiritless, unsociable, stupid, and inhuman;

---

55. Horace, *Satires* 2. 3. Rich as Latin is in such terms Greek can more than match it.

56. See ref. 70. Even more suggestive of schizophrenia are the histories of some patients of Alexander of Tralles and the reports of one or two other observers quoted by Heiberg, J. L.: *loc. cit.*

57. Horace, *Epistles* 2. 2. 128 ff.

58. Juvenal 10. 232-239.

59. Martial, *Epigrams* 12. 56.

60. Rufus of Ephesus, in Brock, A. J., 1929: *Greek Medicine, Being Extracts Illustrative of Medical Writers from Hippocrates to Galen*. Translated and Annotated by A. J. Brock. London and Toronto, Dent. New York, Dutton, pp. 113 ff.

61. Aretaeus The Cappadocian, in Adams, F., 1856: *The Extant Works of Aretaeus, The Cappadocian*. Edited and Translated by Francis Adams. London, Sydenham Society, pp. 3 ff. (Greek), 243 ff. (English).

62. Aretaeus The Cappadocian, in Adams, F.: *op. cit.*, pp. 44 ff. (Gk.), 285 ff. (Eng.).

his utterance is often indistinct, either from the disease or from trauma caused by attacks; "the disease sometimes destroys the understanding, so that the patient becomes altogether fatuous."[63] Of insanity ($μανίη$) he observes: its modes are infinite in species but one in genus; it is a chronic derangement of the mind without fever, and it is thus distinct from the deliriums of drunkenness and from mental disturbance caused by certain edibles such as mandragora and hyoscyamus. Further, "the insanity of which we are speaking has no resemblance to the dotage (*lêrêsis*) which is the calamity of old age;" this never remits and it accompanies the patient to the grave, whereas insanity intermits and with care ceases altogether.[64]

All the medical writers mention melancholia and almost all are preoccupied with the notion of black bile that its name indicates. But to Aretaeus, who supplies a dubious etymological explanation of his attitude, black bile was not a *sine qua non* of melancholia as he understood it. While even Hippocrates, besides laymen, sometimes used the term (or its adjective) loosely, it is evident that Aretaeus was conversant with the characteristics both of the depressive state and of cyclothymia: Melancholia, he says, "is a lowness of spirits (*athymia*) from a single phantasy" and fever is absent. It appears to be the beginning of a part of mania. In those suffering from mania the understanding is sometimes turned to anger, sometimes to joy; but in the melancholics to sorrow and despondency only; they may be suspicious of poisoning (paranoid states are not distinguished from melancholia) and become withdrawn and superstitious; some complain of life and wish to die. If a remission takes place hilarity usually supervenes; but these patients are nevertheless insane.[65]

The following excerpts from Galen's writings bear witness to his experience and sagacity: In the case of a woman who was said to be suffering from insomnia and who refused to be questioned Galen, after excluding fever, formed the opinion that she was affected either by a melancholy dependent on black bile or by trouble about something she was unwilling to confess. A fortunate circumstance enabled him to prove that the cause of her symptoms was a secret love for the dancer Pylades. It matters little that Galen, who liked to report a dramatic success, may have embellished his report or, as he admitted, have had unusual luck: his method and reasoning resembled those of a careful modern clinician. His diagnosis, in the case of a slave, of what we would term an anxiety state proved to be correct and, with the cooperation of the patient's master, he removed its cause.[66]

Writing upon the detection of malingers, he mentions a number of ways by which disease may be simulated and reports two illustrative cases: He found that the cause of a man's ostensibly severe pain was reluctance to go to the assembly, to which he had been summoned by the citizens. Suspecting that a lesion of the skin of another man was factitious, he marshalled the evidence and confirmed the diagnosis in a manner that seems unexceptionable.[67]

---

63. ARETAEUS THE CAPPADOCIAN, in Adams, F.: *op. cit.*, pp. 54 ff. (Gk.), 296 ff. (Eng.).

64. ARETAEUS THE CAPPADOCIAN, in Adams, F.: *op. cit.*, pp. 58 ff. (Gk.), 301 ff. (Eng.).

65. ARETAEUS THE CAPPADOCIAN, in Adams, F.: *op. cit.*, pp. 55 ff. (Gk.), 298 ff. (Eng.). *Athymia* appears in the line: "Ἔστι δὲ ἀθυμίη ἐπὶ φαντασίῃ. Cf. n. 38.

66. GALEN, in Brock, A. J.: *op. cit.*, pp. 213 ff.

67. GALEN, in Brock, A. J.: *op. cit.*, pp. 225 ff.

Caelius Aurelianus lived in the fourth or fifth century and his two treatises, one on acute and the other on chronic diseases, are considered to be translations of a work of Soranus of Ephesus. In the treatise on acute diseases he deals with phrenitis: This never occurs without fever;[68] similar and kindred to it in respect of loss of reason are mania (Caelius defines it as *furor*) "commonly called *insania* (quam vulgo insaniam vocant)", and melancholy. The absence of fever usually serves to distinguish these from phrenitis. But in rare instances mania may at some stages be accompanied by fever; here, however, the loss of reason precedes the fever, while in phrenitis the fever precedes the loss of reason. Caelius adds that it may, from the history, be impossible to determine the beginning of a fever: the state of the pulse may or may not be of help, but *crocidismos*, "a sort of plucking of threads from the covers" and *carphologia*, "a sort of plucking of small pieces of straw from the walls," occur only in phrenitis. Inexperienced doctors may be deceived and think the patient suffering from phrenitis is asleep when he has passed into stupor or coma (*in lethargiam*), and vice versa.[69]

Chronic mental states are discussed in the work devoted to chronic diseases under the heading: *De furore sive insania, quam Graeci manian vocant* (on madness or insanity, called *mania* by the Greeks). After recalling the statement in Plato's Phaedrus that there are two types of *mania* (madness), one from bodily causes and the other divine or inspired, he says that he intends to consider (as becomes a practical physician not concerned with metaphysics) only the type that arises from bodily disease or indisposition. He repeats that mania, as opposed to phrenitis, is a chronic impairment of reason without fever; some cases are mild, some severe. He describes the symptoms of the manic state (in the modern sense of the term) and of depression and phobias; and he mentions a number of patients who had delusions. He does not agree with Apollonius and the followers of Themison that melancholy should be considered a form of mania (he again means in the general sense of madness); but the reasons he gives for distinguishing the two are unsatisfactory: they are based on strange ideas of pathology and the Methodist theories of *strictura* (state of stricture), *solutio* (state of looseness), and *complexio* (mixed state).[70] Caelius has a lengthy passage on treatment: over-enthusiastic medicinal therapy and the use, in most cases, of restraint by bonds are both condemned; the one may be harmful and cannot affect the course of the disease, while the other, if not properly applied, may cause injury and is less effective than the hands of servants.[71]

When the works of writers from Celsus to Caelius are compared, it appears that the writers drew upon wide experience supported by a body of knowledge that, by the time of Caelius, had existed for centuries: their views of some common types of mental disorders are basically similar and reveal an acquaintance with their course and prognosis. Divergence of opinion, depending often on the School to which the practitioner belonged as well as on personal conviction, is naturally most noticeable in speculative theories of pathology and treatment, though all may show common sense in the management of patients.

---

68. CAELIUS AURELIANUS, *On Acute Diseases*, 1. 5.
69. CAELIUS AURELIANUS, *op. cit.*, 1. 42-49.
70. CAELIUS AURELIANUS, *On Chronic Diseases*, 1. 144-152 and 1. 183.
71. CAELIUS AURELIANUS, *op. cit.*, 1. 155-179.

In an account designed in the main to discover what some Greeks and Romans said and thought about mental disorder and the meaning of a few of the terms employed much has had to be omitted: not only have beliefs among other ancient races not been considered—and their systematic study requires special knowledge—but a number of Greek and Roman sources have had to be disregarded; and, in a longer review, those used could have been used more freely than they have been. Nor has it been possible to discuss figures of the Greek and Roman world whose mental state has interested both doctors and historians and sometimes excited controversy: for instance, the personality, constitution, and actions of the early Roman emperors who were savagely attacked by Tacitus and Suetonius are of interest to the physician and the psychiatrist.[72, 73, 74] And other ancient writers have left accounts of psychotic or eccentric persons who created a stir during their lifetime and even after their death; it must suffice to name several authors and their subjects: Menecrates of Syracuse was a physician who, according to Athenaeus (*fl. c.* A.D. 200), had the delusion that he was Zeus;[75] Philostratus (born *c.* A.D. 170) has described the strange life of Apollonius of Tyana, who long after his death was the subject of a controversy between the pagan Hierocles and the Christian bishop Eusebius;[76] Lucian (*c.* A.D. 155-*c.* 200) was a witness of the spectacular suicide of the Cynic Peregrinus, who lit and threw himself on a funeral pyre after the Olympic Games, and has reported it in sarcastic and unsympathetic language.[77] There are numerous instances of suicide in antiquity and all have their modern parallels: some committed it because they were mentally unstable; others, such as Socrates and the victims of the Roman emperors, from necessity; and others, for example, Cato the Younger, because they thought suicide was the only honourable course.[78]

It can be said in conclusion that many types of mental abnormality were known in classical antiquity and that the following are to be identified in the descriptions of the writers who have been cited: hysteria, hypochondriasis, eccentricity, and emotionalism of various kinds; proneness to simulation of disease and self-inflicted injury; the disorder of personality and antisocial behaviour of the criminal psychopath; brutalization in war and revolution; anxiety, phobic, and obsessional states; febrile delirium and intoxication by alcohol and drugs; insanity from or associated with alcoholism; paranoid, delusional and hallucinatory states, including schizophrenia;

---

72. Esser, A., 1958: *Cäsar und die Julisch-Claudischen Kaiser im Biologisch—Ärtzlichen Blickfeld.* Leiden, E. J. Brill.

73. Sandison, A. T., 1958: The Madness of the Emperor Caligula. *Med. Hist.,* 2:202.

74. Moss, G. C., 1963: The Mentality and Personality of the Julio-Claudian Emperors. *Med. Hist.,* 7:165.

75. Athenaeus, *Deipnosophistae* 7. 289A-289F. For a detailed study of Menecrates, his circle, and his times see Weinreich, O. 1933: Menekrates Zeus und Salmoneus. *Tübinger Beiträge zur Altertums-wissenschaft,* 18. Stuttgart, Kohlhammer. The monograph is in three parts: in the second part the case of Menecrates is compared with similar cases of later and modern times; in the third part he is compared with the mythical Salmoneus who, according to the legend, pretended to be Zeus (see Vergil, *Aeneid* 6. 585 ff.).

---

76. Philostratus, *Life of Apollonius of Tyana.* Here magic, superstition, and demonism run riot; it has often been pointed out that they were rife among pagans of the early and later Christian era as well as among Christians and Jews. This work appears in two volumes in the Loeb Classical Library's edition, with English translation by F. C. Conybeare. The second volume also contains The Epistles of Apollonius and The Treatise of Eusebius.

77. Lucian, *The Passing of Peregrinus,* especially 35-45.

78. Plutarch, *Life of Cato Minor,* especially 66-70.

epilepsy and mental deterioration in some epileptics; mania or hypomania, depression and manic-depressive psychosis; dementia.

In the field that has been surveyed there is probably nothing for which a modern parallel could not be found.

REFERENCES AND NOTES

A circumflex accent marks a vowel of a term transliterated from Greek into English as long: such a word has not been latinized, but a latinized form has been preferred for proper names and for some familiar words such as psyche and dysthymia. When Hippocrates is cited the name Jones, followed by the number of the volume and page, refers to the Loeb Classical Library's (1948) edition of Hippocrates, with English translation by W. H. S. Jones; the name Littré, followed by the number of the volume and page, refers to Oeuvres Complètes d'Hippocrate, with French translation by E. Littré, Paris, J. B. Baillière et Fils, 1839-1861; the reference to the Hippocratic treatise cited follows. References to the works of Caelius Aurelianus are from the edition of I. E. Drabkin, Chicago, The University of Chicago Press, 1950. It is acknowledged that translations of occasional lines from the works of some authors are those of the translators of the Loeb Classical Library's editions; a few of these have been slightly altered, mainly for terminological reasons. Other borrowings have, it is hoped, been acknowledged.

Chapter 56

Mental Diseases of Ancient Mesopotamia

J. V. KINNIER WILSON

In a recent Assyriological paper entitled "An introduction to Babylonian psychiatry," a personal attempt has been made to pierce the veil which has long surrounded the whole subject of psychological medicine in Ancient Mesopotamia (Kinnier Wilson, 1965). The present chapter, although philologically less technical, runs closely parallel to this study, and provides a welcome opportunity to share some of the basic ideas with a wider public.

The importance of the subject is that the documents to be considered constitute the very first chapter in the long annals of psychiatry. If one should ask their date the answer is that, although all is preserved only in later copies, it is likely that, as with other medical texts, most of it was first set down in writing in the Old Babylonian period or during the first half of the second millennium B.C. The material is surprisingly extensive and our present endeavours cannot pretend to be more than the first move in the attempt to understand it. Many problems remain which must await the judgements of such authorities as may be stimulated to join the field.

PSYCHOSES

Being free to choose, we may begin by entering the world of delusion. Unlike the delusions of everyday experience those of psychological medicine are beliefs which are stubbornly held, very improbable and impervious to reason. In a first easy text edited by Ebeling (1915 and 1931) we may hear part of the formula of words which a patient was instructed to recite in cases of "rebellious speech." The instruction requires the patient to say: "Having acquired some of the spittle of your mouth, I am assigning the words of . . . [(supply) 'your father,' 'your mother,' 'your sister,' 'love-boy or city prostitute' (as the case may be)] into the all-hiding earth, which opens not her mouth, whose tongue is not rebellious."

What is noticeable about the persons mentioned in this text is that they are all capable of being loved. It may thus be suggested that some frustration or denial of love has converted the subject's love for (one or more of) them into hate, whereupon the mind's "projection" mechanism has turned the thought of "I hate them" into that of "they hate me which justifies me in hating them," and false ideas that they are trying to annoy or provoke him are assisting the patient in personally justifying this new hatred. It is typical of Babylonian psychiatry that the patient often had to perform his own ritual: but the above instruction is not one that any patient is likely to have carried out.

Delusions of another kind are provided by a treatise known as é-gal-TU-ra, a Sumerian term literally meaning "entering the palace." At a first glance this work looks as though it concerned a person who wished to complain to the king of personal injustice: but the associations of the term in other contexts make it clear that a mental state is thus involved, and the whole manual was as certainly compiled by priests and not by any member of the palace administration.

In fact all the evidence points to the conclusion that the work preserves the verbatim utterances of certain schizophrenic patients who believed they were the centre of a conspiracy. Thus when we read (KAR 71 Rev. 20-22):

> It is said that I took over the watch from the guard commander, that I took over the watch from the city magistrate, that I opened the city-gate and let in the enemy,—but before my lord (the king) I swear I have done no such thing,

we may be sure that no one has made any charge at all, and that the patient's fears are but delusional misinterpretations, a fiction of his own mind. A second charge ends with the comment (BM 103385, Rev. vi, 19-20):

> If only female judges could hear my defence I would be [judged pure] as the purest oil, as the purest water.

The comment would have seemed logical to the patient; but since the charge against him was of his own making he would not even have been tried by male judges.

Further point to the diagnosis is provided by the complaint of KAR 71 15 with LKA 107 11, "From one league, two leagues, away you scorch and blaze with wrath," which may convincingly suggest that the patient was suffering from ideas of influence (or passivity), and in the *bēl dabābi* of KAR 71 Obv. 10 and Rev. 3 whose tongue is to be silenced one meets a "conspirator at court," in this instance one of the many persecutors of the ancient psychological scene (see further below). Moreover, in schizophrenia "the patient's attitude to his delusions is typical in that he does not usually act appropriately and thus will go to his doctor where it would have been more logical to have gone to the police" (Curran and Partridge, 1955). This statement explains as perhaps none other can the essential meaning of é-gal-TU-ra. Although going to the doors of the priests the patients nevertheless spoke so much of "going to the palace"—the one place in the land where true justice might be sought—that this was the name the priests eventually came to give to the condition. In fact the palace itself would probably seldom have been involved.

Delusions may be taken a step further by considering the second text of the group published by Ebeling (1949). The passage of our interest translates initially, "If a man has *bēl limutti*'s who persecute him with reb[ellious(?)] tongues, spread rumours, tell tales about him and slander him; if whoever speaks with him [is supposed to] speak nothing but untruths; if, without his knowledge, witchcraft, spells, magic, or other evil machinations of men are turned upon him; and if god, king, superiors, elders, or any officer of the palace household or administration (supposedly) have a grievance(?) against him or are angry with him . . . [perform the following ritual]."

In this text interest centres on the *bēl limutti*. These "mischief-makers" appear to be those rather vaguely sketched persecutors whom the psychotic patient so often refers to simply as "they;" indeed, the delusions of persecution are very clear and carry the hallmark of the technical verb *redû*, "to persecute," itself. What is involved is that the patient believes in an organisation supposedly determined to ruin him, so that in the absence of hallucinosis or affective disorder, and with the evident preservation of coherent thinking, the diagnosis is likely to be paranoia seen in the second, or persecutory, stage according to Magnan's analysis. A secondary interest is that the text introduces two of the most important clichés of Babylonian psychiatry, firstly the "witchcraft, spells, magic, or other evil machinations of men," evidently considered to be the causing agencies of many of the delusions and ideas of psychotic states, and secondly the "god, king, superiors and elders," being the standard phrase for those in authority who are so often the target of the mental patient's outbursts and accusations. With regard to the latter phrase, and the apparent "anger" of these officials with the subject, it should be realized that this anger was almost certainly projected. Not only were the officials *not* angry with the patient, but such being the mental journeys that deluded subjects can make, the chances are that they were not even aware of his existence.

These officials occur again in one of the most inescapable of all Babylonian psychiatric texts. This is AMT 96, 7 with KAR 26, edited, although without analysis, by Ebeling (1915). A translation of the integrated prayer to Marduk, the national god of Babylonia, has been given by von Soden (1953) and by seeing the conditions mentioned as "vielleicht Geisteskrankheiten" von Soden has already paved the way for the following interpretation. "Vielleicht" may in fact be deleted: in the light of its own introduction, *šumma amēlu antašubbû bēl ūri* . . ., the text is to be placed in the first instance under the heading of epileptic psychosis.

The symptoms are given as follows: "If a man is currently suffering from major or minor epileptic attacks . . ., and an *alû*-demon then begins to inflict him with ideas of persecution so that he says—although no one will agree with him that it is so—that the finger of condemnation is being pointed at him behind his back and that god or goddess are angry with him; if he sees horrible, alarming or immoral visions and is (consequently) in a constant state of fear; if he engages in periodic outbursts of anger against god or goddess, is obsessed with delusions of his own mind, evolves(?) his own religion, and says—although (again) they will not allow it—that his family are hostile towards him and that god, king, his superiors and elders treat him unjustly; if all his muscles are subject to paralysis, if his eyes exhibit colours of red (or brown), yellow and black, if he has some speech disorder with spells of forgetfulness, has no desire for female relationships and no inclination to pursue any activity [at all] . . ., (details of the action to be taken follow)."

In this syndrome the delusions (*nullâte*), the suspicions, the vivid hallucinations (translated "visions"), the religiosity, disturbance of memory and loss of interest, will all suit the diagnosis suggested. The symptom of paralysis is taken to indicate that post-convulsive epileptic paralysis which does not long maintain itself. The eye condition is probably not to be interpreted as I have suggested (1965), in terms of visual hallucinations of the "flashes of light" variety, but rather as ecchymoses around the eyes arising from one or more convulsions of a particularly violent nature.

It may be mentioned that line 53 of the text provides another example of the part thought to be played by "witchcraft, spells, magic and other evil machinations" in the production of mental symptoms. The line now becomes our cue to discuss the medical implications of witchcraft, which may be studied through the Babylonian priestly writings in a form possibly more understandable than anywhere else in the ancient world.

The witch is, in part, one of the lost images of psychological medicine. To be sure, despite her massive non-existence, she entered the world at large even as in mediaeval Europe, and undoubtedly even the priests believed in one form of her since, in days before psychology and psychopathology, they could find no other explanation for hostility ($z\bar{\imath}ru$), suicidal tendencies ($zikurud\hat{u}$), and certain other phenomena except in terms of her work. But once a mental subject began to believe in a witch her character changed; she became conditioned by his own illness and his own mind, becoming variously a seducer, snake-charmer, cloud-gatherer, or the like,—and here the priests were not deceived and attempted to give treatment accordingly.

Our concern in this study is with the mental attitudes of the "bewitched," and therefore with the treatise of $Maql\hat{u}$, literally "the Burning," a long work of nine Tablets for which the standard edition is that of Meier (1937). It was principally a collection of incantations concerning witches and persecutors, and its name derives from the fact that, in accompanying rituals, images of such persecutors were made out of certain woods, or otherwise of fat, wax, dough, clay, pitch or the like, and then slowly burnt or melted before the patient in some attempt at psychotherapy. This, at least, is half the story. The other half of it is concerned, firstly, with those witches who were themselves reported as having burnt images of the patient, and then as having buried the ashes in drains, potters' kilns, crossroads, or sundry other unlikely places; secondly, with the grandiose *muštepištu*, the witch who might cause another witch to bewitch the patient; and thirdly, with those again rather recherché witches ($Maql\hat{u}$ IV 119-123) who, being either Elamite, Qutian, Sutian, Lulubian or Hanigalbatian were foreign to Babylon (and as such possibly creations of a mind mischievously determined to confound Babylonian priests or gods supposedly able to control only Babylonian witches)—in short, with witches who had nothing to do with persecution in the proper sense of the term but who were rather the central figures of bizarre confabulation. In many primitive societies the witches belong to social anthropology: in $Maql\hat{u}$ the essential conflicts are mental and personal and belong under the heading of psychosis.

To understand this more clearly it may be said that there is an unmistakable element of illusion in $Maql\hat{u}$. Thus the incantation of III 140ff, "O potsherd of the streets, why are you ever hostile to me?" reveals at once the misinterpretation of something actually being presented to the senses. The potsherd is in fact being interpreted as a witch (*kaššaptu*, line 142) and all that need be said further is that the psychotherapy "turns the tables" on the witch—this being one of the fundamental procedures of Babylonian psychiatry—so that when the patient answers her back (153ff):

I am the spike of a thorn bush, don't you dare step on me;
I am the sting of a scorpion, don't you dare touch me;
I am a jagged mountain, see that your witchcraft, spells, magic and evil machinations don't come near me!

he is being encouraged to think that she in her turn will be suffering from illusions when she looks at him.

For the most part, however, the element of illusion is disguised. Thus it will not be readily apparent in the following (*Maqlû* I, 103-106):

They (my witches) have made me eat bewitched food,
They have made me drink bewitched water;
They have washed me in filthy wash-water,
They have anointed me with salves made from evil drugs.

Nevertheless specific delusions presuppose specific causes, and here are doubtless the delusional elaborations of illusions—the illusion that certain sensations of the skin or body, whether in origin psychical (hallucinations of touch and taste) or physical (paraesthesiae, or fleeting aches and pains) have come to the patient from some source outside of him. It is in this context that the *mušlahhatu* or "female snake-charmer" (*Maqlû* III 43, IV 126, etc.) fits very well, for following subjective feelings or sensations attributed to "poison" the snake-charmer becomes the end result of the patient's theorising as to the cause of the poison. Of many difficult kinds of witch the *mušlahhatu* is, in fact, one of the most understandable. "Two years have gone by," wrote Sadler, 1945, of a certain patient, "and he has not ceased to entertain the idea that a vast number of conspirators are working to poison him." To be sure the witch has gone: but only because nobody, not even the mentally ill, believes in witches any more.

Two further considerations directly connect the witch with psychological medicine. The first concerns those physical complaints which the patient might ascribe to her influences, saying that "they push in my chest, weaken my heart, bind my arms, bend my back, pluck out my hair,"— or do much else besides (*Maqlû* II 31ff., etc). This catalogue of complaints evidently includes observations taken from more than one case, and while they may collectively be interpreted in terms of feelings of passivity, hypochondriasis (of a sort), or even psychomotor hallucinations, it is difficult to think of any normal experience which could easily provide a solution. Of the three suggestions offered I personally prefer the first: for if one should ask what was the main historical antecedent of the rays, wireless waves and electricity which to-day feature in ideas of passivity I believe the answer may be known. It was an influence no less mysterious—the "magic" of witchcraft.

The second point concerns the witch's fellow persecutors. In *Maqlû* I 73ff., II 38ff. and a text published by Lambert, 1957, full lists will be found, headed by "witches," "sorcerers" and "enchanters," and followed by *bēlu*-names of which the *bēl dabābi* and *bēl limutti* we have met before, thus cross-checking with two other psychiatric texts. The *bēl rēdi* is some sort of persecutor by virtue of his very name. The *bēl ikki* which incorporates the word *ikku*, "temper, irritability," is someone who evidently annoys or irritates the patient, but he becomes more understandable when seen as the figure of a projection. For technical reasons which concern a word *egirrû* as meaning "a spoken or overheard phrase of ominous significance to the hearer," the *bēl egirrî* was probably not seen at all, being doubtless that "voice" which may issue short commands or comments, sometimes feared, sometimes respected, in auditory hallucinations. Thus one may tell the witch by the company she keeps.

We may leave the subject of witches and delusions by briefly discussing a condition which is unique to psychiatry. As was noted by Lambert (1957) a term íd-gur-ra literally means "return from the river ordeal acquitted." It occurs in a list giving names of other mental states and it probably means what it says: the patient believes he has returned a free man from the river ordeal in which he and a "witch" have been contending for life. While not a great deal is known about this contest, it is referred to in the second law of Hammurabi's code and in an Old Babylonian letter from Mari on the Middle Euphrates and is undoubtedly to be accepted as an historical fact. The explanation of the term which one may thus suitably offer is that íd-gur-ra was a delusion whereby the patient's fear of death in the ordeal, giving rise to an obsessive-ruminative psychoneurosis not finally to be tolerated, became permanently suppressed into the subconscious.

NEUROSIS AND THE PSYCHOPATHIC STATE

A Babylonian text-book known as *Shurpu*, and running closely parallel to *Maqlû* even to the extent that its material is also presented in nine Tablets, forms the main subject of study for conclusions reached in this section. The work has been well edited by Reiner (1958) and in the following much abbreviated account of it offered translations differ in no major particular from those given in her edition. In the three extracts from the second Tablet cited below the observations were obviously based on a sample of many patients, so that while it might appear that the priest was supplicating on behalf of a single man, the document as a whole probably had no practical application.

I. Be the [mystery?] resolved, in that So-and-so the son of So-and-so does not know it is wrong . . . when he gives with a small measure (37) . . . uses a false balance (42) . . . takes money not lawfully his (43) . . . sets up a false boundary stone (45) . . . enters a friend's house, has intercourse with his friend's wife, sheds his blood and steals his clothes (47-50) . . . when his mouth says "yes" but his heart says "no" and whatever he says is completely untrue (56-7), when he . . . s, shakes and trembles [with rage], destroys [things], throws them out [of the house] or makes them disappear; when he accuses, incriminates, spreads gossip, wrongs, robs or incites others to rob (58-61) . . . when he disarranges the prepared altar-table and angers his god and goddess (79-80). . . .

Here is the Babylonian psychopath—the pathological liar, the swindler, the kleptomaniac, the gossipmonger, the social misfit, the sexual criminal and the murderer—an unmistakable picture.

II. Be the [mystery?] resolved, in that he does not know why he is compelled to take [things], to hide [things] (83-4) . . . to point the finger (of condemnation) at a protecting deity (87) . . . to step in blood or walk about over a place where blood has been shed (93-4) . . . [or why] he has a phobia of meeting an accursed person or of an accursed person meeting him, or of sleeping in the bed, of sitting in the chair, of eating at the table, or drinking from the cup, of an accursed person (98-103). . . .

This section is different and concerns the obsessive-compulsive state. In that phobias are closely related, being irrational obsessional fears, it is perfectly reasonable that the ancient document should have classified the two conditions together. The "accursed person" and the bed, chair, table and cup, belong in fears of contamination (an aspect of phobia), and the patient will be at once recognized by psychologists. If one could have entered his house he would have been found engrossed in a spate of washing, during which the furniture, as also the cup, would have been washed many times over.

III. [Be the mystery? resolved in that he does not know why] he has a morbid fear [or "aversion," *mašâltu*] for beds, chairs, tables, lighted stoves, lamps . . . stables, animals, ploughs, wells, etc. (105 ff.) . . . or for the sanctuaries of [this or that] god or goddess; or why he fears to leave or enter [such and such] city, city-gate or house, or fears (such and such) street, temple or road (122 ff.).

This passage concerns fear of things and fear of places. For lack of information we shall not get far with the first of these groups, but the reason for the fears of the places would be more obvious. In origin they are likely to have been fears for the recurrence of an anxiety attack having an aetiology somehow related to the place, but of which the details have been suppressed into the subconscious together with the guilt, shame, or other emotion originally involved.

If at this point the reader would enquire how it is that, in Babylonian thought, the psychopathic state became so closely linked with obsession, an answer may be attempted here. All hinges on a most difficult, but altogether central, word *māmît*, a derivative from the common verb "to swear," and meaning "oath." However, in psychological contexts this same word evidently means "compulsion," either a compulsion to do, or a compulsion not to do (and, therefore, a fear or phobia for doing) a certain act. The argument must be that such "compulsions" were seen to be so unbreakable as behavior habits—as indeed they are—that, to an outsider, it was as if the subject had sworn an "oath" to do, or not to do, the action involved. Thus interpreted the word *māmît* cuts across all modern schemes of classification, and indeed certain aspects of both schizophrenic and hypomanic behaviour appear to have been described by the use of this term. Even expressive aphasia, the presumed diagnosis of TDP 178 17 which says, "If a man ceaselessly pronounces words of salutation, he is suffering from the *māmît*-illness," is involved in the idea; nor is the logic of it difficult to follow, for if a patient cannot stop crying "Peace be unto you," "Peace be unto you," and so cries *ad infinitum*, one may easily allow that this also, in a sense, is "compulsive" behaviour.

We may conclude this section with a few examples of such compulsions taken from the third Tablet of *Shurpu* which lists over a hundred and seventy of them, selecting the examples so as to provide some secondary headings (given in italics). Thus on the "positive" side of compulsion a patient may be compelled—

35 To slaughter a sheep and touch the death-wound [*blood lust*].
70 To smash doors or door-locks [*malicious conduct*].
16 To implore (stretch out the hands to) the Lamp-god [*pyromania*].
97 To put the breast into the mouth of a child [*sexual perversion*].
144 To lie and blaspheme [*typically psychopathic behaviour*].
84ff. [To play] "Laughing Angels," "Wandering Demons," "Returning Ghosts," "Sneaking Devils;" [to play with] drum or kettle-drum, timbrel or cymbals . . . [to dig] pits and ditches (for the unwary to fall into) [*nerve-racking, noisy, and delinquent behaviour probably to be characterised as childish*].

Similarly, on the negative side of compulsion the patient could have a fear or phobia—

115 Of (certain) days or months . . . [*based on suppressed experience*].
123 Of hunger or hardship [*based on suppressed experience?*].
124 Of having the name of a god invoked in his presence [*guilt complex?*].

131 Of eating an accursed man's food [*fear of contamination*].

It may be emphasized that these entries represent no more than a fraction of the whole catalogue. To do full justice to the thoroughness of Babylonian observational science, *Shurpu* Tablet III, man's first attempt at the classification of "compulsive" behaviour, should be read in its entirety.

DISEASES AND DOCUMENTS

A great deal of work has yet to be done on the proper editing of the text-books of Babylonian psychiatry. In the article referred to at the beginning of this chapter some attempt has been made to indicate what these are, and this may be consulted also for a psychological appraisal of a document thought to be written by a "patient" and hitherto known as "The Poem of the Righteous Sufferer."

Of the priestly writings, three—edin-na-dib-bi-da, šà-zi-ga, and the ér-šà-hun-gá prayers—may be briefly noticed here. All have Sumerian titles, and it will not be immediately obvious why the first of them, literally meaning "crossing the desert," should concern delirium; but as the popular "seeing snakes" so well explains, the hallucinations in delirium involve encounters with sundry hissing or roaring animals, and thus the crossing of the desert is to be taken as a mental journey only. Virtually only one incantation of the work is fully known, but since this has much to do with the roaring of the wild ox and the hissing of the *bašmu*-snake, and the patient's belief that they are trying to kill him, the general line of the interpretation is not in doubt.

As we have suggested, (Kinnier Wilson, 1962) šà-zi-ga will have been concerned with loss of libido, which is what the phrase basically means. Known incantations and procedures, as also a catalogue —LKA 94—giving the first lines of all the incantations in the series, fully bear out this opinion, and in the latter text it is of interest to read something of how the subject was anciently approached. While composed only for use in unobtrusive retreat, such lines as (Obv. i 3) "My stallion, prepare to be ridden," or (*ibid.* 16) "To me, to me, my horse!" or again (*ibid.* 15), "My bolting wild ass" (*sc.* approximately, "I have brought back home"), must nevertheless be allowed their place in the history of psychological thought.

The ér-šà-hun-gá compositions were rather different and concerned the "angry god." Like so much in psychological medicine this deity (*ilu zenû, ilu šabsu*) as well as divine wrath (*kimilti ili*) had no factual existence, and such ideas must be explained either in terms of projection which means that the real, and only, anger was that of the patient, or else in terms of delusions of sin which means that the anger was merely a logical inference of the patient's, and false because based on false premises. It is in fact the latter explanation which best satisfies the evidence of the texts, and since delusions of sin are frequently associated with depression it is not surprising to find this additional element prominently indicated in the compositions. A good example of the type is the much published "Prayer to every god" (see for instance Pritchard, 1950) which, as others like it, has not hitherto been psychologically interpreted; but even the fact that the supplicant—or more accurately the priest acting on his behalf— numbers his sins and wrong-doings as "seven times seven" must arouse suspicion as to the delusional nature of the belief.

But despite its major contribution to the subject Mesopotamia is not alone among the countries of the Near East in providing evidence of mental diseases in ancient times, and it may suitably fall to this chapter to give some notice of our obligation to scribes of other languages. Thus from the land of the Hittites comes especially one most remarkable narrative, first edited under the title *Muršilis Sparachlähmung* by Goetze and Pedersen (1934) and also well analysed from the medical point of view by Oppenheim (1956). The text describes a cultic ritual created to meet the special needs of the Hittite king, Muršili II, who, as the account tells in its opening lines, was caught in a violent storm while making a journey away from his capital, in the course of which, and apparently as the result of shock from a particularly loud thunderclap, he lost to some degree his faculty of speech. This was not serious: but the incident itself, and probably such unexpressed ideas of guilt(?) as later came to be included in the king's understanding of the religious meaning of the event, evidently gave rise to an obsessive-ruminative psychoneurosis which in the course of time—actually after the passage of some years—found repeated expression in troubled dreams. As Oppenheim well argues we may read into the text at this point the fact that the king will have relived the whole episode of the thunderstorm in these dreams, and particularly the central incident of the loss of speech—for finally, in one dream, the "hand of god" struck the king with a motor paralysis which (affected half of his face and) drew his mouth askew(?) (to the opposite side). For the end of the story which concerns the ritual burning in the temple of the Weather-god of the king's chariot, weapons, vestment, and all the paraphernalia associated with the original journey, reference may be made to the authorities stated. From the diagnostic point of view Oppenheim's finding of hysterical aphonia would appear to be altogether sound.

From Anatolia we may turn briefly to Palestine where several of the Hebrew Psalms may be suggested as having origins in psychological medicine, for example Ps. 35 where verse 3 reads, "[O Lord] draw spear and javelin against my persecutors (*rōdepāi*), the delusions being clear from verse 7, "Without cause they hide their nets for me. . . ." In the later verse various suspicions and ideas of reference, indicated by the charge of slander attributed to the persecutors (verse 15), their supposed mocking and gnashing of teeth (verse 16), their supposed hatred and winking of eyes (verse 19), could suggest that the original use of the Psalm was for cases of schizophrenia simplex. For an example of depression with a delusional content one may turn to Psalm 38. The initial pattern of the composition is interesting in that while vs. 1, 4, 6, 8, and 9 appear to indicate the depression (note that "my iniquities have gone over my head" in verse 4 means that they are "more than I can bear"), the intervening vs. 2, 3, 5, 7, and 10 ("Thy arrows have sunk into me . . .," "There is no soundness in my flesh . . .," "My blows grow foul and fester . . .," "My loins are filled with burning . . .," "My heart has been bewitched," —for this meaning see Thomas (1939)— "and my strength fails me") find their best explanation as hypochondriacal ideas. As thus interpreted there would be some variety in the balancing of the couplets. With verse 12, "Those who would seek my life lay snares (for me) . . . and devise treachery" (that is, ways to trick me) "all the day long," the text introduces delusions of another kind. The total of evidence could suggest that, in this instance, the supplication of the priest was for a depressive psychosis.

Our final diagnosis may be made at the Dead Sea. The scrolls which in recent years have been discovered in several caves in this area are world famous, and the strange extra-biblical scrolls have in particular been studied from many different angles—many, that is, but not yet the psychological. They tell of a sect and a sect leader known as the "Teacher of Righteousness" (or "True Teacher," or True Guide"). But when one learns that he was persecuted by a "Wicked Priest," and when one reads also of a "Prince of Lights," an "Angel of Darkness," "Sons of Error," an "Army of the Holy Ones," a "Man of Lies," a "Lion of Wrath," "Periods of Wrath," an "Era of Wickedness," and of a fantastic future War between "Sons of Light" and "Sons of Darkness," it is difficult to think that we are in any sense dealing with a normal religion and a normal person. Since the psychiatrist's word for a "crank" is a paranoiac (Greek παρανοέω, "to think amiss"), and in that "some paranoiacs live at liberty as queer inventors, founders of eccentric sects, or as apostles of peculiar social reforms" (Curran and Partridge, 1955) we may thus find a diagnosis which would appear to suit many aspects of the situation very well. As is typically the case with such a disorder of the intellect (involving φρήν not ψυχή) the paranoiac is easier to recognize than to describe; but at least the weird imagery and ideas of the *War of the Sons of Light against the Sons of Darkness* in which, under the angels of God's dominion, twenty-eight thousand infantry and six thousand horse of the elect of Israel ranged against the Kittim and the army of Belial "shall pursue the enemy to annihilate him in the Battle of God unto his eternal extinction" (ix. 4), should convince anyone of the merits of the suggestion. That in giving the order of battle for this Armageddon the author appears to write as if he were none other than the Almighty's adjutant does not of necessity speak for delusions of grandeur, but the whole account is certainly written in the grand style. The "Kittim," probably more Romanesque than Roman, I would personally think to be the writer's Great Adversary and would not seek to identify them more closely.

Following the initial clue of the (ideas of) persecution it should be that the Teacher of Righteousness, as the suspected patient, was the writer of the document, and the further clue of the names of the "Prince of Lights" variety may suggest that his many writings extended even to documents in which his own name appears. If correctly diagnosed the case will show again to what extent all psychological disease is a child of the times. Whether here, or in a world of witches and angry gods, its outward expressions reflect of necessity the cultural and social background of which it is a part. As the subject belongs to history, so is it to be interpreted in the light of the ideas of history.

REFERENCES

Note: For abbreviations see previous chapter by the author.

CURRAN, D. and PARTRIDGE, M., 1955: *Psychological Medicine.* Edinburgh and London, Livingstone.

EBELING, E., 1915: Assyrische Beschwörungen, Z. Deutsch. Morg. Ges., 69:92 and 96.

———, 1931: Aus dem Tagewerk eines assyrischen Zauberpriesters. Mitt. altorient. Ges., V/3:16.

———, 1949: Beschwörungen gegen den Feind und den bösen Blick aus dem Zweistromlande. Archiv Orientální, 17/1:186.

GOETZE, A. and PEDERSEN, H., 1934: *Muršilis Sprachlähmung.* Copenhagen, Levin and Munksgaard.

KINNIER WILSON, J. V., 1962: Hebrew and Akkadian philological notes. J. Sem. Stud., 7:180.

———, 1965: An introduction to Babylonian psychiatry. Published in the Festschrift to Prof. B. Landsberger. Chicago University Press.

LAMBERT, W. G., 1957: An incantation of the Maqlû type. Arch. Orientforsch., 18:295.

MEIER, G., 1937: Die assyrische Beschwörungssamlung Maqlû. Berlin, *Arch. Orientforsch.*, Beiheft 2.

OPPENHEIM, A. L., 1956: The Interpretation of Dreams in the ancient Near East. Philadelphia, *Trans. Amer. Phil. Soc.*, NS 46/3:230.

PRITCHARD, J. B. (ed.), 1950: *Ancient Near Eastern Texts relating to the Old Testament,* 391. Princeton, University Press.

REINER, E., 1958: Shurpu: A collection of Sumerian and Akkadian incantations. Graz, *Arch. Orientforsch.,* Beiheft 11.

SADLER, W. S., 1945: *Modern Psychiatry,* 491. London, Henry Kimpton.

THOMAS, D. W., 1939: A note on *libbi sěharhar* in Psalm xxxviii 11. *J. Theol. Stud.,* 40:390.

VON SODEN, W. (with Falkenstein, A.), 1953: *Sumerische und akkadische Hymnen und Gebete,* 306. Zürich, Artemis.

Chapter 57

Sexual Behaviour in Ancient Societies

A. T. SANDISON

INTRODUCTION

STORR (1964) has reminded us that Western society has only recently begun to study sexuality in an objective fashion, despite the fact that the sexual impulse is a basic and integral part of human nature. Sexual practices which are acceptable at one time and place may be abhorred as perversions in another and it is probably safe to say that there is no sexual practice which has not somewhere been condemned and elsewhere been accepted.

All human beings carry within themselves the seeds of every sexual deviation. However, it is only when some phenomenon other than heterosexual coitus is compulsorily practised in circumstances where such coitus is available that behaviour should be considered to be deviant. Fellatio and cunnilinctus were, until comparatively recently, classified as perverse but it is now clear that a majority of human males at some time practise these as a part of normal lovemaking (Kinsey et al., 1948, 1953). Similarly, although overt sado-masochistic practices are relatively rare it is clear that sado-masochistic fantasies are common in view of their regular purveyance in modern novels, films and television programmes (Storr, 1964).

Deviant sexual behaviour is the behaviour of those persons who have been, for some reason, unable to form reciprocal equal term relationships with the opposite sex and who cannot, for this reason, give and receive love in a satisfying way. It is not known what proportion of the population is deviant, but the number is likely to be large (Storr, 1964).

The study of sexual deviation is, therefore, very largely the study of sexual activity divorced from love. It is important that we should try to understand the nature of deviation rather than to pass moral judgments upon deviant persons. In this we may be helped by a study of sexual behaviour in ancient societies. Conversely, a knowledge of sexual behaviour in ancient peoples is essential for a fuller understanding of their cultures.

It is always dangerous to generalise from observations on animal behaviour but there is such clear-cut evidence of atypical sexual behaviour in animals that it might be predicted that similar phenomena would occur in primitive man. It is, for example, well known that masturbation occurs in animals, more commonly in the male, that nocturnal dreams are experienced and that homosexual contacts are frequent (Kinsey et al., 1953). Biting occurs in animal coitus (Bloch, 1908). In the primates, masturbation and mouth-genital contacts are not uncommon. Homosexual activity is frequent and in the male anal penetration may be attempted (Altmann 1962). With regard to domesticated animals nymphomania (having an organic cause) occurs in cows, mares and sows and stallions and bulls occasionally "rape" immature females who are not in oestrus (Williams, 1947).

THE HEBREWS

Our knowledge of the sexual behaviour of the Hebrews is largely derived from study of the Bible (Cole, 1960). The Hebrews were essentially an unsophisticated pastoral people with a monotheistically polarised culture surrounded by others mostly in a more advanced state of civilisation and usually having polytheistic fertility or orgiastic cults. For example, in Egypt there were ithyphallic gods and fertility goddesses, in Babylonia the cult of Ishtar had marked sexual overtones, in Canaan there were cults of Baals of phallic type with defaecatory components in the ritual and of Astarte who may be equated with Ishtar. In Greece early fertility goddesses were followed by masculine Olympian deities although temple prostitution still flourished in some areas, while in Rome there was a gradual declension from the morality of the early Republic to the viciousness of the Empire with numerous orgiastic cults derived from many countries.

The Hebrews reacted violently to these cults, although from time to time they were seduced by them. In general, sexuality was regarded as serious, responsible and good. Attempts were made to rejuvenate the aged King David by intercourse with young women. Evidence of strong sexual passion may be found in the Song of Solomon. Every Hebrew was expected to marry and have children. Polygamy and the keeping of concubines was permissible. There were strict hygienic rules which, as Green-Armytage (1945) pointed out, may have favoured conception. Barrenness was tragic and grounds for divorcing a wife. The law of the Levirate marriage required a surviving brother to impregnate the widow of a childless man; it is interesting to note that somewhat similar institutions obtained in a South American tribe (Beals, 1961) and in a Polynesian society in modern times (Danielsson, 1961). During the New Testament period childlessness was less important, divorce was abolished, marriage was monogamous and life-long and celibacy was considered a virtue.

Fornication was frowned upon in Old Testament times but this seems to have been largely related to questions of property rather than to morality since the penalty for fornication with a betrothed woman was death while coitus with an unbetrothed girl resulted only in forced marriage and the payment of a fine to the father. Women sometimes took the initiative in sexual activities, e.g., Lot's daughters (Genesis 19:31), Potiphar's wife (Genesis 39, 7) and Rachel and Leah (Genesis, 30). With the New Testament all extra-marital coitus is condemned. Adultery by the male with an unmarried woman was unimportant since no husband's right was infringed but adultery by the female was a most serious sin since the family blood line could be rendered impure.

Prostitution was regarded in the Old Testament as undesirable but as long as it was of the secular variety condemnation was based on economic rather than moral grounds. Rahab the harlot was not censured; Jephthah was the son of a harlot. Hosea married a harlot and the intercourse of Judah with Tamar acting the harlot evinces no comment. The Hebrews reacted violently to the religious prostitution of surrounding idolatrous fertility cults but at times there appear to have been cult prostitutes (occasionally male as well as female) in Israel, e.g., in the Monarchic period.

Both the Old and New Testaments stigmatise male homosexuality as a sin and a crime: this horror may be related to the idolatrous male prostitution of contiguous fertility cults. As mentioned above, however, male cult prostitutes served in the temple in the time of King Josiah. Male homosexuality is proscribed in Leviticus (*18*, 22 and *20*, 13) and transvestism in Deuteronomy (*22*, 5). Homosexual practices are mentioned in the sins of Sodom (Genesis, *19*:5) and of the men of Gibeah (Judges, *19*:22). There is no evidence of female homosexuality among the Hebrews but the Epistle of Paul to the Romans (*1*, 19-27) indicates that this occurred in the contemporary Gentile Society. Foster (1958) thinks that the love of Ruth for Naomi may have had a sexual component.

Rape is not uncommonly mentioned in the Old Testament, e.g., the rape of Dinah by Schechem (Genesis, *34*, 2), the mass rape of the Levite's concubine by the men of Gibeah which resulted in her death (Judges, *19*:25) and the rape of Tamar by her half-brother Amnon (II Samuel, *13*:14). The penalty for rape of a betrothed woman was death.

Incest among the Ancient Hebrews was proscribed by rules forbidding marriage and coitus among a fairly large circle of relatives by blood and marriage (Leviticus, *18*:6). Nevertheless, those rules were at times disregarded, e.g., in the incestuous relation of Lot and his daughters (Genesis, *19*:33) and of Reuben with his father's concubine (Genesis, *35*:21). The Levirate marriage also cut across the incest law.

Masturbation is not described in the Bible; the sin of Onan was almost certainly coitus interruptus. It is possible that Ezekiel (*16*:17) could be interpreted as female masturbation with an olisbos. The Talmud, however, detailed precautions against masturbation. Nakedness was considered a shameful and humiliating state and priests were required to wear linen breech cloths. Voyeurism and exhibitionism were presumably very rare.

The fact that zoophilia or bestiality is proscribed (Exodus 22:19 and Leviticus 18:23 and 20:15) suggests that as in most pastoral peoples occasional acts of intercourse with animals occurred. The express forbidding of bestiality of women (Leviticus 20:16) with animals may be related to the occurrence of such intercourse in the fertility cults of adjacent societies. The Talmud prohibited widows from keeping pet dogs (Glasner, 1961).

There are no accounts or proscriptions of overtly sexual sadist or masochistic practices nor of necrophilia, coprophilia and paedophilia so it is safe to assume that these must have been absent or extremely rare (Cole, 1960).

THE ANCIENT EGYPTIANS

There is no single monograph which deals with sexual behaviour in the Ancient Egyptians and I am greatly indebted to Cyril Aldred for much help in the preparation of this portion of my essay.

For the great majority of persons marriage was monogamous and bigamy rare; the more wealthy man, however, might have several serf concubines and the Pharaoh was usually polygamous with a harem of queens (one of whom was the great wife), high-born favourites and foreign princesses. Genealogies, however, have shown in non-royal personages that a man might have children by different wives but there is no evidence that such wives were all alive at the same time; in at least one case a surviving childless widow adopted her deceased husband's children by a concubine as her heirs. The status of women was high and if a wife was divorced her dowry apparently had to be returned (Černý, 1945). Early marriage appears to have been the rule, the wife sometimes being as young as twelve to thirteen years of age and the husband fifteen. However, marriage was advised at the age of twenty years. The husband was admonished to feed and clothe the wife well, provide oil for anointing her body, and to make her heart joyful. Widows were free to remarry.

There are injunctions against debauchery but, in general, no stigma was attached to illegitimacy, although possibly among the upper classes female chastity was considered to be important. It has been pointed out elsewhere that Smith and Dawson (1924) interpreted the finding of an unembalmed body of a sixteen-year-old girl who was six months pregnant and who had been violently assaulted as probably the result of discovery by male members of her household of an illegitimate pregnancy. Cyril Aldred, however, in a personal communication indicates that this may simply have been a case of armed robbery.

Certainly Ancient Egyptian love poems have a distinctly erotic element, although statues of husband and wife from Pharaonic times show a rather stiffly affectionate relationship. Incantations and rituals gave back to the dead man his virile force, and in the hereafter provision for felicity included opportunity for sexual expression (Breasted, 1912). There are early hieroglyphic pornographic insults and obscene graffiti were scribbled on ostraca. The Turin Papyrus has never been published in full; it depicts the capers of a bald priest and Theban coquettes and is annotated with ribald remarks (Yoyotte, 1962).

Adultery by the wife was grounds for divorce but, in fact, there are few records of such adultery from the ancient literature. It is possible that the condemnation of adultery proceeded from a sense of preservation of public order. This contrasts with the story in Herodotus. (Bk. II) in which a blind Pharaoh was promised cure if he bathed his eyes with urine from a woman who had never lain with any man but her husband; he had great difficulty in finding a woman whose urine was effective.

There is little documentary evidence of bawdy houses or a licensing system for harlots but since the destitute sometimes had to sell themselves into serfdom for survival, prostitution must have been common. There are admonitions to youths to keep away from harlots; harlots appear to have haunted beer shops.

There is no clear evidence of temple prostitution; indeed, men were deemed impure if they entered a temple precinct after coitus without certain purification ceremonies. There were many temple singers in attendance on various gods of masculine and seminal character but they were usually wives of priests of the same gods (Aldred, personal communication). These masculine gods often took virile forms, e.g., Min as a bull, and many idols were in phallic form. Popular religious stories could be vulgar or even bawdy, e.g., Re recovered his good humour when Hathor lifted her skirt and Seth tried to seduce Horus.

Homosexuality was forbidden in Memphis and in two other nomes. There are no scenes of sodomy from Ancient Egypt. Among the declarations of innocence required for the deceased person to enter the blessed state was a disclaimer that masturbation or sodomy had been committed. Probably, however, homosexuality did occur; according to a popular tale

Pharaoh Pepi IInd was enamoured of one of his generals; the god Seth attempted to violate Horus; and the relationship between Akhenaten and Smenkhkare was somewhat equivocal. Further, from the later period a Coptic love charm is calculated to attract a man to a man. There is no evidence of female homosexuality; there is, however, some evidence of heterosexual anal eroticism in an ostracon in the British Museum, in a XIIth Dynasty statuette and in numerous figures from the Roman period. In the case of the XIIth Dynasty statuette which portrays a male inspecting the anus of a kneeling woman, Sigerist (1951) has suggested that this might also be interpreted as medical inspection of the anus by a physician.

Incest appears to have been restricted to the Royal Family in Pharaonic times, although in the Hellenistic period it was more general. In Ancient Egypt lovers were described as "brother" and "sister," but this was a poetic expression and does not denote incest. Ruffer (1919), in a study of the physical effects on consanguineous marriages in the Royal Families of Ancient Egypt during the XVIIIth Dynasty, XIXth Dynasty and Ptolemaic period showed that no obviously deleterious effects resulted.

As noted above, masturbation was regarded as sinful and the deceased person was required to declare his innocence before entering the blessed state. There was no prudery about dress in Ancient Egypt; women wore a clinging dress held under the bosom by two wide shoulder straps, although in the upper classes the breasts were covered by diaphanous draperies. Boys and girls went naked at all times. Tomb pictures demonstrate pleasure at displays by dancing girls and musicians who wore little or no clothing and the Pharaoh Snefrou was relieved of boredom by an aquatic diversion given by girls clothed only in nets. In Book II of his *History*, Herodotus tells that at festivals at Bubastis women on barges shouted abuse at women on the shore and stood up and hitched up their skirts. The hieroglyphs used recognisable pudendal symbols to indicate "man" and "woman" and "coitus."

A fragment of painted leather from Deir-el-Bahri shows a devotee of Hathor flagellating himself with a whip and producing erection. Hathor is equated with Astarte whose priests also practised flagellation in Classical times (Personal Communication from Cyril Aldred).

We have little knowledge of zoophilia in Ancient Egypt; Isis in the form of a bird pressed herself against the mummified Osiris and restored his vital force. Herodotus, in Book II of his *History*, relates that a male goat had coitus with a woman in Mendes but this may well be a dragoman's story.

There has been debate about the description of necrophilia among the embalmers given by Herodotus (Book II, 89); however, Smith and Dawson (1924) have noted that female mummies of the later period more often show evidence of putrefaction and insect depredation than those of males. I am inclined to believe that this may have been the result of delay in sending such bodies to the embalmers to prevent necrophiliac practices (Sandison, 1963).

As in the case of the Ancient Hebrews, there are no accounts nor proscriptions of overtly sexual sadist or masochist practices nor of paedophilia so it is probable that these were absent or extremely rare.

THE ANCIENT GREEKS

There is a wealth of information concerning sexual behaviour in Ancient

Greece; this has been pieced together from gleanings which have survived from the extensive literature. The sources include the works of philosophers, historians, commentators, epic poets, dramatists (especially those writing domestic comedy), lyrical poets etc. (Licht, 1932; Flacelière, 1962). A further valuable source is derived from study of art-forms—statuary, paintings, medallions, lamps, amulets etc. (Marcadé, 1962).

Since sexual customs and behaviour change over long periods of time, it is only possible in a short space to discuss Greek sexual behaviour in fairly broad terms. Even in the Homeric literature, however, there is little sense of sexual sin. The gods were robustly sexual. This matter of fact attitude, absolute lack of self-consciousness with regard to sex, reason, balance and avoidance of excess persisted throughout the entire Classical and Hellenistic periods. There is evidence of objective sexual study, e.g., Aristotle was aware that anal contractions occurred during orgasm. Sexual behaviour began at an early age. Demosthenes is alleged to have said "Man has the hetairae for erotic enjoyments, concubines for daily use and wives of equal rank to bring up children and to be faithful housewives."

Further, there was clearly a frank phallic element in Greek daily life which was taken as a matter of course. From Archaic times phallic Herms were common; at festivals enormous wooden phalli and phallic birds were taken in procession. Certain persons were recognized to be phallus-makers and perhaps also fashioned olisboi.

The place of the Greek woman was in the home and the function of the Greek wife was to manage the home efficiently and to bring up children. In this respect she was absolute mistress of her house and there is little doubt that most Greek wives led a placid and not unhappy life. She was, however, largely confined to the women's quarters and husbands looked elsewhere for pleasure and intellectual companionship. Spartan girls were more free than those of Athens, and appear to have been visited by their husbands only by stealth in the night; further, a Spartan husband might permit another man to father a child on his wife if the man were of good stock.

Some Greek women probably did not marry and became "old maids;" at marriage the wife was usually younger than the husband and a dowry was paid. If a marriage proved to be barren children might be adopted; surplus children were abandoned by other families and became available for adoption. There was no ritual concept of uncleanness during menstruation in Ancient Greece.

For the Greek husband marriage did not imply renuciation of sexual or aesthetic intercourse with other women, e.g., courtesans and concubines, but wives were expected to remain chaste. Some women undoubtedly committed adultery; indeed, folklore indicated that women were believed to derive more pleasure from coitus than men. Wives could be divorced and otherwise punished for adultery. The man might also be punished if caught in adultery with a married woman or if he seduced an unmarried girl of good character. It was, therefore, safer for men to have recourse to prostitutes and no stigma was attached to such relationships.

Prostitution was first regulated by Solon; prostitutes varied enormously in their physical charms, intellectual ability and price. At one end of the scale came the highly accomplished, well-educated *hetairae*, many of whose names have come down to us. At the other extreme were the brothel whores (for whom there were many witty or coarse synonymous terms) who stood naked or scantily clothed in their bawdy houses and who could be bought for as little as an obol. Brothel keepers paid a tax to the State. An intermediate grade of harlot was the street prostitute—who might advertise in the dust by having 'Follow me" made out in nails on the soles of her shoes—plying her trade in her own or a hired room or copulating in dark corners or at the public baths. A further class of public woman was constituted by the *hieroduli* or temple servants of Corinth, Abydos and other sanctuaries dedicated to Aphrodite Porne. It will be recalled that religious prostitution already existed in the cults of Mylitta at Babylon and of Aphrodite at Byblos.

We have some information on coital techniques. Vase paintings show a variety of postures, some of which appear to require considerable athletic agility. Among the more popular techniques is the male supine with the female straddling but posterior and lateral approaches are also shown. Anal intercourse also occurred; prostitutes are shown accomodating multiple clients by coitus, fellatio, masturbatory techniques and also offering anal intercourse. Both fellatio and cunnilinctus appear in paintings.

Male homosexuality was undoubtedly common in Ancient Greece, although this fact is often glossed over in books written by laymen. The Greeks were decidely ambisexual; the homosexual component was largely paedophilic, being a love of an ephebè or youth. These youths were, however, emphatically *not* children of tender years but adolescents; indeed, sexual relations with immature persons were punished in Ancient Greece. For this reason the practice is sometimes known as ephebephilia rather than paedophilia. Every man of good birth attracted to himself a youth; the attachment was certainly not solely sensual for the older man acted as counsellor, guardian and friend. Ephebephilia was not a feature of lower class life. Boyish beauty was celebrated in Greek literature from the Homeric period onward and was a feature of the myths. It is also immortalized in the plastic arts; indeed, one forms the impression that male beauty influenced the interpretation of the female body as shown by rather immature breast forms in Greek statuary.

Ephebephilia was socially accepted in certain city states of Ancient Greece if based on mutual liking, even if the relationship had carnal as well as aesthetic elements, as was probably inevitable when nakedness, embraces and kisses were common. It was, however, rejected if the boy sold himself for money or presents. It is clear that boys could be hired by contract and could also be had in brothels or houses of accommodation.

Female homosexuality also occurred in Ancient Greece: the poetry of Sappho gives clear indications of love for members of her own sex, although she was probably ambisexual. Philaenis of Leucadia is alleged to have been a notorious tribad and to have written an illustrated book on female homosexual postures. No vase paintings, however, show women in bed together. Lesbos and Sparta were said to be notorious for female homosexuality. *Olisboi* appear to have been used by homosexual pairs.

Masturbation was not regarded as a vice but believed to be a useful method of release; it is mentioned in the literature and depicted on vases. Female masturbation also occurred using manual techniques and also *olisboi* often made of leather. The latter were chiefly made in Miletus and are freely discussed in the 6th mimiambus of Herondas. Vases show *olisboi* in use and it is known that oil was dripped on to them from a small vessel and that women washed after their use.

Nakedness or near nudity was common and clothing was simple. The Minoan female costume left the breasts bare; later the bosom was covered but sometimes by very diaphanous material. Spartan girls wore tunics split up the thigh to the hip and exercised naked; musicians at banquets were often nude. Boys and men frequently went naked in the gymnasium or baths or at games. Voyeurism is, therefore, unlikely: there is a story that King Candaules exhibited his beautiful wife in a state of nudity to his favourite Gyges, but this may have been to satisfy an inordinate vanity in the possession of a beautiful woman. Theophrastus mentions a habit in the character of the "Immodest Man" of lifting up the chiton and exhibiting himself to women. Transvestism was uncommon except in certain religious cults of Cotys but there are accounts of a girl, Dorcion, who preferred male dress and of Amarus who liked to appear in female dress and ornaments (Licht 1932). Agathon, a transvestite, wore a bust bodice and depilated himself like a woman (Flacelière, 1962).

Incest was rejected by public opinion although nowhere were there severe punitive measures. Many of the older myths had a marked incestuous content, but sometimes the myths show that incest was regarded as detestable, e.g., in the case of Oedipus and Jocasta. It appears that even in the archaic period full brother-sister marriage was prohibited. In the Hellenistic period incestuous marriage was the custom in the Greek royal house of Egypt —apparently without any evidence of physical deterioration (Ruffer, 1919). Alcibiades and Axiochus are said to have used a woman in common and also to have both lived with her daughter, although the latter must have been the daughter of one of them.

Greek literature contains several accounts of pygmalionism, i.e., the erotic love of statues or work of arts. Apart from the legendary case of Pygmalion, Lucian describes a youth who fell in love with the statue of the Aphrodite of Cnidus and left seminal stains on the marble. There are similar accounts of youths falling in love with statues of Aphrodite, Agathe Tyche and Eros.

Flagellation occured in Ancient Greece in the religious celebrations of Cybele; so also did castration. There are also accounts of the chastisement of Spartan boys at the altar of Artemis Orthia. Ancient Greek literature is largely free of sadist-masochist material, in contrast to that of Ancient Rome. The story of Heracles who performed female tasks for Omphale who was clad in a lion's skin may have sado-masochist overtones. There is also an account of the bites sustained by Demetrius of Phalerum from the *hetaira* Lamia.

Zoophilia is not infrequent in Greek myths, e.g., the intercourse of Zeus with Leda in the form of a swan and with Persephone as a snake. Pasiphaë had coitus with a bull and bore the Minotaur. Gould and Pyle (1897) cite Plutarch as stating that Aristonymus Ephesius had carnal knowledge of an ass, being tired of women. It seems likely that herdsmen occasionally used their animals as makeshifts as still occurs in Modern Greece and other pastoral regions (Kinsey *et al.*, 1948).

Necrophiliac practices were also rare; Herodotus mentions that Periander of Corinth had coitus with his wife Melissa after he had killed her. Phylarchus tells the story of Dimoetes who had connection with a drowned girl found by him on the seashore until he had to bury her.

Coprophilia is evident in comic and satiric poetry; there are references to micturition and urine, to defaecation, excrement, passing flatus and chamber pots.

Erotic dreams were common in Ancient Greece; these were of heterosexual and homosexual nature, sometimes incestuous, and appear to have been considered worthy of serious consideration as omens. They were experienced by both men and women.

THE ANCIENT ROMANS

Kiefer's (1934) monograph is a useful prime source in the study of the sexual behaviour of the Ancient Romans and depends largely on literary studies. Marcadé (1961) gives much valuable material derived from study of the plastic arts.

It is possible that agamous relationships in early Italy gave way to a matriarchal system in Etruria but in early Republican Rome marriage was monogamous and dominated by the *patria potestas* of the husband. During the early period regular marriage was only possible between patricians and was of the ceremonious *confarreatio* type. Later the plebeians introduced the simpler *coemptio* form, while a third style was by *usus* (custom) according to which one year's continous cohabitation brought a woman into the husband's authority. This transference of authority from the woman's father to her husband could be avoided by an annual separation of three nights.

Customs of the wedding night probably recalled earlier marriage by capture. Phallic songs were sung with obscene jestings and a proportion of brides sat down on the phallus of the fertility god Mutunus Tutunus. Virginity of the bride was esteemed as a token of her probable future marital chastity. In the archaic period the first coitus possibly took place in the presence of witnesses and it is possible that the husband's friends may have had intercourse with the bride in the first instance.

During the Republic the wife was rigidly chaste; love probably played little part in these marriages which were designed for the procreation of children. The mother suckled her children herself in contrast with custom at the later period. The wife was the *domina* of the household but played a part in social life and was not restricted to the women's quarters as in Greece. Divorce was unknown in the earliest period and was probably later instituted in cases of barrenness. During the Imperial period adoption was prevalent. Later, wives could be divorced for adultery or for perverse or disgusting conduct; wives could not, however, divorce husbands. At first there were no statutory penalties for adultery but an adulterous wife might be killed and a male adulterer involved with a married woman might be flogged or castrated. A man might, however, freely commit adultery with slaves or prostitutes. Divorce became commoner in the later Republic and very easy during the Imperial period but Augustus tried to institute reforms, punishing adultery by banishment and loss of property. Augustus's laws recognized concubinage but a man could not have a wife and concubine simultaneously. Laws discriminated against the celibate, but men could get satisfaction from whores and some women became prostitutes in order to be without the reach of the law.

There was a progressive emancipation of Roman women from the middle of the second century B.C.; this was largely political and economic but was accompanied by some sexual emancipation. Some wives were loyal and devoted but evidence of female immorality increases during the Imperial period. Some women enjoyed a sort of free love and might be sexually associated with men without marriage. Even when allowance is made for exaggeration by political enemies, the behaviour of women such as Julia the elder and her daughter and of Messalina appears to have been extremely dissolute, approaching nymphomaniacal states.

Prostitution was common and largely accepted socially; Horace believed that "young men should drop in there, rather than grind some husband's private mill." A register of prostitutes was maintained and patrician women were punished if they prostituted themselves. Some actresses, dancers and musicians approximated to the Greek *hetairae* but the majority of whores were slaves who worked in brothels in certain areas of Rome. Some of these were near the Circus Maximus to cater for those men who were sexually excited by the sadism of the Games. Inns and cookshops also kept girls; other girls—the *scorta erratica*—wandered about. The lowest type were the twopennies or *diobolariae* who practiced at street corners and in graveyards.

Customers were mainly young men but also comprised soldiers, sailors, freedmen, slaves etc. Prostitutes had dyed hair and wore the toga as distinguishing signs and were taxed in the reign of Caligula (Kiefer, 1934).

A well preserved brothel at Pompeii was sited at a street corner and consisted of small rooms opening off a vestibule. A phallus was placed above the door and inside there were erotic paintings. Graffiti on the walls commemorated sexual performances and named *meretrices* and their clients.

Every house bearing a phallic symbol was not, however, a brothel. There was a rich phallic cult in Etruria and Rome, partly concerned with magic. Many persons wore amulets of priapic nature: obscenity was a defence against evil spirits. Priapic bread was distributed at certain festivals. Mutunus Tutunus was an early Roman phallic god who became incorporated into Priapus. Another phallic god, Fascinus, had fertility and harvest significance. The Etruscans placed phalluses on the walls of burial monuments denoting the renewal of life through death and possibly believing in sexual vigour in the tomb. The Tarquinian tomb shows violent sensuality in proximity to representation of virile bulls (Marcadé, 1961).

Male prostitutes were also to be found in Rome. No Italian god had definite homosexual qualities and the Romans, by and large, were suspicious of male nudity which was taken to be synonymous with indecency and impropriety. Cicero believed that homosexuality was a natural product of nakedness. In the later period there was nakedness of men in public baths while women wore only short aprons. At this time also some women wore diaphanous Coan garments. Some prostitutes and certain handsome male slaves kept as sexual favourites were also dressed in thin materials. Women and homosexual men used cosmetics on a large scale to keep their skin fresh and young. Similarly women, effeminate men and homosexuals depilated the body.

It appears that a considerable proportion of prominent Romans were bisexual and admired handsome boys as well as pretty girls. There was not, however, the same idealistic pederastia which occurred in Ancient Greece: Martial denied married men the right to love boys. It has been claimed that the writings of Catullus, Virgil, Tibullus, Petronius, Martial and Juvenal reveal some homosexual tendencies (Kiefer, 1934). Julius Caesar is said to have been loved in his youth by King

Nicomedes; Caligula had homosexual relations with actors (the theatre seems to have been associated with homosexuality): Nero also had homosexual tendencies and Elagabalus sought out men with remarkable penile endowment possibly as part of his religious exercises. The relationship between Hadrian and his favorite Antinous has been much debated. Some authorities have concluded that this included a carnal element and others that the relationship was purely spiritual and aesthetic.

Martial and other authors refer to girls "playing the boy" which presumably refers to posterior anal congress. The *Satyricon* of Petronius shows a marked homosexual or bisexual element in its characters. According to Marcadé sodomy is shown in vase paintings. Juvenal and other writers mention female homosexuality, masturbation and the use of *olisboi*.

As mentioned earlier, the Romans were less at ease with nudity than were the Greeks and we find some evidence of exhibitionism. There are vase paintings of women who are naked except for the strophion or breast band. Caligula is said to have exhibited his wife Caesonia in a naked state. Commodus is alleged to have had some of his concubines debauched before his eyes. In the *Satyricon* a group of persons in the baths gathers around a nude male with a large phallus. Narcissism is described in Ovid's myth. Voyeurism with mirrors is also described (Hirschfeld, n.d.).

Incest does not appear to have been common, although there are allegations that Catullus' mistress Clodia slept with her brother Clodius Pulcher; Caligula is said to have committed incest with his sisters (Sandison 1958) as did Commodus with his sister. Claudius married a niece—a relationship previously considered incestuous. Nero's grandfather is said to have been incestuous and the relation between Nero and his mother appears to have had distinct incestuous overtones.

The greatest blemish on the Roman character was sadism. The earliest Romans were sober, hardworking farmers who came into conflict with their neighbours and, in fighting for survival, became conquerors and empire builders. Their Empire was built on tyranny, blood-shed and murder and after this was attained cruelty became inturned and manifested itself in the sadism of the circus and games. From the start the Romans were cruel to defeated enemies; discipline was strict in Roman homes, schools and armies. Slaves were often treated with appalling cruelty even by Roman women. They were sometimes tortured in a revolting fashion. Penal justice in Rome was excessively harsh and included flogging to death, crucifixion, burning alive, hurling from the Tarpeian Rock etc. Sadism reached its peak at public games in gladatorial conquests and similar spectacles.

As mentioned earlier, prostitutes catered for men sexually inflamed by these sadistic spectacles. Well-born women sometimes had affairs with gladiators, some of whom were infibulated to prevent them wasting their energy in coitus. Masochism is less well-documented but the relationship between Propertius and his prostitute mistress suggests masochism on his part. Elagabalus also seems to have had some masochistic traits.

There is evidence of religious transvestite practices, e.g., in the cult of Elagabalus, and also of flagellation procedures in the cults of Cybele and Dionysius. A painting in the Villa of Mysteries shows female flagellation in a Dionysiac cult celebration. Further, in the *Satyricon* impotence caused by the god Priapus is treated by anal insertions and self-flagellation.

Zoophilia is described in the *Golden Ass* where a well born woman enjoys coitus with an ass and where punitive rape in the arena by an ass is also contemplated. According to Gould and Pyle (1897) male zoophilia also occurred. Marcadé (1961) mentions a realistic series of medallions in which women offer themselves to prancing quadrupeds.

In the *Satyricon* there is a revolting episode of the rape of a seven-year-old girl; a sexual episode in which three persons are implicated is also described. Marcadé illustrates considerable variation in coital position, including the posterior approach in the standing and kneeling positions, straddling by the woman, etc. There are also illustrations of various symplegmatous arrangements, notably in such tombs as the Etruscan Tomb of the Bulls.

THE ANCIENT HINDUS

Meyer (1952) has written a useful monograph on the sexual behaviour of the Ancient Hindus. Our knowledge of life in Ancient India is largely based on the study of the Vedic literature, some dating from the third millenium B.C., and of the great epics—the *Mahabharata* and *Ramayana*. The *Mahabharata* (latter half of 2nd millenium B.C.) is a valuable source of information on sexual relations in Ancient India; it is not a unitary composition but derived from many hands whereas the *Ramayana* (c. 6th century B.C.) has a closer unity despite many interpolations. Written somewhere between the first and fourth century A.D. by Mallanaga the *Kama-Sutra* of *Vatsyayana* is an indispensable compendium on sexual behaviour, probably based on earlier work (Upadhyaya, 1961); much later comes the *Ananga-Ranga* (Burton and Arbuthnot 1963) probably dating from the sixteenth century A.D. A work of similar genre but of Arabian origin is *The Perfumed Garden* (Burton 1963). Sculptural evidence of sexual behaviour in the Mediaeval period is derived from the great temples of Khajuraho and Konarak (Anand 1960).

It must be remembered that our view of sexual behaviour in Ancient India is complicated to some extent by the dual outlook on women compounded of great sensuality and a wistful desire for renunciation of the flesh. By and large, however, prudery had no meaning for the Ancient Indians and erotic questions were widely discussed and written about in the ancient literature; intense eroticism was evident in the Mediaeval period. Chastity appears to have been looked upon as a way of obtaining magical power.

Coitus was regarded as joyful; the ideal of the hero for blissful life was intercourse with thousands of lovely women with long lotus eyes, red lips, rounded arms, firm swelling breasts, lovely waists, great hips and thighs like banana stems. *Jus primae noctis* probably obtained in very early society. Certain heroes were described as 'hands that took away women's girdles, pressed swelling breasts, felt navels and thighs and secret parts and loosened aprons." We are impressed by the athletic ability which must have been required for some of the postural coital variants described in the *Kama-Sutra*.

Contrariwise chastity before marriage was regarded as admirable. Marriage was designed for the begetting of children and for expression of mutual love between husband and wife. Large families were much desired, especially stalwart sons, for religious reasons as well as others. Barren marriages were a great mishap; sometimes a begetter by proxy was obtained and a form of levirate seems to have applied at times.

Girl children were on the whole not

greatly welcomed but later were usually much loved by their parents. Premarital chastity was desirable and after the wedding night the blood-stained *anamandavada* was publicly shown. Before marriage the daughter lived in complete chastity and obedience to the family.

The form of marriage varied according to caste and might be free, by purchase, by capture or by stealing. The father had a duty to provide a husband for his daughter. Polygamy was permissible—a Brahman wife took precedence over all others; polyandry appears to have been rare in the Epic period. It is mentioned occasionally but exclusively in cases of community group marriage of wives among brothers, with a head-wife in common; these probably belonged to one of the aboriginal tribes who have maintained polyandrous habits up to modern times.

Sexual love between married persons was regarded as a great source of joy; the importance of the clitoris in female orgasm was recognized early in India, as well as the occasional necessity for phallic supplements. A woman ought to be married off before the menarche since an embryo was slain each time a maiden had her menses. The post-menstrual period of *ritu* was regarded as proper for conception; a husband had a duty to be with his wife at this time. The menses cleansed a woman of all sexual sin so that even an adulterous wife could demand her rights after the cleansing flow had ceased. Coitus during menstruation was, however, regarded as sinful and in general intercourse during pregnancy was considered undesirable. Coitus was proscribed in the open and required to be performed in private in the evenings or at night. Congress in water was not permitted.

Promiscuous behaviour appears to have been frequent among certain tribes, e.g., the Madras of the North West and others dwelling on the Indus. The Bahika women were described as *great-shelled*, meaning having a large vulva. It was believed that for many women sexual intercourse was the acme of experience and that even old women had a feverish longing for men. Women were believed to derive greater pleasure from coitus than men.

There were penalties for defloration of young women, probably less severe for low caste girls. Rape is described in the Epics and was sometimes punished by loss of property and cutting off the genitalia. Adultery was prohibited in thought, word and deed; however, certain types of women could freely be visited—the public harlot, slave women, wives of actors and singers, and temple dancers. The punishment for adultery varied according to the caste of the woman implicated.

The prostitute might be wealthy, distinguished and well-educated; the Hindu has often sung the praises of the public woman as the embodiment of perfect womanhood. During the Epic period the heroes included *hetairai* in their baggage trains; the strumpet was, however, also an ornament of civic life and women were sometimes kept in large important households for the use of guests. Not surprisingly, there are also admonitions against the snares of harlots, sometimes associated with warnings about gambling.

At this early period temple prostitution does not appear to have been important and during the Epic period *Lingam* worship later associated with Siva was a minor phenomenon. Temple harlots or *deva-dasi* only became important in the second half of the first millenium after Christ and were associated with male deities (Henriques, 1962).

Homosexual sodomy occurred and implied loss of caste if discovered and some homosexual fellatio also went on. Lesbianism among girls was punished by monetary fines and by corporal punishment. Lesbian seduction by married women was more severely punished by shaving the head, amputation of fingers and public display on the back of an ass. Mutual cunnilinctus and *olisboi* appear to have been used in lesbian practices in the harem; bulbs, roots, fruits appear to have been employed.

Olisboi also appear to have been used for female masturbation including the employment of statues with erect *lingam*. Masturbation in the male was disapproved and even nocturnal seminal emissions had to be atoned for.

Men were prohibited from looking at stranger naked women for fear of involuntary pollution; ascetics are said to have sometimes experienced orgasm from simply looking at lovely women. However, certain paintings from Ayunta suggested that the breasts, even of Royal ladies, were sometimes exposed at this time.

As indicated above, homosexual fellatio occurred but it is clear also that heterosexual fellatio and cunnilinctus were not rare but not really socially approved. The women of certain Bengal tribes are said to have been adept at fellatio. Courtesans provided fellatio but married women appear to have eschewed such practices. Some instances of simultaneous heterosexual fellatio and cunnilinctus are described. Anal intercourse and various complex symplegmas are described in the *Kama-sutra* and illustrated in the mediaeval statuary of Khajuraho and Konarak.

Incest appears to have been uncommon and was regarded with great disapproval, only to be wiped out by such severe punishments as death, castration or loss of caste. It is possible, however, that prohibiting of incest among other things were temporarily suspended during certain festivals (Wood, 1961). Zoophilia was proscribed and was probably rare; necrophiliac practices are mentioned in legend only. Coprophiliac and paedophiliac practices are not described.

A somewhat unusual element in Ancient Hindu eroticism is the emphasis on minor sadistic-masochistic practices. Much attention is devoted to such practices as "marking with the nails," the use of love bites, etc. While it is probable that, on the whole, such practices were to be regarded as coming within the range of normal heterosexual love play there do appear to be accounts of overt sadism with killing or injury to the woman during coitus by wedges, scissors and piercing instruments.

THE ANCIENT CHINESE

The study of the sexual behaviour of the Ancient Chinese is the province of the specialist and the layman is largely dependent on Gulik's (1961) monograph.

In Archiac China there may have been a matriarchal system with women initiating sexual relations, but during the period of the Chou Dynasties, i.e., from about 1100-200 B.C., the system became patriarchal. It appears that the Feudal Lords had principal wives, secondary wives and concubines. *Jus primae noctis*, however, was unknown. Marriage was exogamous and the ruler slept with his women in turn. Records were kept of their unions. The women lived in separate quarters and the chastity of girls was protected since only virgins could become principal wives. The daughters of the common people, however, could mix with men. The people held spring festivals with erotic songs and dances; coitus was common and marriages were regularised if the girl became pregnant.

In the later part of the Chou period there was much sexual licence; princes and high officials kept dancing girls and musicians who were used promiscuously by the master, his retinue and guests. These women were slaves and female captives and could be sold and resold. Married women and widows had the opportunity for illicit affairs both in and out of the house; there is a scandalous tale of the widow of a prince who had relations with a palace cook. Wives could be repudiated because of sterility or incurable disease. Coitus was believed to benefit health provided overindulgence did not occur.

Later, in the Han period, *circa*. 200 B.C. to 24 A.D., there came a reaction and lax morals were frowned upon as a menace to stable family life. Great stress was laid on segregation of the sexes, with women living in the inner apartments of the house and going veiled outside. Men and women did not bathe together, nor even share the same clothes racks or chests. The only physical contact was in bed. Sexuality *per se* was not, however, regarded as sinful in Ancient China and women were entitled to satisfaction of their sexual needs. Celibacy was regarded with suspicion. Sexual neglect of women was a grave offence and even a concubine up to the age of fifty years was entitled to coitus every five days. A man might, however, abstain from all coitus for three months after the death of a close relative and his conjugal duties might be abrogated entirely at the age of seventy years.

At about this time marriage manuals with instructions in coital techniques and illustrated by pictures of coital positions were available. Women at this time also utilised cosmetics, powdering the face, neck and shoulders, rougeing the cheeks and painting the lips.

During the period of the Three Kingdoms we have an account of polyandry on the part of Princess Shan-Yin, sister of the Emperor Tze-Yeh, who had no less than thirty consorts. By contrast, the Emperor Huang-ti is said to have enjoyed coitus with twelve hundred women. At about this time Taoist monasteries were said to be the sites of sexual activities *en masse* with promiscuous coupling.

During the Sui period the Emperor Yang-Ti collected erotic pictures and surrounded his bed by mirrors. At this time there appears to have been a predilection for women of prenubile appearance with undeveloped breasts and absent or scanty body hair. There was a reaction in the T'ang period when the desirable woman was sturdy with well-developed breasts, slender waist and heavy hips. Bare bosoms were permissible and dancing girls performed with naked breasts. The Empress Win had strong erotic feelings and surrounded her bed with mirrors: she had numerous lovers and even when nearly seventy years of age required young men.

There was yet another reaction in the Sung period—puritanism became the vogue and ended such unedifying spectacles as those of naked women wrestling. At this period the female foot was the centre of sex appeal and in pictures of women was never shown unshod even if the vulva was clearly depicted. During the Yuan period there are express proscriptions of debauched behaviour. Debauchery was serious if practised with nuns, virgins or widows, was less serious with married women and least of all with prostitutes. Obscene stories were told to titillate the female sexual appetite.

During the Ming period women were again secluded in the home: chastity was valued, divorce was a disgrace and widows could not remarry. The desirable women again had large breasts—pointed, rounded or pendulous—and a prominent belly and buttocks. There are pictures of women having coitus wearing only a sort of brassière and stockings. Later again, however, there was a further revolution of taste and the ideal woman had sloping shoulders and a flat chest.

These facts are given in a little detail since the long period of documentation in Ancient China exemplifies how sexual *mores* may change in a society.

Prostitution appears to have been prevalent at all periods in Chinese history; single and married men might visit prostitutes for amusement and in such a relationship no restrictions or taboos operated as in conjugal coitus. It was believed that harlots, because of frequent coitus with many men, had a very powerful "yin" essence which was very beneficial to the male. Prostitution was regarded as a legitimate occupation: it has been stated that blind women were sometimes driven to prostitution, but this may not be true. Some authorities have seen the use of slaves and female war captives as the forerunner of official prositution. The first historical reference to prostitution dates from the late Chou period. The Emperor Win (140-87 B.C.) provided camp harlots for his armies; at about the same time public brothels were provided and operated by private persons for gain; wealthier persons could purchase brothel girls for their own use. Prostitution continued to flourish through the T'ang and Sung periods till modern times. Venereal disease was known to be associated with prostitution.

Homosexuality is documented from the late Chou period (770-222 B.C.) when some princes kept catamite boys and also had relations with adult males, e.g., ministers of state. Opinion seems to have been neutral in Ancient China on sexual relations between consenting adult persons but denounced homosexuality for profit and praised it if there resulted great art forms, e.g., as between the poets Hsi K'ang and Yuan-Chi in the third century A.D. It was only later in the Han period that male homosexuality flourished and was fashionable; several Han emperors were ambivalent and used both harem women and young boys. The Emperor Wu was a lifelong homosexual and is said to have used a castrated actor. Homosexuality was also common in the Sung period and there appear to have been male prostitutes sometimes dressed as women.

Female homosexuality seems to have been very frequent in Ancient China in the restricted women's quarters; it was thought that when women lived in continuous proximity that lesbianism could hardly be avoided. It was tolerated by the heads of household since the female was believed to possess unlimited "yin" essence so that despite homosexual practices the male was not deprived of it during coitus. Indeed, during the Ming period lesbianism seems to have been encouraged.

For the same reason female masturbation was tolerated since this also did not deprive the male of "yin" essence. There were, however, warnings against excessive use of artificial phalli (*olisboi*) which might damage the woman's internal lining. *Olisboi* of vegetable nature which swelled with the vaginal secretion were sometimes used. Lesbian pairs might use a double *olisbos* provided with two pairs of tapes.

Masturbation by the male was forbidden since this resulted in a complete loss of the "yang" vital male essence. An exception might occasionally be made if a man had long been deprived of female company in case the devitalised semen might clog up his system. By contrast, coitus reservatus was commonly practised to conserve the "yang" male essence and yet obtain essential "yin" female essence at her orgasm. Coitus reservatus appears to have been practised in Ancient China over a period of thousands of years without ill effect.

Precoital cunnilinctus as a preparation of the female for intercourse was approved since there was no loss of "yin" essence. Fellatio was also permitted as a preliminary to coitus, but not to emission since this would result in loss of "yang" essence; the female saliva was believed to contain "yin" essence which was beneficial.

Incest appears to have sometimes occurred. During the Chou period men sometimes had coitus with their father's wives and a woman, Nan-tze (who eventually married Lung, Prince of Wei) was notorious for her incestuous relations with her brothers. This behaviour led to the singing of ribald songs by the field peasants. Relatives of the Emperor Hsiao-ching (c. 156-40 B.C. in the Han dynasty) were degenerate sadists and are said to have had incestuous relationships with sisters and other female relatives. Prince Chien is said to have committed incest with his sisters and Haiyang, son of Ch'ü, also to have had incestuous relationships. These are probably isolated examples, occurring in exalted circles, and in general even "name-incest" (the marrying of a wife or concubine with a surname similar to the male) was avoided.

Zoophilia appears to have been very rare, although during the early Han period Prince Chien used punitive bestiality on harem women who occurred his displeasure and Li Yui mentions a woman who had relations with a dog.

As indicated above, fellatio ("playing the flute") and cunnilinctus appear to have been practised as well as female anal intercourse during the later period but sadism was rare; there are scanty accounts of whipping of prostitutes during the Sung period and of burning incense on the female body. Rare examples of sadistic practises by female upon females are also recorded. There are also rare examples of male masochism.

We are best informed about the Ming period 1368-1644 A.D. because of the availability of erotic novels and plays with an erotic theme and of erotic works of art from this period. Although some novels are markedly scatological there are no references to flagellation, masochism, sadism nor homosexuality. The erotic albums show various, sometimes rather elaborate, heterosexual coital positions, occasional female anal intercourse, cunnilinctus, fellatio and lesbian practices (Gulik, 1961).

OTHER ANCIENT SOCIETIES

For pre-literate peoples our only source of knowledge of sexual behaviour comes from artefacts and art forms. From the Stone Age periods there are representations of animals in coitus and of pregnant buffalo calves denoting an interest in sexuality. A cave relief from Laussel has been interpreted as showing a human couple mating. When men and women are represented their genitalia are often prominent; so-called *Venuses* from Willendorf and Brassempouy show apparent steatopygia, prominent secondary sexual characteristics and careful delineation of, for example, the labia minora (Lewinsohn, 1958). Carved reindeer horn phalli have been discovered in numbers (Abarbanel and Wilbur, 1961). Lewinsohn suggests that a cave drawing shows a sort of voyeurism. Bloch (1908) suggests that coloured pastes were used by these early people as sexual adornment. It seems likely that the effect of castration on domestic animals was known in Neolithic times (Kinsey *et al.* 1953).

From the Bronze age we have evidence

of phallicism, for example, from Corsica where large stone representations of the phallus occur; similar phallic artefacts come from a wide area of the globe, e.g., Babylonia, India, China, Egypt, Palestine, Sudan, Brittany, Argentina, Mexico and North and Central America.

In Ancient Mesopotamia marriage was monogamous although in Sumeria and Babylonia no stigma attached to the keeping of slave concubines. Offspring of such unions were regarded as slaves unless accepted as legitimate by the master. Divorce could be obtained on the grounds of barrenness or ill-health. Further, temple prostitution by *Qadishtu* was taken as a matter of course but it is probable that the story of Herodotus that all Babylonian women prostituted themselves once in a lifetime was a misapprehension of the nature of temple prostitution. The great *Entu* priestesses of the temples were expected to live in chastity; suggestions that these women were sterilised or had recourse to abortion are improbable. It seems likely that male eunuchs played transvestite parts in cult performances similar to those of Cybele. Punitive castration was provided for by laws of Middle Assyria (*c.* 15th cent. B.C.). At other fertility festivals cakes were baked in the representation of the male and female genitalia.

From proverbs and omens we have reason to believe that coitus usually took place in the face to face position with the female lying on her back but plaques show that either coitus *à posteriori* or anal intercourse took place with the male standing and the female leaning forward in an upright position. According to Herodotus Babylonian couples fumigated themselves after coitus. Certain texts suggest that priestesses permitted anal intercourse. Apart from heterosexual coitus it appeared that cunnilinctus, male homosexual sodomy, transvestism and female homosexuality also occurred (Saggs, 1962). There were statutory punishments for rape and seduction (Kinsey *et al.*, 1953).

It is clear that neuroses and psychoses were not infrequent and manifested often by impotence and premature ejaculation. Nocturnal emissions are described and there are accounts of erotic dreams in which persons walked about naked, in which incestuous relations with daughters or mothers-in-law occurred, in which bestiality figured and in which enlarged genitalia appeared (Saggs, 1962). There is said to have been a death penalty for intercourse with cattle although this is a fairly common practice among pastoral peoples.

Monogamy was the rule among the Hittites and only the reigning king had a harem. It is obvious from an edict of Suppiluliama (1380-1346 B.C.) that incest was punished by execution. A man was not allowed to have coitus with sister or cousin. From the context it is obvious that incest was known to be common in other oriental lands (Akurgal, 1962). Euripides in *Andromache* also refers to barbarian incest.

In Ancient Persia according to the *Videvdad* women were respected; girls were expected to be married at the age of fifteen years, preferably to a relative. Menstruation, parturition, ejaculation and death also gave rise to ritual uncleanness. Abortion was a crime and sodomy a major crime punishable by death even for the unconsenting victim (Sigerist, 1961). Herodotus states, however, that at the time of the Invasions the Persians were polygamous and also kept many mistresses; they slept with their women in rotation. They took women with them when they went to war. At this period they are also said to have learned pederasty from the Greeks.

According to Plutarch, Teribazus of Persia decked himself with women's trinkets and was presumably transvestite. Sardanapalus is also said to have been a transvestite (Hirschfeld, n.d.). There are several accounts of incest in the Persian Royal Family—Xerxes with his daughter-in-law, Cambyses with his sisters and Artaxerxes with his sisters also. Herodotus in his *History* also describes reprehensible behaviour by the Persian Army at the time of the Invasion; this varied from handling the breasts of Macedonian women to the repeated rape and death of some women of Phocaea.

Herodotus in his *History* also gives information, possibly sometimes of doubtful value, concerning the sexual behaviour of the barbarian nations. We read that nudity was regarded as indecent in Lydia even for men but that Lydian girls prostituted themselves to provide themselves with a dowry. It is stated that some Caucasian and Indian tribes copulate in the open like cattle and that the latter, like the Ethiopians, have black semen. The latter is obviously of folk-lore origin.

Thracian girls were said to be very promiscuous, as were the women of the Gindanes of Libya who wore an extra leather anklet for each new lover with whom they slept. We are told that the Paeonian lake-dwellers of Thrace were polygamous and that the Massagetae of the Caucasus, the Agathyrsi of Scythia, the Auses and Nasomones of Libya had their women in common with casual intercourse. When a man of the Nasomones married for the first time all the male guests in turn had coitus with the bride and each left her a present. A similar practice was described by Diodorus Siculus as occurring in the Balearics. The king of the Libyan tribe of the Adyrmachidae had *jus primae noctis*.

There is little evidence of sexual deviation in Herodotus but some tales of cruel treatment, for example the sadistic murder by Amestris (wife of Xerxes) of the mother of her daughter-in-law and the cruel behaviour of Pheretima after the siege of Barca when she impaled some of the men of the city and the breasts of their wives upon stakes around the city wall.

The discovery of peat-preserved bodies, probably dating from the early Christian period, in Denmark and North Germany, some of which had been hanged and others suffered throat-cutting, raised the question of whether these were sacrificial victims of a fertility cult or whether they were executed criminals. Two bodies were found close together in Shleswig-Holstein. One was a naked fourteen-year-old girl, the left side of whose head had been shaved and whose eyes were covered by a woven blindfold; nearby was a man aged about forty years who had been strangled with a hazel wand. The question here, of course, is whether the pair had been executed for some sexual offence (Bibby, 1957; Schlabow et al., 1958).

According to Tacitus' *Germania* the Ancient Germans were monogamous and married at mature age. Women were chaste and adultery rare; if discovered adultery was severely punished. There was no mercy for the woman who prostituted her chastity. Means of restricting numbers of children were forbidden and mothers suckled their own infants. In Caesar's *Gaul* it is stated that coitus before the age of twenty was regarded as scandalous. Men and women bathed together, largely unclothed however, in rivers, and naked youths performed a sort of sword dance.

Caesar believed that in Britain a sort of polyandry occurred in which wives were shared between groups of about a dozen men; such groups often comprised brothers or fathers and sons. Brown (1961) states that the ancient Germans drowned transvestites. Bloch (1908) states that homosexuality occurred among the Ancient Scandinavians.

Frumkin (1961) gives some account of Anglo-Saxon England; persons belonged to local kin-ship groups and marriage was designed for procreation. Society was patriarchal and women had menial status. Premarital and extramarital coitus, although proscribed, were common. Male offenders, if detected, were required to pay damages. Adulterous wives might be burned in the earlier period but the laws of Cnut substituted cutting off the nose and ears. Women were regarded as sexual property and sex was neither good nor evil. Prostitution was widespread in England at this time and continued to flourish in the later Mediaeval period. Hirschfeld (n.d.) states that Anglo-Saxon women sometimes flagellated slaves to death.

We have some information concerning sexual behaviour in early American societies. Duran (1963) points out that the art of the Maya presents no sexual expression. The Maya punished adultery with severity and claimed that phallic cults, licence and homosexual behaviour were introduced by the Itza into their region. Voget (1961) states, however, that the Yucatecan Maya and Aztecs assigned wantons to the young men's houses for their pleasure.

The Aztecs appear to have encouraged marriage of men at the age of twenty-five years and of girls at eighteen years. Separation was only possible with great difficulty and expense. Adultery was punishable by death. Sadistic human sacrifice was a notable feature of Aztec religion. Aztec priests were celibate and austere but archaic figurines from Mexico suggest fertility cults of large-breasted goddesses. Beals (1961) states that some Aztec deities were bisexual. Special deities looked after sex activities. No stigma appears to have been attached to male or female homosexuality, nymphomania or prostitution. The Aztecs are said to have made gold statuettes of various forms of copulation including male sodomy. Toltec kings appear to have kept concubines (Burland, 1948).

Our most fruitful source of information about pre-Columbian sexual behaviour is derived from the pottery of the Mochicas of Peru. Much of this has an erotic content; "normal" coitus is rarely shown but many pots show kissing of breasts, fellatio, cunnilinctus, male masturbation, unusual coital postures including posterior approaches and probably anal intercourse. Phallic elements are also common in drinking jars with penile spouts but there are no representations of male homosexuality (Posnansky 1925).

According to Bloch (1908) indiscriminate copulation occurred during foot races at Ancient Peruvian festivals. Voget (1961) believes that it is possible that ceremonial male homosexuality occurred in Coastal areas of Peru and Ecuador at the time of the Spanish Conquest. The Incas included ceremonial flagellation at initiatory rites for young men. The Incas permitted brother-sister and brother-half sister marriage for royal and noble families but incest was otherwise forbidden.

CONCLUSIONS

Gerald Moss writing on mental disorders elsewhere in this volume concludes that "in the field that has been surveyed there is probably nothing for which a modern parallel could not be found." The same might well be said of sexual behaviour in ancient societies.

Kinsey and his associates (1948, 1953) have shown that male and female masturbation, heterosexual petting techniques and mutual manipulations, male and female homosexuality, mouth-genital and animal contacts, premarital and extramarital intercourse, all play a surprisingly large part in contemporary human sexuality in the United States of America. There is good reason to believe that a similar state of affairs obtains in other Occidental societies.

As we have seen, there are known parallels to most of these activities in the sexual behaviour of ancient peoples. There is, perhaps, little in modern times to correspond with the active sadistic practices of Ancient Rome but the Second World War showed that where opportunity offered (for example in concentration camps) individuals gladly undertook sadistic activities. Further, there is much sado-masochistic content in modern literature and drama.

It is perhaps not surprising that patterns of sexual behaviour appear to remain fairly constant and to be influenced only to some extent by the current attitudes of society. Human nature appears to have changed remarkably little over the millenia.

REFERENCES

ABARBANEL, A. and WILBUR, G. B., 1961: Phallicism and Sexual Symbolism. In *The Encylopaedia of Sexual Behaviour*. Ellis, A. and Abarbanel, A. (Eds.) Vol. II, London, Heinemann.

AKURGAL, H., 1962: *Art of the Hittites*. London, Thames and Hudson.

ALTMANN, S. A., 1962: A Field Study of the Sociobiology of Rhesus Monkeys, *Macaca mulatta*. Ann. N.Y. Acad. Sci., 102:338.

ANAND, MULK RAJ., 1960: *Kama Kala: Some Notes on the Philosophical Basis of Hindu Erotic Sculpture*. Geneva, Nagel.

BEALS, C., 1961: Sex Life in Latin America. In *The Encyclopaedia of Sexual Behaviour*. Ellis, A. and Abarbanel, A. (Eds.) Vol. II, London, Heinemann.

BIBBY, G., 1957: *The Testimony of the Spade*. London, Collins.

BLOCH, I., 1908: *The Sexual Life of Our Time*. London, Rebman.

BREASTED, J. H., 1912: *Development of Religion and Thought in Ancient Egypt*. London, Hodder and Stoughton.

BROWN, D. G., 1961: Transvestism and Sex Role Inversion. In *The Encyclopaedia of Sexual Behaviour*. Ellis, A. and Abarbanel, A. (Eds.) Vol. II, London, Heinemann.

BURLAND, C. A., 1948: *Art and Life in Ancient Mexico*. Oxford, Cassirer.

BURTON, R. F., 1963: trans. *The Perfumed Garden of the Shaykh Nefzawi*. London, Spearman.

BURTON, R. F., and ARBUTHNOT, F. A., 1963: trans. *The Ananga-Ranga of Kalyana Malla*. London, Kimber.

ČERNY, J., 1945: The Will of Naunakhte and the Related Documents. J. Egyptn. Archaeol. 31:29.

COLE, W. G., 1960: *Sex and Love in the Bible*. London, Hodder and Stoughton.

DANIELSSON, B., 1961: Sex Life in Polynesia. In *The Encyclopaedia of Sexual Behaviour*. Ed. Ellis, A. and Abarbanel, A. Vol. II, London, Heinemann.

DURAN, C. M., 1963: Surgery of the Mayas. Abbottempo 1(4):14.

FLACELIÈRE, R., 1962: *Love in Ancient Greece*. trans. J. Cleugh. London, Muller.

FOSTER, JEANNETTE H., 1958: *Sex Variant Women in Literature*. London, Muller.

FRUMKIN, R. M., 1961: Early English and American Sex Customs. In *The Encylopaedia of Sexual Behaviour*. Ellis, A. and Abarbanel, A. (Eds.) Vol. I, London, Heinemann.

GLASNER, RABBI S., 1961: Judaism and Sex. In *The Encyclopaedia of Sexual Behaviour*. Ellis, A. and Abarbanel, A. (Eds.) Vol. II, London, Heinemann.

GREEN-ARMYTAGE, V. B., 1945: Some *Obiter Dicta in Obstetrics and Gynaecology*. London, Devonport Press.

GOULD, G. M., and PYLE, W. L., 1897: *Anomalies and Curiosities of Medicine*. London, Rebman.

GULIK, R. H. VAN, 1961: *Sexual Life in Ancient China*. Leiden, Brill.

HENRIQUES, F., 1962: *Prostitution and Society: A Survey*. Vol. I. Primitive, Classical and Oriental. London, MacGibbon and Kee.

HIRSCHFELD, M., no date: *Sexual Anomalies and Perversions*. London, Alder.

KIEFER, O., 1934: *Sexual Life in Ancient Rome*. London, Routledge.

Kinsey, A. C., Pomeroy, W. B., and Martin, C. E., 1948: *Sexual Behavior in the Human Male.* Philadelphia, Saunders.

Kinsey, A. C., Pomeroy, W. B., Martin, C. E., and Gebhard, P. H., 1953: *Sexual Behaviour in the Human Female.* Philadelphia, Saunders.

Lewinsohn, R., 1958: *A History of Sexual Customs.* London, Longmans Green.

Licht, H., 1932: *Sexual Life in Ancient Greece.* London, Routledge.

Marcadé, J., 1961: *Roma-Amor—Essay on Erotic Elements in Etruscan and Roman Art.* Geneva, Nagel.

Marcadé, J., 1962: *Eros-Kalos—Essay on Erotic Elements in Greek Art.* Geneva, Nagel.

Meyer, J. J., 1952: *Sexual Life in Ancient India.* London, Routledge, Kegan Paul.

Posnansky, A., 1925: Die erotischen Keramiken der Mochicas und deren Beziehungen zu occipital deformierten Schädeln. *Abhandl. Anthrop. Ethnol. Urg.* 2:67.

Ruffer, M. A., 1919: On the Physical Effects of Consanguineous Marriages in the Royal Families of Ancient Egypt. *Proc. Roy. Soc. Med.* 12:145.

Saggs, H. W. F., 1962: *The Greatness that was Babylon.* London, Sidgwick and Jackson.

Sandison, A. T., 1958: The Madness of the Emperor Caligula. *Med. Hist.* 2:202.

Sandison, A. T., 1963: The Use of Natron in Mummification in Ancient Egypt. *J. Near East. Stud.* 22:

Schlabow, K., Haage, W., Spatz, H., Klenk, E., Diezel, P. B., Schütrumpf, R., Schäfer, U. and Jankuhn, H., 1958: Zwei Moorleichenfunde aus dem Domslandsmoor Gemarkung Windeby, Kreis Eckernförde. *Praehist Z.* 36:118.

Sigerist, H. E., 1951: *A History of Medicine.* Vol. I. Primitive and Archaic Medicine. New York, Oxford University Press.

Sigerist, H. E., 1961: *A History of Medicine.* Vol. II. Early Greek, Hindu and Persian Medicine. New York, Oxford University Press.

Smith, G. Elliot, and Dawson, W. R., 1924: *Egyptian Mummies.* London, Allen and Unwin.

Storr, A., 1964: *Sexual Deviation.* Harmondsworth, Penguin.

Upadhyaya, S. C., 1961: trans. *Kama Sutra of Vatsyayana.* Bombay, Taraporevala.

Voget, F. W., 1961: Sex Life of the American Indians. In *The Encyclopaedia of Sexual Behaviour.* Ellis, A. and Abarbanel, A. (Eds.) Vol. I. London, Heinemann.

Williams, W. L., 1947: *Diseases of the Genital Organs of Domestic Animals.* 3rd Ed. Worcester, Plimpton.

Wood, R. F., 1961: Sex Life in Ancient Civilisations. In *The Encylopaedia of Sexual Behaviour.* Ellis. A. and Abarbanel, A. (Eds.) Vol. I. London, Heinemann.

Yoyotte, J., 1962: Sexual Behaviour. In *A Dictionary of Egyptian Civilisation.* Posener, G. (Ed.) with assistance of Sauneron, S. and Yoyotte, J. trans. Alix. Macfarlane. London, Methuen.

Index

A

Abscesses, 639-640
  alveolar, 128, 129, 576-582, 625, 702-705
    pseudopathology, 7
    and Harris's lines, 397
    in fossil man, 50, 58
    in primates, 49
  appendix, 496
  Bezold's 464, 468
  breast, 129, 512, 513
  cerebral, 123
  ear, 464-472
  eye, 198, 457
  genital, 506
  liver, 195, 244
  pelvic, 129, 508
  perineal, 515
  peritoneal, 232
  pulmonary, 232
  psoas, 126
  renal, 177, 488, 535
Abortion, 499, 502, 504
Achondroplasia, 432-435, 526-528, 555, 599
Aclasis, diaphyseal, 442
Acne, 450, 453
Acrocephaly, 441-442
Acromegaly, 218, 522-525, 554-555
Actinomycosis, 583
Adenoma, Pituitary, 329, 343, 525
Aeschylus, 711-712
Aetius, 124, 180, 450, 459, 477, 505, 507
Agatharchides of Cnidus, 180
Akhenaten, 343, 429, 525, 738
Alimentary tract, Diseases of, 212-213, 494-497
Alkaptonuria, 20, 22-30
  presumptive, 22-30
Alopecia, 217, 227, 528-530
Amenophis II, 528, 537
  IV, 514
Amenorrhoea, 506
Amentia, 717
Amoebiasis, 195
  hepatic, 182
  pulmonary, 182
Amosis I, 515
Amputation, xii, 64, 67-68, 218, 640-641
Anatomy, in Egypt, 108-109
Anaemia, 178, 194, 421, 506
  haemolytic, 381
  hypochromic, 84, 87
  iron deficiency, 383

Anasarca, 193
Anencephaly, 441, 502
Aneurysms, 475, 477
Angina pectoris, 202, 477
Angioma, 327, 329
Ankylosis of temporo-mandibular joint, 587-588
Anopheles, 85, 117, 118, 144
  A. maculatus, 86
Anoplotherium, 32
Anthracosis, 484, 491, 492
Anthrax, 63, 117, 451
Antihelminthics, 179
Antonine pestilence, 120-121
Anus, Imperforate, 213
Apatosaurus, 42
Aplasia, Partial, 435-436
Apoplexy, 235, 477
Appendicitis, 128, 129, 496
Archigenes of Apameia, 506, 513, 716
Aretaeus, xii, 116, 121, 124, 127, 205, 477, 522, 718-719
Aristophanes, 171, 181, 711-712
Aristotle, 174, 179, 180, 182, 450, 458, 499, 500, 502, 516, 714, 739
Arteriosclerosis 474-475, 481, 482
Arthritis, 102, 126, 255, 352-370, 573, 600-601
  acute 353
  and genital disease, 515-516
  and oral disease, 585-586
  "ankylosing spondylitis," 20, 22, 278, 354, 357-360, 491-492, 601
  chronic, 354-356
  in fossils, 40-42
  in primates, 48
  osteoarthritis, 354-356, 358, 361-362
  pseudopathology, 14
  rheumatoid, 354, 357
  "spondylitis deformans," 585, 600
Artistic representations, 124, 249, 498, 499, 500, 501, 505, 511, 512, 515, 516, 522, 526, 600, 601, 703, 738, 739
  achondroplasia in, 432, 527-528
  blindness in, 457
  circumcision in, 514, 537
  club foot in, 423-424, 600
  dwarfism in, 527-528
  gigantism in, 524
  goitre in, 521
  hare lip in, 17
  hernia in, 445
  hip deformity in, 438, 508
  obesity in, 525, 526, 528, 529

pseudopathology, 15-17
sexual deviation in 501, 738, 739, 744, 745, 747, 749, 750, 751, 753
tuberculosis in, 250-254, 260-263
tumours in, 343
Ascites, 477
Ashurbanipal, 192, 477, 514
Ashurnasirpal II, 194
Asthma, 214, 492
Ataxia, 193
Athens, Plague of, 238-239
Atheroma, 88, 177, 474, 480, 482
Atherosclerosis, 13
Atossa of Persia, Queen, 129, 343, 513
Atresia, oesophageal, 212
of external auditory meatus, 215
Australopithecines, 56, 538, 541, 544
Avicenna, 477

B

Bacteriology, 115-131, 279
bacteria in mummy tissue, 177, 347-348, 490
in fossils, 42
Bilharzia haematobia, 72, 196, 225, 231, 515, 526, 533, 534-535
calcified eggs of, 177, 180, 484, 535
in mummies, 177, 180, 535
Bacillus anthracis, 117
Baghdad boil, 197
Baldness *see* Alopecia
Bartonella bacilliformis, 182
Bed-bugs, 178, 181, 182
Behcet's syndrome, 458
Beri-beri, 72, 73, 227, 229, 417, 418-421
Bering Strait land bridge, 125, 127, 156, 157, 159, 166, 387
Bible, diseases in the, 209-221, 260, 479, 504, 525, 731
blindness in, 457
sexual deviation in, 735-736
skin diseases in, 449-450, 304
Birth injury, 510
Black Death, 116
Blastomycosis, South-American, 182, 451
Blepharitis, 82, 457, 458
Blindness, 215, 228, 457, 458
congenital, 204
day,- 192,
night,- 192, 417, 418, 457, 458
snow,- 459
Blood diseases, 216
Blood-letting, 638-639
Blood-stain on bone, 6
Bog-burials, 13, 115, 185, 484, 752
Boils, 128, 451, 639-640
Borrelia, 235
B. duttoni, 118
B. recurrentis, 118
Boskop skull, 468-469
Brachydactyly, in primates, 53

Breast,
amputation of, 511
diseases of, 511-513
Brodie, Sir Benjamin, viii
Bronchiectasis, 492
Bronchitis, 226, 227, 231
Brucellosis, 63, 72, 117, 353, 357
Bulinus, mollusc, 184, 196, 534

C

Cachexia, 477, 506
malarial, 172,
Caelius Aurelianus, 720
Caesarean section, 203, 505, 641-642
Calcinosis intervertebralis, *see* Alkaptonuria
Calculi, 195, 196, 215, 231, 244, 349-351, 484, 495, 515, 534, 535
gallstones, 495
Camptosaurus, 43
Cancer, 84, 339-341, 646
of breast, 343, 512, 513, 341
of liver, 94
of lung, 88
of ovary, 511
of prostate, 331
of skin, 450, 451
of stomach, 495
Cancrum oris, 243
Carabelli tubercle 543-544
Carbuncles, 128, 129, 244, 449, 451
on eyelids, 459
Carcinoma *see* Cancer
Cardiorenal disease, *see* Vascular disease
Cassius, 450
Castration, 214, 513-514, 741
Cataract, 458, 459, 641
Catheters, Urethral, 533
Celsus, xii, 121, 128, 179, 182, 203, 242,-245, 260, 343, 450-451, 459, 477, 507, 513, 515, 521, 522, 533, 636, 645, 653, 657, 658, 661, 664, 665, 666, 716, 720
Cesspits, residues, xii, 184-188
Cetiosaurus leedsi, 42
Charcot's arthropathy, 356-357
Chickenpox, 119, 121
Chilblain, 244, 450, 451
Childbirth, 203-204, 231, 501-511
death during, 507-508
Chilomastix mesnili, 185
China, ancient,
diseases in, 222-237
Chnoum-hotep, 432, 527
Cholecystitis, 240, 495-496
Cholera, 119, 122, 213, 232
Cholesteatoma, 465, 469, 472
Chondro-osteodystrophy, 20
Chondroma, Ossifying, 327
Cicero, 260, 717
Circumcision, 214, 514-515, 536-537

Cirrhosis, Hepatic, 94, 177, 231, 232, 235, 477
Cleft palate, 551-552
Clostridia, 115-116
   Cl. tetani, 116
   C. botulinum, 116
Club foot *see* Talipes equinovarus
Coloboma, 459
Columella, 180, 181
Confucius, 419, 421
Congenital disorders, 203-204, 423-443, 508
   of the jaws, 551-553
Congenital lacunae, 9, 407
   in primates, 53
Conjunctivitis, 215, 457, 458
Constipation, 497
Consumption, *see* Tuberculosis
Contraception, 102, 498, 499-500
Convulsions, Infantile, 203
Corns, 451
Cranial deformation, Artificial, 556
Cremation, 10, 13-14
Cribra orbitalia, 300, 322, 378, 383, 459
   pseudopathology, 7
Cribra parietalis *see* Osteoporosis symmetrica
Cro-Magnon man, 13, 59, 361, 617, 619
Cryptorchidism, 214, 215
Cushing's syndrome, 525, 526
Cyclothymia, 719
Cystadenocarcinoma, Ovarian, 511
Cystitis, 128, 129, 231, 534
Cysts, 451
   alveolar, 565, 576-582, 584-585
   aneurysmal bone, 325
   eye, 199
   intradiploic epidermoid, 328
   sebaceous, 329, 459

D

Dacryocystitis, 198, 457
Daphoenus, 13
Dead Sea Scrolls, 732
Deafness, 215
Delirium, 200, 730
   febrile (or phrenitis), 715, 716, 720
Dementia, 717, 718
Democedes of Crotona, 343
Dengue, 63, 118
Depopulation 75-78
Depression (mental) 715, 716, 718, 731
Development, 90
Diabetes mellitus, 235, 243, 522
Diarrhoea, 497
Diogenes, 476
Dimetrodon, 37
Diodorus Siculus, 25, 239-240, 752
Dioscorides, 110, 180, 182, 245, 459
Diphtheria, 82, 122, 123, 205, 214, 235
Diplococcus pneumoniae, 123
Diplodocus, 42

Disease,
   bacteriological view, 115-131
   cultural background, 56-68
   definition of, 33-34
   epidemiology of, 78-79
   genetic factors in, 79-81
   in primates, 47-55
   in primitive societies, 69-97
   increase in time, 38-39
   psychological factors in, 77-78
Dislocations, 601, 604-605, 637-638
Distichiasis, 459
Dropsy, 193, 477
   ovarian, ix, 343, 511
Dwarfism, 432-435, 521-522, 526-528, 599
   Chilca skull, 12
Dyschondroplasia, 599
Dysentery, 119, 122, 124, 195, 244
Dysmenorrhoea, 505, 506
Dyspareunia, 506

E

Ear diseases, 198-199, 215, 464-473
Ectropion, 458, 459
Eczema, 217, 244, 451, 454
Electron microscopy, xii, 451
Elephantiasis Arabum, 180
Elephantiasis Graecorum, 245, 305
Empedocles, 476, 502
Empyema, 490, 640
Endocardial diseases *see* Heart disease
Endocrine diseases, 521-531
Entamoeba histolytica, 185
Epidemics, 65-66, 163-164, 213-214, 238-240, 245-246, 279
   in primitive populations, 78-79
   in ancient China, 223-224
Epilepsy, 201-202, 216, 716, 718-719
Epileptic psychosis, 725
Erasistratus of Chios, 476, 515
Ergotism, gangrenous, 478
Erysipelas, 123, 241, 244, 450, 451, 505
Escherichia coli, 129, 185
Eusebius, 451
Excavation hazards, 11-12
Excision, 640-641
Exophthalmos, 199, 457
Exostoses, 552
   in fossils, 43, 325
   osteocartilaginous, 323
Eye diseases, 198-199, 215-216, 457-463
   in mummies, 459-462

F

Facies leprosa *see* Leprosy
Facies nasalis *see* Leprosy
Facies oralis *see* Leprosy
Famine, 63, 89
Fibroids, 506, 507

Fibroma, 329
Fibrous dysplasia, Polyostotic, 371-377
Fistulae,
    alveolar, 563, 573, 580-581
        in fossils, 43
    genital, 506
    lacrymal, 459
    vesico-vaginal, 507
Fleas, 178, 181
Fossil animals: pathological changes in, 31-46
    gigantism in, 525
    palaeozoic, 35-36
    Permian vertebrates, 37, 39
    skull fracture in, 612
Fossil man, 40, 42, 43, 50, 56-59; see also individual groups
Fractures, 7, 53, 67, 491, 511, 601-603, 604-605, 606-622, 623-629, 637-638
    in fossils, 32, 37-38, 43, 606-614
    in primates, 47-55
    post-mortem, 6, 15
Frost-bite, pseudopathology, 7

G

G6PD deficiency, 80, 387
Galen, xii, 110, 121, 123, 127, 171, 172, 173, 174, 179, 180, 182, 193, 260, 343, 459, 476, 513, 516, 522, 719
Gallstones, see calculi
Gangrene, 116, 241, 446
    of the extremities, 477, 522
    of the genitalia, 515
    pseudopathology, 13
Gastro-enteritis, 63
Genital system, Diseases of, 214-215
    female, 498-511
    male, 513-517
Giardia lamblia, 185
Gigantism, 522-525
Glaucoma, 458, 459
Glomerular fibrosis, 480
Glossina flies, 133
    ecology, 135-143
    G. morsitans, 134, 135, 143, 148, 149
    G. pallipides, 138, 142, 143, 148
    G. palpalis, 134, 135, 144, 146, 148
    G. swynnertoni, 138
    G. tachnoides, 135
Goitre, 87, 227, 228, 521, 640
Gonococcal infections, 78, 214
Gonorrhea, 78, 125, 128, 214, 515
Gout, 356, 369, 525, 601
Granville, Augustus Bozzi, viii-ix, 343, 507, 511
Grauballe man, 185
Greece and Rome, Diseases in, 238-246
Growth, 90
    and Harris's lines, 400-403
Gynaecomastia, 218

H

Haematoma, Ossifying, 325, 326, 327
Haematuria, 177, 180, 231, 477, 534, 535
Haemophilia 216
Haemorrhage,
    cerebral, 216, 235, 478, 613, 614
    intestinal, 494
Haemorrhoids, 213, 477, 497
Hammurabi's Code, 198, 259, 457, 511, 728
Hare lip,
    artistic representation, 17
Harris's lines, 15, 390-404
Hatshepsut, 517
Headache, 216
Heart disease, 88, 89, 90, 202, 229, 475
    mitral calcification, 480
Helminths, xiii, 75, 78, 91, 178-180, 184-188, 194-195, 213, 230, 232. see also Bilharzia; hydatid disease
    ancylostomiasis, 195, see Hookworm
    ascaris, 75, 78, 178, 179, 185, 187, 194-195, 231, 232
    Dicrocoelium dendriticum, 185, 187
    Diphyllibothrium latum, 185
    Dirofilaria pongoi, 48
    distosomiasis, hepatic, 180, 231
    Dracunculus medinensis, 180
    eggs, 184-188
    enterobius, 231
    Fasciola hepatica, 185
    filariasis, 48, 81
    herxetef, 178
    hookworm, 84, 88, 178, 179, 195
    in primates, 48-49
    myiasis, 180, 181-182, 461
        urinary, 182
    oxyuris, 180, 185
    roundworm, see Ascaris
    Schistosoma haematobium, 158, 184, 196, 534, 535
    taenia, 178, 179, 180, 185, 196-197
    tapeworm, see Taenia
    Tetrapetalonema digitata, 48
    threadworm, 178, 179
    Toxocara canis, 185
    trichuris, 179, 185, 187
Hemiatrophy, 510
    facial, 556
Hemiplegia, 201, 235, 428, 477, 510, 556
Hepatisation of the lung, 490
Hepatitis, infective, 122, 124, 125, 231
    amoebic, 182
Hermaphrodites, 215, 516-517
Hernia, 640
    in Egypt, 444-446
    inguinal, 445-446
    umbilical 218, 445, 636-637
    in primates, 49
Herod Agrippa, 181
Herod the Great, 181, 479

Herodotus, 25, 99, 181, 182, 260, 450, 502, 513, 515, 516, 522, 536, 630, 678, 710, 712, 737, 738, 741, 751, 752
Herpes Zoster 119, 121
Hip joint, Dysplasia of, 436-440, 508
Hippocratic Collection, xii, 110, 116, 118, 122, 123, 124, 125, 127, 128, 171, 172, 173, 174, 179, 180, 182, 196, 240, 242, 260, 305, 437, 450, 458, 476, 477, 478, 499, 502, 504, 505, 506, 513, 514, 515, 516, 525, 653, 658, 661, 664, 710, 714-715, 716, 719
Home, Sir Everard, viii
Homer, 170, 171, 260, 637, 710-711
Homo habilis, 133
    H. (Pithecanthropus) erectus, vii, 13, 56, 58, 538, 574, 606-611
        femur, 11, 325
    H. sapiens, vii, 133, 538
    H. soloensis, 611-612
Hookworm *see* Helminths
Horace, 260, 717-718, 743
Hunchback, 421, 528. *See also* Tuberculosis
Hutchinson's teeth, 286
Hyaena, Fossil, 32
Hydatid disease, 180, 196-197
Hydatidiform mole, 505, 513
Hydramnion, 505
Hydrocephaly, 428-429, 509, 556
Hydrocoele, 214, 515, 640
Hydrophobia, 116, 229, 712
Hydropic foetus, 505
Hyperemesis, Pregnancy, 505
Hypermetropia, 459
Hyperostosis,
    in fossils, 43-44
    porotic, *see* osteoporosis symmetrica
Hyperplasia, Condylar, 327
Hypertension, 89, 90
Hypertrichosis, 449, 525
Hypochondriasis, 717, 718, 731
Hypoplasia,
    dental, 555, 556
        and Harris's lines, 398-399
    renal, 484
Hypopyon, 459
Hysteria, 711, 715, 716, 718
Hysterical aphonia, 731

I

Imhotep, 105, 604
Impetigo, 127, 451
Immunity, 81
    in invertebrates, 36-37
Infantile paralysis, *see* Poliomyelitis
Infarction, Myocardial *see* Myocardial diseases
Influenza, 119, 120, 122
    pandemic, 78
Inhapy, Queen, 453, 512

J

Jaundice *see* Hepatitis
Jaws, Pathology of, 551-595
    injury to, 623-629
Juvenal, 521, 718

K

Kanam mandible, 326, 330
Keratosis, Solar, 454, 455
Kidney diseases,
    abscesses, 177
    bilharzia haematobia eggs, 177, 180
    congenital atrophy, 177
Klippel-Feil syndrome, 431-432
Kuru, 85
Kwashiorkor, 63, 72, 73, 93

L

Laboratory techniques, 12, 44-45, 187-188, 490
    Harris's' lines, 403-404
    pseudopathology due to, 12
Lathyrism, 478
Leeches, 180-181, 230, 639
Leishmaniasis,
    American, 182
    cutaneous, 197-198, 451
    visceral, 198
Leonidas, 513
Leontiasis ossea, 557
Leprosy, 65, 125, 206-207, 211, 216-217, 226, 229, 233, 236-237 244-245, 281, 291, 295-306, 450, 451, 583, 628
    facies leprosa, 296, 298, 300, 301, 302, 303
    facies nasalis, 297, 298
    facies oralis, 298
    pseudopathology, 7, 17, 162, 181, 216-217, 244, 449
Leptospira, 117
Leucoderma, 127, 217
Leucorrhoea, 506
Lice, 178, 181
Lichen, 451, 453
Limnocyon potens, 44
Literary evidence, *see* Texts, Early
Lithiasis *see* Calculi
Living agents, effect on bones, 9-10
    animal agents, 10-11
    bacteria and moulds, 9
    higher plants, 9-10
Lucretius, 179, 180
Lupus, 244

M

Malaria, 63, 72, 78, 81, 84, 132, 143, 182, 226, 242, 378, 384, 385
    and haemoglobin S, 80
    in ancient Greece, 170-176, 182
    in primates, 48
    in primitive populations, 85-86

Malta fever see Brucellosis
Macrognathia, mandibular, 554-555
Malnutrition, 89, 192-194
Manic state, 720
Martial, 42-44, 717, 718, 743, 744
Mastitis, inflammatory, 343, 513
Mastodon, 32
Mastoid infections, 123, 464-473
   in Rhodesian man, 11, 467-468
   pseudopathology, 7, 15
Measles, 78, 79, 119, 121-122, 451
Meningioma, 326, 327-328
Meningitis, 122, 124, 200-201, 227
Menorrhagia, 506
Mental disorders, 217-218
   congenital defect, 441
   in Greece and Rome, 709-722
   in Mesopotamia, 723-733
   in primitive societies, 91
Merneptah, 446, 474, 479, 528
Mersekha, 433
Meritamon, 407, 409
Mesopotamia, Organic diseases of, 191-208
Metastases, 337, 339-341
Microcephaly, 441
Micrococcus, 42
Micrognathia,
   mandibular, 553-554
   maxillary, 555
Mongolism, 441, 555
Mortality,
   and Harris's lines, 399-400
   in primitive societies, 76, 79
Mummies, vii-ix, 20-30, 108, 111, 115, 120, 123, 126, 127, 129, 177, 178-179, 180, 181, 184, 196, 263-264, 275, 280, 323, 326, 343, 357, 445, 512, 514, 528, 532-537
   ageing of, 532
   clubfoot in, 423-424
   eye diseases in, 459-462
   genital disorders in, 507-517
   Granville's mummy, viii-ix, 343, 507, 511, 588
   hernia in, 446
   intervertebral disc calcification in, 22-30
   parietal thinning in, 407-409
   Peruvian, 181, 185, 255, 481, 507, 512
   respiratory diseases in, 489-492
   skin diseases in, 300-301, 451-452
   smallpox in, 346-348, 451-452
   twins, 502, 503
   vascular disease, 479-484
Mumps, 119, 122, 240
Mutilation, 511, 643-644
   for sexual offences, 746, 747, 753
   hand, 68, 644
   sincipital, 669-670
Mycobacterium leprae, 27
Mycobacterium tuberculosis, 117, 125, 126, 127
Myelomatosis, 286, 337

Myocardial diseases, 88-89, 475
Myopia, 458
Myositis ossificans, 325

N

Naegele's deformity, 508
Naples epidemic, 163, 164
Natron, 25-28, 489, 491
Neandertal man, 59, 523, 538, 541, 545, 557, 612-614
   Dolni Vestonice, 589
   Ehringsdorf, 572, 573, 582, 607, 614
   Gibraltar, 58, 64, 612
   Krapina, 572, 573, 574, 580, 588, 612
   La Chappelle, 58, 352, 361, 572, 573, 588, 612
   La Ferrassie, 572, 578, 588, 612
   La Quina, 588, 612
   oral health, 50, 58, 64, 129, 557, 572-574, 578
   Shanidar, 58, 67
   Skhul, 557, 574, 589, 625
   skull deformity, 13
   Spy, 612
Nebuchadnezzar II, 194
Necroses, in fossils, 43
Neoplasms, see Tumours
Nephretiti, 16
Nesperehân, 263-264
Neuritis, 193
Neurofibroma, 329
Neurosis, 728-730
   psychoneurosis, 731
Nits, 181
Nutrition in primitive populations, 72, 87
Nyctalopia, see Blindness, Night-
Nymphomania, 506, 735
Nystagmus, 192

O

Obesity, 90, 219, 504, 525, 528
Oedema, 193, 203, 235, 420, 477, 505
Ophthalmia 198, 457, 458, 459
   seasonal, 458
Ophthalmoplegia, 192, 206, 457
Orchitis, 515
Oribasius, 174, 507
Oroya Fever, 182, 451
Osteitis, 281, 282, 283, 284, 289, 290, 291, 292, 321, 341, 666
   in fossils 43, 44
   of the jaws, 563, 576-590
   oral, 582
   "osteomyelitis," 126, 129, 282, 293, 357
   "periostitis," 285, 288, 289, 299, 317, 471
   sclerosing osteitis, 582
Osteitis deformans see Paget's disease
"Osteitis fibrosa," 371-377, 525
Osteochondroma, 323, 324, 330
Osteo-dystrophy, in the jaw, 557
Osteogenesis imperfecta, 276-277, 441, 525

Osteoma, 322, 326
  in fossils, 43
  Osteoid, 325, 327
  pseudopathology, 15
Osteomalacia, 288, 508
  in fossils, 44
Osteophytosis, 355, 360, 361, 362-369
Osteoporosis, 407
  osteoporosis circumscripta, 407
  osteoporosis symmetrica, 378-389, 668-669
  pseudopathology, 7
Osteosarcoma, 326, 330-337
  in fossils, 43
Otitis media, 199, 465-467

P

Pachycanthus, 43
Paget's disease of bone, 407
  Gambles cave skull, 15
Palaeobiochemistry, xii
Palaeostomatology, 538-550
Palaeopathology,
  definition of, vii, 31-32
  bibliography of, x
Papilloma, of the skin, 454, 455
Papyri, Egyptian Medical, 98-111, 116, 120, 177, 180, 260, 304, 445-446, 450, 457, 489-492
  Berlin Medical Papyrus, 103, 109, 465, 497, 534, 603
  Carlsberg Papyrus, 104
  Cattaui Papyrus, 105
  Chester Beatty Papyri, 103, 108, 497, 603
  Ebers Papyrus, 98-99, 100-101, 109, 117, 123, 126, 127, 129, 178, 181, 182, 304, 343, 444, 445, 450, 457, 458, 465, 476, 477, 478, 492, 494, 495, 496, 497, 506, 511, 512, 515, 534, 537, 603, 702
  Edwin Smith papyrus, 100, 109, 116, 129, 273-274, 477, 512, 515, 533, 534, 599, 603-605, 702
  Erman Papyrus, 603
  Golenischef Papyrus, 105
  Hearst Papyri, 102-103, 343, 457, 497, 515, 603, 702
  Kahūn Papyrus, 102, 109, 506, 603
  London Medical Papyrus, 103-104, 506, 511, 515, 603
  magical papyri, 104
  Mashâykh Papyrus, 104
  Ramesseum Papyri, 101-102
  Turin Papyrus, 737
  Westcar Papyrus, 502
Paracingular invagination, 542
Paradontal disease *see* periodontal disease
Paralysis, 216
Paranoia, 713, 725, 732
Paraplegia, 272-278
Parasites, xiii, 81, 116, 118, 178-188
  in carboniferous crinoids, 37
  in primates, 48-49
Paratyphoid, 122, 124
Parietal bone, thinning of, 405-412
  biparietal thinning, 413-416
  in ancient Egyptian skulls, 407-409

Paronychia, 129
Parotid gland, enlargement of, 88
Pasteurella pestis, 116
Paulus of Aegineta, xii, 121, 128, 180, 343, 459, 477, 506, 513
Pediculosis, 181, 451
  of the eyelids, 459
Pellagra, 63, 72, 417
Pepi I, 527
  II, 738
Periodontal disease, 129, 212, 558-574, 601
  in fossil animals, 40-41
  in fossil man, 50
Peritonitis, 128, 129
Peru,
  Chilca skull, 12
  pottery, 16, 127, 182, 500, 521, 528, 529, 640, 753
  mummies, 181, 185, 225, 481, 507, 512
  tuberculosis in art, 250-253
Phagocytosis, 36
Philolaus of Tarentum, 515
Photophobia, 458, 459
Phthisis, 492
Physiology, in Egypt, 108-109
Pigeon chest, 421
Piltdown skull, possible pathological diagnosis, 12
Pinta, 125, 152, 153, 156-157, 279
Pityriasis, 451
Placenta,
  abnormalities of, 504-505
  retention of, 505
Plague, 72, 81, 116, 245-246. *See also* Black Death; epidemics.
  bubonic, 245
  pneumonic, 200
Plants, disease in,
  ergot, 63
  fossil plants, 34
  rust, 63
Plasma proteins, 88
Plasmodium, 133, 144
  P. falciparum, 133, 378, 385, 387
  P. malariae, 133, 387
  P. vivax, 133, 387
Plato, 171, 172, 174, 260, 499, 713-714
Pliny, 65, 110, 121, 179, 180, 181, 182, 450, 458, 459, 516, 521
Plutarch, 180, 182
Pneumonia, 119, 122, 123, 129, 177, 199-200, 226, 227, 231, 235, 240, 490, 491, 492, 640, 715
Poisons, 205-206, 232
  alcoholic, 206
Polymastia, 512-513
Poliomyelitis, 122, 124-125, 424, 600
Polydactyly, in primates, 53
Polyneuritis, 417
Polyps, 215, 329, 343, 511, 640
Polythelia, 512-513
Pott's disease *see* Tuberculosis

Presbyopia, 459
Priapism, 515
Primates, Pathology of, 47-55, 129, 133, 150
Primitive societies,
　demography, 75
　depopulation, 75-78
　disease in, 69-97
　epidemiology, 78-79
　isolation, 70-71
　mortality rate, 76, 79
　mutilation, 643-644
　settlement size and mobility, 70-71, 74
　surgery, 635-650
　trephination, 673-701
Procidentia, Uterine, 506
Procopius, 116, 246
Prolapse,
　rectal, 219, 497, 511
　uterine, 506, 511, 637
　vaginal, 511
Propagules, 70-71, 82
Proptosis, 459
Prostatic obstruction, 215
Pseudopathology, 5-19, 321, 325-327
　in mummies, 13, 424
　presumptive alkaptonuria, 22-30
Psoriasis, 217, 227, 244, 354
Psychology, and disease, 77
Psychopathy, 728-730
Psychoses, 723-728
Pterygium, 458, 459, 641
Puerperal fever, 123
　insanity, 505
　sepsis, 240, 502, 505, 508
Punt, Queen of, 438, 528
Pyelitis, 128, 129
Pyorrhea alveolaris, see Periodontal disease

Q
Q. fever, 63

R
Rabies, see Hydrophobia
Radiography, xi-xii, 403-404, 557, 564, 577, 578
　and parietal thinning, 411
　bone pathology, xii
　duplicate X-rays, xiii
　mummies, ix, 22-30
　pseudopathology, resulting from, 15
Radiology, see Radiography
Ramesses II, 479, 485, 515, 528
　III, 528
　IV, 453, 459, 514, 515, 528, 532
　V 446, 452, 515, 528
　VI, 528
Relapsing Fever, 81, 118, 235, 241
Reproductive system, Diseases of, 214-215, 498-520
　see also Genital System
Respiratory diseases, 213-214, 489-493

Rhinitis, 582
Rhodesian Man, vii, 11, 59, 467-468, 523, 613, 614
　oral health, 50, 58-60, 129, 574, 578
Rickets, 6, 228, 278, 379-380, 417, 421-422, 508, 555-556, 601
　fossil, 43
　in animals, 13, 43
　in Pithecanthropus, 13
Rickettsiae, 118, 230
　R. mooseri, 118
　R. prowazeki, 118
Ringworm, see Tinea imbricata
Rogerius Frugardi, 653, 658, 664
Roundworm see Helminths
Rufus of Ephesus, 116, 245, 459, 476, 501, 718

S
Salmonella, 117
　Salmonella typhi, 123
Sarcoma, Osteogenic, 330-337
　periosteal, 341
Scabies, 181, 449, 450, 451
Scalping, 630-634
Scaphocephaly, 6
Scarlet fever, 122, 123
Schistosoma haematobium see Helminths; Bilharzia
Schistosomiasis see Bilharzia
Schizophrenia, 91, 718, 724, 731
Scoliosis, 429-430
Scurvy, 73, 193-194, 212, 417
　pseudopathology, 7
Seknenre, 532
Seneca, 260, 479
Sethes II, 515
Sethos I, 528
Sexual deviation, 734-755
　in ancient Egypt, 736-738
　in ancient Greece, 738-742
　in ancient Rome, 742-745
　in the ancient Chinese, 747-750
　in the ancient Hindus, 745-747
　in the Hebrews, 218, 735-736
Shovel-shape incisor, 451-452
Sickle-cell trait, 133, 143, 150, 378, 381, 383, 384, 385, 387
Sinusitis, 582-583
　in primates, 49
　pseudopathology, 15
Siptah, 124, 424, 444, 528, 599, 603
Skin diseases, 449-456
　in mummies, 451-455
　in the Bible and Talmud, 216-217
Skull perforation, Aetiology of, 15
　in Eskimo skulls, 9
Sleeping sickness see Trypanosomiasis
Smallpox, 119, 120-121, 122, 239, 240, 451
　in a mummy, 346-348
Smith, Sir Grafton Elliot, viii, ix, x

Index

Soil,
   chemical effect on bones, 7-9, 472
      pyrites disintegration, 8
   mechanical effect on bones, 5-9, 321-322
      decalcification, 6
      distortion, 5-6
      mock fracture, 6-7
Soranus of Ephesus, xii, 105, 180, 451, 477, 500, 502, 504, 505, 506, 507, 509, 513, 521, 525
Spina bifida, 277, 278
   in primates, 53
Spirochaetes, 118, 125
   spirochaetal jaundice, 117
Splenomegaly, 172, 182
Splints, 602-603, 637-638
Spondylolisthesis, 601
Sprains, 601, 604-605
Staphylococcus, 353
   Staph. albus, 128
   staph. aureus, 82, 128, 129
Sterility, 504
Streptococcal septicemia, 240
Streptococci, 78, 82, 128, 353
   Str. pyogenes, 123
Styes, 82, 459
Suicide, 93, 721
Surgery, Primitive, 635-650
   dental, in Egypt, 702-705
Sycosis barbae, 451
Syncope, 477
Syphilis, xiv, 125, 244, 279-294, 303, 307, 308, 309, 318, 353, 516, 583-584, 601
   congenital, 286
   endemic, 125, 152, 153, 159-162
   in primitive societies, 86
   lesions in Pre-Columbian bones, 164, 167, 287-293
   pseudopathology, 7, 10
   venereal, 152, 153, 162-167
      geographic distribution, 286-287
Syringomyelia, pseudopathology, 7

T

Talipes equinovarus, 423-428, 599, 600
Talmud, Diseases in, 209-221, 479, 500, 516, 736
Tapeworm, *see* Helminths
Teeth,
   anisodontia, 545
   attrition, 64, 90, 549-550, 564
   calculus, 574-576
   caries, 212, 231, 533, 546-549, 578-579
      and Harris's lines, 396-397
      in fossil animals, 32, 39-40
      in fossil man, 50, 58, 578
      in primates, 49-50
      in primitive societies, 89-90
   delayed dentition, 421
   enamel variations, 542
   extraction, 212
   filling, pseudopathological, 14
   hyperodontia, 540
   hypodontia, 539-540
   impaction, 546, 557
   loss, 578
      in fossil man, 43, 58
   macrodontia, 540-541
   malocclusion, 546, 555-557
   microdontia, 540
   pulp exposure, 58, 64, 549, 576-579, 581
   root variation, 544-545
   taurodontism, 545
   tooth evulsion, 61, 68
Temporo-mandibular joint, 586-590
   ankylosis of, 587-588
Tetanus, 78, 116, 241
Texts, Early
   Arabian, 260, 451
   Assyrian and Babylonian, 191-208, 259-260, 449, 457, 465-467, 477, 723-730
   Chinese, 127, 222-237, 259, 417-422, 654, 655-656
   European, *see* under author
   Hittite, 731, 751
   Indian, 120, 126, 127, 179, 182, 258-259, 451, 458, 477, 502, 504, 505, 522, 525, 654-655, 745, 747
   Maya, 477
   Persian, 181, 450, 751
Textual errors: pseudopathology, 17-18, 98-99, 100, 127, 304
Thalassaemia, 80, 378, 379, 381, 383, 384, 385, 387
Theognis, 711
Theodorides, 182
Threadworm, *see* Helminths
Thrombosis,
   coronary, 202
   septic cerebral, 477
Thucydides, 174, 238-239, 710, 713
Thyrotoxicosis, 521
Ticks, 181
Tinea imbricata, 82, 86, 127, 227
Tissue samples, xiii
Tollund Man, 185
Tonsillitis, 122, 226
   streptococcal, 124
Toxaemia, pregnancy, 203, 505
Trachodon annectens, 13
Trachoma, 125, 128, 198, 215, 226, 457, 458, 459
Tragelaphus scriptus: bushbuck,
   as Trypanosome reservoir, 142, 143, 148
Trauma,
   Cranial, 58, 606-622, 623-629
      in fossil man, 606-614
      pseudopathology, 10
   Post-cranial, 67, 599-605
Trepanation *see* Trephination
Trephination, 9, 10, 14, 15, 429, 430, 467, 642-643, 651-672
   in animals, 667

266562

R
135
B83
HEAL